D1740177

Deep-Time Perspectives on Climate Change: Marrying the Signal from Computer Models and Biological Proxies

The Micropalaeontological Society Special Publications

This collection of papers on ancient climate change and the role of micropalaeontology in its documentation is the second of a series of Special Publications, published by the Geological Society on behalf of The Micropalaeontological Society. TMS has, as one of its aims, the communication of the subject to as wide an audience as possible. With climate change high on the political agenda, this book presents an authoritative account of the Earth's changing environment, especially climate, and the evidence provided by a variety of microfossil groups.

This volume, and that on *Recent Developments in Applied Biostratigraphy*, are published by the Geological Society. Previous volumes that are still in print are published by Springer-Verlag (formerly by Kluwer).

Recent Developments in Applied Biostratigraphy. Powell, A. J. & Riding, J. B. (eds) ISBN 1-86239-187-4, November 2005.

Calcareous Nannofossil Biostratigraphy. Paul R. Bown. ISBN 0-412-78970-1, August 1998.

Early Evolutionary History of Planktonic Foraminifera. M. K. BouDagher-Fadel, F. T. Banner, J. E. Whittaker & M. D. Simmons. ISBN 0-412-75820-2, November 1997.

Micropalaeontology and Hydrocarbon Exploration in the Middle East. M. D. Simmons. ISBN 0-412-42770-2, June 1994.

A Stratigraphic Index of Dinoflagellate Cysts. A. J. Powell. ISBN 0-412-36280-5, November 1991.

Ostracoda and Global Events. R. Whatley & C. Maybury. ISBN 0-412-36300-3, July 1990.

THE MICROPALAEONTOLOGICAL SOCIETY SPECIAL PUBLICATIONS

Deep-Time Perspectives on Climate Change: Marrying the Signal from Computer Models and Biological Proxies

EDITED BY

M. WILLIAMS
University of Leicester, UK

A. M. HAYWOOD
University of Leeds, UK

F. J. GREGORY
PetroStrat, UK

and

D. N. SCHMIDT
University of Bristol, UK

2007
Published for The Micropalaeontological Society by
The Geological Society
London

THE MICROPALAEONTOLOGICAL SOCIETY

The Micropalaeontological Society (TMS) exists 'to advance the education of the public in the study of Micropalaeontology' and is operated 'exclusively for scientific and educational purposes and not for profit'. It was founded as The British Micropalaeontological Group (BMG) in 1970, following a proposal by Professor Leslie Moore of the University of Sheffield and several colleagues who wished to organize a group of palaeontologists with a mutual interest in the study of the microfossils present in British type sections and the provision of a forum for the communication of their results.

Under the guidance of Dr Bob Cummings the Group became the British Micropalaeontological Society (BMS) in 1975 during a period of rapid expansion and the development of the science, particularly in the field of hydrocarbon exploration.

The geographical development of micropalaeontology has resulted in a growth in the international membership of the Society, such that the name was changed to The Micropalaeontological Society in 2001. It is a registered charity in the UK (No. 284013). The Society publishes *The Journal of Micropalaeontology* and from 1973 to 1999 the BMS also published *A Stereo-Atlas of Ostracod Shells*.

The Society comprises six specialist groups which study Foraminifera, Microinvertebrates, Nannofossils, Ostracods, Palynology and Silicofossils. Membership of any of these specialist groups is by personal choice of individual members. The Society organizes an annual general meeting and a number of other scientific meetings and field excursions throughout the year. With the Palaeontological Association, Palaeontographical Society and The Geological Society of London, the TMS assists in the planning of the annual Lyell Meeting of The Geological Society. The Society also manages a website at http://www.nhm.ac.uk/hosted_sites/tms.htm

Published by The Geological Society from:
The Geological Society Publishing House, Unit 7, Brassmill Enterprise Centre, Brassmill Lane, Bath BA1 3JN, UK

(*Orders*:   Tel. +44 (0)1225 445046, Fax +44 (0)1225 442836)
Online bookshop: www.geolsoc.org.uk/bookshop

The publishers make no representation, express or implied, with regard to the accuracy of the information contained in this book and cannot accept any legal responsibility for any errors or omissions that may be made.

British Library Cataloguing in Publication Data

A catalogue record for this book is available from the British Library.

ISBN 978-1-86239-240-3

Typeset by Techset Composition Ltd, Salisbury, UK

Printed by MPG Books Ltd, Bodmin, UK

Distributors

North America
For trade and institutional orders:
The Geological Society, c/o AIDC, 82 Winter Sport Lane, Williston, VT 05495, USA
Orders: Tel + 1 800-972-9892
           Fax + 1 802-864-7626
           E-mail gsl.orders@aidcvt.com

For individual and corporate orders:
AAPG Bookstore, PO Box 979, Tulsa, OK 74101-0979, USA
Orders: Tel + 1 918-584-2555
           Fax + 1 918-560-2652
           E-mail bookstore@aapg.org
           Website http://bookstore.aapg.org

India
Affiliated East-West Press Private Ltd, Marketing Division, G-1/16 Ansari Road, Darya Ganj, New Delhi 110 002, India
Orders: Tel +91 11 2327-9113/2326-4180
           Fax +91 11 2326-0538
           E-mail affiliat@vsnl.com

Contents

Deep-time perspectives on climate change: an introduction

M. WILLIAMS[1], A. M. HAYWOOD[2], F. J. GREGORY[3] & D. N. SCHMIDT[4]

[1]*Department of Geology, University of Leicester, Leicester LE1 7RH, UK*
(e-mail: mri@le.ac.uk.)

[2]*School of Earth & Environment, Environment Building, University of Leeds,*
Leeds LS2 9JT, UK

[3]*PetroStrat Ltd., 33 Royston Road, St Albans, Hertfordshire, AL1 5NF, UK & The Natural*
History Museum, Department of Palaeontology, Cromwell Road, South Kensington,
London SW7 5BD, UK

[4]*Department of Earth Sciences, University of Bristol, Queens Road, Bristol BS8 1RJ, UK*

The geological record provides an excellent archive of climate and environmental change. One of the most important 19th century discoveries in the field of palaeoenvironmental research was the realization that, in the recent past, vast ice-sheets covered areas of the Northern Hemisphere. Initially, the idea that such ice-sheets could wax and wane frequently with regular periodicities was inconceivable. Such behaviour had been suggested in theory; specifically the astronomical theory of James Croll (1864), later embellished by Milutin Milanković (1941), but it was not until relatively recently that the evidence to support such a theory was discovered by palaeoceanographers (Hays *et al.* 1976). The rise of palaeoceanography over the last 50 years was established through the discovery of the link between the chemistry of marine shells and temperature (Urey *et al.* 1951). This relationship was subsequently refined and it was demonstrated that ice volumes have fluctuated periodically over millions of years (Shackleton 1967), just as the astronomical theory had predicted. Geologists soon recognized that the current ice age in which humans evolved was not unique, and that in fact there had been many ice-age periods over the last billion years, in which great ice-sheets frequently covered large areas of the continents. Through palaeontology, so-called 'greenhouse worlds' – intervals that demonstrated clear evidence for prolonged warmth, where little or no ice existed on Earth – were discovered, a prime example being the Cretaceous (Barrera *et al.* 1987). These further enriched the complex tapestry of Earth's environmental and climatic history. Studying the transition periods between greenhouse and icehouse climate has become a popular and intriguing area of research, and several of the papers presented in this volume have sought to illustrate this.

Understanding of the relationships and possible 'tipping points' leading to extreme climates have become more important over the last 20 years or so with the realization that humans are modifying global climate. Our understanding of anthropogenic warming and cooling influences on climate has improved since the Third Assessment Report of the Intergovernmental Panel on Climate Change (Houghton *et al.* 2001), leading to *very high confidence* that the globally averaged net effect of human activities since 1750 has been one of warming (Alley *et al.* 2007, fig. SPM-2). The study of palaeoclimates has sharpened in importance to the point where it is now considered an integral part in informing policy makers on climate change issues, particularly through the work of the Intergovernmental Panel on Climate Change. In an effort to better understand modern climate and future climate change, numerical models of the Earth's climate system have been developed. These climate models (General Circulation Models or 'GCMs') are now in the vanguard of attempts to simulate the effects of greenhouse gas emissions. Although extraordinarily complex, they remain imperfect tools that require validation. In this, the study of past climate records is invaluable. We do not have data for the future but if these models successfully reproduce large-scale climate changes that have occurred in the past, then that may provide us with more confidence in their predictions for the future.

In recent years, the beginning of an integration of geological data with climate modelling studies has advanced both fields tremendously. Data provide essential constraints and validation for model experiments, whilst modelling often suggests alternative interpretations of proxy data and can provide clear indications of where new data collection should take place. Given the large differences in techniques, approaches taken, and language used, this integration represents a challenge to both scientific communities but it is an area in

From: WILLIAMS, M., HAYWOOD, A. M., GREGORY, F. J. & SCHMIDT, D. N. (eds) *Deep-Time Perspectives on Climate Change: Marrying the Signal from Computer Models and Biological Proxies.* The Micropalaeontological Society, Special Publications. The Geological Society, London, 1–3.
1747-602X/07/$15.00 © The Micropalaeontological Society 2007.

which many expect the most significant advances in palaeoclimate studies will emerge. The aim of this book is to bring together data collectors and modellers, and to facilitate the interaction between both groups hopefully leading to a better understanding of how systems evolve and develop.

This book uses the geological timescale as a natural framework around which the papers are organized. It covers a swathe of some 600 million years of Earth history summarized in the first paper of the volume (**Vaughan**) and begins with the climates of the Late Proterozoic through Palaeozoic. This includes times when Earth may have possessed a dominantly greenhouse climate, and times when the world was much cooler, for example during the Neoproterozoic (**Sohl & Chandler**), the Early Palaeozoic Icehouse (**Armstrong**; **Page** *et al.*), or for much of the Late Carboniferous (**Kiehl**) and Early Permian (**Stephenson** *et al.*). This Palaeozoic world was terminated at the Permian–Triassic boundary, a most serious extinction of life on Earth (**Twitchett**).

The Mesozoic world that followed has been modelled in great detail (**Sellwood & Valdes**). In this younger part of the geological record, geographies are well understood and a multitude of fossil and sedimentological data (**Hart; Markwick; Price & Grimes**) can be used to ground truth climate models for this era. The Mesozoic was terminated by another mass extinction of life at the Cretaceous–Tertiary boundary, though this is not reflected in any long-term change in Earth's climate, and may be related to extrinsic (extraterrestrial) factors such as a meteorite impact.

The Cenozoic documents a change from greenhouse to icehouse conditions. The Palaeocene world was warm like its Mesozoic predecessor and was terminated by a brief, sudden and extreme warming (**Sluijs** *et al.*). Cooling, however, has characterized the climate of Earth at least since the Oligocene (**Coxall & Pearson**), terminally leading to large ice-sheets, at both poles (**Hill** *et al.*). The cooler Earth is explored in a number of papers documenting both the terrestrial and marine Neogene world (**Dowsett; Faquette** *et al.*; **Haywood** *et al.*; **Jimenez-Moreno** *et al.*; **Pfuhl & McCave**). The book concludes at the transition from the Pliocene world into the Pleistocene (**Ravelo** *et al.*), in 'deep time', some two million years ago.

Whilst many papers in this book discuss the controlling mechanisms that may have forced major changes in Earth climate, for example, geographical and oceanographic changes during the Late Ordovician, or changing ocean gateways in the Cenozoic (**Schmidt**), other papers detail the biological and chemical proxies that can be used to track ancient climate. These are introduced at the point in the book where they become relevant. Thus, fossils

are introduced early on in the 'Cambrian explosion' of life (**Vannier**), which more than half a billion years ago provided a vast new range of 'organism recorders' to track global oceanographic patterns and climate. Other palaeoceanographic tools, such as the use of the Ca/Mg ratio of foraminifer tests (**Lear**), or alkenones from surface-dwelling marine algae (**Lawrence** *et al.*), and their link to the evolution of the organisms carrying the proxy signal (**Kucera & Schönfeld**) are introduced later in the book, amidst papers dealing with Cenozoic climate, where they find their widest application to palaeoclimate research.

Finally, the strength of using climate models with geological data lies in the iterative process, whereby one discipline can be used to test the other. In this process lies the feasibility to produce, perhaps in the not too distant future, very accurate estimates of Earth's ancient climate. Therein is the key to the future. Perhaps it is fair to say that the great maxim of James Hutton and Charles Lyell, that the *present is the key to the past*, can sometimes be inverted, so that *the past may become the key to the future*.

The editors would like to thank the many reviewers of manuscripts published in this book. We are very grateful to Shell UK (Nick Hogg), the British Antarctic Survey and the British Geological Survey who provided funds to publish the many colour diagrams and figures. We also thank the staff of the Geological Society Publishing House in Bath for their help in producing this volume, in particular Angharad Hills, Helen Floyd-Walker and Sarah Gibbs, and Malcolm Hart as Special Publications Editor of The Micropalaeontological Society (TMS). TMS have been unstinting in their support of this volume. Finally, we thank David Siveter and Jan Zalasiewicz for reading earlier versions of this overview.

References

ALLEY, R. 32 OTHERS 2007. Climate Change 2007: The Physical Science Basis. Summary for Policymakers. Contribution of Working Group I to the Fourth Assessment Report of the Intergovernmental Panel on Climate Change. Available at: http://www.ipcc.ch/SPM2feb07.pdf

BARRERA, E., HUBER, B. T., SAVIN, S. M. & WEBB, P.-N. 1987. Antarctic marine temperatures: Late Campanian through Early Paleocene. *Paleoceanography*, **2**, 21–47.

CROLL, J. 1864. On the physical cause of the change of climate during geological epochs. *Philosophical Magazine*, **28**, 121–137.

HAYS, J. D., IMBRIE, J. & SHACKLETON, N. J. 1976. Variations in the earth's orbit: pacemaker of the ice ages. *Science*, **194**, 1121–1132.

HOUGHTON, J. T. *ET AL.* 2001. Climate Change 2001: The Scientific Basis: Contribution of Working Group I to the Third Assessment Report of the Intergovernmental Panel on Climate Change. Cambridge University Press, Cambridge.

MILANKOVIĆ, M. 1941. Kanon der Erdbestrahlungen und seine Anwendung auf das Eiszeitenproblem., *Special Publication 132*, R. Serbian Academy Belgrade.

SHACKLETON, N. 1967. Oxygen isotope analyses and Pleistocene temperatures re-assessed. *Nature*, **215**, 15–17.

UREY, H. C., LOWENSTAM, H. A., EPSTEIN, S. & MCKINNEY, C. R. 1951. Measurement of paleotemperatures and temperatures of the Upper Cretaceous of England, Denmark, and Southeastern United States. *Bulletin of the Geological Society of America*, **62**, 399–416.

Climate and geology – a Phanerozoic perspective

A. P. M. VAUGHAN

British Antarctic Survey, High Cross, Madingley Rd, Cambridge CB3 0ET UK
(e-mail: a.vaughan@bas.ac.uk)

Abstract: The Phanerozoic is comprised of over 540 million years and, with its defining accompaniment of abundant complex life, provides us with a unique perspective on the extremes of climate change. Understanding these extremes is particularly important if we are to anticipate the possible effects of global warming. The broad sweep of climate change through the Phanerozoic began with relatively cool global temperatures and recovery from late Proterozoic glaciation. This was followed by a mid-Cambrian to Ordovician episode of relatively warm global climate, after which global climate cooled, culminating in the major glaciations of the Carboniferous and Permian Periods. The Triassic and Early Jurassic were warm. The Late Jurassic–Early Cretaceous Period was cool, although without full global glaciation. Global temperatures peaked in the mid-Cretaceous. Since then, global climates have cooled, culminating in Neogene glaciation. These *c.* 100-million-year trends in overall climate show short intense excursions of contrasting climate, many of which have been associated with the mass extinction of life, and with major volcanic and tectonic events. This paper argues that, through the Phanerozoic, two overlapping stable climate regimes appear to have dominated: a high-CO_2 (>1000 ppmv), largely warm climate regime, punctuated by many short-lived episodes of glaciation; and a low-CO_2 (<1000 ppmv), largely cool regime, marked by protracted episodes of superglaciation.

As Hay *et al.* (1997) pointed out: 'the climate of the Holocene is not well-suited to be the baseline climate of the planet'. This is particularly important in light of predicted changes to the global climate anticipated as a result of global warming (Hay *et al.* 1997; IPCC 2001). To understand what may happen, we must examine and try to simulate the warm and cool climates of the more distant past, ideally of the whole Phanerozoic (Fig. 1). The most complete summary of Phanerozoic warm and cool climates continues to be Frakes *et al.* (1992), based on geological evidence from the rock record and oxygen isotopes. They subdivided Phanerozoic time into palaeoclimate 'modes' (Table 1; Fig. 1a), independent of geological time-scale boundaries, improving on the chronostratigraphic system by system treatment by Frakes (1979). These 'modes' are still valid and useful today, although Barron (1993) pointed out that they may represent oversimplifications. For example, Barron (1993) argued that Frakes *et al.* (1992) placed the Pliocene warm interval, currently the focus of much study to understand the possible effects of global warming (e.g. Haywood *et al.* 2005; Barreiro *et al.* 2006), in one of their cold modes. Barron (1993) also contended that they grouped quite dissimilar Late Cretaceous and Early Eocene warm intervals together in the same warm mode. In this chapter, the warm and cool climate modes of Frakes *et al.* (1992) will be reviewed in time-sequential order, and only

developments or changes since their synthesis will be covered in any detail.

Between 1992 and the advent of this book, there have been several reviews of aspects of the Phanerozoic palaeoclimate record, although none on the scale of Frakes *et al.* (1992). The volume edited by Walliser (1995) reviewed global event stratigraphy for the Phanerozoic, which includes many aspects of climate change. Crowell (1999) examined the implications of pre-Mesozoic ice ages on our understanding of the climate system, and Huber *et al.* (2000) specifically treated former warm climates. Boucot & Gray (2001) assessed Phanerozoic climate models based on atmospheric CO_2, and reviewed the many CO_2 proxies, with a particular emphasis on soil and soil-forming organisms. Wallmann (2001) has come closest to the treatment by Frakes *et al.* (1992) and used the record of marine $\delta^{18}O$ values to subdivide the Phanerozoic. He identified three combined icehouse–greenhouse cycles with durations of approximately 127 Myr between the Cambrian and the Triassic, followed by a longer cycle spanning the Jurassic to Cenozoic, which largely correspond to the climate modes proposed by Frakes *et al.* (1992) and discussed in this paper. Subsequently, Wallmann (2004) used a multiproxy approach to model Phanerozoic climates and identify warm and cool climate epochs based on CO_2 levels, proposing a link with the intensity of the galactic cosmic ray flux.

From: WILLIAMS, M., HAYWOOD, A. M., GREGORY, F. J. & SCHMIDT, D. N. (eds) *Deep-Time Perspectives on Climate Change: Marrying the Signal from Computer Models and Biological Proxies.* The Micropalaeontological Society, Special Publications. The Geological Society, London, 5–59.

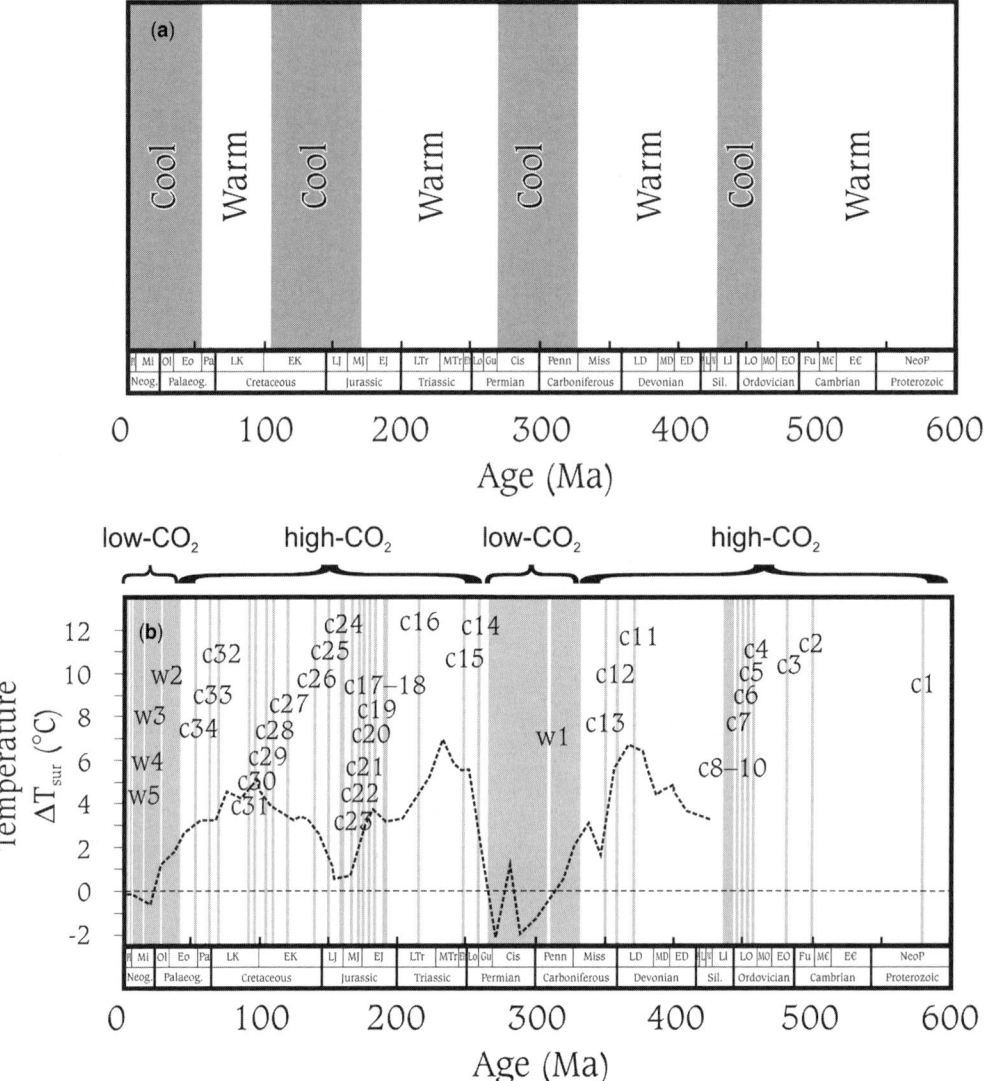

Fig. 1 (a) Late Neoproterozoic and Phanerozoic climate modes of Frakes *et al.* (1992) from Table 1. (b) Intervals of glacial or cool (grey bars) and warm (white areas) climates for the Late Neoproterozoic and Phanerozoic, modified from Fig. 2b of Royer *et al.* (2004) including additional intervals as discussed in the text and listed of Table 1, with the Palaeozoic–Recent palaeotemperature curve of Royer *et al.* (2004) based on $\delta^{18}O$ calibrated with palaeo-pH based on CO_2 proxies and palaeoconcentrations of Ca in seawater; cool intervals labelled c1–c29 (Table 1); warm intervals labelled w1–w5 (Table 1). Brackets above Fig. 1b indicate the durations of high- and low-CO_2 modes for Phanerozoic climate as discussed in the text and listed in Table 1.

Palaeoclimate proxies

We cannot directly observe the climates of the geological past in the way that we have been able to observe climate parameters since the 17th century (e.g. Plaut *et al.* 1995). The key variables, which are both physical and chemical, and include temperature, atmospheric composition and seawater salinity, cannot be measured directly. We depend on chemical and biological systems that have responded, hopefully systematically, to changes in climate, and which have left a record in sediments and ice, which if not pristine, has at least been altered in a systematic way. These so-called proxy data, proxies for short, can be interpreted to yield

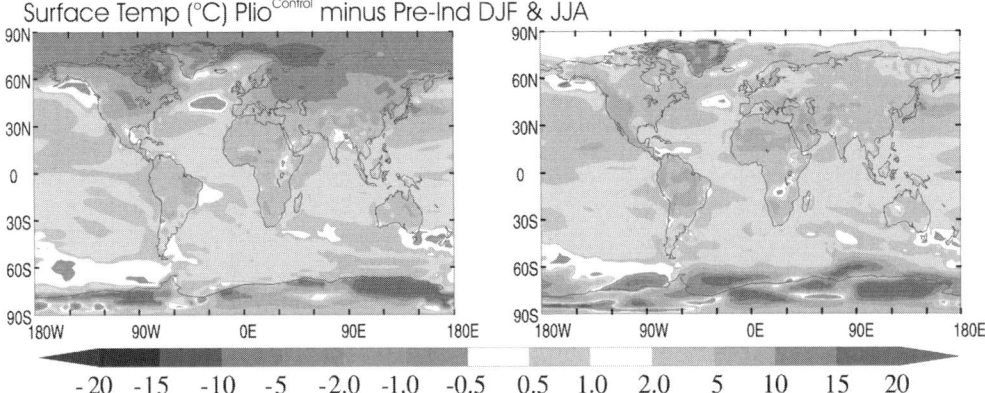

Fig. 2. HadCM3 coupled ocean-atmosphere GCM prediction showing the difference in December, January and February (DJF) & June, July and August (JJA) surface temperatures (°C) between a Mid-Pliocene (*c.* 3 Ma BP) and pre-industrial experiment (redrawn from Haywood *et al.* 2007).

numerical and qualitative data for the key variables of former climates in the geological past (e.g. Wefer *et al.* 1999). Proxies must be evaluated carefully and the level of confidence in a proxy can change as the factors affecting its reliability are better understood (e.g. Elderfield 2002; Williams *et al.* 2005). Geological proxies, i.e. those based on sedimentary facies such as coal or evaporites, can appear relatively straightforward to interpret but may introduce a bias in the signal through incompleteness of the stratigraphic record. There are also parallel factors, such that often only extreme events get preserved, and that preservation of ordinary events often requires peculiarities of tectonic setting (for a discussion of some of these caveats, see Sellwood *et al.* 1993). It is useful to think of the tides in this respect, and when or where today the constant shifting of sand on an average beach might be preserved to record a shoreline. Even in tectonically active areas, such as the western margin of the Americas, earthquakes that shift the base level, such as the 1960 Chile earthquake (e.g. Cisternas *et al.* 2005), or the 1964 Alaska earthquake (e.g. Hamilton *et al.* 2005), happen only once every few hundred years along any particular section of coastline. Therefore, of the order of 150,000 tidal events may go unrecorded, with only the last tides preceding a major earthquake having a chance of preservation. By comparison, chemical and palaeontological methods of estimating Phanerozoic palaeoclimate are complex to interpret, but, (and certainly in the case of oceanic records), they can offer better guarantees of completeness, at least back into the Jurassic.

In the sections below, proxies are reviewed (summarized in Table 2) and organized according to the way that they are grouped within particular

subdisciplines. Chemistry-based proxies are the most abstract and tend to focus on the chemical or isotopic system applied. These are grouped by parameter estimated. Lithological proxies are the most directly related to the parameter estimated and are grouped by rock type, mineralogy or facies interpreted. Palaeontological proxies are subdivided by whether they use taxonomic methods or focus on some morphological aspect of a group of organisms. For clarity and ease of reference, all the proxies reviewed are grouped by parameter estimated in Table 2.

Chemistry-based proxies

Chemistry-based palaeoclimate proxy studies were made possible by the stable isotope work of Harold Urey and others in the 1940s and 1950s (Urey 1947; Epstein *et al.* 1953). Emiliani (1955) published the first quantitative Pleistocene palaeotemperature curve based on the new $\delta^{18}O$ palaeothermometer. This work was substantially built upon by Shackleton and others (Shackleton 1965, 1977 Shackleton & Opdyke 1973; Shackleton & Kennett 1975), culminating in the enormous success of the CLIMAP project in the 1970s (CLIMAP 1981). Table 2 summarizes chemical proxies based on isotope-, element- and biogeochemistry-based systems.

Henderson (2002) has provided a useful framework for the treatment of chemistry-based oceanic proxies, which will form the basis for this section, though here applied to both oceanic and continental chemistry-based proxies. Henderson (2002) distinguished proxies that can be used to reconstruct the physical environment, such as palaeotemperature and mass flows, and those that can be used to

Table 1. *Climate modes and periods of contrasting climate for the Phanerozoic (as shown in Figure 1)*

Climate modes[1] of Frakes et al. (1992)	Warm and cool mode intervals	Absolute age range[2]	New CO_2 modes[3] proposed in this paper	Overall climate	High CO_2 and Low CO_2 mode intervals	Absolute age range[2]	Intervals of contrasting climate[4]	Absolute age[2]
Warm	Ediacaran to Late Ordovician	600–461 Ma	**High**	Warm, fluctuating	Ediacaran–Viséan (late Early Carboniferous)	600–326.4 Ma	mid-Ediacaran (c1)	580 Ma
							Steptoan (mid-Late Cambrian) (c2)	494 Ma
							Tremadocian (Early Ordovician) (c3)	480 Ma
Cool	Late Ordovician to Early Silurian	461–428					early Katian stage, mid-Caradoc (Late Ordovician) (c4)	455 Ma
							mid-Katian stage, Late Caradoc (Late Ordovician) (c5)	452 Ma
							late Katian stage, Early Ashgill (Late Ordovician) (c6)	449 Ma
							Hirnantian (Late Ordovician) (c7)	445 Ma
							Aeronian (Early Silurian) (c8)	439 Ma
							Telychian (Early Silurian) (c9)	436 Ma
							Wenlock (late Early Silurian) (c10)	428 Ma
Warm	mid-Silurian to mid-Early Carboniferous	428–326.4					Frasnian–Famennian boundary (Late Devonian) (c11)	374 Ma
							Devonian–Carboniferous boundary (c12)	360 Ma
							Tournaisian–Viséan boundary (Early Carboniferous) (c13)	345 Ma

Climate	Interval	Age range	Level	Climate detail	Stage interval	Age (Ma)	Event	Age
Cool	Late Early Carboniferous to Late Permian	326.4–270.6	*Low*	Cool, glacial	Serpukhovian (early Late Carboniferous)–Kungurian (latest Early Permian)	326.4–270.6 Ma	Kazimovian (w1)	307 Ma
Warm	Latest Permian to Early Jurassic	270.6–172	*High*	Warm, fluctuating	Roadian (early mid-Permian)–Early Eocene (early mid-Palaeogene)	270.6–49 Ma	Guadalupian–Lopingian boundary (Late Permian) (c14)	260 Ma
							Early Triassic (c15)	250 Ma
							Carnian–Norian boundary (Late Triassic) (c16)	217 Ma
							late Sinemurian (Early Jurassic) (c17)	192 Ma
							Sinemurian-Pliensbachian boundary (Early Jurassic) (c18)	190 Ma
							late Pliensbachian (Early Jurassic) (c19)	184 Ma
							early Toarcian (Early Jurassic) (c20)	182 Ma
							end Toarcian (Early Jurassic) (c21)	176 Ma
Cool	Middle Jurassic to Early Cretaceous	172–105					Aalenian–Bajocian boundary (mid-Jurassic) (c22)	172 Ma
							Bathonian (mid-Jurassic) (c23)	167 Ma
							late Callovian (mid-Jurassic) (c24)	164-161 Ma
							Tithonian (Late Jurassic) (c25)	150 Ma
							Valanginian (Early Cretaceous) (c26)	140 Ma
							late Early Aptian (Early Cretaceous) (c27)	120 Ma
							Early Albian (Early Cretaceous) (c28)	110 Ma

(Continued)

Table 2. Continued

Parameter	Proxy	Effective time range	Recent or key reference
Circulation of oceans	$\delta^{13}C$	Pleistocene–Recent	Lynch-Stieglitz & Fairbanks (1994)
	Foraminiferal Zn/Ca	Pleistocene–Recent	Marchitto et al. (2000)
	Foraminiferal Cd/Ca	Pleistocene–Recent	Boyle (1988)
	$^{9}Be/^{10}Be$ in manganese crusts	Holocene	vonBlanckenburg et al. (1996)
	Nd and Pb isotopes in manganese crusts	Miocene–Recent	Ling et al. (1997)
	Hf isotopes in manganese crusts	Miocene–Recent	David et al. (2001)
	Nd isotopes in foraminifera	Pleistocene–Recent	Vance & Burton (1999)
	Ag/Si ratios	Recent	Zhang et al. (2004)
	^{14}C	Late Pleistocene–Recent	Henderson (2002)
	^{231}Pa and ^{230}Th	Pleistocene–Recent	Henderson (2002)
	Sortable silt	Palaeocene–Recent, but potentially back to the Jurassic	Pfuhl & McCave (2005)
Productivity	Biogenic barite	Pleistocene–Recent	Gingele & Dahmke (1994)
	Solubility of Th, Pa, and Be	Pleistocene–Recent	Chase et al. (2002)
	Sedimentary U concentration	Pleistocene–Recent	Francois et al. (1997)
	$\delta^{66}Zn$ ratio of manganese crusts	Holocene	Maréchal et al. (2000)
	Zn/Si ratios in deep sea hexactinellid sponges	Late Pleistocene–Recent	Ellwood et al. (2004)
	Mo/Al ratios in black shales	Early Ordovician–Recent	Wilde et al. (2004)
	REE in foraminifera	Untested	Haley et al. (2005)
	Transfer function based on carbonate mass accumulation rates of nanoplankton	Palaeocene–Recent	Siesser (1995)
Nutrient utilization	Acritarch diversity analysis	Proterozoic–Permian	Vecoli & Le Herisse (2004)
	Cd/Ca ratio in foraminifera	Pleistocene–Recent	Rickaby & Elderfield (1999)
	Cd/P ratio in phytoplankton particulate matter	Pleistocene–Recent	Cullen & Sherrell (2005)
	$\delta^{15}N$	Early Jurassic–Recent	Robinson et al. (2005)
	$\delta^{30}Si$ of diatom silica	Pleistocene–Recent	Reynolds et al. (2006)
	Transfer functions based on lake diatom assemblages	Early Holocene–Recent, applicable only to lakes	Hausmann & Kienast (2006)
Carbonate alkalinity/ weathering fluxes	Foraminiferal Ba/Ca ratios	Pleistocene–Recent	Lea (1993)
	$^{87}Sr/^{86}Sr$ in carbonate	Cambrian–Recent	Dessert et al.(2001)
	$^{187}Os/^{186}Os$ in mudrocks	Jurassic–Recent	Cohen et al. (2004)
	$^{87}Sr/^{86}Sr$ curve	Cambrian–Recent	McArthur et al. (2001)
	Ge/Si ratios in diatom silica	Pleistocene–Recent	Jones et al. (2002)
	Hf and Nd isotope ratio time trajectories in manganese crusts	Miocene–Recent	van de Flierdt et al. (2002)
pH	Clay mineralogy in ocean sediments	Triassic–Recent	Thiry (2000)
	Mass of individual foraminifera of a particular size	Late Pleistocene–Recent	Lohmann (1995)
	$\delta^{11}B$ of foraminiferal and other carbonate	Neoproterozoic–Recent, but in practice only reliable from Palaeocene–Recent	Pearson & Palmer (2000)

Atmospheric CO₂	calcium-ion concentration of seawater and modelled atmospheric CO$_2$ concentrations	Cambrian–Recent	Royer et al. (2004)
	δ^{13}C of alkenones	Eocene–Recent	Pagani et al. (2005b)
	δ^{13}C of other organic materials	Jurassic–Recent	Hesselbo et al. (2000)
	δ^{13}C of marine, freshwater or pedogenic carbonate	Devonian–Recent	Buggisch (1991); Yemane & Kelts (1996); Royer et al. (2001)
	δ^{11}B and δ^{44}Ca of carbonate calibrated with δ^{18}O and δ^{13}C	Neoproterozoic–Recent	Demicco et al. (2003)
Precipitation/ evaporation	Density of stomata in leaf cuticle	Carboniferous–Recent	Thorn & DeConto (2006)
	Evaporite facies analysis	Archaean–Recent	Ziegler (2003)
	Coal, lignite and peat facies analysis	Devonian–Recent	Sellwood et al. (1993)
	Fusain (fossil charcoal) in clastic sediments	Devonian–Recent	Scott (2000)
	Palaeosols	Cambrian–Recent	Retallack (2001)
	Transfer function from depth to nodular, pedogenic carbonate horizon	Palaeocene–Recent	Retallack (2005)
	Clay mineralogy in ocean sediments	Triassic–Recent	Ahlberg et al. (2003)
	Loess deposits	Palaeogene–Recent	Sun & An (2005)
	Pollen analysis using mutual climatic range and climatic amplitude techniques	Eocene–Recent	Jimenez-Moreno et al. (2005), Klotz et al. (2006)
	Ecological diversity spectra of mammals	Pleistocene–Recent, but potentially applicable back to the Miocene	Fernandez & Pelaez-Campomanes (2005)
Atmospheric circulation	Mammal cenograms	Eocene–Recent	Legendre et al. (2005)
	Tempestite facies analysis	Neoproterozoic–Recent	Agustsdottir et al. (1999)
	Windblown dust in marine sediments, loess deposits	Palaeogene–Recent	Kohfeld & Harrison (2001)
	Dunes, windblown trees	Palaeozoic–Recent	Segalen et al. (2004)
	Aeolianite facies analysis	Devonian–Recent	Le Guern & Davaud (2005)
Glaciation	Glacial sediments	Archaean–Recent	Sellwood et al. (1993)
	Correlation of δ^{18}O and δ^{13}C isotope excursions and sea-level falls	Neoproterozoic–Recent	Miller et al. (2005)
	Glendonite carbonate nodules	Neoproterozoic–Recent	Swainson & Hammond (2001)
	Sub-glacial volcanic deposits	Miocene–Recent	Smellie et al. (2006)
	Gibbsite concentration in soils	Pleistocene–Recent	Ballantyne et al. (2006)
Sea-ice cover	C-25 highly branched isoprenoid alkenes IP25	Holocene	Belt et al. in press
	Transfer functions based on dinoflagellate cyst assemblages	Late Pleistocene–Recent	Peyron & De Vernal (2001)
	Diatom faunal assemblages	Late Pleistocene–Recent	Gersonde & Zielinski (2000)
	Similarity maximum modern-analog techniques on foraminiferal assemblages	Late Pleistocene–Recent	Sarnthein et al. (2003)
	Modern analogue techniques based on diatom assemblages	Late Pleistocene–Recent	Crosta et al. (1998)

See text for explanation

reconstruct aspects of the carbon cycle, such as oceanic pH or atmospheric CO_2 concentrations.

Palaeotemperature. Surface temperature is a reflection of time-averaged net solar energy input. This is energy that is not immediately reradiated back to space or involved in photosynthesis, but which is trapped for a time, either in the atmosphere via, for example, the greenhouse effect and hydrological cycle, e.g. evaporation and ice melting, or in the solid surface, lakes, seas and oceans. Surface temperature is the key variable for the Earth System. In the oceans, it drives the weather and winds, which ultimately power circulation in the atmosphere and oceans. On land and sea, it controls evaporation, which is a major influence on the water cycle and weathering.

As outlined above, $\delta^{18}O$ in carbonate has been used as a surface palaeotemperature proxy since the 1950s (e.g. Emiliani 1955). $\delta^{18}O$ values are determined from skeletal and non-skeletal carbonates (Marshall 1992), with varying degrees of success, depending on whether the carbonate analysed is pristine or secondary. In marine settings, $\delta^{18}O$ ratios are derived from calcareous plankton, such as foraminifera or coccolithophores, as well as other marine invertebrates, such as brachiopods (Korte et al. 2005), corals, and various molluscs (e.g. Gazdzicki et al. 1992; Elorza et al. 1996; Bailey et al. 2003), and also bulk sediments (Tobin & Walker 1997) and cements (Hays & Grossman 1991). Marine records extend back to the Proterozoic (e.g. Samuelsson 1998). In terrestrial settings, speleothems (Winograd et al. 1992) provide a detailed record, as do lake-dwelling ostracods (Schwalb 2003) and gastropods (Grimes et al. 2003), although only extending back to the Pleistocene. Pedogenic carbonate provides a longer time record (e.g. Mack & Cole 2005), although with significant gaps and problems of interpretation. Overall, interpretation of the signal from carbonate is complicated by, for example, the presence of cryptic species, such as in the foraminifera (e.g. Kucera & Darling 2002), biological isotope fractionation during development, so-called 'vital effects' (e.g. Friedrich et al. 2006), and diagenesis (e.g. Wenzel et al. 2000; Williams et al. 2005). $\delta^{18}O$ ratios can also be derived from non-calcareous sources, such as vertebrate tooth enamel (e.g. Bentaleb et al. 2006) and bone (e.g. Barrick et al. 1999), as well as fish scales and otoliths (e.g. Grimes et al. 2003), and diatom silica (e.g. Leng & Barker 2006).

Other carbonate-based palaeotemperature proxies use the ratios of trace elements, such as foraminiferal Mg/Ca (e.g. Elderfield 2002; Elderfield et al. 2002), coralline Sr/Ca (e.g. De Villiers et al. 1994; Wei et al. 2004) and Li/Ca (Marriott

et al. 2004), or Ca isotope ratios such as foraminiferal $\delta^{44}Ca$ (e.g. Gussone et al. 2003). However, Ca isotope ratios have complex species-related variations in sensitivity (e.g. Sime et al. 2005) that complicate their usefulness. Mg/Ca and Sr/Ca from belemnites appears to be effective from the Late Cretaceous back to the Early Jurassic (Rosales et al. 2004).

More recent non-carbonate palaeotemperature proxies include the index of unsaturation of alkenones from the membranes of some Prymnesiophyceae algae (coccolithophores) such as *Emiliania huxleyi* (e.g. Prahl et al. 1988), the so-called UK'37 proxy. This proxy has proved to be very successful (e.g. Bard 2001), is effective back to the Miocene for marine sediments (e.g. Haywood et al. 2005), and has also been applied to lakes (Li et al. 1996). Oxidation effects need to be considered for the older record (e.g. Hoefs et al. 1998). Amino acid racemization epimerization in molluscan calcite is another terrestrial palaeotemperature proxy (e.g. Murray-Wallace et al. 1988), which initially showed promise back to the Pliocene (e.g. Kaufman & Brighamgrette 1993). However, recent studies have shown its application to be limited to the Late Pleistocene (e.g. Andersson et al. 2000). Similar to UK'37, TEX86 is a palaeotemperature proxy based on the tetraether lipids of crenarcheota (Schouten et al. 2002). Initial results suggest that the proxy is robust (e.g. Wuchter et al. 2004). It appears to be resistant to post-depositional oxidation (e.g. Schouten et al. 2004) and is capable of recording palaeotemperatures above the *c.* 29 °C limit of UK37' (e.g. Pelejero & Calvo 2003; Schouten et al. 2003). TEX86 is also applicable to lakes (e.g. Powers et al. 2004), and it has an effective time range extending back to the Mesozoic (e.g. Schouten et al. 2003).

Palaeosalinity. Henderson (2002) highlighted palaeosalinity as a major environmental parameter for which we have no direct marine proxy. Calculations based on $\delta^{18}O$ are prone to error (Schmidt 1999; Rohling 2000), but have been applied back to the Early Triassic (Korte et al. 2005). Recent developments suggest that U/Ca ratios in corals (Ourbak et al. 2006) and alkenone unsaturation ratios (Sikes & Sicre 2002; Blanz et al. 2005; Schouten et al. 2006) show some promise.

Palaeocirculation of oceans. Palaeocirculation proxies either trace the movement of water masses or provide estimates of deep-water flow (Henderson 2002). Of the former, $\delta^{13}C$ (e.g. Lynch-Stieglitz & Fairbanks 1994), Zn/Ca (e.g. Marchitto et al. 2000) and Cd/Ca (Boyle 1988) ratios in foraminiferal carbonate and $^9Be/^{10}Be$ ratios in manganese crusts (Kusakabe et al. 1987; vonBlanckenburg *et al.* 1996) trace water masses by mimicking nutrients. More recent tracers of water masses include isotopes of Nd and Pb (Ling *et al.* 1997), and Hf (David *et al.* 2001) mainly from ferromagnesian crusts on manganese nodules, but also from foraminifera (e.g. Vance & Burton 1999), and Ag/Si ratios (Zhang *et al.* 2004). For deep-water flow, radioactive proxies include ^{14}C, and the U-series daughter products ^{231}Pa and ^{230}Th (Henderson 2002). These have half-lives of less than 100 kyr (e.g. Edwards *et al.* 1997) and are really only applicable to the Pleistocene.

Sea ice. IP25 (Belt *et al.* 2007) is a proxy for palaeo-sea-ice distribution based on the distribution of C25 highly branched isoprenoid alkenes biosynthesized by diatoms living in sea ice, notably genera of Haska and Rhizosolenia (Belt *et al.* 2006). The technique has so far been demonstrated in the Canadian Arctic for the Holocene (Belt *et al.* 2007).

The carbon cycle. Understanding and quantifying the carbon cycle is key to determining the controls on past atmospheric CO_2 (Henderson 2002). For long timescales, i.e. longer than a million years, the main control on atmospheric CO_2 is the balance between volcanogenic gas output and net consumption by silicate weathering and carbonate deposition on the seafloor (Walker *et al.* 1981). For shorter timescales, the oceans (e.g. Henderson 2002), terrestrial vegetation (e.g. Schimel 1995) and soils (e.g. Schlesinger & Andrews 2000) are likely to be the main controls. For example, the oceans contain 50 to 60 times as much CO_2 as the atmosphere (e.g. Henderson 2002; Barker *et al.* 2003). Quantifying, for example, the influence of the oceanic carbon cycle on atmospheric CO_2 must take into account the four species of dissolved inorganic carbon, i.e. dissolved CO_2, undissociated carbonic acid, bicarbonate ion and carbonate ion, by assessing the factors controlling the concentration of these species – in particular, productivity, nutrient utilization, carbonate alkalinity/weathering fluxes, and pH (Henderson 2002). Even in the modern era, the role of soils and terrestrial vegetation remains uncertain (House *et al.* 2003). This paper will deal with the role of soils and vegetation in the section on lithological proxies below.

Palaeoproductivity. Biological productivity has the effect of removing carbon to depth from the surface oceans and the atmosphere. Biogenic barite has traditionally been used as a palaeoproductivity proxy because it sinks through the water column and enters the sedimentary record without much dissolution (e.g. Gingele & Dahmke 1994). New palaeoproductivity proxies for the Pleistocene include the difference in solubility of Th, Pa, and Be (e.g. Chase *et al.* 2002), and sedimentary U concentration (e.g. Francois *et al.* 1997). Isotopic fractionation of transition metal isotopes by biological activity appears to select for lighter values (Zhu *et al.* 2002). The $\delta^{66}Zn$ ratio of manganese crusts shows increasingly positive values with increasing bioproductivity (e.g. Maréchal *et al.* 2000). Fe isotopes (e.g. Beard *et al.* 1999) have proved so far ambiguous (e.g. Zhu *et al.* 2000). Zn/Si ratios in deep-sea hexactinellid sponges increase with increasing rain of carbon as particulate matter from shallower oceanic levels, reflecting increased bioproductivity (Ellwood *et al.* 2004). Mo/Al ratios in black shales provide a palaeoproductivity proxy with a particularly long time range (e.g. Wilde *et al.* 2004). Rare earth element distribution coefficients in planktonic foraminifera also show promise as palaeoproductivity proxies (e.g. Haley *et al.* 2005). However, a recent intercomparison of palaeoproductivity proxies, particularly Ba, has shown some significant interpretation problems (Averyt & Paytan 2004).

Nutrient utilization. Water upwelling from the deep ocean carries with it nutrients and dissolved inorganic carbon. In areas of lower bioproductivity, some of this carbon is returned as CO_2 to the upper oceans and atmosphere. This has been particularly true in the Southern Ocean (e.g. Sigman 2000). The three major biolimiting nutrients are phosphate, nitrate and silicate (Henderson 2002), with an increasing role being recognized for iron (e.g. Kumar *et al.* 1995; Noiri *et al.* 2005). Phosphate is not well-preserved in the sedimentary record, but its behaviour is tracked well by Cd, which substitutes into calcite (Henderson 2002). Cd/Ca ratio in foraminifera allows reconstruction of phosphate utilization (e.g. Rickaby & Elderfield 1999). A recent new proxy for phosphate utilization is Cd/P ratio in phytoplankton particulate matter (Elderfield & Rickaby 2000; Cullen & Sherrell 2005). Nitrate utilization can be tracked by $\delta^{15}N$, which increases in organic matter with increasing nitrogen uptake (Sigman *et al.* 1997; Robinson *et al.* 2005). This proxy has been demonstrated also to work for the Mesozoic (Jenkyns *et al.* 2001). For silicate, biogenic opal (e.g. diatom skeletons) preferentially incorporates the light isotope of silicon, ^{28}Si, resulting in increasingly positive $\delta^{30}Si$ values with increasing silicic acid utilization (De La Rocha *et al.* 1998; Reynolds *et al.* 2006). Rare earth element distribution coefficients in benthic foraminifera also show promise as nutrient utilization proxies (e.g. Haley *et al.* 2005).

Carbonate alkalinity/weathering fluxes. Carbonate alkalinity (the average concentration and distribution of carbonate and bicarbonate ions) determines the chemical speciation of carbon in the upper ocean and therefore the amount of CO_2 that can be drawn into the surface ocean from the atmosphere (Henderson 2002). Rivers constitute the major delivery system for alkalinity (generally bicarbonate ion) and weathering fluxes (most importantly Ca^{2+}) to the oceans (e.g. Raymond & Cole 2003). Foraminiferal Ba/Ca ratios have been used as a proxy for past alkalinity distributions (Lea 1993) although the relationship between the Ba and alkalinity cycles is complex (Rubin et al. 2003). $^{87}Sr/^{86}Sr$ in carbonate (e.g. Dessert et al. 2001) and $^{187}Os/^{186}Os$ ratios in clastic sediments (e.g. Cohen et al. 2004) measure weathering flux from continental sources, but mainly pick up periods of high-rate continental weathering (e.g. Ravizza et al. 2001). From marine sediments, the $^{87}Sr/^{86}Sr$ curve is particularly useful as a weathering proxy, and, with biostratigraphical context, a powerful estimator of numerical age (Francois & Walker 1992; McArthur et al. 2001). Ge/Si ratios in diatom silica (opal) also show some promise as continental weathering proxies (e.g. Munhoven & Francois 1996; Jones et al. 2002), although temperature-dependent effects in the oceans need to be accounted for (e.g. Hammond et al. 2004). Hf and Nd isotope ratio time trajectories in manganese crusts show relative deviations during glacial periods that have been interpreted as a mechanical weathering proxy (van de Flierdt et al. 2002). Overall, good proxies for global oceanic alkalinity and weathering are lacking (Henderson 2002).

pH. Oceanic pH provides insights into how the carbonate chemistry of the oceans, including depth to lysocline, has changed through time (e.g. Sanyal et al. 1995). With increasing oceanic pH, concentration of the chemical species $B(OH)_3$ reduces and $B(OH)_4^-$ increases (e.g. Vengosh et al. 1991; Zeebe 2005). Isotopic fractionation between these two species results in an up to c. 20‰ isotopic shift in $\delta^{11}B$ which is recorded by foraminiferal carbonate (e.g. Vengosh et al. 1991; Zeebe 2005). This proxy has been applied to other carbonate reservoirs (e.g. Honisch et al. 2004) and has been applied over a long timespan, even back to the Silurian (e.g. Joachimski et al. 2005), although reliable estimates only extend back to the Palaeocene (Pearson & Palmer 2000). However, several parameters of the boron/borate/boric acid system need to be better known before long-term estimates can be used with confidence (e.g. Pagani et al. 2005a). Palaeo-pH over Phanerozoic timescales has also been estimated by Royer et al. (2004) based on the calcium-ion concentration of seawater and modelled atmospheric CO_2 concentrations.

Atmospheric CO_2. Atmospheric CO_2 concentrations that pre-date the c. 800 kyr direct record provided by ice cores (e.g. Siegenthaler et al. 2005) are of extreme importance for palaeoenvironmental reconstruction. Of oceanic proxies, $\delta^{13}C$ of organic materials has been particularly successful, and has recently been refined with studies that focus on $\delta^{13}C$ derived from a single group of organisms, for example alkenones (e.g. Pagani 2002), rather than total marine organic material (Henderson 2002). The alkenones record extends back to the Eocene–Oligocene boundary (e.g. Pagani et al. 2005b). Atmospheric CO_2 from $\delta^{13}C$ of other organic materials, including terrestrial materials such as fossil wood (Hesselbo et al. 2000) has been demonstrated back to the Jurassic. $\delta^{13}C$ of carbonate, including both marine (Buggisch 1991), freshwater (Yemane & Kelts 1996) and pedogenic (Royer et al. 2001), has also been used as a proxy for atmospheric and oceanic carbon source and for rates of carbon burial (e.g. Schouten et al. 2000; Strauss & Peters-Kottig 2003). Oceanic pH has also been used as a proxy for atmospheric CO_2 concentration, although this requires assumptions to be made about past dissolved inorganic carbon, Ca flux to the oceans and alkalinity (Henderson 2002; Pearson & Palmer 2002). By this method, Pearson & Palmer (2000) and Demicco et al. (2003) have reconstructed atmospheric CO_2 concentration back 60 Ma. Recently, the technique has been demonstrated for Neoproterozoic carbonates (Kasemann et al. 2005).

Lithological proxies

Sedimentary rocks are the most venerable palaeoclimate proxies in terms of when they were first used (e.g. Lyell 1830; Croll 1867) and they offer palaeoclimatic information from some of the earliest periods of Earth history (e.g. Moore et al. 2001). Köppen & Wegener (1924) made the earliest attempt at quantitative use of sedimentary facies for palaeoclimate reconstruction. The 1960s saw a period of critical appraisal of the value of sedimentary facies as palaeoclimate indicators (e.g. Nairn 1961; Van Houten 1961; Smith 1963). Since the early 1970s, the focus has been increasingly on chemical or isotopic components of the facies rather than interpretation of the facies themselves, most particularly for carbonate sediments, although they continue to be important because of the time range over which they are useful. Sellwood et al. (1993) reviewed the sedimentary facies that are key palaeoclimatic indicators.

Carbonate rocks. Marine carbonate facies are some of the most important palaeoclimate indicators. Sellwood et al. (1993) distinguished between temperature-related facies changes, i.e. related to cool or warm water, and carbonate structures, such as reefs. Modern-day warm-water shelf carbonates form between 30° S and 30° N and include reef-building corals, codiacian algae, and ooids, aggregates and pellets. Temperate-water carbonates, which extended as far as 50° N during the Cretaceous, are characterized by benthic foraminiferans, molluscs, bryozoans, barnacles and calcareous red algae, along with ahermatypic (non-reef-forming) corals. The latitudinal distribution of shelf carbonates has varied through the Phanerozoic, although total area appears to have been relatively invariant (Kiessling *et al.* 2003). Reefs and carbonate build-ups are interpreted as indicating the presence of former warm seaways (Sellwood *et al.* 1993), although the communities of organisms responsible have varied through the Phanerozoic (e.g. Braga & Aguirre 2001; Edinger *et al.* 2002; Leinfelder *et al.* 2005). Sellwood *et al.* (1993) gave the poleward limit of reefs as greater than 30° away from the equator. In the interim, it has been recognized that deep-water, cold-temperate reefs can form at latitudes of up to 65° N (e.g. Freiwald *et al.* 1997; Freiwald & Wilson 1998).

Evaporites. Evaporites are the second major facies treated by Sellwood *et al.* (1993) who defined them as forming anywhere on the Earth's surface where evaporation exceeds precipitation and/or rate of water inflow. Ziegler (2003) added that this must occur under the descending limbs of Hadley cells, today centred between 10° and 40° north or south. However, it is worth noting that hypersaline brines form in Antarctic dry valleys today (Doran *et al.* 2003), at high latitudes and very low temperatures. The earliest evaporites are Archaean in age (e.g. Zharkov 2005).

Storm deposits. Storm deposits are another important palaeoclimatic indicator with high preservation potential (Sellwood *et al.* 1993). Tempestites (e.g. Myrow & Southard 1996; Mohseni & Al-Aasm 2004), other storm deposits (e.g. Hentschke & Milkert 1996; Chaudhuri 2005), and windblown dust (e.g. Dodonov & Baiguzina 1995; Clemens 1998; Qin *et al.* 2005) can indicate changing storminess, and, with modelling, even show the geographical distribution of storm belts through time (Agustsdottir *et al.* 1999). Palaeowind directions are also derivable in some cases (e.g. Allen 1996; Pochat *et al.* 2005).

Glacial sediments. Glacial sediments are archetypal indicators of cold palaeoclimates (Agassiz 1828; Darwin 1842). The direct products of glaciation, deposited in the immediate glacial environment, are generally unsorted diamictons (or, when lithified, diamictite), consisting of rounded granule to large boulder clasts, matrix supported in clay-silt. These are often referred to as boulder clay or till/tillite (Sellwood *et al.* 1993) and are in many cases intercalated with clast-supported conglomerate lenses. In the older record, tillites/glacial diamictites need careful interpretation to avoid confusion with diamictites resulting from meteorite impact (e.g. Mory *et al.* 2000), subaerial debris flows (e.g. Blair 1999), or non-glacially derived sub-aqueous gravity flows (e.g. Eyles 1993), but reliable records extend back to the Archaean (e.g. Young *et al.* 1998). Fluvial systems driven by glacial meltwater produce outwash facies (e.g. Visser 1997; O'Brien *et al.* 1998), which can feed into glaciomarine systems including submarine fans. Glaciers feeding directly into the sea, with or without intervening ice shelfs, also produce distinctive glaciomarine deposits, commonly with dropstones from floating ice or icebergs (e.g. Bennett & Doyle 1996; Visser 1997; Price 1999). More distally, marine sequences can preserve layers of ice-rafted debris, shed from more far-travelled floating ice (e.g. Keany *et al.* 1976; Isbell *et al.* 2001). For periglacial environments, pre-Late Cenozoic evidence is sparse, but there are records from the Neoproterozoic (Williams 1998), and Mesozoic (Constantine *et al.* 1998).

Glendonites. The questions raised by Sellwood *et al.* (1993) over the palaeoclimatic significance of glendonite carbonate nodules have largely been resolved and their status as indicators of cold subaqueous environments confirmed (e.g. Swainson & Hammond 2001). Glendonite carbonate nodules are distinctive pseudomorphs after ikaite, a low-temperature, hydrous form of calcite that may be associated with methane hydrates (Greinert & Derkachev 2004). The precursor ikaite forms under near-freezing conditions at moderately elevated hydrostatic pressure (Swainson & Hammond 2001). Glendonites are characteristic of glaciomarine sediments throughout the Neoproterozoic and Phanerozoic (e.g. Price 1999; McLachlan *et al.* 2001; Alley & Frakes 2003; James *et al.* 2005).

Subglacial volcanic deposits. The deposits of subglacial volcanic eruptions, which include lavas overlying or intercalated with glacial diamicton, and intercalated or laterally associated with hyaloclastite breccia deltas (e.g. Smellie *et al.* 2006), indicate the presence of former ice caps and ice sheets (Smellie 2000). These have been described from Antarctica (e.g. Smellie & Skilling 1994; Smellie *et al.* 2006) and Iceland (e.g. Schopka *et al.* 2006). They constitute a new quantitative proxy for the presence or absence of ice caps (e.g. Schopka *et al.* 2006) and palaeo ice sheet thickness (e.g. Smellie *et al.* 2006), complementing nunatak and trimline studies (e.g. Rae *et al.* 2004; Paus *et al.* 2006). Zeolite compositions from lavas can also be used to distinguish between marine and freshwater/glacial eruptive environments (Johnson & Smellie in press).

Coal, lignite and peat. Coal, lignite and peat deposits are well-recognized indicators of terrestrial humidity (Sellwood *et al.* 1993), i.e. areas where precipitation exceeds evaporation (e.g. Parrish *et al.* 1982; Hallam 1985). A combination of tectonic (e.g. McCabe 1991), biological (e.g. McCabe & Shanley 1992), eustatic (e.g. Staub 2002) and climatic (e.g. McCabe & Parrish 1992) factors is required for the preservation of coal- or lignite-forming mires. There are two main types (Moore 1995): rain-fed mires, which form in humid maritime climates at high–mid-latitudes today; and flow-fed mires, which currently form at low latitudes. The latitudes of coal formation may, however, have been different in the past (Crowley & North 1991), although, for the Palaeozoic, for example, both low-latitude flow-fed mires (e.g. Edwards 1998) and high-latitude, rain- and flow-fed mires (e.g. Michaelsen & Henderson 2000) have been recognized.

Fusain. Wildfires (Harris 1981; Sellwood *et al.* 1993; Glasspool 2000) are represented in the geological record by the presence of fossil charcoal, predominantly fusain (Scott 1989; Scott *et al.* 2000), in sedimentary rocks, and are indicators of a rapid growing season punctuated by periods of drought (Finkelstein *et al.* 2005) terminated with thunderstorms (Edwards 1984). The lack of a more complete treatment of fire through the Phanerozoic, as highlighted by Sellwood *et al.* (1993), has since been addressed by the review of Scott (2000). Late Carboniferous wildfires in tropical lowland peats (Scott 2000) may reflect elevated levels of atmospheric O_2 at that time (Berner *et al.* 2003).

Palaeosols. Palaeosols provide important palaeoclimatic information, in particular palaeoprecipitation (e.g. Sellwood *et al.* 1993; Retallack 2001). Palaeosols represent intervals or areas of low or no sedimentation. Their mineral and chemical composition reflects the interaction between their source terrigenous clastic sediments and the processes of weathering, which can be physical, chemical and biological. Sellwood *et al.* (1993) identified five different, major, climatically significant palaeosol types. Laterites and bauxites they linked to humid tropical climate with a long dry season. They distinguished between pedogenic and groundwater laterites, the latter of which occur in low land and

coastal settings, deriving their iron from groundwater. Sellwood *et al.* (1993) pointed out that although most bauxites occur with laterites, some may form in depressions in karst. Price *et al.* (1997a) modelled the occurrence of bauxites and showed that they require mean annual temperatures greater than 22 °C or 23 °C and high precipitation for at least six months of the year. Clay-rich vertisols, with characteristic internal features and microstructures, were linked by Sellwood *et al.* (1993) with exclusively warm temperate to tropical climates with four to eight dry months each year, in some cases associated with playas in otherwise arid or semi-arid regions. However, recent data suggest that they can also form in humid climates (e.g. Nordt *et al.* 2004). Chemical analyses of vertisol profiles provide palaeoprecipitation proxies (e.g. Stiles *et al.* 2001; Driese *et al.* 2005).

Calcretes and dolocretes, calcium or magnesium carbonate-rich soils respectively, are also a marker for arid or semi-arid areas, both cold (e.g. Lauriol & Clark 1999; Rowe & Maher 2000) and warm (e.g. Jimenez-Espinosa & Jimenez-Millan 2003), and are often found in association with zones of gypsum precipitation (gypcretes). Alonso-Zarza (2003) has pointed out that there is a continuum between calcretes and palustrine (swamp or marsh) carbonates, and that the latter can form rapidly and under much more humid conditions than calcretes, requiring care when making palaeoenvironmental interpretations. This is particularly important when determining palaeoprecipitation via transfer functions derived from the depth to the nodular, pedogenic carbonate horizon (Retallack 2005).

Cementation of soil, or the underlying saprolite, by secondary silica forms silcretes (e.g. Webb & Golding 1998). Since the work by Sellwood *et al.* (1993), silcretes have been identified forming today in central Australia. Although Webb & Golding (1998) suggested an association with long-term aridity and seasonally high evaporation, allowing silica to precipitate from highly saline groundwaters, they argue that silcretes still have no clear climatic association. Recent work by Ullyott & Nash (2006) and Alexandre *et al.* (2004) suggested that silcretes may also form in cool and wet climates.

The final category of palaeosol identified by Sellwood *et al.* (1993) are podzols (spodosols of Mack *et al.* 1993). These form in cool wet environments where humus accumulates, promoting acidic conditions (e.g. Bonifacio *et al.* 2006). Sellwood *et al.* (1993) pointed out, however, that podzols may form under warmer climatic regimes in well-drained siliceous substrates (e.g. Van Niekerk *et al.* 1999).

Mack & James (1994) proposed a simpler scheme using the classification scheme of Mack *et al.* (1993), and based on a theoretical model of four palaeoclimatic palaeosol associations. They defined: (1) a wet equatorial zone, characterized by oxisols (laterites and bauxites), with secondary clay-rich argillisols and humic histosols (including coal); (2) a subtropical dry zone containing calcisols (calcretes and dolocretes) with subsidiary gypsisols (gypcretes) and vertisols; (3) a moist mid-latitude zone with argillisols, spodosols (rich in organic matter and iron/aluminium oxides; common in coniferous forests today and equivalent to podzols), gleysols (waterlogged soils) and histosols; and (4) a polar zone with gleysols and protosols (regolithic soils).

Clay mineralogy.   Of parallel importance to soils, the distribution in oceanic sediments of clay minerals will reflect weathering processes and soil development in adjacent continents, providing an indirect record of terrestrial climate (e.g. Chamley 1989; Sellwood *et al.* 1993; Chamley 1998; Thiry 2000). A recent review by Thiry (2000) highlights the difficulties in extracting a climate signal, in particular the time lag introduced by the time it takes for soils to form, and biases in the signal that result from, for example, the longevity of kaolinite deposits (indicative of tropical wet climates). Differential settling of clay particles between proximal and distal settings may also introduce a bias in the signal (e.g. Simkevicius *et al.* 2003). Nevertheless, for example, the declining proportions of crystallized smectite and chlorite, and increasing illite in marine sediments, have been used to trace the transition from humid to subpolar and polar conditions in the high southern latitudes (e.g. Ehrmann *et al.* 2005), and characteristic proportions of kaolinite, smectite, chlorite and illlite indicate warm equable Late Triassic and humid Early Jurassic climates at mid-northern latitudes (Ahlberg *et al.* 2003). In terrestrial settings, Ballantyne (1994) suggested that low gibbsite concentration of soils in formerly glaciated areas is a proxy for glacial erosion, which has recently been confirmed by cosmogenic isotope dating (Ballantyne *et al.* 2006).

Sortable silt.   This is a relatively new proxy for palaeocurrent speed and rates of deep-water flow (e.g. McCave *et al.* 1995). It uses the 10–63 μm fraction of marine sediment because this fraction is most sensitive to sorting in response to hydrodynamic processes and its properties can be used to infer current speed (McCave *et al.* 1995). The proxy works well for the Cenozoic (e.g. Pfuhl & McCave 2005), where contourite drifts have been identified from seafloor topography and geographic association, but should, in theory be applicable to the entire interrogateable seafloor record.

Aeolianites.   The final class of palaeoclimate indicator facies treated by Sellwood *et al.* (1993) was aeolianites (wind-blown sediments). As Sellwood *et al.* (1993) noted, the simple association between these deposits and hot arid climates has been seriously questioned and their main palaeoclimatic significance is as palaeowind indicators (e.g. Allen 1993; Adams 2003; Segalen *et al.* 2004; Le Guern & Davaud 2005). Fine-grained aeolianites, such as loess deposits and dust (for review see Kohfeld & Harrison 2001), are important in that they provide the only proxy for atmosphere palaeocirculation (e.g. Henderson 2002). Palaeocirculation information can be derived from interpretation of temporal variations in source region from the chemical and mineralogical composition of dust (e.g. Nakai *et al.* 1993). In addition to providing palaeowind and palaeocirculation data (e.g. Sun *et al.* 2004), loess deposits also indicate cold, arid periglacial climates (Lagroix & Banerjee 2002; Sun & An 2005).

Palaeontological proxies

The empirical use of fossils and fossil assemblages as palaeoclimate indicators can be traced back at least as far as Lyell (1830) and their value was even recognized in classical times (Imbrie & Newell 1964). Quantifying former climates using biological data is based on three key assumptions (Woodcock 1992): (1) that climatic factors limit the occurrences or associations of taxa; (2) that taxa have characteristics (such as leaf shape or stomatal density) that respond to climate; and (3) that climate causes variance in relative frequency of taxa. It is useful to subdivide studies that use biological data into taxonomic and morphological types (Wing & Greenwood 1993) based on whether or not they address assumption (2) above. Quantitative, proxy-based palaeoclimate reconstruction based on fossil data began for terrestrial environments in the 1950s, based on pollen analysis (Faegri & Iverson 1950), and for the marine environment in the early 1970s with the development of micropalaeontology-based transfer functions by Imbrie & Kipp (1971). Quantitative techniques relate biological parameters to changes in environment variables such as temperature, pH or salinity. These parameters can either be measured of individuals (morphological techniques), or of a faunal assemblage (taxonomic techniques).

Taxonomic methods.   For the terrestrial realm, the primary taxonomic methods for palaeoclimate reconstruction use assemblages of pollen (e.g. Birks 1981) or other palynomorphs (e.g. Stanley 1970; Loboziak *et al.* 1989). Pollen analysis, using climatic amplitude (e.g. Jimenez-Moreno *et al.* 2005) and mutual climatic range (e.g. Klotz *et al.* 2006) techniques, provides data on palaeotemperature and palaeoprecipitation (e.g. Fauquette & Bertini 2003) as far back as the Eocene (e.g. Harrington 2004). Taxonomic methods based on plant macrofossils (e.g. Baghai & Jorstad 1995), vertebrates (Markwick 1998; Friedman *et al.* 2003), molluscs (e.g. Moine & Rousseau 2002) and arthropods (e.g. Pilny *et al.* 1987; Williams & Eyles 1995) provide palaeotemperature data, and have been demonstrated for the Cretaceous. Taxon analysis of chironomid midges provides lake palaeosalinity data back to the Late Pleistocene (e.g. Walker 1991) and insect assemblages have been proposed as a marker for low salinity for the Early Cretaceous (Coram & Jarzembowski 2002). More sophisticated palaeoclimatic treatments of terrestrial faunal assemblages use herbivorous mammals and their feeding habits. These include ecological diversity spectra (e.g. Andrews *et al.* 1979; Fernandez & Pelaez-Campomanes 2005), which quantify a range of parameters including temperature, humidity/aridity and annual precipitation potentially back to the Miocene, and cenograms (e.g. Legendre 1986; Rodriguez 1999; Legendre *et al.* 2005), which provide information on temperature and humidity/aridity back to the Eocene. The nearest living relative method (e.g. MacGinitie 1941), which blurs the boundary between taxonomic and morphological techniques, makes palaeoclimatic inferences based on fossil groups and assemblages by using the ranges and climatic responses of their modern descendants or sister groups (Wing & Greenwood 1993), and is particularly good for the Cenozoic. The nearest living relative method provides palaeotemperature data from both plants (Poole *et al.* 2005; Wang *et al.* 2005) back to the Jurassic, and animals (e.g. Hutchison 1982; Moe & Smith 2005) back to the Palaeocene.

For the oceanic realm, taxonomic methods of deriving palaeoclimate parameters, such as temperature and salinity, are most commonly derived via transfer functions, which were first developed by Imbrie & Kipp (1971). These can be calculated from marine (e.g. Sejrup *et al.* 2004) and lacustrine (e.g. Hausmann & Kienast 2006) microfossil assemblages, and are commonly applied to Pliocene or younger assemblages (e.g. Andersson 1997), although for palaeoproductivity they extend back to the Palaeocene (Siesser 1995). Transfer functions have evolved since the principal components regression work of Imbrie & Kipp (1971) and now include weighted averaging and weighted averaging least squares methods (Sejrup *et al.* 2004; Hausmann & Kienast 2006), as well as artificial neural networks (Malmgren & Nordlund 1997; Peyron & De Vernal 2001) and maximum likelihood and Bayesian techniques (e.g. Robertson

et al. 1999). The applications of transfer function techniques have broadened away from the planktonic foraminifera and radiolaria of Imbrie & Kipp (1971). They now include, in the oceans, benthic foraminifera (e.g. Sejrup *et al.* 2004), diatoms (e.g. Zielinski *et al.* 1998), ostracods (e.g. Brouwers *et al.* 1991), dinoflagellates (e.g. Peyron & De Vernal 2001), and in lakes, chironomid midges (e.g. Korhola *et al.* 2001) and diatoms (e.g. Roberts & McMinn 1999). Parameters estimated include, oceanic palaeoproductivity (e.g. Siesser 1995), ocean bottom-water palaeotemperatures (e.g. Brouwers *et al.* 1991), palaeosalinity (e.g. Sejrup *et al.* 2004), sea-surface palaeotemperatures (e.g. Andersson 1997; Malmgren & Nordlund 1997; Zielinski *et al.* 1998), and seasonal extent of former sea-ice cover (e.g. Peyron & De Vernal 2001). In lakes, palaeonutrient utilization (e.g. Hausmann & Kienast 2006), surface palaeotemperature (Korhola *et al.* 2001) and palaeosalinity (Roberts & McMinn 1999) have been estimated. Non-transfer-function-based techniques on diatoms have used faunal assemblages (Gersonde & Zielinski 2000) and modern analogue techniques (Crosta *et al.* 1998) as proxies for sea-ice extent back to the Pleistocene. Similarity maximum modern-analogue techniques on foraminiferal assemblages from core tops have also been used as a proxy for sea-ice extent (Sarnthein *et al.* 2003), again back to the Late Pleistocene. For the older, pre-Cenozoic, record, other non-transfer-function-based methods use faunal assemblages of marine molluscan macrofossils to estimate palaeotemperature (e.g. Kafanov & Volvenko 1997) and palynomorphs, such as acritarchs, for palaeoproductivity (e.g. Vecoli & Le Herisse 2004) with applications back to Ordovician times.

Morphological methods. Morphological methods generally centre around leaf margin analysis, beginning with the work of Bailey & Sinnott (1916) and culminating in the work of Wolfe (1993) on the climate-leaf analysis multivariate program, or CLAMP. This is widely used as a palaeotemperature proxy (e.g. Gregory-Wodzicki 2000; Kennedy 2003; Spicer *et al.* 2005) and has been extended back to the Permian (Glasspool *et al.* 2004). The width of growth rings in fossil wood is also an important morphological technique for palaeotemperature and palaeoprecipitation (e.g. Fritts 1976; Creber & Chaloner 1985; Francis 1986; Poole *et al.* 2005). Although results need to be assessed in terms of taxonomic analysis (Brison *et al.* 2001), the technique has been demonstrated to be effective from Carboniferous times to Recent (e.g. Falcon-Lang 1999; Morgans 1999; Francis & Poole 2002). Measuring maximum latewood density of tree rings gives the ability to measure hemispheric variations in temperature (Briffa *et al.* 2004), although this has only been applied to historically recent material. The density of stomata in leaf cuticle has been used as a proxy for palaeo-CO_2 concentration (Woodward 1987; Woodward & Bazzaz 1988). This has been shown to be effective back to the Carboniferous (e.g. Otto-Bliesner & Upchurch 1997; McElwain *et al.* 1999; Retallack 2002; Thorn & DeConto 2006). For the oceans, palaeo-carbonate ion concentration, which determines the water depth at which all calcite has dissolved from the sediment or lysocline (e.g. Ridgwell 2005), has been determined by estimating the degree of dissolution of individual foraminifera of a particular size (e.g. Lohmann 1995). However, the situation is more complicated in glacial periods (Broecker & Clark 2001) and dissolution effects at the sediment–water interface cannot be discounted (e.g. De Villiers 2005).

Modelling

Palaeoclimate modelling falls into two main categories. The main, and most significant, category is by the generation and interpretation of climate analogues. These can take the form of conceptual models (Parrish 1993), or be carried out using numerical, computer-based global circulation models (GCMs) or Earth models of intermediate complexity (EMICs) (for a recent review, see McGuffie & Henderson-Sellers 2001). These models attempt to generate analogues for climate over a range of spatial and temporal scales up to and including the complete globe. The second category is what are called geochemical (Royer *et al.* 2001), mass-balance (e.g. Beerling 1999), or trend models (e.g. Richards 1998). These use palaeoclimate proxies or model outputs to reconstruct the evolution of a particular atmospheric or oceanic chemical species, such as oxygen or carbon dioxide, over a range of geological time, up to and including the entire geological record.

Climate analogue modelling. It was recognized in the early 19th century that former global climates differed from today's (Agassiz 1828; Darwin 1842). Following the work of Tyndall (1861) and Arrhenius (1896), a key role for CO_2 was suspected from the beginning (Chamberlin 1897, 1899). The first empirical palaeoclimatic reconstructions for the Phanerozoic were carried out by Köppen & Wegener (1924), who used a simple zonal climate model and the distribution of palaeoclimatic indicator facies such as coal or glacial sediments in a continental drift framework to refine the fit of the continents. These applications of palaeoclimate studies based on simple zonal schemes continued into the 1960s (e.g. Nairn 1961). It was only with

the advent of palaeomagnetism in the 1950s and plate tectonics in the 1960s (summarized in Irving 2005) that continental palaeolatitudes could be independently fixed. This freed the geological record to provide ground truth for increasingly empirical, conceptual (Parrish 1993) and numerical (Schneider & Dickinson 1974) palaeoclimate models. The earliest numerical models were relatively simple analogues of Late Pleistocene glacial climates (Alyea 1972; Gates 1974). These quickly moved to modelling of palaeoclimates with global circulation models (GCMs) initially of Late Pleistocene glacial climates (Gates 1976), and at the end of the decade the CLIMAP project saw the first attempts to provide model boundary conditions and to evaluate climate models with proxy data (CLIMAP 1981). From the 1980s, palaeoclimate modelling was extended first to the Mesozoic (Barron *et al.* 1981; Kutzbach & Gallimore 1989; Chandler *et al.* 1992; Valdes & Sellwood 1992) and then the Palaeozoic (Crowley & Baum 1992; Kutzbach & Ziegler 1993; Otto-Bliesner 1995; Gibbs *et al.* 1997).

Although the first coupled ocean-atmosphere GCM was realized in the 1960s (Manabe & Bryan 1969), it was not until the 1990s that these were applied to palaeoclimate studies (Stocker *et al.* 1992; Cubasch *et al.* 1997; Schiller *et al.* 1997) initially for the early Holocene and the last interglacial (Eemian) (Texier *et al.* 1997; Montoya *et al.* 1998). Since 2000, these GCMs have been increasingly applied to the older Cenozoic and Mesozoic (Otto-Bliesner *et al.* 2002; Haywood & Valdes 2004; Haywood *et al.* 2004) (Fig. 2). In the 1990s, multiple or ensemble runs (e.g. Cubasch *et al.* 1994; Hoar *et al.* 2004) and asynchronous coupling (e.g. Liu *et al.* 1999; Dutton & Barron 2000), where the outputs of one EMIC or GCM is used to force climate parameters in a second EMIC or GCM, were used to model climate behaviour.

Since the late 1990s, the emphasis has been on model 'validation' or 'evaluation' (the second term is preferred because it does not confer any sense of 'approval' (Kohfeld & Harrison 2000). This uses multiproxy datasets (e.g. Price *et al.* 1995, 1997*b*; Sellwood & Valdes 1997; Sellwood *et al.* 2000; Felzer & Thompson 2001; Haywood *et al.* 2004) to minimize the uncertainties inherent in numerical models (Dickinson 1989). A second trend is that models have become more sophisticated with coupling of ice-sheet (e.g. Ridley *et al.* 2005) and biome (e.g. Haywood *et al.* 2002*a*; Snyder *et al.* 2004; Brentnall *et al.* 2005) models to the coupled atmosphere–ocean GCMs.

Geochemical or mass-balance modelling. Geochemical or mass-balance modelling uses calculations of the transfer of material between the

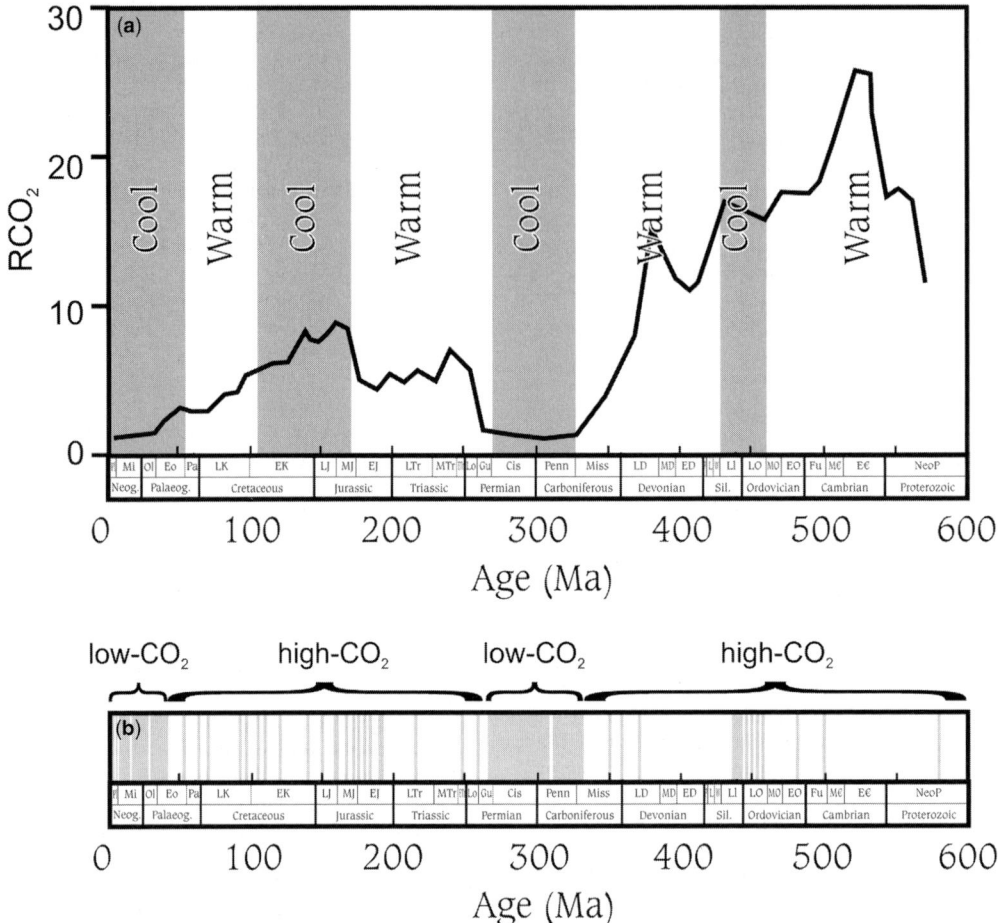

Fig. 3. (a) Palaeo-CO_2 atmospheric concentration as a multiple 'R' of pre-industrial values using the curve of Berner (2003a) superimposed on the climate modes of Frakes *et al.* (1992). (b) Intervals of glacial or cool (grey bars) and warm (white areas) climates and durations of high- and low-CO_2 modes for the Late Neoproterozoic and Phanerozoic modified from Fig. 1b shown for reference.

lithosphere, hydrosphere and atmosphere (Li 1972; Garrels & Lerman 1984; Lasaga & Berner 1998), to assess the long-term evolution of atmospheric and oceanic composition. It is by necessity generalized and the geographical patchiness and temporal incompleteness of the terrestrial and particularly the pre-Jurassic marine record must qualify its interpretation. Modelling of the long-term carbon cycle is one of the keys to understanding Phanerozoic climate change (e.g. Berner & Barron 1984; Gifford 1994; Berner 1998; Berner & Kothavala 2001). Changes in the carbon cycle can be measured directly, with varying degrees of precision, from organic carbon in sedimentary rocks or carbonate in limestone (this latter may take the form of sedimentary rock or carbonate cements in, for example, soils) (e.g. Berner 1998).

Of secondary importance is the long-term oxygen cycle (e.g. Berner & Canfield 1989; Berner 1999, 2001; Berner *et al.* 2000). Some authors argue that oxygen concentration changed substantially during the Phanerozoic, possibly peaking in the late Carboniferous (e.g. Berner & Canfield 1989; Beerling *et al.* 2002) although there are strong counterarguments that it was relatively unvarying (e.g. Lenton & Watson 2000). The long-term oxygen cycle is indirectly modelled from the carbon cycle and the sulphur cycle (e.g. Berner 2005) (the sulphur cycle is measured directly from evaporitic rocks and iron pyrites in sedimentary rocks (e.g. Francois & Gerard 1986; Railsback 1992)).

Three other cycles that are important for understanding atmospheric evolution are the nitrogen cycle (e.g. Falkowski 1997), the phosphorus cycle (Guidry & Mackenzie 2000; Lenton 2001) and the silicon cycle in the oceans (Ragueneau *et al.* 2000; Matsumoto *et al.* 2002). All three are implicated in the drawdown of carbon and its incorporation in marine sediments (e.g. Falkowski 1997; Lenton 2001; Matsumoto *et al.* 2002). One of the most spectacular successes of mass-balance modelling is in providing support for the extreme atmospheric changes associated with the Neoproterozoic 'Snowball Earth' model (Hoffman *et al.* 1998). Recent mass-balance models are sophisticated, incorporating multiple elements and both major oceanic and atmospheric cycles (e.g. Hansen & Wallmann 2003; Bergman *et al.* 2004).

Phanerozoic climate modes

Frakes *et al.* (1992) looked at climates back to 600 million years ago and identified what they termed 'climate modes' with alternating episodes of cool or warm climates (Table 1; Fig. 1a). They defined 'cool modes' as 'times of global refrigeration during which ice was present on earth'. Their 'warm modes' they defined as times 'when climates were globally warm, as indicated by the abundance of evaporites, geochemical data, faunal distributions, etc, and with little or no polar ice'. Frakes *et al.* (1992) argued that warm–cool climate mode pairs spanned intervals of approximately 150 million years: about half a galactic year. Superimposed upon these climate modes were brief intervals of contrasting climates (Fig. 1b; Table 1), which they did not consider sufficiently long to merit climate modes of their own. These are important, however, and the geological background to some will be dealt with in slightly more detail in this chapter, as they give insights into the extremes of climate change. All numerical ages quoted for chronostratigraphic units and boundaries are according to Gradstein *et al.* (2004).

Frakes *et al.* (1992) proposed eight climate modes for the Phanerozoic (Table 1; Fig. 1a). They grouped these into four warm: (i) Cambrian to Middle Ordovician; (ii) Mid-Silurian to Early Carboniferous; (iii) latest Permian to Early Jurassic; (iv) Late Cretaceous to Early Palaeogene, inclusive; and four cool: (i) Late Ordovician to Early Silurian; (ii) Early Carboniferous to Late Permian; (iii) Late Jurassic to Early Cretaceous; (iv) Mid-Palaeogene to Recent, inclusive. These are treated alternately in time-sequential order below.

Climate modes

Earliest Cambrian to Middle Ordovician warm mode. Frakes *et al.* (1992) placed the beginning of this warm mode at the end of the latest Neoproterozoic glaciation, which they assumed to be at the Precambrian–Cambrian boundary. Since then, the latest Precambrian, Ediacaran Period has been added (Knoll *et al.* 2006) post-dating the last Neoproterozoic glacial deposits, this period starting between 635 and 600 million years ago (Knoll *et al.* 2006). The Ediacaran terminates at the earliest Cambrian, now 540 million years ago (Gradstein *et al.* 2004), 30 million years later than the dates used by Frakes *et al.* (1992) from the Decade of North American Geology (DNAG) timescale of Palmer (1983). Ediacaran stratigraphic (Knoll *et al.* 2006) and stable isotope (Le Guerroue *et al.* 2006) data suggest that it post-dates major Neoproterozoic glaciation and should be included in the warm mode. Frakes *et al.* (1992) suggested that this warm mode ended prior to the beginning of the Ordovician Katian stage which is now dated at about 461 million years ago, an interpretation endorsed by Page *et al.* (2007) in this volume.

This warm mode is marked by substantial deposition of evaporites and moderately high rates of carbonate formation by Phanerozoic standards (Frakes *et al.* 1992). One notable biogeochemical event during this warm mode is the deposition of substantial phosphorite deposits around the time of the Ediacaran–Cambrian boundary (Cook 1992). This accumulation of phosphorite differs from other Phanerozoic accumulations in that it does not appear to be associated with oceanic upwelling and may be related to a change in ocean chemistry associated with a rise in atmospheric oxygen and the formation of the first hard skeletons.

At least three brief intervals of probable cooler climate punctuate this warm mode. A brief glacial episode is recognized in the Ediacaran at about 580 Ma (c1, Fig. 1b) from $\delta^{13}C$ values in marine carbonates (Le Guerroue *et al.* 2006). Another short-lived episode of cool, possibly glacial climate is identified in the Steptoan of the Late Cambrian (e.g. Saltzman *et al.* 2000), at *c.* 494 Ma (c2, Fig. 1b), again from a positive excursion of up to 5‰ in brachiopod carbonate (Cowan *et al.* 2005) immediately followed by a major global sea-level drop (Glumac & Spivak-Bimdorf 2002). Although Frakes *et al.* (1992) observed that unequivocal Cambrian or Early Ordovician glacial sediments are rare, Early Ordovician, Tremadocian, *c.* 480 Ma (c3, Fig. 1b), glacial deposits have been identified in the Meguma Zone of Nova Scotia (Schenk 1995).

Late Ordovician to Early Silurian cool mode. Suggestions for the duration of this cool mode have varied since Frakes *et al.* (1992). From a review of Phanerozoic glacial deposits, Evans (2003) estimated that it lasted of the order of 17 million years, from the latest Ordovician (Hirnantian), *c.* 445 Ma until the Llandovery–Wenlock boundary at the end of the Early Silurian, 428 Ma. Royer *et al.* (2004), using pH-corrected Phanerozoic $\delta^{18}O$ trends, estimate the duration at less than 10 million years, and Brenchley *et al.* (2003) estimate that widespread ice sheets lasted less than 1 Myr. However, Page *et al.* (2007) in this volume present evidence that support the assessment of Frakes *et al.* (1992) with a beginning at *c.* 461 Ma.

Frakes *et al.* (1992) documented widespread glacial sediments over a large area of the African and South American parts of Gondwana, and in terranes probably sourced from there. Page *et al.* (2007) in this volume present sea-level data and lithological evidence from West Africa, Canada and France for two middle Caradoc and one early Ashgill cool or glacial episodes in the Late Ordovician (Katian stage) (c4–c6, Table 1 and Fig. 1b) and there is some evidence in Baltica for cool phases at these times (Kaljo *et al.* 2003). These do not appear to represent glaciation on the scale of Hirnantian and Early Silurian episodes and may indicate a gradual transition from the preceding warm mode. Data from Baltica restricts full glaciation to the latest Ordovician and Early Silurian (Jeppsson & Calner 2003; Kaljo *et al.* 2003; Legrand 2003). Four glacial episodes are recognized (Kaljo *et al.* 2003), including the latest Ordovician Hirnantian episode, at *c.* 445 Ma (c7, Fig. 1b), and Silurian glacial episodes in the Aeronian, *c.* 439 Ma (c8, Fig. 1b), Telychian, *c.* 436 Ma (c9, Fig. 1b), and earliest Wenlock (Sheinwoodian), *c.* 428 Ma (c10, Fig. 1b)). The Early Silurian is marked by a series of global oceanic changes, termed Primo and Secundo states by Aldridge *et al.* (1993), inferred to have been determined by whether deeper oceanic circulation is driven by cold, or saline, density currents (e.g. Jeppsson & Aldridge 2000). Given the relative brevity of this cool mode, it is not marked by any notable deviations from cool climate, although Secundo states in the Llandovery to earliest Wenlock interval are marked by warm high latitudes (Aldridge *et al.* 1993).

Mid-Silurian to Early Carboniferous warm mode. The latest clear evidence for Silurian glaciation comes from isotopic and sedimentological evidence in the Early Wenlock (Azmy *et al.* 1998; Kaljo *et al.* 2003), placing the start of this warm mode slightly earlier than the end-Wenlock suggested by Frakes *et al.* (1992). The end of this warm mode is open to interpretation. Frakes *et al.* (1992) placed it at the Viséan–Namurian boundary (the Namurian is now called the Serpukhovian (Gradstein *et al.* 2004)), *c.* 326 Myr ago according to Gradstein *et al.* (2004). However, Evans (2003) included the high-latitude Late Devonian,

Famennian and Early Carboniferous, Tournaisian/ Viséan glaciations in the following cool mode, which would terminate this warm mode at *c.* 375 Ma.

Based on the distribution of carbonates, evaporites and coal deposits, Frakes *et al.* (1992) argued that this warm mode demonstrated a progressive global warming to Late Devonian times and tentatively placed its end at the beginning of the Serpukhovian. Carbonates show a progressive expansion to higher and higher latitudes during this period. However, within this warm mode, there is evidence for Late Devonian glaciation at the Frasnian–Famennian boundary (Evans 2003), *c.* 375 Ma (c11, Fig. 1b), a Gondwana 'mini-glaciation' at the Devonian–Carboniferous boundary (Caplan & Bustin 1999), *c.* 360 Ma (c12, Fig. 1b) and glaciation in the Early Carboniferous (Tournaisian–Viséan boundary), *c.* 345 Ma (c13, Fig. 1b). Frakes *et al.* (1992) explained these glaciations by the passage of parts of Gondwana over the South Pole, largely confirmed by Evans (2003), i.e. high latitude glaciations not requiring any major perturbation of the Earth System. However, the pH-corrected CO_2 and palaeotemperature curve of Royer *et al.* (2004) suggest that following a palaeotemperature peak at *c.* 380 Ma (early Late Devonian), there is a steady decline into the Carboniferous, with notable steeper drops between around 375 and 360 Ma, which coincide with the Famennian (Streel *et al.* 2000), Devonian–Carboniferous boundary (Caplan & Bustin 1999) and Early Carboniferous (Evans 2003) glaciations.

Early Carboniferous to Late Permian cool mode. Frakes *et al.* (1992) placed the beginning of this cool mode in the early Serpukhovian, at *c.* 325 Ma. Evidence from South Africa and Australia supports an early glacial phase in this cool mode between 325 and 315 Ma (Evans 2003). The end of this cool mode was estimated by Frakes *et al.* (1992) to be in the Late Permian at what was the Kazanian–Tatarian boundary (now the Wuchiapingian–Changhsingian) at *c.* 254 Ma. The most recent data (Eyles *et al.* 2002, 2006) place the latest glacially influenced sedimentation in the latest Early Permian, Kungurian, at *c.* 271 Ma.

The latitudinal distribution of glacial sediments summarized by Frakes *et al.* (1992) suggests that low-latitude glaciation (35–40° from the equator) during this cool mode occurred in two main phases: the first in the late Serpukhovian to end-Moscovian (previously called late Namurian and Westphalian) and the second in the Asselian and Sakmarian of the earliest Permian (Frakes *et al.* 1992). Evans (2003) argues that this second phase was at its most intense in a brief period around 297 Ma. In the intervening Kasimovian/Gzelian

(Stephanian), glacial deposits extended no farther towards the equator than 50° (Frakes *et al.* 1992). Interpretation is made more complex because Gondwana was moving rapidly across the high southern latitudes during this cool mode (Frakes *et al.* 1992). Frakes *et al.* (1992) argued that post-Sakmarian glaciation was probably restricted to Antarctica, possibly providing a source of dropstone material to Australian and South African sedimentary basins until late in the Early Permian (e.g. Eyles *et al.* 2002, 2006; Evans 2003). Carbonates and evaporites are restricted to low latitudes during this cool mode (Frakes *et al.* 1992), with evaporites showing a retreat to more equatorial latitudes from the Viséan onwards, with carbonates showing a similar change in the mid-Serpukhovian. Both carbonates and evaporites show a recovery from the mid Early Permian onwards (Frakes *et al.* 1992).

The short-lived climatic amelioration in the Kasimovian/Gzelian (Stephanian) (e.g. Frakes *et al.* 1992; Bruckschen *et al.* 1999) merits some discussion. Pennsylvanian cycles of terrestrial to fluvial to submarine sedimentation, called cyclothems and often associated with coal formation (e.g. Heckel 1986), which are an expression of glacioeustatic sea-level changes (Veevers & Powell 1987), are linked to Milankovitch cyclicity, although there is some discussion over what periodicity may be the primary controlling one (e.g. Ross & Ross 1985; Heckel 1986; Crowley *et al.* 1993; Algeo *et al.* 2004; Driese & Ober 2005). In this scheme, the Kasimovian/Gzelian showed a shortening of duration of the major controlling cycles (Ross & Ross 1985). However, with a reduction in duration of the Pennsylvanian from 34 to 19 million years (Klein 1990; Gradstein *et al.* 2004), this simple explanation for Kasimovian/Gzelian warming, and a link to Milankovitch-paced cyclicity, was called into question. An examination of $\delta^{18}O$ records from Carboniferous brachiopods in western Europe and the former USSR (Bruckschen *et al.* 1999), coupled with the $^{87}Sr/^{86}Sr$ and $\delta^{13}C$ record, shows drops in $\delta^{18}O$ and $^{87}Sr/^{86}Sr$ during the Kasimovian/Gzelian consistent with global warming and suggests that this was quite short-lived and restricted to the Kasimovian at *c.* 307 Ma (w1, Fig. 1b).

Latest Permian to Early Jurassic warm mode. Based on the evidence for the latest ice-rafting in Australian basins (Eyles *et al.* 2002, 2006), the start of this warm mode fits best at the beginning of the Roadian at *c.* 271 Ma, which is slightly earlier than the end-Wuchiapingian (Kazanian), *c.* 254 Ma, proposed by Frakes *et al.* (1992). Frakes *et al.* (1992) suggested that this warm mode ended with the first evidence for ice-rafted debris in Jurassic sediments. These are recorded in the Middle Jurassic of northern Asia, at the Aalenian–Bajocian boundary (Chumakov & Frakes 1997), now dated at *c.* 172 Ma (Fig. 1b), although Price (1999) summarized evidence for possible glendonites in the Early Jurassic in Pliensbachian times at *c.* 190 Ma.

This warm mode is singular in that it contains the extinctions just before and at the Permian–Triassic boundary, the latter of which is the most severe in the Phanerozoic (e.g. Knoll *et al.* 1996; Hallam & Wignall 1999). These major changes in the biosphere have implications for lithology-based interpretation of early Triassic climates, for example, the 'coal gap' (Retallack *et al.* 1996), when no coals formed anywhere on the globe even in appropriate sedimentary settings, and interpretations based on palaeosol formation, i.e. spodosols were absent (Retallack 1997). Based on relatively low-precision $\delta^{18}O$ data and a restricted set of lithological indicators (including palaeosols and reef limestones), Frakes *et al.* (1992) argued that this warm mode demonstrated an early rapid rise in temperatures in the latest Permian and Early Triassic, followed by a slow cooling trend through the Triassic and Early Jurassic. The more recent, pH-calibrated $\delta^{18}O$-based palaeotemperature curve of Royer *et al.* (2004) indicates a very similar overall pattern, with a sharp rise in temperatures from late Early Permian times. Palaeotemperatures reached a peak in the early Middle Triassic, at *c.* 245 Ma, followed by a slow decline towards the Mid-Jurassic, only interrupted by a short-lived rise around the Pliensbachian–Toarcian boundary, at *c.* 183 Ma (Hesselbo *et al.* 2000; Royer *et al.* 2004).

Cooler contrasting climates punctuated this warm mode, with, in some cases, combined evidence for glaciation from isotope excursions and rapid sea-level changes. The earliest-documented probable glacial event in this warm mode occurred across the Guadalupian–Lopingian boundary in the Late Permian (Isozaki *et al.* 2006), at *c.* 260 Ma (c14 Fig. 1b). Although the Triassic appears to be devoid of cool intervals, stable isotope studies by Korte *et al.* (2005) indicated that short-lived, high-amplitude $\delta^{13}O$ excursions occurred in the Early Triassic, at *c.* 250 Ma (c15, Fig. 1b), and a positive $\delta^{18}O$ excursion occurred at the Carnian–Norian boundary, at *c.* 217 Ma (c16, Fig. 1b). In the Early Jurassic, rapid sea-level changes provide evidence of short-lived glaciation in the Late Sinemurian, *c.* 192 Ma (c17, Fig. 1b), at the Sinemurian–Pliensbachian boundary, *c.* 190 Ma (c18, Fig. 1b), in the late Pliensbachian, *c.* 184 Ma (c19, Fig. 1b), and at the end Toarcian, *c.* 176 Ma (c21, Fig. 1b), (e.g. Price 1999; Immenhauser 2005). This Early Jurassic interval is also marked by rapid temperature rise around the Pliensbachian–Toarcian boundary, *c.* 183 Ma (e.g. Pálfy & Smith 2000; Bailey *et al.* 2003), which appears to have been followed by a brief early Toarcian cool or possibly glacial episode, notionally at *c.* 182 Ma (c20, Fig. 1b) (Morard *et al.* 2003; Wignall *et al.* 2005). The steep temperature rise deviates from the long-term cooling trend demonstrated by Royer *et al.* (2004). Belemnite Mg/Ca, Sr/Ca, and Na/Ca ratios increase by a factor of between 1.7 and 2, coincident with a 3‰ negative shift in $\delta^{18}O$ and indicating an abrupt warming in northwest Europe of 6–7 °C and a more active hydrological cycle in this interval (Bailey *et al.* 2003).

Late Jurassic to early Cretaceous cool mode. According to Frakes *et al.* (1992), the start of this cool mode is relatively clearly marked by the appearance of dropstones in high-latitude Mid-Jurassic marine deposits from the Aalenian–Bajocian boundary (e.g. Chumakov & Frakes 1997), at *c.* 172 Ma (c22, Fig. 1b). However, more recent evidence could place the beginning as early as the Pliensbachian (e.g. Price 1999; Immenhauser 2005). Compared with the evidence for its initiation, the termination of this cool mode is more difficult to define. Frakes *et al.* (1992) used the latest appearance of ice-rafted debris in Australian marine sequences of the Eromanga basin in the early Albian of the Late Cretaceous, which would approximately place it at *c.* 110 Ma (Fig. 1b). However, a continuation of short-lived glacial episodes from the Mid-Jurassic through the Late Cretaceous is supported by the pH-calibrated $\delta^{18}O$-based palaeotemperature curve of Royer *et al.* (2004), who argue that the 'Late Jurassic to Early Cretaceous' cool mode is fundamentally different from the 'Early Carboniferous to late Permian' and 'Mid-Palaeogene to Recent' cool modes and is more akin to the 'Late Ordovician to early Silurian' one.

Frakes *et al.* (1992) argued that this mode was cooler than the preceding and following ones, based on the evidence for seasonal ice at high latitudes in both hemispheres, although not as cold as Palaeozoic cool modes. Based on a variety of indicators, including evidence of ice, palaeobotanical data, oxygen isotopes, clay minerals and the distribution of coals, evaporites and carbonates, they suggested that the intensity of cool climates varied through the mode. Frakes *et al.* (1992) proposed a cool Mid-Jurassic, with seasonal ice in the Bajocian and Bathonian, *c.* 167 Ma (c23, Fig. 1b), a conclusion supported by Price (1999). Frakes *et al.* (1992) suggested warming during the end-Mid-Jurassic Callovian, although new oxygen isotope data from shark-tooth enamel and ammonite migration data suggest that the end-Mid-Jurassic (late Callovian) was actually a time of glaciation,

lasting *c.* 2.6 million years (Dromart *et al.* 2003), at *c.* 161–164 Ma (c24, Fig. 1b). Glaciation at this time is also suggested by the pH-calibrated $\delta^{18}O$-based palaeotemperature curve of Royer *et al.* (2004). For the Oxfordian, Kimmeridgian and Tithonian of the Late Jurassic, Frakes *et al.* (1992) suggested the warmest climates of the mode, although Price (1999) gave evidence for cold or sub-freezing polar climates in the Tithonian, at *c.* 150 Ma (c25, Fig. 1b). The suggestion of a cooler and more humid earliest Cretaceous (Valanginian and Hauterivian) (Frakes *et al.* 1992) is consistent with the evidence for cool Valanginian climates, *c.* 140 Ma (c26 Fig. 1b), presented by Price (1999). Frakes *et al.* (1992) proposed a gradual warming trend in the later Early Cretaceous following further evidence for ice in the late Early Aptian, *c.* 120 Ma (c27, Fig. 1b), (e.g. Price 1999) towards a peak in temperatures in the Albian at the beginning of the following warm mode.

Given the variation in climates during this cool mode, it is particularly difficult to pick out climates that deviate significantly from any chosen trend. However, the Late Callovian cooling identified by Dromart *et al.* (2003) is particularly sharp (1–3 °C for lower to middle latitudes) and associated with increased drawdown of organic carbon and an abrupt fall in global sea-levels.

Late Cretaceous to Early Palaeogene warm mode. Frakes *et al.* (1992) place the beginning of this warm mode at the start of the Late Cretaceous in the mid-Albian at *c.* 105 Ma. The end of this warm mode is defined by the onset of global cooling as documented in the $\delta^{18}O$ record of marine foraminifera from about *c.* 55 Ma onwards (Frakes *et al.* 1992), which is the Palaeocene–Eocene boundary.

The status of this warm mode has been seriously questioned by Royer *et al.* (2004). Like the preceding cool mode, this warm mode shows an oscillating climate with shorter cool and warm periods, although the warm periods are substantially warmer than are those in the preceding cool mode. The mode is punctuated by a series of glacial episodes, with evidence of these from sea-level records (e.g. Miller *et al.* 2005), oxygen isotopes (e.g. Miller *et al.* 2003) and strontium (e.g. Stoll & Schrag 1996). Eustatic sea-level data (e.g. Immenhauser 2005), from the Russian and Arabian platforms, show sharp short-lived falls in the late Early and Late Albian, at *c.* 110 and *c.* 105 Ma (c28, c29, Fig. 1b), which may indicate high-latitude ice caps at that time. Similarly, eustatic sea-level data from offshore New Jersey, combined with the dates of positive excursions in the marine foraminiferal $\delta^{18}O$ record (e.g. Miller *et al.* 2005) indicates short-lived Antarctic ice-caps of the order of

40–50% Neogene volumes. The best evidence for these (e.g. Miller *et al.* 2005) is in the mid-Cenomanian, *c.* 96 Ma (c30, Fig. 1b), mid-Turonian, 93–92 Ma (c31, Fig. 1b), and at the Campanian–Maastrichtian boundary, 70.6 Ma (c32, Fig. 1b).

The temporal distribution of warm and cool or glacial episodes during this warm mode has changed substantially since the synthesis by Frakes *et al.* (1992). Frakes *et al.* (1992) suggested that the Albian was one of the warmest periods of this warm mode, along with the Coniacian and Campanian. New data suggest that the warmest period of this warm mode, and one of the warmest periods of the Phanerozoic, was during the Turonian (Wilson *et al.* 2002; Bice *et al.* 2006), and that the Albian was possibly quite cool with at least two short-lived high-latitude glaciations (e.g. Immenhauser 2005). Additionally, the Turonian episode of maximum warmth is bracketed by short-lived glacial periods in the Cenomanian and Coniacian according to Royer *et al.* (2004), and punctuated by a glacial episode according to Miller *et al.* (2005).

Carbon dioxide levels modelled to generate sea-surface temperatures based on foraminiferal $\delta^{18}O$ and Mg/Ca data suggest variation by a factor of four (600–2400 ppmv) in the Albian to Turonian interval (Bice & Norris 2002; Bice *et al.* 2006). Benthic foraminifera assemblage data from Venezuela suggest that the following Santonian interval was warm, with some cooling towards the Campanian boundary (Rey *et al.* 2004). Frakes *et al.* (1992) suggested that the Campanian was also warm and this is supported by more recent calcareous nannofossil transfer function data from the Indian Ocean (Lees 2002).

As suggested by Frakes *et al.* (1992), the Maastrichtian was a time of cooler climate. Cooling began at the Campanian–Maastrichtian boundary, *c.* 71 Ma, with evidence of cooler ocean temperatures from benthic foraminiferal and calcareous nannofossil assemblages (Friedrich *et al.* 2005), and eustatic sea-level drop consistent with substantial, short-lived glaciation at high latitudes (e.g. Miller *et al.* 2005). Indian Ocean calcareous nannofossil data suggest that the late Maastrichtian was a period of warming (Lees 2002), which is also suggested by Canadian palaeosol $\delta^{13}C$ data (Nordt *et al.* 2002).

The very latest Maastrichtian to Early Danian (Early Palaeocene) saw a period of rapid changes in climate, associated with rapid changes in sea-level as indicated by planktonic foraminiferal assemblages (e.g. Keller *et al.* 2002; Keller 2004). Palaeobotanical data suggest that the Early Palaeocene, *c.* 65 Ma (c33, Fig. 1b), was cool at high latitudes with warming by the end of the Danian (Poole *et al.* 2005). New data suggest that the later

Palaeocene was subtropical at high northern latitudes, with a peak across the Palaeocene–Eocene boundary (Moran *et al.* 2006; Sluijs *et al.* 2006) at *c.* 55 Ma, the so-called Palaeocene–Eocene thermal maximum (e.g. Zachos *et al.* 1993; Farley & Eltgroth 2003). Cooling occurred rapidly in the Early Eocene, *c.* 55 Ma (c34, Fig. 1b), heralding the end of this warm mode, with freshening of Arctic Ocean surface waters in the latest Early Eocene, at *c.* 49 Ma, and the first glacially rafted sediments evident in Arctic Ocean sequences in the early Mid-Eocene at *c.* 45 Ma (Moran *et al.* 2006).

Given the range of climate variability during this warm mode summarized above, it could be argued that the deviations from warm climates were more the norm. The short-lived Albian and Late Cretaceous glaciations are of particular interest. They are associated with sharp (<1 Myr) drops in sea-level of up to 80 m in the Albian (e.g. Immenhauser 2005) and up to 40 m in the Late Cretaceous (e.g. Miller *et al.* 2005). The Late Cretaceous examples are also associated with positive $\delta^{18}O$ excursions of the order of 1‰. As Miller *et al.* (2005) point out, these are comparable to pre-Pleistocene sea-level and $\delta^{18}O$ changes linked with substantial Antarctic ice caps. Given the recent evidence for large variations in Late Cretaceous atmospheric CO_2 concentrations (e.g. Bice *et al.* 2006), these brief glacial excursions are less surprising than they would have been at the time of writing of Frakes *et al.* (1992). Another transient climatic episode in this warm period, the Palaeocene–Eocene thermal maximum (Zachos *et al.* 1993), at 55.8 Ma, may at face value seem an odd choice, given that it represents a period of extreme warmth. However, like the Pliensbachian–Toarcian event (e.g. Pálfy & Smith 2000; Bailey *et al.* 2003), it occurs towards the end of the warm mode (Zachos *et al.* 2001) and is followed by a short-lived episode of cooling, in the earliest Eocene (Wing *et al.* 2000). It is associated with a global strong positive excursion in $\delta^{13}C$ of 2.5‰ (Zachos *et al.* 1993; Dickens 2000) and subtropical Arctic Ocean temperatures (Sluijs *et al.* 2006). Extraterrestrial ^3He data suggest that it had a rapid onset (<few thousand years) and lasted *c.* 120,000 years with possible accelerated drawdown of CO_2 from the atmosphere and oceans (Farley & Eltgroth 2003).

Mid-Palaeogene to Recent cool mode

This cool mode is perhaps the most intensively studied of all the climate modes described by Frakes *et al.* (1992), which is reflected by their devoting as many pages to it as to all of the other climate modes combined. It differs from preceding climate modes in that continuous and high-resolution proxy records track its changes, e.g. $\delta^{18}O$ for palaeotemperature (e.g. Zachos *et al.* 2001; Royer *et al.* 2004), $\delta^{11}B$ for palaeo-pH and palaeo-CO_2 (Pearson & Palmer 2000), Sr/Ca from benthic foraminiferal calcite for weathering fluxes (Lear *et al.* 2003*a*), etc. Its beginning is marked by cooling in the early Middle Eocene (Moran *et al.* 2006), at *c.* 49 Ma, following the Palaeocene–Eocene thermal maximum (Zachos *et al.* 1993). This is slightly later than the Early Eocene proposed by Frakes *et al.* (1992). In principle, we have not yet reached the end of this cool mode (e.g. Loutre & Berger 2000), but whether or not we have done so may be dependent on human activities in the near future (e.g. IPCC 2001).

The present cool mode is marked by an overall cooling trend, reflected in the $\delta^{18}O$ and fossil plant assemblages, punctuated by a series of sharp changes, culminating in Pleistocene bihemispheric glaciation (Frakes *et al.* 1992). The earliest evidence of cooling, that marks the beginning of this cool mode, comes from new marine core data from the Arctic Ocean that suggest substantial freshening of ocean waters close to the Early–Mid-Eocene boundary (Moran *et al.* 2006), at *c.* 49 Ma. These new data support the onset of Arctic ice-rafting from Mid-Eocene times onwards, at *c.* 45 Ma, 35 million years earlier than previously assumed (Moran *et al.* 2006). This earlier evidence for Arctic ice is supported by recent sedimentary and foraminiferal geochemistry data from the Pacific, which suggest that bipolar glaciation was episodically present in the late Mid-Eocene from *c.* 42 Ma onwards (Tripati *et al.* 2005).

The first sharp change in this cool mode, represented by major positive excursions of *c.* 1.5‰ in $\delta^{18}O$ and *c.* 1.4‰ in benthic $\delta^{13}C$, and sea-level drops of up to 125 m, occurred at the Eocene–Oligocene boundary (e.g. Tripati *et al.* 2005), *c.* 34 Ma. These changes are interpreted to mark the formation of the first large permanent ice sheets in Antarctica (e.g. Zachos & Kump 2005), although terrestrial palaeotemperature proxy data suggest that the positive excursion in $\delta^{18}O$ was more likely to be related to changes in Antarctic ice volume rather than a lowering of palaeotemperatures (e.g. Grimes *et al.* 2005). Alkenones data suggest that atmospheric CO_2 levels were still relatively high at this time, but that they continued to drop though the Oligocene, from 1000–1500 ppmv in the Late Eocene, reaching modern levels by latest Oligocene times (Pagani *et al.* 2005*b*).

The Early and early Late Oligocene is marked by cool climates with a series of short-lived positive excursions in the $\delta^{18}O$ curve (Miller *et al.* 1991), labelled Oi-1, 1a and 1b, at *c.* 34 Ma, *c.* 33 Ma and *c.* 32 Ma (e.g. Zachos *et al.* 2001; Pekar *et al.* 2002), and Oi-2, 2*, 2a, 2b and 2c at *c.* 30 Ma,

c. 29 Ma, *c.* 28 Ma, *c.* 27 Ma and *c.* 26 Ma (Pekar *et al.* 2002; Wade & Palike 2004). These were associated with sea-level drops of up to 65 m (Wade & Palike 2004) and are interpreted to represent oscillating, large-scale Antarctic ice sheets (Zachos *et al.* 2001). The later Late Oligocene, *c.* 25 Ma (w2, Fig. 1b), shows a brief return to warmer climates in both hemispheres (Barreda & Palamarczuk 2000; De Man & Van Simaeys 2004; Villa & Persico 2006), followed by a return to glacial conditions at the Oligocene–Miocene boundary (e.g. Zachos *et al.* 2001).

Like the Oligocene, the Miocene is a time of cool climates with a series of short-lived positive excursions in the $\delta^{18}O$ curve associated with short-lived drops in sea-level (Miller *et al.* 1991). The first and most major of these, Mi-1 (Miller *et al.* 1991), is at the Oligocene–Miocene boundary, at *c.* 23 Ma. This is followed by at least five further $\delta^{18}O$ stages, Mi-2 to Mi-6 (and substages), at *c.* 16 Ma, *c.* 15 Ma, *c.* 13 Ma, *c.* 12 Ma and *c.* 10 Ma (Miller *et al.* 1991; Flower 1999). A significant overall positive step in $\delta^{18}O$ occurred between Mi-2 and Mi-4, between *c.* 16 Ma and *c.* 12 Ma (e.g. Flower 1999), the so-called 'mid-Miocene cooling' event, associated with the establishment of a semi-permanent East Antarctic ice sheet and changes in oceanic circulation (e.g. Westerhold *et al.* 2005). From Mi-6 onwards, sedimentological evidence suggests the presence of glaciation outside Antarctica in both hemispheres (e.g. Denton & Armstrong 1969; Mercer & Sutter 1982; Duncan & Helgason 1998).

As suggested by Frakes *et al.* (1992), the early Tortonian of the Late Miocene, *c.* 11 Ma (w3, Fig. 1b), shows a slight warming of global climates, from, for example, Mg/Ca ratios in benthic foraminifera (Lear *et al.* 2003*b*). The Late Miocene saw a resumption of significant cooling, from $\delta^{18}O$ in benthic foraminifera and accompanying sea-level falls, that resulted in glaciation during the Messinian of the latest Miocene, between *c.* 6.26 and *c.* 5.50 Ma, although terminating prior to the Miocene–Pliocene boundary (e.g. Hodell *et al.* 2001), at *c.* 5.3 Ma. Extreme desiccation of the Mediterranean basin occurred at this time (e.g. Hodell *et al.* 1986; Krijgsman *et al.* 1999). Although glacio-eustatic sea-level falls and more arid global climate are contributory factors, the primary driver for isolation of the Mediterranean appears to be tectonic (e.g. Krijgsman *et al.* 1999; Hodell *et al.* 2001), particularly as similar tectonically driven desiccation episodes affected basins at its western margin shortly before in the Tortonian (Krijgsman *et al.* 2000).

The very latest Miocene and Early Pliocene, *c.* 5.3 Ma, show a return to milder climates (w4, Fig. 1b), as indicated by changes in the $\delta^{18}O$ of benthic and planktonic foraminifera from the Atlantic (Vidal *et al.* 2002; Reuning *et al.* 2006), thermohaline circulation modelling (Ravelo & Andreasen 2000), and Antarctic glacial sedimentology (Prentice & Krusic 2005). Climate warming peaked in the mid-Pliocene, between *c.* 3.3 Ma and *c.* 3 Ma (w5, Fig. 1b), and as described by Haywood *et al.* (2002b), 'Earth experienced a significant sustained period of greater global warmth spanning a time frame longer than any interglacial of the Quaternary period'. Evidence for this warmth comes from Australian ostracod faunal data (Warne 2005), alkenones palaeothermometry (Haywood *et al.* 2005), multiproxy analysis of Arctic Ocean ODP core data (Knies *et al.* 2002) and climate modelling (e.g. Haywood *et al.* 2000, 2002a).

In the Late Pliocene, at some point around 2.7 Ma, probably at marine isotope stage G6 (*c.* 2.74 Ma) (Bartoli *et al.* 2005), global climate cooled. Ocean-bottom nitrogen isotope data (Sigman *et al.* 2004), alkenone unsaturation ratios and diatom oxygen isotope ratios (Haug *et al.* 2005), Ge/Si ratios (Lin & Chen 2002), and ostracod assemblages (Yamada *et al.* 2005), suggest that this was the beginning of full-scale Northern Hemisphere glaciation (e.g. Klotz *et al.* 2006). It marks the onset of relatively continuous intensifying glacial-to-interglacial cycles to the present day (e.g. Bartoli *et al.* 2005). Late Pliocene and Early Pleistocene glacial–interglacial cycles show a 41 kyr periodicity, probably controlled by the astronomical obliquity cycle (e.g. Park & Maasch 1993). Between *c.* 1.2 and *c.* 0.6 Ma, centred on *c.* 0.95 Ma, the so-called mid-Pleistocene climate transition or climate revolution, the periodicity shifted to 100 kyr (e.g. Tziperman & Gildor 2003). This is consistent, temporally, with control by the precessional cycle, although the variation in insolation associated with this cycle appears to be too small compared with the associated significant increase in the intensity of Northern Hemisphere glaciation and increase in amplitude of glacial–interglacial intervals (e.g. Winckler *et al.* 2004). Evidence for this change comes from positive excursions in oceanic $\delta^{13}C$ (Raymo *et al.* 1997; Wang *et al.* 2004), global eustatic sea-level fall (Kitamura & Kawagoe 2006), mass extinctions of benthic foraminifera (Hayward 2001; Kawagata *et al.* 2005, 2006), and the onset of major glaciation in the European Alps (Muttoni *et al.* 2003). The final stages of this transition are associated with unusual deposition of diatomaceous, carbonaceous and carbonate sediments in the South Atlantic and Mediterranean (Schmieder *et al.* 2000; Gingele & Schmieder 2001).

No particularly notable intervals of contrasting climate mark this cool mode. The Late Oligocene, *c.* 25 Ma, mid-Miocene, *c.* 11 Ma, and latest Miocene–Early Pliocene, *c.* 5.3 Ma milder intervals (w2–4, Fig. 1b) are mainly of note because they briefly offset the overall cooling trend). The 'mid-Miocene cooling' event between Mi-2 and Mi-4, *c.* 16 Ma–12 Ma (e.g. Flower 1999) is interesting because the early stages coincide with eruption of the Columbia River basalts between *c.* 16.1 and *c.* 15 Ma (Hooper *et al.* 2002), which is the most recent large igneous province eruption known. The only interval comparable to the Palaeozoic Kazimovian warm interval is, perhaps, the mid-Pliocene warm interval (w5, Fig. 1b), which had sea-levels higher than today and East Antarctic climates similar to modern-day Chile (Haywood *et al.* 2002b).

Major Phanerozoic controls on climate

Long-term secular variation in the carbon cycle

The carbon cycle, with its links to biological and tectonomagmatic activity, is extremely important for the long-term evolution of climate in the Phanerozoic. The geochemically modelled atmospheric CO_2 curves of Berner (e.g. Berner & Kothavala 2001; Berner 2003a) (Fig. 3a) indicate that pCO_2^{atm} was high in the Early Palaeozoic, low during Permo-Carboniferous times, relatively high again in the Mesozoic, and that it has been in decline since then. Although the relationship between palaeotemperature and pCO_2^{atm} is less clear at the very highest concentrations seen during the Mesozoic (Bice *et al.* 2006), there is general agreement of a positive correlation between pCO_2^{atm} and global average temperature (e.g. Kump 2002; Retallack 2002; Royer *et al.* 2004; Siegenthaler *et al.* 2005). Examination of the curve of Berner (2003a), (Fig. 3a) and comparison with the palaeotemperature analysis by Royer *et al.* (2004) (Fig. 3b) suggests that Phanerozoic climate can be subdivided into long-term, fluctuating high CO_2 modes, interleaved with shorter-term, low CO_2 modes (Fig. 1b). Similarly, Raymond & Metz (2004) suggested that for Phanerozoic glaciation there are short-term, overall high CO_2 glacial modes, such as during the Late Ordovician and Cretaceous (Fig. 1b), and longer-term, low CO_2 glacial modes, such as in the Permo-Carboniferous and Late Cenozoic (Fig. 1b).

Of course, something that is not always expressed is what constitutes high or low pCO_2^{atm} levels and where the threshold lies. Royer *et al.* (2004) suggested that levels above 1000 ppmv should be considered as high and those below that should be considered as low, with a transitional range between about 600 and 1000 ppmv. A brief examination of the geological record suggests that this is reasonable. The earliest Cenozoic Antarctic glaciation, Oi-1 at the beginning of the Oligocene (e.g. Tripati *et al.* 2005), began when atmospheric CO_2 levels were at about 1000 parts per million and falling (Pagani *et al.* 2005*b*); recent data suggest that CO_2 levels in the cool Early Cretaceous were of the order of 560 to 1200 parts per million (Haworth *et al.* 2005), compared with up to 2400 ppmv for the warmer Late Cretaceous (Bice *et al.* 2006).

Tectonism

A key driver of long-term variability in atmospheric CO_2 is the amalgamation and fragmentation of continents, the so-called supercontinent cycle (e.g. Condie 2002; Murphy & Nance 2005). Major orogenic events (Fig. 4a), which exhume deep levels of the continental crust, have been proposed as drivers of major draw-down of atmospheric CO_2 (Raymo 1991; Hay 1996). These are associated with changes in palaeoelevation (e.g. Ruddiman & Kutzbach 1991), weathering rates (Raymo 1991), atmospheric circulation (Gunnell 1998) and atmospheric dustiness (e.g. Kohfeld & Harrison 2000). Silicate weathering products added to the ocean from the current orogenic belt that extends from Papua New Guinea to the Pyrenees, and includes the Himalayas, have been implicated in the overall decline of atmospheric CO_2 levels during the Late Palaeogene and Neogene and the onset of global glaciation (Raymo 1991), although modelling suggests that the net effect is not sufficient to explain the degree of cooling observed (Kerrick & Caldeira 1999).

The current phase of the supercontinent cycle is one of amalgamation (Fig. 4a), which is a likely time of major orogenesis. Similarly, in the Permo-Carboniferous, Pangaea was amalgamating (e.g. Veevers 2004), resulting in the Hercynian–Alleghenian–Uralian orogeny (Fig. 4a), comparable in scale, duration and rate of exhumation, to the modern-day collisional orogenic belt described above. Although the situation is slightly complicated by the rapid evolution of land plants in the Permo-Carboniferous, the addition of silicate weathering products to the Palaeo-Tethyan and Panthallassan oceans was likely to have been key in the large-scale decline of atmospheric CO_2 levels. Conversely, the rifting phase of Pangaea in the Jurassic was associated with elevated rates of CO_2 degassing (Kerrick 2001). Figure 3a and 3b suggest that there is a relatively good correlation between orogenesis that results in the amalgamation of supercontinents and low-CO_2 climate modes. Other orogenic events do not seem to correlate in time with cool episodes. The correlation with 'aragonite' and 'calcite' seas (Hardie 1996) (Fig. 4b) suggests that this is more closely linked to the supercontinental cycle with only very indirect influence on climate.

Another important aspect of tectonic changes over shorter timescales is the opening and closing of oceanic gateways (e.g. Smith & Pickering 2003) (Fig. 5) and change in the configuration of land and sea (e.g. Barron *et al.* 1980; Ramstein *et al.* 1997). The opening of the Drake Passage and Tasman Gateways in the earliest Oligocene (Fig. 5) has been implicated in the onset of Antarctic isolation through initiation of the Antarctic Circumpolar Current (Livermore *et al.* 2004; Mackensen 2004). Very shortly afterwards, in the Early Oligocene, subsidence of the Greenland–Scotland ridge is suggested to be the trigger for the onset of North Atlantic Deep Water formation and the beginning of the modern thermohaline circulation (Via & Thomas 2006). In the middle Miocene, closure of the Panamanian Gateway between North and South America appears to have triggered the major cooling step at that time (e.g. von der Heydt & Dijkstra 2006). Full-scale Northern Hemisphere glaciation may have followed as a result of advection of warmer waters northwards, with increased delivery of moisture to high latitudes (e.g. Lear *et al.* 2003*b*). Further back in time, plate configurations are much less certain (though see Armstrong this volume). Nevertheless, opening and closing of gateways during the final amalgamation of Pangaea in the Permian have been linked to climate changes at that time (Saltzman 2003).

Magmatism

A major source of variation in the rate of addition of juvenile, mantle-derived CO_2 to the atmosphere and oceans is the rate of eruption of magmatic large igneous provinces (LIPs), volcanic eruption cycles with volumes greater than $100,000$ km^3 of magma, which appear to have a *c.* 170-million-year cyclicity over the past 1500 million years (Prokoph *et al.* 2004). McElwain *et al.* (2005) have suggested that LIP eruptions also add substantial CO_2 by intruding into coal seams. Enhanced rates of magmatism, particularly associated with superplume episodes, such as in the Cretaceous (Larson 1995), are implicated in intervals of high CO_2 (Caldeira & Rampino 1991). The LIP record for the Phanerozoic is biased towards the record of terrestrial LIP eruptions (Fig. 6). Consequently, the record is much more complete for the period after 170 million years ago, when the presently interrogateable seafloor record began. The Jurassic–Cretaceous, Mesozoic peak in LIP emplacement (Fig. 6), the most recent peak in the superplume

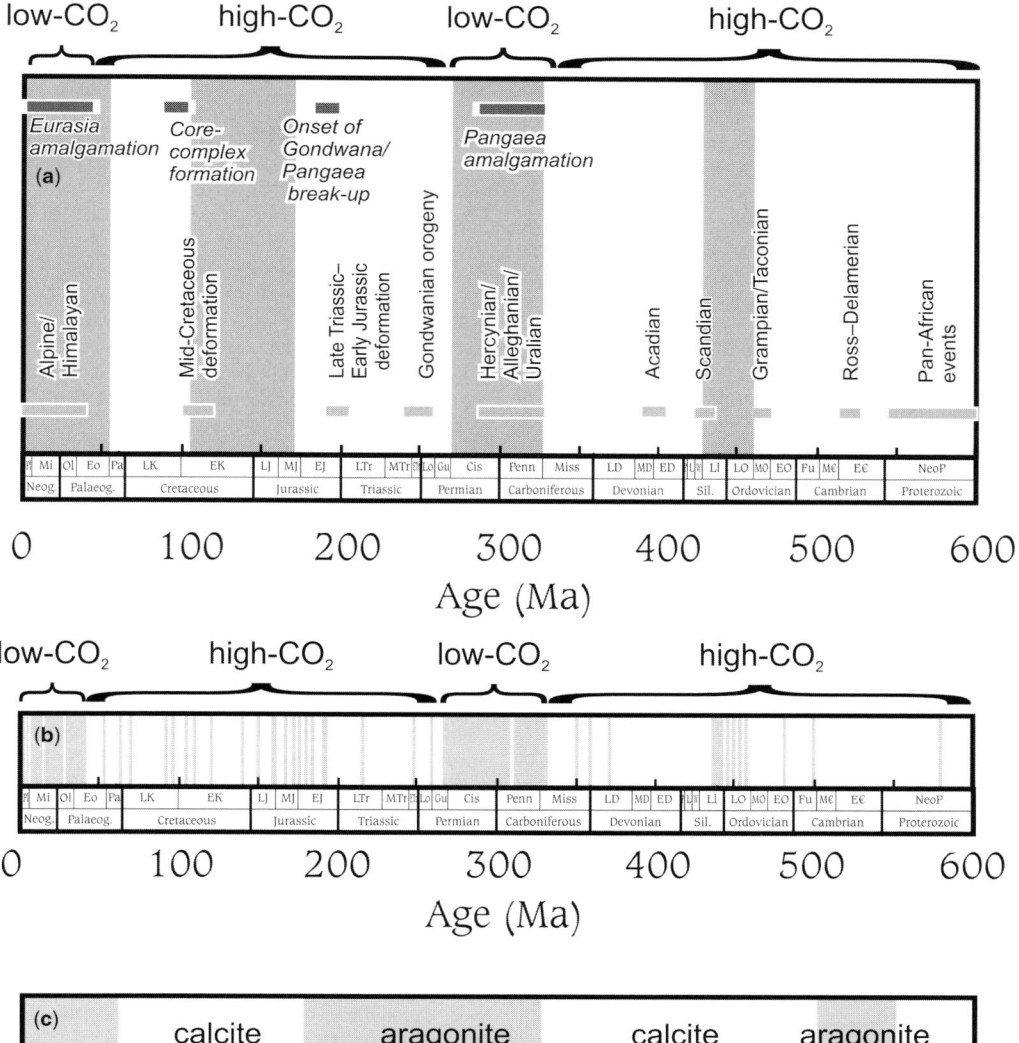

Fig. 4. Figure 1a superimposed with the following. (**a**) Late Neoproterozoic and Phanerozoic tectonic events modified after Vaughan & Livermore (2005 fig. 3), except age of Rodinia break-up (Murphy *et al.* 2004), Pan-African events (Thomas *et al.* 2002*b*; Heilbron & Machado 2003; Meert 2003), Ross–Delamerian (Boger & Miller 2004), Grampian/Taconian (Pincivy *et al.* 2003; Dewey 2005), Scandian (Kirkland *et al.* 2006) and Acadian (Sherlock *et al.* 2003). (**b**) Intervals of glacial or cool (grey bars) and warm (white areas) climates and durations of high- and low-CO_2 modes for the Late Neoproterozoic and Phanerozoic modified from Fig. 1b shown for reference (**c**) Phanerozoic intervals of 'calcite' and 'aragonite' seas for the Phanerozoic as suggested by Hardie (1996).

cycle, is well represented in the geological record and coincides with a relative peak in atmospheric CO_2 and some of the highest global temperatures in the Phanerozoic (Vaughan & Storey 2007).

The record for LIP emplacement pre-170 million years ago is sketchy. Some Late Triassic and Early Jurassic oceanic LIPs have been recognized (e.g. Vaughan & Storey 2007), but that is as far back as the record of oceanic LIPs extends. Earlier LIPs are all terrestrial, and largely

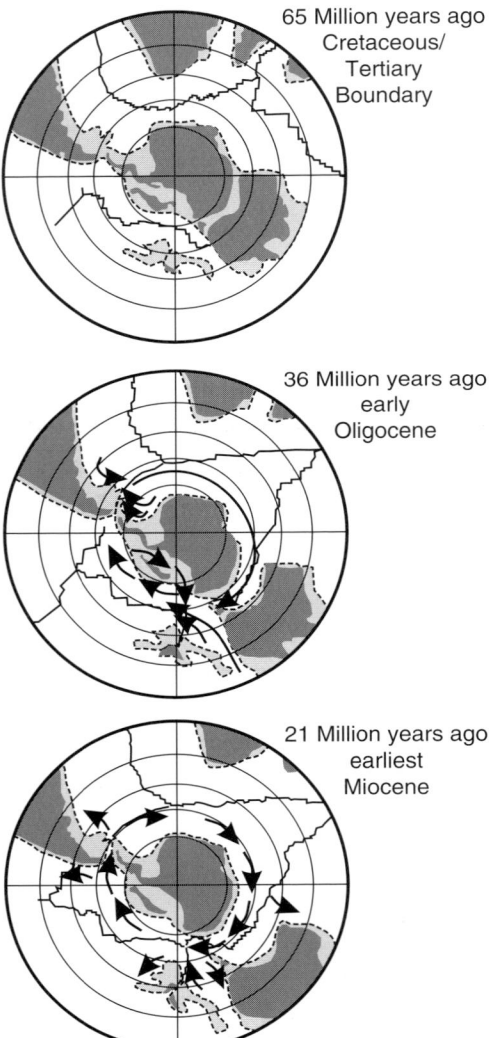

65 Million years ago
Cretaceous/
Tertiary
Boundary

36 Million years ago
early
Oligocene

21 Million years ago
earliest
Miocene

Fig. 5. Reconstruction of Cenozoic opening of the Drake Passage and Tasman oceanic gateways and onset of the circum Antarctic current (redrawn after Kennett 1977).

recognized from dyke swarms rather than basalt or rhyolite lavas (Ernst *et al.* 2005). In Figure 6, LIP eruptions through the Phanerozoic are plotted by area (Fig. 6a) and by volume (Fig. 6b). Volume is one of the key parameters required to assess the impact of a LIP, but volumes are difficult to obtain for LIPs older than Early Jurassic. Dyke swarm extent gives a proxy for area for older LIPs, so for Phanerozoic-length timescales it is useful to examine LIPs by erupted area. For the Mesozoic, there is a striking time coincidence in some cases between LIP eruptions and cool or glacial episodes, e.g. the Siberian Traps, Karoo magmatism and Shatsky Rise LIP (Fig. 6a).

If basaltic LIP eruptions are possible drivers of elevated global temperature, silicic LIPs and super-eruptions appear to be associated with global cooling. For example, Briffa *et al.* (1998) showed a strong correspondence between large silicic eruptions over the past 600 years and severe Northern Hemisphere winters based on maximum latewood density in tree rings. Over longer timescales, Rampino & Self (1992) suggested that the Toba super-eruption 73,500 years ago accelerated the Oxygen Isotope Stage 5a to 4 interglacial to glacial transition, and Prueher & Rea (1998) suggested that a 10-fold increase in volcanicity in the North Pacific *c.* 2.67 million years ago coincided with an intensification of Late Pliocene continental glaciation. Farther back in time, the Early and early Middle Jurassic period of Gondwana break-up silicic LIP eruptions (Fig. 6a) coincide with a period of four glacial episodes over that interval (Fig. 1b). Modelling by Jones *et al.* (2005) suggests that the silicic magmatism is insufficient on its own to trigger glaciation, although deviation from normal temperatures may last decades. The evidence suggests that cooling associated with silicic eruptions may act as an intensifier in the onset of glaciation where other conditions are favourable.

The rate of obduction of ophiolites (Fig. 7a), a slice of ocean floor added to continental margins rather than being subducted, has also been suggested as a proxy for peaks in the superplume cycle (Vaughan & Scarrow 2003). Again, as with LIP eruptions, from Fig. 7a, there is a striking positive time correlation between ophiolite obduction pulses and Palaeozoic and Mesozoic cool intervals, particularly for the Steptoan in the Cambrian (c2, Table 1), the Tremadocian in the Early Ordovician (c3, Table 1), mid- and Late Jurassic intervals, and the Late Cretaceous and Palaeogene.

The rate of eruption of kimberlites (Fig. 7b), also a significant source of juvenile CO_2, appears to correlate positively with the rate of basaltic LIP emplacement. At face value, kimberlite volumes appear too small to perturb the CO_2 balance of the

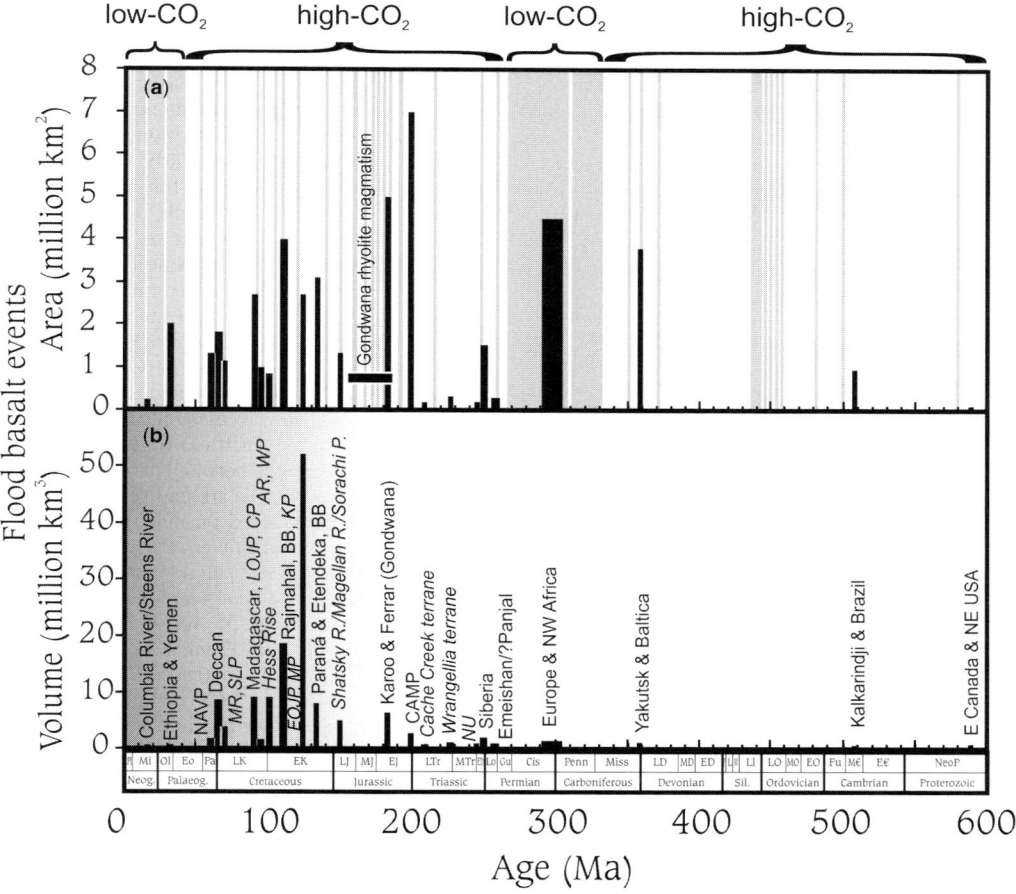

Fig. 6. (**a**) Figure 1b intervals of glacial or cool (grey bars) and warm (white areas) climates for the Late Neoproterozoic and Phanerozoic overlain by the timing, and approximate, estimated areas of large igneous province flood basalt events (including Gondwana rhyolite magmatism; duration from Riley *et al.* (2001)) for the Neoproterozoic to Neogene modified from Ernst & Buchan (2001) and on-line database at http://www.largeigneousprovinces.org/ downloads.html (accessed 9/7/2006), except Ethiopia (Kieffer *et al.* 2004), North Atlantic volcanic province (Eldholm & Grue 1994), Bunbury Basalts (Ingle *et al.* 2004), CAMP (McHone 2002), Nilufer unit (Genc 2004), Brazil (Santos *et al.* 2002) and Kalkarindji LIP (Glass & Phillips 2006); Gondwana rhyolite magmatism shown as horizontal grey bar. (**b**) Data as for (a) but showing large igneous province volumes instead. Names of continental flood basalts indicated in plain type; oceanic flood basalts indicated in italic type. Gradationally shaded area shows the geological interval for which sea floor is preserved.

atmosphere and oceans for more than a few years. However, there is a good general positive correlation, visible on Fig. 7b, between peaks in kimberlite eruption rate and high-CO_2 climate intervals and lulls in kimberlite eruption rates and low-CO_2 climate modes, which suggests that they may have a more significant and longer-term effect.

Overall, these proxies, in conjunction with the CO_2 and the seafloor record, suggest a major peak in oceanic LIP emplacement during the Cambro-Ordovician, even in the absence of surviving

direct evidence. These LIPs were probably oceanic, as were most of the Cretaceous LIPs, and constitute the Early Palaeozoic superplume event proposed by Larson (1991) and Barnes (2004). Significant LIP emplacement intervals, also probably representing superplume events, although on a smaller scale than the Mesozoic and Cambro-Ordovician events, have also been suggested for the Late Carboniferous–Early Permian (e.g. Larson 1991; Doblas *et al.* 1998) and Late Permian (e.g. Medvedev *et al.* 2003).

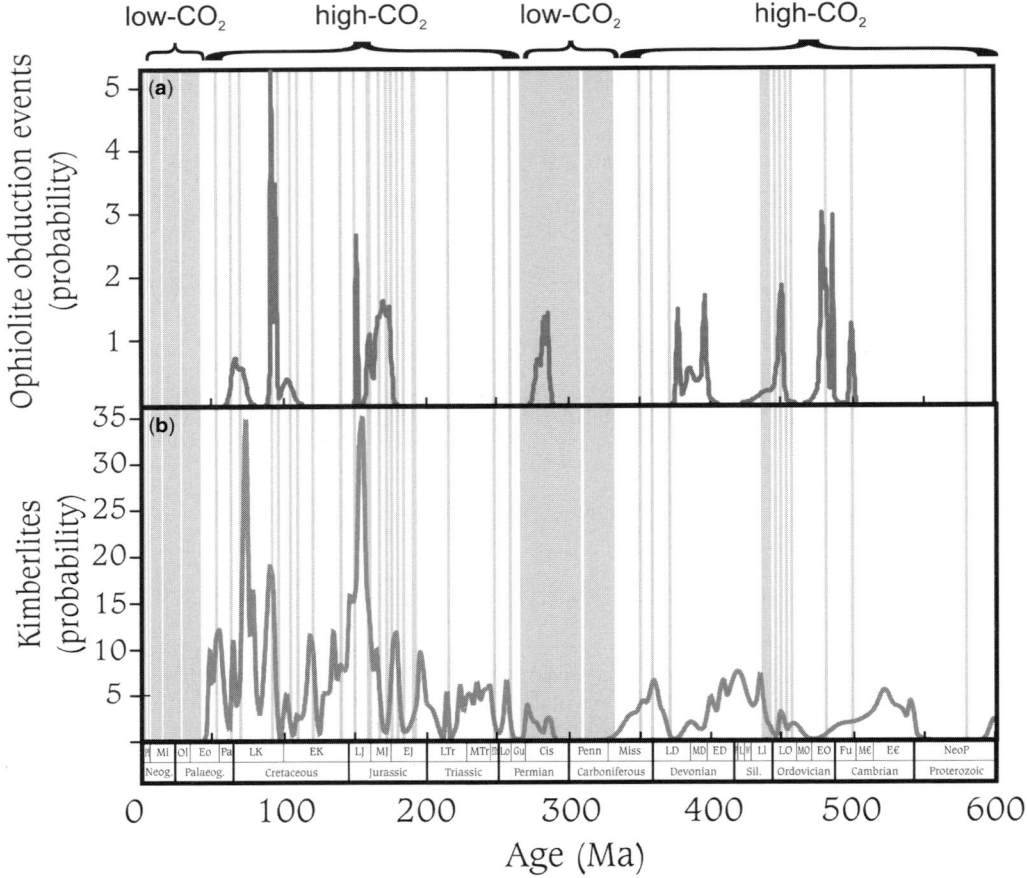

Fig. 7. Fig 1b intervals of glacial or cool (grey bars) and warm (white areas) climates for the Late Neoproterozoic and Phanerozoic overlain by the following. (**a**) Cumulative frequency plot (Ludwig 1999) of Phanerozoic ophiolite obduction events modified from Vaughan & Scarrow (2003). (**b**) Cumulative frequency plot (Ludwig 1999) of Phanerozoic kimberlite age data replotted from Griffin *et al.* (1999), Heaman & Kjarsgaard (2000), Belousova *et al.* (2001), Davis & Miller (2001), Heaman *et al.* (2003), Westerlund *et al.* (2004) and Jelsma *et al.* (2004).

Biological evolution

The evolution of land plants with complex root systems caused a major change in atmospheric composition (e.g. Berner 1998) and the evolutionary rise in land plants has been strongly implicated in the severity of the long-term CO_2 reduction during the Permo-Carboniferous low-CO_2 interval and global glaciation (e.g. Berner 2003*b*; Igamberdiev & Lea 2006). Plants increased the rate of continental silicate weathering rates, enhancing the drawdown of CO_2 in the oceans as carbonate; organic acids generated by plant roots are significant agents of silicate weathering (Caldeira 2006). Aided by actively subsiding basins and many cycles of glacio-eustatic sea-level rise and fall, or cycles of humidity–aridity, plants also directly

contributed to the removal of CO_2 from the atmosphere by forming peat, which was subsequently turned to coal (e.g. Heckel 1996). Similar suggestions for the Cenozoic have been made with regard to the rise of the angiosperms (Igamberdiev & Lea 2006).

Another major change happened in the Late Triassic with the evolution of plankton capable of fixing carbonate to form hard skeletons, allowing the deep oceans to become a significant sink for carbonate (e.g. Ridgwell 2005; Erba 2006). This was initiated by the change from dominance in the oceans of phytoplankton with green algal plastids, e.g. acritarchs, to those with red algal plastids, e.g. coccolithophorids, at the Permian–Triassic boundary (Grzebyk *et al.* 2003) culminating in the Late Triassic (e.g. Ridgwell 2005). Prior to this, carbonate fixing was carried out in continental shelf areas by largely benthic and infaunal organisms (e.g. Ridgwell 2005). Major drops in sea-level in pre-Mesozoic time, by diminishing shelf area, drastically reduced the capacity of the Earth System to sequester carbon as carbonate, a regulated and buffered process by comparison with the more rapid drawdown of CO_2 as organic carbon, saturating the oceans (Ridgwell 2005). Since the Late Triassic, only at times of probable ocean acidification do we see a return to the Palaeozoic and Early Mesozoic mechanism, and, in addition to the swings in climate, sea-level, and isotopic systems described above, this is commonly associated with mass extinction of organisms, for example, at the Cretaceous–Palaeogene boundary (e.g. Hollander *et al.* 1993). However, the relationship between carbonate and organic carbon formation and atmospheric CO_2 is complex and still not well understood (see Barker *et al.* 2003 for a recent review).

Astronomical controls

Changes in insolation driven by the Earth's orbital parameters of precession, obliquity and eccentricity, have long been recognized as important drivers of climatic change, if only widely accepted in the last 30 years (e.g. Croll 1867; Milankovitch 1941; Hays *et al.* 1976; Zachos *et al.* 2001; Berger & Loutre 2004). These occur on 19 kyr, 41 kyr, and 100 kyr frequencies, respectively. Additionally, there are so-called 'grand' cycles related to the orbits of the major planets with periodicities at the present day of 400 kyr, and 1.25, 2.35 and 4.6 Myr (e.g. Olsen 2001). Modelling based on Milankovitch theory has been applied with varying degrees of success to explanation of cyclical climate variations for much of the Neoproterozoic (e.g. Lewis *et al.* 2003) and Phanerozoic (e.g. Herrmann *et al.* 2003; Wade & Palike 2004; Holbourn *et al.* 2005; Huynh & Poulsen 2005). These studies suggest that insolation minima of the orbital cycles are the likely trigger for glaciation at low CO_2 levels at long and short timescales. The effects of long-term changes in the intensity of insolation at the peaks of the Milankovitch cycle are not known, but may be a significant factor in longer-term climate change.

Although the Sun is recognized as a short-term variable (Solanki & Krivova 2004), the influence of variation in solar output at longer timescales, although potentially an extremely important driver of climate change (e.g. Walker *et al.* 1981), is currently difficult to model or quantify. For example, Foukal *et al.* (2006) have shown that variations in the Sun's brightness (solar luminosity) caused by changes in sunspot area are insufficient to account for warming since the Little Ice Age; however,

they point out that additional climate forcing by variations in solar ultraviolet output or the solar wind cannot be ruled out. Apart from the obvious effect of variation in insolation, solar activity, via accompanying changes in the strength of the Sun's magnetic field, may also inversely affect the galactic cosmic ray flux (Usoskin *et al.* 2005), with possible implications for low-level cloudiness (e.g. Yu 2002; Harrison & Stephenson 2006). For example, Ramirez *et al.* (2006) have suggested that cosmic ray modulated variation in low-level cloud cover may be responsible for temperature changes up to 50% that predicted for doubling of atmospheric CO_2. In terms of variations in solar output and insolation at the Earth's surface, a study of sunspot activity for the Holocene, using dendrochronologically dated variations in ^{14}C, suggests that current solar activity is at its highest for 11,000 years (Solanki *et al.* 2004).

Variations in the galactic cosmic ray flux as a function of the location of the Solar System relative to the spiral arms of the Galaxy has been implicated as a driver of long-term climate change (Wallmann 2004; Wendler 2004; Shaviv 2005). Figure 8 shows a summary of the warm and cool modes of Frakes *et al.* (1992) for the last 600 million years, overlain by blocks showing likely periods of enhanced galactic cosmic ray flux of Wallmann (2004). Interestingly, there appears to be a loose time correlation between peaks in the galactic cosmic ray flux and protracted periods of Phanerozoic glaciation (Wallmann 2004). However, this is less clear for high- and low-CO_2 climate modes (Fig. 8b). Mechanisms proposed to explain links between galactic cosmic ray flux and climate include the reflective and cooling effects of enhanced low-level cloudiness nucleated by cosmic rays penetrating the atmosphere (Shaviv 2005). However, Royer *et al.* (2004) showed that a pH-corrected Phanerozoic palaeotemperature curve predicts proxy-based cool and warm epochs more accurately than predicted variation in galactic cosmic ray flux (Shaviv & Veizer 2003).

Figure 8 suggests that, stochastically, the periodicity of changes in the galactic cosmic ray flux and the more closely coupled supercontinent and superplume cycles have been partially in step, although not in any way coupled, during the Phanerozoic. Galactic cosmic ray flux mechanisms for global cooling (e.g. Shaviv & Veizer 2003) may have at times intensified major glaciations, but are unlikely to be directly responsible for them. Interestingly, Evans (2003) has pointed out that, taking into account preservational bias, sedimentological or isotopic evidence for major global glaciation is absent between the 2.5–2.2 Ga Palaeoproterozoic Huronian glaciation and the Neoproterozoic 'Snowball Earth' glaciations of *c.* 750 million years ago

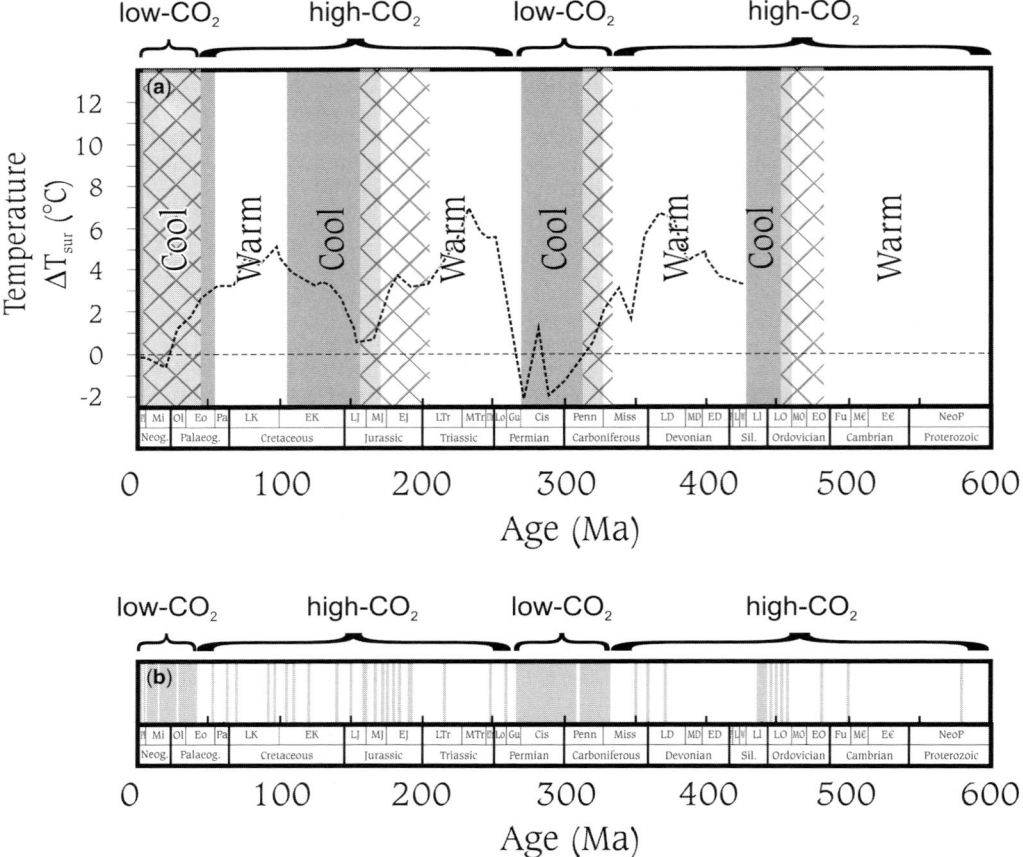

Fig. 8. (**a**) Figure 1a superimposed with intervals of high galactic cosmic ray flux for the Phanerozoic from Wallmann (2004). (**b**) Intervals of glacial or cool (black bars) and warm (white areas) climates and durations of high- and low-CO_2 modes for the Late Neoproterozoic and Phanerozoic modified from Figure 1b shown for reference.

(Evans 2000). One outside possibility is that peaks in the galactic cosmic ray flux and periods of low-CO_2 (lows in the superplume cycle and times of supercontinental amalgamation) were out of phase during the later Palaeoproterozoic to early Neoproterozoic interval, moderating the severity of low-CO_2 mode glaciation.

Other mechanisms

Studies of the chemical composition of evaporite deposits and of fluid inclusions in halite (e.g. Hardie 1996; Lowenstein *et al.* 2001) suggest that the carbonate composition of seawater has varied systematically through the Phanerozoic, alternating between 'aragonite seas', as at the present day, and 'calcite seas' (e.g. Hardie 1996; Berner 2004) (Fig. 4c), such as in the Cretaceous. It has been argued that these changes in composition are variously related to rates of magmatic or hydrothermal

activity at spreading ridges, changes in Mg/Ca ratio, and pCO_2 (summarized in Adabi 2004). As summarized by Erba (2006), a broad general relationship is evident with aragonite seas during periods of supercontinental amalgamation and relatively low hydrothermal activity at spreading ridges, i.e. relatively low-CO_2 intervals, and calcite seas during periods of high hydrothermal activity at spreading ridges and at times of supercontinental dispersal, i.e. relatively high-CO_2 intervals.

Dissociation of seabed methane hydrate deposits (e.g. Nisbet 1992; Hesselbo *et al.* 2000), or high-latitude permafrost-hosted methane hydrate deposits (e.g. Nisbet 2002), is implicated in generating peaks in atmospheric CO_2, with accompanying palaeotemperature rises, and paired negative–positive spikes in $\delta^{13}C$ (e.g. van de Schootbrugge *et al.* 2005). The proposal is that, destabilized by warming of ocean waters (e.g. Thomas *et al.* 2002*a*), or decompression caused by a drop in sea-level (e.g. Maslin *et al.* 2005), methane hydrate deposits in shelf sediments catastrophically decrepitate. This transfers large volumes of isotopically light, up to $-60‰$ $\delta^{13}C$, carbon as methane to the atmosphere (e.g. Zachos *et al.* 2005). Methane is a powerful greenhouse gas in its own right (Badr *et al.* 1992), so this initial release may cause a pulse of global warming, but atmospheric chemistry models suggest that the methane rapidly oxidizes to CO_2 (<10 years, Nisbet pers. comm. 2006), prolonging the elevation of temperatures. The increase in atmospheric CO_2 triggers a rise in ocean productivity, initially resulting in a negative spike in $\delta^{13}C$, which progresses to more positive values as drawdown of CO_2 continues (Jenkyns 2003).

Methane hydrate dissociation has been proposed as the mechanism for events at the boundaries of the Permian–Triassic (e.g. Winguth & Maier-Reimer 2005), Pliensbachian–Toarcian (Hesselbo *et al.* 2000) and Palaeocene–Eocene (Dickens 2000). The Permian–Triassic event coincided with the eruption of the Siberian Traps (a major basaltic and silicic LIP) (Wignall 2001) and the Early Jurassic methane hydrate release was also coincident with major LIP eruptions 183 Ma (Fig. 6a) associated with the onset of Gondwana break-up (Vaughan & Storey 2007).

Discussion

Examination of the curves of Berner & Kothavala (2001), review of the geological data presented above, and comparison with the palaeotemperature analysis by Royer *et al.* (2004) suggests an alternative to the warm and cool modes of Frakes *et al.* (1992). As outlined in Table 1 and Figure 1b, it is proposed that there are long-term high-CO_2 warm modes to Phanerozoic climate, punctuated by short-lived contrasting cool or glacial intervals, and shorter-term low-CO_2 cool modes, such as the current Oligocene–Recent interval, with rarer contrasting short-lived warm intervals, that culminate in times of major global glaciation or superglaciations.

Examination of the geological record of cool and warm climates, as summarized in Table 1, Figure 1b, and Figures 3, 4, 6 & 7, would suggest that high-CO_2 modes, which have a predominance of warm climates, coincide with peaks in the superplume cycle, i.e. with high rates of emplacement of LIPs, and coincide with periods of continental dispersal in the supercontinent cycle. These are periods of high rates of magmatic activity and elevated hydrothermal activity at mid-ocean ridges. Although the correlation is not perfect, high-CO_2 modes also loosely coincide with 'calcite' seas. Overall, these modes are times of high-CO_2 flux and relatively low rates of continental weathering.

Low-CO_2 modes, which are generally of much shorter duration, appear to coincide with lulls in the superplume cycle, i.e. with very low rates of LIP emplacement (Fig. 6). They coincide with periods of amalgamation in the supercontinent cycle with very large-scale orogenesis and high rates of crustal exhumation (Fig. 4). These are periods of relatively low rates of magmatic activity and low rates of hydrothermal activity at mid-ocean ridges. Again, although the correlation is not perfect, low-CO_2 modes approximate with 'aragonite' seas. Overall, these modes are times of relatively low-CO_2 flux and elevated rates of continental weathering.

Of particular interest, are the periods of contrasting climate (summarized in Table 1). High-CO_2 modes are punctuated by frequent short-lived cool or glacial episodes. Low-CO_2 modes, on the other hand, only rarely experience short-lived warming episodes, such as that in the mid-Pliocene. The short-lived cool or glacial episodes in high-CO_2 modes have some common characteristics. They generally occur at times of eustatically high sea-level, but are marked by a rapid drop in sea-level, such as in the Late Callovian (Dromart *et al.* 2003). They are preceded by or overlap with episodes of deposition of carbonaceous sediments, such as black shales or sapropels as in the Turonian (Wilson *et al.* 2002; Bice *et al.* 2006). Where reliable isotope data can be extracted, such as for the Mesozoic and early Cenozoic, they are also associated with positive excursions in $\delta^{13}C$ and $\delta^{18}O$, again seen in the Late Callovian (Dromart *et al.* 2003).

A modified scheme of climate modes for the Phanerozoic

The Earliest Cambrian to mid-Ordovician, Late Ordovician to Early Silurian and Late Silurian to mid–Early Carboniferous climate modes of Frakes *et al.* (1992) (warm–cool–warm respectively) coincided with the early Palaeozoic high-CO_2, climate mode (Table 1, Fig. 1b), which was largely warm with CO_2 concentrations fluctuating on short timescales. As suggested above, this high-CO_2 mode can probably be extended back into the Ediacaran of the late Neoproterozoic. This is based on the analysis of Royer *et al.* (2004), and the CO_2 curve of Berner (2003a). It largely overlaps with the early Palaeozoic 'calcite' sea (e.g. Holland 2004), indicating a time of high hydrothermal activity at spreading ridges and continental dispersal. As discussed above, proxy data suggest that this is an interval of a peak in the superplume cycle with elevated rates of LIP eruptions.

Of the short-lived cool, possibly glacial, phases during this high-CO_2 mode, the Ediacaran short-lived glacial interval at *c.* 580 Ma (c1, Fig. 1b; Table 1) coincides with the inception of the Middle Iapetus LIP of Ernst & Buchan (2001). Eruption of the Kalkarindji LIP in Australia (Glass & Phillips 2006) overlaps with the Middle Cambrian, Steptoan positive $\delta^{13}C$ excursion, major sea-level drop and probable short-lived glacial episode (e.g. Saltzman *et al.* 2000) (c2, Fig. 1b; Table 1).

For the short-lived cool or glacial periods in the early Ordovician, late Ordovician, and Early Silurian (c3–c10, Fig. 1b; Table 1), there are no recorded episodes of basaltic LIP magmatism, although indirect proxy evidence suggests that they were likely. The late Ordovician Hirnantian glaciation (c7, Fig. 1b; Table 1) was probably the most severe and is associated, although not simultaneously, with black shale deposition and a mass extinction event (Marshall *et al.* 1997). The three Early Silurian glaciations (Kaljo *et al.* 2003) (c8–c10, Fig. 1b; Table 1) do coincide with an inferred period of elevated global silicic volcanism, mainly represented by thick bentonite horizons, representing individual explosive eruptions of up to 1000 cubic kilometres each (Fortey *et al.* 1996). These may have resulted in short periods of global cooling, as suggested for explosive eruptions during the past 600 years (e.g. Briffa *et al.* 1998), or have acted as a trigger for glaciation (e.g. Rampino & Self 1992; Jones *et al.* 2005).

During the latter part of this high-CO_2 mode, short-lived glaciation is associated with black shale deposition, sea-level changes and mass extinction in the Late Devonian, the Kellwasser event of the Frasnian–Famennian (Buggisch 1991) (c11, Fig. 1b; Table 1), and at the Devonian–Carboniferous boundary, the Hangenberg event (Caplan & Bustin 1999) c12, Fig. 1b; Table 1). This latter event coincides with the eruption of the Yakutsk–Baltica LIP (Fig. 6a), one of the largest-known LIPs of the Palaeozoic. A final phase of short-lived glaciation occurred in the Viséan of the Early Carboniferous (e.g. Eyles 1993) (c13, Fig. 1b; Table 1).

The following late Palaeozoic cool, low-CO_2 mode (Table 1, Fig. 1b) coincides with the Permo-Carboniferous cool mode of Frakes *et al.* (1992) and was a period of superglaciation. It coincided with a peak of orogenesis in the supercontinent cycle, with activity along the Alleghenian–Hercynian–Uralian orogeny as Pangaea finished amalgamating (e.g. Veevers 2004). It coincided with the late Palaeozoic 'aragonite' sea (e.g. Holland 2004), indicating a time of low hydrothermal activity at spreading ridges and continental amalgamation. This low-CO_2 mode occurs at a low in the superplume cycle, with only one LIP recognized in the Late Carboniferous and Early Permian (Kazimovian–Artinskian) (Doblas *et al.*

1998). This low-CO_2 cool mode is punctuated by only one short-lived warm interval, in the Kazimovian (Bruckschen *et al.* 1999) (w1, Fig. 1b; Table 1), which precedes the peak in Permo-Carboniferous superplume magmatism (Doblas *et al.* 1998) (Fig. 6a). Warming in the Kazimovian was associated with mass extinction of land plants (Edwards 1998).

The mid-Phanerozoic warm, high-CO_2 mode (Table 1, Fig. 1b) incorporates the Latest Permian to Early Jurassic, Middle Jurassic to Early Cretaceous, and Late Cretaceous to Palaeocene climate modes of Frakes *et al.* (1992) (warm–cool–warm respectively). Like the early Palaeozoic high-CO_2 mode, this mode was largely warm with CO_2 concentrations fluctuating on short timescales. This mode occurs during a period of continental dispersal following the break-up of Pangaea and Gondwana from Early Jurassic times onwards. It overlaps with the Mesozoic to Palaeogene 'calcite' sea (e.g. Holland 2004), indicating a time of high hydrothermal activity at spreading ridges and continental dispersal. Direct magmatic evidence suggests that this is an interval of a peak in the superplume cycle with elevated rates of continental and oceanic LIP eruptions (Vaughan & Storey in press). At least 25 LIPs erupted between early Jurassic times and the Late Palaeocene, many of which coincided with cool or glacial intervals (Fig. 6a). The short-lived cooler episodes in this warm, high-CO_2 mode (Table 1, Fig. 1b) are associated with black shale deposition, oxygen and carbon isotope excursions, sea-level changes and mass extinctions.

Notable events occurred in the Late Permian (c14, Fig. 1b; Table 1), associated with the eruption of the Emeishan Traps and the Guadalupian–Lopingian mass extinction event (Wignall 2001), in the earliest Triassic (c15, Fig 1b, Table 1) associated with the eruption of the Siberian Traps and the largest mass extinction of the Phanerozoic (e.g. Reichow *et al.* 2002), and in the Late Triassic and Early Jurassic (c16–c21, Fig. 1b; Table 1), associated with emplacement of the Central Atlantic Magmatic Province and Gondwana Large Igneous Province and several mass extinction events (summarized in Vaughan & Livermore 2005). Further events occurred throughout the Jurassic and Cretaceous and in the Late Palaeocene–Early Eocene (c22–c34, Fig. 1b; Table 1). Mid-Cretaceous LIP eruptions are not particularly noted for mass extinction events, although these LIPs were largely submarine and oceanic, which may have reduced their impact on the biosphere (Wignall 2001).

The Cenozoic cool, low-CO_2 mode (Table 1, Fig. 1b) is equivalent to the Early Eocene to present cool mode of Frakes *et al.* (1992), and like the Late Palaeozoic low-CO_2 mode is also a time of superglaciation. It coincides with a peak of orogenesis in the supercontinent cycle, with Cenozoic activity along the orogenic belt that extends from the Pyrenees to Papua New Guinea, and includes the Himalayas, as a new supercontinent forms. Orogenesis is also active along the western margin of the Americas (Dickinson 2004; Sobolev & Babeyko 2005). This cool mode occurs during the current 'aragonite' sea (e.g. Holland 2004), indicating a time of low hydrothermal activity at spreading ridges. This low-CO_2 mode occurs at a low in the superplume cycle, with only two small LIPs having erupted in the last 50 million years: the Ethiopian traps, which erupted *c.* 30 million years ago (Hofmann *et al.* 1997), and the Columbia River basalts *c.* 16 million years ago (Hooper *et al.* 2002). This low-CO_2 cool mode is punctuated by four short-lived warm intervals (w2–w5, Fig. 1b; Table 1), in the Late Oligocene, Late Miocene, latest Miocene–Early Pliocene and mid-Pliocene (mid-Neogene) (Table 1, Fig. 1b).

Conclusions

Emanuel (2002) observed that although climate appears to be sensitive to small orbitally driven changes in insolation, often with abrupt consequences, it is on another level stable, in that it has shown only relatively small variations despite an almost 30% increase in solar output since formation of the Earth 4560 million years ago. Emanuel (2002) suggested that the climate may have a limited series of overlapping stable regimes that result in more than one equilibrium state for the same insolation conditions. This has also been modelled for the thermohaline circulation (Marotzke & Willebrand 1991), and was suggested by Chamberlin (1906) in the early 20th century. A review of the palaeoclimatic evidence suggests that through the Phanerozoic two overlapping stable climate regimes appear to have dominated: a high-CO_2, largely warm climate regime, punctuated by short-lived episodes of glaciation driven by negative-feedback processes in the carbon cycle; and a low-CO_2, largely cool regime, marked by protracted episodes of superglaciation. Other stable regimes are obviously possible, the most recent of which were seen during the low-latitude, 'Snowball Earth' global glaciations of the Neoproterozoic (e.g. Hoffman *et al.* 1998).

The future

From a proxy point of view, a major gap in the arsenal of techniques available to scientists is a simple and reliable one for marine salinity, particularly given its importance for the density of ocean waters and the control on deep-ocean circulation (Henderson 2002). Given the success of the crenarchaeotal-lipid-membrane-based Tex86 proxy for palaeotemperature (Schouten *et al.* 2002), perhaps the halophile euryarchaeota, which are widely present in terrestrial and marine environments (e.g. Elshahed *et al.* 2004; Herndl *et al.* 2005), may provide a palaeosalinity equivalent. A gap also exists for good proxies of global oceanic alkalinity and weathering (Henderson 2002), which exert a key control on CO_2 drawdown. From a modelling point of view, the way forward, given the steady increase in computer processing power, is almost certainly towards increasingly coupled GCMs. The most ambitious plans, at the time of writing, aim to couple biome and ice-sheet models with coupled ocean–atmosphere GCMs (e.g. Haywood *et al.* 2002*b*). Model evaluation with the highest-quality proxy data from the geological record is the only way that we can increase our confidence in model outputs, an essential for anticipating future climates, as outlined by Kump (2002). From a geological point of view, the recent success of the Integrated Ocean Drilling Program Arctic Coring Expedition to the Lomonosov Ridge (Brinkhuis *et al.* 2006; Moran *et al.* 2006; Sluijs *et al.* 2006) illustrates the continued importance of coring the ocean basins. Drilling programmes are planned in both polar regions and the importance of this cannot be overstated.

Euan Nisbet and Jan Zalasiewicz are thanked for constructive and insightful reviews that helped refine many of the topics presented here.

References

ADABI, M. H. 2004. A re-evaluation of aragonite versus calcite seas. *Carbonates and Evaporites*, **19**, 133–141.

ADAMS, K. D. 2003. Estimating palaeowind strength from beach deposits. *Sedimentology*, **50**, 565–577.

AGASSIZ, L. 1828. On the erratic blocks of the Jura. *Edinburgh New Philosophical Journal*, **24**, 176–179.

AGUSTSDOTTIR, A. M., BARRON, E. J., BICE, K. L. ET AL. 1999. Storm activity in ancient climates 2. An analysis using climate simulations and sedimentary structures. *Journal of Geophysical Research-Atmospheres*, **104**, 27295–27320.

AHLBERG, A., OLSSON, I. & SIMKEVICIUS, P. 2003. Triassic–Jurassic weathering and clay mineral dispersal in basement areas and sedimentary basins of southern Sweden. *Sedimentary Geology*, **161**, 15–29.

ALDRIDGE, R. J., JEPPSSON, L. & DORNING, K. J. 1993. Early Silurian oceanic episodes and events. *Journal of the Geological Society*, **150**, 501–513.

ALEXANDRE, A., MEUNIER, J. D., LLORENS, E., HILL, S. M. & SAVIN, S. M. 2004. Methodological improvements for investigating silcrete formation: petrography, FT-IR and oxygen isotope ratio of silcrete quartz cement, Lake Eyre Basin (Australia). *Chemical Geology*, **211**, 261–274.

ALGEO, T. J., SCHWARK, L. & HOWER, J. C. 2004. High-resolution geochemistry and sequence stratigraphy of the Hushpuckney Shale (Swope Formation, eastern Kansas): implications for climato-environmental dynamics of the Late Pennsylvanian Midcontinent Seaway. *Chemical Geology*, **206**, 259–288.

ALLEN, J. R. L. 1993. Palaeowind: geological criteria for direction and strength. *Philosophical Transactions of the Royal Society of London Series B-Biological Sciences*, **341**, 235–242.

ALLEN, J. R. L. 1996. Windblown trees as a palaeoclimate indicator: The character and role of gusts. *Palaeogeography, Palaeoclimatology, Palaeoecology*, **121**, 1–12.

ALLEY, N. F. & FRAKES, L. A. 2003. First known Cretaceous glaciation: Livingston Tillite Member of the Cadna-owie Formation, South Australia. *Australian Journal of Earth Sciences*, **50**, 139–144.

ALONSO-ZARZA, A. M. 2003. Palaeoenvironmental significance of palustrine carbonates and calcretes in the geological record. *Earth-Science Reviews*, **60**, 261–298.

ALYEA, F. N. 1972. Numerical simulation of an ice age paleoclimate. PhD thesis, Fort Collins, Colorado State University.

ANDERSSON, C. 1997. Transfer function vs. modern analog technique for estimating Pliocene sea-surface temperatures based on planktic foraminiferal data, western equatorial Pacific Ocean. *Journal of Foraminiferal Research*, **27**, 123–132.

ANDERSSON, T., FORMAN, S. L., INGOLFSSON, O. & MANLEY, W. F. 2000. Stratigraphic and morphologic constraints on the Weichselian glacial history of northern Prins Karls Forland, western Svalbard. *Geografiska Annaler Series A-Physical Geography*, **82A**, 455–470.

ANDREWS, P., LORD, J. M. & EVANS, E. M. N. 1979. Patterns of ecological diversity in fossil and modern mammalian faunas. *Biological Journal of the Linnean Society*, **11**, 177–205.

ARMSTRONG, H. A. 2007. On the cause of the Ordovician glaciation. *In*: WILLIAMS, M., HAYWOOD, A., GREGORY, J. & SCHMIDT, D. W. (eds) *Deep-time perspectives on climate change: marrying the signal from computer models and biological proxies*. The Micropalaeontological Society, Special Publication. Geological Society, London, 101–122.

ARRHENIUS, S. 1896. On the influence of carbonic acid in the air upon the temperature of the ground. *Philosophical Magazine Series 5*, **4**, 237–276.

AVERYT, K. B. & PAYTAN, A. 2004. A comparison of multiple proxies for export production in the equatorial Pacific. *Paleoceanography*, **19**.

AZMY, K., VEIZER, J., BASSETT, M. G. & COPPER, P. 1998. Oxygen and carbon isotopic composition of Silurian brachiopods: Implications for coeval seawater and glaciations. *Geological Society of America Bulletin*, **110**, 1499–1512.

BADR, O., PROBERT, S. D. & OCALLAGHAN, P. W. 1992. Methane – a Greenhouse Gas in the Earth's, Atmosphere. *Applied Energy*, **41**, 95–113.

BAGHAI, N. L. & JORSTAD, R. B. 1995. Paleontology, paleoclimatology and paleoecology of the Late Middle Miocene Musselshell Creek flora, Clearwater County, Idaho: a preliminary study of a new fossil flora. *Palaios*, **10**, 424–436.

BAILEY, I. W. & SINNOTT, E. W. 1916. The climatic distribution of certain types of angiosperm leaves. *American Journal of Botany*, **3**, 24–39.

BAILEY, T. R., ROSENTHAL, Y., MCARTHUR, J. M., VAN DE SCHOOTBRUGGE, B. & THIRLWALL, M. F. 2003. Paleoceanographic changes of the Late Pliensbachian–Early Toarcian interval: a possible link to the genesis of an Oceanic Anoxic Event. *Earth and Planetary Science Letters*, **212**, 307–320.

BALLANTYNE, C. K. 1994. Gibbsitic soils on former nunataks: implications for ice sheet reconstruction. *Journal of Quaternary Science*, **9**, 73–80.

BALLANTYNE, C. K., MCCARROLL, D. & STONE, J. O. 2006. Vertical dimensions and age of the Wicklow Mountains ice dome, Eastern Ireland, and implications for the extent of the last Irish Ice Sheet. *Quaternary Science Reviews*, **25**, 2048–2058.

BARD, E. 2001. Comparison of alkenone estimates with other paleotemperature proxies. *Geochemistry, Geophysics, Geosystems*, **2**.

BARKER, S., HIGGINS, J. A. & ELDERFIELD, H. 2003. The future of the carbon cycle: review, calcification response, ballast and feedback on atmospheric CO_2. *Philosophical Transactions of the Royal Society of London Series A-Mathematical Physical and Engineering Sciences*, **361**, 1977–1998.

BARNES, C. R. 2004. Was there an Ordovician superplume? *In*: WEBBY, B. D., DROSER, M. L., PARIS, F. & PERCIVAL, I. G. (eds) *The Great Ordovician Biodiversification Event*. Columbia University Press, New York, 77–80.

BARREDA, V. & PALAMARCZUK, S. 2000. Palynostratigraphy of Late Oligocene–Miocene deposits in the southern San Jorge Gulf area, Santa Cruz Province, Argentina. *Ameghiniana*, **37**, 103–118.

BARREIRO, M., PHILANDER, G., PACANOWSKI, R. & FEDOROV, A. 2006. Simulations of warm tropical conditions with application to middle Pliocene atmospheres. *Climate Dynamics*, **26**, 349–365.

BARRICK, R. E., FISCHER, A. G. & SHOWERS, W. J. 1999. Oxygen isotopes from turtle bone: Applications for terrestrial paleoclimates? *Palaios*, **14**, 186–191.

BARRON, E. J. 1993. Book Review: *Climate Modes of the Phanerozoic* – FRAKES, L. A., FRANCIS, J. E. & SYKTUS, J. I. *Nature*, **365**, 26–26.

BARRON, E. J., SLOAN, J. L. & HARRISON, C. G. A. 1980. Potential significance of land-sea distribution and surface albedo variations as a climatic forcing factor: 180 My to the present. *Palaeogeography, Palaeoclimatology, Palaeoecology*, **30**, 17–40.

BARRON, E. J., THOMPSON, S. L. & SCHNEIDER, S. H. 1981. An ice-free Cretaceous: results from climate model simulations. *Science*, **212**, 501–508.

BARTOLI, G., SARNTHEIN, M., WEINELT, M., ERLENKEUSER, H., GARBE-SCHONBERG, D. & LEA, D. W. 2005. Final closure of Panama and the onset of northern hemisphere glaciation. *Earth and Planetary Science Letters*, **237**, 33–44.

BEARD, B. L., JOHNSON, C. M., COX, L., SUN, H., NEALSON, K. H. & AGUILAR, C. 1999. Iron isotope biosignatures. *Science*, **285**, 1889–1892.

BEERLING, D. J. 1999. Quantitative estimates of changes in marine and terrestrial primary productivity over the past 300 million years. *Proceedings of the Royal Society of London Series B-Biological Sciences*, **266**, 1821–1827.

BEERLING, D. J., LAKE, J. A., BERNER, R. A., HICKEY, L. J., TAYLOR, D. W. & ROYER, D. L. 2002. Carbon isotope evidence implying high O_2/CO_2 ratios in the Permo-Carboniferous atmosphere. *Geochimica et Cosmochimica Acta*, **66**, 3757–3767.

BELOUSOVA, E. A., GRIFFIN, W. L., SHEE, S. R., JACKSON, S. E. & O'REILLY, S. Y. 2001. Two age populations of zircons from the Timber Creek kimberlites, Northern Territory, as determined by laser-ablation ICP-MS analysis. *Australian Journal of Earth Sciences*, **48**, 757–765.

BELT, S. T., MASSÉ, G., ROWLAND, S. J., LeBLANC, B., POULIN, M. & MICHEL, C. 2007. A novel chemical fossil of palaeo sea ice: IP$_{25}$. *Organic Geochemistry* **38**, 16–27.

BELT, S. T., MASSE, G., ROWLAND, S. J. & ROHMER, M. 2006. Highly branched isoprenoid alcohols and epoxides in the diatom *Haslea ostrearia* Simonsen. *Organic Geochemistry*, **37**, 133–145.

BENNETT, M. R. & DOYLE, P. 1996. Global cooling inferred from dropstones in the Cretaceous: Fact or wishful thinking? *Terra Nova*, **8**, 182–185.

BENTALEB, I., LANGLOIS, C., MARTIN, C. ET AL. 2006. Rhinocerotid tooth enamel $^{18}O/^{16}O$ variability between 23 and 12 Ma in southwestern France. *Comptes Rendus Geoscience*, **338**, 172–179.

BERGER, A. & LOUTRE, M. F. 2004. Astronomical theory of climate change. *Journal de Physique Iv*, **121**, 1–35.

BERGMAN, N. M., LENTON, T. M. & WATSON, A. J. 2004. COPSE: A new model of biogeochemical cycling over Phanerozoic time. *American Journal of Science*, **304**, 397–437.

BERNER, R. A. 1998. The carbon cycle and CO_2 over Phanerozoic time: the role of land plants. *Philosophical Transactions of the Royal Society of London Series B-Biological Sciences*, **353**, 75–81.

BERNER, R. A. 1999. Atmospheric oxygen over Phanerozoic time. Proceedings of the National Academy of Sciences, **96**, 10955–10957.

BERNER, R. A. 2001. Modeling atmospheric O_2 over Phanerozoic time. *Geochimica et Cosmochimica Acta*, **65**, 685–694.

BERNER, R. A. 2003a. The long-term carbon cycle, fossil fuels and atmospheric composition. *Nature*, **426**, 323–326, 10.1038/nature02131.

BERNER, R. A. 2003b. The rise of trees and their effects on Paleozoic atmospheric CO_2 and O_2. *Comptes Rendus Geoscience*, **335**, 1173–1177.

BERNER, R. A. 2004. A model for calcium, magnesium and sulfate in seawater over Phanerozoic time. *American Journal of Science*, **304**, 438–453.

BERNER, R. A. 2005. The carbon and sulfur cycles and atmospheric oxygen from middle Permian to middle Triassic. *Geochimica et Cosmochimica Acta*, **69**, 3211–3217.

BERNER, R. A. & BARRON, E. J. 1984. Factors affecting atmospheric CO_2 and temperature over the past 100 million years. *American Journal of Science*, **284**, 1183–1192.

BERNER, R. A., BEERLING, D. J., DUDLEY, R., ROBINSON, J. M. & WILDMAN, R. A. 2003. Phanerozoic atmospheric oxygen. Annual Review of Earth and Planetary Sciences, **31**, 105–134.

BERNER, R. A. & CANFIELD, D. E. 1989. A new model for atmospheric oxygen over Phanerozoic time. *American Journal of Science*, **289**, 333–361.

BERNER, R. A. & KOTHAVALA, Z. 2001. GEOCARB III: A revised model of atmospheric CO_2 over Phanerozoic time. *American Journal of Science*, **301**, 182–204.

BERNER, R. A., PETSCH, S. T., LAKE, J. A. ET AL. 2000. Isotope fractionation and atmospheric oxygen: implications for Phanerozoic O_2 evolution. *Science*, **287**, 1630–1633.

BICE, K. L., BIRGEL, D., MEYERS, P. A., DAHL, K. A., HINRICHS, K. U. & NORRIS, R. D. 2006. A multiple proxy and model study of Cretaceous upper ocean temperatures and atmospheric CO_2 concentrations. *Paleoceanography*, **21**.

BICE, K. L. & NORRIS, R. D. 2002. Possible atmospheric CO_2 extremes of the Middle Cretaceous (late Albian–Turonian). *Paleoceanography*, **17**.

BIRKS, H. 1981. The use of pollen analysis in the reconstruction of past climates: a review. *In*: WIGLEY, T. M. L., INGRAM, M. J. & FARMER, G. (eds) *Climate and History*. Cambridge University Press, Cambridge, 111–138.

BLAIR, T. C. 1999. Sedimentology of the debris flow dominated Warm Spring Canyon alluvial fan, Death Valley, California. *Sedimentology*, **46**, 941–965.

BLANZ, T., EMEIS, K. C. & SIEGEL, H. 2005. Controls on alkenone unsaturation ratios along the salinity gradient between the open ocean and the Baltic Sea. *Geochimica et Cosmochimica Acta*, **69**, 3589–3600.

BOGER, S. D. & MILLER, J. M. 2004. Terminal suturing of Gondwana and the onset of the Ross–Delamerian Orogeny: the cause and effect of an Early Cambrian reconfiguration of plate motions. *Earth and Planetary Science Letters*, **219**, 35–48.

BONIFACIO, E., SANTONI, S., CELI, L. & ZANINI, E. 2006. Spodosol-Histosol evolution in the Krkonose National Park (CZ). *Geoderma*, **131**, 237–250.

BOUCOT, A. J. & GRAY, J. 2001. A critique of Phanerozoic climatic models involving changes in the CO_2 content of the atmosphere. *Earth-Science Reviews*, **56**, 1–159.

BOYLE, E. A. 1988. Cadmium: chemical tracer of deep-water paleoceanography. *Paleoceanography*, **3**, 471–489.

BRAGA, J. C. & AGUIRRE, J. 2001. Coralline algal assemblages in upper Neogene reef and temperate carbonates in Southern Spain. *Palaeogeography, Palaeoclimatology, Palaeoecology*, **175**, 27–41.

BRENCHLEY, P. J., CARDEN, G. A., HINTS, L. ET AL. 2003. High-resolution stable isotope stratigraphy of Upper Ordovician sequences: Constraints on the timing of bioevents and environmental changes associated with mass extinction and glaciation. *Geological Society of America Bulletin*, **115**, 89–104.

BRENTNALL, S. J., BEERLING, D. J., OSBORNE, C. P., HARLAND, W. M., FRANCIS, J. E., VALDES, P. J. & WITTIG, V. E. 2005. Climatic and ecological determinants of leaf lifespan in polar forests of the high CO_2 Cretaceous 'greenhouse' world. *Global Change Biology*, **11**, 2177–2195.

BRIFFA, K. R., JONES, P. D., SCHWEINGRUBER, F. H. & OSBORN, T. J. 1998. Influence of volcanic eruptions on Northern Hemisphere summer temperature over the past 600 years. *Nature*, **393**, 450–455.

BRIFFA, K. R., OSBORN, T. J. & SCHWEINGRUBER, F. H. 2004. Large-scale temperature inferences from tree rings: a review. *Global and Planetary Change*, **40**, 11–26.

BRINKHUIS, H., SCHOUTEN, S., COLLINSON, M. E. *ET AL*. 2006. Episodic fresh surface waters in the Eocene Arctic Ocean. *Nature*, **441**, 606–609, 10.1038/nature04692.

BRISON, A. L., PHILIPPE, M. & THEVENARD, F. 2001. Are Mesozoic wood growth rings climate-induced? *Paleobiology*, **27**, 531–538.

BROECKER, W. S. & CLARK, E. 2001. Reevaluation of the $CaCO_3$ size index paleocarbonate ion proxy. *Paleoceanography*, **16**, 669–671.

BROUWERS, E. M., JORGENSEN, N. O. & CRONIN, T. M. 1991. Climatic significance of the ostracode fauna from the Pliocene Kap Kobenhavn Formation, north Greenland. *Micropaleontology*, **37**, 245–267.

BRUCKSCHEN, P., OESMANN, S. & VEIZER, J. 1999. Isotope stratigraphy of the European Carboniferous: proxy signals for ocean chemistry, climate and tectonics. *Chemical Geology*, **161**, 127–163.

BUGGISCH, W. 1991. The global Frasnian–Famennian Kellwasser Event. *Geologische Rundschau*, **80**, 49–72.

CALDEIRA, K. 2006. Forests, climate, and silicate rock weathering. *Journal of Geochemical Exploration*, **88**, 419–422.

CALDEIRA, K. & RAMPINO, M. R. 1991. The mid-Cretaceous super plume, carbon dioxide, and global warming. *Geophysical Research Letters*, **18**, 987–990.

CAPLAN, M. L. & BUSTIN, R. M. 1999. Devonian–Carboniferous Hangenberg mass extinction event, widespread organic-rich mudrock and anoxia: causes and consequences. *Palaeogeography, Palaeoclimatology, Palaeoecology*, **148**, 187–207.

CHAMBERLIN, T. C. 1897. A group of hypotheses bearing on climatic changes. *Journal of Geology*, **5**, 653–683.

CHAMBERLIN, T. C. 1899. An attempt to frame a working hypothesis of the cause of glacial periods on an atmospheric basis. *Journal of Geology*, **7**, 545–584, 667–685, 751–787.

CHAMBERLIN, T. C. 1906. On a possible reversal of deep-sea circulation and its influence on geologic climates. *Proceedings of the American Philosophical Society*, **45**, 33–43.

CHAMLEY, H. 1989. *Clay Sedimentology*. Springer, Berlin.

CHAMLEY, H. 1998. Clay mineral sedimentation in the ocean. *In*: PAQUET, H. & CLAUER, N. (eds) *Soils and Sediments, Mineralogy and Geochemistry*. Springer, Heidelberg, 269–302.

CHANDLER, M. A., RIND, D. & RUEDY, R. 1992. Pangaean climate during the Early Jurassic: GCM simulations and the sedimentary record of paleoclimate. *Geological Society of America Bulletin*, **104**, 543–559.

CHASE, Z., ANDERSON, R. F., FLEISHER, M. Q. & KUBIK, P. 2002. The influence of particle composition on scavenging of Th, Pa and Be in the ocean. *Earth and Planetary Science Letters*, **204**, 215–229, 10.1016/S0012–821X(02)00984–6.

CHAUDHURI, A. K. 2005. Climbing ripple structure and associated storm-lamination from a Proterozoic carbonate platform succession: Their environmental and petrogenetic significance. *Journal of Earth System Science*, **114**, 199–209.

CHUMAKOV, N. M. & FRAKES, L. A. 1997. Mode of origin of dispersed clasts in Jurassic shales; Southern part of the Yana-Kolyma fold belt, North East Asia. *Palaeogeography, Palaeoclimatology, Palaeoecology*, **128**, 77–85.

CISTERNAS, M., ATWATER, B. F., TORREJON, F. *ET AL*. 2005. Predecessors of the giant 1960 Chile earthquake. *Nature*, **437**, 404–407.

CLEMENS, S. C. 1998. Dust response to seasonal atmospheric forcing: Proxy evaluation and calibration. *Paleoceanography*, **13**, 471–490.

CLIMAP 1981. *Seasonal reconstruction of the Earth's surface at the last glacial maximum*. Map and Chart Series MC-36.1–18. Geological Society of America.

COHEN, A. S., COE, A. L., HARDING, S. M. & SCHWARK, L. 2004. Osmium isotope evidence for the regulation of atmospheric CO_2 by continental weathering. *Geology*, **32**, 157–160.

CONDIE, K. C. 2002. The supercontinent cycle: are there two patterns of cyclicity? *Journal of African Earth Sciences*, **35**, 179–183.

CONSTANTINE, A., CHINSAMY, A., VICKERS-RICH, P. & RICH, T. H. 1998. Periglacial environments and polar dinosaurs. *South African Journal of Science*, **94**, 137–141.

COOK, P. J. 1992. Phosphogenesis around the Proterozoic–Phanerozoic transition. *Journal of the Geological Society*, **149**, 615–620.

CORAM, R. A. & JARZEMBOWSKI, E. A. 2002. Diversity and ecology of fossil insects in the Dorset Purbeck succession, southern England. *Life and Environments in Purbeck Times*. Special Papers in Palaeontology Series. Palaeontological Association, Aberystwyth, **68**, 257–268.

COWAN, C. A., FOX, D. L., RUNKEL, A. C. & SALTZMAN, M. R. 2005. Terrestrial-marine carbon cycle coupling in similar to 500-m.y.-old phosphatic brachiopods. *Geology*, **33**, 661–664.

CREBER, G. T. & CHALONER, W. G. 1985. Tree growth in the Mesozoic and early Tertiary and the reconstruction of paleoclimates. *Palaeogeography, Palaeoclimatology, Palaeoecology*, **52**, 35–59.

CROLL, J. 1867. On the change in the obliquity of the ecliptic, its influence on the climate of the polar regions and the level of the sea. *Philosophical Magazine Series 4*, **33**, 426–445.

CROSTA, X., PICHON, J. J. & BURCKLE, L. H. 1998. Application of modern analog technique to marine Antarctic diatoms: Reconstruction of maximum sea-ice extent at the Last Glacial Maximum. *Paleoceanography*, **13**, 284–297.

CROWELL, J. C. 1999. Pre-Mesozoic ice ages: their bearing on understanding the climate system, *Geological Society of America Memoirs*, **145**, 112.

CROWLEY, T. J. & BAUM, S. K. 1992. Modeling Late Paleozoic glaciation. *Geology*, **20**, 507–510.

CROWLEY, T. J. & NORTH, G. R. 1991. *Paleoclimatology*. Oxford University Press, Oxford.

CROWLEY, T. J., YIP, K. J. J. & BAUM, S. K. 1993. Milankovitch Cycles and Carboniferous Climate. *Geophysical Research Letters*, **20**, 1175–1178.

CUBASCH, U., SANTER, B. D., HELLBACH, A. ET AL. 1994. Monte-Carlo Climate-Change Forecasts with a Global Coupled Ocean-Atmosphere Model. *Climate Dynamics*, **10**, 1–19.

CUBASCH, U., VOSS, R., HEGERL, G. C., WASZKEWITZ, J. & CROWLEY, T. J. 1997. Simulation of the influence of solar radiation variations on the global climate with an ocean-atmosphere general circulation model. *Climate Dynamics*, **13**, 757–767.

CULLEN, J. T. & SHERRELL, R. M. 2005. Effects of dissolved carbon dioxide, zinc, and manganese on the cadmium to phosphorus ratio in natural phytoplankton assemblages. *Limnology and Oceanography*, **50**, 1193–1204.

DARWIN, C. R. 1842. Notes on the effects produced by ancient glaciers of Caernarvonshire, and on the boulders transported by floating ice. *London, Edinburgh, and Dublin Philosophical Magazine and Journal of Science*, **21**, 180–188.

DAVID, K., FRANK, M., O'NIONS, R. K., BELSHAW, N. S. & ARDEN, J. W. 2001. The Hf isotope composition of global seawater and the evolution of Hf isotopes in the deep Pacific Ocean from Fe–Mn crusts. *Chemical Geology*, **178**, 23–42.

DAVIS, W. J. & MILLER, A. R. 2001. *A Late Triassic Rb-Sr phlogopite isochron age for a kimberlite dyke from the Rankin Inlet area, Nunavut*. Radiogenic Age and Isotope Studies: Report 14, **Current Research 2001-F3**.

DE LA ROCHA, C. L., BRZEZINSKI, M. A., DENIRO, M. J. & SHEMESH, A. 1998. Silicon-isotope composition of diatoms as an indicator of past oceanic change. *Nature*, **395**, 680–683.

DE MAN, E. & VAN SIMAEYS, S. 2004. Late Oligocene Warming Event in the southern North Sea Basin: benthic foraminifera as paleotemperature proxies. *Netherlands Journal of Geosciences-Geologie En Mijnbouw*, **83**, 227–239.

DEMICCO, R. V., LOWENSTEIN, T. K. & HARDIE, L. A. 2003. Atmospheric pCO_2 since 60 Ma from records of seawater pH, calcium, and primary carbonate mineralogy. *Geology*, **31**, 793–796.

DENTON, G. H. & ARMSTRONG, R. L. 1969. Miocene–Pliocene glaciations in southern Alaska. *American Journal of Science*, **267**, 1121–1142.

DESSERT, C., DUPRE, B., FRANCOIS, L. M. ET AL. 2001. Erosion of Deccan Traps determined by river geochemistry: impact on the global climate and the $^{87}Sr/^{86}Sr$ ratio of seawater. *Earth and Planetary Science Letters*, **188**, 459–474.

DE VILLIERS, S. 2005. Foraminiferal shell-weight evidence for sedimentary calcite dissolution above the lysocline. *Deep-Sea Research Part I-Oceanographic Research Papers*, **52**, 671–680.

DE VILLIERS, S., SHEN, G. T. & NELSON, B. K. 1994. The Sr/Ca temperature relationship in coralline aragonite: influence of variability in (Sr/Ca) seawater and skeletal growth parameters. *Geochimica et Cosmochimica Acta*, **58**, 197–208.

DEWEY, J. F. 2005. Orogeny can be very short. *Proceedings of the National Academy of Sciences of the United States of America*, **102**, 15286–15293.

DICKENS, G. R. 2000. Methane oxidation during the Late Palaeocene Thermal Maximum. *Bulletin de la Société Geologique de France*, **171**, 37–49.

DICKINSON, R. E. 1989. Uncertainties of estimates of climatic change: a review. *Climatic Change*, **15**, 5–13.

DICKINSON, W. R. 2004. Evolution of the North American Cordillera. Annual Review of Earth and Planetary Sciences, **32**, 13–45.

DOBLAS, M., OYARZUN, R., LOPEZ-RUIZ, J. ET AL. 1998. Permo-Carboniferous volcanism in Europe and northwest Africa: a superplume exhaust valve in the centre of Pangaea? *Journal of African Earth Sciences*, **26**, 89–99.

DODONOV, A. E. & BAIGUZINA, L. L. 1995. Loess stratigraphy of Central Asia: Palaeoclimatic and palaeoenvironmental aspects. *Quaternary Science Reviews*, **14**, 707–720.

DORAN, P. T., FRITSEN, C. H., McKAY, C. P., PRISCU, J. C. & ADAMS, E. E. 2003. Formation and character of an ancient 19-m ice cover and underlying trapped brine in an 'ice-sealed' east Antarctic lake. *Proceedings of the National Academy of Sciences of the United States of America*, **100**, 26–31.

DRIESE, S. G., NORDT, L. C., LYNN, W. C., STILES, C. A., MORA, C. I. & WILDING, L. P. 2005. Distinguishing climate in the soil record using chemical trends in a vertisol climosequence from the Texas Coast Prairie, and application to interpreting Paleozoic paleosols in the Appalachian basin, USA. *Journal of Sedimentary Research*, **75**, 339–349.

DRIESE, S. G. & OBER, E. G. 2005. Paleopedologic and paleohydrologic records of precipitation seasonality from early Pennsylvanian 'underclay' Paleosols, USA. *Journal of Sedimentary Research*, **75**, 997–1010.

DROMART, G., GARCIA, J. P., PICARD, S., ATROPS, F., LECUYER, C. & SHEPPARD, S. M. F. 2003. Ice age at the Middle-Late Jurassic transition? *Earth and Planetary Science Letters*, **213**, 205–220.

DUNCAN, R. A. & HELGASON, J. 1998. Precise dating of the Holmatindur cooling event in eastern Iceland: Evidence for mid-Miocene bipolar glaciation. *Journal of Geophysical Research-Solid Earth*, **103**, 12397–12404.

DUTTON, J. F. & BARRON, E. J. 2000. Intra-annual and interannual ensemble forcing of a regional climate model. *Journal of Geophysical Research-Atmospheres*, **105**, 29523–29538.

EDINGER, E. N., COPPER, P., RISK, M. J. & ATMOJO, W. 2002. Oceanography and reefs of recent and Paleozoic tropical epeiric seas. *Facies*, **47**, 127–149.

EDWARDS, D. 1984. Fire regimes in the biomes of South Africa. *In*: BOOYSEN, P. D. V. & TAINTON, N. M.

(eds) *Ecological effects of fire*. Ecological Studies, **48**, 19–38.

EDWARDS, D. 1998. Climate signals in Palaeozoic land plants. *Philosophical Transactions of the Royal Society of London Series B-Biological Sciences*, **353**, 141–156.

EDWARDS, R. L., CHENG, H., MURRELL, M. T. & GOLDSTEIN, S. J. 1997. Protactinium-231 dating of carbonates by thermal ionization mass spectrometry: implications for Quaternary climate change. *Science*, **276**, 782–786, 10.1126/science.276.5313.782.

EHRMANN, W., SETTI, M. & MARINONI, L. 2005. Clay minerals in Cenozoic sediments off Cape Roberts (McMurdo Sound, Antarctica) reveal palaeoclimatic history. *Palaeogeography, Palaeoclimatology, Palaeoecology*, **229**, 187–211.

ELDERFIELD, H. 2002. Foraminiferal Mg/Ca paleothermometry: expected advances and unexpected consequences. *Geochimica et Cosmochimica Acta*, **66**, A213.

ELDERFIELD, H. & RICKABY, R. E. M. 2000. Oceanic Cd/P ratio and nutrient utilization in the glacial Southern Ocean. *Nature*, **405**, 305–310.

ELDERFIELD, H., VAUTRAVERS, M. & COOPER, M. 2002. The relationship between shell size and Mg/Ca, Sr/Ca, δO^{18}, and δC^{13} of species of planktonic foraminifera. *Geochemistry, Geophysics, Geosystems*, **3**.

ELDHOLM, O. & GRUE, K. 1994. North Atlantic volcanic margins: dimensions and production rates. *Journal of Geophysical Research-Solid Earth*, **99**, 2955–2968.

ELLWOOD, M. J., KELLY, M., NODDER, S. D. & CARTER, L. 2004. Zinc/silicon ratios of sponges: A proxy for carbon export to the seafloor. *Geophysical Research Letters*, **31**.

ELORZA, J., GARCIA-GARMILLA, F. & JAGT, J. W. M. 1996. Diagenesis-related differences in isotopic and elemental composition of late Campanian and early Maastrichtian inoceramids and belemnites from NE Belgium: Palaeoenvironmental implications. *Geologie en Mijnbouw*, **75**, 349–360.

ELSHAHED, M. S., NAJAR, F. Z., ROE, B. A., OREN, A., DEWERS, T. A. & KRUMHOLZ, L. R. 2004. Survey of archaeal diversity reveals an abundance of halophilic Archaea in a low-salt, sulfide- and sulfur-rich spring. *Applied and Environmental Microbiology*, **70**, 2230–2239.

EMANUEL, K. 2002. A simple model of multiple climate regimes. *Journal of Geophysical Research-Atmospheres*, **107**.

EMILIANI, C. 1955. Pleistocene temperatures. *Journal of Geology*, **63**, 538–578.

EPSTEIN, S., BUCHSBAUM, R., LOWENSTAM, H. A. & UREY, H. C. 1953. Revised carbonate-water isotopic temperature scale. *Geological Society of America Bulletin*, **64**, 1315–1325.

ERBA, E. 2006. The first 150 million years history of calcareous nannoplankton: Biosphere-geosphere interactions. *Palaeogeography, Palaeoclimatology, Palaeoecology*, **232**, 237–250.

ERNST, R. E. & BUCHAN, K. L. 2001. Large mafic magmatic events through time and links to mantle plume heads. *In*: ERNST, R. E. & BUCHAN, K. L. (eds) *Mantle Plumes: Their Identification Through Time*, **352**, 483–575.

ERNST, R. E., BUCHAN, K. L. & CAMPBELL, I. H. 2005. Frontiers in Large Igneous Province research. *Lithos*, **79**, 271–297, 10.1016/j.lithos.2004.09.004.

EVANS, D. A. D. 2000. Stratigraphic, geochronological, and paleomagnetic constraints upon the neoproterozoic climatic paradox. *American Journal of Science*, **300**, 347–433.

EVANS, D. A. D. 2003. A fundamental Precambrian–Phanerozoic shift in Earth's glacial style? *Tectonophysics*, **375**, 353–385.

EYLES, N. 1993. Earth's, glacial record and its tectonic setting. *Earth-Science Reviews*, **35**, 1–248.

EYLES, N., MORY, A. J. & BACKHOUSE, J. 2002. Carboniferous-Permian palynostratigraphy of west Australian marine rift basins: resolving tectonic and eustatic controls during Gondwanan glaciations. *Palaeogeography, Palaeoclimatology, Palaeoecology*, **184**, 305–319.

EYLES, N., MORY, A. J. & EYLES, C. H. 2006. 50-million-year-long record of glacial to postglacial marine environments preserved in a Carboniferous-Lower Permian graben, Northern Perth Basin, Western Australia. *Journal of Sedimentary Research*, **76**, 618–632.

FAEGRI, K. & IVERSON, J. 1950. *Textbook of Modern Pollen Analysis*. Ejnar Munksgaard, Copenhagen.

FALCON-LANG, H. J. 1999. The Early Carboniferous (Courceyan-Arundian) monsoonal climate of the British Isles: evidence from growth rings in fossil woods. *Geological Magazine*, **136**, 177–187.

FALKOWSKI, P. G. 1997. Evolution of the nitrogen cycle and its influence on the biological sequestration of CO_2 in the ocean. *Nature*, **387**, 272–275, 10.1038/387272a0.

FARLEY, K. A. & ELTGROTH, S. F. 2003. An alternative age model for the Paleocene-Eocene thermal maximum using extraterrestrial ^{3}He. *Earth and Planetary Science Letters*, **208**, 135–148.

FAUQUETTE, S. & BERTINI, A. 2003. Quantification of the northern Italy Pliocene climate from pollen data: evidence for a very peculiar climate pattern. *Boreas*, **32**, 361–369.

FELZER, B. A. & THOMPSON, S. L. 2001. Evaluation of a regional climate model for paleoclimate applications in the Arctic. *Journal of Geophysical Research-Atmospheres*, **106**, 27407–27424.

FERNANDEZ, M. H. & PELAEZ-CAMPOMANES, P. 2005. Quantitative palaeoclimatic inference based on terrestrial mammal faunas. *Global Ecology and Biogeography*, **14**, 39–56.

FINKELSTEIN, D. B., PRATT, L. M., CURTIN, T. M. & BRASSELL, S. C. 2005. Wildfires and seasonal aridity recorded in Late Cretaceous strata from south-eastern Arizona, USA. *Sedimentology*, **52**, 587–599.

FLOWER, B. P. 1999. Cenozoic deep-sea temperatures and polar glaciation: the oxygen isotope record. *Terra Antartica Reports*, **3**, 27–42.

FORTEY, N. J., MERRIMAN, R. J. & HUFF, W. D. 1996. Silurian and late Ordovician K-bentonites as a record of late Caledonian volcanism in the British Isles. *Transactions of the Royal Society of Edinburgh-Earth Sciences*, **86**, 167–180.

FOUKAL, P., FROHLICH, C., SPRUIT, H. & WIGLEY, T. M. L. 2006. Variations in solar luminosity and their effect on the Earth's climate. *Nature*, **443**, 161–166.

FRAKES, L. A. 1979. *Climates through geologic time.* Elsevier, Amsterdam.

FRAKES, L. A., FRANCIS, J. E. & SYKTUS, J. I. 1992. *Climate modes of the Phanerozoic – The history of the Earth's climate over the past 600 million years.* Cambridge University Press, Cambridge.

FRANCIS, J. E. 1986. Growth rings in Cretaceous and Tertiary wood from Antarctica and their paleoclimatic implications. *Palaeontology*, **29**, 665–684.

FRANCIS, J. E. & POOLE, I. 2002. Cretaceous and early Tertiary climates of Antarctica: evidence from fossil wood. *Palaeogeography, Palaeoclimatology, Palaeoecology*, **182**, 47–64.

FRANCOIS, L. M. & GERARD, J. C. 1986. A Numerical-Model of the Evolution of Ocean Sulfate and Sedimentary Sulfur During the Last 800 Million Years. *Geochimica et Cosmochimica Acta*, **50**, 2289–2302.

FRANCOIS, L. M. & WALKER, J. C. G. 1992. Modeling the Phanerozoic carbon cycle and climate: constraints from the ^{87}Sr-^{86}Sr isotopic ratio of seawater. *American Journal of Science*, **292**, 81–135.

FRANCOIS, R., ALTABET, M. A., YU, E.-F. *ET AL.* 1997. Contribution of Southern Ocean surface-water stratification to low atmospheric CO_2 concentrations during the last glacial period. *Nature*, **389**, 929–935.

FREIWALD, A., HENRICH, R. & PÄTZOLD, J. 1997. Anatomy of a deep-water coral reef mound from Stjernsund, West Finnmark, northern Norway. *In*: JAMES, N. P. & CLARKE, J. A. D. (eds) *Cool-water carbonates.* SEPM Special Volume. SEPM, Tulsa, Oklahoma, **56**, 741–762.

FREIWALD, A. & WILSON, J. B. 1998. Taphonomy of modern deep, cold-temperate water coral reefs. *Historical Biology*, **13**, 37–52.

FRIEDMAN, M., TARDUNO, J. A. & BRINKMAN, D. B. 2003. Fossil fishes from the high Canadian Arctic: further palaeobiological evidence for extreme climatic warmth during the Late Cretaceous (Turonian-Coniacian). *Cretaceous Research*, **24**, 615–632.

FRIEDRICH, O., HERRLE, J. O. & HEMLEBEN, C. 2005. Climatic changes in the Late Campanian-Early Maastrichtian: Micropaleontological and stable isotopic evidence from an epicontinental sea. *Journal of Foraminiferal Research*, **35**, 228–247.

FRIEDRICH, O., SCHMIEDL, G. & ERLENKEUSER, H. 2006. Stable isotope composition of Late Cretaceous benthic foraminifera from the southern South Atlantic: Biological and environmental effects. *Marine Micropaleontology*, **58**, 135–157.

FRITTS, H. C. 1976. *Tree rings and climate.* Academic Press, London.

GARRELS, R. M. & LERMAN, A. 1984. Coupling of the sedimentary sulfur and carbon cycles: an improved model. *American Journal of Science*, **284**, 989–1007.

GATES, W. L. 1974. Numerical simulation of Ice Age climate. *Transactions-American Geophysical Union*, **55**, 259–259.

GATES, W. L. 1976. Numerical simulation of Ice Age climate with a global general circulation model. *Journal of the Atmospheric Sciences*, **33**, 1844–1873.

GAZDZICKI, A., GRUSZCZYNSKI, M., HOFFMAN, A., MALKOWSKI, K., MARENSSI, S. A., HALAS, S. & TATUR, A. 1992. Stable carbon and oxygen isotope record in the Paleogene La Meseta Formation, Seymour Island, Antarctica. *Antarctic Science*, **4**, 461–468.

GENC, S. C. 2004. A Triassic large igneous province in the Pontides, northern Turkey: geochemical data for its tectonic setting. *Journal of Asian Earth Sciences*, **22**, 503–516.

GERSONDE, R. & ZIELINSKI, U. 2000. The reconstruction of late Quaternary Antarctic sea-ice distribution – the use of diatoms as a proxy for sea-ice. *Palaeogeography, Palaeoclimatology, Palaeoecology*, **162**, 263–286.

GIBBS, M. T., BARRON, E. J. & KUMP, L. R. 1997. An atmospheric pCO_2 threshold for glaciation in the Late Ordovician. *Geology*, **25**, 447–450.

GIFFORD, R. 1994. The global carbon cycle: a viewpoint on the missing sink. *Australian Journal of Plant Physiology*, **21**, 1–15, 10.1071/PP9940001.

GINGELE, F. & DAHMKE, A. 1994. Discrete barite particles and barium as tracers of paleoproductivity in South Atlantic sediments. *Paleoceanography*, **9**, 151–168, 10.1029/93PA02559.

GINGELE, F. X. & SCHMIEDER, F. 2001. Anomalous South Atlantic lithologies confirm global scale of unusual mid-Pleistocene climate excursion. *Earth and Planetary Science Letters*, **186**, 93–101.

GLASS, L. M. & PHILLIPS, D. 2006. The Kalkarindji continental flood basalt province: A new Cambrian large igneous province in Australia with possible links to faunal extinctions. *Geology*, **34**, 461–464, 10.1130/G22122.1.

GLASSPOOL, I. 2000. A major fire event recorded in the mesofossils and petrology of the Late Permian, Lower Whybrow coal seam, Sydney Basin, Australia. *Palaeogeography, Palaeoclimatology, Palaeoecology*, **164**, 357–380.

GLASSPOOL, I. J., HILTON, J., COLLINSON, M. E., WANG, S. J. & LI CHENG, S. 2004. Foliar physiognomy in Cathaysian gigantopterids and the potential to track Palaeozoic climates using an extinct plant group. *Palaeogeography, Palaeoclimatology, Palaeoecology*, **205**, 69–110.

GLUMAC, B. & SPIVAK-BIRNDORF, M. L. 2002. Stable isotopes of carbon as an invaluable stratigraphic tool: An example from the Cambrian of the northern Appalachians, USA. *Geology*, **30**, 563–566.

GRADSTEIN, F. M., OGG, J. G., SMITH, A. G. *ET AL.* 2004. *A Geologic Time Scale 2004.* Cambridge University Press, Cambridge.

GREGORY-WODZICKI, K. M. 2000. Relationships between leaf morphology and climate, Bolivia: implications for estimating paleoclimate from fossil floras. *Paleobiology*, **26**, 668–688.

GREINERT, J. & DERKACHEV, A. 2004. Glendonites and methane-derived Mg-calcites in the Sea of Okhotsk, Eastern Siberia: implications of a venting-related ikaite/glendonite formation. *Marine Geology*, **204**, 129–144.

GRIFFIN, W. L., RYAN, C. G., KAMINSKY, F. V. *ET AL.* 1999. The Siberian lithosphere traverse: mantle terranes and the assembly of the Siberian Craton. *Tectonophysics*, **310**, 1–35.

GRIMES, S. T., HOOKER, J. J., COLLINSON, M. E. & MATTEY, D. P. 2005. Summer temperatures of late Eocene to early Oligocene freshwaters. *Geology*, **33**, 189–192.

GRIMES, S. T., MATTEY, D. P., HOOKER, J. J. & COLLINSON, M. E. 2003. Paleogene paleoclimate reconstruction using oxygen isotopes from land and freshwater organisms: the use of multiple paleo-proxies. *Geochimica et Cosmochimica Acta*, **67**, 4033–4047.

GRZEBYK, D., SCHOFIELD, O., VETRIANI, C. & FALKOWSKI, P. G. 2003. The Mesozoic radiation of eukaryotic algae: the portable plastid hypothesis. *Journal of Phycology*, **39**, 259–267.

GUIDRY, M. W. & MACKENZIE, F. T. 2000. Apatite weathering and the Phanerozoic phosphorus cycle. *Geology*, **28**, 631–634.

GUNNELL, Y. 1998. Passive margin uplifts and their influence on climatic change and weathering patterns of tropical shield regions. *Global and Planetary Change*, **18**, 47–57.

GUSSONE, N., EISENHAUER, A., HEUSER, A. ET AL. 2003. Model for kinetic effects on calcium isotope fractionation (δCa-44) in inorganic aragonite and cultured planktonic foraminifera. *Geochimica et Cosmochimica Acta*, **67**, 1375–1382.

HALEY, B. A., KLINKHAMMER, G. P. & MIX, A. C. 2005. Revisiting the rare earth elements in foraminiferal tests. *Earth and Planetary Science Letters*, **239**, 79–97.

HALLAM, A. 1985. A review of Mesozoic climates. *Journal of the Geological Society, London*, **142**, 433–445.

HALLAM, A. & WIGNALL, P. B. 1999. Mass extinctions and sea-level changes. *Earth-Science Reviews*, **48**, 217–250.

HAMILTON, S., SHENNAN, I., COMBELLICK, R., MULHOLLAND, J. & NOBLE, C. 2005. Evidence for two great earthquakes at Anchorage, Alaska and implications for multiple great earthquakes through the Holocene. *Quaternary Science Reviews*, **24**, 2050–2068.

HAMMOND, D. E., MCMANUS, J. & BERELSON, W. M. 2004. Oceanic germanium/silicon ratios: Evaluation of the potential overprint of temperature on weathering signals. *Paleoceanography*, **19**.

HANSEN, K. W. & WALLMANN, K. 2003. Cretaceous and Cenozoic evolution of seawater composition, atmospheric O_2 and CO_2: A model perspective. *American Journal of Science*, **303**, 94–148.

HARDIE, L. A. 1996. Secular variation in seawater chemistry: An explanation for the coupled secular variation in the mineralogies of marine limestones and potash evaporites over the past 600 m.y. *Geology*, **24**, 279–283.

HARRINGTON, G. J. 2004. Structure of the North American vegetation gradient during the late Paleocene/early Eocene warm climate. *Evolutionary Ecology Research*, **6**, 33–48.

HARRIS, T. M. 1981. Burnt ferns from the English Wealden. *Proceedings of the Geologists' Association*, **92**, 47–58.

HARRISON, R. G. & STEPHENSON, D. B. 2006. Empirical evidence for a nonlinear effect of galactic cosmic rays on clouds. *Proceedings of the Royal Society A-Mathematical Physical and Engineering Sciences*, **462**, 1221–1233.

HAUG, G. H., GANOPOLSKI, A., SIGMAN, D. M. ET AL. 2005. North Pacific seasonality and the glaciation of North America 2.7 million years ago. *Nature*, **433**, 821–825.

HAUSMANN, S. & KIENAST, F. 2006. A diatom-inference model for nutrients screened to reduce the influence of background variables: Application to varved sediments of Greifensee and evaluation with measured data. *Palaeogeography, Palaeoclimatology, Palaeoecology*, **233**, 96–112.

HAWORTH, M., HESSELBO, S. P., MCELWAIN, J. C., ROBINSON, S. A. & BRUNT, J. W. 2005. Mid-Cretaceous pCO2 based on stomata of the extinct conifer *Pseudofrenelopsis* (Cheirolepidiaceae). *Geology*, **33**, 749–752.

HAY, W. W. 1996. Tectonics and climate. *Geologische Rundschau*, **85**, 409–437.

HAY, W. W., DECONTO, R. M. & WOLD, C. N. 1997. Climate: Is the past the key to the future? *Geologische Rundschau*, **86**, 471–491.

HAYS, J. D., IMBRIE, J. & SHACKLETON, N. J. 1976. Variations in the Earth's orbit: pacemaker of the ice ages. *Science*, **194**, 1121–1132.

HAYS, P. D. & GROSSMAN, E. L. 1991. Oxygen isotopes in meteoric calcite cements as indicators of continental paleoclimate. *Geology*, **19**, 441–444.

HAYWARD, B. W. 2001. Global deep-sea extinctions during the Pleistocene ice ages. *Geology*, **29**, 599–602.

HAYWOOD, A. M., DEKENS, P., RAVELO, A. C. & WILLIAMS, M. 2005. Warmer tropics during the mid-Pliocene? Evidence from alkenone paleothermometry and a fully coupled ocean-atmosphere GCM. *Geochemistry, Geophysics, Geosystems*, **6**, 1–20, 10.1029/2004GC000799.

HAYWOOD, A. M., SELLWOOD, B. W. & VALDES, P. J. 2000. Regional warming: Pliocene (3 Ma) paleoclimate of Europe and the Mediterranean. *Geology*, **28**, 1063–1066.

HAYWOOD, A. M. & VALDES, P. J. 2004. Modelling Pliocene warmth: contribution of atmosphere, oceans and cryosphere. *Earth and Planetary Science Letters*, **218**, 363–377.

HAYWOOD, A. M., VALDES, P. J., FRANCIS, J. E. & SELLWOOD, B. W. 2002a. Global middle Pliocene biome reconstruction: A data/model synthesis. *Geochemistry, Geophysics, Geosystems*, **3**.

HAYWOOD, A. M., VALDES, P. J., SELLWOOD, B. W. & KAPLAN, J. O. 2002b. Antarctic climate during the middle Pliocene: model sensitivity to ice sheet variation. *Palaeogeography, Palaeoclimatology, Palaeoecology*, **182**, 93–115.

HAYWOOD, A. M., VALDES, P. J. & MARKWICK, P. J. 2004. Cretaceous (Wealden) climates: a modelling perspective. *Cretaceous Research*, **25**, 303–311.

HAYWOOD, A. M., VALDES, P. J. & PECK, V. L. 2007. A permanent El Niño-like state during the Pliocene? *Paleoceanography*, **22**, pa 1213.

HEAMAN, L. M. & KJARSGAARD, B. A. 2000. Timing of eastern North American kimberlite magmatism: continental extension of the Great Meteor hotspot track? *Earth and Planetary Science Letters*, **178**, 253–268.

HEAMAN, L. M., KJARSGAARD, B. A. & CREASER, R. A. 2003. The timing of kimberlite magmatism in North America: implications for global kimberlite genesis and diamond exploration. *Lithos*, **71**, 153–184.

HECKEL, P. H. 1986. Sea-level curve for Pennsylvanian eustatic marine transgressive–regressive depositional cycles along midcontinent outcrop belt, North America. *Geology*, **14**, 330–334.

HECKEL, P. H. 1996. Glacial-eustatic base-level climatic model for late Middle to Late Pennsylvanian coal-bed formation in the Appalachian basin. *Journal of Sedimentary Research Section B-Stratigraphy and Global Studies*, **65**, 348–356.

HEILBRON, M. & MACHADO, N. 2003. Timing of terrane accretion in the Neoproterozoic-Eopaleozoic Ribeira orogen (SE Brazil). *Precambrian Research*, **125**, 87–112.

HENDERSON, G. M. 2002. New oceanic proxies for paleoclimate. *Earth and Planetary Science Letters*, **203**, 1–13.

HENTSCHKE, U. & MILKERT, D. 1996. Power spectrum analyses of storm layers in marine silty sediments: A tool for a paleoclimatic reconstruction? *Journal of Coastal Research*, **12**, 898–906.

HERNDL, G. J., REINTHALER, T., TEIRA, E., VAN AKEN, H., VETH, C., PERNTHALER, A. & PERNTHALER, J. 2005. Contribution of Archaea to total prokaryotic production in the deep Atlantic Ocean. *Applied and Environmental Microbiology*, **71**, 2303–2309.

HERRMANN, A. D., PATZKOWSKY, M. E. & POLLARD, D. 2003. Obliquity forcing with 8–12 times preindustrial levels of atmospheric pCO$_2$ during the Late Ordovician glaciation. *Geology*, **31**, 485–488.

HESSELBO, S. P., GROCKE, D. R., JENKYNS, H. C., BJERRUM, C. J., FARRIMOND, P., BELL, H. S. M. & GREEN, O. R. 2000. Massive dissociation of gas hydrate during a Jurassic oceanic anoxic event. *Nature*, **406**, 392–395.

HOAR, M. R., PALUTIKOF, J. P. & THORNE, M. C. 2004. Model intercomparison for the present day, the mid-Holocene, and the Last Glacial Maximum over western Europe. *Journal of Geophysical Research-Atmospheres*, **109**.

HODELL, D. A., CURTIS, J. H., SIERRO, F. J. & RAYMO, M. E. 2001. Correlation of late Miocene to early Pliocene sequences between the Mediterranean and North Atlantic. *Paleoceanography*, **16**, 164–178.

HODELL, D. A., ELMSTROM, K. M. & KENNETT, J. P. 1986. Latest Miocene benthic δ^{18}O changes, global ice volume, sea-level and the Messinian salinity crisis. *Nature*, **320**, 411–414.

HOEFS, M. J. L., VERSTEEGH, G. J. M., RIJPSTRA, W. I. C., DE LEEUW, J. W. & DAMSTE, J. S. S. 1998. Postdepositional oxic degradation of alkenones: Implications for the measurement of palaeo sea surface temperatures. *Paleoceanography*, **13**, 42–49.

HOFFMAN, P. F., KAUFMAN, A. J., HALVERSON, G. P. & SCHRAG, D. P. 1998. A Neoproterozoic snowball Earth. *Science*, **281**, 1342–1346.

HOFMANN, C., COURTILLOT, V., FERAUD, G., ROCHETTE, P., YIRGU, G., KETEFO, E. & PIK, R. 1997. Timing of the Ethiopian flood basalt event and implications for plume birth and global change. *Nature*, **389**, 838–841.

HOLBOURN, A., KUHNT, W., SCHULZ, M. & ERLENKEUSER, H. 2005. Impacts of orbital forcing and atmospheric carbon dioxide on Miocene ice-sheet expansion. *Nature*, **438**, 483–487.

HOLLAND, H. D. 2004. Sea level, sediments, and the composition of Phanerozoic seawater. *Geochimica et Cosmochimica Acta*, **68**, A332–A332.

HOLLANDER, D. J., MCKENZIE, J. A. & HSU, K. J. 1993. Carbon isotope evidence for unusual plankton blooms and fluctuations of surface water CO$_2$ in Strangelove Ocean after terminal Cretaceous event. *Palaeogeography, Palaeoclimatology, Palaeoecology*, **104**, 229–237.

HONISCH, B., HEMMING, N. G., GROTTOLI, A. G., AMAT, A., HANSON, G. N. & BUMA, J. 2004. Assessing scleractinian corals as recorders for paleo-pH: Empirical calibration and vital effects. *Geochimica et Cosmochimica Acta*, **68**, 3675–3685.

HOOPER, P. R., BINGER, G. B. & LEES, K. R. 2002. Ages of the Steens and Columbia River flood basalts and their relationship to extension-related calc-alkalic volcanism in eastern Oregon. *Geological Society of America Bulletin*, **114**, 43–50.

HOUSE, J. I., PRENTICE, I. C., RAMANKUTTY, N., HOUGHTON, R. A. & HEIMANN, M. 2003. Reconciling apparent inconsistencies in estimates of terrestrial CO$_2$ sources and sinks. *Tellus Series B-Chemical and Physical Meteorology*, **55**, 345–363.

HUBER, B. T., MACLEOD, K. G. & WING, S. L. 2000. *Warm Climates in Earth History*, Cambridge University Press, Cambridge, **xi**, 455.

HUTCHISON, J. H. 1982. Turtle, crocodilian, and champsosaur diversity changes in the Cenozoic of the north-central region of western United States. *Palaeogeography, Palaeoclimatology, Palaeoecology*, **37**, 149–164.

HUYNH, T. T. & POULSEN, C. J. 2005. Rising atmospheric CO$_2$ as a possible trigger for the end-Triassic mass extinction. *Palaeogeography, Palaeoclimatology, Palaeoecology*, **217**, 223–242.

IGAMBERDIEV, A. U. & LEA, P. J. 2006. Land plants equilibrate O$_2$ and CO$_2$ concentrations in the atmosphere. *Photosynthesis Research*, **87**, 177–194.

IMBRIE, J. & KIPP, N. 1971. A new micropalaeontological method for quantitative paleoclimatology: application to late Pleistocene Caribbean core V28–238. *In*: TUREKIAN, K. K. (ed.) *The late Cenozoic Glacial Ages*. Yale University Press, New Haven, 77–181.

IMBRIE, J. & NEWELL, N. 1964. *Approaches to Paleoecology*. John Wiley & Sons Inc., New York.

IMMENHAUSER, A. 2005. High-rate sea-level change during the Mesozoic: New approaches to an old problem. *Sedimentary Geology*, **175**, 277–296.

INGLE, S., SCOATES, J. S., WEIS, D., BRUGMANN, G. & KENT, R. W. 2004. Origin of Cretaceous continental tholefites in southwestern Australia and eastern India: insights from Hf and Os isotopes. *Chemical Geology*, **209**, 83–106.

IPCC 2001. *Climate Change 2001: the scientific basis – contribution of working group I to the third assessment report of IPCC.*

IRVING, E. 2005. The role of latitude in mobilism debates. *Proceedings of the National Academy of Sciences of the United States of America*, **102**, 1821–1828.

ISBELL, J. L., MILLER, M. F., BABCOCK, L. E. & HASIOTIS, S. T. 2001. Ice-marginal environment and ecosystem prior to initial advance of the late Palaeozoic ice sheet in the Mount Butters area of the central Transantarctic Mountains, Antarctica. *Sedimentology*, **48**, 953–970.

ISOZAKI, Y., KAWAHATA, H. & OTA, A. 2006. A unique carbon isotope record across the Guadalupian-Lopingian (Permian) boundary in Panthalassan paleoatoll carbonates: the high-productivity 'Kamura event' and its collapse. *Geophysical Research Abstracts*, **8**, 5365.

JAMES, N. P., NARBONNE, G. M., DALRYMPLE, R. W. & KYSER, T. K. 2005. Glendonites in Neoproterozoic low-latitude, interglacial, sedimentary rocks, northwest Canada: Insights into the Cryogenian ocean and Precambrian cold-water carbonates. *Geology*, **33**, 9–12.

JELSMA, H. A., DE WIT, M. J., THIART, C., DIRKS, P., VIOLA, G., BASSON, I. J. & ANCKAR, E. 2004. Preferential distribution along transcontinental corridors of kimberlites and related rocks of Southern Africa. *South African Journal of Geology*, **107**, 301–324.

JENKYNS, H. C. 2003. Evidence for rapid climate change in the Mesozoic–Palaeogene greenhouse world. *Philosophical Transactions of the Royal Society of London Series A-Mathematical Physical and Engineering Sciences*, **361**, 1885–1916.

JENKYNS, H. C., GROCKE, D. R. & HESSELBO, S. P. 2001. Nitrogen isotope evidence for water mass denitrification during the early Toarcian (Jurassic) oceanic anoxic event. *Paleoceanography*, **16**, 593–603.

JEPPSSON, L. & ALDRIDGE, R. J. 2000. Ludlow (late Silurian) oceanic episodes and events. *Journal of the Geological Society*, **157**, 1137–1148.

JEPPSSON, L. & CALNER, M. 2003. The Silurian Mulde Event and a scenario for secundo-secundo events. *Transactions of the Royal Society of Edinburgh-Earth Sciences*, **93**, 135–154.

JIMENEZ-ESPINOSA, R. & JIMENEZ-MILLAN, J. 2003. Calcrete development in Mediterranean colluvial carbonate systems from SE Spain. *Journal of Arid Environments*, **53**, 479–489.

JIMENEZ-MORENO, G., RODRIGUEZ-TOVAR, F. J., PARDO-IGUZQUIZA, E., FAUQUETTE, S., SUC, J. P. & MULLER, P. 2005. High-resolution palynological analysis in late early-middle Miocene core from the Pannonian Basin, Hungary: climatic changes, astronomical forcing and eustatic fluctuations in the Central Paratethys. *Palaeogeography, Palaeoclimatology, Palaeoecology*, **216**, 73–97.

JOACHIMSKI, M. M., SIMON, L., VAN GELDERN, R. & LECUYER, C. 2005. Boron isotope geochemistry of Paleozoic brachiopod calcite: Implications for a secular change in the boron isotope geochemistry of seawater over the Phanerozoic. *Geochimica et Cosmochimica Acta*, **69**, 4035–4044.

JOHNSON, J. S. & SMELLIE, J. L. 2007. Zeolite compositions as proxies for eruptive palaeoenvironment. *Geochemistry, Geophysics, Geosystems*, **8**, Q03009.

JONES, G. S., GREGORY, J. M., STOTT, P. A., TETT, S. F. B. & THORPE, R. B. 2005. An AOGCM simulation of the climate response to a volcanic super-eruption. *Climate Dynamics*, **25**, 725–738.

JONES, I. W., MUNHOVEN, G., TRANTER, M., HUYBRECHTS, P. & SHARP, M. J. 2002. Modelled glacial and non-glacial HCO_3^-, Si and Ge fluxes since the LGM: little potential for impact on atmospheric CO_2 concentrations and a potential proxy of continental chemical erosion, the marine Ge/Si ratio. *Global and Planetary Change*, **33**, 139–153.

KAFANOV, A. I. & VOLVENKO, I. V. 1997. Bivalve molluscs and Cenozoic paleoclimatic events in the northwestern Pacific Ocean. *Palaeogeography, Palaeoclimatology, Palaeoecology*, **129**, 119–153.

KALJO, D., MARTMA, T., MANNIK, P. & VIIRA, V. 2003. Implications of Gondwana glaciations in the Baltic late Ordovician and Silurian and a carbon isotopic test of environmental cyclicity. *Bulletin de la Société Géologique de France*, **174**, 59–66.

KASEMANN, S. A., HAWKESWORTH, C. J., PRAVE, A. R., FALLICK, A. E. & PEARSON, P. N. 2005. Boron and calcium isotope composition in Neoproterozoic carbonate rocks from Namibia: evidence for extreme environmental change. *Earth and Planetary Science Letters*, **231**, 73–86.

KAUFMAN, D. S. & BRIGHAMGRETTE, J. 1993. Aminostratigraphic correlations and paleotemperature implications, Pliocene–Pleistocene high-sea-level deposits, northwestern Alaska. *Quaternary Science Reviews*, **12**, 21–33.

KAWAGATA, S., HAYWARD, B. W., GRENFELL, H. R. & SABAA, A. 2005. Mid-Pleistocene extinction of deep-sea foraminifera in the North Atlantic Gateway (ODP sites 980 and 982). *Palaeogeography, Palaeoclimatology, Palaeoecology*, **221**, 267–291.

KAWAGATA, S., HAYWARD, B. W. & GUPTA, A. K. 2006. Benthic foraminiferal extinctions linked to late Pliocene-Pleistocene deep-sea circulation changes in the northern Indian Ocean (ODP Sites 722 and 758). *Marine Micropaleontology*, **58**, 219–242.

KEANY, J., LEDBETTER, M., WATKINS, N. & HUANG, T. C. 1976. Diachronous deposition of ice-rafted debris in sub-Antarctic deep sea sediments. *Geological Society of America Bulletin*, **87**, 873–882.

KELLER, G. 2004. Low-diversity, late Maastrichtian and early Danian planktic foraminiferal assemblages of the eastern Tethys. *Journal of Foraminiferal Research*, **34**, 49–73.

KELLER, G., ADATTE, T., STINNESBECK, W., LUCIANI, V., KAROUI-YAAKOUB, N. & ZAGHBIB-TURKI, D. 2002. Paleoecology of the Cretaceous-Tertiary mass extinction in planktonic foraminifera. *Palaeogeography, Palaeoclimatology, Palaeoecology*, **178**, 257–297.

KENNEDY, E. M. 2003. Late Cretaceous and Paleocene terrestrial climates of New Zealand: leaf fossil evidence from South Island assemblages. *New Zealand Journal of Geology and Geophysics*, **46**, 295–306.

KENNETT, J. P. 1977. Cenozoic evolution of Antarctic glaciation, the circum-Antarctic Ocean, and their impact on global paleoceanography. *Journal of Geophysical Research*, **82**, 3843–3860.

KERRICK, D. M. 2001. Present and past nonanthropogenic CO_2 degassing from the solid Earth. *Reviews of Geophysics*, **39**, 565–585.

KERRICK, D. M. & CALDEIRA, K. 1999. Was the Himalayan orogen a climatically significant coupled source and sink for atmospheric CO_2 during the Cenozoic? *Earth and Planetary Science Letters*, **173**, 195–203.

KIEFFER, B., ARNDT, N., LAPIERRE, H. *ET AL.* 2004. Flood and shield basalts from Ethiopia: magmas from the African Superswell. *Journal of Petrology*, **45**, 793–834, doi:10.1093/petrology/egg108.

KIESSLING, W., FLUGEL, E. & GOLONKA, J. 2003. Patterns of Phanerozoic carbonate platform sedimentation. *Lethaia*, **36**, 195–225.

KIRKLAND, C. L., DALY, J. S., EIDE, E. A. & WHITEHOUSE, M. J. 2006. The structure and timing of lateral escape during the Scandian Orogeny: A combined strain and geochronological investigation in Finnmark, Arctic Norwegian Caledonides. *Tectonophysics*, **425**, 159–189.

KITAMURA, A. & KAWAGOE, T. 2006. Eustatic sea-level change at the Mid-Pleistocene climate transition: new evidence from the shallow-marine sediment record of Japan. *Quaternary Science Reviews*, **25**, 323–335.

KLEIN, G. D. 1990. Pennsylvanian time scales and cycle periods. *Geology*, **18**, 455–457, doi:10.1130/0091–7613(1990)018<0455:PTSACP>2.3.CO;2.

KLOTZ, S., FAUQUETTE, S., COMBOURIEU-NEBOUT, N., UHL, D., SUC, J. P. & MOSBRUGGER, V. 2006. Seasonality intensification and long-term winter cooling as a part of the Late Pliocene climate development. *Earth and Planetary Science Letters*, **241**, 174–187.

KNIES, J., MATTHIESSEN, J., VOGT, C. & STEIN, R. 2002. Evidence of 'Mid-Pliocene (similar to 3 Ma) global warmth' in the eastern Arctic Ocean and implications for the Svalbard/Barents Sea ice sheet during the late Pliocene and early Pleistocene (similar to 3–1.7 Ma). *Boreas*, **31**, 82–93.

KNOLL, A. H., BAMBACH, R. K., CANFIELD, D. E. & GROTZINGER, J. P. 1996. Comparative Earth history and Late Permian mass extinction. *Science*, **273**, 452–457.

KNOLL, A. H., WALTER, M. R., NARBONNE, G. M. & CHRISTIE-BLICK, N. 2006. The Ediacaran Period: a new addition to the geologic time scale. *Lethaia*, **39**, 13–30.

KOHFELD, K. E. & HARRISON, S. P. 2000. How well can we simulate past climates? Evaluating the models using global palaeoenvironmental datasets. *Quaternary Science Reviews*, **19**, 321–346.

KOHFELD, K. E. & HARRISON, S. P. 2001. DIRTMAP: the geological record of dust. *Earth-Science Reviews*, **54**, 81–114.

KÖPPEN, V. P. & WEGENER, A. 1924. *Die Klimate der Geologischen Vorzeit*. Gebrüder Bornträger, Berlin.

KORHOLA, A., BIRKS, H. J. B., OLANDER, H. & BLOM, T. 2001. Chironomids, temperature and numerical models: a reply to Seppala. *Holocene*, **11**, 615–622.

KORTE, C., KOZUR, H. W. & VEIZER, J. 2005. Delta C-13 and delta O-18 values of Triassic brachiopods and carbonate rocks as proxies for coeval seawater

and palaeotemperature. *Palaeogeography, Palaeoclimatology, Palaeoecology*, **226**, 287–306.

KRIJGSMAN, W., GARCES, M., AGUSTI, J., RAFFI, I., TABERNER, C. & ZACHARIASSE, W. J. 2000. The 'Tortonian salinity crisis' of the eastern Betics (Spain). *Earth and Planetary Science Letters*, **181**, 497–511.

KRIJGSMAN, W., HILGEN, F. J., RAFFI, I., SIERRO, F. J. & WILSON, D. S. 1999. Chronology, causes and progression of the Messinian salinity crisis. *Nature*, **400**, 652–655.

KUCERA, M. & DARLING, K. F. 2002. Cryptic species of planktonic foraminifera: their effect on palaeoceanographic reconstructions. *Philosophical Transactions of the Royal Society of London Series A-Mathematical Physical and Engineering Sciences*, **360**, 695–718.

KUMAR, N., ANDERSON, R. F., MORTLOCK, R. A., FROELICH, P. N., KUBIK, P., DITTRICHHANNEN, B. & SUTER, M. 1995. Increased biological productivity and export production in the glacial Southern Ocean. *Nature*, **378**, 675–680.

KUMP, L. R. 2002. Reducing uncertainty about carbon dioxide as a climate driver. *Nature*, **419**, 188–190, 10.1038/nature01087.

KUSAKABE, M., KU, T. L., SOUTHON, J. R., VOGEL, J. S., NELSON, D. E., MEASURES, C. I. & NOZAKI, Y. 1987. Distribution of ^{10}Be and ^{9}Be in the Pacific Ocean. *Earth and Planetary Science Letters*, **82**, 231–240, doi:10.1016/0012–821X(87)90198–1.

KUTZBACH, J. E. & GALLIMORE, R. G. 1989. Pangaean climates: megamonsoons of the megacontinent. *Journal of Geophysical Research-Atmospheres*, **94**, 3341–3357.

KUTZBACH, J. E. & ZIEGLER, A. M. 1993. Simulation of Late Permian climate and biomes with an atmosphere–ocean model: comparisons with observations. *Philosophical Transactions of the Royal Society of London Series B-Biological Sciences*, **341**, 327–340.

LAGROIX, F. & BANERJEE, S. K. 2002. Paleowind directions from the magnetic fabric of loess profiles in central Alaska. *Earth and Planetary Science Letters*, **195**, 99–112.

LARSON, R. L. 1991. Geological consequences of superplumes. *Geology*, **19**, 963–966.

LARSON, R. L. 1995. The mid-Cretaceous superplume episode. *Scientific American*, **272**, 82–86.

LASAGA, A. C. & BERNER, R. A. 1998. Fundamental aspects of quantitative models for geochemical cycles. *Chemical Geology*, **145**, 161–175.

LAURIOL, B. & CLARK, I. 1999. Fissure calcretes in the arctic: a paleohydrologic indicator. *Applied Geochemistry*, **14**, 775–785.

LE GUERN, P. & DAVAUD, E. 2005. Recognition of ancient carbonate wind deposits: lessons from a modern analogue, Chrissi Island, Crete. *Sedimentology*, **52**, 915–926.

LE GUERROUE, E., ALLEN, P. A., COZZI, A., ETIENNE, J. L. & FANNING, M. 2006. 50 Myr recovery from the largest negative δC^{13} excursion in the Ediacaran ocean. *Terra Nova*, **18**, 147–153.

LEA, D. W. 1993. Constraints on the alkalinity and circulation of glacial circumpolar deep water from benthic foraminiferal barium. *Global Biogeochemical Cycles*, **7**, 695–710.

LEAR, C. H., ELDERFIELD, H. & WILSON, P. A. 2003*a*. A Cenozoic seawater Sr/Ca record from benthic foraminiferal calcite and its application in determining global weathering fluxes. *Earth and Planetary Science Letters*, **208**, 69–84.

LEAR, C. H., ROSENTHAL, Y. & WRIGHT, J. D. 2003*b*. The closing of a seaway: ocean water masses and global climate change. *Earth and Planetary Science Letters*, **210**, 425–436.

LEES, J. A. 2002. Calcareous nannofossil biogeography illustrates palaeoclimate change in the Late Cretaceous Indian Ocean. *Cretaceous Research*, **23**, 537–633.

LEGENDRE, S. 1986. Analysis of mammalian communities from the late Eocene and Oligocene of southern France. *Palaeovertebrata*, **16**, 191–212.

LEGENDRE, S., MONTUIRE, S., MARIDET, O. & ESCARGUEL, G. 2005. Rodents and climate: A new model for estimating past temperatures. *Earth and Planetary Science Letters*, **235**, 408–420.

LEGRAND, P. 2003. Late Ordovician-early Silurian paleogeography of the Algerian Sahara. *Bulletin de la Société Geologique de France*, **174**, 19–32.

LEINFELDER, R. R., SCHLAGINTWEIT, F., WERNER, W., EBLI, O., NOSE, M., SCHMID, D. U. & HUGHES, G. W. 2005. Significance of stromatoporoids in Jurassic reefs and carbonate platforms – concepts and implications. *Facies*, **51**, 299–337.

LENG, M. J. & BARKER, P. A. 2006. A review of the oxygen isotope composition of lacustrine diatom silica for palaeoclimate reconstruction. *Earth-Science Reviews*, **75**, 5–27.

LENTON, T. M. 2001. The role of land plants, phosphorus weathering and fire in the rise and regulation of atmospheric oxygen. *Global Change Biology*, **7**, 613–629.

LENTON, T. M. & WATSON, A. J. 2000. Redfield revisited 2. What regulates the oxygen content of the atmosphere? *Global Biogeochemical Cycles*, **14**, 249–268.

LEWIS, J. P., WEAVER, A. J., JOHNSTON, S. T. & EBY, M. 2003. Neoproterozoic "snowball Earth": Dynamic sea ice over a quiescent ocean. *Paleoceanography*, **18**, art. no.–1092.

LI, J. G., PHILP, R. P., PU, F. & ALLEN, J. 1996. Long-chain alkenones in Qinghai Lake sediments. *Geochimica et Cosmochimica Acta*, **60**, 235–241.

LI, Y.-H. 1972. Geochemical mass balance among lithosphere, hydrosphere, and atmosphere. *American Journal of Science*, **272**, 119–137.

LIN, H. L. & CHEN, C. J. 2002. A late Pliocene diatom Ge/Si record from the Southeast Atlantic. *Marine Geology*, **180**, 151–161.

LING, H. F., BURTON, K. W., O'NIONS, R. K., KAMBER, B. S., VON BLANCKENBURG, F., GIBB, A. J. & HEIN, J. R. 1997. Evolution of Nd and Pb isotopes in Central Pacific seawater from ferromanganese crusts. *Earth and Planetary Science Letters*, **146**, 1–12, 10.1016/S0012–821X(96)00224–5.

LIU, Z., GALLIMORE, R. G., KUTZBACH, J. E., XU, W., GOLUBEV, Y., BEHLING, P. & SELIN, R. 1999. Modeling long-term climate changes with equilibrium asynchronous coupling. *Climate Dynamics*, **15**, 325–340.

LIVERMORE, R. A., EAGLES, G., MORRIS, P. & MALDONADO, A. 2004. Shackleton Fracture Zone: no barrier to early circumpolar ocean circulation. *Geology*, **32**, 797–800.

LOBOZIAK, S., STREEL, M. & DEALMEIDABURJACK, M. I. 1989. Paleoclimatic Conclusions from a Comparison between Middle-Upper Devonian Miospores Assemblages from Libya and Brazil. *Geobios*, **22**, 247–251.

LOHMANN, G. P. 1995. A model for variation in the chemistry of planktonic foraminifera due to secondary calcification and selective dissolution. *Paleoceanography*, **10**, 445–457.

LOUTRE, M. F. & BERGER, A. 2000. Future climatic changes: are we entering an exceptionally long interglacial? *Climatic Change*, **46**, 61–90, 10.1023/A:1005559827189.

LOWENSTEIN, T. K., TIMOFEEFF, M. N., BRENNAN, S. T., HARDIE, L. A. & DEMICCO, R. V. 2001. Oscillations in Phanerozoic seawater chemistry: Evidence from fluid inclusions. *Science*, **294**, 1086–1088.

LUDWIG, K. R. 1999. Isoplot Ex version 2, Berkeley Geochronology Center, Berkeley.

LYELL, C. 1830. *Principles of Geology*.

LYNCH-STIEGLITZ, J. & FAIRBANKS, R. G. 1994. A conservative tracer for glacial ocean circulation from carbon isotope and palaeo-nutrient measurements in benthic foraminifera. *Nature*, **369**, 308–310, 10.1038/369308a0.

MACGINITIE, H. D. 1941. A Middle Eocene flora from the central Sierra Nevada. *Carnegie Institute of Washington Publications*, **534**, 1–94.

MACK, G. H. & COLE, D. R. 2005. Geochemical model of delta O-18 of pedogenic calcite versus latitude and its application to Cretaceous palaeoclimate. *Sedimentary Geology*, **174**, 115–122.

MACK, G. H. & JAMES, W. C. 1994. Paleoclimate and the global distribution of paleosols. *Journal of Geology*, **102**, 360–366.

MACK, G. H., JAMES, W. C. & MONGER, H. C. 1993. Classification of paleosols. *Geological Society of America Bulletin*, **105**, 129–136.

MACKENSEN, A. 2004. Changing Southern Ocean paleocirculation and effects on global climate. *Antarctic Science*, **16**, 369–386.

MALMGREN, B. A. & NORDLUND, U. 1997. Application of artificial neural networks to paleoceanographic data. *Palaeogeography, Palaeoclimatology, Palaeoecology*, **136**, 359–373.

MANABE, S. & BRYAN, K. 1969. Climate calculations with a combined ocean-atmosphere model. *Journal of the Atmospheric Sciences*, **26**, 786–789.

MARCHITTO, T. M., CURRY, W. B. & OPPO, D. W. 2000. Zinc concentrations in benthic foraminifera reflect seawater chemistry. *Paleoceanography*, **15**, 299–306.

MARÉCHAL, C. N., NICOLAS, E., DOUCHET, C. & ALBARÈDE, F. 2000. Abundance of zinc isotopes as a marine biogeochemical tracer. *Geochemistry, Geophysics, Geosystems*, **1**, 10.1029/1999GC000029.

MARKWICK, P. J. 1998. Fossil crocodilians as indicators of Late Cretaceous and Cenozoic climates: implications for using palaeontological data in reconstructing palaeoclimate. *Palaeogeography, Palaeoclimatology, Palaeoecology*, **137**, 205–271.

MAROTZKE, J. & WILLEBRAND, J. 1991. Multiple equilibria of the global thermohaline circulation. *Journal of Physical Oceanography*, **21**, 1372–1385.

MARRIOTT, C. S., HENDERSON, G. M., BELSHAW, N. S. & TUDHOPE, A. W. 2004. Temperature dependence of delta Li-7, delta Ca-44 and Li/Ca during growth of calcium carbonate. *Earth and Planetary Science Letters*, **222**, 615–624.

MARSHALL, J. D. 1992. Climatic and oceanographic isotopic signals from the carbonate rock record and their preservation. *Geological Magazine*, **129**, 143–160.

MARSHALL, J. D., BRENCHLEY, P. J., MASON, P., WOLFF, G. A., ASTINI, R. A., HINTS, L. & MEIDLA, T. 1997. Global carbon isotopic events associated with mass extinction and glaciation in the late Ordovician. *Palaeogeography, Palaeoclimatology, Palaeoecology*, **132**, 195–210.

MASLIN, M., VILELA, C., MIKKELSEN, N. & GROOTES, P. 2005. Causes of catastrophic sediment failures of the Amazon Fan. *Quaternary Science Reviews*, **24**, 2180–2193.

MATSUMOTO, K., SARMIENTO, J. L. & BRZEZINSKI, M. A. 2002. Silicic acid leakage from the Southern Ocean: A possible explanation for glacial atmospheric pCO_2. *Global Biogeochemical Cycles*, **16**.

MCARTHUR, J. M., HOWARTH, R. J. & BAILEY, T. R. 2001. Strontium isotope stratigraphy: LOWESS Version 3: best fit to the marine Sr-Isotope curve for 0–509 Ma and accompanying look-up table for deriving numerical age. *Journal of Geology*, **109**, 155–170.

MCCABE, P. J. 1991. Tectonic controls on coal accumulation. *Bulletin de la Société Géologique de France*, **162**, 277–282.

MCCABE, P. J. & PARRISH, J. T. 1992. Tectonic and climatic controls on the distribution and quality of Cretaceous coals. *In*: MCCABE, P. J. & PARRISH, J. T. (eds) *Controls on the distribution and quality of Cretaceous coals*. Geological Society of America Special Papers, **267**, 1–15.

MCCABE, P. J. & SHANLEY, K. W. 1992. Organic control on shoreface stacking patterns: bogged down in the mire. *Geology*, **20**, 741–744.

MCCAVE, I. N., MANIGHETTI, B. & ROBINSON, S. G. 1995. Sortable silt and fine sediment size composition slicing: parameters for paleocurrent speed and paleoceanography. *Paleoceanography*, **10**, 593–610.

MCELWAIN, J. C., BEERLING, D. J. & WOODWARD, F. I. 1999. Fossil plants and global warming at the Triassic-Jurassic boundary. *Science*, **285**, 1386–1390.

MCELWAIN, J. C., WADE-MURPHY, J. & HESSELBO, S. P. 2005. Changes in carbon dioxide during an oceanic anoxic event linked to intrusion into Gondwana coals. *Nature*, **435**, 479–482.

MCGUFFIE, K. & HENDERSON-SELLERS, A. 2001. Forty years of numerical climate modelling. *International Journal of Climatology*, **21**, 1067–1109.

MCHONE, J. G. 2002. Volatile emissions of Central Atlantic Magmatic Province basalts: mass assumptions and environmental consequences. *In*: HAMES, W. E., MCHONE, J. G., RENNE, P. R. & RUPPEL, C. (eds) *The Central Atlantic Magmatic Province*. American Geophysical Union Geophysical Monographs, **136**, 241–254.

MCLACHLAN, I. R., TSIKOS, H. & CAIRNCROSS, B. 2001. Glendonites (pseudomorphs after ikaite) in Late Carboniferous marine Dwyka beds in southern Africa. *South African Journal of Geology*, **104**, 265–272.

MEDVEDEV, A. Y., AL'MUKHAMEDOV, A. I. & KIRDA, N. P. 2003. Geochemistry of Permo–Triassic volcanic rocks of West Siberia. *Geologiya I Geofizika*, **44**, 86–100.

MEERT, J. G. 2003. A synopsis of events related to the assembly of eastern Gondwana. *Tectonophysics*, **362**, 1–40.

MERCER, J. H. & SUTTER, J. F. 1982. Late Miocene earliest–Pliocene glaciation in southern Argentina: implications for global icesheet history. *Palaeogeography, Palaeoclimatology, Palaeoecology*, **38**, 185–206.

MICHAELSEN, P. & HENDERSON, R. A. 2000. Facies relationships and cyclicity of high-latitude, Late Permian coal measures, Bowen Basin, Australia. *International Journal of Coal Geology*, **44**, 19–48.

MILANKOVITCH, M. 1941. Canon of insolation and the ice-age problem, Royal Serbian Academy, Special Publications, Vol. **132**, Translated from German by Israel Program for Scientific Translations, Jerusalem, 633.

MILLER, K. G., SUGARMAN, P. J., BROWNING, J. V. ET AL. 2003. Late Cretaceous chronology of large, rapid sea-level changes: glacioeustasy during the greenhouse world. *Geology*, **31**, 585–588.

MILLER, K. G., WRIGHT, J. D. & BROWNING, J. V. 2005. Visions of ice sheets in a greenhouse world. *Marine Geology*, **217**, 215–231.

MILLER, K. G., WRIGHT, J. D. & FAIRBANKS, R. G. 1991. Unlocking the ice house: Oligocene–Miocene oxygen isotopes, eustasy, and margin erosion. *Journal of Geophysical Research-Solid Earth and Planets*, **96**, 6829–6848.

MOE, A. P. & SMITH, D. M. 2005. Using pre-Quaternary Diptera as indicators of paleoclimate. *Palaeogeography, Palaeoclimatology, Palaeoecology*, **221**, 203–214.

MOHSENI, H. & AL-AASM, I. S. 2004. Tempestite deposits on a storm-influenced carbonate ramp: An example from the Pabdeh formation (Paleogene), Zagros Basin, SW. *Journal of Petroleum Geology*, **27**, 163–178.

MOINE, O. & ROUSSEAU, D. D. 2002. Terrestrial molluscs and temperature: a new quantitative transfer function. *Comptes Rendus Palevol*, **1**, 145–151.

MONTOYA, M., CROWLEY, T. J. & VON STORCH, H. 1998. Temperatures at the last interglacial simulated by a coupled ocean-atmosphere climate model. *Paleoceanography*, **13**, 170–177.

MOORE, J. M., TSIKOS, H. & POLTEAU, S. 2001. Deconstructing the Transvaal Supergroup, South Africa: implications for Palaeoproterozoic palaeoclimate models. *Journal of African Earth Sciences*, **33**, 437–444.

MOORE, P. D. 1995. Biological processes controlling the development of modern peat-forming ecosystems. *International Journal of Coal Geology*, **28**, 99–110.

MORAN, K., BACKMAN, J., BRINKHUIS, H. ET AL. 2006. The Cenozoic palaeoenvironment of the Arctic Ocean. *Nature*, **441**, 601–605, doi:10.1038/nature04800.

MORARD, A., GUEX, J., BARTOLINI, A., MORETTINI, E. & DE WEVER, P. 2003. A new scenario for the Domerian – Toarcian transition. *Bulletin de la Société Geologique de France*, **174**, 351–356.

MORGANS, H. S. 1999. Lower and Middle Jurassic woods of the Cleveland Basin (North Yorkshire), England. *Palaeontology*, **42**, 303–328.

MORY, A. J., IASKY, R. P., GLIKSON, A. Y. & PIRAJNO, F. 2000. Woodleigh, Carnarvon Basin, Western Australia: a new 120 km diameter impact structure. *Earth and Planetary Science Letters*, **177**, 119–128.

MUNHOVEN, G. & FRANCOIS, L. M. 1996. Glacial-interglacial variability of atmospheric CO_2 due to changing continental silicate rock weathering: A model study. *Journal of Geophysical Research-Atmospheres*, **101**, 21423–21437.

MURPHY, J. B. & NANCE, R. D. 2005. Do supercontinents turn inside-in or inside-out? *International Geology Review*, **47**, 591–619.

MURPHY, J. B., PISAREVSKY, S. A., NANCE, R. D. & KEPPIE, J. D. 2004. Neoproterozoic–Early Paleozoic evolution of peri-Gondwanan terranes: implications for Laurentia–Gondwana connections. *International Journal of Earth Sciences*, **93**, 659–682.

MURRAY-WALLACE, C. V., KIMBER, R. W. L. & BELPERIO, A. P. 1988. Holocene paleotemperature studies using amino acid racemization reactions. *Australian Journal of Earth Sciences*, **35**, 575–577.

MUTTONI, G., CARCANO, C., GARZANTI, E. *ET AL*. 2003. Onset of major Pleistocene glaciations in the Alps. *Geology*, **31**, 989–992.

MYROW, P. M. & SOUTHARD, J. B. 1996. Tempestite deposition. *Journal of Sedimentary Research*, **66**, 875–887.

NAIRN, A. E. M. 1961. *Descriptive Palaeoclimatology*. Interscience Publishers, New York.

NAKAI, S., HALLIDAY, A. N. & REA, D. K. 1993. Provenance of dust in the Pacific Ocean. *Earth and Planetary Science Letters*, **119**, 143–157.

NISBET, E. G. 1992. Sources of atmospheric CH_4 in early postglacial time. *Journal of Geophysical Research-Atmospheres*, **97**, 12859–12867.

NISBET, E. G. 2002. Have sudden large releases of methane from geological reservoirs occurred since the Last Glacial Maximum, and could such releases occur again? *Philosophical Transactions of the Royal Society of London Series A-Mathematical Physical and Engineering Sciences*, **360**, 581–607.

NOIRI, Y., KUDO, I., KIYOSAWA, H., NISHIOKA, J. & TSUDA, A. 2005. Influence of iron and temperature on growth, nutrient utilization ratios and phytoplankton species composition in the western subarctic Pacific Ocean during the SEEDS experiment. *Progress in Oceanography*, **64**, 149–166.

NORDT, L., ATCHLEY, S. & DWORKIN, S. I. 2002. Paleosol barometer indicates extreme fluctuations in atmospheric CO_2 across the Cretaceous-Tertiary boundary. *Geology*, **30**, 703–706.

NORDT, L. C., WILDING, L. P., LYNN, W. C. & CRAWFORD, C. C. 2004. Vertisol genesis in a humid climate of the coastal plain of Texas, USA. *Geoderma*, **122**, 83–102.

O'BRIEN, P. E., LINDSAY, J. F., KNAUER, K. & SEXTON, M. J. 1998. Sequence stratigraphy of a sandstone-rich Permian glacial succession, Fitzroy Trough, Canning Basin, Western Australia. *Australian Journal of Earth Sciences*, **45**, 533–545.

OLSEN, P. E. 2001. Grand cycles of the Milankovitch band. *Eos Transactions*, **82**, F2.

OTTO-BLIESNER, B. L. 1995. Continental drift, runoff, and weathering feedbacks: implications from climate model experiments. *Journal of Geophysical Research-Atmospheres*, **100**, 11537–11548.

OTTO-BLIESNER, B. L. & UPCHURCH, G. R. 1997. Vegetation-induced warming of high-latitude regions during the late Cretaceous period. *Nature*, **385**, 804–807.

OTTO-BLIESNER, B. L., BRADY, E. C. & SHIELDS, C. 2002. Late Cretaceous ocean: coupled simulations with the National Center for Atmospheric Research climate system model. *Journal of Geophysical Research-Atmospheres*, **107**.

OURBAK, T., CORREGE, T., MALAIZE, B., LE CORNEC, F., CHARLIER, K. & PEYPOUQUET, J. P. 2006. A high-resolution investigation of temperature, salinity, and upwelling activity proxies in corals. *Geochemistry, Geophysics, Geosystems*, **7**.

PAGANI, M. 2002. The alkenone-CO_2 proxy and ancient atmospheric carbon dioxide. *Philosophical Transactions of the Royal Society of London Series A-Mathematical Physical and Engineering Sciences*, **360**, 609–632.

PAGANI, M., LEMARCHAND, D., SPIVACK, A. & GAILLARDET, J. 2005*a*. A critical evaluation of the boron isotope-pH proxy: The accuracy of ancient ocean pH estimates. *Geochimica et Cosmochimica Acta*, **69**, 953–961.

PAGANI, M., ZACHOS, J. C., FREEMAN, K. H., TIPPLE, B. & BOHATY, S. 2005*b*. Marked decline in atmospheric carbon dioxide concentrations during the Paleogene. *Science*, **309**, 600–603.

PAGE, A., ZALASIEWICZ, J., WILLIAMS, M. & POPOV, L. 2007. Were transgressive black shales a negative feedback modulating glacioeustasy in the Early Palaeozoic Icehouse? *In*: WILLIAMS, M., HAYWOOD, A. M., GREGORY, F. J. & SCHMIDT, D. M. (eds) *Deep-Time Perspectives on Climate Change: Marrying the Signal from Computer Models and Biological Proxies*. The Micropalaeontological Society, Special Publication. Geological Society, London. 123–157.

PÁLFY, J. & SMITH, P. L. 2000. Synchrony between Early Jurassic extinction, oceanic anoxic event, and the Karoo–Ferrar flood basalt volcanism. *Geology*, **28**, 747–750.

PALMER, A. R. 1983. The Decade of North-American Geology – 1983 Geologic Time Scale. *Geology*, **11**, 503–504.

PARK, J. & MAASCH, K. A. 1993. Plio-Pleistocene time evolution of the 100-kyr cycle in marine paleoclimate records. *Journal of Geophysical Research-Solid Earth*, **98**, 447–461.

PARRISH, J. T. 1993. A brief discussion of the history, strengths and limitations of conceptual climate models for pre-Quaternary time. *Philosophical Transactions of the Royal Society of London Series B-Biological Sciences*, **341**, 263–266.

PARRISH, J. T., ZIEGLER, A. M. & SCOTESE, C. R. 1982. Rainfall patterns and the distribution of coals and evaporites in the Mesozoic and Cenozoic. *Palaeogeography, Palaeoclimatology, Palaeoecology,* **40**, 67–101.

PAUS, A., VELLE, G., LARSEN, J., NESJE, A. & LIE, O. 2006. Lateglacial nunataks in central Scandinavia: Biostratigraphical evidence for ice thickness from Lake Flafattjonn, Tynset, Norway. *Quaternary Science Reviews,* **25**, 1228–1246.

PEARSON, P. N. & PALMER, M. R. 2000. Atmospheric carbon dioxide concentrations over the past 60 million years. *Nature,* **406**, 695–699.

PEARSON, P. N. & PALMER, M. R. 2002. The boron isotope approach to paleo-pCO$_2$ estimation. *Geochimica et Cosmochimica Acta,* **66**, A586–A586.

PEKAR, S. F., CHRISTIE-BLICK, N., KOMINZ, M. A. & MILLER, K. G. 2002. Calibration between eustatic estimates from backstripping and oxygen isotopic records for the Oligocene. *Geology,* **30**, 903–906.

PELEJERO, C. & CALVO, E. 2003. The upper end of the U37K temperature calibration revisited. *Geochemistry, Geophysics, Geosystems,* **4**.

PEYRON, O. & DE VERNAL, A. 2001. Application of artificial neural networks (ANN) to high-latitude dinocyst assemblages for the reconstruction of past sea-surface conditions in Arctic and sub-Arctic seas. *Journal of Quaternary Science,* **16**, 699–709.

PFUHL, H. A. & MCCAVE, I. N. 2005. Evidence for late Oligocene establishment of the Antarctic Circumpolar Current. *Earth and Planetary Science Letters,* **235**, 715–728.

PILNY, J. J., MORGAN, A. V. & MORGAN, A. 1987. Paleoclimatic implications of a Late Wisconsinan insect assemblage from Rostock, southwestern Ontario. *Canadian Journal of Earth Sciences,* **24**, 617–630.

PINCIVY, A., MALO, M., RUFFET, G., TREMBLAY, A. & SACKS, P. E. 2003. Regional metamorphism of the Appalachian Humber zone of Gaspe Peninsula: Ar-40/Ar-39 evidence for crustal thickening during the Taconian orogeny. *Canadian Journal of Earth Sciences,* **40**, 301–315.

PLAUT, G., GHIL, M. & VAUTARD, R. 1995. Interannual and interdecadal variability in 335 years of Central England Temperatures. *Science,* **268**, 710–713.

POCHAT, S. P., VAN DEN DRIESSCHE, J., MOUTON, V. & GUILLOCHEAU, F. 2005. Identification of Permian palaeowind direction from wave-dominated lacustrine sediments (Lodeve Basin, France). *Sedimentology,* **52**, 809–825.

POOLE, I., CANTRILL, D. & UTESCHER, T. 2005. A multi-proxy approach to determine Antarctic terrestrial palaeoclimate during the Late Cretaceous and Early Tertiary. *Palaeogeography, Palaeoclimatology, Palaeoecology,* **222**, 95–121.

POWERS, L. A., WERNE, J. P., JOHNSON, T. C., HOPMANS, E. C., DAMSTE, J. S. S. & SCHOUTEN, S. 2004. Crenarchaeotal membrane lipids in lake sediments: A new paleotemperature proxy for continental paleoclimate reconstruction? *Geology,* **32**, 613–616.

PRAHL, F. G., MUEHLHAUSEN, L. A. & ZAHNLE, D. L. 1988. Further evaluation of long-chain alkenones as indicators of paleoceanographic conditions. *Geochimica et Cosmochimica Acta,* **52**, 2303–2310, 10.1016/0016-7037(88)90132-9.

PRENTICE, M. L. & KRUSIC, A. G. 2005. Early Pliocene Alpine glaciation in Antarctica: Terrestrial versus tidewater glaciers in Wright Valley. *Geografiska Annaler Series a-Physical Geography,* **87A**, 87–109.

PRICE, G. D. 1999. The evidence and implications of polar ice during the Mesozoic. *Earth-Science Reviews,* **48**, 183–210.

PRICE, G. D., SELLWOOD, B. W. & VALDES, P. J. 1995. Sedimentological evaluation of general circulation model simulations for the 'greenhouse' 'earth: Cretaceous and Jurassic. *Sedimentary Geology,* **100**, 159–180.

PRICE, G. D., VALDES, P. J. & SELLWOOD, B. W. 1997a. Prediction of modern bauxite occurrence: Implications for climate reconstruction. *Palaeogeography, Palaeoclimatology, Palaeoecology,* **131**, 1–13.

PRICE, G. D., VALDES, P. J. & SELLWOOD, B. W. 1997b. Quantitative palaeoclimate GCM validation: Late Jurassic and mid-Cretaceous case studies. *Journal of the Geological Society,* **154**, 769–772.

PROKOPH, A., ERNST, R. E. & BUCHAN, K. L. 2004. Time-series analysis of large igneous provinces: 3500 Ma to present. *Journal of Geology,* **112**, 1–22.

PRUEHER, L. M. & REA, D. K. 1998. Rapid onset of glacial conditions in the subarctic North Pacific region at 2.67 Ma: Clues to causality. *Geology,* **26**, 1027–1030.

QIN, X. G., CAI, B. G. & LIU, T. S. 2005. Loess record of the aerodynamic environment in the east Asia monsoon area since 60,000 years before present. *Journal of Geophysical Research-Solid Earth,* **110**.

RAE, A. C., HARRISON, S., MIGHALL, T. & DAWSON, A. G. 2004. Periglacial trimlines and nunataks of the Last Glacial Maximum: the Gap of Dunloe, southwest Ireland. *Journal of Quaternary Science,* **19**, 87–97.

RAGUENEAU, O., TREGUER, P., LEYNAERT, A. *ET AL.* 2000. A review of the Si cycle in the modern ocean: recent progress and missing gaps in the application of biogenic opal as a paleoproductivity proxy. *Global and Planetary Change,* **26**, 317–365.

RAILSBACK, L. B. 1992. A geological numerical-model for paleozoic global evaporite deposition. *Journal of Geology,* **100**, 261–277.

RAMIREZ, J., MENDOZA, B., MENDOZA, V. & ADEM, J. 2006. Effects of an assumed cosmic ray-modulated low global cloud cover on the Earth's temperature. *Atmosfera,* **19**, 169–179.

RAMPINO, M. R. & SELF, S. 1992. Volcanic winter and accelerated glaciation following the Toba supereruption. *Nature,* **359**, 50–52.

RAMSTEIN, G., FLUTEAU, F., BESSE, J. & JOUSSAUME, S. 1997. Effect of orogeny, plate motion and land sea distribution on Eurasian climate change over the past 30 million years. *Nature,* **386**, 788–795.

RAVELO, A. C. & ANDREASEN, D. H. 2000. Enhanced circulation during a warm period. *Geophysical Research Letters,* **27**, 1001–1004.

RAVIZZA, G., NORRIS, R. N., BLUSZTAJN, J. & AUBRY, M. P. 2001. An osmium isotope excursion associated with the late Paleocene thermal maximum: evidence of intensified chemical weathering. *Paleoceanography,* **16**, 155–163.

RAYMO, M. E. 1991. Geochemical evidence supporting CHAMBERLIN, T. C., theory of glaciation. *Geology*, **19**, 344–347.

RAYMO, M. E., OPPO, D. W. & CURRY, W. 1997. The mid-Pleistocene climate transition: A deep sea carbon isotopic perspective. *Paleoceanography*, **12**, 546–559.

RAYMOND, A. & METZ, C. 2004. Ice and its consequences: Glaciation in the Late Ordovician, Late Devonian, Pennsylvanian-Permian, and Cenozoic compared. *Journal of Geology*, **112**, 655–670.

RAYMOND, P. A. & COLE, J. J. 2003. Increase in the export of alkalinity from North America's largest river. *Science*, **301**, 88–91, 10.1126/science.1083788.

REICHOW, M. K., SAUNDERS, A. D., WHITE, R. V., PRINGLE, M. S., AL'MUKHAMEDOV, A. I., MEDVEDEV, A. I. & KIRDA, N. P. 2002. $^{40}Ar/^{39}Ar$ dates from the West Siberian Basin: Siberian flood basalt province doubled. *Science*, **296**, 1846–1849.

RETALLACK, G. J. 1997. Palaeosols in the upper Narrabeen Group of New South Wales as evidence of early Triassic palaeoenvironments without exact modern analogues. *Australian Journal of Earth Sciences*, **44**, 185–201.

RETALLACK, G. J. 2001. *Soils of the Past: An Introduction to Paleopedology*. Blackwell Science, Oxford.

RETALLACK, G. J. 2002. Carbon dioxide and climate over the past 300 Myr. *Philosophical Transactions of the Royal Society of London Series A*, **360**, 659–673.

RETALLACK, G. J. 2005. Pedogenic carbonate proxies for amount and seasonality of precipitation in paleosols. *Geology*, **33**, 333–336.

RETALLACK, G. J., VEEVERS, J. J. & MORANTE, R. 1996. Global coal gap between Permian-Triassic extinction and Middle Triassic recovery of peat-forming plants. *Geological Society of America Bulletin*, **108**, 195–207.

REUNING, L., REIJMER, J. J. G., BETZLER, C., TIMMERMANN, A. & STEPH, S. 2006. Sub-Milankovitch cycles in periplatform carbonates from the early Pliocene Great Bahama Bank. *Paleoceanography*, **21**.

REY, O., SIMO, J. A. & LORENTE, M. A. 2004. A record of long- and short-term environmental and climatic change during OAE3: La Luna Formation, late Cretaceous (Santonian-early Campanian), Venezuela. *Sedimentary Geology*, **170**, 85–105.

REYNOLDS, B. C., FRANK, M. & HALLIDAY, A. N. 2006. Silicon isotope fractionation during nutrient utilization in the North Pacific. *Earth and Planetary Science Letters*, **244**, 431–443.

RICHARDS, G. R. 1998. Identifying trends in climate: An application to the Cenozoic. *International Journal of Climatology*, **18**, 583–594.

RICKABY, R. E. M. & ELDERFIELD, H. 1999. Planktonic foraminiferal Cd/Ca: Paleonutrients or paleotemperature? *Paleoceanography*, **14**, 293–303.

RIDGWELL, A. 2005. A Mid Mesozoic Revolution in the regulation of ocean chemistry. *Marine Geology*, **217**, 339–357.

RIDLEY, J. K., HUYBRECHTS, P., GREGORY, J. M. & LOWE, J. A. 2005. Elimination of the Greenland ice sheet in a high CO_2 climate. *Journal of Climate*, **18**, 3409–3427.

RILEY, T. R., LEAT, P. T., PANKHURST, R. J. & HARRIS, C. 2001. Origins of large volume rhyolitic volcanism in the Antarctic Peninsula and Patagonia by crustal melting. *Journal of Petrology*, **42**, 1043–1065.

ROBERTS, D. & MCMINN, A. 1999. A diatom-based palaeosalinity history of Ace Lake, Vestfold Hills, Antarctica. *Holocene*, **9**, 401–408.

ROBERTSON, I., LUCY, D., BAXTER, L. ET AL. 1999. A kernel-based Bayesian approach to climatic reconstruction. *Holocene*, **9**, 495–500.

ROBINSON, R. S., SIGMAN, D. M., DIFIORE, P. J., ROHDE, M. M., MASHIOTTA, T. A. & LEA, D. W. 2005. Diatom-bound N-15/N-14: New support for enhanced nutrient consumption in the Ice Age subantarctic. *Paleoceanography*, **20**.

RODRIGUEZ, J. 1999. Use of cenograms in mammalian palaeoecology. A critical review. *Lethaia*, **32**, 331–347.

ROHLING, E. J. 2000. Paleosalinity: confidence limits and future applications. *Marine Geology*, **163**, 1–11.

ROSALES, I., QUESADA, S. & ROBLES, S. 2004. Paleotemperature variations of Early Jurassic seawater recorded in geochemical trends of belemnites from the Basque-Cantabrian basin, northern Spain. *Palaeogeography, Palaeoclimatology, Palaeoecology*, **203**, 253–275.

ROSS, C. A. & ROSS, J. R. P. 1985. Late Paleozoic depositional sequences are synchronous and worldwide. *Geology*, **13**, 194–197.

ROWE, P. J. & MAHER, B. A. 2000. 'Cold' stage formation of calcrete nodules in the Chinese Loess Plateau: evidence from U-series dating and stable isotope analysis. *Palaeogeography, Palaeoclimatology, Palaeoecology*, **157**, 109–125.

ROYER, D. L., BERNER, R. A. & BEERLING, D. J. 2001. Phanerozoic atmospheric CO_2 change: evaluating geochemical and paleobiological approaches. *Earth-Science Reviews*, **54**, 349–392.

ROYER, D. L., BERNER, R. A., MONTAÑEZ, I. P., TABOR, N. J. & BEERLING, D. J. 2004. CO_2 as a primary driver of Phanerozoic climate. *GSA Today*, **14**, 4–10, 10.1130/1052-5173(2004)014 <4:CAAPDO>2.0.CO;2.

RUBIN, S. I., KING, S. L., JAHNKE, R. A. & FROELICH, P. N. 2003. Benthic barium and alkalinity fluxes: is Ba an oceanic paleo-alkalinity proxy for glacial atmospheric CO_2? *Geophysical Research Letters*, **30**.

RUDDIMAN, W. F. & KUTZBACH, J. E. 1991. Plateau uplift and climatic change. *Scientific American*, **264**, 66.

SALTZMAN, M. R. 2003. Late Paleozoic ice age: oceanic gateway or pCO₂? *Geology*, **31**, 151–154.

SALTZMAN, M. R., RIPPERDAN, R. L., BRASIER, M. D. ET AL. 2000. A global carbon isotope excursion (SPICE) during the Late Cambrian: relation to trilobite extinctions, organic-matter burial and sea level. *Palaeogeography, Palaeoclimatology, Palaeoecology*, **162**, 211–223.

SAMUELSSON, J. 1998. Carbon and oxygen isotope geochemistry of Early Neoproterozoic successions on the Kola Peninsula, northwest Russia. *Norsk Geologisk Tidsskrift*, **78**, 291–303.

SANTOS, J. O. S., HARTMANN, L. A., MCNAUGHTON, N. J. & FLETCHER, I. R. 2002. Timing of mafic magmatism in the Tapajo's Province (Brazil) and implications for the evolution of the Amazon Craton: evidence from baddeleyite and zircon U–Pb SHRIMP geochronology. *Journal of South American Earth Sciences*, **15**, 409–429.

SANYAL, A., HEMMING, N. G., HANSON, G. N. & BROECKER, W. S. 1995. Evidence for a higher pH in the glacial ocean from boron isotopes in foraminifera. *Nature*, **372**, 234–236, 10.1038/373234a0.

SARNTHEIN, M., PFLAUMANN, U. & WEINELT, M. 2003. Past extent of sea ice in the northern North Atlantic inferred from foraminiferal paleotemperature estimates. *Paleoceanography*, **18**.

SCHENK, P. E. 1995. Meguma Zone. *In*: WILLIAMS, H. (ed.) *Geology of the Appalachian – Caledonian Orogen in Canada and Greenland*. Geological Survey of Canada, Geology of Canada, **6**, 261–277.

SCHILLER, A., MIKOLAJEWICZ, U. & VOSS, R. 1997. The stability of the North Atlantic thermohaline circulation in a coupled ocean-atmosphere general circulation model. *Climate Dynamics*, **13**, 325–347.

SCHIMEL, D. S. 1995. Terrestrial ecosystems and the carbon cycle. *Global Change Biology*, **1**, 77–91.

SCHLESINGER, W. H. & ANDREWS, J. A. 2000. Soil respiration and the global carbon cycle. *Biogeochemistry*, **48**, 7–20.

SCHMIDT, G. A. 1999. Error analysis of paleosalinity calculations. *Paleoceanography*, **14**, 422–429.

SCHMIEDER, F., VON DOBENECK, T. & BLEIL, U. 2000. The Mid-Pleistocene climate transition as documented in the deep South Atlantic Ocean: initiation, interim state and terminal event. *Earth and Planetary Science Letters*, **179**, 539–549.

SCHNEIDER, S. H. & DICKINSON, R. E. 1974. Climate modeling. *Reviews of Geophysics*, **12**, 447–493.

SCHOPKA, H. H., GUDMUNDSSON, M. T. & TUFFEN, H. 2006. The formation of Helgafell, southwest Iceland a monogenetic subglacial hyaloclastite ridge: sedimentology, volcano–ice interaction hydrology and volcano–ice interaction. *Journal of Volcanology and Geothermal Research*, **152**, 359–377.

SCHOUTEN, S., HOPMANS, E. C. & DAMSTE, J. S. S. 2004. The effect of maturity and depositional redox conditions on archaeal tetraether lipid palaeothermometry. *Organic Geochemistry*, **35**, 567–571.

SCHOUTEN, S., HOPMANS, E. C., FORSTER, A., VAN BREUGEL, Y., KUYPERS, M. M. M. & DAMSTE, J. S. S. 2003. Extremely high sea-surface temperatures at low latitudes during the middle Cretaceous as revealed by archaeal membrane lipids. *Geology*, **31**, 1069–1072.

SCHOUTEN, S., HOPMANS, E. C., SCHEFUSS, E. & DAMSTE, J. S. S. 2002. Distributional variations in marine crenarchaeotal membrane lipids: a new tool for reconstructing ancient sea water temperatures? *Earth and Planetary Science Letters*, **204**, 265–274.

SCHOUTEN, S., OSSEBAAR, J., SCHREIBER, K., KIENHUIS, M. V. M., LANGER, G., BENTHIEN, A. & BIJMA, J. 2006. The effect of temperature, salinity and growth rate on the stable hydrogen isotopic composition of long chain alkenones produced by *Emiliania huxleyi* and *Gephyrocapsa oceanica*. *Biogeosciences*, **3**, 113–119.

SCHOUTEN, S., VAN KAAM-PETERS, H. M. E., RIJPSTRA, W. I. C., SCHOELL, M. & DAMSTE, J. S. S. 2000. Effects of an oceanic anoxic event on the stable carbon isotopic composition of Early Toarcian carbon. *American Journal of Science*, **300**, 1–22.

SCHWALB, A. 2003. Lacustrine ostracodes as stable isotope recorders of late-glacial and Holocene environmental dynamics and climate. *Journal of Paleolimnology*, **29**, 267–351.

SCOTT, A. C. 1989. Observations on the nature and origin of fusain. *International Journal of Coal Geology*, **12**, 443–475.

SCOTT, A. C. 2000. The Pre-Quaternary history of fire. *Palaeogeography, Palaeoclimatology, Palaeoecology*, **164**, 281–329.

SCOTT, A. C., LOMAX, B. H., COLLINSON, M. E., UPCHURCH, G. R. & BEERLING, D. J. 2000. Fire across the K-T boundary: initial results from the Sugarite Coal, New Mexico, USA. *Palaeogeography, Palaeoclimatology, Palaeoecology*, **164**, 381–395.

SEGALEN, L., ROGNON, P., PICKFORD, M., SENUT, B., EMMANUEL, L., RENARD, M. & WARD, J. 2004. Reconstitution of dune morphologies and palaeowind regimes in the Proto-Namib since the Miocene. *Bulletin de la Société Géologique de France*, **175**, 537–546.

SEJRUP, H. P., BIRKS, H. J. B., KRISTENSEN, D. K. & MADSEN, H. 2004. Benthonic foraminiferal distributions and quantitative transfer functions for the northwest European continental margin. *Marine Micropaleontology*, **53**, 197–226.

SELLWOOD, B. W., PRICE, G. D., SHACKLETON, N. J. & FRANCIS, J. E. 1993. Sedimentary facies as indicators of Mesozoic palaeoclimate [and discussion]. *Philosophical Transactions of the Royal Society of London Series B-Biological Sciences*, **341**, 225–233.

SELLWOOD, B. W. & VALDES, P. J. 1997. Geological evaluation of climate general circulation models and model implications for Mesozoic cloud cover. *Terra Nova*, **9**, 75–78.

SELLWOOD, B. W., VALDES, P. J. & PRICE, G. D. 2000. Geological evaluation of multiple general circulation model simulations of Late Jurassic palaeoclimate. *Palaeogeography, Palaeoclimatology, Palaeoecology*, **156**, 147–160.

SHACKLETON, N. J. 1965. The high-precision isotopic analysis of oxygen and carbon in carbon dioxide. *Journal of Scientific Instruments*, **42**, 689–692, 10.1088/0950-7671/42/9/306.

SHACKLETON, N. J. 1977. Oxygen isotope stratigraphic record of the Late Pleistocene. *Philosophical Transactions of the Royal Society of London Series B-Biological Sciences*, **280**, 169–182.

SHACKLETON, N. J. & KENNETT, J. P. 1975. Late Cenozoic oxygen and carbon isotopic changes at DSDP site 284: implication for glacial history of the Northern Hemisphere and Antarctica. *Initial Reports of the Deep Sea Drilling Program*, **29**, 801–807.

SHACKLETON, N. J. & OPDYKE, N. D. 1973. Oxygen isotope and paleomagnetic stratigraphy of Equatorial Pacific core V28–238: oxygen isotope temperatures and ice volumes on a 10^5 and 10^6 year scale. *Quaternary Research*, **3**, 39–55.

SHAVIV, N. J. 2005. On the link between cosmic rays and terrestrial climate. *International Journal of Modern Physics A*, **20**, 6662–6665.

SHAVIV, N. J. & VEIZER, J. 2003. Celestial driver of Phanerozoic climate? *GSA Today*, **13**, 4–10.

SHERLOCK, S. C., KELLEY, S. P., ZALASIEWICZ, J. A., SCHOFIELD, D. I., EVANS, J. A., MERRIMAN, R. J. & KEMP, S. J. 2003. Precise dating of low-temperature deformation: Strain-fringe analysis by Ar-40–Ar-39 laser microprobe. *Geology*, **31**, 219–222.

SIEGENTHALER, U., STOCKER, T. F., MONNIN, E. *ET AL.* 2005. Stable carbon cycle-climate relationship during the late Pleistocene. *Science*, **310**, 1313–1317.

SIESSER, W. G. 1995. Paleoproductivity of the Indian Ocean during the Tertiary period. *Global and Planetary Change*, **11**, 71–88.

SIGMAN, D. M. 2000. Global carbon cycle and its links to carbon fixation by oceanic phytoplankton. *Abstracts of Papers of the American Chemical Society*, **219**, U839–U839.

SIGMAN, D. M., ALTABET, M. A., FRANCOIS, R. & WHELAN, J. 1997. Diatom microfossil N isotopes support the hypothesis of higher nitrate utilization in the Southern Ocean during the last ice age. *Abstracts of Papers of the American Chemical Society*, **214**, 69–GEOC.

SIGMAN, D. M., JACCARD, S. L. & HAUG, G. H. 2004. Polar ocean stratification in a cold climate. *Nature*, **428**, 59–63.

SIKES, E. L. & SICRE, M. A. 2002. Relationship of the tetra-unsaturated C-37 alkenone to salinity and temperature: Implications for paleoproxy applications. *Geochemistry Geophysics, Geosystems*, **3**.

SIME, N. G., DE LA ROCHA, C. L. & GALY, A. 2005. Negligible temperature dependence of calcium isotope fractionation in 12 species of planktonic foraminifera. *Earth and Planetary Science Letters*, **232**, 51–66.

SIMKEVICIUS, P., AHLBERG, A. & GRIGELIS, A. 2003. Jurassic smectite and kaolinite trends of the East European Platform: implications for palaeobathymetry and palaeoclimate. *Terra Nova*, **15**, 225–229.

SLUIJS, A., SCHOUTEN, S., PAGANI, M. *ET AL.* 2006. Subtropical Arctic Ocean temperatures during the Palaeocene/Eocene thermal maximum. *Nature*, **441**, 610–613, 10.1038/nature04668.

SMELLIE, J. L. 2000. Subglacial eruptions. *In*: SIGURDSSON, H. (ed.) *Encyclopaedia of Volcanoes*. Academic Press, San Diego, 403–418.

SMELLIE, J. L., MCINTOSH, W. C. & ESSER, R. 2006. Eruptive environment of volcanism on Brabant Island: Evidence for thin wet-based ice in northern Antarctic Peninsula during the Late Quaternary. *Palaeogeography, Palaeoclimatology, Palaeoecology*, **231**, 233–252.

SMELLIE, J. L. & SKILLING, I. P. 1994. Products of subglacial volcanic eruptions under different ice thicknesses: Two examples from Antarctica. *Sedimentary Geology*, **91**, 115–129.

SMITH, A. G. & PICKERING, K. T. 2003. Oceanic gateways as a critical factor to initiate icehouse Earth. *Journal of the Geological Society, London*, **160**, 337–340.

SMITH, A. J. 1963. A striated pavement beneath the basal Gondwana sediments on the Ajay River, Bihar, India. *Nature*, **198**, 880, 10.1038/198880b0.

SNYDER, P. K., DELIRE, C. & FOLEY, J. A. 2004. Evaluating the influence of different vegetation biomes on the global climate. *Climate Dynamics*, **23**, 279–302.

SOBOLEV, S. V. & BABEYKO, A. Y. 2005. What drives orogeny in the Andes? *Geology*, **33**, 617–620.

SOLANKI, S. K. & KRIVOVA, N. A. 2004. Solar irradiance variations: From current measurements to long-term estimates. *Solar Physics*, **224**, 197–208.

SOLANKI, S. K., USOSKIN, I. G., KROMER, B., SCHUSSLER, M. & BEER, J. 2004. Unusual activity of the Sun during recent decades compared to the previous 11,000 years. *Nature*, **431**, 1084–1087.

SPICER, R. A., HERMAN, A. B. & KENNEDY, E. M. 2005. The sensitivity of CLAMP to taphonomic loss of foliar physiognomic characters. *Palaios*, **20**, 429–438.

STANLEY, E. A. 1970. Maastrichtian plant microfossil assemblages and their paleogeographic and paleoclimatic implications for latest Mesozoic time. *American Association of Petroleum Geologists Bulletin*, **54**, 2506.

STAUB, J. R. 2002. Marine flooding events and coal bed sequence architecture in southern West Virginia. *International Journal of Coal Geology*, **49**, 123–145.

STILES, C. A., MORA, C. I. & DRIESE, S. G. 2001. Pedogenic iron-manganese nodules in Vertisols: A new proxy for paleoprecipitation? *Geology*, **29**, 943–946.

STOCKER, T. F., WRIGHT, D. G. & MYSAK, L. A. 1992. A zonally averaged, coupled ocean–atmosphere model for paleoclimate studies. *Journal of Climate*, **5**, 773–797.

STOLL, H. M. & SCHRAG, D. P. 1996. Evidence for glacial control of rapid sea level changes in the Early Cretaceous. *Science*, **272**, 1771–1774.

STRAUSS, H. & PETERS-KOTTIG, W. 2003. The Paleozoic to Mesozoic carbon cycle revisited: the carbon isotopic composition of terrestrial organic matter. *Geochemistry, Geophysics, Geosystems*, **4**, 1–15, 1083, 10.1029/2003GC000555.

STREEL, M., CAPUTO, M. V., LOBOZIAK, S. & MELO, J. H. G. 2000. Late Frasnian-Famennian climates based on palynomorph analyses and the question of the Late Devonian glaciations. *Earth-Science Reviews*, **52**, 121–173.

SUN, D. H., BLOEMENDAL, J., REA, D. K. *ET AL.* 2004. Bimodal grain-size distribution of Chinese loess, and its palaeoclimatic implications. *Catena*, **55**, 325–340.

SUN, Y. B. & AN, Z. S. 2005. Late Pliocene–Pleistocene changes in mass accumulation rates of eolian deposits on the central Chinese Loess Plateau. *Journal of Geophysical Research-Atmospheres*, **110**.

SWAINSON, I. P. & HAMMOND, R. P. 2001. Ikaite, $CaCO_3 \cdot 6H_2O$: Cold comfort for glendonites as paleothermometers. *American Mineralogist*, **86**, 1530–1533.

TEXIER, D., DE NOBLET, N., HARRISON, S. P. *ET AL.* 1997. Quantifying the role of biosphere-atmosphere feedbacks in climate change: coupled model simulations for 6000 years BP and comparison with palaeodata for northern Eurasia and northern Africa. *Climate Dynamics*, **13**, 865–882.

THIRY, M. 2000. Palaeoclimatic interpretation of clay minerals in marine deposits: an outlook from the continental origin. *Earth-Science Reviews*, **49**, 201–221.

THOMAS, D. J., ZACHOS, J. C., BRALOWER, T. J., THOMAS, E. & BOHATY, S. 2002*a*. Warming the fuel for the fire: Evidence for the thermal dissociation of methane hydrate during the Paleocene–Eocene thermal maximum. *Geology*, **30**, 1067–1070.

THOMAS, R. J., CHEVALLIER, L. P., GRESSE, P. G. ET AL. 2002*b*. Precambrian evolution of the Sirwa Window, Anti-Atlas Orogen, Morocco. *Precambrian Research*, **118**, 1–57.

THORN, V. C. & DECONTO, R. 2006. Antarctic climate at the Eocene/Oligocene boundary – climate model sensitivity to high latitude vegetation type and comparisons with the palaeobotanical record. *Palaeogeography, Palaeoclimatology, Palaeoecology*, **231**, 134–157.

TOBIN, K. J. & WALKER, K. R. 1997. Ordovician oxygen isotopes and paleotemperatures. *Palaeogeography, Palaeoclimatology, Palaeoecology*, **129**, 269–290.

TRIPATI, A., BACKMAN, J., ELDERFIELD, H. & FERRETTI, P. 2005. Eocene bipolar glaciation associated with global carbon cycle changes. *Nature*, **436**, 341–346.

TYNDALL, J. 1861. On the absorption and radiation of heat by gases and vapours, and on the physical connection of radiation, absorption, and conduction. *Philosophical Magazine Series 4*, **22**, 169–94, 273–285.

TZIPERMAN, E. & GILDOR, H. 2003. On the mid-Pleistocene transition to 100-kyr glacial cycles and the asymmetry between glaciation and deglaciation times. *Paleoceanography*, **18**.

ULLYOTT, J. S. & NASH, D. J. 2006. Micromorphology and geochemistry of groundwater silcretes in the eastern South Downs, UK. *Sedimentology*, **53**, 387–412.

UREY, H. C. 1947. The thermodynamic properties of isotopic substances. *Journal of the Chemical Society*, **1947**, 562–581, 10.1039/JR9470000562.

USOSKIN, I. G., SCHUSSLER, M., SOLANKI, S. K. & MURSULA, K. 2005. Solar activity, cosmic rays, and Earth's temperature: A millennium-scale comparison. *Journal of Geophysical Research-Space Physics*, **110**.

VALDES, P. J. & SELLWOOD, B. W. 1992. A paleoclimate model for the Kimmeridgian. *Palaeogeography, Palaeoclimatology, Palaeoecology*, **95**, 47–72.

VAN DE FLIERDT, T., FRANK, M., LEE, D. C. & HALLIDAY, A. N. 2002. Glacial weathering and the hafnium isotope composition of seawater. *Earth and Planetary Science Letters*, **198**, 167–175.

VAN DE SCHOOTBRUGGE, B., MCARTHUR, J. M., BAILEY, T. R., ROSENTHAL, Y., WRIGHT, J. D. & MILLER, K. G. 2005. Toarcian oceanic anoxic event: An assessment of global causes using belemnite C isotope records. *Paleoceanography*, **20**.

VAN HOUTEN, F. B. 1961. Climatic significance of red beds. *In*: NAIRN, A. E. M. (ed.) *Descriptive Palaeoclimatology*. Interscience, New York, 89–139.

VAN NIEKERK, H. S., BEUKES, N. J. & GUTZMER, J. 1999. Post-Gondwana pedogenic ferromanganese deposits, ancient soil profiles, African land surfaces and palaeoclimatic change on the Highveld of South Africa. *Journal of African Earth Sciences*, **29**, 761–781.

VANCE, D. & BURTON, K. 1999. Neodymium isotopes in planktonic foraminifera: a record of the response of continental weathering and ocean circulation rates to climate change. *Earth and Planetary Science Letters*, **173**, 365–379, 10.1016/S0012–821X(99)00244-7.

VAUGHAN, A. P. M. & LIVERMORE, R. A. 2005. Episodicity of Mesozoic terrane accretion along the Pacific margin of Gondwana: implications for superplume-plate interactions. *In*: VAUGHAN, A. P. M., LEAT, P. T. & PANKHURST, R. J. (eds) *Terrane Processes at the Margins of Gondwana*. Geological Society, London, Special Publications, **246**, 143–178.

VAUGHAN, A. P. M. & SCARROW, J. H. 2003. Ophiolite obduction pulses as a proxy indicator of superplume events? *Earth and Planetary Science Letters*, **213**, 407–416.

VAUGHAN, A. P. M. & STOREY, B. C. 2007. A new supercontinent self-destruct mechanism: evidence from the Late Triassic–Early Jurassic. *Journal of the Geological Society, London*, **164**, 382–392.

VECOLI, M. & LE HERISSE, A. 2004. Biostratigraphy, taxonomic diversity and patterns of morphological evolution of Ordovician acritarchs (organic-walled microphytoplankton) from the northern Gondwana margin in relation to palaeoclimatic and palaeogeographic changes. *Earth-Science Reviews*, **67**, 267–311.

VEEVERS, J. J. 2004. Gondwanaland from 650–500 Ma assembly through 320 Ma merger in Pangea to 185–100 Ma breakup: supercontinental tectonics via stratigraphy and radiometric dating. *Earth Science Reviews*, **68**, 1–132, 10.1016/j.earscirev.2004.05.002.

VEEVERS, J. J. & POWELL, C. M. 1987. Late Paleozoic glacial episodes in Gondwanaland reflected in transgressive–regressive depositional sequences in Euramerica. *Geological Society of America Bulletin*, **98**, 475–487.

VENGOSH, A., KOLODNY, Y., STARINSKY, A., CHIVAS, A. R. & MCCULLOCH, M. T. 1991. Coprecipitation and isotopic fractionation of boron in modern biogenic carbonates. *Geochimica et Cosmochimica Acta*, **55**, 2901–2910.

VIA, R. K. & THOMAS, D. J. 2006. Evolution of Atlantic thermohaline circulation: Early Oligocene onset of deep-water production in the North Atlantic. *Geology*, **34**, 441–444, 10.1130/G22545.1.

VIDAL, L., BICKERT, T., WEFER, G. & ROHL, U. 2002. Late Miocene stable isotope stratigraphy of SE Atlantic ODP Site 1085: Relation to Messinian events. *Marine Geology*, **180**, 71–85.

VILLA, G. & PERSICO, D. 2006. Late Oligocene climatic changes: Evidence from calcareous nannofossils at Kerguelen Plateau Site 748 (Southern Ocean). *Palaeogeography, Palaeoclimatology, Palaeoecology*, **231**, 110–119.

VISSER, J. N. J. 1997. Deglaciation sequences in the Permo-Carboniferous Karoo and Kalahari basins of southern Africa: A tool in the analysis of cyclic glaciomarine basin fills. *Sedimentology*, **44**, 507–521.

VON DER HEYDT, A. & DIJKSTRA, H. A. 2006. Effect of ocean gateways on the global ocean circulation in the late Oligocene and early Miocene. *Paleoceanography*, **21**.

VONBLANCKENBURG, F., ONIONS, R. K., BELSHAW, N. S., GIBB, A. & HEIN, J. R. 1996. Global distribution of beryllium isotopes in deep ocean water as derived from Fe-Mn crusts. *Earth and Planetary Science Letters*, **141**, 213–226.

WADE, B. S. & PALIKE, H. 2004. Oligocene climate dynamics. *Paleoceanography*, **19**.

WALKER, I. R. 1991. Modern Assemblages of Arctic and Alpine Chironomidae as Analogs for Late-Glacial Communities. *Hydrobiologia*, **214**, 223–227.

WALKER, J. C. G., HAYS, P. B. & KASTING, J. F. 1981. A negative feedback mechanism for the long-term stabilization of Earth's surface-temperature. *Journal of Geophysical Research-Oceans and Atmospheres*, **86**, 9776–9782.

WALLISER, O. H. 1995. *Global Events and Event Stratigraphy in the Phanerozoic*. Springer-Verlag, Berlin.

WALLMANN, K. 2001. The geological water cycle and the evolution of marine δO^{18} values. *Geochimica et Cosmochimica Acta*, **65**, 2469–2485.

WALLMANN, K. 2004. Impact of atmospheric CO_2 and galactic cosmic radiation on Phanerozoic climate change and the marine δO^{18} record. *Geochemistry, Geophysics, Geosystems*, **5**.

WANG, P. X., TIAN, J., CHENG, X. R., LIU, C. L. & XU, J. 2004. Major Pleistocene stages in a carbon perspective: The South China Sea record and its global comparison. *Paleoceanography*, **19**.

WANG, Y. D., MOSBRUGGER, V. & ZHANG, H. 2005. Early to Middle Jurassic vegetation and climatic events in the Qaidam Basin, northwest China. *Palaeogeography, Palaeoclimatology, Palaeoecology*, **224**, 200–216.

WARNE, M. T. 2005. The global Mio-Pliocene climatic equability and coastal ostracod faunas of southeast Australia. *Palaeogeography, Palaeoclimatology, Palaeoecology*, **225**, 248–265.

WEBB, J. A. & GOLDING, S. D. 1998. Geochemical mass-balance and oxygen-isotope constraints on silcrete formation and its paleoclimatic implications in Southern Australia. *Journal of Sedimentary Research*, **68**, 981–993.

WEFER, G., BERGER, W. H., BIJMA, J. & FISCHER, G. 1999. Clues to ocean history: a brief overview of proxies. *In*: FISCHER, G. & WEFER, G. (eds) *Use of Proxies in Paleoceanography: Examples from the South Atlantic*. Springer, Berlin, 1–68.

WEI, G. J., YU, K. F. & ZHAO, J. X. 2004. Sea surface temperature variations recorded on coralline Sr/Ca ratios during Mid-Late Holocene in Leizhou Peninsula. *Chinese Science Bulletin*, **49**, 1876–1881.

WENDLER, J. 2004. External forcing of the geomagnetic field? Implications for the cosmic ray flux – climate variability. *Journal of Atmospheric and Solar-Terrestrial Physics*, **66**, 1195–1203.

WENZEL, B., LECUYER, C. & JOACHIMSKI, M. M. 2000. Comparing oxygen isotope records of Silurian calcite and phosphate – δO^{18} compositions of brachiopods and conodonts. *Geochimica et Cosmochimica Acta*, **64**, 1859–1872.

WESTERHOLD, T., BICKERT, T. & ROHL, U. 2005. Middle to late Miocene oxygen isotope stratigraphy of ODP site 1085 (SE Atlantic): new constraints on Miocene climate variability and sea-level fluctuations. *Palaeogeography, Palaeoclimatology, Palaeoecology*, **217**, 205–222.

WESTERLUND, K. J., GURNEY, J. J., CARLSON, R. W., SHIREY, S. B., HAURI, E. H. & RICHARDSON, S. H. 2004. A metasomatic origin for late Archean eclogitic diamonds: implications from internal morphology of diamonds and Re–Os and S isotope characteristics

of their sulfide inclusions from the late Jurassic Klipspringer kimberlites. *South African Journal of Geology*, **107**, 119–130.

WIGNALL, P. B. 2001. Large igneous provinces and mass extinctions. *Earth-Science Reviews*, **53**, 1–33.

WIGNALL, P. B., NEWTON, R. J. & LITTLE, C. T. S. 2005. The timing of paleoenvironmental change and cause-and-effect relationships during the early Jurassic mass extinction in Europe. *American Journal of Science*, **305**, 1014–1032.

WILDE, P., LYONS, T. W. & QUINBY-HUNT, M. S. 2004. Organic carbon proxies in black shales: molybdenum. *Chemical Geology*, **206**, 167–176.

WILLIAMS, G. E. 1998. Late Neoproterozoic periglacial aeolian sand sheet, Stuart Shelf, South Australia. *Australian Journal of Earth Sciences*, **45**, 733–741.

WILLIAMS, M., HAYWOOD, A. M., TAYLOR, S. P., VALDES, P. J., SELLWOOD, B. W. & HILLENBRAND, C. D. 2005. Evaluating the efficacy of planktonic foraminifer calcite δO^{18} data for sea surface temperature reconstruction for the Late Miocene. *Geobios*, **38**, 843–863.

WILLIAMS, N. E. & EYLES, N. 1995. Sedimentary and Paleoclimatic Controls on Caddisfly (Insecta, Trichoptera) Assemblages During the Last Interglacial-to-Glacial Transition in Southern Ontario. *Quaternary Research*, **43**, 90–105.

WILSON, P. A., NORRIS, R. D. & COOPER, M. J. 2002. Testing the Cretaceous greenhouse hypothesis using glassy foraminiferal calcite from the core of the Turonian tropics on Demerara Rise. *Geology*, **30**, 607–610.

WINCKLER, G., ANDERSON, R. F., STUTE, M. & SCHLOSSER, P. 2004. Does interplanetary dust control 100 kyr glacial cycles? *Quaternary Science Reviews*, **23**, 1873–1878.

WING, S. L., BAO, H. & KOCH, P. L. 2000. An early Eocene cool period? Evidence for continental cooling during the warmest part of the Cenozoic. *In*: HUBER, B. T., MACLEOD, K. S. & WING, S. C. (eds) *Warm Climates in Earth History*. Cambridge University Press, Cambridge, 197–237.

WING, S. L. & GREENWOOD, D. R. 1993. Fossils and fossil climate: the case for equable continental interiors in the Eocene. *Philosophical Transactions of the Royal Society of London Series B-Biological Sciences*, **341**, 243–252.

WINGUTH, A. M. E. & MAIER-REIMER, E. 2005. Causes of the marine productivity and oxygen changes associated with the Permian-Triassic boundary: A reevaluation with ocean general circulation models. *Marine Geology*, **217**, 283–304.

WINOGRAD, I. J., COPLEN, T. B., LANDWEHR, J. M. ET AL. 1992. Continuous 500,000-year climate record from vein calcite in Devils Hole, Nevada. *Science*, **258**, 255–260.

WOLFE, J. A. 1993. A method of obtaining climatic parameters from leaf assemblages. *US Geological Survey Bulletin*, **2040**, 1–73.

WOODCOCK, D. W. 1992. Climate reconstruction based on biological indicators. *Quarterly Review of Biology*, **67**, 457–477.

WOODWARD, F. I. 1987. Stomatal numbers are sensitive to increases in CO_2 from pre-industrial levels. *Nature*, **327**, 617–618.

WOODWARD, F. I. & BAZZAZ, F. A. 1988. The responses of stomatal density to CO_2 partial-pressure. *Journal of Experimental Botany*, **39**, 1771–1781.

WUCHTER, C., SCHOUTEN, S., COOLEN, M. J. L. & DAMSTE, J. S. S. 2004. Temperature-dependent variation in the distribution of tetraether membrane lipids of marine Crenarchaeota: Implications for TEX86 paleothermometry. *Paleoceanography*, **19**.

YAMADA, K., TANAKA, Y. & IRIZUKI, T. 2005. Paleoceanographic shifts and global events recorded in late Pliocene shallow marine deposits (2.80–2.55 Ma) of the Sea of Japan. *Palaeogeography, Palaeoclimatology, Palaeoecology*, **220**, 255–271.

YEMANE, K. & KELTS, K. 1996. Isotope geochemistry of Upper Permian early diagenetic calcite concretions: Implications for Late Permian waters and surface temperatures in continental Gondwana. *Palaeogeography, Palaeoclimatology, Palaeoecology*, **125**, 51–73.

YOUNG, G. M., VON BRUNN, V., GOLD, D. J. C. & MINTER, W. E. L. 1998. Earth's oldest reported glaciation: Physical and chemical evidence from the Archean Mozaan Group (similar to 2.9 Ga) of South Africa. *Journal of Geology*, **106**, 523–538.

YU, F. Q. 2002. Altitude variations of cosmic ray induced production of aerosols: Implications for global cloudiness and climate. *Journal of Geophysical Research-Space Physics*, **107**.

ZACHOS, J., PAGANI, M., SLOAN, L., THOMAS, E. & BILLUPS, K. 2001. Trends, rhythms, and aberrations in global climate 65 Ma to present. *Science*, **292**, 686–693.

ZACHOS, J. C. & KUMP, L. R. 2005. Carbon cycle feedbacks and the initiation of Antarctic glaciation in the earliest Oligocene. *Global and Planetary Change*, **47**, 51–66.

ZACHOS, J. C., LOHMANN, K. C., WALKER, J. C. G. & WISE, S. W. 1993. Abrupt climate change and transient climates during the Paleogene: a marine perspective. *Journal of Geology*, **101**, 191–213.

ZACHOS, J. C., ROHL, U., SCHELLENBERG, S. A. *ET AL.* 2005. Rapid acidification of the ocean during the Paleocene–Eocene thermal maximum. *Science*, **308**, 1611–1615.

ZEEBE, R. E. 2005. Stable boron isotope fractionation between dissolved $B(OH)_3$ and $B(OH)_4^{mm}$. *Geochimica et Cosmochimica Acta*, **69**, 2753–2766.

ZHANG, Y., OBATA, H. & NOZAKI, Y. 2004. Silver in the Pacific Ocean and the Bering Sea. *Geochemical Journal*, **38**, 623–633.

ZHARKOV, M. A. 2005. Evaporite sedimentation in the Precambrian as related to changes in biosphere and seawater chemistry, article 1: evaporites of the Archean and lower Proterozoic. *Stratigraphy and Geological Correlation*, **13**, 134–142.

ZHU, X. K., GUO, Y., WILLIAMS, R. J. P. *ET AL.* 2002. Mass fractionation processes of transition metal isotopes. *Earth and Planetary Science Letters*, **200**, 47–62.

ZHU, X. K., O'NIONS, R. K., GUO, Y. L. & REYNOLDS, B. C. 2000. Secular variation of iron isotopes in North Atlantic Deep Water. *Science*, **287**, 2000–2002.

ZIEGLER, A. M., ESHEL, G., REES, P. M., ROTHFUS, T. A., ROWLEY, D. B. & SUNDERLIN, D. 2003. Tracing the tropics across land and sea: Permian to present. *Lethaia*, **36**, 227–254.

ZIELINSKI, U., GERSONDE, R., SIEGER, R. & FUTTERER, D. 1998. Quaternary surface water temperature estimations: Calibration of a diatom transfer function for the Southern Ocean. *Paleoceanography*, **13**, 365–383.

Reconstructing Neoproterozoic palaeoclimates using a combined data/modelling approach

L. E. SOHL & M. A. CHANDLER

Center for Climate Systems Research at Columbia University, 2880 Broadway, New York, New York 10025, USA. (e-mail: les14@columbia.edu; tel: (212) 678-5550/fax: (212) 678-5648)

Abstract: Climate reconstructions of the Neoproterozoic Era (1000–542 Ma) face special challenges because many proxies used to constrain younger palaeoclimates are not available/applicable in Precambrian time. Given the few available proxies, deep time climate simulations are best viewed as a means to address more fundamental questions about the nature of climate change and to address disparities in data interpretation by examining phenomena from a process-related perspective. The Global Climate Model (GCM) simulations presented here were aimed at determining what combination of forcings might have permitted the initiation of low- to mid-latitude continental ice-sheets during the Sturtian glacial interval, *c.* 750 Ma. However, despite the formation of extensive extratropical ice cover, tropical regions in these experiments remain too warm for the initiation of large ice-sheets. The enhanced precipitation along the leading edge of icy regions suggests that the addition of topographic relief and dynamic ice flow could make ice-sheets viable into subtropical regions. However, these simulations suggest that 'hard' snowball Earth solutions are only likely for much earlier intervals in Earth history, and are certainly not viable in combination with large accumulations of greenhouse gases.

The Neoproterozoic Era (1000 to 542 Ma) is a remarkable interval in Earth history from the standpoint both of climatic change and biological innovation. On the climate front, we have evidence for tremendous swings between conditions for which we have as yet no adequate explanation. Parts of the Neoproterozoic Era are marked by an equable climate that is perhaps warmer than we would have expected, given a less-luminous Sun (a smaller-scale version of the 'faint young Sun' paradox afflicting early Earth palaeoclimate; see e.g. Tajika 2003; Chumakov 2004). However, we also have two of the most severe glaciations in Earth history, the so-called 'snowball Earth' events, in which continental-scale ice-sheets were able to exist, at sea level, within 20° of the equator (Park 1997; Schmidt & Williams 1995; Sohl *et al.* 1999). On the biological front, we have the first appearance of macroscopic multicellular organisms (the Mistaken Point and Ediacaran faunal assemblages) toward the end of the Neoproterozoic Era (Narbonne 2005), a prelude to the great expansion in biodiversity known as the Cambrian explosion. Perhaps more importantly, there is the evolution of the first shelly marine animals such as *Cloudina* (e.g. Hofmann & Mountjoy 2001), whose ability to create their own shells may have forever altered ocean chemistry and helped established the marine carbon cycle as we now know it (Ridgwell *et al.* 2003; Bartley & Kah 2004).

As the macroscopic fossil record appears almost entirely limited to a time some 60 million years after the snowball Earth glaciations occurred (Narbonne 2005), there has been considerable speculation about the extent to which the extreme climate changes could have driven mass extinctions and/or biological innovations. Naturally, we would like to explore these potential impacts by better quantifying characteristics of Neoproterozoic climate, such as surface air temperature, snow cover, and sea-ice extent through the use of proxies and computer models. However, reconstructing the palaeoclimate of an interval so far in Earth's past presents special challenges not faced by geoscientists and climate modellers who work on younger time intervals. The palaeoclimate proxies available to constrain climate forcings, such as greenhouse gas levels and ocean circulation, are not as quantitative or readily interpreted as we might like, while boundary conditions for Global Climate Models, such as continent/ocean distribution, topography, ground cover, and even solar luminosity, are not well constrained.

The purpose of Neoproterozoic climate modelling is therefore largely limited to addressing more fundamental questions about the nature of climate change and its potential relationship to biological and ecosystem changes. Since our understanding of Neoproterozoic environments is limited compared to those of the Phanerozoic Era, the need to address the big picture questions first is not a handicap. In fact, Neoproterozoic climate change has the potential to shed light on the large-scale forcings involved in creating the broadest range of natural

From: WILLIAMS, M., HAYWOOD, A. M., GREGORY, F. J. & SCHMIDT, D. N. (eds) *Deep-Time Perspectives on Climate Change: Marrying the Signal from Computer Models and Biological Proxies.* The Micropalaeontological Society, Special Publications. The Geological Society, London, 61–80.
1747-602X/07/$15.00 © The Micropalaeontological Society 2007.

climate variability that the Earth has ever displayed. It is also clear that with such extreme environmental ranges, many first-order physical processes – such as moisture fluxes, dynamic transports of mass and momentum, detailed parameterizations of hydrologic processes, and to the extent possible, ocean processes – are critical to an accurate portrayal and understanding of the climates of this time. The Neoproterozoic Era thus presents us with astounding, non-hypothetical climate scenarios that both captivate geoscientists and modellers, and require a high degree of communication and cooperation amongst a wide range of research specializations.

One cautionary note must be kept in mind when modelling deep time palaeoclimates such as the snowball Earth intervals, for which boundary constraints are scarce: it is very easy to fall into the trap of setting up simulations that explore interesting theoretical considerations, but do not address any questions regarding actual geological events in a manner that is truly useful for interpreting the past. In planning and evaluating deep time palaeoclimate simulations, climate modellers need to be aware not only of their own model limitations, but also the need to be careful about trying too hard to achieve a specific goal rather than seeking to understand the results that models deliver based on available geological inputs. Our approach to climate modelling has focused on creating an ensemble of experiments that define a range of possible climate forcings and boundary conditions, and allowing the GCM to tell us how the forcings, feedbacks and other processes interact to yield various climate solutions.

Overview of climate trends and chronology of events for the Neoproterozoic Era

The Neoproterozoic Era can be divided into three broad climate intervals, roughly equivalent to chronostratigraphic divisions and defined by the inferred presence of little or no glacial deposits in the geological record (the Tonian and Ediacaran Periods) vs. the presence of widely distributed glacial deposits that suggest that an ice age was in progress (the Cryogenian Period) (see Fig. 1). Firm age constraints are hard to come by for any climatically significant changes reflected in sedimentation patterns; indeed, this is a general problem for the entire era. For the sake of convenience, we define the boundaries of the glacial intervals by the approximate age ranges of the glacial deposits, although global cooling trends probably began at an earlier date and proceeded gradually

into the more extreme state. The Neoproterozoic climatic intervals can be described as follows:

The *pre-Cryogenian warm interval* (informal name, Tonian Period; 1000–850 Ma) is defined on the basis of a general abundance of substantial, presumably warm-water carbonate deposits that are sometimes associated with evaporite rocks (e.g. the Mackenzie Mountains Supergroup and Coates Lake Group, northwestern Canada, Jefferson & Parrish 1989; the Callana and Burra Groups, southern Australia, Preiss 1987). This interval is assumed to represent a continuation of the warm, equable conditions believed to have existed for the previous 1200 million years (Hambrey & Harland 1981, 1985; Lowe 1992; Buick *et al.* 1995).

The *Cryogenian interval* (850–635 Ma) is marked by glacial deposition on nearly every modern landmass except Antarctica (Hambrey & Harland 1981, 1985) during two episodes commonly referred to as the Sturtian and Marinoan (or Varanger) glaciations. What makes these two glaciations unusual is their severity compared with the Pleistocene glaciation, leading to the nickname 'snowball Earth' glaciations. Reliable palaeomagnetic evidence for ice-sheets having existed at low latitudes near sea level is strongest for the Marinoan glacial deposits of Australia (e.g. Sohl *et al.* 1999). Less well constrained are the positions of other glaciated continents, but existing palaeomagnetic data suggest that concurrent glaciation may have also existed at mid- to high latitudes (Torsvik *et al.* 1995; Meert *et al.* 1994). There is also evidence of low-latitude glaciation in northwestern Canada during the Sturtian glaciation (Park 1997), with other possible Sturtian glacial deposits at low latitudes in Australia (McWilliams & McElhinny 1980), India (e.g. Unrug 1992), and Namibia (Meert *et al.* 1995). Between these two glaciations is an interglacial interval of unknown duration, marked by the presence of largely siliciclastic successions (e.g. sandstones and shales) that contain varying proportions of carbonate rocks, such as in Australia (Preiss 1987) and northwestern Canada (Narbonne & Aitken 1995). The sedimentology and mineralogy of the interglacial carbonate rocks in south Australia suggest that subtropical to tropical conditions existed within 15–20° of the equator, and imply that latitudinal temperature gradients may have been comparable to the present day (Sohl 2000; cf. James *et al.* 2005).

Until recently, the commonly accepted timing of Cryogenian glaciations assumed a separation of over 100 million years between the Sturtian and Marinoan episodes, with the approximate age of the Sturtian set at 750–725 Ma and the age of the Marinoan at *c.* 600 Ma. New radiometric age constraints (Kendall *et al.* 2006) now suggest that glacial deposits associated with the Sturtian

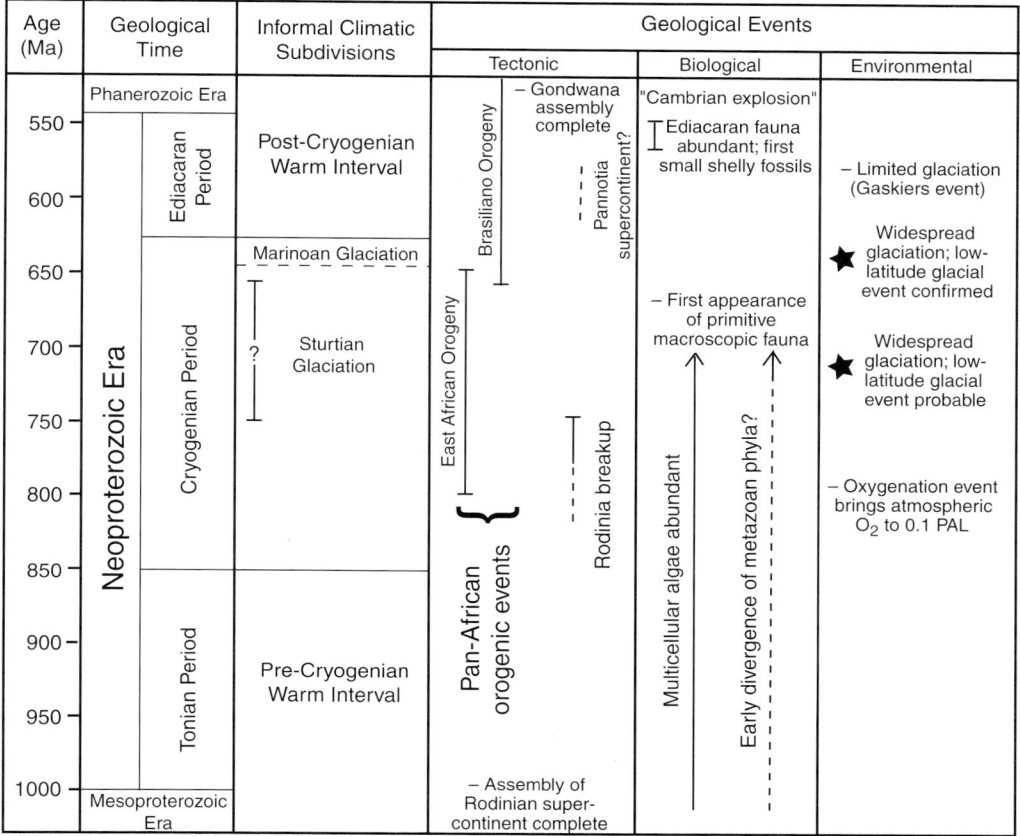

Fig. 1. Earth history context for climatic intervals of the Neoproterozoic Era (adapted from Chandler & Sohl 2000). Stratigraphic subdivisions follow Cowie *et al.* (1989) and Knoll *et al.* (2004). Ages and age ranges for geological and climatic events are approximate, given sparse radiometric age constraints in most cases. Tectonic events are after Hoffman (1991); Powell *et al.* (1993); Meert & van der Voo (1997); Dalziel (1997). Biological events follow Hofmann *et al.* (1990); Bromham *et al.* (1998); Grotzinger *et al.* (1995); Narbonne (2005). Environmental events according to Schmidt & Williams (1995); Canfield & Teske (1996); Park (1997); Sohl *et al.* (1999); Thompson & Bowring (2000); Kendall *et al.* (2006).

glaciation encompass such a broad timespan – perhaps 750 Ma to 650 Ma – that our view of Cryogenian glacial episodes is in serious need of revision. For the purposes of this paper, we assume a 'classic' Sturtian glacial scenario set at *c.* 750 Ma.

The *post-Cryogenian warm interval* (635–542 Ma, essentially coincident with the recently ratified Ediacaran Period; Knoll *et al.* 2004) is defined based upon the development of several substantial shallow-water carbonate platform successions, implying warm-water depositional conditions in a number of widespread locations, e.g. India (Shankar *et al.* 1993; Jiang *et al.* 2002), Siberia (Pelechaty *et al.* 1996), and south China (Jiang *et al.* 2003). The Gaskiers glaciation has recently been shown to occur during this otherwise warm interval (Thompson & Bowring 2000; Condon *et al.* 2005), although it does not appear to have been as severe or as long-lived as the previous snowball Earth-type glaciations. There have been suggestions that additional minor glacial events occurred based in part upon sedimentological observations and fluctuations in the global δ^{13}C isotopic curve (e.g. DiBona 1991; Kaufman *et al.* 1997), but debate continues over these purported events (Zhang *et al.* 2005).

Challenges in deep-time palaeoclimate reconstruction

As amply illustrated by the other chapters in this volume, geochemical and biological proxies are among the most common tools for indirectly extracting climate and primary productivity information from the geological record. These proxies include stable isotope values primarily measured in sediments and faunal assemblages in both the marine and terrestrial realms. The proxies can be related to climate system variables such as ocean temperature and circulation patterns, ocean productivity, seawater alkalinity, and atmospheric CO_2 levels. The utility of proxies has been demonstrated repeatedly in Pleistocene and younger sediments, while their use in older intervals of the Phanerozoic Era has met with varying levels of success (see reviews in Bradley 1994; Henderson 2002).

However, a review of the previous descriptions of the Neoproterozoic climate intervals will show that mention of geochemical and biological proxies is conspicuously lacking. That is because the deeper one goes into Earth's past, the more problematic the use of these proxies becomes, since the sources of the proxy data are either unavailable or not readily interpreted owing to various factors. A principal problem is that many of the common organic sources for the geochemical proxies, and indeed all the climate-sensitive faunal assemblages typically used for the Cenozoic Era (e.g. incidence of *Pachyderma* left- *vs.* right-coiling foraminifera; occurrences of bryozoans *vs.* corals), simply do not exist in the Neoproterozoic Era: the organisms had not yet evolved. Organic-walled microfossils called acritarchs did exist throughout the Neoproterozoic Era, but their fossil record is remarkably dull for much of that time. There are few distinctive index forms until after the Marinoan glaciation, so there are limited biostratigraphic constraints to associate with the proxy data measured. Small shelly fossils (SSFs), the first organisms with calcite shells, only make their first appearance near the close of the Neoproterozoic Era (*c.* 550 Ma; Narbonne 2005), and so they too are of little help as sources of isotopic proxy data for palaeoclimate reconstructions. In any case, we have insufficient data to determine whether the SSFs interacted with their ecosystem in a manner similar enough to foraminifera, such that we could draw the same environmental conclusions from the proxies measured. The TEX_{86} proxy for deriving sea surface temperatures (SSTs) from organic lipids in marine crenarchaeota (Schouten *et al.* 2003) may hold some promise, but the proxy is fairly new and remains untested for this time period.

There are alternative sources to purely organic sources of geochemical proxy data – the Neoproterozoic rock record does include a fair proportion of limestones and dolomites – but these sources also have severe limitations. Diagenetic (postdepositional chemical) changes can alter the values of geochemical proxies substantially.

Oxygen isotope ($\delta^{18}O$) values are perhaps the most vulnerable to alteration (Killingley 1983; Schrag *et al.* 1995), but none of the other geochemical proxies is completely immune to alteration after burial (e.g. Lehmann *et al.* 2002). Even with great care in sample collection, it is not always possible to determine whether the values measured in the laboratory reflect the original geochemical conditions in the samples' depositional environment (Jiang 2002; Kaufman *et al.* 2006). Mineralogical compositions and textures are also frequently altered through diagenesis (e.g. aragonite dissolves and is replaced by calcite; carbonate mud recrystallizes during neomorphism; Tucker 1990), and it may not be possible to determine the original state of the sample.

Despite these difficulties, we are not left without any means to get a sense of Neoproterozoic climatic conditions. The occurrence and distribution of 'climate-sensitive' sediments becomes the main proxy for palaeoclimate, through simple analogy to modern occurrences of the same types of rocks (Briden & Irving 1964; Parrish 1998). Glacial sediments require the action of glaciers, and thus a cold climate. Shallow-water limestones require warm water and supersaturation of carbonate in marine waters, which is typical of tropical marine shelf environments (assuming, perhaps erroneously, an ocean composition similar to present). Evaporite deposits require a hot, dry climate to evaporate seawater in restricted basins. The global climate picture must then be pieced together by making correlations between stratigraphic successions in different regions, and available palaeolatitude data, to determine the palaeogeographic extent of approximately age- and climatically equivalent rocks.

As a prime example of this methodology, the arguments in favour of the two Neoproterozoic snowball Earth glaciations depend principally on the sheer number and widespread palaeogeographical extent of presumed age-equivalent glacial deposits, which in some cases lay well down into low latitudes (Fig. 2; Hambrey & Harland 1981, 1985). There are additional, smaller-scale sedimentary features that also argue for cold climates across a range of latitudes during the glacial intervals. For example, there are examples of fossil permafrost features called sand wedges (e.g. Nystuen 1976; Deynoux 1982; Williams & Tonkin 1985; Deynoux *et al.* 1989; Zhang 1994), vertically foliated sand-filled cracks in frozen ground that are known to have formed during the Pleistocene ice age under arid conditions, when the annual average air temperature was below freezing but seasonally above freezing (Black 1976). There is localized disruption of sediments underlying glacial diamictites, which can be interpreted as ice-contact deformation at a shoreline. There is also the probable occurrence of glendonites, metastable

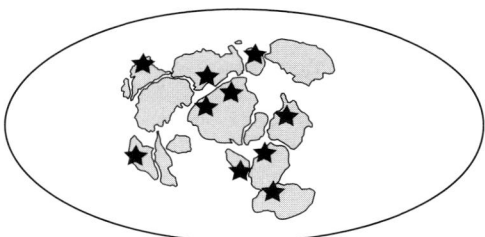

Fig. 2. Reconstruction of the continental distribution during the Sturtian glacial interval used in this study, with the locations of Neoproterozoic glacial deposits dated to this interval marked by stars. The locations of individual landmasses are based upon the 'classic' Rodinia reconstructions suggested by Hoffman (1991), Powell *et al.* (1993), Torsvik *et al.* (1996), and Weil *et al.* (1998). Glacial deposits are noted as per Hambrey & Harland (1981) and Evans (2000).

minerals such as ikaite (since replaced by calcite) that are characteristic of colder saline-rich environments, within some of the shallow marine deposits (Sohl 2000; James *et al.* 2005).

The puzzle of the extreme nature of the Sturtian and Marinoan glaciations is counterbalanced by the puzzle of why the Neoproterozoic Era as a whole was not uniformly colder, given that the Sun was approximately 4.5 to 8% less luminous over the course of the era (Gough 1981). Again, the evidence we have in favour of warm climatic conditions comes from geological terrestrial/shallow marine proxy sources. Outside of the glacial intervals, there are widespread deposits of shallow marine carbonate rocks (limestones and dolomites) that are deposited today mainly in tropical settings, where the warmth of the seawater encourages the chemical reactions that precipitate carbonate out of the water (e.g. Preiss 1987; Jefferson & Parrish 1989; Jiang *et al.* 2002). Prior to the Sturtian glaciation, there are also regionally extensive mid-latitude deposits of evaporitic rocks (salts such as gypsum and halite) that must have accumulated in hot, arid environments (e.g. Hill *et al.* 2000), a depositional setting also more commonly associated today with tropical regions. The overall portrait of the non-glacial intervals of the Neoproterozoic Era is one of an equable climate, not unlike that described for the Mesozoic Era.

Of these two 'end-member' climate conditions we interpret from the Neoproterozoic geological record, the snowball Earth glaciations have attracted most of the attention in recent years, in large part because of the discussion over the possible climatic influence on the evolution of macroscopic life. Previous iterations of the snowball Earth hypothesis (Hoffman *et al.* 1998; Hoffman & Schrag 2002) have taken the geological

evidence for widespread cold climates and extrapolated a vision of a world practically entombed in ice, with oceans frozen over or nearly so for millions of years. Such a condition would obviously have had an enormous impact on the ability of life to survive in abundance, had the freeze-over actually happened – and therein lays a point of contention. Proponents of the 'hard' snowball Earth have suggested that total or near-total sea-ice cover is necessary to explain both an interpreted rapid transition (a few hundreds to thousands of years) from the glacial to non-glacial state, as well as unusual ^{13}C signatures, as low as $-5\permil$, in carbonate rocks ('cap carbonates') directly overlying the glacial deposits (Hoffman *et al.* 1998; Hoffman & Schrag 2002). Those in favour of a slightly less extreme scenario, the 'slushball' Earth, point to sedimentary deposits (ice-rafted debris in deep marine settings) that can be used to argue in favour of more open ocean rather than less (McMechan 2000; Condon *et al.* 2002; Kellerhals & Matter 2003). Icebergs carrying the debris would have needed room to drift and melt as they dropped their sediment load, and active, wet-based continental glaciers probably needed a significant open-water source of moisture for precipitation, in order to be maintained. The problem here is that there are no data available that can *directly* support or refute the existence of total or near-total global sea-ice cover – there is no geological proxy in the Neoproterozoic Era for sea-ice cover at all.

The only tool we then have available to evaluate the likelihood of nearly, or totally, frozen oceans during either of the Neoproterozoic snowball Earth glaciations is computer climate simulation. To illustrate one possible modelling approach, we use here a version of the GISS Global Climate Model (GCM) to simulate the 'classic' Sturtian glaciation, *c.* 750 Ma, which has slightly better geological constraints available for the model boundary conditions. Our principal goal is to attempt to duplicate, as closely as possible, the surface conditions that would have permitted the existence of large-scale ice-sheets in mid- to low latitudes as indicated by the distribution of glacial deposits in the geological record. We emphasize that the GCM is used here to assess whether known forcings are consistent with direct geological evidence for the extreme glacial episodes and, additionally, to supply the physical process information required to unravel the mechanisms that led to climate change. Tests of the hypotheses for the exact nature of these snowball Earth glaciations ('hard' snowball *vs.* 'slushball') must ultimately take into account a combination of geochemical data, sedimentary features, depositional settings, and numerical model behaviour.

Choosing a climate model for palaeoclimate simulations

There are several classes of climate models that are employed in exploring deep-time palaeoclimate. Energy balance models (EBMs) are well suited for calculating the radiative effects of changes in greenhouse gas concentrations, especially when those concentrations are quite large with respect to the modern atmosphere. This circumstance typically arises in simulations of early Earth climate, when carbon dioxide and/or methane concentrations are thought to have been several orders of magnitude higher than at present (e.g. Caldeira & Kasting 1992). Earth models of intermediate complexity (EMICs) come in a wide variety of forms and typically incorporate a much broader range of parameters than EBMs, including a hydrologic cycle (Claussen *et al.* 2002). EMIC simulations can be run longer than GCM simulations, but computing efficiency is achieved in part by lowering the grid resolution of the model, as well as prescribing, to varying degrees, atmospheric and oceanic physics. Global climate models (GCMs) provide the most detailed representation of the climate system and its dynamics. GCMs include not only explicit calculation of energy balance, but also incorporate the conservation of mass, moisture and momentum. Although GCMs are much more computationally expensive than the simpler models, their higher geographic resolution makes them most appropriate for examining the effects of altered boundary conditions, and they are most directly compared with spatially delimited palaeoclimate proxy data (e.g. the distribution of glacial deposits). Large-scale climate change scenarios that directly related to tropical feedback mechanisms (such as snowball Earth climates) would be difficult to simulate accurately without inclusion of moisture and momentum physics.

For the study presented here, we employ an atmospheric GCM that incorporates an approximation of the Neoproterozoic palaeogeography (*c.* 750 Ma) in addition to other boundary conditions drawn from palaeoclimate proxy data.

Global climate model description

The experimental results presented here were produced using a newer version of the GISS Model II global climate model (GCM). The GCM is a three-dimensional Cartesian grid-point model, which solves numerically the equations of conservation for energy, mass, momentum and moisture, as well as the equation of state. It uses a horizontal grid resolution of 8° latitude by 10° longitude, with nine layers in the atmosphere, and has two ground layers for hydrological storage (which is minimal in our Neoproterozoic simulations, since all land is defined as desert). The model accounts for seasonal and diurnal solar variations and contains parameterizations for large-scale and convective clouds, background aerosols, and all radiatively important trace gases. Gases explicitly incorporated into the radiation scheme include carbon dioxide, methane and nitrous oxides (anthropogenic chlorofluorocarbons, or CFCs, were of course zeroed out in these simulations). The model generates precipitation whenever supersaturated conditions occur, and snow depth is based on the balance between snowfall, melting and sublimation.

Sea surface temperatures (SSTs) are calculated using model-derived surface energy fluxes and specified ocean heat convergences. The ocean heat convergences vary both seasonally and regionally, but are otherwise fixed. This is the primary mixed-layer 'Q-flux' ocean model developed for use with the GISS GCM, full details of which are described in Russell *et al.* (1985), and in appendix A of Hansen *et al.* (1988). We modified the original method of Russell *et al.* (1985) by using five harmonics instead of two in defining the seasonally varying energy flux and upper ocean energy storage, which improves accuracy in regions of seasonal sea ice formation. By deriving vertical fluxes and upper ocean heat storage from a run with appropriate palaeogeography for the Sturtian interval, the model provides a more self-consistent method for obtaining ocean heat transports.

The role of the ice/albedo feedback will be critical in experiments that produce significant changes in sea ice and snow cover. Newly fallen snow in the model has an albedo of 0.85 and ages within 50 days to a lower limit of 0.5. The sea-ice parameterization is thermodynamic; below $-1.6\,°C$, ice of 0.5 m thickness forms over a fractional area of the grid box and henceforth grows horizontally as needed to maintain energy balance. Surface fluxes change the ocean water and sea-ice temperature in proportion to the area of a grid cell they cover. Conductive cooling occurs at the ocean/ice interface, with sea ice thickening if the water temperature remains at $-1.6\,°C$. Sea ice melts when the ocean warms to $0\,°C$, and the SST in a grid box remains at $0\,°C$ until all ice has melted in that cell. The albedo of snow-free sea ice is independent of thickness; it is assigned a value of 0.55 in the visible spectrum and 0.3 in the near infrared, for a spectrally weighted value of 0.45.

Setting up boundary conditions and forcings for a snowball Earth simulation

The scarcity of any palaeoclimate proxy data for the Neoproterozoic Era beyond the distribution and characteristics of climate-sensitive sediments leaves rather more degrees of freedom in setting up simulations than would be the case for more recent palaeoclimates. For certain boundary conditions and initial forcings, we have no choice but to make educated guesses as to appropriate values based upon the information we do have. At the same time, the considerable age of the Neoproterozoic Era also introduces new considerations, such as significant changes in solar luminosity and palaeogeography, which would typically not be of interest to Pleistocene ice age or future climate modellers.

The key boundary conditions and forcings we altered for this series of Sturtian simulations, and the rationale behind the values selected, are as follows:

Solar luminosity

According to current astrophysical theory, a G2-type yellow star such as the Sun grows brighter as it ages; various standard stellar evolution models suggest that the Sun would have been 25 to 30% less luminous when the Solar System formed. Gough (1981) gives the luminosity change as

$$L(t) = \left[1 + \frac{2}{5}(1 - t/t_\odot)\right]^{-1} L_\odot \qquad (1)$$

where $t \leq t_\odot$, and $t_\odot$ is the age of the Earth. Using Gough's equation, with $t_\odot = 4550$ Ma (Dalrymple 1991), the solar luminosity value for 750 Ma is 6.19% less than the modern value of 1366.619 W m^{-2}, or 1282.026 W m^{-2}. Incoming solar radiation is reduced in GISS Model II by decreasing the total amount of shortwave radiation entering the top of the atmosphere, and it is proportionally reduced at all wavelengths.

Palaeogeography/topography

The supercontinent of Rodinia, which had formed by the beginning of the Neoproterozoic Era (Hoffman 1991; Dalziel 1995), began to break apart again c. 800–780 Ma (e.g. Preiss 1987; Narbonne & Aitken 1995), although widespread rifting leading to the final break-up apparently did not occur until approximately 750 Ma, commensurate with the approximate onset of global cooling and the Sturtian glacial interval (e.g. Powell et al. 1993; Borg & DePaolo 1994). For our Sturtian experiments, we developed a palaeocontinental reconstruction based upon the available palaeomagnetic data (see Evans 2000 for a summary of palaeomagnetic constraints on glacial deposits) and other geological constraints (e.g. Dalziel 1997) used to define what Rodinia may have looked like just prior to break-up. Such reconstructions are necessarily tentative, as reliable palaeomagnetic data are not abundant and age constraints on the relevant rocks are not well defined (see e.g. Wingate et al. 2002 for a discussion of alternative reconstructions). However, a radically different continental configuration for the Sturtian interval, as compared with modern geography, does provide an opportunity to examine the possible effects of varying land distribution on climate, and specifically whether the low to mid-latitude landmass distribution could have played a key role in triggering glaciation by increasing the surface albedo of tropical regions (as per Kirschvink 1992).

Since the true relief for the Sturtian interval is unknown, we set the topography in all Sturtian simulations at a uniform 50-metre height for all land areas. Some of our previous unpublished Marinoan experiments have suggested that orography may be an important influence on the initiation of low-latitude glaciation. Topographic estimates can be made for presumed elevated regions based upon the age and distribution of orogenic belts and rift zones (Ziegler et al. 1985), but we do not pursue that issue in this study.

Vegetation

The GCM requires the assignment of vegetation categories to each grid cell containing land; these assignments are integral to the ground hydrology parameterization of our GCM (Hansen et al. 1983). Since the earliest-known fossils of vascular land plants are Ordovician in age (Wellman et al. 2003), it is generally assumed that Neoproterozoic continents were not vegetated in the modern sense (cf. Heckman et al. 2001; Kennedy et al. 2006). We mimic the lack of land vegetation by assigning desert conditions to all land area (Matthews 1983).

Greenhouse gases

The level of atmospheric CO_2 during the Neoproterozoic Era is of prime importance for understanding palaeoclimate conditions, but proxy data are few, so varying viewpoints exist as to appropriate levels. Some researchers suggest that atmospheric CO_2 levels needed to be high, perhaps 10× modern, in order to counteract the decreased solar luminosity (e.g. Kasting 1987). Others have suggested that CO_2 should have been at levels close to the modern value, given that the size of the dissolved inorganic carbon (DIC) reservoir in the oceans, which can be correlated with atmospheric pCO_2, was not significantly different from the modern DIC reservoir (Bartley & Kah 2004). Our previous simulations run for the younger Marinoan

(Varanger) glaciation (Chandler & Sohl 2000) showed that modern and 4× modern levels of CO_2 produced climatic conditions much too warm to support low-latitude glaciation, despite a less-luminous Sun. Even when we reduced CO_2 to a level of just 40 ppmv, less than a quarter of the minimum value during the late Pleistocene glaciation (Bartola *et al.* 1987), it was still not quite sufficient to make the Earth cold enough to support continental-scale ice-sheets in tropical latitudes.

For this series of Sturtian simulations then, we opted to modify combinations of both CO_2 and methane (CH_4). Although methane is present in the modern atmosphere in much lower concentrations than CO_2, it is a considerably more powerful greenhouse gas (as much as 32× as effective as CO_2 in trapping heat), and we anticipated an increased cooling through the reduction of both gases in our simulations. Greenhouse gas levels employed for CO_2 included 315 ppmv (the 'modern' value measured in 1958), 140 ppmv (half the accepted pre-industrial value) and 40 ppmv (an extreme value we carry over from previous experiments). For methane, gas levels included 1.224 ppmv (again, the 'modern' value measured in 1958), and 0.612 (half the modern value) and 0.306 ppmv (25% of the modern level, and slightly less than the Late Pleistocene [145 ka] minimum value of *c.* 0.325 ppmv; Chappellaz *et al.* 1990).

Ocean heat transports

The transport of heat by ocean circulation is critical to the distribution of temperatures on the planet. Today, the oceans transport heat, on an annually averaged global basis, away from tropical and subtropical areas and into the middle and high latitudes. Previous simulations (e.g. Chandler & Sohl 2000) have shown that reducing ocean heat transports allows the polar regions to cool to a greater extent but sequesters heat in tropical regions. Although these changes are a simple redistribution of energy, the global climate impact is not negligible because altered distributions of ocean–atmosphere energy exchange directly affect certain key climate feedback mechanisms (e.g. ice albedo feedback, tropical moisture feedback). Thus, the ultimate impact of changes in ocean heat transport can be to alter the global climate. For this series of Sturtian experiments, we simulated the potential effects of poleward ocean heat transports that were 100%, 50% and 10% of the modern value. The geographic distribution of the ocean transports are necessarily modified since the Neoproterozoic ocean basin configurations are considerably different from modern configurations.

Ocean heat storage

As described above, the GCM uses a mixed-layer ocean model with a Q-flux parameterization to simulate horizontal transport of energy in the oceans. For these experiments, we adjusted the maximum mixed-layer depth (MLD) to be the full depth of the ocean (up to 4000 m) so the ocean heat storage capacity is equal to that of the full ocean. The GISS GCM does have an option to use a simple diffusion into the deeper ocean to mimic the sequestration of heat below the relatively shallow modern mixed layer. However, the diffusion rates are parameterized based on the geography of modern deep water production and are not appropriate for use with an altered ocean basin configuration. A fully dynamic coupled ocean is a desirable option, and some researchers have already employed one in their simulations of a snowball Earth-type ice age (e.g. Poulsen *et al.* 2001); however, with no Neoproterozoic bathymetry information available, even a dynamic ocean model does not provide definitive answers regarding the state of ocean circulation.

Experimental design

In order to determine just how much change we are introducing to a simulation by modifying the forcings, we have run two types of experimental control runs. In these, the forcings such as solar luminosity, atmospheric greenhouse gases and ocean heat transports are set at their modern levels (specifically, our control runs use 1958 forcing values, since that was the first year direct, continuous measurements of greenhouse gas levels were made). Our baseline reference for the model is a control run with modern geography, appropriately labelled 'Modern' for this series of experiments. For palaeoclimate experiments, we also conducted a control run (labelled S001R) using the Sturtian-age Rodinian continental distribution, but with all other forcings identical to the modern control. This 'palaeo-control' run allows us to make more meaningful assessments of the effects of solar and greenhouse gas forcings on the latitude by longitude climate differences for the Sturtian interval, which are otherwise dominated by the presence/absence of land at any given geographic location. The effects of the palaeogeographic changes on climate are shown separately as well.

All of the Sturtian experiments and control runs were run for 100 simulated years, and results analysed based on averages from the final five years of each simulation. A list of the specific setup of each simulation is shown in Table 1, and key

Table 1. *Palaeoclimate simulations discussed in this paper*

RunID	Forcings			
	Solar luminosity (% reduction)	Atmospheric CO_2 (ppm)	Atmospheric CH_4 (ppm)	OHT (% modern)
Modern	100	315	1.224	100
S001R	100	315	1.224	100
S002	−6.19	315	1.224	100
S003	−6.19	140	1.224	100
S004	−6.19	140	1.224	50
S005	−6.19	140	0.612	100
S006	−6.19	140	0.612	50
S007	−6.19	40	1.224	100
S008	−6.19	40	0.612	50
S009	−6.19	40	0.306	50
S010	−6.19	40	0.306	10

climate feedback quantities for each simulation are listed in Table 2.

There are ramifications to using a very deep maximum mixed-layer depth, and these are covered in the discussion section below. Our value of 4000 m is far greater than the average mixed-layer depth of 127 m in the modern ocean, but modern seasonal values in the cold polar regions are in fact substantially deeper (Kara *et al.* 2001) and are not out of the question for a very cold Neoproterozoic ocean during extreme glaciation events (D. Martinson, pers. comm. 2006). Additional examination of the effects of mixed-layer depth assignments in GCMs is certainly warranted.

Simulation results

Individual forcings

Palaeogeography. Our first comparison, between Modern and Sturtian (S001R) control runs, is a simple test of the effects of a change in landmass distribution. Kirschvink (1992) and Hoffman *et al.* (1998) had previously suggested that a concentration of landmass in mid- to low-latitude regions would lead to increased global cooling, on the assumption that land's generally higher albedo would reflect more incoming solar radiation back out to space than the equivalent area of (initially) ice-free ocean. However, we found that the Sturtian control run was notably warmer than the Modern control, in both tropical areas (by 1.6 °C) and in terms of global average temperature (by 2.0 °C) (see Table 3), clearly precluding the possibility that concentrating land at low latitudes would, on its own, induce a snowball Earth glaciation (see also Fig. 3a, b). It appears that while the centring

of the Rodinia supercontinent on the equator does increase ground albedo, the concentration of land in one area (rather than the broken land distribution in the Modern) also permits higher net absorption of radiation over a broader area of the ocean, primarily in tropical and mid-latitude regions. At the same time, ocean ice cover was also higher for S001R (Table 3); this is not a contradiction of the warmer global average Sturtian temperature, but rather a reflection of having less land and more ocean area available for sea-ice growth in the polar regions.

Solar luminosity. A comparison between S002 and the control run S001R reveals the significant cooling impact of a 6.19% reduction in incident solar radiation, as per Gough's (1981) equation. Simulation S002 is nearly 4 °C colder than the Sturtian control run, both in the global average and in the tropical average temperatures (Table 3). Total snow and ice cover, snow depth and ocean ice cover are all nearly double that of S001R, although the increase is fairly tightly focused in zonal bands between 60° and 75° latitude in both hemispheres, along the edge of the maximum annual sea-ice extent in S002. The increase in snow and ice cover is coincident with the greatest decreases in temperature, with the equator-to-pole temperature gradient steepening in S002 as anticipated. However, despite the dramatic decrease in global and tropical average temperatures, a significant portion of the planet – from the mid-latitudes to tropical regions – remains above 0 °C on an annual average basis, with zonal average temperatures in excess of 10 °C extending as far as 38° north and south of the equator. Such conditions rule out the possibility of annually persistent snow and ice cover that could develop into ice sheets at low latitudes.

Table 2. *Values for key climate feedback quantities**

Run ID	Altered forcings			Surface air temperature		Snow and ice			Cloud cover		Albedo		Atm. water vapour
	CO$_2$ (ppm)	CH$_4$ (ppm)	OHT (% mod)	Global mean (°C)	Tropical (°C at 4° N)	Total cover (%)	Snow depth (mm)	Ocean ice cover (%)	Low level (%)	High level (%)	Planetary (%)	Ground (%)	(mm H$_2$O)
Modern	315	1.224	100	13.11	24.8	12.24	7.3	5.1	36.95	28.68	30.46	12.03	23.5
S001R	315	1.224	100	15.11	26.4	7.52	4.1	6.4	33.55	28.67	29.78	13.8	23.1
S002	315	1.224	100	11.03	22.6	13.35	7.8	11.4	33.73	28.29	30.68	15.78	17.7
S003	140	1.224	100	9.11	21.5	16.63	10.0	14.2	34.66	27.69	31.59	17.22	16.2
S004	140	1.224	50	7.98	22.9	19.07	12.0	16.2	37.22	27.77	33.06	18.28	16.8
S005	140	0.612	100	8.77	21.4	17.53	10.6	15.0	34.92	27.77	31.82	17.64	16.1
S006	140	0.612	50	7.76	22.8	19.26	12.2	16.4	37.40	27.78	33.16	18.37	16.6
S007	40	1.224	100	6.29	19.8	20.70	13.3	17.4	36.44	27.29	32.93	19.23	14.4
S008	40	0.612	50	4.68	21.1	23.85	18.0	20.0	39.30	26.86	34.69	21.00	14.5
S009	40	0.306	50	4.45	21.0	24.16	20.0	20.1	39.38	26.95	34.75	21.15	14.4
S010	40	0.306	10	3.36	22.1	26.50	28.5	22.0	41.78	27.48	36.20	22.71	14.8

* Values shown are global annual averages unless otherwise specified. All simulations except the Modern control run use Sturtian palaeogeography, and all runs except the Modern run and S001R use a 6.19% reduction in solar luminosity.

Table 3. *Differences in pairs of simulations for changes to a single forcing**

Run ID	Altered forcings	Surface air temperature		Snow and ice			Cloud cover		Albedo		Atm. water vapour
		Global mean (°C)	Tropical (°C at 4° N)	Total cover (%)	Snow depth (mm)	Ocean ice cover (%)	Low level (%)	High level (%)	Planetary (%)	Ground (%)	(mm H₂O)
Geography only											
Modern control–S001R		**2**	**1.6**	*− 4.72*	*− 3.2*	**1.3**	*− 3.4*	*− 0.01*	*− 0.68*	**1.77**	*− 0.4*
Solar radiation only											
S001R–S002		*− 4.08*	*− 3.8*	**5.83**	**3.7**	**5**	**0.18**	*− 0.38*	**0.9**	**1.98**	*− 5.4*
Atmospheric CO₂											
S002–S003	315 > 140 ppm	*− 1.92*	*− 1.1*	**3.28**	**2.2**	**2.8**	**0.93**	*− 0.6*	**0.91**	**1.44**	*− 1.5*
(CH₄ = 1.224 ppm, OHT = 100)											
S002–S007	315 > 40 ppm	*− 4.74*	*− 2.8*	**7.35**	**5.5**	**6.0**	**2.71**	*− 1.0*	**2.25**	**3.45**	*− 3.3*
(CH₄ = 1.224 ppm, OHT = 100)											
S003–S007	140 > 40 ppm	*− 2.82*	*− 1.7*	**4.07**	**3.3**	**3.2**	**1.78**	*− 0.4*	**1.34**	**2.01**	*− 1.8*
(CH₄ = 1.224 ppm, OHT = 100)											
S006–S008	140 > 40 ppm	*− 3.08*	*− 1.7*	**4.59**	**5.8**	**3.6**	**1.9**	*− 0.92*	**1.53**	**2.63**	*− 2.1*
(CH₄ = 0.612 ppm, OHT = 50)											
Atmospheric CH₄											
S003–S005	1.224 > 0.612 ppm	*− 0.34*	*− 0.1*	**0.9**	**0.6**	**0.8**	**0.26**	**0.08**	**0.23**	**0.42**	*− 0.1*
(CO₂ = 140 ppm, OHT = 100)											
S008–S009	0.612 > 0.306 ppm	*− 0.23*	*− 0.1*	**0.31**	**2**	**0.1**	**0.08**	**0.09**	**0.06**	**0.15**	*− 0.1*
(CO₂ = 40 ppm, OHT = 50)											
Ocean heat transports											
S003–S004	100% > 50%	*− 1.13*	**1.4**	**2.44**	**2**	**2**	**2.56**	**0.08**	**1.47**	**1.06**	**0.6**
(CO₂ = 140 ppm, CH₄ = 1.224 ppm)											
S005–S006	100% > 50%	*− 1.01*	**1.4**	**1.73**	**1.6**	**1.4**	**2.48**	**0.01**	**1.34**	**0.73**	**0.5**
(CO₂ = 140 ppm, CH₄ = 0.612 ppm)											
S009–S010	50% > 10%	*− 1.09*	**1.1**	**2.34**	**8.5**	**1.9**	**2.4**	**0.53**	**1.45**	**1.56**	**0.4**
(CO₂ = 40 ppm, CH₄ = 0.306 ppm)											

* For comparisons based on geography and solar radiation, modern forcings for greenhouse gases and ocean heat transports apply. For other comparisons, values of 'unchanged' forcings may deviate from modern but are consistent within the pairs of simulations examined; specific forcings are indicated for each pair listed. Increases in climate feedback quantities are indicated in bold face; decreases are shown in italics.

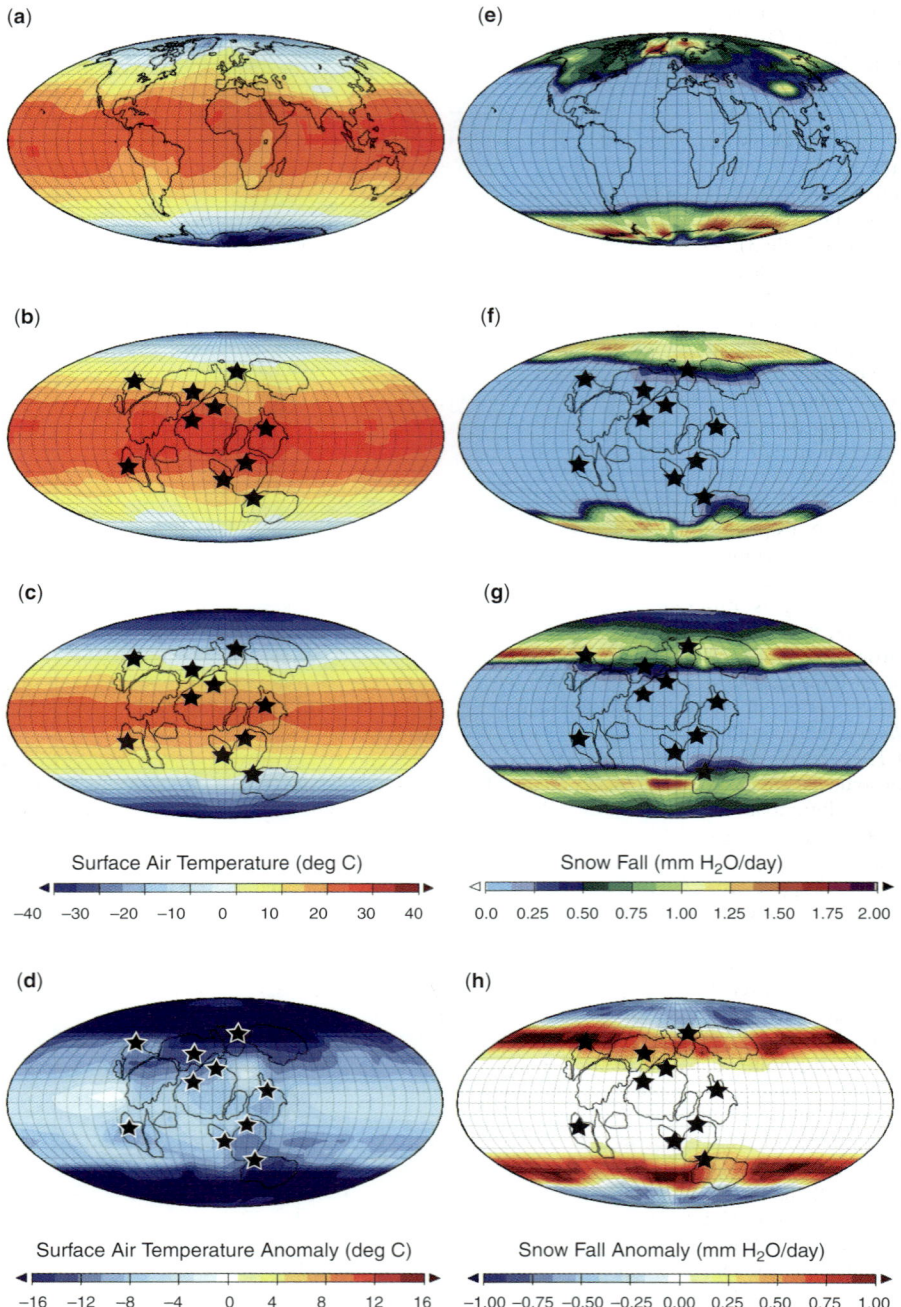

Fig. 3. Comparison of the Modern control run and Sturtian control run S001R with simulation S010, the coldest of the series of experiments discussed in this chapter. The locations of known Sturtian-age glacial deposits are indicated by stars in the plots for S001R and S010, as per Hambrey & Harland (1981) and Evans (2000). Annual average surface air temperatures are shown for the Modern control run, S001R and S010 in (**a**), (**b**), and (**c**), respectively. The significant decrease in surface temperatures in S010 compared to S001R is highlighted in (**d**). Similarly, annual average snowfall is shown for the Modern control run, S001R and S010 in (**e**), (**f**), and (**g**), respectively, with the contrast in snowfall between S010 and S001R illustrated in (**h**).

Carbon dioxide. There are four pairs of simulations in which CO_2 is changed while other forcings were held constant at various values (see Table 1) under a less-luminous Sun: for S002/S003, CO_2 is decreased from 315 to 140 ppmv; for S002/S007, CO_2 is decreased from 315 to 40 ppmv; and for S003/S007 and S006/S008, CO_2 is decreased from 140 to 40 ppmv (the difference between the latter pairs being the CH_4 and OHT levels; see also Table 3). As expected, each successive decrease in CO_2 leads to cooling, both in tropical regions and on the global average; increases in total snow and ice cover and albedo; and decreases in atmospheric water vapour as the atmosphere becomes progressively colder and less capable of retaining moisture. The greatest cooling among these simulation pairs is, not surprisingly, induced by the maximum (275 ppmv) decrease in CO_2 (illustrated through a differencing of S002 and S007), is $-4.74\,°C$ for the global annual average and $-2.8\,°C$ in tropical regions (Table 3). The extent of cooling is comparable to that induced by decreased solar radiation alone, although the pattern of cooling is different: decreasing CO_2 creates a larger drop in global average temperatures than in tropical regions, reflecting the 'blanketing' effect of an atmospheric gas concentration versus the 'spotlight' effect of incident solar radiation.

Methane. There are three pairs of simulations in which CH_4 is changed while other forcings were held constant at various values (see Table 1) under a less-luminous Sun: for S003/S005 and S004/S006, CH_4 is decreased from 1.224 to 0.612 ppmv (half of the modern level); and for S008/S009, CH_4 is dropped from 0.612 to 0.306 ppmv. Given methane's greenhouse warming potential, we had anticipated a noticeable enhancement to cooling effects produced by reductions in the other forcings. However, the reductions in atmospheric methane produced relatively little change across the board for the variables examined (Table 3). Temperatures did not drop more than a few tenths of a degree C; snow and ice cover, cloud cover, and albedo all increased by less than 1%; and atmospheric water vapour was virtually unchanged. The reason for this relative lack of response to methane forcing lies in the atmospheric levels used for these simulations. Although the warming potential of methane increases rapidly with concentration, the changes to the low concentrations we employed simply do not evoke significant impacts.

Altered ocean heat transports. There were three pairs of simulations in which OHT was reduced while other forcings were held constant at various values (see Table 1) under a less-luminous Sun: for S003/S004 and S005/S006, OHT was reduced to 50% of modern, and for S009/S010, OHT was reduced to just 10% of the modern level. As anticipated from previous experiments (Chandler & Sohl 2000), reducing OHT increased the average annual temperature of tropical regions, since less heat was shifted poleward; the average temperatures here never dipped below a warm $20\,°C$ (see Table 3). The concentration of heat in tropical areas did not, however, prevent a reduction in the global average temperature and, in fact, total snow and ice cover as well as snow depth increased with each decrease in OHT (Table 3). Much of the increase in snowfall was over areas newly covered by ocean ice, which acted as a platform for snow accumulation, although some of the increase did occur over portions of land in the Northern Hemisphere.

Combined forcings

As with our previous snowball Earth experiments (Chandler & Sohl 2000), the combined effects of alterations to multiple forcings were more effective in both reducing surface temperatures and increasing snow and ice cover than individual forcings alone. Since the pattern of climatic change is similar for runs S002 through S010, varying chiefly by degree, we focus here on the results of our coldest simulation, S010. This simulation had the greatest reductions in CO_2, CH_4 and OHT, in addition to solar luminosity (see Tables 1 and 3), compared to the Sturtian control run S001.

With a global annual average surface temperature of just $3.36\,°C$, S010 is $11.75\,°C$ colder than S001, and $9.75\,°C$ colder than the Modern control run. Snow and ice coverage in S010 expands to more than a quarter of the planet as well (26.5%), with a notable increase in both planetary and ground albedo as a consequence (6.42% and 8.91% over S001R, respectively). The decrease of atmospheric water vapour by more than a third from S001R is both a reflection of the increasingly cold and dry atmosphere as well as a feedback in and of itself, since water vapour is the most powerful greenhouse gas of all. However, despite the profound overall cooling observed in S010, the reduced OHT sequestered sufficient heat in equatorial regions to make annual average tropical temperatures warmer in S010 than in experiments with greater ocean heat transports (S007, S008 and S009) (see Tables 2 and 3). The tropical temperature in S010 is still colder than the Modern control, but only by $2.7\,°C$, while it was $4.3\,°C$ cooler than the Sturtian control run S001R.

Perhaps the best way to appreciate the extent of cooling in S010 is to compare it visually to both the Modern and Sturtian control runs. Maps of the annual average surface air temperature show that the contrast between the modern and palaeo control runs is small (Fig. 3a and Fig. 3b), which is reasonable given that the only difference between the two is geography. The annual average surface air temperature for S010 (Fig. 3c), however, is clearly colder than its palaeo control, S001R (Fig. 3b), with above-freezing temperatures confined to a narrower band in tropical and lower mid-latitude regions. The anomaly plot (Fig. 3d) illustrates the temperature differential in S010 compared to S001R and underscores the considerable extent of cooling associated with the maximum decreases in solar luminosity, greenhouse gases and ocean heat transport. A similar pattern is evident in maps of the annual average snowfall, expressed here as millimetres of water equivalent per day (1 mm H_2O equivalent is roughly equal to 10 mm freshly fallen snow). The Modern (Fig. 3e) and S001R (Fig. 3f) runs are similar to each other, while S010 shows snowfall rates that are considerably increased (Fig. 3g) in middle latitudes (Fig. 3h). Note that the maximum snowfall (as well as differential increase) is highest along the leading edges of the sea-ice cover of both hemispheres in the two Sturtian simulations (Fig. 4b–c), relating directly to the availability of moisture for precipitation along the sea ice–open ocean boundary, despite the low atmospheric water vapour content for S010 in particular (see Table 2). Note also that our choice of low topography everywhere over the continents has a substantial influence on possible sites of low-latitude snow accumulation. There is no equivalent in our Sturtian simulations to the modern Tibetan Plateau or Western Cordillera, which today can shift annual average snowfall contours equatorward by 10 to 20 degrees of latitude.

Another series of maps that emphasize the extent of global cooling in S010 are shown in Figure 4, which contrasts the Sturtian vs. Modern annual snow and ice coverage in the Northern and Southern Hemispheres. The polar ice caps in S001R (Fig. 4b) are even smaller than those of the modern climate (Fig. 4a), especially in the south polar region, illustrating the regional warming effect of the open polar oceans in the Sturtian as contrasted with the land-covered (Antarctica) and land-locked (Arctic) poles in present-day geography. Both northern and southern polar ice caps in S010 are vast (Fig. 4c), with the 50% snow and ice cover contour extending to roughly 45° latitude and impinging upon the mid-latitude landmasses in both hemispheres. The zonal average temperatures for each of these simulations (Fig. 5) illustrates the temperature control on the snow and ice cover extent quite well; the upper 'shoulders' of each profile coincide with the latitude at which the annual average surface air temperature reaches the freezing mark. For the Modern control run, this point is around 60° latitude; for S001R,

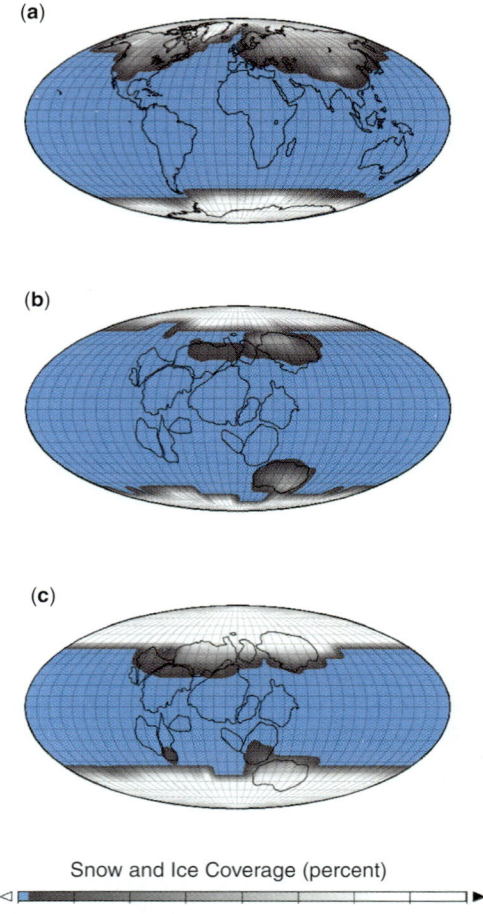

(a)

(b)

(c)

Snow and Ice Coverage (percent)

◁ ▶

0.0   12.5   25.0   37.5   50.0   62.5   75.0   87.5   100.0

Fig. 4. Contrasts in annual average snow and ice coverage between the Modern control run, S001R and S010 (shown in (**a**), (**b**) and (**c**), respectively).

the freezing mark coincides with the Modern control run in the Northern Hemisphere, but is at 70° latitude in the Southern Hemisphere (reflecting the warming produced by the removal of Antarctica from a polar position); and for S010, the freezing mark is at a remarkably low 45° latitude.

Discussion

Having identified S010 as our coldest simulation, we need to assess whether the surface conditions described by S010 are favourable to what we know of the distribution of Sturtian glacial deposits. The surface air temperature and snowfall maps for S010 (Fig. 3c, g), which include the marked positions of known Sturtian-age glacial rocks, show

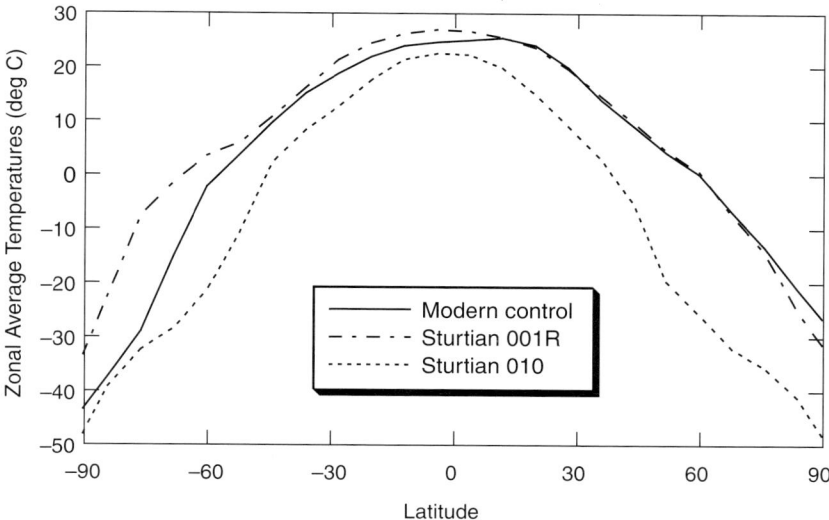

Fig. 5. Zonal average temperature plots for the Modern control run, S001R and S010. The curve for S001R shows considerable warming in the Southern Hemisphere compared to the Modern control, a reflection of the warming created by the removal of a polar landmass (Antarctica). The steep temperature gradients in S010, especially between 40 and 60 degrees latitude in both hemispheres, shows the impact of drastically reduced ocean heat transports on the mid- to high-latitude regions.

significant inconsistencies with the geographic locations of low-latitude glacial deposits. Most of those deposits fall within a tropical zone that is actually quite warm in this simulation, with an annual average temperature in excess of 20 °C (Fig. 3c). Not surprisingly, such warmth is not conducive to precipitation as snowfall, much less the existence of annually persistent snow and ice cover that could lead to ice sheet formation. This enhanced tropical warmth despite an overall much colder planet is the direct result of reducing ocean heat transports to just 10% of the modern level, which sequesters heat in the tropical oceans. The zonal temperature gradient for S010 thus ends up fairly steep compared to the Modern and Sturtian control runs (Fig. 5). This steep gradient acts as a barrier to land and sea-ice encroachment into tropical latitudes, in contrast to other climate studies which use diffusive rather than dynamic means to transport heat out of tropical regions (e.g. Donnadieu *et al.* 2004). While the lack of tropical sea ice is not a problem for 'soft' snowball Earth scenarios, the lack of tropical land ice is, as these simulations do not reproduce surface conditions necessary for the onset and persistence of low-latitude ice-sheet formation.

Our previous simulations of the younger Varanger (Marinoan) glaciation (Chandler & Sohl 2000) did produce snow and ice coverage on land in tropical regions. The Sturtian simulations are considerably warmer than our original Varanger

simulations, by as much as 15 °C and more in tropical regions, despite similar combinations of solar and greenhouse gas forcings. This contrast is probably the result of two key factors. First, our previous Varanger palaeogeographic reconstruction was dominated by a landmass configuration that covered the south polar region and extended northward across all latitudes of the Southern Hemisphere. Relative to the Sturtian interval, there was little land at tropical latitudes. The higher concentration of mid- to high-latitude landmass would allow Varanger ice-sheets to initiate more readily, and the ice-sheets themselves could then contribute to climate change through elevated surface albedo and topographic influence on wind and precipitation. Such local climate effects would allow for some equatorward expansion of snow cover. In the real world, where ice dynamics play an important role, we surmise that stable accumulations of snow in the mid-latitudes would probably lead to flow of glaciers into lower latitudes. In this regard, Sturtian palaeogeography is somewhat 'handicapped' by the difficulty of initiating as well as maintaining ice-sheets at lower latitudes.

Secondly, the maximum mixed-layer depth for our previous Varanger simulations was only set to 127 m, compared to the 4000 m used in the Sturtian experiments. As a result, the Varanger mixed-layer ocean had less heat storage available than the Sturtian ocean and was more susceptible to rapid cooling. Indeed, this rapid cooling sometimes causes instabilities in the model as the ocean freezes through the entire depth of the shallower mixed layer. Our use of a 4000 m maximum for the simulations discussed here allowed us to avoid instabilities, but the greater heat capacity of such a thick mixed layer depth means that the simulations do not achieve equilibrium within 100 years. We can assume that the simulations would cool further, but the final results are likely to be no cooler than our initial Varanger experiments; it would simply have taken longer to reach that point. The role of the maximum mixed-layer depth and the correct assignment of geographic and seasonal values do 'however' need to be explored in much greater detail, and additional study with fully coupled ocean–atmosphere GCMs is warranted.

Still, our results from simulation S010, the coldest of the Sturtian runs completed, bring to light some additional interesting points for future investigations. One point is that although overall atmospheric water vapour levels are considerably reduced, the relatively warm tropical regions act as a heat engine driving enhanced snowfall along the margins of the annually persistent snow and ice cover (Fig. 3g). The contrast between S001R and S010 in these marginal areas is in excess of 4 mm H_2O per day, a difference not unlike the precipitation contrast between the modern Sahara Desert and northeastern United States. It may be that a tropical heat engine is required, at least initially, for there to be sufficient snow and ice build-up in mid-latitude areas, thus overcoming the Sturtian palaeogeographic handicap. Ice-sheet growth on land at these locations, rather than sea ice growth, may be needed to provoke the additional cooling that would ultimately allow continent-scale ice-sheets to spread across the fragments of Rodinia.

Recall also that the reconstruction for these simulations did not include topographic relief. The uniform 50-metre height of our Sturtian continents is low compared to the 236-metre average elevation for our Modern control run. In the Modern control run, the extension of snow and ice cover into mid-latitude regions has also clearly been influenced by the presence of the Tibetan Plateau, the Western Cordillera and the Andes Mountains (Fig. 3e). The possibility that orographic effects influenced climate during the Sturtian glaciation cannot be excluded.

Irrespective of such additional influences, which may indeed have made low-latitude continental icesheets viable, the expansion of sea ice across all latitudes does not seem to be consistent with any understanding of the likely range of climate forcings that existed during the Neoproterozoic Era. Earlier in Earth history, extreme glacial episodes may very well have resulted in 'hard' snowball

Earth scenarios. For example, the Huronian glaciation of the Palaeoproterozoic Era (c. 2.1 Ga) would have occurred under solar luminosities that were as much as 15% below modern – or more than twice the decrease experienced by the Sturtian glaciation. Three-dimensional climate models, like the GISS GCM, have sensitivities that suggest a runaway icehouse can be achieved at modern levels of greenhouse gases if solar luminosities are only 85% to 90% of modern (J. Hansen, pers. comm, 2001). Thus the palaeoclimates of early Earth are generally thought to require higher amounts of atmospheric greenhouse gases to explain why the planet was not continuously in a 'hard snowball' state.

Conclusions

In this effort to replicate climatic conditions amenable to the development of large ice-sheets in low to mid-latitude continental regions during the Neoproterozoic Sturtian glaciation, no single climate forcing is sufficient to cool the planet adequately. Reductions in solar luminosity, carbon dioxide, methane and ocean heat transports all induce some cooling, albeit in different ways, with solar luminosity and carbon dioxide having the greatest impact on temperature. The effects of the solar luminosity reduction and maximum decrease in carbon dioxide to 40 ppmv are similar in magnitude, although CO_2 reduction has a greater global impact while solar luminosity reduction has a greater impact on tropical regions. Decreases in methane lead to only minor enhancement of cooling brought on by other forcing changes; this result perhaps runs counter to what might be expected, given methane's high greenhouse warming potential, but is in fact in line with methane's nonlinear response to changes in concentration. Reductions in ocean heat transports led to global cooling, with dramatic cooling at higher latitudes, but slight warming in some tropical regions, as heat is less efficiently moved away from low latitude areas as ocean heat transport declines.

Combinations of altered forcings did increase the amount of global cooling in the model, with the most extreme combination of forcings in simulation S010 producing a world with sea ice extending to 40° latitude and a global average surface temperature of just 3.36 °C. Despite this considerable cooling with respect to the Modern and Sturtian control runs, most continental areas never became cold enough to support annually persistent snow and ice cover, which is required for the large ice-sheets inferred from the geological record of glacial deposits to develop. Taken together, however, the tropical warming induced by reduced ocean heat transports, coincident increase in precipitation, and the steep zonal average temperature gradient, suggests that the model need not produce much more cooling before larger portions of land would be capable of supporting year-round snow and ice cover. Coupled dynamic ice-sheet modelling (a field that still needs to see considerable advances even for modern ice-sheets) may be important to accurately portraying the land-ice masses of the Neoproterozoic Era.

While this series of experiments was not intended to address specifically any of the tenets of the 'hard' snowball Earth hypothesis (Hoffman *et al.* 1998; Hoffman & Schrag 2002), our results do have some implications for two key assumptions not directly based upon data in the geological record. The first assumption is that concentration of landmass in low to mid-latitude regions would increase the planet's albedo and lead to global cooling. We find that a low- to mid-latitude landmass concentration does increase ground albedo, but planetary albedo itself decreases and the planet is warmer than anticipated. A contributing factor appears to be the absence of a landmass like modern Antarctica in the polar regions, which is isolated from heat exchange with lower latitudes by the oceanic Circumpolar Current.

A second assumption of the 'hard' snowball Earth hypothesis is that global cooling would have lead to sea ice encroachment into ever-lower latitudes, until an ice-albedo feedback effect set in and rapidly produced total or near-total sea ice cover in tropical regions. We find that reducing ocean heat transports to as little as 10% of the modern value does permit sea ice to extend further equatorward, but incoming solar radiation is then sequestered within a smaller area of the tropical ocean. The resulting tropical temperatures are even higher than the Modern control run, and act as an absolute barrier to sea ice development over a sizeable swathe of ocean. Our results in this case highlight the utility of computer models in 'checking' conceptual models of climatic change, even if the computer model itself does not fully reproduce the features preserved in the geological record.

The authors thank Jeff Jonas and Michael Shopsin for programming assistance. This work has been supported by NSF grants ATM-9907640 (to N. Christie-Blick, LS & MC) and ATM-0214400, ATM-0231400 and ATM-03 23516 (to MC).

References

BARTLEY, J. K. & KAH, L. C. 2004. Marine carbon reservoir, C_{org}-C_{carb} coupling, and the evolution of the Proterozoic carbon cycle. *Geology*, **32**, 129–132, doi: 10.1130/G19939.1.

BARTOLA, J. M., RAYNAUD, D., KOROTKEVICH, Y. S. & LORIUS, C. 1987. Vostok ice core provides 160,000-year record of atmospheric CO_2. *Nature*, **329**, 408–414.

BLACK, R. F. 1976. Periglacial features indicative of permafrost: ice and soil wedges. *Quaternary Research*, **6**, 3–26.

BORG, S. G. & DePAOLO, D. J. 1994. Laurentia, Australia, and Antarctica as a Late Proterozoic supercontinent: Constraints from isotopic mapping. *Geology*, **22**, 307–310.

BRADLEY, R. S. 1994. *Quaternary Paleoclimatology* (4th edn.). Chapman & Hall, London.

BRIDEN, J. C. & IRVING, E. 1964. Palaeolatitude spectra of sedimentary palaeoclimatic indicators. *In*: NAIRN, A. E. M. (ed.) *Problems in Palaeoclimatology*. Interscience Publishers, London & New York, 199–224.

BROMHAM, L., RAMBAUT, A., FORTEY, R., COOPER, A. & PENNY, D. 1998. Testing the Cambrian explosion hypothesis by using a molecular dating technique. *Proceedings of the National Academy of Science (USA)*, **95**, 12386–12389.

BUICK, R., DES MARAIS, D. J. & KNOLL, A. H. 1995. Stable isotopic compositions of carbonates from the Mesoproterozoic Bangemall Group, northwestern Australia. *Chemical Geology*, **123**, 153–171.

CALDEIRA, K. & KASTING, J. F. 1992. Susceptibility of the early Earth to irreversible glaciation caused by carbon dioxide clouds. *Nature*, **359**, 226–228.

CANFIELD, D. E. & TESKE, A. 1996. Late Proterozoic rise in atmospheric oxygen concentration inferred from phylogenetic and sulphur-isotope studies. *Nature*, **382**, 127–132.

CHANDLER, M. A. & SOHL, L. E. 2000. Climate forcings and the initiation of low-latitude ice sheets during the Neoproterozoic Varanger glacial interval. *Journal of Geophysical Research*, **105**, 20737–20756.

CHAPPELLAZ, J., BARNOLA, J. M., RAYNAUD, D., KOROTKEVICH, Y. S. & LORIUS, C. 1990. Ice-core record of atmospheric methane over the past 160,000 years. *Nature*, **345**, 127–131.

CHUMAKOV, N. M. 2004. Trends in global climate changes inferred from geological data. *Stratigraphy and Geological Correlation*, **12**, 117–138.

CLAUSSEN, M., MYSAK, L. A., WEAVER, A. J. ET AL. 2002. Earth system models of intermediate complexity: closing the gap in the spectrum of climate system models. *Climate Dynamics*, **18**, 579–586.

CONDON, D. J., PRAVE, A. R. & BENN, D. I. 2002. Neoproterozoic glacial-rainout intervals: Observations and implications. *Geology*, **30**, 35–38.

CONDON, D., ZHU, M. Y., BOWRING, S., WANG, W., YANG, A. H. & JIN, Y. G. 2005. U-Pb ages from the Neoproterozoic Doushantuo Formation, China. *Science*, **308**, 95–98.

COWIE, J. W., ZIEGLER, W. & REMANE, J. 1989. Stratigraphic Commission accelerates progress, 1984 to 1989. *Episodes*, **12**, 79–83.

DALRYMPLE, G. B. 1991. *The Age of the Earth*. Stanford University Press, Stanford.

DALZIEL, I. W. D. 1995. Earth before Pangea. *Scientific American*, **272**, 58–63.

DALZIEL, I. W. D. 1997. Neoproterozoic–Paleozoic geography and tectonics: Review, hypothesis, environmental speculation. *Geological Society of America Bulletin*, **109**, 16–42.

DEYNOUX, M. 1982. Periglacial polygonal structures and sand wedges in the late Precambrian glacial formations of the Taoudeni Basin in Adrar of Mauritania (West Africa). *Palaeogeography, Palaeoclimatology, Palaeoecology*, **39**, 55–70.

DEYNOUX, M., KOCUREK, G. & PROUST, J. N. 1989. Late Proterozoic periglacial eolian deposits on the west African platform, Taoudeni Basin, western Mali. *Sedimentology*, **36**, 531–549.

DIBONA, P. A. 1991. A previously unrecognized Late Proterozoic succession: Upper Wilpena Group, northern Flinders Ranges, South Australia. *Quarterly Geological Notes of the Geological Society of South Australia*, **117**, 2–9.

DONNADIEU, Y., RAMSTEIN, G., FLUTEAU, F., ROCHE, D. & GANOPOLSKI, A. 2004. The impact of atmospheric and oceanic heat transports on the sea-ice-albedo instability during the Neoproterozoic. *Climate Dynamics*, **22**, 293–306.

EVANS, D. A. D. 2000. Stratigraphic, geochronological, and paleomagnetic constraints upon the Neoproterozoic climatic paradox. *American Journal of Science*, **300**, 347–433.

GOUGH, D. O. 1981. Solar interior structure and luminosity variations. *Solar Physics*, **74**, 21–34.

GROTZINGER, J. P., BOWRING, S. A., SAYLOR, B. Z. & KAUFMAN, A. J. 1995. Biostratigraphic and geochronologic constraints on early animal evolution. *Science*, **270**, 598–604.

HAMBREY, M. J. & HARLAND, W. B. (eds) 1981. *Earth's Pre-Pleistocene Glacial Record*. Cambridge University Press, New York.

HAMBREY, M. J. & HARLAND, W. B. 1985. The late Proterozoic glacial era. *Palaeogeography, Palaeoclimatology, Palaeoecology*, **51**, 255–272.

HANSEN, J. E., RUSSELL, G. & RIND, D. 1983. ET AL. Efficient three-dimensional global models for climate studies: Models I and II. *Monthly Weather Review*, **111**, 609–662, doi: 10.1175/1520-0493(1983)111 < 0609:ETDGMF > 2.0.CO;2.

HANSEN, J., FUNG, I., LACIS, A., RIND, D., LEBEDEFF, R., RUEDY, R., RUSSELL, G. & STONE, P. 1988. Global climate changes as forecast by Goddard Institute for Space Studies three-dimensional model. *Journal of Geophysical Research*, **93**, 9341–9364.

HECKMAN, D. S., GEISER, D. M., EIDELL, B. R., STAUFFER, R. L., KARDOS, N. L. & HEDGES, S. B. 2001. Molecular evidence for the early colonization of land by fungi and plants. *Science*, **293**, 1129–1133.

HENDERSON, G. M. 2002. New oceanic proxies for paleoclimate. *Earth and Planetary Science Letters*, **203**, 1–13.

HILL, A. C., AROURI, K., GORJAN, P. & WALTER, M. R. 2000. Geochemistry of marine and non-marine environments of a Neoproterozoic cratonic carbonate/evaporite: the Bitter Springs Formation, central Australia. *In*: GROTZINGER, J. P. & JAMES, N. P. (eds) *Carbonate Sedimentation and Diagenesis in the Evolving Precambrian World*. Society for Sedimentary Geology, Special Publication, **67**, 327–344.

HOFFMAN, P. F. 1991. Did the breakout of Laurentia turn Gondwanaland inside-out? *Science*, **252**, 1409–1412.

HOFFMAN, P. F., KAUFMAN, A. J., HALVERSON, G. P. & SCHRAG, D. P. 1998. Snowball Earth. *Science*, **281**, 1342–1346.

HOFFMAN, P. F. & SCHRAG, D. P. 2002. The snowball Earth hypothesis: Testing the limits of global change. *Terra Nova*, **14**, 129–155.

HOFMANN, H. J., NARBONNE, G. M. & AITKEN, J. D. 1990. Ediacaran remains from intertillite beds in northwestern Canada. *Geology*, **18**, 1199–1202.

HOFMANN, H. J. & MOUNTJOY, E. W. 2001. *Namacalathus–Cloudina* assemblage in Neoproterozoic Miette Group (Byng Formation), British Columbia: Canada's oldest shelly fossils. *Geology*, **29**, 1091–1094.

JAMES, N. P., NARBONNE, G. M., DALRYMPLE, R. W. & KYSER, T. K. 2005. Glendonites in low-latitude, interglacial, sedimentary rocks, northwest Canada: Insights into the Cryogenian ocean and Precambrian cold-water carbonates. *Geology*, **33**, 9–12.

JEFFERSON, C. W. & PARRISH, R. R. 1989. Late Proterozoic stratigraphy, U-Pb ages, and rift tectonics, Mackenzie Mountains, northwestern Canada. *Canadian Journal of Earth Sciences*, **26**, 1784–1801.

JIANG, G. Q. 2002. *Neoproterozoic sequence and chemostratigraphy*. PhD thesis, Columbia University.

JIANG, G. Q., CHRISTIE-BLICK, N., KAUFMAN, A. J., BANERJEE, D. M. & RAI, V. 2002. Sequence stratigraphy of the Neoproterozoic Infra Krol Formation and Krol Group, Lesser Himalaya, India. *Journal of Sedimentary Research*, **72**, 524–542.

JIANG, G. Q., SOHL, L. E. & CHRISTIE-BLICK, N. 2003. Neoproterozoic stratigraphic comparison of the Lesser Himalaya (India) and Yangtze block (south China): Paleogeographic implications. *Geology*, **31**, 917–920.

KARA, A. B., ROCHFORD, P. A. & HURLBURT, H. E. 2001. *Naval Research Laboratory Mixed Layer Depth (NMLD) Climatologies*. Naval Research Laboratory, Washington, D.C., NRL Report: NRL/FR/7330-01-9995.

KASTING, J. F. 1987. Theoretical constraints on oxygen and carbon dioxide concentrations in the Precambrian atmosphere. *Precambrian Research*, **34**, 205–229.

KAUFMAN, A. J., KNOLL, A. H. & NARBONNE, G. M. 1997. Isotopes, ice ages, and terminal Proterozoic earth history. *Proceedings of the National Academy of Sciences* (USA), **95**, 6600–6605.

KAUFMAN, A. J., JIANG, G., CHRISTIE-BLICK, N., BANERJEE, D. M. & RAI, V. 2006. Stable isotope record of the terminal Neoproterozoic Krol platform in the Lesser Himalayas of northern India. *Precambrian Research*, **147**, 156–185.

KELLERHALS, P. & MATTER, A. 2003. Facies analysis of a glaciomarine sequence, the Neoproterozoic Mirbat Sandstone Formation, Sultanate of Oman. *Eclogae Geologicae Helvetiae*, **96**, 49–70.

KENDALL, B., CREASER, R. A. & SELBY, D. 2006. Re-Os geochronology of post-glacial black shales in Australia: Constraints on the timing of 'Sturtian' glaciation. *Geology*, **34**, 729–732.

KENNEDY, M. J., DROSER, M., MAYER, L. M., PEVEAR, D. & MROFKA, D. 2006. Late Precambrian oxygenation; inception of the clay mineral factory. *Science*, **311**, 1446–1449, doi: 10.1126/science.1118929.

KILLINGLEY, J. S. 1983. Effects of diagenetic recrystallization on $^{18}O/^{16}O$ values of deep sea sediments. *Nature*, **301**, 594–596.

KIRSCHVINK, J. L. 1992. Late Proterozoic low-latitude global glaciation: The snowball earth. *In*: SCHOPF, J. W. & KLEIN, C. (eds) *The Proterozoic Biosphere, A Multidisciplinary Study*. Cambridge University Press, New York, 51–52.

KNOLL, A. H., WALTER, M. R., NARBONNE, G. M. & CHRISTIE-BLICK, N. 2004. A new period for the geologic time scale. *Science*, **305**, 621–622, doi: 10.1126/science.1098803.

LEHMANN, M. F., BERNASCONI, S. M., BARBIERI, A. & McKENZIE, J. A. 2002. Preservation of organic matter and alteration of its carbon and nitrogen isotope composition during simulated and *in situ* early sedimentary diagenesis. *Geochimica et Cosmochimica Acta*, **66**, 3573–3584.

LOWE, D. R. 1992. Major events in the geological development of the Precambrian Earth. *In*: SCHOPF, J. W. & KLEIN, C. (eds) *The Proterozoic Biosphere: A Multidisciplinary Study*. Cambridge University Press, Cambridge, UK, 67–75.

MATTHEWS, E. 1983. Global vegetation and land-use: New high-resolution data-bases for climate studies. *Journal of Climate and Applied Meteorology*, **22**, 474–487.

McMECHAN, M. E. 2000. Vreeland Diamictites – Neoproterozoic glaciogenic slope deposits, Rocky Mountains, northeast British Columbia. *Bulletin of Canadian Petroleum Geology*, **48**, 246–261.

McWILLIAMS, M. O. & McELHINNY, M. W. 1980. Late Precambrian paleomagnetism of Australia: The Adelaide Geosyncline. *Journal of Geology*, **88**, 1–26.

MEERT, J. G. & VAN DER VOO, R. 1997. The assembly of Gondwana 800–550 Ma. *Journal of Geodynamics*, **23**, 223–235.

MEERT, J. G., VAN DER VOO, R. & AYUB, S. 1995. Paleomagnetic investigation of the Neoproterozoic Gagwe lavas and Mbozi complex, Tanzania and the assembly of Gondwana. *Precambrian Research*, **74**, 225–244.

MEERT, J. G., VAN DER VOO, R. & PAYNE, T. W. 1994. Paleomagnetism of the Catoctin Volcanic Province: A new Vendian-Cambrian apparent polar wander path for North America. *Journal of Geophysical Research*, **99**, 4625–4641.

NARBONNE, G. M. 2005. The Ediacara biota: Neoproterozoic origin of animals and their ecosystems. *Annual Review of Earth and Planetary Sciences*, **33**, 421–442.

NARBONNE, G. M. & AITKEN, J. D. 1995. Neoproterozoic of the Mackenzie Mountains, northwestern Canada. *Precambrian Research*, **73**, 101–121.

NYSTUEN, J. P. 1976. Late Precambrian Moelv Tillite deposited on a discontinuity surface associated with a fossil ice wedge, Rendalen, southern Norway. *Norsk Geologisk Tidsskrift*, **56**, 29–50.

PARK, J. K. 1997. Paleomagnetic evidence for low-latitude glaciation during deposition of the Neoproterozoic Rapitan Group, Mackenzie Mountains, N. W. T., Canada. *Canadian Journal of Earth Sciences*, **34**, 34–49.

PARRISH, J. T. 1998. *Interpreting Pre-Quaternary Climate from the Geologic Record: Perspectives in Paleobiology and Earth History*. Columbia University Press, New York. World Wide Web Address: http://www.earthscape.org/r3/parrish/index.html

PELECHATY, S. M., KAUFMAN, A. J. & GROTZINGER, J. P. 1996. Evaluation of $\delta^{13}C$ chemostratigraphy for intrabasinal correlation: Vendian strata of Northeast Siberia. *Geological Society of America Bulletin*, **108**, 992–1003.

POULSEN, C. J., PIERREHUMBERT, R. T. & JACOB, R. L. 2001. Impact of ocean dynamics on the simulation of the Neoproterozoic 'snowball Earth'. *Geophysical Research Letters*, **28**, 1575–1578.

POWELL, C. MCA., LI, Z. X., McELHINNY, M. W., MEERT, J. G. & PARK, J. K. 1993. Paleomagnetic constraints on timing of the Neoproterozoic breakup of Rodinia and the Cambrian formation of Gondwana. *Geology*, **21**, 889–892.

PREISS, W. V. (compiler) 1987. *The Adelaide Geosyncline: Late Proterozoic Stratigraphy, Sedimentation, Palaeontology and Tectonics*. Geological Survey of South Australia Bulletin **53**.

RIDGWELL, A. J., KENNEDY, M. J. & CALDEIRA, K., 2003. Carbonate deposition, climate stability, and Neoproterozoic ice ages. *Science*, **302**, 859–862, doi: 10.1126/science.1088342.

RUSSELL, G. L., MILLER, J. R. & TSANG, L.-C. 1985. Seasonal oceanic heat transports computed from an atmospheric model. *Dynamics of Atmospheres and Oceans*, **9**, 253–271.

SCHMIDT, P. W. & WILLIAMS, G. E. 1995. The Neoproterozoic climatic paradox: Equatorial palaeolatitude for Marinoan glaciation near sea level in South Australia. *Earth and Planetary Science Letters*, **134**, 107–124.

SCHOUTEN, S., HOPMANS, E. C., FORSTER, A., VAN BREUGEL, Y., KUYPERS, M. M. M. & SINNINGHE DAMSTÉ, J. S. 2003. Extremely high sea surface temperatures at low latitudes during the middle Cretaceous as revealed by archaeal membrane lipids. *Geology*, **31**, 1069–1072.

SCHRAG, D. P., DEPAOLO, D. J. & RICHTER, F. M. 1995. Reconstructing past sea surface temperatures: Correcting for diagenesis of bulk marine carbonate. *Geochimica et Cosmochimica Acta*, **59**, 2265–2278.

SHANKAR, R., KUMAR, G., MATHUR, V. K. & JOSHI, A. 1993. Stratigraphy of Blaini, Infra Krol, Krol and Tal succession, Krol belt, Lesser Himalaya. *Indian Journal of Petroleum Geology*, **2**, 99–136.

SOHL, L. E. 2000. *Paleoclimatology of the Neoproterozoic interglacial to Marinoan glacial succession (~650–575 Ma), central Flinders Ranges, South Australia*. PhD thesis, Columbia University.

SOHL, L. E., CHRISTIE-BLICK, N. & KENT, D. V. 1999. Paleomagnetic polarity reversals in Marinoan (ca. 600 Ma) glacial deposits of Australia: Implications for the duration of low-latitude glaciation in Neoproterozoic time. *Geological Society of America Bulletin*, **111**, 1120–1139.

TAJIKA, E. 2003. Faint young sun and the carbon cycle: implications for the Proterozoic global glaciations. *Earth and Planetary Science Letters*, **213**, 443–453.

THOMPSON, M. D. & BOWRING, S. A. 2000. Age of the Squantum 'Tillite,' Boston Basin, Massachusetts: U-Pb zircon constraints on terminal Neoproterozoic glaciation. *American Journal of Science*, **300**, 630–655.

TORSVIK, T. H., LOHMANN, K. C. & STURT, B. A. 1995. Vendian glaciations and their relation to the dispersal of Rodinia: Paleomagnetic constraints. *Geology*, **23**, 727–730.

TORSVIK, T. H., SMETHURST, M. A., MEERT, J. G., VAN DER VOO, R., MCKERROW, W. S., BRASIER, M. D., STURT, B. A. & WALDERHAUG, H. J. 1996. Continental break-up and collision in the Neoproterozoic and Palaeozoic – A tale of Baltica and Laurentia. *Earth-Science Reviews*, **40**, 229–258.

TUCKER, M. E. 1990. Diagenetic processes, products and environments. *In*: TUCKER, M. E. & WRIGHT, V. P. (eds) *Carbonate Sedimentology*. Blackwell Scientific Publications, Oxford, 314–364.

UNRUG, R. 1992. The supercontinent cycle and Gondwanaland assembly: Component cratons and the timing of suturing events. *Journal of Geodynamics*, **16**, 215–240.

WEIL, A. B., VAN DER VOO, R., MAC NIOCAILL, C. & MEERT, J. G. 1998. The Proterozoic supercontinent Rodinia: paleomagnetically derived reconstructions for 1100 to 800 Ma. *Earth and Planetary Science Letters*, **154**, 13–24.

WELLMAN, C. H., OSTERLOFF, P. L. & MOHIUDDIN, U. 2003. Fragments of the earliest land plants. *Nature*, **425**, 282–285.

WILLIAMS, G. E. & TONKIN, D. G. 1985. Periglacial structures and palaeoclimatic significance of a late Precambrian block field in the Cattle Grid copper mine, Mount Gunson, South Australia. *Australian Journal of Earth Sciences*, **32**, 287–300.

WINGATE, M. T. D., PISAREVSKY, S. A. & EVANS, D. A. D. 2002. Rodinia connections between Australia and Laurentia: no SWEAT, no AUSWUS? *Terra Nova*, **14**, 121–128.

ZHANG, S. H., JIANG, G. Q., ZHANG, J. M., SONG, B., KENNEDY, M. J. & CHRISTIE-BLICK, N. 2005. U-Pb sensitive high-resolution ion microprobe ages from the Doushantuo Formation in south China: Constraints on late Neoproterozoic glaciations. *Geology*, **33**, 473–476.

ZHANG, Q. R. 1994. Environmental evolution during the early phase of Late Proterozoic glaciation, Hunan, China. *In*: DEYNOUX, M., MILLER, J. M. G., DOMACK, E. W., EYLES, N., FAIRCHILD, I. J. & YOUNG, G. M. (eds) *Earth's Glacial Record*. Cambridge University Press, New York, 260–266.

ZIEGLER, A. M., ROWLEY, D. B., LOTTES, A. L., SAHAGIAN, D. L., HULVER, M. L. & GIERLOWSKI, T. C. 1985. Paleogeographic interpretation: with an example from the Mid-Cretaceous. *Annual Review of Earth and Planetary Sciences*, **13**, 385–425.

Early Cambrian origin of complex marine ecosystems

J. VANNIER

*UMR 5125 PEPS, CNRS, Université Lyon 1, Campus de la Doua, Bâtiment Géode, 69622
Villeurbanne Cedex, France (e-mail: jean.vannier@univ-lyon1.fr)*

Abstract: Marine ecosystems with complex trophic structure began to develop in the early Cambrian. Evidence from exceptional biotas (e.g. Chengjiang in S. China) and other fossil sites indicates a high level of biological interactivity (e.g. prey–predator relationships) and the colonization of a wide range of pelagic and benthic niches by predators, scavengers, detritus feeders and filter-feeders. A chain of biotic innovations and events seems to have played a crucial role in the construction of modern trophic webs: (1) the achievement of complex nervous systems, visual organs and motor functions; (2) the development of predation (e.g. by arthropods and infaunal worms); (3) the colonization of new niches (e.g. water column); and (4) the introduction of new tiers such as the mesozooplankton (e.g. chaetognaths, arthropods). All of them changed the dynamics of the ecosystem (energy flow, transfer of biomass, food source), introduced new selection pressures, triggered a cascade of adaptive behavioural and morphofunctional responses in the trophic layers, and catalysed biological diversification. The role of non-biological factors (e.g. oxygen, water chemistry, carbon cycle, climates) may have been important in the early stages of metazoan evolution and ecospace colonization, but these were probably negligible in the ecosystem build-up process itself. The early Cambrian global re-organization of the marine ecosystem resulted in more complex trophic links and stronger interactions between plankton and benthos. This unprecedented increase of interdependence between organisms, whose dynamics through the Precambrian–Cambrian transition still needs detailed explanation, made the marine ecosystems as a whole, and for the first time in their history, highly vulnerable to environmental perturbations such as climatic variations, changes in ocean chemistry and ultimately human action.

The Precambrian–Cambrian transition represents a key period in the evolution of Life on Earth, during which most animal phyla appear in the fossil record. The wealth of fossil evidence obtained over recent years from exceptional biotas such as that of Chengjiang in China (Hou *et al.* 2004) clearly demonstrates the existence of diverse marine communities in the early Cambrian (*c.* 540–520 Ma). At least 50% of the modern animal phyla, among them the major components of the present-day marine diversity (arthropods, chordates), are already represented in these communities. The faunal elements which dominate the early Cambrian fauna in terms of species number and numerical abundance are, by far, the arthropods, followed by sponges and worms. The so-called 'Cambrian explosion' of metazoans is accompanied by a burst of functional innovations in all vital aspects of animal life such as locomotion, feeding and vision (Parker 2003). These novelties undoubtedly generated an unprecedented behavioural complexity which in turn allowed animals to colonize new horizons of the marine ecospace (Knoll & Caroll 1999). The Precambrian–Cambrian transition indeed witnessed unprecedented and profound ecological changes. The spectrum of ecological niches occupied by animals expanded considerably with a variety of crawlers, burrowers and swimmers

invading both the sea-floor and the water column. A new type of biological system that transferred mass and energy via a complex food chain and through a cascade of trophic levels started to build up, laying the foundations of modern-style ecosystems. Over a relatively short period of time, possibly a few tens of million years, the marine realm metamorphosed, passing from a relatively simple organization dominated by microorganisms and sessile organisms (e.g. 'Ediacara' biota, sponges) to highly complex ecosystems where a myriad of new ecological interactions such as predation occupied a central position and exerted feedback effects on the environment and evolutionary processes.

This paper focuses here on the early stages of the construction of modern ecosystems and the early colonization of new marine ecospace. It does not present a comprehensive compilation of fossil data from all Cambrian biota. The goal is to provide the reader with an updated (and as vivid as possible) picture of the composition and functioning of the early Cambrian ecosystem, based on one particular example, that of the Chengjiang biota in South China. Two reasons explain this choice of fossil assemblage. The Chengjiang biota is by far the best-documented of all early Cambrian exceptional fossil sites known to date. The completeness of its fossil record is exceptionally high,

From: WILLIAMS, M., HAYWOOD, A. M., GREGORY, F. J. & SCHMIDT, D. N. (eds) *Deep-Time Perspectives on Climate Change: Marrying the Signal from Computer Models and Biological Proxies.* The Micropalaeontological Society,
Special Publications. The Geological Society, London, 81–100.
1747-602X/07/$15.00 © The Micropalaeontological Society 2007.

covering a vast array of skeletonized and soft-bodied organisms. The only biota that rivals that of Chengjiang in terms of diversity and complete-ness is that of the middle Cambrian Burgess Shale (Briggs *et al.* 1994; Conway Morris 1986; and more recently Caron 2005 and Caron & Jackson 2006). The 'ecological inventory' presented here should be viewed as the initial and necessary step to further studies on the evolution of marine ecology through time. Indeed, important unan-swered issues need to be tackled, such as the dynamics of trophic structures across the Precam-brian–Cambrian transition. However, the fossil record of metazoans before the Cambrian is patchy (e.g. Ediacaran organisms, embryos, trace fossils) and links are still extremely difficult to establish between late Precambrian marine life and the fully-fledged animal communities of the early Cambrian. The 'post-early Cambrian' evol-ution of trophic structures is also a puzzling ques-tion. Does the complexity of present-day ecosystems find its origin in a unique event that occurred over a relatively short period of time half a billion years ago or, on the contrary, result from successive changes throughout the Phanero-zoic? In this paper, the crucial issue of biological and non-biological triggers is discussed, especially whether the 'Cambrian explosion' and its associ-ated ecological turnover may have been catalysed or not by environmental or intrinsic factors.

Biodiversity of marine life in the early Cambrian

Our perception of the global diversity of marine life across the Precambrian–Cambrian transition is extremely patchy and, to some extent, biased by preservational factors. However, several major Lagerstätten such as Chengjiang in China (Hou *et al.* 2004), Sirius Passet in Greenland (Conway Morris & Peel 1990, 1995; Budd 1993, 1998), Emu Bay in Australia (Nedin 1999), Comley in England (Siveter *et al.* 2001), and Sinsk in Siberia (Ivantsov *et al.* 2005; Ponamarenko 2005), that all preserve an array of soft-bodied and sclerotized organisms, provide precious windows into the animal richness of the early Cambrian commu-nities. The Chengjiang biota discovered in 1984 and currently studied by several teams of Chinese scientists and their foreign collaborators, provides a relatively accurate picture of early Cambrian (*c.* 525 Ma) marine life in a shallow-water environ-ment and under tropical latitudes. At least 40% of the present-day phyla are represented in the early Cambrian Chengjiang fauna, among them the top components of the Recent marine diversity such as arthropods and chordates (Fig. 1). Most of the

'missing' groups are microscopic soft-bodied organisms such as nematodes and platyhelminthes that have an extremely low potential to be fossi-lized. Arthropods dominate the Chengjiang fauna (Fig. 1) with *c.* 40% of species, followed by sponges (12.6%) and priapulid worms (8.2%). Various extinct phyla and groups with uncertain affinities such as lobopods, hyolithids, vetulicolans and anomalocaridids account for *c.* 20% of species. However, this 525 Ma-old marine fauna, by its composition, remains surprisingly familiar with an overwhelming dominance of arthropods.

To what degree taphonomic conditions such as transport and decay altered the original composition of the biota is an important issue that should not be neglected. Detailed qualitative and quantitative studies of a succession of fossil assemblages from the middle Cambrian Burgess Shale (Caron & Jackson 2006) have recently brought some clues and indicate that all organisms present at the time of burial were preserved independently of their original tissue composition. Similar methods applied to the Chengjiang Lagerstätte (Zhao *et al.* 2006) deposited under similar conditions (most probably successive storm-generated deposits) lead to similar conclusions that fossil assemblages from event layers do represent original commu-nities. Perhaps the only missing elements of these communities are pelagic ones (e.g. small planktonic organisms if present) that may have decayed before they could reach the sea-floor.

Colonization of early Cambrian marine ecospace

Pelagic invaders and pioneer zooplankton

Present-day pelagic ecosystems cover more than 70% of the surface of the Earth and are strongly influ-enced by atmosphere–ocean (coupling) interactions related to hydrodynamic processes. They are inhab-ited by an extraordinary variety of organisms from picoplankton (below 0.2–2 micrometres) to large fishes and mammals (e.g. whales). All of them interact via a complex food chain and a cascade of trophic levels with autotrophic primary producers (phytoplankton such as dinoflagellates and cyano-bacteria) as the first level, heterotrophic consumers such as herbivores (e.g. crustacean copepods) as the second, and carnivores (e.g. chaetognaths) as a third. Zooplankton play a central role in the pelagic food web both in terms of biomass and energy fluxes. By exploiting and recycling microscopic phytoplankton, the zooplankton produces massive quantities of nutrient-rich particles that constitute an exploitable resource for benthic communities, thus linking benthic and pelagic realms. This

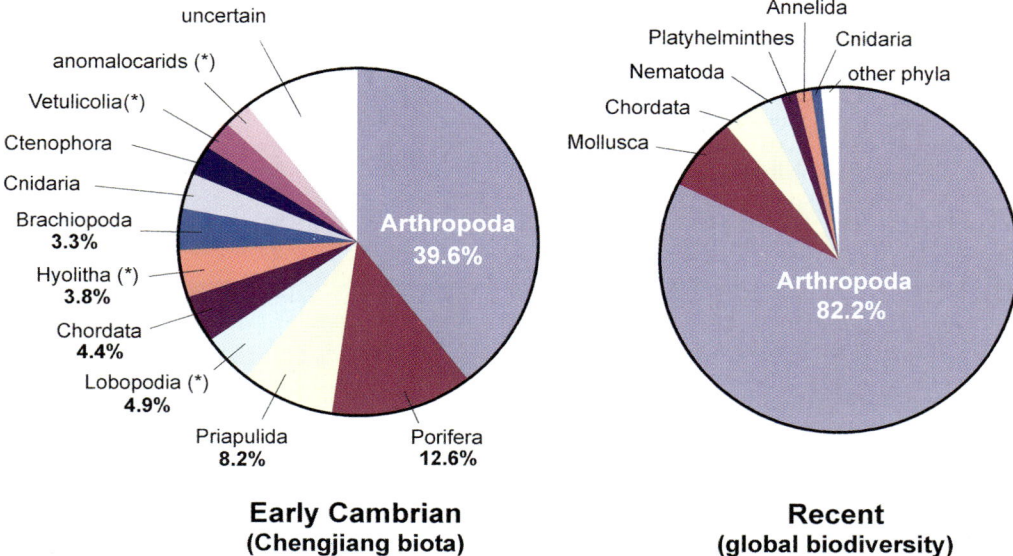

Fig. 1. Diversity of the early Cambrian Chengjiang fauna ($N = 182$ species) compared with Recent global animal biodiversity (based on Brusca & Brusca 2003). These diagrams are meant to show the general characteristics of the Chengjiang fauna and the Recent fauna (e.g. dominance of arthropods). Major differences in sample scale do not allow reliable comparisons.

complex process also contributes to the drawdown of atmospheric carbon into the bodies of marine organisms and its burial in ocean sediments. It is particularly efficient in greenhouse worlds where oceans are strongly stratified and ocean bottoms are anoxic. Thus, the establishment of new modes of mass and energy transfers within the plankton may have been an important event in the development of Earth's carbon cycle.

The origin and construction of this highly complex pelagic ecosystem, crucial to the functioning of the biosphere, is still an enigma. The animal invasion of the water column probably started early in the evolutionary history of several phyla (Rigby & Milsom 1996; Vannier *et al.* 2003). Potential pelagic animals such as ctenophores, eldoniids that superficially resemble jellyfish, and bivalved arthropods such as *Isoxys*, *Tuzoia* and *Zhenghecaris* do occur in the early Cambrian Chengjiang biota (Fig. 2). Their assumed pelagic lifestyle is based on combined evidence from their functional morphology, distribution patterns, depositional environments and also on analogies with Recent forms. Considering their overall size that typically exceeds 20 mm, most of these swimmers fall within the category of nekton – i.e. organisms with abilities to control their movement. Nektonic animals were probably diverse and may have been particularly abundant in the lowermost levels of the water column, among them the enigmatic

vetulicolids, some streamlined fish-like chordates and the celebrated anomalocaridids which were undoubtely the largest predators (exceeding 1 m long; Chen *et al.* 1994) of early Cambrian time. However, considerable uncertainties remain concerning the exact ecological niches, bathymetric range and migration patterns of these assumed 'pelagic' colonizers.

Whether true zooplankton – by definition organisms smaller than 20 mm and unable to overcome the dispersive effect of water current – were present or not in the early Cambrian oceans is another challenging question for palaeobiologists that has an important bearing on the structure of the early trophic web. Again, fossil evidence is sparse. Small bivalved arthropods resembling Recent pelagic ostracods occur in the early Cambrian black shales of South China that were deposited along the distal shelf margins and slope environments of the Yangtze Platform (Steiner *et al.* 1993, 2001; Zhu *et al.* 2004; Vannier *et al.* 2007). These dysoxic/anoxic deposits typically lack benthic faunas and have great potential to preserve organisms that lived in the water column (e.g. graptolites in the Ordovician). Fragments of well-preserved appendages found in the lower Cambrian of Canada (Mount Cap) that show some resemblances with the filter-feeding apparatus of Recent pelagic crustaceans are also serious indicators of the presence of arthropod zooplankton (Butterfield 1994, 2001*a*, *b*). Other zooplanktonic candidates are

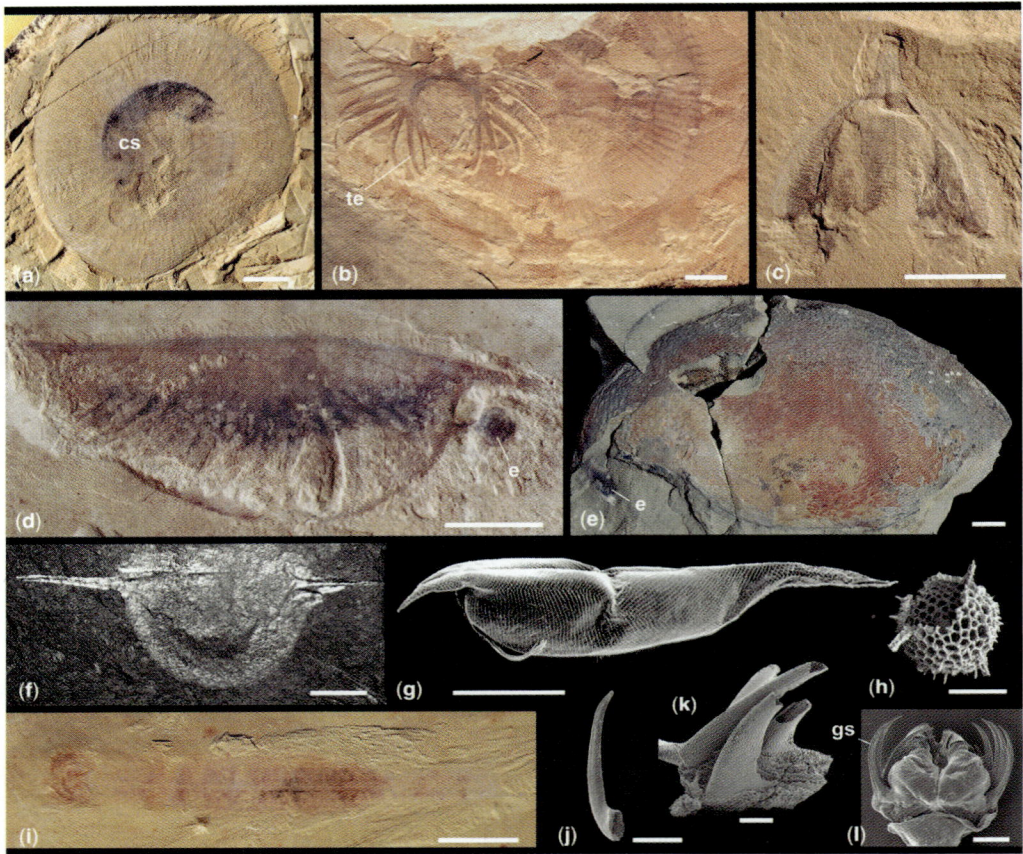

Fig. 2. Early Cambrian pelagic invaders and pioneer zooplankton. (**a**), (**b**) The jellyfish-like eldoniid *Stellostomites eumorphus* Sun & Hou 1987, showing disc, coiled sac and tentacles (courtesy M.-Y. Zhu; see Zhu *et al.* 2002). (**c**) The ctenophore *Maotianoascus octonarius* (see Hu 2005). (**d**), (**e**) The nektonic arthropods *Isoxys curvirostratus* Vannier & Chen 2001, and *Zhenghecaris shankouensis* Vannier *et al.* 2006. (**f**), (**g**) The bivalved arthropod *Sunella bispinata* Cui & Hou in Huo *et al.* 1991 and its possible Recent analogue *Conchoecia daphnoides* (Claus 1890) (Crustacea, Ostracoda) from Japan (see Vannier & Chen 2001). (**h**) Radiolarian (courtesy A. Braun; see Braun *et al.* 2004). (**i**), (**l**) The Cambrian chaetognath *Protosagitta spinosa* Hu in Chen *et al.* 2002, complete specimen, and details of the head structure of the Recent *Sagitta nagae* from Japan; note grasping spines (Vannier *et al.* 2006). (**j**), (**k**) The protoconodonts *Protoherzina unguliformis* Missarzhevsky 1973 and *Mongolodus longispinus* (Yang & He 1984) respectively, isolated and naturally clustered (assumed grasping apparatus) elements interpreted as the grasping spines of chaetognaths. cs, coiled sac; e, eye; gs, grasping spines; te, tentacles. Fossil specimens all from the lower. Cambrian Chengjiang biota except (f) from the bituminous concretions of the Hetang Formation, lower Cambrian Zhejiang, S. China, (h) from lower Cambrian black shales at MaoPing, Hubei, S. China (Vannier *et al.* in preparation), (j) and (k) from the lowermost Cambrian of S. China (*Anabarites trisulcatus–Protoherzina anabarica* Biozone; see Steiner *et al.* 2004; Vannier *et al.* 2005). Scale bars 10 mm in (a)–(e), 5 mm in (f) and (i), 1 mm in (g), 500 μm in (j), 200 μm in (k), (l) and 100 μm in (h).

the chaetognaths or 'arrow worms' (Shinn 1997). These tiny (adults *c.* 2–120 mm long), extremely prolific animals are often second to copepod crustaceans in terms of abundance and biomass in the world's oceans (5–15% biomass). Chaetognaths play an important role in the marine trophic web both as consumers and as a food source for larger animals. They have been firmly recognized in the

Chengjiang biota (Fig. 2i; Chen & Huang 2002; Chen *et al.* 2002; Hu 2005; Vannier *et al.* 2005 and 2007) but the group as a whole may have had a much wider distribution if we consider that protoconodonts (Fig. 2j, k) were true chaetognaths or closely related to them. Indeed, there is a set of convincing evidence (e.g. external morphology, ultrastructure, natural clusters; see Szaniawski 2002 for complete references) that strongly supports the chaetognath affinities of protoconodonts. Their almost global distribution (China, India, Europe, Kazakhstan, North America) and very early occurrence close to the Precambrian–Cambrian boundary suggest that chaetognaths *sensu lato* were among the first zooplanktonic invaders of the water column. The obvious raptorial function of their circumoral apparatuses (grasping spines) also indicates that they counted amongst the earliest active predator metazoans. However, puzzling questions remain concerning this early predator zooplankton. Was it exploiting epipelagic and mesopelagic niches or other levels in the water column? Abundant food suitable for chaetognaths to prey upon such as tiny arthropods or meroplanktonic larvae is likely to have proliferated close to the water–sediment interface, thus possibly driving chaetognaths to occupy hyperbenthic niches (*c.* 1–10 m above bottom). This hypothesis finds strong support in that Recent hyperbenthic communities are effectively intensively exploited for food by chaetognaths (Choe & Deibel 2000).

The oldest radiolarians were recently described from black bituminous limestone concretions found from the lower Cambrian Hetang Formation of South China (Fig. 2 h; Braun *et al.* 2004). Radiolarians are holoplanktonic protozoans that are widely distributed in the present-day oceans and ancient oceanic settings, and occur throughout the water column from the near surface to hundreds of metres depth. Nutrition of radiolarians involves a large variety of materials, including many zooplankton groups such as copepods, crustacean larvae, ciliates and flagellates, and such phytoplankton groups as diatoms and coccolithophores. They often share relationships with dinoflagellate symbionts. Clearly, radiolarians, adding to chaetognaths *sensu lato*, and bivalved arthropods appear as another possibly zooplanktonic component of the early Cambrian oceans. Whether Cambrian radiolarians were particle feeders, scavenging nutrients randomly from the water, or predators (e.g. acritarchs?) is unknown.

Infaunal colonizers

The Precambrian–Cambrian transition is characterized by a remarkable increase of animal activity within the sediment, attested by evidence of deeper and more intense bioturbation. As a consequence, the substrates on which marine benthic animals lived changed from being relatively firm with a sharp sediment–water interface and the dominance of microbial mats to having high water content and a blurry sediment–water interface. This change in substrates evidenced by detailed sedimentological studies has been termed the

'Cambrian Substrate Revolution' (Bottjer *et al.* 2000; Dornbos *et al.* 2004, 2005) and is thought to have had profound ecological implications – first of all that of providing shelter and new niches to diverse benthic metazoans. Traces of infaunal and epifaunal activities such as scratch marks, tracks and burrows of all kinds become more complex through the Precambrian–Cambrian transition (e.g. Jensen 1997; Zhu 1997). However, two crucial elements of the animal interaction with sediment are still poorly documented: (1) the precise identity of the organisms which produced the traces imprinted in sediment; and (2) the exact locomotory and feeding strategies used by the pioneer colonizers of the benthos. Potential tracemakers in the Chengjiang biota are numerous and comprise a wide range of superficial crawlers and diggers such as epibenthic and nektobenthic arthropods (Fig. 3). However, less abundant are the animals that were potentially capable of penetrating deeper into the sediment and to adopt true infaunal habits. These are the inarticulate brachiopods and several groups of worm *sensu lato*. The lingulid brachiopods *Lingulellotreta* and *Lingulella* are very common constituents of the Chengjiang biota (Hou *et al.* 2004) and both have an extremely long and flexible pedicle, often approaching ten times the length of the shell. Although modern lingulid brachiopods live in burrows, permanently attached by a comparable contractile pedicle, the assumed infaunal habit of these early lingulids has been questioned by some authors who merely favour epibenthic or even epiplanktonic lifestyles (e.g. attached to floating organisms such as eldoniids; Chen & Zhou 1997).

The worm fauna from Chengjiang is overwhelmingly dominated by priapulids with at least 13 different species and diverse anatomies (Han *et al.* 2004; Huang *et al.* 2004a, b; Huang 2005). Recent priapulids form a relict group with only 16 extant species, most of them being infaunal burrowers in fine and often poorly oxygenated muddy sediments. Other representatives of the group construct tubes or live interstitially among sediment particles. Recent community analyses reveal that priapulids, at least locally, were probably the most abundant animals of the Chengjiang biota (Dornbos & Chen 2006) comprising up to 45% of all fossil specimens. By contrast with their present-day situation, they may have exerted a major influence on the energy flow of the trophic web. The remarkable morphological similarities between the early Cambrian priapulids and their Recent counterparts have important evolutionary and ecological implications. For example, no major anatomical change seems to have occurred within the priapuliid lineage, which suggests evolutionary stasis over the last 525 million years. The detailed morphology of the

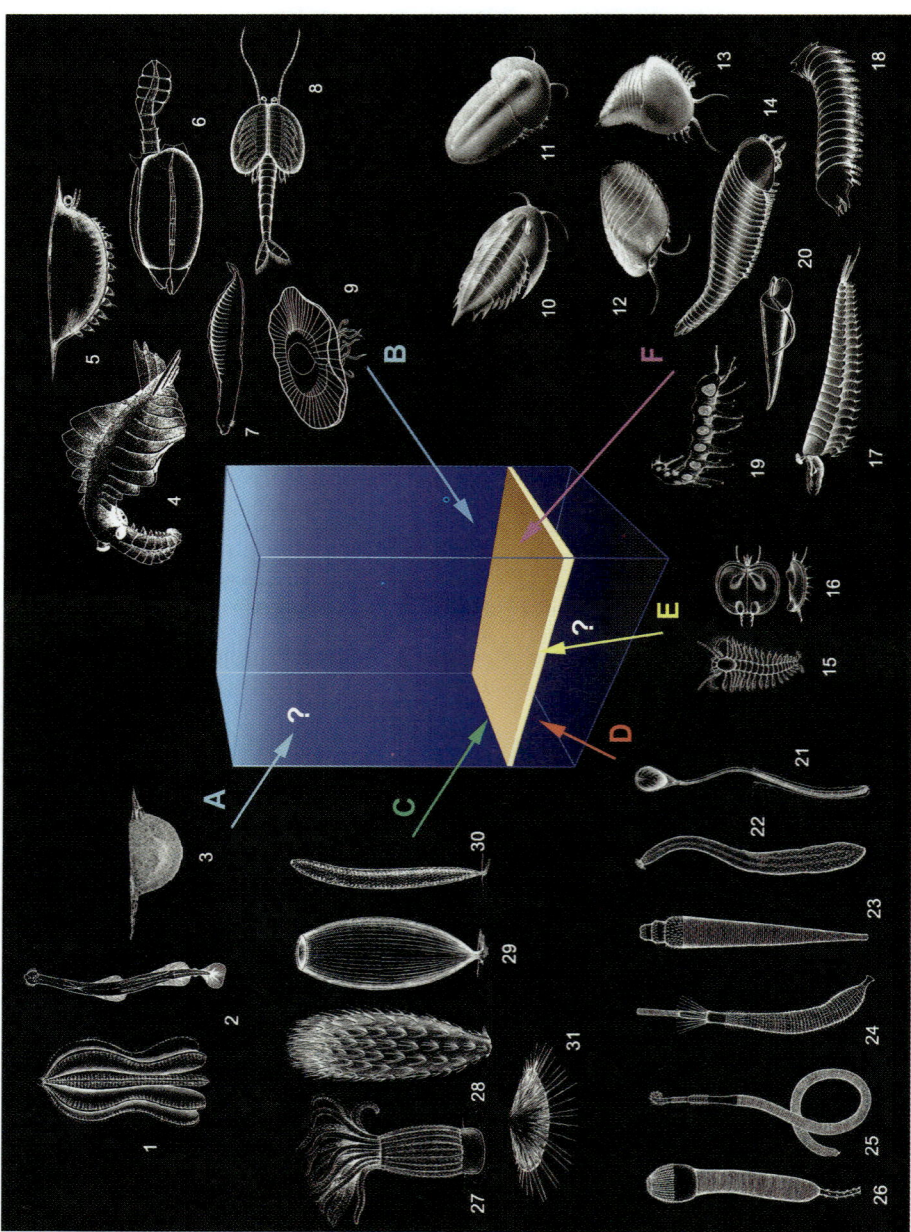

Fig. 3. Early Cambrian occupancy of marine ecospace in the water column, at the water–sediment interface and within the sediment. A pelagic niches (bathymetrical range unknown); B nektonic/nektobenthic niches; C, F epibenthic niches inhabited by sessile and vagrant organisms, respectively; D infaunal niches; E possible meiofaunal niches. 1, *Maotianoascus* (ctenophore); 2, chaetognath; 3, *Sunella* (bradoriid arthropod); 4, anomalocaridid; 5, *Isoxys* (arthropod); 6, vetulicolid (uncertain affinities); 7, *Haikouella* (chordate); 8, waptiid (arthropod); 9, eldoniid (jellyfish-like organism); 10–18, the arthropods *Sinoburius, Naraoia, Kuamaia, Cindarella, Fuxianhuia, Ercaia, Kunmingella, Fortiforceps* and *Urokodia*; 19, *Microdictyon* (lobopod); 20, hyolithid; 21, *Lingulellotreta* (brachiopod); 22, *Cambrosipunculus* (sipunculan worm); 23–26, the priapulid worms *Paraselkirkia, Corynetis, Palaeoscolex* and *Xiaoheiqingella*; 27, *Xianguangia* (cnidarian); 28, *Allonia* (chancelloriid); 29–31, the sponges *Quadrolaminiella, Paraleptomitella* and *Choia*. Animal reconstructions after Hou *et al.* 2004 (modified), Huang 2005 and Collins 1996 (anomalocaridid). Not to scale.

feeding apparatus (e.g. eversible introvert, pharynx, scalid rows; Fig. 4b), body design (e.g. indicating hydrostatic skeleton of the body cavity, body muscle action) and external ornament (e.g. annulations, spines, hooks, plates) clearly indicate that most lower Cambrian priapulids were predators and had capabilities to move through the sediment by peristaltic burrowing in a similar way as does the Recent *Priapulus caudatus* (Fig. 4h). External ornament seems to have acted as anchoring features. Recently, the discovery at Chengjiang of complete worms, preserved inside their lined burrows (Huang 2005; Zhang *et al.* 2006) confirms that some, if not all, the early Cambrian priapulids were able to make dwelling tubes. However, the depth, orientation and geometry of their burrows are less clear. Were their tubes U-shaped, vertical, and deep or on the contrary parallel or slightly oblique to the sediment surface? The fact that burrows were made in the topmost water-saturated layer of sediment where storm events probably remobilized this particular layer, and that compaction obliterated many features, makes the question particularly difficult to answer. Preliminary observations of live specimens (Kristineberg Marine Station, Sweden) indicate capabilities for penetrating relatively deep into organic-rich muds (several tens of cm in laboratory conditions; unpublished information). More detailed experiments and *in-situ* submarine observations are expected to provide new information on the exact locomotory behaviour of these animals on and within the sediment.

Although classically used as biostratigraphical indicators of the Precambrian–Cambrian boundary (Narbonne *et al.* 1987), trace fossils such as *Treptichnus* have long been an enigma for palaeobiologists. Recent observations of live specimens of *Priapulus* from Sweden (Huang *et al.* 2004*c*) indicate that these traces were most probably produced by the crawling action of priapulids over the substratum. These comparative studies reveal astonishing behavioural similarities between the Recent and the earliest ancestors of priapulid worms.

Other worms such as sipunculans occur in the Chengjiang biota (Huang *et al.* 2004*d*), although much rarer than priapulids. Details of their digestive tract (U-shaped, anus location) and, again, strong analogies with extant species (Fig. 4i) suggest that early Cambrian sipunculans were selective or non-selective detritivores, possibly direct deposit feeders, and shallow burrowers. Annelids are by far the most successful Recent worms with more than 16 500 described species (Brusca & Brusca 2003). *Facivermis* (Hou *et al.* 2004; Huang 2005) with circumoral tentacles recalling the cirri of modern annelids may indicate that the group was present in the early Cambrian benthic communities, although extremely minor compared with priapulids. Because bioturbation enhances organic decomposition, nutrient cycling, redistribution of organic material, and oxygenation of sediment, the appearance of bioturbators in the ecological scenery, at least in shallow-water environments, may have had large-scale repercussions, for example that of liberating carbon back to the atmosphere.

Epibenthic dwellers

The water–sediment interface is by far the environment where most Recent marine faunas attain a peak of numerical abundance and species diversity. This tendency is confirmed in the early Cambrian Chengjiang fauna where more than 90% of the described species are epibenthic dwellers (Fig. 3). Sponges are a diverse component of the fauna (Fig. 3), most of them being demosponges, a group in which the collagen skeleton is reinforced by siliceous spicules. Most Chengjiang sponges were anchored to the substrate or more rarely rested on the sediment. Comparisons with Recent species suggest that they were sessile filter feeders upon microscopic particles present in the water column and, most probably size-selective particle feeders. In Recent sponges, internal particle capture occurs in the 2 to 5 μm range, the most common food items being bacteria, small protists, unicellular algae and organic detritus (Brusca & Brusca 2003). The large variety of shape (from rounded to tubular) and size of the Chengjiang sponges may indicate niche partitioning such as the selective exploitation of particulate food source via different strategies of capture. Their tiering level was relatively low (below 15 cm above sediment–water interface) but 50 cm to possibly 1 m-tier hexactinellid sponges have been reported elsewhere on the Yangtze Platform, in deeper-water black shale environments (lowermost Cambrian Hetang Formation; Yuan *et al.* 2002). More enigmatic are the chancellorids (Fig. 5a; Bengtson & Hou 2001; Janussen *et al.* 2002). Although sac-like and sclerite-bearing animals similar to sponges, their ecology is unresolved. Suspension feeders such as tunicates resembling Recent solitary ascidians were also present (Chen *et al.* 2003) and, along with sponges and other sessile metazoans, contributed to extracting plankton and organic detritus from the water column.

The diversity of epibenthic arthropods is huge (Figs 3, 5) with countless dorsoventrally flattened and multisegmented forms that possessed biramous appendages clearly adapted to crawling (typically an articulated endopod and a leaf-life exopod with a respiratory function; Fig. 6i). The arthropods also comprised abundant 'bivalved' forms – i.e.

J. VANNIER

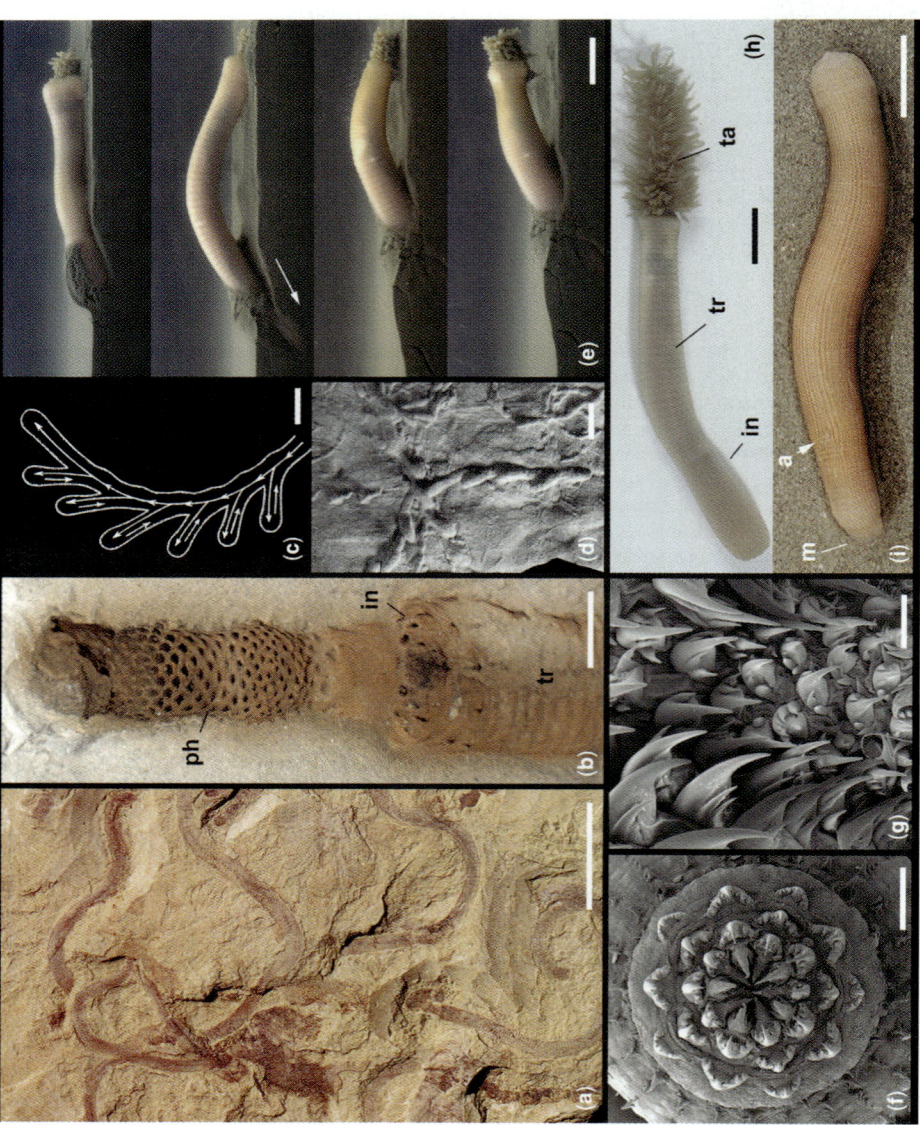

Fig. 4. Infaunal colonizers. (**a**), (**b**) Priapulid worms from the Chengjiang biota; accumulation of *Maotianshania* on the surface of a bedding plane and exceptional 3D preservation of the front part of *Cricoscomia jinningensis* Hou & Sun 1988 (see Huang 2005) showing pharynx. (**c**), (**d**) Loop-shaped traces of the Recent *Priapulus caudatus* (from unpublished experiments; see Huang *et al.* 2004*d*; Huang 2005) and assumed fossil analogues from the lower Cambrian Buen Formation of Greenland (*Trichophycus pedum*; Bryant & Pickerill 1990). (**e**)–(**h**) Morphology and behaviour of the Recent priapulid *Priapulus caudatus* from Sweden. (**e**) Burrowing in mud (J. Vannier and D.-Y. Huang, unpublished observations). (**f**), (**g**) Scanning electron micrographs of the eversible pharynx showing strong teeth in relation with predator feeding mode (compare teeth with (b)). (**h**) Body division with introvert, trunk and tail. (**i**) Recent sipunculan worm from Brittany, France with location of mouth and anus. Fossil specimens all from the lower Cambrian Chengjiang biota. A = anus; in = introvert; m = mouth; ph = pharynx; ta = tail; tr = trunk. Scale bars 10 mm in (a), (c)–(e), and (h), and (i); 1 mm in (b), 500 µm in (f), and 200 µm in (g).

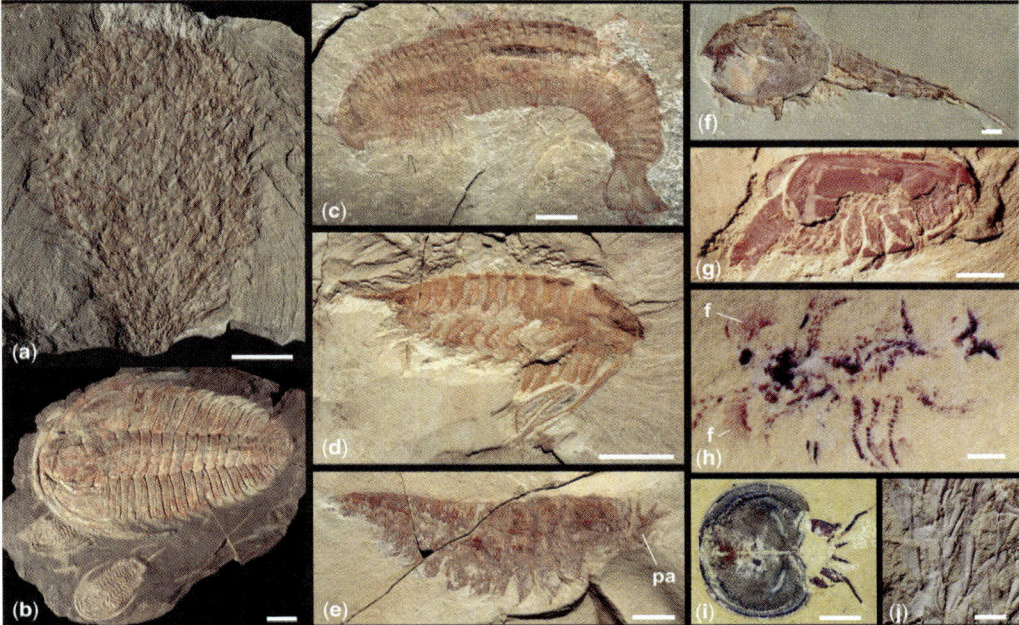

Fig. 5. Epibenthic dwellers, (**a**) Sessile cup-like chancelloriid with triradiate sclerites. (**b**)–(**i**) Arthropods, (**b**) The trilobite *Redlichia takooensis* (see Nedin 1999). (**c**) The multisegmented *Shankouia zkenghei* Chen *et al.* 2004. (**d**) *Leanchoilia illecebrosa* (Hou 1987). (**e**) *Haikoucaris ercaienis* Chen *et al.* 2004 with prehensile appendage (**f**) Waptiid. (**g**) *Canadaspis laevigata* Hou & Bergström 1991. (**h**) *Ercaia minuscula* Chen *et al.* 2001 with specialized fan-like antennae. (**i**) The bradoriid *Kunmingella douvillei* (Mansuy 1912) preserved in articulated 'butterfly' orientation and showing posterior appendages (courtesy Derek Siveter), (**j**) Accumulation of conical hyolith shells. f, Fan-like antenna; pa, prehensile appendage. All from the L. Cambrian Chengjiang biota except (b) from the lower Cambrian Emu Shale, Kangaroo Island, SW Australia (courtesy C. Nedin). Scale bars 10 mm in (a), (b), 5 mm in (c), (d), (e), (g) and (j), 2 mm in (f), 1 mm in (i), and 500 μm in (h) and (j).

characterized by a flexible, dorsally folded carapace which enveloped the cephalon and most of the trunk of the animal. Typical examples of these arthropods are the waptiids (Fig. 5f) which recall Recent phyllocarid crustaceans such as *Nebalia* (Vannier *et al.* 1997). Their flexible limbless abdomen and paddle-like tail indicate capabilities for both swimming in the vicinity of the bottom and presumably stirring up the sediment for food search or protection. Bradoriids also belong to this informal category of 'bivalved athropods' (Fig. 5i; Hou *et al.* 1996, 2002; Shu *et al.* 1999). These tiny animals were capped by a folded dorsal shield that, in life attitude, is supposed to have been held widely open in a so-called 'butterfly position'. Their preserved ventral anatomy and external resemblance to modern crustaceans support the view that bradoriids were motile epibenthic dwellers of the sediment–water interface, and, possibly, had a lifestyle and ecological niche similar to Recent crustacean ostracods (Vannier & Chen 2005). Bradoriids may have been detritus feeders, or, similarly to some Recent ostracods (Vannier *et al.* 1998), microscavengers

or micropredators of unknown non-mineralized animals (e.g. meiofaunal organisms or larvae). Bradoriids were among the most prolific animals of the Chengjiang biota, occurring in numerous localities and horizons, and in different sedimentary facies of the Yangtze Platform (Hou *et al.* 2002). Like trilobites, bradoriids diversified rapidly and had a global distribution both at high and low latitude during the early and middle Cambrian (Williams *et al.* 2007). Their high numerical abundance at Chengjiang suggests that bradoriids, along with other invertebrates, may have been important recyclers on the early Cambrian sea-bed (Fig. 3). The general prevalence of epibenthic arthropods (predators, scavengers and detritivores; e.g. Chengjiang biota) may well be explained by the availibility of food (abundant epibenthic prey, carcasses of both benthic and pelagic animals, detritus) at the water–sediment interface.

The meiofauna is a major ecological category of the present-day water–sediment interface, that encompasses animals living within the uppermost millimetre or centimetre of sediment and ranging in

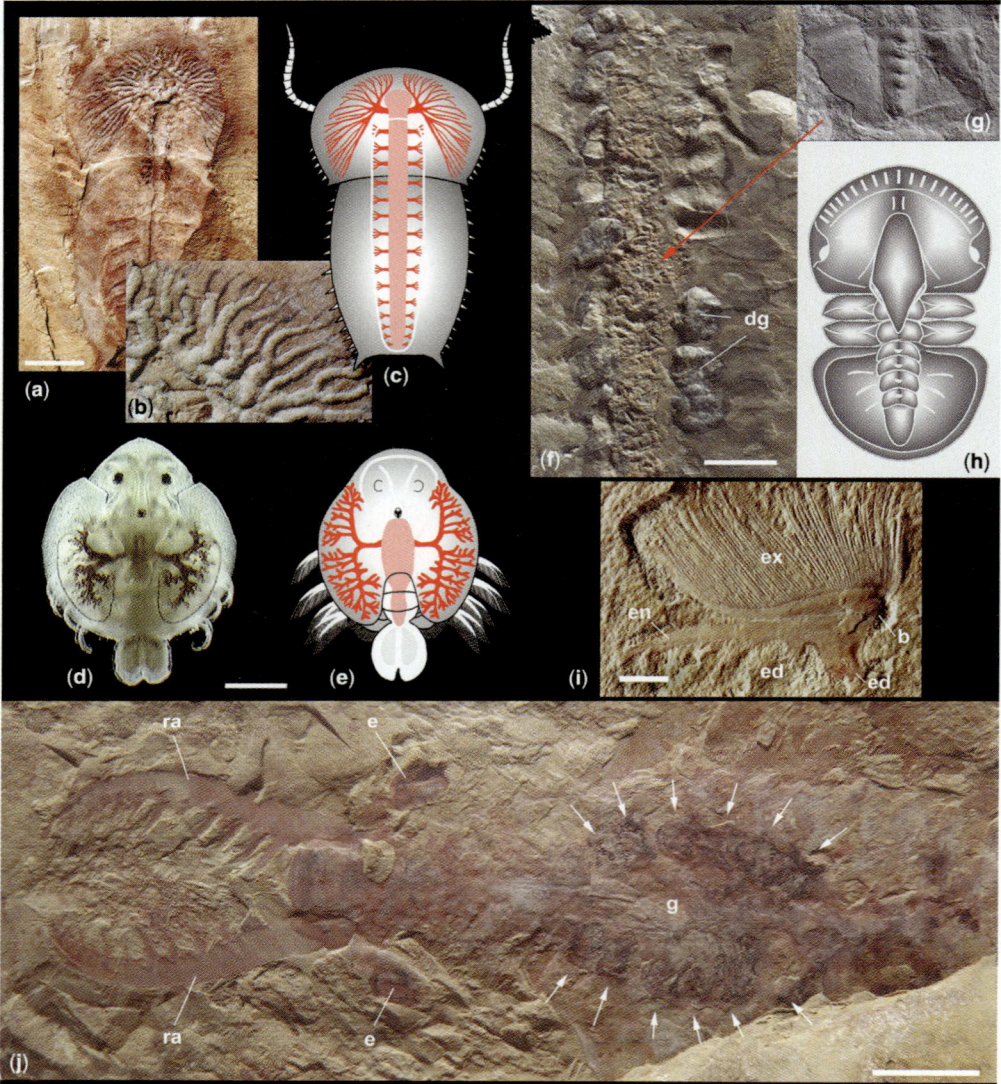

Fig. 6. Reconstructing the feeding modes and strategies of early Cambrian animals. (**a**)–(**e**) Digestive system of naraoiid arthropods (*Naraoia spinosa* Zhang & Hou 1985) from the Chengjiang biota; digestive and storage functions of diverticles (red) inferred from Recent branchiuran crustaceans (ectoparasites of fish). (**f**)–(**h**) Preserved gut contents and digestive glands in an arthropod from the Kaili biota (see Zhu *et al.* 2004), with details of prey (tiny eodiscoid trilobites). (**i**) Appendage of *Misszhouia longicaudata* Zhang & Hou 1985 from Chengjiang showing strong spinous endites near the base of endopod and directed inwards (courtesy Derek Siveter). (**j**) Juvenile anomalocaridid from Chengjiang showing its frontal raptorial appendages, stalked eyes and digestive system (white arrows indicate assumed lobe-like digestive glands on both sides of the gut). b, base of appendage; dg, digestive glands; e, eye; ed, endites; en, endopod; ex, exopod; g, gut; ra, raptorial appendages. Scale bars 10 mm in (a) and (j), 5 mm in (f), 2 mm in (d) and (i), and 1 mm in (d) and (e).

size from approximately 0.1 mm to 1 m (nematodes, copepods, foraminiferans, ostracods, loriciferans, etc.). A series of remarkably well-documented studies by D. Waloszek and his collaborators (see Waloszek 1993 and Maas *et al.* 2003 for complete key references) attest to the presence of a diverse meiofauna in the upper Cambrian 'Orsten' fauna of Sweden and other regions. Whether a meiofauna *sensu stricto* already existed in the early Cambrian remains an open issue. Tiny arthropods such as *Ercaia* (2–4 mm long; Fig. 5g and Chen *et al.* 2001) with its fan-like setose antennae recalling those of Recent copepods, *Primicaris* (Zhang *et al.* 2003) and possible loriciferans (Huang 2005), all found in the Chengjiang biota, are early Cambrian meiofaunal candidates. However, these organisms do not conform with the strict definition of the meiofauna in terms of body size.

Although molluscs are among the most prolific invertebrates in Recent marine ecosystems and have an extremely well-documented fossil record (e.g. gastropods, ammonites), their early evolutionary history has long remained obscure. It is only recently that the re-interpretation of *Odontogriphus* and *Wiwaxia* from the Burgess Shale (Caron *et al.* 2006) provided convincing evidence that soft-bodied molluscs, fully equipped with a radula and mantle cavity, were present in the Cambrian seas as epibenthic grazers of microbial mats (*Odontogriphus* associated with cyanobacterium *Morania*) and biofilms. Although trace fossils indicate that mat-grazing was probably the most common feeding type of the late Precambrian vagile metazoans (Seilacher *et al.* 2005), we have here detailed evidence of the feeding mechanism (radular scraping) of the early molluscs. Molluscs comparable with those from the Burgess Shale and resembling *Vetustovermis* (Glaessner 1979; Emu Bay Shale Lagerstätte) were also present in the early Cambrian Chengjiang biota (Huang 2005; unpublished description).

Reconstructing the early Cambrian marine food chain

The innumerable fossil evidence obtained from exceptionally preserved biotas such as those of Chengjiang (see above), the Burgess Shale, Sirius Passet, Emu Bay, Comley and Sinsk clearly demonstrates the existence of diverse marine ecosystems in the early Cambrian, but vital aspects of the functioning of these early biological systems are still obscure; for example, the nature of the primary production, available food sources, feeding strategies of animals, prey/predator relationships and the structure of the trophic web in general (e.g. food-chain length). Relatively few attempts have been made to tackle these important issues (Conway Morris 1986; Butterfield 2001*a*, *b*; Vannier & Chen 2005; Hu 2005; Caron 2005). The trophic analysis made by Conway Morris (1986) in the Phyllopod Bed fauna from the middle Cambrian Burgess Shale documents deposit feeders, suspension feeders, predators and scavengers, and reconstructs a possible food chain. Estimates of

numbers of individuals are used in this pilot study to infer biomass and the relative importance of feeding types within the ecosystem. However, the use of the Phyllopod Bed as a time-averaged assemblage of unknown duration reduces the reliability of the results obtained by these quantitative methods (Caron & Jackson 2006). Concerning the early Cambrian Chengjiang Lagerstätte and comparable sites, detailed analysis of the fossil content at the level of each depositional event (e.g. storm-induced deposits) would be the only way to reconstruct quantitative aspects of the trophic structure. The information given here is qualititative and seen as the initial step for future detailed synecological studies.

Fossil evidence for reconstruction of Cambrian food chains is obtainable in different ways, including: (1) the analysis of preserved gut contents, (2) the functional morphology of fossil animals that may be used to infer feeding mechanisms (e.g. predatorial habits) and trophic types, (3) the assessment of coprolite content and feeding traces, and (4) the study of particular structures such as bite marks and drill holes or various predation damage that may testify to interactions between animals. However, the diet and feeding methods of Cambrian animals can rarely be well constrained and have been enlightened by only very few cases, most of them from the middle Cambrian Burgess Shale (Butterfield 2001*b*; Vannier & Chen 2005 for references). Another potential but rarely exploited source of information is provided by fossil aggregates interpreted by authors (e.g. Conway Morris & Robison 1988; Chen & Zhou 1997; Nedin 1999; Shu *et al.* 1999; Babcock 2003 ; Vannier & Chen 2005) as possible coprolites.

Gut contents

This is the most valuable direct evidence for deciphering animal feeding interactions. Several animals from the Burgess Shale have skeletal and organic remains preserved *in situ* within their alimentary tract. For example, the priapulid worm *Ottoia* (Conway Morris 1977) shows hyolith shells aligned in its digestive tract, and the arthropod *Sidneyia* (Bruton 1981) has gut contents with exoskeletal fragments of hyoliths, trilobites and much-comminuted shelly material including possible bradoriids. Similarly, accurate information is available on the gut contents of an arthropod (dorsoventrally flattened; *c.* 10 cm long) from the lower middle Cambrian Kaili Lagerstätte of South China (Fig. 6f–h; Zhu *et al.* 2004). Skeletal fragments of tiny eodiscoid trilobites were discovered within the cylindrical gut of this predator arthropod possibly related to *Fuxianhuia* (Fig. 3, no. 14) which also possessed paired digestive glands (Vannier & Chen 2002; Butterfield 2002).

Unfortunately, no such direct evidence is available in early Cambrian animals.

Functional morphology

The exceptional preservation of digestive systems, especially in arthropods and worms, also provides key information on the feeding strategies of Cambrian animals. For example, the trilobite-like naraoiid arthropods present in the Chengjiang biota have most peculiar ramifying diverticles in the head and trunk regions (Fig. 6a–c). Comparisons with Recent crustaceans revealed the function of these enigmatic distensible diverticles (Vannier & Chen 2002). They played an important role in food storage and digestion and seem to have been adapted to intermittent scavenging or predation.

The structure of the appendages is often a good indicator of animal feeding methods. Naraoiids had appendages (endopods; Fig. 6i) with strong spinose endites projected inwards with the obvious function of squeezing and shredding food. Here, functional morphology clearly supports the hypothesis that naraoiids were predators or scavengers. Recent studies (Chen *et al.* 2004; Maas *et al.* 2004) provide similar examples and emphasize the prehensile function of the 'great-appendage' in a variety of arthropods from Chengjiang. However, it is uncertain whether these organs had a true raptorial (predation) function or were simply used by the animal in food handling (e.g. scavenging on carcasses). Although anomalocaridids (Fig. 6f) do not resemble any extant marine animal, their raptorial habits are not questioned. The design of their huge frontal appendages armed with strong spines and the structure of its jaw apparatus (Whittington & Briggs 1985; Collins 1996; Nedin 1999; Vannier & Chen 2005) make anomalocaridids the most fearful predators of the Cambrian seas.

Coprolites

Several types of fossil aggregates from the Chengjiang biota have been recently interpreted as coprolites (Vannier & Chen 2005). Elliptical ones contain randomly distributed exoskeletal remains of typically small-to-medium-sized bivalved arthropods (e.g. bradoriids, waptiids) and were possibly produced by anomalocaridids or other unknown predators. The ribbon-like ones with typically small hyolith shells are interpreted as the faeces of infaunal carnivorous worms such as priapulids. Coprolite-like assemblages also occur in the early Cambrian Emu Bay biota (Nedin 1999). This new coprolite data set adds to morphofunctional information obtained from fossil organisms (see above) and indicates that predation occurred at different levels of the water column with: (1) endobenthic

predators (diverse priapulid fauna) feeding near the sediment–water interface; (2) epibenthic predators/scavengers (almost exclusively arthropods); (3) predators living in the lower levels of the water column (e.g. anomalocaridids).

Bite marks and drill holes

These features provide additional indirect evidence of animal interactions (Babcock 2003). For example, healed injuries on trilobites from the Emu Shale biota have been interpreted as the traces of predation damage by anomalocaridids (Nedin 1999) and suggest that at least some anomalocaridids were durophagous predators feeding upon epibenthic prey. However, it is not clear from the morphology of anomalocaridid mouthpieces how these bites may have been inflicted on their prey. It seems that the plates of the circular jaws had the main function to engulf and push the prey into the mouth, and to break apart the hard exoskeletons of prey such as trilobites. Finer mechanical processing of food was most probably achieved by additional teeth that lined the inner wall of the mouth and, possibly, through the action of the peristaltic contractions of the oesophagus. These interpretations remain hypothetical because no Recent marine animals possess such feeding methods.

Borings on small Cambrian brachiopod valves (Conway Morris & Bengtson 1994) indicate that predation possibly occurred at a microscopic level and that specialized adaptations for penetrating shelled prey had already been acquired by unidentified predators.

Complexity of early Cambrian trophic structure

The fossil evidence for diverse animal interactions presented here is still fragmentary and does not allow generalizations concerning the structure and functioning of the early Cambrian marine ecosystem as a whole. However, it does provide new information to resolve some of the ecological puzzles and to help our understanding of the main biological transfers of mass and energy between the different living components and ecological categories. Our tentative reconstruction (Fig. 7) aims to show that some of the key principles of the functioning of present-day ecosystems were effectively operating in the early Cambrian.

Primary production

In the modern pelagic realm, phytoplankton primary production is *c.* 90% due to photosynthetic prokaryotic bacteria (e.g. *Synechococcus* and

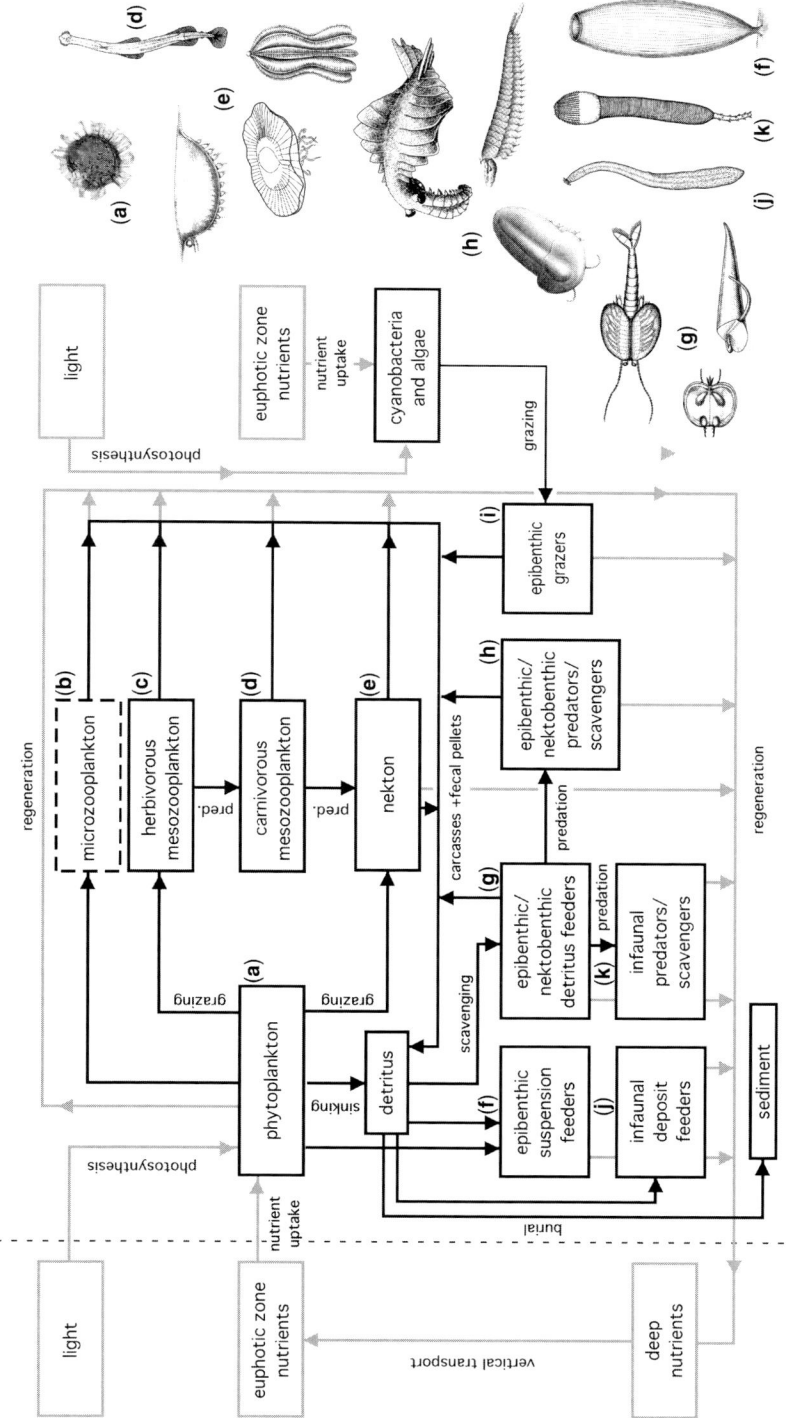

Fig. 7. Assumed general organization of the early Cambrian marine ecosystem based on fossil evidence. Physical forcing and input functions in grey. Transfers of mass and energy are shown by black arrows and solid lines (dotted frame = hypothetical). Ecological categories lettered (**a**) to (**k**) with typical forms illustrated aside. (**a**) Primary phytoplankton producers (e.g. acritarchs). (**d**) Predator zooplankton (e.g. chaetognaths). (**e**) Nekton (e.g. bivalved arthropods, eldoniids and ctenophores). (**f**) Sessile suspension feeders (e.g. sponges). (**g**) Epibenthic/nektobenthic detritus feeders (waptiids, hyoliths and bradoriids). (**h**) Epibenthic/nektobenthic predators/scavengers (e.g. anomalocaridids, *Fortiforceps* and naraoiid). (**j**) Infaunal deposit feeders (sipunculan worms). (**k**) Infaunal predators (e.g. priapulid worms). Microzooplankton between 20 and 200 μm, mesozooplankton between 0.2 and 20 mm, nekton >20 mm.

Prochlorococcus), the rest being provided by eukaryotes such as dinoflagellates and diatoms. The smallest of these components have no potential to be registered in the fossil record because they simply do not sink and decay within the water column unless they aggregate to larger and denser particles (marine snow). Acritarchs (Fig. 7a) that are commonly interpreted as the degradation-resistant cysts of dinoflagellates are abundant and diverse in the lower Cambrian rocks and the most tangible indicators of primary productivity in the water column. The shift from a relatively uniform leiosphaerid phytoplankton through the Proterozoic to a much more diverse and rapidly evolving acanthomorphic phytoplankton in the Cambrian is seen by some authors (e.g. Butterfield 2001*a*) as a response to the introduction of mesoplanktonic grazers (e.g. arthropods). The fact that this radiation through the Proterozoic–Cambrian transition parallels the biodiversification of metazoans supports the view that major changes in primary production may have been a possible ecological trigger of the 'Cambrian Explosion'. In addition, such qualitative and quantitative changes in primary production through the Precambrian–Cambrian transition may have had a direct effect on the carbon cycle, possibly a substantial increase of CO_2 extraction from the atmosphere. These hypotheses need to be tested by accurate quantative data. In addition to microscopic phytoplankton, megascopic filament-like algae are present in the Chengjiang biota (Hou *et al.* 2004) and obviously contributed to the primary production of shallow-water environments.

Primary zooplanktonic consumers

The presence of herbivorous mesozooplankton (Fig. 7c) in the early Cambrian is currently supported by only one set of strong fossil evidence: the remains of filter-feeding apparatuses of assumed crustacean origin (Butterfield 1994, 2001*b*). The hypothesis that these filter feeders may have played the same role as copepods in modern oceans is interesting but needs to be tested by additional fossil data.

Secondary planktonic and nektonic consumers

The chaetognaths and protoconodont animals that were possibly both adapted to mid-water predation (Fig. 2) are serious candidates for a secondary level of zooplanktonic consumers in the early Cambrian ocean (Fig. 7d). The widespread introduction of prey–predator relationships into the pelagic realm

during the Precambrian–Cambrian transition via the chaetognaths seems to represent another important innovation in the tiering and functioning of early marine ecosystems that may have laid the foundations of modern-style marine food chains. However, there remains the question of their food source (herbivorous zooplankton?, larval stages of benthic animals?) that lack fossil evidence. Other potential pelagic consumers such as macrozooplankton (e.g. ctenophores, eldoniids) and nekton (e.g. early chordates, vetulicolids, arthropods), were certainly inhabiting the water column, although very little is known of their exact lifestyles, diet and feeding strategies.

Suspension feeders

Sinking detritus derived from the biological activity of phyto-and zooplankton, added to microorganisms, most probably generated a major food source for sponges and other sessile suspension feeders (Fig. 7f). This assumption is based on the feeding methods of Recent sponges.

Detritivores and scavengers

This ecological category which encompassed a huge variety of vagrant animals from worms (e.g. sipunculans) to arthropods (from millimetre to large size), and hyoliths (Fig. 7g, i) undoubtedly exploited: (1) the flux of nutrient-rich particles that was sinking out from the water column (including the 'repackaged food' such as faecal pellets from zooplankton) and (2) carcasses of various types deposited on the bottom. Detritivores and scavengers along with microbial activity were powerful recyclers of organic matter at the water–sediment interface.

Benthic–nektobenthic primary carnivores

Predation occurred at different levels of the water column with: (1) endobenthic predators (diverse priapulid fauna; Fig. 7j) feeding near the sediment–water interface; (2) epibenthic predators (almost exclusively arthropods); (3) predators living in the lower levels of the water column (e.g. anomalocaridids). Communities living at or close to the water-sediment interface (epibenthic *sensu stricto*, meiobenthic and demersal animals) were exposed to a multidirectional predatorial pressure from infaunal, epifaunal and midwater predators. These early Cambrian predators had already acquired complex behaviour such as hunting (e.g. anomalocaridids, priapulids) and predator avoidance in which advanced sensory systems were most likely involved.

Secondary and tertiary carnivores?

Although predation was obviously diverse, nothing indicates that the food chain extended beyond the level of primary carnivores. For example, the association of arthropod hunting traces with their possible prey (priapulid worms) in the lower Cambrian *Mickwitzia* Sandstone of Sweden (Jensen 1997; Butterfield 2001*b*) remains questionable. That the diet of anomalocaridids may have included other carnivores is possible (trilobites; priapulid worms?; see evidence from coprolites Vannier & Chen 2005) but still lacks direct evidence from gut contents.

The Cambrian 'ecological revolution': biotic *vs.* environmental triggers

The unprecedented high trophic complexity of the early Cambrian ecosystem (Fig. 7) addresses the puzzling question of what factors, environmental or otherwise, may have triggered this 'ecological revolution'. One key aspect of the problem is the remarkable diversity of animal interactions that characterizes the beginning of the Cambrian and is well exemplified by prey–predator relationships. Indeed, nothing in the fossil record indicates that such strong animal interactions and complex trophic links existed before the Cambrian (e.g. Ediacaran assemblages in the late Proterozoic; early metazoan embryos of the Doushantuo Formation, China dated to *c.* 580 Ma or younger, see Xiao *et al.* 1998). According to the 'Light Switch' theory of Parker (2003), it is the introduction of vision among early Cambrian marine communities that was the main trigger of the ecological turnover. Precambrian animals may have possessed chemical or mechano-receptors, even simple light receptors, but no visual organ that could form an image (Parker 2003). According to this theory, selection pressure in the Precambrian was relatively low. But with the introduction of eyes, mainly in arthropods, the visual appearance of animals became important. This event is considered to have been the starting point of a totally new mode of animal interaction and is likely to have triggered active predation and its cascade of feedback effects (e.g. anti-predatorial responses from prey such as exoskeletal protection and biomineralization, and new behaviour). Indeed, arthropods bearing well-developed eyes, some being pedunculate and mobile, are numerous among early Cambrian communities. One may argue that there is no reason to favour vision and that other anatomical innovations (e.g. digestive system, appendages, other senses) may have had a comparable triggering effect on the whole system. However, vision has an enormous

potential, that of promoting active and selective predation and enabling the multidirectional exploration of environment (e.g. colonization of new niches). For these reasons, vision may have played a leading role. Vision can be an efficient tool if linked with a nervous system that allows the animal to process images and give an adequate response to visual stimuli. Major improvements in the functioning of the nervous system of early metazoans may actually be the crucial innovation that gave rise to vision, but also boosted other vital activities such as limb coordination, locomotion (epibenthic and off-bottom), and feeding (e.g. active predation). The increase of neural complexity during the Precambrian–Cambrian transition may have given access to a variety of new behaviour that altogether changed the ecological rules.

The pioneer invasion of the water column by mesozooplankton from originally benthic organisms (e.g. possibly via meroplanktonic larvae) may have been one of the key events that had profound effects on the dynamics of the early Cambrian food web (Butterfield 1994, 2001*a*; this paper). Why and when the first ecological shifts from a benthic to pelagic lifestyle occurred remains an enigma, but must have been a response to some biotic (benthic predation?) or environmental pressure. The addition of this new tier is likely to have provoked major changes in the energy flow and biomass transfer through the system: (1) by 'repackaging' phytoplankton as larger sinking, nutrient-rich particles (e.g. faecal pellets) for both pelagic and benthic consumers (Butterfield 2001*a*, *b*) and (2) by creating a new source of food for the nekton. Such changes would have triggered a cascade of adaptive responses in the underlying trophic layers and catalysed biological diversification. Butterfield (2001*a*) also put forward that major changes in the phytoplankton composition occurred across the Precambrian–Cambrian transition and suggested a possible linkage between the radiation of acanthotrophic phytoplankton and the explosive radiation of animals. The question of whether the invasion of mesozooplanktic grazers in the water column introduced new selection pressure and boosted the evolution of phytoplankton, or that changes in phytoplankton composition promoted the advent of midwater grazers, is still unresolved.

In addition to these possible biotic factors (mesozooplankton, phytoplankton turnover) and equally important to us, is the introduction of both active and passive predation in several major animal groups (e.g. arthropods, worms).

A most important event with profound repercussions on the functioning of early ecosystems is the animal biomineralization that took place between *c.* 550 and 530 Ma across the Precambrian–Cambrian transition in the form of a huge variety of shells, spicules and sclerites (see late Precambrian *Cloudina*, e.g. Hua *et al.* 2005 and early Cambrian Small Shelly Fossils (SSF), e.g. Conway Morris 2001). Prior to this date, fossil skeletons are virtually absent. These biomineralization processes involve silica (e.g. sponges), calcium carbonate and phosphate. There seems to be a correlation between water chemistry and carbonate biomineralization (Harper *et al.* 1997). That changes in water chemistry enabled or facilitated biomineralization is possible but may not explain the explosive radiation of biomineralization, exemplified by the diversity of SSF in the Tommotian (early Cambrian). Some authors (Conway Morris 2001) expressed the view that the appearance of skeletons in general, mineralized or not, is ecological, chiefly a response to the increase of predation. Supporting evidence for this hypothesis is that the earliest predators, although armed with powerful skeletonized raptorial apparatuses, are devoid of mineralized parts, for example priapulids, anomalocaridids and chaetognaths. Biomineralized skeletons obviously had not only the function of protecting animals against the attacks of predators. The earliest skeletons such as those of sponges (siliceous spicules) seem to have been developed for support rather than protection and may have been secreted under conditions of unusual supersaturation. Experimental and biochemical studies (e.g. functionality of the organic matrix) using recent models seem to be promising approaches to the question of early biomineralization and its relation to water chemistry.

Combined geological, sedimentological and geochemical evidence clearly indicate that several major disturbances affected the global ecosytem in the late Precambrian (Neoproterozoic). Changes in carbon isotopes correlated with glacial deposits are good indicators of several glacial episodes before the Cambrian (e.g. Kaufman *et al.* 1997; and Condon *et al.* 2005 for chronology). For example, two major Neoproterozoic glaciations (Marinoan, *c.* 635 Ma and Gaskiers, *c.* 580 Ma) are recognized in Namibia, Oman, China and Newfoundland. They are correlated with negative $\delta^{13}C$ excursions and their temporal framework is well constrained by U–Pb ages (Myrow & Kaufman 1999; Condon *et al.* 2005). However, the links between these climatic and environmental changes and the biotic events that took place during the Precambrian–Cambrian transition are not easy to establish. These glacial episodes occur before the early Cambrian bioradiation and before the first occurrence of metazoan embryos (e.g. from Doushantuo Formation, China) and can hardly be seen as the direct cause of any trophic

reorganization. However, a sharp $\delta^{13}C$ negative excursion at around 550 Ma and apparently unrelated to glacial episodes is far more enigmatic. This isotopic anomaly seems to be synchronous with the appearance of larger and more complex mobile metazoans (e.g. evidence from horizontal trace fossils). But again what may suggest a possible feedback relationship (Condon *et al.* 2005) between evolutionary developments (increase of benthic activity) and seawater chemistry, requires precise explanations. Global perturbations in the carbon cycle near the Precambrian–Cambrian boundary (short-lived negative excursion in the carbon-isotopic composition of seawater; Zhu *et al.* 2004 for China) would suggest an extinction event (Knoll & Carroll 1999) affecting the global ocean ecosystem. If it was confirmed that such an event effectively took place, then it may have accelerated the disapperance of the Ediacaran biota and created unique ecological opportunities for metazoans to colonize a full range of vacant niches (Fig. 3).

The hypothesis that an increase of atmospheric oxygen may have removed physical barriers to the evolution of large size (Knoll & Carroll 1999) and higher metabolism is plausible. But there is a lack of firm geochemical evidence that a rise of oxygen levels coincided with the early Cambrian bioradiation and more precisely the construction of a complex ecosystem. However, evidence from high-resolution carbon and sulphur isotope records obtained from Oman (Fike *et al.* 2006) indicates that the ocean became increasingly oxygenated between the end of the Marinoan glaciation (Precambrian, *c.* 635 Ma) and before the extinction and subsequent evolutionary radiation across the Precambrian–Cambrian boundary. This major oceanic change may have had a key role in the establishment of a new type of oceanic trophic web and in the evolution of marine animals as a whole. Rather than a unique, initial trigger, it may be a chain of biotic events probably concentrated within a relatively short time interval that eventually modified animal interactions, and laid the foundations of modern marine ecosystems. Among these events were the achievement of more complex nervous systems, visual organs and motor functions, the increase of predation pressure (both passive and active), exoskeletal biomineralization, the colonization of new niches and the addition of new tiers. The early Cambrian global re-organization of the marine ecosystem resulted in more complex trophic links and also in a more closely dependent plankton and benthos (Butterfield 2001*b*). This high level of interdependence made marine ecosystems highly vulnerable to environmental perturbations, in the first instance those induced by climatic changes.

The author thanks the French Ministries of Foreign Affairs, Education and Research (DRIC), and the Chinese Ministry of Science and Technology (MOST), for funding via the PRA T03-04 Program, Mark Williams (Leicester) for reading the manuscript, Derek Siveter (Oxford) for photographs of Chengjiang specimens and David Siveter (Leicester) and Martin Brasier (Oxford) for their helpful reviews. Contribution UMR5125-07.045.

References

BABCOCK, L. E. 2003. Trilobites in Paleozoic predator–prey systems, and their role in reorganization of Early Paleozoic ecosystems. *In*: KELLEY, P. H., KOWALEWSKI, M. & HANSEN, T. A. (eds) *Predator–Prey Interactions in the Fossil Record*. Kluwer Academic/Plenum Publishers, New York, 317pp.

BENGTSON, S. & HOU, X.-G. 2001. The integument of Cambrian chancelloriids. *Acta Palaeontologica Polonica*, **46**, 1–22.

BOTTJER, D. J., HAGADORN, J. W. & DORNBOS, S. Q. 2000. The Cambrian substrate revolution. *Geological Society of America Today*, **10**, 1–7.

BRAUN, A., CHEN, J.-Y., MAAS, A. & WALOSZEK, D. 2004. Plankton from Early Cambrian black shale series on the YangTze Platform and its influences on lithologies. *Progress in Natural Science*, **14**, 7–12.

BRIGGS, D. E. G., ERWIN, D. H. & COLLIER, F. J. 1994. *The Fossils of the Burgess Shale*. Smithsonian Institution Press, Washington, DC, 238pp.

BRUSCA, R. C. & BRUSCA, G. J. 2003. *Invertebrates*, 2nd. edn. Sinauer Associates, Sunderland, Mass, 936pp.

BRUTON, D. L. 1981. The arthropod *Sidneyia inexpectans*, Middle Cambrian, Burgess Shale, British Columbia. *Philosophical Transactions of the Royal Society, London*, **B300**, 553–585.

BRYANT, I. D. & PICKERILL, R. K. 1990. Lower Cambrian trace fossils from the Buen Formation of Central North Greenland. Preliminary observations. *Gronlands Geologiske Undersolgelse*, **147**, 44–62.

BUDD, G. E. 1993. A Cambrian gilled lobopod. *Nature*, **364**, 709–711.

BUDD, G. E. 1998. Stem group arthropods from the Lower Cambrian Sirius Passet fauna of Northern Greenland. *In*: FORTEY, R. A. & THOMAS, R. (eds) *Arthropod Relationships*. Systematics Association Special Volume, **55**, Chapman & Hall, London, 125–138.

BUTTERFIELD, N. J. 1994. Burgess Shale-type fossils from the lower Cambrian shallow shelf sequence in northwestern Canada. *Nature*, **369**, 477–479.

BUTTERFIELD, N. J. 2001a. Ecology and Evolution of Cambrian Plankton. *In*: ZHURAVLEV, A. Y. & RIDING, R. (eds) *The Ecology of the Cambrian Radiation*. Columbia University Press, 200–216.

BUTTERFIELD, N. J. 2001b. Cambrian food webs. *In*: BRIGGS, D. E. G. & CROWTHER, P. R. (eds) *Palaeobiology II*, Blackwell Science, London, 40–43.

BUTTERFIELD, N. J. 2002. *Leanchoilia* guts and the interpretation of three-dimensional structure in Burgess Shale-type fossils. *Paleobiology*, **28**, 155–171.

CARON, J. B. 2005. Taphonomy and community analysis of the Middle Cambrian Greater Phyllopod Bed, Burgess Shale. Unpublished PhD dissertation, University of Toronto, Toronto, 316pp.

CARON, J.-B. & JACKSON, D. A. 2006. Taphonomy of the Greater Phyllopod Bed community, Burgess Shale. *Palaios*, **21**, 451–465.

CARON, J.-B., SCHELTEMA, A., SCHANDER, C. & RUDKIN, D. 2006. A soft-bodied mollusc with radula from the Middle Cambrian Burgess Shale. *Nature*, **442**, 159–163.

CHEN, J.-Y. & HUANG, D.-Y. 2002. A possible Lower Cambrian chaetognath (arrow worm). *Science*, **298**, 187.

CHEN, J.-Y., HUANG, D.-Y., PENG, Q.-Q., CHI, H.-M., WANG, X.-Q. & FENG, M. 2003. The first tunicate from the Early Cambrian of South China. *Proceedings of the National Academy of Sciences of the United States of America*, **100**, 8314–8318.

CHEN, L.-Z., LUO, H.-L., HU, S.-X., YIN, J.-Y., JIANG, Z.-W., WU, Z.-L., LI, F. & CHEN, A.-L. 2002. *Early Cambrian Chengjiang Fauna in Eastern Yunnan, China*. Yunnan Science and Technology Press, Kunming.

CHEN, J.-Y., RAMSKÖLD, L. & ZHOU, G.-Q. 1994. Evidence for monophyly and arthropod affinity of Cambrian giant predators. *Science*, **264**, 1304–1308.

CHEN, J.-Y., VANNIER, J. & HUANG, D.-Y. 2001. The origin of crustaceans: new evidence from the Early Cambrian of China. *Proceedings of the Royal Society, London, Biological Sciences*, **1482**, 2181–2187.

CHEN, J.-Y., WALOSZEK, D. & MAAS, A. 2004. A new 'great appendage' arthropod from the lower Cambrian Maotianshan Shale fauna, South China. *Lethaia*, **37**, 1–12.

CHEN, J.-Y. & ZHOU, G.-Q. 1997. Biology of the Chengjiang fauna. *Bulletin of the National Museum of Natural Science*, **10**, 11–105.

CHOE, N. & DEIBEL, D. 2000. Seasonal vertical distribution and population dynamics of the chaetognath *Parasagitta elegans* in the water column and hyperbenthic zone of Conception Bay, Newfoundland. *Marine Biology*, **137**, 847–885.

CLAUS, C. 1890. Die Gattungen und Arten der Mediteranen und atlantischen Halocypriden nebst Bemerkungen über die Organisation derselben. *Arbeiten aus dem zoologischen Institute der Universität Wien und der zoologischen Station in Triest*, **9**, 1–34.

COLLINS, D. 1996. The "evolution" of *Anomalocaris* and its classification in the arthropod class Dinocarida (nov.) and Order Radiodonta (nov.). *Journal of Paleontology*, **70**, 280–293.

CONDON, D., ZHU, M.-Y., BOWRING, S., WANG, W., YANG, A. & JIN, Y.-G. 2005. U-Pb ages from the Neoproterozoic Doushantuo Formation, China. *Science*, **308**, 95–98.

CONWAY MORRIS, S. 1977. Fossil priapulid worms. *Special Papers in Palaeontology*, **20**, 1–95.

CONWAY MORRIS, S. 1986. The community structure of the Middle Cambrian Phyllopod Bed (Burgess Shale). *Palaeontology*, **29**, 423–467.

CONWAY MORRIS, S. 2001. Significance of early shells. *In*: BRIGGS, D. E. G. & CROWTHER, P. R. (eds) *Palaeobiology II*. Blackwell Science, London, 31–40.

CONWAY MORRIS, S. & BENGTSON, S. 1994. Cambrian predators: possible evidence from boreholes. *Journal of Paleontology*, **68**, 1–23.

CONWAY MORRIS, S. & PEEL, J. S. 1990. Articulated halkieriids from the Lower Cambrian of North Greenland. *Nature*, **345**, 802–804.

CONWAY MORRIS, S. & PEEL, J. S. 1995. Articulated halkieriids from the Lower Cambrian of North Greenland and their role in early Cambrian protostome evolution. *Philosophical Transactions of the Royal Society, London*, **B347**, 305–358.

CONWAY MORRIS, S. & ROBISON, R. A. 1988. Middle Cambrian priapulids and other soft-bodied fossils from Utah and Spain. *The University of Kansas Paleontological Contributions*, **117**, 1–22.

DORNBOS, S., BOTTJER, D. & CHEN, J.-Y. 2004. Evidence for seafloor microbial mats and associated metazoan lifestyles in Lower Cambrian phosphorites of Southwest China. *Lethaia*, **37**, 127–137.

DORNBOS, S. Q., BOTTJER, D. J. & CHEN, J.-Y. 2005. Paleoecology of benthic metazoans in the Early Cambrian Maotianshan Shale biota and the Middle Cambrian Burgess Shale biota: evidence for the Cambrian substrate revolution. *Palaeogeography, Palaeoclimatology, Palaeoecology*, **220**, 47–67.

DORNBOS, S. Q. & CHEN, J.-Y. 2006. Community palaeoecology of the Early Cambrian Maotianshan Shale biota: ecologic dominance of priapulid worms. *Abstracts of the 2nd International Palaeontological Congress, Beijing*, 303.

FIKE, D. A., GROTZINGER, J. P., PRATT, L. M. & SUMMONS, R. E. 2006. Oxidation of the Ediacaran Ocean. *Nature*, **444**, 744–747.

GLAESSNER, M. F. 1979. Lower Cambrian Crustacea and annelid worms from Kangaroo Island, South Australia. *Alcheringa*, **3**, 21–31.

HAN, J., SHU, D.-G., ZHANG, Z.-F. & LIU, J.-N. 2004. The earliest-known ancestors of Recent Priapulomorpha from the Early Cambrian Chengjiang Lagerstätte. *Chinese Science Bulletin*, **49**, 1860–1868.

HARPER, E. M., PALMER, T. J. & ALPHEY, J. R. 1997. Evolutionary response by bivalves to changing Phanerozoic sea-water chemistry. *Geological Magazine*, **134**, 403–407.

HOU, X.-G. 1987. Two new arthropods from Lower Cambrian, Chengjiang, eastern Yunnan. *Acta Palaeontologica Sinica*, **26**, 236–256. [In Chinese, with English summary.]

HOU, X.-G. & BERGSTRÖM, J. 1991. The arthropods of the Lower Cambrian Chengjiang fauna, with relationships and evolutionary significance. *In*: SIMONETTA, A. M. & CONWAY MORRIS, S. (eds) *The Early Evolution of Metazoa and the Significance of Problematic Taxa*, 179–187. Cambridge University, Press, Cambridge.

HOU, X.-G., ALDRIDGE, R. J., BERGSTRÖM, I., SIVETER, DAVID, J., SIVETER, DEREK, J. & FENG, X.-H. 2004. *The Cambrian Fossils of Chengjiang, China*. Blackwell Science, Oxford, 233pp.

HOU, X.-G., SIVETER, D. J., WILLIAMS, M. & FENG, X.-H. 2002. A monograph of the bradoriid arthropods from the Lower Cambrian of SW China. *Transactions of the Royal Society of Edinburgh, Earth Sciences*, **92**, 347–409.

HOU, X.-G., SIVETER, D. J., WILLIAMS, M., WALOSZEK, D. & BERGSTRÖM, J. 1996. Preserved appendages in the arthropod *Kunmingella* from the early Cambrian of China: its bearing on the systematic position of the Bradoriida and the fossil record of the Ostracoda. *Philosophical Transactions of the Royal Society*, **B351**, 1131–1145.

HU, S.-X. 2005. Taphonomy and Palaeoecology of the Early Cambrian Chengjiang Biota from Eastern Yunnan, China. *Berliner Paläobiologische Abhandlungen*, **7**, 1–197.

HUA, H., CHEN, Z., YUAN, X.-L., ZHANG, L. & XIAO, S.-H. 2005. Skeletogenesis and asexual reproduction in the earliest biomineralizing animal *Cloudina*. *Geology*, **33**, 277–280.

HUANG, D.-Y. 2005. Early Cambrian worms from SW China: morphology, systematics, lifestyles and evolutionary significance. Unpublished PhD thesis, Université Claude Bernard Lyon 1. 247pp.

HUANG, D.-Y., VANNIER, J. & CHEN, J.-Y. 2004a. Anatomy and lifestyles of Early Cambrian priapulid worms exemplified by *Corynetis* and *Anningella* from the Early Cambrian Maotianshan Shale (SW China). *Lethaia*, **37**, 21–33.

HUANG, D.-Y., VANNIER, J. & CHEN, J.-Y. 2004b. Recent Priapulidae and their early Cambrian ancestors: comparisons and evolutionary significance. *Geobios*, **37**, 217–228.

HUANG, D.-Y., ZHU, M.-Y., VANNIER, J. & CHEN, J.-Y. 2004c. Priapulid worms are the possible trace makers of the Early Cambrian *Trichophycus pedum*. 48th Annual Meeting of the Palaeontological Association, Lille.

HUANG, D.-Y., CHEN, J.-Y., VANNIER, J. & SAIZ-SALINAS, J. I. 2004d. Early Cambrian Sipunculan worms from southwest China. *Proceedings of the Royal Society London*, **B271**, 1671–1676.

HUO, S.-C., SHU, D.-G. & CUI, Z.-L. 1991. *Cambrian Bradoriida of China*. Geological Publishing House, Beijing. 249pp.

IVANTSOV, A. Y., ZHURAVLEV, A. Y., LEGUTA, A. V., KRASSILOV, V. A., MELNIKOVA, L. M. & USHATINSKAYA, G. T. 2005. Palaeoecology of the Early Cambrian Sinsk biota from the Siberian Platform. *Palaeogeography, Palaeoclimatology, Palaeoecology*, **220**, 69–88.

JANUSSEN, D., STEINER, M. & ZHU, M.-Y. 2002. New well-preserved scleritomes of Chancelloriidae from the Early Cambrian Yu'anshan Formation (Chengjiang, China) and the Middle Cambrian Wheeler Shale (Utah, USA) and paleobiological implications. *Journal of Paleontology*, **76**, 596–606.

JENSEN, S. 1997. Trace fossils from the Lower Cambrian *Mickwitzia* sandstone, south Central Sweden. *Fossils and Strata*, **42**, 1–111.

KAUFMAN, A. J., KNOLL, A. H. & NARBONNE, G. M. 1997. Isotopes, ice ages and terminal Proterozoic earth history. *Proceedings of the National Academy of Sciences, USA*, **94**, 6600–6605.

KNOLL, A. H. & CARROLL, S. B. 1999. Early animal evolution: emerging views from comparative biology and geology. *Science*, **284**, 2129–2137.

MAAS, A., WALOSZEK, D., CHEN, J.-Y., BRAUN, A., WANG, X.-Q. & HUANG, D.-Y. 2004. Phylogeny and life habits of Early arthropods – Predation in the

Early Cambrian Sea. *Progress in Natural Science* (Special Issue), 124–132.

MAAS, A., WALOSZEK, D. & MÜLLER, K. J. 2003. Morphology, ontogeny and phylogeny of the Phosphatocopina (Crustacea) from the Upper Cambrian 'Orsten' of Sweden. *Fossils and Strata*, **49**, 1–238.

MANSUY, H. 1912. Partie 2, Paléontologie. *In*: DEPRAT, J. & MANSUY, H. E'tude géologique du Yun-Nan oriental. *Mémoire du Service Géologique de l'Indochine*, **1**, 1–146.

MISSARZHEVSKY, V. V. 1973. Conodont-shaped organisms from Precambrian–Cambrian boundary beds of the Siberian Platform and Kazakhstan. *In*: ZHURAVLEVA, I. T. (ed.) *Problemy Paleontologii i Biostratygrafii nizhnego kembriya Sibiri i Dal'nego Vostoka.* Trudy Instituta Geologii i Geofiziki SO ANSSSR, **49**, 53–59.

MYROW, P. M. & KAUFMAN, A. J. 1999. A newly discovered cap carbonate above Varanger-age glacial deposits in Newfoundland. *Journal of Sedimentary Research*, **69**, 784–793.

NARBONNE, G. M., MYROW, P. M., LANDING, E. & ANDERSON, M. M. 1987. A candidate stratotype for the Precambrian-Cambrian boundary, Fortune Head, Burin Peninsula, southeastern Newfoundland. *Canadian Journal of Earth Sciences*, **34**, 1277–1293.

NEDIN, C. 1999. *Anomalocaris* predation on nonmineralized and mineralized trilobites. *Geology*, **27**, 987–990.

PARKER, A. 2003. In the blink of an eye: how vision kickstarted the big bang of evolution. Free Press, 316pp.

PONAMARENKO, A. G. (ed.) 2005. Unique Sinsk localities of Early Cambrian organisms (Siberian Platform). *Transactions of the Palaeontological Institute, NAUKA, Moscow*, **284**, 1–143.

RIGBY, S. & MILSOM, C. 1996. Benthic origin of zooplankton: an environmentally determined macroevolutionary effect. *Geology*, **24**, 52–54.

SEILACHER, A., BUATOIS, L. & MANGANO, L. G. 2005. Trace fossils in the Ediacaran–Cambrian transition: behavioral diversification, ecological turnover and environmental shift. *Palaeogeography, Palaeoclimatology, Palaeoecology*, **227**, 323–356.

SHINN, G. L. 1997. Chaetognatha. *In*: HARRISON, F. W. & RUPPERT, E. E. (eds) *Microscopic Anatomy of Invertebrates*, Volume 15: Hemichordata, Chaetognatha and the invertebrate chordates, Wiley-Liss, New York, 103–220.

SHU, D.-G., VANNIER, J., LUO, H.-L., CHEN, L.-Z., ZHANG, X.-L. & HU, S.-X. 1999. Anatomy and lifestyle of *Kunmingella* (Arthropoda, Bradoriida) from the Chengjiang fossil Lagerstätte (lower Cambrian, Southwest China). *Lethaia*, **32**, 279–298.

SIVETER, D. J., WILLIAMS, M. & WALOSZEK, D. 2001. A phosphatocopid crustacean with appendages from the Lower Cambrian. *Science*, **293**, 479–481.

STEINER, M., LI, G.-X., QIAN, Y. & ZHU, M.-Y. 2004. Lower Cambrian Small Shelly Fossils of northern Sichuan and southern Shaanxi (China), and their biostratigraphic importance. *Geobios*, **37**, 259–275.

STEINER, M., MEHL, D., REITNER, J. & ERDTMANN, B.-D. 1993. Oldest entirely preserved sponges and other fossils from the lowermost Cambrian and a new facies reconstruction of the Yangtze Platform (China). *Berliner Geowissenschaftlichen Abhandlungen*, **9**, 293–329.

STEINER, M., WALLIS, E., ERDTMANN, B.-D., ZHAO, Y.-L. & YANG, R.-D. 2001. Submarine-hydrothermal exhalative ore layers in black shales from South China and associated fossils–insights into a Lower Cambrian facies and bio-evolution. *Palaeogeography, Palaeoclimatology, Palaeoecology*, **169**, 165–191.

SUN, W.-G. & HOU, X.-G. 1987. Early Cambrian medusae from Chengjiang, Yunnan, China. *Acta Palaeontologica Sinica*, **26**, 300–305. [In Chinese, with English summary.]

SZANIAWSKI, H. 2002. New evidence for the protoconodont origin of chaetognaths. *Acta Palaeontologica Polonica*, **47**, 405–419.

VANNIER, J., ABE, K. & IKUTA, K. 1998. Feeding in myodocopid ostracods: functional morphology and laboratory observations from videos. *Marine Biology*, **132**, 391–408.

VANNIER, J., BOISSY, P. & RACHEBOEUF, P. R. 1997. Locomotion in *Nebalia bipes*: a model for Palaeozoic phyllocarid crustaceans. *Lethaia*, **30**, 89–104.

VANNIER, J. & CHEN, J.-Y. 2001. The early Cambrian colonization of pelagic niches exemplified by *Isoxys* (Arthropoda). *Lethaia*, **33**, 295–311.

VANNIER, J. & CHEN, J.-Y. 2002. Digestive system and feeding mode in Cambrian naraoiid arthropods. *Lethaia*, **35**, 107–120.

VANNIER, J. & CHEN, J.-Y. 2005. Early Cambrian food chain: New evidence from fossil aggregates in the Maotianshan Shale Biota, SW China. *Palaios*, **20**, 3–26.

VANNIER, J., RACHEBOEUF, P., BRUSSA, E., WILLIAMS, M., RUSHTON, A. W. A., SERVAIS, TH., & SIVETER, D. 2003. Cosmopolitan arthropod zooplankton in the Ordovician seas. *Palaeogeography, Palaeoclimatology, Palaeoecology*, **195**, 173–191.

VANNIER, J., STEINER, M., RENVOISÉ, E., HU, S.-X. & CASANOVA, J.-P. 2005. Arrow worms: small marine predators from 'deep time'. *Acta Micropalaeontologica Sinica*, **22** (Supplement), 189–190.

VANNIER, J., CHEN, J.-Y., HUANG, D.-Y., CHARBONNIER, S. & WANG, X.-Q. 2006. Thylacocephalan arthropods: their Early Cambrian origin and evolutionary significance. *Acta Paleontologica Polonica*, **51**, 1–14.

VANNIER, J., STEINER, M., RENVOISÉ, E., HU, S.-X. & CASANOVA, J.-P. 2007. Early Cambrian origin of modern food webs: evidence from predator arrow worms. *Proceedings of the Royal Society London, Biological Sciences*.

WALOSZEK, D. 1993. The Upper Cambrian *Rehbachiella* and the phylogeny of Brachiopoda and Crustacea. *Fossils and Strata*, **32**, 1–202.

WHITTINGTON, H. B. & BRIGGS, D. E. G. 1985. The largest Cambrian animal, *Anomalocaris, Burgess Shale, British Columbia. Philosophical Transactions of the Royal Society London*, **B309**, 569–609.

WILLIAMS, M., SIVETER, D. J., POPOV, L. E. & VANNIER, J. M. C. 2007. Biogeography and affinities of the bradoriid arthropods: cosmopolitan microbenthos of the Cambrian seas. *Palaeogeography, Palaeoclimatology, Palaeoecology*, **248**, 202–232.

XIAO, S. H., ZHANG, Y. & KNOLL, A. H. 1998. Three-dimensional preservation of algae and animal embryos in a Neoproterozoic phosphorite. *Nature*, **391**, 553–558.

YANG, X. & HE, T. 1984. New Small Shelly Fossils from Lower Cambrian Meishucun Stage of Nanjiang Area, northern Sichuan. *Professional Papers in Stratigraphy and Palaeontology*, **13**, 35–47.

YUAN, X.-L., XIAO, S.-H., PARSLEY, R. L., ZHOU, C.-M., CHEN, Z. & HU, J. 2002. Towering sponges in an Early Cambrian Lagerstätte: disparity between nonbilaterian and bilaterian epifaunal tierers at the Neoproterozoic–Cambrian transition. *Geology*, **30**, 363–366.

ZHANG, W.-T. & HOU, X.-G. 1985. Preliminary notes on the occurrence of the unusual trilobite *Naraoia* in Asia. *Acta Palaeontologica Sinica*, **24**, 591–595. [In Chinese, with English summary.]

ZHANG, X.-G., HOU, X.-G. & BERGSTRÖM, J. 2006. Early Cambrian priapulid worms buried with their lined burrows. *Geological Magazine*, **143**, 743–748.

ZHANG, X.-L., HAN, J., ZHANG, Z.-F., LIU, H.-Q. & SHU, D.-G. 2003. Reconsideration of the supposed naraoiid larva from the Early Cambrian Chengjiang Lagerstätte, South China. *Palaeontology*, **46**, 447–465.

ZHAO, F.-C., ZHU, M.-Y. & CARON, J.-B. 2006. Taphonomy and community of the Early Cambrian Chengjiang Lagerstätte from the Mafang Section, SW China. *Abstracts of the 2nd International Palaeontological Congress, Beijing*, 313–314.

ZHU, M.-Y. 1997. Precambrian–Cambrian trace fossils from Eastern Yunnan, China: implications for Cambrian explosion. *Bulletin of the National Museum of Natural Science*, **10**, 275–312.

ZHU, M.-Y., VANNIER, J., VAN ITEN, H. & ZHAO, Y.-L. 2004. Direct evidence for predation on trilobites in the Cambrian. *Biology Letters*, **271**, 277–280.

ZHU, M.-Y., ZHANG, J.-M., STEINER, M., YANG, A., LI, G.-X. & ERDTMANN, B. D. 2004. Sinian–Cambrian stratigraphic framework for shallow- to deep-water environments of the YangTze Platform: an integrated approach. *Progress in Natural Science Special Issue*, 75–84.

ZHU, M.-Y., ZHAO, Y.-L. & CHEN, J.-Y. 2002. Revision of the Cambrian medusiform animals *Stellostomites eumorphus* Sun and Hou and *Pararotadiscus guizhouensis* (Zhao and Zhu) from South China. *Geobios*, **35**, 165–185.

On the cause of the Ordovician glaciation

H. A. ARMSTRONG

*Durham University, Department of Earth Sciences, Palaeozoic Environments Group,
South Road, Durham DH1 3LE, UK (e-mail: h.a.armstrong@durham.ac.uk)*

Abstract: Commonality of patterns and processes, identified from geological proxy data, occurs in the sequence of events leading to Cenozoic and Ordovician glaciations. Both glaciations were set against a backdrop of long-term declining pCO_2 likely initiated by changes in plate configuration that resulted in increased weathering and nutrient cycling into the oceans. Rapid expansion of ice volume was triggered by the redirection of warm, circumequatorial currents into high latitudes to provide a source of warm moist air and high levels of snowfall. Once ice-sheets were large enough to survive successive precession and obliquity maxima, eccentricity pacing of ice-margin processes embedded in obliquity and the monsoonal climate system embedded in precession largely controlled their size. Changes in family and generic diversity in conodonts, ostracods and graptolites reflect a longer glaciation scenario for the Ordovician. The rising diversity trajectory to the Llanvirn was terminated in the Caradoc. This was followed by a slight rise in diversity in all groups into the Ashgill as taxa adapted to the new environmental conditions. A decline in diversity, of varying severity, into the Llandovery reflects the impact of mass extinction. Commonality in the sequence of events and pattern of environmental change leads to the rejection of the Ordovician glaciation being unique in Earth history.

For much of the Phanerozoic, the Earth experienced a mild greenhouse climate, punctuated by severe glaciations, resulting in fundamental re-organization of the Earth's ocean-climate system. The glaciation at the end of the Ordovician was closely associated with the second largest of the three great Phanerozoic mass extinction events (Bambach *et al.* 2004; Sepkoski Jr 1996) and was thus an agent of major change in global biodiversity.

The Ordovician world was significantly different from that in the Cenozoic (Barnes 2004*a*): the land was largely devoid of vegetation, atmospheric oxygen levels were *c.* 50% present atmospheric levels (PAL) (Berner 2001). Ordovician sea levels were the highest in the Phanerozoic (Frakes *et al.* 1992; Hallam 1992) and large areas of the continents were submerged. Greenhouse climate conditions were maintained by higher (8 to 18×PAL) pCO_2 levels (Berner 1992, 1994; Yapp & Poths 1992; Berner & Kothavala 2001). Though there is some doubt over these very high estimates (Boucot & Gray 2001; Herrmann *et al.* 2003, 2004). These are not the conditions that typify Cenozoic glaciation and this has led some workers to view the Ordovician glaciation as short, intense and 'unique in Earth history'. No consensus exists as to the cause of this glaciation.

The focus of this contribution is to reject the axiom of uniqueness and apply the principle of uniformitarianism ('the present is the key to the past') to assess, within the limits of the geological data, the extent to which explanatory hypotheses for late Cenozoic glaciation can be used to understand the end-Ordovician glaciation.

Stratigraphical evidence for Ordovician glaciation

Recognizing the timing of onset and duration of a glaciation is critical to elucidating any causal hypothesis. A glaciation (onset of cooling to Termination) and glacial maximum (peak glaciation) are identified from proxy data for the Ordovician event. The Guttenburg Positive Carbon Isotope Excursion (GICE) during the Chatfieldian (mid-Caradoc; Fig. 1) has been proposed as the start of the Ordovician glaciation approximately 10 Ma before the Hirnantian glacial maximum (Saltzman 2005; Saltzman & Young 2005; Tobin *et al.* 2005).

However, direct evidence of glaciation from high palaeolatitudes is poorly constrained and the early record is largely incomplete. Glaciogenic deposits are known from the late Caradoc of Bolivia, Peru and Argentina (Crowell *et al.* 1980), Morocco (Hamoumi 1999) and the Sahara (Beuf *et al.* 1977; Biju-Duval *et al.* 1981; Legrand 1988, 1993, 1995) and are used as evidence for a Caradoc to Wenlock glaciation (Frakes *et al.* 1992); though some of the earlier records have been redated as late Ordovician (Paris *et al.* 1995). In the Argentine Precordillera, there are at least three separate glacial advances that appear to pre-date the *Hirnantia* Fauna and the coeval $\delta^{13}C$ isotope excursion (Astini 1999; Peralta & Carter 1999; Sheehan 2001). In South Africa, the Pakhuis Tillite is overlain by the Soom Shale, which in turn is overlain by the Hirnantian-aged Disa Siltstone Member (of the Cedarberg Formation) which contains a cold-water

From: WILLIAMS, M., HAYWOOD, A. M., GREGORY, F. J. & SCHMIDT, D. N. (eds) *Deep-Time Perspectives on Climate Change: Marrying the Signal from Computer Models and Biological Proxies.* The Micropalaeontological Society, Special Publications. The Geological Society, London, 101–121.

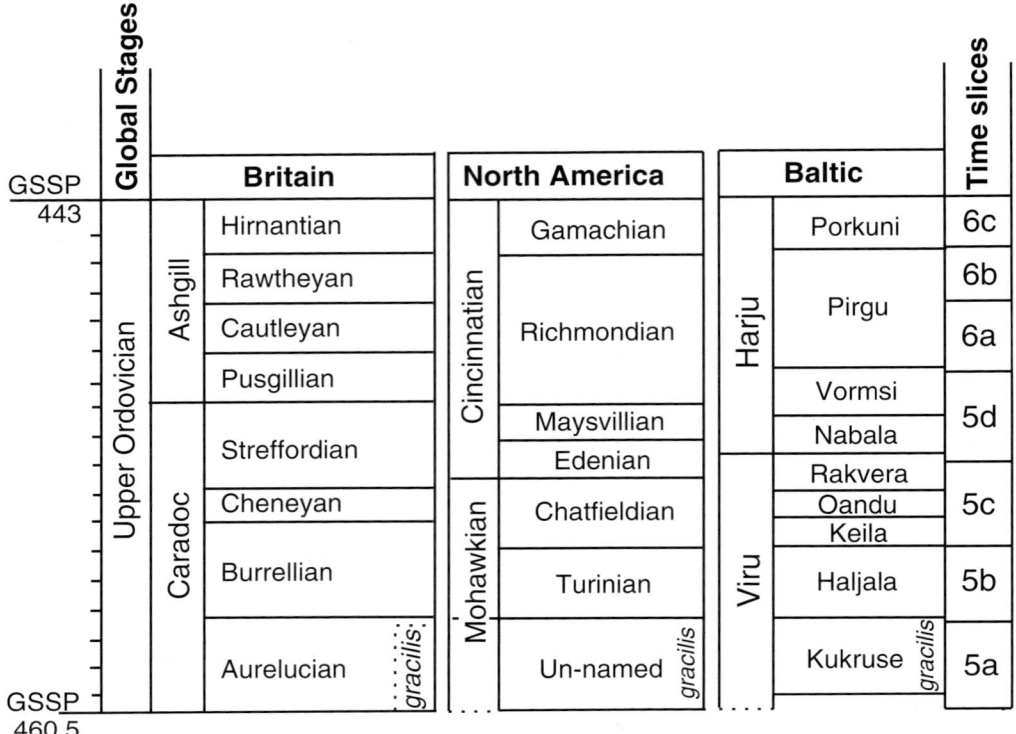

Fig 1. Upper Ordovician chronostratigraphy showing the correlation of schemes for Britain, North America and Balto-Scandinavia. Time slices and radiometric timescale are as defined in Webby *et al.* 2004.

Mediterranean Province fauna. The Pakhuis Tillite has been correlated with other pre-Hirnantian tillites in North Africa (Sutcliffe *et al.* 2000).

The best-constrained proxy records of glaciation are from 'far field' sections in low latitudes. Prior to the Hirnantian glacial maximum, deep-sea fan deposits, such as at Dob's Linn, Scotland, were dominated by anoxic black shales. Ashgill deposits at Dob's Linn are essentially oxic mudstones, indicating ventilation at least to bathyal water depths and a thermally stratified Iapetus Ocean (Armstrong & Coe 1997). The deepest parts of the oceans could have remained anoxic (Brenchley *et al.* 1994; Armstrong & Coe 1997).

The prominent HICE carbon isotopic excursion (Underwood *et al.* 1997) is regarded as coincident with the glacial maximum (Brenchley *et al.* 1994). Recent revision of the graptolite biostratigraphy at Dob's Linn (Melchin *et al.* 2003) places the base of the isotope excursion in the *pacificus* Sub-biozone (late *anceps* Biozone; Rawtheyan) as predicted from sedimentary analysis (Armstrong & Coe 1997). The glacial maximum lies within the *persculptus* Biozone (Hirnantian; Fig. 2). In many other low latitude areas, a major unconformity occurs in shallow

marine settings, coincident with the Hirnantian glacial maximum (Brenchley *et al.* 1994). Sequence stratigraphical analysis of low latitude peritidal carbonates (e.g. Armstrong & Lane 1981) and ice-proximal sedimentary records suggests at least four glacial advances (Moreau *et al.* 2004; Turner *et al.* 2005).

Defining the end of the glaciation proves more difficult. The start of the post-glacial transgression (the end of the glacial maximum) is a global eustatic event marking a return of anoxic black shale deposition (Armstrong *et al.* 2005; Sutcliffe *et al.* 2000). At Dob's Linn, the base of the black Birkhill Shales occurs within the mid-*persculptus* Biozone. However, glacial deposits are known into the Silurian (Frakes *et al.* 1992; Grahn & Caputo 1992). Biotic cycles in the early Silurian that correlate broadly with postulated sea-level changes, are explained by a model that involves episodic changes in oceanic state (Aldridge *et al.* 1993). Primo episodes were characterized by cool high-latitude climates, cold oceanic bottom waters, and high nutrient supply which supported abundant and diverse planktonic communities. Secundo episodes were characterized by warmer high-latitude climates, salinity-dense oceanic bottom waters, low diversity

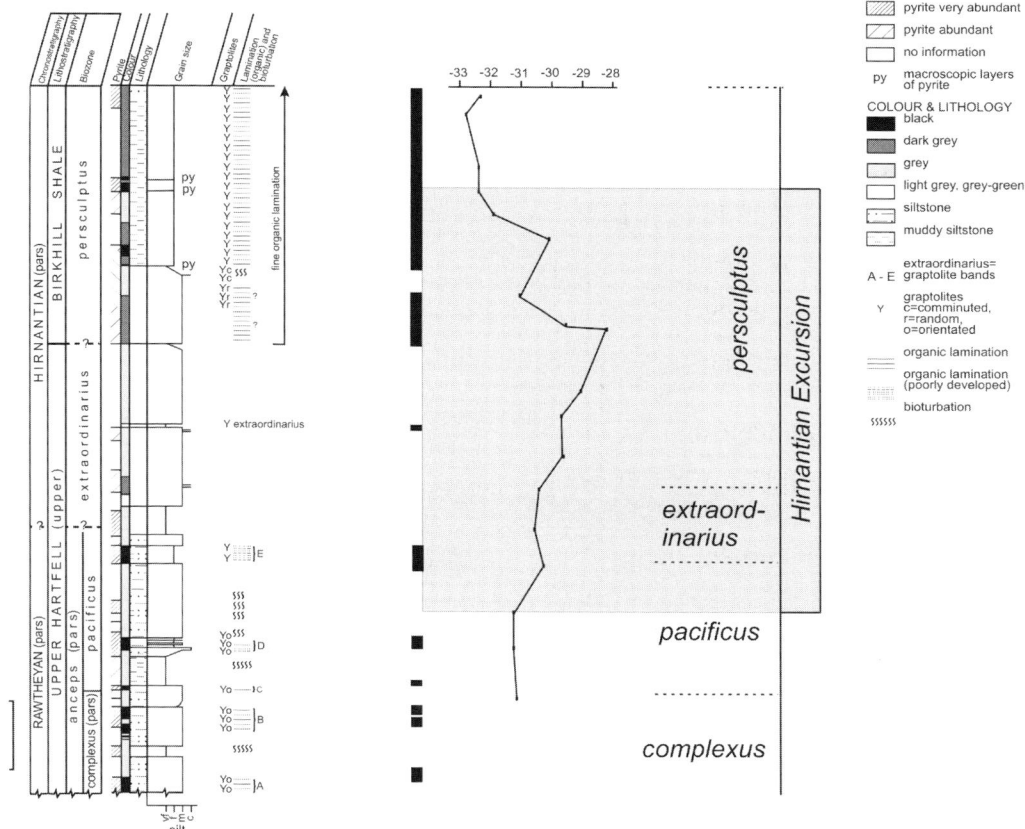

Fig. 2. Composite lithostratigraphy, biostratigraphy and chemostratigraphy of the Upper Hartfell and lower Birkhill Shales at Dob's Linn (GSSP for the base of the Silurian). Sedimentary log after Armstrong & Coe 1997, δ^{13}C curve from Underwood *et al.* 1997 and graptolite zonation from Melchin *et al.* 2003. The revised biostratigraphy now places the base of the *persculptus* Biozone below the *extraordinarius* band and the *anceps* bands A to D are pre-isotope excursion and hence pre-glacial maximum. The return of basin anoxia at the base of the Birkhill Shale in the mid-*persculptus* Biozone coincides with the shift to more negative δ^{13}C values and the end of the glacial maximum.

planktonic communities, and carbonate formation in shallow waters. The pattern of turnover shown by conodont faunas, together with sedimentological information and data from other fossil groups, permit the identification of two cycles in the Llandovery to earliest Wenlock interval (Aldridge *et al.* 1993). Primo episodes, with an approximate periodicity of 2.5 Ma, likely correlate with orbitally forced ice re-advances at high latitudes, suggesting the glaciation continued to the Wenlock (Page *et al.* 2007).

Causative hypotheses for Ordovician glaciation

To date, a longer Ordovician glacial scenario has not been considered in the development of causative hypotheses, and computer simulations have largely focused on the Hirnantian glacial maximum. Two general hypotheses have emerged to explain the Hirnantian glacial maximum.

A 'Monterey mechanism' of increased primary marine productivity and organic carbon burial, stimulated by newly available nutrients brought to the surface by the initiation of the thermohaline circulation (THC; Berry & Wilde 1978; Wilde 1991). Evidence of coastal upwelling along the southern margin of Laurentia, typically a chert-phosphate association, is found in shelfal settings, from the Chatfieldian and has been cited as evidence for the existence of vigorous THC (Pope & Steffen 2003). This is inconsistent with the evidence for the initiation of the THC in the Ashgill and alternately these deposits reflect shallow coastal upwelling generated by offshore winds. Climate modelling indicates much of southern Laurentia lay beneath the zone of northeasterly trade winds during the Upper Ordovician, where offshore winds would have predominated (Parrish 1982; Herrmann *et al.* 2004, 2005). Cold surface ocean currents can also feed cold, nutrient-rich waters into low palaeo-latitudes. Computer simulations, incorporating Caradoc palaeogeography and high sea levels, indicate the southwest and southerly margin of Laurentia was affected by the cold South Gondwana Current (Herrmann *et al.* 2004; Fig. 3).

The 'Monterey hypothesis' also predicts large-scale carbon burial to drive a fall in pCO_2

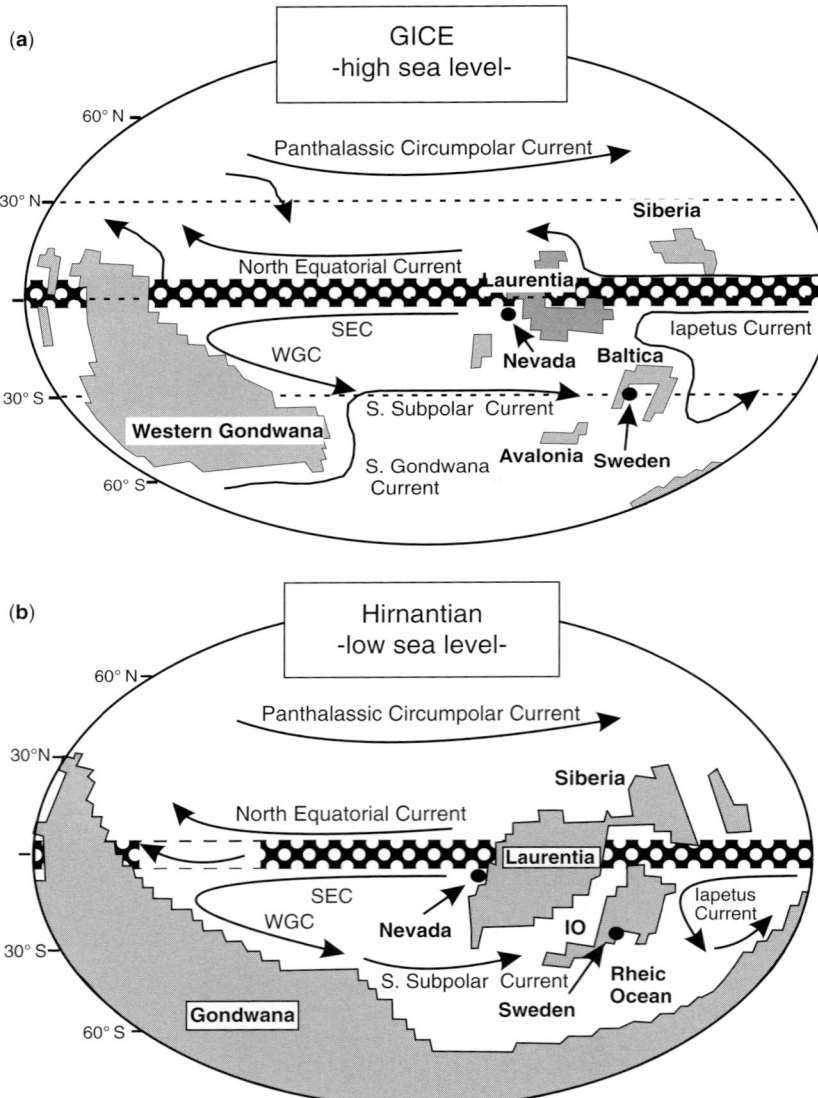

Fig. 3. Palaeogeographical reconstructions and surface ocean circulation patterns for the Early and Late Ordovician (simplified from Herrmann *et al.* 2005). Sketch maps showing palaeogeography, palaeoceanography and ITCZ position during the early part of the Late Ordovician (3a, Caradoc; 3b Ashgill). Maps are simplified from OGCM model outputs (from Herrmann *et al.* 2005). Current patterns are generalized to bold arrows centred on currents with velocities >5 cm sec^{-1}; the dotted zone is the ITCZ. Abbreviations: SEC South Equatorial Current; WGC West Gondwana Current (new name proposed to avoid confusion with Antarctica which did not come into existence until the Cretaceous); IC Iapetus Current; IO Iapetus Ocean.

(Vincent & Berger 1985). During the Ashgill there is no evidence, at least to bathyl depths, of widespread deposition of organic carbon, only short periods of black shale deposition and ocean anoxia (Armstrong & Coe 1997; fig. 1). $\delta^{13}C$ values from central Nevada suggest rising pCO_2 during the Hirnantian glacial maximum (Kump *et al.* 1995, 1999).

The silicate weathering hypothesis was proposed to counter these problems (Kump *et al.* 1999). In this hypothesis, the rate of silicate weathering is the dominant control on pCO_2. As ice-sheets build and spread across Gondwana, they cover areas previously undergoing silicate weathering. Global silicate weathering rates decline and pCO_2 rises, providing the positive feedback that moderated climate and finally bringing the glaciation to an end. This provides an explanation for rising atmospheric pCO_2 levels during the glacial maximum, but this phenomenon could be equally well explained by increased exposure and hydrolysis of low latitude carbonate platforms during the glacial lowstand; a process occurring in the absence of calcifying plankton and with a declining, calcifying shelf benthos during mass extinction. The net result would have been an increase in ocean and atmosphere $CO_2.$

At the present day, only hydrolysis affects atmospheric CO_2 balance and this would have been much reduced by the low temperatures during the Ashgill. A further criticism of the silicate weathering hypothesis has also been raised: the hypothesis is only based on the observation of rising atmospheric pCO_2 levels in a single section and needs to be corroborated (Sheehan 2001). Long-term (starting at GICE), more positive trends in stable isotopes of carbon and oxygen from Nevada suggest increasing global marine productivity and cooling. These trends are consistent with seawater $^{87}Sr/^{86}Sr$ values, used as a proxy for continental erosion. These values decreased from 0.709 to 0.7078 during the early Caradoc but then rose from the mid-Caradoc into the Silurian (Fig. 4). Both the proposed hypotheses for late Ordovician glaciation therefore remain plausible but flawed and no glacial scenario has yet been developed for a longer glaciation.

Cenozoic glaciations – key stages and hypotheses for understanding the Ordovician

Stage 1: onset of Antarctic glaciation

The established paradigm for the sudden and widespread increase in ice volume in Antarctica and

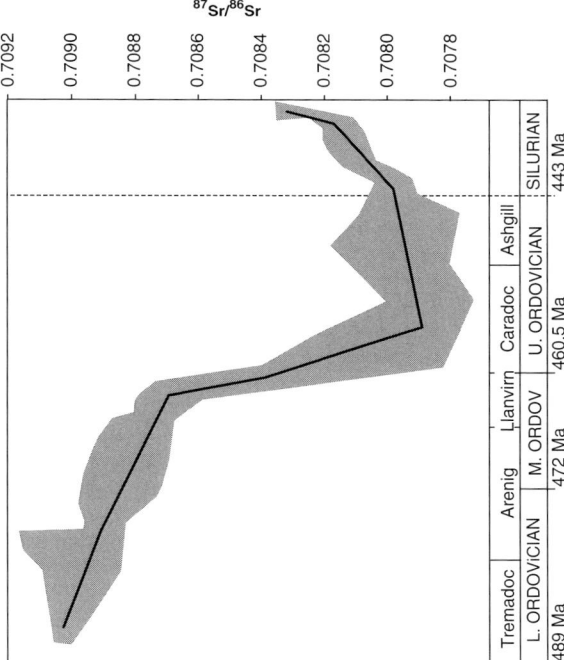

Fig. 4. Strontium isotope variations in seawater during the Ordovician incorporating all currently available, well-preserved brachiopod and carbonate component data (redrawn from Shields *et al.* 2003). Envelope corresponds to a best case overall reproducibility of 25 ± 10^{-6} (Veizer *et al.* 1999; Ebneth *et al.* 2001).

the associated shift toward colder temperatures near the Eocene–Oliocene boundary (*c.* 34 Ma) centres on the opening of Southern Ocean gateways, the initiation of the Antarctic Circumpolar Current and the thermal isolation of the continent (Kennett 1977). However, global climate model (GCM) studies, using prescribed sea-surface temperatures (SSTs) to represent open versus closed Drake Passage conditions, show a modest role for changes in ocean heat transport in the formation of the Antarctic ice-sheet, and suggest the possibility that warmer SSTs are more favourable for ice-sheet growth by increasing snowfall in the continental interior (Oglesby 1989; Prentice & Mathews 1991). Instead, a coupled global climate–dynamical ice-sheet model, accounting for palaeogeography, greenhouse gas concentrations, changing orbital parameters and varying ocean heat transport simulations, shows that declining Cenozoic pCO_2 played a dominant role in causing the glaciation (DeConto & Pollard 2003). Estimates of Early Cenozoic pCO_2 are between 2 and $5 \times$ PAL, declining until reaching near-modern values in the early Neogene (Pearson & Palmer 2000).

The DeConto and Pollard model shows the progressive enlargement of the East Antarctic Ice Sheet (EAIS) that at first comprised small, isolated ice-caps rapidly formed on the highest Antarctic plateaus. As pCO_2 declined and the Earth cooled, these isolated ice-caps rapidly merged into permanent ice-sheets (DeConto & Pollard 2003). Paleogene Antarctic ice-sheets are viewed as being temperate (Barrett *et al.* 1987; Denton *et al.* 1991; Hambrey *et al.* 1991; Zachos *et al.* 1992) and paced by Milankovitch orbital parameters (Zachos *et al.* 1996; Naish *et al.* 2001) in much the same way as the Late Cenozoic ice-sheets of the Northern Hemisphere.

At pCO_2 levels between *c.* $3 \times$ and $2 \times$ PAL, height mass balance feedbacks were initiated during orbital periods (high eccentricity, low obliquity and aphelion during austral summer) producing cold austral summers and triggering much larger, highly dynamic terrestrial ice-sheets (Birchfield *et al.* 1982; Abe-Ouchi & Blatter 1993; Crowley *et al.* 1994; Macqueda *et al.* 1998). Within a single 40-ka obliquity cycle, ice-sheets reached sea level around most of the continental margin of Antarctica with the potential for producing ice-rafted debris (DeConto & Pollard 2003), significantly enhancing THC and impacting ice-margin processes (Ruddiman 2003). Two questions arise from this model: what were the causes of declining pCO_2 during the Cenozoic, and what sequence of events led to increased snowfall at high southerly latitudes necessary to trigger ice-sheet growth?

An erratic decline in pCO_2 occurred between 55 and 40 Ma and may have been caused by reduced

CO_2 outgassing from ocean ridges, volcanoes and metamorphic belts and increased carbon burial (Pearson & Palmer 2000). Since the early Miocene (about *c.* 24 Ma), pCO_2 levels appear to have remained below 500 ppm and were more stable than before, although transient intervals of CO_2 reduction may have occurred during periods of rapid cooling during the mid-Miocene and Pliocene (Pearson & Palmer 2000). Both these later events have been associated with the formation and/or intensification of the Northern Hemisphere glaciation.

Stage 2: Mid-Miocene cooling, expansion of the EAIS

The mid-Miocene Monterey carbon isotope excursion remains an enigmatic event. The base of the *c.* 3‰ shift in $\delta^{13}C$ coincides with a positive shift in Sr isotope values, correlated with the rise of the Himalayas (particularly Tibet) to present-day elevation (Garizone *et al.* 2000; Dettman *et al.* 2003; Spicer *et al.* 2003). Global cooling, recorded in the marine $\delta^{18}O$ record, started *c.* 2 million years after the start of the Monterey excursion (Pagani *et al.* 1999). The most likely scenario to explain this event is that the increased supply of nutrients to the oceans from the Himalayas (Clift & Gaedicke 2002; Clift *et al.* 2002) forced a carbon cycle response leading to drawdown of CO_2 into the oceans and geosphere that eventually produced a negative greenhouse effect and cooling. This event was short-lived and by the late Miocene $\delta^{13}C$ and $\delta^{18}O$ had recovered to pre-event levels.

Stage 3: Onset of Laurentide glaciation

Models for the initiation of the Northern Hemisphere glaciation (*c.* 4 Ma–present) provide insight into the question of increasing supply of moisture to high latitudes and increased precipitation. This is encapsulated in the 'snow gun hypothesis' (originally proposed for the Pacific: Haug & Tiedemann 1998; Billups 2005; Haug *et al.* 2005). Current hypotheses include:

(a) The closure of the Central Atlantic Seaway at the Isthmus of Panama isolated the North Atlantic circulation from the Pacific. This was shortly followed by a change in the mean latitude of the Atlantic Intertropical Convergence Zone (ITCZ; Peterson & Haug 2006). Prior to 4.4 Ma, a northern position of the ITCZ limited warm surface-water advection into the subtropical region of the Atlantic. Instead, warm water flow was directed eastwards in the North Equatorial Counter Current, reducing the poleward heat flux (Richardson & Reverdin 1987). By 4.3 Ma, the ITCZ had shifted south (i.e. equatorially) diverting warm water into the subtropics via the North Brazil Coastal Current. Poleward oceanic transport of warm salty surface water increased snowfall at high northern latitudes (the 'snow gun hypothesis' (Billups 2005; Haug *et al.* 2005) and increased the rate of North Atlantic deep-water formation thus increasing THC (Billups *et al.* 1999).

(b) Downwind changes in Sub-arctic Pacific Ocean stratification (Haug *et al.* 2005) resulting from increasing seasonal temperature contrast. Warmer summer SSTs result in a warmer atmosphere that can hold more moisture and this is blown onto continental North America, falling as snow.

Stage 4: The Mid-Pleistocene Transition

The Laurentide ice-sheet reached a critical size at the Mid-Pleistocene Transition (*c.* 1 million years ago) when ice volume became more sensitive to orbital forcing. Eccentricity modulation of ice-margin processes (entrained by obliquity) and the monsoon system and boreal wetlands (entrained by precession) forced changes in pCO_2 at a 100 000-year timescale. These feedbacks drove the system deeper into glaciation, and initiated the glacial-to-interglacial cycles, quasi-persistent until today (Ruddiman 2003).

Evidence of climate change during the Caradoc and Ashgill

Geological proxy data provide the opportunity to identify this sequence of stages for the Ordovician glaciation.

Stage 1

The exact timing of the onset of glaciation during the Ordovician is difficult to constrain. Stable isotope evidence provides the key proxy data for understanding the pattern of climate change and carbon cycle dynamics. The $\delta^{13}C$ and $\delta^{18}O$ stable isotopic record from Nevada suggests stable greenhouse climates until sustained cooling was initiated at the GICE (Chatfieldian, mid-Caradoc; Fig. 5). Energy balance models (Crowley & Baum 1995) and atmospheric circulation models (Gibbs & Kump 1994; Crowley & Baum 1995; Gibbs *et al.* 1997, 2000), assuming high levels of atmospheric pCO_2, have revealed that the accumulation of late Ordovician permanent polar snow was highly sensitive to atmospheric pCO_2. The ocean general circulation model for the Caradoc and Ashgill indicates that polar temperatures would have been low enough to sustain ice-sheets once pCO_2 levels fell to $8 \times$ PAL and that falling sea level reinforced this possibility (Hermann *et al.* 2005). GICE

coincides with the onset of low latitude coastal upwelling (Pope & Steffen 2003) and an increase in frequency of glacial deposits (Frakes *et al.* 1992). In most areas of the world, GICE appears to correlate with regressive or lowstand deposits (Saltzman & Young 2005; Tobin *et al.* 2005) and the appearance of cold oceanic water in the epeiric sea of Laurentia (Kolata *et al.* 2001). These coincidences imply that a pCO_2 threshold was crossed at GICE that allowed the formation of a polar ice-sheet.

Stage 2: GICE as an equivalent to the Monterey Event

The mid-Caradoc is a critical period of Earth system change. $^{87}Sr/^{86}Sr$ ratios show an intriguing switch to a positive trend that is sustained into the Silurian (Fig. 4), GICE indicates a major perturbation in the carbon cycle and marks the onset of a long-term cooling trend (see below); events comparable to those during the mid-Miocene.

The high fidelity stable oxygen and carbon isotope record (Cambrian to Silurian) from central Nevada (Saltzman 2005; Saltzman & Young 2005; Fig. 5) provides evidence for changing temperature–salinity conditions in tropical latitudes during GICE and the later Ordovician. In this record, GICE is marked by a positive shift in $\delta^{13}C_{calcite}$ values of *c.* 5‰ (-2.37 to 3.55‰). Above the Eureka Quartzite, $\delta^{13}C_{calcite}$ values become increasingly positive from the base of HICE (Fig. 5).

Two populations of $\delta^{18}O_{calcite}$ values can also be identified clustering around -10‰ and -6‰, the more negative values range from the base of the Ordovician to low in the Upper Ordovician (in Fig. 5 from 380 to 752 m) and recur in the lower Maysvillian (at 950 m, low Katian in terms of the newly named global stages) and in the Richmondian (1100 to 1150 m, high Katian). The last of these corresponds to the Boda Warm Event postulated on faunal and sedimentological grounds (Boucot *et al.* 2003; Fortey & Cocks 2005).

$\delta^{18}O_{calcite}$ values spanning GICE have also been recovered from the Kullsberg Limestone, Sweden. Values from brachiopods in this section (Marshall & Middleton 1990) and marine cements (Tobin *et al.* 2005) range from -5.5 for pre-GICE Excursion, brachiopods to -3.5 to -1.8‰ for marine cements during GICE (Fig. 6). A similar scale-positive shift for GICE has also been reported from several sections in China (Wang *et al.* 1997) and the Baltic region (Kaljo *et al.* 2004).

Plotting $\delta^{18}O$ values from the Nevada and Sweden records in temperature–salinity–isotope space allows characterization of hydrographical

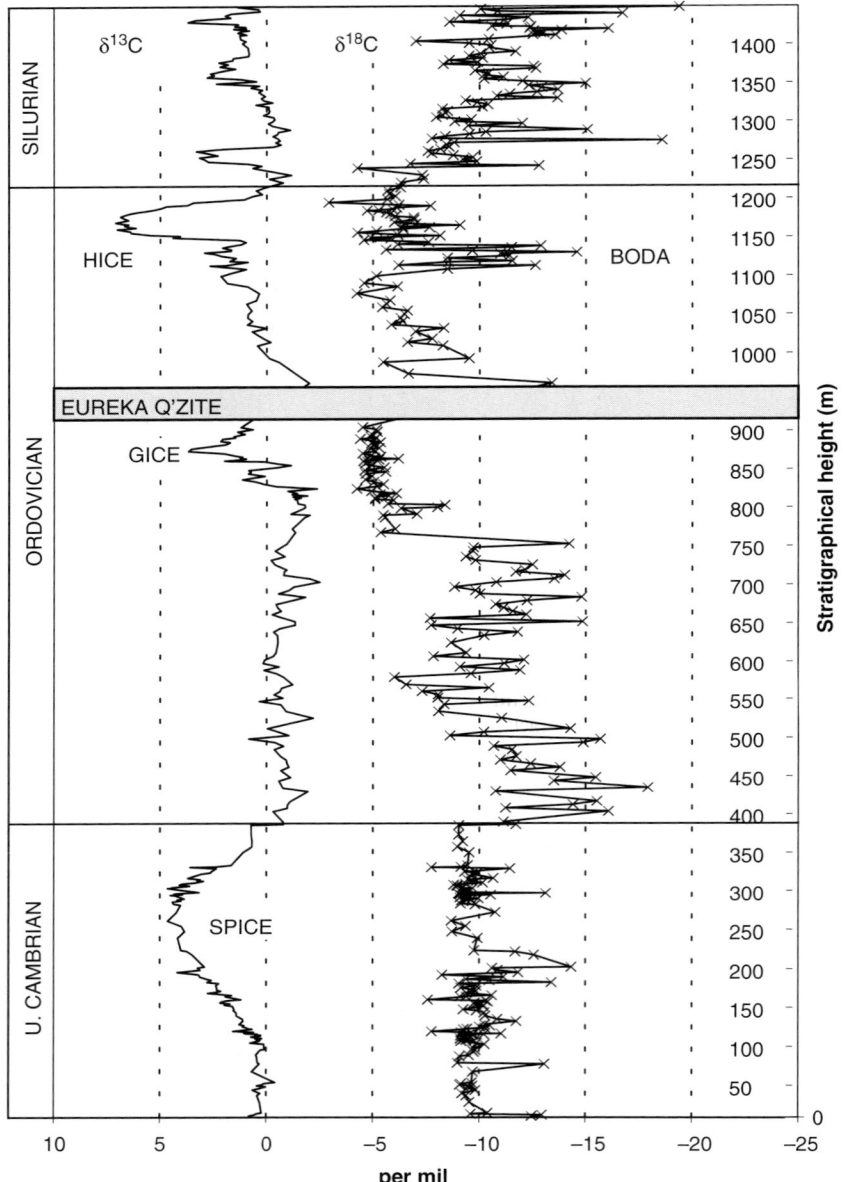

Fig. 5. Middle to Upper Ordovician low latitude stable carbon and oxygen isotope curves from central Nevada (data from Saltzman & Young 2005); oxygen isotope data plotted with permission. These show a positive shift in $\delta^{13}C$ in the Chatfieldian (GICE excursion) and a negative shift in $\delta^{18}O$ (warming equivalent to the Boda Event) that precedes the Hirnantian glacial maximum. Radiometric ages for stage boundaries taken from Webby *et al.* (2004). Saltzman & Young 2005 and Tobin *et al.* 2005 have argued the case for the fidelity of their data as a primary seawater record. Anomalously low stable oxygen isotope values for the Early Palaeozoic have been interpreted to be indicative of diagenetic alteration (Land 1995), high seawater temperatures (Karhu & Epstein 1986), salinity stratification (Railsback *et al.* 1990), high seawater pH (Wenzel *et al.* 2000), low seawater $\delta^{18}O$ (Veizer *et al.* 1986) or a combination of high tropical temperatures and low $\delta^{18}O$ seawater (Shields *et al.* 2003). Depleted $\delta^{18}O$ values are a common feature of the Early Palaeozoic. Values in excess of $-10‰$ are also recorded from the Upper Cambrian – Lower Ordovician and the Upper Silurian (Wenlock–Ludlow) in the Nevada section. 'Megacycles' of highly depleted ($> -10‰$) to more

conditions at these localities (Fig. 6). In Nevada, prior to the Guttenburg Carbon Isotope Excursion, there was a change from high temperature, low salinity and isotopically depleted ($\delta^{18}O_{calcite} = -10‰$) equatorial water to cooler, higher salinity, isotopically heavier subtropical water ($\delta^{18}O_{calcite} = -5‰$) is equivalent to a fall in temperature from c. 40 to 20 °C. $\delta^{18}O_{calcite}$ values from the Kullsberg Limestone indicate a change from subtropical to temperate or transitional water (Fig. 6; a mixture of subtropical and high latitude deep water as defined by Wilde & Berry 1984); and equivalent to a fall in temperature from c. 22 to 10 °C (Tobin et al. 2005).

Marine faunal diversity and provinciality also reflect changes in global temperature and oceanography associated with the onset of glaciation. The pattern of provinciality in conodonts, trilobites and graptolites was dramatically modified during the early Caradoc and this has been previously related to global transgression (Barnes et al. 1995). Alternately, this could reflect the transition from warm- to cool-water conditions in the early Chatfieldian (Brookfield 1988; Brookfield & Brett 1998; Lavoie 1995; Lavoie & Asselin 1998; Patzkowsky & Holland 1993; Patzkowsky et al. 1997). During the later Chatfieldian in Sweden, there were brief migrations of tropical ('Midcontinent' Province, warm equatorial) brachiopods, trilobites, tabulate corals, conodonts, stromatoporoids, and other organisms into mid–high latitudes (Saltzman & Young 2005; Tobin et al. 2005). Faunal evidence suggests there was a brief warming at the end of GICE that was either directly or indirectly linked to the end of the carbon isotopic excursion.

Stage 3: the Boda Event – a record of ocean gateway closure and changing ocean circulation

The Boda warm event coincides approximately with the *A. ordovicicus* conodont Biozone and ceases with the advent of the Hirnantian (Boucot

et al. 2003). In the Nevada record, the Boda Event is characterized by $\delta^{18}O_{calcite}$ values of -10 to $-15‰$ and $\delta^{13}C_{calcite}$ values of c. 2‰.

Closure of ocean gateways and concomitant changes in ocean circulation are an essential precursor of the snow gun hypothesis. Barnes (2004a, b) suggested a causal link between a superplume breakout during the mid-Ordovician (early Darriwilian) and Late Ordovician glaciation, but did not elaborate on the nature of this link. During the Mid-Ordovician, Gondwana drifted southward, towards the South Pole. In high southern latitudes, the Rheic Ocean opened with the continued drift northwards of Avalonia. The Iapetus Ocean and Proto-Tethys were connected though a seaway north of Baltica and currents may have flowed between them (Wilde 1991; Fig. 3a).

During the Ashgill, Gondwana continued to move southward with South American, African and parts of the Antarctic in high southern latitudes astride the South Pole. The Tournquist Sea closed, linking Avalonia and Baltica. The Iapetus Ocean narrowed with a seaway at the equator separating Laurentia–Siberia from Avalonia–Baltica. The Rheic Ocean widened, which acted as the major water communication between the widening south Palaeo-Pacific and Proto-Tethys (Fig. 3b).

There is general agreement between several generations of OGCM modelled circulation patterns (Wilde 1991; Poussart et al. 1999; Herrmann et al. 2004, 2005). The ocean circulation in the Northern Hemisphere was predominantly zonal due to an absence of large landmasses. During the Caradoc, the North Equatorial Counter-Current flowed westward, relatively uninterrupted from the west coast of Laurentia to Siberia (Fig. 3a). The South Equatorial Counter-Current was blocked in the west by the Australian–Antarctic area of Gondwana and was redirected into the poleward-flowing 'West Gondwana Current' to join with the South Gondwana Current at the south Panthalassic convergence (Fig. 3a). Southward movement of Gondwana in the Ashgill caused a reduction in the strength of the South Gondwana Current and, at low sea levels, this current disappeared. The northward

Fig. 5. (*continued*) positive ($-5‰$) $\delta^{18}O$ values, occur through the Nevada record with a frequency c. 30 Ma. Carboniferous $\delta^{18}O$ values, for the period affected by glacial and interglacial climate episodes, also fall within the range $-10‰$ to $-1‰$ (Bruckschen et al. 1999). Tropical values fall as low as $-16‰$ (Tucker et al. 2003). The most depleted $\delta^{18}O_{calcite}$ values are interpreted to occur during Namurian and Visean greenhouse climate phases (Bruckschen et al. 1999). Carboniferous $\delta^{13}C_{calcite}$ values rise from 1 to 5‰ between interglacial and glacial periods, similar to those from the Ordovician. Highly depleted $\delta^{18}O$ values are also reported at the present day in low tropical and subtropical latitudes (34° N to 34° S) where fluvial discharge is high. $\delta^{18}O$ values of -8 to $-12.8‰$ are recorded in surface ocean waters during winter months (October to April) close to the Ganges–Brahmaputra and the Indus River discharges and at the mouth of the Orinoco River. During Southern Hemisphere summer months, similar, highly depleted waters occur along the western seaboard of South America (http://www.giss.nasa.gov/data/o18data). Highly depleted stable isotopic values in the Nevada and Swedish records are therefore regarded as a primary oceanographic signal.

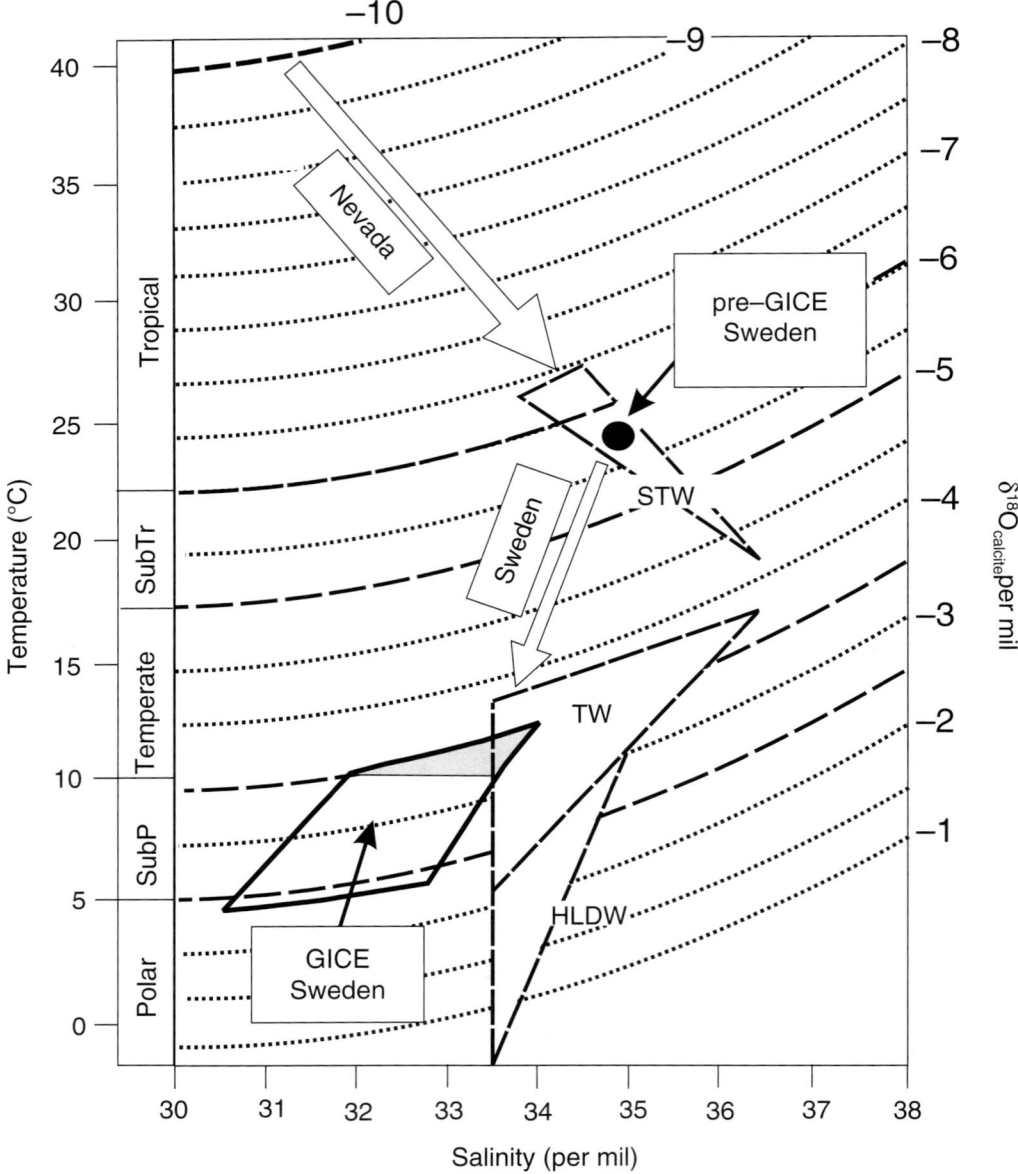

Fig. 6. Temperature–salinity– $\delta^{18}O$ plot based on that first published by Wilde & Berry (1984) and modified by Tobin *et al.* (2005). Published data are from Nevada (Saltzman & Young 2005) and Sweden (Marshall & Middleton 1990; Tobin *et al.* 2005; $\delta^{18}O_{calcite}$ values shown by dotted lines; dashed lines define tropical through polar notation; James 1997). The grey field is the subset of the Sweden data that lie above 10 °C, the minimum temperature required for marine calcite cement precipitation as described by Tobin *et al.* (2005). Modelled seawater fields HLDW (high latitude deep, low salinity water); TW (transitional water); STW (subtropical) are after Wilde & Berry (1984).

movement of Avalonia in the Ashgill narrowed the Iapetus Ocean, which in turn narrowed and increased the strength of the Iapetus Current (Fig. 3b).

Marine benthic megafossils from approximately the mid-Ashgill of Central and Southern Europe and North Africa track changes in palaeoceanography, in particular the expansion of a warm climate marine fauna (Fig. 7), and bioclastic limestones, into the Mediterranean cool-water realm (Boucot *et al.* 2003; Fortey & Cocks 2005).

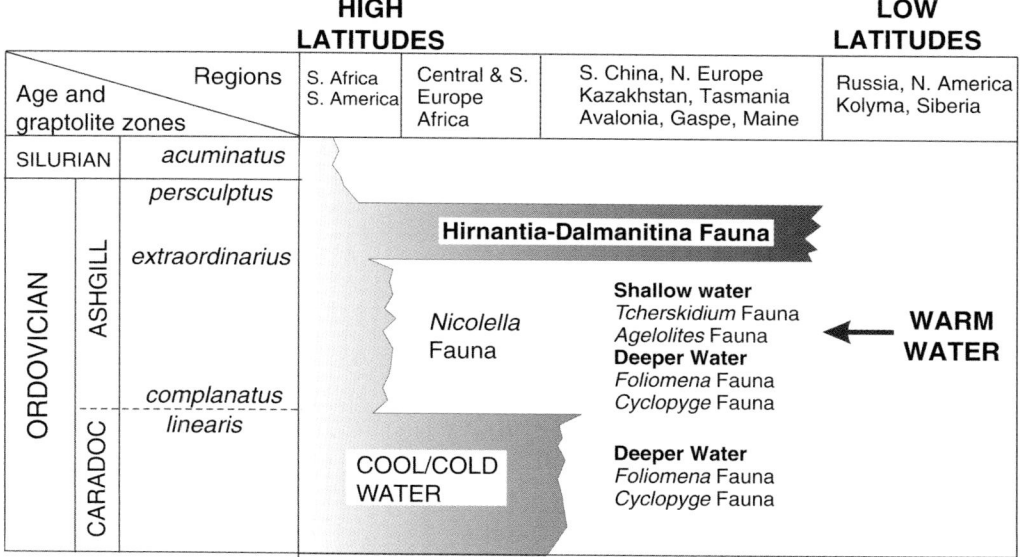

Fig. 7. Diagram depicting the latitudinal distribution of later Ordovician and earliest Silurian cool/cold water brachiopod faunas. The *Agelolites* fauna only occurs in the mid-Ashgill in low latitudes (after Boucot *et al.* 2003).

The snow gun hypothesis predicts poleward flow of warm surface currents, leading to a mixing of Laurentian and Gondwanan water masses. $^{143}Nd/^{144}Nd$ data from brachiopod and conodonts appear to represent the isotopic values of early Palaeozoic seawater and can be used to characterize different water masses (Wright *et al.* 1984; Keto & Jacobsen 1987; Wright & Barnes 2002). Keto & Jacobsen (1987) characterized two distinct water masses in the Iapetus Ocean from the Cambrian on the basis of $^{143}Nd/^{144}Nd$ values and suggested that either a physical barrier or thermal stratification kept these water masses separate. The geochemical distinction between the northern and southern Iapetus water masses breaks down in the Ashgill with δ_{Nd} converging on an average value of -9 (Keto & Jacobsen 1987; Holmden *et al.* 1996; Wright & Barnes 2002; Fig. 8).

Stage 4: Orbital forcing during the Hirnantian glacial maximum

HICE is marked by a positive shift in $\delta^{13}C_{calcite}$ of *c*. 6‰ (1.25 to 7.1‰). Many studies have identified a short interval of high $\delta^{13}C$ (2 to 7‰ positive shift) and $\delta^{18}O$ (2 to 3‰ positive shift) during the Rawtheyan to Hirnantian (Marshall & Middleton 1990; Underwood *et al.* 1997; Finney *et al.* 1999; Kump *et al.* 1999; Saltzman 2005; Saltzman & Young 2005).

Growing ice-sheets reach a critical size when they become sensitive to orbital forcing (Ruddiman 2003). Using analogy with the Pleistocene, it has been proposed that the Hirnantian glacial maximum represents two eccentricity-paced glacial–interglacial cycles (Armstrong *et al.* 2005; Sutcliffe *et al.* 2000) and that biotic extinction occurred at these timescales (Armstrong 1995). However, no direct evidence has been provided to support this hypothesis.

In order to test this hypothesis, a model of synthetic orbital variations has been constructed and January solar insolation, values for the top of the atmosphere at 60° and 70° S have been calculated (Fig. 9). The model spans two eccentricity cycles consistent with the proposed duration of the Hirnantian glacial maximum. It is not possible to constrain the period and amplitude of obliquity (and precession) beyond a few tens of millions of years (Laskar 1989; Laskar *et al.* 1993) and Pleistocene values have been used in the model. Periods of maximum solar insolation occur when obliquity is high, eccentricity reduces and perihelion is approached during austral summer (orbits 2, 4, 6, 8, 10, 12, 14, 16 in Fig. 9b, c). Cooler austral summers are predicted at intermediate summer solar insolation, during orbital periods of low eccentricity and low obliquity, and aphelion occurs during austral summer (orbits 3, 5, 7, 9, 11, 13, 15 in Fig. 9b). The coldest austral summers (minimal solar insolation during the

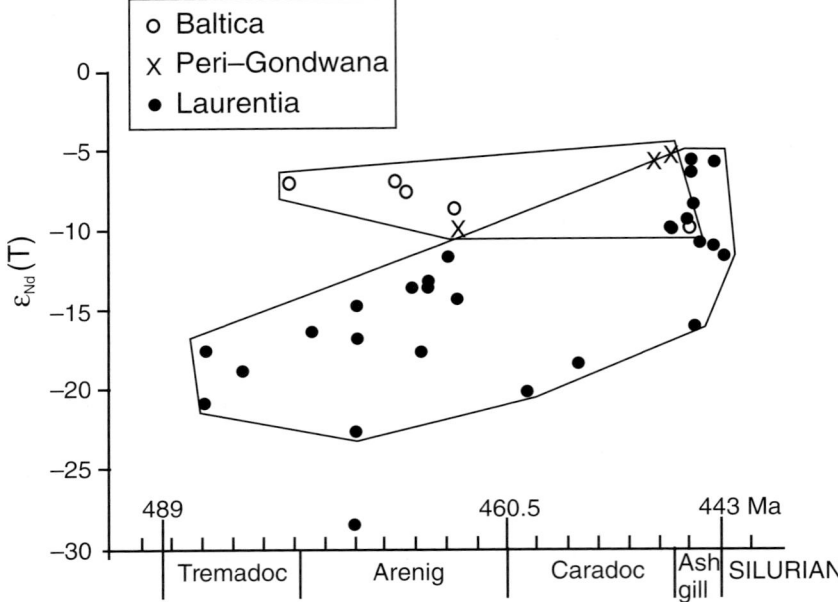

Fig. 8. Plot of Nd isotopic values from Ordovician conodonts. These show the Early Ordovician separation of Laurentian circumequatorial water from peri-Gondwana water. By the Caradoc–Ashgill, the convergence of Laurentia–Baltica and peri-Gondwana values indicates a mixing of these water masses (from Wright & Barnes 2002).

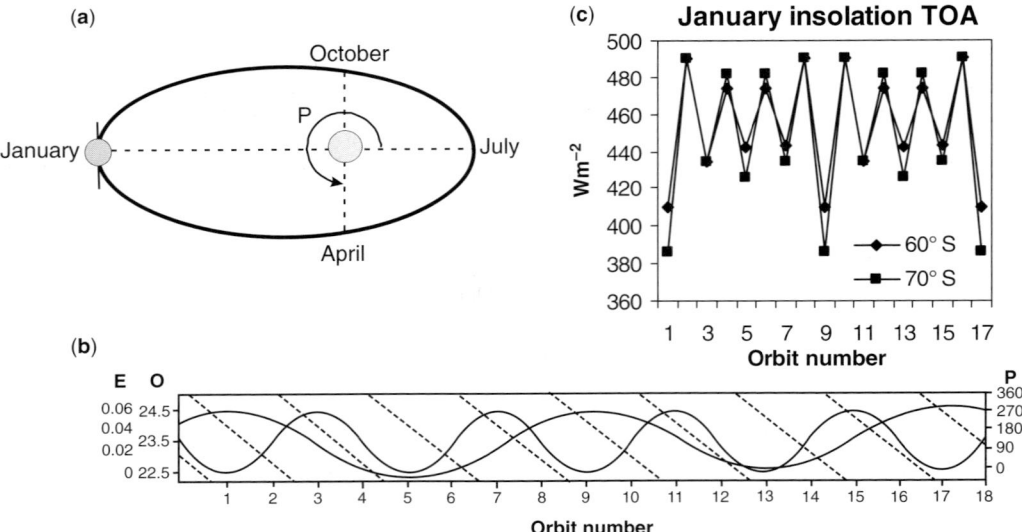

Fig. 9. Synthetic orbital variation used in the GCM–ice-sheet model for Antarctic glaciation (DeConto & Pollard 2003). (**a**) Schematic representation of the orbital configuration at the beginning of each 40 000-year sequence, yielding the coldest possible austral summers (Orbit 1 in other figures). P is the prograde angle between the perihelion and Northern Hemisphere vernal equinox, and is related to the precession cycle. This determines the month when Earth is at aphelion. (**b**) Synthetic orbital sequence. Numbers 1–18 correspond to orbit time steps of 10,000 years, orbit 1 corresponds to the diagram in (a). With maximum eccentricity (0.05), minimum obliquity (22.5°) and P = 270°. Precession, obliquity and eccentricity have sinusoidal periodicities of 20, 40, 80 ka respectively, so that eccentricity completes half a cycle during an obliquity cycle. (**c**) Austral summer (January) insolation at the top of the atmosphere for orbital steps 1–17. Values are for 60° and 70° south, and bracket the proposed palaeolatitude of southern Jordan.

austral summer) occur when eccentricity is high, obliquity low and aphelion occurs during austral summer (Fig. 9a; orbits 1, 9 and 17 in Fig. 9b, c).

Assuming summer temperature (ablation) is the most important limitation on ice-sheet size, major ice advances will occur during the coldest possible austral summers with an eccentricity period. Smaller-scale ice advances will occur during intermediate insolation conditions. Ice retreats would occur during the warmest austral summers and have an obliquity period.

If the size of the Hirnantian ice-sheet was orbitally forced, then glaciogenic sections at high latitudes should show a pattern of major (equivalent to glacials) and intermediate ice advances (equivalent to stadials), distinguished on the degree of ice incision. The Hirnantian section in southern Jordan records two ice-related, unconformity-bounded palaeovalley sequences, with up to 150 m of erosion at each unconformity. The upper sequence contains a glacially striated pavement with two cross-cutting striation directions (Turner et al. 2005). This record has been interpreted to indicate two major ice advances that can be correlated to Saudi Arabia (Vaslet 1990) and Libya (Turner et al. 2005). Minor and intermediate ice advances appear to have a more regional extent (Turner et al. 2005). From the synthetic orbital model, this record would span two eccentricity cycles as proposed for the Hirnantian glacial maximum.

Well-constrained ice-proximal successions in northern Gondwana record two ice-related depositional sequences, separated by major erosional unconformities. Proxy evidence of orbital pacing of ice-margin processes has also been reported from the early post-glacial section in southern Jordan. Here, cycles of increasing marine organic productivity have been related to ice melting and interpreted as obliquity-paced (Armstrong et al. 2005). Evidence for low latitude, precession-scale climate responses is found in cycle periodicities within evaporites from Australia (Williams 1991).

Implications for global biodiversity

The most profound impact on the marine biosphere occurred during the glacial maximum and associated end-Ordovician mass extinction (Bambach et al. 2004). During this event, the relics of the Cambrian Evolutionary Fauna suffered major taxic loss (44% families), the Palaeozoic Evolutionary Fauna c. 30% loss and newly evolved elements of the Modern Evolutionary Fauna decline by only 4% (Sepkoski 1996). Extrapolating from these values suggests that 85% of species went extinct (Jablonski 1991). A first strike of extinction coincides with the onset of the glacial maximum,

when sea levels fell dramatically, draining shallow epicontinental seas, intensifying thermohaline circulation, and depressing the thermocline, oxycline and picnocline (Brenchley et al. 1994, 1995; Brenchley & Marshall 1999). Following the first strike of extinction, surviving faunas and diversity stabilised, organisms had adapted to the new ecological situation (Sheehan 2001). A second strike of mass extinction coincided with the end of the glacial maximum, when sea levels rose and oceans returned rapidly to pre-Hirnantian conditions (Brenchley et al. 1994, 1995; Brenchley & Marshall 1999). In broad terms, the first strike of extinction eliminated communities from the epicontinental seas and new Hirnantian shelf and slope communities rapidly evolved to replace them. The second strike eliminated members of the newly evolved Hirnantian communities. Glacially induced ecological changes were most marked in the epicontinental seas (Sheehan 2001).

With a 'long' glaciation scenario a fall in familial diversity would be expected to occur during the later Caradoc (associated with GICE) and with ongoing cooling to the Hirnantian. A general pattern of sustained diversity increase until the mid-Caradoc followed by a general decline through the late Ordovician is seen in many elements of the Palaeozoic Evolutionary Fauna (Webby et al. 2004).

A more detailed analysis of single clade responses in the ecologically diverse groups – conodonts, ostracods and the planktonic graptolites – indicates parallel patterns of diversity change through the Ordovician (Fig. 10). Family and generic diversity data (calculated as d_{norm} as recommended by Sepkoski 1975) and corrected for stage length duration) show parallel patterns: a rising trajectory into the Llanvirn, followed by a decline in the Caradoc, a slight rise in groups in the Ashgill and declines of varying severity by the Llandovery (reflecting the mass extinction event).

Interpretation

Was GICE the start of the Ordovician glaciation? The mid-Caradoc was a critical period of Earth system change. The coincidence of more positive $^{87}Sr/^{86}Sr$ isotopes and a major perturbation in the carbon cycle during GICE marks a brief but severe interval of cooling followed by a long-term cooling trend. This would be equivalent to a 12 to 20 °C fall in SST if this was explained entirely in terms of temperature, and similar to the magnitude of fall in deep ocean waters at the start of the Antarctic glaciation (Ruddiman 2000). Faunal provinciality and migration patterns support cooling during GICE followed by warming. Events associated with GICE require more detailed analysis and

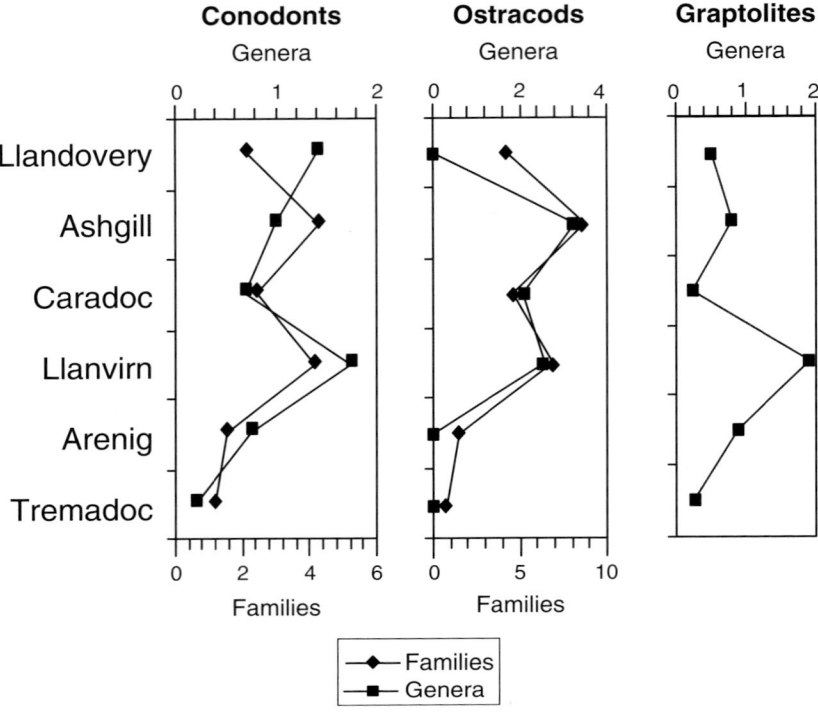

Fig. 10. Normalized diversity plots for conodonts, ostracods and graptolites. Family diversities for conodonts and ostracods are taken from range charts in Benton (1995). Conodont genera ranges were taken from Sweet (1988) as the most recent compilation. Ostracod generic ranges (Llanvirn–Ashgill) are from Williams *et al.* (2003). Normalized diversity ($d_{norm} = a + 0.5 \times b + 0.5 \times c + 0.5 \times d$, where a is the number of range through taxa within the stage; b is the number of taxa originating in and ranging beyond the stage; c is the number of taxa ranging into but going extinct within the stage; d is the number of taxa whose range is restricted to the stage (after Sepkoski 1975). This metric compensates for zonal range chart data, which over-represent true ranges as few taxa will completely span the zones in which they first or last appear, or to which they are confirmed. d_{norm} has been corrected for stage duration. Plots show the normalized diversity trajectories for both family and generic data.

particularly more detailed correlation of geological proxies; however, at our current state of knowledge there appears to be a sequence of events, comparable to those during the mid-Miocene that resulted in a short global cooling and initiation of polar ice-sheets. Glacial deposits and glacial features provide evidence for limited Caradoc ice-sheets in South America, and ice-volume changes have been implicated as a cause of relative sea level change in the Caradoc succession of Morocco (Hamoumi 1999) and Laurentia (Pope & Read 1997, 1998). The patchy sedimentary record suggests ice-caps may have been more like the temperate ice-sheets of Antarctica during the Paleogene. Both GICE and the Monterey excursions were short-lived, suggesting either ocean nutrients declined (unlikely due to the continuing Caledonian Orogen) or ice-sheets remained small, not reaching the critical size to provide cooling feedbacks.

Following GICE, there is a long-term increase in $\delta^{13}C$ values. The mechanism for this trend remains unclear. Increased silicate weathering, Monterey effects associated directly or indirectly with the developing Taconic/Caledonian Orogeny, or gradually increasing ice-sheets might be implicated.

The Boda Event is interpreted as a period of increased flow of warm tropical and subtropical water into high latitudes, the trigger to rapid ice-volume increase. This was due to changing palaeogeography, ocean gateway closure and sea level fall, producing a zonal warm-water circulation in the Southern Hemisphere bringing warm water and moist air across large areas of the Gondwana margin. There is a general match (within current biostratigraphical resolution) between the start of the Boda Event and the switch from anoxic to oxic conditions in the low latitude Iapetus Ocean, suggesting increased THC at this time.

The Caradoc palaeogeographical configuration and computed surface currents result in an increase of southward ocean heat transport of up to 42% compared to the present day and also results in an increased meridional overturning (Wilde 1991; Poussart *et al.* 1999; Herrmann *et al.* 2004, 2005). The late Ashgill plate configuration and low sea levels reduced the Iapetus Ocean to an embayment and created a meridional barrier in the tropical regions. This change and the cessation of the South Gondwana Current resulted in a more zonal ocean circulation and warm-water flow to high southerly latitudes. The importance of a shift in the position of the ITCZ remains to be demonstrated.

This scenario is comparable with that at the initiation of the Gulf Stream in the Northern Hemisphere at 4.3 Ma (Bartoli *et al.* 2005). Herrmann *et al.* (2004) showed that surface circulation is sensitive to sea level change. As ice-sheets grew during the Ashgill and they reduced relative sea level, this resulted in a decline in poleward ocean heat transport. Concomitant increasing ice–snow albedo would have fed back into the climate system, further enhancing global cooling.

Computer simulations show that cooling during the Late Ordovician can be attributed to changes in atmospheric pCO_2, palaeogeography and sea level (Crowley *et al.* 1994; Crowley & Baum 1995; Gibbs *et al.* 1997, 2000; Herrmann *et al.* 2003, 2004). Threshold pCO_2 values for the build-up of ice may have been crossed during GICE. During the late Rawtheyan to early Hirnantian, the ice-sheet reached a critical size (equivalent to the Mid-Pleistocene transition, approximately one million years ago), it became more sensitive to orbital forcing. Assuming summer temperature (ablation) is the most important limitation on ice-sheet size, then ice-sheet growth occurred during the coldest possible austral summers (low obliquity, high eccentricity and aphelion corresponding to the austral summer). During these orbital periods, eccentricity-paced highly dynamic ice-sheets expanded and incised the continental shelves. Subsidiary ice advances occurred between these major events, during orbital periods of low eccentricity, low obliquity and aphelion in the austral winter. The Hirnantian glacial maximum, as documented in ice-proximal settings, lasted two eccentricity cycles.

During the Plio-Pleistocene, the monsoon is an important and rapid response, low latitude climate system that, through changes in the nutrient flux to the oceans, feeds back into the global carbon cycle. Intensification of the monsoon system results in the drawdown of atmospheric CO_2, causing cooling on a precession timescale (Kutzbach & Webb III 1991; Ruddiman 2000). During the late Llanvirn to early Caradoc Northern Hemisphere summer (Fig. 11, bottom panel), the atmospheric circulation pattern may have been characterized by stable high-pressure zones and the development of intense low-pressure zones in the circumequatorial zone. This produced a monsoonal pressure pattern over eastern Laurentia, Baltica and probably Siberia and western Gondwana (Wilde 1991). Monsoonal circulation in the Indo-Australian-Antarctic regions of Gondwana is likely to have intensified during glaciation (Wilde 1991) and equatorial upwelling along the ITCZ would have been enhanced along the west coast of Laurentia (Pope & Steffen 2003).

During the Hirnantian, eccentricity modulation of processes entrained by precession (e.g. monsoons) and obliquity (e.g. ice-margin processes) led to cyclic changes in ocean nutrient flux. This forced changes in pCO_2 that initiated quasi-glacial-to-interglacial cycles; these impacted the tropics through changes in global temperature and relative sea level change.

Limited evidence from well-documented clades suggests biotic diversity declined from the late Caradoc, reflecting changing oceanographic and climatic conditions at this time. The two-stage mass extinction, at the base of the Hirnantian and during the mid-Hirnantian, can be explained in terms of temperature changes, major oceanic re-organization and ecosystem disruption. Despite the high taxonomic impact, the mass extinction did not have a lasting effect on the Palaeozoic Evolutionary Fauna, and by middle Silurian times marine family diversity had rebounded to pre-extinction levels (Droser *et al.* 1997, 2000; Sheehan 2001).

The rapid termination of glacial maximum conditions in the late Hirnantian remains problematic. This may have been the result of a coincidence of rising atmospheric CO_2 levels, sourced from the weathering of exposed low latitude carbonate platforms and warm austral summers. If the melting Hirnantian ice-sheet remained at a size to be sensitive to orbital effects, postulated further small-scale ice advances during the Early Silurian might then be explained in terms of Milankovitch forcing.

Conclusions

This review of geological proxies demonstrates a commonality in the pattern and sequence of events leading to the Cenozoic and Ordovician glaciations, and supports a commonality of cause. Though pCO_2 levels were different, both glaciations were set against a backdrop of long-term declining pCO_2, likely initiated by changes in plate configuration that resulted in increased silicate weathering and nutrient cycling into the

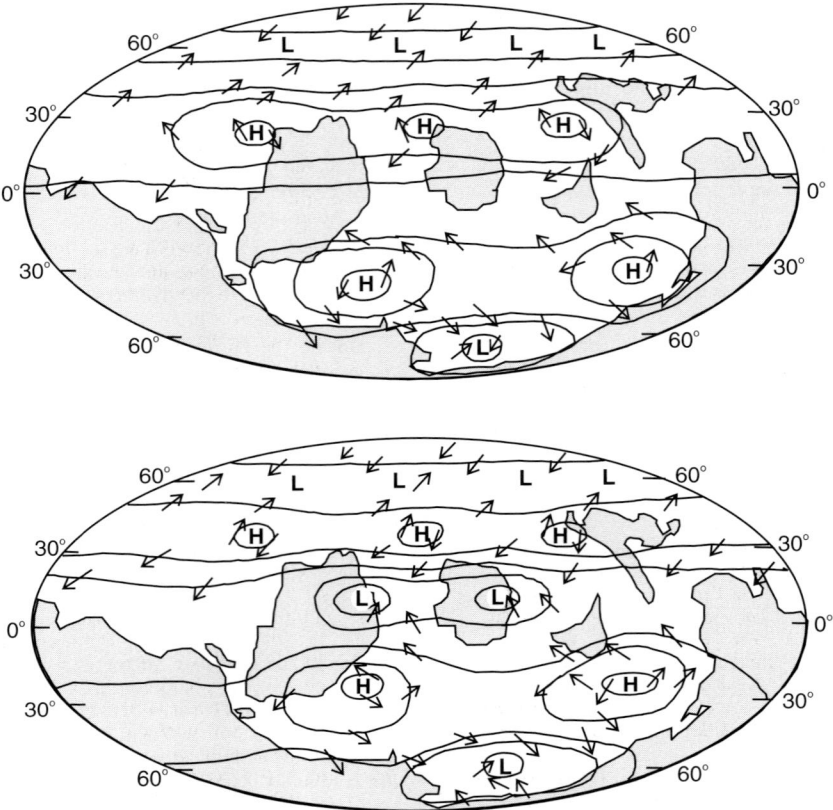

Fig. 11. Atmospheric circulation patterns during the Late Ordovician (late Llanvirn to early Caradoc). Top, Northern Hemisphere winter; bottom, Northern Hemisphere summer. Base map is generalized (redrawn after Parrish 1982).

oceans. Rapid expansion of ice volume was triggered by the redirection of warm, circumequatorial currents into high latitudes to provide a source of warm moist air and high levels of snowfall. Once ice-sheets were large enough to survive successive precession and obliquity maxima, eccentricity pacing of processes embedded in obliquity and precession largely controlled their size. This supports a unified theory for glaciation which integrates plate tectonics, changing ocean circulation and orbital/insolation effects, and a rejection of the axiom that the Ordovician glaciation and associated mass extinction were unique in Earth history.

The editors are thanked for the opportunity to produce this review, which has benefited from discussions with colleagues, particularly Dr A. W. Owen and members of IGCP Project 503. Some of the ideas were developed during the tenure of NERC grant (NER/B/2000/000068). An earlier draft of the manuscript was commented on by Prof. M. E. Tucker, and this paper has further benefited from the comments of the referees Dr P. Brenchley (University of Liverpool) and Prof. T. Meidla (University of Tartu) and the editorial assistance of Dr M. Williams. Dr M. Saltzmann is thanked for granting permission to publish his $\delta^{18}O$ data, stored in the GSA data repository.

References

ABE-OUCHI, A. & BLATTER, H. 1993. On the initiation of ice sheets. *Glaciology*, **18**, 203–207.

ALDRIDGE, R. J., JEPPSSON, L. & DORNING, K. J. 1993. Early Silurian oceanic episodes and events. *Journal of the Geological Society, London*, **150**, 501–513.

ARMSTRONG, H. A. 1995. High-resolution biostratigraphy (conodonts and graptolites) of the Upper Ordovician and Lower Silurian – a chronometric framework in which to evaluate the causes of the mass extinction. *Modern Geology*, **20**, 1–28.

ARMSTRONG, H. A. & COE, A. L. 1997. Deep sea sediments record the geophysiology of the end Ordovician glaciation. *Journal of the Geological Society, London*, **154**, 929–934.

ARMSTRONG, H. A. & LANE, P. D. 1981. The Un-named Silurian (?) dolomite formation of Børglum Elv, central Peary Land. *Rapport Groenlands Geologiske Undersøgelse*, **106**, 29–34.

ARMSTRONG, H. A., TURNER, B. R., MAKHLOUF, I. A., WILLIAMS, M., AL SMADI, A. & ABU SALAH, A. 2005. Origin, sequence stratigraphy and depositional environment of an Upper Ordovician (Hirnantian), peri-glacial black shale, Jordan. *Palaeogeography Palaeoclimatology Palaeoecology*, **220**, 273–289.

ASTINI, R. A. 1999. The Late Ordovician glaciation in the Proto-Andean margin of Gondwana revisited: geodynamic implications. *Acta Universitatis Carolinae, Geologica*, **43**, 171–173.

BAMBACH, R. K., KNOLL, A. H. & WANG, S. 2004. Origination, extinction, and mass depletions of marine diversity. *Paleobiology*, **30**, 522–542.

BARNES, C. R. 2004a. Ordovician oceans and climate. *In*: WEBBY, B. D., PARIS, F., DROSER, M. L. & PERCIVAL, I. G. (eds) *The Great Ordovician Biodiversification Event. Critical Moments and Perspectives in Earth History and Paleobiology*. Columbia University Press, New York, 72–77.

BARNES, C. R. 2004b. Was there an Ordovician superplume event? *In*: WEBBY, B. D., PARIS, F., DROSER, M. L. & PERCIVAL, I. G. (eds) *The Great Ordovician Biodiversification Event* Columbia University Press, New York, 77–80.

BARNES, C. R., FORTEY, R. A. & WILLIAMS, S. H. 1995. The pattern of global bio-events during the Ordovician Period. *In*: WALLIER, O. H. (ed.) *Global Events and Event Stratigraphy in the Phanerozoic*. Springer, 141–172.

BARRETT, P. J., ELSTON, D. P., HARWOOD, D. M., McKELVEY, B. C. & WEBB, P.-N. 1987. Mid-Cenozoic record of glaciation and sea-level change on the margin of Victoria Land basin, Antarctica. *Geology*, **15**, 634–637.

BARTOLI, G., SARNTHEIN, M., WEINELT, M., ERLENKEUSER, H., GARBE-SCHÖNBERG, D. & LEA, D. W. 2005. Final closure of Panama and the onset of northern hemisphere glaciation. *Earth and Planetary Science Letters*, **237**, 33–44.

BENTON, M. J. 1995. *The Fossil Record 2*. Chapman & Hall, London.

BERNER, R. A. 1992. Palaeo-CO_2 and climate. *Nature*, **358**, 114.

BERNER, R. A. 1994. GEOCARB II: a revised model for atmospheric CO_2 over Phanerozoic time. *American Journal of Science*, **294**, 56–91.

BERNER, R. A. 2001. Modelling atmospheric O_2 over Phanerozoic time. *Geochimica et Cosmochimica Acta*, **65**, 685–694.

BERNER, R. A. & KOTHAVALA, Z. 2001. Geocard III; a revised model of atmospheric CO_2 over Phanerozoic time. *American Journal of Science*, **301**, 182–204.

BERRY, W. B. N. & WILDE, P. 1978. Progressive ventilation of the oceans-an explanation for the distribution of the Lower Paleozoic black shales. *American Journal of Science*, **278**, 27–75.

BEUF, S., BIJU-DUVAL, B., DE CHAARPAL, D., ROGON, P., GARIEL, O. & BENNACEF, A. 1977. Les gres du Paléozoique inférior au Sahara. Sédimentation et discontinuites. Evolution structurale d'un craton. *Publ. Inst. Fr. Pétr., Collection Sc. & Tech. du Pétrol.*, **19**.

BIJU-DUVAL, B., DEYNOUX, M. & ROGNON, P. 1981. Late Ordovician tillites of the central Sahara. *In*: HAMBREY, M. J. & HARLAND, W. B. (eds) *Earth's pre-Pleistocene Glacial Record*. Cambridge University Press, Cambridge, 99–107.

BILLUPS, K. 2005. Snow maker for the ice ages. *Nature*, **433**, 809–810.

BILLUPS, K., RAVELO, A. C., ZACHOS, J. C. & NORRIS, R. D. 1999. Link between oceanic heat transport, thermohaline circulation, and the Intertropical Convergence Zone in the early Pliocene Atlantic. *Geology*, **27**, 319–322.

BIRCHFIELD, G. E., WEERTMAN, J. & LUNDE, A. T. 1982. A model study of the role of high latitude topography in the climate response to orbital insolation anomalies. *Journal of Atmospheric Science*, **39**, 71–87.

BOUCOT, A. J. & GRAY, J. 2001. A critique of Phanerozoic climatic models involving changes in the CO_2 content of the atmosphere. *Earth Science Reviews*, **56**, 1–159.

BOUCOT, A. J., RONG, J.-Y., CHEN, X. & SCOTESE, C. R. 2003. Pre-Hirnantian Ashgill climatically warm event in the Mediterranean region. *Lethaia*, **36**, 119–132.

BRENCHLEY, P. J., CARDEN, G. A. F. & MARSHALL, J. D. 1995. Environmental changes associated with the 'first strike' of the Late Ordovician mass extinction. *Modern Geology*, **20**, 69–82.

BRENCHLEY, P. J. & MARSHALL, J. D. 1999. Relative timing of critical events during the late Ordovician mass extinction-new data from Oslo. *Acta Universitatis Carolinae, Geologica*, **43**, 187–191.

BRENCHLEY, P. J., MARSHALL, J. D., CARDEN, G. A. F. ET AL. 1994. Bathymetric and isotopic evidence for a short lived Late Ordovician glaciation in a greenhouse period. *Geology*, **22**, 295–298.

BROOKFIELD, M. E. 1988. A mid-Ordovician temperate carbonate shelf – the Black River and Trenton Limestone Groups of southern Ontario, Canada. *Sedimentary Geology*, **60**, 137–153.

BROOKFIELD, M. E. & BRETT, C. E. 1998. Paleoenvironments of the Mid-Ordovician (Upper Caradocian) Trenton Limestones of southern Ontario, Canada: storm sedimentation on a shoal-basin shelf model. *Sedimentary Geology*, **57**, 75–105.

BRUCKSCHEN, P., OESMANN, S. & VEIZER, J. 1999. Isotope stratigraphy of the European Carboniferous: proxy signals for ocean chemistry, climate and tectonics. *Chemical Geology*, **161**, 127–163.

CLIFT, P. & GAEDICKE, C. 2002. Accelerated mass flux to the Arabian Sea during the middle to late Miocene. *Geology*, **30**, 207–210.

CLIFT, P., LEE, J. I., CLARK, M. K. & BLASZTAJN, J. 2002. Erosional response of South China to arc rifting and monsoonal strengthening: a record from South China Sea. *Marine Geology*, **184**, 207–226.

CROWELL, J. C., SUÁREZ-SORUCO, R. & ROCHA-CAMPOS, A. C. 1980. Silurian glaciation in central South America. *In*: CRESSWELL, M. M. & VELLA, P. (eds) *Gondwana Five: Selected Papers and Abstracts of Papers presented at the Fifth International Gondwana Symposium, Wellington, New Zealand*. Balkema, Rotterdam, 105–110.

CROWLEY, T. J. & BAUM, S. K. 1995. Toward reconciling Late Ordovician (~ 440 Ma) glaciation with very high CO_2 levels. *Journal of Geophysical Research*, **96**, 22597–22610.

CROWLEY, T. J., YIP, K.-J. & BAUM, S. K. 1994. Snowline instability in a general circulation model: Application to the Carboniferous glaciation. *Climate Dynamics*, **10**, 363–376.

DECONTO, R. M. & POLLARD, D. 2003. A coupled climate–ice sheet modelling approach to the Early Cenozoic history of the Antarctic ice sheet. *Palaeogeography Palaeoclimatology Palaeoecology*, **198**, 39–52.

DENTON, G. H., PRENTICE, M. L. & BURCKLE, L. H. 1991. Cenozoic history of the Antarctica Ice sheet. *In*: (TINGEY, R. J.) (ed). *The Geology of Antarctica* Oxford University Press, Oxford, 365–433.

DETTMAN, D. L., XIAOMIN, F., CARMALA, N., GARIZONE, C. N. & JIJUN, L. 2003. Uplift-driven climate change at 12 Ma: a long $\delta^{18}O$ record from the NE margin of the Tibetan plateau. *Earth and Planetary Science Letters*, **214**, 267–277.

DROSER, M. L., BOTTJER, D. J. & SHEEHAN, P. M. 1997. Evaluating the ecological architecture of major events in the Phanerozoic history of marine invertebrate life. *Geology*, **25**, 167–170.

DROSER, M. L., BOTTJER, D. J., SHEEHAN, P. M. & MCGHEE JR, G. R. 2000. Decoupling of taxonomic and ecologic severity of Phanerozoic marine mass extinctions. *Geology*, **28**, 675–678.

EBNETH, S., SHIELDS, G. A., VEIZER, J., MILLER, J. F. & SHERGOLD, J. H. 2001. High resolution strontium isotope stratigraphy across the Cambrian–Ordovician transition. *Geochemica et Cosmochimica Acta*, **65**, 2273–2292.

FINNEY, S., BERRY, W. B. N., COOPER, J. D. *ET AL.* 1999. Late Ordovician mass extinction: A new perspective from stratigraphic sections in central Nevada. *Geology*, **27**, 215–218.

FORTEY, R. A. & COCKS, L. R. M. 2005. Late Ordovician global warming – The Boda event. *Geology*, **33**, 405–408.

FRAKES, L. A., FRANCIS, J. E. & SYKES, J. I. 1992. *Climate Modes of the Phanerozoic*. Cambridge University Press, Cambridge.

GARIZONE, C. N., QUADE, J., DECELLES, P. G. & ENGLISH, N. B. 2000. Predicting palaeoelevation of Tibet and the Himalaya from $\delta^{18}O$ vs. altitude gradients in meteoric water across the Nepal Himalaya. *Earth and Planetary Science Letters*, **183**, 215–229.

GIBBS, M. T., BARRON, E. J. & KUMP, L. R. 1997. An atmospheric pCO_2 threshold for glaciation in the Late Ordovician. *Geology*, **25**, 447–450.

GIBBS, M. T., BICE, K. L., BARRON, E. J. & KUMP, L. R. 2000. Glaciation in the Early Palaeozoic greenhouse: the roles of paleogeography and atmospheric CO_2. *In*: HUBER, B. T., MACLEOD, K. G. & WING, S. L. (eds) *Warm Climates in Earth History*. Cambridge University Press, Cambridge, 386–422.

GIBBS, M. T. & KUMP, L. R. 1994. Global chemical weathering at the last glacial maximum and the present: sensitivity to changes in lithology and hydrology. *Paleoceanography*, **9**, 529–543.

GRAHN, Y. & CAPUTO, M. V. 1992. Early Silurian glaciations in Brazil. *Palaeogeography Palaeoclimatology Palaeoecology*, **51**, 9–15.

HALLAM, A. 1992. *Phanerozoic Sea-Level Changes*. Columbia University Press, New York.

HAMBREY, M. J., LARSEN, B. & EHRMANN, W. U. 1991. The glacial record from the Prydz Bay continental shelf, East Antarctica. *In*: BARRON, J. & LARSEN, B. (eds) *Ocean Drilling Program Scientific Results*. Ocean Drilling Program, College Station, **119**, 77–132.

HAMOUMI, N. 1999. Upper Ordovician glaciation spreading and its sedimentary record in Moroccan North Gondwana margin. *Acta Universitatis Carolinae, Geologica*, **43**, 109–112.

HAUG, G. H., GANOPOLSKI, A., SIGMAN, D. M. *ET AL.* 2005. North Pacific seasonality and the glaciation of North America 2.7 million years ago. *Nature*, **433**, 821–825.

HAUG, G. H. & TIEDEMANN, R. 1998. Effect of the formation of the Isthmus of Panama on Atlantic Ocean thermohaline circulation. *Nature*, **393**, 673–676.

HERRMANN, A. D., HAUPT, B. J., PATZKOWSKY, M. E., SEIDOV, D. & SLINGERLAND, R. L. 2005. Response of Late Ordovician paleoceanography to changes in sea level, continental drift, and atmospheric pCO_2: potential causes for long-term cooling and glaciation. *Palaeogeography Palaeoclimatology Palaeoecology*, **210**, 385–401.

HERRMANN, A. D., PATZKOWSKY, M. E. & POLLARD, D. 2003. Obliquity forcing with 8–12 times preindustrial levels of atmospheric pCO_2 during the Late Ordovician glaciation. *Geology*, **31**, 485–488.

HERRMANN, A. D., PATZKOWSKY, M. E. & POLLARD, D. 2004. The impact of paleogeography, pCO_2, poleward ocean heat transport and sea level change on global cooling during the Late Ordovician. *Palaeogeography Palaeoclimatology Palaeoecology*, **206**, 59–74.

HOLMDEN, C., CREASER, R. A., MUEHLENBACHS, K., BERGSTROM, S. M. & LESLIE, S. A. 1996. Isotopic and elemental systematics of Sr and Nd in 454 Ma biogenic apatites: Implications for paleoseawater studies. *Earth and Planetary Science Letters*, **142**, 425–437.

JABLONSKI, D. 1991. Extinctions: a paleontological perspective. *Science*, **253**, 754–757.

JAMES, N. P. 1997. The cool-water carbonate depositional realm. *Special Publications*, SEPM, 1–20.

KALJO, D. L., HINTS, L., MARTMA, T., NOLVAK, J. & ORASPOLD, A. 2004. Late Ordovician carbon isotope trend in Estonia, its significance in stratigraphy and environmental analysis. *Palaeogeography Palaeoclimatology Palaeoecology*, **210**, 165–185.

KARHU, J. & EPSTEIN, S. 1986. The implication of the oxygen isotope records in coexisting cherts and phosphates. *Geochimica et Cosmochimica Acta*, **50**, 1745–1756.

KENNETT, J. P. 1977. Cenozoic evolution of Antarctic glaciation, the circum-Antarctic oceans and their impact on global paleoceanography. *Journal of Geophysical Research*, **82**, 3843–3859.

KETO, L. S. & JACOBSEN, S. B. 1987. Nd and Sr isotopic variations of Early Palaeozoic oceans. *Earth and Planetary Science Letters*, **84**, 7–41.

KOLATA, D. R., HUFF, W. D. & BERGSTRÖM, S. M. 2001. The Ordovician Sebree Trough: An oceanic passage to the Midcontinent United States. *Geological Society of America, Bulletin*, **113**, 1067–1078.

KUMP, L. R., ARTHUR, M. A., PATZKOWSKY, M. E., GIBBS, M. T., PINKUS, D. S. & SHEEHAN, P. M. 1999. A weatherinng hypothesis for glaciation at high atmospheric pCO_2 during the Late Ordovician. *Palaeogeography Palaeoclimatology Palaeoecology*, **152**, 173–187.

KUMP, L. R., GIBBS, M. T., ARTHUR, M. A., PATZKOWSKY, M. E. & SHEEHAN, P. M. 1995. Hirnantian glaciation and the carbon cycle. *In*: COOPER, J. D. (ed.) *Ordovician Odyssey: Short Papers for the 7th Int. Symp. Ordovician System. Pacific Section*. Society for Sedimentary Geology, Fullerton, 299–302.

KUTZBACH, J. E. & WEBB, III, T. 1991. Late Quaternary climatic and vegetational change in Eastern North America: Concepts, Models and Data. *In*: SHANE, L. C. K. & CUSHING, E. J. (eds) *Quaternary Landscapes*. University of Minnesota Press, Minneapolis.

LAND, L. S. 1995. Oxygen and carbon isotopic composition of Ordovician brachiopods: implications for coeval sea water: discussion. *Geochemica et Cosmochimica Acta*, **59**, 2843–2844.

LASKAR, J. 1989. A numerical experiment on the chaotic behaviour of the Solar System. *Nature*, **338**, 237–238.

LASKAR, J., JOUTEL, F. & BOUDIN, F. 1993. Orbital, precessional, and insolation quantities for the Earth from -20 MYR to $+10$ MYR. *Astronomy and Astrophysics*, **270**, 522–533.

LAVOIE, D. 1995. A Late Ordovician high-energy temperate-water carbonate ramp, southern Quebec, Canada: implications for Late Ordovician oceanography. *Sedimentology*, **42**, 95–116.

LAVOIE, D. & ASSELIN, E. 1998. Upper Ordovician facies in the Lac Saint-Jean outlier, Quebec (eastern Canada): palaeoenvironmental significance for Late Ordovician oceanography. *Sedimentology*, **45**, 817–832.

LEGRAND, P. 1988. The Ordovician-Silurian boundary in the Algerian Sahara. *Bulletin of the British Museum (Natural History), Geology*, **43**, 171–176.

LEGRAND, P. 1993. Ashgillian Graptolites of the Chirfa region. *Bulletin Centres de Recherches Exploration – Production ELF (Essences et Lubrifants de France), Aquitaine*, **17**, 435–441.

LEGRAND, P. 1995. Evidence and concerns with regard to the late Ordovician glaciation in North Africa. *In*: COOPER, J. D., DROSER, M. L. & FINNEY, S. (eds) *Ordovician Odyssey: Short Papers for the Seventh International Symposium on the Ordovician System*. SEPM, Las Vegas, 165–169.

MACQUEDA, M., WILLMOTT, A. J., MAMBER, J. L. & DARBY, M. S. 1998. An investigation of the small ice cap instability in the Southern Hemisphere with coupled atmosphere-sea-ice-ocean-terrestrial ice model. *Climate Dynamics*, **14**, 329–352.

MARSHALL, J. D. & MIDDLETON, P. D. 1990. Changes in marine isotopic composition and the Late Ordovician glaciation. *Journal of the Geological Society, London*, **147**, 1–4.

MELCHIN, M. J., HOLMDEN, C. & WILLIAMS, S. H. 2003. Correlation of graptolite biozones, chitinozoan biozones and carbon isotope curves through the Hirnantian. *In*: ALBANESI, G. L., BERESI, M. S. & PERALTA, S. H. (eds) *Ordovician from the Andes INSUGEO, Serie Correlacion Geologica*, **17**, 101–104.

MOREAU, J., GHIENNE, J.-F., DEYNOUX, M. & RUBINO, J.-L. 2004. Ice-proximal sedimentary records of the Late Ordovician glacial cycles. *In*: *IGCP Project 503, Erlangen. Programme with Abstracts*, Erlanger geologische Abhandlungen–Sonderband, **5**, 55.

NAISH, T. R. *ET AL*. 2001. Orbitally induced oscillations in the East Antarctic ice sheet at the Oligocene-Miocene boundary. *Nature*, **413**, 719–729.

OGLESBY, R. J. 1989. A GCM study of Antarctic Glaciation. *Climate Dynamics*, **3**, 135–156.

PAGANI, M., ARTHUR, M. A. & FREEMAN, K. H. 1999. Miocene evolution of atmospheric carbon dioxide. *Paleoceanography*, **14**, 273–292.

PAGE, A., ZALASIEWICZ, J. A., WILLIAMS, M. & POPOV, L. 2007. Were transgressive black shales a negative feedback modulating glacioeustacy in the Early Palaeozoic Icehouse? *In*: WILLIAMS, M., HAYWOOD, A. M., GREGORY, F. J. & SCHMIDT, D. N. (eds) *Deep-Time Perspective on Climate Change. Marrying the Signal from Computer Models and Biological Proxies*. The Micropalaeontology Society, Special Publications. The Geological Society, London, 123–158.

PARIS, F., ELAOUAD-DEBBAJ, Z., JAGLIN, J. C., MASSA, D. & OULEBSIR, L. 1995. Chitinozoa and late Ordovician glacial events on Gondwana. *In*: COOPER, C., DROSER, M. L. & FINNEY, S. (eds) *Ordovician Odyssey*. Las Vegas, 171–176.

PARRISH, J. T. 1982. Upwelling and petroleum source beds, with reference to the Palaeozoic. *AAPG Bulletin*, **66**, 750–774.

PATZKOWSKY, M. E. & HOLLAND, S. M. 1993. Biotic response to a Middle Ordovician paleoceanographic event in eastern North America. *Geology*, **21**, 619–622.

PATZKOWSKY, M. E., SLUPIK, L. M., ARTHUR, M. A., PANCOST, R. D. & FREEMAN, K. H. 1997. Late Middle Ordovician environmental change and extinction: Harbinger of the Late Ordovician or continuation of Cambrian patterns? *Geology*, **25**, 911–914.

PEARSON, P. N. & PALMER, M. R. 2000. Atmospheric carbon dioxide over the past 60 million years. *Nature*, **406**, 695–699.

PERALTA, S. H. & CARTER, C. H. 1999. Don Braulio Formation (late Ashgillian-early Llandoverian, San Juan Precordillera, Argentina): stratigraphic remarks and paleoenvironmental significance. *Acta Universitatis Carolinae, Geologica*, **43**, 225–28.

PETERSON, L. C. & HAUG, G. H. 2006. Variability in the mean latitude of the Atlantic Intertropical Convergence Zone as recorded by riverine input of sediments to the Cariaco Basin (Venezuela). *Palaeogeography Palaeoclimatology Palaeoecology*, **234**, 97–113.

POPE, M. C. & READ, J. F. 1997. High-resolution surface and subsurface sequence stratigraphy of Middle and Late Ordovician (Late Mohawkian to Cincinnatian) foreland basin rocks, Kentucky and Virginia. *American Association of Petroleum Geologists, Bulletin*, **81**, 1866–1893.

POPE, M. C. & READ, J. F. 1998. Ordovician meter-scale cycles: Implications for climate and eustatic fluctuations in the central Appalachians during a global greenhouse, non-glacial to glacial transition. *Palaeogeography Palaeoclimatology Palaeoecology*, **138**, 27–42.

POPE, M. C. & STEFFEN, J. B. 2003. Widespread, prolonged late Middle to Late Ordovician upwelling in North America: A proxy record of glaciation? *Geology*, **31**, 63–66.

POUSSART, F., WEAVER, A. J. & BARNES, C. R. 1999. Late Ordovician glaciation under high atmospheric CO_2: A coupled model analysis. *Paleooceanography*, **14**, 542–558.

PRENTICE, M. L. & MATHEWS, R. K. 1991. Tertiary ice sheet dynamics: The snow gun hypothesis. *Journal of Geophysical Research*, **96**, 6811–6827.

RAILSBACK, L. B., ACKERLY, S. C., ANDERSON, T. F. & CISNE, J. L. 1990. Palaeontological and isotope evidence for warm saline deep waters in Ordovician oceans. *Nature*, **343**, 156–159.

RICHARDSON, P. L. & REVERDIN, G. L. 1987. Seasonal cycle of velocity in the Atlantic North Equatorial countercurrent as measured by surface drifters, current meters and ship drifts. *Journal of Geophysical Research–Oceans*, **92**, 3691–3708.

RUDDIMAN, W. F. 2000. *Earth's Climate: Past and Future*. W. H. Freeman and Co. New York.

RUDDIMAN, W. F. 2003. Orbital insolation, ice volume, and greenhouse gases. *Quaternary Science Reviews*, **22**, 1597–1629.

SALTZMAN, M. R. 2005. Phosphorus, nitrogen and the redox evolution of the Paleozoic oceans. *Geology*, **33**, 573–576.

SALTZMAN, M. R. & YOUNG, S. A. 2005. Long-lived glaciation in the Late Ordovician? Isotopic and sequence-stratigraphic evidence from western Laurentia. *Geology*, **33**, 109–112.

SEPKOSKI JR, J. J. 1975. Stratigraphic biases in the analysis of taxonomic survivorship. *Paleobiology*, **1**, 343–355.

SEPKOSKI JR, J. J. 1996. Patterns of Phanerozic extinction: a perspective from global databases. *In*: WALLISER, O. H. (ed.) *Global Events and Event Stratigraphy in the Phanerozoic*, Springer, Berlin, 35–51.

SHEEHAN, P. M. 2001. The Late Ordovician mass extinction. *Annual Review of Earth and Planetary Sciences*, **29**, 331–364.

SHIELDS, G. A., CARDEN, G. A. F., VEIZER, J., MEIDLA, T., RONG, J.-Y. & RONG, Y.-L. 2003. Sr, C, and oxygen isotope geochemistry of Ordovician brachiopods: a major isotopic event around the Middle–Late Ordovician transition. *Geochemica et Cosmochimica Acta*, **67**, 2005–2025.

SPICER, R. A., HARRIS, N. B. W., WIDDOWSON, M. *ET AL.* 2003. Constant elevation of southern Tibet over the past 15 million years. *Nature*, **421**, 622–624.

SUTCLIFFE, O. E., DOWDESWELL, J. A., WHITTINGTON, R. J., THERON, J. N. & CRAIG, J. 2000. Calibrating the Late Ordovician glaciation and mass extinction by the eccentricity of Earth's orbit. *Geology*, **28**, 967–970.

SWEET, W. C. 1988. The conodonts: morphology, taxonomy, paleoecology, and evolutionary history of a long extinct animal phylum. Clarendon Press, Oxford.

TOBIN, K. J., BERGSTRÖM, S. M. & DE LA GARZA, P. 2005. A mid-Caradocian (453 Ma) drawdown in atmospheric pCO_2 without ice sheet development? *Palaeogeography Palaeoclimatology Palaeoecology*, **226**, 187–204.

TUCKER, M. E., GALLAGHER, J., LEMON, K. & LENG, M. J. 2003. The Yoredale Cycles of Northumbria: High-frequency clastic-carbonate sequences of the mid-Carboniferous icehouse world. *Open University Geological Society Journal*, **24**, 1–6.

TURNER, B. R., MAKHLOUF, I. M. & ARMSTRONG, H. A. 2005. Late Ordovician (Ashgillian) glacial deposits in southern Jordan. *Sedimentary Geology*, **181**, 73–91.

UNDERWOOD, C. J., CROWLEY, S. F., MARSHALL, J. D. & BRENCHLEY, P. J. 1997. High resolution carbon isotope stratigraphy of the basal Silurian stratotype (Dob's Linn, Scotland) and its global correlation. *Journal of the Geological Society, London*, **154**, 709–718.

VASLET, D. 1990. Upper Ordovician glacial deposits in Saudi Arabia. *Episodes*, **13**, 147–161.

VEIZER, J., ALA, D., AZMY, K. *ET AL.* 1999. $^{87}Sr/^{86}Sr$, $\delta^{13}C$ and $\delta^{18}O$ evolution of Phanerozoic seawater. *Chemical Geology*, **161**, 59–88.

VEIZER, J., FRITZ, P. & JONES, B. 1986. Geochemistry of brachiopods: oxygen and carbon isotopic records of Paleozoic oceans. *Geochemica et Cosmochimica Acta*, **50**, 1679–1696.

VINCENT, E. & BERGER, W. H. 1985. Carbon dioxide and polar cooling in the Miocene: the Monterey hypothesis. *In*: BROECKER, W. S. & SUNDQUIST, E. T. (eds) *The Carbon Cycle and Atmospheric CO_2: Natural Variations Archean to Present*. AGU Geophysical Monograph, **32**, 455–468.

WANG, K., CHATTERTON, B. D. E. & WANG, Y. 1997. An organic carbon isotope record of the Late Ordovician to Early Silurian marine sedimentary rocks, Yangtze Sea. South China: implications for CO_2 changes during the Hirnantian glaciation. *Palaeogeography Palaeoclimatology Palaeoecology*, **132**, 147–158.

WEBBY, B. D., PARIS, F., DROSER, M. L. & PERCIVAL, I. G. 2004. *The Great Ordovician Biodiversification Event*. Columbia University Press, New York.

WENZEL, B., LÉCUYER, C. & JOACHIMSKI, M. M. 2000. Comparing oxygen isotope records of Silurian calcite and phosphate – $\delta^{18}O$ compositions of brachiopods and conodonts. *Geochemica et Cosmochimica Acta*, **64**, 1859–1872.

WILDE, P. 1991. Oceanography in the Ordovician. *In*: BARNES, C. R. & WILLIAMS, S. H. (eds) *Advances in Ordovician Geology*. Paper 90–9. Geological Survey of Canada, 283–298.

WILDE, P. & BERRY, W. B. N. 1984. Destabilization of the oceanic density structure and its significance to marine "extinction" events. *Palaeogeography Palaeoclimatology Palaeoecology*, **48**, 143–162.

WILLIAMS, G. E. 1991. Milankovitch-band cyclicity in bedded halite deposits contemporaneous with Late Ordovician – Early Silurian glaciation, Canning Basin, Western Australia. *Earth and Planetary Science Letters*, **103**, 143–155.

WILLIAMS, M., FLOYD, J. D., SALAS, M. J., SIVETER, D. J., STONE, P. & VANNIER, J. M. C. 2003. Patterns of ostracod migration for the 'North Atlantic' region during the Ordovician. *Palaeogeography Palaeoclimatology Palaeoecology*, **195**, 193–228.

WRIGHT, C. A. & BARNES, C. R. 2002. Neodymium isotopic composition of Ordovician conodonts as a seawater proxy: Testing paleogeography. *Geochemistry Geophysics Geosystems*, **3**, doi: 10.1029/2001GC000195.

WRIGHT, J., SEYMOUR, R. S. & SHAW, H. F. 1984. REE and Nd isotopes in conodont apatite; variations with geological age and depositional environment. *In*: CLARK, D. L. (ed.) *Conodont Biofacies and Provincialism Special Paper* **196**. Geological Society of America, 325–340.

YAPP, C. J. & POTHS, H. 1992. Ancient atmospheric CO_2 pressures inferred from natural goethites. *Nature*, **353**, 342–344.

ZACHOS, J. C., BREZA, J. R. & WISE, S. W. 1992. Early Oligocene ice sheet expansion on Antarctica: Stable isotope and sedimentological evidence from Kerguelen Plateau, southern Indian Ocean. *Geology*, **20**, 569–573.

ZACHOS, J. C., QUINN, T. M. & SALAMY, K. A. 1996. High-resolution (10^4 years) deep-sea foraminiferal stable isotope records of the Eocene–Oligocene climate transition. *Paleooceanography*, **11**, 251–266.

Were transgressive black shales a negative feedback modulating glacioeustasy in the Early Palaeozoic Icehouse?

A. A. PAGE[1,2], J. A. ZALASIEWICZ[1], M. WILLIAMS[1] & L. E. POPOV[3]

[1]*Department of Geology, University of Leicester, Leicester, LE1 7RH, UK*
(e-mail: aap30@esc.cam.ac.uk)

[2]*Department of Earth Sciences, University of Cambridge, Downing Street,*
Cambridge, CB2 3EQ, UK

[3]*National Museum of Wales, Department of Geology, Cathays Park,*
Cardiff CF10 3NP, UK

Abstract: The Early Palaeozoic Icehouse (Late Ordovician–Early Silurian, *c.* 455–425 Ma) was a remarkable event in the Earth's climatic history, marked by extensive glaciations occurring at a time of elevated atmospheric CO_2. The oceanography of the Early Palaeozoic Icehouse was markedly different from that of modern oceans, with frequent episodes of oceanic anoxia and high concentrations of CO_2 which may have acidified the oceans and restricted carbonate burial. Thus, the marine organic carbon reservoir may have more strongly influenced long-term changes in atmospheric CO_2 than at present. We suggest that deposition of black shales represented a major sink for atmospheric carbon. Sequence stratigraphy reveals that widespread black shale deposition occurred in transgressions, whereas regressions are characterized by deposition of bioturbated facies, allowing changes in lithofacies and deep-water redox conditions to be related to the Early Palaeozoic carbon cycle. Assuming increased temperature is a function of increased atmospheric CO_2, and that glacioeustatic sea-level can serve as a proxy for temperature due to changing ice volume, we infer that the deposition of transgressive black shales may have acted as a negative feedback mechanism, drawing down CO_2 and preventing the onset of runaway greenhouse conditions.

The Early Palaeozoic represents an important interval in Earth biosphere evolution. It post-dated the origin of large metazoans and complex, tiered marine food webs (Butterfield 1997), and is succeeded by the radiation of land plants (Berner 1998; Gensel & Edwards 2001). It therefore represents an intermediate state between the oxygen-poor Proterozoic palaeoenvironment and the well-oxygenated world of the Late Palaeozoic and post-Palaeozoic (Berner 2003; Catling & Claire 2005). This interval marks a non-actualistic solution to the Earth's carbon budget. Though generally considered an interval of long-lived, stable greenhouse conditions (e.g. Gibbs *et al.* 2000; Montañez 2002; Church & Coe 2003, fig. 5.4), major glaciations nonetheless occurred in the late Ordovician and early Silurian. These glaciations occurred at elevated atmospheric CO_2 (Royer 2006) and transitions between oxic and anoxic marine conditions were frequent (Figs 1d & 2). In a time before the evolution of a complex land biota, most of the organic carbon reservoir must have existed in oceans, where it was buried as black shale (Fig. 1). Despite recent advances in general circulation models (GCMs) and the application of climatically sensitive stable isotopes to infer palaeoenvironmental change, Early Palaeozoic climate remains somewhat enigmatic, no doubt in part due to its lack of analogue in the modern world.

Instead, the Early Palaeozoic needs to be understood in its own terms. Much as Charles Lapworth, Adam Sedgwick and Roderick Murchison carefully unpicked the undifferentiated 'greywacke' successions mapped by William Smith and Charles Lyell in the 19th century, and established the stratigraphic divisions of the Lower Palaeozoic (Rudwick 1985; Secord 1986; Oldroyd 1990), the 21st century sees the need for Early Palaeozoic workers to return to its stratigraphy and establish how global lithostratigraphic patterns of continental weathering and carbonate and black shale burial relate to its palaeoclimate, thereby determining the large-scale controls on the carbon cycle at this time.

The Early Palaeozoic carbon cycle and climate

Understanding chemical oceanography and carbon cycling in the Early Palaeozoic is difficult. The precise magnitude of atmospheric CO_2 at this time

From: WILLIAMS, M., HAYWOOD, A. M., GREGORY, F. J. & SCHMIDT, D. N. (eds) *Deep-Time Perspectives on Climate Change: Marrying the Signal from Computer Models and Biological Proxies.* The Micropalaeontological Society, Special Publications. The Geological Society, London, 123–156.
1747-602X/07/$15.00 © The Micropalaeontological Society 2007.

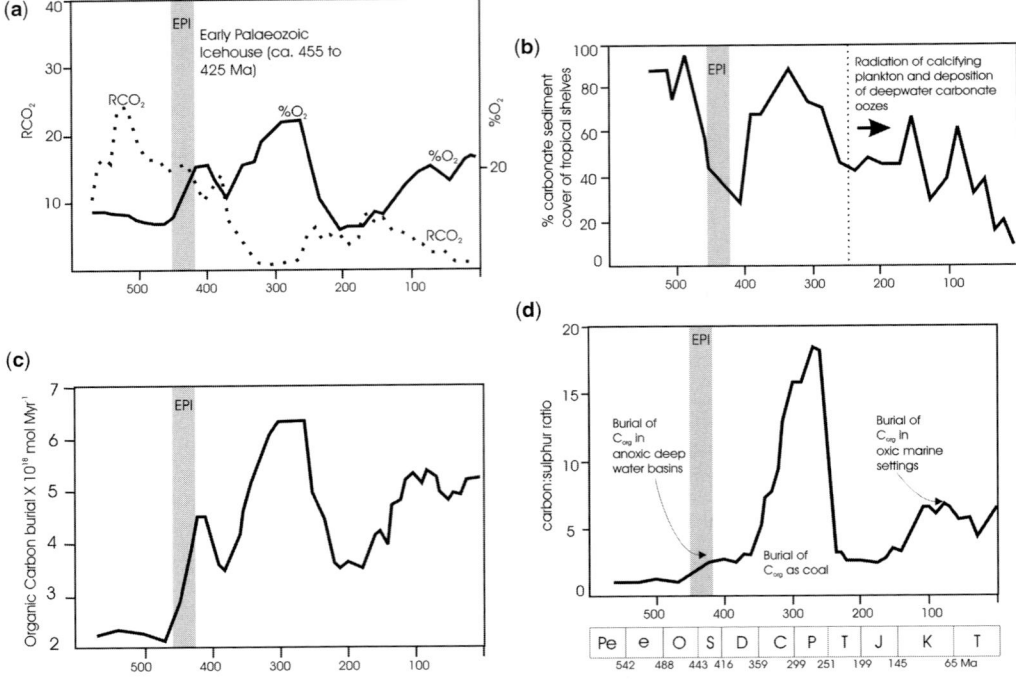

Fig. 1. Graphs illustrating changes in the nature of the carbon cycle through Phanerozoic time, showing the Early Palaeozoic dominated by organic-carbon burial in deep-marine anoxic waters under conditions of elevated atmospheric CO_2. (**a**) Temporal changes in atmospheric CO_2 relative to pre-industrial levels and partial pressure of atmospheric O_2 (after Berner 2001; Berner & Kothavala 2001). (**b**) Reduced carbonate deposition during the EPI as recorded in the relative proportion of low-latitude ($<30°$ N/S) shelf area occupied by carbonate sediments with time (data from Walker *et al.* 2002). (**c**) Temporal changes in organic carbon burial flux, showing a notable high in the EPI when organic carbon was predominantly buried in black shales (after Berner 2003). (**d**) Ratio of the accumulation rate of organic carbon and pyrite–sulphur (C/S) in sediments versus time (after Berner 2003): low C/S values reflect deposition of organic carbon in euxinic basins, high values correspond to burial in terrestrial freshwater swamps, and intermediate values are found in normal marine sediments (Berner & Raiswell 1983).

is uncertain, and the relation between CO_2 regulation and Early Palaeozoic climate is not fully resolved (see discussions in Ridgwell 2005; Royer 2006). However, available proxy data agree well with Berner and Kothavala's (2001) GEOCARB III model of atmospheric CO_2 levels over Phanerozoic time (Crowley & Berner 2001; Royer *et al.* 2004; Royer 2006), providing support for extremely elevated CO_2 levels in this interval (Fig. 1a). This provides support for key assumptions of the GEOCARB model and its descendants. Among these assumptions is that long-term drawdown of atmospheric CO_2 into the oceans was a consequence of (a) continental silicate weathering and burial in carbonates, and (b) photosynthesis and burial of organic carbon (Berner 1991, 1994, 2006; Berner & Kothavala 2001). CO_2 regulation must have been reflected in the specific pattern of organic and inorganic carbon burial in the Early

Palaeozoic (Fig. 1), which differs notably from that of the Neoproterozoic (Rothman *et al.* 2003) and the rest of the Palaeozoic (Berner 2003).

In the Early Palaeozoic, carbonate burial was restricted to the continental shelves (Walker *et al.* 2002), whilst organic carbon was predominantly buried in deep-water anoxic environments (Fig. 1d). The advent of biomineralization in the Cambrian explosion facilitated carbonate burial relative to the Neoproterozoic (Rothman *et al.* 2003; Ridgwell 2005). However, the Early Palaeozoic may have lacked a well-developed marine carbonate buffer, which is highly sensitive to increased atmospheric CO_2 (Barker *et al.* 2003). The radiation of the calcifying plankton in the Triassic profoundly affected the pattern of marine carbonate deposition (Martin 1995), and a modern ocean analogue for carbonate burial may not be applicable before this (cf. Ridgwell 2005). Likewise, the advent of digestion, bioturbation and a macroplankton faecal express in the Cambrian explosion (Butterfield 1997) no doubt significantly affected the cycling of the organic carbon reservoir, which lacked the sustained, large-scale fluctuations witnessed in the Neoproterozoic (Hayes et al. 1999; Rothman et al. 2003). However, prior to the greening of the continents in the Late Palaeozoic (Berner 1998; Gensel & Edwards 2001), the marine organic carbon reservoir dominated the burial of organic carbon (Berner 2003), and the frequent intervals of marine anoxia that typify the Early Palaeozoic are probably a consequence of this (Figs 1d & 3).

Constraining CO_2 drawdown in this interval requires good estimates of the burial fluxes of carbonates and organic carbon (Fig. 1b, c). There are two ways of achieving such estimates. The first approach takes the known volume of carbon held in preserved strata and applies a correction to account for the progressive volume loss due to erosion, subduction and metamorphism (e.g. Berner & Canfield 1989; Walker et al. 2002). The second approach applies GCMs and/or geochemically appropriate mass-balance models to proxy and/or mass flux curves (e.g. Berner 2003; Locklair & Lermann 2005; Ridgwell 2005). Neither of these methods is without problems: the former depending heavily on the dataset and the latter depending on the assumptions of the model. Though sophisticated approaches such as GCMs or multifactor box modelling allow palaeoclimatic hypotheses to be quantitatively established and/or tested, they may be extremely sensitive to certain parameters and differing algorithms can produce different results based on similar datasets (see Haywood et al. 2005). Moreover, GCMs are highly dependent on changes in palaeogeography, ocean bathymetry, pCO_2, insolation and albedo. So, unless these factors are well constrained, their results should be considered conservatively before universally accepting their applicability in non-actualistic environments (cf. Ridgwell 2005).

The Early Palaeozoic climate includes the seemingly paradoxical occurrence of extensive glaciations (the Early Palaeozoic Icehouse or EPI as defined below) at elevated atmospheric CO_2 (Royer 2006). Given the long-recognized coupling of CO_2 and temperature (Arrhenius 1896; Chamberlin 1899) and more recent affirmations of the sensitivity of temperature to CO_2 (e.g. Shackleton 2000; Zachos et al. 2001; Kump 2002; Siegenthaler et al. 2005), a link between CO_2 and temperature in the Early Palaeozoic seems reasonable. Decreased cosmic ray flux also may have contributed to globally cooler temperatures during the EPI (Veizer et al. 2000; Shaviv 2002; Shaviv & Veizer 2003), but this was insufficient to induce glaciation alone (Royer 2006).

Most models suggest that atmospheric CO_2 was the key control on temperature and ice formation in the Ordovician and Silurian, predicting a pCO_2-ice threshold around 3000 ppm (Kump et al. 1999; Hermann et al. 2003, 2004a, b; Royer 2006). These values are significantly lower than the GEOCARB III or GEOCARBSULF estimate for Hirnantian CO_2 levels at c. 4000 ppm (Berner & Kothavala 2001; Berner 2006). The GEOCARB/ GEOCARBSULF estimates are consistent with estimates from goethite (Yapp & Poths 1992) and significantly lower than the single palaeosol-based estimate of CO_2 for the Ashgill at c. 5600 ppm (Royer 2006). However, these models operate on longer timescales than the duration of the short-lived Hirnantian glacial maximum (cf. Sutcliffe et al. 2000), so the discrepancy between the estimated CO_2-ice threshold and estimates of atmospheric CO_2 may not be inconsistent (cf. Royer 2006). That is, glacial events could have been too rapid to be captured by either these models or the sparse proxy record. Individual glaciations in the EPI may have been short-lived events related to rapid CO_2 drawdown and cooling (e.g. Kump et al. 1999).

A stratigraphic approach to Early Palaeozoic glaciations

Lithostratigraphic correlation may establish a link between deposition of CO_2 sinks and glaciations during the Early Palaeozoic. CO_2 may be drawn down from the atmosphere and sequestered in rocks by (a) photosynthesis and burial of organic carbon, or (b) continental silicate weathering and carbonate deposition. In glacial intervals, changes in ice-volume allow changes in glacioeustatic sea-level to serve as a proxy for atmospheric CO_2 (assuming that ice volume was a decreasing function of temperature and that temperature was an increasing function of atmospheric CO_2). The stratigraphic occurrence of carbonates and argillaceous sediments has been well documented in the identification of Primo/Secundo or Humid/Arid episodes (e.g. Jeppsson 1990, 1997; Aldridge et al. 1993; Jeppsson et al. 1995; Bickert et al. 1997; Cramer & Saltzman 2007; see also discussion). We adopt a complementary approach by comparing the stratigraphic distributions of black shale with glacioeustatic sea-level curves, isotopic data and evidence of glaciations.

This approach depends on the selection of high quality datasets with well-resolved stratigraphies and accurate correlations. These are discussed in the Appendix. We have been cautious in assigning evidence of ice formation to the glacial maxima, as discussed below.

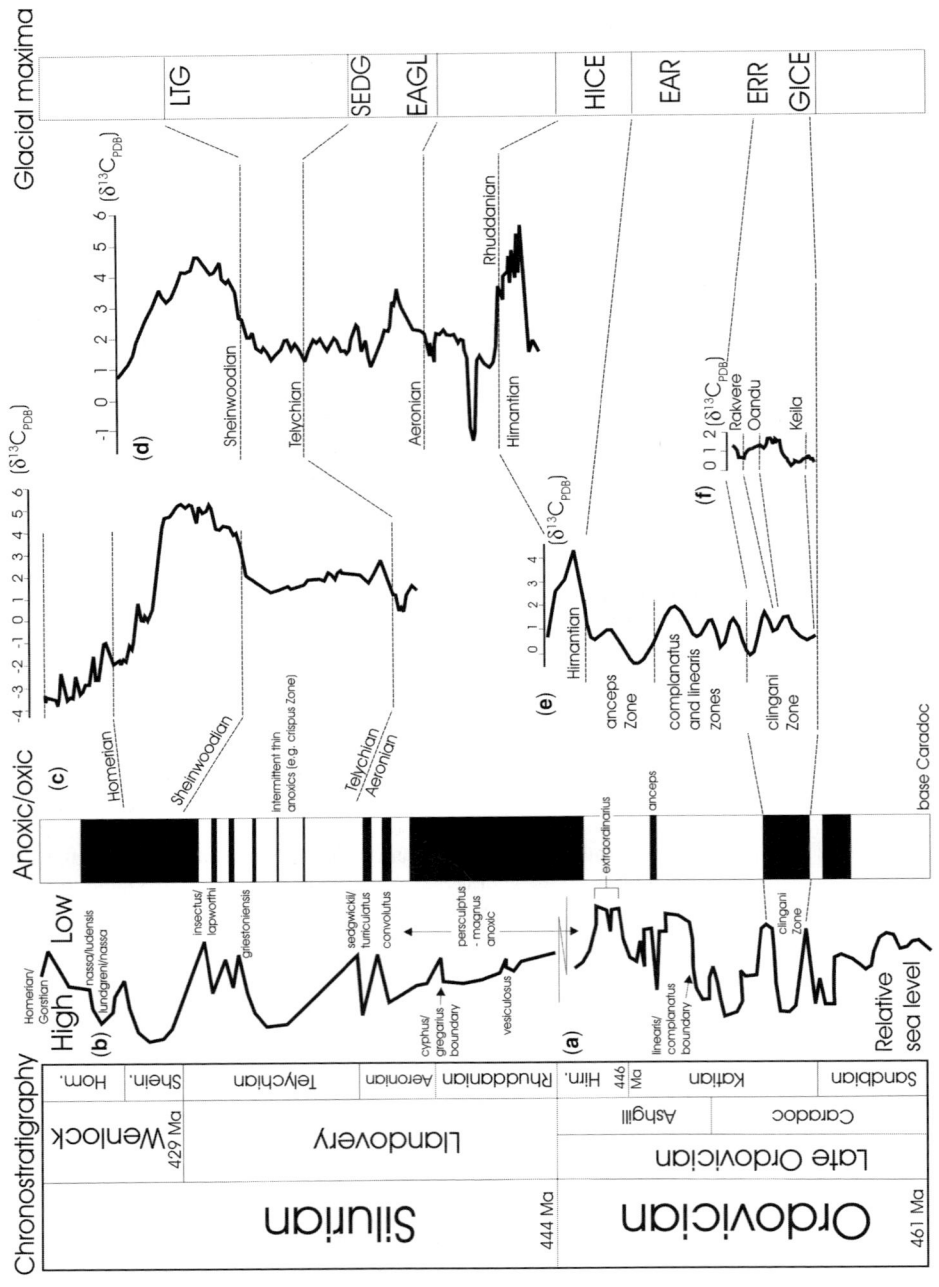

The Early Palaeozoic Icehouse

The Early Palaeozoic Icehouse (EPI) was an approximately 30-million-year interval comprising seven currently recognized glacial maxima (Table 1; Fig. 2). We propose that the EPI began with the Guttenberg Limestone carbon isotope excursion (GICE) in the Caradoc (Ordovician), and ended with Ireviken event deglacial transgression of the earliest Wenlock (Silurian). The EPI reached its greatest extent in the short-lived Hirnantian event identified by Brenchley *et al.* (1994), but there is a good evidence for extensive ice formation and significant glacioeustatic change throughout the EPI (Tables 1 & 2; Fig. 2). Several other authors have argued for an extended period of glaciation (e.g. Frakes *et al.* 1992; Eyles 1993; Evans 2003; Ghienne 2003; Kaljo *et al.* 2003; Nielsen 2003*a*; Saltzman & Young 2005).

The sedimentary record of glaciations in the EPI is predominantly held on Gondwana, with Ordovician deposits generally found in Africa, and Silurian deposits in South America (Table 2; Eyles 1993; Díaz-Martínez & Grahn 2007). This continental-scale diachronism may reflect the movement of Gondwana across the South Pole (cf. Fortey & Cocks 2003). The Tamadjert Fm of Saharan Africa displays a 200 m sequence of glacial deposits. These include extensive tillites and diamictites and abundant evidence of glacial erosion stretching from the Caradoc to the late Llandovery or even the earliest Wenlock (Beuf *et al.* 1971; Biju-Dival *et al.* 1981 and references therein). Three periods of continent-wide diamictite deposition are recognized in the early Silurian of South America (Caputo 1998; Díaz-Martínez 2007). After the latest Llandovery, the South American record of glaciation becomes ambiguous. Glaciogenic diamictites in the San Gabán–Cancañiri–Zapla and Nhamundá Fms (Table 2) are overlain by strata yielding early Wenlock conodonts and late Telychian–early Wenlock chitinozoans respectively (Díaz-Martínez 2007; Grahn in Cramer & Saltzman 2007). The most recent works on these glacial deposits consider them as having

an entirely Llandovery age (Díaz-Martínez 2007; Díaz-Martínez & Grahn 2007). Though the Kirusillas Fm of Bolivia contains diamictites of early Wenlock age (Merino 1991; Díaz-Martínez 2007), these lack glacially abraded clasts and are considered to be sediment gravity flows (Díaz-Martínez 2007; Díaz-Martínez & Grahn 2007).

The coupled, rapid variation in the isotopic and glacioeustatic records (Fig. 2) that continues throughout the EPI clearly marks a genetic change in the Earth system. The strong co-variation in these records may arise from the interval being a prolonged icehouse event (Kaljo *et al.* 2003; Nielsen 2003*b*). The good correspondence of third-order variation in eustatic sea-level curves (Ross & Ross 1996; Nielsen 2003*a*, *b*), along with evidence of ice advance and retreat (Table 2) argues for ice-volume controlling sea-level during the EPI. Likewise, there is strong co-variation in $\delta^{13}C$ and $\delta^{18}O$ data throughout the EPI (cf. Azmy *et al.* 1998; Shields *et al.* 2003). During the EPI, we interpret positive $\delta^{13}C$ excursions as being due to increased weathering of shallow carbonates exposed in regressions (cf. Kump *et al.* 1999; Melchin & Holmden 2006; see also the Appendix). Whilst in the EPI, positive $\delta^{18}O$ excursions are interpreted as due to cooling rather than increased salinity alone (Azmy *et al.* 1998). Therefore, coupled, positive $\delta^{13}C$ and $\delta^{18}O$ excursions may indicate glaciations if consistent with other evidence.

The synchronous onset of significant, coupled isotopic and eustatic fluctuations at the onset of the Katian in the Ordovician, mark the beginning of the EPI (Fig. 2; Patzkowsky *et al.* 1997; Kaljo *et al.* 2003; Nielsen 2003*a*, *b*). This corresponds to the mid-Caradoc *clingani* graptolite Zone (Cooper & Sadler 2004; Goldman *et al.* 2005). There is, though, also evidence of glacial erosion and a regression in the earliest Caradoc (Hamoumi 1999; Nielsen 2003*a*, *b*). The termination of the EPI is marked by the decoupling of isotopic and glacioeustatic variation in the earliest Wenlock (Fig. 2). In the latest Telychian, there is good evidence of ice (Table 2). Prior to the early

---

Fig. 2. (Opposite) The relationship between extent of marine anoxia and sea-level during the EPI based on the chronostratigraphy of Cooper & Sadler (2004) and Melchin *et al.* (2004), but placing the Llandovery–Wenlock Boundary at Ireviken datum 2 after the recommendations of Loydell *et al.* (2003) and Calner *et al.* (2004). Transgressive anoxia is apparent by comparing (**a**) summary sea-level curves to (**b**) the oxic/anoxic stratigraphy (dark bands, anoxic; white, oxic). Sea-level curves drawn after Ross & Ross (1996) and Nielsen (2003*a*); the oxic/anoxic stratigraphy is compiled from Figure 3. The Ross & Ross (1996) curve has been modified in accordance with its recorrelation in Loydell (1998). (**c–f**) Stable carbon-isotope curves: (c, d) redrawn from Kaljo *et al.* (2003); (e) redrawn from Kaljo *et al.* (2004); (f) redrawn from Ainsaar *et al.* (1999). (**g**) Glacial maxima recognized in Table 1: GICE, Guttenberg regression; ERR, early Rakvere regression; EAR, early Ashgill regression; HICE, Hirnantian glaciation; EAGL, early Aeronian glaciation; SEDG, sedgwickii Zone glaciation; LTG, late Telychian glaciation.

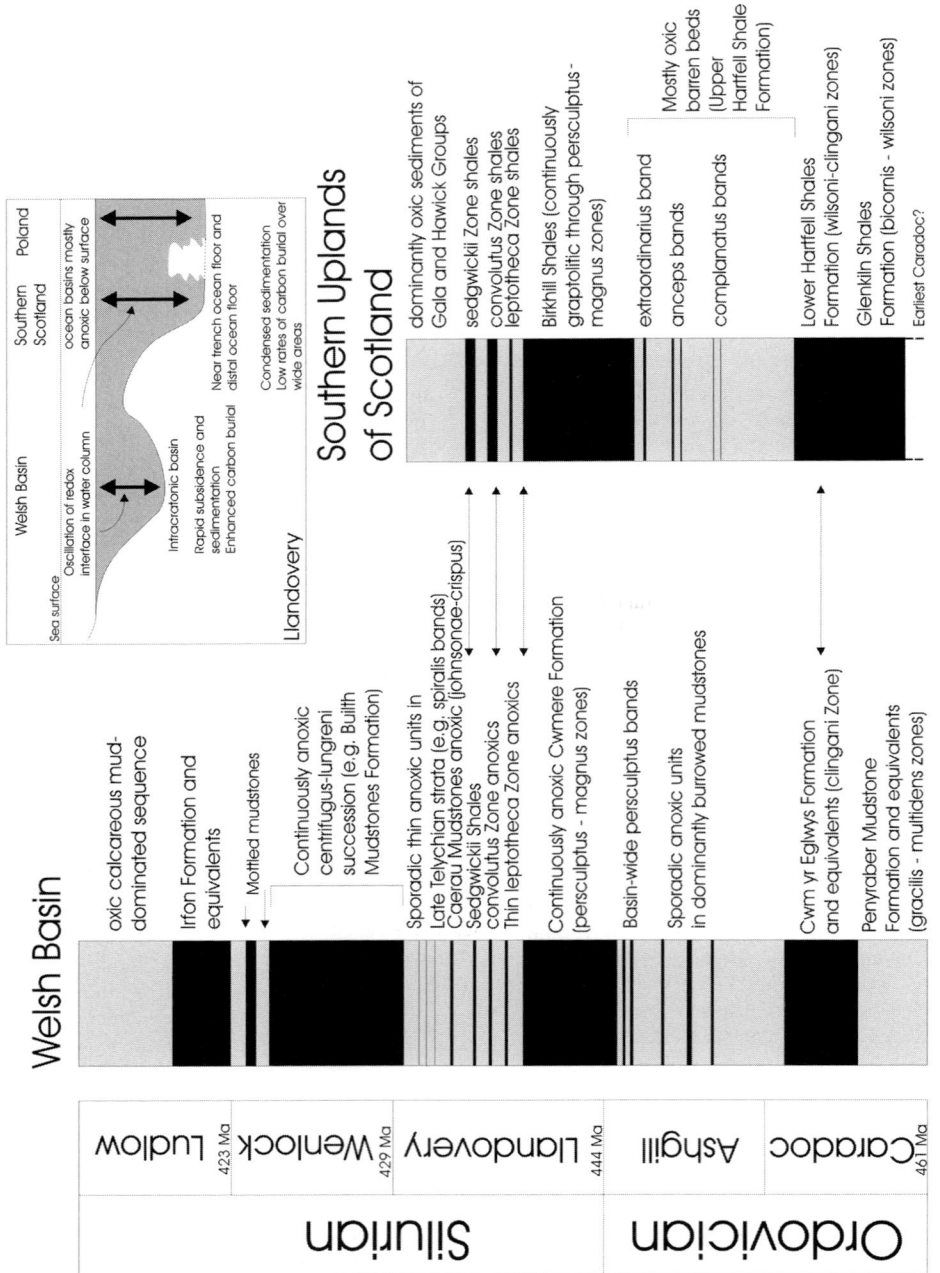

Wenlock Ireviken excursion *sensu* Cramer & Saltzman (2005, 2007), there are coupled, positive excursions in $\delta^{13}C$ and $\delta^{18}O$ (Bickert *et al.* 1997; Azmy *et al.* 1998) during a regression (cf. Loydell 1998). The Ireviken excursion itself occurred at a time of climatic amelioration (Cramer & Saltzman 2005, 2007). It witnessed a positive $\delta^{13}C$ excursion at globally high sea-level (Cramer & Saltzman 2005, 2007). This positive $\delta^{13}C$ excursion was accompanied by a slight negative $\delta^{18}O$ excursion, which may indicate warming (Bickert *et al.* 1997; Azmy *et al.* 1998). The decoupling of isotopic and glacioeustatic co-variation in the early Wenlock (Fig. 2) marks a genetic change in the Earth system. It is unlike any interval in the EPI itself, and accompanied climatic amelioration witnessed in the development of extensive limestone reefs (e.g. Copper 1994; Brunton *et al.* 1998).

Most recent work on Early Palaeozoic glaciations has focused on the Hirnantian (e.g. Sutcliffe *et al.* 2001; Hermann *et al.* 2004*a*, *b*; Armstrong *et al.* 2005; Le Heron *et al.* 2005) since Brenchley *et al.* (1994) argued for a short-lived Late Ordovician glaciation. This may partly reflect its coincidence with a mass extinction at a major stratigraphic division (e.g. Chen *et al.* 2000, 2005), as well as the deposition of major hydrocarbon source rocks in overlying strata (Lüning *et al.* 2000; Berner 2003). However, we have used the criteria of Brenchley *et al.* (1994) to recognize six more glacial maxima in the EPI. Namely, that sequence stratigraphic evidence for large, global lowstands, coincident with positive $\delta^{13}C$ and $\delta^{18}O$ excursions and evidence of ice formation, indicates a glacial maximum (Table 1; Fig. 2).

Within the EPI, individual glacial maxima were separated by warmer intervals (e.g. Fortey & Cocks 2005), and recent research has highlighted clear variability in the Silurian palaeoenvironment (e.g. Jeppsson 1990; Aldridge *et al.* 1993; Bickert *et al.* 1997; Johnson 2006; Calner & Eriksson 2006). The frequent evidence of ice formation and rapid, third-order variation in eustatic sea-level (e.g. Azmy *et al.* 1998; Caputo 1998; Loydell 1998; Hamoumi 1999; Ghienne 2003; Nielsen 2003*a*, *b*; Johnson 2006) suggests that ice-sheets may have dynamically expanded and retreated during the EPI.

Glacial maxima in the EPI

The seven glacial maxima of the EPI have been either assigned to existing named events or given stratigraphically descriptive names based on their nature. Thus, some are referred to as regressions and others are called glaciations, depending on the weight of evidence. We deal with each glacial maximum in turn below.

The Guttenberg regression (*GICE*) as defined in Table 1 and Figure 2 is named after the Guttenberg Limestone Member of the Decorah Fm in the Upper Mississippi Valley, USA, where the positive $\delta^{13}C$ excursion was first recognized (Hatch *et al.* 1987). This carbon isotope excursion has subsequently been recognized elsewhere and is considered to represent a global event (e.g. Patzkowsky *et al.* 1997; Ainsaar *et al.* 1999; Kaljo *et al.* 2004). It was accompanied by a synchronous positive $\delta^{18}O$ excursion of earliest *clingani* graptolite Zone age (Shields *et al.* 2003; Tobin *et al.* 2005). The high-resolution chemostratigraphy of Ludvigson *et al.* (2004) shows that the GICE occurred after three other smaller positive $\delta^{13}C$ excursions, and that it took place in the *P. tenuis* conodont Zone and *americanus* graptolite Zone (equivalent to the earliest *clingani* Zone). In the Katian GSSP (Black Knob Ridge, Oklahoma, USA), there is a seemingly synchronous $\delta^{13}C$ excursion just above the Sandbian–Katian boundary (Goldman *et al.* 2005). This is coincident with the late Keila age regression noted by Kaljo *et al.* (2003) and Nielsen (2003*a*); and Ludvigson *et al.* (2004) also noted that the GICE occurred in a time of stratigraphical downlap. Due to the lack of well-dated glacial deposits in this interval (Table 2), there is no unambiguous link with ice formation, but it may represent a period of cooling as continental ice-sheets were beginning to expand (cf. Patzkowsky *et al.* 1997).

The early Rakvere regression (*ERR*) is named after the stage in Estonia, where it is recognized in the carbon isotope and sequence stratigraphic records (Table 1; Fig. 2). This regression took place within the latest *clingani* graptolite Zone (Nielsen 2003*a*), equivalent to the *A. superbus* conodont Zone (Nõlvak *et al.* 2006). Though there is evidence for glacial erosion at this time (Table 2), there is no evidence for extensive tillite formation.

Fig. 3. (Opposite) Correlation of marine anoxia between UK depositional basins based on the widespread occurrence of graptolite-bearing anoxic mudrock facies, with inset schematic showing the differing depositional environments associated with this facies. Stratigraphies and inset compiled from original work of Toghill (1968); Cave (1979); Baker (1981); White *et al.* (1991). Davies *et al.* (1997, 2003); Pratt *et al.* (1995); Zalasiewicz *et al.* (1995); Porębska & Sawłowicz (1997); Schofield *et al.* (2004); and Verniers & Vandenbroucke (2006). Note that the Hirnantian and Llandovery oxic–anoxic stratigraphy of the Welsh Basin bears close resemblance to the Lake District and Howgill Fells succession of Northern England (Rickards 1970; Hutt 1974; Rickards & Woodcock 2005).

Table 1. *Name, age and evidence for each of the seven glacial maxima in the EPI; $\delta^{13}C_{PDB}$ = most positive value of carbon isotopes in positive excursion based on regional isotopic compilations; $\Delta\delta^{13}C_{PDB}$ = total change in carbon isotope values during positive isotope excursions based on regional isotopic compilations; $\delta^{18}O_{PDB}$ = most positive value of oxygen isotopes in positive excursion recorded in brachiopod shells; $\Delta\delta^{18}O_{PDB}$ = total change in oxygen isotope values during positive isotope excursions recorded in brachiopod shells; oxygen isotope data taken from Tobin et al. (2005) for the Guttenberg regression, from Brenchley et al. (1994) for the Hirnantian Glaciation in the Baltic region, and from Azmy et al. (1998) for all Silurian events; all other data compiled from Table 2, Figure 2, Kaljo et al. (2007), and references therein*

Glacial maxima	Evidence for ice	$\Delta\delta^{13}C_{PDB}$	$\delta^{13}C_{PDB}$	$\Delta\delta^{18}O_{PDB}$	$\delta^{18}O_{PDB}$	Extent of regression	Timing
Guttenberg regression	poorly dated glacial-erosive features in North Africa	~ + 1.0‰	+ 1.7‰	~ + 2.3‰	−2.7‰	medium	early *caudatus* graptolite Zone
early Rakvere regression	poorly dated glacial-erosive features in North Africa	~ + 0.8‰	+ 1.8‰	–	–	medium	late *clingani* graptolite Zone
early Ashgill regression	poorly dated Ashgill tillites and glacial erosive features in North Africa	~ + 1.2‰	+ 2.0‰	–	–	large	end *linearis* graptolite Zone
Hirnantian glaciation	Pan-Gondwanan tillites & diamictites containing *Hirnantia* fauna	~ + 4.0‰	+ 4.3‰	~ + 4.2‰	0‰	large	*extraordinarius–early persculptus* graptolite Zones
early Aeronian glaciation	*gregarius* Zone diamictite in South America	~ + 2.0‰	+ 3.0‰	~ + 0.6‰	−4.5‰	small	*gregarius* (?*magnus*) graptolite Zone
sedgwickii s.l. glaciation	well-dated diamicties in South America	~ + 0.5‰	+ 2.0‰	~ + 0.7‰	−4.4‰	medium	*sedgwickii* graptolite Zone
late Telychian glaciation	well-dated diamicties in South America	~ + 1.5‰*	+ 2‰*	~ + 1.0‰*	−4.6‰*	large	?*insectus–lapworthi* graptolite Zones

*These excursions occur in the late Telychian significantly preceding the Ireviken excursion *sensu* Cramer & Saltzman (2005, 2007).

This, along with its expression in the sea-level and isotopic record, may indicate that it was a relatively small, short-lived event.

The early Ashgill regression (EAR) appears to be a more significant event based on both its sea-level and isotopic record (Table 1; Fig. 2). Nielsen (2003a) noted significant regression in the early Ashgill *complanatus* graptolite Zone. This is synchronous with regression and a positive $\delta^{13}C$ excursion noted by Kaljo *et al.* (2004) at the basal Pirgu stage in Estonia (Nõlvak *et al.* 2006). There is good evidence for glacial sediments being deposited in the Ashgill of North Africa (which was close to the palaeomagnetic South Pole). However, this regression cannot be tied to any one particular high-latitude event, due to imprecision in the biostratigraphy of these successions (Table 2).

The Hirnantian glaciation (HICE) occurs as two pulses of glaciation within the *extraordinarius–persculptus* graptolite Zones (Sutcliffe *et al.* 2000). It represents the glacial maximum of the EPI and has received extensive study. It is well constrained and clearly globally extensive. The extent of the eustatic and isotopic variations associated with this are shown in Table 1 and Figure 2, with more detailed treatment of this event being found in Brenchley *et al.* (1994, 2003), Marshall *et al.* (1997), Sutcliffe *et al.* (2001), and Armstrong (this volume).

The early Aeronian glaciation (EAGL) has a significant expression in marine carbon isotope values, but this may also represent an increased effect of carbonate weathering relative to the Ordovician glaciations. There is clear evidence for ice formation during the *gregarius* graptolite Zone (Tables 1 & 2), but this is a long interval, which can be subdivided into the *triangulatus*, *magnus* and *argenteus* zones (see Hutt 1974). Precisely how ice formation in this event correlates with sea-level is unclear: the Rhuddanian–Aeronian boundary regression seen in the Ross & Ross (1996) sea-level curve is not recognized in other records, which show marked regressions at or around the *magnus–argenteus* graptolite Zone boundary (Johnson *et al.* 1991; Loydell 1998; Johnson 2006). Azmy *et al.* (1998) show the onset of a positive $\delta^{13}C$ and $\delta^{18}O$ excursion in the *triangulatus* Zone, but do not present data for the *magnus* and *argenteus* zones. Kaljo *et al.* (2003) illustrate a longer $\delta^{13}C$ excursion, which both they and Johnson (2006) correlate to an early Aeronian glaciation.

The sedgwickii graptolite Zone glaciation (SEDG) is discussed at length below. Loydell (1998) and Johnson (2006) both show major regressions at this time. Azmy *et al.* (1998) and Johnson (2006) argued for glaciations based on isotopic data, which we note correspond to well-dated diamictites (Table 2).

The late Telychian glaciation (LTG) is the final glacial maximum of the EPI. As noted above, the final pulse of diamictite deposition in South America can be assigned to the late Telychian, though cannot be constrained to any particular zone (Table 2). The late Telychian is coincident with regressions: Figure 2 illustrates a rapid sea-level fall in the late *insectus* graptolite Zone (see also Appendix), while Loydell (1998) shows a rapid regression in the *lapworthi* graptolite Zone with a lowstand throughout the *lapworthi–insectus* interval, also recognized in SW Siberia by Yolkin *et al.* (1997). During this interval, Azmy *et al.* (1998) show a positive $\delta^{13}C$ and $\delta^{18}O$ excursion, beginning in the *crenulata* graptolite Zone and reaching a maximum in the *centrifugus* graptolite Zone. Likewise, Kaljo *et al.* (1998, 2003) show that the onset of this positive $\delta^{13}C$ transition occurred in the late Llandovery, with subsequent studies placing this in the *Pt. amorphognathoides* conodont Zone in Estonia (Kaljo & Martma 2006) and possibly towards its base (see Cramer & Saltzman 2005). As such, we suggest this glaciation occurred within the *lapworthi–insectus* graptolite Zone interval.

Oxic–anoxic stratigraphy and sea-level in the EPI

Correlating the UK oxic–anoxic stratigraphy against glacioeustatic sea-level curves reveals a repeated relation between black shale deposition and deglacial transgressions throughout the EPI (Figs 2 and 3). Conversely, glaciations themselves correspond to well-oxygenated deep waters. Comparison with deep-water successions influenced by the Rheic and Palaeotethys oceans suggests these oxic–anoxic transitions may have had global extent. The glacioeustatic sea-level and oxic–anoxic stratigraphy are discussed briefly below, before the post-Hirnantian transgression and *sedgwickii* Zone regressions are dealt with in detail.

All the oxic to anoxic transitions of the EPI represent maximum flooding surface transgressive black shales (*sensu* Wignall & Maynard 1993; Wignall 1994). They are regionally extensive and post-date glacial maxima (Fig. 2; Woodcock *et al.* 1996; Armstrong *et al.* 2005). The GICE regression precedes the development of extensive anoxia within the Welsh Basin during the *clingani* graptolite Zone. At this time, widespread, organic carbon-rich and phosphatic black shales, such as the Nod Glas and Cwm-yr-Eglwys Mudstone Fms (Cave & Dixon 1993; Davies *et al.* 2003), were laid down upon oxic mudrocks and limestones, such as the Carswyn and Penyraber Mudstone Fms (Fig. 2; Pratt *et al.* 1995; Davies *et al.* 2003). In the

Platteville–Decorah formations of eastern Iowa (USA), the Guttenberg Member is brown shale (Ludvigson *et al.* 2004). This occurs above a well-laminated shale and widespread phosphatic bed of the Spects Ferry Member and below the well-laminated shales and blackened/phosphatic hardgrounds of the Ion Member (Ludvigson *et al.* 2004). Likewise, pyritic graptolite shales of *clingani* Zone age Nakkholmen Fm in the Oslo area (Norway) overlie limestones, as do coeval graptolite shales in the lower Mossen Fm of Västergötland (Nielsen 2003*a*). This represents an oxic–anoxic transition that reflects a profound drowning event (Nielsen 2003*a, b*).

The early Ashgill regression preceded the *anceps* graptolite Zone transgression, with the UK record witnessing simultaneous intervals of black shale deposition in an otherwise oxic succession (Fig. 3). For example, the 'Red Vein' in Wales, a thin unit of anoxic mudstones bearing graptolites of probable *anceps* graptolite Zone age (e.g. Schofield *et al.* 2004), appears synchronous with the *anceps* bands in Scotland (Williams 1982). The Hirnantian glaciation, with an acme in the *extraordinarius* graptolite Zone (Sutcliffe *et al.* 2000), preceded the deposition of globally extensive transgressive black shales in the *persculptus–acuminatus* graptolite zones (Fig. 4 & text below). Likewise, the *sedgwickii* graptolite Zone glacial event is characterized by the deposition of oxic facies during a lowstand, which is sandwiched between transgressive black shales (Fig. 5 & text below). The *convolutus* graptolite Zone represents a major transgression and global highstand following the early Aeronian glaciation, with widespread black shale deposition noted in Loydell (1998), synchronous with the UK *convolutus* bands of graptolitic shale (Fig. 3). Similarly, the deglacial transgression in the latest Telychian is characterized by the onset of anoxia and the deposition of marine black shales in the *centrifugus* Zone in the UK. These include the Builth Mudstones in the Welsh Basin (Woodcock *et al.* 1996; Zalasiewicz & Williams 1999) and the Brathay Mudstones in northern England (Rickards 1970; Rickards & Woodcock 2005). These were both deposited on essentially oxic, late Llandovery successions (Davies *et al.* 1997; Rickards & Woodcock 2005). In Baltoscandia, this event also sees the deposition of graptolitic shales on greenish-grey marlstones in the Ohesaare core from Estonia (Loydell *et al.* 1998) and on oolitic limestone/grey-green shale interbeds in Bornholm (Bjerreskov 1975; AAP unpublished observations April 2006). Similarly, Lüning *et al.* (2005) noted the deposition of 'hot shales' on the North African/Arabian margin during the *centrifugus* to *firmus* graptolite zones. The relation between deglaciation, transgression

and anoxia in the late Telychian–early Wenlock has been reviewed in depth by Cramer and Saltzman (2007). Further examples of transgressive black shales deposited after the Llandovery glaciations may be found in Loydell (1998).

During the EPI, the deposition of bioturbated facies, representing conditions of deep-marine oxygenation, occurred during regressions and glacial maxima (Brenchley 1988; Loydell 1998; Fig. 2). The Hirnantian and *sedgwickii* Zone glacial maxima correspond to intervals of grey shale and deposition of mottled (i.e. bioturbated) mudstones (Figs 4 & 5). Likewise, the positive carbon isotope excursion in the mid-*gregarius* graptolite Zone that marks the early Aeronian glaciation may correspond to the onset of oxic deposition in the mid-*magnus* graptolite Zone of the Welsh Basin (Fig. 3). In Black Knob Ridge, Oklahoma, the maximum $\delta^{13}C$ excursion corresponding to GICE occurs in an interval with extremely diminished C_{org} content in an otherwise organic-rich sequence (Goldman *et al.* 2005). This perhaps represents an interval of increased deep-water ventilation (cf. Ludvigson *et al.* 2004). The early Rakvere regression (latest *clingani* graptolite Zone) may possibly correlate with the transition from black shales to limestone at Whitland, South Wales. Also, the Fjäcka Shales of Sweden are deposited on the well-oxygenated facies of the Slandrom Limestone of Sweden and Bestorp Limestone of Västergötland, representing transgressive black shales deposited after the early Rakvere event (Männil & Meidla 1994).

Transgressive anoxia: post-Hirnantian glaciation oceanic anoxic event

The late *persculptus* and *acuminatus* graptolite Zones are characterized by deposition of globally extensive transgressive black shales (Fig. 4) immediately following the *extraordinarius*–early *persculptus* graptolite Zone acme of the Hirnantian glaciation (Sutcliffe *et al.* 2000, 2001), suggesting a fundamentally deglacial origin for the onset of global marine anoxia. This followed the enhanced deep ocean circulation and oxygenation that characterized the Hirnantian glaciation (Brenchley 1988; Brenchley *et al.* 1994; Armstrong & Coe 1997). High-palaeolatitude sedimentary successions typically consist of Hirnantian glacial deposits immediately overlain by black shales (e.g. Sutcliffe *et al.* 2001; Armstrong *et al.* 2005, 2006). Low-palaeolatitude settings see unambiguously oxic facies such as deep-water bioturbated mudstones overlain by deglacial black shales (e.g. Mu 1988; Armstrong & Coe 1997; Davies *et al.* 1997; Chen *et al.* 2000, 2005; Verniers &

Vandenbroucke 2006). Shallow-water shelly faunas in low- to mid-palaeolatitudes may also be buried below *persculptus* graptolite Zone black shales (e.g. Bjerreskov 1975; Mu 1988; Davies *et al.* 1997; Chen *et al.* 2000, 2005), though some shallow successions in the palaeotropics may see limestone deposition going on uninterrupted (e.g. Barnes & Bolton 1988). These deglacial black shales are widely palaeogeographically distributed and represent a global event (Fig. 4) perhaps comparable to the Mesozoic oceanic anoxic events (cf. Cohen *et al.* 2004).

Deposition of regionally extensive black shales on maximum flooding surfaces (*sensu* Wignall & Maynard 1993; Wignall 1994) in the *persculptus* graptolite Zone provides strong evidence for their onset in the end Ordovician–early Silurian transgression (Ross & Ross 1995, 1996; Loydell 1998; Lüning *et al.* 2000; Nielsen 2003*a*). However, precise biostratigraphic dating of high-latitude black shales is hindered by the relative scarcity of graptolites in these settings (Skevington 1974; Zalasiewicz 2001) and the prevalence of non-

diagnostic taxa with long ranges (Lüning *et al.* 2000). Nevertheless, where high-latitude, postglacial black shales contain a sufficient fauna to permit dating, the onset of anoxia can be assigned to the *persculptus* graptolite Zone: e.g. the black shales of the Cedarberg Fm, South Africa, and the Don Braulio Fm, Argentina (Sutcliffe *et al.* 2000, 2001); Batra Fm, Jordan (Armstrong *et al.* 2005); and Murzuq Basin, Libya (Lüning *et al.* 2000). Though Lüning *et al.* (2006) noted that the base of the Batra Fm is diachronous, with its lower member yielding an *acuminatus* graptolite Zone fauna towards the (present-day) North (Armstrong *et al.* 2005, 2006), this is neither inconsistent with the onset of anoxia occurring in the *persculptus* graptolite Zone, nor is it inconsistent with deposition of the Batra Fm as a maximum flooding surface black shale. As the transgression continued into the early Silurian, the oxygen minimum zone would have shoaled further up the shelf (Armstrong *et al.* 2006). The formation of early Silurian black shales, such as at the base of the Qusaiba Shale, Saudi Arabia (Aoudeh & Al-Hajri 1995), is

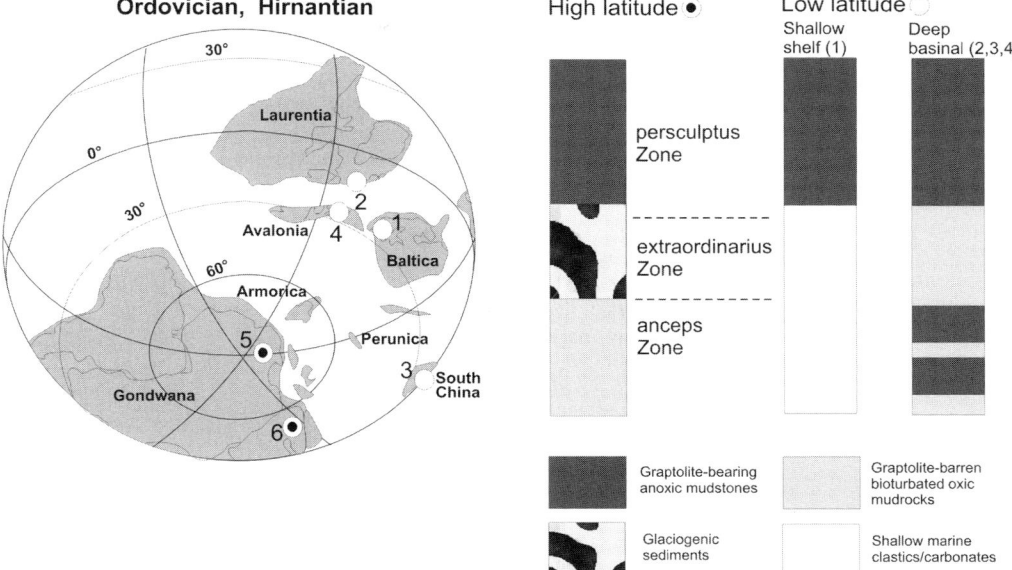

Fig. 4. Schematic lithographic logs showing the onset of transgressive anoxia in different settings during the *persculptus* graptolite Zone anoxic event. Palaeogeographical reconstruction showing the position of continents and associated terranes in the Hirnantian, annotated with localities where black shale is deposited on a maximum flooding surface: 1. Bavnegård Well, Bornholm, Denmark (Bjerreskov 1975); 2. Dob's Linn, Scotland UK (Toghill 1968; Armstrong & Coe 1997; Verniers & Vandenbroucke 2006); 3. Fenxiang, Yingang, Yangtze region, southern China (Mu 1988; Chen *et al.* 2000, 2005); 4. Cwmere Fm, central Welsh Basin, UK (Woodcock *et al.* 1996; Davies *et al.* 1997); 5. Murzuq Basin, Libya (Lüning *et al.* 2000); 6. Batra Fm, Jordan (Armstrong *et al.* 2005). Palaeogeographic reconstructions mainly after Torsvik *et al.* (1998). The relative position of Gondwana, Armorica, Baltica, Avalonia, Laurentia and Siberia are largely unmodified.

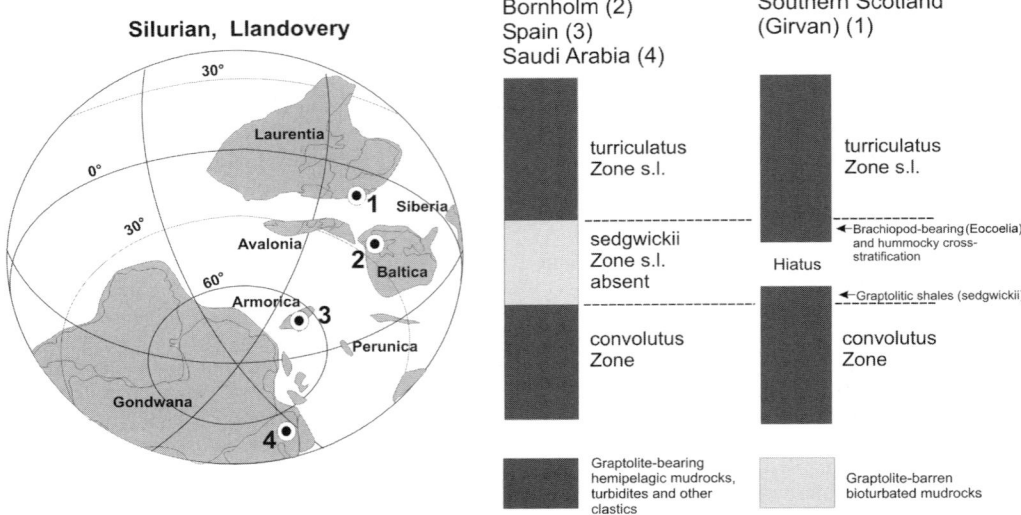

Fig. 5. Schematic lithographic logs showing evidence of regressive oxygenation during the *sedgwickii* graptolite Zone glaciation. Palaeogeographical reconstruction showing the position of continents and associated terranes in the mid-Llandovery, showing global event of deep-water oxygenation: 1. Girvan Group, Scotland, UK (Floyd & Williams 2003); 2. Øleå, Bornholm, Denmark (Bjerreskov 1975); 3. Western Iberian Cordillera, NE Spain (Gutiérrez-Marco & Štorch 1998); 4. Qusaiba Shale, Qalibah Fm, Saudi Arabia (Miller & Melvin 2005; AAP/JAZ/MW unpublished observations). Palaeogeographic reconstructions again mainly after Torsvik *et al.* (1998), with amendments as noted in Figure 4. Also, by the beginning of the Silurian, the amalgamation of some Kazakh crustal terranes probably led to the formation of a substantial landmass north of Tarim and South China (Koren *et al.* 2003). The position of Annamia close to South China and the Karakum–Tajik plate is mainly based on the strong affinities of the late Silurian (Ludlow–Pridoli) brachiopod faunas (Rong *et al.* 1995; Thong-Dzuy *et al.* 2001).

evidence that oceanic anoxia and deposition of maximum flooding surface black shales continued as the transgression continued.

The influx of deglacial meltwater in the oceans of the *persculptus* graptolite Zone may have been critical to the onset of transgressive anoxia: it may have increased marine stratification through the formation of low-salinity surface waters and, by providing a source of nutrients via continental weathering, stimulated marine productivity. This may be analogous to the formation of sapropels in the Neogene Mediterranean Basin (Rohling & Gieskes 1989; Rohling 1994; Scrivner *et al.* 2004). Buoyant, low-salinity surface waters, strengthening the pycnocline in the deglacial Hirnantian Ocean, may have precluded deep-water thermohaline circulation to sufficiently maintain a well-oxygenated sea-floor. Periglacial outwash may have carried sufficient nutrients to fuel the deposition of the 'hot shales' of North Africa and Arabia (cf. Meybeck 1982), which are characterized by a total organic-carbon content of up to 17% (Lüning *et al.* 2000), well above that found in normal black shales.

Some authors have declared that 'hot shale' deposition may be a result of upwelling (Lüning *et al.* 2000, 2005, 2006), but this disaccords with both their widespread, synchronous deposition and with GCMs of Hirnantian circulation (Hermann *et al.* 2003, 2004*a*, *b*). Upwelling is a regionally localized phenomenon and the oxygen minimum zone associated with upwelling zones is only stable on decadal timescales (Wignall 1994). Moreover, meridional and monsoonal coastal upwelling is restricted to low- to mid-latitudes (Parrish 1982), so are unlikely to apply to the 'hot shales', which occur at high palaeolatitudes. Meanwhile, end-Ordovician continental configuration is inconsistent with widespread zonal coastal upwelling (Armstrong *et al.* 2006). Zonal upwelling occurs when north or south continental margins lie adjacent to the major zonal wind systems (Parrish 1982). Comparing modern palaeogeographical reconstructions (Scotese & McKerrow 1991; Cocks & Torsvik 2002) with the high-latitude zonal wind predicted in the late Ordovician atmospheric simulations of Parrish (1982) shows that major winds were primarily orthogonal to the

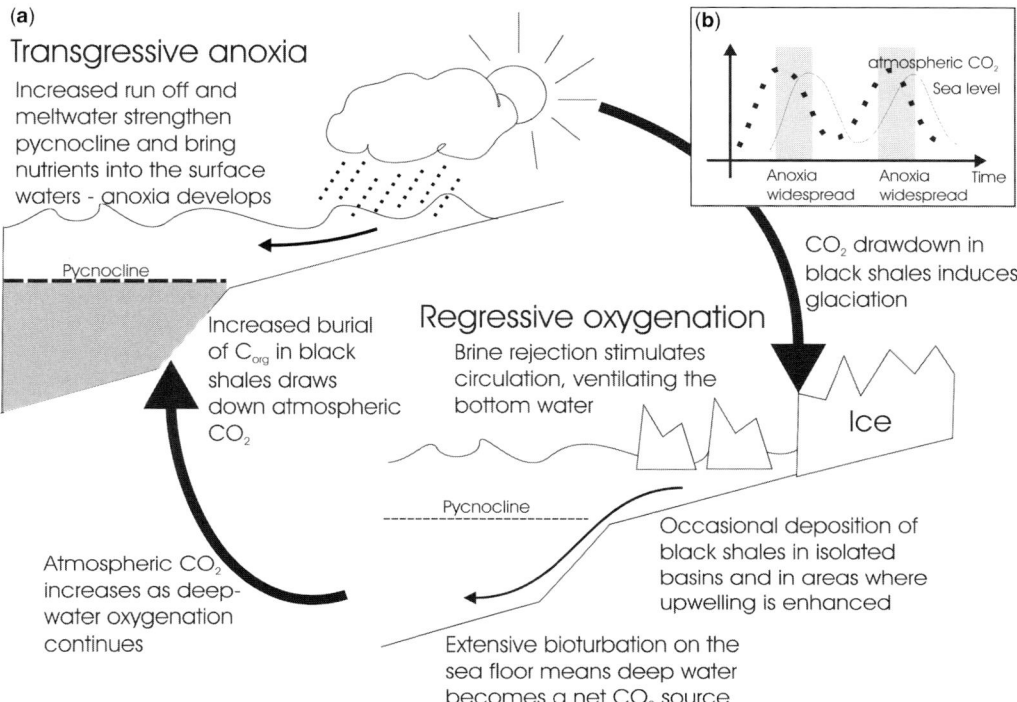

Fig. 6. (**a**) Summary cartoon showing end-member transgressive and regressive oceans in the EPI, outlining a model in which transgressive black shale deposition may serve as a negative-feedback mechanism modulating glacioeustasy (**b**) Graphs showing the postulated relationship between CO_2 temperature and sea-level, given that intervals of transgressive black shale deposition may draw-down significant atmospheric CO_2.

Gondwanan margin, making widespread zonal upwelling untenable. This is borne out by recent, more sophisticated GCMs for the late Ordovician, that show predominantly onshore ocean currents around the North African and Arabian margins of Gondwana (Hermann *et al.* 2003, 2004a, b). As such, upwelling alone can neither account for the simultaneous onset of globally widespread anoxia in the *persculptus* graptolite Zone nor for the widespread high-latitude occurrence of the most organic-rich shales.

Furthermore, deglacial melting would most likely be to the detriment of the increased thermohaline circulation needed to sustain upwelling. In the Quaternary of the North Atlantic, intervals of meltwater outwash are associated with increased ocean stratification and more sluggish deep-water circulation (Rahmstorf 2002). Instead, an alternative nutrient source may lie in increased continental weathering during deglaciation, providing a major source of both major and trace nutrients (Broecker 1982; Meybeck 1982) and stimulating productivity (as modelled by Kump & Arthur 1999. Given that the locus of glacial melting is

focused on the Gondwana palaeocontinent, this nutrient source could explain the relatively increased organic content of high-palaeolatitude black shales compared to those in low palaeolatitude black shales. This would leave anoxia at low palaeolatitudes as a consequence of increased ocean stratification and decreased thermohaline circulation due to high-palaeolatitude ice melting. This, coupled with sediment starvation (cf. Wignall 1991), would lead to an expanded oxygen-minimum zone. Hence, periglacial run-off provides a mechanism for anoxia that would have simultaneously increased global seawater stratification as well as stimulating productivity and export production (see schematic in Fig. 6). This highlights the fundamentally deglacial nature of transgressive black shales in the EPI.

Regressive oxygenation: the *sedgwickii* graptolite Zone event

The *sedgwickii* graptolite Zone is marked by global regression of plausibly glacial origin during which sediments were deposited in well-oxygenated deep water overlying earlier black shales (Table 1; Figs 2 & 5). Graptolites are rarely preserved in extensively bioturbated facies, though graptolites could clearly exist in well-oxygenated waters (Armstrong & Coe 1997; Floyd & Williams 2003; AAP/JAZ/ MW unpublished observations). Therefore, bioturbated graptolite-free intervals in otherwise continually graptolitic successions may be taken as evidence of increased sedimentary ventilation, seen in both high- and low-latitude successions in the Iapetus, Rheic and Palaeotethys Oceans (Fig. 5 & references therein). It is clear that the boundary between the preceding *convolutus* graptolite Zone and the *sedgwickii* graptolite Zone is characterized by the development of oxic strata, and that widespread anoxia is again developed by the onset of the *guerichi* graptolite Zone (Loydell 1998).

However, the *sedgwickii* graptolite Zone also contains a distinctive interval of black shale, which is only seen in certain successions from the Iapetus and Rheic Oceans. In UK successions in Wales and Dob's Linn, this thin black shale is sandwiched between a sequence of grey shales and mottled mudstones (e.g. Toghill 1968; Cave 1979). The *sedgwickii* Zone of the Kullatorp core (Västergötland, Sweden) comprises greyish-green mudstones in a succession of otherwise graptolitic black shales. However, the Silvberg and Gulleråsen-Sanden sections of Dalarna contain a finely laminated shale yielding *St. sedgwickii* (Loydell 1998). In Bornholm, Denmark, however, there is an abrupt transition between the highly graptolitic black shale of the *cometa* Band that characterizes the top of the *convolutus* graptolite Zone and a heavily bioturbated, non-graptolitic silty, grey mudstone (Bjerreskov 1975; AAP unpublished observations on Øleå section and collections in the Geological Museum, University of Copenhagen). The *sedgwickii* shale band in the Welsh Basin is generally thought to represent a transgressive black shale (Cave 1979; Baker 1981; Davies & Waters 1995; Woodcock *et al.* 1996). However, it is unlike the transgressive black shales of the *persculptus* graptolite Zone and early Wenlock (Woodcock *et al.* 1996; Lüning *et al.* 2005; Armstrong *et al.* 2005, 2006) in which organic-carbon burial is more profoundly expressed at high palaeolatitudes. Instead, this brief black shale event in the *sedgwickii* Zone is only observed at certain low-latitude settings. There is no major graptolitic shale burial associated with the *sedgwickii* Zone age strata in the Qusaiba Shale, Saudi Arabia (Zalasiewicz *et al.* 2007). This may reflect increased isolation of semi-enclosed basins at low sea-level, suggesting local rather than global significance (e.g. Loydell 1994).

There is good evidence of a regression within the lithofacies deposited during the *sedgwickii* graptolite Zone, and no graptolites have been recovered from the low-latitude successions of the Western Iberian Cordillera, Spain (Gutiérrez-Marco & Štorch 1998), the Prague Basin, Czech Republic (Štorch 1986, 1994), and the Canadian Arctic (Melchin 1989). In Girvan, southern Scotland, UK, the late *sedgwickii* Zone contains shallow-water brachiopods and deposits with hummocky cross-stratification in the Lower Camregan Grits Fm overlying the black shales of the Pencleuch Shale Fm (Floyd & Williams 2003), and in Spengill, Howgill Fells, UK, the *sedgwickii* graptolite Zone contains the only occurrence of limestones and grits in its Llandovery succession (Rickards 1970). The evidence for a regression in UK deep-water strata may correlate with the regressions recognized in the *sedgwickii* s.l. graptolite Zone in sequence stratigraphic analyses of shallow-water facies in Baltoscandia, North America and Siberia (Johnson 1996; Ross & Ross 1996; Yolkin *et al.* 1997), and coincides with the formation of diamictite in Gondwana (Caputo 1998); whilst the overlying *guerichi* graptolite Zone itself appears to mark the onset of extensive marine anoxia (Loydell 1998), which may be linked to a transgression (Fig. 2).

Increased oxygenation of the marine realm in the *sedgwickii* graptolite Zone may be a consequence of high-latitude ice formation stimulating more rapid thermohaline circulation, as is seen in the modern-day Atlantic (Rahmstorf 2002). The *sedgwickii* Zone regression is coincident with diamictite deposition in South America (Caputo 1998; Table 1; Fig. 2), suggesting glacioeustatic control. The formation of marine ice would have resulted in brine rejection, creating cold, dense waters at high latitudes. On sinking, these may have driven a more vigorous deep-water circulation, providing a greater flux of oxygen to the deep oceans (see Fig. 6), consistent with the well-ventilated deep-water facies seen in glacial maxima during the EPI (Fig. 2; Brenchley 1988). The global cessation of anoxic facies at the end of the *convolutus* graptolite Zone and their return in the *guerichi* graptolite Zone reflects a third-order change in depositional style, consistent with a glacioeustatic control on oceanic redox state (Church & Coe 2003).

A simple model for carbon burial and glacioeustasy in the EPI

During the EPI, there was a fundamental link between glacioeustatic sea-level change and the burial of organic carbon in deglacially transgressive black shales, which may have represented a negative feedback mechanism that served to stabilize climate (Fig. 6), and prevented the onset of runaway greenhouse conditions. Given that sea-level may represent a proxy for atmospheric temperature, which itself is a function of pCO_2, the deposition of globally extensive black shales on maximum flooding surfaces may have served to slow the initial warming after the glacial maxima by drawing down CO_2 from the ocean–atmosphere system. This would sustain the EPI and prevent onset of greenhouse conditions that characterized most of the Early Palaeozoic (Frakes *et al.* 1992; Gibbs *et al.* 2000; Montañez 2002; Church & Coe 2003).

This model for deposition of black shales due to periglacial meltwater increasing both productivity and ocean stratification predicates that oceanic anoxia was intimately linked to glacioeustasy. Black shale deposition in transgressions may have been significant to influence cooling, and therefore a regression, by drawing down atmospheric carbon (see the schematic graph in Fig. 6). Likewise, the onset of well-oxygenated oceans due to brine rejection driving deep-water circulation may have served to prevent organic carbon burial to sustain fully glacial conditions, which might have led to another transgression. With global marine oxygenation linked to ice formation and melting, we infer a strong link between organic carbon burial in black shales, sea level and atmospheric CO_2 during the EPI. This model now needs to be rigorously tested against both the sedimentary record of the EPI and by developing theoretical models of the carbon budget.

Black shale deposition and the EPI carbon cycle

Though we have clearly shown a strong link between black shale deposition and deglacial transgressions, relating anoxia and carbon-burial flux is not straightforward. Models of the carbon cycle show a significant increase in carbon burial as black shales close to the Ordovician–Silurian boundary (Fig. 1c). Meanwhile, Ronov *et al.* (1980) estimated that the amount of organic carbon buried in Ordovician and Silurian black shales is comparable to that in Permo-Carboniferous strata, a time when the organic reservoir may have exerted a significant influence on atmospheric CO_2 (e.g. Bruckschen *et al.* 1999). As the EPI corresponds to a major low in shelf-carbonate deposition, organic carbon burial in black shales may have been critical to regulating the carbon cycle (Fig. 1; Patzkowsky *et al.* 1997). The increase in atmospheric O_2 during this interval is consistent with increased organic-carbon burial (Fig. 1a, c; cf. Berner 2003; Catling & Clare

2005), empirically linking black shale deposition and atmospheric change. Though we have shown that black shale burial occurs in transgressions, the lateral extent of black shale is unclear, making it hard to estimate the carbon burial flux from rock volume.

The oxygenation–depth profile of EPI oceans is largely unknown. Oxic–anoxic transitions occur in open-ocean settings, demonstrating that this is not a phenomenon exclusive to restricted basins or epicontinental seas (cf. Porębska & Sawłowicz 1997). Similarly, anoxia may be developed in shelf settings relatively close to the storm wave-base, such as in the Qusaiba Shale of Saudi Arabia, where anoxic or dysoxic facies alternate with bioturbated mudstones and facies yielding benthic faunas (Miller & Melvin 2005). Precisely how shallow anoxic conditions may extend in the EPI is unclear (cf. Pancost *et al.* 1998). Whether these transgressive anoxic events represent shoaling of an expanded oxygen minimum zone onto the shelf, or whether there was widespread whole oceanic anoxia in the EPI, remains uncertain.

We concur with Cramer & Saltzman (2005, 2007) that the deposition of organic carbon in epicontinental black shales cannot account for positive $\delta^{13}C$ excursions in the Silurian. As they noted, widespread deposition of shales in epicontinental seas does not coincide with these excursions. Neither do our data show evidence for increased carbon burial in deep-water settings with strong oceanic influence during any of the positive $\delta^{13}C$ excursions (Table 1; Fig. 2). We also contest the suggestion of upwelling and increased carbon burial as an explanation for positive $\delta^{13}C$ excursions in the EPI (e.g. Young *et al.* 2005). All positive $\delta^{13}C$ excursions in the EPI are best explained by increased weathering of shallow marine carbonates in regressions (e.g. Kump *et al.* 1999; Melchin & Holmden 2006) as noted in the Appendix. Young *et al.* (2005) argued that the Guttenberg regression was synchronous with biomarker evidence for photic-zone anoxia based on the data of Pancost *et al.* (1998). This however represents a miscorrelation. The high-resolution stratigraphy of Ludvigson *et al.* (2004) demonstrates that the interval of photic zone anoxia identified by Pancost *et al.* (1998) corresponds to the Spects Ferry Member of the Platteville Fm. The Spects Ferry Member underlies the Guttenberg Limestone, preceding the EPI, whilst the Guttenberg regression corresponds to lithological evidence of a more-oxic interval (cf. Ludvigson *et al.* 2004). As noted above, there is no direct evidence for increased organic carbon preservation or productivity during the positive $\delta^{13}C$ excursions of the EPI (cf. Chen *et al.* 2005). Because of this, and the inability of upwelling to explain the synchronous onset of global anoxia, we attribute anoxia to increased deglacial outwash causing oceanic stratification (Fig. 6).

It is still unclear how black shale burial varies with regard to the burial of other facies associated with atmospheric CO_2 drawdown. On timescales greater than 1000 years, carbonate burial provides a sink for CO_2 (Elderfield 2002). Continental silicate weathering draws down CO_2 over geological timescales and may have been the key driver of changes in atmospheric CO_2 over geologically long timescales (Holland 1978; Berner 1991; Raymo 1991; Kump *et al.* 1999; Cohen *et al.* 2004). As continental weathering is thought to increase with greater temperatures (Berner *et al.* 1983; Velbel 1993; Hovius 1998), and the rate of erosion appears to be greater in unstable (transitional) climates than in stable greenhouse or icehouse climates (e.g. Shuster *et al.* 2005), one would expect a greater CO_2 drawdown due to weathering in transgressions. Our model for drawdown of CO_2 in transgressive anoxia due to increased freshwater runoff from the continents would augment this, complementing drawdown of CO_2 to the oceans by silicate weathering and limestone burial (e.g. Kump *et al.* 1999). Improved constraints on the relation between weathering, carbonate deposition, black shale burial and changes in global temperature would further improve our understanding of EPI climate.

Comparison of black shale and carbonate burial in the EPI

Black shale burial can be compared to carbonate deposition and changes in global temperature by relating the oxic/anoxic stratigraphy of Figure 2 to Primo/Secundo (P/S) and Humid/Arid (H/A) cycles. These cycles relate changes in carbonate deposition to changes in temperature and deep-water circulation. P/S and H/A cycles were recognized in the Silurian by Jeppsson (1990, 1997) and Bickert *et al.* (1997) respectively. These cycles have recently been reviewed and synthesized by Cramer & Saltzman (2005; 2007), who viewed Silurian climate as alternating between two end-member oceanographic regimes. The first regime is characterized by cooler (P), wetter (H) climates with high sea-levels; argillaceous limestone deposition took place in shallower successions and black shales were deposited on the continental shelf, with oxic deep waters. The other regime is characterized by warmer (S), more arid (A) climates with lower sea-level; reefs formed in shallow successions and limestones were deposited on the shelf with black shales deposited in anoxic open oceans. It should, however, be noted that Jeppsson (1990, 1997) and Bickert *et al.* (1997) differ in their views on where and when anoxia occur.

Bickert *et al.* (1997) suggested the open oceans were anoxic throughout both H and A episodes with anoxia shoaling onto the shelf in H episodes. Jeppsson (1990, 1997) proposed oxic deep waters occurred in P episodes with anoxia occurring on the shelf due to high productivity in these episodes; conversely, he proposed that S episodes were characterized by well-oxygenated shelf conditions and anoxia in the open oceans.

There is no simple relationship between the oxic–anoxic stratigraphy of the British Isles and either P/S or H/A cycles. In the Ordovician, Kaljo *et al.* (1999; 2004) recognized humid and arid episodes in both the mid–late Caradoc, which was predominantly anoxic, and the Ashgill, which was predominantly oxic (Fig. 3). Likewise, observations of glacial maxima and anoxia do not accord well with the P/S model. For example, the Spirodden Secundo episode lasts from the *persculptus* graptolite Zone until the *argenteus* graptolite Zone (Aldridge *et al.* 1993), a time recognized as being a warmer deglacial interval. This episode has a similar duration to the anoxic conditions in the UK (*persculptus–magnus* graptolite zones interval) as shown in Figure 2. In contrast, the Malmøykalven Secundo episode (Aldridge *et al.* 1993) corresponds to the *sedgwickii* graptolite Zone, a period of globally extensive oxic conditions and glaciation, as discussed above. Moreover, the P/S episodes recognized in the Llandovery (Aldridge *et al.* 1993) do not correspond well with glacioeustatic sea-level changes or episodes of organic carbon burial in black shales (Fig. 2; Loydell 1998). There was a relatively low rate of limestone burial during the EPI (Fig. 1b), and glacial maxima show a clearer link between changes in oceanic redox state than they do with either P/S or H/A cycles. So, we suggest that ocean anoxia seems to reflect changes in atmospheric CO_2 and temperature more clearly than does carbonate deposition.

Discussion

The model works well for the clearly deglacial anoxic events in the EPI, especially those in the *clingani*, *persculptus* and *centrifugus* graptolite zones (Fig. 2). Likewise, there is clear evidence for increased deep-water ventilation and the deposition of well-oxygenated mudstones in glaciations themselves (Figs 2 & 5). However, it only addresses how black shale deposition may serve as a negative feedback to stabilize EPI climate rather than providing a mechanism for the onset and termination of ice formation in the EPI. Recognizing repeated glacial maxima within the EPI, rather than just focusing on the Hirnantian event, significantly furthers our understanding of Early

Palaeozoic climate. Characterizing the lithostratigraphic patterns associated with each glacial maximum has established a consistent set of events associated with ice-sheet formation and retreat. The factors associated with ice formation within individual EPI glacial maxima can be potentially inferred from these 'event stratigraphies'. The differences in scale between each of the glacial maxima may be used to infer which factors were most important for ice formation in the EPI. That is, the most important factors controlling ice formation are likely to be most strongly expressed in large events such as the Hirnantian glacial maximum, but only weakly expressed in smaller events. Once the factors which control ice formation in glacial maxima have been established, it may be possible to infer what was responsible for the onset and termination of the EPI.

We have employed stratigraphic correlation of sub-graptolite zone resolution to recognize individual glacial maxima within the EPI and their associated lithostratigraphic changes. By combining isotopic data with glacioeustatic curves and lithostratigraphic patterns, there is considerable potential for developing a highly resolved Ordovician–Silurian global event stratigraphy. By comparison, biostratigraphy alone offers comparatively poor resolution and global coverage. The first appearances of many graptolites are diachronous (cf. Williams *et al.* 2003, 2004; Cooper & Sadler 2004). Likewise, the relative scarcity of graptolites in cool waters and in shallow environments (Skevington 1974; Finney & Berry 1997; Loydell 1998; Zalasiewicz 2001) may preclude accurate correlation between high and low latitudes (e.g. Lüning *et al.* 2000), and also between graptolite and conodont or shelly-fossil biostratigraphies (e.g. Mullins & Aldridge 2004). Contrastingly, carbon isotope stratigraphies appear to show good correlations between differing facies and palaeogeographical settings, allowing global correlation (e.g. Underwood *et al.* 1997; Melchin & Holmden 2006; Kaljo *et al.* 2007). Moreover, the strong coupling of both positive carbon isotope excursions with changes in deep-water redox conditions and glacioeustasy (Figs 2 & 6) represents a method of correlating third-order sequence stratigraphic changes, potentially providing a new tool for Early Palaeozoic stratigraphy. Likewise, this model for formation of transgressive black shales and their role in carbon burial (Fig. 6) may extend to other intervals. For example, the Toarcian oceanic anoxic event in the Jurassic saw continental run-off corresponding to the onset of anoxia, which induced CO_2 drawdown and subsequent cooling (Cohen *et al.* 2004).

In the EPI, neither the occurrence of glacial maxima nor oxic–anoxic transitions are regularly spaced (Fig. 2) and these events may therefore have been externally forced. The lack of observed cyclicity in the occurrence of glacial maxima suggests that the negative feedback due to CO_2 drawdown in transgressive black shale deposition was insufficient to create regular, self-sustaining glacial–interglacial cycles. The long intervals of anoxic marine conditions (e.g. late Caradoc, early Silurian and late Telychian–early Wenlock) and predominantly oxic marine conditions (e.g. early Caradoc, Ashgill, Telychian) may suggest that there was a second-order, possibly tectonic control on ocean redox conditions (see also Leggett 1980; Leggett *et al.* 1981). Likewise, the distribution of the currently recognized glacial maxima may be of second-order rather than third-order periodicity. Such second-order forcing, possibly by the opening and closing of oceanic gateways (e.g. Smith & Pickering 2003; Armstrong this volume), or drawdown of CO_2 due to increased silicate weathering in orogenies (Kump *et al.* 1999), may ultimately be responsible for the conditions that facilitated ice formation in the EPI.

Conclusions

Recognizing the EPI as a long-lived interval revises our understanding of Early Palaeozoic climate, showing a long-lived icehouse in an interval previously thought to be dominated by greenhouse conditions (cf. Brenchley *et al.* 1994; Gibbs *et al.* 2000; Montañez 2002; Church & Coe 2003; Royer 2006). Its *c.* 30-million-year duration makes it comparable to other long-lived icehouses such as those in the Cenozoic, Permo-Carboniferous and even the Neoproterozoic. All of these events are characterized by waxing and waning of ice-sheets on different timescales with potentially different forcing and feedback mechanisms dependent on timescale. Moreover, each of these icehouses represents a markedly different solution to the Earth's carbon budget, and the locus and importance of organic burial varied considerably between them (cf. Fig. 1). Though atmospheric CO_2 levels appear to have controlled global temperature over geological time (Royer 2006), each of these icehouses is characterized by a markedly different carbon cycle, which needs to be understood in its own terms.

Though there are many uncertainties shrouding our understanding of glaciation in this interval, the EPI is notably different from those of the modern oceans: it may have represented a time when the deposition of black shales in conditions of marine anoxia played a significant role in mediating the carbon cycle and climate.

We wish to thank David Harper and an anonymous reviewer for their helpful and insightful comments. AAP wishes to thank Eddie Blackett, Andrea Snelling (both Leicester & BGS), Nigel Woodcock & Barrie Rickards (Cambridge) and Jakob Vinther (Copenhagen) for showing him around *persculptus* and *sedgwickii* graptolite Zone sections in Wales, Sedbergh and Bornholm, and for providing access to their collections from these areas. Likewise, we thank Merrell Miller (Saudi Aramco) for providing access to boreholes of late Aeronian–early Telychian strata from Saudi Arabia. Cambridge Earth Science contribution 8775.

Appendix: Choice/interpretation of datasets and correlations

Details of individual datasets used in this study and our interpretations of them are discussed in turn below.

Stratigraphic framework: our correlations follow those of Cooper & Sadler (2004) for the Ordovician, and Melchin *et al.* (2004) for the Silurian, although we correlate the Llandovery–Wenlock boundary to Ireviken datum 2 (see next paragraph). In addition, we refer to Nõlvak *et al.* (2006) for correlation of the Ordovician timescale in Estonia to the global standard. We refer to other stratigraphic compilations in the text as necessary and assume that all major isotopic excursions are globally synchronous.

The International Commission on Stratigraphy website notes that the correlation of the Llandovery–Wenlock boundary GSSP is 'imprecise' (www.stratigraphy.org). To resolve this, the stratigraphy of the late Telychian–early Wenlock interval has received much attention of late (e.g. Jeppsson 1997; Loydell *et al.* 2003; Munnecke *et al.* 2003; Mullins & Aldridge 2004; Cramer & Saltzman 2005, 2007). The absence of taxonomically identifiable graptolites in the GSSP (Mullins & Aldridge 2004) makes correlation to graptolite zones uncertain (Melchin *et al.* 2004). However, the GSSP has a good conodont and microfossil stratigraphy, although the position of the golden spike does not correspond to the base of any particular biozone (Mabillard & Aldridge 1985; Mullins & Aldridge 2004). Instead, recent works have correlated this boundary at a slightly younger level, namely Ireviken datum 2 (e.g. Loydell *et al.* 2003; Calner *et al.* 2004). This represents the boundary between the lower and upper *Ps. bicornis* conodont zonal levels (Jeppsson 1997), which is close to the base of the *murchisoni* graptolite Zone (Loydell *et al.* 2003). This level has been suggested as a correlatable level for the boundary on the International Commission on Stratigraphy website (www.stratigraphy.org).

Most original works on Ordovician deposits (Table 2) correlate these strata to the British stages (cf. Fortey *et al.* 1995, 2000). As such, we refer to these stages throughout this paper (the relations between the British stages and the international stages of Cooper & Sadler [2004] are shown in Fig. 2). However, we use the term Hirnantian *sensu*

Cooper & Sadler (2004) as this stage is well-defined with good global correlation. We note that the definition and correlation of the British Ordovician stages is not unproblematic (Fortey *et al.* 1995, 2000; Cooper & Sadler 2004). The recent placement of the GSSPs for these stages reflects improved Ordovician biostratigraphy. The recently named Sandbian and Katian Stages of the Ordovician are defined on the well-correlated first appearances of the graptolites *Nemagraptus gracilis* and *Ensigraptus caudatus*, even though the first appearance of these graptolites is locally diachronous (cf. Williams *et al.* 2003, 2004). Therefore, we feel that the historical correlation of glacial deposits to the British stages, and our use of the UK oxic–anoxic stratigraphy, justify our reference to these 'old-fashioned' terms and we wish to highlight that use of old terms does not necessarily reflect the employment of outmoded correlations.

Criteria for recognizing glacial maxima: The glacial maxima identified in the EPI (Table 1) are recognized using an argument similar to that employed by Brenchley *et al.* (1994). Namely, glacial maxima occur when rapid regressions are accompanied by synchronous oxygen and carbon evidence of cooling, if there are contemporaneous glacial deposits. Glaciations may be recognized from either the deposition of diamictites containing glaciogenic clasts, ice-rafted debris in distal marine settings, or from glacial erosive features (Eyles 1993). However, evidence of ice may not necessarily be evidence of glacial maxima, as ice-sheets may persist through interglacials. Also, an extensive unconformity of Caradoc–Hirnantian age persists through much of Africa and Arabia (Destombes *et al.* 1985; Sutcliffe 2001), potentially removing evidence of glaciations in this interval. Likewise, correlation between high-latitude glacial deposits and equivalent low-latitude strata may be hindered by (a) the low-abundance of graptolites in high-latitude environments (Lüning *et al.* 2000; Zalasiewicz 2001), and (b) the general absence of limestone-hosted shelly faunas in these settings (e.g. Walker *et al.* 2002). However, if the co-occurrence of ice formation with regressions and positive $\delta^{13}C$ and $\delta^{18}O$ excursions is not contradicted by biostratigraphic data, then it seems more parsimonious to consider them to be related to a glaciation rather than being caused by separate events.

Sea-level curves: The sea-level curves illustrated in Figure 2 are based on sequence stratigraphic analyses of shallow-water facies. These may be 'calibrated' to the depths of such facies in modern environments (Ross & Ross 1996), though such estimates may also vary according to sediment flux or local topography (Orr 2001). Nevertheless, other methods for estimating sea-level change possess inherent uncertainties. Faunally derived sea-level curves may represent changes in palaeoenvironment rather than deepening *per se* (Orr 2001). For example, 'quantitative' sea-level curves based on conodont assemblages do not produce consistent results in different environments (e.g. Zhang *et al.* 2006). Glacial maxima within the EPI may be associated with

Table 2. *Evidence for ice formation during the EPI; palaeolatitudes inferred from palaeogeographical reconstructions for the Caradoc, Hirnantian and Llandovery–Wenlock boundary, quoted to within ±5° N/S; stratigraphic nomenclature reflects the original author's usage*

Stage	Age	Location	Palaeolat.	Evidence of ice	Evidence for age
CARADOC *sensu* Fortey *et al.* (1995, 2000)	'Llandeilo-Caradoc boundary' Hamoumi (1999)	Lower Ktaoua Formation, Zagora, Morocco	80° S	glacial pavement with polished surface displaying *roche moutonnée*-like forms, undulating surfaces, graze, score & nail-shaped groove joints (Beuf *et al.* 1971; Hamoumi 1999)	surface forms base of L. Ktaoua Fm, which has ?*gracilis–clingani* graptolite Zone age (Destombes *et al.* 1985) based on comparison of trilobite & brachiopod fauna with UK and Bohemia; overlies 1st Bani Formation of 'Llandeilo' age (Destombes *et al.* 1985) with erosive contact
	'Lower Caradoc' Hamoumi (1999)	Lower Ktaoua Formation, Zagora, central Anti-Atlas, Morocco	80° S	surface remnants at corrie heads displaying battered & scoured surfaces, nivation hollows and thermokarst, on top of glaciotectonised and jointed glaciomarine sediments (Beuf *et al.* 1971; Hamoumi 1999)	within L. Ktaoua Fm, which has ?*gracilis–clingani* graptolite Zone age (Destombes *et al.* 1985), based on comparison of trilobite & brachiopod fauna with UK and Bohemia
	Caradoc (Pickerill *et al.* 1979)	Gander Bay Tillites, Davidsville Group, Newfoundland, Canada	65° S[†]	diamictites and locally abundant dropstones (Pickerill *et al.* 1979)	oldest strata in the Davidsville Group contain conodonts of Llanvirn–Llandeilo age; diamictites immediately underlie graptolitic slate of Caradoc age (Pickerill *et al.* 1979)
	Caradoc (Schenck & Lane 1981)	White Rock Fm, Nova Scotia, Canada	65° S[†]	marine-rafted tillite with clasts in small lenses and dropstone fabrics in varvites (Schenck & Lane 1981), quartz meta-arenites showing polished striations and groves on facets (Schenck 1972)	tillite overlain by 'poorly preserved, limited [graptolite] fauna of Caradoc or younger age' (Schenck 1972; Schenck & Lane 1981)
CARADOC or ASHGILL	'middle Caradoc–Ashgill boundary' Hamoumi (1999)	Touririne Fm, Eastern Anti-Atlas, Morocco	75° S	'frost dominated cold tidal [surface]... display (*sic*) ice wedges, desiccation polygons and karstification' Hamoumi (1999)	Lower Touririne Sandstone member contains trilobite fauna comparable to '*peltifer* graptolite Zone' fauna of Letná Fm, Bohemia; Upper Touririne sandstone member coeval with strata of *clingani–? complanatus* graptolite zone age; Touririne Fm unconformably overlain by Hirnantian tillites; stratigraphy in Destombes *et al.* (1985)

(Continued)

Table 2. *Continued*

Stage	Age	Location	Palaeolat.	Evidence of ice	Evidence for age
ASHGILL (excluding Hirnantian *sensu* Cooper & Sadler 2004)	'Upper Ordovician' Legrand (1985)	Tamadjert Fm*, Western Hoggar, Algeria	60° S	buried landscapes show glacial landforms, with striations, drumlins, vestiges of moraines, traces of solifluction, ?pingos; terrestrial and marine tillite deposits (Beuf *et al.* 1971; Biju-Duval *et al.* 1981; Legrand 1985)	unconformably overlies strata of Caradoc age (Biju-Duval *et al.* 1981); marine facies may contain late Caradoc trilobites, brachiopods and graptolites (Gatinskiy *et al.* 1966); upper part of formation contains middle–late Llandovery graptolites (Legrand 1970)
	'Upper Ordovician' Tucker & Reid (1973)	Sierra Leone	50° S	ice-drop tillites with carbonate boulders (Tucker & Reid 1973)	lithological similarity to and correlation with similar deposits in Guinea which are overlain by Llandovery graptolite shales (Tucker & Reid 1973)
	Ashgill (Doré 1981)	Tillite de Feuguerolles, Normandy, France	40° S	ice-drop tillites and diamictites with glacially striated clasts (Doré & Le Gall 1973; Doré 1981)	conformably underlain by strata containing Caradoc age trilobites and other fossils (Robardet *et al.* 1972; Doré & Le Gall 1973; Doré 1981); conformably overlain by strata yielding latest Ordovician–earliest Silurian graptolite fauna (Phillipot & Robardet 1971; Doré 1981)
	'Late Ordovician' Deynoux & Trompette (1981)	Taoudeni Basin, West Africa	70° S	terrestrial and marine tillites in the area near the Hodh with striated boulders; glacially reworked deposits with outwash fans 'similar to the Icelandic sandur' (Deynoux & Trompette 1981); 'micro-cordons' that probably represent subglacial eskers in englacial tunnels; structures similar to *fentes minces* (Deynoux & Trompette 1981); glacial pavements and *roches moutonées* with striations, furrows and crescentic fractures in the Hodr; glaciotectonic features including ice-push ridges and *fractures en gradin* (Biju-Dival *et al.* 1974)	glacial deposits have erosive discomformity at their base (Deynoux & Trompette 1981; Deynoux, Sougy & Trompette 1985) overlying the upper part of Supergroup 2, which has an age near the Cambro-Ordovician boundary based on inarticulate brachiopods and trace fossils (Legrand 1969); base comparable with Caradoc–Ashgill unconformity in the Hoggar, Tassilis & Anti-Atlas (Deynoux & Trompette 1981); glacial deposits conformably but ?diachronously overlain by graptolite faunas of Upper Ashgill–middle Llandovery age (Underwood *et al.* 1998)

HIRNANTIAN *sensu* Cooper & Sadler (2004) *extraordinarius* graptolite Zone acme (Sutcliffe *et al.* 2000, 2001)	Northern Africa: Upper 2nd Bani Fm, Anti-Atlas Mts, Morocco; Djebel Serraf Fm, Ougarta Mts, Algeria	75° S	synchronous, large-scale tillite and diamictite deposition; two phases of regionally extensive glaciomarine shelf sequences and subglacial erosive surfaces; ice-contact fans, ice-rafted debris (Sutcliffe *et al.* 2000, 2001; see also Hamoumi 1999)	*Hirnantia* fauna within glaciogenic deposits, and in underlying formations (erosive contacts); disconformably overlain by strata containing Rhuddanian graptolites (Destombes *et al.* 1985; Sutcliffe *et al.* 2000, 2001)
extraordinarius graptolite Zone acme (Sutcliffe *et al.* 2000, 2001)	Melez Chograne & Memouniat Frms, Libya	75° S	glaciomarine shelf deposits, ice-rafted debris and erosive surfaces covered by ice-contact deposits (Havlíček & Massa 1973; Sutcliffe *et al.* 2001)	*Hirnantia* fauna throughout succession, overlain by Llandovery graptolites (Havlíček & Massa 1973; Sutcliffe *et al.* 2001)
extraordinarius graptolite Zone acme (Sutcliffe *et al.* 2000, 2001)	Tichit glacial group, the Hodh, Mauritania	70° S	glacially striated dropstones, diamictites (Deynoux & Trompette 1981; Ghienne 2003)	dropstones coexist with graptolites of Upper Ashgill age (Underwood *et al.* 1998)
extraordinarius graptolite Zone acme (Sutcliffe *et al.* 2000, 2001)	Pakhuis Fm, South Africa	15° S	tillites & diamictites, two subglacial erosive surfaces with striated pavements and boulders, ice-rafted debris (Rust 1981; Sutcliffe *et al.* 2000, 2001)	*Hirnantia* fauna in conformably overlying Cedarberg Fm (Sutcliffe *et al.* 2001)
?Hirnantian	Tabuk Fm, Arabian peninsula	50° S	diamictites & tillites with some striated, faceted and polished clasts; boulder pavements with striations (McClure 1978)	tillite interfingers with late Caradoc age graptolite shale also containing trilobites of late Caradoc or early Ashgill age (Young 1981); overlying Qusaiba Shale contains a Rhuddanian age graptolite fauna (Lüning *et al.* 2000)
Hirnantian (Armstrong *et al.* 2005)	Ammar Fm, southern Jordan	50° S	tillite; glacial unconformity at base and two episodes of glacial incision; conglomerates with glacially faceted and striated clasts (Abed *et al.* 1993)	conformably overlain by *persculptus* graptolite Zone fauna (Armstrong *et al.* 2005)
Hirnantian (Caputo 1998)	Don Braulio Fm, Argentina	15° S‡	diamictite some striated, faceted and polished pebbles and cobbles (Büggish & Astini 1993)	overlain by *Hirnantia* fauna (Sutcliffe *et al.* 2001)

(Continued)

Table 2. *Continued*

Stage	Age	Location	Palaeolat.	Evidence of ice	Evidence for age
	?Hirnantian (Caputo 1998)	Iapó Fm, Paraná Basin, Brazil	25° S	diamictites with faceted and striated clasts (Maack 1957; Rocha-Campos 1981)	lithological comparison with South African and Argentinian tillites (Caputo 1998); Iapó Fm discordantly overlies rhyolites of the Castro Group dated at 450 ± 25 Ma (Bigarella 1970); correlation with interfingering Vila Maria Fm suggests diamictite overlain by early Llandovery palynomorph and shelly fauna (Caputo & Crowell 1985)
	Hirnantian	Prague Basin, Czech Republic	55° S	two intervals of diamictite deposition (Brenchley & Štorch 1989)	conformably underlain by *Mucronaspis* fauna and *anceps* Zone graptolites; conformably overlain by Hirnantian fauna (Štorch & Mergl 1989)
LLANDOVERY (Rhuddanian)	?Rhuddanian	San Gabán–Cancañiri–Zapla Fms, Bolivia, Argentina & Peru	55° S	widely extensive diamictites with striated and faceted clasts (Crowell *et al.* 1981; Caputo & Crowell 1985; Díaz-Martínez & Grahn 2007)	oldest diamictite horizon overlain by Aeronian chitinozoan fauna and underlain by Rhuddanian chitinozoan; (Díaz-Martínez & Grahn 2007); these formations unconformably overlie Caradoc strata showing evidence of Ashgillian deformation (Crowell *et al.* 1981)
	'Upper Ordovician or Lower Silurian' (Kennedy 1981)	Stoneville & Beaver Cove Fms, Newfoundland, Canada	60° S[†]	diamictite beds; dropstones probably derived by iceberg rafting (Kennedy 1981)	Stoneville & Beaver Cove Fms are coeval (Kennedy 1981); former underlain by poorly preserved corals of Upper Ordovician or Lower Silurian age (McCann & Kennedy 1974) and lithological correlatives of its upper part have yielded Llandovery age fossils (Eastler 1969; Kennedy 1981)
LLANDOVERY (Aeronian)	*gregarius* graptolite Zone (Caputo 1998)	Nhamundá Fm, Amazonas Basin Brazil	60° S	diamictite; ice-push & ice-shear deformation structures (Carozzi *et al.* 1973; Caputo 1998)	diamictite immediately overlain by *gregarius* Zone graptolite fauna and chitinozoan fauna (Grahn & Paris 1992; Caputo 1998)

Age	Formation/location	Latitude	Lithology	Notes
'late Aeronian–early Telychian' (Caputo 1998)	Nhamundá Fm, Amazonas Basin, Brazil	60° S	diamictite; ice-push & ice-shear deformation structures (Carozzi et al. 1973; Caputo 1998)	overlies *gregarius* Zone fauna; shales lateral to tillites yield an early Telychian chitinozoan fauna (Caputo 1998)
?Aeronian	San Gabán–Cancañiri–Zapla Fms, Bolivia, Argentina & Peru	60° S	widely extensive diamictites with striated and faceted clasts (Crowell et al. 1981; Caputo & Crowell 1985; Díaz-Martínez & Grahn 2007)	overlies shales yielding Aeronian chitinozoans (Díaz-Martínez & Grahn 2007), overlain by shales and a younger diamictite horizon
Llandovery (Caputo 1998)	Ipu Fm, Parnaíba Basin & Cariri Valley, Brazil	70° S	three diamictite layers (Caputo & Crowell 1985; Grahn & Caputo 1992); faceted pebbles (Kegel 1953)	interfingers with Tianguá Fm, which contains Early Silurian chitinozoans and acritarchs (Caputo & Lima 1984; individual diamictites may correlate with the better-dated diamictites in the Nhamundá Fm (Grahn & Caputo 1992)
late Telychian (Grahn in Cramer & Saltzman 2007)	Nhamundá Fm, Amazonas Basin, Brazil	70° S	diamictites and tillites; ice-push & ice-shear deformation structures (Carozzi et al. 1973; Caputo 1998)	Late Telychian–early Wenlock chitinozoan fauna in interfingering shales (Caputo 1998; first appearance of chitinozoa *M. magaritana* above the youngest tillite (Grahn in Cramer & Saltzman 2007)
LLANDOVERY (Telychian, including *centrifugus* Zone) Late Telychian: (Díaz-Martínez 2007)	San Gabán–Cancañiri–Zapla Fms, Bolivia, Argentina & Peru	65° S	widely extensive diamictites with striated and faceted clasts (Crowell et al. 1981; Caputo & Crowell 1985; Díaz-Martínez & Grahn 2007)	youngest diamictite conformably overlain by Sacla limestone, which has early Wenlock age based on occurrence of the conodont *O. sagitta rhenana* (Díaz-Martínez 2007); however, acritarchs and chitinozoans in intercalated shale horizons suggest a Llandovery–Wenlock boundary age (Suárez-Soruco 1995; this diagnosis may suggest an older age than Ireviken Datum 2 (see Appendix)

*Synonym of Felar Felar Fm.
†This part of Nova Scotia is thought to have been on the margin of West Africa at this time (Kennedy 1981).
‡Position of the Argentine Precordillera poorly constrained at this interval.

faunal turnover and changes in oceanic temperature and oxygenation. As such, faunally derived sea-level curves do not necessarily provide independent evidence of glacioeustasy in this interval. We only make passing reference to Loydell (1998), even though this offers well-defined evidence of Silurian sea-level change with good stratigraphic control. However, there is no curve defined using a comparable method for the Ordovician. And, as Loydell's (1998) sea-level curve uses deposition of graptolite shale as a criterion for establishing sea-level change, employing it would preclude an independent test of the relationship between sea-level and black shale distribution during the EPI.

Though we have used sea-level curves for the Ordovician and Silurian by different authors (Ross & Ross 1996; Nielsen 2003*a*), both are compiled using the same method and correlate sequences in North America and Baltoscandia. As the Iapetus Ocean closed, there may have been an increased local-tectonic component in the Laurentian record of sea-level change during the EPI (e.g. McKerrow *et al.* 2000). Nevertheless, correlation between two palaeocontinents reduces the chance of conflating relative tectonic changes with global changes in sea-level, and significant global events should register above local noise. The Nielsen (2003*a*) sea-level curve for the Ordovician shows a strong correlation with the equivalent sea-level curve by Ross & Ross (1995), but offers better stratigraphic resolution. Meanwhile, the Ross & Ross (1996) curve for the Silurian employs a method consistent with that used by Nielsen (2003*a*) in the Ordovician. The original Ross & Ross (1996) sea-level curve has poor biostratigraphic control in the late Telychian (Loydell 1998), with the authors referring to an undifferentiated *crenulata* Zone between the *griestoniensis* and *centrifugus* zones. This interval can be differentiated into four graptolite zones (e.g. Loydell *et al.* 1998). As such, Figure 2 illustrates an amended version of the Ross & Ross (1996) curve based on the recorrelation of their original stratigraphy by Loydell (1998).

The sequence stratigraphic patterns observed by Ross & Ross (1996) may be consistent with 20–60 m changes in sea-level in less than 1–2 Ma. The rates and frequency of sea-level change during the EPI (Ross & Ross 1996; Nielsen 2003*a*, *b*) are consistent with third-order sequence stratigraphic cycles. They are therefore more likely glacially than tectonically forced (Church & Coe 2003).

Oxygen isotopes: The oxygen isotope data presented in Table 1 are obtained from the shells of brachiopods. These were selected on the basis that they showed no trace element or microstructural evidence of significant diagenetic alteration (Brenchley *et al.* 1994; Marshall *et al.* 1997; Azmy *et al.* 1998; Tobin *et al.* 2005). Modern brachiopods secrete a low-Mg calcite shell at or near isotopic equilibrium with seawater (Lowenstam 1961; Carpenter & Lohmann 1995; James *et al.* 1997), which tends to resist diagenesis (Marshall *et al.* 1997; Azmy *et al.* 1998). The isotopic composition of their shells shows little deviation due to vital effects at the present day (Carpenter &

Lohmann 1995). Analysis of multi-taxa assemblages in the Silurian suggests that vital effects may not have been significant in the Palaeozoic (Samtleben *et al.* 2001). Hence, these fossil data have been regarded as representative of the isotopic composition of Early Palaeozoic seawater (e.g. Veizer *et al.* 1997; Samtleben *et al.* 2001).

Positive $\delta^{18}O$ excursions can be achieved by decreases in temperature or increases in salinity (e.g. Hays & Grossman 1991). The latter can be achieved due to increased ice volume or decreased freshwater input, and both may occur along with decreased temperature in glaciations (e.g. Azmy *et al.* 1998; Tobin *et al.* 2005). Though changes in salinity alone have been argued to account for the positive isotope excursions observed in this interval (e.g. Samtleben *et al.* 1996; Bickert *et al.* 1997), such changes represent salinity change of *c.* 14‰, which is an implausibly large change that cannot be tolerated by brachiopods (Azmy *et al.* 1998). As such, we interpret these positive excursions as representing decreases in temperature, which may be accompanied by smaller increases in salinity. Therefore, positive $\delta^{18}O$ excursions can be related to glaciations if consistent with other evidence (e.g. Brenchley *et al.* 1994; Azmy *et al.* 1998).

Carbon isotopes: The isotopic curves illustrated in Figure 2 represent a consistent record of the $\delta^{13}C$ of carbonates throughout the EPI. These curves were compiled using bulk rock carbonates in the Baltic region. These strata have a well-resolved stratigraphy correlated to graptolite, conodont and chitinozoan zones (e.g. Loydell *et al.* 1998, 2003; Loydell & Nestor 2005; Nõlvak *et al.* 2006). In this interval, the Estonian chronostratigraphic nomenclature is well-established across Baltoscandia and widely used for correlation of carbonate-hosted assemblages (Cooper & Sadler 2004; Melchin *et al.* 2004). This allows precise correlation between sea-level curves and isotope data.

The $\delta^{13}C$ curves we illustrate were compiled by a single research group using consistent sampling strategies and analytic techniques on bulk rock carbonates (for methods, see Kaljo *et al.* 1997, 1998, 2001). These data show little evidence for diagenetic alteration (Kaljo *et al.* 1997, 1998, 2001). The curves in Figure 2 are consistent with each other. Individual excursions show good accord with both organic and inorganic carbon-isotope records derived in different regions (e.g. Underwood *et al.* 1997; Kump *et al.* 1999; Melchin & Holmden 2006). The major positive excursions highlighted in Table 1 and Figure 2 are recognized in the more-or-less continuous record of carbon isotopes from the shells of Silurian brachiopods (e.g. Azmy *et al.* 1998). They also occur in the bulk rock data from the Ordovician of Laurentia (e.g. Patzkowsky *et al.* 1997; Saltzman & Young 2005) and the compilation for the Palaeozoic of the Great Basin, USA (Saltzman 2005). The similarity of the $\delta^{13}C$ curves in Figure 2 to these other $\delta^{13}C$ curves highlights that the bulk-rock curves of Baltica record global perturbations in the carbon cycle.

Kump & Arthur (1999) and Kump *et al.* (1999) show that positive carbon-isotope excursions can be achieved by (a) increasing productivity, (b) increasing the burial flux of organic carbon, or (c) by positive excursions in the $\delta^{13}C$ value of riverine input into the marine carbon reservoir due to increased terrestrial carbonate weathering. These alternatives can be distinguished by analysis of coupled patterns of organic and inorganic carbon isotopes (cf. Kump & Arthur 1999; Kump *et al.* 1999). For example, coupled organic and inorganic $\delta^{13}C$ data are available for the Hirnantian glaciation, and these positive $\delta^{13}C$ excursions have been interpreted as representing changes in weathering (Kump *et al.* 1999; Melchin & Holmden 2006). This excursion occurs during a major regression (e.g. Brenchley *et al.* 2004; Melchin & Holmden 2006), which exposed shallow marine carbonates to terrestrial weathering, resulting in a more positive $\delta^{13}C$ value of river waters (Kump *et al.* 1999). Given that all the positive $\delta^{13}C$ excursions of the EPI correspond to lowstands in cooler intervals with decreased organic carbon burial in deep-water settings (Table 1; Fig. 2), all positive $\delta^{13}C$ excursions may reasonably be interpreted as being due to increased weathering of shallow carbonates exposed in regressions.

The positive $\delta^{13}C$ excursion isotope excursion associated with the Hirnantian glaciation is hard to reconcile with increased productivity or organic preservation (cf. Brenchley *et al.* 1994). This event is coincident with a mass extinction (e.g. Sutcliffe *et al.* 2000; Chen *et al.* 2005), when there is no evidence of increased organic burial in even the deepest-water facies (e.g. Armstrong & Coe 1997). We therefore believe it is best to consider the positive carbon-isotope excursions of the EPI to represent cooler events if they are coincident with regression (e.g. Patzkowsky *et al.* 1997; Azmy *et al.* 1998; Kaljo *et al.* 2003; Tobin *et al.* 2005; Johnson 2006), and that these excursions may be related to glaciations if consistent with other evidence.

Oxic–anoxic stratigraphy: Anoxic intervals are recognized from the occurrence of laminated hemipelagic mudrocks in deep-water settings (i.e. below storm wave base), which often contain graptolites. Graptolites are organic walled macrozooplankton and the majority of their fossil record comes from distal, anoxic mudrocks (Chapman 1991; Underwood 1992; Finney & Berry 1997). They may also be sporadically found in oxic facies, should they have undergone rapid burial. For example, *Stimulograptus sedgwickii* occurs in well-ventilated sandstones and siltstones in the Girvan area, UK (Floyd & Williams 2003), and in the shelf successions of the Llandovery area, Wales, UK (Cocks *et al.* 1984). Likewise, tiny graptolite fragments have been reported from rocks showing evidence of bioturbation (e.g. Armstrong & Coe 1997; Mullins & Aldridge 2004). Thus, the presence of graptolites in well-laminated, dark-grey or black mudrocks is evidence of anoxia rather than graptolite palaeoecology (cf. Berry *et al.* 1987). Similarly, the absence of macrofossil graptolites in poorly laminated

and/or burrowed paler grey shales is more typical of oxygenated bottom waters and sediments.

The oxic–anoxic stratigraphy for the EPI presented here (Fig. 3 and refs therein) is derived from correlating anoxic intervals in the deep-water record of UK successions. These successions are located in the Welsh Basin and Southern Uplands of Scotland, which occur on the Iapetus margins of Avalonia and Laurentia respectively (Zalasiewicz 2001). They have a well-established, high-resolution stratigraphy that allows such oxic–anoxic transitions to be recognized at a sub-graptolite zone resolution (e.g. Verniers & Vandenbroucke 2006). Thus, if anoxic facies are deposited simultaneously in the Welsh Basin and Southern Uplands, they represent at least an Iapetus-wide anoxic event (Fig. 2). Where intervals of anoxia in the Southern Uplands and Welsh Basin do not correlate, it may be that sediment redox conditions represent local rather than oceanic events. Leggett (1980) made a similar compilation for the Early Palaeozoic of the UK, employing stage level correlations. However, our higher resolution oxic–anoxic stratigraphy allows individual events to be correlated at a biozone level (e.g. Fig. 3).

The basis of the UK record as a reliable record of global marine anoxia requires consideration of the changing depositional settings of both the Welsh Basin and Southern Uplands. We select data from intervals where these strata were deposited in shelf and/or deep-basin environments. To establish the global extent, we also correlate this stratigraphy with redox changes recognized in the deep-water record of the Rheic and/or Palaeothethys Oceans. As far as we are aware, the record of marine anoxia in the deep-water facies of the UK represents the only well-dated, continuous succession where the oxic–anoxic stratigraphy has been sufficiently documented to assemble a composite oxic–anoxic stratigraphy for the EPI.

The Welsh Basin was a restricted basin on the eastern margin of Avalonia. Its depositional setting and sedimentary history are reviewed by Woodcock (2000) and Zalasiewicz (2001). Local changes in freshwater run-off, nutrient input or upwelling may have induced localized anoxic events by altering productivity and stratification. Its sedimentary record stretches from the Cambrian to latest Silurian. Nevertheless, the widespread volcanism prior to the mid-Caradoc significantly disrupted patterns of marine topography, subsidence and deposition (Woodcock 1990). This resulted in a more ambiguous and locally variable pattern of basin redox conditions at this time (cf. Leggett 1980). Once sediment input outpaced subsidence in the late Silurian (King 1994; Woodcock 2000), the basin became increasingly shallow and it rapidly filled with sediment. The oxic–anoxic stratigraphy of the Welsh Basin correlates extremely well with similar stratigraphies where available for the deep-water successions of the Howgill Fells and Lake District, Northern England (cf. Rickards 1970; Hutt 1974; Rickards & Woodcock 2005). Thus, the Welsh Basin provides a well-resolved record of the oxic–anoxic stratigraphy of eastern Avalonia throughout the EPI.

In contrast, the Moffat Shale Group of the Southern Uplands is commonly thought to represent an ocean floor environment approaching a trench (Leggett *et al.* 1979; Leggett 1987). It may have been susceptible to changes in upwelling, which could have induced localized anoxia by increasing export production (e.g. Finney & Berry 1997). The Southern Uplands record of the EPI is contained within an accretionary prism formed from Iapetus thrust-slices during the Caledonian Orogeny (Leggett *et al.* 1979; Leggett 1987; Strachan 2000). The succession has an early Caradoc to late Llandovery age (Leggett 1980, 1987; Strachan 2000). Subsequently, the Moffat Shale Group was overlain by the massive flysch deposits of the Gala and Hawick groups of late Llandovery to Wenlock age (White *et al.* 1991). At a larger scale, the mudstones of the Moffat Shale Group comprise a generally distal, condensed succession of Ordovician and early Llandovery age, with slightly more expanded and proximal facies in the mid- and late Llandovery. So, rather than being a restricted basin, the Southern Uplands provide a well-resolved record of open marine conditions throughout all but the terminal part of the EPI.

The major anoxic events of the Welsh Basin correlate well with those from the Southern Uplands where data are available (Fig. 3). Hence, there is no reason to believe that the late Llandovery–Wenlock of the Welsh Basin is unrepresentative of the Iapetus Ocean redox conditions, especially as synchronous changes are seen in northern England (cf. Rickards 1970; Rickards & Woodcock 2005). Thus, the oxic–anoxic stratigraphies of the UK basins (Fig. 3), from which we compiled the summary oxic–anoxic stratigraphy (Fig. 2), probably represent the best record available of oxic–anoxic transitions in low-latitude deep waters during the EPI.

References

ABED, A. M., MAKHLOUF, I. M., AMIREH, B. S. & KHALIL, B. 1993. Upper Ordovician glacial deposits in southern Jordan. *Episodes*, **16**, 316–328.

AINSAAR, L., MEIDLA, T. & MARTMA, T. 1999. Evidence for a widespread carbon isotopic event associated with late Middle Ordovician sedimentological and faunal changes in Estonia. *Geological Magazine*, **136**, 49–62.

ALDRIDGE, R. J., JEPPSSON, L. & DORNING, K. J. 1993. Early Silurian oceanic episodes and events. *Journal of the Geological Society, London*, **150**, 501–513.

AOUDEH, S. M. & AL-HAJRI, S. A. 1995. Regional distribution and chronostratigraphy of the Qusaiba Member of the Qalibah Formation in the Nafud Basin, northwestern Saudi Arabia. *In*: AL-HUSSEINI, M. I. (ed.) *Geo '94, The Middle East Petroleum Geosciences*, vol. 1. Gulf PetroLink, Manama, Bahrain, 143–154.

ARMSTRONG, H. A. & COE, A. L. 1997. Deep sea sediments record the geophysiology of the end Ordovician glaciation. *Journal of the Geological Society, London*, **154**, 929–934.

ARMSTRONG, H. A., TURNER, B. R., MAKHLOUF, I. M., WEEDON, G. P., WILLIAMS, M., AL SMADI, A. & ABU SALAH, A. 2005. Origin, sequence stratigraphy and depositional environment of an Upper Ordovician (Hirnantian) deglacial black shale, Jordan. *Palaeogeography, Palaeoclimatology, Palaeoecology*, **220**, 273–289.

ARMSTRONG, H. A., TURNER, B. R., MAKHLOUF, I. M., WEEDON, G. P., WILLIAMS, M., AL SMADI, A. & ABU SALAH, A. 2006. Reply to 'Origin, sequence stratigraphy and depositional environment of an upper Ordovician (Hirnantian) deglacial black shale, Jordan'. *Palaeogeography, Palaeoclimatology, Palaeoecology*, **230**, 356–360.

ARRHENIUS, S. 1896. On the influence of carbonic acid in the air upon the temperature on the ground. *Philosophical Magazine and Journal of Science*, **41**, 237–275.

AZMY, K., VEIZER, J., BASSETT, M. G. & COPPER, P. 1998. Oxygen and carbon isotopic composition of Silurian brachiopods: Implications for coeval seawater and glaciations. *Geological Society of America Bulletin*, **110**, 1499–1512.

BAKER, S. J. 1981. The graptolite biostratigraphy of a Llandovery outlier near Llanystumdwy, Gwynedd, North Wales. *Geological Magazine*, **118**, 355–365.

BARKER, S., HIGGINS, J. A. & ELDERFIELD, H. 2003. The future of the carbon cycle: review, calcification response, ballast and feedback on atmospheric CO_2. *Philosophical Transactions of the Royal Society of London, series A*, **361**, 1977–1999.

BARNES, C. R. & BOLTON, T. E. 1998. The Ordovician–Silurian boundary on Manitoulin Island, Ontario, Canada. *Bulletin of the British Museum, Natural History, Geology*, **43**, 247–253.

BERNER, R. A. 1991. A model for atmospheric CO_2 over Phanerozoic time. *American Journal of Science*, **291**, 339–376.

BERNER, R. A. 1994. GEOCARB II: A revised model of atmospheric CO_2 over Phanerozoic time. *American Journal of Science*, **294**, 56–91.

BERNER, R. A. 1998. The carbon cycle and CO_2 over Phanerozoic time: the role of land plants. *Philosophical Transactions of the Royal Society, series B*, **353**, 75–82.

BERNER, R. A. 2001. Modelling atmospheric O_2 over Phanerozoic time. *Geochimica et Cosmochimica Acta*, **65**, 685–694.

BERNER, R. A. 2003. The long-term carbon cycle, fossil fuels and atmospheric composition. *Nature*, **426**, 323–326.

BERNER, R. A. 2006. GEOCARBSULF: A combined model for Phanerozoic atmospheric O_2 and CO_2 over Phanerozoic time. *Geochimica et Cosmochimica Acta*, **70**, 5653–5664.

BERNER, R. A. & CANFIELD, D. E. 1989. A new model for atmospheric oxygen over Phanerozoic time. *American Journal of Science*, **289**, 333–361.

BERNER, R. A. & KOTHAVALA, Z. 2001. Geocarb III: A revised model of atmospheric CO_2 over Phanerozoic time. *American Journal of Science*, **301**, 182–204.

BERNER, R. A. & RAISWELL, R. 1983. Burial of organic carbon and pyrite sulfur in sediments over Phanerozoic time: a new theory. *Geochimica et Cosmochimica Acta*, **47**, 855–862.

BERNER, R. A., LASAGA, A. C. & GARRELS, R. M. 1983. The carbonate–silicate geochemical cycle and its effect on atmospheric carbon dioxide over the past 100 million years. *American Journal of Science*, **283**, 641–683.

BERRY, W. B. N., WILDE, P. & QUINBY-HUNT, M. S. 1987. The oceanic non-sulfidic oxygen minimum zone; a habitat for graptolites? *Bulletin of the Geological Society of Denmark (Meddelelser fra Dansk Geologisk Forening)*, **35**, 103–114.

BEUF, S., BIJU-DUVAL, B., DE CHARPAL, O., ROGNON, P., GARIEL, O. & BENNACEF, A. 1971. *Les grès du Paléozoïque inférieur au Sahara (Sédimentation et discontinuities, evolution structurale d'un craton)*. Publication de l'Institut Français Pétrole, **18**.

BICKERT, T., PÄTZOLD, J., SAMTLEBEN, C. & MUNNECKE, A. 1997. Paleoenvironmental changes in the Silurian indicated by stable isotopes in brachiopod shells from Gotland, Sweden. *Geochimica et Cosmochimica Acta*, **61**, 2717–2730.

BIGARELLA, J. J. 1970. Continental drift and palaeocurrent analysis. Second Gondwana Symposium, Proceedings and Papers. Council for Scientific and Industrial Research, South Africa, 73–97.

BIJU-DUVAL, B., DEYNOUX, M. & ROGNON, P. 1974. Essai d'interprétation de 'fractures en gradins' observées dans les formations glaciaires Précambriennes et Ordoviciennes du Sahara. *Revue de Geographie Physique et de Geologie Dynamique*, **26**, 503–512.

BIJU-DIVAL, B., DEYNOUX, M. & ROGNON, P. 1981. Late Ordovician tillites of the Central Sahara. *In*: HAMBREY, M. J. & HARLAND, W. B. (eds) *Earth's Pre-Pleistocene Glacial Record*. Cambridge University Press, Cambridge, 99–107.

BJERRESKOV, M. 1975. Llandoverian and Wenlockian graptolites from Bornholm. *Fossils and Strata*, **8**, 1–94.

BRENCHLEY, P. J. 1988. Environmental changes close to the Ordovician–Silurian boundary. *Bulletin of the British Museum of Natural History (Geology)*, **43**, 377–385.

BRENCHLEY, P. J. & ŠTORCH, P. 1989. Environmental changes in the Hirnantian (upper Ordovician) of the Prague Basin, Czechoslovakia. *Geological Journal*, **24**, 165–181.

BRENCHLEY, P. J., MARSHALL, J. D. & CARDEN, G. A. F. *ET AL*. 1994. Bathymetric and isotopic evidence for a short-lived Late Ordovician glaciation in a greenhouse period. *Geology*, **22**, 295–298.

BRENCHLEY, P. J., CARDEN, G. A., HINTS, L. *ET AL*. 2003. High-resolution isotope stratigraphy of Late Ordovician sequences: constraints on the timing of bio-events and environmental changes associated with mass extinction and glaciation. *Geological Society of America, Bulletin*, **115**, 89–104.

BROECKER, W. S. 1982. Glacial to interglacial changes in ocean chemistry. *Progress in Oceanography*, **11**, 151–197.

BRUCKSCHEN, P., OESMANN, S. & VEIZER, J. 1999. Isotope stratigraphy of the European Carboniferous: proxy signals for ocean chemistry, climate and tectonics. *Chemical Geology*, **161**, 127–163.

BRUNTON, F. R., SMITH, L., DIXON, O. A., COPPER, P., NESTOR, H. & KERSHAW, S. 1998. Silurian reef episodes, changing seascapes, and paleobiogeography. *In*: LANDING, E. & JOHNSON, M. E. (eds) *Silurian Cycles, Linkages of Dynamic Stratigraphy with Atmospheric, Oceanic, and Tectonic Changes*. New York State Museum Bulletin, **491**, 265–282.

BÜGGISH, W. & ASTINI, R. 1993. The Late Ordovician ice age: new evidence from the Argentine Precordillera. *In*: FINDLAY, R. H., UNRUG, R., BANKS, M. R. & VEEVERS, J. J. (eds) *Gondwana Eight – Assembly, Evolution and Dispersal*. Balkema, Rotterdam, 439–447.

BUTTERFIELD, N. J. 1997. Plankton ecology and the Proterozoic–Phanerozoic transition. *Paleobiology*, **23**, 247–262.

CALNER, M. & ERIKSSON, M. E. 2006. Silurian research at the crossroads. *GFF*, **128**, 73–74.

CALNER, M., JEPPSSON, L. & MUNNECKE, A. 2004. The Silurian of Gotland – Part I: Review of the stratigraphic framework, event stratigraphy, and stable carbon and oxygen isotope development. *Erlanger Geologische Abhandlungen, Sonderband*, **5**, 113–131.

CAPUTO, M. V. 1998. Ordovician–Silurian glaciations and global sea-level changes. *In*: LANDING, E. & JOHNSON, M. E. (eds) *Silurian Cycles, Linkages of Dynamic Stratigraphy with Atmospheric, Oceanic, and Tectonic Changes*. New York State Museum Bulletin, **491**, 15–25.

CAPUTO, M. V. & CROWELL, J. C. 1985. Migration of glacial centers across Gondwana during Paleozoic Era. *Geological Society of America, Bulletin*, **96**, 1020–1036.

CAPUTO, M. V. & LIMA, E. C. 1984. Estratigrafia, idade e correlação do Grupo Serra Grande – Bacia do Parnaíba. *Congresso Brasileiro de Geologia*, **33**, 740–753.

CAROZZI, A. V., PAMPLONA, H. R. P., CASTRO, J. C. & CONTREIRAS, C. J. A. 1973. Ambientes deposicionais e evolução tecto-sedimentar da seção elástica paleozóica da Bacia do Médio Amazonas. *Congresso Brasileiro de Geologia*, **27**, 279–314.

CARPENTER, S. J. & LOHMANN, K. C. 1995. $\delta^{18}O$ and $\delta^{13}C$ values of modern brachiopods. *Geochimica et Cosmochimica Acta*, **59**, 3749–3764.

CATLING, D. C. & CLAIRE, M. 2005. How Earth's atmosphere evolved to an oxic state: A status report. *Earth and Planetary Science Letters*, **237**, 1–20.

CAVE, R. 1979. Sedimentary environments of the basinal Llandovery of mid-Wales. *In*: HARRIS, A. L., HOLLAND, C. H. & LEAKE, B. E. (eds) *The Caledonides of the British Isles – Reviewed*. Special Publication of the Geological Society, London, **8**, 517–26.

CAVE, R. & DIXON, R. J. 1993. The Ordovician and Silurian of the Welshpool area. *In*: WOODCOCK, N. H. & BASSETT, M. G. (eds) *Geological Excursions in Powys, Central Wales*. Geological Series – National Museum of Wales, **14**, 51–84

CHAMBERLIN, T. C. 1899. An attempt to frame a working hypothesis of the cause of glacial periods on an atmospheric basis. *Journal of Geology*, **7**, 545–584.

CHAPMAN, A. J. 1991. How are they preserved? *In*: PALMER, D. & RICKARDS, R. B. (eds) *Graptolites:*

Writing in the Rocks. Fossils illustrated, vol. 1. The Boydell Press, Woodbridge, UK, 6–11.

CHEN X., RONG JIAYU, C. E. MITCHELL, D. A. T. HARPER, FAN, JUNXUAN, ZHAN, RENBIN, ZHANG, YUANDONG, LI, RONGYU & WANG, YI 2000. Late Ordovician to earliest Silurian graptolite and brachiopod biozonation from the Yangtze region, South China, with a global correlation. *Geological Magazine*, **137**, 623–650.

CHEN, X., MELCHIN, M. J., SHEETS, H. D., MITCHELL, C. E. & FAN, J.-X. 2005. Patterns and Processes of latest Ordovician graptolite extinction and recovery based on data from South China. *Journal of Paleontology*, **79**, 842–861.

CHURCH, K. D. & COE, A. L. 2003. Processes controlling relative sea-level change and sediment supply. *In:* COE, A. L. (ed.) *The Sedimentary Record of Sea-Level Change.* Cambridge University Press, Cambridge, 99–117.

COCKS, L. R. M. & TORSVIK, T. H. 2002. Earth geography from 500 to 400 million years ago: A faunal and palaeomagnetic review. *Journal of the Geological Society of London*, **159**, 631–644.

COCKS, L. R. M., WOODCOCK, N. H., RICKARDS, R. B., TEMPLE, J. T. & LANE, P. D. 1984. The Llandovery Series of the type area. *Bulletin of the British Museum (Natural History): Geology*, **38**, 131–182.

COHEN, A. S., COE, A. L., HARDING, S. M. & SCHWARZ, L. 2004. Osmium isotope evidence for the regulation of atmospheric CO_2 by continental weathering. *Geology*, **32**, 157–160.

COOPER, R. A. & SADLER, P. M. 2004. The Ordovician Period. *In:* GRADSTEIN, F. M., OGG, J. G. & SMITH, A. G. (eds) *A Geologic Time Scale 2004.* Cambridge University Press, Cambridge, 165–187.

COPPER, P. 1994. Ancient reef ecosystem expansion and collapse. *Coral Reefs*, **13**, 3–11.

CRAMER, B. D. & SALTZMAN, M. R. 2005. Sequestration of ^{12}C in the deep ocean during the early Wenlock (Silurian) positive carbon isotope excursion. *Palaeogeography, Palaeoclimatology, Palaeoecology*, **219**, 333–349.

CRAMER, B. D. & SALTZMAN, M. R. 2007. Fluctuations in epeiric sea carbonate production during Silurian positive carbon isotope excursions: A review of proposed paleoceanographic models: *Palaeogeography, Palaeoclimatology, Palaeoecology*, in press.

CROWELL, J. C., SUÁREZ-SORUCO, R. & ROCHA-CAMPOS, A. C. 1981. The Silurian Cancañiri (Zapla) Formation of Bolivia, Argentina and Peru. *In:* HAMBREY, M. J. & HARLAND, W. B. (eds) *Earth's Pre-Pleistocene Glacial Record.* Cambridge University Press, Cambridge, 902–907.

CROWLEY, T. J. & BERNER, R. A. 2001. CO_2 and climate change. *Science*, **292**, 870–872.

DAVIES, J. R. & WATERS, R. A. 1995. The Caban Conglomerate and Ystrad Meurig Grits Formation – nested channels and lobe switching on a mud dominated latest Ashgill to Llandovery lope-apron, Welsh Basin, UK. *In:* PICKERING, K. T., HISCOTT, R. N., KENYON, N. H., RICCI LUCCHI, F. & SMITH, R. D. A. (eds) *Atlas of Deep Water Environments: Architectural Style in Turbidite Systems.* London: Chapman & Hall, 182–93.

DAVIES, J. R., FLETCHER, C. J. N., WATERS, R. A., WILSON, D., WOODHALL, D. G. & ZALASIEWICZ, J. A. 1997. Geology of the country around Llanilar and Rhayader. *Memoir of the British Geological Survey*, Sheets 178 and 179 (England and Wales). HMSO, London.

DAVIES, J. R., WATERS, R. A., WILBY, P. R., WILLIAMS, M. & WILSON, D. 2003. *Geology of Cardigan and Dinas Island district – a brief explanation of the geological map. Sheet explanation of the British Geological Survey.* 1: 50000 sheet 193 (including part of sheet 210) Cardigan & Dinas Island (England & Wales). Keyworth, Nottingham, British Geological Survey.

DESTOMBES, J., HOLLAND, H. & WILLEFERT, S. 1985. Lower Palaeozoic rocks of Morocco. *In:* HOLLAND, C. H. (ed.) *Lower Palaeozoic of north-western and west-central Africa.* Lower Palaeozoic Rocks of the World, **4**, 91–337, John Wiley & Sons, Chichester.

DEYNOUX, M. & TROMPETTE, R. 1981. Late Ordovician tillites of the Taodeni Basin, West Africa. *In:* HAMBREY, M. J. & HARLAND, W. B. (eds) *Earth's pre-Pleistocene Glacial Record.* Cambridge University Press, 89–96.

DEYNOUX, M., SOUGY, J. & TROMPETTE, R. 1985. Lower Palaeozoic rocks of west Africa and the western part of central Africa. *In:* HOLLAND, C. H. (ed.) *Lower Palaeozoic of North-Western and West-Central Africa.* Lower Palaeozoic Rocks of the World, **4**, 337–497, John Wiley & Sons, Chichester.

DIAZ-MARTINEZ, E. 2007. The Sacta Limestone Member (early Wenlock): cool-water, temperate carbonate deposition at the distal foreland of Gondwana's active margin, Bolivia. *Palaeogeography, Palaeoclimatology, Palaeoecology*, in press.

DIAZ-MARTINEZ, E. & GRAHN, Y. 2007. Early Silurian glaciation along the western margin of Gondwana (Peru, Bolivia, and northern Argentina): Palaeogeographic and geodynamic setting. *Palaeogeography, Palaeoclimatology, Palaeoecology*, in press.

DORÉ, F. & LE GALL, J. 1973. Presence et position stratigraphique de la tillite Ordovicienne dans le Maine (E. du Massif Armoricain). *Bulletin de la Société Géologique de France*, **15**, 32–33.

DORÉ, F. 1981. The late Ordovician tillite in Normandy (Armorican Massif) *In:* HAMBREY, M. J. & HARLAND, W. B. (eds) *Earth's Pre-Pleistocene Glacial Record.* Cambridge University Press, Cambridge, 582–584.

EASTLER, T. E. 1969. Silurian geology of the Change Islands and eastern Notre Dame Bay, Newfoundland. *Memoir of the American Association of Petroleum Geologists*, **12**, 425–432.

ELDERFIELD, H. 2002. Climate change: Carbonate mysteries. *Science*, **296**, 1618–1621.

EVANS, D. A. D. 2003. A fundamental Precambrian–Phanerozoic shift in Earth's glacial style? *Tectonophysics*, **375**, 353–385.

EYLES, N. 1993. Earth's glacial record and its tectonic setting. *Earth Science Reviews*, **35**, 1–248.

FINNEY, S. C. & BERRY, W. B. N. 1997. New perspectives on graptolite distributions and their use as indicators of platform margin dynamics. *Geology*, **25**, 919–922.

FLOYD, J. D. & WILLIAMS, M. 2003 (for 2002). A revised correlation of Silurian rocks in the Girvan district, SW

Scotland. *Transactions of the Royal Society of Edinburgh, Earth Sciences*, **25**, 383–392.

FORTEY, R. A. & COCKS, L. R. M. 2003. Palaeontological evidence bearing on global Ordovician–Silurian continental reconstructions. *Earth Science Reviews*, **61**, 245–307.

FORTEY, R. A. & COCKS, L. R. M. 2005. Late Ordovician global warming – The Boda Event. *Geology*, **33**, 405–408.

FORTEY, R. A., HARPER, D. A. T., INGHAM, J. K., OWEN, A. W. & RUSHTON, A. W. A. 1995. A revision of Ordovician series and stages from the historical type area. *Geological Magazine*, **132**, 15–30.

FORTEY, R. A., HARPER, D. A. T., INGHAM, J. K., OWEN, A. W., PARKES, M. A., RUSHTON, A. W. A. & WOODCOCK, N. H. 2000. A revised correlation of Ordovician rocks in the British Isles. *Geological Society of London Special Report*, **24**, 1–83.

FRAKES, L. A., FRANCIS, J. E. & SYKTUS, J. I. 1992. *Climate Modes of the Phanerozoic*. Cambridge University Press, Cambridge.

GATINSKIY, Y. G., KLOCHKO, V. P., ROZMAN, K. S. & TROFIMOV, D. M. 1966. Novyye dannyye po stratigrafii paleozoyskikh otlozheniy yuzhnoy Sakhary. *Akadwmii Nauk SSSR, Doklady*, **170**, 1154–1157.

GENSEL, P. G. & EDWARDS, D. 2001. *Plants Invade the Land: Evolutionary and Environmental Perspectives*. Columbia University Press, New York.

GHIENNE, J.-F. 2003. Late Ordovician sedimentary environments, glacial cycles, and post-glacial transgression in the Taoudeni Basin, West Africa. *Palaeogeography, Palaeoclimatology, Palaeoecology*, **189**, 117–145.

GIBBS, M. T., BICE, K. L., BARRON, E. J. & KUMP, L. R. 2000. Glaciation in the early Paleozoic 'greenhouse'; the roles of paleogeography and atmospheric CO_2. *In*: HUBER, B. T., MACLEOD, K. G. & WING, S. L. (eds) *Warm Climates in Earth History*. Cambridge University Press, Cambridge, 386–422.

GOLDMAN, D., LESLIE, S. A., NÕLVAK, J. & YOUNG, S. 2005. The Black Knob Ridge section, southeastern Oklahoma, USA: a possible Global Stratotype-Section and Point (GSSP) for the base of the Middle Stage of the Upper Ordovician Series. www.stratigraphy.org/BKR.pdf.

GRAHN, Y. & CAPUTO, M. V. 1992. Early Silurian glaciations in Brazil. *Palaeogeography, Palaeoclimatology, Palaeoecology*, **99**, 9–15.

GRAHN, Y. & PARIS, F. 1992. Age and correlation of the Trombetas Group, Amazonas Basin, Brazil. *Revue de Micropaléontologie*, **35**, 197–209.

GUTIÉRREZ-MARCO, J. C. & ŠTORCH, P. 1998. Graptolite biostratigraphy of the Lower Silurian (Llandovery) shelf deposits of the Western Iberian Cordillera, Spain. *Geological Magazine*, **135**, 71–92.

HAMOUMI, N. 1999. Upper Ordovician glaciation spreading and its sedimentary record in Moroccan North Gondwana margin. *In*: KRAFT, P. & FJÄCKA, O. (eds) *Quo Vadis Ordovician?* Acta Universitatis Carolinae, Geologica, **43**, 111–114.

HATCH, J. R., JACOBSON, S. R., WITZKE, B. J. *ET AL.* 1987. Possible late Middle Ordovician organic carbon isotope excursion: evidence from Ordovician oils and hydrocarbon source rocks, mid-

continent and east-central United States. *American Association of Petroleum Geologists, Bulletin*, **71**, 1342–1354.

HAVLÍČEK, V. & MASSA, D. 1973. Brachiopodes de l'Ordovicien Supérieur de Libye occidental: implications stratigraphiques regionales. *Geobios*, **6**, 267–290.

HAYES, J. M., STRAUSS, H. & KAUFMAN, A. J. 1999. The abundance of ^{13}C in marine organic matter and isotopic fractionation in the global biogeochemical cycle of carbon during the past 800 Ma. *Chemical Geology*, **161**, 103–125.

HAYS, P. D. & GROSSMAN, E. L. 1991. Oxygen isotopes in meteoric calcite cements as indicators of continental paleoclimate. *Geology*, **19**, 441–444.

HAYWOOD, A. M., DEKENS, P., RAVELO, A. C. & WILLIAMS, M. 2005. Warmer tropics during the mid-Pliocene? Evidence from alkenone paleothermometry and a fully coupled ocean-atmosphere GCM. *Geochemistry, Geophysics, Geosystems*, **6**, Q03010.

HERRMANN, A. D., PATZKOWSKY, M. E. & POLLARD, D. 2003. Obliquity forcing with 8–12 times preindustrial levels of atmospheric pCO_2 atmospheric CO_2 during the Late Ordovician glaciation. *Geology*, **31**, 485–488.

HERRMANN, A. D., PATZKOWSKY, M. E. & POLLARD, D. 2004*a*. The impact of paleogeography, pCO_2, poleward ocean heat transport and sea level change on global cooling during the Late Ordovician. *Palaeogeography, Palaeoclimatology, Palaeoecology*, **206**, 59–74.

HERRMANN, A. D., HAUPT, B. J., PATZKOWSKY, M. E., SEIDOV, D. & SLINGERLAND, R. L. 2004*b*. Response of Late Ordovician paleoceanography to changes in sea level, continental drift, and atmospheric pCO_2: potential causes for long-term cooling and glaciation. *Palaeogeography, Palaeoclimatology, Palaeoecology*, **210**, 385–401.

HOLLAND, H. D. 1978. *The Chemistry of the Atmosphere and the Ocean*. Wiley Interscience, New York.

HOVIUS, N. 1998. Controls on sediment supply by large rivers. *In*: SHANLEY, K. W. & MCCABE, P. J. (eds) *Relative Role of Eustasy, Climate, and Tectonism in Continental Rocks*. SEPM Special Publication, **59**, 3–16.

HUTT, J. E. 1974. The Llandovery graptolites of the English Lake District. Part 1. *Monograph of the Palaeontographical Society, London*, **128**, 1–56.

JAMES, N. P., BONE, Y. & KYSER, T. K. 1997. Brachiopod $\delta^{18}O$ values do reflect ambient oceanography: Lacepede Shelf, southern Australia. *Geology*, **25**, 551–554.

JEPPSSON, L. 1990. An oceanic model for lithological and faunal changes tested on the Silurian record. *Journal of the Geological Society, London*, **147**, 663–674.

JEPPSSON, L. 1997. The anatomy of the Mid–Early Silurian Ireviken Event and a scenario for P–S events. *In*: BRETT, C. E. & BAIRD, G. C. (eds) *Paleontological Events: Stratigraphic, Ecological, and Evolutionary Implications*. Columbia University Press, New York, 451–492.

JEPPSSON, L., ALDRIDGE, R. J. & DORNING, K. J. 1995. Wenlock (Silurian) oceanic episodes and events.

Journal of the Geological Society, London, **152**, 487–498.

JOHNSON, M. E. 1996. Stable cratonic sequences and a standard for Silurian eustasy. *In*: WITZKE, B. J., LUDVIGSON, G. A. & DAY, J. (eds) *Paleozoic Sequence Stratigraphy: Views from the North American Craton.* Geological Society of America Special Paper, **306**, 203–211.

JOHNSON, M. E. 2006. Relationship of Silurian sea-level fluctuations to oceanic episodes and events. *GFF*, **128**, 115–122.

JOHNSON, M. E., KALJO, D. L. & RONG, J.-Y. 1991. Silurian eustasy. *In*: BASSETT, M. G., LANE, P. D. & EDWARDS, D. (eds) *The Murchison Symposium.* Special Papers in Palaeontology, **44**, 145–163.

KALJO, D. & MARTMA, T. 2006. Application of carbon isotope stratigraphy to dating the Baltic Silurian rocks, *GFF*, **128**, 123–129.

KALJO, D., MARTMA, T., MANNIK, P. & VIIRA, V. 2003. Implications of Gondwana glaciations in the Baltic Late Ordovician and Silurian and a carbon isotopic test of environmental cyclicity. *Bulletin de la Société Géologique de France*, **174**, 59–66.

KALJO, D., HINTS, L., HINTS, O., MARTMA, T. & NÕLVAK, J. 1999. Carbon isotope excursions and coeval biotic-environmental changes in the late Caradoc and Ashgill of Estonia. *In*: KRAFT, P. & FJAKA, O. (eds) *Quo Vadis Ordovician?* Acta Universitatis Carolinae, Geologica, **43**, 507–510.

KALJO, D., HINTS, L., MARTMA, T. & NÕLVAK, J. 2001. Carbon isotope stratigraphy in the latest Ordovician of Estonia. *Chemical Geology*, **175**, 49–59.

KALJO, D., HINTS, L., MARTMA, T., NÕLVAK, J. & ORASPÕLD, A. 2004. Late Ordovician carbon isotope trend in Estonia, its significance in stratigraphy and environmental analysis. *Palaeogeography, Palaeoclimatology, Palaeoecology*, **210**, 165–185.

KALJO, D., KIIPLI, T. & MARTMA, T. 1997. Carbon isotope event markers through the Wenlock–Pridoli sequence in Ohesaare (Estonia) and Priekule (Latvia). *Palaeogeography, Palaeoclimatology, Palaeoecology*, **132**, 211–224.

KALJO, D., KIIPLI, T. & MARTMA, T. 1998. Correlation of carbon isotope events and environmental cyclicity in the East Baltic Silurian. *In*: LANDING, E. & JOHNSON, M. E. (eds) *Silurian Cycles: Linkages of Dynamic Stratigraphy with Atmospheric, Oceanic and Tectonic Changes.* New York State Museum Bulletin, **491**, 297–312.

KALJO, D., MARTMA, T. & SAADRE, T. 2007. Post-Hunnebergian Ordovician carbon isotope trend in Baltoscandia, its environmental implications and some similarities with that of Nevada. *Palaeogeography, Palaeoclimatology, Palaeoecology*, in press.

KEGEL, W. 1953. Contribuição para o estudo do. Devoniano da Bacia do Parnaíba. *Boletim da Divisão de Geologia e Mineralogia, Rio de Janeiro*, **141**, 1–48.

KENNEDY, M. J. 1981. The early Palaeozoic Stoneville Formation, northern Newfoundland. *In:* HAMBREY, M. J. & HARLAND, W. B. (eds) *Earth's pre-Pleistocene Glacial Record.* Cambridge University Press, Cambridge, 713–716.

KING, L. M. 1994. Subsidence analysis of eastern Avalonian sequences; implications for Iapetus closure.

Journal of the Geological Society, London, **151**, 647–657.

KOREN, T. L., POPOV, L. E., DEGTJAREV, K. E., KOVALEVSKY, O. P. & MODZALEVSKAYA, T. L. 2003. Kazakhstan in the Silurian. *In*: LANDING, E. & JOHNSON, M. E. (eds) *Silurian Lands and Seas: Palaeogeography Outside of Laurentia.* New York State Museum, Bulletin, **493**, 323–344.

KUMP, L. R. 2002. Reducing uncertainty about carbon dioxide as a climate driver. *Nature*, **419**, 188–190.

KUMP, L. R. & ARTHUR, M. A. 1999. Interpreting carbon-isotope excursions: carbonates and organic matter. *Chemical Geology*, **161**, 181–198.

KUMP, L. R., ARTHUR, M. A., PATZKOWSKY, M. E., GIBBS, M. T., PINKUS, D. S. & SHEEHAN, P. M. 1999. A weathering hypothesis for glaciation at high atmospheric pCO_2 during the Late Ordovician. *Palaeogeography, Palaeoclimatology, Palaeoecology*, **152**, 173–187.

LE HERON, D. P., SUTCLIFFE, O. E., WHITTINGTON, R. J. & CRAIG, J. 2005. The origins of glacially related soft-sediment deformation structures in Upper Ordovician glaciogenic rocks: implication for ice sheet dynamics. *Palaeogeography, Palaeoclimatology, Palaeoecology*, **218**, 75–103.

LEGGETT, J. K. 1980. British Lower Paleozoic black shales and their palaeo-oceanographic significance. *Journal of the Geological Society, London*, **137**, 139–156.

LEGGETT, J. K. 1987. The Southern Uplands as an accretionary prism; the importance of analogues in reconstructing palaeogeography. *Journal of the Geological Society, London*, **144**, 737–752.

LEGGETT, J. K., MCKERROW, W. S., COCKS, L. R. M. & RICKARDS, R. B. 1981. Periodicity in the early Palaeozoic marine realm. *Journal of the Geological Society*, **138**, 167–176.

LEGGETT, J. K., MCKERROW, W. S., MORRIS, J. H., OLIVER, G. J. H. & PHILLIPS, W. E. A. 1979. The north-western margin of the Iapetus Ocean. *In*: HARRIS, A. L., HOLLAND, C. H. & LEAKE, B. E. (eds) *The Caledonides of the British Isles; reviewed.* Special Publication of the Geological Society, London, **8**, 499–512.

LEGRAND, P. 1969. Description de *Westonia chudeaui* sp. nov. Brachiopode inarticulé de l'Adrar mauritanien (Sahara occidental). *Bulletin de la Société Géologique de France*, **11**, 251–256.

LEGRAND, P. 1970. Les couches à *Diplograptus* du Tassili de Tarit (Ahmet, Sahara algérien). *Bulletin de la Société d'Histoire Naturelle de Afrique du Nord*, **60**, 3–58.

LEGRAND, P. 1985. Lower Palaeozoic rocks of Algeria. *In*: HOLLAND, C. H. (ed.) *Lower Palaeozoic of North-Western and West-Central Africa. Lower Palaeozoic Rocks of the World*, **4**, 5–89, John Wiley & Sons, Chichester.

LOCKLAIR, R. E. & LERMANN, A. 2005. A model of Phanerozoic cycles of carbon and calcium in the global ocean: Evaluation and constraints on ocean chemistry and input fluxes. *Chemical Geology*, **217**, 113–126.

LOWENSTAM, H. A. 1961, Mineralogy, $^{18}O/^{13}C$ ratios, and strontium and magnesium contents of recent and fossil brachiopods and their bearing on the history of oceans. *Journal of Geology*, **69**, 241–260.

LOYDELL, D. K. 1994. Early Telychian changes in graptoloid diversity and sea level. *Geological Journal*, **29**, 355–368.

LOYDELL, D. K. 1998. Early Silurian sea-level changes. *Geological Magazine*, **135**, 447–471.

LOYDELL, D. K. & NESTOR, V. 2005. Integrated graptolite and chitinozoan biostratigraphy of the upper Telychian (Llandovery, Silurian) of the Ventspils D-3 core, Latvia. *Geological Magazine*, **142**, 369–376.

LOYDELL, D. K., KALJO, D. & MÄNNIK, P. 1998. Integrated biostratigraphy of the lower Silurian of the Ohesaare core, Saaremaa, Estonia. *Geological Magazine*, **135**, 769–783.

LOYDELL, D. K., MÄNNIK, P. & NESTOR, V. 2003. Integrated biostratigraphy of the lower Silurian of the Aizpute-41 core, Latvia. *Geological Magazine*, **140**, 205–229.

LUDVIGSON, G. A., WITZKE, B. J., GONZALEZ, L. A., CARPENTER, S. J., SCHNEIDER, C. L. & HASIUK, F. H. 2004. Late Ordovician (Turinian–Chatfieldian) carbon isotope excursions and their paleoceanographic significance. *Palaeogeography, Palaeoclimatology, Palaeoecology*, **210**, 187–214.

LÜNING, S., CRAIG, J., LOYDELL, D. K., STORCH, P. & FITCHES, B. 2000. Lower Silurian 'hot shales' in North Africa and Arabia: regional distribution and depositional model. *Earth Science Reviews*, **49**, 21–200.

LÜNING, S., LOYDELL, D. K., ŠTORCH, P., SHAHIN, Y. & CRAIG, J. 2006. Origin, sequence stratigraphy and depositional environment of an upper Ordovician (Hirnantian) deglacial black shale, Jordan – Discussion. *Palaeogeography, Palaeoclimatology, Palaeoecology*, **230**, 352–355.

LÜNING, S., SHAHIN, Y. M., LOYDELL, D. K., AL-RABI, H. T., MASRI, A., TARAWNEH, B. & KOLONIC, S. 2005. Anatomy of a world-class source rock: distribution and depositional model of Silurian organic-rich shales in Jordan and implications for hydrocarbon potential. *AAPG Bulletin*, **89**, 1397–1427.

MAACK, R. 1957. Über Vereisungsperioden und Vereisungsspuren in Brasilien. *Geologische Rundschau*, **45**, 547–595.

MABILLARD, J. E. & ALDRIDGE, R. J. 1985. Microfossil distribution across the base of the Wenlock Series in the type area. *Palaeontology*, **28**, 89–100.

MÄNNIL, R. & MEIDLA, T. 1994. The Ordovician System of the East European Platform (Estonia, Latvia, Lithuania, Belorussia, parts of Russia, Ukraine and Moldova). *In*: WEBBY, B. D., ROSS, R. J. & ZHEN, Y. Y. (eds) *The Ordovician System of the East European Platform and Tuva (southeastern Russia)*. IUGS Publication, **28**, 1–52.

MARSHALL, J. D., BRENCHLEY, P. J., MASON, P., WOLFF, G. A., ASTINI, R. A., HINTS, L. & MEIDLA, T. 1997. Global carbon isotopic events associated with mass extinction and glaciation in the late Ordovician. *Palaeogeography, Palaeoclimatology, Palaeoecology*, **132**, 195–210.

MARTIN, R. E. 1995. Cyclic and secular variation in microfossil biomineralization – clues to the biogeochemical evolution of Phanerozoic oceans. *Global and Planetary Change*, **11**, 1–23.

MCCANN, A. M. & KENNEDY, M. J. 1974. A probable glacio-marine deposit of Late Ordovician–Early Silurian age from the north central Newfoundland Appalachian Belt. *Geological Magazine*, **111**, 549–564.

MCCLURE, H. A. 1978. Early Palaeozoic glaciation in Arabia. *Palaeogeography, Palaeoclimatology, Palaeoecology*, **25**, 315–326.

MCKERROW, W. S., MAC NIOCIALL, C. & DEWEY, J. F. 2000. The Caledonian Orogeny redefined. *Journal of the Geological Society, London*, **157**, 1149–1154.

MELCHIN, M. J. 1989. Llandovery graptolite biostratigraphy and paleobiogeography, Cape Phillips Formation, Canadian Arctic Islands. *Canadian Journal of Earth Sciences*, **26**, 1726–1746.

MELCHIN, M. J. & HOLMDEN, C. 2006. Carbon isotope chemostratigraphy in Arctic Canada: Sea-level forcing of carbonate platform weathering and implications for Hirnantian global correlation. *Palaeogeography, Palaeoclimatology, Palaeoecology*, **234**, 186–200.

MELCHIN, M. J., COOPER, R. A. & SADLER, P. M. 2004. The Silurian Period. *In*: GRADSTEIN, F. M., OGG, J. G. & SMITH, A. G. (eds) *A Geologic Time Scale 2004*. Cambridge University Press, Cambridge, 165–187.

MERINO, D. 1991. Primer registro de conodontos siluricos en Bolivia. *Revisita Tecnica de YPFB*, **12**, 271–274.

MEYBECK, M. 1982. Carbon, nitrogen, and phosphorus transport by world rivers. *American Journal of Science*, **282**, 401–450.

MILLER, M. A. & MELVIN, J. 2005. Significant new biostratigraphic horizons in the Qusaiba member of the Silurian Qalibah formation of central Saudi Arabia, and their sedimentologic expression in a sequence stratigraphic context. *GeoArabia*, **10**, 49–92.

MONTAÑEZ, I. P. 2002. Biological skeletal carbonate records changes in major-ion chemistry of paleo-oceans. *PNAS*, **99**, 15852–15854.

MU, E.-Z. 1988. The Ordovician–Silurian boundary in China. *Bulletin of the British Museum (Natural History)*, **43**, 117–131.

MULLINS, G. L. & ALDRIDGE, R. J. 2004. Chitinozoan biostratigraphy of the basal Wenlock Series (Silurian) global stratotype section and point. *Palaeontology*, **47**, 745–773.

MUNNECKE, A., SAMTLEBEN, C. & BICKERT, T. 2003. The Ireviken event in the Lower Silurian of Gotland, Sweden – relation to similar Palaeozoic and Proterozoic events. *Palaeogeography, Palaeoclimatology, Palaeoecology*, **195**, 99–124.

NIELSEN, A. T. 2003a. Ordovician sea level changes: a Baltoscandian perspective. *In*: WEBBY, B. D., PARIS, F., DROSER, M. & PERCIVAL, I. G. (eds) *The Great Ordovician Biodiversification Event*. Columbia University Press, New York, 84–93.

NIELSEN, A. T. 2003b. Late Ordovician sea level changes: evidence of Caradoc glaciations? *Geophysical Research Abstracts, European Geosciences Union*, **5**.

NÕLVAK, J., HINTS, O. & MÄNNIK, P. 2006. Ordovician timescale in Estonia: recent developments. *Proceedings of the Estonian Academy of Sciences, Geology*, **55**, 95–108.

OLDROYD, D. R. 1990. *The Highland Controversy: Constructing Geological Knowledge through Fieldwork in*

Nineteenth-Century Britain. Science and Its Conceptual Foundations series. Chicago University Press, Chicago.

ORR, P. J. 2001. Bathymetric Indicators. *In*: BRIGGS, D. E. G. & CROWTHER, P. R. (eds) *Palaeobiology II*. Blackwell Science, Oxford, 479–482.

PANCOST, R. D., FREEMAN, K. H., PATZKOWSKY, M. E., WAVREK, D. & COLLISTER, J. W. 1998. Molecular indicators of redox and marine photoautotroph composition in the late Middle Ordovician of Iowa, USA. *Organic Geochemistry*, **29**, 1649–1662.

PARRISH, J. T. 1982. Upwelling and petroleum source beds, with reference to the Palaeozoic. *AAPG Bulletin*, **66**, 750–774.

PATZKOWSKY, M. E., SLUPIK, L. M., ARTHUR, M. A., PANCOST, R. D. & FREEMAN, K. H. 1997. Late Middle Ordovician environmental change and extinction: harbinger of the Late Ordovician or continuation of Cambrian patterns? *Geology*, **25**, 911–914.

PHILLIPOT, A. & ROBARDET, M. 1971. Nouvelles donnees sur les formations siluriennes de Domfront (Orne). *Bulletin de la Société Géologique et Mineralogique de Bretagne, Serie C*, **3**, 41–47.

PICKERILL, R. K., PAJARI, G. E. & CURRIE, K. L. 1979. Evidence of Caradocian glaciation in the Davidsville Group of north-eastern Newfoundland. *Current Research, Part C, Papers of the Canadian Geological Survey*, **79**, 67–72.

PORĘBSKA, E. & SAWŁOWICZ, Z. 1997. Palaeoceanographic linkage of geochemical and graptolite events across the Silurian–Devonian boundary in Bardzkie Mountains (Southwest Poland). *Palaeogeography, Palaeoclimatology, Palaeoecology*, **132**, 343–354.

PRATT, W. T., WOODHALL, D. G. & HOWELLS, M. F. 1995. Geology of the country around Cadair Idris. Memoir for 1:50,000 Geological Sheet 149 (England and Wales). London, HMSO.

RAHMSTORF, S. 2002. Ocean circulation and climate during the past 120,000 years, *Nature*, **419**, 207–214.

RAYMO, M. E. 1991. Geochemical evidence supporting T.C. Chamberlin's theory of glaciation. *Geology*, **19**, 344–347.

RICKARDS, R. B. 1970. The Llandovery (Silurian) graptolites of the Howgill Fells, Northern England. *Palaeontographical Society Monographs*, **123**, 1–108.

RICKARDS, R. B. & WOODCOCK, N. H. 2005. Stratigraphical revision of the Windermere Supergroup (Late Ordovician–Silurian) in the southern Howgill Fells, NW England. *Proceedings of the Yorkshire Geological Society*, **55**, 263–285.

RIDGWELL, A. 2005. A Mid Mesozoic Revolution in the regulation of ocean chemistry. *Marine Geology*, **217**, 339–357.

ROBARDET, M., HENRY, J. L., NION, J., PARIS, F. & PILLET, J. 1972. La Formation du Pont-de-Caer (Caradocien) dans les synclinaux de Domfront et de Sees (Normandie). *Société Géologique du Nord, Annales*, **92**, 117–137.

ROCHA-CAMPOS, A. C. 1981. Early Palaeozoic Iapo Formation of Paraná, Brazil. *In*: HAMBREY, M. J. & HARLAND, W. B. (eds) *Earth's pre-Pleistocene Glacial Record*. Cambridge University Press, Cambridge, 908–909.

ROHLING, E. J. & GIESKES, W. W. C. 1989. Late Quaternary changes in Mediterranean Intermediate Water density and formation rate. *Paleoceanography*, **4**, 531–545.

ROHLING, E. J. 1994. Review and new aspects concerning the formation of Mediterranean sapropels, *Marine Geology*, **122**, 1–28.

RONG, J.-Y., BOUCOT, A. J., SU, Y.-Z. & STRUSZ, D. L. 1995. Biogeographical analysis of Late Silurian brachiopod faunas, chiefly from Asia and Australia. *Senckenberg Lethaea*, **28**, 39–60.

RONOV, A. B., KHAIN, V. E., BALUKHOVSKY, A. N. & SESLAVINSKY, K. B. 1980. Quantitative analysis of Phanerozoic sedimentation. *Sedimentary Geology*, **25**, 311–325.

ROSS, C. A. & ROSS, J. R. P. 1995. North American Ordovician depositional sequences and correlations. *In*: COOPER, J. D. & FINNEY, S. C. (eds) *Ordovician Odyssey. SEPM, Pacific Section*, Fullerton California, 309–313.

ROSS, C. A. & ROSS, R. P. 1996. Silurian sea-level fluctuations. *In*: WITZKE, B. J., LUDVIGSON, G. A. & DAY, J. (eds) *Paleozoic Sequence Stratigraphy: Views from the North American Craton*. Geological Society of America, Special Paper, **306**, 187–192.

ROTHMAN, D. H., HAYES, J. M. & SUMMONS, R. E. (2003) Dynamics of the Neoproterozoic carbon cycle. *Proceedings of the National Academy of Science*, **100**, 8124–8129.

ROYER, D. L. 2006. CO_2-forced climate thresholds during the Phanerozoic. *Geochimica et Cosmochimica Acta*, **70**, 566–567.

ROYER, D. L., BERNER, R. A., MONTAÑEZ, I. P., TABOR, N. J. & BEERLING, D. J. 2004. CO_2 as a primary driver of Phanerozoic climate. *GSA Today*, **14**, 4–10.

RUDWICK, M. J. S. 1985. *The great Devonian controversy: The shaping of scientific knowledge among gentlemanly specialists*. Science and its conceptual foundations series. Chicago University Press, Chicago.

RUST, I. C. 1981. Early Palaeozoic Pakhuis Tillite, South Africa. *In*: HAMBREY, M. J. & HARLAND, W. B. (eds) *Earth's pre-Pleistocene Glacial Record*. Cambridge University Press, Cambridge, 113–117.

SALTZMAN, M. R. 2005. Phosphorus, nitrogen, and the redox evolution of the Paleozoic oceans. *Geology*, **33**, 573–576.

SALTZMAN, M. R. & YOUNG, S. A. 2005. A long-lived glaciation in the Late Ordovician? Isotopic and sequence stratigraphic evidence from western Laurentia. *Geology*, **33**, 109–112.

SAMTLEBEN, C., MUNNECKE, A., BICKERT, T. & PÄTZOLD, J. 2001. Shell succession, assemblage and species dependent effects on the C/O-isotopic composition of brachiopods: examples from the Silurian of Gotland. *Chemical Geology*, **175**, 61–107.

SAMTLEBEN, C., MUNNECKE, A., BICKERT, T. & PÄTZOLD, J. 1996. The Silurian of Gotland (Sweden): Facies interpretation based on stable isotopes in brachiopod shells. *Geologische Rundschau*, **85**, 278–292.

SCHENCK, P. E. 1972. Possible Late Ordovician glaciation of Nova Scotia. *Canadian Journal of Earth Science*, **9**, 95–107.

SCHENCK, P. E. & LANE, T. E. 1981. Early Paleozoic tillite of Nova Scotia, Canada. *In*: HAMBREY, M. J. & HARLAND, W. B. (eds) *Earth's pre-Pleistocene Glacial Record*. Cambridge University Press, Cambridge, 707–710.

SCHOFIELD, D. I., DAVIES, J. R, WATERS, R. A., WILBY, P. R., WILLIAMS, M. & WILSON, D. 2004. Geology of the Builth Wells district – a brief explanation of the geological map. Sheet explanation of the British Geological Survey. 1:50000 sheet 196 Builth Wells (England & Wales). Keyworth, Nottingham, British Geological Survey.

SCOTESE, C. R. & MCKERROW, W. S. 1991. Ordovician plate tectonic reconstructions. *Canadian Geological Survey Paper*, **90–99**, 271–282.

SCRIVNER, A. E., VANCE, D. & ROHLING, E. J. 2004. New neodymium isotope data quantify Nile involvement in Mediterranean anoxic episodes. *Geology*, **32**, 565–568.

SECORD, J. A. 1986. *Controversy in Victorian geology: the Cambrian–Silurian dispute*. Princeton University Press, Princeton.

SHACKLETON, N. J. 2000. The 100,000-year ice-age cycle identified and found to lag temperature, carbon dioxide, and orbital eccentricity. *Science*, **289**, 1897–1902.

SHAVIV, N. J. 2002. Cosmic ray diffusion from the galactic spiral arms, iron meteorites, and a possible climatic connection. *Physical Review Letters*, **89**, art. no. 051102.

SHAVIV, N. J. & VEIZER, J. 2003. Celestial driver of Phanerozoic climate?, *GSA Today*, **13**, 4–10.

SHIELDS, G. A., CARDEN, G. A. F., VEIZER, J., MEIDLA, T., RONG, J.-Y. & LI, R.-Y. 2003. Sr, C and O isotope geochemistry of Ordovician brachiopods: a major isotopic event around the Middle–Late Ordovician transition. *Geochimica et Cosmochimica Acta*, **67**, 2005–2025.

SHUSTER, D. L., EHLERS, T. A., RUSMOREN, M. E. & FARLEY, K. A. 2005. Rapid glacial erosion at 1.8 Ma revealed by ^{4}He/^{3}He thermochronometry. *Science*, **310**, 1668–1670.

SIEGENTHALER, U., STOCKER, T., MONNIN, E. *ET AL.* 2005. Stable carbon cycle-climate relationship during the late Pleistocene. *Science*, **310**, 1313–1317.

SKEVINGTON, D. 1974. Controls influencing the composition and distribution of Ordovician graptolite provinces. *In*: RICKARDS, R. B., JACKSON, D. E. & HUGHES, C. P. (eds) *Graptolite Studies in Honour of O. M. B. Bulman*. Special Papers in Palaeontology, **13**, 59–73.

SMITH, A. G. & PICKERING, K. T. 2003. Oceanic gateways as a critical factor to initiate icehouse Earth. *Journal of the Geological Society*, **160**, 337–340.

ŠTORCH, P. 1986. Ordovician–Silurian boundary in the Prague Basin (Barrandian area, Bohemia). *Sborník Geologických věd, Geologie*, **41**, 69–103.

ŠTORCH, P. 1994. Graptolite biostratigraphy of the Lower Silurian (Llandovery and Wenlock) of Bohemia. *Geological Journal*, **29**, 137–165.

ŠTORCH, P. & MERGL, M. 1989. Králodvor/Kosov boundary and the late Ordovician environmental changes in the Prague Basin (Barrandian area, Bohemia). *Sborník Geologických Věd, Geologie*, **44**, 117–153.

STRACHAN, R. A. 2000. Mid-Ordovician to Silurian sedimentation and tectonics on the northern active margin of Iapetus. *In*: WOODCOCK, N. H. & STRACHAN, R. A. (eds) *Geological History of Britain Ireland*. Blackwell, Oxford, 107–123.

SUÁREZ-SORUCO, R. 1995. Comentarios sobre la edad de la Formacion Cancañiri. *Revista Tecnica de Yacamientos Petroliferos Fisicales Bolivianos*, **16**, 51–54.

SUTCLIFFE, O. E., DOWDESWELL, J. A., WHITTINGTON, R. J., THERON, J. N. & CRAIG, J. 2000. Calibrating the Late Ordovician glaciation and mass extinction by the eccentricity of Earth's orbit. *Geology*, **28**, 967–970.

SUTCLIFFE, O. E., HARPER, D. A. T., SALEM, A. A., WHITTINGTON, R. J. & CRAIG, J. 2001. The development of an atypical *Hirnantia* brachiopod fauna and the onset of glaciation in the late Ordovician of Gondwana. *Transactions of the Royal Society of Edinburgh, Earth Sciences*, **92**, 1–14.

THONG-DZUY, T., BOUCOT, A. J., RONG, J.-Y. & FANF, Z.-J. 2001. Late Silurian marine shelly fauna of Central and Northern Vietnam. *Geobios*, **34**, 315–338.

TOBIN, K. J., BERGSTRÖM, S. M. & DE LA GARZA, P. 2005. A mid-Caradocian (453 Ma) drawdown in atmospheric pCO$_2$ atmospheric CO$_2$ without ice sheet development? *Palaeogeography, Palaeoclimatology, Palaeoecology*, **226**, 187–204.

TOGHILL, P. 1968. The graptolite assemblages and zones of the Birkhill Shales (Lower Silurian) at Dobb's Linn. *Palaeontology*, **11**, 654–678.

TORSVIK, T. H. 1998. Palaeozoic palaeogeography: A North Atlantic viewpoint. *GFF*, **120**, 109–118.

TUCKER, M. E. & REID, P. C. 1973. The sedimentology and context of late Ordovician glacial marine sediments from Sierra Leone, West Africa. *Palaeogeography, Palaeoclimatology, Palaeoecology*, **13**, 289–307.

UNDERWOOD, C. J. 1992. Graptolite Preservation and Deformation. *Palaios*, **7**, 178–186.

UNDERWOOD, C. J., CROWLEY, S. F., MARSHALL, J. D. & BRENCHLEY, P. J. 1997. High resolution carbon isotope stratigraphy of the basal Silurian stratotype (Dob's Linn, Scotland) and its global correlation. *Journal of the Geological Society, London*, **154**, 709–718.

UNDERWOOD, C. J., DEYNOUX, M. & GHIENNE, J.-F. 1998. High Palaeolatitude (Hodh, Mauritania) recovery of graptolite fauna after the Hirnantian (end Ordovician) extinction event. *Palaeogeography, Palaeoclimatology, Palaeoecology*, **142**, 91–105.

VEIZER, J., BRUCKSCHEN, P. & PAWELLEK, F. *ET AL.* 1997. Oxygen isotope evolution of Phanerozoic seawater. *Palaeogeography, Palaeoclimatology, Palaeoecology*, **132**, 159–172.

VEIZER, J., GODDERIS, Y. & FRANÇOIS, L. M. 2000. Evidence for decoupling of atmospheric CO$_2$ and global climate during the Phanerozoic eon. *Nature*, **408**, 698–701.

VELBEL, M. A. 1993. Temperature dependence of silicate weathering in nature: How strong a negative feedback on long-term accumulation of atmospheric CO_2 and global greenhouse warming? *Geology*, **21**, 1059–1062.

VERNIERS, J. & VANDENBROUCKE, T. R. A. 2006. Chitinozoan biostratigraphy in the Dob's Linn Ordovician–Silurian GSSP, Southern Uplands, Scotland. *GFF*, **128**, 195–202.

WALKER, L. J., WILKINSON, B. H. & IVANY, L. C. 2002. Continental drift and Phanerozoic carbonate accumulation in shallow-shelf and deep-marine settings. *Journal of Geology*, **110**, 75–87.

WHITE, D. E., BARRON, H. F., BARNES, R. P. & LINTERN, B. C. 1991. Biostratigraphy of late Llandovery (Telychian) and Wenlock turbiditic sequences in the SW Southern Uplands, Scotland. *Transactions of the Royal Society of Edinburgh, Earth Sciences*, **82**, 297–322.

WIGNALL, P. B. 1991. Model for transgressive black shales? *Geology*, **19**, 167–170.

WIGNALL, P. B. 1994. *Black Shales*. Clarendon Press, Oxford.

WIGNALL, P. B. & MAYNARD, J. R. 1993. The sequence stratigraphy of transgressive black shales. *In*: KATZ, B. J. & PRATT, L. (eds) *Source Rocks within a Sequence Stratigraphic Framework*. American Association of Petroleum Geologists Studies in Geology, **37**, 35–47.

WILLIAMS, M., DAVIES, J. R., WATERS, R. A., RUSHTON, A. W. A. & WILBY, P. R. 2003. Stratigraphical and palaeoecological importance of Caradoc (Upper Ordovician) graptolites from the Cardigan area, southwest Wales. *Geological Magazine*, **140**, 549–571.

WILLIAMS, M., RUSHTON, A. W. A., WOOD, B., FLOYD, J. D., SMITH, R. & WHEATLEY, C. 2004. A revised graptolite biostratigraphy for the lower Caradoc (Upper Ordovician) of southern Scotland. *Scottish Journal of Geology*, **40**, 97–114.

WILLIAMS, S. H. 1982. The Late Ordovician fauna of the *anceps* bands at Dob's Linn, southern Scotland. *Geologica et Palaeontologica*, **16**, 29–56.

WOODCOCK, N. H. 1990. Sequence stratigraphy of the Palaeozoic Welsh Basin. *Journal of the Geological Society, London*, **147**, 537–547.

WOODCOCK, N. H. 2000. Late Ordovician to Silurian evolution of Eastern Avalonia during convergence with Laurentia. *In*: WOODCOCK, N. H. & STRACHAN, R. A. (eds) *Geological History of Britain and Ireland*. Blackwell, Oxford.

WOODCOCK, N. H., BUTLER, A. J., DAVIES, J. R. & WATERS, R. A. 1996. Sequence stratigraphical analysis of late Ordovician and early Silurian depositional systems in the Welsh Basin: a critical assessment. *In*: HESSELBO, S. P. & PARKINSON, D. N. (eds) *Sequence Stratigraphy in British Geology*. Special Publication of the Geological Society of London, **103**, 197–208.

YAPP, C. J. & POTHS, H. 1992. Ancient atmospheric CO_2 pressures inferred from natural goethites. *Nature*, **355**, 342–344.

YOLKIN, E. A., SENNIKOV, N. V., BAKHAREV, N. K., IXOKH, N. G. & YAZIKOV, A. YU. 1997. Periodicity of deposition in the Silurian and relationships of global geological events in the middle Paleozoic of the southwestern margin of the Siberian continent. *Geologiya i Geofizika*, **38**, 596–607.

YOUNG, G. M. 1981. Early Palaeozoic tillites of the northern Arabian Peninsula. *In*: HAMBREY, M. J. & HARLAND, W. B. (eds) *Earth's Pre-Pleistocene Glacial Record*. Cambridge University Press, Cambridge, 338–340.

YOUNG, S. A., SALTZMAN, M. R. & BERGSTROM, S. A. 2005. Upper Ordovician (Mohawkian) carbon isotope stratigraphy in Eastern and Central North America: regional expression of a perturbation of the global carbon cycle. *Palaeogeography, Palaeoceanography, Palaeoclimatology*, **222**, 53–76.

ZACHOS, J., PAGANI, M., SLOAN, L., THOMAS, E. & BILLUPS, K. 2001. Trends, rhythms, and aberrations in global climate 65 Ma to present. *Science*, **292**, 686–693.

ZALASIEWICZ, J. A. 2001. Graptolites as constraints on models of sedimentation across Iapetus: a review. *Proceedings of the Geologists Association*, **112**, 237–251.

ZALASIEWICZ, J. A. & WILLIAMS, M. 1999. Graptolite biozonation of the Wenlock Series (Silurian) of the Builth Wells district, central Wales. *Geological Magazine*, **136**, 263–283.

ZALASIEWICZ, J. A., RUSHTON, A. W. A. & OWEN, A. W. 1995. Late Caradoc graptolitic faunal gradients across the Iapetus Ocean. *Geological Magazine*, **132**, 611–617.

ZALASIEWICZ, J., WILLIAMS, M., MILLER, M., PAGE, A. & BLACKETT, E. 2007. Early Silurian (Llandovery) graptolites from central Saudi Arabia: first documented record of Telychian faunas from the Arabian Peninsula. *GeoArabia*, **12**, in press.

ZHANG, S., BARNES, C. R. & JOWETT, D. M. S. 2006. The paradox of the global standard Late Ordovician–Early Silurian sea level curve: Evidence from conodont community analysis from both Canadian Arctic and Appalachian margins. *Palaeogeography, Palaeoclimatology, Palaeoecology*, **236**, 246–271.

Modelling climates of the Late Palaeozoic

J. T. KIEHL

National Center for Atmospheric Research, Earth Sun System Laboratory, 1850 Table Mesa Drive, Boulder, CO 80305 USA (e-mail: jtkon@ucar.edu)

Abstract: Climate models are comprehensive tools for studying Earth's past, present and future climate conditions. A spectrum of climate models is described with special focus on three-dimensional coupled atmosphere, ocean, land and cryosphere models. At present, these are the most sophisticated models to look at Earth's climate. Three time periods of the Late Palaeozoic are considered: the Carboniferous, the middle Permian and the latest Permian. Modelling the climate of these periods is reviewed to illustrate how models are applied to understanding past climates. Finally, the future development of climate modelling of the Late Palaeozoic is discussed.

Why model the past? This question may arise in considering studies of deep time. Geological and geochemical data provide a rich observational understanding of how Earth evolved through time. By carrying out careful analysis of field data, a picture of Earth develops for specific regions of the planet. However, these types of data are site specific, since it is impossible to carry out such studies for large areas of Earth over many time periods. These studies also provide a picture that is partially developed, because missing information always permits the possibility of multiple explanations for an event in geological history. For example, the existence of mass extinctions at certain time periods may be explained in many different ways (e.g. Hallam & Wignall 1997; Benton 2003).

Model simulations provide another tool for looking at Earth's past. They are not sufficient to provide all the information needed to understand a particular time period, but they may provide supplemental information on processes that played a role in determining Earth's past. These models, for the most part, simulate a time-averaged picture of temperature, precipitation, ocean circulation and the state of the cryosphere. They are global in coverage and, thus, can provide information on regions that may lie between sites where geological data exist. Fully coupled climate models are the most comprehensive and 'self consistent' tools available to study Earth's climate. As such, these models are valuable aids in understanding the physical mechanisms that shape Earth's climate.

Finally, these models also are used to look at the future state of Earth's climate. For this reason, it is important to evaluate the accuracy of these models against past climates. If these models can successfully replicate climates of the past, especially under diverse conditions, then there is greater confidence that the models can be used for future climate simulations. This is especially true in modelling past hothouse climates where CO_2 levels were greater than pre-industrial levels (e.g. Sloan & Barron 1990; Huber *et al.* 2000), for this is where our climate is projected to be in the future.

The Late Palaeozoic offers a unique challenge for climate models. The concentration of atmospheric CO_2 is believed to have ranged from present-day levels up to ten times present concentrations (e.g. Retallack 2002; Berner 2004; Royer *et al.* 2004; Royer 2006). The positions of the continents ranged from the single megacontinent, Pangaea, to one where major continents existed in the Northern and Southern Hemispheres, allowing for circumglobal ocean circulation in the tropics (Ziegler *et al.* 1979; Stampfli & Borel 2002). These diverse conditions provide a good test of the robustness of climate models.

This paper is organized in the following way: the following section considers the spectrum of models that exist to study past climates, and this is followed by a description of fully coupled global climate models as well as the input data required by these global models. The next section describes results from a few applications of models to different Late Palaeozoic climates. Finally, the future development of climate model capabilities for deep-time research is briefly discussed.

Spectrum of climate models

Climate models have been developed for many applications. Models that do not explicitly include all the physical and dynamical processes of the climate system are referred to as low-order climate models. Perhaps the most popular low-order model is the energy-balance model. This model is often two-dimensional (latitude and longitude) and assumes that at each point on Earth there is a balance between the incoming

From: WILLIAMS, M., HAYWOOD, A. M., GREGORY, F. J. & SCHMIDT, D. N. (eds) *Deep-Time Perspectives on Climate Change: Marrying the Signal from Computer Models and Biological Proxies.* The Micropalaeontological Society, Special Publications. The Geological Society, London, 157–167.

shortwave energy, the outgoing longwave energy and the horizontal transport of energy by atmospheric motions (for more information, see North & Stevens 2006).

The amount of shortwave energy available to the climate system depends upon the incoming solar radiation, the reflectivity of the region (or albedo), and the amount of solar energy absorbed by the surface–atmosphere system (Kiehl 1992). The surface temperature, the amount of cloud and the gaseous absorbers in the atmosphere determine the outgoing longwave energy (Kiehl 1992). The horizontal transport of energy is generally assumed to be diffusive, i.e. the transport is related to the temperature difference across a given point. These types of models have been used to look at many time periods in Earth's history (e.g. Crowley *et al.* 1991; Hyde *et al.* 1999).

Models of the ocean also vary in their level of complexity. The low-order models assume fixed ocean heat transport, but allow for the heat capacity of the ocean mixed layer. These so-called slab-ocean models, or mixed-layer models are often coupled to energy-balance models of the atmosphere or atmospheric general circulation models. The next level of complexity is to consider the zonally symmetric circulation of the oceans. These models calculate the circulation as a function of latitude and depth, and convey information about the overturning circulation of the oceans. The most complex level of ocean models are three-dimensional in extent and explicitly solve the equations of motion for the entire ocean system. These models allow for various mixing processes in the ocean and usually include a model for the ocean boundary layer. These models (e.g. Winguth *et al.* 2002) have been coupled to energy-balance models to form climate models of intermediate complexity. Their advantage is that they can be run for long time integrations.

The cryosphere can also be included in the low-order climate models (Hyde *et al.* 1999). For sea-ice, a thermodynamic model can be used to simulate the seasonal evolution of sea-ice amount and thickness. For glacial ice, thermodynamic and dynamic models of various levels of sophistication are used.

The exact configuration of the model is dictated by a number of factors. First, the particular problem under investigation may require lengthy computational integrations. For example, modelling a period of glacial growth and demise may require simulations of tens of thousands of years, which would necessitate a model of intermediate complexity. Secondly, computational limitations may force one to use a simpler model for a problem.

Global climate models

Global climate models (e.g. Trenberth 1992; Washington & Parkinson 2005) are the most comprehensive models used to look at Earth's climate. They are three-dimensional numerical representations of the climate system. The system contains separate numerical models for the atmosphere, ocean, sea-ice and land components. These component models are coupled together to represent the physical aspects of Earth's climate system. Each component model solves a system of physical equations representing conservation properties, e.g. conservation of momentum, energy and mass. The mathematical solution of these equations uses discrete forms of these expressions. Thus, a discrete grid is created for each component in terms of latitude, longitude and vertical extent (or pressure). An example of a discrete grid for the atmosphere is shown in Figure 1. The equations are also discretized in time, as well.

Any feature that is solved on the *grid scale* and *time scale* of the component model is called a resolved scale of the climate system. Typical spatial grids for climate system models are of the order of 200 kilometres, while the time step is around 10 to 20 minutes. This means that any physical property (e.g. clouds, ice polynyas, river systems, mountainous terrain or ocean eddies) that is of a spatial scale less than the grid size is not explicitly resolved in the model. These features, many of which are critical aspects of the climate system, are parametrically represented, or *parameterized*, in models.

How does a parameterization work? The process that is parameterized must be formulated in terms of the resolved scale properties. For example, the fractional amount of cloud in a grid box may be related to the average amount of moisture and temperature in that grid box. The specific relationship between the sub-grid-scale process (e.g. fractional cloud amount) and the large-scale resolved quantities may be based on empirical data or a finer-scale model combined with theoretical considerations (e.g. conservation of mass). At present, there are many different ways for parameterization of any given sub-grid-scale process, which leads to a diversity of climate models (e.g. Schlesinger 1988; Kiehl 1992). It is also apparent that changes in these parameterizations can lead to significant differences in climate simulations (Kiehl *et al.* 2006). The dependence of the simulated climate to the choice of parameterization is a concern in applying models to specific climate problems. However, it is found that diverse models tend to agree in many features despite the differences in parameterizations and numerical solution techniques (e.g. McAveny *et al.* 2001). It is worth

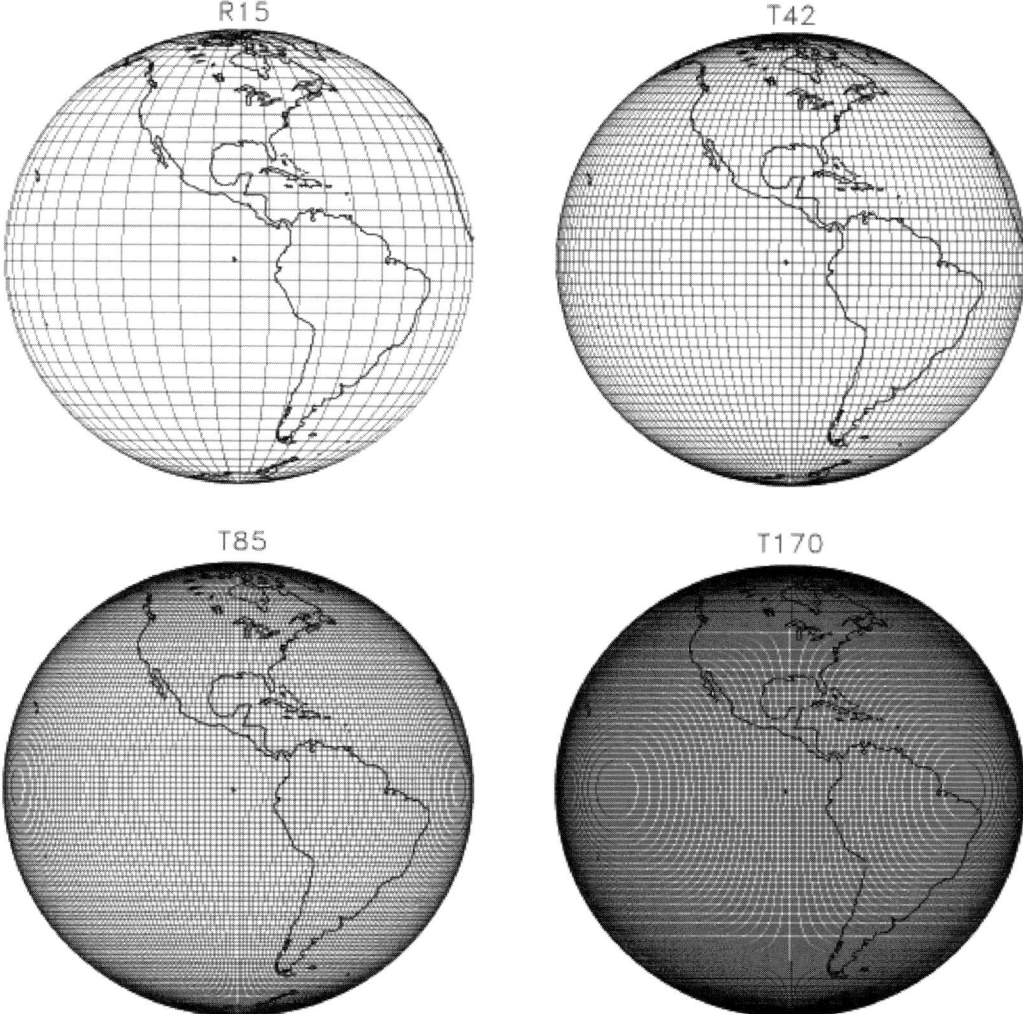

Fig. 1. Four grid resolutions for atmospheric climate models from Washington & Parkinson (2005). The R15 resolution is equivalent to 800 km by 500 km grid box, the T42 resolution is equivalent to 300 km by 300 km grid box, the T85 resolution is equivalent to 150 km by 150 km grid box, and the T170 resolution is equivalent to 75 km by 75 km grid box.

mentioning that the range in climate response to increased CO_2, the so-called range in climate sensitivity, mainly arises from the range in parameterizations employed in models (e.g. Randall *et al.* 2006).

Although the three-dimensional models are quite sophisticated in their formulation, the price one pays is that they are computationally expensive, namely, the finer the spatial grid scale, the smaller the time step, the more expensive the model. Thus, if a model were run at 1 or 2 kilometre spatial resolution, the cost would be computationally prohibitive. Parameterization is therefore a necessity and not a convenience, given current computational technology. Given the time scales of

geological processes (100,000 to millions of years), it is impossible to simulate time-varying, or transient, climate processes with these types of models. This is one reason that models of intermediate complexity are still attractive, for these models can simulate long time scale events. However, these models also are more dependent on parameterizations than the fully coupled three-dimensional models. The researcher must look at the specific question under study and decide which tool is best for their problem.

The three-dimensional models provide the user with a vast array of climate variables. Beyond the standard state variables of atmospheric temperatures, winds, precipitation, ocean temperatures, salinity and ocean circulation, the models currently predict cloud information, ocean tracers, soil properties, sea-ice properties, to name just a few. At present there are more predicted variables than measured, even for present-day conditions. For palaeoclimate simulations, it is often best to relate the model variable to the palaeoproxy in some manner (e.g. Rees *et al.* 1999; Gibbs *et al.* 2002).

Boundary conditions

Climate models require certain boundary conditions. For palaeoclimate simulations, the solar luminosity must be specified for the appropriate time period. At the time of the Late Palaeozoic, the solar output was *c.* 2–3% less than present. The solar output has monotonically increased over geologic time. The orbital configuration of the Sun–Earth geometry also varies in time. Thus, the declination of Earth and eccentricity of the Earth's orbit about the Sun is also required in a simulation. The spatial location of the continental landmasses, palaeogeography, and the ocean bathymetry are essential boundary conditions. Continental topography is also an essential boundary condition that is required for the three-dimensional climate models. If continental ice-masses were present, then the location and topography of these features are also necessary climate boundary conditions. The chemical composition of the palaeo-atmosphere is also specified in these models. For example, the mixing ratio of CO_2 is usually prescribed, along with the mixing ratios of methane and other greenhouse gases. The amount and distribution of water vapour is predicted by the model, and not specified.

Once the boundary conditions are specified, then an initial state is specified for the various components: atmosphere, ocean, land, sea-ice. Ideally, one could start with initial isothermal and rest (no atmospheric, ocean or ice motions) conditions. However, climate models typically start with conditions that are more realistic, for example present-day conditions. The climate model, if run for a sufficiently long time, will produce a new steady-state climate. In other words, for the given boundary conditions, the model will arrive at a solution of the climate system where the global annual mean amount of incoming absorbed solar radiation is balanced by the outgoing longwave radiation, or the system is in energy balance. To achieve this type of steady state requires the system to be integrated for a few thousand years, which is the steady-state time scale of the ocean component. To date there is only one such Palaeozoic simulation that meets this criterion (Kiehl & Shields 2005). Most

climate simulations are run for a few hundred years, which ensures that the upper ocean has reached a quasi-steady state.

Note that if the boundary conditions change in time, e.g. the amount of CO_2 varies, or the position of the continents changes, then the climate is in a transient state. Thus, global annual energy balance is not obtained at any given time in the simulation.

Climate simulations of the Late Palaeozoic

Three geological time periods are considered to illustrate the application of models in the Late Palaeozoic: the Carboniferous, the middle Permian and the latest Permian.

Carboniferous

There are a number of modelling studies of the Carboniferous time period, where both energy-balance models and global atmospheric climate models were used. Crowley *et al.* (1991) used a two-dimensional (latitude versus longitude) energy-balance model to investigate the role of luminosity and orbital variations to Gondwanan glaciation. They used three different ice-sheet assumptions and considered present and reduced (3%) solar luminosity. They also considered a present orbital configuration and one which minimized summer insolation in the Southern Hemisphere. They compared their Carboniferous simulations to Pleistocene simulations, which was a time of large glacial ice extent. They found that neither a reduced luminosity nor a cold summer orbital configuration would lead to glacial conditions for the Carboniferous. However, the combination of lower luminosity and cold summer insolation did lead to temperatures that could sustain a large Gondwanan ice-sheet. Crowley (1994) further summarizes a wide range of aspects of Pangaean climates from various model studies.

Crowley & Baum (1994) used the GENESIS atmospheric General Circulation Model (GCM) with prescribed sea-surface temperatures to simulate the climate of the Westphalian period (*c.* 305 Ma). The model simulated climate in terms of precipitation, and evaporation patterns agreed quite well with the distribution of coals and evaporites for this time period. The model also simulated southern Gondwanaland temperatures that were well below freezing, which are not supported by data.

Crowley *et al.* (1996) used the atmospheric GCM with specified sea-surface temperatures to look at conditions for coal formation in the Carboniferous. The Carboniferous was a time of large carbon sequestration, which occurred in regions where peat bogs formed. These peat formations are indicative of regions of high precipitation. Crowley *et al.* (1996) again assumed a 3% reduction in solar luminosity and carried out two simulations assuming a cold summer and warm summer orbital configuration, relative to the Southern Hemisphere. Their model results show a significant correlation between the location of maximum precipitation and coal formation. This indicates that the climate model is doing reasonably well in capturing the location of precipitation during this time period. Valdes & Crowley (1998) carried out an intercomparison of two GCM simulations of the Carboniferous (305 Ma). Both of the climate models used the same boundary conditions. Valdes & Crowley (1998) found significant differences in simulated southern polar temperatures and in precipitation features. Some of these differences are related to horizontal resolution.

Crowley & Baum (1994) employed a two-dimensional energy-balance model to look at the relative importance of changes in geography, solar luminosity and CO_2 levels to glaciation during the Carboniferous. They found that combined changes in both palaeogeography and CO_2 concentration gave the best explanation for glacial formation at this time. Baum & Crowley (1991) and Crowley *et al.* (1994) studied the stability of the snowline during the Carboniferous in both an energy-balance model and a global climate model. They found that both models exhibited instability in the snowline due to changes in solar luminosity, which may help to explain the rapidity of glacial formation in the climate system. Crowley *et al.* (1993) using the GENESIS global climate model found that orbital variations can explain glacial to interglacial climates observed in the mid-Carboniferous.

Otto-Bliesner (1995, 1998) has carried out simulations of the Carboniferous climate using an atmospheric GCM to look at the effects of mountain topography on location of precipitation. This is an important question when considering modelling the location of coal deposits during this time period. Otto-Bliesner (1995) found that using realistic mountain topography was critical for capturing the location of maximum precipitation in the tropics at this time. Otto-Bliesner (1996) used the GENESIS climate model coupled to a 50-metre ocean mixed layer model to explore the initiation of a continental ice-sheet in the Late Carboniferous, and carried out a series of sensitivity studies with the model that included changes to land elevation, changes in solar luminosity, and a prescribed ice-sheet. In this study, regions of persistent summer snow cover were used as an indicator of ice-sheet initiation and indicated that changes in solar luminosity were the dominant factor in persistence of summer snow cover.

Middle–Late Permian

There have been a number of studies of Permian climates (Kutzbach *et al.* 1990; Kutzbach & Ziegler 1993; Fluteau *et al.* 2001). Kutzbach & Gallimore (1989) used a low-resolution atmospheric GCM to investigate the existence of megamonsoons on the Pangaean continent. Their model used a 50-metre-deep ocean mixed layer with a thermodynamic sea-ice model. The model simulated the seasonal migration of a large monsoon pattern in the northern and southern Tethys Sea. Kutzbach (1994) extended his study of megamonsoons to look at the role of orbital variations in modulating the strength of precipitation patterns for the Pangaean megacontinent.

Crowley *et al.* (1989) used a two-dimensional energy-balance model to look at the seasonal cycle of the supercontinent Pangaea. They found large seasonal temperature ranges in the Southern Hemisphere of greater than 35 °C. These large seasonal changes would be difficult to rectify with equable climates for such a continental configuration.

Recently, Winguth *et al.* (2002) have used a two-dimensional energy-balance model coupled to a three-dimensional ocean GCM to study the climate of the Middle Permian. The wind forcing for the ocean model is obtained by running an atmospheric GCM with the surface temperature conditions from the energy-balance model. The system also included a thermodynamic sea-ice model. The combination of these four component models creates a system model of intermediate complexity. Winguth *et al.* (2002) applied this model to study the climate of the Wordian (*c.* 265 Ma).

They assume a 2.1% reduction in solar luminosity for this time period, and performed simulations with three different orbital configurations, one with a circular orbit and an axial tilt of 23.5°, and simulations with the cold and warm summer orbits employed by Crowley *et al.* (1996). They carried out four simulations assuming 1, 2, 4, and 8 times pre-industrial levels of atmospheric CO_2. The strength of the Panthalassic overturning circulation is shown in Figure 2. They find efficient formation of deep water in both hemispheres for all levels of CO_2.

Rees *et al.* (2002) used a diverse range of data to classify terrestrial environments into regional biomes. They then compared these data-derived biome patterns to biomes derived from climate model simulations for the Permian. Model simulations were carried out for 1, 4 and 8 times pre-industrial CO_2 levels. They found that the best overall agreement between model and data is for 4 and 8 times pre-industrial CO_2 levels. However, none of the climate simulations produced biomes in high southern latitudes that match the data.

MERIDIONAL CIRCULATION [SV]

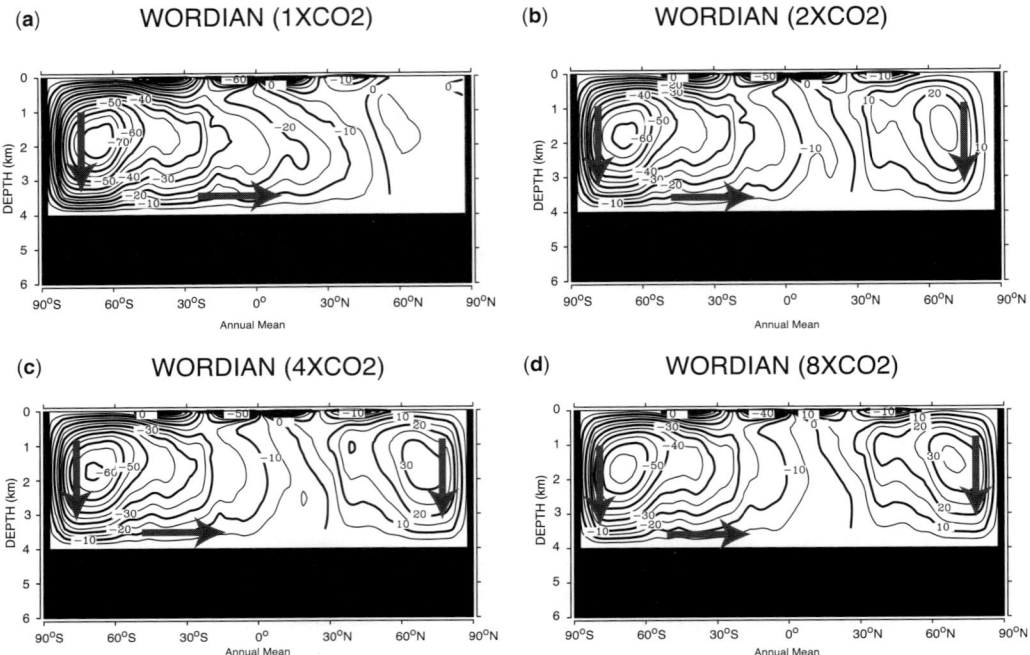

Fig. 2. Meridional overturning circulation for the middle Permian ocean simulation from Winguth *et al.* (2002). Shown is the zonal average of the rate of overturning of water mass for four different levels of atmospheric carbon dioxide. The arrows indicate the direction of the circulation of water mass. The magnitude of the flow of water is in units of Sverdrups (Sv), where $1 \text{ Sv} = 10^6 \text{ m}^3 \text{ s}^{-1}$.

Latest Permian

The latest Permian is a very interesting period in Earth history. The largest-known extinction of life occurred at the boundary of the Permian–Triassic (*c.* 251 Ma), where approximately 95% of marine life and 75% of terrestrial life went extinct in perhaps as little as a few thousand years (Twitchett 2007, this volume). A number of hypotheses have been put forward to explain this event (e.g. Erwin 2005; Benton 2003), including global warming, bolide impact, ocean anoxia, and acid rain. This was a time of massive flood basalt activity that lasted for *c.* 700,000 years. Isotopic and geologic data indicate that the oceans were anoxic at this time (Wignall & Twitchett 2002), whilst terrestrial palaeosol data indicate that the poles were very warm (Taylor *et al.* 1992).

Studies of the latest Permian ocean using three-dimensional global ocean models forced with pre-scribed sea-surface boundary conditions (Hotinski *et al.* 2001; Zhang *et al.* 2001) indicate that a slug-gish ocean circulation may have existed, which would support anoxic or dysoxic conditions.

However, Zhang *et al.* (2001) found that this state does not maintain itself for long periods. Although ocean-only studies are important for understanding sensitivity of the ocean state to different climate forcings, in reality the ocean is strongly coupled with the atmosphere and land (through runoff) pro-cesses. This coupling allows for feedback processes between these various components.

Kiehl & Shields (2005) carried out a fully coupled three-dimensional study of the latest Permian climate, which included coupled atmos-pheric, ocean, land and sea-ice components. They employed the Permian palaeogeography and palaeotopography of Ziegler and colleagues as shown in Figure 3. The model assumed a ten-fold increase in CO_2 compared to present (e.g. Kidder & Worsley 2004). The initial climate state for the ocean was taken from the end of a Cretaceous simu-lation, which assumed elevated CO_2. The fully coupled simulation was run for 2700 years, suffi-ciently long for the entire system to come into a steady state. No ocean acceleration techniques were used in this simulation. This means that the deep ocean circulation is also in a steady state,

Permian T31 Topography

Topography   m

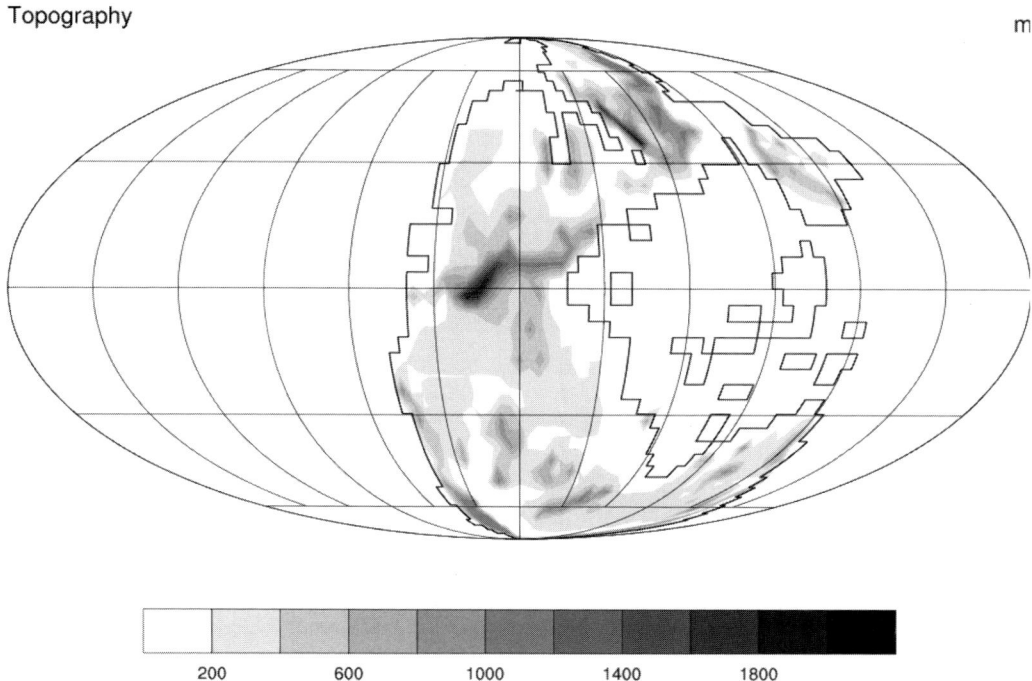

200   600   1000   1400   1800

Fig. 3. Latest Permian topography used in the study of Kiehl & Shields (2005). Topography height in metres.

which is required to look at the implications of ocean mixing for anoxic conditions.

The simulated land surface temperatures for winter and summer conditions are shown in Figure 4. Compared to a present-day simulation, the poles have warmed by 35 to 40 °C, while the tropics have warmed by 5 °C. The polar amplification is due to two factors: (1) the presence of more water vapour at the higher latitudes leads to a larger greenhouse effect, (2) the presence of open ocean to the poles allows for ocean heat transport to the poles. Land temperatures in the subtropical desert regions are quite high (35 °C).

As discussed, a cause for ocean anoxia is a very sluggish ocean circulation. One way to look at the efficiency of ocean turnover is a quantity called ideal age. This is a passive tracer in the ocean model that is a measure of the length of time (in years) that a parcel of water at a given depth has last been exposed to the ocean surface. Thus, long ideal ages (thousand years) indicate a very slow ocean circulation. Figure 5 shows the ideal age at 2000 metre depth in a present-day simulation and the latest Permian simulation. Note that for the present day the ideal age in the North Atlantic is very short, i.e. less than 50 years. For the Permian

simulation, the ideal age at this depth is very long (*c*. 1000 years) indicating a very inefficient connection between water at depth and the surface. This slow circulation, coupled with very warm sea-surface temperatures, which reduces the solubility of oxygen in the ocean, leads to anoxic or dysoxic conditions in the Panthalassic Ocean, and hence marine extinction. A parallel simulation of the latest Permian has been carried out with present levels of CO_2, and in this simulation the ocean circulation is very vigorous and much colder. This indicates that the primary driver to changes in ocean circulation (and anoxic conditions) is the amount of atmospheric CO_2, not the palaeogeographic conditions of Pangaea.

Future directions in palaeoclimate modelling

The descriptions in the previous section have focused on models of the physical climate system. However, Earth's climate involves far more than the physical behaviour of the atmosphere, ocean, land and cryosphere. Biogeochemical processes play a very important role in determining the

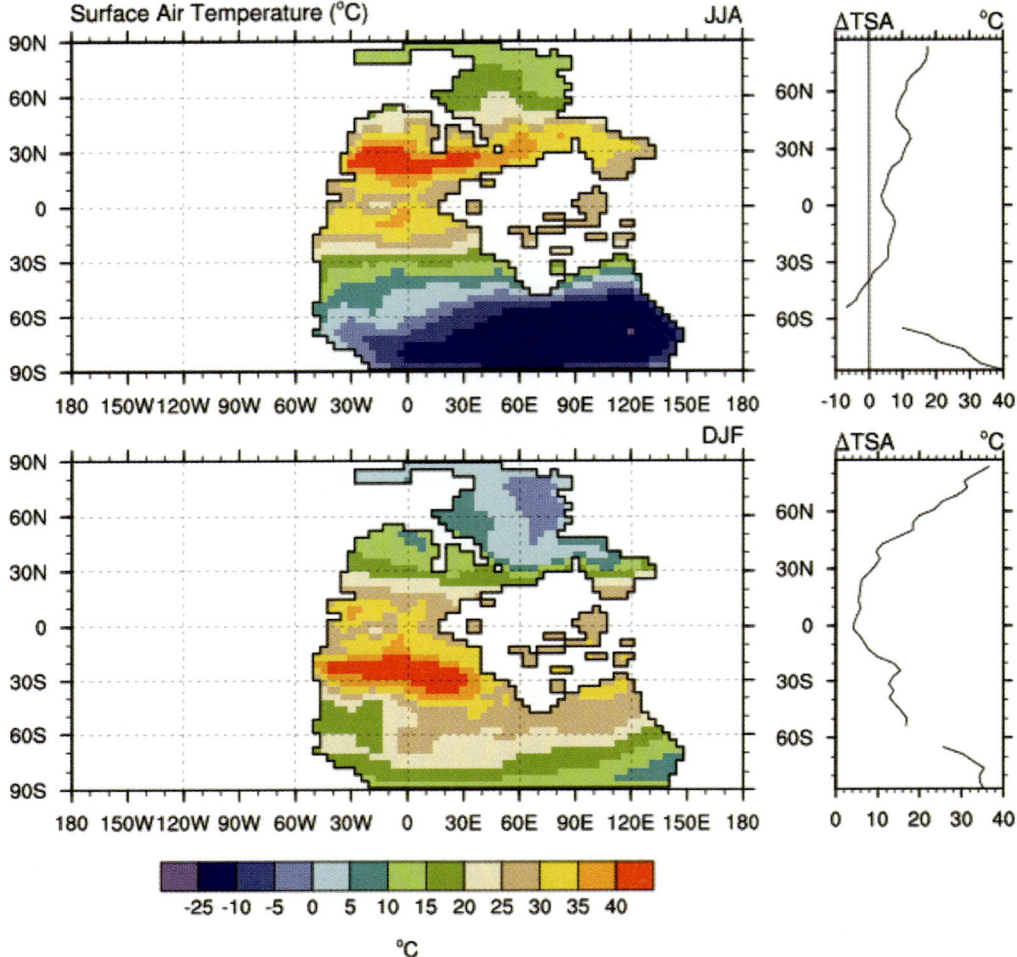

Fig. 4. Latest Permian surface temperatures (°C) for June–July–August (top) and December–January–February (bottom). Differences of zonal mean land surface temperatures from the present-day simulation are shown on the right (from Kiehl & Shields 2005).

levels of atmospheric greenhouse gases, e.g. CO_2 and CH_4. Biogeochemistry also involves the fractionation of various isotopes, which provide important signatures of Earth's climate system. Current studies of these interactions employ energy-balance models coupled to box-geochemistry models (e.g. Godderis & Joachimski 2004).

Atmospheric chemical processes that form ozone, acid rain and aerosols also play a key role in the climate system. Thus, to truly model Earth's climate requires the development and application of comprehensive Earth System Models. Many climate centres are actively involved in developing these types of models, and there is

much interest in applying these to important geological questions. Recent studies by Kump *et al.* (2005) and Lamarque *et al.* (2006) have applied atmospheric chemistry models to the latest Permian. Lamarque *et al.* (2006) use a full three-dimensional atmospheric chemistry model to study the effects of elevated methane levels on atmospheric ozone. They find that with sufficiently high methane levels, atmospheric ozone collapses.

Dynamic vegetation models are also currently being used in models of deep-time climates (e.g. Haywood & Valdes 2006; Shellito & Sloan 2006). These models allow for feedbacks between the physical climate system and regional ecosystems.

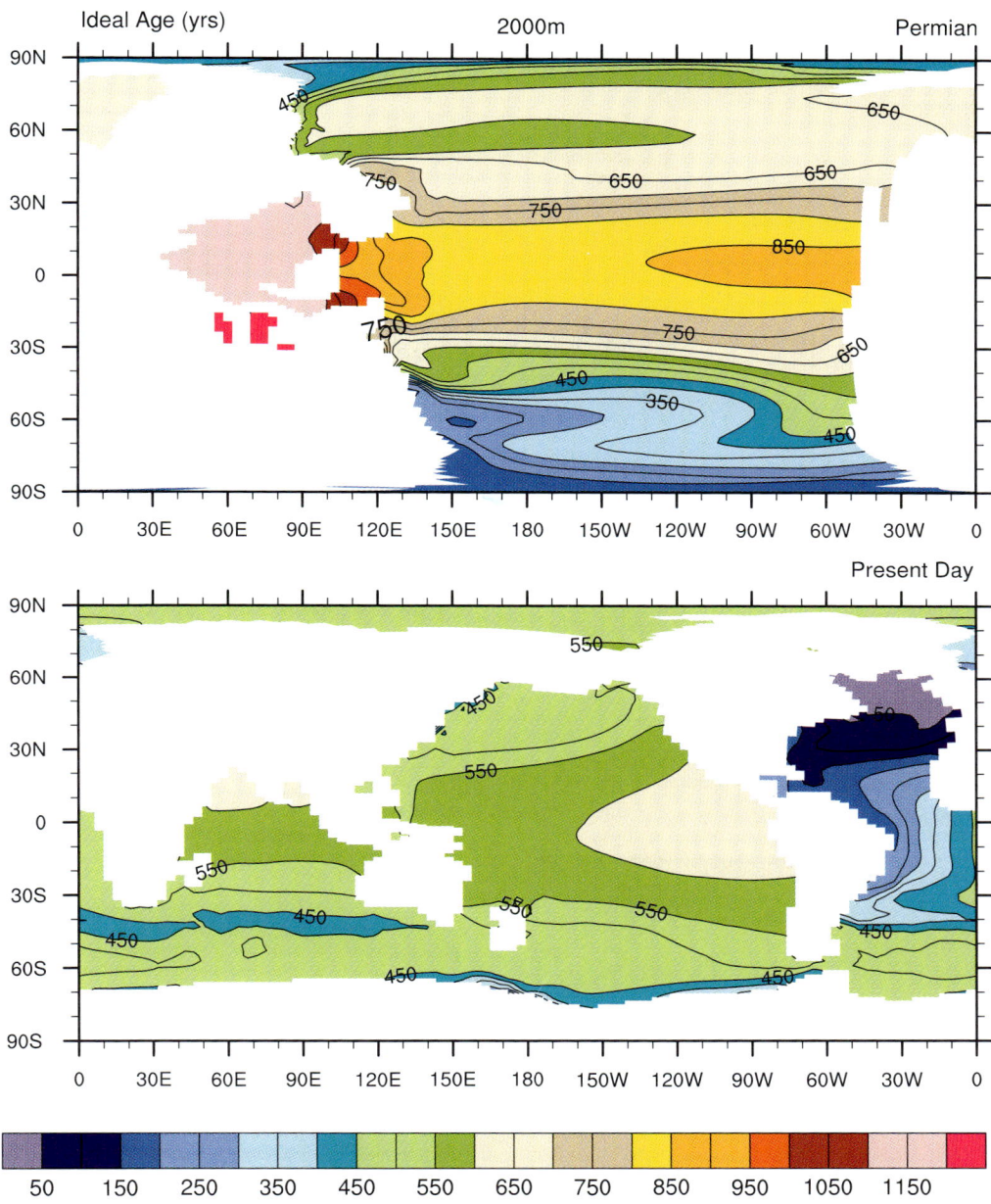

Fig. 5. Ideal age of ocean water (years) at 2000 metres depth (from the simulation of Kiehl & Shields 2005). Top figure is a Permian reconstruction and the bottom is present day.

Results from the predicted vegetation distributions can be evaluated against site-specific palaeodata. Additional coupling between vegetation and atmospheric chemistry through the emission of precursor gases like isoprene leads to additional feedback processes in the Earth system.

References

BAUM, S. K. & CROWLEY, T. J. 1991. Seasonal snowline instability in a climate model with realistic geography: application to Carboniferous (~300 Ma) glaciation. *Geophysical Research Letters*, **18**, 1719–1722.

BENTON, M. J. 2003. *When Life Nearly Died, The Greatest Mass Extinction of All Time*, Thames & Hudson.

BERNER, R. A. 2004. *The Phanerozoic Carbon Cycle*. Oxford University Press.

CROWLEY, T. J. 1994. Pangean climates. *In*: KLEIN, G. D. (ed.) *Pangea: Paleoclimate, Tectonics, and Sediment during accretion, zenith and breakup of a supercontinent*, Boulder, CO, Geological Society of America Special Paper **288**, 25–39.

CROWLEY, T. J. & BAUM, S. K. 1994. General circulation model study of late Carboniferous interglacial climates. *Paleoclimates*, **1**, 3–21.

CROWLEY, T. J., HYDE, W. T. & SHORT, D. A. 1989. Seasonal cycle variations on the supercontinent of Pangea. *Geology*, **17**, 457–460.

CROWLEY, T. J., BAUM, S. K. & HYDE, W. T. 1991. Climate model comparison of Gondwanan and Laurentide glaciations. *Journal of Geophysical Research*, **96**, 9217–9226.

CROWLEY, T. J., YIP, K.-J. & BAUM, S. K. 1993. Milankovitch cycles and Carboniferous climate. *Geophysical Research Letters*, **20**, 1175–1178.

CROWLEY, T. J., YIP, K.-J. & BAUM, S. K. 1994. Snowline instability in a general circulation model: application to Carboniferous glaciation. *Climate Dynamics*, **10**, 363–376.

CROWLEY, T. J., YIP, K.-J., BAUM, S. K. & MOORE, S. B. 1996. Modelling carboniferous coal formation. *Paleoclimates*, **2**, 159–177.

ERWIN, D. H. 2005. *Extinction: How Life on Earth Nearly Ended 250 Million Years Ago*. Princeton University Press.

FLUTEAU, F., BESSE, J., BROUTIN, J. & RAMSTEIN, G. 2001. The late Permian climate. What can be inferred from climate modeling concerning Pangea scenarios and Hercynian range altitude. *Palaeogeography, Palaeoclimatology, Palaeoecology*, **167**, 39–71.

GIBBS, M. T., REES, P. McA., KUTZBACH, J. E., ZIEGLER, A. M., BEHLING, P. J. & ROWLEY, D. B. 2002. Simulations of Permian climate and comparisons with climate-sensitive sediments. *The Journal of Geology*, **110**, 33–55.

GODDERIS, Y. & JOACHIMSKI, M. M. 2004. Global change in Late Devonian: modeling the Frasnian–Famennian short-term carbon isotope excursions. *Palaeogeography, Palaeoclimatology, Palaeoecology*, **202**, 309–329.

HALLAM, A. & WIGNALL, P. 1997. *Mass Extinctions and Their Aftermath*. Oxford University Press.

HAYWOOD, A. M. & VALDES, P. J. 2006. Vegetation cover in a warmer world simulated using a dynamic global vegetation model for the Mid-Pliocene. *Palaeogeography, Palaeoclimatology, Palaeoecology*, **237**, 412–427.

HOTINSKI, R. M., BICE, K. L., KUMP, L. R., NAJJAR, R. G. & ARTHUR, M. A. 2001. Ocean stagnation and end-Permian anoxia. *Geology*, **29**, 7–10.

HUBER, B. T., MACLEOD, K. G. & WING, S. L. 2000. *Warm Climate in Earth History*. Cambridge University Press.

HYDE, W. T., CROWLEY, T. J., TARASOV, L. & PELTIER, W. R. 1999. The Pangean ice age: studies with a coupled climate-ice sheet model. *Climate Dynamics*, **15**, 619–629.

KIDDER, D. L. & WORSLEY, T. R. 2004. Causes and consequences of extreme Permo-Triassic warming to globally equable climate and relation to Permo-Triassic extinction and recovery. *Palaeogeography, Palaeoclimatology, Palaeoecology*, **203**, 207–237.

KIEHL, J. T. 1992. Atmospheric general circulation modeling. *In*: TRENBERTH, K. E. (ed.) *Climate System Modeling*, Cambridge University Press, 319–370.

KIEHL, J. T. & SHIELDS, C. A. 2005. Climate simulation of the latest Permian: Implications for mass extinction. *Geology*, **33**, 757–760.

KIEHL, J. T., SHIELDS, C. A., HACK, J. J. & COLLINS, W. D. 2006. The climate sensitivity of the Community Climate System Model version 3 (CCSM3). *Journal of Climate*, **19**, 2584–2596.

KUMP, L. R., PAVLOV, A. & ARTHUR, M. A. 2005. Massive release of hydrogen sulfide to the surface ocean and atmosphere during intervals of oceanic anoxia. *Geology*, **33**, 397–400.

KUTZBACH, J. E. 1994. Idealized Pangean climates: sensitivity to orbital change. *In*: KLEIN, G. D. (ed.) *Pangea: Paleoclimate, Tectonics, and Sediment during accretion, zenith and breakup of a supercontinent*, Boulder, CO, Geological Society of America Special Paper **288**, 41–55.

KUTZBACH, J. E. & GALLIMORE, R. G. 1989. Pangean climates: Megamonsoons of the megacontinent. *Journal of Geophysical Research*, **94**, 3341–3357.

KUTZBACH, J. E., GUETTER, P. J. & WASHINGTON, W. M. 1990. Simulated circulation of an idealized ocean for Pangean time. *Paleoceanography*, **5**, 299–317.

KUTZBACH, J. E. & ZIEGLER, A. M. 1993. Simulation of Late Permian climate and biomes with an atmosphere–ocean model: comparison with observations. *Philosophical Transactions of the Royal Society of London*, **B341**, 327–340.

LAMARQUE, J.-F., KIEHL, J. T., SHIELDS, C. A., BOVILLE, B. A. & KINNISON, D. E. 2006. Modeling the response to changes in tropospheric methane concentration: application to the Permian–Triassic boundary. *Paleoceanography*, **21**, PA3006, doi:10.1029/2006PA001276.

MCAVANEY, B. J., COVEY, C., JOUSSAUME, S. *ET AL.* 2001. Model Evaluation. *In*: HOUGHTON, J. T. (ed.) *Climate Change 2001, The Scientific Basis*. Cambridge University Press, 471–523.

NORTH, G. R. & STEVENS, M. J. 2006. Energy-balance climate models. *In*: KIEHL, J. T. & RAMANATHAN, V. (eds) *Frontiers of Climate Modeling*, Cambridge University Press, 52–72.

OTTO-BLIESNER, B. L. 1995. Continental drift, runoff and weathering feedbacks: Implications from climate model experiments. *Journal of Geophysical Research*, **100**, 11537–11548.

OTTO-BLIESNER, B. L. 1996. Initiation of a continental ice sheet in a global climate model (GENESIS). *Journal of Geophysical Research*, **101**, 16909–16920.

OTTO-BLIESNER, B. L. 1998. Effects of tropical mountain elevations on the climate during the past: Climate model experiments. *In*: CROWLEY, T. J. & BURKE, K. C. (eds) *Tectonic Boundary Conditions for Climate Reconstructions*, Oxford Monographs on Geology and Geophysics, 100–115.

RANDALL, D. A., SCHLESINGER, M. E., MELESHKO, V., GALIN, V., MORCRETTE, J.-J. & WETHERALD, R. 2006. Cloud feedbacks. *In*: KIEHL, J. T. & RAMANATHAN, V. (eds) *Frontiers of Climate Modeling*, Cambridge University Press, 217–250.

REES, P., GIBBS, M. T., ZIEGLER, A. M., KUTZBACH, J. E. & BEHLING, P. J. 1999. Permian climates: Evaluating model predictions using global paleobotanical data. *Geology*, **27**, 891–894.

REES, P. M., ZIEGLER, A. M., GIBBS, M. T., KUTZBACH, J. E., BEHLING, P. J. & ROWLEY, D. B. 2002. Permian phytogeographic patterns and climate data/model comparisons. *Journal of Geology*, **110**, 1–31.

RETALLACK, G. J. 2002. Carbon dioxide and climate over the past 300 Myr. *Philosophical Transactions of the Royal Society of London*, **A360**, 659–673.

ROYER, D. L., BERNER, R. A., MONTAÑEZ, I. P., TABOR, N. J. & BEERLING, D. J. 2004. CO_2 as a primary driver of Phanerozoic climate. *GSA Today*, **14**, 4–10.

ROYER, D. L. 2006. CO_2-forced climate thresholds during the Phanerozoic. *Geochemica et Cosmochemica Acta*, **70**, 5665–5675.

SCHLESINGER, M. E. (ed.) 1988. *Physically Based Modelling and Simulation of Climate and Climate Change – Part I*. Kluwer.

SHELLITO, C. J. & SLOAN, L. C. 2006. Reconstructing a lost Eocene paradise, part II: on the utility of dynamic vegetation models in pre-Quaternary climate studies. *Global and Planetary Change*, **50**, 18–32.

SLOAN, L. C. & BARRON, E. J. 1990. Equable climates during Earth history? *Geology*, **18**, 489–492.

STAMPFLI, G. M. & BOREL, G. D. 2002. A plate tectonic model for the Paleozoic and Mesozoic constrained by dynamic plate boundaries and restored synthetic ocean isochrones. *Earth and Planetary Science Letters*, **196**, 17–33.

TAYLOR, E. L., TAYLOR, T. N. & CUNEO, N. R. 1992. The present is not the key to the past: A polar forest from the Permian of Antarctica. *Science*, **257**, 1675–1677.

TRENBERTH, K. E. 1992. *Climate System Modeling*, Cambridge University Press.

TWITCHETT, R. J. 2007. Climate change across the Permian/Triassic boundary. *In*: WILLIAMS, M., HAYWOOD, A. M., GREGORY, J. & SCHMIDT, D. (eds) *Deep-Time Perspectives on Climate Change: Marrying the Signal from Computer Models and Biological Proxies*, The Micropalaeontological Society, Special Publications. The Geological Society, London.

VALDES, P. J. & CROWLEY, T. J. 1998. A climate model intercomparison for the Carboniferous. *Paleoclimates*, **2**, 219–238.

WASHINGTON, W. M. & PARKINSON, C. L. 2005. *An Introduction to Three-Dimensional Climate Modeling* (2nd edn), University Science Books.

WIGNALL, P. B. & TWITCHETT, R. J. 2002. Extent, duration, and nature of the Permian–Triassic superanoxic event. *In*: KOEBERL, C. & MACLEOD, K. C. (eds) *Catastrophic Events and Mass Extinctions: Impacts and Beyond*, Geological Society of America Special Paper **356**, 395–413.

WINGUTH, A. M. E., HEINZE, C., KUTZBACH, J. E. *ET AL.* 2002. Simulated warm polar currents during the middle Permian. *Paleoceanography*, **17**, doi:10.1029/2001PA000646.

ZHANG, R., FOLLOWS, M. J., GROTZINGER, J. P. & MARSHALL, J. 2001. Could the Late Permian deep ocean have been anoxic? *Paleoceanography*, **16**, 317–329.

ZIEGLER, A. M., SCOTESE, C. R., MCKERROW, W. S., JOHNSON, M. E. & BAMBACH, R. K. 1979. Paleozoic paleogeography. *Annual Reviews of Earth and Planetary Science*, **7**, 473–502.

The Early Permian fossil record of Gondwana and its relationship to deglaciation: a review

M. H. STEPHENSON[1], L. ANGIOLINI[2] & M. J. LENG[3]

[1]*British Geological Survey, Keyworth, Nottingham, NG12 5GG, UK (e-mail: mhste@bgs.ac.uk)*

[2]*Dipartimento di Scienze della Terra 'A. Desio', Università degli Studi di Milano, Via Mangiagalli 34, Milano, 20133, Italy*

[3]*NERC Isotope Geosciences Laboratory, Keyworth, Nottingham, NG12 5GG, UK*

Abstract: Deglaciation sequences of Early Permian age in Gondwana have until now been distinguished mainly on lithological criteria by reference to climate-sensitive lithologies and associated geochemistry; whereas identification on biotic criteria such as vegetational or faunal change has not been employed. Present palaeontological data, which are widely- scattered geographically and of different stratigraphic scales and resolutions, show diversity increase from glacial conditions to post-glacial conditions. Amongst the marine fauna, a cold-water fauna consisting of bivalves such as *Eurydesma* and *Deltopecten*, and brachiopods such as *Lyonia* and *Trigonotreta*, were established in the earliest post-glacial marine transgressions. Above this is a more warmer, increasingly warmer, temperate fauna, including brachiopods, bryozoans, bivalves, cephalopods, gastropods, conularids, fusulinids, small foraminifers, asterozoans, blastoids and crinoids. The polynomorph succession shows change from monosaccate pollen assemblages, associated with fern spores, to more diverse assemblages with common non-taeniate bisaccate pollen. In Oman, where this has been studied in greatest detail, the upland saw changes from a glacial monosaccate pollen-producing flora to a warmer climate bisaccate pollen-producing flora; while in the terrestrial lowlands, a parallel change occurred from a glacial fern flora to a warmer climate colpate pollen-producing and lycopsid lowland flora. The sedimentary organic matter of the associated clastic rocks shows a decreasing $\delta^{13}C$ trend believed to reflect palaeoatmospheric change due to post-glacial global warming.

Early Permian farfield isotope studies, compiled by other workers, from brachiopods from the southern Urals, show a $\delta^{18}O$ decline of 2.5‰ in the Early Permian (Asselian to Artinskian) and stable $\delta^{13}C$ values of around +4.3‰ in the same period. This farfield evidence is in part consistent with palaeontological data since the most likely cause for the decline in $\delta^{18}O$ is the return of isotopically light waters to the oceans from melting of glaciers at high latitudes.

At present, we are living in an interglacial stage of the Late Cenozoic ice age that began around the earliest Oligocene (Crowell 1995) and has not yet ended. However, the Carboniferous–Permian ice age is the most widespread and geologically well-represented ice age of the Phanerozoic and has the benefit that it ended more than 270 million years ago, and therefore can be studied in its entirety. By studying the waning glaciation of the Early Permian, we may be able to learn more of the interplay of factors influencing glaciation, and hence of the effects that might be expected in the near future in our own era.

The Late Palaeozoic glaciation of Gondwana comprised three distinct non-overlapping episodes. Though there is some controversy about their precise timing (see e.g. Veevers & Powell 1987; Isbell *et al.* 2003; Jones & Fielding 2004), it appears that two short glaciations occurred in the Frasnian (latest Devonian) and Mid-Carboniferous respectively, and that the glacigene rocks associated

with them were deposited mainly by alpine glaciers. The third and latest glaciation of Pennsylvanian to Early Permian age is thought to have been more widespread as there is evidence for continental ice-sheets covering a total area of 17.9 to 22.6 × 10^6 km^2 (Veevers *et al.* 1994; Veevers & Tewari 1995; López-Gamundí 1997; Wopfner & Casshyap 1997; Isbell *et al.* 2003). The sedimentary rocks deposited in this last major glacial episode occur in cratonic basins (e.g. Kalahari, Rubh al Khali and Paraná basins; Levell *et al.* 1988; Veevers *et al.* 1994) rift basins (e.g. Damodar, Son, Mahanadi basins; Veevers & Tewari 1995); foreland basins (Veevers *et al.* 1994; Stollhofen *et al.* 2000; Jones & Fielding 2004; e.g. Sydney and Karoo basins) and successor basins (central Transantarctic Mountains; Isbell 1999) that are now exposed in South America, the Falkland Islands, Africa, Arabia, Madagascar, Antarctica, India and Australia (Fig. 1).

Two main facies have been recognized (Isbell *et al.* 2003). The first consists of massive diamictite

From: WILLIAMS, M., HAYWOOD, A. M., GREGORY, F. J. & SCHMIDT, D. N. (eds) *Deep-Time Perspectives on Climate Change: Marrying the Signal from Computer Models and Biological Proxies*. The Micropalaeontological Society, Special Publications. The Geological Society, London, 169–189.
1747-602X/07/$15.00 © The Micropalaeontological Society 2007.

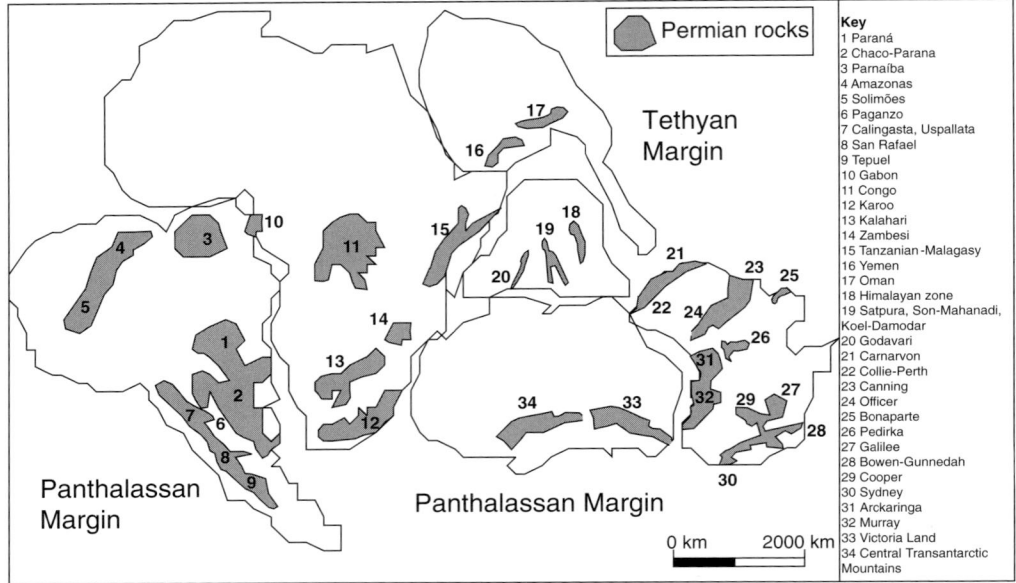

Fig. 1. Chief basins in the Early Permian of Gondwana.

resting on striated surfaces, sheared diamictite and sandstone and shale. These rocks were deposited subglacially as lodgement and deformation till, and in glaciofluvial and glaciolacustrine palaeoenvironments. The second facies association consists of massive diamictite overlying gradational and sharp contacts, stratified diamictite, dropstonebearing shales and shales without dropstones. These rocks represent diverse styles of glaciomarine deposition including deposition at or near an ice front, rain-out debris from ice shelves or ice tongues, deposition from rafts of ice, and open marine sedimentation.

Within Lower Permian strata, an abrupt change from glacial to post-glacial deposits records the rapid withdrawal of ice from depositional basins throughout Gondwana (Isbell *et al.* 2003). In most basins, a distinct contact between glacial facies below and fluvial, lacustrine and marine clastic facies above occurs. The marine facies may be associated with the *Eurydesma* fauna (Dickins 1984) or with other cold- to temperate-water faunas such as that of the Manendragarh Limestone in India (Chandra 1994; Frakes *et al.* 1975; Shah & Sastry 1975; Venkatachala & Tiwari 1987), the Callytharra Formation in Western Australia (Dickins 1992) and the Saiwan Formation/Haushi limestone in Oman (Angiolini *et al.* 2003). Waning of Early Permian glaciation in the Gondwana continent was also manifested in farfield events of Euramerica and Tethys in terms of lithology and plant evolution

(Ziegler *et al.* 2002), macropalaeontology (Angiolini *et al.* 2005), cyclic sedimentation related to eustatic change (Isbell *et al.* 2003), oxygen and hydrogen isotope ratios in pedogenic phyllosilicates (Tabor & Montañez 2002, 2005) and seawater isotope composition (Veizer *et al.* 1999; Korte *et al.* 2005).

The biostratigraphy of Gondwanan Early Permian post-glacial sections is relatively well known in isolation, but the detailed evolution of biota after the glaciation, in the context purely of palaeoclimate rather than in terms of biostratigraphy, is not well studied. The advantages of developing an Early Permian biotic deglaciation model would be in understanding in detail the response of life to increasing temperatures and other climate change. Potentially, measures of extinction rate versus time and versus rate of palaeoenvironmental change (perhaps quantified in terms of per mil $\delta^{13}C$ change) could be developed if an Early Permian biotic deglaciation model could be scaled against time. Such measures would be very useful in the study of modern biotic change during global warming. In addition, understanding some of the palaeoclimatic controls on deglaciation may be important in geological environments that are well known for accumulation of coal, gas and oil (see Potter *et al.* 1995), and in understanding the distribution of so-called postglacial 'hot shales' which can be important source rocks of hydrocarbons.

The purpose of this chapter is to summarize and synthesize palaeontological and isotopic data relating to post-glacial Early Permian climate change and to discuss the progress and challenges associated with developing an Early Permian biotic deglaciation model.

Timing of Early Permian deglaciation

Biostratigraphic correlation of facies of most Early Permian deglaciation within the northern Gondwanan basins and in the Karoo, Falkland Island and Antarctic basins indicates that deglaciation took place in the late Asselian or early Sakmarian (Wopfner & Kreuser 1986; Veevers & Powell 1987; Tiwari 1994; Vijaya 1994; Crowell 1995; Wopfner 1999; Eyles *et al.* 2002; Stephenson & Osterloff 2002; Isbell *et al.* 2003; Stephenson *et al.* 2005) in or around the *Granulatisporites confluens* palynomorph biozone of Foster & Waterhouse (1988). In basins at lower latitudes during the Permian (see Isbell *et al.* 2003), for example the Solimões, Amazonas and Parnaíba basins in present-day northern South America, and along the Panthalassan margin of South America (e.g. the Tepuel, Calingasta, Uspallata and Paganzo basins), glaciation was restricted to the Mississippian and Pennsylvanian (González Bonorino 1992; López-Gamundí *et al.* 1992; Isbell *et al.* 2003). In eastern Australia and Tasmania, there is evidence of post-Sakmarian glaciation in the form of lonestones, and the eastern Australian post-Sakmarian succession as a whole has been regarded as being indicative of a cool climate (Dickins 1978, 1996; Martini & Glooschenko 1985; Martini & Banks 1989). Lonestones have been reported to occur at three levels in the Bowen Basin (Dickins 1996), and in the Sydney Basin (Dickins 1996; Eyles *et al.* 1998) and Tasmania (Banks & Clark 1987). Dickins (1996) discussed lonestone occurrences and noted their coincidence with glendonites, concluding that there is evidence for post-Sakmarian glaciation in the area. Crowell (1995) considered the lonestones in the Sydney Basin were large enough to suggest nearby glaciation, whereas those in the Bowen Basin were of rafted iceberg origin and therefore not necessarily suggestive of nearby glaciation. Recent work (Jones & Fielding 2004) also casts doubt on post-Sakmarian glaciation in the Bowen and Galilee basins, and in the New England fold belt of Queensland. Noting the persistence of glaciation, at least in the extreme SE of Australia and in Tasmania, and distribution of deglaciation sediments in other areas of Gondwana, Crowell (1995) concluded that glaciation persisted longer on the polar side of the Gondwana margin than along the Tethyan margin, and Frank *et al.*

(2005) suggested persistent glacial conditions may have been maintained by upwelling of cold water during sea-level highstands.

Aside from SE Australia and the northerly basins of South America, the weight of sedimentological and geochemical evidence therefore indicates that most Gondwanan deglaciation took place in the late Asselian or early Sakmarian (Crowell 1995; Wopfner 1999; Isbell *et al.* 2003). However, the evidence for deglaciation is mainly sedimentological, and biotic change has not, as a rule, been considered. Thus the relationship between sedimentologically defined deglaciation and biotic change during post-glacial warming needs to be examined.

Sedimentologically defined deglaciation

Most Carboniferous–Permian deglaciation sequences are identified on sedimentological and associated geochemical data (e.g. Theron & Blignault 1975; Charrier 1986; Visser & Young 1990; Bonorino 1992; López-Gamundí *et al.* 1992; Crowell 1995; Visser 1995, 1996, 1997; Dickins 1996; Diekmann & Wopfner 1996; González Isbell & Cúneo 1996; Santos *et al.* 1996; Wopfner & Diekmann 1996; Eyles *et al.* 1997; Gonzalez 1997; Faure & Cole 1999; Apak & Backhouse 1999; Wopfner 1999; Stollhofen *et al.* 2000; Eyles & de Broekert 2001; Scheffler *et al.* 2003). Some studies (e.g. Kneller *et al.* 2004) revealed local detail of glacial melting, changing sediment supply and incidences of local-scale catastrophic sedimentation events. However, a distinction should be made between deglaciation in the sense of glacial melting related to climate warming and attendant biotic response, and sedimentological concepts of deglaciation, which primarily describe the sedimentological (usually fairly local) effects of the demise of glaciers. Visser (1997) defined a deglaciation sequence as '... an upward thinning sediment body (or package) deposited seaward of the ice-grounding line during a major recessional phase of a marine ice margin ...' Other authors (e.g. Wopfner 1999) have used much broader definitions not confined to sedimentological criteria or to the marine realm, but encompassing mineralogical and geochemical aspects as well as considering aspects of climate. It seems unlikely that all deglaciation sequences identified purely on sedimentological grounds, for example employing Visser's definition, would have a biotic response. Studies of deglaciation by Visser (1996, 1997) showed that in close similarity to Holocene marine ice-sheets, the ice-sheets of the Karoo and Kalahari basins of southern Africa were unstable during sea-level fluctuations. During sea-level rise, the ice-sheet margin loses its points of contact with the ground and therefore the ice-sheet retreats to a stable upglacial position. Thus, Visser (1997) showed that in the Karoo, the primary reason for glacial melting in some parts of the succession is not warming but sea-level change. This has also been established for modern deglaciation, for example in the Irish Sea (Eyles & McCabe 1989). Visser's observations of sea-level change and deglaciation were primarily concerned with explaining sedimentary cycles of varying glacial influence, but these cycles, because they were controlled by sea level, are not the same as climatic 'glacials' and 'interglacials' nor are they likely to relate to coherent parallel changes in biota. Thus, care is needed when identifying deglaciation sequences that might be studied for biotic change.

Eyles *et al.* (2002) demonstrated a similar independence of sequences interpreted to be of deglacial origin from simple climate warming. They commented that models like those developed by Visser that link deglaciation to sea-level change, and which assume therefore that events or key lithologies have sequence-stratigraphic global correlation significance, may be flawed because of local tectonic controls. Eyles *et al.* (2002) applied a new Early Permian palynological biozonation across six western Australian basins (Bonaparte, Canning, Carnarvon, Collie, Gunbarrel and Perth basins), and concluded that shales previously considered to be of a common deglacial eustatic origin were of different ages. They concluded that some 'post-glacial shales' were present due to local tectonic subsidence rather than deglacial eustatic sea-level rise.

Thus, the key point in identifying successions for study of biotic response to deglaciation is to be able to distinguish between sequences interpreted to be deglacial in origin, caused primarily by local tectonics, sea-level rise or climate change. Clearly, these are linked casually but for an understanding of biotic change in relation to deglaciation only those sequences identified as being related to climate change are worthy of detailed further study.

Previous studies of biotic change associated with Asselian–Sakmarian Gondwana deglaciation

Gondwana glacial deposits are well known for the almost complete lack of macrofauna between the Namurian–Westphalian and late Asselian (Kemp *et al.* 1977; Truswell 1980; Jones & Truswell 1992); however, palynomorphs in the form of spores and pollen, and other non-terrestrial propagules, are quite common in the same deposits. The post-glacial (post-mid-Asselian) macrofauna and macroflora is also more diverse, reflecting climatic amelioration, and includes brachiopods, bivalves, gastropods, conularids, bryozoans, echinoderms, barnacles, foraminifera, ostracods, fishes, asterozoans, nautiloids, ammonoids, insects and plants (see e.g. Dickins 1957; Hudson & Sudbury 1959; Cousminer 1965; Closs 1970; Frakes *et al.* 1975; Shah & Sastry 1975; Venkatachala & Tiwari 1987; Foster & Waterhouse 1988; Archangelsky 1990; Chandra 1994; Vijaya 1994; Archbold & Dickins 1996; Pant 1996; Angiolini *et al.* 1997, 2003, 2004, 2006; Gonzalez 1997; Waterhouse 1997; Briggs 1998; Leonova 1998; Archbold 1999, 2000; Foster & Archbold 2001; Larghi 2005). Similarly, palynomorphs indicate a general increase in diversity in post-glacial Early Permian sediments (Bharadwaj 1969; Truswell, 1980; Stephenson *et al.* 2003). However, most studies of both palynology, fauna and flora have been concerned with taxonomy or biostratigraphy, rather than been aimed at detailed study of climate change.

Because of their abundance throughout, palynomorphs are the dominant method of biostratigraphic correlation in Gondwanan Lower Permian studies. There are too many studies to be listed here, but a number of publications summarize data for the different regions of Gondwana and form the basis for initial attempts at pan-Gondwanan correlation. These include for South America: Russo *et al.* (1980) and Vergel (1993) for the Chacoparana Basin in Argentina; Césari and Gutiérrez 2000 and Azcuy and Jelin (1980) for the Paganzo Basin, Argentina; Marques-Toigo (1991) and Souza (2003) for the Paraná Basin of Brasil. Archangelsky *et al.* (1980) and Azcuy (1979) also summarized data for the whole of South America. Attempts at pan-African or pan-southern African palynological correlation schemes include those of Falcon *et al.* (1984), Anderson (1977), Falcon (1975), MacRae (1988), Hart (1967) and Millsteed (1999). In Arabia, palynological schemes were summarized and presented as a unified preliminary scheme by Stephenson *et al.* (2003). In India, summaries of schemes include those of Tiwari (1994), Venkatachala *et al.* (1995) and Vijaya *et al.* (2001). Early Permian Australian palynostratigraphy was summarized by Kemp *et al.* (1977), Price (1983, 1997) and Foster (1983) but detailed basinal studies have also been published (e.g. Galilee Basin, Jones & Truswell (1992); Collie Basin, Backhouse (1991); Canning Basin, Powis (1979, 1984); Foster & Waterhouse (1988)). In Antarctica, some palynostratigraphy is available for the Asselian–Sakmarian (e.g. Kyle 1977; Farabee *et al.* 1991; Lindström 1995). A few authors have attempted to summarize and correlate across the Gondwana region as a whole using palynology (e.g. Bharadwaj 1969; Kemp 1975; Truswell 1980).

Considering the undoubted wealth of palaeontological, and particularly palynological, data from glacial and deglacial Lower Permian rocks, it is surprising that most studies of biotic change in relation to deglaciation have been rather general in character and have not attempted detailed bed-by-bed or metre-by-metre reconstruction of biotic change. Most have been concerned with generalities reconstructing climate change over long stratigraphic intervals with rather low sampling densities and using data from a wide geographical area (e.g. Dickins 1985, 1992, 1996).

South America

There have been a number of general studies examining faunal and floral changes through the Upper Palaeozoic in South America. González (1997, p. 235; and references therein) working on the basins of Argentina concluded that '... the origins and extinctions of the major Carboniferous and Early Permian faunal groups are closely associated with sedimentological evidence of climatic alterations due to temperature ranges ...'. González (1997) recognized two major faunal groups in the Early Permian of Argentina, the Uspallatian and the Bonetian. The first, characterized by *Cancrinella* cf. *farleyensis* (Etheridge & Dun), was considered by González (1997) to be of Asselian age and to represent glacial conditions, while the second (considered of Sakmarian age) was associated with cold-water transgression and characterized by *Eurydesma*, *Merismopteria*, *Megadesmus* and *Deltopecten*. González (1997) considered that the demise of the Bonetian fauna corresponded with the last Permian glaciation in Argentine basins.

However, the faunal changes described by González (1997) are based on erroneous age attribution (Simanauskas & Cisterna 2001) since his Uspallatian glacial fauna has been shown to be younger (Sakmarian) by Cisterna & Simanauskas (2000). Also, the *Tivertonia jachalensis–Streptorhynchus inaequiornatus* Biozone of the Tupe Formation of the Paganzo Basin, considered to be Late Carboniferous by González (1997), is given a middle–late Asselian age by Archbold *et al.* (2004).

Cisterna and Simanauskas (2000), Simanauskas Cisterna (2001), Cisterna *et al.* (2002*a*, *b*) and Archbold *et al.* (2004) have recently established a revised biostratigraphy and correlation of Upper Carboniferous–Lower Permian brachiopods of the Pre-Cordilleran basins of Argentina (Rio Blanco, Calingasta–Uspallata, Paganzo, Tepuel–Genoa basins). Simanauskas and Cisterna (2001) described an uppermost Carboniferous–Lower Permian macrofaunal assemblage from the El Paso

Formation of the Calingasta–Uspallata Basin. This fauna, previously considered to be Early Carboniferous in age, is recorded in shales between diamictites and shows a gradual increase in biodiversity.

The most complete faunal succession has been described from the Rio Blanco Basin (lower and middle members of the Rio del Penon Formation) as consisting of three brachiopod associations spanning the Asselian–early Sakmarian (Cisterna & Simanauskas 2000). The lower association (Assemblage I) comes from black shales above sandstones, conglomerates and paraconglomerates (diamictites?). The authors do not interpret the faunal succession in terms of biotic change after the Gondwanan glaciation, but they indicate its affinity to the Asselian–Tastubian faunas of Peninsular India and of the Lyons Group and Callytharra Formation of Western Australia. *Trigonotreta* sp. from their Assemblage I is similar to the primitive species of the *Trigonotreta* group such as *Trigonotreta hesdoensis* (Sahni & Dutt) from the Asselian–early Tastubian of Peninsular India. This fauna is regarded as early Asselian by Archbold *et al.* (2004). Interestingly, Associations II and III do not show a significant increase in biodiversity with respect to the lower assemblage, even if some of their characterizing species are similar to species of the Sakmarian of Western Australia (species of *Costatumulus*, *Neochonetes* and *Tivertonia*). According to Archbold *et al.* (2004), assemblages II and III span the middle Asselian–early Sakmarian time interval.

A general study of South American floral change in relation to climate change is that of Goldberg (2004), who studied macro- and microfloral diversity throughout the Permian in the Paraná Basin in southern Brasil and compared this with results from nine Australian Gondwanan basins. Her methodology was to count the number of taxa (genera/species) and graph these against lithostratigraphic and chronostratigraphic units as well as climate-sensitive lithologies. Data points corresponded nominally to the centre of formations so essentially this is a broad-based study too coarse to allow detailed study of post-glacial climate change. The study, however, showed a great increase in microfloral diversity from the Asselian–early Sakmarian Itararé Group (dominantly glacial) to the post-glacial coal-bearing late Sakmarian–early Kungurian Rio Bonito Formation, probably related to climatic amelioration. An accompanying decrease in floral diversity was attributed by Goldberg (2004) to taphonomic and local factors. Australian basins, examined using the same methodology, showed rather inconsistent results with no obvious amelioration even from the Asselian to the Sakmarian.

Africa

Although this area has been well-studied from the point of view of sedimentological and geochemical aspects of deglaciation (e.g. Visser & Young 1990; Visser 1995, 1996, 1997; Diekmann & Wopfner 1996; Faure & Cole 1999; Wopfner 1999; Stollhofen *et al.* 2000; Scheffler *et al.* 2003), there have been few attempts to relate biotic change to deglaciation, apart from brief references to overall changes in palynomorph assemblages in relation to climate throughout the Permian by Oesterlen & Millsteed (1994) and Nyambe & Utting (1997), and a study of the five Ecca Group coal seams of the Witbank Basin, Transvaal, South Africa (Falcon *et al.* 1984). Though Falcon *et al.* (1984) did not give a chronostratigraphic age for the Ecca Group, recent radioisotopic dates for the Dwyka and Ecca Groups in the Cape Province and Namibia (Bangert *et al.* 1999) suggest that the Ecca Group is Asselian–Sakmarian in age, and is therefore of equivalent age to deglacial sequences studied in, for example, Oman. Falcon *et al.* (1984) suggested that the oldest Witbank seams (Nos 1 and 2), were formed soon after the most '. . . intensive glaciation, characteristic of the Dwyka period'. In similarity with Arabian palynological assemblages of this time, Seam No. 1 is characterized by, amongst others, *Microbaculispora tentula* Tiwari, and abundant radially and bilaterally symmetrical monosaccate pollen. Seam No. 2 is similar but contains greater numbers of non-taeniate bisaccate pollen, and changes between Seams No. 1 and 2 may therefore be similar to those occurring below the *Alisporites indarraensis* Biozone of Oman, and the *Striatopodocarpites fusus* Biozone of the Collie Basin, Western Australia (Backhouse 1991; see later discussion). Falcon *et al.* (1984) suggested that the floral development of the Witbank area was influenced by the proximity of the sea, and by eustatic fall and rise. They suggested that coal-forming periods were associated with regressive periods where large flat coastal lowland areas were available for plant colonization, though this is inconsistent with more recent concepts of sequence stratigraphy that relate coal formation to sea-level rise and high water table levels. Relating regression with glacioeustatic sea-level fall in this way, they therefore considered coal-forming periods to correspond to short-lived glacial advances during Ecca times, though they considered the main trend through the Ecca to be one of overall climatic amelioration and increase in floral diversity.

There are relatively few studies of Early Permian fauna from southern Africa. Non-marine arthropod ichnotaxa attributed to crustaceans are reported from the shallow lacustrine Ecca Group that overlies the Upper Dwyka shales (Braddy & Briggs 2002), and Dickins (1961) reported *Eurydesma* from the Dwyka beds.

Arabia

It has long been known that deposits of the Permo-Carboniferous glaciation extended to Oman in the Arabian Peninsula in the form of the Al Khlata Formation. This formation is present in the subsurface in much of the southern part of Oman south of the Oman Mountains and extends northwestward beneath the Rub 'al Khali into Saudi Arabia (Braakman *et al.* 1982; Levell *et al.* 1988; Al-Belushi *et al.* 1996). The distribution of climate-sensitive lithologies within the overlying lower Gharif member of the Gharif Formation *sensu* Hughes Clarke (1988) indicates that post-glacial climate change must have occurred in Oman (Wopfner & Diekmann 1996; Wopfner 1999). Much of the Al Khlata Formation is composed of diamictites. These are succeeded by the Rahab member, which is considered to be a post-glacial lacustrine unit (Hughes Clarke 1988; Levell *et al.* 1988; Wopfner 1999). The lowest lower Gharif member sediments are sandstones and mudstones of probable fluvial origin, but above these is a restricted marine interval marked by the acritarch *Ulanisphaeridium omanensis* Stephenson & Osterloff, which is termed the 'maximum flooding shale' (Guit *et al.* 1995) and represents a possible post-glacial eustatic flooding event (Sharland *et al.* 2001; Stephenson & Osterloff 2002).

Above the 'maximum flooding shale', in the subsurface of South Oman, are beds correlative with the outcropping Saiwan Formation of north central Oman (Guit *et al.* 1995; Sharland *et al.* 2001). Angiolini *et al.* (2003) described faunal evidence for warming within the Saiwan Formation, by observing the development of a pioneer cold-water brachiopod community (*Pachycyrtella* palaeocommunity) at its base (late early Sakmarian), followed by a more mature secondary ecological community of a wider marine biota of late Sakmarian age. The *Pachycyrtella* pioneer palaeocommunity records the evolution from glacial to cool temperate climate conditions between the early and late Sakmarian, and colonized the unoccupied environment provided by the sharp transgression at maximum deglaciation (Angiolini *et al.* 2003). This pioneer palaeocommunity is characterized by (1) random distribution pattern over a limited area, (2) clustering in groups, (3) numerical dominance (>85%), (4) suspension feeding, (5) rapid rates of reproduction and growth (*r*-strategy), (6) early maturity and (7) high mortality rates in the juveniles. The disappearance of the basal palaeocommunity is related to a drastic change in the physical environment, recording the interplay of final Gondwanan deglaciation and initial Neotethys opening (Angiolini *et al.* 2003). The secondary ecological succession developed rapidly, indicating significant climatic amelioration and more stable environmental conditions (Angiolini *et al.* 2003). Biodiversity indices significantly increase from the pioneer palaeocommunity (Margalef index 0.2–1.9; Shannon–Wiener index 0.1–1.5; equitability generally <0.5) to the overlying more mature palaeocommunity (Margalef index 1–2.8; Shannon–Wiener index 0.5–2.4; equitability 0.8–1). Climate change continued into middle Gharif member times since that unit contains common calcrete horizons and redbeds (Guit *et al.* 1995).

The palynological succession within the Rahab and lower Gharif members in well sections Thuleilat (TL)-16 and -42 in south Oman allows detailed metre-by-metre reconstruction of floral changes within the deglaciation period (Stephenson & Osterloff 2002; Stephenson *et al.* 2005). The reconstruction relies on palynomorphs in the sedimentary rocks accurately representing the hinterland vascular plant communities. In broad terms, the numbers and variety of palynomorphs are probably an adequate guide to the numbers and variety of plants on the hinterland, but factors such as differing yields of spores and pollen by different plants, and processes such as sorting, may cause taphonomic biases.

In the lowest part of the section (below 940 m in TL-16 and 944 m in TL-42; Figs 2 and 3), a cold climate fern wetland community was present on lowland outwash alluvial plains while on the surrounding uplands or better-drained ground, a primitive conifer community developed (Stephenson & Osterloff 2002; Stephenson *et al.* 2005). There is also evidence that a specialized autochronous freshwater algal palaeoecology occupied small ponds in the fern lowland wetland environment. Later in the post-glacial period (above 940 m in TL-16 and 944 m in TL-42), these communities were replaced. In the lowland alluvial plains, a cycad-like and lycopsid vegetation developed, while in the uplands or better-drained areas a taeniate- and non-taeniate bisaccate pollen-producing glossopterid or other gymnospermous flora was established. Later (above 925 m in TL-16 and 915 m in TL-42), restricted marine conditions occurred in parts of the sedimentary basin; within the bodies of brackish or salt water, an ephemeral microflora and fauna (indicated by rare acritarchs and microforaminiferal linings) developed. This restricted marine transgression probably advanced from the northeast (Konert *et al.* 2001) and may represent the first major deglacial eustatic sea-level rise of the Permian in the Arabian region (Stephenson & Osterloff 2002; Stephenson *et al.* 2005).

A carbon isotope record from organic matter within the same sections of the Al Khlata and lower Gharif formations of TL-16 and -42 shows a $\delta^{13}C_{organic}$ trend from -21 to $-24\%o$ (Stephenson *et al.* 2005; Figs 2 and 3). The trend may be due to changes in palaeoatmospheric CO_2 since they cannot be accounted for by changes in organic matter type or preservation. Changing $^{13}C/^{12}C$ ratios in atmospheric CO_2 are known to occur in response to changes in the proportions of carbonate and organic carbon 'burial' (Erwin 1993), and between glacial and interglacial environments due to changes in biomass and productivity (Arens *et al.* 2000). The decreasing trend seen in TL-16 and TL-42 may therefore represent change in the carbon cycle although it is impossible to identify a specific mechanism.

Australia

The Early Permian of Australia is perhaps the most intensively studied of the former parts of Gondwana, but comparatively few studies address the biotic response to Asselian–Sakmarian deglaciation. After his pioneering work in the 1970s and 1980s, Dickins (1993) summarized the faunal succession of Western Australia in relation to climatic warming in broad terms. Glacial features are present in the Nangetty Formation of the Perth Basin, the Lyons Group of the Carnarvon Basin, and the Grant Group and Paterson Formation of the Canning Basin. Intercalated with, and overlying these, are rocks containing a cold-water marine fauna consisting of the bivalves *Eurydesma* and *Deltopecten*, the gastropod *Keeneia*, and the brachiopods *Lyonia lyoni* (Prendergast), *Rhynchopora australasica* Archbold, *Kiangsiella* sp., *Grumantia* cf. *costellata* Clarke, *Cyrtella australis* Thomas, *Ambikella notoplicata* Archbold and Thomas and *Trigonotreta lyonsensis* Archbold and Thomas. Dickins (1985) recognized five glacial/interglacial phases in the Lyons Group, with the intercalated marine fauna remaining relatively uniform throughout its deposition and indicative of cold water. Above this, in the Fossil Cliff Member of the Holmwood Shale (Perth Basin), the Callythara Formation (Carnarvon Basin) and the Nura Nura Member of the Poole Sandstone (Canning Basin), there is a more diverse, slightly warmer-water fauna, including brachiopods, bryozoans, blastoids and crinoids. Amongst the bivalves, Dickins (1993) considered the bivalve *Edmondia* to indicate warm conditions.

Archbold (1998) revised the faunal succession of Western Australia and showed an increase in biodiversity from the Asselian *Lyonia lyoni* Biozone (7 species), through the Tastubian (early Sakmarian) *Trigonotreta occidentalis* Biozone (10 species) to the Sterlitamakian (late Sakmarian) *Coronalosia*

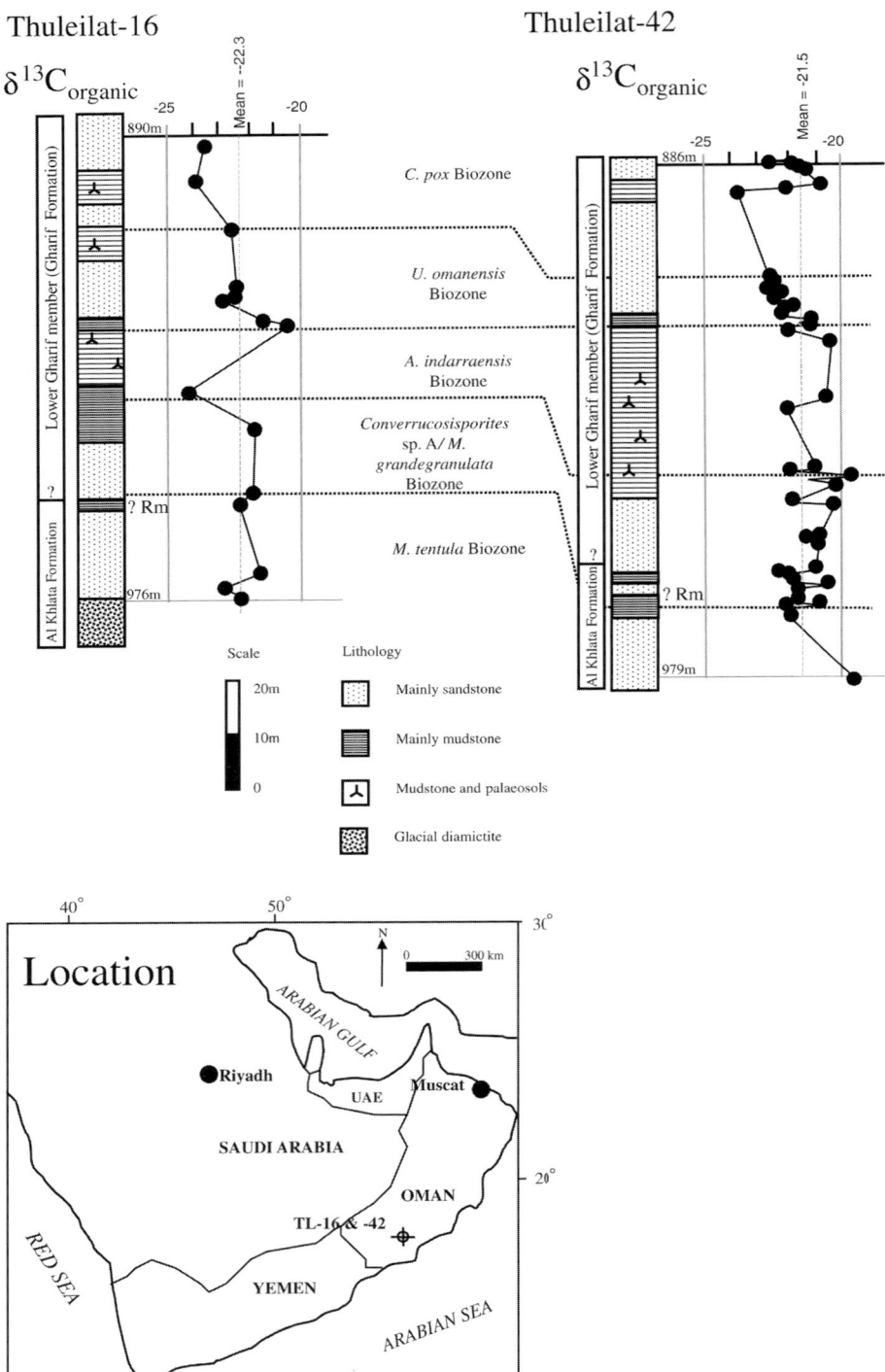

Fig. 2. Correlation of Thuleilat (TL) well sections of Oman based on palynology, and flattened on the base of the *U. omanensis* Biozone. Note that the boundaries between the Rahab member, the Al Khlata Formation and the Gharif Formation are not defined precisely in the Thuleilat wells. ? Rm, possible position of Rahab member. Reproduced with permission from Stephenson *et al.* (2005). Inset map shows location.

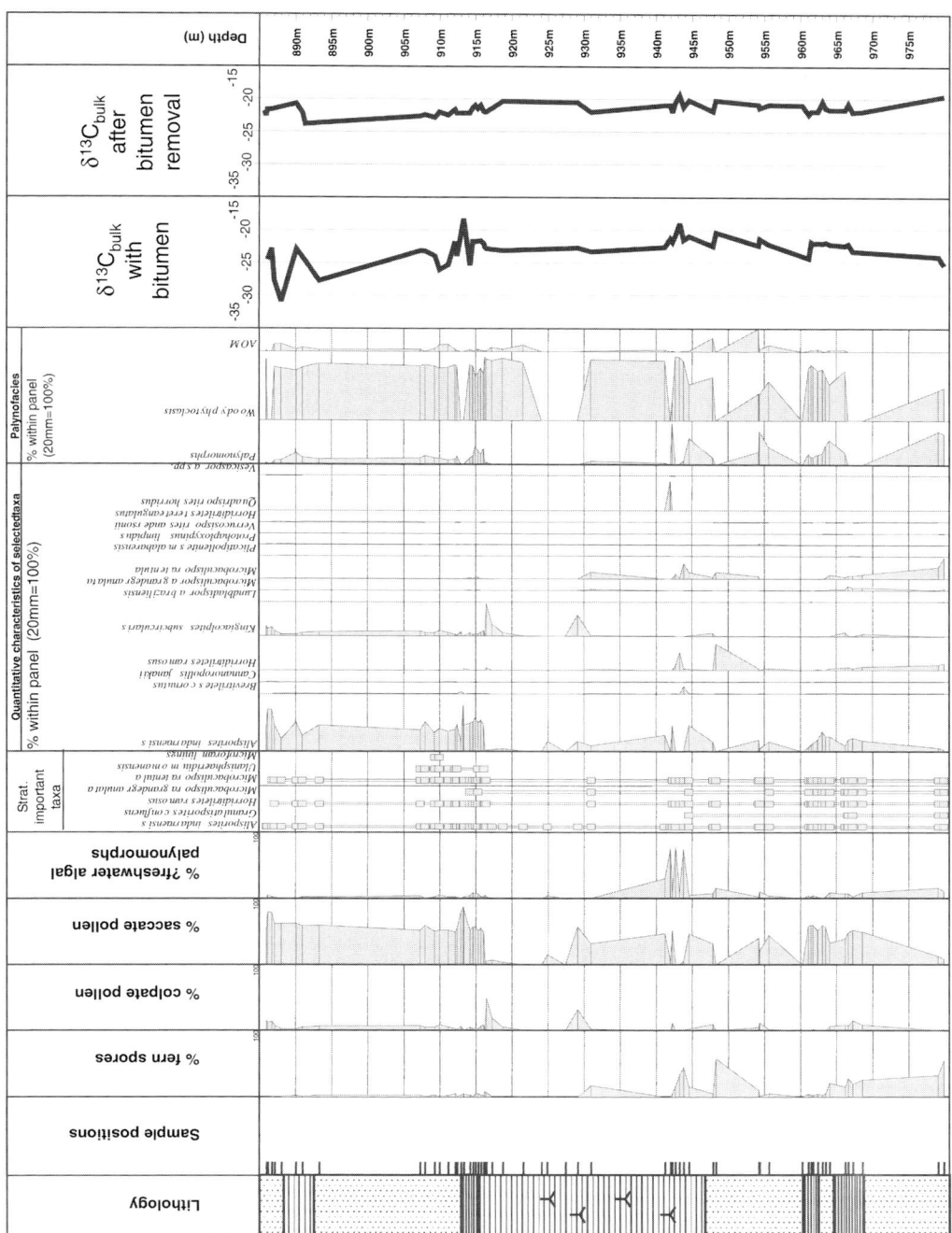

Fig. 3. Gross quantitative palynological character, ranges of selected palynomorphs, palynofacies, and $\delta^{13}C_{org}$ values for the Thuleilat-42 well section. AOM, amorphous organic matter. Reproduced with permission from Stephenson *et al.* (2005).

irwinensis (formerly *Strophalosia irwinensis*) Biozone (40 species). There is a slow trend in increase of biodiversity between the first two biozones (Asselian to Tastubian in age), but a sudden shift occurs in the Sterlitamakian *Coronalosia irwinensis* Biozone. This biotic change is consistent with the water-temperature curve established by Archbold & Shi (1995) and based on the presence/absence of Tethyan brachiopod genera, which are considered indicators of warm water. According to this curve, cold-water temperatures (5 °C) in the *Lyonia lyoni* Biozone evolved to warmer temperatures (20 °C) in the *Coronalosia irwinensis* Biozone, which is characterized by high biodiversity and by the invasion of a significant number of Tethyan genera (about 14). Also oxygen isotope data available for Lower Permian Australian spiriferids (Compston 1960; Lowenstam 1961, 1964) indicate the potential for a significant temperature change during Asselian–Tastubian times and late Artinskian times.

In recent reviews of the palaeobiogeography of Australasian brachiopod faunas, Archbold (2000, 2001) demonstrated the striking difference in generic and specific composition of brachiopods on either side of the continent. The eastern side of Australasia was affected by cold-water currents: eastern Australia marine faunas have a strong link with New Zealand and New Caledonia and a lower diversity than coeval Western Australian (westralian) faunas. They show an increase in biodiversity from south (Tasmania) to north (Bowen Basin, Queensland) which is interpreted as reflecting an increase in temperature from cold to cool temperate. Clarke (1990) illustrated brachiopod and bivalve assemblages from glendonite-bearing siltstones above the glacigene sequence in Tasmania and interpreted these as cold-water low-diversity faunas.

The eastern Australian brachiopod biozonation of Briggs (1998) shows a slight increase in the number of taxa from the Asselian *Lyonia bourkei* Biozone to the Tastubian *Strophalosia concentrica* and *Strophalosia subcircularis* biozones with a reverse trend in the Sterlitamakian *Bandoproductus walkomi* Biozone. A significant increase in diversity is recorded later, in the lower Artinskian *Echinalosia curtosa* Biozone. This biotic change is consistent with persistence of cold climate and glaciation in eastern Australia during the late Sakmarian (Frank *et al.* 2005).

In palynological terms, few deglaciation sequences have been studied to the same level of detail as the Rahab and Lower Gharif members of Oman, hence close comparisons are not possible. However, sequences in Western Australia studied by Backhouse (1991) allow some comparison. Backhouse (1991) noted decreases in monosaccate

pollen and later increases in small bisaccate pollen within and just above a section assignable to the *Granulatisporites confluens* Biozone of Foster & Waterhouse (1988). Backhouse (pers. comm., 2000) considered the bisaccate pollen to be closely similar to *Alisporites indarraensis* Segroves. No *Ulanisphaeridium omanensis*-bearing horizon was reported by Backhouse (1991) though the low sample frequency of Backhouse's study may not have been sufficient to detect such a narrow horizon.

Lithostratigraphic similarities between the deglaciation sequences in Oman and the Canning Basin, Western Australia have been noted by Wopfner (1999). The scarcity of palynological information from the Canning Basin sequences does not allow close comparison, but it is notable that the acritarch *Ulanisphaeridium berryense* McMinn, a taxon closely similar to *U. omanensis*, was recorded by Foster & Waterhouse (1988) from the upper parts of the sequence assigned to the *G. confluens* Biozone in Calytrix No. 1 well in the Canning Basin.

India

It has long been known that the dominantly continental Early Permian strata of the lower Talchir Formation and overlying Karharbari Formation (Indian Peninsula) contain both glacigene diamictites (Ghosh & Mitra 1972; Frakes *et al.* 1975) and thin marine intercalations (Sastry & Shah 1964). Shah and Sastry (1975) considered the marine beds at Manendragarh and Rajhara to be within the lower and middle parts of the Talchir Formation, and those at Umaria and Badhaura to be in the basal part of the Karharbari Formation. They further considered the two sets of marine beds to have been deposited by separate transgressions, the earlier in the Asselian, the later in the Sakmarian. However, Frakes *et al.* (1975) considered there to be no difference in the faunas at Umaria and Manendragarh and assumed them to be of the same age, being characterized chiefly by the shelly fauna *Eurydesma*, *Peruvispira* and *Trigonotreta*. Dickins & Shah (1979) supported the view of Shah & Sastry (1975) and reported a list of foraminifers, bryozoans, brachiopods, bivalves, gastropods and crinoids from Manendragarh and Umaria suggesting correlation to the Lyons Group marine faunas of Western Australia. The more diversified marine fauna from Badhaura, with more than 17 brachiopod species, is considered younger and correlateable to the Callytharra Formation of Western Australia. According to Archbold *et al.* (1996), the Manendragarh brachiopod fauna is characterized by *Trigonotreta hesdoensis* (Sahni & Dutt), which is one of the most primitive species of the genus (Angiolini *et al.* 2005). Slightly younger (late Tastubian) is *Trigonotreta narsarhensis* (Reed); a later evolutionary form of *T. hesdoensis* from Umaria which is associated with *Bandoproductus umariensis* (Reed), *Arctitreta* sp., *Tomiopsis barakerensis* (Reed) (= *Ambikella*) and *Cleiothyridina* sp. The Badhaura fauna is more diversified, containing species of the genera *Semilingula*, *Arctitreta*, *Derbyia*, *Strophalosia*, *Aulosteges*, *Cyrtella*, *Neospirifer*, *Crassispirifer*, *Ambikella*, *Hoskingia*, *Fletcherithyris* and *Gilledia*, and is considered to be at least Sterlitamakian in age by Archbold *et al.* (1996). Apart from the discussion of Dickins (1985), who recognized the Umaria beds as representing the Tastubian post-glacial eustatic rise, no attempt has been made to link the biotic change with the post-glacial climatic change in India.

Venkatachala and Tiwari (1987) suggested that other marine intervals could be identified within the Talchir Formation on the basis primarily of palynomorphs such as acritarchs and leiospheres. Although no detailed attempts have been made to link climate change with macrofaunal and macrofloral change through the Talchir and Karharbari formations, palynomorphs have been studied with this aspect in mind (e.g. Banerjee & D'Rozario 1990; Tiwari 1994; Vijaya 1994; Venkatachala *et al.* 1995; Pant 1996; Vijaya *et al.* 2001). The most detailed of these studies in relation to the Early Permian is that of Vijaya (1994). Within the Talchir Formation, Vijaya (1994) recognized three phases recording an upsection increase in microfloral diversity. The earliest of these (Phase I, considered of Asselian age) is characterized by common radially and bilaterally symmetrical monosaccate pollen and simple trilete spores. Phase II is characterized by sculptured monosaccate pollen, taeniate bisaccate pollen and more common sculptured spores. Finally, Phase III, considered of early Sakmarian age, is characterized by a maximum diversity of forms. Vijaya (1994) considered Phase III to correlate with the *Granulatisporites confluens* Biozone. These palynological changes cannot be directly related to facies changes in a single section, since Vijaya (1994) presents the data in a generalized way, but clearly the sequence is deglacial in form and has some similarity with palynological changes recorded elsewhere in Gondwana. Similar phases attributed to climatic warming were also suggested by Banerjee & D'Rozario (1990).

Antarctica

Late Palaeozoic rocks of glacial origin outcrop in the central Transantarctic Mountains, Victoria Land and the Ellsworth Mountains of west Antarctica. Isbell and Cúneo (1996) demonstrated rapid change from glacigene to sediments of post-glacial origin within the Weller Coal Measures of southern Victoria Land. In an investigation of the palaeoenvironment of the Weller Coal Measures, Francis *et al.* (1993) considered that winter temperatures were low during deposition, suggesting that adjacent uplands were probably still glaciated. They also suggested that summer temperatures were nevertheless adequate for a peat-forming flora and forest vegetation (evidenced from very large *in situ* *Glossopteris* stumps). The forests may have been deciduous (Taylor *et al.* 1992), reflecting such seasonality, and the leaves of *Glossopteris* are often found in thick mats (Francis *et al.* 1993). The age of the Weller Coal Measures and underlying diamictites, and therefore of the date of this rapid deglaciation, are uncertain however, and can only be placed between Eastern Australian Palynological Stage 2 (see Evans 1969; Barrett & Kyle 1975) and Stage 4 (Kyle 1977). Thus, the deglaciation cannot be directly correlated with others in for example Oman and the Collie Basin, Western Australia.

Lindström (1995), however, showed detailed palynological changes occurring within the *Granulatisporites confluens* Biozone in rocks in western Dronning Maud Land. She correlated these changes with the lower Weller Coal Measures and the underlying Darwin Tillite and considered the rocks to have been deposited in a cold climate, not far from active glaciation. The sections Lindström (1995) studied were stratigraphically rather short (both less than 10 m) and so probably represent a 'snapshot' of climate change, but some of the changes are consistent with those noted in Oman, Peninsular India and Western Australia. At Lindström's Locality A, for example, the highest samples are dominated by non-taeniate bisaccate pollen, whilst monosaccate pollen dominate the lower part of the section. *Granulatisporites confluens* occurs throughout, apart from two sample levels toward the centre of the section.

Few Early Permian macrofossils have been described from Antarctica. Kelly *et al.* (2001) reported *Cancrinella* sp. in bioturbated mudstones from locality B at Alexander Island (Antarctic Peninsula) which according to Archbold *et al.* (2004) is comparable to *Costatumulus* sp. B. of the Rio Blanco basin Association II (Argentina) and given a middle–late Asselian age.

Farfield isotope and atmospheric studies

There are relatively few studies using isotopes which are specific to the Early Permian; but there are low-resolution isotope curves which record seawater composition for most of the Phanerozoic based on compilations of the carbon and oxygen isotope composition of brachiopods (e.g. Veizer *et al.* 1999). Through the Phanerozoic, changes in marine $\delta^{13}C$ are interpreted in terms of global changes in burial and re-oxidation of organic matter; although other processes such as release of low $\delta^{13}C$ volcanic or mantle CO_2 into the atmosphere, discharges of oceanic methane and overturn of low $\delta^{13}C$ anoxic bottom waters are all described (Korte *et al.* 2005). Variation in $\delta^{18}O$ in brachiopod shells is thought to be largely due to the growth and retreat of continental glaciers, although multimillion-year variations in $\delta^{18}O$ may also be a function of interactions of the oceans with the lithosphere driven by tectonics (e.g. Veizer *et al.* 1997). The Phanerozoic seawater curves show that in the Early Permian there was a rapid decline in $\delta^{18}O$ despite the curve for $\delta^{13}C$ being rather flat (Veizer *et al.* 1999). The only dedicated Permian isotope seawater curves are based on a compilation of data from brachiopods from a series of discrete basins. The $\delta^{18}O$ data show this decline of 2.5‰ in the Early Permian (Asselian to late Artinskian) (Fig. 4, see also Korte *et al.* 2005), over a timespan of 10 million years. The study of Korte *et al.* (2005) is also one of few that describes rigorous screening of brachiopod carbonate data for diagenetic alteration using a combination of microtextural and trace element content.

However, the Early Permian brachiopod data given in Korte *et al.* (2005) are entirely derived from the southern Urals (sections from Usolka, Sakmara and Dalnij Tyulkas; references in Korte *et al.* 2005). The southern Urals were palaeoequatorial at this time, and therefore climatically distinct from the cold-climate seas in which most postglacial Gondwanan brachiopods lived. However, Korte *et al.* (2005) assumed that southern Urals brachiopods lived in basins that were open to the Palaeotethys Ocean, and so were capable of reflecting a global signal. Korte *et al.* (2005) considered a 2‰ spread in both $\delta^{13}C$ and $\delta^{18}O$ in coeval brachiopods as sufficient to allow for global differences including variations in seawater temperature with latitude (James *et al.* 1997; Brand *et al.* 2003). Korte *et al.* (2005) therefore assumed that the decline in $\delta^{18}O$ in the Early Permian Urals brachiopods is a global signature and not a geographical artefact, so that the data can represent far-field evidence to help decipher changing environment around Gondwana. In general, the $\delta^{13}C$ data in the Early Permian are stable around +4.3‰. In contrast, the $\delta^{18}O$ data from the Asselian and lower Sakmarian brachiopods have values between +0.1 to −2.3‰ (the latter being the highest of all the values throughout the Permian); values drop to −3.5‰ in the late Artinskian. Korte *et al.* (2005) considered that the most likely cause for this decline in $\delta^{18}O$ is the

return of isotopically light waters to the oceans from melting of Permian glaciers at high latitudes (Fig. 5). This explanation is not entirely consistent with evidence from brachiopods and palynology (see earlier discussion), because that suggests deglaciation was confined to the Asselian and Sakmarian. Given that the Late Palaeozoic and Pleistocene glaciations were probably of similar magnitude (Crowley & Baum 1992), it has been calculated that we can ascribe an approximate 2.3‰ increase in seawater $\delta^{18}O$ to the 'ice volume effect' (Korte *et al.* 2005). The maximum range in data we see over this period is 3.6‰ (+0.1‰ to −3.5‰ shown on fig. 7 of Korte *et al.* 2005). Therefore, the remaining 1.3‰ in $\delta^{18}O$ could be caused by an increase in temperature of approximately 5 °C in the Urals. Korte *et al.* (2005) concluded therefore that the decline of the Gondwana ice-sheet was due to global warming because of the requirement for both temperature and ice volume to explain the $\delta^{18}O$ variation.

Another Early Permian farfield temperature change reconstruction from a study of oxygen and hydrogen isotope ratios in pedogenic phyllosilicates is that of Tabor & Montañez (2005). The Midland Basin of Texas and the southern Anadarko Basin of Oklahoma were equatorial during the Early Permian. The phyllosilicates have δD and $\delta^{18}O$ values that are thought to represent Early Permian meteoric waters at a specific temperature. Using the mineral-water temperature-dependent fractions equations, it was shown that the palaeotemperature of phyllosilicate crystallization ranged from 22 ± 3 °C to 35 ± 3 °C (calcite $\delta^{18}O$ data) across the Carboniferous–Permian boundary in the East Midland basin.

One of the few open marine isotope records from the Early Permian has been reported from the Karoo Basin, South Africa (Scheffler *et al.* 2003). The basin contains organic matter thought to be almost entirely of marine origin, and this was measured for $\delta^{13}C$. The $\delta^{13}C$ composition of marine organic matter is often thought to reflect the $\delta^{13}C$ signature of atmospheric CO_2 and therefore reflect changes in pCO_2 (Scheffler *et al.* 2003). The lowest $\delta^{13}C$ values therefore represent full glacial conditions, while deglaciation is represented by higher $\delta^{13}C$. The Karoo Basin has $\delta^{13}C_{organic}$ which increases (<−25‰ to −22‰) from 308 to 288 Ma, a trend that Scheffler *et al.* (2003) compared to Veizer *et al.*'s (1999) $\delta^{13}C$ data from equatorial shelly carbonates for the same time period. The trend to increasing $\delta^{13}C$ values in both the Karoo organic matter and equatorial carbonates is thought to represent the same global pCO_2 shift. The Ecca Group which has the highest $\delta^{13}C_{organic}$ values is thought to have been, at least in part, deposited during the earliest Asselian.

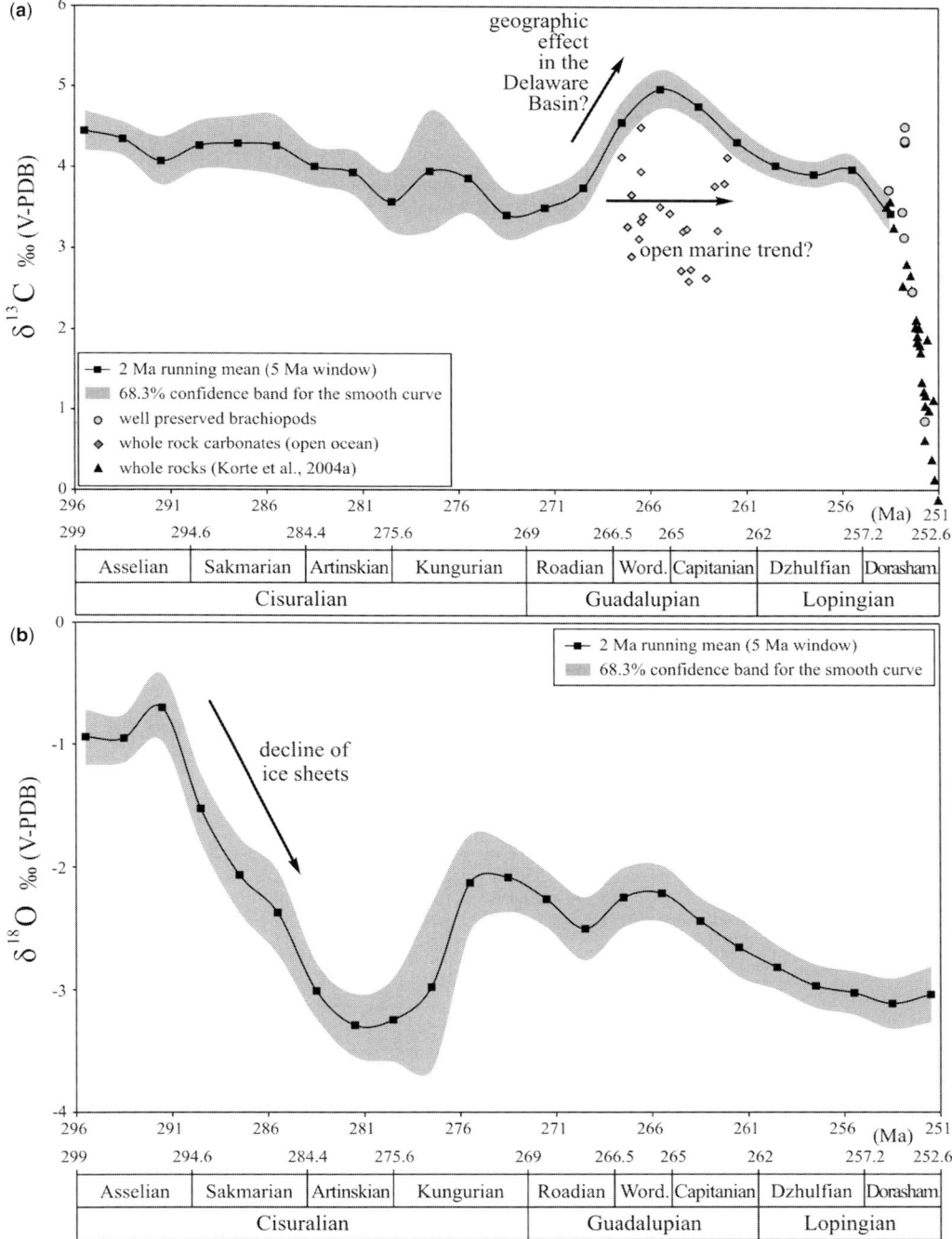

Fig. 4. Running $\delta^{13}C$ mean (**a**) and $\delta^{18}O$ mean (**b**) (2 Ma step, 5 Ma window), based on well-preserved brachiopod shells for the interval 296–253 Ma. The Early Permian brachiopod data are entirely derived from the Southern Urals, which were in a palaeo-equatorial position, although it is assumed that the brachiopods were living in basins which were open to the Tethys Ocean. Reproduced with permission from Korte *et al.* (2005).

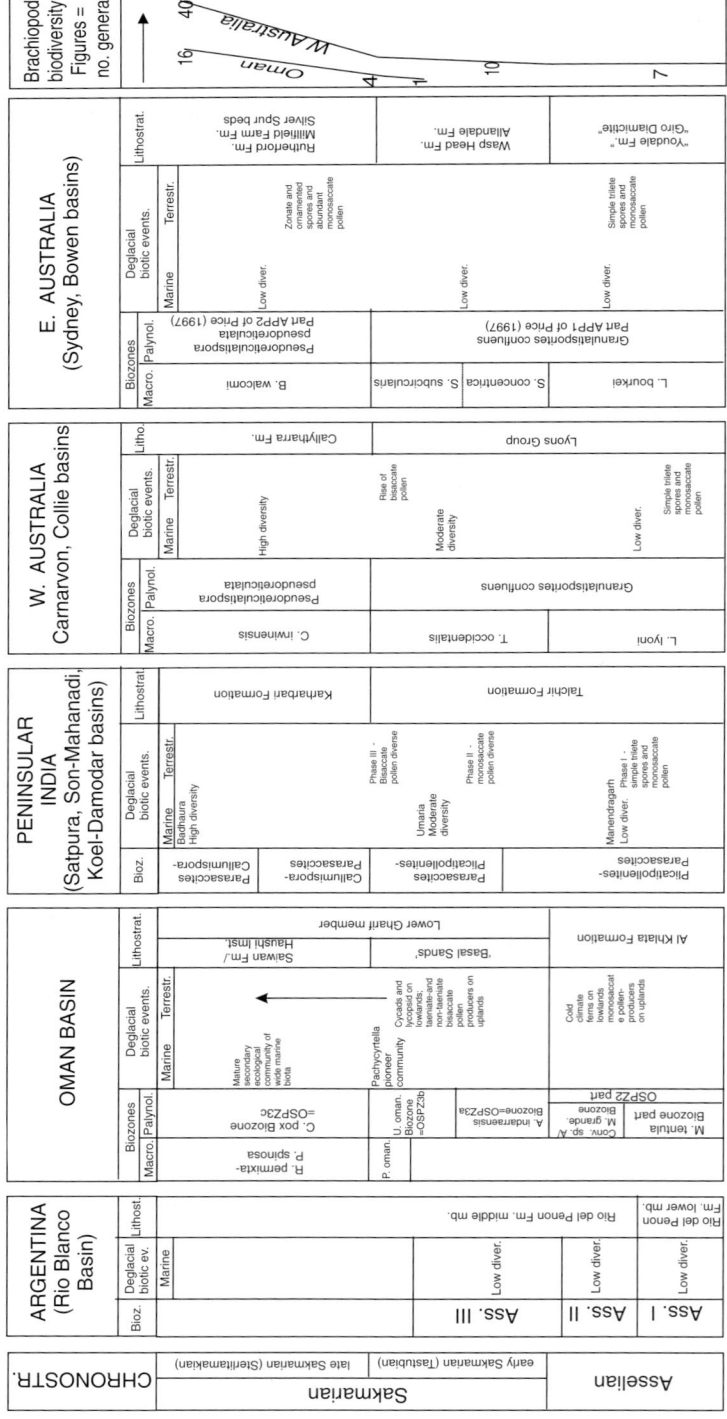

Fig. 5. Distribution of Asselian–Sakmarian palynological and macrofaunal assemblages in selected basins across Gondwana. Note that only sections with established faunal successions are included, thus Africa and Antarctica are excluded. As discussed in the text, it is not possible to correlate precisely in this time interval, thus this diagram is **not** primarily aimed at correlation but at comparing biotic responses to deglaciation. Argentina, Rio Blanco Basin biozones after Cisterna and Simanauskas (2000); no palynological data specific to Rio Blanco Basin. Oman Basin biozones after Angiolini *et al.* (2003), Stephenson and Osterloff (2002) and Stephenson *et al.* (2003). Indian basins biozones after Vijaya *et al.* (2001). Western Australian biozones after Archbold (2001, Carnarvon Basin) and Backhouse (1991, Collie Basin). Eastern Australian biozones after Archbold (2001) and Price (1997).

Further study

The Early Permian rocks of Gondwana that can be related to climate-forced deglaciation are well-represented sequences of pre-Cenozoic glaciation, and therefore form an invaluable resource to develop a Gondwana biotic deglaciation model. The advantages of developing such a deglaciation model would be in understanding the response of life to increasing temperatures and other climate change, and might be useful in the study of modern biotic change during global warming. However, to achieve such a model, a number of factors need to be considered:

1. More detailed bed-by-bed interdisciplinary palaeontological studies of measured sections demonstrably related to climate-forced deglaciation must be carried out. Clearly, such studies are limited to sections with well-preserved fossils and more or less continuous stratigraphical fossil occurrence. While this may be possible for palynology, it is unlikely to be possible for macrofossils, which in post-glacial Early Permian sequences are stratigraphically and geographically rather isolated. Palynological assemblages that are defined quantitatively would be important in such studies, but such assemblages at least in glacigene sediments, have been shown to be influenced by facies (Stephenson 1998), and reworking (Osterloff et al. 2004). Similarly, factors related to differential spore or pollen production by different plants can influence the quantitative characteristics of palynological assemblages (Wilson 1963), as can over-representation of natural diversity due to excessive taxonomic splitting. Thus, study of biotic change needs to compensate for such factors.

2. Sections for study must be coeval so that similarities (due to general climate change) and differences (presumably due to local responses to deglaciation) can be distinguished. However, establishing isochroneity is very difficult, and it is likely that palynological data will be required to correlate sequences – while also defining climate change – so there will be attendant problems of circularity of argument. An advantage of establishing a Gondwana-wide palynostratigraphy, uniting the four or five schemes presently in existence for the former continents of Gondwana, would also be in allowing other palaeontological data, of which there are considerable amounts, to be used. It is also possible that final glacial events may be of slightly different ages in different places. This has already been established on a large scale, since glaciation in southeast Australia can be shown to have persisted into the Artinskian, but also there may be variations in periods of maximum rate of biotic change from region to region, due to differing palaeolatitudes, for example. Finally, many of the post-glacial sequences for which only palynological information is available are mainly non-marine, for example in India and Arabia. Non-marine glacial basins are known for numerous disconformities and non-sequences, and for large-scale cyclical reworking of older sediments. It may not always be possible, even under the best biostratigraphical circumstances, to be able to correlate such rocks across a large continent.

Conclusions

1. The Early Permian rocks of Gondwana that can be related to climate-forced deglaciation are perhaps the best-known and best-represented sequences of pre-Cenozoic glaciation, and therefore form an invaluable resource for future research, to develop a Gondwana biotic deglaciation model.

2. Present data, which are widely scattered geographically and of different stratigraphic scales and resolutions, clearly show diversity increase from glacial conditions to post-glacial conditions.

3. Amongst the marine fauna, a cold-water marine fauna consisting of bivalves such as *Eurydesma* and *Deltopecten*, and brachiopods such as *Lyonia*, were established in the earliest post-glacial marine transgressions that did not affect all of Gondwana. Above this is a more diverse, slightly warmer, cold temperate fauna, including brachiopods, bryozoans, blastoids and crinoids like those of the Saiwan Formation of Oman. A summary of marine faunal and palynological changes in Gondwanan Early Permian deglaciation sequences is shown in Figure 5.

4. The palynomorph succession, which covers the deglaciation period more fully, shows some consistency across Gondwana in Asselian–Sakmarian rocks. Very broadly, a change from monosaccate pollen assemblages, associated with fern spores, to more diverse assemblages with common non-taeniate bisaccate pollen, occurs through the deglaciation period (Fig. 5). In Oman, where this has been studied in greatest detail, the upland saw changes from a glacial monosaccate pollen-producing flora to a warmer climate bisaccate pollen-producing flora; in the lowlands, a parallel change occurred from a glacial fern flora to a warmer-climate colpate pollen-producing and lycopsid lowland flora. The sedimentary organic matter of the clastic rocks of the sequence records a corresponding $\delta^{13}C$ trend (from approximately -21 to -24%) believed to reflect palaeoatmospheric change due to post-glacial global warming.

5. Of the few Early Permian farfield isotope studies, the most detailed (Korte et al. 2005), compiled from brachiopods from the southern Urals, shows a $\delta^{18}O$ decline of 2.5% in the Early

Permian (Asselian to Artinskian) and stable $\delta^{13}C$ values of around +4.3‰ in the same period. This farfield evidence is in part consistent with Gondwanan palaeontological data since the most likely cause for the decline in $\delta^{18}O$ is the return of isotopically light waters to the oceans from melting of glaciers at high latitudes.

M. H. Stephenson and M. J. Leng publish this chapter with permission of the Director of the British Geological Survey. The authors thank Elsevier Publishing for permission to reproduce figures 6 and 8 from Korte *et al.* (2005) and the Geological Society for permission to reproduce figures 4 and 5 from Stephenson *et al.* (2005).

References

AL-BELUSHI, J. D., GLENNIE, K. W. & WILLIAMS, B. P. J. 1996. Permo-Carboniferous Glaciogenic Al Khlata Formation, Oman: a new hypothesis for origin of its glaciation. *GeoArabia*, **1**, 389–404.

ANDERSON, J. M. 1977. The biostratigraphy of the Permian and Triassic. Part 3. A review of Gondwana Permian palynology with particular reference to the northern Karroo Basin of South Africa. *Memoirs of the Botanical Survey of South Africa*, **41**, 1–133.

ANGIOLINI, L., BALINI, M., GARZANTI, E., NICORA, A. & TINTORI, A. 2003. Gondwanan glaciation and opening of Neotethys: the Al Khlata and Saiwan formations of Interior Oman. *Palaeogeography, Palaeoclimatology, Palaeoecology*, **196**, 99–123.

ANGIOLINI, L., CRASQUIN-SOLEAU, S., PLATEL, J.-P., ROGER, J., VACHARD, D., VASLET, D. & AL HUSSEINI, M. 2004. Saiwan, Gharif and Khuff formations, Haushi–Huqf Uplift, Oman. *In*: AL-HUSSEINI, M. I. (ed.) *Carboniferous, Permian and Triassic Arabian Stratigraphy*. GeoArabia Special Publication 3, Gulf PetroLink, Manama, Bahrain, 149–183.

ANGIOLINI, L., BRUNTON, H. & GAETANI, M. 2005. Early Permian (Asselian) brachiopods from Karakorum (Pakistan) and their palaeobiogeographical significance. *Palaeontology*, **48**, 69–86.

ANGIOLINI, L., BUCHER, H., PILLEVUIT, A. *ET AL.* 1997. Early Permian (Sakmarian) brachiopods from southeastern Oman. *Geobios*, **30**, 379–405.

ANGIOLINI, L., STEPHENSON, M. H. & LEVEN, Y. 2006. Correlation of the Lower Permian surface Saiwan Formation and subsurface Haushi Limestone, central Oman. *GeoArabia*, **11**, 17–38.

APAK, S. N. & BACKHOUSE, J. 1999. Stratigraphy and petroleum exploration objectives of the Permo-Carboniferous sucession on the Barbwire Terrace and adjacent areas, northwest Canning Basin, Western Australia. *Geological Survey of Western Australia, Report*, **68**.

ARCHANGELSKY, S. 1990. Plant distribution in Gondwana during the Late Palaeozoic. *In*: TAYLOR, T. N. & TAYLOR, E. L. (eds) *Antarctic Paleobiology*. Springer Verlag, New York, pp. 102–117.

ARCHANGELSKY, S., AZCUY, C. L., PINTO, I. D., GONZÁLEZ, C. R., MARQUES-TOIGO, M., RÖSLER,

O. & WAGNER, R. H. 1980. The Carboniferous and Early Permian of the South American Gondwana area: a summary of the biostratigraphic information. *Actas del II Congreso Argentino de Paleontología y Bioestratigrafía y I Congreso Latinoamericano de Paleontología 1978*, **4**, 257–269.

ARCHBOLD, N. W. 1998. Marine biostratigraphy and correlation of the west Australian Permian basins. *In*: PURCELL, P. G. & PURCELL, R. R. (eds) *The Sedimentary Basins of Western Australia 2*, Proceedings of the Petroleum Exploration Society of Australia Symposium, 141–151.

ARCHBOLD, N. W. 1999. Permian Gondwana correlations: the significance of the western Australian marine Permian. *Journal of African Earth Sciences*, **29**, 63–75.

ARCHBOLD, N. W. 2000. Palaeobiogeography of the Australasian Permian. *Memoir of the Association of Australasian Palaeontologists*, **23**, 287–310.

ARCHBOLD, N. W. 2001. Pan-Gondwanan, Early Permian (Asselian-Sakmarian-Aktastinian) correlations. *In*: WEISS, R. E. (ed.) *Contributions to Geology and Paleontology of Gondwana – in Honour of Helmut Hopfner*. Geological Institute, University of Cologne, 29–39.

ARCHBOLD, N. W. & DICKINS, J. M. 1996. Permian (Chart 6). *In*: YOUNG, G. C. & LAURIE, J. R. (eds) *An Australian Phanerozoic Timescale*. OUP, Melbourne, 127–135.

ARCHBOLD, N. W. & SHI, G. R. 1995. Permian brachiopod faunas of Western Australia: Gondwanan-Asian relationships and Permian climate. *Journal of Southeast Asian Earth Sciences*, **11**, 207–215.

ARCHBOLD, N. W., CISTERNA, G. A. & SIMANAUSKAS, T. 2004. The Gondwanan Carboniferous–Permian boundary revisited: new data from Australia and Argentina. *Gondwana Research*, **7**, 125–133.

ARCHBOLD, N. W., SHAH, S. C. & DICKINS, J. M. 1996. Early Permian brachiopod faunas from Peninsular India: their Gondwanan relationships. *Historical Biology*, **11**, 125–135.

ARENS, N. C., HOPE JAHREN, A. & AMUNDSON, R. 2000. Can C3 plants faithfully record the carbon isotope composition of atmospheric carbon dioxide? *Paleobiology*, **26**, 137–164.

AZCUY, C. L. 1979. A review of the early Gondwana palynology of Argentina and South America. IV International Palynological Conference, 1976, Birbal Sahni Institute Lucknow, India, **2**, 175–185.

AZCUY, C. L. & JELIN, R. 1980. Las palinozonas del limite Carbónico–Pérmico en la Cuenca Paganzo. *Actas delII Congreso Argentina de Palaeontologia y Bioestratigrafia y I Congreso Latinoamericano de Paleontologia*, **4**, 51–67.

BACKHOUSE, J. 1991. Permian palynostratigraphy of the Collie Basin, Western Australia. *Review of Palaeobotany and Palynology*, **67**, 237–314.

BANERJEE, M. & D'ROZARIO, A. 1990. Palynostratigraphic correlation of Lower Gondwana sediments in the Chuparbhita and Hura basins, Rajmahal Hills, Eastern India. *Review of Palaeobotany and Palynology*, **65**, 239–255.

BANGERT, B., STOLLHOFEN, H., LORENZ, V. & ARMSTRONG, R. 1999. The geochronology and significance of ash-fall tuffs in the glaciogenic Carboniferous-Permian Dwyka Group of Namibia and South Africa. *Journal of African Earth Sciences*, **29**, 33–49.

BANKS, M. R. & CLARKE, M. J. 1987. Changes in the geography of the Tasmania Basin on the Late Paleozoic. *In*: MCKENZIE, G. D. (ed.) *Gondwana Six: Stratigraphy, Sedimentology and Paleontology*, Washington DC, American Geophysical Union, 1–14.

BARRETT, P. J. & KYLE, R. A. 1975. The Early Permian glacial beds of South Victoria Land and the Darwin Mountains, Antarctica. *In*: CAMPBELL, K. W. (ed.) *Gondwana Geology*, Australian National University Press, Canberra, 333–346.

BHARADWAJ, D. C. 1969. Lower Gondwana Formations. 6th Congress of the Stratigraphy and Geology of the Carboniferous, Sheffield 1967. *Compte Rendu*, **1**, 255–274.

BRAAKMAN, J., LEVELL, B., MARTIN, J., POTTER, T. L. & VAN VLIET, A. 1982. Late Palaeozoic Gondwana glaciation in Oman. *Nature*, **299**, 48–50.

BRADDY, S. J. & BRIGGS, D. E. G. 2002. New Lower Permian nonmarine arthropod trace fossils from New Mexico and South Africa. *Journal of Paleontology*, **76**, 546–557.

BRAND, U., LOGAN, A., HILLER, N. & RICHARDSON, J. 2003. Geochemistry of modern brachiopods: applications and implications for oceanography and paleoceanography. *Chemical Geology*, **198**, 305–334.

BRIGGS, D. J. C. 1998. Permian Productidina and Strophalosiidina from the Sydney-Bowen Basin and New England Orogen: systematics and biostratigraphic significance. *Memoir of the Association of Australasian Palaeontologists*, **19**, 1–258.

CÉSARI, S. N. & GUTIÉRREZ, P. R. 2000. Palynostratigraphy of Upper Palaeozoic sequences in central-western Argentina. *Palynology*, **24**, 113–146.

CHANDRA, S. K. 1994. *Marine signatures in the Gondwanas of Peninsular India and Permian palaeogeography*. Ninth International Gondwana Symposium, Hyderabad, India, January, 529–538.

CHARRIER, R. 1986. The Gondwana glaciation in Chile: description of alleged glacial deposits and palaeogeographic conditions bearing on the extension of the ice cover in southern South America. *Palaeogeography, Palaeoclimatology, Palaeoecology*, **56**, 151–175.

CISTERNA, G. A. & SIMANAUSKAS, T. 2000. Brachiopods from the Rio del Penon Formation, Rio Blanco Basin, Upper Palaeozoic of Argentina. *Revista Española de Paleontología*, **15**, 129–151.

CISTERNA, G. A., SIMANAUSKAS, T. & ARCHBOLD, N. W. 2002a. Permian brachiopods from the Tupe Formation, San Juan Province, Precordillera, Argentina. *Alcheringa*, **26**, 177–200.

CISTERNA, G. A., SIMANAUSKAS, T. & ARCHBOLD, N. W. 2002b. The Permian brachiopod genus *Trigonotreta* Koening and its occurrence in Argentina. *Ameghiniana*, **39**, 213–220.

CLARKE, M. J. 1990. Late Palaeozoic (Tamarian; Late Carboniferous–Early Permian) cold water brachiopods from Tasmania. *Alcheringa*, **14**, 53–76.

CLOSS, D. 1970. Intercalation of goniatites in the Gondwanic glacial beds of Uruguay. Gondwana Stratigraphy, 1st Gondwana Symposium, Buenos Aires, 1967, *UNESCO, Earth Sciences*, **2**, 197–202.

COMPSTON, W. 1960. The carbon isotopic composition of certain marine invertebrates and coals from the Australian Permian. *Geochimica et Cosmochimica Acta*, **18**, 1–22.

COUSMINER, H. L. 1965. Permian spores from Apillapampa, Bolivia. *Journal of Paleontology*, **39**, 1097–1111.

CROWELL, J. C. 1995. The ending of the Late Paleozoic ice age during the Permian paper. *In*: SCHOLLE, P. A., PERYT, T. M. & ULMER-SCHOLLE, D. S. (eds) *The Permian of Northern Pangea Vol. 1. Palaeogeography, Palaeoclimates, Stratigraphy*. Springer Verlag, 62–74.

CROWLEY, T. J. & BAUM, S. K. 1992. Modelling late Paleozoic glaciation. *Geology*, **20**, 507–510.

DICKINS, J. M. 1957. Lower Permian pelecypods and gastropods from the Carnarvon Basin, Western Australia. *Australian Bureau of Mines and Mineral Resources Bulletin*, **41**, 75pp.

DICKINS, J. M. 1961. Permian pelecypods newly recorded from eastern Australia. *Palaeontology*, **4**, 119–130.

DICKINS, J. M. 1978. Climate of the Permian in Australia: the invertebrate faunas. *Palaeogeography, Palaeoclimatology, Palaeoecology*, **23**, 33–46.

DICKINS, J. M. 1984. Evolution and climate in the Upper Palaeozoic. *In*: BRENCHLEY, P. (ed.) *Fossils and Climate*. John Wiley & Sons, Chichester, 317–327.

DICKINS, J. M. 1985. Late Palaeozoic glaciation. *Bureau of Mines and Mineral Resources Journal of Australian Geology and Geophysics*, **9**, 163–169.

DICKINS, J. M. 1992. Permian geology of Gondwana countries: a review. *International Geology Reviews*, **34**, 986–1000.

DICKINS, J. M. 1993. Palaeoclimate. *In*: SKWARKO, S. K. (ed.) *Palaeontology of Western Australia*. Geological Survey of Western Australia Bulletin, **135**, 7–9.

DICKINS, J. M. 1996. Problems of a Late Palaeozoic glaciation in Australia and subsequent climate in the Permian. *Palaeogeography, Palaeoclimatology, Palaeoecology*, **125**, 185–197.

DICKINS, J. M. & SHAH, S. C. 1979. Correlation of the Permian marine sequence of India and Western Australia. *In*: LASKAR, B. & RAJA RAO, C. S. (eds) *Fourth International Gondwana Symposium, Papers, Vol. II*. Delhi, India, 387–408.

DIEKMANN, B. & WOPFNER, H. 1996. Petrographic and diagenetic signatures of climate change in peri- and postglacial Karoo sediments of SW Tanzania. *Palaeogeography, Palaeoclimatology, Palaeoecology*, **125**, 5–25.

ERWIN, D. H. 1993. *The Great Paleozoic Crisis: life and death in the Permian*. Columbia University Press, New York, 327pp.

EVANS, P. R. 1969. Upper Carboniferous and Permian palynological stages and their distribution in eastern Australia. *In*: *Gondwana Stratigraphy*, IUGS Symposium Argentina (1967), UNESCO, Earth Sciences, **2**, 41–54.

EYLES, N. & DE BROEKERT, P. 2001. Glacial tunnel valleys in the Eastern Goldfields of Western Australia cut below the Late Paleozoic Pilbara ice sheet. *Palaeogeography, Palaeoclimatology, Palaeoecology*, **171**, 29–40.

EYLES, N. & MCCABE, A. M. 1989. The Late Devensian (<22000 YBP) Irish Sea Basin: the sedimentary record of a collapsed ice sheet margin. *Quaternary Science Reviews*, **8**, 307–351.

EYLES, N., EYLES, C. H. & GOSTIN, V. A. 1997. Iceberg rafting and scouring in the Early Permian Shoalhaven Group of New South Wales, Australia: evidence of Heinrich-like events? *Palaeogeography, Palaeoclimatology, Palaeoecology*, **136**, 1–17.

EYLES, C. H., EYLES, N. & GOSTIN, V. A. 1998. Facies and allostratigraphy of high-latitude, glacially influenced, marine strata of the Early Permian southern Sydney Basin, Australia. *Sedimentology*, **45**, 121–161.

EYLES, N., MORY, A. J. & BACKHOUSE, J. 2002. Carboniferous-Permian palynostratigraphy of west Australian rift basins: resolving tectonic and eustatic controls during Gondwanan glaciations. *Palaeogeography, Palaeoclimatology, Palaeoecology*, **184**, 305–319.

FALCON, R. M. S. 1975. Palynostratigraphy of the Lower Karroo sequence in Sebungwe District, Mid Zambezi Basin, Rhodesia. *Palaeontologia Africana*, **18**, 1–29.

FALCON, R. M. S., PINHEIRO, H. & SHEPERD, P. 1984. The palynobiostratigraphy of the major coal seams in the Witbank Basin with lithostratigraphic, chronostratigraphic and palaeoclimatic implications. *Comunicações dos Serviços Geológicos de Portugal*, **70**, 215–243.

FARABEE, M. J., TAYLOR, E. L. & TAYLOR, T. N. 1991. Late Permian palynomorphs from the Buckly Formation, central Transantarctic Mountains, Antarctica. *Review of Palaeobotany and Palynology*, **69**, 353–368.

FAURE, K. & COLE, D. 1999. Geochemical evidence for lacustrine microbial blooms in the vast Permian main Karoo, Paraná, Falkland Islands and Huab basins of southwestern Gondwana. *Palaeogeography, Palaeoclimatology, Palaeoecology*, **152**, 189–213.

FOSTER, C. B. 1983. Review of the time frame for the Permian of Queensland. *In: Symposium on Permian Geology of Queensland*, Proceedings of the Geological Society of Australia Queensland Division, 107–120.

FOSTER, C. B. & ARCHBOLD, N. W. 2001. Chronologic anchor points for the Permian and Early Triassic of the eastern Australian basins. *In:* WEISS, R. E. (ed.) *Contributions to Geology and Paleontology of Gondwana – in Honour of Helmut Hopfner*, Geological Institute, University of Cologne, 175–199.

FOSTER, C. B. & WATERHOUSE, J. 1988. The *Granulatisporites confluens* Oppel Zone and Early Permian marine faunas from the Grant Formation on the Barbwire Terrace, Canning Basin, Australia. *Australian Journal of Earth Sciences*, **35**, 135–157.

FRAKES, L. A., KEMP, E. M. & CROWELL, J. C. 1975. Late Paleozoic glaciation: Part IV, Asia. *Geological Society of America Bulletin*, **86**, 454–464.

FRANCIS, J. E., WOOLFE, K. J., ARNOT, M. J. & BARRETT, P. J. 1993: *In: Permian Forests of Allan Hills Antarctica: the Palaeoclimate of Gondwanan High Latitudes*. Special Papers in Palaeontology, **49**, 75–83.

FRANK, T. D., JONES, A., FIELDING, C. & THOMAS, S. 2005. Protracted glacial conditions in eastern Gondwana maintained by strengthened upwelling during marine highstands. *Geological Society of America Abstracts with Programs*, **37**, No. 7, p.256; Salt Lake City Annual Meeting (Oct. 16–19, 2005*).*

GHOSH, P. K. & MITRA, N. D. 1972. Sedimentary framework of glacial and periglacial deposits of the Talchir Formation of India. *In:* HOUGHTON, S. H. (ed.), *Proceedings and Papers of the 2nd Gondwana Symposium*, South Africa, 213–224.

GOLDBERG, K. 2004. Floral diversity in the assessment of paleoclimate in the Paraná Basin, southern Brasil. *Journal of Geology*, **112**, 719–727.

GONZALEZ, C. R. 1997. Upper Palaeozoic glaciation and Carboniferous and Permian faunal changes in Argentina. *In:* DICKINS, J. M. (ed.) *Late Palaeozoic and Early Mesozoic Circum-Pacific Events and Their Global Correlation*, Cambridge University Press, 235–243.

GONZALEZ-BONORINO, G. 1992. Carboniferous glaciation in Gondwana. Evidence for grounded marine ice and continental glaciation in southwestern Argentina. *Palaeogeography, Palaeoclimatology, Palaeoecology*, **91**, 363–375.

GUIT, F. A., AL-LAWATI, M. H. & NEDERLOF, P. J. R. 1995. Seeking new potential in the Early-Late Permian Gharif play, West Central Oman. *In:* AL-HUSSEINI, M. I. (ed.), *Geo '94 The Middle East Petroleum Geosciences*, **2**, Gulf Petrolink, Bahrain, 447–462.

HART, G. F. 1967. The stratigraphic subdivision and equivalents of the Karroo sequence as suggested by palynology. *IUGS Gondwana Symposium Proceedings*, **1**, Buenos Aires, 23–35.

HUDSON, R. G. S. & SUDBURY, M. 1959. Permian brachiopoda from south-east Arabia. *Notes et Mémoires sur le Moyen-Orient*, Muséum National d'Histoire Naturelle, Paris, **7**, 19–55.

HUGHES CLARK, M. W. 1988. Stratigraphy and rock unit nomenclature in the oil producing area of interior Oman. *Journal of Petroleum Geology*, **11**, 5–60.

ISBELL, J. L. 1999. The Kukri Erosion Surface; a reassessment of its relationship to rocks of the Beacon Supergroup in the central Transantarctic Mountains, Antarctica. *Antarctic Science*, **11**, 228–238.

ISBELL, J. L. & CÚNEO, N. R. 1996. Depositional framework of Permian coal-bearing strata, southern Victoria Land, Antarctica. *Palaeogeography, Palaeoclimatology, Palaeoecology*, **125**, 217–238.

ISBELL, J. L., MILLER, M. F., WOLFE, K. L. & LENAKER, P. A. 2003. Timing of late Paleozoic glaciation in Gondwana: was glaciation responsible for the development of northern hemisphere cyclothems? *Geological Society of America Special Paper*, **370**, 5–24.

JAMES, N. P., BONE, Y. & KYSER, T. K. 1997. Brachiopod $\delta^{18}O$ values do reflect ambient oceanography: Lacepede Shelf, southern Australia. *Geology*, **25**, 551–554.

JONES, A. T. & FIELDING, C. R. 2004. Sedimentological record of the late Paleozoic glaciation in Queensland, Australia. *Geology*, **32**, 153–156.

JONES, M. J. & TRUSWELL, E. M. 1992. Late Carboniferous and Early Permian palynostratigraphy of the Joe Joe Group, southern Galilee Basin, Queensland, and implications for Gondwana Stratigraphy. *Bureau of Mines and Mineral Resources Journal of Australian Geology and Geophysics*, **13**, 143–185.

KELLY, S. R. A., DOUBLEDAY, P. A., BRUNTON, C. H. C., DICKINS, J. M., SEVASTOPULO, G. D. & TAYLOR, P. D. 2001. First Carboniferous and ?Permian marine macrofaunas from Antarctica and their tectonic implications. *Journal of the Geological Society, London*, **158**, 219–232.

KEMP, E. M. 1975. Palynology of Late Palaeozoic glacial deposits of Gondwanaland. *In*: CAMPBELL, K. W. (ed.) *Gondwana Geology*, Australian National University Press, Canberra, 125–134.

KEMP, E. M., BALME, B. E., HELBY, R. J., KYLE, R. A., PLAYFORD, G. & PRICE, P. L. 1977. Carboniferous and Permian palynostratigraphy in Australia and Antarctica: a review. *Bureau of Mines and Mineral Resources Journal of Australian Geology and Geophysics*, **2**, 177–208.

KNELLER, B., MILANA, J. P., BUCKEE, C. & AL JA'AIDI, O. 2004. A depositional record of deglaciation in a paleofjord (Late Carboniferous [Pennsylvanian] of San Juan Province, Argentina): the role of catastrophic sedimentation. *Geological Society of America Bulletin*, **116**, 348–367.

KONERT, G., ABDULKADER, M. A., AL-HAJRI, A. A. & DROSTE, H. J. 2001. Paleozoic stratigraphy and hydrocarbon habitat of the Arabian Plate. *GeoArabia*, **6**, 407–442.

KORTE, C., JASPER, T., KOZUR, H. W. & VEIZER, J. 2005. $\delta^{18}O$ and $\delta^{13}C$ of Permian brachiopods: A record of seawater evolution and continental glaciation. *Palaeogeography, Palaeoclimatology, Palaeoecology*, **224**, 333–351.

KYLE, R. A. 1977. Palynostratigraphy of the Victoria Group of south Victoria Land, Antarctica. *New Zealand Journal of Geology and Geophysics*, **20**, 1081–1102.

LARGHI, C. 2005. *Dickinsartella* fauna from the Saiwan Formation (Oman): a bivalve fauna testifying to the Late Sakmarian (Early Permian) climatic amelioration along the north-eastern Gondwanan fringe. *Rivista Italiana di Paleontologia e Stratigrafia*, **111**, 353–375.

LEONOVA, T. B. 1998. Permian ammonoids of Russia and Australia. *Proceedings of the Royal Society of Victoria*, **110**, 157–162.

LEVELL, B. K., BRAAKMAN, J. H. & RUTTEN, K. W. 1988. Oil-bearing sediments of Gondwana glaciation in Oman. *American Association of Petroleum Geologists Bulletin*, **72**, 775–796.

LINDSTRÖM, S. 1995. Early Permian palynostratigraphy of the Northern Heimefrontfjella Mountain Range, Dronning Maud Land, Antarctica. *Review of Palaeobotany and Palynology*, **89**, 359–415.

LÓPEZ-GAMUNDÍ, O. R. 1997. Glacial-postglacial transition in the late Paleozoic basins of southern South America. *In*: MARTINI, I. P. (ed.) *Late Glacial and Postglacial Environmental Changes*, OUP, Oxford 147–168.

LÓPEZ-GAMUNDÍ, O. R., LIMARINO, C. O. & CÉSARI, S. 1992. Late Paleozoic paleoclimatology of central western Argentina. *Palaeogeography, Palaeoclimatology, Palaeoecology*, **91**, 305–329.

LOWENSTAM, H. A. 1961. Mineralogy, $^{18}O/^{16}O$ ratios and strontium and magnesium contents of recent and fossil brachiopods and their bearing on the history of the oceans. *Journal of Geology*, **69**, 241–260.

LOWENSTAM, H. A. 1964. Palaeotemperatures of the Permian and Cretaceous Periods. *In*: NAIRN, A. E. M. (ed.) *Problems in Palaeoclimatology*, Interscience, London, 227–248.

MACRAE, C. S. 1988. Palynostratigraphical correlation between the Lower Karoo sequence of the Waterburg and Pafuri coal basins and the Hammanskraal plant macrofossil locality, RSA. *Memoirs of the Geological Survey of South Africa*, **75**, 1–217.

MARQUES-TOIGO, M. 1991. Palynobiostratigraphy of the southern Brasilian Neopalaeozoic Gondwana sequence. *Proceedings of the Seventh International Gondwana Symposium*, Instituto de Geosciencias, University of Sao Paulo, 503–516.

MARTINI, I. P. & BANKS, M. R. 1989. Sedimentology of the cold-climate, coal-bearing, Lower Permian, Lower Freshwater Sequence of Tasmania. *Sedimentary Geology*, **64**, 25–41.

MARTINI, I. P. & GLOOSCHENKO, W. A. 1985. Cold climate peat formation in Canada and its relevance to Lower Permian Coal Measures of Australia. *Earth Science Reviews*, **22**, 107–140.

MILLSTEED, B. D. 1999. Palynology of the Early Permian coal-bearing deposits near Vereeniging, Free State, South Africa. *Bulletin of the Council for Geoscience*, **124**, 1–77.

NYAMBE, I. A. & UTTING, J. 1997. Stratigraphy and palynostratigraphy, Karoo Supergroup (Permian and Triassic), mid-Zambezi Valley, southern Zambia. *Journal of African Earth Sciences*, **24**, 563–583.

OESTERLEN, P. & MILLSTEED, B. 1994. Lithostratigraphy palaeontology and sedimentary environments of the western Cabora Bassa Basin, lower Zambezi Valley, Zimbabwe. *South African Journal of Geology*, **97**, 205–224.

OSTERLOFF, P. L., AL-HARTHY, A., PENNEY, R. *ET AL.* 2004. Gharif and Khuff formations, subsurface Interior Oman. *In*: AL-HUSSEINI, M. (ed.) *Carboniferous, Permian and Early Triassic Arabian stratigraphy*, GeoArabia Special Publication **3**, Gulf PetroLink, Manama, Bahrain, 83–147.

PANT, D. D. 1996. The biogeography of the late Paleozoic floras of India. *Review of Palaeobotany and Palynology*, **90**, 79–98.

POTTER, P. E., FRANCA, A. B., SPENCER, C. W. & CAPUTO, M. V. 1995. Petroleum in glacially-related sandstones of Gondwana: a review. *Journal of Petroleum Geology*, **18**, 397–420.

POWIS, G. D. 1979. *Palynology of the Late Palaeozoic glacial sequence, Canning Basin, Western Australia*. Unpublished PhD. thesis, University of Western Australia.

POWIS, G. D. 1984. Palynostratigraphy of the Late Carboniferous sequence, Canning Basin, W. A. *In*: PURCELL, P. G. (ed.) *The Canning Basin, W. A.: Proceedings of the Geological Society of Australia/Petroleum Geological Society of Australia Symposium*. Perth, 1984, 429–438.

PRICE, P. L. 1983. A Permian palynostratigraphy for Queensland. *In*: FOSTER, C. B. (ed.) *Permian Geology of Queensland, Geological Society of Australia, Brisbane*, 155–211.

PRICE, P. L. 1997. Permian to Jurassic palynostratigraphic nomenclature of the Bowen and Surat basins.

In: GREEN, P. (ed.) *The Surat and Bowen Basins, Southeast Queensland*, Queensland Department of Mines and Energy, Brisbane, 137–178.

RUSSO, A., ARCHANGELSKY, S. & GAMERRO, J. C. 1980. Los depósitas supra Palaeozoicos en el subsuelo de la Llanura Chaco-Pampeana, Argentina. *Actas del II Congreso Argentino de Paleontología y Bioestratigrafía y I Congreso Latinoamericano de Paleontología 1978*, **4**, 157–173.

SANTOS, P. R., DOS ROCHA-CAMPOS, A. C. & CANUTO, J. R. 1996. Patterns of late Palaeozoic deglaciation in the Paraná Basin, Brazil. *Palaeogeography, Palaeoclimatology, Palaeoecology*, **125**, 165–184.

SASTRY, M. V. A. & SHAH, S. C. 1964. Permian marine transgression in peninsular India. *22nd International Geological Congress Proceedings*, **3**, 139–150.

SCHEFFLER, K., HOERNES, S. & SCHWARK, L. 2003. Global changes during Carboniferous–Permian glaciation of Gondwana: linking polar and equatorial climate evolution by chemical proxies. *Geology*, **31**, 605–608.

SHAH, S. C. & SASTRY, M. V. A. 1975. Significance of Early Permian marine faunas of Peninsular India. *In*: CAMPBELL, K. W. (ed.) *Gondwana Geology*. Australian National University Press, Canberra, 391–395.

SHARLAND, P. R., ARCHER, R., CASEY, D. M. *ET AL.* 2001. *Arabian Plate Sequence Stratigraphy*, GeoArabia Special Publication, Gulf Petrolink, Bahrain, **2**, 371pp.

SIMANAUSKAS, T. & CISTERNA, G. A. 2001. Braquiopodos articulados de la formacion El Paso, Paleozoico superior, Precordillera Argentina. *Revista Española de Paleontología*, **16**, 209–222.

SOUZA, P. A., PETRI, S. & DINO, R. 2003. Late Carboniferous palynology from the Itararé Subgroup (Parana Basin) at Araçoiba da Serra, São Paulo State, Brazil. *Palynology*, **27**, 39–74.

STEPHENSON, M. H. 1998. *Stratigraphic and systematic palynology of Permian and Permo-Carboniferous rocks of Oman and Saudi Arabia*. Unpublished PhD thesis, University of Sheffield.

STEPHENSON, M. H. & OSTERLOFF, P. L. 2002. Palynology of the deglaciation sequence represented by the Lower Permian Rahab and Lower Gharif members, Oman. *American Association of Stratigraphic Palynologists Contribution Series*, **40**, 1–32.

STEPHENSON, M. H., OSTERLOFF, P. L. & FILATOFF, J. 2003. Integrated palynological biozonation of the Permian of Saudi Arabia and Oman: progress and problems. *GeoArabia*, **8**, 467–496.

STEPHENSON, M. H., LENG, M. J., VANE, C. H., OSTERLOFF, P. L. & ARROWSMITH, C. 2005. Investigating the record of Permian climate change from argillaceous sediments, Oman. *Journal of the Geological Society, London*, **162**, 1–11.

STOLLHOFEN, H., STANISTREET, I. G., BANGERT, B. & GRILL, H. 2000. Tuffs, tectonism and glacially related sea-level changes, Carboniferous-Permian, southern Namibia. *Palaeogeography, Palaeoclimatology, Palaeoecology*, **161**, 127–150.

TABOR, N. J. & MONTAÑEZ, I. P. 2002. Shifts in late Paleozoic atmospheric circulation over western equatorial Pangea: insights from pedogenic mineral $\delta^{18}O$ compositions. *Geology*, **30**, 1127–1130.

TABOR, N. J. & MONTAÑEZ, I. P. 2005. Oxygen and hydrogen isotope compositions of Permian pedogenic phyllosilicates: Development of modern surface domain arrays and implications for palaeotemperature reconstruction. *Palaeogeography, Palaeoclimatology, Palaeoecology*, **223**, 127–146.

TAYLOR, T. N., TAYLOR, E. L. & CÚNEO, N. 1992. The present is not the key to the past: a polar forest from the Permian of Antarctica. *Science*, **257**, 1675–1677.

THERON, J. N. & BLIGNAULT, H. J. 1975. A model for the sedimentation of the Dwyka glacials in the southwestern Cape. *In*: CAMPBELL, K. W. (ed.) *Gondwana Geology*, Australian National University Press, Canberra, 347–356.

TIWARI, R. S. 1994. Palynoevent stratigraphy in Gondwana sequence of India. *Ninth International Gondwana Symposium*, Hyderabad, India, January 1994, 3–19.

TRUSWELL, E. M. 1980. Permo-Carboniferous palynology of Gondwanaland: progress and problems in the decade of 1980. *Bureau of Mines and Mineral Resources Journal of Australian Geology and Geophysics*, **5**, 95–111.

VEEVERS, J. J. & POWELL, C. MCA. 1987. Late Paleozoic glacial episodes in Gondwanaland reflected in transgressive-regressive depositional sequences in Euramerica. *Geological Society of America Bulletin*, **98**, 475–487.

VEEVERS, J. J. & TEWARI, R. C. 1995. Gondwana master basin of Peninsular India between Tethys and the interior of the Gondwanaland Province of Pangea: Boulder, Colorado, *Geological Society of America Memoir*, **187**, 72pp.

VEEVERS, J. J., POWELL, C. M., COLLINSON, J. W. & LÓPEZ-GAMUNDÍ, O. R. 1994. Synthesis. *In*: VEEVERS, J. J. & POWELL, C. M. (eds) *Permian-Triassic Pangean basins and foldbelts along the Panthalassan Margin of Gondwanaland*. Boulder, Colorado, Geological Society of America Memoir, **184**, 331–353.

VEIZER, J., BRUCKSCHEN, P. & PAWELLEK, F. *ET AL.* 1997. Oxygen isotope evolution of Phanerozoic seawater. *Palaeogeography, Palaeoclimatology, Palaeoecology*, **132**, 159–172.

VEIZER, J., ALA, D., AZMY, K. *ET AL.* 1999. $^{87}Sr/^{86}Sr$, $\delta^{13}C$ and $\delta^{18}O$ evolution of Phanerozoic seawater. *Chemical Geology*, **161**, 59–88.

VENKATACHALA, B. S. & TIWARI, R. S. 1987. Lower Gondwana marine incursions: periods and pathways. *Palaeobotanist*, **36**, 24–29.

VENKATACHALA, B. S., TIWARI, R. S. & VIJAYA, 1995. Diversification of spore-pollen character states in the Indian Permian. *Review of Palaeobotany and Palynology*, **85**, 319–340.

VERGEL, M. 1993. Palinoestratigrafía de la secuencia Neopalaeozoica de la Cuenca Chacoparanense, Argentina. *12th International Congress of the Stratigraphy and Geology of the Carboniferous and Permian, Buenos Aires 1993*, **1**, 201–212.

VIJAYA, 1994. Advent of Gondwanan deposition on Indian Peninsula: a palynological reflection and relationship. *Ninth International Gondwana Symposium, Hyderabad, India*, 283–298.

VIJAYA & RAM, AWATAR, 2001. Vertical distribution of spore and pollen index species in the Permian sequence on Peninsular India. *In*: WEISS, R. E. (ed.) *Contributions to geology and paleontology of Gondwana – in honour of Helmut Hopfner*, Geological Institute, University of Cologne, 475–495.

VISSER, J. N. J. & YOUNG, G. M. 1990. Major element geochemistry and paleoclimatology of the Permo–Carboniferous glacigene Dwyka Formation and post–glacial mudrocks in southern Africa. *Palaeogeography, Palaeoclimatology, Palaeoecology*, **81**, 49–57.

VISSER, J. N. J. 1995. Post-glacial Permian stratigraphy and geography of southern and central Africa: boundary conditions for climatic modelling. *Palaeogeography, Palaeoclimatology, Palaeoecology*, **118**, 213–243.

VISSER, J. N. J. 1996. Controls on Early Permian shelf deglaciation in the Karoo Basin of South Africa. *Palaeogeography, Palaeoclimatology, Palaeoecology*, **125**, 129–139.

VISSER, J. N. J. 1997. Deglaciation sequences in the Permo-Carboniferous Karoo and Kalahari basins of southern Africa: a tool in the analysis of cyclic glaciomarine basin fills. *Sedimentology*, **44**, 507–521.

WATERHOUSE, J. B. 1997. The Permian time-scale. *Permophiles*, **30**, 6–8.

WILSON, L. R. 1963. Teratological forms in pollen of *Pinus flexilis* James. *Journal of Palynology*, **1**, 106–110.

WOPFNER, H. 1999. The Early Permian deglaciation event between East Africa and northwestern Australia. *Journal of African Earth Sciences*, **29**, 77–90.

WOPFNER, H. & KREUSER, T. 1986. Evidence for Late Palaeozoic glaciation in southern Tanzania. *Palaeogeography, Palaeoclimatology, Palaeoecology*, **56**, 259–275.

WOPFNER, H. & DIEKMANN, B. 1996. The Late Palaeozoic Idusi Formation of southwest Tanzania: a record of change from glacial to postglacial conditions. *Journal of African Earth Sciences*, **22**, 575–595.

WOPFNER, H. & CASSHYAP, S. M. 1997. Transition from freezing to subtropical climates in the Permo-Carboniferous of Afro-Arabia and India. *In*: MARTINI, I. P. (ed.) *Late Glacial and Postglacial Environmental Changes*, OUP, Oxford, 192–212.

ZIEGLER, A. M., REES, P. MCA. & NAUGOLNYKH, S. V. 2002. The Early Permian floras of Prince Edward Island, Canada: differentiating global from local effects of climate change. *Canadian Journal of Earth Sciences*, **39**, 223–238.

Climate change across the Permian/Triassic boundary

R. J. TWITCHETT

School of Earth, Ocean and Environmental Sciences, University of Plymouth, Drake Circus, Plymouth, PL4 8AA, UK (e-mail: richard.twitchett@plymouth.ac.uk)

Abstract: The Permian–Triassic interval was a critical period of time for most of Earth's biosphere and witnessed the most severe extinction event of the Phanerozoic. Present evidence suggests that extraterrestrial influences had little, if anything, to do with these dramatic changes in the biota. Instead, evidence is growing that climate change was a major factor. Late Permian and Early Triassic glacial deposits are unknown and there is no geological evidence for even a brief episode of global cooling at this time. Studies of high latitude palaeosols and data from isotope analyses of biogenic and abiogenic marine carbonates suggest that the Permian–Triassic was a time of global warming, although some of the isotope data are derived from samples that have clearly been altered during diagenesis. Many authors have attributed Permian–Triassic temperature rise to a runaway greenhouse driven by elevated atmospheric CO_2 levels, causing temperature rise and breakdown of methane hydrates, which in turn led to methane release and oxidation, thus further increasing atmospheric CO_2. Apart from a negative $\delta^{13}C$ isotope excursion, which could have been the result of many other factors, no evidence exists for methane release. In some sections, the isotope shift post-dates the extinction level. There is excellent geological evidence for oceanographic changes such as ocean stagnation, widespread development of anoxia, loss of productivity and sea-level rise. These have all been attributed to rising global temperatures, a conclusion supported by results from recent climate simulation models. Re-analysis of the Sr isotope data demonstrates that seawater $^{87}Sr/^{86}Sr$ ratio varied little through the end-Permian event and immediate aftermath, although values were rising prior to the event and again from the late Induan. This may indicate a temporary decrease in riverine influx to the oceans, which may be climate-related. Overall, the increasing correspondence between computer-generated predictions of Permian–Triassic climate and geological evidence bodes well for future studies on the palaeoclimate of this critical interval in Earth history.

Rocks spanning the Permian–Triassic (P–Tr) interval contain a record of the most severe extinction event of the Phanerozoic and the subsequent recovery of global biodiversity (e.g. Erwin 1993; Benton 2003). One of the most important crises in the history of life on Earth, the end-Permian event is ranked first for both the magnitude of diversity loss and the severity of the ecological upheaval (McGhee *et al.* 2004). It is estimated that globally, in the marine realm, some 50% of families died out, which scales to a loss of some 80–96% of species. Terrestrial biodiversity loss was similar or slightly higher (Benton 1995). Studies on a local and regional scale tend to support the higher estimates of species-level extinction (Jin *et al.* 2000). It is generally accepted that two crises occurred in the latter part of the Permian: one near the Capitanian–Wuchiapingian boundary (termed the end-Guadalupian event) (e.g. Stanley & Yang 1994), dated at 260 Ma, and the major end-Permian event, which occurred near the end of the Changhsingian stage. This latter event has attracted most study and has been dated at 252.6 Ma (Mundil *et al.* 2004).

Most research focus has been on the larger of the two events, yet still there are many unresolved debates, especially concerning the fundamentals of timing and cause. Recent studies have concluded that the extinction event was very rapid indeed, taking on the order of 10^4 years (Rampino *et al.* 2000; Twitchett *et al.* 2001), although several animal groups declined in diversity through the latter 20 million years of the Permian. Other authors argue that each region suffered rapid extinction, coincident with the onset of oceanic anoxia, but that the onset of anoxia occurred in different regions at different times leading to a globally diachronous event spanning 1–2 million years (Wignall & Hallam 1992; Wignall & Newton 2003). However, the data marshalled in support of this view have been challenged (e.g. Retallack 2004), and the debate is still open. In a comprehensive review published 14 years ago, Erwin (1993) discussed the many potential causes of the end-Permian event, including sea-level change, flood basalt volcanism, climate change, anoxia, oceanic overturn and methane release, as well as possible extraterrestrial impact. Presently, the main debate revolves around the relative merits of extraterrestrial impact *vs.* global warming triggered by the continental flood basalt eruptions of the Siberian Traps (Benton & Twitchett 2003). Evidence for

From: WILLIAMS, M., HAYWOOD, A. M., GREGORY, F. J. & SCHMIDT, D. N. (eds) *Deep-Time Perspectives on Climate Change: Marrying the Signal from Computer Models and Biological Proxies.* The Micropalaeontological Society, Special Publications. The Geological Society, London, 191–200.
1747-602X/07/$15.00 © The Micropalaeontological Society 2007.

impact has been recently presented (e.g. Becker *et al.* 2001, 2004), but the quality, inferences and interpretations of these data have been heavily criticized (e.g. Renne *et al.* 2004). Presently, the most widely accepted model for the end-Permian extinction is a chain of events following repeated eruptions of the Siberian Trap basalts over a period of perhaps 500,000 years that culminated in a runaway greenhouse effect (e.g. Wignall 2001; Kidder & Worsley 2004).

Recovery from the extinction took time. Global marine diversity, at the family level, apparently took some 100 million years to reach pre-extinction levels (e.g. Benton 1995), although ecological recovery was somewhat quicker, with complex structures such as reefs reappearing in the Middle Triassic (some 10–15 million years after the event). Initial post-extinction ecosystems are characterized by low diversity, low complexity assemblages of abundant but small-sized organisms, often dominated by one or two taxa (Schubert & Bottjer 1995; Twitchett 1999), interpreted to be the result of environmental stress. Recent work on the marine record has supported earlier hypotheses that continuing environmental stress, specifically the presence of marine anoxia, was important in preventing rapid recovery (Twitchett *et al.* 2004).

Understanding P–Tr climate change is important for several reasons. First, knowledge of the climate changes that were associated with past episodes of major extinction may be critical for placing the present-day biotic crisis and climate change in historical perspective (cf. Smith 2001). Secondly, climate change (specifically rapid global warming) is presently implicated as one of the leading causes of the P–Tr extinction event.

Evidence of global cooling?

In the late 1980s, Stanley (1988) invoked global cooling in order to explain the loss of the low palaeolatitude reef ecosystems, and further postulated that cooling was a common cause of all Phanerozoic extinctions. His hypothesis that severe global cooling occurred at the P/Tr boundary and caused the extinction event was largely discredited after revisions in the stratigraphic position of the glacial deposits that provided the core of the evidence. These are now considered to be no younger than Middle Permian in age (Erwin 1993).

However, despite the fact that no glacial deposits are known from the latest Permian Changhsingian stage, a few authors continue to invoke cooling as a (or the) cause of the extinction.

For example, Kozur (1998) suggested that the worldwide spread of 'cold-water' conodont taxa (e.g. *Clarkina carinata*) is proof of cooling during the P–Tr interval. Unfortunately, there is no evidence that the taxa identified by Kozur really were cold-water forms, and no oxygen isotope measurements have been published for these taxa. Recently, Wignall (2001) postulated that the Siberian Trap volcanism may have led to a brief cooling episode, possibly with a short sharp glaciation, prior to the onset of a runaway greenhouse. In the absence of glacial deposits, the only possible evidence of this supposed cooling in the rock record is the (eustatic?) sea-level fall recorded in most P–Tr sections. However, as pointed out by Wignall (2001) and others, the extinction horizon in most sections occurs during the subsequent transgression and so even if sea-level fall was caused by cooling (which has not been demonstrated), it seems highly unlikely that cooling caused the mass extinction.

Permian–Triassic climate change: the present model

The leading present hypothesis is that rapid and severe global warming was responsible for the end-Permian mass extinction event (e.g. Benton & Twitchett 2003; Kidder & Worsley 2004). This model (Fig. 1) has evolved since the early 1990s, and incorporates a number of potential extinction mechanisms that were at one time (Erwin 1993) considered to be independent, though not necessarily mutually exclusive, possibilities.

The present hypothesis is that the Siberian Trap flood basalt eruptions vented large amounts of CO_2 into the atmosphere over a relatively short period of time. This resulted in rising global temperatures. Warming then led to the destabilization and disassociation of shallow (marine and/or terrestrial) gas hydrate deposits, which vented large volumes of CH_4 into the oceans and atmosphere. This CH_4, although rapidly oxidized to CO_2, then caused more warming, which in turn would have caused the dissociation of further gas hydrate reservoirs. During this positive feedback loop, some sort of threshold was probably reached, beyond which the natural systems that normally reduce carbon dioxide levels could not operate and a 'runaway greenhouse' ensued. Global warming and elevated atmospheric CO_2 levels would have had devastating effects on terrestrial ecosystems, and also in the marine realm, where it is believed to have caused a rise in sea level, stagnation, oceanic anoxia, possible acidification and a decrease in surface primary productivity.

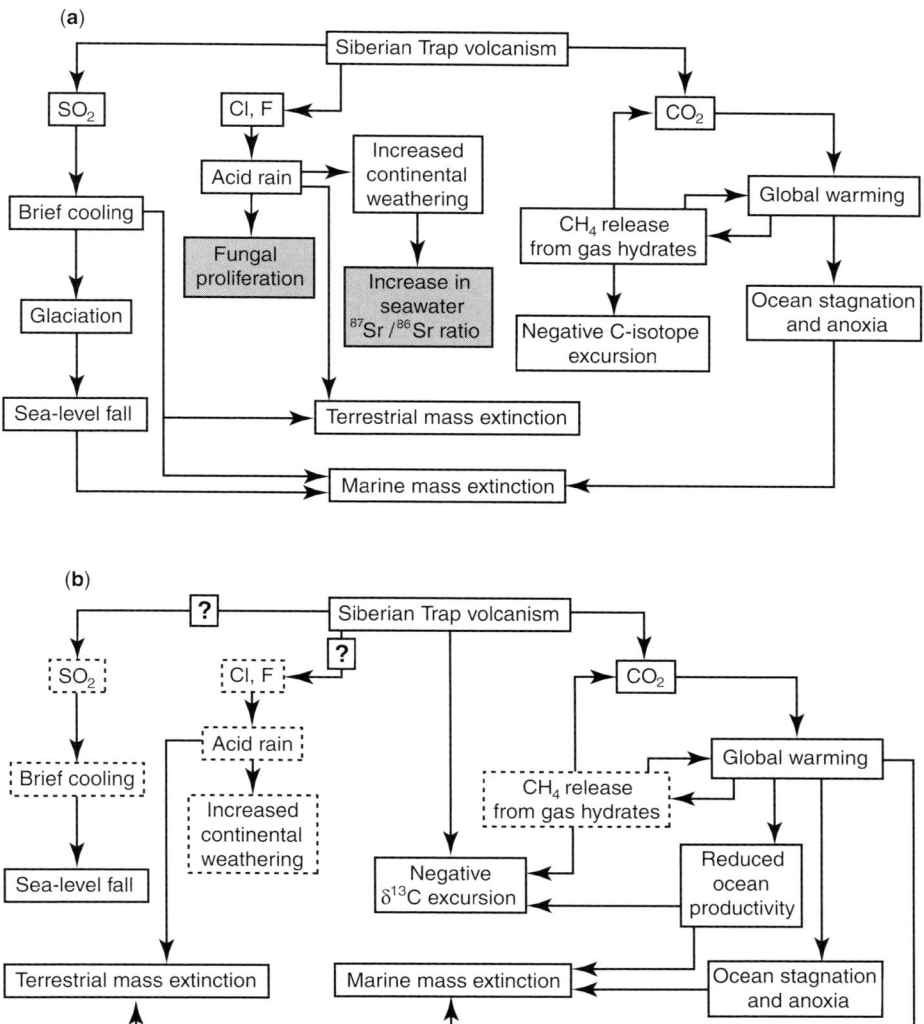

Fig. 1. Flood basalt-induced climate change and the end-Permian extinction event. (**a**) Model proposed by Wignall (2001). Grey boxes indicate those aspects which have since been shown to be incorrect, namely the supposed fungal spike (see Foster *et al.* 2002) and the increase in seawater $^{87}Sr/^{86}Sr$ ratio (see text). (**b**) Refined model from discussions in the text. Boxes with dashed lines are those for which there is no supporting geological evidence, or for which the evidence presented is considered weak or equivocal. Arrows arising from boxes with dashed lines should be considered dubious. Boxes with solid lines are those for which there is reasonable to good supporting geological evidence. Question marks indicate arrows arising from boxes with solid lines that represent unsupported inferences. See text for details.

However, despite the present widespread acceptance of this model, it is worth considering the veracity of some of the evidence cited in support. In particular, the network of cause-and-effect links proposed by e.g. Wignall (2001, p. 13) are in need of objective re-analysis (Fig. 1b).

Evidence of Permian–Triassic warming

$\delta^{18}O$ record of marine carbonates

In the late 1980s, the first detailed multidisciplinary study of a completely cored P/Tr boundary section (the Gartnerkofel-1 core of southern Austria) was conducted. Holser *et al.* (1989) analysed the oxygen isotopes from bulk carbonate samples and demonstrated the presence of a large decrease in the proportion of heavy oxygen isotopes (δ^{18}O) in carbonates spanning the P/Tr boundary. Values declined from about $-0.5\%_o$ prior to the extinction horizon, *c.* $-4\%_o$ around the extinction level, through to a maximum of about $-8\%_o$ in the earliest Triassic. The authors interpreted these data as indicating a 5–6 °C increase in temperature. A similar δ^{18}O shift has been observed at numerous other P/Tr boundary sections, and the quoted figure of 5–6 °C is often repeated in the literature (e.g. Erwin 1993; Benton 2003).

However, interpretation of the oxygen isotope record is problematic as isotope values from bulk carbonates are very sensitive to alteration during burial and diagenesis. Certainly, the limestones of Gartnerkofel-1 have been heavily recrystallized, especially around the boundary interval (contra Holser *et al.* 1989), and thus the oxygen isotope data should be viewed with caution. Despite these potential problems, Kidder and Worsley (2004) noted that the temperature rise inferred by Holser *et al.* (1989) fits precisely with their own estimate based on an assumed fourfold increase in atmospheric CO_2 (although this does not constitute real evidence that either or both inferences are correct).

It would be useful to have oxygen isotope data from independent archives. Biogenic marine carbonate is one potential, but hitherto underutilized, resource. Although one of the hardest-hit groups during the end-Permian crisis, articulate brachiopods span the extinction horizon in many sections. Oxygen isotope data from P–Tr articulate brachiopod shells from Spitsbergen have been published (Gruszczynski *et al.* 1989), but re-analysis of the samples showed evidence of diagenetic alteration (Mii *et al.* 1997) and so they should be viewed with caution. No other P–Tr articulate brachiopod archives have yet been studied. One of the most common brachiopods in the immediate aftermath of the event is the inarticulate *Lingula* (e.g. Rodland & Bottjer 2001). Recently, Rodland *et al.* (2003) have demonstrated, using modern and Early Triassic *Lingula*, that the oxygen isotopes of the organophosphate portion of the lingulid shell are highly variable, and do not reflect secretion in equilibrium with seawater. However, the carbonate fraction of modern *Lingula* shell apparently is secreted in equilibrium with seawater (Rodland *et al.* 2003). Analyses of the carbonate fraction of P–Tr lingulids may therefore provide a reliable palaeotemperature proxy.

Another potential archive of palaeotemperature data that has yet to be explored is the P–Tr conodont record. Conodonts are present in all marine P–Tr sections, are the biostratigraphic tool of choice for P–Tr global correlations and are even used to define the base of the Triassic (e.g. Yin *et al.* 2001). Several studies have shown that conodont elements can preserve palaeoclimate data and it is likely that future studies will exploit this presently underutilized resource.

Analysis of fossil soils

Early Triassic fossil soils (palaeosols) of high southerly palaeolatitudes (up to 85° S), which were analysed by Retallack (1999), record morphological characteristics that, at the present day, typify low temperate latitudes, and which formed under warmer conditions than soils of the Late Permian from the same localities. This may be qualitative evidence of global warming. However, as Pangaea as a whole was drifting northwards through the P–Tr interval, any locality in the Southern Hemisphere might be expected to show evidence of a warming trend even if global climate remained unaltered, and until the results are replicated by Northern Hemisphere data they should be viewed with a degree of caution.

The analysis of isotope changes in soil carbonates, which form in direct contact with the atmosphere, may provide further evidence of temperature change during the P–Tr interval, provided they are diagenetically unaltered. Ghosh *et al.* (2001) inferred a temperature rise of 5 °C from the early Guadalupian to the Middle Triassic following analyses of soil carbonates from the Satpura Basin of central India, but the time resolution is far too coarse to shed any light on temperature changes of the extinction–recovery interval. Pedogenic carbonate nodules from Karoo are believed to have been altered by diagenesis and so the higher resolution oxygen isotope record from South Africa does not preserve a primary signal (MacLeod *et al.* 2000).

Evidence for elevated CO_2 levels

If the evidence for warming across the P–Tr event is accepted, then one might expect evidence for a rapid and dramatic rise in greenhouse gases in the palaeo-atmosphere. However, such evidence remains equivocal. The best evidence provided for an increase in CO_2 derives from analysis of the stomatal density of plant cuticles by Retallack (2001), although the time resolution is somewhat coarse. Certainly, the Early Permian to Late Triassic interval has been considered by many authors (e.g. Berner 1997) to have been a time of rising atmospheric CO_2 levels, based on geochemical modelling. This hypothesis is supported by estimates of palaeo-pCO_2 based on analyses of the isotopic compositions of pedogenic carbonates (e.g. Ghosh *et al.* 2001). Studies of pedogenic carbonates from Karoo have yielded a high-resolution dataset of carbon isotope measurements (MacLeod *et al.* 2000; Ward *et al.* 2005), but the values have never been converted to palaeo-pCO_2 estimates and there are some concerns about whether these data reflect primary values (MacLeod *et al.* 2000).

Evidence for gas hydrate dissociation and CH_4 release

Regarding methane venting, the most often-cited piece of 'evidence' for this cornerstone of the P–Tr global warming model is the large negative excursion in $\delta^{13}C$ recorded in shallow marine carbonates (e.g. Erwin 1993; Kidder & Worsley 2004). The *c.* 3–4‰ negative shift in $\delta^{13}C$ (locally up to 8‰) appears to be too large to be explained by any other mechanism, such as volcanic emissions, and methane venting from gas hydrate deposits is regarded as the only viable alternative (e.g. Erwin 1993). A negative shift is recorded in marine carbonates, terrestrial soil carbonates (e.g. Retallack 2001), bulk organic matter (e.g. Twitchett *et al.* 2001), and biomarker molecules (Grice *et al.* 2005) and so appears to reflect a real atmospheric change.

However, some caution is required as large negative shifts could have many other causes. A simple productivity crash may be all that is required to produce a sharp negative shift in marine carbonates (e.g. Kump 1991). The P–Tr negative shift in bulk organic matter is likely due to a change in organic matter source, such as a reduction in the relative contribution of material from higher plants (Foster *et al.* 1997). The negative shift in individual biomarker molecules may likewise reflect an undetected change in source. Volcanically vented CO_2 is also enriched in ^{12}C, with the Siberian Traps and other volcanic centres, such as South China, providing potential sources. At the conclusion of a recent comprehensive study, Berner (2002) noted that it is not possible to reject all of these other possible factors, and the end-Permian negative shift in $\delta^{13}C$ was likely driven by methane release associated with mass mortality and volcanic degassing. Thus, methane venting is but one possible explanation for the observed negative shift in $\delta^{13}C$. One should not regard the negative shift as unequivocal evidence of methane flux to the atmosphere. Until independent evidence for methane venting is found, the methane-induced runaway part of the 'runaway greenhouse' model will remain open to question. In addition, geochemical modelling suggests that oxidation of the released methane would not in any case have produced enough CO_2 to trigger catastrophic warming (Berner 2002). Finally, in some localities the $\delta^{13}C$ shift clearly occurred after the extinction crisis (Twitchett *et al.*

2001), and therefore was more probably related to the consequences, rather than cause(s), of the catastrophe.

Evidence of warming-related kill mechanisms?

The P/Tr warming episode is supposed to have resulted in marine extinction by causing sea-level rise, productivity crash, oceanic stagnation and anoxia (e.g. Wignall 2001; Benton & Twitchett 2003) (Fig. 1). In addition, temperature rise itself, if high enough, could be lethal, especially in shallow water at low palaeolatitudes (Kidder & Worsley 2004). Geological evidence for these environmental changes is good, and data from modelling experiments suggest that warming could have provided a trigger. For example, evidence from the most complete sections indicates that disappearance of the Permian taxa occurred during a time of global sea-level rise (e.g. Wignall *et al.* 1996). Modelling results indicate that an average whole-ocean temperature rise of 15 °C could raise sea level by approximately 20 m simply through thermal expansion (Kidder & Worsley 2004).

Regarding ocean stagnation and anoxia, a substantial body of data has accumulated to show that marine ecosystems of the Early Triassic, even those in very shallow water (above storm wave base), were less well oxygenated than during Late Permian, pre-extinction times. Evidence derives from a variety of independent sources such as facies analysis, trace fossil studies, palaeoecology, geochemical data, isotopic analyses, and biomarker distributions (e.g. Wignall & Hallam 1992; Wignall & Twitchett 1996, 2002; Twitchett 1999; Grice *et al.* 2005). The deepest parts of the world's oceans were apparently oxygen-restricted from the latest Changhsingian to the Middle Triassic: the 'Superanoxic Event' of Isozaki (1997). During the early Induan (Griesbachian), most shelf settings experienced episodic development of euxinic conditions, comparable to the present-day Black Sea (e.g. Grice *et al.* 2005). These euxinic intervals alternated with intervals of slightly elevated, but still subnormal, oxygen concentrations, allowing a limited, depauperate benthos of typically small-sized organisms to colonize (e.g. Twitchett 1999). Only the shallowest settings of Neotethys appeared to have escaped, at least temporarily (e.g. Krystyn *et al.* 2003). Following this early Induan peak in oxygen-poor conditions, oxygenation of the marine shelf improved somewhat, and only the deeper basins remained oxygen-restricted (Wignall & Twitchett 2002). Individual regions seem to have responded differently and both re-oxygenation and ecological recovery were diachronous globally (e.g. Twitchett *et al.* 2004).

Although one contributory factor to the development of widespread anoxic and euxinic conditions of the P–Tr interval was the low atmospheric oxygen levels at this time (e.g. Berner 2005), climate also played a role. Wignall & Twitchett (1996) inferred that global warming and a reduction in the pole-to-equator temperature gradient would result in ocean stagnation and development of oxygen-poor bottom waters. Computer simulations of the oceans and climate of the Late Permian have confirmed that this is plausible (e.g. Hotinski *et al.* 2001). Finally, warmer water also holds less dissolved oxygen than cooler water, which, coupled with low atmospheric oxygen levels, explains the development of oxygen-poor conditions above storm wave base in equatorial regions (Wignall & Twitchett 1996). Thus, evidence for ocean stagnation and anoxia is excellent, and it is a plausible consequence of global warming. Although it clearly affected post-extinction recovery (Twitchett *et al.* 2004), whether it contributed directly to the extinction is still debated (e.g. Retallack 2004).

Regarding productivity crash, the geological evidence is more circumstantial. Given that most shelf settings of the Early Triassic experienced low oxygen conditions, which should promote the preservation of organic matter, the total organic carbon (TOC) content of nearly all Lower Triassic shelf sedimentary rocks, both carbonates and siliciclastics, is surprisingly low (Twitchett 2001). Only one, very localized, Lower Triassic petroleum source rock is known: the basal Kockatea Shale of the Perth basin, Western Australia (Grice *et al.* 2005). In most cases, both low palaeolatitude carbonates and mid-palaeolatitude siliciclastics (e.g. East Greenland), TOC content actually decreases from the oxygenated, well-bioturbated Changhsingian sediments to the overlying, unbioturbated Griesbachian sediments (Twitchett *et al.* 2001); difficult to explain without invoking a decline in productivity levels. As surface productivity relies on efficient nutrient recycling, which itself depends on ocean circulation, the sluggish warm water oceans of the Early Triassic would be expected to support much lower levels of primary production (Wignall & Twitchett 1996; Kidder & Worsley 2004). Other factors that may potentially have affected oceanic productivity include elevated atmospheric CO_2, which may have lowered the pH of surface waters. However, no direct evidence of this exists. Finally, the negative shift in $\delta^{13}C$ of marine carbonates may have been caused, at least in part, by a collapse in primary productivity (Kump 1991; Berner 2002).

In contrast, high original TOC levels have been inferred for sedimentary rocks of the Panthalassa deep ocean floor, preserved in the accreted terranes of Japan (e.g. Kakuwa 1996). This may reflect a real difference between productivity in the shelf seas and the open ocean; i.e. the shelf seas mostly experienced a drop in productivity, the open oceans a rise. However, there is a significant lithological change in the terrane sequences, from cherts below the P/Tr boundary to claystones above. Thus, the apparent rise in TOC may simply be due to the disappearance of the chert, which would otherwise dilute the organic matter in the rocks, and enhanced preservation of organic matter on the anoxic ocean floor (Isozaki 1997). Lack of chert resulted from near-total disappearance of the radiolaria. Clearly, productivity in the Permian–Triassic marine realm was patchy, as it is at the present day, and while most sections record a local decrease in TOC, others apparently do not.

Evidence for increased seawater $^{87}Sr/^{86}Sr$ ratios?

It has long been recognized that the $^{87}Sr/^{86}Sr$ ratio of seawater decreased through the Middle Permian, to a Phanerozoic minimum in the early–middle Lopingian. Using data from conodont elements, Martin & Macdougall (1995) suggested that the high rate of decrease is only explicable through decreased river input into the world's oceans, which they postulated was the result of the formation of Pangaea and an increase in internal drainage. Subsequently, the $^{87}Sr/^{86}Sr$ ratio increased dramatically through the remaining Late Permian and into the Middle Triassic.

This rise in seawater $^{87}Sr/^{86}Sr$ ratio began long before the extinction horizon and P/Tr boundary, and continued long after. Thus, contrary to the suggestions of many authors (e.g. Erwin 1993; Wignall 2001; Korte *et al.* 2003), it cannot have been due to increased continental weathering triggered by Siberian Trap volcanism or some other effect of the extinction event, such as the supposed widespread loss of terrestrial vegetation (e.g. Ward 2000 and Ward *et al.* 2005). Indeed, the Palaeozoic–Mesozoic rise in seawater $^{87}Sr/^{86}Sr$ ratio was entirely independent of the causes and/or consequences of the end-Permian extinction event. Following analyses of neodymium isotopes over the same period, Martin & Macdougall (1995) concluded that it was the result of increases in both the riverine flux to the oceans and in the $^{87}Sr/^{86}Sr$ ratio of river water over this period. Korte *et al.* (2003) reached similar conclusions and inferred episodes of global humidity as an explanation for increasing river flux to the oceans.

The long-term Late Permian to Middle Triassic trend of rising seawater $^{87}Sr/^{86}Sr$ revealed by large-scale studies has been assumed to continue uninterrupted across the P/Tr boundary and crisis level (Martin & Macdougall 1995; Korte *et al.* 2003). However, when individual data are plotted at the (conodont) zonal resolution, supplemented by data from other sources, a different picture emerges (Fig. 2). The Late Permian to Middle Triassic trend of rising ^{87}Sr/^{86}Sr ratios is temporarily interrupted from around the extinction horizon (latest Changhsingian, *chanxingensis* Zone) until the latest Griesbachian *carinata* Zone, whereupon ratios begin to rise once more (Fig. 2). This brief apparent interruption to the long-term rise is not simply an artefact of the duration of conodont zones near the boundary, or the nature of the

plot in Figure 2: the span of time represented is approximately a million years (e.g. Jin *et al.* 2000). Taken against a background of rising seawater ^{87}Sr/^{86}Sr, this interval of little change, or even slight decline, around the P–Tr interval implies a temporary increase in mid-ocean ridge activity, a decline in riverine flux to the oceans, a decline in the ^{87}Sr/^{86}Sr ratio of river water or a combination of any of these.

Geological evidence exists for each of these possibilities. Rifting events occurred in many regions during the P–Tr interval and the presence of a major marine transgression suggests that the mid-ocean ridge systems were very active during this period. Reduction of the strontium isotope

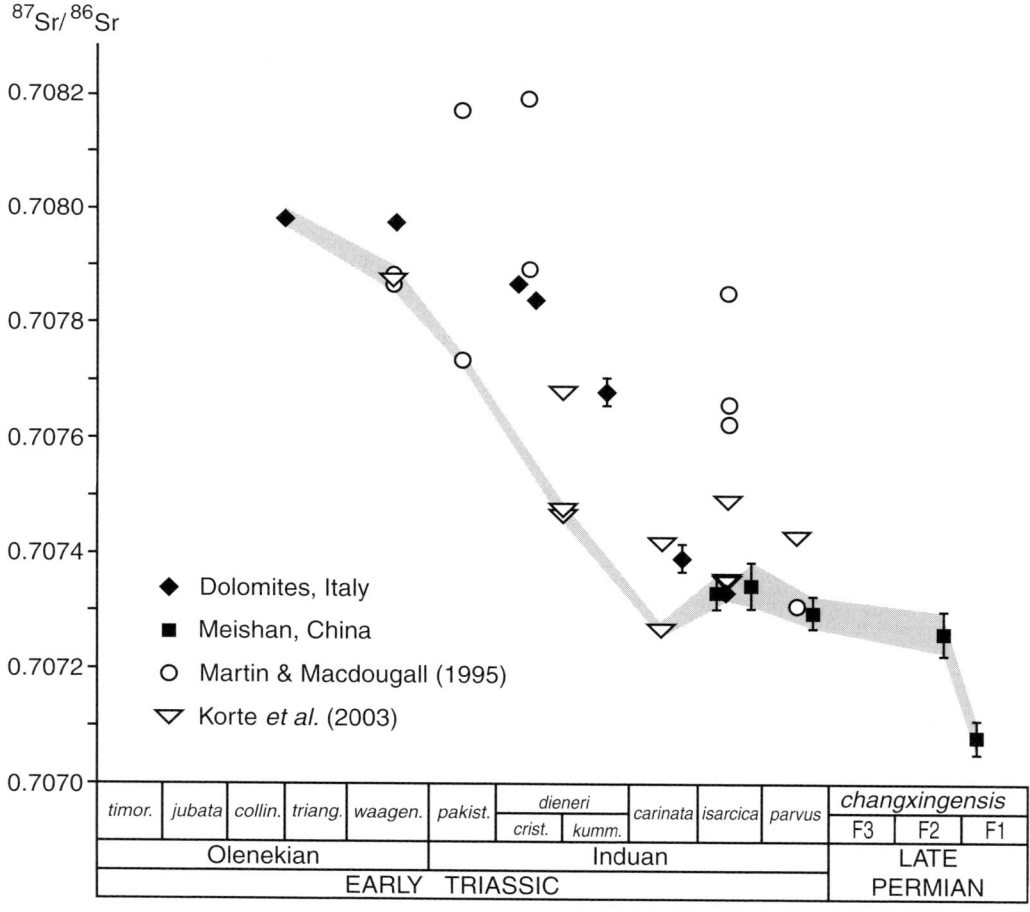

Fig. 2. Late Permian and Early Triassic seawater ^{87}Sr/^{86}Sr data from analyses of conodont elements and plotted at a zone-by-zone scale. Open symbols are from previously published data. Solid symbols are unpublished data (from Twitchett 1997). Grey shaded area highlights the lowest ^{87}Sr/^{86}Sr ratios for each conodont zone (i.e. those data least affected by diagenesis). Note that through the P/Tr boundary (from the *changxingensis* Zone to the *carinata* Zone), values do not rise, but remain more-or-less constant. Samples from widely separate localities show similar values, suggesting that the pattern truly reflects a global seawater signal.

ratio of this river water may be the result of the presence of evaporites or limestones in the catchment area. The global Upper Permian rock record comprises numerous evaporite deposits (Zharkov 1981), and weathering of these during the latest Permian to earliest Triassic regression and transgression may have greatly increased the amount of low radiogenic strontium entering the world's oceans. Sedimentary and palaeobotanical evidence of widespread aridity around the P/Tr boundary (e.g. Smith 1995) may have resulted in a reduction in overall river flux to the oceans. Finally, it is perhaps noteworthy that the duration of this temporary break in the long term rise of seawater $^{87}Sr/^{86}Sr$ also corresponds to the peak of the P–Tr Superanoxic Event, when the world's oceans were most stratified (e.g Isozaki 1997; Wignall & Twitchett 2002). Clearly, further high-resolution analyses of the P–Tr $^{87}Sr/^{86}Sr$ record are required to fully resolve these issues.

The published $^{87}Sr/^{86}Sr$ data appear at odds with the suggestions, from sedimentological evidence (e.g. Newell et al. 1999; Ward et al. 2000) and biomarkers (Sephton et al. 2005) that there was a peak in run-off around the extinction horizon caused by the supposed catastrophic loss of vegetation at this time. Clearly, the $^{87}Sr/^{86}Sr$ data do not support this hypothesis; increased riverine flux to the oceans began earlier (Martin & Macdougall 1995) and there is no additional peak near the extinction horizon. Although more work is needed to resolve this, one possibility is that the P–Tr sedimentological record, showing a change in depositional style to more braided rivers carrying a coarser load (e.g. Newell et al. 1999), reflects a more seasonal rainfall pattern under semi-arid conditions but not a sudden increase in overall (annual) riverine flux to the oceans. The apparent catastrophic increase in soil products entering the shelf sea, from biomarker evidence from northern Italy (Sephton et al. 2005), is perhaps not unexpected in very proximal settings at a time of major sea-level transgression and does not represent a sudden increase in runoff.

Climate models and simulations

As technology has advanced and climate models have become more sophisticated, there has been an improvement in agreement between simulations of the latest Permian–earliest Triassic palaeoclimate and the inferences made from analyses of the rock and fossil records. One of the early conclusions from worldwide studies of independent facies, fossil and geochemical evidence was that the P–Tr interval was characterized by widespread and long-lasting low oxygen conditions (e.g. Wignall & Twitchett 1996; Isozaki 1997). Most authors interpreted the inferred anoxia and euxinia as being the

result of global warming leading to sluggish ocean circulation. However, it was not until the work of Hotinski et al. (2001) that a computer simulation was able to recreate low oxygen conditions in the late Permian oceans using temperature rise as a simple forcing mechanism.

The most recent and sophisticated computer simulation, that employs a linked ocean and atmosphere model, gives the closest agreement yet to the empirical geological and palaeontolgical data. The simulation of Kiehl & Shields (2005) used an atmospheric CO_2 concentration of 10×present-day levels and confirmed Hotinski et al.'s (2001) conclusions that ocean circulation would be greatly curtailed. In addition, the Kiehl & Shields (2005) simulation reproduced ice-free poles (high latitude sea-surface temperatures of 8 °C), which is in agreement with palaeontological data for warm high latitudes (e.g. Beauchamp & Baud 2002), and which contrasts strongly with the results of early climate models that predicted high latitude continental ice (e.g. Golonka et al. 1994). The modelled palaeotropical sea-surface temperatures were only some 2 degrees higher than present day, whereas the modelled palaeoequatorial land temperatures reached extreme highs of 51 °C. As further geological data are collected, using presently untapped palaeotemperature archives such as Northern Hemisphere P–Tr palaeosols and conodont and brachiopod oxygen isotopes, the predictions of this latest model can be further tested.

Conclusions

The Permian–Triassic interval is characterized by rising global temperatures that were possibly triggered by CO_2 release from the Siberian Trap flood basalt eruptions, although atmospheric CO_2 had been rising throughout the latter part of the Permian and may simply have reached a critical threshold at this time. This warming event is associated with, and probably caused, sea-level rise, a reduction in ocean circulation and the expansion of low oxygen conditions globally and to very shallow depths. The development of shallow water anoxia was aided by low atmospheric concentrations of O_2. Early Triassic seas were not only sluggish, but also probably supported very low levels of surface productivity. Global warming, ocean stagnation and loss of productivity probably contributed to the major biotic crisis of the latest Permian. No geological evidence exists for an episode of global cooling at this time. Several aspects of the volcanically triggered runaway greenhouse model for the P–Tr interval (e.g. Wignall 2001) do not stand up to objective analysis nor are they supported by empirical geological data. In particular, there is a lack of independent evidence of methane release from gas hydrates. Also, seawater $^{87}Sr/^{86}Sr$ ratios did not rise through the

P/Tr boundary interval as has commonly been assumed, and rather than reflect widespread continental weathering and runoff more probably indicate a brief episode of decreased river influx to the world's oceans at this time (a sign of global aridity?). Future high-resolution analyses of palaeoclimate archives such as soil carbonates, conodont elements and brachiopod shells are likely to improve our understanding of P–Tr climate change.

The author thanks P. Wignall and M. Benton for their critical reviews.

References

BEAUCHAMP, B. & BAUD, A. 2002. Growth and demise of Permian biogenic chert along northwest Pangea: evidence for end-Permian collapse of thermohaline circulation. *Palaeogeography, Palaeoclimatology, Palaeoecology*, **184**, 37–63.

BECKER, L., POREDA, R. J., HUNT, A. G., BUNCH, T. E. & RAMPINO, M. 2001. Impact event at the Permian–Triassic boundary: evidence from extraterrestrial noble gases in fullerenes. *Science*, **291**, 1530–1533.

BECKER, L., POREDA, R. J., BASU, A. R., POPE, K. O., HARRISON, T. M., NICHOLSON, C. & IASKY, R. 2004. Bedout: a possible end-Permian impact crater offshore of northwestern Australia. *Science*, **304**, 1469–1476.

BENTON, M. J. 1995. Diversification and extinction in the history of life. *Science*, **268**, 52–58.

BENTON, M. J. 2003. *When Life Nearly Died: the Greatest Mass Extinction of All Time*. Thames & Hudson, London, 336pp.

BENTON, M. J. & TWITCHETT, R. J. 2003. How to kill (almost) all life: the end-Permian extinction event. *Trends in Ecology and Evolution*, **18**, 358–365.

BERNER, R. A. 1997. Paleoclimate – the rise of plants and their effect on weathering and atmospheric CO_2. *Science*, **276**, 544–546.

BERNER, R. A. 2002. Examination of hypotheses for the Permo-Triassic boundary extinction by carbon cycle modelling. *Proceedings of the National Academy of Sciences*, **99**, 4172–4177.

BERNER, R. A. 2005. The Carbon and Sulfur Cycles and atmospheric oxygen from middle Permian to middle Triassic. *Geochimica et Cosmochimica Acta*, **69**, 3211–3217.

ERWIN, D. H. 1993. *The great Paleozoic crisis: life and death in the Permian*. Columbia University Press, New York.

FOSTER, C. B., LOGAN, G. A., SUMMONS, R. E., GORTER, J. D. & EDWARDS, D. S. 1997. Carbon isotopes, kerogen types and the Permian–Triassic boundary in Australia: implications for exploration. *Australian Petroleum Production and Exploration Association Journal*, **37**, 472–489.

FOSTER, C. B., STEPHENSON, M. H., MARSHALL, C., LOGAN, G. A. & GREENWOOD, P. F. 2002. Revision of *Reduviasporonites* Wilson 1962: description, illustration, comparison and biological affinities. *Palynology*, **26**, 35–58.

GHOSH, P., GHOSH, P. & BHATTACHARYA, S. K. 2001. CO_2 levels in the Late Palaeozoic and Mesozoic atmosphere from soil carbonate and organic matter, Satpura Basin, central India. *Palaeogeography, Palaeoclimatology, Palaeoecology*, **170**, 219–236.

GOLONKA, J., ROSS, M. I. & SCOTESE, C. R. 1994. Phanerozoic paleogeographic and paleoclimatic modeling maps. *In*: EMBRY, A. F., BEAUCHAMP, B. & GLASS, D. J. *Pangea: Global Environments and Resources*, Canadian Society of Petroleum Geologists, Memoir **17**, pp. 1–47.

GRICE, K., CAO, C., LOVE, G. D. *ET AL.* 2005. Photic zone euxinia during the Permian–Triassic Superanoxic Event. *Science*, **307**, 706–709.

GRUSZCZYNSKI, M., HALAS, S., HOFFMAN, A. & MALKOWSKI, K. 1989. A brachiopod calcite record of the oceanic carbon and oxygen isotope shifts at the Permian/Triassic transition. *Nature*, **337**, 64–68.

HOLSER, W. P., SCHÖNLAUB, H. P., ATTREP, M., BOEKELMANN, K., KLEIN, P., MAGARITZ, M. & ORTH, C. J. 1989. A unique geochemical record at the Permian–Triassic boundary. *Nature*, **337**, 39–44.

HOTINSKI, R. M., BICE, K. L., KUMP, L. R., NAJJAR, R. G. & ARTHUR, M. A. 2001. Ocean stagnation and end-Permian anoxia. *Geology*, **29**, 7–10.

ISOZAKI, Y. 1997. Permo-Triassic boundary superanoxia and stratified superocean: records from lost deep sea. *Science*, **276**, 235–238.

JIN, Y. G., WANG, Y., WANG, W. & ERWIN, D. H. 2000. Pattern of marine mass extinction near the Permian–Triassic boundary in South China. *Science*, **289**, 432–436.

KAKUWA, Y. 1996. The Permian–Triassic mass extinction event recorded in bedded chert sequences in southwest Japan. *Palaeogeography, Palaeoclimatology, Palaeoecology*, **121**, 35–51.

KIDDER, D. L. & WORSLEY, T. R. 2004. Causes and consequences of extreme Permo–Triassic warming to globally equable climate and relation to the Permo–Triassic extinction and recovery. *Palaeogeography, Palaeoclimatology, Palaeoecology*, **203**, 207–237.

KIEHL, J. T. & SHIELDS, C. A. 2005. Climate simulation of the latest Permian: implications for mass extinction. *Geology*, **33**, 757–760.

KORTE, C., KOZUR, H. W., BRUCKSHEN, P. & VEIZER, J. 2003. Strontium isotope evolution of Late Permian and Triassic seawater. *Geochimica et Cosmochimica Acta*, **67**, 47–62.

KOZUR, H. W. 1998. Some aspects of the Permian–Triassic boundary (PTB) and of the possible causes for the biotic crisis around this boundary. *Palaeogeography, Palaeoclimatology, Palaeoecology*, **143**, 227–272.

KRYSTYN, L., BAUD, A., RICHOZ, S. & TWITCHETT, R. J. 2003. A unique Permian–Triassic boundary section from Oman. *Palaeogeography, Palaeoclimatology, Palaeoecology*, **191**, 329–344.

KUMP, L. R. 1991. Interpreting carbon-isotope excursions: Strangelove oceans. *Geology*, **19**, 299–302.

MACLEOD, K. G., SMITH, R. M. H., KOCH, P. L. & WARD, P. D. 2000. Timing of mammal-like reptile extinctions across the Permian–Triassic boundary in South Africa. *Geology*, **28**, 227–230.

MARTIN, E. E. & MACDOUGALL, J. D. 1995. Sr and Nd isotopes at the Permian–Triassic boundary: a record of climate change. *Chemical Geology*, **125**, 73–99.

MCGHEE, G. R., SHEEHAN, P. M., BOTTJER, D. J. & DROSER, M. L. 2004. Ecological ranking of Phanerozoic biodiversity crises: ecological and taxonomic severities are decoupled. *Palaeogeography, Palaeoclimatology, Palaeoecology*, **211**, 289–297.

MII, H. S., GROSSMAN, E. L. & YANCEY, T. E. 1997. Stable carbon and oxygen isotopic shifts in Permian seas of West Spitsbergen: global change or diagenetic artifact. *Geology*, **25**, 227–230.

MUNDIL, R., LUDWIG, K. R., METCALFE, I. & RENNE, P. R. 2004. Age and timing of the Permian mass extinctions: U/Pb dating of closed-system zircons. *Science*, **305**, 1760–1763.

NEWELL, A. J., TVERDOKHLEBOV, V. P. & BENTON, M. J. 1999. Interplay of tectonics and climate on a transverse fluvial system, Upper Permian, Southern Uralian Foreland Basin, Russia. *Sedimentary Geology*, **127**, 11–29.

RAMPINO, M. R., PROKOPH, A. & ADLER, A. 2000. Tempo of the end-Permian event: high resolution cyclostratigraphy at the Permian–Triassic boundary. *Geology*, **28**, 643–646.

RENNE, P. R., MELOSH, H. J., FARLEY, K. A. *ET AL.* 2004. Is Bedout an impact structure? Take 2. *Science*, **306**, 610–611.

RETALLACK, G. J. 1999. Postapocalyptic greenhouse paleoclimate revealed by earliest Triassic paleosols in the Sydney Basin, Australia. *Bulletin of the Geological Society of America*, **111**, 52–70.

RETALLACK, G. J. 2001. A 300 million year record of atmospheric carbon dioxide from fossil plant cuticles. *Nature*, **411**, 287–290.

RETALLACK, G. J. 2004. Comment – Contrasting deep-water records from the Upper Permian and Lower Triassic of South Tibet and British Columbia: evidence for a diachronous mass extinction (Wignall & Newton 2003). *Palaios*, **19**, 101–102.

RODLAND, D. L. & BOTTJER, D. J. 2001. Biotic recovery from the end-Permian mass extinction: behavior of the inarticulate brachiopod *Lingula* as a disaster taxon. *Palaios*, **16**, 95–101.

RODLAND, D. L., KOWALEWSKI, M, DETTMAN, D. L., FLESSA, K. W., ATUDOREI, V. & SHARP, Z. D. 2003. High resolution analysis of the $\delta^{18}O$ in the biogenic phosphate of modern and fossil lingulid brachiopods. *Journal of Geology*, **111**, 441–453.

SCHUBERT, J. K. & BOTTJER, D. J. 1995. Aftermath of the Permian–Triassic mass extinction event: paleoecology of Lower Triassic carbonates in the western USA. *Palaeogeography, Palaeoclimatology, Palaeoecology*, **116**, 1–39.

SEPHTON, M. A., LOOY, C., BRINKHUIS, H., WIGNALL, P. B., DE LEEUW, J. W. & VISSCHER, H. 2005. Catastrophic soil erosion during the end-Permian biotic crisis. *Geology*, **33**, 941–944.

SMITH, R. M. H. 1995. Changing fluvial environments across the Permian–Triassic boundary in the Karoo Basin, South Africa, and possible causes of tetrapod extinctions. *Palaeogeography, Palaeoclimatology, Palaeoecology*, **117**, 81–104.

SMITH, A. B. 2001. Large scale heterogeneity of the fossil record: implications for Phanerozoic biodiversity studies. *Philosophical Transactions of the Royal Society, London, Series B*, **356**, 351–367.

STANLEY, S. M. 1988. Climatic cooling and mass extinction of Paleozoic reef communities. *Palaios*, **3**, 228–232.

STANLEY, S. M. & YANG, X. 1994. A double mass extinction at the end of the Paleozoic Era. *Science*, **266**, 1340–1344.

TWITCHETT, R. J. 1997. *Palaeoenvironments of the Lower Triassic of the Dolomites, northern Italy.* Unpublished PhD thesis, University of Leeds.

TWITCHETT, R. J. 1999. Palaeoenvironments and faunal recovery after the end-Permian mass extinction. *Palaeogeography, Palaeoclimatology, Palaeoecology*, **154**, 27–37.

TWITCHETT, R. J. 2001. Incompleteness of the Permian–Triassic fossil record: a consequence of productivity decline? *Geological Journal*, **36**, 341–353.

TWITCHETT, R. J., LOOY, C. V., MORANTE, R., VISSCHER, H. & WIGNALL, P. B. 2001. Rapid and synchronous collapse of marine and terrestrial ecosystems during the end-Permian mass extinction event. *Geology*, **29**, 351–354.

TWITCHETT, R. J., KRYSTYN, L., BAUD, A., WHEELEY, J. R. & RICHOZ, S. 2004. Rapid marine recovery after the end-Permian mass extinction event in the absence of marine anoxia. *Geology*, **32**, 805–808.

WARD, P. D., MONTGOMERY, D. R. & SMITH, R. 2000. Altered river morphology in South Africa related to the Permian–Triassic extinction. *Science*, **289**, 1740–1743.

WARD, P. D., BOTHA, J., BUICK, R. *ET AL.* 2005. Abrupt and gradual extinction among Late Permian land vertebrates in the Karoo Basin, South Africa. *Science*, **307**, 709–714.

WIGNALL, P. B. 2001. Large igneous provinces and mass extinction. *Earth Science Reviews*, **53**, 1–33.

WIGNALL, P. B. & HALLAM, A. 1992. Anoxia as a cause of the Permian–Triassic mass extinction: facies evidence from northern Italy and the western United States. *Palaeogeography, Palaeoclimatology, Palaeoecology*, **93**, 21–46.

WIGNALL, P. B. & NEWTON, R. 2003. Contrasting deep-water records from the Upper Permian and Lower Triassic of South Tibet and British Columbia: evidence for a diachronous mass extinction. *Palaios*, **18**, 153–167.

WIGNALL, P. B. & TWITCHETT, R. J. 1996. Oceanic anoxia and the end-Permian mass extinction. *Science*, **272**, 1155–1158.

WIGNALL, P. B. & TWITCHETT, R. J. 2002. Extent, duration and nature of the Permian–Triassic superanoxic event. *In*: KOEBERL, C. & MACLEOD, K. G. (eds) *Catastrophic Events and Mass Extinctions: Impacts and Beyond*. Geological Society of America Special Papers, **356**, 395–413.

WIGNALL, P. B., KOZUR, H. & HALLAM, A. 1996. The timing of palaeoenvironmental changes at the Permian/Triassic (P/Tr) boundary using conodont biostratigraphy. *Historical Biology*, **12**, 39–62.

YIN, H. F., ZHANG, K. X., TONG, J. N., YANG, Z. Y. & WU, S. B. 2001. The Global Stratotype Section and Point (GSSP) of the Permian–Triassic boundary. *Episodes*, **24**, 102–114.

ZHARKOV, M. A. 1981. *History of Paleozoic Salt Accumulation*. Springer–Verlag, Berlin.

Mesozoic climates

B. W. SELLWOOD[1] & P. J. VALDES[2]

[1]*Geography Department, School of Human and Environmental Sciences, The University,
Whiteknights, Reading, RG6 6AB, UK (e-mail: b.w.sellwood@reading.ac.uk)*

[2]*School of Geographical Sciences, University of Bristol,
University Road, Bristol, BS8 1SS, UK*

Abstract: Compared to the present, the Mesozoic Earth was an alien world with a greenhouse climate, as we illustrate here by reference to Upper Triassic, Upper Jurassic, Lower and Upper Cretaceous palaeoclimate reconstructions generated on a General Circulation Model. Throughout the Mesozoic, dense forests grew close to both poles but experienced months-long daylight in warm summers, and months-long darkness in cold, sometimes snowy, winters. Ocean depths were warm (8 °C or more to the ocean floor) and reefs, with corals, grew 10° of latitude further north and south than at the present time. The whole Earth was warmer than now by several degrees centigrade, generating higher atmospheric humidity and a greatly enhanced hydrological cycle. Modelling suggests that much of the rainfall was predominantly convective in character, and often focused over the oceans and leaving major desert expanses on the continental areas. Polar ice-sheets are unlikely to have been present for most of the time because of the high summer temperatures achieved. An absence of major terrestrial ice-caps storing water enriched with ^{16}O is reflected in oxygen isotopic compositions in marine shells more negative than during later times when ice began to accumulate around the poles. Localized mountain glaciers cannot be ruled out, particularly over Antarctica during the Cretaceous. There is no convincing evidence of the sort of short-term, and large-scale, eustatic changes associated with major glaciation and deglaciation comparable with that of the Neogene. Model output does however suggest the possibility of sea-ice, particularly in the nearly enclosed Arctic seaway through some of the year, notably during the Cretaceous. Model results for the Jurassic suggest the possibility of upland ice-sheets during times corresponding with orbitally induced climatic minima. During the Triassic and Jurassic, the world was predominantly warm with at least four times present atmospheric CO_2 and model outputs for evaporation and precipitation conform well with the known distributions of evaporites, calcretes and other climatically sensitive facies.

In recent years, General Circulation Models (GCMs) have been used to evaluate ancient climates. These models use the laws of physics and an understanding of past geography to simulate climatic responses. They are objective in character and it is now possible to compare results from different GCMs for a range of times and over a wide range of parameterizations for past, present and future (e.g. in terms of predictions of surface air temperature, precipitation, surface moisture (precipitation minus evaporation). GCMs are currently producing simulated climate predictions for the Mesozoic which compare favourably with the distributions of climatically sensitive facies (e.g. coals, evaporites and palaeosols) and faunal distributions (e.g. coral reefs and crocodilians). They can be used effectively in the prediction of oceanic upwelling sites and the distribution of petroleum source rocks and phosphorites. Models also produce evaluations of other parameters that do not leave a geological record (e.g. cloud cover, snow cover) and equivocal phenomena such as storminess. Parameterization of sub-grid-scale processes is the main weakness in GCMs (e.g. land surfaces, convection, cloud behaviour) and model output for continental interiors is apparently too cold, in winter, by comparison with palaeontological data. The sedimentary, palaeontological and geochemical record provides an important way that GCMs may themselves be evaluated and this is important because the same GCMs are being used currently to predict possible changes in future climate.

The message from the geological record is clear. Through the Phanerozoic, the Earth's climate has changed significantly, both on a variety of timescales and over a range of climatic states, usually baldly referred to, and oversimplified, as 'greenhouse' and 'icehouse'. These terms disguise many subtle states between these extremes.

Broad general documentations of the Earth's climate state through geological time, based largely on the evidence provided by climate proxy data (sedimentary facies, fossil biota, geochemical information), have been made by many authors (e.g. Frakes 1979; Crowley & North 1991; Frakes *et al.* 1992; Parrish 1994; Valdes *et al.* 1999;

From: WILLIAMS, M., HAYWOOD, A. M., GREGORY, F. J. & SCHMIDT, D. N. (eds) *Deep-Time Perspectives on Climate Change: Marrying the Signal from Computer Models and Biological Proxies.* The Micropalaeontological Society, Special Publications. The Geological Society, London, 201–224.
1747-602X/07/$15.00 © The Micropalaeontological Society 2007.

Huber *et al.* 2000; Skelton *et al.* 2003). Frakes (1979) suggested on the basis of such data that, over the past 500 Ma, the Earth has been generally warmer than at present, but with significant cool phases in the Upper Ordovician, Permo-Carboniferous (Gondwana Glaciation) and the Quaternary to Holocene.

The predominance of episodes warmer than present ('greenhouse phases'), and relative rarity of intervals such as present, with significant ice-caps ('icehouse phases'), is particularly striking. Direct evidence for glaciation (from the distribution of facies such as tillites, ice-rafted dropstones, striated pavements, glendonites etc.) outside of the Oligocene–Recent, Permo-Carboniferous and Upper Ordovician, has led to the idea that equable conditions have prevailed in the past when, by implication, the Earth is believed to have been generally ice-free. Based on negative evidence, the Mesozoic is believed by many to be an essentially ice-free time interval (e.g. Hallam 1985, 1994), a warm interval within which possible cooler 'snaps' existed (e.g. Price 1999). Evidence for the existence of Mesozoic ice comes from a combination of sea-level data and the record of strontium content in marine shells which suggest brief (*C.* 2 million year) icy intervals within an otherwise warm Mesozoic (reviewed in Royer *et al.* 2004). It has been recognized in recent years that the onset and growth of significant glaciation in Antarctica took place over a time-frame far greater than was originally thought. Critical evidence comes from the isotopic record in the oceans, recording a progressive and stepwise increase in the $\delta^{18}O$ content of the ocean waters (e.g. Zachos *et al.* 1992, 2002). Recently, Tripati *et al.* (2005) have suggested that Antarctic glaciation could have even started in the mid-Eocene. There is now direct sedimentary evidence from the Antarctic area itself of iceberg dropstone activity since the late Eocene and of ice-cap advances & retreats since around the Oligocene (e.g. Billups & Schrag 2002; DeConto & Pollard 2003; Ivany *et al.* 2006) with the East Antarctic ice-sheet rapidly growing to about its present size at about 15 Ma (summarized in Hall *et al.* 2003).

Rocks of the Mesozoic era (Fig. 1; see also Fig. 2) span a 186 Ma time interval sandwiched between two mass extinction events recorded in the fossil biota (Benton 2003), but they have yet to provide unequivocal evidence of glacial processes. There is, on the other hand, a wealth of information reflecting global average temperatures throughout the Mesozoic, and these seem to have been around 6° to 9 °C warmer than at present. The data include thermophilic organisms such as reefs, and tropical carbonate belts that extended at least 10° of latitude poleward of their present

distributions (Sellwood & Price 1994; Huber & Watkins 1992; Deconto *et al.* 2000; and review by Johnson *et al.* 2002), widespread bauxites and other palaeosols (Price *et al.* 1997), extensive distributions of evaporites and desert deposits, ocean-derived palaeotemperature data (Sellwood & Price 1994; Crowley & Zachos 2000; Bice *et al.* 2003) and information from temperature-sensitive terrestrial biota such as crocodilians and plants (e.g. Spicer *et al.* 1994; Markwick 1998). Pearson *et al.* (2001) and Williams *et al.* (2005) have, however, raised serious concerns about the interpretation of the isotope record from foraminiferally derived isotopic data because diagenetic alteration can bias them towards cooler values. Despite this problem, climate proxies suggest the Mesozoic Earth was warmer than present. This is somewhat paradoxical because Earth would have received, from the Sun, around 1% less incident solar flux at the top of the atmosphere than it does today (at the start of the Jurassic it was about almost 2% less), these differences being due to the Sun's evolution (Kump *et al.* 1999).

Despite suggestions that changes in cosmic ray flux might explain cooler and warmer phases in Earth history, a better explanation appears to derive from the correlation between high levels of atmospheric CO_2 and warmer episodes, and of low levels of CO_2 with the occurrence of well-documented, long-lived and aerially extensive continental glaciations (as argued by Royer *et al.* 2004).

Following a sea-level low at the start of the Triassic, and despite significant lowstands in the early Jurassic and in the early Cretaceous, it has been well established that global sea-levels throughout much of the Mesozoic were generally much higher than at present (reviewed in Hallam 1992). The apparent absence of major ice-caps accounts for some of this eustatic rise (by comparison with the present), but absence of ice and the thermal expansion of water can only contribute around 100 m or so to the rise. In addition, high rates of sea-floor spreading (on average 75% greater than at present) will have led, it is assumed, to the elevation of oceanic ridges and the displacement of ocean waters over what were, for most of the time interval, relatively subdued or subsident continental masses (e.g. Skelton *et al.* 2003).

A recent approach to understanding past climate regimes on Earth has been through the application of complex computer models, specifically: Atmospheric General Circulation Models (AGCMs), Ocean General Circulation Models (OGCMs) and recently with even more complex coupled ocean–atmosphere GCMs (OAGCMs). There are now many contributions in this field

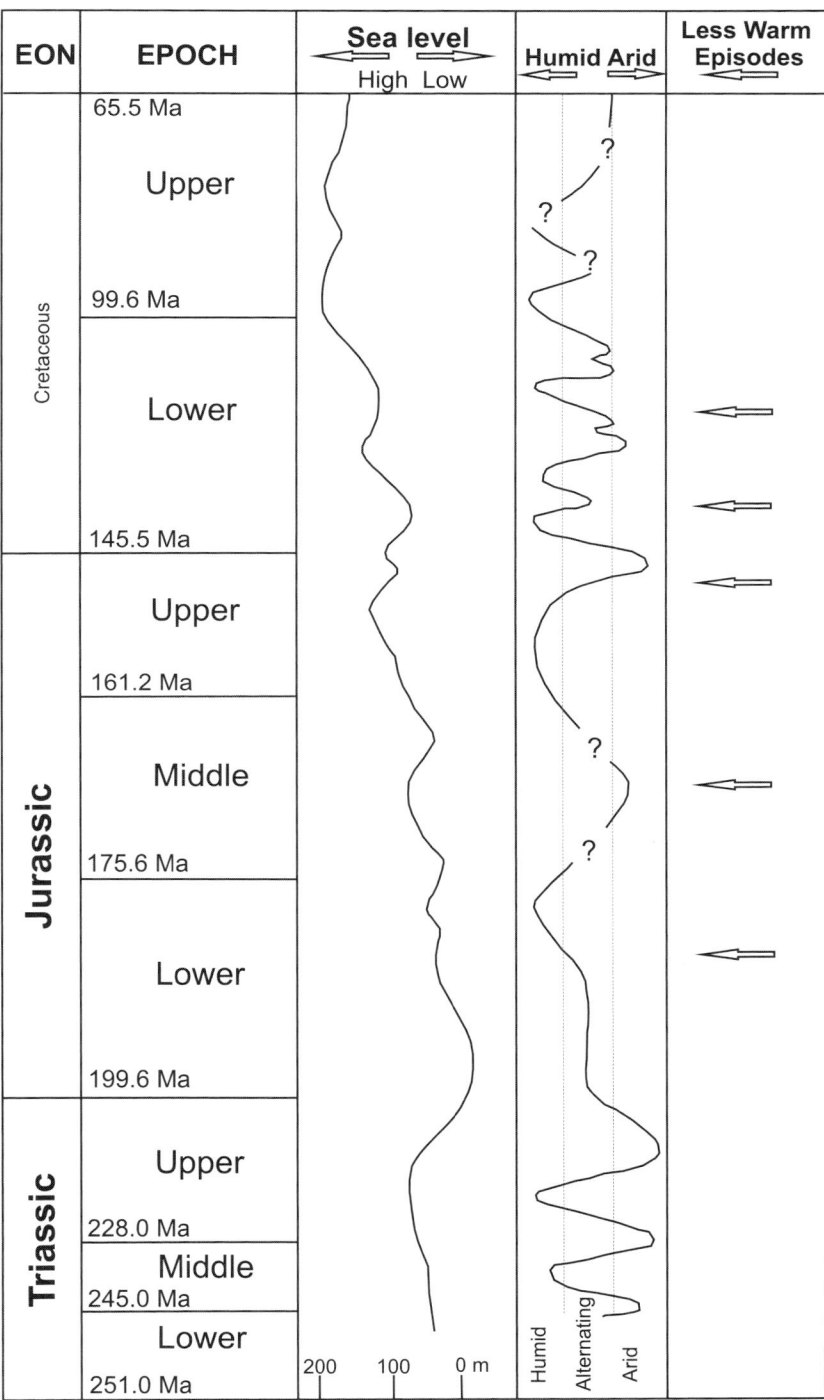

Fig. 1. Sea-level, aridity–humidity cycles and timing of less warm episodes during the Mesozoic (after Price 1999) and the Mesozoic timescale (after Gradstein *et al.* 2004).

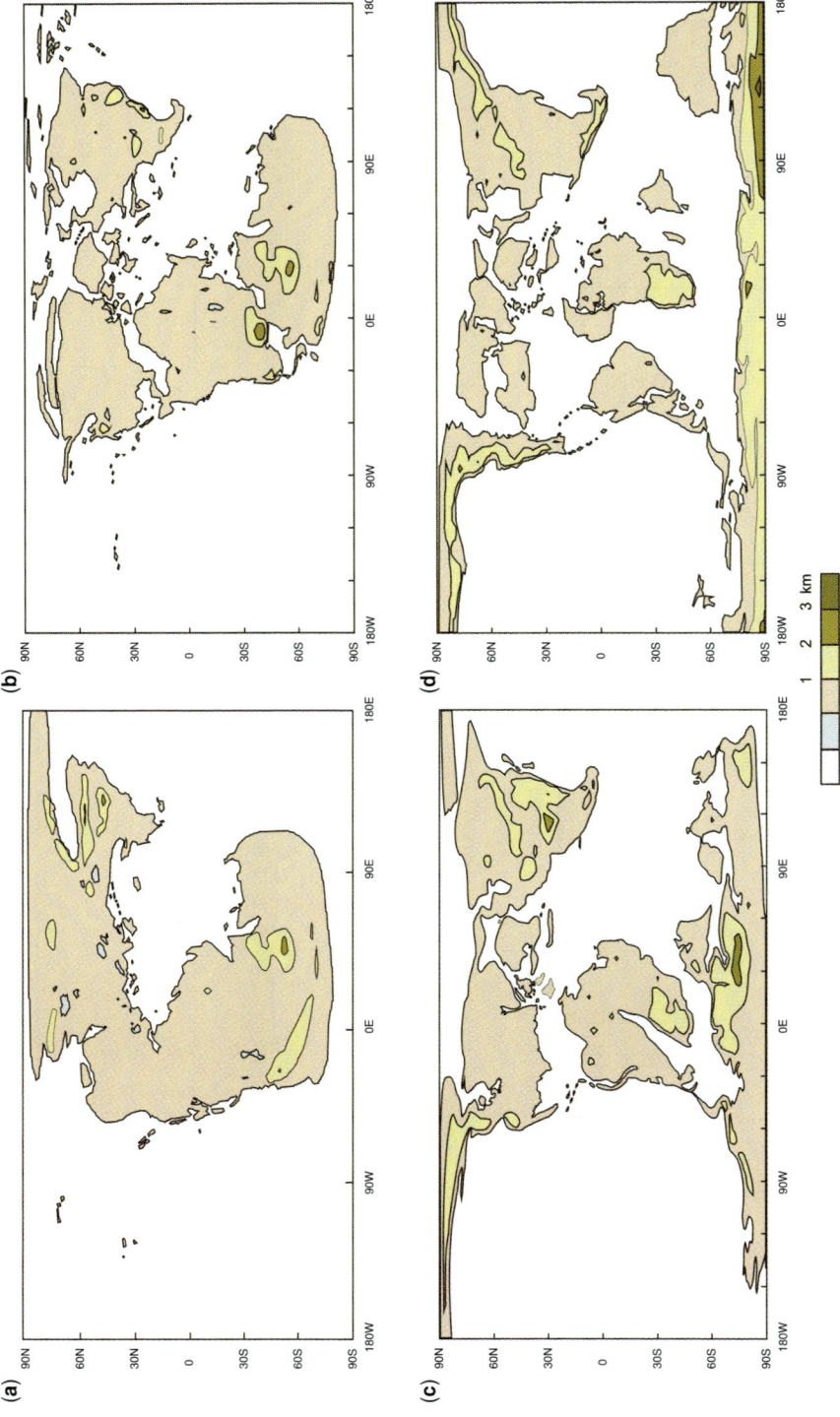

Fig. 2. Palaeogeographies and palaeo-orographies for the Upper Triassic (**a**), Upper Jurassic (**b**), Lower Cretaceous (**c**) and Upper Cretaceous (**d**), after Smith *et al.* (1994).

with the following papers and reviews reflecting something of the evolution in this approach on both sides of the Atlantic: Barron (1983, 1987); Barron & Washington (1985); Kutzbach & Gallimore (1989); Chandler *et al.* (1992); Moore *et al.* (1992*a*, *b*); Valdes & Sellwood (1992); Barron *et al.* (1994); Kutzbach & Ziegler (1994); Price *et al.* (1995); Valdes *et al.* (1995); Handoh *et al.* (1999); Valdes *et al.* (1999); Bjerrum *et al.* (2001); Poulsen *et al.* (2001); and Sellwood & Valdes (2006). Mesozoic greenhouse climates have even been considered as possible analogues for future climate (Crowley 1996), thus heightening interest in model outputs on the Mesozoic.

We present here the output from new model simulations generated using the HadAM3 version of the UKMO atmosphere GCM (Figs 3–11), and we review geological data, and the output from other models, to throw light on the climate of the Mesozoic Earth and on the validity of the models themselves. The model simulations are based on an atmosphere-only model, but the results are similar to fully coupled model simulations which are currently underway. The palaeogeographies and palaeo-orographies (Fig. 2a–d) used in the modelling are based on those of Smith *et al.* (1994). Model outputs for the Upper Triassic (Figs 3, 4), Upper Jurassic (Figs 5, 6), Lower Cretaceous (Figs 7, 8) and Upper Cretaceous (Figs 9, 10) can be compared with interpreted climate zonations for these time periods based on geological data (Fig. 11, simplified from Scotese 2000).

Model description

The model was developed at the Hadley Centre for Climate Prediction and Research, which is part of the UK Meteorological Office. The GCM consists of a linked atmospheric model and sea-ice model. The horizontal resolution of the atmospheric model is 2.5° in latitude and 3.75° in longitude. This provides a grid spacing at the equator of 278 km north to south and 417 km east to west. The atmospheric model consists of 19 layers. It also includes a radiation scheme that can represent the effects of minor trace gases. Its land surface scheme includes the representation of the freezing and melting of soil moisture. The representation of evaporation includes the dependence of stomatal resistance on temperature, vapour pressure and CO_2 concentration. There is an adiabatic diffusion scheme, to simulate the horizontal mixing of tracers.

Other aspects of the workings of this model have been described in Pope *et al.* (2000) and its use in palaeoclimate studies is illustrated in Haywood *et al.* (2002).

Triassic: model/proxy data comparison

For the Triassic (Figs 3, 4, 11), we present simulations from the atmospheric GCM, using a simple zonally symmetric sea-surface temperature. A significant feature of the Triassic Earth is that landmasses are generally low-lying and almost symmetrically distributed in a broad arc about the equator (Fig. 2a).

Temperatures, sea-ice and snow depth (Fig. 3)

A major aspect of the modelled Earth is its overall warmth. Despite temperatures plunging to -20 °C and below over Siberia in Northern Hemisphere winter, and to similarly low values over southernmost Gondwana in Southern Hemisphere winter, the annual temperature average of the Earth is subdued in these high latitude areas because of the high summer values achieved there (*c.* 24 °C). These high summer values preclude the possibility of year-round ice and snow. Freezing temperatures are modelled to come on over Siberia during October, plunging to their lowest during February and then rapidly rising with the onset of spring, in April. May to September are warm months with those regions experiencing the coldest winters achieving summer temperatures, which rise to around 25 °C. Continental regions between about 40° N and S are generally warm (>20 °C) throughout the year but have sustained highs above 30 °C and even above 40 °C during most of the year. Air temperatures over the oceans are largely latitudinal and move, as expected, with the movement of the ITCZ. There is no sea-ice but a significant proportion of the precipitation falls as snow during the winter months in each hemisphere poleward of about 60° (Fig. 3).

Temperature-limited facies

From the model, low temperatures would not have been a significant inhibitor to coral reef development through the Tethys with a maximum range of 35° S to 33° N in the Upper Triassic (Flügel 2002). These sponge-coral and coral-dominated reefs are interpreted to have had ranges close to those of their modern counterparts and their higher latitude range limits are very close to the 20 °C isotherm in the model. Prolific reefs and associated platform carbonates are found throughout the Tethyan region. On the Pacific (Panthalassa) margin of Pangaea, coral reefs extend from Oregon in the north to northern Chile in the south, localities which also lay close to the simulated 20 °C mean annual isotherm.

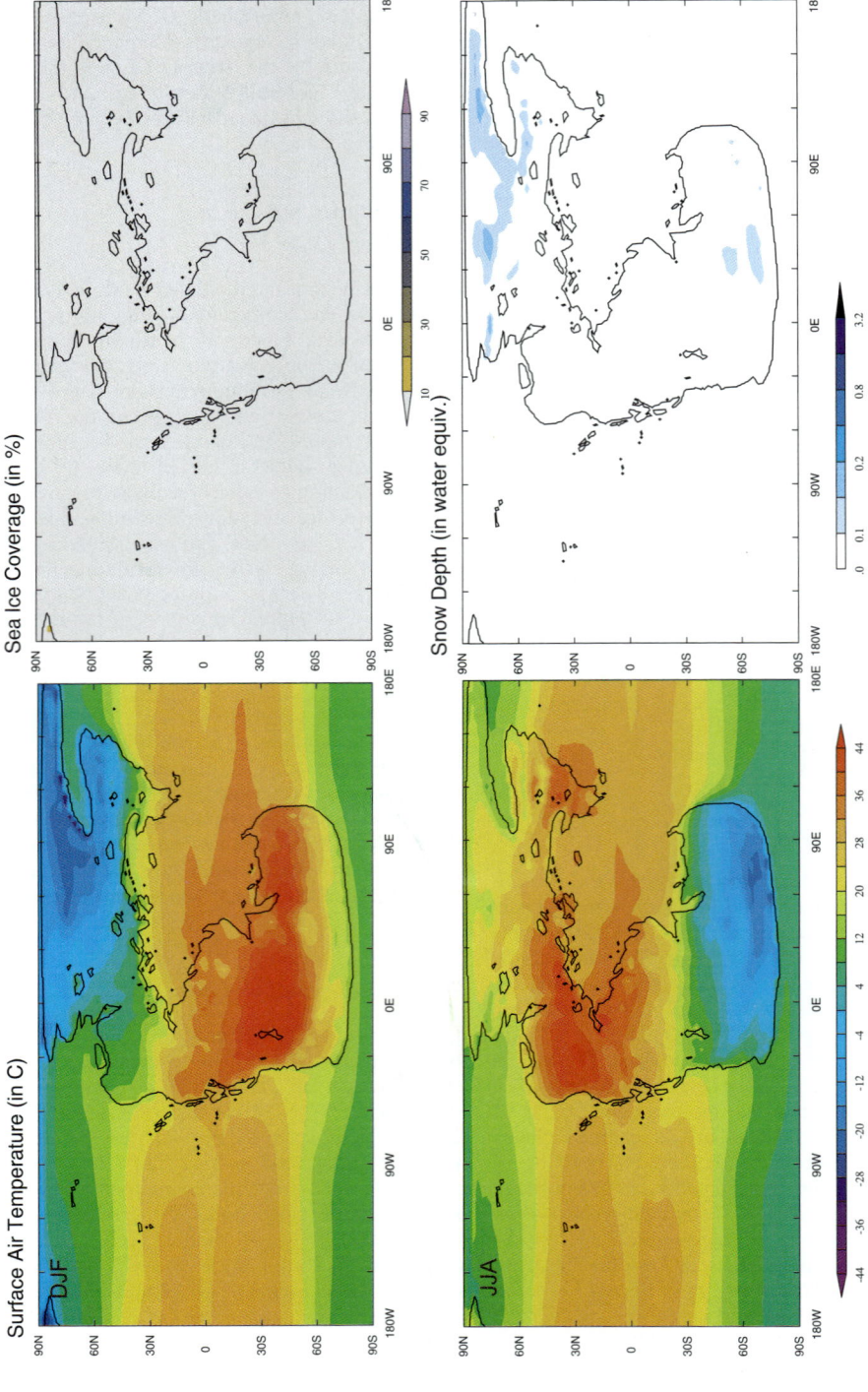

Fig. 3. Model simulated mean seasonal temperatures for the Upper Triassic, for (DJF) December–January–February season, and June–July–August (JJA) season. Units are in °C and the contour interval is every 4 °C. Modelled Average Upper Triassic annual sea-ice coverage (%). Modelled Upper Triassic annual average snow thickness (in metres of water equivalent).

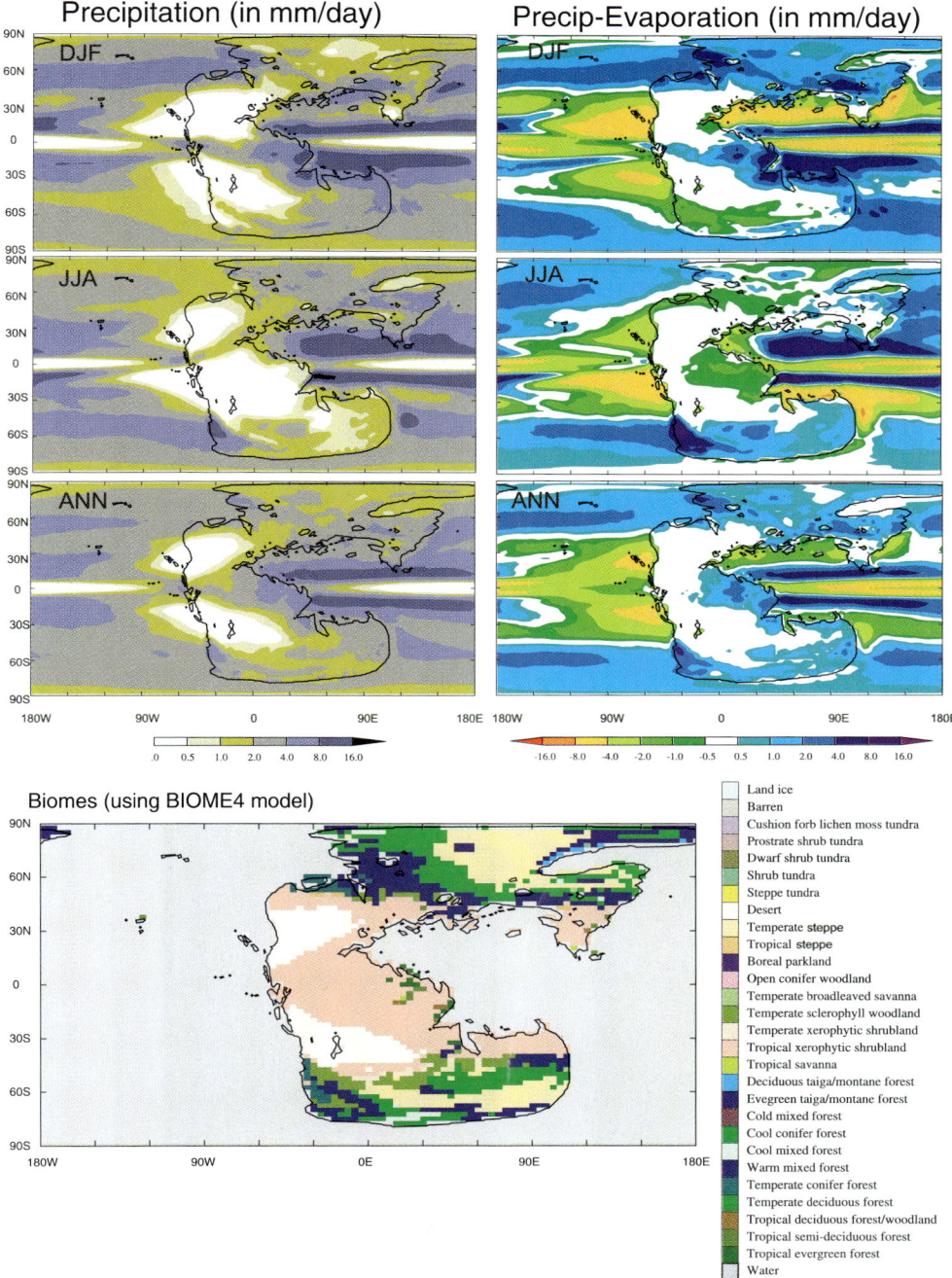

Fig. 4. Modelled simulated mean seasonal precipitation for the Upper Triassic, for (DJF) December–January–February season, (JJA) June–July–August season and Annual average (ANN). Units are in mm/day and the contour interval is not regular. Modelled precipitation minus evaporation (P–E) in mm per day for the Upper Triassic (negative values indicate excess evaporation over precipitation), for (DJF) December–January–February season, (JJA) June–July–August season and Annual average (ANN). Biome zones for the Upper Triassic based on the model predicted temperatures and precipitation (*note*: zones are named as present-day equivalents).

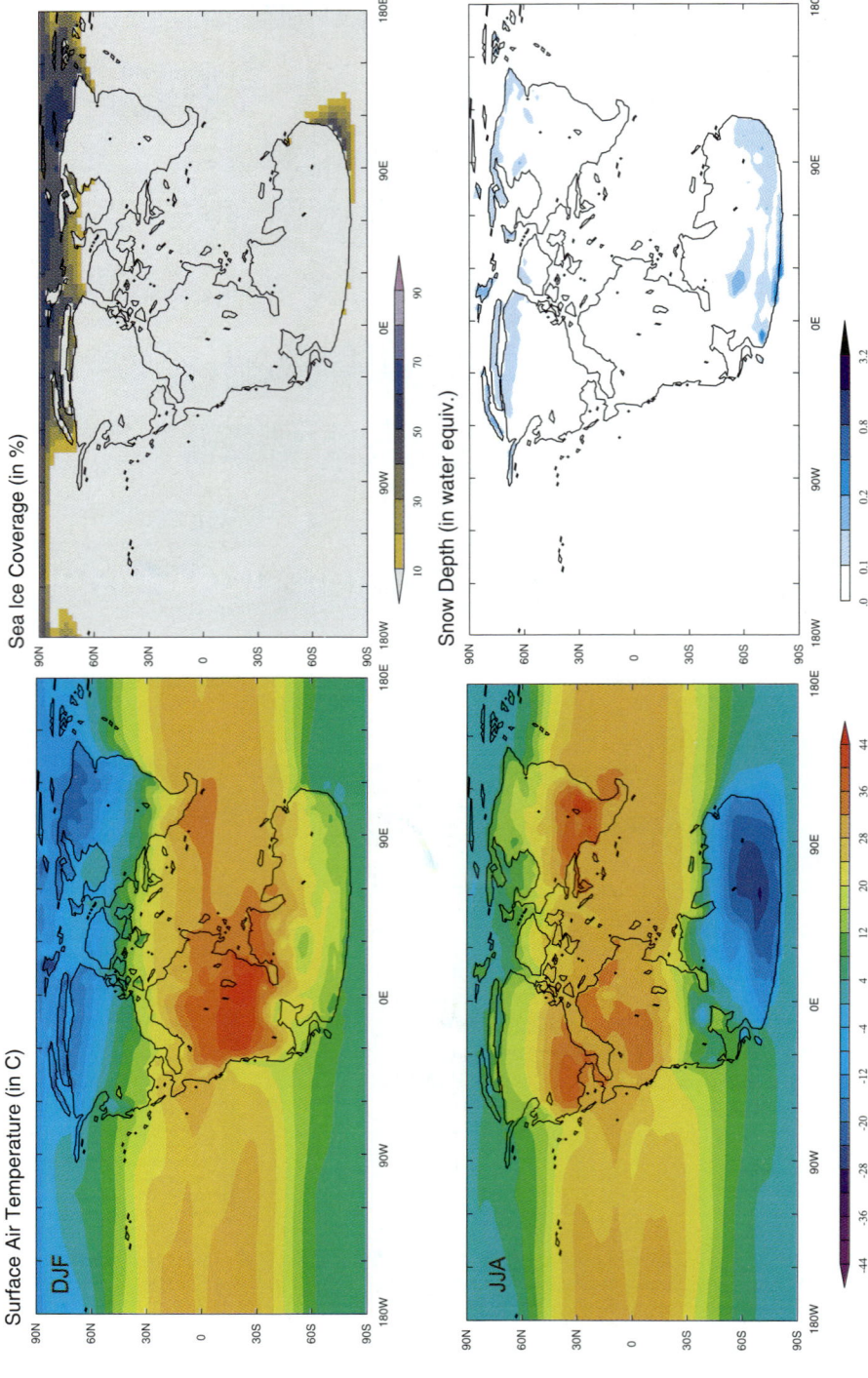

Fig. 5. Model simulated mean seasonal temperatures for the Upper Jurassic, for (DJF) December–January–February season, and June–July–August (JJA) season. Units are in °C and the contour interval is every 4 °C. Modelled Average Upper Jurassic annual sea-ice coverage (%). Modelled annual average Upper Jurassic snow thickness (in metres of water equivalent).

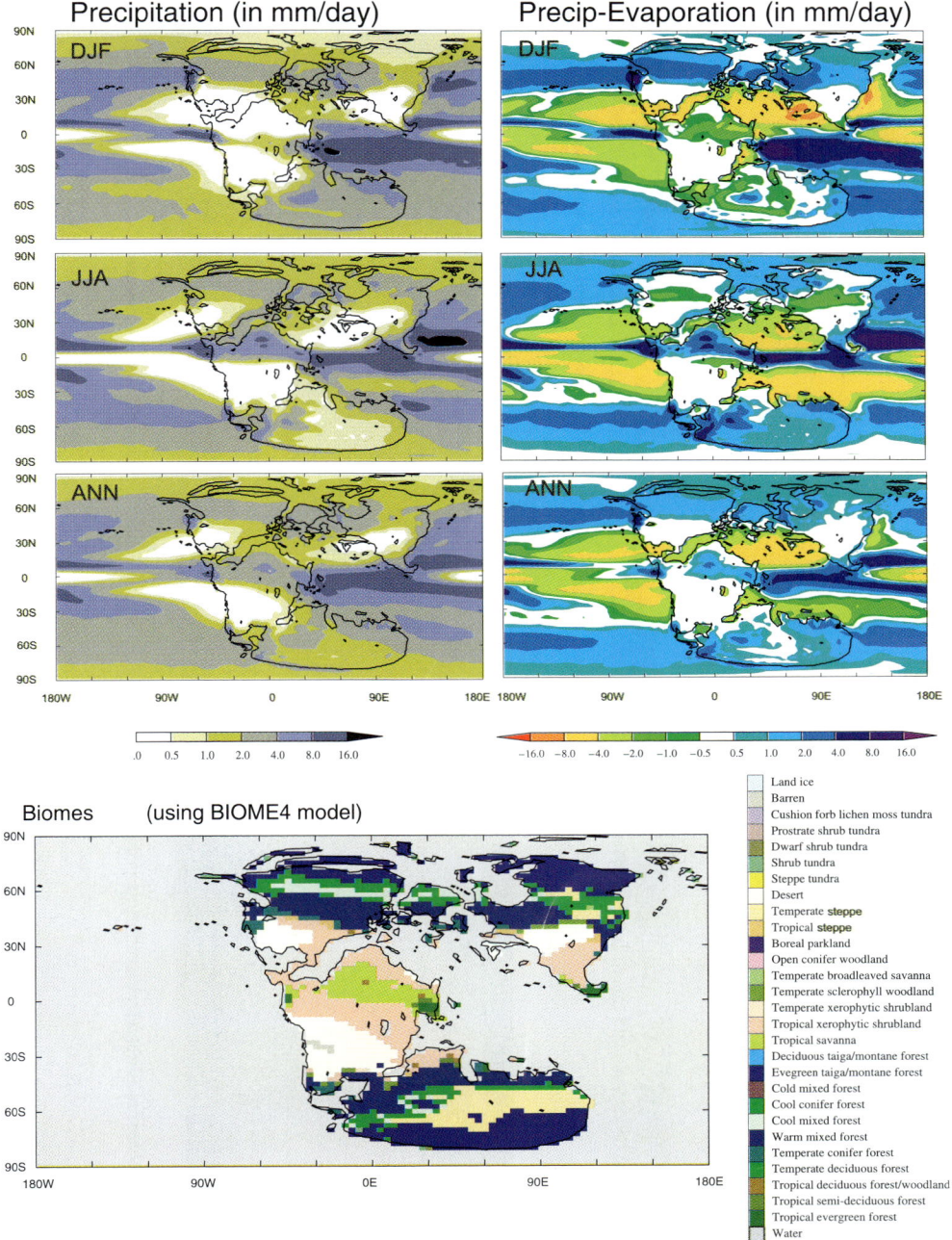

Fig. 6. Model simulated mean seasonal precipitation for the Upper Jurassic, for (DJF) December–January–February season, (JJA) June–July–August season and Annual average (ANN). Units are in mm/day and the contour interval is not regular. Modelled precipitation minus evaporation (P–E) in mm per day for the Upper Jurassic (negative values indicate excess evaporation over precipitation), for (DJF) December–January–February season, (JJA) June–July–August season and Annual average (ANN). Biome zones for the Upper Jurassic based on the model predicted temperatures and precipitation (*note*: zones are named as present-day equivalents).

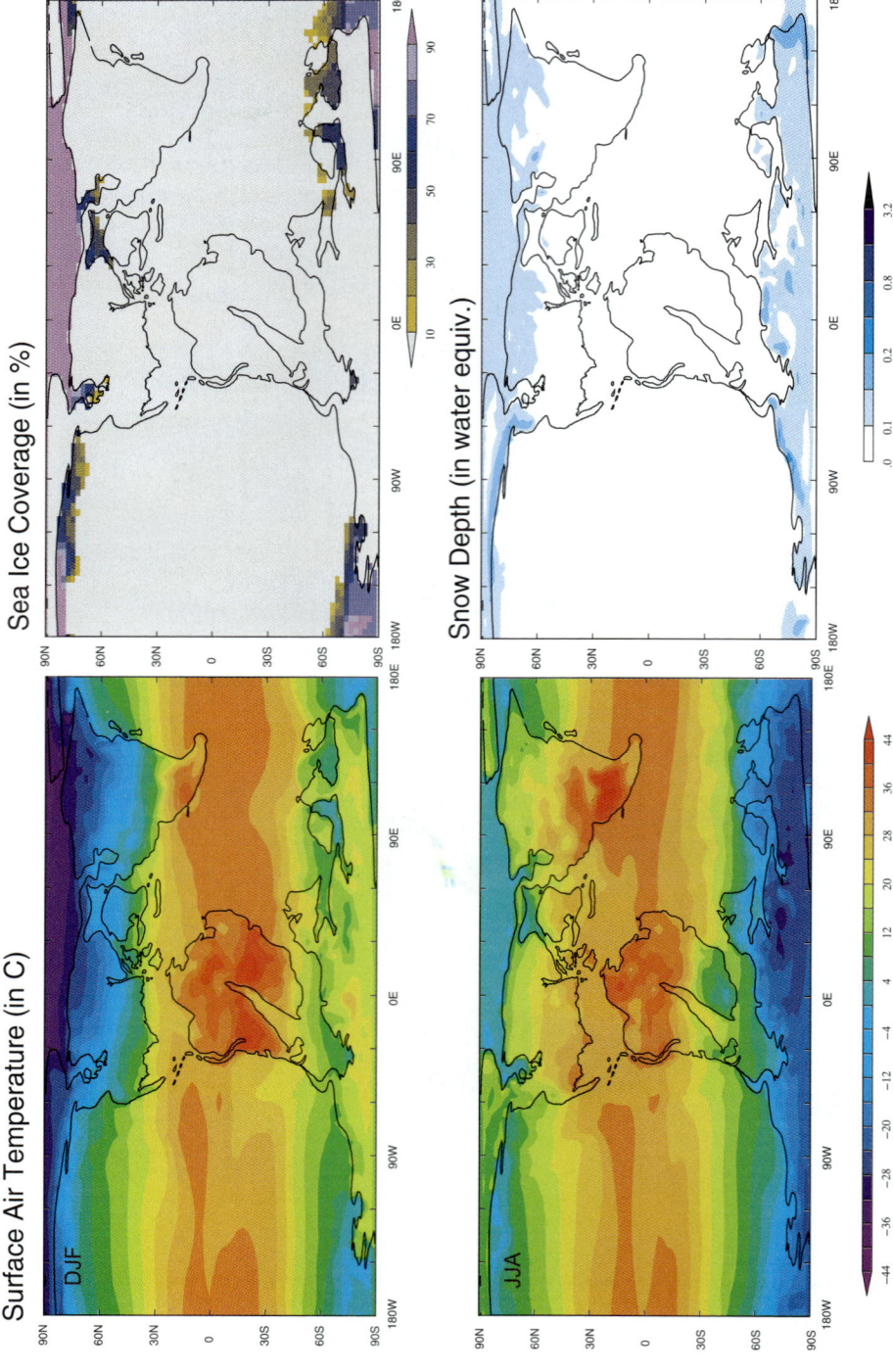

Fig. 7. Model simulated mean seasonal temperatures for the Lower Cretaceous, for (DJF) December–January–February season, and June–July–August (JJA) season. Units are in °C and the contour interval is every 4 °C. Modelled Average Lower Cretaceous annual sea-ice coverage (%). Modelled annual average Lower Cretaceous snow thickness (in metres of water equivalent).

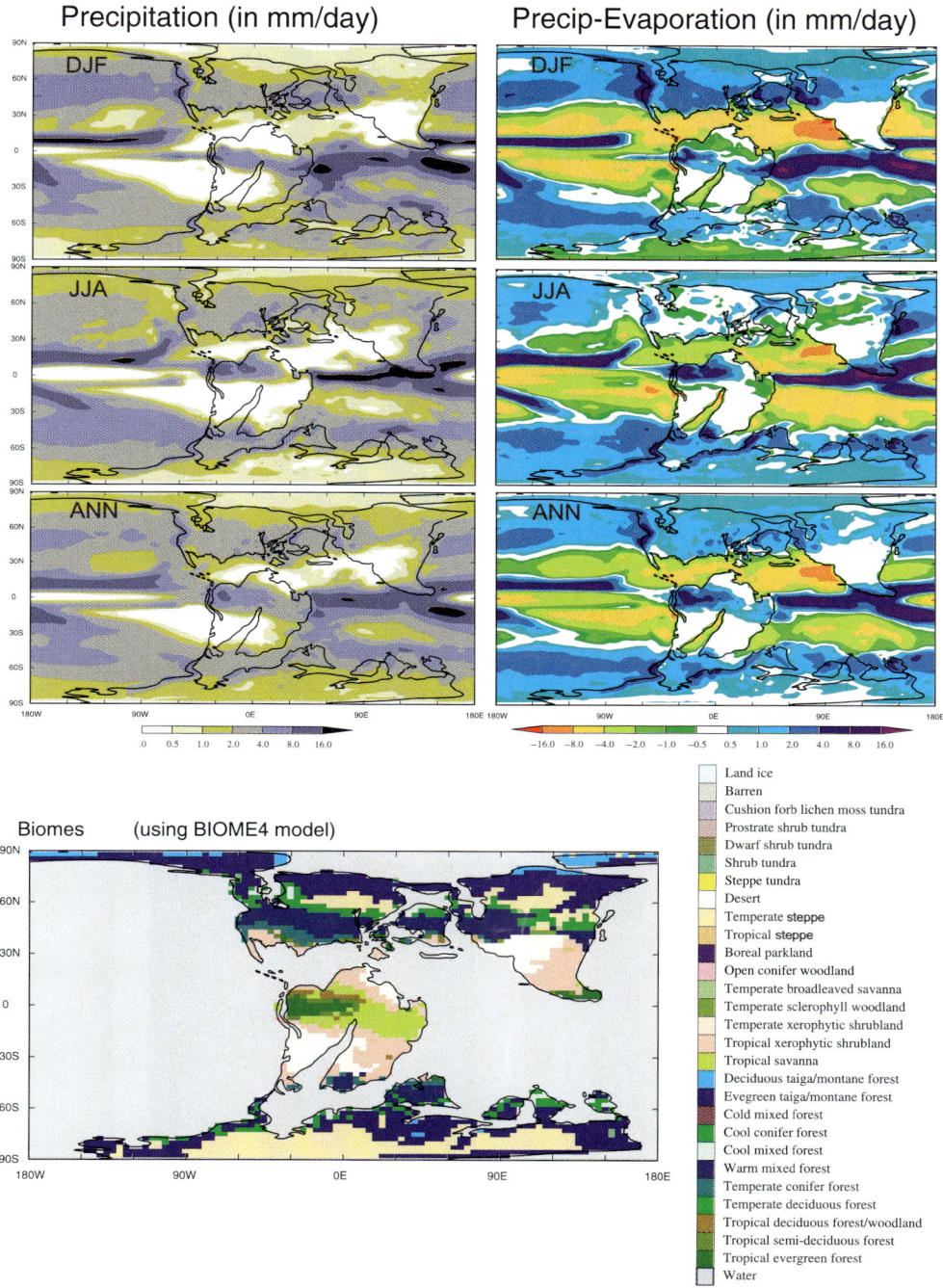

Fig. 8. Model simulated mean seasonal precipitation for the Lower Cretaceous, for (DJF) December–January–February season, (JJA) June–July–August season and Annual average (ANN). Units are in mm/day and the contour interval is not regular. Modelled precipitation minus evaporation (P–E) in mm per day for the Lower Cretaceous (negative values indicate excess evaporation over precipitation), for (DJF) December–January–February season, (JJA) June–July–August season and Annual average (ANN). Biome zones for the Lower Cretaceous based on the model predicted temperatures and precipitation (*note*: zones are named as present-day equivalents).

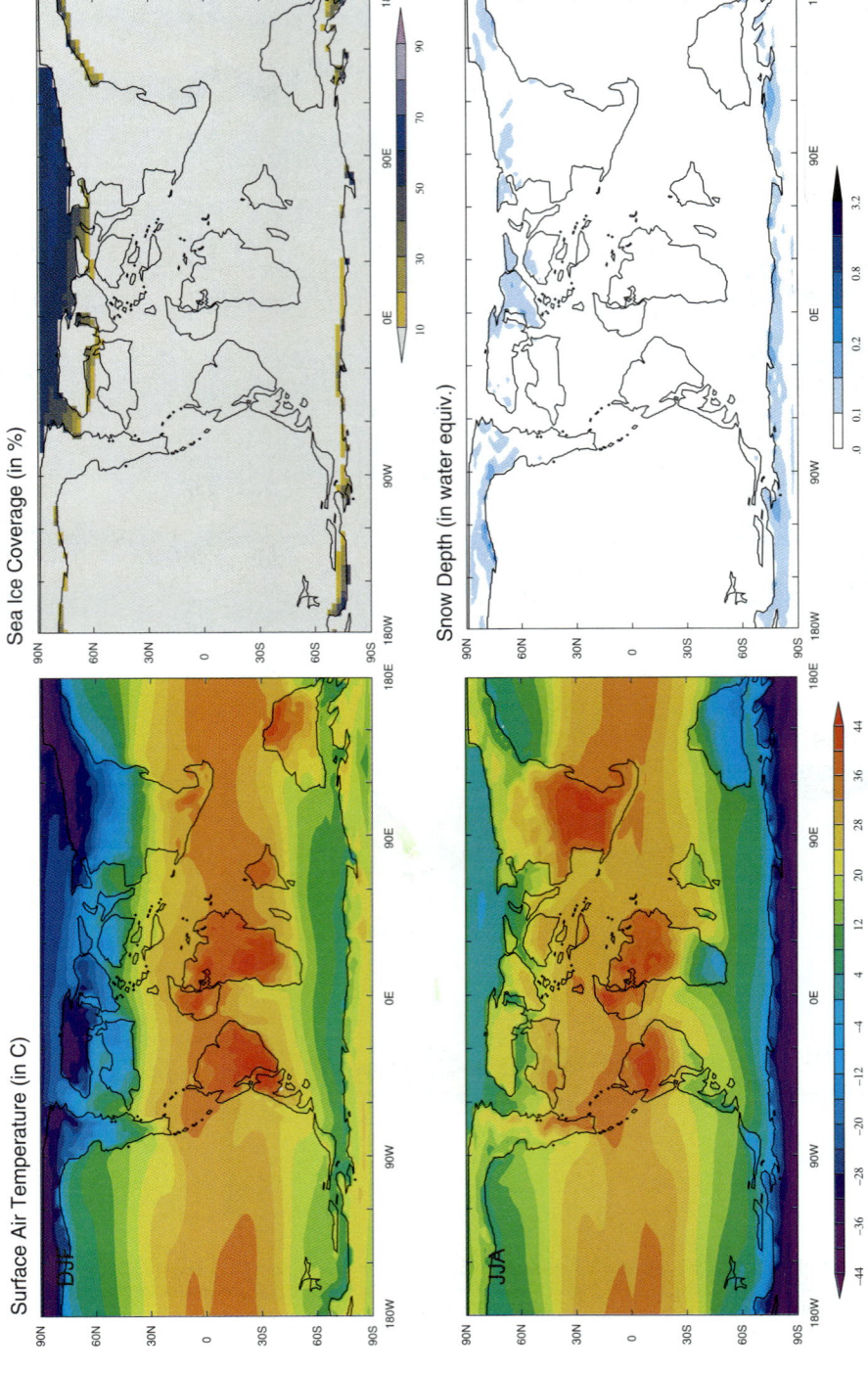

Fig. 9. Model simulated mean seasonal temperatures for the Upper Cretaceous, for (DJF) December–January–February season and (JJA) June–July–August season. Units are in °C and the contour interval is every 4°C. Modelled Average Upper Cretaceous annual see-ice coverage (%). Modelled annual average Upper Cretaceous snow thickness (in metres of water equivalent).

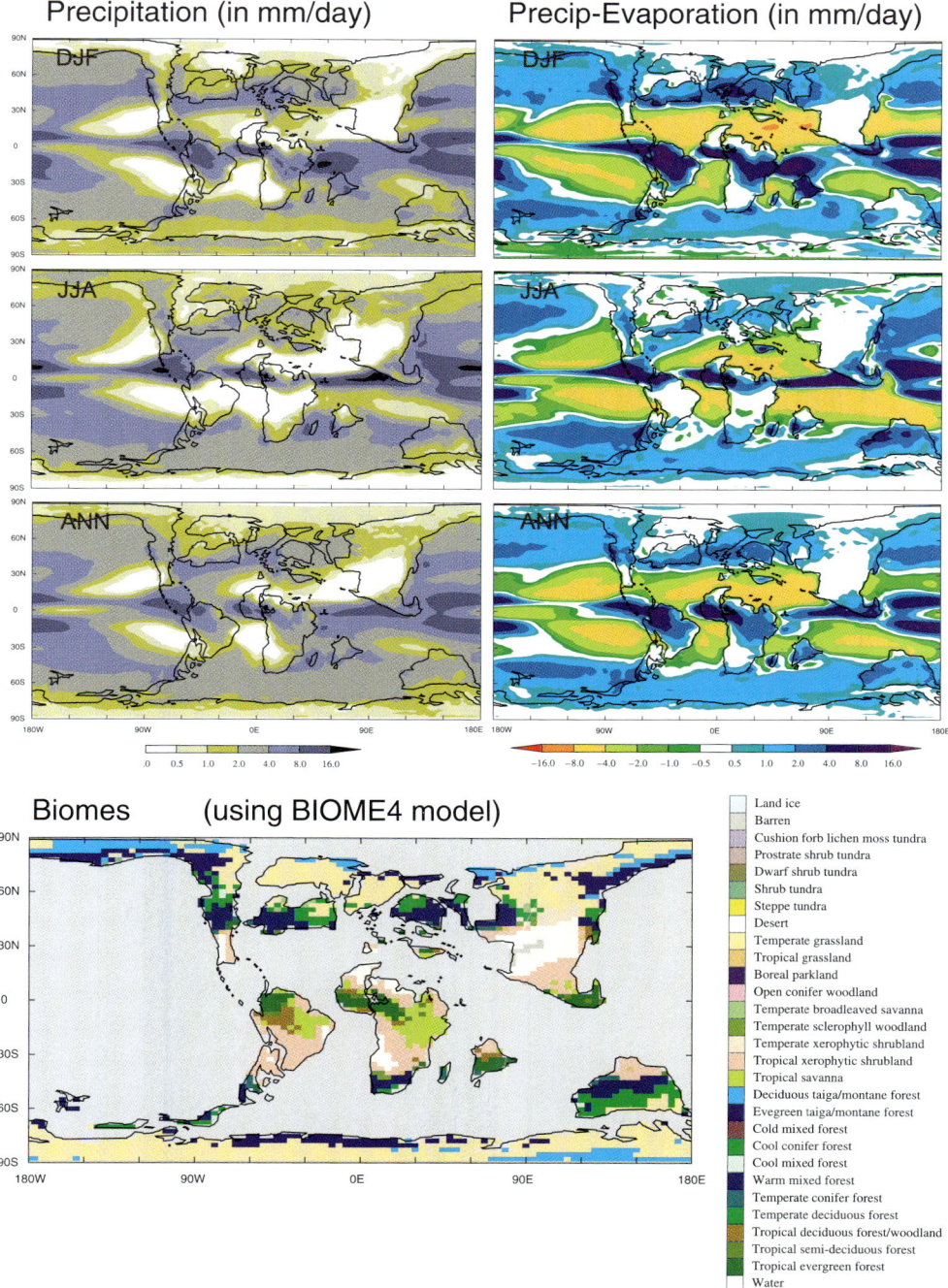

Fig. 10. Model simulated mean seasonal precipitation for the Upper Cretaceous, for (DJF) December–January–February season, (JJA) June–July–August season and Annual average (ANN). Units are in mm/day and the contour interval is not regular. Modelled precipitation minus evaporation (P–E) in mm per day for the Upper Cretaceous (negative values indicate excess evaporation over precipitation), for (DJF) December–January–February season, (JJA) June–July–August season and Annual average (ANN). Biome zones for the Upper Cretaceous based on the model predicted temperatures and precipitation (*note*: zones are named as present-day equivalents).

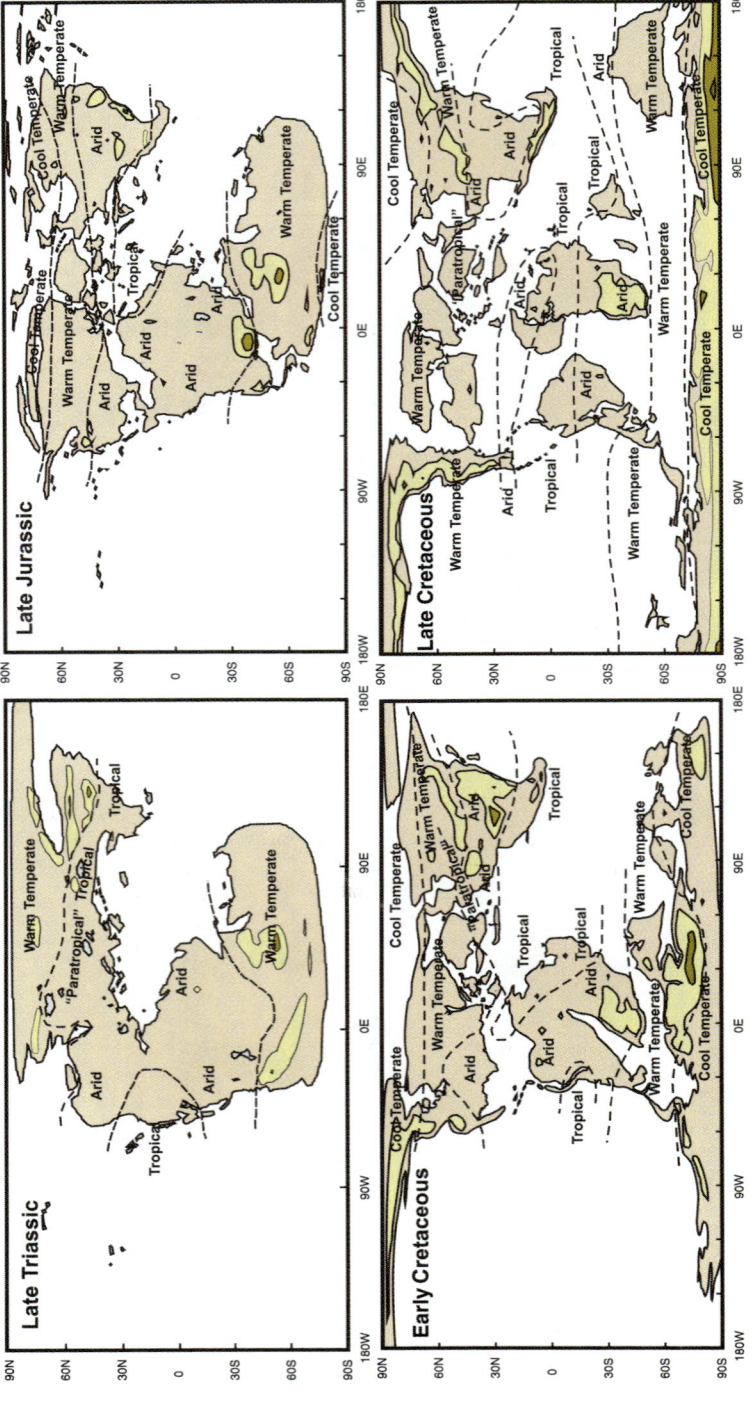

Fig. 11. Environmental zones for the Upper Triassic, Upper Jurassic, Lower Cretaceous and Upper Cretaceous interpreted from geological proxy data (simplified from Scotese 2000 PALEOMAP project).

Precipitation minus evaporation (Fig. 4)

Large tracts of Pangaea between about 40° N and S are modelled to receive very little rainfall. Much of the planet's rainfall is over the oceans, being convective in character, with the main zone of rainfall migrating north and south through the year with the movement of the ITCZ. The southern and western margin of the Tethys is modelled to have a monsoonal climate with rains that started to peak in December through April, and evaporation exceeding precipitation for the rest of the year. Much of Pangaea has an excess of evaporation over precipitation and highly seasonal rainfall, with only high continental latitudes being areas where precipitation exceeds evaporation. It is important to note, in Figure 4 and subsequent Precipitation–Evaporation figures, that once modelled soil moisture reaches zero no further desiccation can occur, so much of the land in white (i.e. with Precip–Evap between −0.5 and +0.5 mm day) is dry.

Over the seaways, the relationship between evaporation and precipitation is expressed in ocean salinities raised above those of modern normal marine (35‰ NaCl), in particular the western end of the Tethys where precipitation minus evaporation is generally in excess of 2 mm per day (expressed as −2 mm on Fig. 2a). These values are modelled to give open sea salinities that exceed 40‰ and inshore values in excess of 45‰ NaCl.

Triassic facies and floras (Fig. 2c)

Walter (1985) defined a series of climate zones: (10) ice; (9) polar; (8) cold temperate; (7) cool temperate; (6) cold temperate arid (steppe/desert); (5) warm temperate, humid; (4) winterwet (warm temperate, dry summer); (3) desert (subtropical arid); (2) summer wet (tropical, humid summer); (1) tropical humid (see also Scotese 2000 Paleomap Project http://www.scotese.com and Fig. 11). Of these, zones 10 to 7 are unrecognized in the Triassic by reference to geological data (Frakes et al. 1992; Ziegler et al. 1994; Scotese 2000), an observation compatible with model output. We present here the results from BIOME 4, which evaluates the outputs for a range of parameterizations and converts these into a prediction of terrestrial biomes based on modern-day systems. The biomes predicted are what would be expected, given the climate parameters, but for modern-day equivalents. Although there are no direct comparisons in the Triassic for modern-day plant communities such as grasslands (termed 'steppe' in Fig. 4), it is instructive to see where, for example, comparable climate zones may have occurred in the past.

We present these outputs for all the time periods we consider in this paper.

In the early Triassic, the highest latitude floras contain the lycopod *Pleuromeia*, which has a world-wide distribution and reflects a global climate largely devoid of frosts, at least in coastal settings (Ziegler et al. 1994). From the global distribution of lithologic indicators of climate, Scotese (see above and Fig. 11) considers the interior of Pangaea to have been hot and dry, and notes that warm temperate climates extended to the poles. To permit these floras to extend to 75° in both hemispheres, as is observed, would require poles far warmer than at present, which is consistent with modelled output and supports Scotese's (2000) interpretation that this may have been one of the hottest times in Earth history. The distribution of extensive calcretes throughout Europe and the western Tethyan margin, with associated evaporites during the Middle Triassic, is in agreement with the modelled climate: hot and predominantly arid but with a short wet season.

In the Middle Triassic, warm temperate floras extended to about 70° N in NE Siberia and coals formed (Ziegler et al. 1994). Southern Hemisphere coal swamps are found in S. Argentina, S. Chile, and Antarctica, again compatible with modelled zones with precipitation exceeding evaporation for much of the year. Biomes and facies suggest that the equatorial zone is a predominantly arid belt, again lending support to the validity of the model. Taylor et al. (2000) have studied fossil wood from the Middle Triassic of Victoria Land and the Transantarctic mountains of Antarctica. The original trees grew at 70–75° S palaeolatitude and exhibit growth ring patterns that suggest a highly seasonal habitat. Light, rather than temperature, was the limiting factor which controlled plant growth, with long periods of winter darkness occurring in these near-polar habitats. Although some trees show evidence of frost damage (from an unseasonal frost early in the fall or late in the spring), the authors believe that the palaeobotanical evidence indicates that these areas were far warmer than predicted on the basis of either model or physical data alone.

In the late Trias, a warm temperate climate was dominant between 30° to 50° of latitude in both hemispheres (Fig. 11). Floras typical of this biome are mixed with dry subtropical floras in the south, in S. Europe, and with cool temperate floras to the north in northern China. There is not much in the way of climatic zonality and Meyen (1997) suspected that the whole Earth may have been essentially frost-free. Ziegler et al. (1994) think this an overstated case. The model generally supports this scenario, but lack of precipitation typifies many continental interiors between these latitudes, especially away from the coastal zones.

Thus, the model output shows a good correlation with globally derived data particularly that for terrestrial plant biomes and terrigenous proxy facies (Ziegler *et al.* 1994; Scotese 2000), and also with the distribution of reefs (Flügel 2002). Terrestrial reptile distributions (e.g. Tucker & Benton 1984) can be interpreted to be broadly in line with model output.

Jurassic (Kimmeridgian): model/proxy data comparison

The model output presented here is for the Upper Jurassic (Figs 5, 6, 11), by which time Pangaea was breaking up (Fig. 2b), leading to a major new low- to mid-latitude seaway (the Central Atlantic) which extended Tethyan influences westwards as far as the present-day Caribbean. In the Southern Hemisphere Seaways were opening between what will become Madagascar–India and East Africa, and western Antarctica and South America–South Africa. Land bridges still connected large tracts of palaeocontinental Laurentia and Gondwana. Europe comprised an archipelago of low islands separated by shallow seaways.

Jurassic Climate System (Figs 5 and 6)

Late Jurassic models, largely based on Kimmeridgian or Tithonian times, are well represented in the literature (reviewed in Valdes *et al.* 1999; and Rees *et al.* 2000), and we present model outputs for temperature, sea-ice, snow depth, precipitation, precipitation–evaporation, and BIOME 4 as Figures 5 and 6.

Poleward of about 50°, the continents in both hemispheres experienced similar seasonal temperature regimes, with cool cold winters but generally warm summers. Modelled winter temperatures permit the extensive development of sea-ice (Fig. 5), but summer temperatures are too high to permit low altitude glaciation even though significant snow accumulation had occurred in the winter months (Fig. 5). Model simulations have suggested that during times comparable with orbital conditions that triggered Quaternary glaciations, upland sites could have accumulated ice-sheets over the southernmost parts of Gondwana. Such accumulations are likely to have been dynamic in character (rapid waxing and waning) but could have promoted rapid small-scale (a few metres) of sea-level change (Valdes *et al.* 1995; Price *et al.* 1997). Low winter temperatures over Siberia and China would have surely favoured feathered, migratory dinosaurs. Low latitude continental areas remained hot throughout the year and air temperatures over the oceans are largely

longitudinal in character, moving with the movement of the ITCZ.

Bjerrum *et al.* (2001) suggest that marine provincial boundaries, associated in the literature with temperature (e.g. Arkell 1956), were controlled by density differences in ocean waters, southward-flowing currents (from the Arctic into Europe), possibly being related to thermohaline circulation. When reduced-salinity Arctic waters spread southwards towards the Tethys, influxes of boreal faunas occurred, the converse occurring during times of predominantly northerly flow. In the South Atlantic (Falklands area, Deep Sea Drilling Project sites 330 and 511, Price & Sellwood 1997), apparently anomalously high seawater palaeotemperatures (17°–18 °C at around palaeolatitude 60 °S), derived from $\delta^{18}O$ values, may have resulted from freshwater run-off depleting surface waters with respect to $\delta^{18}O$. This emphasizes the necessity for isotopic values to be interpreted with caution, the apparent warmth at these latitudes being perhaps more apparent than real. A similar situation has recently also been reported from one of these sites (DSDP Site 511 for the Turonian Cretaceous: Bice *et al.* 2003).

Precipitation minus Evaporation (Fig. 6)

As in the Triassic, much of the modelled precipitation is convective in character, falling over the oceans, with the main zone of rainfall migrating with the movement of the ITCZ. Major continental tracts between about 40° N and S of the equator were dry for much of the year, reflected in biome output as desert or near-desert (modern equivalent being tropical xerophytic shrubland in Fig. 6). Seasonal rains over equatorial Africa and coastal South America generate moister biomes. The Earth's wetlands occurred in mid- to high latitudes favoured by modest rainfall combined with lower rates of evaporation, the warm temperate zones where most Jurassic coals occur (Scotese 2000; Fig. 11).

Jurassic facies, floras and climate zones (Figs 6 and 11)

Jurassic plant productivity and maximum diversity were concentrated at mid- to high latitudes, reflecting a migration of the zone of peak productivity from low to higher latitudes during greenhouse times. Forests in mid-latitudes were dominated by a mixture of ferns, cycadophytes, sphenophytes, pteridophytes and conifers (reviewed and modelled in Rees *et al.* 2000). Low latitude vegetation was predominantly xeromorphic and only patchily forested, with small-leafed forms of conifers and cycadophytes. Polar latitudes were dominated by broad-leafed conifers and ginkgophytes which have been interpreted as deciduous. Tropical evergreen (rainforest-type) biomes are rare or absent, as are tundra and glacial biomes. The Rees *et al.* (2000) compilation represents a considerable refinement on the approach of Hallam (1985, 1994), and the model simulations of Moore *et al.* (1992*a*, *b*), Valdes & Sellwood (1992), Valdes *et al.* (1995), Price *et al.* (1995) and references in these works. The symmetrical arrangement of Jurassic coals and evaporites about the equator suggests that precipitation exceeded evaporation in mid- to high northern latitudes. Between about 30° N and the equator, evaporation is greater than precipitation whereas evaporation and precipitation are approximately in balance (E = P) around the equator. Between about 10° and 30° S, evaporation once again exceeds precipitation and, although evidence in the far south is sparse, there is a return to excess precipitation in the mid- to high southern latitudes. The patterns obtained from the models are in broad agreement with the geological data presented and published previously (and reviewed in Valdes *et al.* 1999 and Scotese 2000 [generalized in Fig. 11]).

In the Late Jurassic, modelling described in Rees *et al.* (2000), and Valdes & Sellwood (1992), atmospheric CO_2 concentrations of four times pre-industrial values were used, values wholly compatible with GEOCARB model results (Royer *et al.* 2004), and evidence from fossil bryophytes in comparison with experimental results on modern equivalents (Fletcher *et al.* 2005). No major changes of climate are indicated for the Jurassic as a whole, and Rees *et al.* (2000) suspect that the climate zones they reported remained more-or-less constant but that because of plate movements continental areas moved across them. Such an interpretation requires support from palaeomagnetic data, which are, as yet, lacking. Hallam (1985), in an alternative hypothesis, suggested that the Late Jurassic spread of aridity in S. Eurasia might have been caused by regional tectonics, creating a rain shadow effect.

Cretaceous: model/proxy data comparison

During the Cretaceous, a series of major palaeogeographic changes took place, which is why we give results for both the Lower and Upper Cretaceous (Figs 7–11). In the Lower Cretaceous, the southern continents (Gondwanaland) were separated from the northern continents. In the Lower Cretaceous South America and Africa began to separate as the northward-opening South Atlantic formed, and by the Upper Cretaceous this separation was complete. Part of Antarctica lay over the South Pole, but the continent was still connected to South America, India and Australia. In the Northern Hemisphere, the Arctic was a nearly enclosed seaway (Fig. 2). Another significant change was the increased orography of the Earth, with uplands over both the northern and southern continents (Fig. 2), reflecting enhanced plate tectonic activity. By the Upper Cretaceous, Antarctica (now fully separated from island India, island Madagascar and island Australia), lay within the South Polar region. The South Atlantic was open, and Eastern Siberia and Alaska were conjoined, close to the North Pole. A broad seaway stretched north–south through North America, and Europe remained an archipelago (Fig. 2).

According to Poulsen *et al.* (2001) and Skelton *et al.* (2003), the mid-Cretaceous was a time of unusually active tectonism and sea-floor spreading, associated with enhanced volcanic outgassing. In the marine realm, this is a time of episodic Oceanic Anoxic Events (OAEvent 1 – Aptian–Early Albian; OAEvent 2 – Cenomanian/Turonian boundary). Outgassing caused enhanced CO_2 levels in the atmosphere, and plate movements caused palaeogeographic changes that in turn caused major global oceanic circulation, in particular the presence or absence of a marine connection between the North and South Atlantic. Reporting model experiments employing the NCAR Parallel Ocean Climate Model, Poulsen *et al.* (2001) suggest that, during the Turonian, a gateway or sill opened in the Atlantic allowing Antarctic Bottom Water to ventilate the Atlantic basins. Handoh *et al.* (1991) have modelled paleo-upwelling and organic-rich sediments in the Cenomanian Atlantic and confirm that a critical factor controlling North Atlantic ventilation was the establishment, for the first time, of connection between the North and South Atlantic Oceans in the Turonian. A series of global changes in climate and oceanography at the Cenomanian/Turonian boundary were probably triggered by the initiation of this connection between the formerly separated Atlantic Oceans. Beckmann *et al.* (2005), using ODP data from offshore Ivory Coast (W. Africa) and GCM simulations, have argued for precession-driven fluctuations in river discharge during the Coniacian–Campanian. Phases of enhanced outflow can here be linked with episodes of oceanic anoxia.

The model results shown use the atmosphere model with prescribed sea-surface temperatures which are based on provisional results from a fully coupled atmosphere–ocean (HadCM3L). These latter simulations have not reached complete equilibrium in the deep ocean and therefore we choose to show an atmosphere-only simulation. However, we believe (based on evidence from previous work with this model) that the simulations are

sufficiently close to the final coupled results and
thus justify further analysis.

Modelled temperature (Figs 7, 9)

In both the Lower and Upper Cretaceous, continental areas north and south of around 60° latitude
show rapid seasonal gradients towards cold winter
temperatures, low latitudes remaining warm year-
round. Both these conditions led to the development
of extensive winter sea-ice, particularly in the virtually enclosed Arctic seaway, but also around the
Antarctic perimeter. Winter precipitation was as
snow in these boreal areas, both more abundant
and more extensive in the Lower Cretaceous
model than in that for the Upper Cretaceous.

The large mid-west seaway over North America
in the Upper Cretaceous helps to ameliorate both
summer and winter temperatures there, as had
already been demonstrated by Valdes et al.
(1996). Lowlands bordering the southern shores of
this seaway are warm whereas the rising cordillera
to the west of this seaway is a cool zone.

Low latitude zones are uniformly hot year-long
throughout the Cretaceous. But in the Southern
Hemisphere the changing palaeogeographic configuration impacts significantly on India and
Australia, as they both move northwards, from the
Lower to the Upper Cretaceous (Figs 7 and 9).
Antarctica generally has low sub-zero values
through the winter, but is modelled to have summer
temperatures too warm to sustain an ice-cap.

Precipitation (Figs 8, 10) and
climate zones (Fig. 11)

Because of the high sea-surface temperatures in the
Cretaceous, there is a greatly enhanced hydrological cycle by comparison with the present. Much
convective rainfall is over the equatorial regions
and, as with both the Triassic and Jurassic, this is
concentrated in the oceans.

In both hemispheres, and in both the Lower and
Upper Cretaceous, the zone of highest rainfall,
where precipitation also far exceeds evaporation,
moves with the ITCZ. The distribution of Cretaceous coals and kaolinite typifies the areas modelled to have nearly year-round moisture whereas
evaporites and calcretes are recorded from the
more arid areas in the north (Scotese 2000;
Fig. 11), so in both cases the models appear to be
reasonably compatible with proxy data.

Throughout the Cretaceous, during the winter
months, the northern polar lands of Greenland,
North America, although receiving modest precipitation, have much less evaporation. Significant precipitation, both here and in Siberia, is modelled to

occur as winter snow. Northern mid-latitudes have
an excess of winter moisture in a zone extending
almost as far south as 30° N. Evaporation predomi-
nates from here to just north of the equator, embra-
cing northern and much of western Africa which, in
the Lower Cretaceous, experiences a marked
summer wet season during the June–July–August
period (Fig. 8). By comparison with the preceding
Jurassic, the humid zone has moved south, the
ITCZ bringing monsoonal rains to large tracts of
present-day Brazil, Central Africa and southern
Arabia. In the Arabian Gulf, the Lower Cretaceous
Lafan Formation marks a phase of intense karstifi-
cation and kaolinite development which is wholly
compatible with the model output.

Except in its far northern more arid zone,
Australia is modelled to have had sufficient moisture to be extensively forested, warm in the north
and temperate in the south. Antarctica is a dry continent but just about in balance year-round with
respect to evaporation and precipitation, with
much of the winter precipitation falling as snow.

Evaporation and precipitation (coupled with
local run-off) have an impact on the salinities of
ocean waters (derived as results from the models
but not figured here). In the open Pacific and
Indian Oceans, these are predicted to be at their
most saline at around 30° N and 30° S, with equatorial zones also having relatively high salinities
(>36‰). Marine salinities are modelled to be
highest over areas now occupied by N Africa
(38–40‰) and Iran (38–45‰). Water salinities
are also high in the Argentina and Chile embay-
ments. Throughout the American mid-west
seaway, salinities are low (approaching 30‰), and
remain low into the enclosed Arctic Basin, promoting there the formation, locally, of winter sea-ice.

Cretaceous vegetation and
biomes (Figs 8, 10)

Using the GENESIS model to evaluate the coupled
climate-vegetation system, applying a predictive
vegetation model (EVE), DeConto et al. (2000)
simulated climate, vegetation and ocean interactions for the Upper Cretaceous. Vegetation simulations were justifiably tuned by the elimination of
grasses and sedges (which had yet to appear).
They found that broadleaf herbaceous angiosperms
(referred to as 'forbs'), ferns and shrubs dominate
the cover in the niche today occupied by grasslands,
generating what they termed a 'forb–fern prairie'
(equivalent to 'steppe' and 'grassland' in the
BIOME 4 output shown here). To maintain the
types of high latitude coastal and continental climates warm enough to sustain the plant and
animal communities now well documented from

very high latitudes, in both hemispheres they needed to run the model with 1680 ppm CO_2 in the atmosphere (comparable with the evaluation of Royer *et al.* 2004). Under these simulated conditions, the Antarctic Peninsula became forested by evergreen linear-leaf trees (*Araucaria* and *Podocarpus*), a model prediction that is supported by known fossil occurrences. In the Arctic, boreal forests were fringed by forb–fern prairie. Tropical ocean temperatures are modelled to have reached 30–33 °C which is warmer than at present and a little warmer than most values calculated by reference to oxygen isotope studies (e.g. Crowley & Zachos 2000). The equator-to-pole thermal gradient is much reduced by comparison with present-day values (Mean Annual equatorial SST of 32 °C and polar seawater value of 8 °C), but despite this the simulation has ocean surface circulations almost as vigorous as those of the present. This ocean heat transport provides a mechanism for enhancing the warming of coastal zones in high latitudes. The absence of a circumAntarctic ocean precludes the development of an Antarctic Circumpolar Current. Persistent low pressure over continents generated convergence zones and the advection of warm moist air from the oceans into continental interiors. High latitude forests maintained a relatively low albedo in these areas and this, coupled with warm summer temperatures, helped to maintain polar warmth in a self-perpetuating system. Deep waters within the ocean are simulated to have formed in high southern latitudes with temperatures of 10–12 °C and salinities of around 35‰.

Temperature-limited facies

Johnson *et al.* (2002) record the distribution of Cretaceous reefs. Coral and rudistid reefs are generally considered to have been warm-water (Beauvais 1992), rudist bioherms being restricted to generally lower latitudes than were coral buildups. Nonetheless, through the Cenomanian to Santonian, coral reefs extended to nearly 40° N (in Europe and New Mexico), and to 40° S (Madagascar). The main focus of reef growth at this time was through the Tethyan belt from the Caribbean, through Mediterranean Europe, and the Middle East. In the latest Cretaceous, corals are more restricted, ranging from scattered sites along the equator to about 40° N in northern France. Rudist reefs extend from just south of the equator to about 35° N. It is interesting to note that the high salinity/high temperature zones in the model correspond generally with the distribution of rudist-dominated reef communities (cf. Johnson *et al.* 2002). By reference to cold month means temperatures in our model, it is noticeable that the 24 °C isotherm lies much closer to the equator in the

Southern Hemisphere than it does in the Northern Hemisphere (approaching 20° S and 30° N). Johnson *et al.* (2002) show a similarly asymmetric distribution of reef belts.

By reference to facies and other climate proxies, the mid-Cretaceous (*c.* 105 to *c.* 89 Ma) appears to have been one of the warmer time intervals of the Phanerozoic (Veizer *et al.* 2000, reviewed in Johnson *et al.* 2002), and maybe rivalling the Triassic. Warm and cool temperate climates extended into high polar latitudes (Askin & Spicer 1995; Herman & Spicer 1996; Tarduno *et al.* 1998). The physiognomy of fossil leaves from mid- to late Cretaceous plants from the Antarctic Peninsula (then at around 65° S) suggests mean annual temperatures and 17 °C to 19 °C (Hayes *et al.* 2006). Isotopic data from planktic foraminifera suggest SSTs as high as 35 °C (Norris *et al.* 2002; Wilson *et al.* 2002) and high latitude SSTs at 20 °C or above (Huber *et al.* 1995; Bice *et al.* 2002). Bottom water temperatures, derived from benthic foraminifera, suggest that palaeodepths greater than 1000 m ranged between 11 °C and 19 °C (Huber *et al.* 1999, 2002).

Estimates of atmospheric CO_2 concentrations range widely between <900 ppm and >4000 ppm, being derived from fossil leaf stomatal indices. Ginkgo species in Siberia have been interpreted to suggest values between 4000–5500 ppm (Retallack 2001), but these are probably too high. Carbon isotopic values derived from marine organic compounds from the Western Interior Seaway of North America and North Atlantic have been interpreted to indicate values between 830–1100 ppm (Freeman & Hayes 1992). A wide range of other values have also been obtained based on analyses of fossil pedogenic carbonates (e.g. Cerling 1991, who estimates 1500–3000 ppm). Berner & Kothavala (2001) suggest values of around 1500 ppm at the start of the Cretaceous, falling gradually towards 1000 ppm by the mid-Campanian and then dropping towards *c.* 800 ppm approaching the K–T boundary. Ekart *et al.* (1999) suggest a similar trend in values but a much broader range (from >2500 ppm at the start of the Cretaceous declining to around 500 ppm at its close). All these authors indicate a wide measure of uncertainty, which typifies such evaluations, with many researchers listing error bars exceeding 1500 ppm. Nonetheless, as Bice & Norris (2002) point out, to have SSTs in tropical and temperate latitudes as high as 30 °C may require CO_2 concentrations as high as 4500 ppm or more. They furthermore speculate that the CO_2 content of the atmosphere may have been highly dynamic, exhibiting wide fluctuations and over a variety of timescales. Recent research, combining experiments with modern bryophytes and detailed investigation of fossil counterparts from Antarctica, are interpreted to indicate mid-Cretaceous values in the range 1000–1400 ppm, a range supported by geochemical modelling (Fletcher *et al.* 2005).

Saltzman & Barron (1982) reported evidence of saline deep water masses within the Campanian/Maastrichtian South Atlantic, based on the $\delta^{18}O$ values derived from inoceramid shells. Such data have been refined more recently and D'Hondt & Arthur (2002) are now able to recognize, on the basis of stable isotopic data from benthic foraminifera, at least three deep water masses in the Late Maastrichtian oceans. The coolest intermediate-depth water gives temperatures of c. 6 °C and originated in a high latitude Southern Ocean. The deepest waters give 10 °C and probably originated in the northern Atlantic. The warmest intermediate waters were 13–15 °C and have an unknown source. Much of this deep water, they suggest, was preconditioned for winter sinking by low- or middle-latitude evaporation (much as are the Mediterranean, Red Sea and North Atlantic waters today), but only a small proportion (c. 10%) of it could have come directly from such a source. Thus low- or mid-latitude evaporation is likely to have played a significant role in preconditioning some surface waters for sinking, but the sinking being activated by seasonal cooling of such waters. Warm waters would have generally restricted the oceanic storage capacity for both CO_2 and methane hydrates, by comparison with icehouse times, such as the present and later Cenozoic.

Based on isotopic data derived from planktonic foraminifera, there have been several claims that oceanic equatorial waters were only as warm as, or cooler, during the Cretaceous than at present, despite the general equability indicated by strong evidence of polar warmth (e.g. Sellwood *et al.* 1994 [c. 25–26 °C]; D'Hondt & Arthur 1996 [c. 20–21 °C]). Also using $\delta^{18}O$ values, but from pristine shell material, Wilson and Opdyke (1996) have shown that rudist reefs grew at 27–32 °C (i.e. at least as warm as at present). So the apparently cooler temperatures reported can now be explained by cryptic diagenetic alteration of foraminiferan shells in cool oceanic bottom waters (Pearson *et al.* 2001) with equatorial values of 28–32 °C being confirmed. Jenkyns *et al.* (2004) have recently interpreted Maastrichtian Arctic marine palaeotemperatures using TEX_{86}, a palaeothermometer based on the membrane lipids from marine planktonic Crenarchaeota. They inferred SSTs of c. 15 °C at 80° N (Alpha Ridge, Arctic Ocean) but this probably reflects summer temperatures and that mean annual temperatures are much lower, with the possibility of seasonal sea-ice. This value, when compared with the equatorial temperatures cited above, gives an equator to

North Pole gradient of a mere 15 °C. Huber *et al.* (1995) had previously inferred an equator to South Pole gradient of c. 14 °C based on oxygen-isotopic ratios from planktonic foraminifera. Nonetheless, there appears to be a well-documented trend towards cooling global oceanic temperatures, and cooling and less saline bottom waters, approaching the end of the Cretaceous (reviewed in Macleod *et al.* 2000).

By reference to the palaeo-distribution of fossil crocodilians, Markwick (1998, and this volume) has evaluated the cooling in Earth's climate over the past 100 Ma. He shows that temperature is the principal influence on crocodilian global distributions, with the coldest month mean temperature (CMM) of 5.5 °C marking the minimum thermal limit for the group (corresponding to a modern mean annual temperature (MAT) of c. 14.2 °C). He suggests that during the late Cretaceous and early Palaeogene, MATs in excess of 14.4 °C (CMMs > 5.5 °C) permeated throughout midlatitudes and coastal regions in high latitudes. Crocodiles are recorded just N of 30° N in N America and 43° S in S America, in the latter case associated with turtles. Turtles are also recorded from India at 35° S. Crocodiles extend to 40° N in China and 30° N in Europe. These distributions are close to a modelled CMM of 6 °C in China, >8 °C in Europe, and >8 °C in Patagonia. But they occur in areas modelled to have a CMM of −8 °C in North America, clearly presenting a major discrepancy between model output and geological reality. The remarkable recent fossil evidence that several groups of dinosaurs were feather-insulated (e.g. in China) is quite compatible with the low winter temperatures modelled in these areas. If modelled temperatures are correct, then many dinosaur groups living in mid- to high latitudes are likely to have been feathered.

Conclusions

Model simulations for the Mesozoic throw important light on climatic processes, and the behaviour of Earth Systems, under greenhouse conditions. The model output generally compares favourably with climate proxy data, but discrepancies between data and models reveal some serious modelling problems (or possibly incorrect interpretation of the proxies). In continental interiors, model simulations generate winter conditions considered to be too cold (by 15 °C or more) by reference to palaeontological data.

Ocean temperatures could have been much warmer than today, not only at the surface but also to great depths (8 °C to the ocean floor). Such temperatures would have greatly reduced the oceanic storage capacity both for CO_2 (in ocean waters) and methane hydrates (in ocean-floor sediments) by comparison with the present day. The simulations shown in this paper are with atmosphere–only models which require the specification of surface ocean conditions. However, we based these sea–surface temperatures on preliminary results from fully coupled model simulations for these periods. Thus the close agreement between model output and proxy climate data in coastal and open sea areas suggests that the models are performing well.

In its greenhouse mode, the Earth has greatly enhanced evaporation and precipitation by comparison with the present, but much of the rainfall is convective in character, falling over the oceans. There is no convincing evidence to indicate that short-term and large-scale glacio-eustatic changes occurred during the Mesozoic, even though models suggest the possibility of seasonal sea-ice. To replicate climatic conditions similar to those indicated by proxy facies, a very large increase in atmospheric CO_2 is required (at least $4\times$ present-day values).

We gratefully acknowledge the helpful corrections and suggestions concerning the first draft of this paper provided by Professors Tony Hallam and Jane Francis, and the editors.

References

ARKELL, W. J. 1956. *Jurassic geology of the world.* Oliver & Boyd, Edinburgh.

ASKIN, R. A. & SPICER, R. A. 1995. The Late Cretaceous and Cenozoic history of vegetation and climate of northern and southern high latitudes: a comparison. *In: Effects of Past Global Change on Life.* Studies in Geophysics, National Research Council, National Academy Press, Washington, 156–173.

BARRON, E. J. 1983. A warm equable Cretaceous: the nature of the problem. *Earth-Science Reviews*, **19**, 305–338.

BARRON, E. J. 1987. Eocene equator-to-pole surface ocean temperatures: a significant climate problem. *Palaeoclimatology, Palaeoecology, Palaeogeography*, **2**, 729–740.

BARRON, E. J. & WASHINGTON, W. M. 1985. Warm Cretaceous climates: high atmospheric CO_2 as a plausible mechanism. *In:* SUNDQUIST, E. T. & BROEKER, W. S. (eds) *The Carbon Cycle and Atmospheric CO_2: Natural Variations Archean to Present.* Geophysical Monograph Series, Washington, DC, AGU, 546–553.

BARRON, E. J., FAWCETT, P. J., POLLARD, D. & THOMPSON, S. 1994. Model simulations of Cretaceous climates: the role of geography and carbon dioxide. *In:* ALLEN, J. R. L., HOSKINS, B. J., SELLWOOD, B. W., SPICER, R. S. & VALDES, P. J. (eds) *Palaeoclimates and their Modelling: with Special Reference to the Mesozoic Era.* Chapman & Hall, 99–108.

BEAUVAIS, L. 1992. Palaeobiogeography of the Early Cretaceous corals. *Palaeogeography, Palaeoclimatology, Palaeoecology*, **92**, 233–247.

BECKMANN, B, FLOGEL, S., HOFMANN, P., SCHULZ, M. & WAGNER, T. 2005. Orbital forcing of river discharge in tropical Africa and ocean response. *Nature*, **437**, 241–244.

BENTON, M. J. 2003. *When Life Almost Died.* Thames & Hudson Ltd, London, 336pp.

BERNER, R. A. & KOTHAVALA, Z. 2001. GEOCARB III: A revised model of atmospheric CO_2 over Phanerozoic time. *American Journal of Science*, **301**, 182–204.

BICE, K. L. & NORRIS, R. D. 2002. Possible atmospheric extremes of the Middle Cretaceous (Albian–Turonian). *Paleoceanography*, **17**, 1070, doi: 10.1029/2002PA000778, 2002.

BICE, K. L., HUBER, B. T. & NORRIS, R. 2003. Extreme polar warmth during the Cretaceous greenhouse? Paradox of late Turonian ^{18}O record at Deep Sea Drilling Project Site 511. *Paleoceanography*, **18**, 1031, doi:10.1029/2002PA000848, 2003.

BILLUPS, K. & SCHRAG, D. P. 2002. Paleotemperatures and ice volume of the past 27 Myr revisited with paired Mg/Ca and $^{18}O/^{16}O$ measurements from benthic foraminifera. *Paleoceanography*, **17**, 11pp.

BJERRUM, C., SURLYK, F., CALLOMON, J. H. & SLINGERLAND, R. L. 2001. Numerical paleoceanographic study of the Early Jurassic transcontinental Laurasian Seaway. *Paleoceanography*, **16**, 390–404.

CERLING, T. E. 1991. Carbon dioxide in the atmosphere: evidence from Cenozoic and Mesozoic paleosols. *American Journal of Science*, **291**, 377–400.

CHANDLER, M. A., RIND, D. & RUEDY, R. 1992. Pangaean climate during the Early Jurassic: GCM simulations and the sedimentary record of palaeoclimate. *Geological Society of America Bulletin*, **104**, 543–559.

CROWLEY, T. J. 1996. Remembrance of things past: Greenhouse lessons from the geological record, *Consequences*, 2, U.S. GCRIO [Online], (available at http://www.gcrio.org/CONSEQUENCES/winter96/ geoclimate.html)

CROWLEY, T. J. & NORTH, G. R. 1991. *Paleoclimatology*, Oxford University Press. 349pp.

CROWLEY, T. J. & ZACHOS, J. C. 2000. Comparison of zonal temperature profiles for past warm time periods. *In:* HUBER, B. T., MACLEOD, K. G. & WING, S. T. (eds) *Warm Climates in Earth History.* Cambridge University Press, 50–76.

DECONTO, R. M. & POLLARD, D. 2003. Rapid Cenozoic glaciation of Antarctica induced by declining atmospheric CO_2. *Nature*, **421**, 245–249.

DECONTO, R. M., BRADY, E. C., BERGENGREN, J. & HAY, W. W. 2000. Late Cretaceous climate, vegetation and ocean interactions. *In:* HUBER, B. T., MACLEOD, K. G. & WING, S. T. (eds) *Warm Climates in Earth History.* Cambridge University Press, 275–296.

D'HONDT, S. & ARTHUR, M. A. 1996. Late Cretaceous oceans and the cool tropics paradox. *Science*, **271**, 1838–1841.

D'HONDT, S. & ARTHUR, M. A. 2002. Deep water in the late Maastrichtian ocean. *Paleoceanography*, **17**, doi:10.1029/1999PA000486, 2002.

EKART, D. D., CERLING, T. E., MONTANEX, I. P. & TABOR, N. J. 1999. A 400 million year carbon isotope record of pedogenic carbonate: implications for paleoatmospheric carbon dioxide. *American Journal of Science*, **299**, 805–827.

FLETCHER, B. J., BEERLING, D. J., BRENTALL, S. J. & ROYER, D. L. 2005. Fossil bryophytes as recorders of ancient CO_2 levels: experimental evidence and a Cretaceous case study. *Global Biogeochemical Cycles*, **19**, GB3012, doi:10:1029/2005GB002495.

FLÜGEL, E. 2002. Triassic reef patterns. *In*: KIESSLING, W., FLÜGEL, E. & GOLONKA, J. (eds) *Phanerozoic Reef Patterns*. SEPM, Special Publication, **72**, 391–463, Tulsa.

FRAKES, L. A. 1979. *Climates Throughout Geologic Time*. Elsevier, Amsterdam, 310pp.

FRAKES, L. A., FRANCIS, J. E. & SYKTUS, J. I. 1992. *Climate Modes of the Phanerozoic*. Cambridge University Press, 274pp.

FREEMAN, K. H. & HAYES, J. M. 1992. Fractionation of carbon isotopes by phytoplankton and estimates of ancient CO_2 levels. *Global Bigeochemical Cycles*, **6**, 185–198.

GRADSTEIN, F. M., OGG, J. G, SMITH, A. G., BLEEKER, W. & LOURENS, L. J. 2004. A new Geologic Time Scale with special reference to the Precambrian and Neogene. *Episodes*, **27**, 83–100.

HALL, I. R., MCCAVE, I. N., ZAHN, R., CARTER, L., KNUTZ, P. C. & WEEDON, G. P. 2003. Paleocurrent reconstruction of the deep Pacific inflow during the middle Miocene: reflections of East Antarctic ice sheet growth. *Paleoceanography*, **18**, 1040, doi:10.1029/2002PA000817, 2003.

HALLAM, A. 1975. *Jurassic Environments*, Cambridge University Press, 269pp.

HALLAM, A. 1985. A review of Mesozoic climates. *Journal of the Geological Society, London*, **142**, 433–445.

HALLAM, A. 1992. Phanerozoic sea-level changes. Columbia University Press, New York, 266pp.

HALLAM, A. 1994. Jurassic climates as inferred from the sedimentary and fossil record. *In*: ALLEN, J. R. L., HOSKINS, B. J., SELLWOOD, B. W., SPICER, R. S. & VALDES, P. J. (eds) *Palaeoclimates and Their Modelling: with Special Reference to the Mesozoic Era*. Chapman & Hall, 79–88.

HANDOH, I. C., BIGG, G. R., JONES, E. J. W. & MASAMICHI, I. 1999. An ocean modelling study of the Cenomanian Atlantic: equatorial paleo-upwelling, organic-rich sediments and the consequences for a connection between the proto-North and South Atlantic. *Geophysical Research Letters*, **26**, 223–226.

HAQ, B. U., HARDENBOL, J. & VAIL, P. R. 1987. Chronology of fluctuating sea levels since the Triassic. *Science*, **235**, 1156–1166.

HAYES, P. A., FRANCIS, J. E., CANTRILL, D. J. & CRAME, J. A. 2006. Palaeoclimate analysis of Late Cretaceous angiosperm leaf floras, James Ross Island, Antarctica. *In*: FRANCIS, J. E., PIRRIE, D. & CRAME, J. A. (eds) *Cretaceous–Tertiary High Latitude Palaeoenvironments, James Ross Basin, Antarctica*. Geological Society, London, Special Publications, **258**, 49–62.

HAYWOOD, A. M., VALDES, P. J., SELLWOOD, B. W. & KAPLAN, J. O. 2002. Antarctic climate during the middle Pliocene: model sensitivity to ice sheet variation. *Palaeogeography, Palaeoclimatology, Palaeoecology*, **182**, 93–115.

HERMAN, A. B. & SPICER, R. A. 1996. Palaeobotanical evidence for a warm Cretaceous Arctic Ocean. *Nature*, **380**, 330–333.

HUBER, B. T. & WATKINS, D. K. 1992. Biogeography of Campanian–Maastrichtian calcareous plankton in the region of the Southern Ocean: paleogeographic and palaeoclimatic implications. *In*: KENNETT, J. P. & WARNKE, D. A. (eds) *The Antarctic Paleoenvironment: A Perspective on Global Change*. American Geophysical union, Antarctic Research Series, **56**, 31–60.

HUBER, B. T., HODELL, D. A. & HAMILTON, C. P. 1995. Mid- to Late Cretaceous climate of the southern high latitudes: stable isotopic evidence for minimal equator-to-pole thermal gradients. *Geological Society of America Bulletin*, **107**, 1164–1191.

HUBER, B. T., LECKIE, R. M., NORRIS, R. D., BRALOWER, T. J. & CO BABE 1999. Foraminiferal assemblage and stable isotopic change across the Cenomanian-Turonian boundary in the subtropical North Atlantic. *Journal of Foraminiferal Research*, **29**, 392–417.

HUBER, B. T., MACLEOD, K. G. & WING, S. T. 2000. *Warm Climates in Earth History*. Cambridge University Press, 462pp.

HUBER, B. T., NORRIS, R. D. & MACLEOD, K. G. 2002. Deep-sea paleotemperature record of extreme warmth during the Cretaceous. *Geology*, **30**, 123–126.

IVANY, L. C., VAN SIMAEYS, S., DOMACK, E. W. & SAMSON, S. D. 2006. Evidence for an earliest Oligocene ice sheet on the Antarctic Peninsula. *Geology*, **34**, 377–380.

JENKYNS, H. C., FOSTER, A., SCHOUTEN, S. & DAMSTE, J. S. S. 2004. High temperatures in the Late Cretaceous Arctic Ocean. *Nature*, **432**, 888–892.

JOHNSON, C. C., SANDERS, D., KAUFFMAN, E. G. & HAY, W. 2002. Patterns and processes influencing Upper Cretaceous reefs. *In*: KIESSLING, W., FLÜGEL, E. & GOLONKA, J. (eds) *Phanerozoic Reef Patterns*. SEPM, Special Publication, **72**, 549–585, Tulsa.

KUMP, L. R., KASTING, J. F. & CRANE, R. G. 1999. *The Earth System*. Prentice Hall, 351pp.

KUTZBACH, J. E. & GALLIMORE, R. G. 1989. Pangaean climates – Megamonsoons of the megacontinent. *Journal of Geophysical Research – Atmospheres*, **94**, 3341–3357.

KUTZBACH, J. E. & ZIEGLER, A. M. 1994. Simulation of Late Permian climate and biomes with an atmosphere–ocean model – comparisons with observations. *Philosophical Transactions of the Royal Society of London*, **B 341**, 327–340.

MACLEOD, K. G., HUBER, B. T. & DUCHARME, M. L., 2000. Paleontological and geochemical constraints on the deep oceans during the Cretaceous greenhouse interval. *In*: HUBER, B. T., MACLEOD, K. G. & WING, S. T. (eds) *Warm Climates in Earth History*. Cambridge University Press, 241–274.

MARKWICK, P. J. 1998. Fossil crocodilians as indicators of Late Cretaceous and Cenozoic climates: implications for using palaeontological data in reconstructing palaeoclimate. *Palaeogeography, Palaeoclimatology, Palaeoecology*, **137**, 205–271.

MARKWICK, P. J. 2007. Examination of palaeographic and palaeoclimatic significance of selected climate proxies, and their use in quantitatively evaluating mesozoic and Cenozoic climate model results. *In*: WILLIAMS, M., HAYWOOD, A. M., GREGORY, F. J. & SCHMIDT, D. N. (eds) *Deep-Time Perspectives on Climate Change: Marrying the Signal from Computer Models and Biological Proxies*. The Micropalaeontological Society, Special Publications. The Geological Society, London, 251–312.

MEYEN, S. V. 1997. Permian conifers of Western Angaraland. *Review of Palaeobotany and Palynology*, **96**, 351–447.

MOORE, G. T., HAYASHIDA, D. N., ROSS, C. A. & JACOBSON, S. R. 1992a. Paleoclimate of the Kimmeridgian/Tithonian (Late Jurassic) world, I. Results using a general- circulation model. *Palaeogeography, Palaeoclimatology, Palaeoecology*, **93**, 113–150.

MOORE, G. T., SLOAN, L. C., HAYASHIDA, D. N., ROSS, C. A. & UMRIGAR, N. P. 1992b. Paleoclimate of the Kimmeridgian/Tithonian (Late Jurassic) world, II: sensitivity tests comparing 3 different paleotopographic settings. *Palaeogeography, Palaeoclimatology, Palaeoecology*, **95**, 229–252.

NORRIS, R. D., BICE, K. L., MAGNO, E. A. & WILSON, P. A. 2002. Jiggling the tropical thermostat in the Cretaceous hothouse. *Geology*, **30**, 299–302.

PARRISH, J. T. 1994. A brief discussion of the history, strengths and limitations of conceptual climate models for pre-Quaternary time. *In*: ALLEN, J. R. L., HOSKINS, B. J., SELLWOOD, B. W., SPICER, R. S. & VALDES, P. J. (eds) *Palaeoclimates and Their Modelling: with Special Reference to the Mesozoic Era*. Chapman & Hall, 55–58.

PEARSON, P. N., DITCHFIELD, P. W., SINGANO, J., HARCOURT-BROWN, K. G., NICHOLAS, C. J., OLSSON, R. K., SHACKLETON, N. J. & HALL, M. A. 2001. Warm tropical sea surface temperatures in the Late Cretaceous and Eocene epochs. *Nature*, **413**, 481–487.

POE, V. D., GALLANI, M. L., ROWNTREE, P. R. & STRATTON, R. A. 2000. The impact of new physical parametrizations in the Hadley Centre climate model: HadAM3: *Climate Dynamics*, **16**, 123–146.

POPE, V. D., GALLANI, M. L., ROWNTREE, P. R. & STRATTON, R. A. 2000. The impact of new physical parametrizations in the Hadley Centre climate model: HadAM3: *Climate Dynamics*, **16**, 123–146.

POULSEN, C. J., BARRON, E. J, ARTHUR, M. A. & PETERSON, W. H. 2001. Response of the mid-Cretaceous global oceanic circulation to tectonic and CO_2 forcings. *Paleoceanography*, **16**, 576–592.

PRICE, G. D. 1999. The evidence and implications of polar ice during the Mesozoic. *Earth–Science Reviews*, **48**, 183–210.

PRICE, G. D. & SELLWOOD, B. W. 1997. "Warm" palaeotemperatures from high Late Jurassic palaeolatitudes (Falkland Plateau): Ecological, environmental or diagenetic controls? *Palaeogeography, Palaeoclimatology, Palaeoecology*, **129**, 315–327.

PRICE, G. D., SELLWOOD, B. W. & VALDES, P. J. 1995. Sedimentological evaluation of general circulation model simulations for the "greenhouse" Earth: Cretaceous and Jurassic case studies. *Sedimentary Geology*, **100**, 159–180.

PRICE, G. D., VALDES, P. J. & SELLWOOD, B. W. 1997. Quantitative palaeoclimate GCM validation: Late Jurassic and mid-Cretaceous case studies. *Journal of the Geological Society of London*, **154**, 769–772.

REES, P. M., ZIEGLER, A. M. & VALDES, P. J. 2000. Jurassic phytogeography and climates: new data and model comparisons. *In*: HUBER, B. T., MACLEOD, K. G. & WING, S. T. (eds) *Warm Climates in Earth History*. Cambridge University Press, 297–318.

RETALLACK, G. J. 2001. A 300 million-year record of atmospheric carbon dioxide from fossil plant cuticles. *Nature*, **411**, 287–290.

ROYER, D., BERNER, R. A., MONTAÑEZ, I. P., TABOR, N. J. & BEERLING, D. J. 2004. CO_2 as a primary driver of Phanerozoic climate. *GSA Today*, **14**, 4–10.

SALTZMAN, E. S & BARRON, E. J. 1982. Deep circulation in the Late Cretaceous – oxygen isotope paleotemperatures from *Inoceramus* remains in DSDP cores. *Palaeogeography, Palaeoclimatology, Palaeoecology*, **40**, 167–181.

SCOTESE, C. R. 2000. Paleomap Project, http://www.scotese.com

SELLWOOD, B. W. & PRICE, G. D. 1994. Sedimentary facies as indicators of Mesozoic palaeoclimate. *In*: ALLEN, J. R. L., HOSKINS, B. J., SELLWOOD, B. W., SPICER, R. S. & VALDES, P. J. (eds) *Palaeoclimates and Their Modelling: with Special Reference to the Mesozoic Era*. Chapman & Hall, 17–25.

SELLWOOD, B. W. & VALDES, P. J. 2006. Mesozoic climates: general circulation models and the rock record. *Sedimentary Geology*, **190**, 269–287.

SELLWOOD, B. W., PRICE, G. D. & VALDES, P. J. 1994. Cooler estimates of Cretaceous temperatures. *Nature*, **370**, 453–455.

SKELTON, P., SPICER, R. A., KELLEY, S. P. & GILMOUR, I. 2003. *The Cretaceous World*. The Open University Cambridge University Press, 360pp.

SMITH, A. G., SMITH, D. G. & FUNNELL, B. M. 1994. *Atlas of Mesozoic and Cenozoic Coastlines*. Cambridge University Press, 99pp.

SPICER, R. A., REES, P. M. & CHAPMAN, J. 1994. Cretaceous phytogeography and climate signals. *In*: ALLEN, J. R. L., HOSKINS, B. J., SELLWOOD, B. W., SPICER, R. S. & VALDES, P. J. (eds) *Palaeoclimates and Their Modelling: with Special Reference to the Mesozoic Era*. Chapman & Hall, 69–78.

TARDUNO, J. A., BRINKMAN, D. B., RENNE, P. R., COTTRELL, R. D., SCHER, H. & CASTILLO, P. 1998. Evidence for extreme climatic warmth from Late Cretaceous Arctic vertebrates. *Science*, **282**, 2241–2244.

TAYLOR, E. L., TAYLOR, T. N. & CUNÉO, R. 2000. Permian and Triassic high latitude paleoclimates: evidence from fossil biotas. *In*: HUBER, B. T., MACLEOD, K. G. & WING, S. T. (eds) *Warm*

Climates in Earth History. Cambridge University Press, 321–350.

TRIPATI, A., BACKMAN, J., ELDERFIELD, H. & FERRETTI, P. 2005. Eocene bipolar glaciation associated with global carbon cycle changes. *Nature*, **436**, 341–346.

TUCKER, M. E. & BENTON, M. J. 1984. Triassic environments, climates and reptile evolution. *Palaeogeography, Palaeoclimatology, Palaeoecology*, **40**, 361–379.

VALDES, P. J. & SELLWOOD, B. W. 1992. A palaeoclimate model for the Kimmeridgian. *Palaeogeography, Palaeoclimaotlogy, Palaeoecology*, **95**, 47–72.

VALDES, P. J., SELLWOOD, B. W. & PRICE, G. D. 1995. Modelling Late Jurassic Milankovitch climate variations. *In*: HOUSE, M. R. & GALE, A. S. (eds) *Orbital Forcing Timescales and Cyclostratigraphy*. Geological Society Special Publication, **85**, 115–132.

VALDES, P. J., SELLWOOD, B. W. & PRICE, G. D. 1996. Evaluating concepts of Cretaceous equability. *Palaeoclimates*, **1**, 139–158.

VALDES, P. J., SPICER, R. A., SELLWOOD, B. W. & PALMER, D. C. 1999. *Understanding Past Climates: Modelling Ancient Weather*. CD–ROM, Routledge.

VEIZER, J., GODDERIS, Y. & FRANCOIS, L. M. 2000. Evidence for decoupling of atmospheric CO_2 and global climate during the Phanerozoic eon. *Nature*, **408**, 698–701.

WALTER, H. 1985. Vegetation of the Earth. Springer-Verlag, Berlin, 318pp.

WILLIAMS, M., HAYWOOD, A. M., TAYLOR, S. P., VALDES, P. J., SELLWOOD, B. W. & HILLENBRAND, C-D. 2005. Evaluating the efficacy of planktonic foraminifer calcite d^{18}O data for sea surface temperature reconstruction for the Late Miocene. *Geobios*, **38**, 843–863.

WILSON, P. A. & OPDYKE, B. N. 1996. Equatorial sea-surface temperatures for the Maastrichtian revealed through remarkable preservation of metastable carbonate. *Geology*, **24**, 555–558.

WILSON, P. A., NORRIS, R. D. & COOPER, M. J. 2002. Testing the Cretaceous greenhouse hypothesis using glassy foraminiferal calcite from the core of the Turonian tropics on Demerara Rise. *Geology*, **30**, 607–610.

ZACHOS, J. C., ARTHUR, M. A., BRALOWER, T. J. & SPERO, H. J. 2002. Palaeoclimatology – Tropical temperatures in greenhouse episodes. *Nature*, **419**, 897–898.

ZACHOS, J. C., BREZA, J. R. & WISE, S. W. 1992. Early Oligocene ice-sheet expansion on Antarctica – stable isotope and sedimentological evidence from Kerguelen Plateau, southern Indian Ocean. *Geology*, **20**, 569–573.

ZIEGLER, A. M., PARRISH, J. M., JIPING, Y. *ET AL*. 1994. Early Mesozoic phytogeography and climate. *In*: ALLEN, J. R. L., HOSKINS, B. J., SELLWOOD, B. W., SPICER, R. S. & VALDES, P. J. (eds)` *Palaeoclimates and Their Modelling: with Special Reference to the Mesozoic Era*. Chapman & Hall, 89–97.

New approaches for quantifying the Cretaceous terrestrial climate record

G. D. PRICE & S. T. GRIMES

School of Earth, Ocean and Environmental Sciences, University of Plymouth, Drake Circus, Plymouth, PL4 8AA, UK (email: g.price@plymouth.ac.uk)

Abstract: Climatic variability is reviewed, as recorded in the terrestrial record of the Cretaceous and with particular emphasis of newer techniques and advances. Physiognomic analysis of flora from the high latitude sites of the Northern and Southern Hemispheres has provided extensive evidence for warm polar climates and for a gradual decline in terrestrial mean annual temperatures as the Cretaceous drew to a close. These data are largely consistent with the marine record although the limited data available for the Arctic Ocean may require a fuller evaluation of its validity and significance. Climate-related geochemical variations in sediments and phosphate oxygen isotope analysis of continental vertebrates are also consistent with Mid-Cretaceous polar warmth followed by Late Cretaceous cooling. Considerably less quantitative data have been derived from tropical palaeolatitudes. Nevertheless, available data indicate a warm to hot, semi-arid low latitude Cretaceous climate. An apparent lack of a tropical everwet zone results in limited tropical rainforest biomes, which are restricted to a narrow palaeoequatorial belt.

The Cretaceous is considered to have been warmer than today on the basis of a variety of palaeoclimate data and model studies. Among the more compelling pieces of evidence for a Cretaceous greenhouse are isotopic signatures consistent with high sea-surface temperatures preserved by planktonic foraminifera inhabiting the equatorial zone (e.g. Wilson *et al.* 2002) and evidence of temperate forests inside the Arctic and Antarctic circles (e.g. Spicer & Parrish 1990; Falcon-Lang & Cantrill 2001; Francis & Poole 2002). Although quantifying terrestrial palaeotemperatures has proven to be a challenging endeavour, terrestrial sediments and biota are a rich and informative source of data providing much information regarding the different climatic states of the Cretaceous (e.g. Frakes *et al.* 1992; Spicer & Corfield 1992; Chumakov *et al.* 1995; Francis & Poole 2002; Russell & Paesler 2003). It is the aim of this paper to explore the climatic variability as recorded in the terrestrial record of the Cretaceous, with particular emphasis on newer techniques and advances. For example, by using reconstructed vegetation types and angiosperm leaf physiognomic studies, it is possible to construct Cretaceous temperature curves (e.g. Kennedy *et al.* 2002; Spicer *et al.* 2002; Craggs 2005) with internally and externally consistent results. Likewise, phosphate oxygen isotope analysis of tooth enamel from mammals has become a routine tool for the reconstruction of terrestrial palaeoclimates during the Tertiary (e.g. Koch *et al.* 1995; Fricke *et al.* 1998; Grimes *et al.* 2005) and the Quaternary (e.g. Delgado-Huertas *et al.* 1997; Stephan 2000). The principles behind these new palaeoclimate reconstruction techniques are now starting to be applied to continental vertebrates (e.g. Amiot *et al.* 2004), including turtle bone, fish scales and dinosaur teeth and bone of Cretaceous age.

Palaeobotanical evidence

High latitude sites

The morphology-based approach to reconstructing the palaeoclimate of the Cretaceous has been used with great success, particularly with respect to palaeobotanical reconstructions (e.g. Wolfe & Upchurch 1987; Spicer & Parrish 1990; Kennedy *et al.* 2002; Spicer *et al.* 2002). Furthermore, as a result of the continued development of Wolfe's (1979) pioneering work, which resulted in the first quantitative relationship between leaf margins and mean annual temperature (MAT), absolute palaeotemperatures are now routinely applied to palaeobotanical reconstructions of the Cretaceous climate. For example, leaf margin analysis yields a Cenomanian MAT of 10 ± 3 °C on the coastal plain of northern Alaska (Spicer & Parrish 1990), whilst analysis of the leaf physiognomy of the Albian–Cenomanian Krivorechenskaya Formation of the Grebenka area, NE Russia (Spicer *et al.* 2002), suggests that the plants experienced a MAT of 13.0 ± 1.8 °C (Figs 1, 2). Further data of Spicer & Parrish (1990) and Craggs (2005) from Alaska and NE Russia, also based on physiognomic analysis, reveals a trend suggesting a gradual decline in MATs as the Cretaceous drew to a

From: WILLIAMS, M., HAYWOOD, A. M., GREGORY, F. J. & SCHMIDT, D. N. (eds) *Deep-Time Perspectives on Climate Change: Marrying the Signal from Computer Models and Biological Proxies.* The Micropalaeontological Society, Special Publications. The Geological Society, London, 225–233. 1747-602X/07/$15.00 © The Micropalaeontological Society 2007.

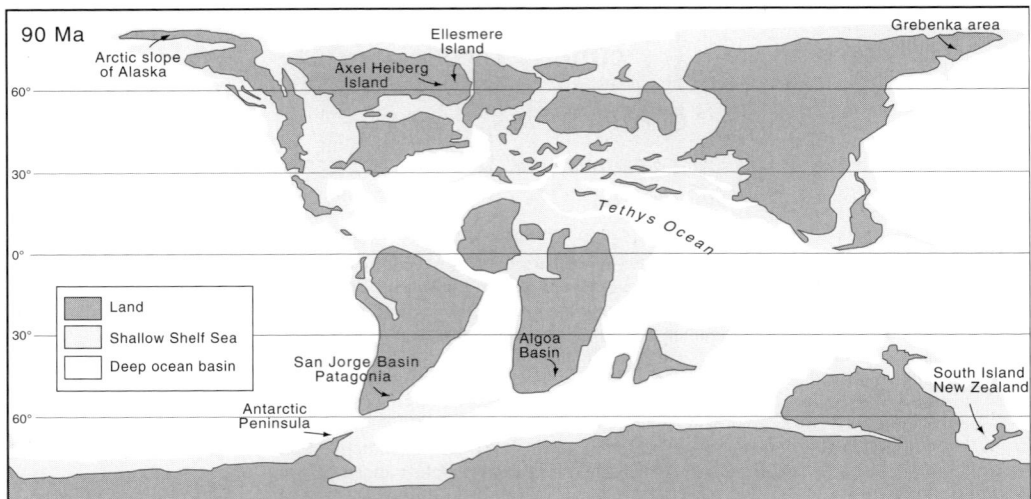

Fig. 1. Upper Cretaceous global palaeogeography (*c.* 90 Ma) showing locations discussed within the text (modified from Blakey 2006).

close. For the Maastrichtian, MATs in the region of 5 °C (Spicer & Parrish 1990) are described (Fig. 2), consistent with only occasional winter freezing. A late Cretaceous (Campanian–Maastrichtian) polar forest from NW Ellesmere Island, Canada (Fig. 1) is thought to represent an extinct biome, which was subjected to a polar light regime and elevated atmospheric CO_2 concentrations (Falcon-Lang

et al. 2004). Their results compare closely with those from coeval circum-Arctic fossil sites and support the occurrence of a latest Cretaceous cool mode, demonstrating the existence of cool temperate forests at high palaeolatitudes in the Canadian Arctic.

A comparison between these terrestrial palaeo-temperatures from Alaska and NE Russia and

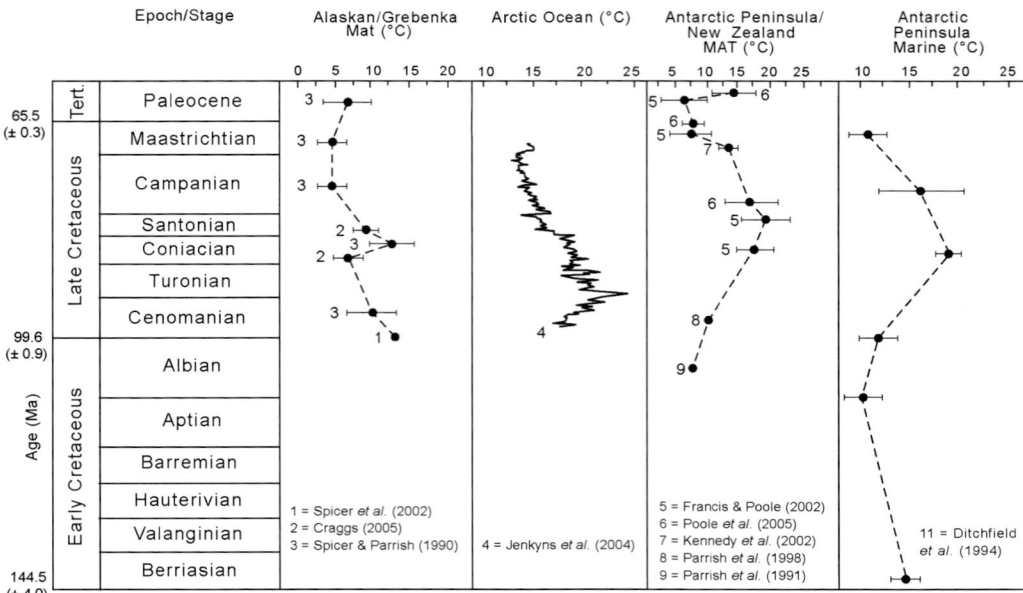

Fig. 2. Compilation of Cretaceous terrestrial palaeotemperature estimates for high latitude Northern and Southern Hemispheres, compared with estimates of marine palaeotemperatures. Ages from Gradstein *et al.* (2004).

temperature estimates for the Arctic Ocean (e.g. Jenkyns *et al.* 2004) are illustrated in Figure 2. The data in this figure clearly show that a decline in Arctic Ocean temperatures from the Mid-Cretaceous through into the Maastrichtian is mirrored in the local terrestrial environment. However, there is a clear difference in the absolute palaeotemperatures, with Jenkyns *et al.* (2004) suggesting very warm (subtropical) Arctic seas, which are certainly inconsistent with the much lower values based upon the palaeobotanical estimates for the local terrestrial climate. This inconsistency may be because the new data, provided by Jenkyns *et al.* (2004), are an extrapolation of isotopic data derived from the English Chalk using TEX86, a new palaeothermometer, the validity of which may require a fuller evaluation. The temperature estimates derived from palaeobotanical evaluations do, however, compare well with General Circulation Model (GCM) simulations of temperature. For example, Upchurch *et al.* (1999) simulate high latitude land temperatures for the Maastrichtian ranging from *c.* −14 °C in the winter to 10 °C in the summer (see also Sellwood & Valdes 2006). Coeval marine temperatures (Fig. 3) reached a maximum of *c.* 4 °C in the summer.

The Northern Hemisphere, high latitude, palaeobotanical palaeotemperatures compare well with those derived from leaf-margin analysis and vegetational physiognomy of Cretaceous floras from Antarctica and the South Island of New Zealand (e.g. Parrish *et al.* 1998; Poole *et al.* 2005, fig. 2). Furthermore, the Southern Hemisphere, low latitude, palaeobotanical palaeotemperatures for the Late Cretaceous are also consistent with GCM data for the Campanian (e.g. DeConto *et al.* 1999). These climate simulations require 1500 ppm of atmospheric CO_2 and increased polar ocean heat transport to replicate the overall polar warmth, along with warm winters within continental interiors (see Upchurch *et al.* 1999 and Fig. 3). In contrast, seasonally cold polar temperatures, and more controversially limited polar ice-caps, have been postulated to have existed during the Early Cretaceous (e.g. Price 1999). Many studies have recorded large and globally synchronous Milankovitch-influenced oscillations of Cretaceous sea level, and fluctuating polar ice volumes have sometimes been invoked as a primary mechanism to account for such changes. Furthermore, Frakes *et al.* (1992) reported tillites and dropstones, providing some equivocal evidence for the presence of ice-sheets. However, the distribution of these ice proxies within Cretaceous sediments is limited (see Chumakov *et al.* 1995).

The vascular tissues of plants display a variety of anatomical features that can be correlated with climate, especially the analyses of growth rings in

fossil wood, providing important information regarding temperature, rainfall, seasonality and climate trends (e.g. Francis & Poole 2002; Philippe *et al.* 2004; Poole *et al.* 2005). For example, the studies of the northern Antarctic Peninsula region by Francis & Poole (2002) and Poole *et al.* (2005) drew on a large dataset of Antarctic conifer and angiosperm woods in order to derive an insight into the palaeoclimate of the region. Climate signals from fossil wood, supported by sedimentary and geochemical evidence, indicate warm and relatively wet conditions during the Coniacian to Early Campanian (Francis & Poole 2002). However, narrower growth rings suggest that the climate cooled during the Maastrichtian and Palaeocene. Furthermore, a study of dicotyledonous wood anatomy indicates that, in addition to cooling, a wet and possibly seasonal climate also prevailed at this

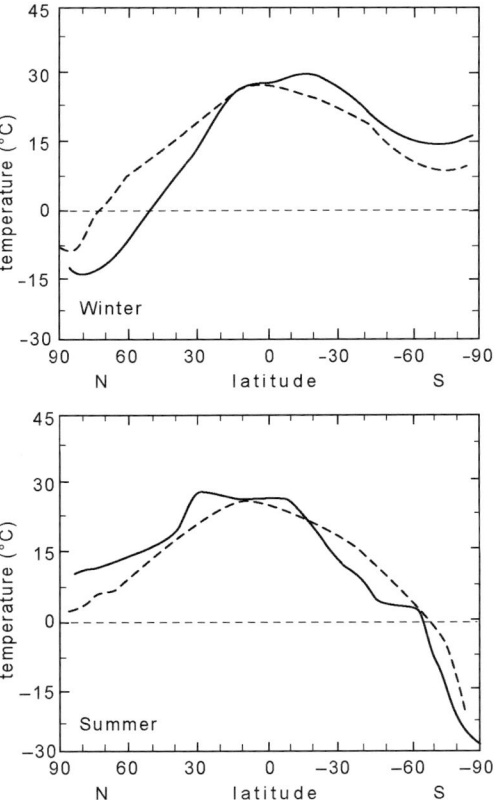

Fig. 3. General Circulation Model (GCM) simulated Late Cretaceous (Maastrichtian) zonally averaged surface palaeotemperatures (°C) for winter (January) and summer (July). Modified from Upchurch *et al.* (1999) from their EVERGREEN TREE GCM experiment. Solid lines are estimates for land palaeotemperatures and dashed lines are for the ocean.

time, with tentatively estimated MATs falling from 7 to 4 °C (Francis & Poole 2002 and Fig. 2). At similar latitudes, plant macrofossil assemblages collected from the Late Cretaceous of South Island, New Zealand (Kennedy *et al.* 2002) indicate, via multivariate leaf morphological analysis (CLAMP), MATs ranging between 12 and 15 °C. These data are consistent with those derived from the Antarctic Peninsula (Fig. 2). As with the Northern Hemisphere high latitude data, the decline in terrestrial Arctic MATs is again consistent with palaeotemperatures derived from the marine realm (e.g. Ditchfield *et al.* 1994). But, unlike with the Northern Hemisphere data, there is good agreement of MATs between the marine and terrestrial realms in the Southern Hemisphere.

Mid-latitude and tropical sites

For the mid-latitudes of the Cretaceous, detailed physiognomic analysis of leaf floras, and analysis of the vascular tissues of plants, are less abundant than those from the high and low latitudes. However, a number of key studies have been undertaken (e.g. Wolfe & Upchurch 1987; Falcon-Lang 2003; Philippe *et al.* 2004). For example, a study of conifer woods from the mid-palaeolatitude Upper Cretaceous (Campanian) Two Medicine Formation, Montana, USA (Falcon-Lang 2003) revealed the complete absence of true growth rings. This, Falcon-Lang (2003) argues, indicated that these trees grew under an equable, megathermal climate that was otherwise favourable for continuous, year-round cambial activity. These results support the results of earlier Cretaceous GCM simulations which suggest that mean winter temperatures for the Western Interior of the USA were 16–20 °C, and summer mean temperatures were 20–24 °C (Valdes *et al.* 1996; Sellwood & Valdes 2006). Leaf physiognomic data suggest that NW Montana lay in the transition zone between a non-seasonal megathermal climate and a weakly seasonal mesothermal climate, a region with a MAT close to 20 °C, and a MAT range of only 8 °C (Wolfe & Upchurch 1987). Wolfe & Upchurch (1987) also provide further data for the Cretaceous of this region in which MATs ranged from 21 °C in the Albian, rising to 25 °C in the Turonian and maintaining similar values in the Campanian and Maastrichtian. Similar values have been derived from the physiognomic analysis of Mid-Cretaceous leaf floras of the Czech Republic (Spicer 2003).

Although Africa was an extensive landmass throughout the Cretaceous, relatively little quantitative data presently exist for its palaeoclimatic reconstruction. An extensive deposit of fossil conifer wood from the Late Jurassic and the Early Cretaceous of Mali, southern Sahara, has however been described by Bamford *et al.* (2002). Their work also reveals indistinct tree ring structures in the fossil conifers, supporting previous suggestions that West Africa had an equable climate during this part of the Mesozoic. Likewise, plant remains from the Early Cretaceous (Valanginian) Kirkwood Formation of the Algoa Basin, South Africa (Fig. 1) indicate a warm to hot, semi-arid climate (Gomez *et al.* 2002). In Saharan Algeria and Tunisia, an abundance of ferns, cycad-like plants and primitive conifers, suggests a moist climate characterized the Early Cretaceous. An abrupt change to a more semi-arid climate is indicated by the influx of xeric conifers during the Barremian. These plants continued as the dominant flora into Cenomanian times (Russell & Paesler 2003; Ziegler *et al.* 2003). Angiosperm leaves from Lebanon, southern Egypt and Sudan also indicate a change to warm and semi-arid climate during Cenomanian times (Dilcher & Basson 1990). This interpretation is supported by Chumakov *et al.* (1995) and Ziegler *et al.* (2003) who show the dominance of sedimentary aridity proxies, such as evaporite deposits, occurring in the mid-latitudes during the Mid-Cretaceous. A similar pattern is also seen in Tertiary and Recent evaporite deposits, which demonstrate that Hadley circulation patterns dominated the tropics as dry areas (Ziegler *et al.* 2003). A true tropical rainforest biome is restricted to a narrow equatorial belt in northern South America and Nigeria and Somalia (e.g. Upchurch *et al.* 1999), perhaps due to the lack of a tropic everwet zone related to the broad spread of available precipitation and imposed high rates of evaporation (Ziegler *et al.* 2003).

Palaeobiological and isotopic evidence

With respect to the Cretaceous fauna, a number of methods have been employed in order to make inferences about terrestrial palaeoclimate. First, taxonomic analogy: a method that is based upon morphological analogy. This, together with a suite of geochemical and isotopic approaches, has been successfully applied to Cretaceous faunas. The taxonomy-based approach, using both fauna and flora, has been widely employed in order to reconstruct climatic zones throughout the Mesozoic (e.g. Barnard 1973). However, the major disadvantage of this approach is that the climatic preference of the nearest living relative may not accurately reflect those of extinct forms because of evolutionary change, extinction, or geographic range restriction related to non-climatic factors (Wing & Greenwood 1993; Markwick 1998). For example, the occurrence of dinosaurs and large amphibians at high latitudes has frequently been taken to reflect warm and equable polar climates. Cretaceous polar dinosaurs are known from New Zealand, Australia, Siberia, Svalbard, Canada, Antarctica, and Alaska (e.g. Edwards *et al.* 1978; Tarduno *et al.* 1998; Rich *et al.* 2002; Fiorillo & Parrish 2004). However, if some dinosaurs were warm blooded (Bakker 1980; Seebacher 2003; Amiot *et al.* 2006), they may have been able to tolerate colder winters than that suggested by thermal tolerance studies (Paul 1988). The discovery of remarkably preserved early Cretaceous dinosaurs covered with primitive feathers (e.g. Ji *et al.* 2001; Norell *et al.* 2002) also provides evidence that these animals may well have developed feathers to maintain warmth before they were used for flight. Alternatively, dinosaurs may have seasonally migrated to warmer climates or hibernated during seasonally cold winters (e.g. Axelrod 1984; Parrish *et al.* 1987; Kear 2005).

Recently, Fricke & Rogers (2000) examined the oxygen isotope ratios of tooth enamel from Late Cretaceous crocodiles and coexisting theropod dinosaur taxa from several different localities in order to investigate thermoregulation. They concluded that for theropods the isotopic variation was consistent with that of homeothermic animals having constant body temperatures independent of their environments, i.e. warm blooded (see Barrick & Kohn 2000 for qualified comments). Hence, it is possible that dinosaurs are not strictly bound to tropical climates, like modern giant reptiles (Buffetaut 2004), and consequently Cretaceous palaeoclimate reconstructions using thermal tolerance studies should be considered in this context.

Using the climatic preference of the nearest living relative of champsosaurs and crocodilians has also been undertaken to evaluate the characteristics of polar climates. The discovery of champsosaurs on Axel Heiberg Island (72° N) (Tarduno *et al.* 1998) contrasts with the low latitude distribution of their nearest living relatives, the crocodiles, which are considered as their morphological and ecological analogues. It is therefore considered likely that champsosaurs could not have tolerated prolonged exposure to subfreezing conditions and Tarduno *et al.* (1998) and Friedman *et al.* (2003) estimated a MAT in excess of 14 °C for the high Canadian Arctic based on the thermal tolerances of extant crocodilians. Likewise, the detailed study of Markwick (1998) also recorded fossil crocodilians occurring up to at least 60° N for the Campanian to Mid-Eocene and as a consequence inferred MATs for those regions of >14.2 °C.

Further evidence of Cretaceous (Aptian–Albian) polar temperatures has been provided by oxygen isotope data from carbonate cements in concretions from Victoria, Australia. Gregory *et al.* (1989) and Ferguson *et al.* (1999) proposed that meteoric fluids with $\delta^{18}O$ values as low as $-20‰$ were involved in the precipitation of the early calcites

and are comparable with $\delta^{18}O$ of calcites formed beneath modern glaciers. Gregory *et al.* (1989) suggest their data indicate MATs of less than 5 °C and perhaps as low as -6 °C and should be taken to be indicative of the presence of seasonal ice and possibly permanent ice at high elevations or in the interior of the Antarctic continent itself. Palaeotemperature estimates based upon these oxygen isotope ratios of calcite derived from concretions are still, however, controversial.

The analysis of phosphate oxygen isotope ratios of tooth enamel from mammals has become a routine tool for the reconstruction of terrestrial palaeoclimates in the Cenozoic. These techniques are now starting to be applied to continental vertebrate remains (including turtle bone, fish scales and dinosaur teeth and bone) of Cretaceous age (Fricke & Rogers 2000). For example, Amiot *et al.* (2004) have used these techniques to demonstrate that during the Late Campanian–Mid-Maastrichtian, the temperature gradient at palaeolatitudes 83 °N to 32 °S was more steep (0.4 ± 0.1 °C/°latitude) than the present-day one (0.6 °C/°latitude), with temperatures that decreased from about 30 °C near the equator to about -5 °C at the poles. Such estimates of terrestrial temperature for the Late Cretaceous are certainly in line with those constructed from physiognomy and the characteristics of the vascular tissues of plants (e.g. Wolfe and Upchurch 1987; Spicer and Parrish 1990, see Fig. 2). Above 30° of palaeolatitude, Amiot *et al.* (2004) also noted that their data were consistent with air temperatures that were also a little higher than at present.

However, such isotopic reconstructions of ancient palaeoclimates may be affected by diagenesis. For example, the analysis of the stable isotope composition and the rare earth element content of vertebrate remains from the Late Cretaceous of northern Spain (Lano) by Lécuyer *et al.* (2003) indicates that the geochemical compositions of the biogenic apatites have most likely been acquired during multi-stage diagenesis involving complete recrystallization of the apatite in the presence of aqueous fluids. Despite this diagenetic alteration, they argued that the narrow range of phosphate $\delta^{18}O$ values measured in the closely associated marine (sharks and rays) and freshwater vertebrate species (crocodiles and turtles) reflects the temperature and isotopic composition of groundwaters that prevailed in the burial environment. Furthermore, if it is assumed that all oxygen in the phosphate has exchanged with groundwaters, the oxygen isotope composition of the vertebrate remains suggests that a warm climate (MAT = 20–25 °C) existed in the Late Cretaceous of northern Spain.

The oxygen isotopic composition of pedogenic carbonates also provides a potentially powerful tool for palaeoclimate reconstruction. For example, recent geochemical modelling of the $\delta^{18}O$ of pedogenic calcite against latitude in modern palaeosol environments has allowed oxygen isotope data from the Early (Aptian) and Late (Maastrichtian) Cretaceous of the Western interior of North America to be evaluated (Mack & Cole 2005). The results indicate that Cretaceous palaeosol calcite, as a function of palaeolatitude, plot in the modern winter-precipitation field, which, along with vertic features, may suggest a wet-winter palaeoclimate. Steeper, positive slopes for the Cretaceous curves compared with the modern curves may also be attributable to greater latitudinal gradients in MATs and soil temperatures in the Cretaceous by comparison to today, and/or to a greater degree of evaporation in Cretaceous soils at low latitudes compared to today (Mack & Cole 2005). Similarly, Nordt et al. (2003) and Dworkin et al. (2005) evaluated climate change during the Maastrichtian by analysing the oxygen isotopic composition of pedogenic carbonates from a series of stacked palaeosols in west Texas. Their data yielded MATs ranging from 15–22 °C, entirely consistent with those data derived from palaeobotanical studies (see Fig. 2). Nordt et al. (2003) also note a strong coupling between Maastrichtian terrestrial climates and ocean temperatures that were possibly forced by Deccan trap volcanism.

A less conventional method of reconstructing the palaeoclimate in the Cretaceous involves using insect trace fossil associations in palaeosols (the Coprinisphaera ichnofacies of Genise et al. 2000) from the Late Cretaceous Laguna Palacios Formation in Central Patagonia (San Jorge Basin), southern South America (Genise et al. 2002). Genise et al. (2002) demonstrated that the ecological preferences of trace fossils from bees of the subfamily Halictinae, as well as some features of the nests, suggest a temperate, seasonal climate and an environment dominated by low vegetation for the Laguna Palacios Formation, which is also compatible with sedimentological and pedogenic evidence.

Summary

A number of new techniques and advances have provided data characterizing the climatic variability as recorded in the terrestrial record of the Cretaceous. The physiognomic analysis of flora from the high latitude sites of the Northern Hemisphere for example (Spicer & Parrish 1990; Craggs 2005) has proved useful in providing evidence for warm polar climates and a gradual decline in terrestrial MATs as the Cretaceous drew to a close. These data are consistent with the findings of Amiot et al. (2004) who undertook phosphate oxygen isotope analysis of continental vertebrates to demonstrate that during the Late Campanian–Mid-Maastrichtian, temperatures ranged from about 30 °C near the equator to about −5 °C at the poles. Likewise, stable isotope data gathered from molluscs (ammonites, belemnites and bivalves) from the Southern Hemisphere (Ditchfield et al. 1994) and microfossils (planktonic and benthic foraminifers) (Barrera et al. 1987), along with studies of growth-rings in fossil wood (Francis & Poole 2002; Poole et al. 2005) and climate-related geochemical variations in sediments (Dingle & Lavelle 1998) are all consistent with Mid-Cretaceous polar warmth followed by a Late Cretaceous cooling trend both in the marine and terrestrial realms. The exact timing of peak warmth does, however differ between sites. Considerably less quantitative data have been derived from the Cretaceous tropics but those studies of terrestrial flora, fauna and climatically- sensitive sediments that do exist show the dominance of sedimentary aridity proxies such as evaporite deposits demonstrating the dominance of Hadley cells on tropical circulations patterns (e.g. Ziegler et al. 2003). The primary forcing factor for warm temperatures in the Cretaceous is usually considered to have been an increase in the concentration of atmospheric CO_2. The maintenance of polar warmth requires the export of sufficient heat from the low latitudes (Herman & Spicer 1996). Thus far, even comprehensive models of climate have yielded anomalously low winter temperatures in continental interiors inhabited by large herbivorous dinosaurs (e.g. DeConto et al. 1999). Recent studies have reduced much of this anomaly by way of factors such as increased CO_2 in concert with greater oceanic poleward heat flux (to prevent the tropical oceans overheating as a consequence of higher CO_2 levels). Furthermore, high-latitude forests are believed to have played an important role in maintaining polar warmth during the Cretaceous (Upchurch et al. 1999).

Early versions of this paper benefited from critical comments and discussions provided by Dr. Richard Twitchett (University of Plymouth). The paper was also improved by helpful reviews by James Riding (British Geological Survey) and Bruce Sellwood (University of Reading). This paper forms part of the output from a NERC grant (NE/C507237/1) awarded to Stephen Grimes.

References

AMIOT, R., LECUYER, C., BUFFETAUT, E., FLUTEAU, F., LEGENDRE, S. & MARTINEAU, F. 2004. Latitudinal temperature gradient during the Cretaceous Upper Campanian-Middle Maastrichtian: $\delta^{18}O$ record of continental vertebrates. Earth and Planetary Science Letters, 226, 255–272.

AMIOT, R., LECUYER, C., BUFFETAUT, E., ESCARGUEL, G., FLUTEAU, F. & MARTINEAU, F. 2006. Oxygen isotopes from biogenic apatites suggest widespread endothermy in Cretaceous dinosaurs. *Earth and Planetary Science Letters*, **246**, 41–54.

AXELROD, D. I. 1984. An interpretation of Cretaceous and Tertiary biota in polar regions. *Palaeogeography, Palaeoclimatology, Palaeoecology*, **45**, 105–147.

BAKKER, R. T. 1980. Dinosaur heresy-dinosaur renaissance: Why we need endothermic archosaurs for a comprehensive theory of bioenergentic evolution. *In*: THOMAS, R. D. K. & OLSON, E. C. (eds) *A Cold look at the Warm Blooded Dinosaurs*. American Association of the Advancement of Science Select Symposium **28**, 351–462.

BAMFORD, M. K., ROBERTS, E. M., SISSOKO, F., BOUARE, M. L. & O'LEARY, M. A. 2002. An extensive deposit of fossil conifer wood from the Mesozoic of Mali, southern Sahara. *Palaeogeography, Palaeoclimatology, Palaeoecology*, **186**, 115–126.

BARNARD, P. D. W. 1973. Mesozoic floras. *In*: HUGHES, N. F. (ed.) *Organisms and Continents through Time*. Special Papers in Palaeontology, **12**, 175–188.

BARRERA, E., HUBER, B. T., SAVIN, S. M. & WEBB, P. N. 1987. Antarctic marine temperatures: late Campanian through early Paleocene. *Paleoceanography*, **2**, 21–47.

BARRICK, R. E. & KOHN, M. J. 2000. Multiple taxon–multiple locality approach to providing oxygen isotope evidence for warm-blooded theropod dinosaurs: Comment. *Geology*, **27**, 565–566.

BLAKEY, R. 2006 Global Plate Tectonics and Paleogeography. Northern Arizona University, URL: http://www2.nau.edu/~rcb7/090Marect.jpg

BUFFETAUT, E. 2004. Polar dinosaurs and the question of dinosaur extinction: a brief review. *Palaeogeography, Palaeoclimatology, Palaeoecology*, **214**, 225–231.

CHUMAKOV, N. M., ZHARKOV, M. A., HERMAN, A. B. *ET AL*. 1995. Climatic belts of the mid Cretaceous time. *Stratigraphy and Geological Correlation*, **3**, 241–260.

CRAGGS, H. J. 2005. Late Cretaceous climate signal of the Northern Pekulney Range Flora of northeastern Russia. *Palaeogeography, Palaeoclimatology, Palaeoecology*, **217**, 25–46.

DECONTO, R. M., HAY, W. W., THOMPSON, S. L. & BERGENGREN, J. 1999. Late Cretaceous climate and vegetation interactions: The cold continental interior paradox. *In*: BARRERA, E. & JOHNSON, C. (eds) *Evolution of the Cretaceous Ocean/Climate System*. Geological Society of America Special Paper, **332**, 391–406.

DELGADO-HUERTAS, A., LACUMIN, P. & LONGINELLI, A. 1997. A stable isotope study of fossil mammal remains from the Paglicci Cave, southern Italy, 13 to 33 ka BP: palaeoclimatological considerations. *Chemical Geology*, **141**, 211–223.

DILCHER, D. I. & BASSON, P. W. 1990. Mid Cretaceous angiosperm leaves from a new fossil locality in Lebanon. *Botanical Gazette*, **151**, 538–547.

DINGLE, R. V. & LAVELLE, M. 1998. Late Cretaceous–Cenozoic climatic variations of the northern Antarctic Peninsula: new geochemical evidence and review. *Palaeogeography, Palaeoclimatology, Palaeoecology*, **141**, 215–232.

DITCHFIELD, P. W., MARSHALL, J. D. & PIRRIE, D. 1994. High latitude palaeotemperature variation: New data from the Tithonian to Eocene of James Ross island, Antarctica. *Palaeogeography, Palaeoclimatology, Palaeoecology*, **107**, 79–101.

DWORKIN, S. I., NORDT, L. & ATCHLEY, S. 2005. Determining terrestrial paleotemperatures using the oxygen isotopic composition of pedogenic carbonate. *Earth and Planetary Science Letters*, **237**, 56–68.

EDWARDS, M. B., EDWARDS, R. & COLBERT, E. H. 1978. Carnosaurian footprints in the Lower Cretaceous of eastern Spitsbergen. *Journal of Palaeontology*, **52**, 940–941.

FALCON-LANG, H. J. 2003. Growth interruptions in silicified conifer woods from the Upper Cretaceous Two Medicine Formation, Montana, USA: implications for palaeoclimate and dinosaur palaeoecology. *Palaeogeography, Palaeoclimatology, Palaeoecology*, **199**, 299–314.

FALCON-LANG, H. J. & CANTRILL, D. J. 2001. Leaf phenology of some mid-Cretaceous polar forests, Alexander Island, Antarctica. *Geological Magazine*, **138**, 39–52.

FALCON-LANG, H. J., MACRAE, R. A. & CSANK, A. Z. 2004. Palaeoecology of late Cretaceous polar vegetation preserved in the Hansen point volcanics, NW Ellesmere Island, Canada. *Palaeogeography, Palaeoclimatology, Palaeoecology*, **212**, 45–64.

FERGUSON, K. M., GREGORY, R. T. & CONSTANTINE, A. 1999. Lower Cretaceous (Aptian Albian) secular changes in the oxygen and carbon isotope record from high paleolatitude, fluvial sediments, southeast Australia: Comparisons to the marine record. *In*: BARRERA, E. & JOHNSON, C. C. (eds) *Evolution of the Cretaceous Ocean-Climate System*. Geological Society of America Special Paper, **332**, 59–72.

FIORILLO, A. R. & PARRISH, J. T. 2004. The first record of a Cretaceous dinosaur from southwestern Alaska. *Cretaceous Research*, **25**, 453–458.

FRAKES, L. A., FRANCIS, J. E. & SYKTUS, J. I. 1992. Climate Modes of the Phanerozoic. Cambridge University Press, Cambridge, 274pp.

FRANCIS, J. E. & POOLE, I. 2002: Cretaceous and early Tertiary climates of Antarctica: evidence from fossil wood. *Palaeogeography, Palaeoclimatology, Palaeoecology*, **182**, 47–64.

FRICKE, H. C. & ROGERS, R. R. 2000. Multiple taxon–multiple locality approach to providing oxygen isotope evidence for warm-blooded theropod dinosaurs. *Geology*, **28**, 799–802.

FRICKE, H. C., CLYDE, W. C., O'NEIL, J. R. & GINGERICH, P. D. 1998. Evidence for rapid climate change in North America during the latest Palaeocene thermal maximum: oxygen isotope compositions of biogenic phosphate from the Bighorn Basin (Wyoming). *Earth and Planetary Science Letters*, **160**, 193–208.

FRIEDMAN, M., TARDUNO, J. A. & BRINKMAN, D. B. 2003. Fossil fishes from the high Canadian Arctic: further palaeobiological evidence for extreme climatic warmth during the Late Cretaceous (Turonian–Coniacian). *Cretaceous Research*, **24**, 615–632.

GENISE, J. F., MANGANO, M. G., BUATOIS, L. A., LAZA, J. H. & VERDE, M. 2000. Insect trace fossil associations in paleosols: The Coprinisphaera ichnofacies. *Palaios*, **15**, 49–64.

GENISE, J. F., SCIUTTO, J. C., LAZA, J. H., GONZALEZ, M. G. & BELLOSI, E. S. 2002. Fossil bee nests, coleopteran pupal chambers and tuffaceous paleosols from the Late Cretaceous Laguna Palacios Formation, Central Patagonia (Argentina). *Palaeogeography, Palaeoclimatology, Palaeoecology*, **177**, 215–235.

GOMEZ, B., MARTINEZ-DELCLOS, X., BAMFORD, M. & PHILIPPE, M. 2002. Taphonomy and palaeoecology of plant remains from the oldest African Early Cretaceous amber locality. *Lethaia*, **35**, 300–308.

GRADSTEIN, F. M., OGG, J. G. & SMITH, A. G. 2004. *A Geologic Time Scale 2004*. Cambridge University Press, Cambridge, 589pp.

GREGORY, R. T., DOUTHITT, C. B., DUDDY, I. R., RICH, P. & RICH, T. H. 1989. Oxygen isotope composition of carbonate concretions from the Lower Cretaceous of Victoria, Australia: implications for the evolution of meteoric waters on the Australian continent in a paleopolar environment. *Earth and Planetary Science Letters*, **92**, 27–42.

GRIMES, S. T., HOOKER, J. J., COLLINSON, M. E. & MATTEY, D. P. 2005. Summer temperatures of late Eocene to early Oligocene freshwaters. *Geology*, **33**, 189–192.

HERMAN, A. B. & SPICER, R. A. 1996. Palaeobotanical evidence for a warm Cretaceous Arctic Ocean. *Nature*, **380**, 330–333.

JENKYNS, H. C., FORSTER, A., SCHOUTEN, S. & DAMSTE, J. S. S. 2004. High temperatures in the Late Cretaceous Arctic Ocean. *Nature*, **432**, 888–892.

JI, Q., NORELL, M. A., GAO, K. Q., JI, S. A. & REN, D. 2001. The distribution of integumentary structures in a feathered dinosaur. *Nature*, **410**, 1084–1088.

KEAR, B. P. 2005. Marine reptiles from the Lower Cretaceous (Aptian) deposits of White Cliffs, southeastern Australia: implications of a high latitude, cold water assemblage. *Cretaceous Research*, **26**, 769–782.

KENNEDY, E. M., SPICER, R. A. & REES, P. M. 2002. Quantitative palaeoclimate estimates from late cretaceous and Paleocene leaf floras in the northwest of the South Island, New Zealand. *Palaeogeography, Palaeoclimatology, Palaeoecology*, **184**, 321–345.

KOCH, P. L., ZACHOS, J. C. & DETTMAN, D. L. 1995. Stable isotope stratigraphy and paleoclimatology of the Paleogene Bighorn Basin (Wyoming, USA). *Palaeogeography, Palaeoclimatology, Palaeoecology*, **115**, 61–89.

LÉCUYER, C., BOGEY, C., GARCIA, J. P. ET AL. 2003. Stable isotope composition and rare earth element content of vertebrate remains from the Late Cretaceous of northern Spain (Lano): did the environmental record survive? *Palaeogeography, Palaeoclimatology, Palaeoecology*, **193**, 457–471.

MACK, G. H. & COLE, D. R. 2005. Geochemical model of $\delta^{18}O$ of pedogenic calcite versus latitude and its application to Cretaceous palaeoclimate. *Sedimentary Geology*, **174**, 115–122.

MARKWICK, P. J. 1998. Fossil crocodilians as indicators of Late Cretaceous and Cenozoic climates: implications for using palaeontological data in reconstructing palaeoclimate. *Palaeogeography, Palaeoclimatology, Palaeoecology*, **137**, 205–271.

NORDT, L., ATCHLEY, S. & DWORKIN, S. 2003. Terrestrial evidence for two greenhouse events in the Latest Cretaceous. *GSA Today*, **13**, 4–9.

NORELL, M., JI, Q., GAO, K. Q., YUAN, C. X., ZHAO, Y. B. & WANG, L. X. 2002. 'Modern' feathers on a non-avian dinosaur. *Nature*, **416**, 36–37.

PARRISH, J. M., PARRISH, J. T., HUTCHSON, J. H. & SPICER, R. A. 1987. Late Cretaceous vertebrate fossils from the North Slope of Alaska and implications for dinosaur ecology. *Palaios*, **2**, 377–389.

PARRISH, J. T., SPICER, R. A., DOUGLAS, J. G., RICH, T. H. & VICKERS-RICH, P. 1991. Continental climate near the Albian South Pole and comparison with the climate near the North Pole. *Geological Society of America Abstracts with Programs*, **23**, A302.

PARRISH, J. T., DANIEL, I. L., KENNEDY, E. M. & SPICER, R. A. 1998. Paleoclimatic significance of mid-Cretaceous floras from the middle Clarence Valley, New Zealand. *Palaios*, **13**, 149–159.

PAUL, G. S. 1998. Physiological, migratorial, climatological, geophysical, survival, and evolutionary implications of Cretaceous polar dinosaurs. *Journal of Paleontology*, **62**, 640–652.

PHILIPPE, M., BAMFORD, M., MCLOUGHLIN, S. ET AL. 2004. Biogeographic analysis of Jurassic–early Cretaceous wood assemblages from Gondwana. *Review of Palaeobotany and Palynology*, **129**, 141–173.

POOLE, I., CANTRILL, D. & UTESCHER, T. 2005. A multi-proxy approach to determine Antarctic terrestrial palaeoclimate during the Late Cretaceous and Early Tertiary. *Palaeogeography, Palaeoclimatology, Palaeoecology*, **222**, 95–121.

PRICE, G. D. 1999. The evidence and implications of polar-ice during the Mesozoic. *Earth Science Reviews*, **48**, 183–210.

RICH, T. H., VICKERS-RICH, P. & GANGLOFF, R. A. 2002. Polar dinosaurs. *Science*, **295**, 979–980.

RUSSELL, D. A. & PAESLER, M. A. 2003. Environments of Mid-Cretaceous Saharan dinosaurs. *Cretaceous Research*, **24**, 569–588.

SEEBACHER, F. 2003. Dinosaur body temperatures: the occurrence of endothermy and ectothermy. *Paleobiology*, **29**, 105–122.

SELLWOOD, B. W. & VALDES, P. J. 2006. Mesozoic climates: general circulation models and the rock record. *Sedimentary Geology*, **190**, 269–287.

SPICER, R. A. 2003. Changing climate and biota. In: SKELTON, P. W. (ed.) *The Cretaceous World*. Cambridge University Press, pp. 85–162.

SPICER, R. A., AHLBERG, A., HERMAN, A. B., KELLEY, S. P., RAIKEVICH, M. I. & REES, P. M. 2002. Palaeoenvironment and ecology of the middle Cretaceous Grebenka flora of northeastern Asia. *Palaeogeography, Palaeoclimatology, Palaeoecology*, **184**, 65–105.

SPICER, R. A. & CORFIELD, R. M. 1992. A review of terrestrial and marine climates in the Cretaceous with implications for modelling the 'Greenhouse earth'. *Geological Magazine*, **129**, 169–180.

SPICER, R. A. & PARRISH, J. T. 1990. Late Cretaceous–early Tertiary palaeoclimates of the northern high latitudes: a quantitative view. *Journal of the Geological Society, London*, **147**, 329–341.

STEPHAN, E. 2000. Oxygen isotope analysis of animal bone phosphate: method refinement, influence of consolidants, and reconstruction of palaeotemperatures from Holocene sites. *Journal of Archaeological Science*, **27**, 523–535.

TARDUNO, J. A., BRINKMAN, D. B., RENNE, P. R., COTTRELL, R. D., SCHER, H. & CASTILLO, P. 1998. *Evidence for Extreme Climatic Warmth from Late Cretaceous Arctic Vertebrates. Science*, **282**, 2241–2244.

UPCHURCH, G. R., OTTO-BLIESNER, B. L. & SCOTESE, C. 1999. Terrestrial vegetation and its effects on climate during the Cretaceous. *In*: BARRERA, E. & JOHNSON, C. C. (eds) Evolution of the Cretaceous Ocean-Climate System. *Geological Society of America Special Paper*, **332**, 407–426.

VALDES, P. J., SELLWOOD, B. W. & PRICE, G. D. 1996. Evaluating Concepts of Cretaceous equability. *Palaeoclimates*, **1**, 139–158.

WILSON, P. A., NORRIS, R. D. & COOPER, M. J. 2002. Testing the Cretaceous greenhouse hypothesis using glassy foraminiferal calcite from the core of the Turonian tropics on Demerara Rise. *Geology*, **30**, 607–610.

WING, S. L. & GREENWOOD, D. R. 1993. Fossils and fossil climate: the case for equable continental interiors in the Eocene. *Philosophical Transactions of the Royal Society, London*, **B341**, 243–252.

WOLFE, J. A. 1979. Temperature parameters of humid to mesic forests of eastern Asia and relation to forests of other regions of the northern hemisphere and Australasia. *U. S. Geological Survey Professional Paper*, **1106**, 37pp.

WOLFE, J. A. & UPCHURCH, G. R. 1987. North American nonmarine climates and vegetation during the Late Cretaceous. *Palaeogeography, Palaeoclimatology, Palaeoecology*, **61**, 33–77.

ZIEGLER, A. M., ESHEL, G., REES, P. M., ROTHFUS, T. A., ROWLEY, D. B. & SUNDERLIN, D. 2003. Tracing the tropics across land and sea: Permian to present. *Lethaia*, **36**, 227–254.

Late Cretaceous climates and foraminiferid distributions

M. B. HART

School of Earth, Ocean & Environmental Sciences, University of Plymouth, Drake Circus,
Plymouth PL4 8AA, UK (e-mail: mhart@plymouth.ac.uk)

Abstract: The late Cretaceous records a warm, greenhouse, period of Earth history with a distinct 'hot greenhouse' interval in the latest Cenomanian and earliest Turonian. This was coupled with the highest sea levels of the Mesozoic and very low N–S temperature gradients. Polar regions are thought to have been temperate, rather than cold, although there is a body of opinion that regards some of the late Cretaceous sea-level changes as being glacio-eustatic. Towards the end of the Cretaceous, there was a brief period of warming and poleward migration of planktic foraminifera that was followed by a near end-Cretaceous cooling event.

The late Cretaceous is widely regarded as having been one of the warmest periods of Earth history. Between the icehouse conditions of the Permo-Carboniferous and the late Cenozoic glaciations, the Mesozoic is often described as recording a 'greenhouse' condition, with the late Cretaceous recognized as a 'hot greenhouse' (Huber *et al.* 2002; Norris *et al.* 2002). Evidence for this view comes from the high sea-surface temperatures of equatorial regions (Wilson *et al.* 2002) and the presence of temperate vegetation within the Arctic and Antarctic circles (Spicer & Parrish 1990; Falcon-Lang & Cantrill 2001; Falcon-Lang *et al.* 2004).

Late Cretaceous palaeogeography

The second phase in the break-up of Pangaea began in the early Cretaceous at about 120 Ma (Hay *et al.* 1999). This increased rate of oceanic crust production (Larson 1991; Kauffman & Hart 1996, Fig. 1) began the separation of S. America and Africa, India and Africa, and India and Antarctica/Australia (Hart *et al.* 2001, fig. 6). At the same time, the Andes were beginning to be uplifted and, with the arrival of exotic terranes (Wrangellia, Stikinia, etc.), the Rocky Mountains were also emergent. The latter separated the Pacific Ocean from the Western Interior Seaway of North America (Fig. 1; Dean & Arthur 1998). The enhanced rate of oceanic crust production, and the broad profile of the expanded mid-oceanic ridges, displaced significant amounts of oceanic water onto the expanded shelves as sea level is estimated to have been 100–200 metres (or more) higher than the present time (Hancock 1976, 1989; Hancock & Kauffman 1979). Coupled with this expanded ridge volume were the developing oceanic plateaux of Kerguelen (in the Indian Ocean) and Ontong–Java (in the Pacific Ocean) which must have also had a significant impact on sea level (Larson

1991). The high temperatures recorded from the marine environment (Frakes 1999; Bice & Norris 2002; Huber *et al.* 2002; Norris *et al.* 2002; Bice *et al.* 2003, 2006) and on land (Tarduno *et al.* 1998; Falcon-Lang *et al.* 2004; Price & Grimes, this volume) suggest that the Earth may have been ice-free during the late Cretaceous. This view has been challenged using a number of lines of evidence, most noticeably rapidly fluctuating sea levels (Miller *et al.* 1999; Gale *et al.* 2002) and other isotopic evidence (Ditchfield *et al.* 1994; Stoll & Schrag 1996, 2000).

By the end of the late Cretaceous, this high rate of sea-floor spreading had reduced (Larson 1991) and, as a result, sea levels had fallen slightly. At the time of the Cretaceous/Tertiary (K/T) boundary, the North and South Atlantic Oceans had joined, the Eastern and Western Indian Oceans had opened significantly and Australia was beginning to rift from Antarctica (Fig. 2). The Andes and Rocky Mountains were continuing to form and a series of ophiolites had been emplaced in Oman (Glennie *et al.* 1974, 1990), Cyprus (Robertson & Hudson 1974; Robertson 1975) and Syria (Knipper *et al.* 1990; Al-Riyami & Robertson 2002). Elsewhere in the Mediterranean area there was ongoing tectonic activity in the Cretaceous (Jongsma *et al.* 1985).

Distribution of Cretaceous planktic foraminifera

Despite all these major global changes, the Upper Cretaceous succession of Britain is almost totally within the chalk facies which represents over 30 million years of near-uniform pelagic sedimentation. Within this succession in the Anglo-Paris Basin (Southern England), the proportion of planktic foraminifera fluctuates markedly. Figure 3 shows a schematic graphical representation of the relative percentage of planktic foraminifera (of the

From: WILLIAMS, M., HAYWOOD, A. M., GREGORY, F. J. & SCHMIDT, D. N. (eds) *Deep-Time Perspectives on Climate Change: Marrying the Signal from Computer Models and Biological Proxies.* The Micropalaeontological Society, Special Publications. The Geological Society, London, 235–250.

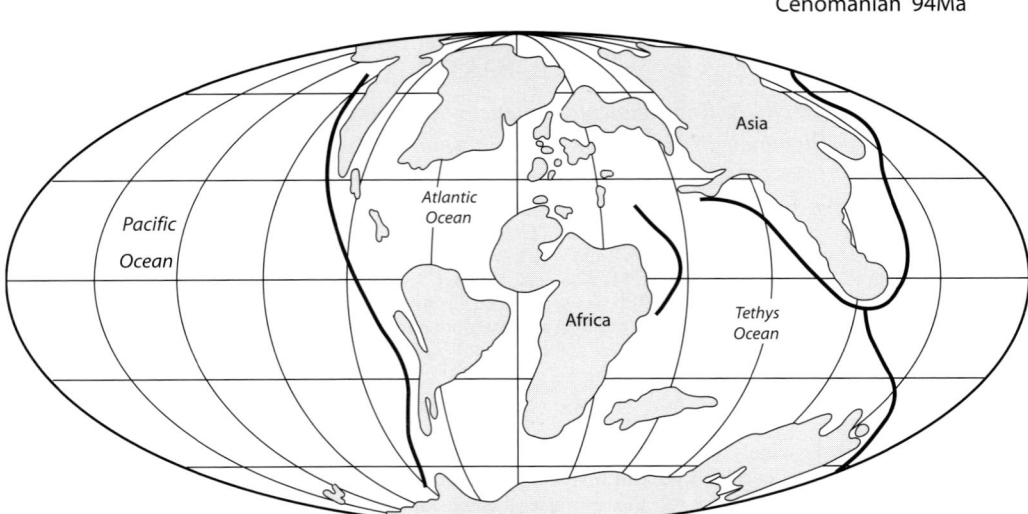

Fig. 1. Cenomanian (94 Ma) palaeogeographical map; base map derived from Smith *et al.* (1994), with subduction zones based on the palaeogeographical map of Scotese (www.scotese.com/cretaceo.htm).

total fauna) in the 250–500 μm size fraction (although counts based on other size fractions give a slightly different value inspite of having essentially the same pattern). In the early Cenomanian, planktic foraminifera are relatively rare with keeled species (e.g. *Rotalipora, Thalmaninella, Praeglobotruncana*) only occurring in relatively small numbers at restricted levels. Such horizons are, however, laterally persistent and can be used for correlation (including for the route of the Channel Tunnel; see Hart 2004, figs 4, 5). In the mid-Cenomanian, there is an increase in the percentage of planktic taxa (Carter & Hart 1977; Hart & Bailey 1979; Hart 2004), although the percentage

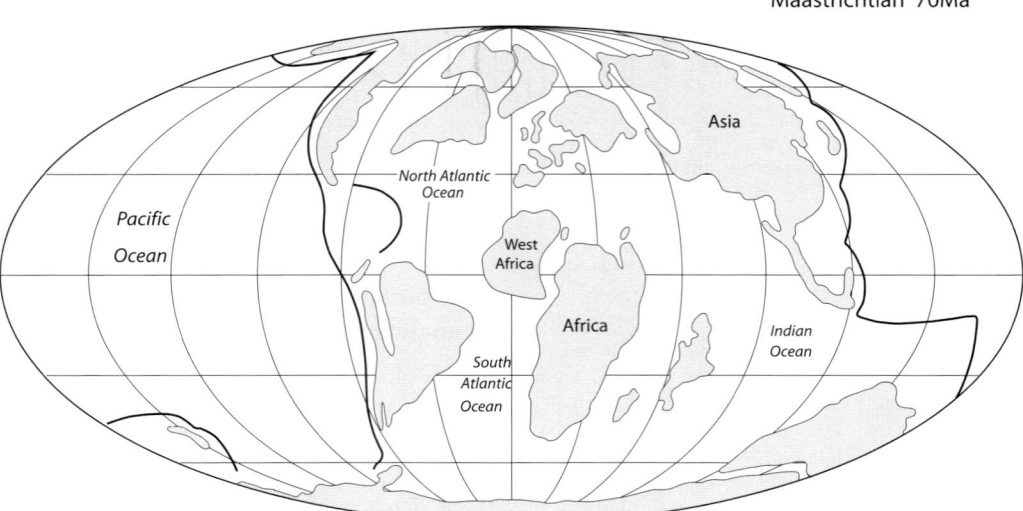

Fig. 2. Maastrichtian (70 Ma) palaeogeographical map; base map derived from Smith *et al.* (1994), with subduction zones based on the palaeogeographical map of Scotese (www.scotese.com/k/t.htm).

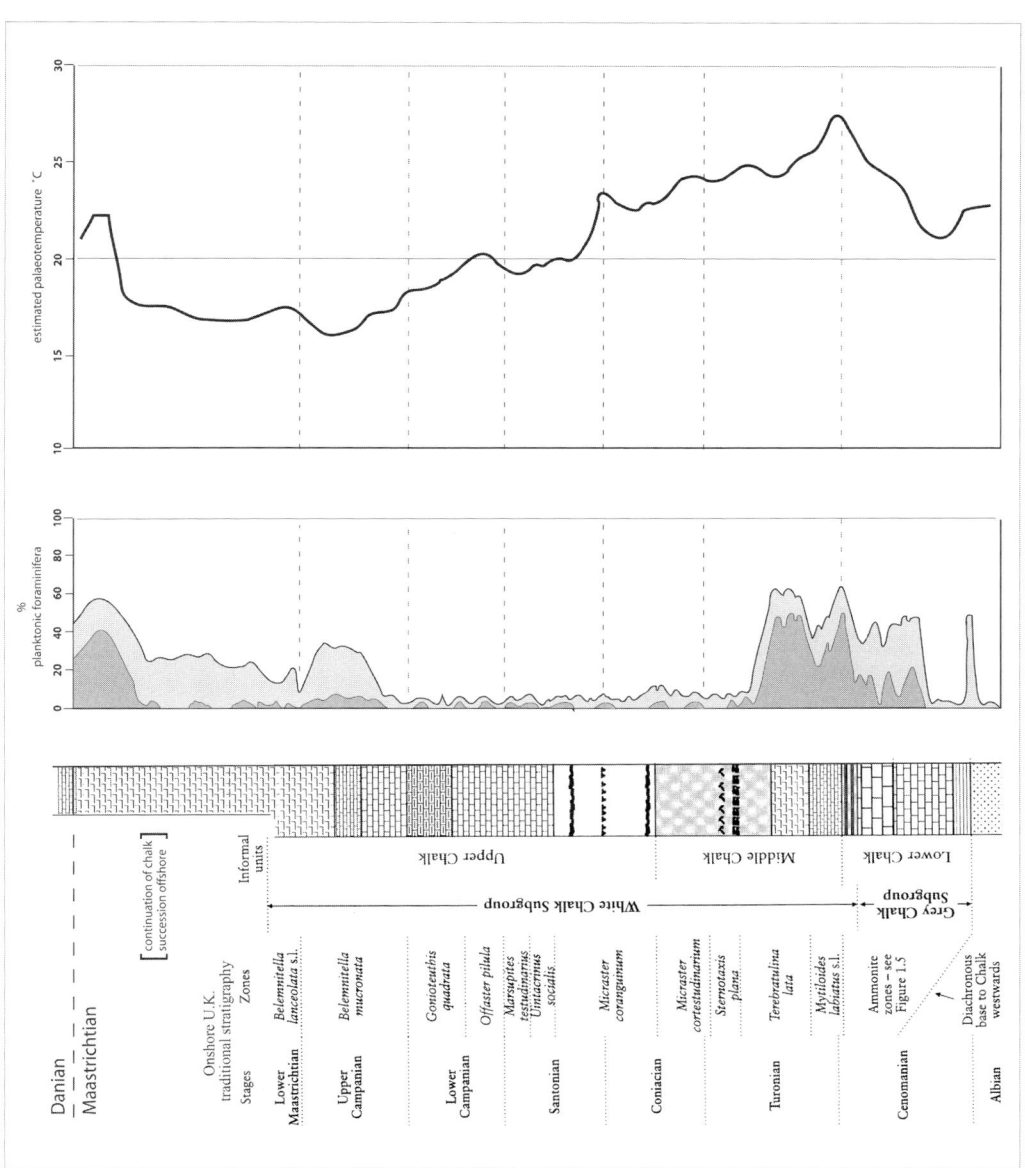

Fig. 3. UK succession (based on Mortimore *et al.* 2001), schematic distribution of planktic foraminifera (based on counts of the 250 μm fraction and schematic temperature curve based on Jenkyns *et al.* (1994).

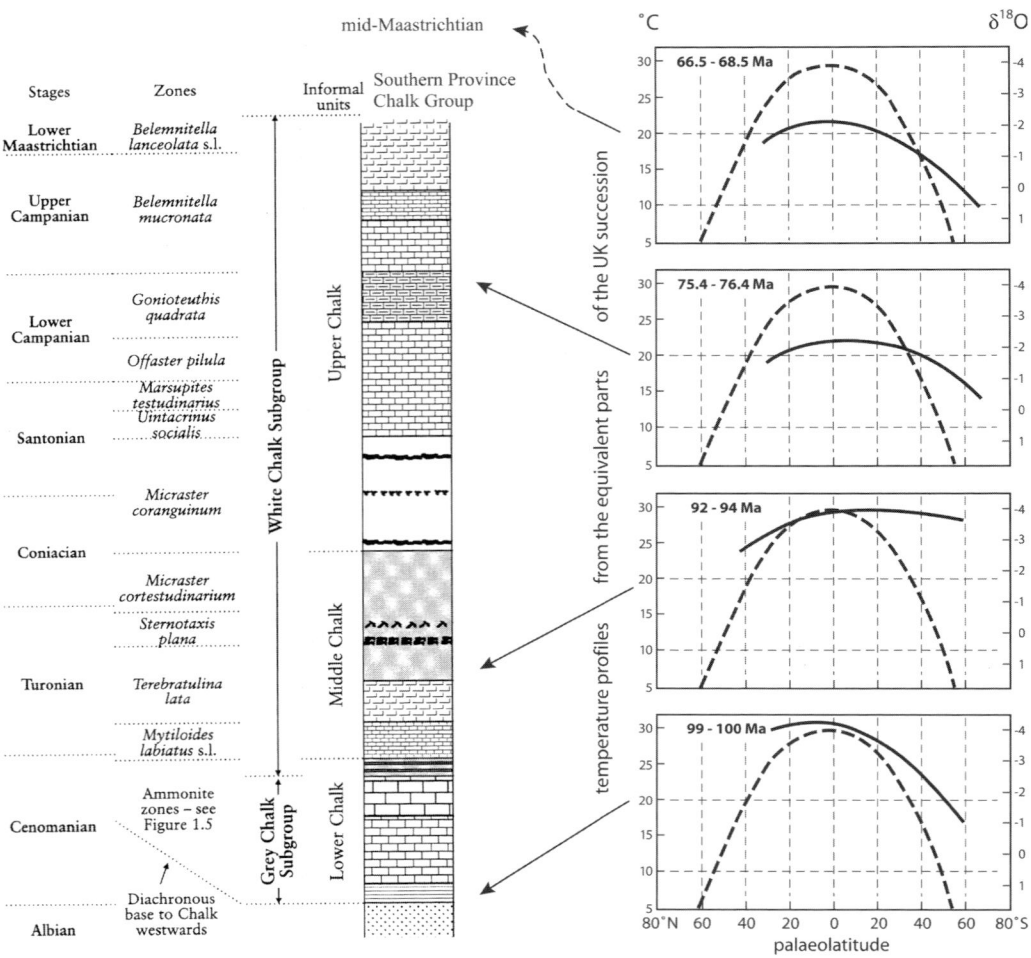

Fig. 4. North–South temperature profiles for the Late Cretaceous succession based on data in Huber *et al.* (2002). Each of the profiles has been located (approximately) against the log of the UK succession. The Demerara Rise data (Bice *et al.* 2006) come from the equatorial region and in almost all cases (especially the data derived from Mg/Ca ratios) would be higher than the values shown in the graphs. If correct, then these data would make the profiles 'peak' in the Tropics, rather than appearing rather 'flat'.

of keeled forms remains relatively low, aside from 'pulses' with highs of 25–30%. The early to mid-Turonian is a little different with a planktic percentage up to 60–70% and a dominance of keeled, Tethyan forms such as *Helvetoglobotruncana helvetica, Dicarinella* spp. and *Marginotruncana* spp. (Hart & Weaver 1977; Hart 1997). In the mid-Turonian, there is a major change associated with a global sea-level fall that is recorded in Britain. (Hancock 1976, 1989; Hancock & Kauffman 1979; Hart 1990, 1997, in press), Western Interior Seaway of the USA (West *et al.* 1998; Scott *et al.* 1998) and India (Tewari *et al.* 1998).

In the mid–late Turonian, Voigt & Wiese (2000) have reported synchronous changes in the δ¹⁸O

isotope record in a number of sections in Europe. Despite the problems of diagenesis in these carbonate-rich sediments, the synchroneity of the changes implies a regional signal. This cooling is associated with a number of palaeobiogeographical changes. The relationship to the mid–late Turonian sea-level fall and the effect of changes in the circulation patterns caused by this event are as yet unclear. Keeled planktonic foraminifera become less numerous through this event, as reported by Hart & Weaver (1977) and Hart (1990, 1997). In Eastern and Northern England, and especially in the North Sea Basin, the abundance of planktic foraminifera is greater than the levels recorded in the Anglo-Paris Basin (Bailey 1978; Swiecicki 1980;

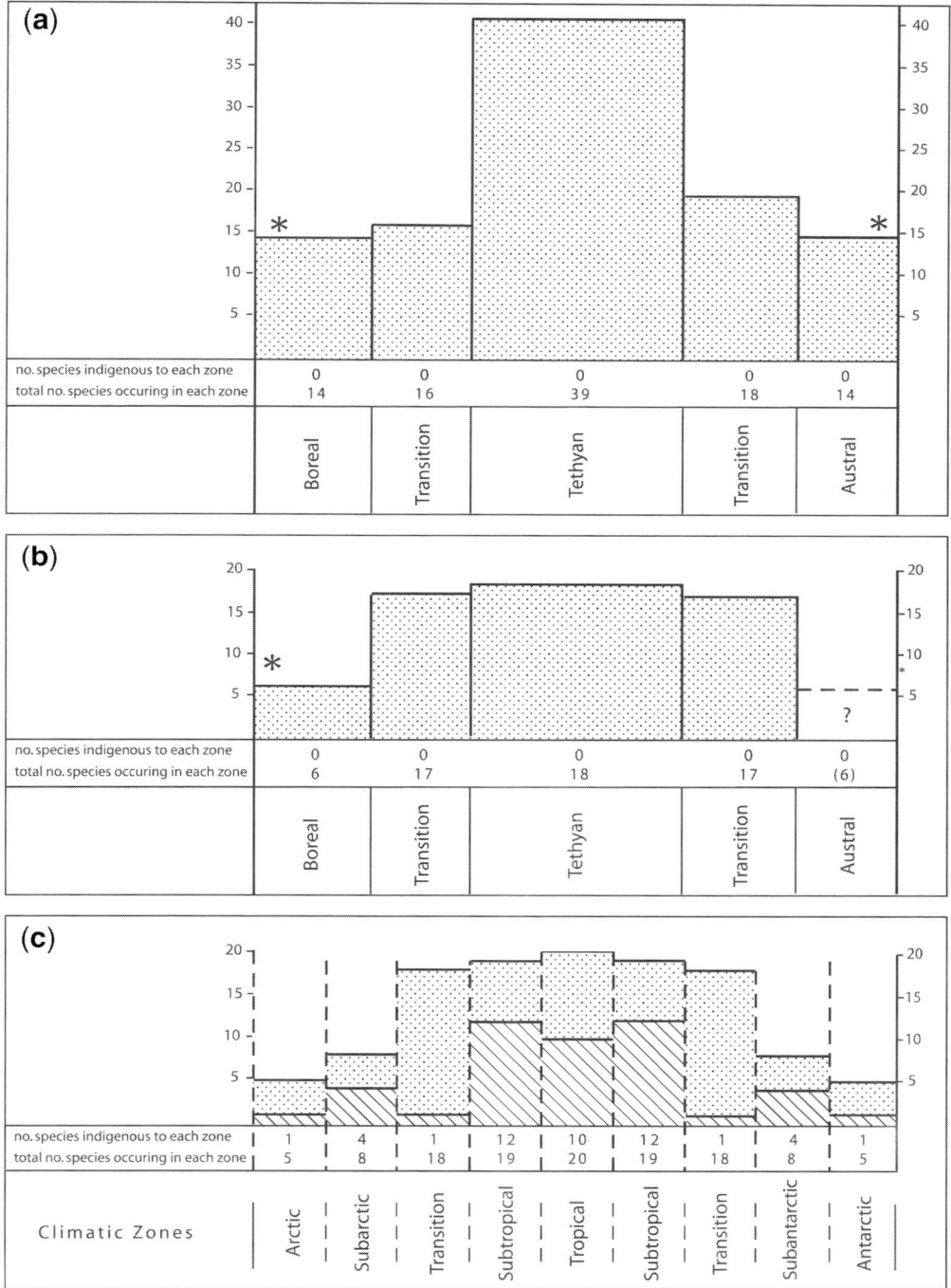

Fig. 5. See caption on p. 240.

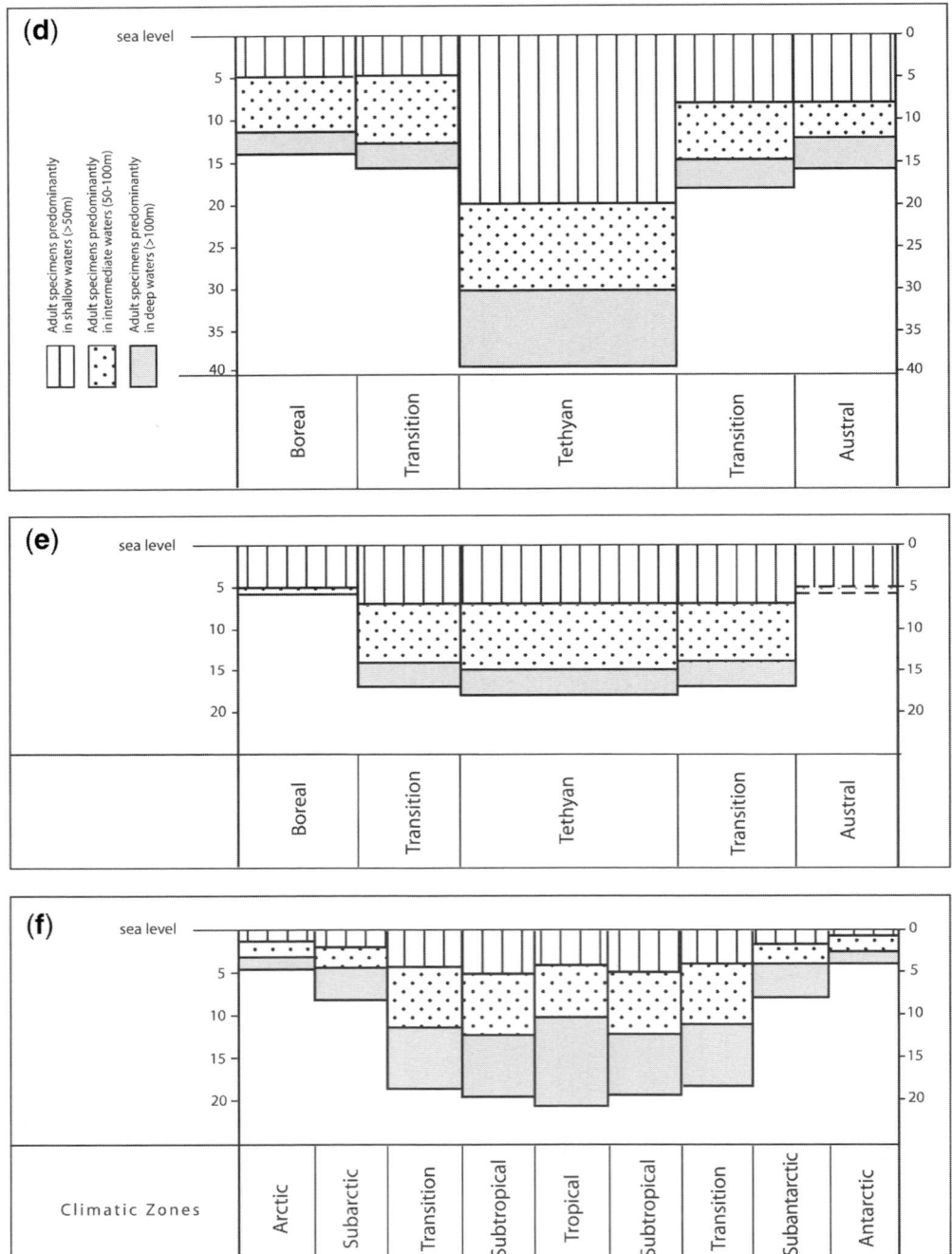

Fig 5. Distribution of planktic foraminifera and morphotypes in the early Maastrichtian, latest Albian and Recent (based on data in Bé 1977). (**a**) Distribution of early Maastrichtian planktic foraminifera. The * marks the occurrence of high latitude assemblages that are dominated by agglutinated foraminifera; (**b**) Distribution of planktic foraminifera in the latest Albian; (**c**) distribution of Recent planktic foraminifera (based on Bé 1977); (**d**) distribution of planktic morphotypes in the early Maastrichtian; (**e**) distribution of planktic morphotypes in the latest Albian; and (**f**) distribution of Recent planktic foraminifera (based on Bé 1977). Diagram modified from that presented in Hart (2000).

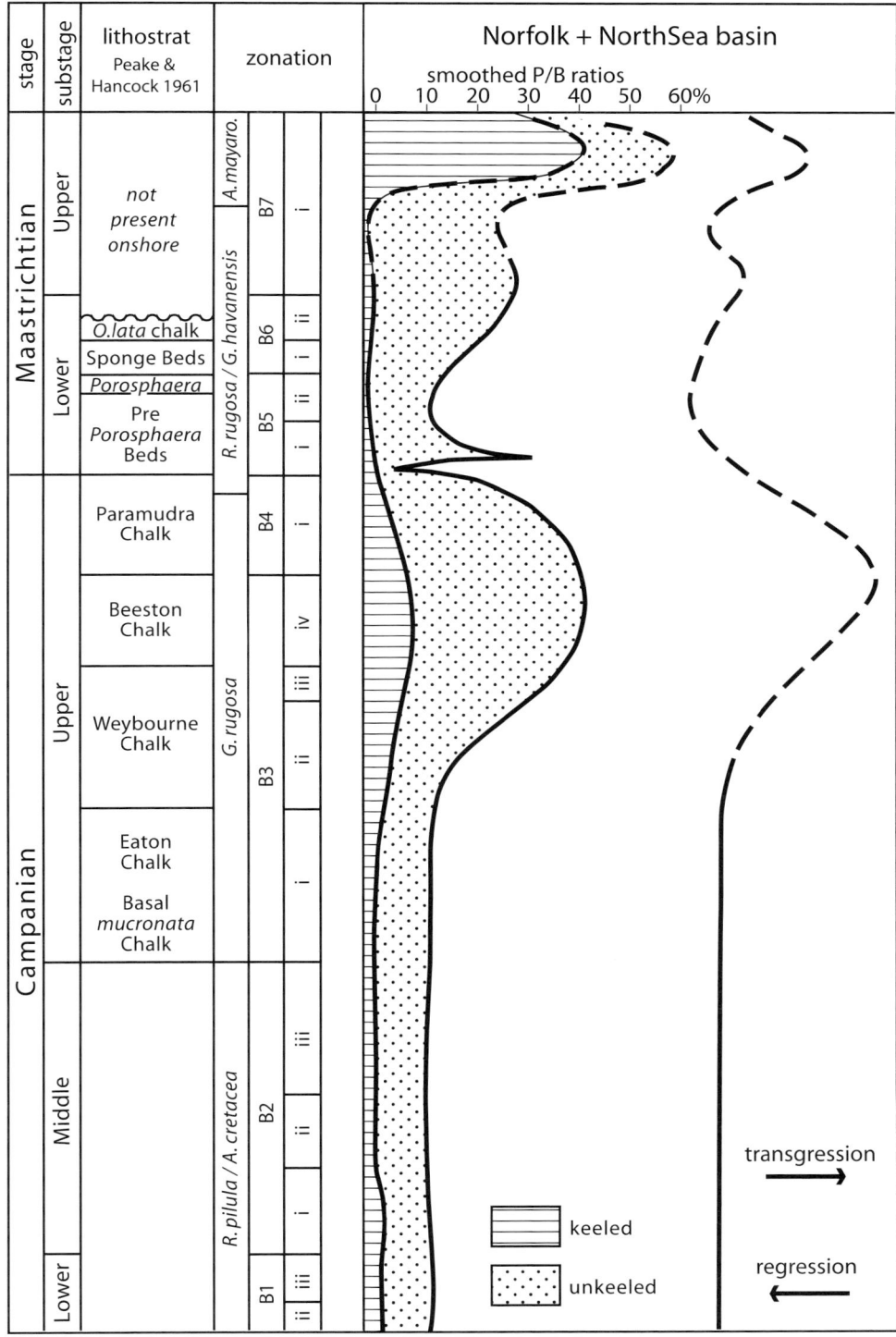

Fig. 6. Schematic distribution of planktic foraminifera in the Campanian–Maastrichtian successions of Norfolk and the Central North Sea Basin (after Swiecicki 1980) and (on the right) a tentative interpretation of relative sea-level changes.

Bergen & Sikora 1999; King *et al.* 1989). While water depth is clearly one of the controls on this distribution, migration – and barriers to migration (such as the London–Brabant High) – are also important. The full story of the distributional controls operating on the planktic foraminifera in this complex palaeogeographic setting is still to be determined.

Bandy (1967) was the first to suggest that the northward migration of keeled foraminifera (*Rotalipora, Globotruncana*, etc.) could be used as a palaeoclimatic signal and his work has formed the basis of many subsequent investigations (recently reviewed by Hart 2000). Since the time of Bandy's original work, the use of stable isotopes in the construction of climate models has not only developed but become standard practice. In the chalk succession of S.E. England, Jenkyns *et al.* (1994) have recorded a $\delta^{18}O$ curve which has been used as a standard reference for late Cretaceous palaeotemperatures in the mid-latitudes of the Northern Hemisphere (Fig. 3). It is noticeable that abundant keeled planktic foraminifera are recorded where the Jenkyns *et al.* data suggest surface water palaeotemperatures of *c.* 24 °C. In the Coniacian to Campanian succession of southern Britian, while some keeled taxa are recorded (e.g. *Dicarinella primitiva/concavata, Globotruncana arca, G. bulloides, G. ventricosa, Marginotruncana* spp.), these are not the Tethyan (=warm water) forms used in the standard zonation of the late Cretaceous (e.g. *Dicarinella asymetrica, Globotruncana aegyptiaca, G. elevata, Globotruncanita calcarata*). An additional problem in the chalk facies is the preservation of planktic foraminifera. Studies of flints and flint meal (soft powder preserved inside hollow flints) have shown that the proportions of planktic taxa are usually greater in the flint meal; and one can sometimes find taxa that are not recorded in the body of the chalk outside the flints. The mechanism for such selective preservation is also poorly understood (Curry 1982; Hart *et al.* 1986; Bailey & Clayton, in press).

In recent years, there has been a remarkable increase in the stable isotope and palaeotemperature data available for the late Cretaceous. Of particular importance has been the work on the well-preserved faunas of the Blake Nose (30 °N palaeolatitude, North Atlantic Ocean) transect (Huber *et al.* 2002) and the Demerara Rise (Norris *et al.* 2002; Wilson *et al.* 2002; Bice *et al.* 2006). These data indicate a 'warm greenhouse' condition in the late Albian to mid–late Cenomanian, a 'hot greenhouse' in the latest Cenomanian to early Campanian and a 'cool greenhouse' in the remainder of the Campanian and Maastrichtian. Huber *et al.* (2002, figs 2, 3) used these data from the Blake Plateau, together with data from Southern High Latitudes (ODP Sites 511 and 690), to reconstruct cross-latitude

profiles for the late Cretaceous interval (Fig. 4). All of these cross-latitude profiles are quite 'flat', with the tropics at, or even below, present levels while the Southern High Latitudes were significantly warmer than the present day. Their data for the latest Albian are quite compatible with the distribution of planktic foraminifera and the morphotype analysis presented by Hart (2000, figs 1, 2). The cross-latitude profiles shown in Figure 4 should be compared to the morphotype analysis in Figure 5.

In the early Campanian, the isotope data of Jenkyns *et al.* (1994) suggest that, in the chalk seas of N.W. Europe, the temperature fell below 20 °C and this is quite compatible with the relatively low diversity, generally non-keeled planktic fauna (although the depth of water must also have been a limiting factor) seen at this stratigraphical level. The fauna is dominated by *Archaeoglobigerina* spp. and keeled generalists such as *Globotruncana bulloides* and *G. linneiana*. As indicated above, none of the zonally important planktic taxa are present although they have been reported from North Atlantic DSDP/ODP/IPOD cores (Schönfeld & Burnett 1991; Hart & Aplin 2002). As many of these taxa appear to have inhabited deeper-water environments, it is difficult to separate the restriction as being due to temperature or water depth. Huber *et al.* (2002) record that the early Campanian marks the boundary between the 'hot greenhouse' and 'cool greenhouse' regimes even though the N–S temperature profile of the mid-Campanian (Huber *et al.* 2002, fig. 3; Fig. 4) suggests a cooler (20–22 °C) equatorial region but a temperature of 15 °C at palaeolatitude 70 °S. Recently, Jenkyns *et al.* (2004, fig. 3) have used Arctic Ocean data to propose mid-Campanian temperatures close to the North Pole of *c.* 15 °C. This suggestion is certainly inconsistent with much lower palaeobotanical estimates for the region (Falcon-Lang *et al.* 2004). This discrepancy may be methodological, as Jenkyns *et al.* (2004, fig. 3) have used TEX$_{86}$ data from the Arctic Ocean cores to extrapolate their British temperature curve into the Arctic Ocean.

Other mid-Campanian data suggest that mean annual temperatures were 7–14 °C warmer than today (Brady *et al.* 1998). There is no direct evidence of any significant ice at high latitudes (Frakes *et al.* 1992) and thermal gradients are regarded as having been low (Barrera *et al.* 1987; Huber *et al.* 1995, 2002; Brady *et al.* 1998). Modelling of the mid-Campanian (Thompson & Pollard 1997; Brady *et al.* 1998) suggests that warm salty bottom water (WSBW) can be formed in the high latitudes of the Southern Hemisphere by cooling warm salty water. These models appear to confirm that poleward heat transport was able to maintain global warm conditions throughout the Campanian.

This view has been challenged by Miller *et al.* (1999) who have suggested that rapid sea-level changes of 30–40 m in the New Jersey (USA) succession indicate a glacio-eustatic control. In support of these New Jersey data, Miller *et al.* (1999) point to a drop in temperature based on Site 690 in the Southern Ocean. It is clear from the Miller *et al.* data, as well as the $\delta^{18}O$ record of Huber *et al.* (2002) and Jenkyns *et al.* (1994), that the Maastrichtian does record a time of greater climatic variability in terms of both the palaeotemperature record and the migration events of planktic foraminifera.

In the Norfolk succession of Britain, Swiecicki (1980) recorded an increase in numbers of planktic foraminifera in the latest Campanian and in the mid-Maastrichtian (Fig. 6). In the North Sea Basin and Faeroe–Shetland Basin, commercial wells also record quite dramatic fluctuations in planktic numbers (King *et al.* 1989; Akker *et al.* 2000) in the latest Campanian and Maastrichtian (Fig. 7). The Akker *et al.* data, from well 205/10-2B in the Faeroe–Shetland Basin, records an increase in the planktic fauna in the latest Campanian (up to 40% of the fauna) comparable to the data from onshore Norfolk (Fig. 6) which records a maximum level of *c.* 30%. The Faeroe–Shetland Basin data (Akker *et al.* 2000, fig. 3) also indicate that, below this level, in the early to mid-Campanian and in the Santonian, planktic forms are rare and only occasionally record <5%. In Norfolk, the

comparable values for the Campanian are >10% and in the rest of Southern England the Santonian values rarely exceed 10% of the fauna.

The succession in Norfolk ends in the Lower Maastrichtian (Hart *et al.* 1989; Hart & Swiecicki 2003) but, up to that level, the relative percentage of planktic foraminifera is comparable to that recorded by Akker *et al.* (2000, fig. 3). The late Campanian increase in planktic foraminifera is also recorded in the Northern Viking Graben and Outer Moray Firth (King *et al.* 1989, fig. 8.6) although it is not recorded in the Southern North Sea Basin (which is directly offshore from the Norfolk coast). Akker *et al.* (2000, fig. 3) also record further 'floods' of planktic foraminifera in the mid- and late Maastrichtian. A very similar pattern is also reported from the Northern Viking Graben (King *et al.* 1989, fig. 8.6), where Zone FCN21b of the latest Maastrichtian is characterized by a flood of *Pseudotextularia elegans*. King *et al.* identify Zone FCN21 as having two subzones (FCN21b–*P. elegans* Subzone and a lower FCN21a–*Rugoglobigerina* spp. Subzone). Zone FCN21b of the uppermost Maastrichtian is characterized by an assemblage that is 80–90% planktic foraminifera, including the nominate taxon and *Abathomphalus, Globotruncanella, Globigerinelloides, Heterohelix, Racemiguembelina, Contusotruncana* and *Rugoglobigerina*. Of particular note is the abundance of *P. elegans* and

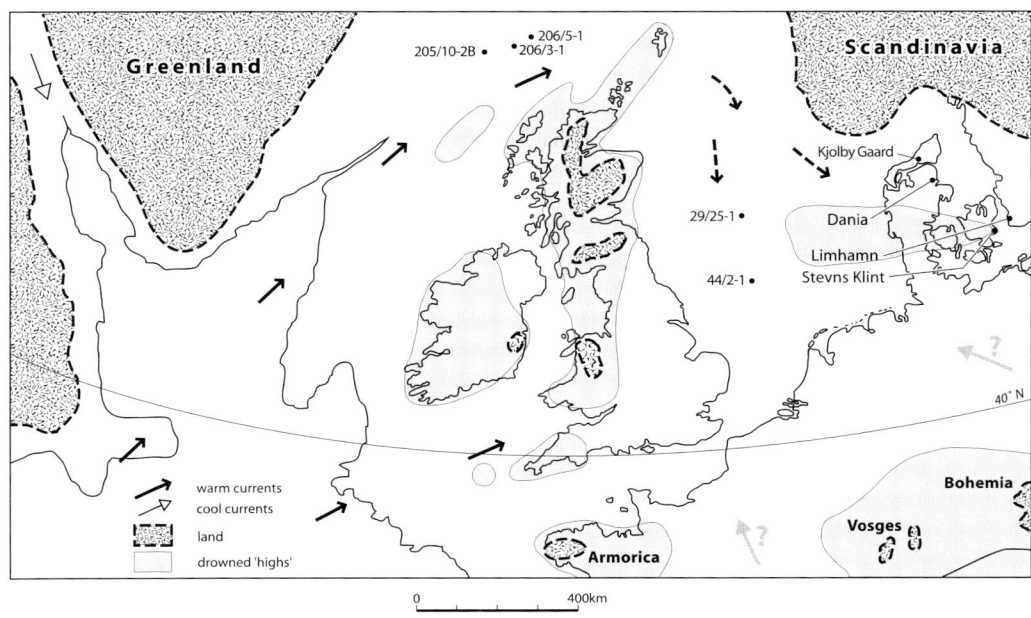

Fig. 7. Location map of the North Atlantic Ocean, North Sea Basin, UK and Denmark with potential land areas and submarine highs indicated. Based on data in Ziegler (1982) and Hancock & Rawson (1992).

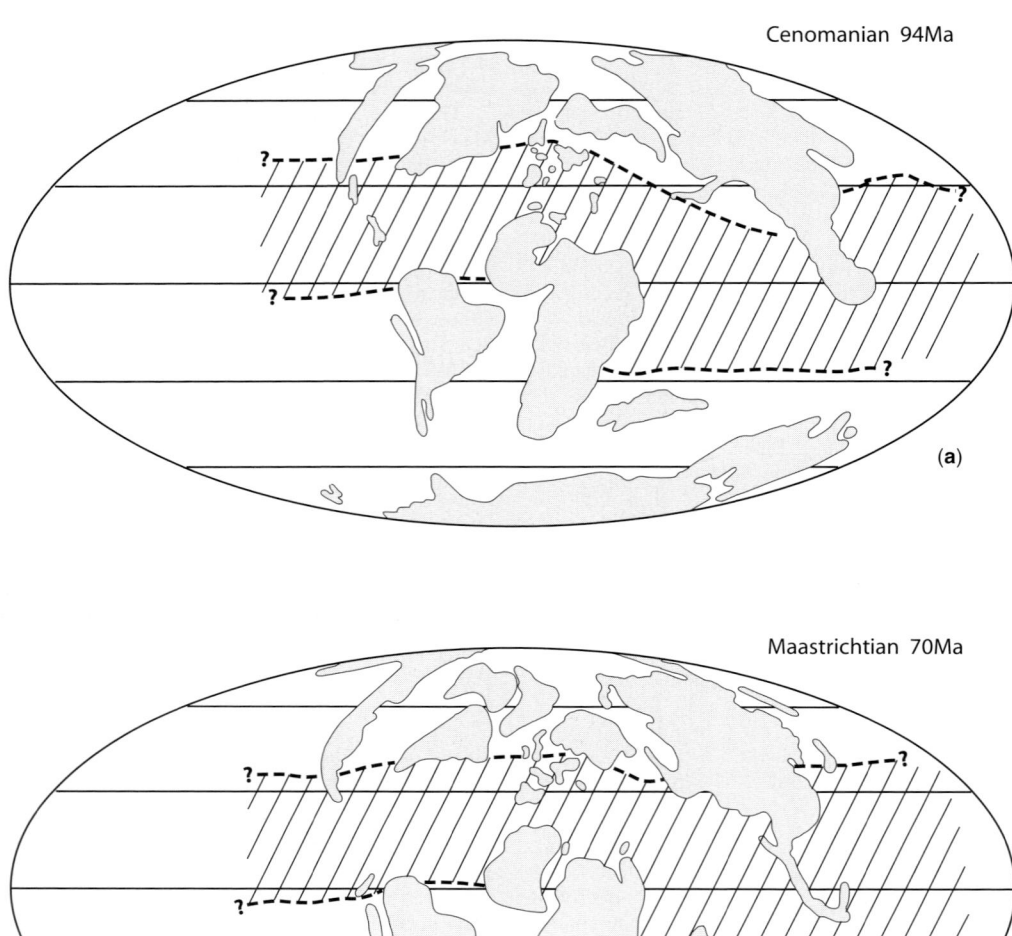

Fig. 8. Distribution maps for 'larger' foraminifera in (**a**) the Cenomanian (based on Fig. 1) and (**b**) the Campanian–Maastrichtian (based on Fig. 2) intervals. In many cases, the boundaries of the areas in which larger foraminifera are to be found are conjectural–based on very limited data. Larger foraminifera require shallow, oxygenated environments with clear water allowing light penetration, as well as warm temperatures. Everywhere within the boundaries cannot, therefore, support such faunal assemblages.

large *Contusotruncana contusa.* This assemblage was also described from Shell/Esso wells 29/25-1 and 44/2-1 in the Central North Sea Basin by Swiecicki (1980).

In Northern Denmark, this assemblage has also been described from marly chalk at Kjølby Gaard (Troelsen 1955 and unpublished work of the author) and in Dania Quarry (no longer accessible

as a result of landscaping). At Linhamn, Sweden (Malmgren 1982) and Stevns Klint (Denmark), *P. elegans/Racemiguembelina fructicosa* transitional forms are found in significant (6–12% of total fauna) numbers up to the level of the twin hardgrounds, between the white coccolith chalk and the 'Grey Chalk' (see Hart *et al.* 2005*a*, fig. 9). Hart *et al.* (2005*a*) noted that the full effect of the 'warming event' seemed to be absent (based on the isotope data), while – above the twin hardgrounds – there was an indication of a very latest Maastrichtian cooling event. In the 'Grey Chalk' above the twin hardgrounds, *P. elegans*/*Racemiguembelina fructicosa* transitional forms are only recorded in very low numbers (<1% of the total assemblage). *C. contusa* and other keeled species are missing from the successions at Limhamn and Stevns Klint (Malmgren 1982; Feist 2001; Hart *et al.* 2005*a*), probably as a direct result of the more restricted water depth in that area which is caused by the edge of the Fennoscandian Border Zone (Hart *et al.* 2005*a*, fig. 2). The shallow nature of the environment is confirmed by the presence of the well-known bryozoan 'mounds' (Surlyk 1997 and references therein). In the northern part of the Viking Graben, the same marl-rich chalks (with abundant planktic foraminifera) as were previously exposed in Dania Quarry are recorded as the Jorsalfare Formation (Isaksen & Tonstad 1989). In the Southern North Sea Basin, in the Flamborough Outlier, Stewart & Bailey (1996, p. 166, fig. 4) record the presence of *P. elegans* (and associated benthic taxa) in Maastrichtian chalks that are overlain by floods of bryozoan debris. This suggests a latest Maastrichtian shallowing event which may be related to other events close to the K/T boundary in the region of Stevns Klint (see Hart *et al.* 2005*a*) and also recorded in the Maastricht area of the Netherlands (Schiøler *et al.* 1997).

In a detailed analysis of circum-Antarctic plankton assemblages, Huber & Watkins (1992) have shown a number of poleward and equatorward migrations which they attribute to climate change. Using data from Sites 689 and 690, Huber & Watkins (1992, fig. 23) record a peak of keeled planktic diversity at the base of the late Maastrichtian, followed by an equatorward migration of a number of taxa (including *Abathomphalus mayaroensis*) followed, in the very latest Maastrichtian, by a poleward migration of *Pseudotextularia elegans*. This late Maastrichtian, high latitude, warming event, and especially the poleward migration of a *P. elegans* dominated fauna, has also been described from the Atlantic Ocean by Olsson *et al.* (2001). Their data, and interpretation, are confirmed by the above information from the Faeroe–Shetland Basin and the North Sea Basin. It appears likely that the pulse of warming, and the associated faunas, came from the Atlantic Ocean, through the Faeroe–Shetland Basin, into the Viking Graben and on into the North Sea Basin from the north, rather than from the southeast through the Danish–Polish Furrow (Fig. 7). Olsson *et al.* (2001) suggest that this latest Maastrichtian warming event began approximately

450,000 years before the K/T boundary and was over 22,000 years before the boundary, although this timing and how it correlates with events in Denmark (Hart *et al.* 2005*a*) and the Netherlands (Schiøler *et al.* 1997) is still to be fully investigated. Olsson *et al.* (2001) link the warming event to temperate/warm floras in Dakota (Johnson 1999) and Greenland (Wolfe & Upchurch 1987). The warming has also been linked to the eruption of the Deccan Traps in Chron 29R (Courtillot *et al.* 1986; Hansen *et al.* 1996; Olsson *et al.* 2002).

Distribution of Cretaceous benthic foraminifera

Cretaceous benthic foraminifera can be used in ecological modelling (e.g. Sliter & Baker 1972; Jones & Charnock 1985; Koutsoukos & Hart 1990) and there is an enormous database with which to reconstruct latitudinal (=climate) variations. The 'larger' foraminifera are quite restricted in their distribution and are often used to delimit tropical regions. It must be remembered that the term 'larger' foraminifera does not signify a taxonomic category but is an umbrella term for a range of families and genera. Generally, such taxa are normally in excess of 1 mm maximum diameter, have a complex internal structure and are often identified to the species level only by the use of thin sections. Modern 'larger' foraminifera often contain algal symbionts (Murray 1991, pp. 10–13; Hallock *et al.* 1991, pp. 44–45; ter Kuile 1991) and are, therefore, limited to shallow, warm and clear water. In the late Cretaceous, there are two quite distinct assemblages, one in the Cenomanian and the other in the Campanian–Maastrichtian interval. The Cenomanian taxa include *Orbitolina* and the praealveolinids (including *Praealveolina, Ovalvulina*, etc.) as well as other forms such as *Chrysalidina*, which although quite large are often left out of this grouping. Orbitolinids have a long history in the early Cretaceous, but are still quite abundant in the Cenomanian. They often occur in carbonate-rich environments (Oman Mountains; Simmons & Hart 1987) although they are also known from sandy, often glauconite-rich, environments (e.g. South-West England and the western part of the Paris Basin; Carter & Hart 1977). The praealveolinids that tend to be more abundant in carbonate-rich environments are less well known in clastic depositional settings (Hart *et al.* 2005*b*).

Dilley (1971, 1972, 1973) has summarized the distribution of the orbitolinids and his data have only been modified slightly in the last 30 years (Fig. 8). The praealveolinids are more restricted to the Tethyan region, being most abundant from Portugal in the west through to the Middle East.

The orbitolinids and praealveolinids disappeared in the latest Cenomanian, either terminated by the latest Cenomanian OAE ll extinction event (Kennedy & Simmons 1991; Caus *et al.* 1997; Hart *et al.* 2005*b*) or 'drowned' by the very high sea levels close to the Cenomanian–Turonian boundary (Hancock 1976, 1989).

In the Campanian–Maastrichtian interval, a second assemblage of 'larger' foraminifera is recorded. This includes the genera *Orbitoides, Lepidorbitoides, Siderolites* and *Omphalocyclus.* This fauna, or elements of it, is known from almost the same area as that occupied by the orbitolinids and praealveolinids with, in addition, the Cauvery Basin in South India (Blanford 1862; Vrendenburg 1908: Ravindran 1980; Tewari 1996; Hart *et al.* 2000, 2001) and a part of Madagascar (Abramovich *et al.* 2002). Both the Cenomanian and Campanian–Maastrichtian distributions clearly pick out the tropical belt with a maximum distribution from 42° N–40° S. This pattern is a precursor of the Cenozoic and Recent distribution of 'larger' foraminifera quite closely. These distributions should be considered in conjunction with the north–south temperature profiles (Fig. 4).

The Cretaceous/Tertiary boundary

As indicated above, there is a growing body of evidence of a marked warming event in the latest Maastrichtian. This is followed, quite abruptly, by a cooling event that has been estimated – from $\delta^{18}O$ benthic foraminifera data – to be *c.* 1 °C (Keller *et al.* 1993, 2002; Barrera & Keller 1994; Corfield 1994; Brinkhuis *et al.* 1998; Olsson *et al.* 2001; Hart *et al.* 2004, 2005*a*, fig. 9). This cooling was relatively short-lived and temperatures are thought to have recovered early in the Danian, even within the Cerithium Limestone according to Schmitz *et al.* (1992). At El Kef, temperatures recovered within zone P1a, according to Galeotti *et al.* (2004).

Conclusions

The evidence from the distribution of planktic and benthic foraminifera suggests that the late Cretaceous was warm, with low north–south temperature gradients. The latest Cenomanian and earliest Turonian has been identified as an interval of 'hot greenhouse' conditions and this was followed by a generally warm, though cooling, interval in the Turonian through Campanian. In the Maastrichtian, there were quite marked temperature fluctuations, with a distinct warming phase in the late Maastrichtian that was followed by a rapid, end-Maastrichtian cooling.

The author would like to thank several colleagues for fruitful discussions on Cretaceous climates, including Michèle Caron, Jodie Fisher, Brian Huber, Gerta Keller, Gregory Price and Matthew Watkinson. Haydon Bailey is particularly thanked for his advice and guidance on the offshore data, especially North Sea Basin. The two reviewers (Haydon Bailey and Andy Henderson) are thanked for their sound advice as both helped in the improvement of the paper. John Abraham is thanked for preparing the figures.

References

ABRAMOVICH, S., KELLER, G., ADATTE, T. *ET AL.* 2002. Age and paleoenvironment of the Maastrichtian to Paleocene of the Mahajanga Basin, Madagascar: a multidisciplinary approach. *Marine Micropaleontology,* **47,** 17–70.

AKKER, T. J. H. A., VAN DEN KAMINSKI, M. A., GRADSTEIN, F. M. & WOOD, J. 2000. Campanian to Palaeocene biostratigraphy and palaeoenvironments in the Foula Sub-basin, west of Shetland Islands, UK. *Journal of Micropalaeontology,* **19,** 23–43.

AL-RIYAMI, K. & ROBERTSON, A. 2002. Mesozoic sedimentary and magmatic evolution of the Arabian continental margin, northern Syria: evidence from the Baer-Bassit Melange. *Geological Magazine,* **139,** 395–420.

BAILEY, H. W. 1978. *A foraminiferal biostratigraphy of the Lower Senonian of Southern England.* Unpublished PhD Thesis, Plymouth Polytechnic [now University of Plymouth].

BAILEY, H. W. & CLAYTON, C. J. in press. Foraminiferids from flint meals and 'rotten' flints – the choice of an eclectic. *In*: WHITTAKER, J. E. & HART, M. B. (eds) *Micropalaeontology, Sedimentary Environments and Stratigraphy: A tribute to Dennis Curry,* The Micropalaeontological Society, Special Publications. The Geological Society, London.

BANDY, O. L. 1967. Cretaceous planktonic foraminiferal zonation. *Micropaleontology,* **13,** 1–31.

BARRERA, E., HUBER, B., SAVIN, S. & WEBB, P. 1987. Antarctic marine temperatures: Late Campanian though early Paleocene. *Paleoceanography,* **2,** 21–47.

BARRERA, E. & KELLER, G. 1994. Productivity across the Cretaceous–Tertiary boundary in high-latitudes. *Geological Society of America, Bulletin,* **106,** 1254–1266.

BÉ, A. W. H. 1977. An ecological, zoogeographic and taxonomic review of Recent planktonic Foraminifera. *In*: RAMSAY, A. T. S. (ed.) *Oceanic Micropalaeontology,* Academic Press, London, **Vol. 1,** 1–100.

BERGEN, J. A. & SIKORA, P. J. 1999. Microfossil diachronism in southern Norwegian North Sea chalks: Valhall and Hod fields. *In*: JONES, R. W. & SIMMONS, M. D. (eds) *Biostratigraphy in Production and Development Geology,* Geological Society, London, Special Publications, **152,** 85–111.

BICE, K. L. & NORRIS, R. D. 2002. Possible atmospheric CO_2 extremes of the Middle Cretaceous (late Albian–Turonian). *Paleoceanography,* **17,** 1070.

BICE, K. L., HUBER, B.T. & NORRIS, R. D. 2003. Extreme polar warmth during the Cretaceous greenhouse? Paradox of the late Turonian $\delta^{18}O$ record at Deep Sea Drilling Project Site 511. *Paleoceanography*, **18**, 1031.

BICE, K. L., BIRGEL, D., MEYERS, P. A., DAHL, K. A., HINRICHS, K.-U. & NORRIS, R. D. 2006. A multiple proxy and model study of cretaceous upper ocean temperatures and atmospheric CO_2 concentrations. *Paleoceanography*, **21**, PA2002.

BLANFORD, H. F. 1862. On the Cretaceous and other rocks of the South Arcot and Trichinopoly districts. *Geological Survey of India, Memoir*, **4**, 217pp.

BRADY, E. C., DeCONTO, R. M. & THOMPSON, S. L. 1998. Deep water formation and poleward ocean heat transport in the warm climate extreme of the Cretaceous (80 Ma). *Geophysical Research Letters*, **25**, 4205–4208.

BRINKHUIS, H., BUJAK, J. P., SMIT, J., VERSTEEGH, G. J. M. & VISSCHER, H. 1998. Dinoflagellate-based sea surface temperature reconstructions across the Cretaceous–Tertiary boundary. *Palaeogeography, Palaeoclimatology, Palaeoecology*, **141**, 67–83.

CARTER, D. J. & HART, M. B. 1977. Aspects of mid-Cretaceous stratigraphical micropalaeontology. *Bulletin of the British Museum, Natural History (Geology)*, **29**, 1–135.

CAUS, E., TEIXELL, A. & BERNAUS, J. M. 1997. Depositional model of a Cenomanian–Turonian extensional basin (Sopeira Basin, NE Spain): interplay between tectonics, eustasy and biological productivity. *Palaeogeography, Palaeoclimatology, Palaeoecology*, **129**, 23–36.

CORFIELD, R. 1994. Palaeocene oceans and climate: an isotopic perspective. *Earth-Science Reviews*, **37**, 225–252.

COURTILLOT, V., BESSE, J., VANDAMME, D., MONTIGNY, R., JAEGER, J.-J. & CAPPETTA, H. 1986. Deccan flood basalts at the Cretaceous/Tertiary boundary? *Earth and Planetary Science Letters*, **80**, 361–374.

CURRY, D. 1982. Differential preservation of foraminiferids in the English Upper Cretaceous – consequential observations. *In*: BANNER, F. T. & LORD, A. R. (eds) *Aspects of Micropalaeontology*, George Allen & Unwin, London, 240–261.

DEAN, W. E. & ARTHUR, M. A. (eds) 1998. Stratigraphy and Paleoenvironments of the Cretaceous Western Interior Seaway, USA. *SEPM Concepts in Sedimentology and Paleontology*, **6**, 255pp.

DILLEY, F. C. 1971. Cretaceous Foraminiferal Biogeography (& Bibliography). *In*: MIDDLEMISS, F. A., RAWSON, P. F. & NEWALL, G. (eds) *Faunal Provinces in Space and Time*, Geological Journal, Special Issue, **4**, 169–190.

DILLEY, F. C. 1972. Cretaceous Larger Foraminifera. *In*: HALLAM, A. (ed.) *Atlas of Palaeobiogeography*, Elsevier, Amsterdam, Ch. 37.

DILLEY, F. C. 1973. Larger Foraminifera and seas through time. *In*: HUGHES, N. F. (ed.) *Organisms and continents through time*, Special Papers in Palaeontology, The Palaeontological Association, London, **12**, 155–168.

DITCHFIELD, P. W., MARSHALL, J. D. & PIRRIE, D. 1994. High latitude palaeotemperature variation: New data from the Tithonian to Eocene of James Ross Island, Antarctica. *Palaeogeography, Palaeoclimatology Palaeoecology*, **107**, 79–101.

FALCON-LANG, H. J. & CANTRILL, D. J. 2001. Leaf phenology of some mid-Cretaceous polar forests, Alexander Island, Antarctica. *Geological Magazine*, **138**, 39–52.

FALCON-LANG, H. J., MacRAE, R. A. & CSANK, A. Z. 2004. Palaeoecology of late Cretaceous polar vegetation preserved in the Hansen Point volcanics, NW llesmere Island, Canada. *Palaeogeography, Palaeoecology, Palaeoclimatology*, **212**, 45–64.

FEIST, S. E. 2001. *Benthic foraminifera across the K/T Boundary, Denmark: A signal of global change*. Unpublished MRes Thesis, University of Plymouth.

FRAKES, L. A. 1999. Estimating the global thermal state from Cretaceous sea surface and continental temperature data. *In*: BARRERA, E. & JOHNSON, C. C. (eds) *Evolution of the Cretaceous Ocean–Climate System*, Geological Society of America, Special Paper, **332**, 238–300.

FRAKES, L. A., FRANCIS, J. E. & SYKTUS, J. I. 1992. *Climate modes of the Phanerozoic*. Cambridge University Press, Cambridge.

GALE, A. S., HARDENBOL, J., HATHWAY, B., KENNEDY, W. J., YOUNG, J. R. & PHANSALKER, V. 2002. Global correlation of Cenomanian (Upper Cretaceous) sequences: Evidence for Milankovitch control on sea level. *Geology*, **30**, 291–294.

GALEOTTI, S., BRINKHUIS, H. & HUBER, M. 2004. Records of post-Cretaceous-Tertiary boundary millennial-scale cooling from the Western Tethys: A smoking gun for the impact-winter hypothesis. *Geology*, **32**, 529–532.

GLENNIE, K. W., BOEUF, M. G. A., HUGHES CLARK, M. W., MOODY-STUART, M., PILAAR, W. F. H. & REINHARDT, B. M. 1974. The Geology of the Oman Mountains. *Verhandelingen van het Koninklijk Nederlands Geologish-Mijnbouwkundig Genootschaap*, **31**, 423pp.

GLENNIE, K. W., HUGHES-CLARKE, M. W., BOEUF, M. G. A., PILAAR, W. F. H. & REINHARDT, B. M. 1990. Inter-relationship of Makran–Oman Mountains belts of convergence. *In*: ROBERTSON, A. H. F., SEARLE, M. P. & RIES, A. C. (eds) *The Geology and Tectonics of the Oman Region*, Geological Society, Special Publications, **49**, 773–786.

HALLOCK, P., RÖTTGER, R. & WETMORE, K. 1991. Hypotheses on form and function in foraminifera. *In*: LEE, J. J. & ANDERSON, O. R. (eds) *Biology of Foraminifera*, Academic Press Ltd, London, 41–72.

HANCOCK, J. M. 1976. The petrology of the Chalk. *Proceedings of the Geologists' Association*, **86**, 499–535 [for 1975].

HANCOCK, J. M. 1989. Sea-level changes in the British region during the Late Cretaceous. *Proceedings of the Geologists' Association*, **100**, 565–594.

HANCOCK, J. M. & KAUFFMAN, E. G. 1979. The great transgressions of the Late Cretaceous. *Journal of the Geological Society, London*, **136**, 175–186.

HANCOCK, J. M. & RAWSON, P. F. 1992. Cretaceous. *In*: COPE, J. C. W., INGHAM, J. K. & RAWSON, P. F. (eds) *Atlas of Palaeogeography and Lithofacies*, Geological Society, London, Memoir, **13**, 131–139.

HANSEN, H. J., TOFT, P., MOHABEY, D. M. & SURKAR, A. 1996. Lameta age: Dating the main pulse of the Deccan Traps volcanism. In National Symposium Deccan Flood Basalts, India. *Gondwana Geology Magazine*, **2**, 365–374.

HART, M. B. 1990. Cretaceous sea level changes and global eustatic curves; evidence from SW England. *Proceedings of the Ussher Society*, **7**, 268–272.

HART, M. B. 1997. The application of micropalaeontology to sequence stratigraphy; an example from the chalk succession of South-West England. *Proceedings of the Ussher Society*, **9**, 158–163.

HART, M. B. 2000. Climatic modelling in the Cretaceous using the distribution of planktonic Foraminiferida. *In*: HART, M. B. (ed.) *Climates: Past and Present*, Geological Society, London, Special Publications, **181**, 33–41.

HART, M. B. 2004. The mid-Cenomanian non-sequence: micropalaeontological detective story. *In*: BEAUDOIN, A. B. & HEAD, M. J. (eds) *The Palynology and Micropalaeontology of Boundaries*, Geological Society, London, Special Publications, **230**, 187–206.

HART, M. B. in press. Cretaceous foraminifera from the Turonian succession at Beer, South East Devon. *Cretaceous Research.*

HART, M. B. & APLIN, K. C. 2002. Correlation of the Campanian/Maastrichtian boundary in N.W. Europe; can the foraminifera from the Goban Spur provide some answers? *Geoscience in South-West England*, **10**, 304–311.

HART, M. B. & BAILEY, H. W. 1979. The distribution of the planktonic Foraminiferida in the mid-Cretaceous of NW Europe. *Aspekte der Kreide Europas, IUGS Series A*, **6**, 527–542.

HART, M. B., BAILEY, H. W., SWIECICKI, A. & LAKEY, B. R. 1986. Upper Cretaceous Flint meal faunas from Southern England. *In*: SIEVEKING, G., DE, G. & HART, M. B. (eds) *The Scientific Study of Flint and Chert*, Cambridge University press, Cambridge, 89–97.

HART, M. B., BAILEY, H. W., CRITTENDEN, S., FLETCHER, B. N., PRICE, R. J. & SWIECICKI, A. 1989. Cretaceous. *In*: JENKINS, D. G. & MURRAY, J. W. (eds) *Stratigraphical Index of Fossil Foraminifera*, Ellis Horwood Ltd, Chichester, 273–371.

HART, M. B., BHASKAR, A. & WATKINSON, M. P. 2000. Larger foraminifera from the Upper Cretaceous of the Cauvery Basin, S. E. India. *Geological Society of India, Memoir*, **46**, 159–171.

HART, M. B., JOSHI, A. & WATKINSON, M. P. 2001. Mid-Late Cretaceous stratigraphy of the Cauvery Basin and the development of the Eastern Indian Ocean. *Journal of the Geological Society of India*, **58**, 217–229.

HART, M. B. & SWIECICKI, A. 2003. The Maastrichtian of Norfolk. *Netherlands Journal of Geosciences (Geologie en Mijnbouw)*, **82**, 233–245.

HART, M. B. & WEAVER, P. P. E. 1977. Turonian microbiostratigraphy of Beer, S. E. Devon. *Proceedings of the Ussher Society*, **4**, 86–93.

HART, M. B., FEIST, S. E., PRICE, G. D. & LENG, M. J. 2004. Re-appraisal of the K/T boundary succession at Stevns Klint, Denmark. *Journal of the Geological Society, London*, **161**, 885–892.

HART, M. B., FEIST, S. E., HÅKANSSON, E. ET AL. 2005a. The Cretaceous–Palaeogene boundary succession at Stevens Klint, Denmark: Foraminifers and stable isotope stratigraphy. *Palaeogeography, Palaeoclimatology, Palaeoecology*, **224**, 6–26.

HART, M. B., CALLAPEZ, P. M., FISHER, J. K. ET AL. 2005b. Micropalaeontology and stratigraphy of the Cenomanian/Turonian boundary in the Lusitanian Basin, Portugal. *Journal of Iberian Geology*, **31**, 311–326.

HAY, W. W., DECONTO, R., WOLD, C. N. ET AL. 1999. Alternative global Cretaceous palaeogeography. *In*: BARRERA, E. & JOHNSON, C. C. (eds) *The evolution of Cretaceous ocean–climate system*, Geological Society of America, Special Paper, **332**, 1–47.

HUBER, B. T. & WATKINS, D. K. 1992. Biogeography of Campanian–Maastrichtian calcareous plankton in the region of the Southern Ocean: paleogeographic and paleoclimatic implications. *The Antarctic Paleoenvironment: A Perspective on Global Change, Antarctic Research Series*, **56**, 31–60.

HUBER, B. T., HODELL, D. A. & HAMILTON, C. P. 1995. Middle–Late Cretaceous climate of the southern high latitudes: Stable isotopic evidence for minimal equator-to-pole thermal gradients. *Geological Society of America, Bulletin*, **107**, 1164–1191.

HUBER, B. T., NORRIS, R. D. & MACLEOD, K. G. 2002. Deep-sea paleotemperature record of extreme warmth during Cretaceous. *Geology*, **30**, 123–126.

ISAKSEN, D. & TONSTAD, K. 1989. A revised Cretaceous and Tertiary lithostratigraphic nomenclature for the Norwegian North Sea. *Norwegian Petroleum Directorate, Bulletin*, **5**, 59pp.

JENKYNS, H. C., GALE, A. S. & CORFIELD, R. M. 1994. Carbon- and oxygen-isotope stratigraphy of the English Chalk and Italian Scaglia and its palaeoclimatic significance. *Geological Magazine*, **131**, 1–34.

JENKYNS, H. C., FORSTER, A., SCHOUTEN, S. & DAMSTÉ, J. S. S. 2004. High Temperatures in the Late Cretaceous Arctic Ocean. *Nature*, **432**, 888–892.

JOHNSON, K. 1999. The megaflora of the Hell Creek Formation, south-western Dakota: biostratigraphy and paleoecology of the end-Cretaceous terrestrial vegetation. *Annual Meetings of the Geological Society of America, Denver, Colorado, abstracts with program*, 72.

JONES, R. W. & CHARNOCK, M. A. 1985. "Morphogroups" of agglutinating foraminifera. Their life positions and feeding habits and potential applicability in (paleo)ecological studies. *Revue de Paléobiologie*, **4**, 311–320.

JONGSMA, D., VAN HINTE, J. & WOODSIDE, J. M. 1985. Geologic structure and Neotectonics of the North African continental margin south of Sicily. *Marine & Petroleum Geology*, **2**, 156–179.

KAUFFMAN, E. G. & HART, M. B. 1996. Cretaceous Bio-Events. *In*: WALLISER, O. H. (ed.) *Global Events and Event Stratigraphy*, Springer, Berlin, Heidelberg, 285–312.

KELLER, G., BARRERA, E., SCHMITZ, B. & MATTSON, E. 1993. Gradual mass extinction, species survivorship, and long-term environmental changes across the Cretaceous–Tertiary boundary in high latitudes. *Geological Society of America, Bulletin*, **105**, 979–997.

KELLER, G., ADATTE, T., BURNS, S. J. & TANTAWY, A. A. 2002. High-stress paleoenvironment during the late Maastrichtian to early Paleocene in Central Egypt. *Palaeogeography, Palaeoclimatology, Palaeoecology*, **187**, 35–60.

KENNEDY, W. J. & SIMMONS, M. D. 1991, Mid-Cretaceous ammonites and associated microfossils from the central Oman Mountains. *Newsletters in Stratigraphy*, **25**, 127–154.

KING, C., BAILEY, H. W., BURTON, C. A. & KING, A. D. 1989. Cretaceous of the North Sea. *In*: JENKINS, D. G. & MURRAY, J. W. (eds) *Stratigraphical Atlas of Fossil Foraminifera*, Ellis Horwood Ltd, Chichester, 372–417.

KNIPPER, A. L., KOPAEVITCH, L. F. & RUKIEH, M. 1990. Age and origin of ophicalcites from the Ba'r–Bassit ophiolite massif, Syria. *Ofioliti*, **15**, 79–86.

KOUTSOUKOS, E. A. M. & HART, M. B. 1990. Cretaceous foraminiferal morphogroup distribution patterns, palaeocommunities and trophic structures: a case study from the Sergipe Basin, Brazil. *Transactions of the Royal Society of Edinburgh: Earth Sciences*, **81**, 221–246.

LARSON, R. L. 1991. Latest pulse of the Earth: evidence for a mid-Cretaceous superplume. *Geology*, **19**, 547–550.

MALMGREN, B. A. 1982. Biostratigraphy of planktic Foraminifera from the Maastrichtian white chalk of Sweden. *Geologiska Föreningens I Stockholm Föhandlingar*, **103**, 357–375.

MILLER, K. G., BARRERA, E., OLSSON, R. K., SUGARMAN, P. J. & SAVIN, S. M. 1999. Does ice drive early Maastrichtian eustasy? *Geology*, **27**, 783–786.

MORTIMORE, R. N., WOOD, C. J. & GALLOIS, R. W. 2001. *British Upper Cretaceous Stratigraphy*, Geological Conservation Review Series, No. 23, Joint Nature Conservation Committee, Peterborough.

MURRAY, J. W. 1991. *Ecology and Palaeoecology of Benthic Foraminifera*. Longman Scientific & Technical, Harlow, 397pp.

NORRIS, R. D., BICE, K. L., MAGNO, E. A. & WILSON, P. A. 2002. Jiggling the tropical thermostat in the Cretaceous hothouse. *Geology*, **30**, 299–302.

OLSSON, R. K., WRIGHT, J. D. & MILLER, K. G. 2001. Paleobiogeography of *Pseudotextularia elegans* during the latest Maastrichtian global warming event. *Journal of Foraminiferal Research*, **31**, 275–282.

OLSSON, R. K., MILLER, K. G, BROWNING, J. D., WRIGHT, J. D. & CRAMER, B. S. 2002. Sequence stratigraphy and sea level change across the Cretaceous–Tertiary boundary on the New Jersey passive margin. *In*: KOEBERL, C. & MACLEOD, K. G. (eds) *Catastrophic Events and Mass Extinctions: Impacts and Beyond*. Geological Society of America, Special Paper, **356**, 97–108.

PRICE, G. D. & GRIMES, S. T. 2007. New approaches for quantifying the Cretaceous terrestrial climate record. *In*: WILLIAMS, M., HAYWOOD, A. M., GREGORY, F. J. & SCHMIDT, D. N. (eds) *Deep-Time Perspectives on Climate Change: Marrying the Signal from Computer Signal from computer models and Biological proxies*. The Micropalaeontological Society, Special Publications. The Geological Society, London. 225–234.

RAVINDRAN, C. N. 1980. *Foraminiferal biostratigraphical studies of the Ariyalur Group of Tiruchirapalli Cretaceous rocks of Tamil Nadu State*. Unpublished PhD Thesis, University of Madras, Chennai, India.

ROBERTSON, A. H. F. 1975. Cyprus umbers: basalt-sediment relationships on a Mesozoic ocean ridge. *Journal of the Geological Society, London*, **131**, 511–531.

ROBERTSON, A. H. F. & HUDSON, J. 1974. Pelagic sediments in the Cretaceous and Tertiary history of the Troodos Massif, Cyprus. *Special Publication of the International Association of Sedimentologists*, **1**, 403–436.

SCHIØLER, P., BRINKHUIS, H., RONCAGLIA, L. & WILSON, G. J. 1997. Dinoflagellate biostratigraphy and sequence stratigraphy of the Type Maastrichtian (Upper Cretaceous), ENCI Quarry, The Netherlands. *Marine Micropaleontology*, **31**, 65–95.

SCHMITZ, B., KELLER, G. & STENVALL, O. 1992. Stable isotope and foraminiferal changes across the Cretaceous–Tertiary boundary at Stevns klint, Denmark: Arguments for long-term oceanic instability before and after Bolide-impact event. *Palaeogeography, Palaeoclimatology, Palaeoecology*, **96**, 233–260.

SCHÖNFELD, J. & BURNETT, J. 1991. Biostratigraphical correlation of the Campanian–Maastrichtian boundary: Lägerdorf-Hemmoor (northwestern Germany), DSDP Sites 548A, 549 and 551 (eastern North Atlantic) with palaeobiogeographical and palaeoceanographical implications. *Geological Magazine*, **128**, 479–503.

SCOTT, R. W., FRANKS, P. C., EVETTS, M. J., BERGEN, J. A. & STEIN, J. A. 1998. Timing of Mid-Cretaceous relative sea level changes in the Western Interior: Amoco No. 1 Bounds Core. *In*: DEAN, W. E. & ARTHUR, M. A. (eds) *Stratigraphy and Paleoenvironments of the Cretaceous Western Interior Seaway, USA*, SEPM Concepts in Sedimentology and Paleontology, **6**, 11–34.

SIMMONS, M. D. & HART, M. B. 1987. The biostratigraphy and microfacies of the Early to Mid-Cretaceous carbonates of Wadi Mi'aidin, Central Oman Mountains. *In*: HART, M. B. (ed.) *Micropalaeontology of Carbonate Environments*, Ellis Horwood Ltd, Chichester, 176–207.

SLITER, W. V. & BAKER, R. A. 1972. Cretaceous bathymetric distribution of benthic foraminiferids. *Journal of Foraminiferal Research*, **2**, 167–183.

SMITH, A. G., SMITH, D. G. & FUNNELL, B. M. 1994. *Atlas of Mesozoic and Cenozoic Coastlines*, Cambridge University Press, Cambridge.

SPICER, R. A. & PARRISH, J. T. 1990. Late Cretaceous – early Tertiary palaeoclimates of the northern high latitudes: a quantitative view. *Journal of the Geological Society, London*, **147**, 329–341.

STEWART, S. A. & BAILEY, H. W. 1996. The Flamborough Tertiary outlier, southern North Sea. *Journal of the Geological Society, London*, **153**, 163–173.

STOLL, H. M. & SCHRAG, D. P. 1996. Evidence for glacial control of rapid sea level changes in the Early Cretaceous. *Science*, **272**, 1771–1774.

STOLL, H. M. & SCHRAG, D. P. 2000. High-resolution stable isotope records from the Upper Cretaceous rocks of Italy and Spain: Glacial episodes in a greenhouse planet? *Geological Society of America, Bulletin*, **112**, 308–319.

SURLYK, F. 1997. A cool-water carbonate ramp with bryozoan mounds: Late Cretaceous-Danian of the Danish Basin. *In*: JAMES, N. P. & CLARKE, J. D. A. (eds) *Cool Water Carbonates*, SEPM Special Publication, **56**, 293–307.

SWIECICKI, A. 1980. *A foraminiferal biostratigraphy of the Campanian and Maastrichtian chalks of the United Kingdom*. Unpublished PhD Thesis, Plymouth Polytechnic [now University of Plymouth]; 2 vols.

TARDUNO, J. A., BRINKMAN, D. B., RENNE, P. R., COTTRELL, R. D., SCHER, H. & CASTILLO, P. 1998. Evidence for extreme climatic warmth from Late Cretaceous Arctic vertebrates. *Science*, **282**, 2241–2244.

TER KUILE, B. 1991. Mechanisms for calcification and carbon cycling in algal symbiont-bearing foraminifera. *In*: LEE, J. J. & ANDERSON, O. R. (eds) *Biology of Foraminifera*, Academic Press Ltd, London, 73–89.

TEWARI, A. 1996. *The Middle to Late Cretaceous microbiostratigraphy (foraminifera) and lithostratigraphy of the Cauvery Basin, southeast India*. Unpublished PhD Thesis, University of Plymouth.

TEWARI, A., HART, M. B. & WATKINSON, M. P. 1998. *Teredolites* from the Gardudamangalam Sandstone Formation (late Turonian–Coniacian), Cauvery Basin, Southeast India. *Ichnos*, **6**, 75–98.

THOMPSON, S. L. & POLLARD, D. 1997. Greenland and Antarctica mass balances for present and doubled CO_2 from the GENESIS Version 2 Global Climate Model. *Journal of Climate*, **10**, 158.

TROELSEN, J. C. 1955. *Globotruncana contusa* in the white chalk of Denmark. *Micropaleontology*, **1**, 76–82.

VOIGT, S. & WIESE, F. 2000. Evidence of Late Cretaceous (Late Turonian) climate cooling from oxygen-isotope variations and palaeobiogeographic changes in Western and Central Europe. *Journal of the Geological Society, London*, **157**, 737–743.

VRENDENBURG, W. 1908. The Cretaceous *Orbitoides* of India. *Records of the Geological Survey of India*, **36**, 172–213.

WEST, O. L. O., LECKIE, R. M. & SCHMIDT, M. 1998. Foraminiferal paleoecology and paleoceanography of the Greenhorn Cycle along the southwestern margin of the Western Interior Sea. *In*: DEAN, W. E. & ARTHUR, M. A. (eds) *Stratigraphy and Paleoenvironments of the Cretaceous Western Interior Seaway, USA*, SEPM Concepts in Sedimentology and Paleontology, No. **6**, 79–99.

WILSON, P. A., NORRIS, R. D. & COOPER, M. J. 2002. Testing the Cretaceous greenhouse hypothesis using glassy foraminiferal calcite from the core of the Turonian tropics on Demerara Rise. *Geology*, **30**, 607–610.

WOLFE, J. A. & UPCHURCH, G. R. JR 1987. North American non-marine climates and vegetation during the Late Cretaceous. *Palaeogeography, Palaeoclimatology, Palaeoecology*, **61**, 33–77.

ZIEGLER, P. A. 1982. *Geological Atlas of Western and Central Europe*. Shell Internationale Petroleum Maatschappij B. V. The Hague, +40 enclosures. 130pp.

The palaeogeographic and palaeoclimatic significance of climate proxies for data-model comparisons

P. J. MARKWICK

Petroleum Systems Evaluation Group, GETECH Group plc., Kitson House, Elmete Hall, Elmete Lane, Leeds, LS8 2LJ, UK (e-mail: pjm@getech.com)

Abstract: Palaeoclimate interpretations based on geological proxies of climate are fundamental to our understanding of climate change in the geological record. Most proxy definitions depend upon analogy with modern-day relationships, and the validity of this has long been questioned, especially for biological climate proxies. In the early 19th century the solution was to assume that if multiple proxies indicated the same climate then this increased the probability that the interpretation was correct. This probabilistic approach is advocated here. A further criticism has been that climate interpretations based on proxies are mainly qualitative. This is a problem when such data are used to constrain the output of computer-based numerical climate models, which are inherently quantitative.

In this study, I examine the climatic and palaeoclimatic significance of a selection of climate proxies (including crocodilians, turtles, amphibians, coals and evaporites), illustrating how each can be quantitatively defined in the modern using the concept of 'climate space'. 'Climate space' is a concept taken from ecology, analogous to petrological phase space, that defines the (usually multidimensional) environmental limits in which an organism can survive. This concept can be also used for any climate proxy, including sedimentological features ('facies' or 'depositional' space). I then show how these derived climate interpretations are applied to geological climate proxy occurrences, which are then used to evaluate climate model output directly using Geographical Information Systems (GIS). GIS combines the storage and querying functionality of relational databases with the spatial context provided by maps. Unfortunately, the geological record is neither a complete nor impartial witness to the past, and even quantitative interpretations include inherent uncertainties that must be recognized and stated: *viz.*, temporal and spatial heterogeneities and imprecision. This requires the careful compilation of large databases, and the use of 'control' groups to constrain significant absences of climate proxies. These spatial heterogeneities are examined using a series of palaeogeographic maps for the Cretaceous to Recent, which are provided here. Such maps are also important because they also act as one of the main boundary conditions for palaeoclimate models.

The interpretation of past climates is largely derived from geological phenomena, usually fossil or sedimentary, by analogy with the phenomena's modern climate limits. In most cases, these climate proxies offer only the broadest guides to contemporary climates: for example, evaporites imply net evaporation during deposition (P < E: precipitation is less than evaporation), but do not relate to a specific value; coals and peats have often been interpreted to represent 'wet' conditions, without any reference to what 'wet' means, and such generalizations can be misleading. For example, peats are more an indication of contemporary productivity and local hydrological conditions, which may or may not reflect precipitation (Lottes & Ziegler 1994). Similarly, red beds are frequently associated with arid conditions; but while this is often the case, it is not always so, and the description of 'red beds' in the literature (without any other sedimentological information) cannot be assumed to indicate past

arid conditions (Dubiel & Smoot 1994). Part of the problem is that there have been few studies that have examined any particular proxy in detail in terms of its palaeoclimatic significance (Bárdossy 1982; Gyllenhaal 1991; Lottes & Ziegler 1994; Markwick 1994, 1996; Price *et al.* 1997). With the advent of GCMs (General Circulation Models), the need for more quantitative climate proxies has become critical and the lack of well-constrained climate proxies has become readily apparent.

The nature of the problem

Computer-based climate models, specifically GCMs, are the key means of investigating the dynamics and pattern of future (Houghton *et al.* 2001) and past climate change. They also provide a powerful tool in frontier oil and gas exploration by modelling many of the environmental processes needed to predict the past distribution of source,

From: WILLIAMS, M., HAYWOOD, A. M., GREGORY, F. J. & SCHMIDT, D. N. (eds) *Deep-Time Perspectives on Climate Change: Marrying the Signal from Computer Models and Biological Proxies*. The Micropalaeontological Society, Special Publications. The Geological Society, London, 251–312.

reservoir and seal facies (Barron 1985; Huc 1990; Kruijs & Barron 1990; Barron & Moore 1994; Burggraf *et al.* 2006).

But, models are experiments that must be constantly tested against observations to assess their veracity (Markwick 1998). For palaeoclimatology, these observations are represented by geological climate proxies: indirect indicators of the contemporary climate, which include fossil, sedimentary, mineralogical and geochemical evidence (Wolfe 1971, 1978, 1993, 1994; Ziegler *et al.* 1987, 2003; Price *et al.* 1995; Parrish 2001; Zachos *et al.* 2001).

Models are also quantitative and therefore the proxies used to evaluate them must also be defined quantitatively. Numerical climate ranges have been assigned to many geological climate proxies (Bárdossy 1982; Gyllenhaal 1991; Lottes & Ziegler 1994; Markwick 1994, 1996; Price *et al.* 1997), but invariably these relationships have been defined in the modern, either through comparison of the climate represented by the geographic limits of each proxy's modern distribution (Markwick 1996; Price *et al.* 1997), or through experiment (Colbert *et al.* 1946). This also applies to physiographic methods such as CLAMP (Wolfe 1993, 1994) or stomatal analysis (Beerling *et al.* 2001; Royer *et al.* 2001; Beerling 2002; Beerling & Royer 2002). It also pertains to demonstrably numerical methods, such as stable isotope geochemistry (Price & Sellwood 1997; Pearson *et al.* 2001; MacLeod & Bergen 2004), which is not without caveats, including the influence of diagenesis, 'vital' effects, spatial and depth-related isotope heterogeneity, and inter-species variations. In the early 19th century, Fleming (1829, 1830) questioned the validity of using modern-day climate relationships as the basis for interpreting past climate through analogy, especially for biological climate proxies (today often referred to as NLR, nearest living relative, or NLE, nearest living equivalent), and argued that there could be no way of proving beyond reasonable doubt that an organism in the past, no matter how similar it resembled a living form, had the same climate tolerances. This is a criticism that has been raised by numerous subsequent authors (Sloan & Barron 1992).

Fleming's (1829, 1830) objections to using climate proxies were countered by Conybeare (1829) who argued that geologists do not rely on any one climate proxy, but multiple proxies, and that the combined force of these analogues increases the probability that a palaeoclimate interpretation is correct. This probabilistic approach is still the most compelling (Markwick 1996, 1998; Sinka & Atkinson 1999) and is that advocated here (Fig. 1). But even accepting this, defining the climate represented by each proxy is problematic

and a number of questions must be addressed, especially when using fossils. These include the following:

(1) To what extent does the modern distribution of a proxy represent its full potential range, and therefore define the actual physical climate limits that can then be applied to the past? For biological systems, this can include the consequences of biogeographical and ecological limits independent of the immediate climate.

(2) Which climate/environmental variables are the important limiting factors for any proxy? Most proxies are the consequence of a number of environmental factors and there is always the possibility of autocorrelation.

(3) To what extent is an observed proxy a consequence of local conditions rather than the general state of the atmosphere at a resolution comparable to model output?

(4) What is the consequence of temporal climate variability? This relates to the problem of equating climate in the meteorological sense (the state of the atmosphere over decades) to palaeoclimate, which on a regional or global scale can be a heterogeneic pastiche of various 'climate' states that existed over millions of years (Markwick & Lupia 2001).

In order to more rigorously define modern climate tolerances, Markwick (1996, 1998) used

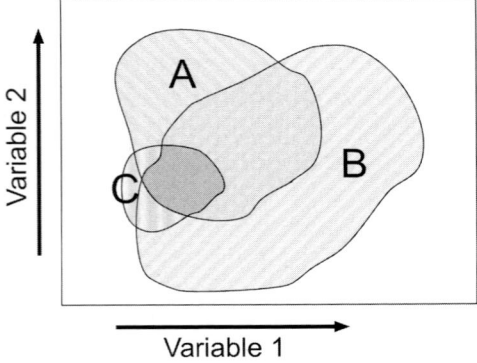

Fig. 1. Schematic representation of 2D climate space (Markwick 1996, 1998) occupied by three different climate proxies, A, B and C. The union of the three (darker shaded area) is the mutual climate range (Sinka & Atkinson 1999). In reality, the climate space occupied by each individual proxy is multivariate, and like petrological phase space, the addition of more variables can change the shape of that space. The more climate proxies that can be used to refine the interpretation of palaeoclimate, the greater the probability that the interpretation is true.

the concept of climate space taken from ecological studies (Porter & Gates 1969) to define geological proxies (Fig. 1). This concept is analogous to phase space in igneous and metamorphic petrology; like mineral phase space, climate proxies are rarely defined by a single environmental variable: for example, crocodilians are not just limited by coldest-month temperatures, but also by the presence of standing water, which provides a buffer against temperature extremes (Smith 1979; Gans 1989). The intersection in climate space of multiple proxies should therefore provide the most probable interpretation of palaeoclimate for an interval/location. This idea was developed by Sinka & Atkinson (1999) and named by them Mutual Climate Range analysis (MCR; see Fig. 1), which they applied to climatic interpretations of Quaternary fossil floras.

With the climate of each proxy defined, and accepting the associated caveats, interpretations can be assigned to geological occurrences. But how significant is each occurrence? The geological record is neither a complete nor an impartial witness to the past such that any palaeoclimate conclusion must also take into account spatial and temporal heterogeneities and imprecisions (Lupia & Markwick 1998; Markwick & Lupia 2001; Markwick & Valdes 2004). For any studied time-slice, the distribution of rocks of that age is limited by the distribution of accommodation space, which in turn is dictated by tectonics, sediment supply and base-level. The distribution of this outcrop and subcrop (notwithstanding issues of accessibility and collection logistics; Markwick 1996, 1998) represents the maximum extent of possible occurrences that have any significance for comparing with model results. This should be obvious, but the consequences are important since it means that over large parts of the globe there is no way of testing model veracity (i.e. the areas of pre-time-slice outcrop). Within the area of potential record (subcrop and outcrop), two assumptions must be made: 1 that unless there is clear evidence of transport, the presence of a climate proxy must be assumed to represent contemporary conditions in the location found (Lyell 1830); 2 the presence of a biological climate proxy (fossil) must mean the existence of a viable standing population rather than an errant individual (Markwick 1996, 1998). Thus, the presence of a biological climate proxy is significant, but an absence need not be unless the absence can be constrained by the presence of a taphonomic 'control group', a concept developed by Bottjer & Jablonski (1988). This can be applied to all proxies, but as I have already stated interpretations based on the presence of these proxies become more probable the more proxies that are available to support them.

Palaeogeography provides the spatial context in which to understand the distribution of climate proxy data and the associated uncertainties discussed above, as well as providing the framework within which to make data-model comparisons through GIS. Palaeogeography is also the principal boundary condition for the models themselves, and since geography has long been known to influence the distribution and nature of climate (Humboldt 1828; Lyell 1830), and thereby climate model results (Barron 1981; Barron & Washington 1984; Barron et al. 1993; Sloan 1994; Bice et al. 1998; Poulsen et al. 1998), it must be explicitly defined. In this study, I will also show how the interplay of geography and model resolution can have an important role in affecting data-model comparisons, especially in areas of rapid changes in relief, which unfortunately are usually typical of terrestrial depositional systems (viz., foreland or extensional basins, Markwick 1996).

The geological record is further complicated by time. A time-slice represented by a palaeoclimate simulation may be based on palaeogeographic boundary conditions that stratigraphically occupy millions of years of time (e.g. an 'Early Eocene' simulation chronostratigraphically encompasses some 7.2 million years (Gradstein et al. 2004)), with all the potential climatic, environmental and tectonic variability that can occur during that time. Because the data used to compile the map (and the climate proxy data itself used to ground-truth the consequent model results) are unlikely to represent the same 'moment' (horizon) within the time-slice, any derived map may not represent a world that ever existed contemporaneously, but is in reality a patchwork of possible local or regional geographies (Markwick & Valdes 2004). Correlation limitations over long distances mean that this dilemma is currently irresolvable for regional and global studies, such that any definitive representation of the palaeogeography of a time-slice must therefore be viewed with care. In terms of modelling, the solution is to run sensitivity experiments that represent the possible range, or end-member states, in the boundary conditions, whether this be the atmospheric chemistry, palaeogeography or other input variables. Ziegler et al. (1983, 1985) set their palaeogeographies to represent the maximum transgression within an interval, and this would represent one possible extreme, although with the caveat that the maximum transgression need not be coeval around the world due to local tectonic effects.

Consequently, the matrix of possible boundary condition variations is quite large. The consequence of this temporal heterogeneity for data-model comparisons is that the distribution of climate proxies represents the maximum preserved extent of all the range variations during the time-slice. This can best be thought of by analogy with a tidal strandline on a beach, with the maximum landward strandline marking the spring tide rather than the smaller daily variations. Thus, the northward extent of Eocene warm climate indicators such as crocodilians or palms into Ellesmere Island (West *et al.* 1977; Estes & Hutchison 1980; McKenna 1983) could potentially represent the consequence of a single warm spike such as the PETM (Palaeocene–Eocene Thermal Maximum) (Hooker & Dashzeveg 2003; Wing *et al.* 2004), rather than the longer-term state of the Eocene climate. The crucial thing is to consider such heterogeneities when using climate proxy data to evaluate model success or failure.

The net consequence of addressing these issues is the need for robust datasets, both for defining the climate tolerances of proxies in the modern day, and reconstructing the palaeoclimatology of the past. Because such datasets have to be global to be robust, at least for the modern day when defining climate limits, they are consequently large and this in turn requires the careful design of computer-based databases to store and analyse this information (Peuquet 1988; Markwick & Lupia 2001). It is particularly important to ensure that data provenance and quality are attributed correctly, because they impinge on quantifications of uncertainty, which must be defined in probabilistic approaches to palaeoclimatology.

This study

This paper presents a summary of some of the major climate proxies, in particular fossil terrestrial vertebrates, indicating their general climatic significance, a description of the datasets upon which these interpretations are based (as used in this study), and examples of how some of the climate limits are defined (with associated uncertainties). This is followed by a survey of the spatial distribution of each proxy from a polar perspective by using a series of Cretaceous and Tertiary palaeogeographic maps. These also illustrate the spatial and temporal heterogeneity of the data. I then examine and discuss several methods of data-model comparison using GIS and how this is a powerful tool with which to quantitatively compare and analyse data and model results, and a means of identifying sources of uncertainty.

Data and interpretation

Data-model comparisons for palaeoclimate simulations require the compilation and analysis of two broadly defined datasets: 1 the distribution of climate proxies in the modern (with the associated collection of climate data) with which to define

the climate space occupied by a proxy; 2 the distribution of proxies for each studied time-slice in the geological record. This in turn requires the design of large global databases with which to compile and store the data (since GCMs are themselves global and therefore it is essential to 'ground truth' them with global datasets that allow assessment over different and geographically disparate areas). The logistics of these large databases have been discussed by many authors (Peuquet 1988; Markwick & Lupia 2001); I draw on over 20 years of published experience including the design, compilation and analysis of the lithological, palaeogeographic and palaeoecological databases at the University of Chicago (Raup 1972; Sepkoski 1982; Ziegler *et al.* 1985; Crane & Lidgard 1989; Markwick 1996; Lupia 1997).

The whole process of data compilation, visualization and analysis has been greatly facilitated over the last ten years with the increasing availability and development of desktop GIS. Throughout this study we have used ESRI's ArcGIS software. GIS is a computer-based spatial, relational database system, in which geographic features are represented by points, lines, polygons and rasters (grids and images). Each feature is an individual record, which has associated data attributable to it. These records can be linked ('joined') to other records, based on spatial relationships or common attributes, and can be thoroughly interrogated by way of queries. A crucial strength of GIS is the ability to query across different data types, for instance point data (such as fossil occurrences) with gridded data (climate model output).

Datasets of the modern

The definition of modern climate space attributed to each proxy used in this study is based on a survey of published biogeographical, geographical, ecological and climate datasets compiled over the last 15 years (Markwick 1996). Data types include polygons (e.g. geographic range maps), rasters (e.g. climate grids) and points. Of these, point data are the most precise and carry the most confidence because they can be tied to a documented occurrence at a particular time and place. However, this spatial and temporal precision must be balanced with the need to define the climate: the general state of the atmosphere over several decades. To be of use, climate data representing a statistically significant time-series must co-occur with the climate proxy occurrence, and this is rare. Terrestrial vertebrates in particular, can respond to short-term, synoptic changes in the environment by moving, sometimes over great distances, and so care must be taken in using such precise occurrence data. For floras and sedimentological proxies, these issues are less important because such proxies can be considered 'static' within the time-frame of the climate being defined.

To try to account for such heterogeneities, Markwick (1998, 2001) generated 'synthetic' faunal lists by assigning faunas to 1060 global distributed climate stations compiled by Muller (1982) for vegetation studies. Each of these stations includes significantly long time-series of monthly data for major climate variables including mean daily temperature, mean precipitation, radiation and potential evapotranspiration (PET). To these have been added additional parameters, including annual metrics, combinations of variables, biomes (Walter 1985) and satellite-derived Normalised Difference Vegetation Index, NDVI (Goward et al. 1985; Cihlar et al. 1991; NOAA-EPA Global Ecosystems Database Project 1992); details are given in Markwick (1998). Faunal lists are assigned to each station by overlaying climate station distributions onto published species distribution maps (Hall & Kelson 1959; Van der Brink 1967; Little 1977; Arnold & Burton 1978; Cook 1984; Stebbins 1985; Uhl & Dransfield 1987; Branch 1988; Bouchardy & Moutou 1989; Eisenberg 1989; Kingdon 1990; Conant & Collins 1991; Grenard 1991; Cogger 1992; Iverson 1992a; Redford & Eisenberg 1992; Strahan 1992). A 50 km radius circle is drawn around each station, and an occurrence registered where the taxon's distribution intersects this circle. 50 km was chosen as the radial limit as it represents a typical de-correlation distance for precipitation, which is the most sensitive climate parameter to spatial heterogeneity. Ecologically, this approach removes local, small-scale faunal and floral heterogeneities, and thereby emulates the spatial and temporal time-averaging in the fossil record, with which this modern dataset can thereby be directly compared. 50 km also approximately equates with the scale of regional General Circulation Models $(0.5° \times 0.5°)$.

A detailed survey of all geological climate proxies is beyond the scope of this paper, so the aim here is to provide a brief discussion of a series of examples to illustrate the methods that we then apply to data-model comparisons (summarized in Table 1). Although the emphasis here is on terrestrial vertebrate climate proxies, we also include a brief analysis of the climate significance of palms (Uhl & Dransfield 1987), evaporites and coals, which includes a survey of published gazetteers and the compilations of Gyllenhaal (1991), Lottes and Ziegler (1994) and Ziegler et al. (2003). These datasets have been spatially interrogated using GIS with digital climatic, oceanographic and elevational datasets including the following: recorded monthly climate data for 1060 climate stations taken from Müller's (1982) compilation for vegetation studies; gridded climatologies (New et al. 1999); salinity and temperature (NOAA-EPA Global Ecosystems Database Project 1992).

For biological climate proxies, biodiversity has been used extensively to indicate the environmental conditions given modern relationships between species richness (diversity) and climate (Ostrom 1970; Pianka 1977; Currie 1991; Currie & Fritz 1993; Rosenzweig 1995; Markwick 2000, 2001). This remains equivocal because the exact mechanism by which climate and diversity are linked is unclear; however, as is shown below, the diversity (richness) of a biological climate proxy increases as the environment/climate becomes more conducive to survival, which equates with increasing distance in climate space from the climatic limits. Therefore, whilst the presence or absence of a proxy defines the proxy's climate space extent (as represented in the modern), increasing diversity can be used as a measure of increasing probability that the occurrence lies within the defined climate space.

Reptiles

'Reptiles' have long been interpreted to represent 'warm' climates (Lyell 1830; Owen 1850a, b, 1851; Colbert 1964b; Ostrom 1970), given that most modern members of the group are physiologically ectothermic and thereby dependent on ambient thermal conditions (Gans 1982). For palaeoclimatology, they also have the advantage that they are a common component of fossil vertebrate faunas, with specimens that are easily identifiable, at least at a generic or higher level. Although snakes and lizards have been used as climate indicators (Head 2005) and are included in the database, this paper concentrates on the two most common reptile climate proxies, crocodilians and turtles.

Crocodilians

A comprehensive study of crocodilians as climate proxies (Markwick 1996, 1998) concluded that modern crocodilians are limited to those areas with coldest-month mean temperatures of at least 5 °C, and the presence of standing water to act as a thermal buffer against extremes of temperature, based on published physiological experiments (Colbert et al. 1946; Smith 1975) and biogeographical analysis (Markwick 1998) (Fig. 3). 5 °C is the critical thermal minimum found by Brisbin et al. (1982). The critical thermal maxima and minima are the limiting temperatures at which ectotherms cease to be able to act independently and at which

Table 1. *A summary of selected climate proxies and their climatic interpretation*

Proxy	Climate	Notes and references
Biological proxies		
Crocodilians	CMM >5.0 °C (today = MAT ≥ 16.0 °C) standing water (modern distribution coincides with annual p >500 mm)	Crocodilians are dependent on local hydrology (the presence of permanent water), which need not correspond to areas of high precipitation; e.g. Nile River. Markwick (1998)
Turtles	Dependence on temperature and precipitation, but details vary according to family	This study
Tortoises (Testudinidae)	Non-freezing conditions (especially giant tortoises, genus *Geochelone*)	Giant tortoises have been used extensively for palaeoclimate studies in the Cenozoic (Hibbard 1960; Brattstrom 1961) (Markwick 2001; Böhme *et al.* 2006)
Amphibians	Biodiversiy linked with temperature and especially precipitation	This study
Palms	Although some palms can tolerate freezing conditions, arborescent forms are typical of subtropical and warmer climates	
Coral reefs	Hermatypic corals: water $T > c.$ 20 °C; clear water (light is very important)	Need to qualify what is meant by 'reefs'. Importance of light (Sellwood Price 1994; Kiessling 2001)
Peats/Coals	Productivity: mos. $T > 10°$ & $p > 40$ mm. Peat Prediction: percentage of months with $T > 10°$ that also have $p > 40$ mm	Need to also consider local hydrology, evolution of plants necessary, tectonics, subsidence (Lottes Ziegler 1994)
Sedimentary proxies		
Evaporites	$P < E$ (at least seasonally) usually indicative of low rainfall (the modern distribution of evaporites corresponds to areas with annual $p < c.$ 500 mm)	Need to know type and nature of evaporites (composition, extent, thickness: isolated pseudomorphs after halite compared with for example, Zechstein salts) The limit of 500 mm corresponds to that for semi-arid regions (Glennie 1970; Sonnenfeld 1984); deserts, $< c.$ 250 mm annually
Bauxites	MAT ≥ 22 °C p >1200 mm/yr, 9 mos p ≥ 40 mm	Difficult to date, climate interpretation is based on comparisons of modern climate to distribution of modern bauxites (Bárdossy & Aleva 1990; Price *et al.* 1997)
Red beds	$P < E$ (but with major caveats); indicates pH and Eh conditions	Need to know sedimentological characteristics of red beds (Dubiel & Smoot 1994)
Phosphorite	Ocean upwelling: concentration of nutrients leading to high productivity	To concentrate the phosphate requires that the organic matter be removed (either through physical movement or more often through oxidation
Carbonates	Depends on type of carbonate	Dictated as much by light as temperature (Ziegler *et al.* 1984). Sellwood and Price distinguished between warm water carbonates dominated by Mg-calcite and aragonite, and temperate carbonates dominated by low Mg calcite (Sellwood & Price 1994)
Tillites	Contemporary glaciers	
Glacial drop-stones	Calving glaciers at sea-level or sea-ice; generally of local origin	Differentiating ice-rafted dropstones from those rafted by other mechanisms requires some care, Markwick & Rowley (1998)

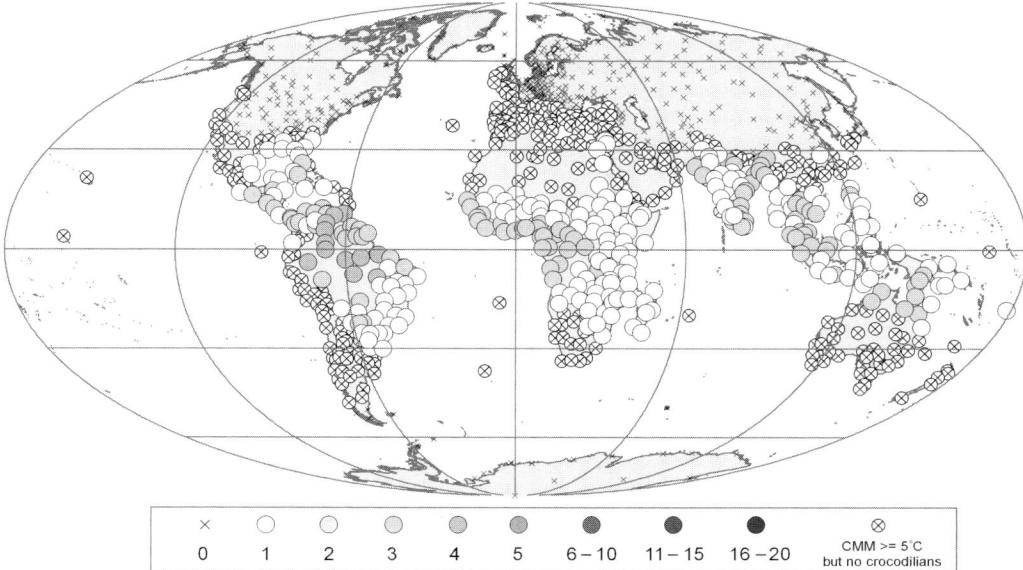

Fig. 2. Distribution of modern crocodilians showing their species diversity. The distribution of areas with coldest-month mean temperatures (CMM) of greater than 5 °C, the critical thermal minimum for crocodilians is also shown to indicate areas that fall within crocodilian climate space but do not have crocodilians.

if they are exceeded the animals will die (Cowles & Bogert 1944; Brattstrom 1965). Not all of the thermal crocodilian climate space is occupied as shown in Figure 2, which reflects this need for water, as well as the effects of human–crocodilian interactions and habitat destruction (Mediterranean coasts). This is an important consequence of this biogeographic method of defining climate space limits since it allows for apparent unfilled parts of climate space to be tested for additional limiting factors (Markwick 1998).

The value of 5 °C CCM also generally equates to areas below which waters freeze for a sufficient duration such that airholes through the ice can no

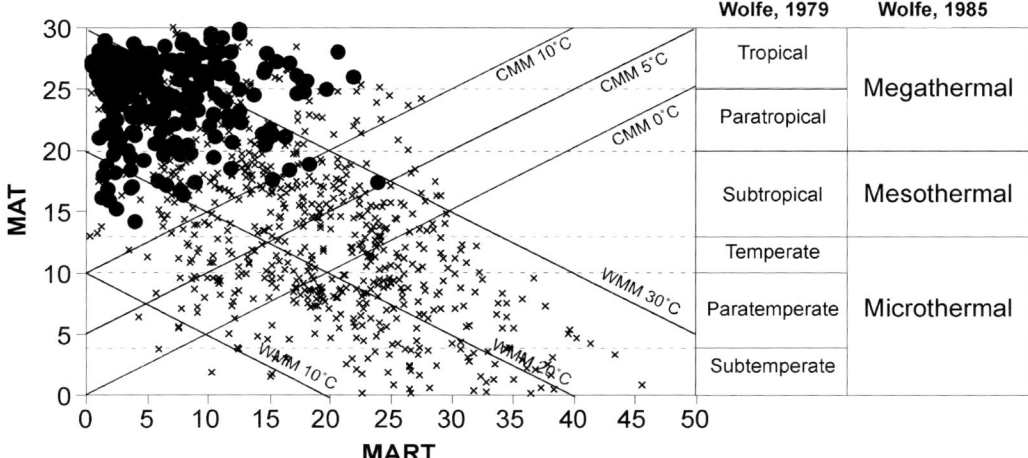

Fig. 3. Distribution of modern crocodilians in bivariate thermal climate space. Small crosses represent distribution of all climate stations and therefore significant crocodilian absences. MAT-mean annual temperature; MART-mean annual range of temperature.

longer by maintained by individual crocodilians and they drown (Brisbin *et al.* 1982). This physical limit must apply to fossil forms, if (like their modern relatives) they are ectotherms and ecologically used water as a thermal buffer. For members of the crocodilian crown group (Benton & Clark 1988), this is probably a reasonable assumption, but more disparately related groups may have had different physiological and ecological preferences. Willis & Stilwell (2000) claimed that crocodilians have changed their physiology through time, but provided no evidence to support this assertion.

Turtles

Fossil turtles, as ectotherms, are also commonly associated with 'warm' climates (Lyell 1830); but today, their range extends north into Scandinavia and southern Alaska, and south into Argentina and New Zealand (Iverson 1992*a*). The most poleward occurrences are dominated by marine turtles that migrate with the prevailing warm ocean currents, for which oceanographic temperatures and salinities are more appropriate than climate *per se*. Terrestrial turtles also extend into relatively high latitudes (Fig. 4) and occupy a variety of climate regimes, which led Broin (1984) to recommend caution when using turtles in isolation for reconstructing palaeoclimate, especially for older time intervals. Most turtle species are aquatic and, like crocodilians, freshwater turtles use water as a thermal buffer against temperature extremes. Consequently,

winter freezing and consequent drowning are important physical limits to turtle range, although the use of burrows is also common (Brattstrom 1965). Storey (1990) has documented freezing of small turtles (Storey & Storey 1988; Storey 1990), but this is size dependent and documented cases of turtle specimens that are subject to persistently low winter temperatures show that they tend to die the following year. Measurements of critical thermal minimum temperatures are sparse: Brattstrom (1965) cites several approximate values of about 5 °C for both chelydrid turtles and small tortoises, although the 5 °C CMM limit observed for crocodilians does not seem to apply to turtles (Fig. 5); in general, aquatic turtles appear to be able to withstand colder climates than terrestrial tortoises (Figs 6 and 7). Tolerance to extreme warmth is also moderated by water: Hutchison *et al.* (1966) found that aquatic turtles had lower critical thermal maxima than the fully terrestrial testudinids (tortoises).

The importance of both water and temperature is manifest in climatic relationships with turtle species diversity (species richness) (Fig. 4), although published results show how this is not an unequivocal relationship, and does vary with geography. Studies by Schall & Pianka (1977, 1978) of turtle species diversity in North America and the Iberian Peninsula indicated a strong positive correlation with temperature; but a subsequent 'global' study by Iverson (1992*b*) found that the strongest correlation globally was with annual precipitation and

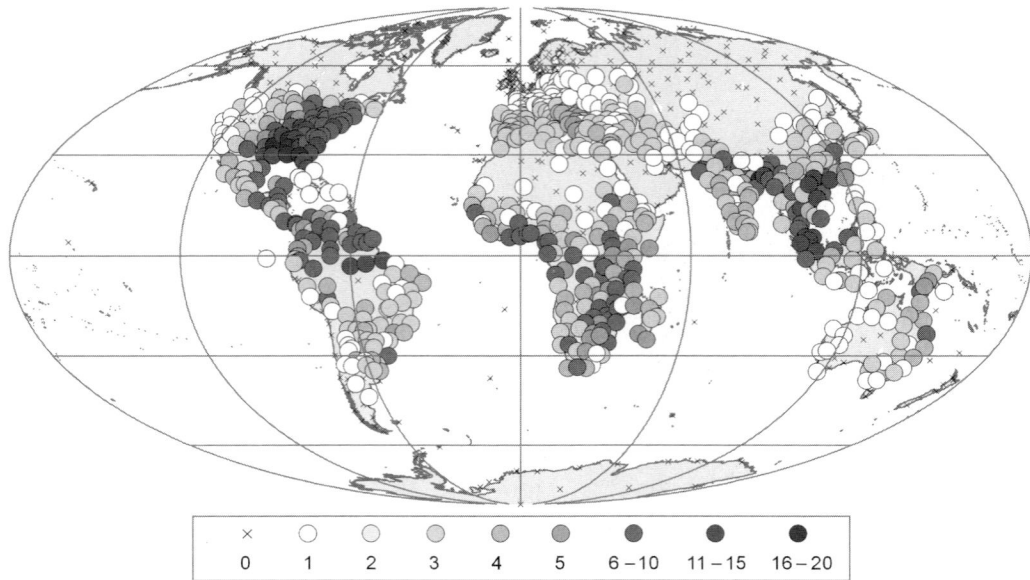

Fig. 4. Distribution of modern non-marine turtles with their species diversity.

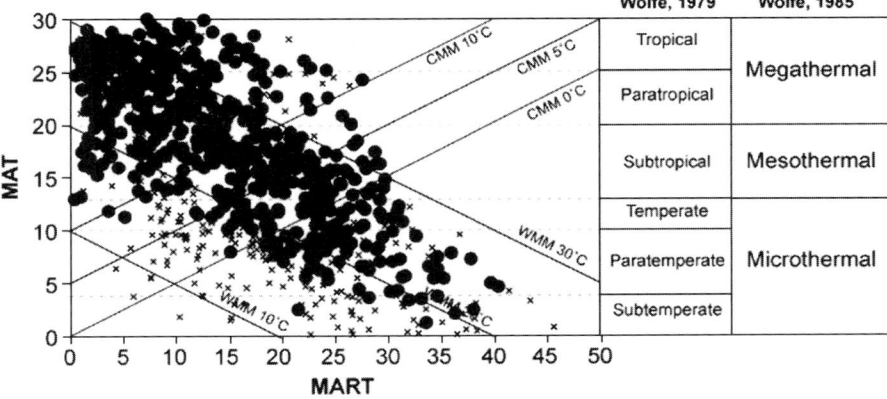

Fig. 5. Distribution of modern non-marine turtles in bivariate thermal climate space. Small crosses represent distribution of all climate stations and therefore significant turtle absences.

not temperature. Correlations varied by turtle group. Our own studies suggest that both precipitation and temperature are important, although temperature dominates. Using the unrotated scores of the Principal Components Analysis of 16 climate parameters (used by Markwick (1996, 1998) in his analysis of crocodilians), a Mann-Whitney test of turtle distributions shows a significant distribution along both of the first two axes (Fig. 8).

Tortoises (members of the family Testudinidae) appear to be more robust indicators of climate since today they are exclusively terrestrial (Fig. 9). However, most tortoises use burrows as buffers against temperature extremes, and as long as these dens remain in the range of 10–20 °C, tortoises

can survive in areas with short-term freezing temperatures (Brattstrom 1961). Giant tortoises (*Geochelone* spp.) are more sensitive to temperature, given their size. In his study of Plio-Pleistocene climates, Hibbard (1960) used giant tortoises to indicate a lack of freezing conditions, noting that today there is no evidence that this genus burrow to avoid temperature extremes. This is borne out by their modern limited extent in thermal climate space (Fig. 7), although like other turtles there is also a strong link with precipitation; however, the results of the Mann-Whitney test would suggest that this relationship is not as strong as for aquatic turtles (Fig. 8). Similarities between *Geochelone* and crocodilian climate space may be significant.

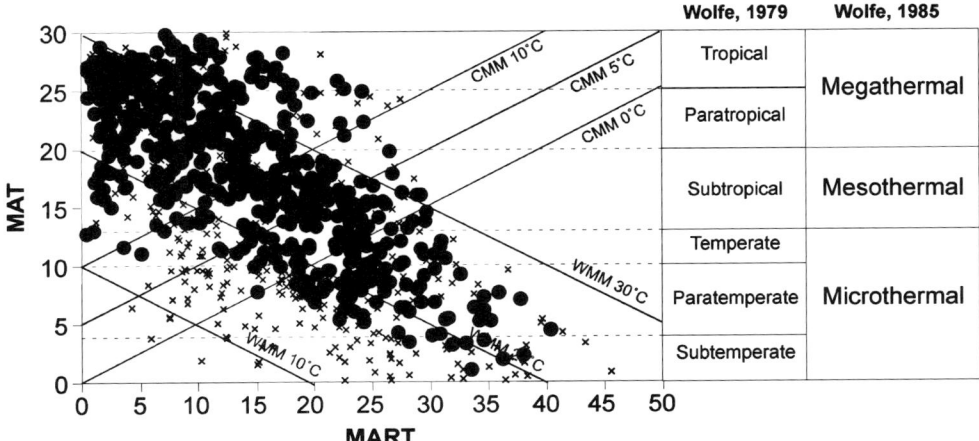

Fig. 6. Distribution of modern freshwater turtles in bivariate thermal climate space.

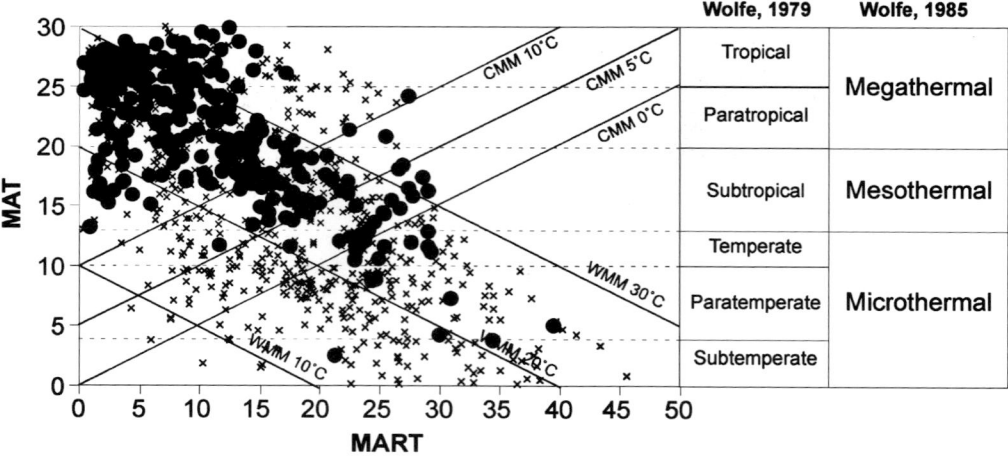

Fig. 7. Distribution of modern tortoises (Family Testudinidae) in bivariate thermal climate space.

Amphibians

Today, amphibians occur in many environments, but their general distribution, as measured by their species diversity (Fig. 10), indicates a strong dependence on water and temperature as would be expected from their physiology (Currie 1991; Markwick 2001). However, amphibians have a broad latitudinal range (Colbert 1964*b*) and are also known to aestivate for prolonged periods to escape drought periods (in the order of years in the Australian deserts), which may complicate any use of precipitation or temperature cut-offs. Nonetheless, broad palaeoclimate conclusions have been made using amphibians (e.g. Colbert 1964*a*), with their presence generally equated with 'water availability'. Böhme *et al.* (2006) have recently used amphibian assemblages to reconstruct palaeo-precipitation in the Neogene, based on different ecological groupings. Preservation may partly explain the limited number of studies using amphibians as climate proxies: although common in microvertebrate assemblages, the fragility of amphibian bones means that for most vertebrate localities diagnostic elements have been rarely preserved and documented (albeit at the level of 'amphibians present').

Dinosaurians

The use of non-avian dinosaurians as a climate indicator is problematic, given the fact that there are no living members, and consequently the uncertainties persist about their exact physiology (Ostrom 1970;

Hopson 1976; Paul 1992; Barrick & Showers 1999). Markwick *et al.* (2000*c*) found no clear link between dinosaurian generic assemblages and modelled climate (either temperatures or precipitation) for the Maastrichtian, with the geographic disparity accounting for most of the variance in the data. If, as many suggest, dinosaurians are endothermic or at least gigantothermic, then their biodiversity and distribution should show more similarities with modern mammals than reptiles. For the time being, we have included them on our maps, as a control of taphonomic biases, but with the provision that they (or at least some members of the Dinosauria) may have some climate significance.

Mammals

There have been a number of excellent studies linking mammal assemblages with their contemporary climate, usually through various mammalian feeding strategies (Janis 1989, 1993; Collinson & Hooker 2003). The endothermic physiology of mammals means that they are relatively insulated from direct correspondence with ambient climate, as can be seen by the lack of a clear relationship between mammalian biodiversity and climate or latitude, except through the filter of physiology and feeding strategies (Markwick 2001). This is, in essence, the basis of the Cenogram method of Legendre (1989; also Legendre & Hartenberger 1992), which uses the relation between the body mass distribution of all mammal species in a community and their environment, and includes the vegetation structure and climate. In this study, we use

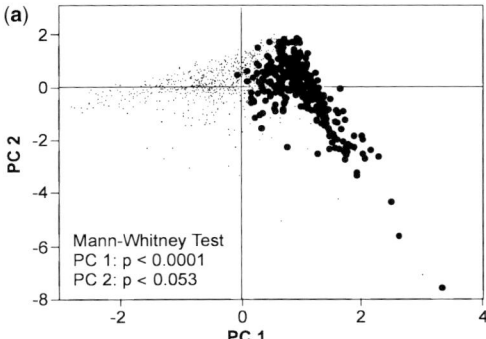

(a)

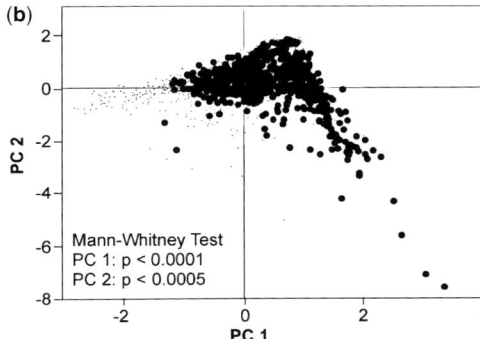

(b)

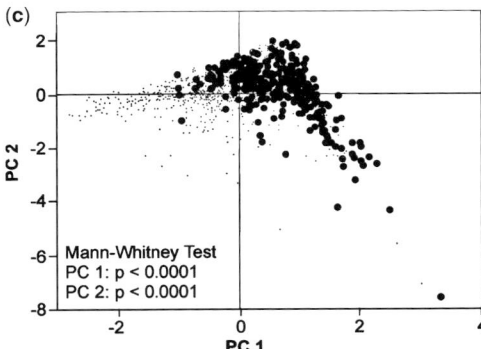

(c)

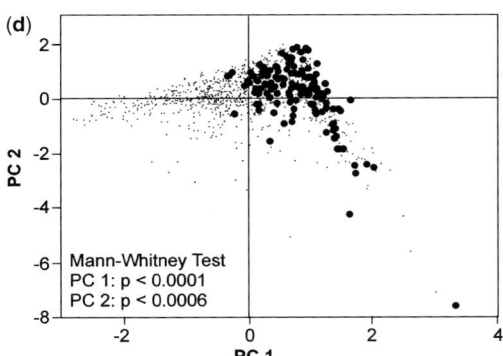

(d)

Fig. 8. (*Continued*).

Fig. 8. Location of modern crocodilian and turtle localities (large circles) in relation to the population of all available stations (small dots) plotted against components 1 and 2 of a Principal Components Analysis of 16 climate parameters (Markwick 1996, 1998). PC 1 is dominated by temperature variables and accounts for 52.7% of the scatter; PC 2 comprises mostly Precipitation variables and accounts for 25.7% of the distribution. The Mann-Whitney non-parametric test is used to examine whether these crocodilian and turtle localities are significantly distributed along either axis: (**a**) crocodilians, (**b**) aquatic turtles, (**c**) *Geochelone*, (**d**) tortoises.

fossil mammals as a taphonomic control group for more climate-sensitive vertebrates, but draw attention to their potential use in their own right.

Palms

Palms, like crocodilians and turtles, have long been associated with 'warm' climates, and have been extensively used in palaeoclimatology (Wing & Greenwood 1993; Greenwood & Wing 1995). A survey of the present-day distribution of 'palms' is shown in Figure 11, based on Uhl & Dransfield (1987), which shows the broad latitudinal range. Figure 12 shows palms in thermal climate space. Numerous studies have suggested a cellular intolerance to sustained freezing conditions, but the survival of ornamental palms in many parts of northern Europe and North America would suggest a greater tolerance to cold, and this has long been recognized by horticulturalists (Francko 2003). Palms such as the Needle Palm (*Rhapidophyllum hystrix*) can tolerate sub-zero temperatures. Arborescent forms are more sensitive to cold as indicated by published lists of minimum temperatures (e.g. Noblick 1998; Walters 1998; Francko & Wilson 2001). However, as with all plants, the survival of an individual to temperature extremes (especially when mediated by humans) need not indicate the viability of a population, which is also dependent on reproductive processes: flowering, and the ripening and germination of seeds. Larcher (1980) points out that these latter processes generally require warmer temperatures than is needed for plant growth, with the rate of germination in most plant species rising with increasing temperature. This may account for why palm biodiversity, as for vertebrates, also increases with increasing temperatures. Their position in climate space would

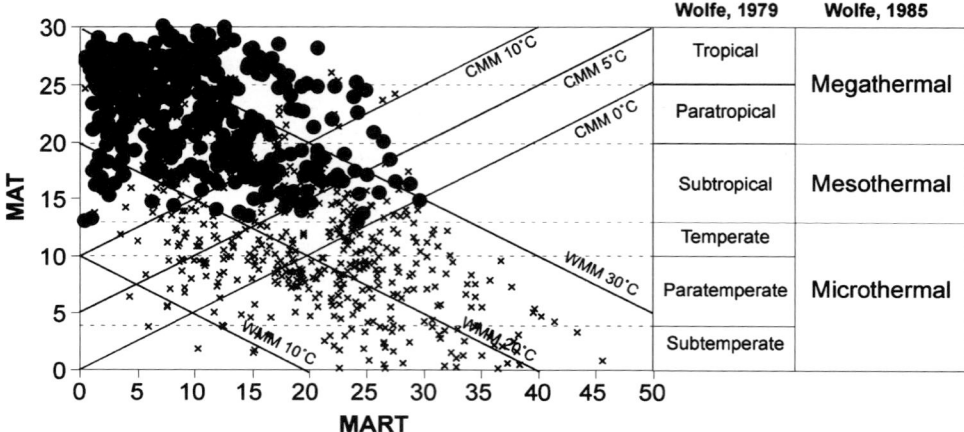

Fig. 9. Distribution of modern giant tortoises (*Geochelone* spp.) in bivariate thermal climate space. Small crosses represent distribution of all climate stations and therefore significant turtle absences.

suggest that temperature is a major limiting factor, although again the presence of water is critical; this includes groundwater within reach of root penetration (Francko 2003). Today, palms are generally a minor element in most floras, and rarely dominate except in specialist environments such as mangrove swamps (e.g. *Nypa*). Consequently, the palaeoclimatic significance must be in

agreement with other proxies. An additional problem for palms is the taxonomic uncertainty.

Biomes

Biomes (or 'zonobiomes') are areas of the Earth defined by a similar climate (originally based on an analysis of 8000 climate diagrams from

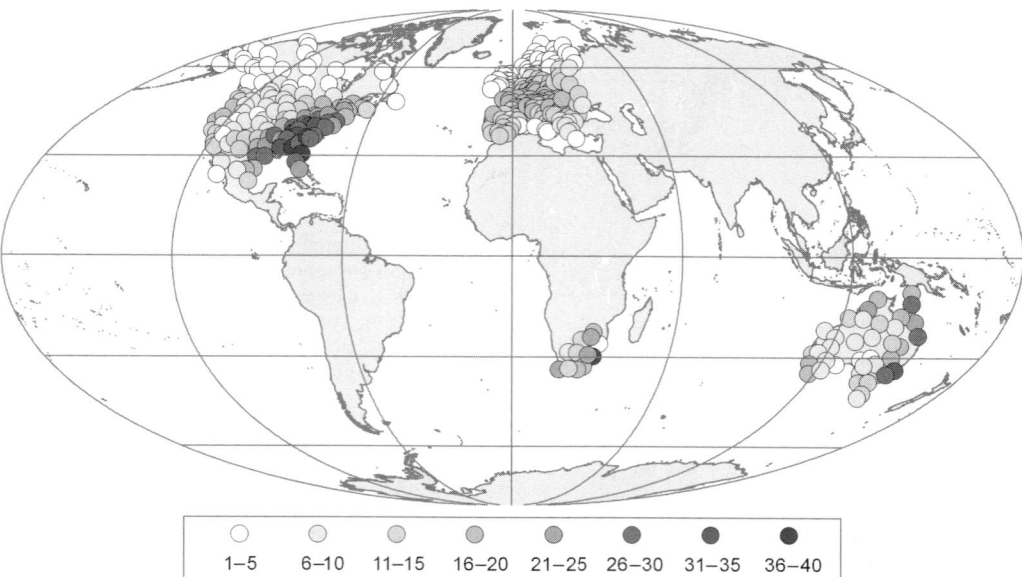

Fig. 10. Distribution of modern amphibian species diversity for North America, Europe, Australia and southern Africa showing correspondence to both temperature and precipitation. Markwick (2001) showed how the relative importance of temperature and/or precipitation varied geographically: for Australia, the strongest correlation is with precipitation (the limiting factor); in Europe, the strongest correlation is with temperature; for North America, taken as a whole, amphibian diversity correlates with both temperature and precipitation (see Markwick 2001 for data sources).

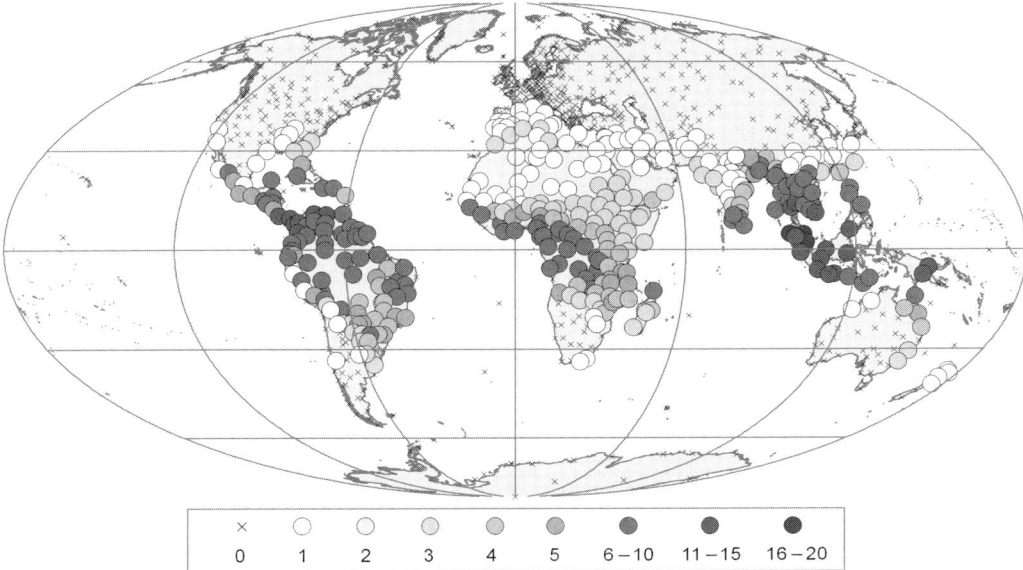

Fig. 11. Distribution of modern palms with their familial diversity using the taxonomy given in Uhl & Dransfield (1987).

around the world, including monthly rainfall and temperature; Walter & Lieth 1967) and a homogeneous vegetation type (Fig. 13; Table 2). The method assumes that the present strong dependence of vegetation on climate applies throughout time, such that similar biomes should be recognizable from fossil floras in the geological record. The limitation of such assumptions has already been discussed. Fossil floras are assigned to biomes in two principal ways: 1 using the biome assignment of the closest living relative for each taxon in a fossil list and then assigning the biome that scores the highest (Horrell 1991); 2 using multivariate statistical techniques (specifically correspondence analysis) to order the floras, and then look for appropriate biome breaks from resulting gradients (Ziegler 1990). However, while biomes, in the sense of climatically delimited floral

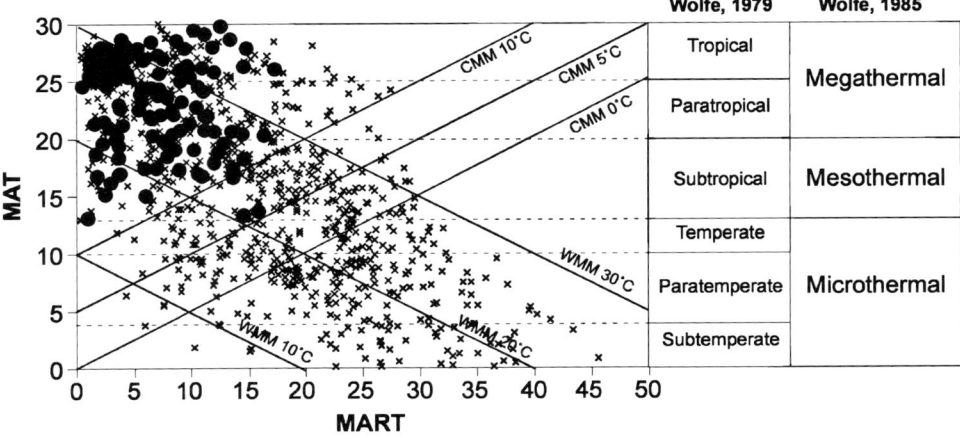

Fig. 12. Distribution of modern palms in bivariate thermal climate space. Small crosses represent distribution of all climate stations and therefore significant absences.

Table 2. *Biome definitions based on Ziegler 1990*

BIOME	Present vegetation	Climate descriptor	Precipitation systems	Poleward boundary	Sedimentary proxies
10	none	ice	orographic	none	tillites
9	tundra (treeless)	arctic	some summer frontal	0 mos $T > 0\,°C$	periglacial features
8	boreal coniferous forests (taiga)	cold temperate	summer frontal	<1 mo $T \geq 10\,°C$	podozol soils, well-developed peats
7	steppe to desert with cold winters	mid-latitude desert	none due to distance or rain shadows	<4 mos $T \geq 10\,°C$	chemozem or sierozem soils, evaporites
6	nemoral broadleaf deciduous forests	cool temperate	winter frontal, summer convective	<4 mos $T \geq 10\,°C$	occasional peats, grey and brown forest soils
5	temperate evergreen forests	warm temperate	winter frontal, summer convective	winter $T < 0\,°C$	yellow and red podzolic soils, peats locally
4	sclerophyllous woody plants	winterwet	winter frontal	≥ 10 mos $p > 20$mm	brown earth soils, very few peats
3	subtropical desert vegetation	subtropical desert	coastal: none due to cold upwelling offshore inland: none, descending limb of Hadley cell	coastal: limit of winter rains inland: winter $T < 0\,°C$	coastal: phosphorites, ORM, cherts, diatomites, sabkhas inland: aeolian dune sands, evaporites
2	tropical deciduous forests or savannas	summerwet	subtropical: summer monsoon tropical: summer extension of ITCZ	<3 mos $p > 20$mm	
1	evergreen tropical rainforest	everwet	tropical: coastal diurnal equatorial: ITCZ	<11 mos $p > 20$mm	bauxites, laterites, peats

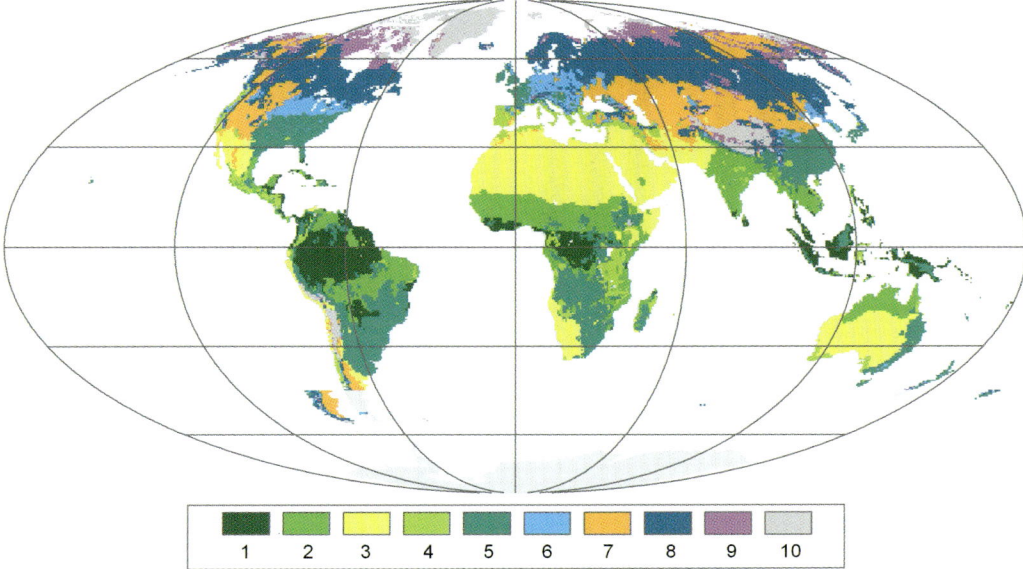

Fig. 13. Distribution of modern biomes based on Walter (1985). See Table 1 for details.

assemblages, may be discernible in the fossil record (pattern), it is less clear whether they can be assumed to directly relate to Recent biomes and thereby their climate. Nonetheless, this method has been applied to the geological past with some success: Maastrichtian (Horrell 1991) and Permian (Kutzbach & Ziegler 1993).

Here, we have largely ignored the use of fossil floral data, despite its importance, since it is covered in other chapters. However, biomes provide a convenient and familiar context with which to understand the climatic significance of vertebrate climate proxies, coals and evaporites (Figs 14–19).

Coals and peats

Coals, as the lithified, compressed remains of plant material, are commonly associated with warm, humid vegetated environments (Moore & Bellamy 1974; Stach 1982); but today, peats (the precursors of coal) can occur in a wide range of environments (Fig. 20), where a high local water table prevents desiccation and oxidation of accumulating plant material, and acidic conditions limit the rate of bacterial degradation (Moore & Bellamy 1974).

Most predictors of peats start with the prediction of plant productivity (Lottes & Ziegler 1994). At the simplest physiological level, plants require light, heat and water, and plant productivity is closely tied to these factors, especially their persistence through the year (the growing season). As a

precursor to predicting coal distribution, Lottes & Ziegler (1994) used the number of months with mean temperatures greater than 10 °C and precipitation greater than 40 mm to emulate this growing season, which they found matched well with the satellite-derived Normalized Difference Vegetation Index (NDVI), which is used as a proxy for terrestrial productivity (Goward & Dye 1987; Cihlar *et al.* 1991; Weng *et al.* 2004). Temperature affects metabolic processes and the 10 °C cut-off used by Lottes & Ziegler (1994) is the approximate limit used by Walter (1985) to define the general temperature above which plant growth occurs; Larcher (1980) discusses this in greater detail, noting that although most plant activity occurs between 5 and 40 °C (Larcher 1980, p. 27), there is some geographic and species variation with growth in tropical plants occurring above 12–15 °C, whilst in the Arctic some activity can occur down to 0 °C (Larcher 1980, p. 29). Water is of course important, and peats are often used to contrast with evaporites as a measure of the contemporary hydrological regime in the geological record (Parrish *et al.* 1982; Ziegler *et al.* 1987; Gyllenhaal 1991; Ziegler *et al.* 2003). The two proxies do dominate different parts of precipitation climate space, as shown in Figures 21–22, and occur in areas with more positive values of P–E (Precipitation minus Evaporation) (Fig. 23). Amphibian physiology has a similar dependence and Markwick (2001) found a very strong correlation between North American angiosperm

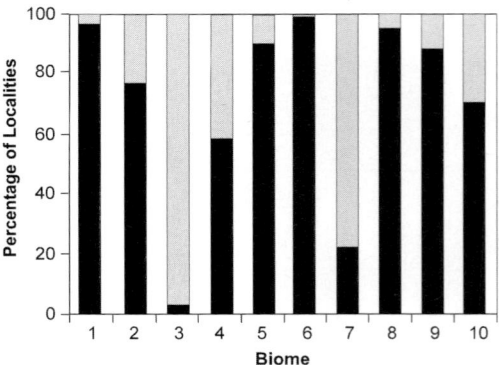

Fig. 14. Relative percentage of modern peats (black) and evaporites (grey) in each Walter biome. Dominance of evaporites in the desert biomes (3 and 7) is clearly shown.

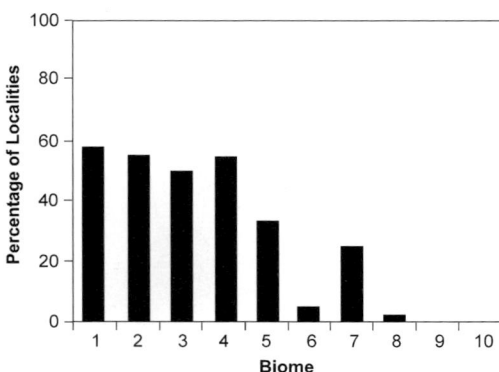

Fig. 17. Percentage of climate stations from Markwick (1996, 1998) in each biome that falls within the range of terrestrial tortoises.

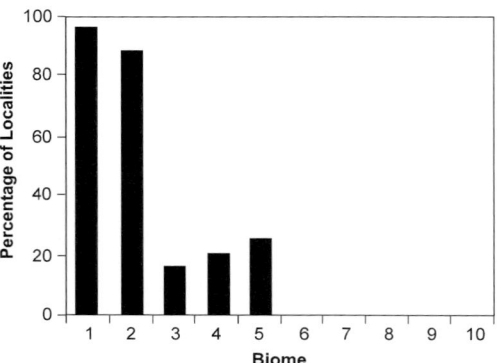

Fig. 15. Percentage of climate stations from Markwick (1996, 1998) in each biome that falls within the range of crocodilians.

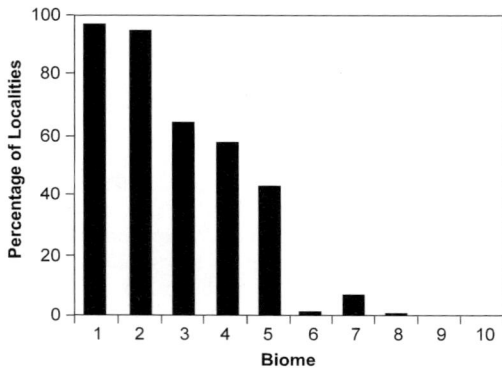

Fig. 18. Percentage of climate stations from Markwick (1996, 1998) in each biome that falls within the range of palms.

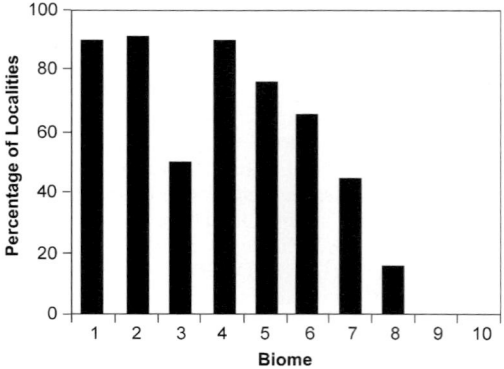

Fig. 16. Percentage of climate stations from Markwick (1996, 1998) in each biome that falls within the range of terrestrial aquatic turtles.

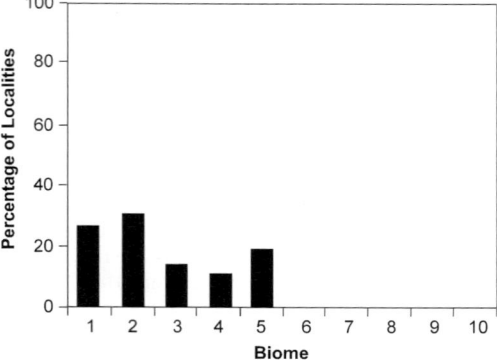

Fig. 19. Percentage of climate stations from Markwick (1996, 1998) in each biome that falls within the range of the giant tortoise, *Geochelone* spp.

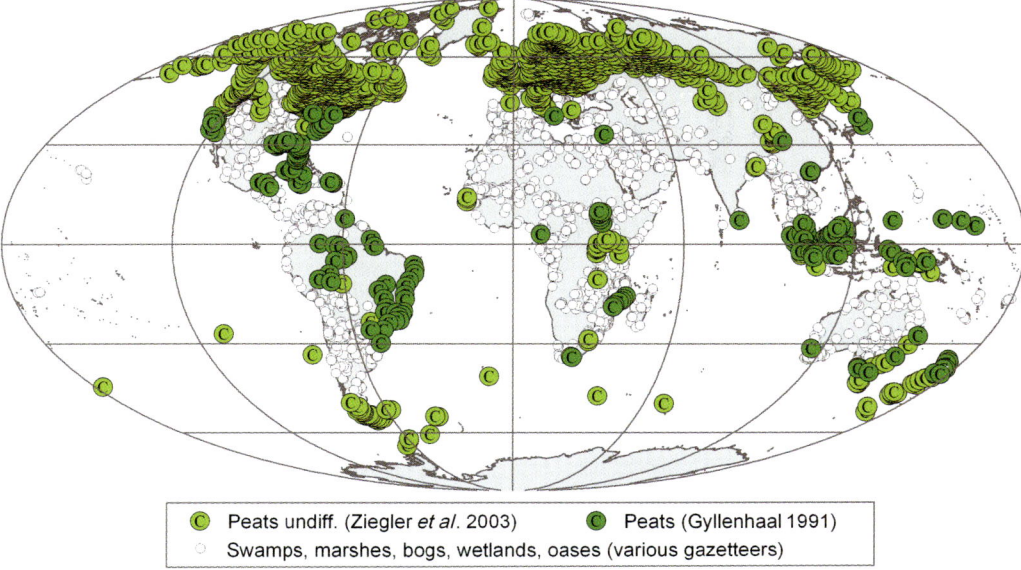

Fig. 20. Distribution of modern peats from the Paleogeographic Atlas Project's lithological databases (Ziegler *et al.* 1987, 2003; Lottes & Ziegler 1994) and Gyllenhaal (1991).

species diversity and amphibian species diversity. Mammal biodiversity also equates with primary plant productivity through the filter of feeding strategies (Markwick 2001).

Productivity does not necessarily guarantee peat formation; high temperatures equate with increased production, but also increased degradation rates. High latitudes, in contrast, may have only short growing seasons, but preservation may be enhanced by low temperatures. Lottes & Ziegler (1994) modified their 'production' metric to reflect this preservational angle, by only considering those months with precipitation greater than 40 mm and temperatures greater than 10 °C as a percentage of those months with temperatures above 10 °C. This formed their 'peat prediction map', which could account for over 80% of their peat localities (Fig. 24). This provides a useful starting point for

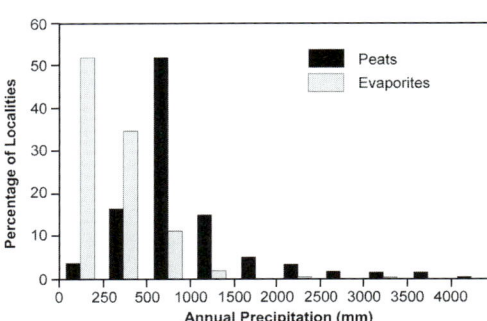

Fig. 21. Relative frequency diagram showing percentage of all peat or evaporite localities in the modern database that fall into each annual precipitation category. 500 mm is commonly used as the maximum limit for 'deserts' (Glennie 1970). Precipitation data from the 1961–1990 CRU climatology (New *et al.* 1999). See also Gyllenhaal (1991).

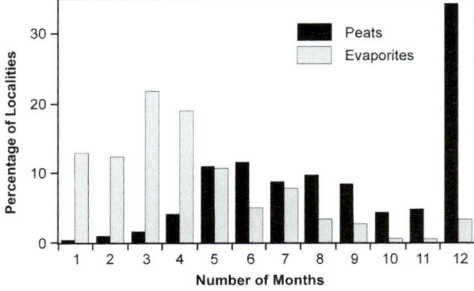

Fig. 22. Relative frequency diagram showing percentage of all peat or evaporite localities in the modern database as a function of the number of months with mean precipitation >40 mm. Precipitation data from the 1961–1990 CRU climatology (New *et al.* 1999). This suggests that a measure of precipitation continuity through the year (seasonality) can be distinguished by the presence of evaporites dominant, mixed evaporites and peats, peats dominant.

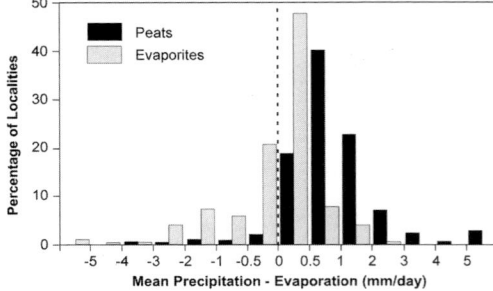

Fig. 23. Relative frequency diagram showing percentage of all peat or evaporite localities as a function of specified intervals of Precipitation minus Evaporation. As can be expected, evaporites generally occur to the left of this diagram (more-negative values of P–E) and coals to the right (more-positive values). In the absence of reliable observations of evaporation, this graph is derived from GCM results for the modern day (unpublished results Paul Valdes, University of Bristol).

interpreting coal-forming environments in the geological record, but Lottes & Ziegler (1994) did acknowledge that their peat dataset is heavily biased by peats in cold temperate areas (Biome 8, Walter 1985), which are largely an artefact of Pleistocene glacial retreat. In addition (and this is a problem that applies to many studies quantifying climate proxies in the modern), their method could not account for major peat depositional systems in otherwise arid areas, such as the Sudd of Sudan, where water is the consequence of precipitation in far-field upstream areas. This is a also a weakness in many GCMs, which represent runoff as a consequence of vertical filling of soil

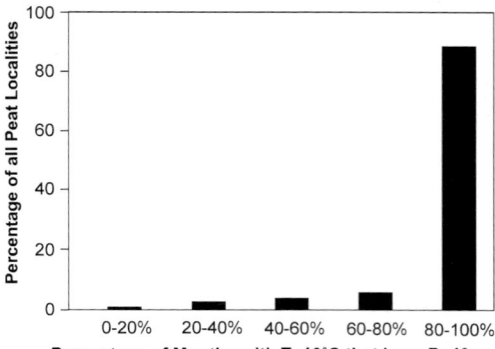

Fig. 24. Percentage of peat localities in the modern database that occur as a function of percentage of months with mean temperatures >10 °C that also have precipitation >40mm. This was used to generate the peat prediction map of Lottes & Ziegler (1994).

grid cells, with no lateral transport (either as groundwater or fluvial flow).

Evaporites

Sedimentary evaporites (Fig. 25) form through salt saturation of salt-bearing waters, and indicate hydrological conditions in which net water loss (through evaporation) is greater than net water gain (via groundwater and/or surface flow and direct precipitation). Consequently, evaporites have been frequently used to identify arid areas in the geological record, usually defined by limited precipitation; today, arid 'deserts' (represented by Biomes 3 and 7 of Walter (1985); Figs 13–14) are restricted to regions with less than 500 mm of precipitation a year (Glennie 1970), with 250 mm being a convenient division between semi-arid and arid conditions. This is supported by the results shown here (Fig. 21).

Evaporites are not simply a consequence of low precipitation *per se*, but low relative humidity and high net evaporation (Sonnenfeld 1984). Unfortunately, neither variable (especially evaporation) is unequivocally recorded at modern climate stations. Consequently, the definition of climate space for arid indicators is problematic. Because of this, numerous indirect methods of defining evaporation have been derived, such as that of Köppen (1931) or Thornthwaite (1948), which have then been applied to various climate proxies (Gyllenhaal 1991). Equations with high temperatures also need to be tempered, with the highest evaporative losses in arid and semi-arid areas today occurring in the winter season, especially in continental areas (Sonnenfeld 1984).

Evaporites do require water to form and Sonnenfeld (1984) noted that the majority of evaporites today occur in semi-arid areas in which there is at least a short wet season, as long as this does not balance out the annual net evaporative loss. However, periods of rainfall can lead to dissolution of developing evaporite minerals and this needs to be considered. Coastal evaporites, such as in sabkhas, can form under extreme arid conditions, given water replenishment from marine sources. This juxtaposition of water and high evaporation leads to the frequent co-occurrence of both 'wet' and 'dry' climate proxies in the geological record (Sonnenfeld 1984), for example the Late Cretaceous–Palaeogene lake basins of South China (Wu *et al.* 1997).

Gyllenhaal (1991) analysed a series of hydrological climate proxies in order to reconstruct precipitation gradients in the Palaeozoic. This multiple proxy approach was summarized into a series of proxy 'assemblages' (Fig. 26). These are in broad agreement with my own analyses using a larger

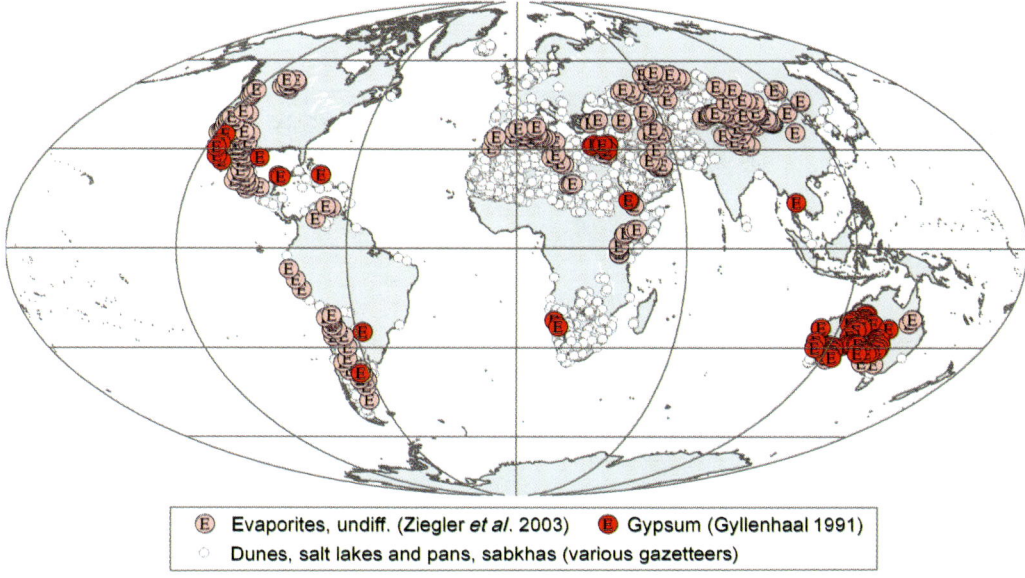

Fig. 25. Distribution of modern evaporites and other indicators of aridity used in this study.

modern-day dataset (Fig. 27). The crucial issue again is not to rely on a single climate proxy. Glennie (1970) provides a useful summary and discussion of other 'desert' indicators, with an emphasis on sedimentary factors, such as the presence of detrital clays and 'fresh' feldspars indicative of low rates of chemical weathering. Other aridity proxies used by Gyllenhaal (1991) include sand seas (dune fields over large areas, hundreds of kilometres, to differentiate them from coastal dunes), which indicate the absence of vegetation, and low annual precipitation values (<500 mm and generally < four months of precipitation greater than 40 mm), but require a sand supply. Palaeosols, such as caliches and calcretes, occupy a broader precipitation range (Gyllenhaal 1991), and are dependent on the nature of the parent material (see also Retallack 1990); and vertisols, which most importantly provide evidence of hydrological seasonality. Seasonality can also be indicated by sedimentary laminae in arid settings, especially when evaporites, such as hopper crystals in the Green River Formation, can be demonstrated to occur in only one part of annual couplets. 'Red-beds' are commonly used to indicate aridity, but are more an indication of Eh conditions, with climatic interpretations depending on supporting sedimentological evidence (Dubiel & Smoot 1994), and are not used further here. On a smaller scale, mud-cracks and rain-pits indicate some

measure of evaporation but also explicitly need some water to form.

Palaeontological evidence for aridity is generally negative, with a few exceptions such as *Classopolis* pollen, which is interpreted as evidence of 'dry' conditions (Vakhrameev 1981). In general, however, it is the absence of fossil climate proxies that is used (e.g. crocodilians and turtles require standing water as a buffer against extremes of temperature, while amphibians need water for reproduction); but to be significant, such absences must be qualified by the presence of taphonomic 'control groups', as discussed earlier in this paper. However, the presence of water in a region may not reflect the ambient precipitation, but rather conditions upstream (e.g. Nile Valley). This potential

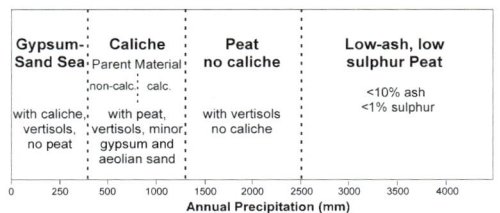

Fig. 26. Climate proxy assemblages defined by Gyllenhaal (1991) based on his detailed survey of modern climate proxies used to define precipitation gradients in the geological record.

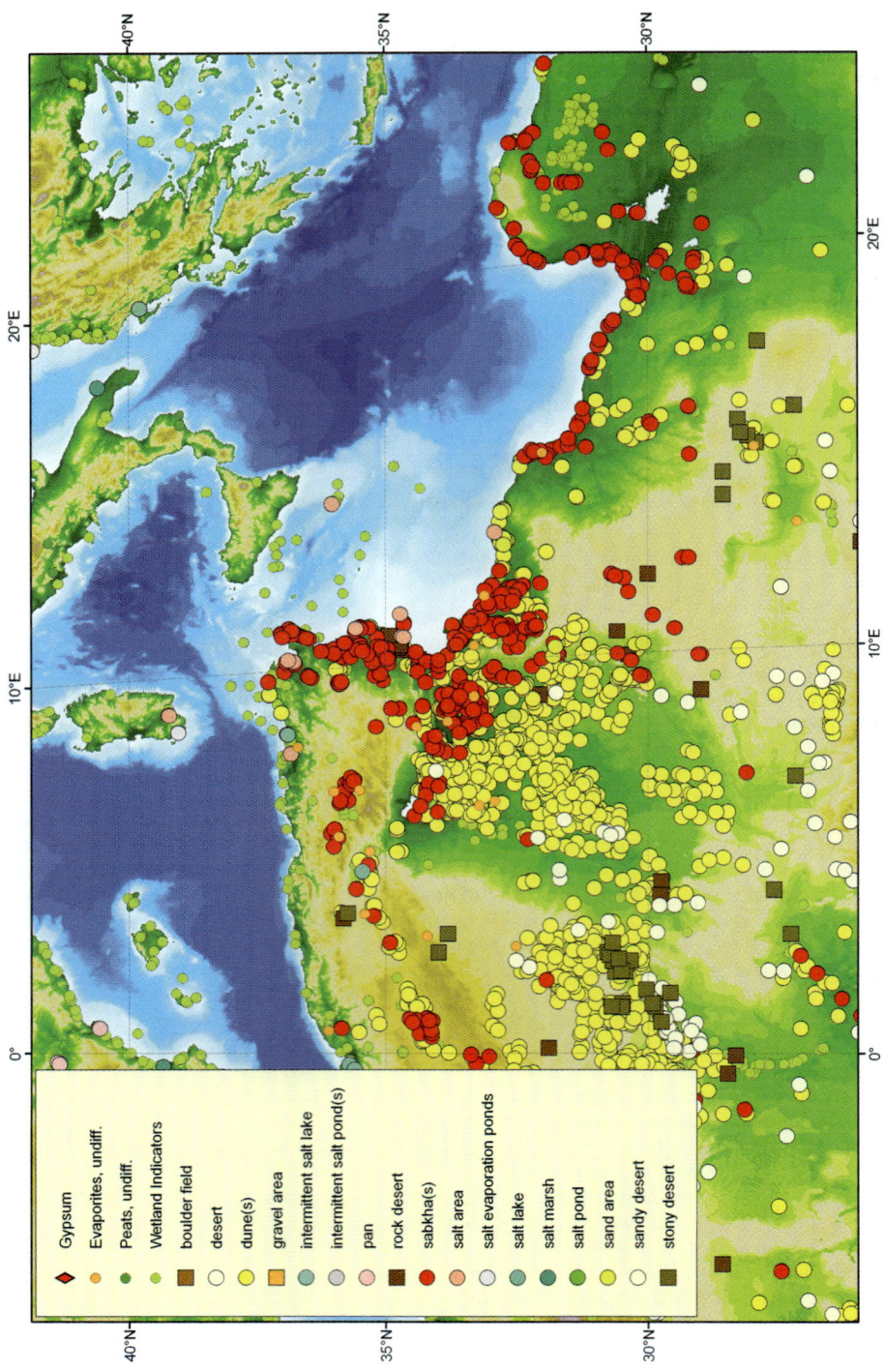

Fig. 27. A North African example of the modern hydrological indicator used in this study, showing the density of data. This is part of ongoing detailed study examining hydrological proxies and hydrological results form GCM experiments. Of the parameters shown, only man-made 'salt ponds' and generic 'salt-areas' fall outside of the aridity limit of 500mm annual precipitation (Markwick *et al.* 2005).

complication is compounded by preservational biases in fossil vertebrates, which appear to be preferentially (but not always) preserved in semi-arid environments, with the acidity of many wetter climates (conducive to peat and floral preservation, due to the concentration of especially humic acids) facilitating bone loss (Markwick 1996, 1998).

Datasets of the past

Having defined the 'climate space' represented by different climate proxies, climate interpretations can be applied to geological occurrences. As previously stated, such applications are not without caveats, whether it be uncertainties in age assignments, or the consequences of time-averaging (Behrensmeyer & Chapman 1993; Markwick 1998; Markwick & Lupia 2001). Although a site-by-site assessment of multiple proxy interpretations is recommended, which also take account of such uncertainties, as well as site-specified sedimentological data that might refine palaeoclimate interpretations, the general pattern of climate proxies provides useful information that can help constrain the general validity of climate model experiments. In this section, we show the distribution of fossil crocodilians, turtles, amphibians, non-avian dinosaurians and mammals from the Aptian to the Present. These are derived from a fossil vertebrate database compiled by Markwick (1996), which includes detailed specimen, environmental, lithological and stratigraphic information on about 6000 Cretaceous and Cenozoic fossil vertebrate localities, with taxonomic and ecological data for 22,000 extant and fossil vertebrate and floral taxa (including habitat, size and diet). These are shown together with the contemporary distribution of peats, coals, evaporites and tillites, ostensibly from the lithological databases of the Paleogeographic Atlas Project (Ziegler *et al.* 1985, 1987, 2003; Lottes & Ziegler 1994). Marine phosphorites and reefs are also included from this dataset, as a general indication of the coeval state of the ocean, which plays such an important part in regulating atmospheric conditions. These data have been supplemented with phosphorite data from various sources (BSC 1964; Notholt *et al.* 1989; Burnett & Riggs 1990; Orris & Chernoff 2002*a*, *b*).

Palaeogeography – defining the spatial context

The global palaeogeographic maps (Figs 28–49) are based on a series of more limited time-slices constructed by the Palaeogeographic Atlas Project at the University of Chicago (Ziegler *et al.* 1983, 1985; Hulver *et al.* 1993; Markwick *et al.* 2000*a*). These were modified in 1996–1997 to include additional data and intervening time-slices in order to provide a complete suite of 22 Late Cretaceous to Recent Sub-epoch (Cenozoic) and Stage (Cretaceous) maps, including sensitivity tests for the modern (+100 and −100m) (Markwick *et al.* 2000*a*). The methods used to construct these maps are described in Ziegler *et al.* (1985) and Markwick and Valdes (2004), in which the observed relationship between modern elevation and tectono-physiographic and environmental settings is used to assign values to similar settings in the geological record. The geological data used are derived entirely from published literature, utilizing the large lithological and reference databases of the Paleogeographic Atlas Project (Ziegler *et al.* 1985; Hulver *et al.* 1993), the databases of Markwick (see Markwick 1996), and the libraries of the universities of Chicago, Reading and Leeds. Although only the terrestrial topography is shown on these maps, palaeobathymetry is now being systematically added to each as part of an extensive update programme (Markwick & Valdes 2004).

The original maps were drawn at Chicago onto A3-sized basemaps (approximately 1:100,000,000) on which were printed reconstructed plate outlines (the plate reconstructions are those of Rowley, 1995 unpublished work), a simplified present coastline, and a modern five-degree grid (all rotated to their appropriate past position). They were all subsequently redrawn at a scale of 1:45,000,000 (Mollweide) and 1:30,000,000 (polar orthographic). Contours were then drawn, of which the first is the shoreline, which of all the levels of elevation is perhaps the least controversial. The mapping of the other topographic contours assumes the following: 1 that the modern relationship between elevation and tectonics is an acceptable assumption for reconstructing past topography (uniformitarianism); and 2 that mountain belts are generally long-lived features. Additional information, including sedimentological and palaeobiogeographical evidence, was used where available.

These maps were then digitized as ArcInfo coverages (a file format used in ESRI's ArcInfo GIS software) using polar projections for areas poleward of about 50–60°, and the Mollweide Projection for the tropical and equatorial regions. Consequently, these maps may be considered to be precise at about 1:30,000,000, but it is not recommended that they be enlarged above this scale. Since 1998, the palaeogeographies have been converted to ArcGIS shapefiles.

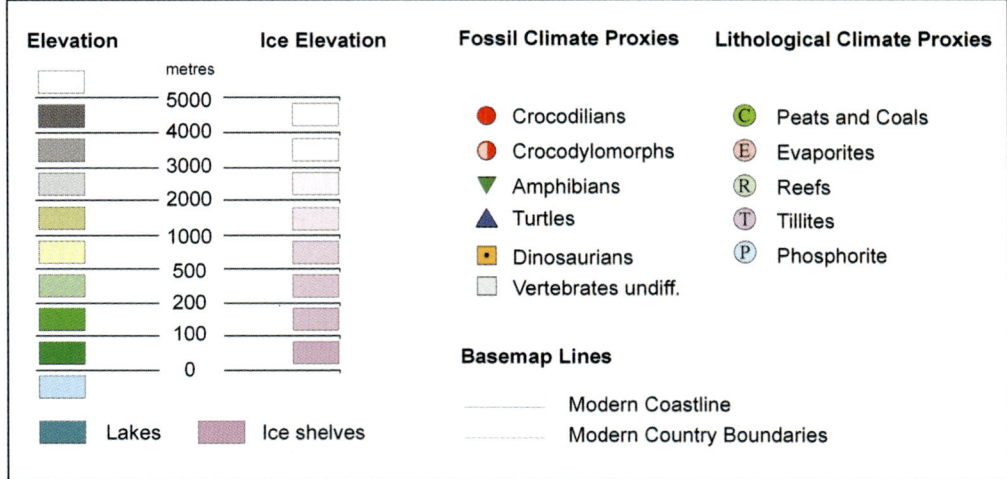

Fig. 28. The map legend for Figures 29–49 which show the distribution of climate proxies and palaeogeography for each time interval in the study. Maps are shown using north and south polar orthographic projections.

The resulting maps provide the opportunity for better understanding not only the interaction between geography, topography and climate, but also the role of geography in delimiting past biogeographic and biodiversity patterns. They highlight the coincidence of orographic changes, with the known palaeoclimatic cooling of the Middle and Late Eocene (e.g. the uplift of the Transantarctic Mountains, Rocky Mountains, and those of central Asia, coincident with increasing isolation of the Antarctic continent). Changes in the distribution of major mid-latitude seaways are also suggestive of a link with at least regional climate changes. The presence of highland in East Antarctica (Gamburtsev subglacial Mountains) continues to be problematic, since their great height might be assumed to be the locus of ice accumulation, even during 'hot-house' intervals. However, Markwick & Rowley (1998) have shown that even were these mountains glaciated above 1500 or 2000 m it would only accommodate an ice-sheet comparable to a eustatic sea-level change today of about 8 or 2 m, respectively, which would probably not be resolvable on a global scale in the geological record. It is also interesting to note that the initiation of glaciation of Antarctica in the Middle Eocene coincides with the continent moving slightly off of the pole.

Data-model comparisons

Model evaluation takes two forms:

1. *Data-model comparisons*: the assessment of model veracity through the comparison of results with observations, which can include modern-day weather station reports, satellite data, present-day occurrences of climate proxies, historical records, and geological occurrences of climate proxies.

2. *Sensitivity experiments*: the investigation of model behaviour in which input parameters are varied systematically, usually through a series of experiments that form the matrix of possible scenarios. Sensitivity tests for palaeoclimatology usually focus on varying boundary conditions: alternative palaeogeographic reconstructions, tectonic models, orbital parameterization, and atmospheric compositions. They can also be used to assess potential intra-time-slice variations (Burggraf *et al.* 2006), as well as intrinsic variability due to changing process representation in models, inter-model variability (Harrison 2004; Otto-Bliesner *et al.* 2004), and potential analytical error (factors such as inter-computer processing variability, interaction of different model modules/components, coding issues, and missing or incorrectly represented processes).

In this section, I provide some examples of the logistics of data-model comparisons using the climate proxy datasets described above and the results from model simulations run in collaboration with Professor Paul Valdes, University of Bristol (Markwick *et al.* 2000*b*, 2002, 2005; Markwick & Valdes 2003; Haywood *et al.* 2004; Burggraf *et al.* 2006).

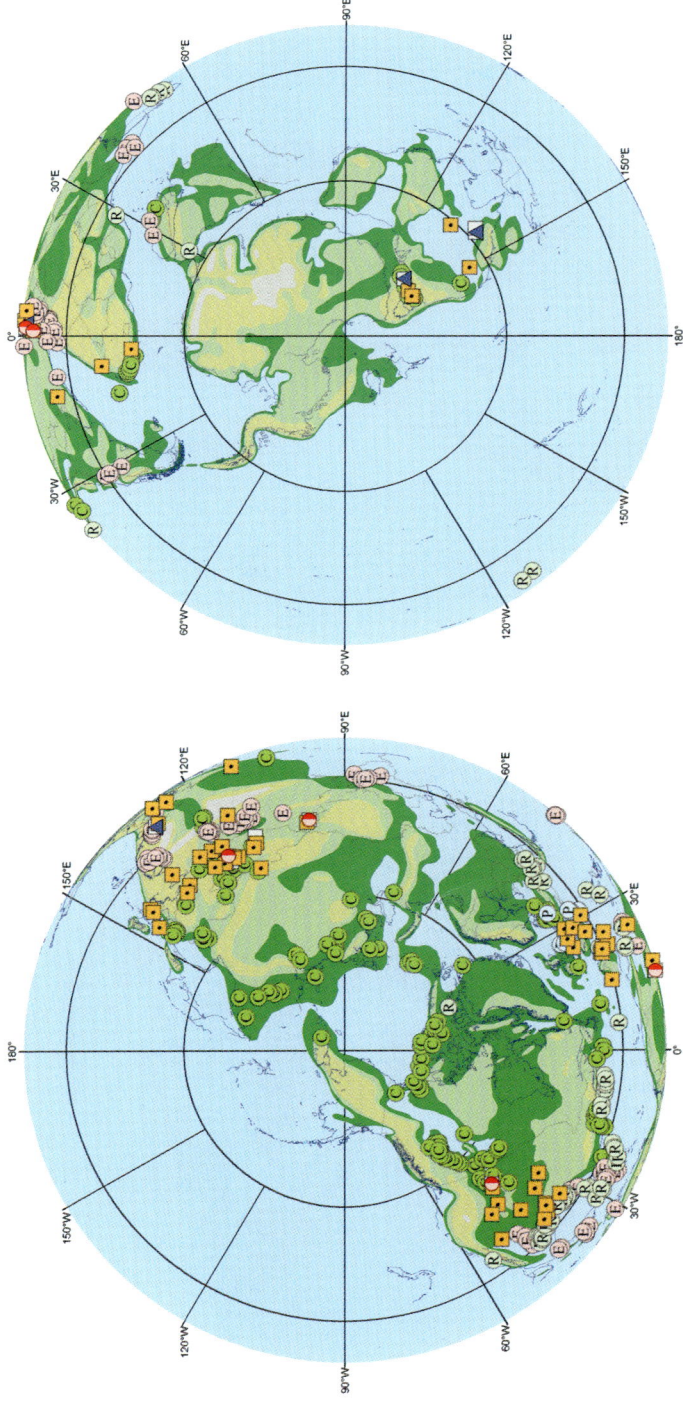

Fig. 29. Palaeogeographic map of the Aptian. See Figure 28 for legend. Left, northern hemisphere; right, southern hemisphere.

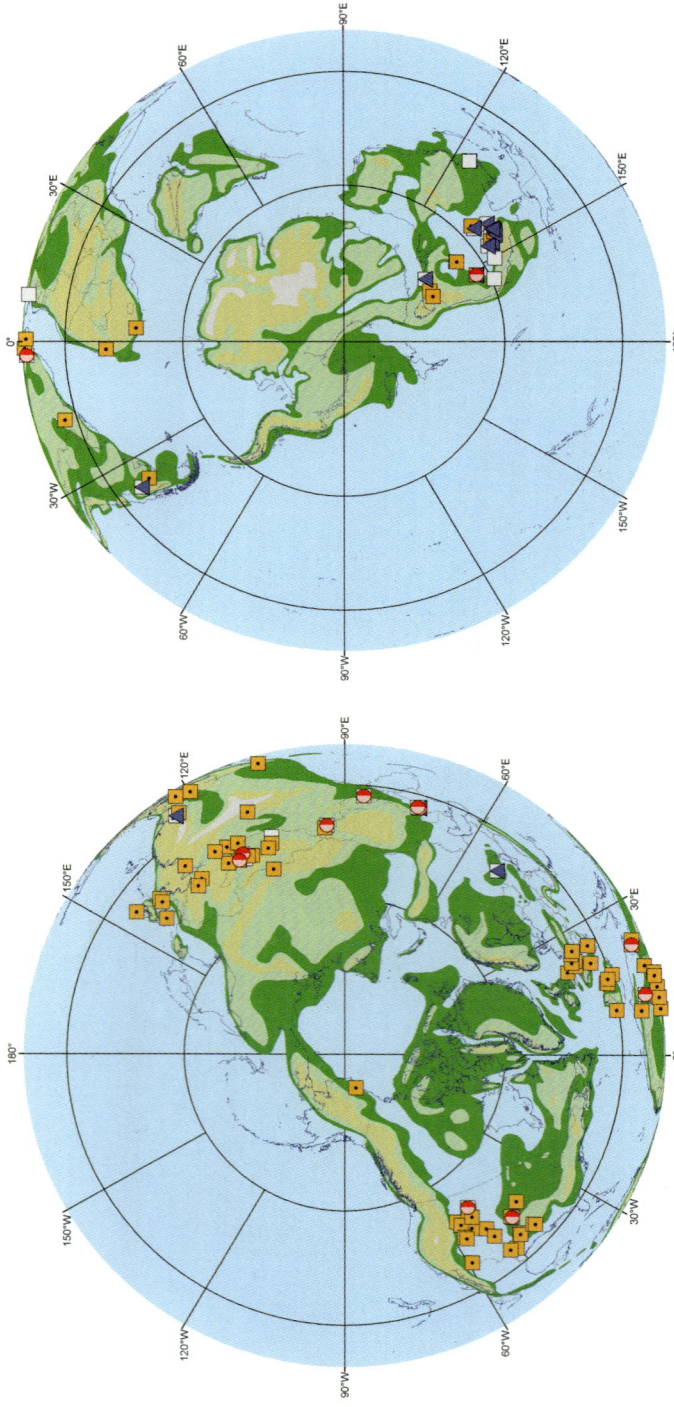

Fig. 30. Palaeogeographic map of the Albian. See Figure 28 for legend. Left, northern hemisphere; right, southern hemisphere.

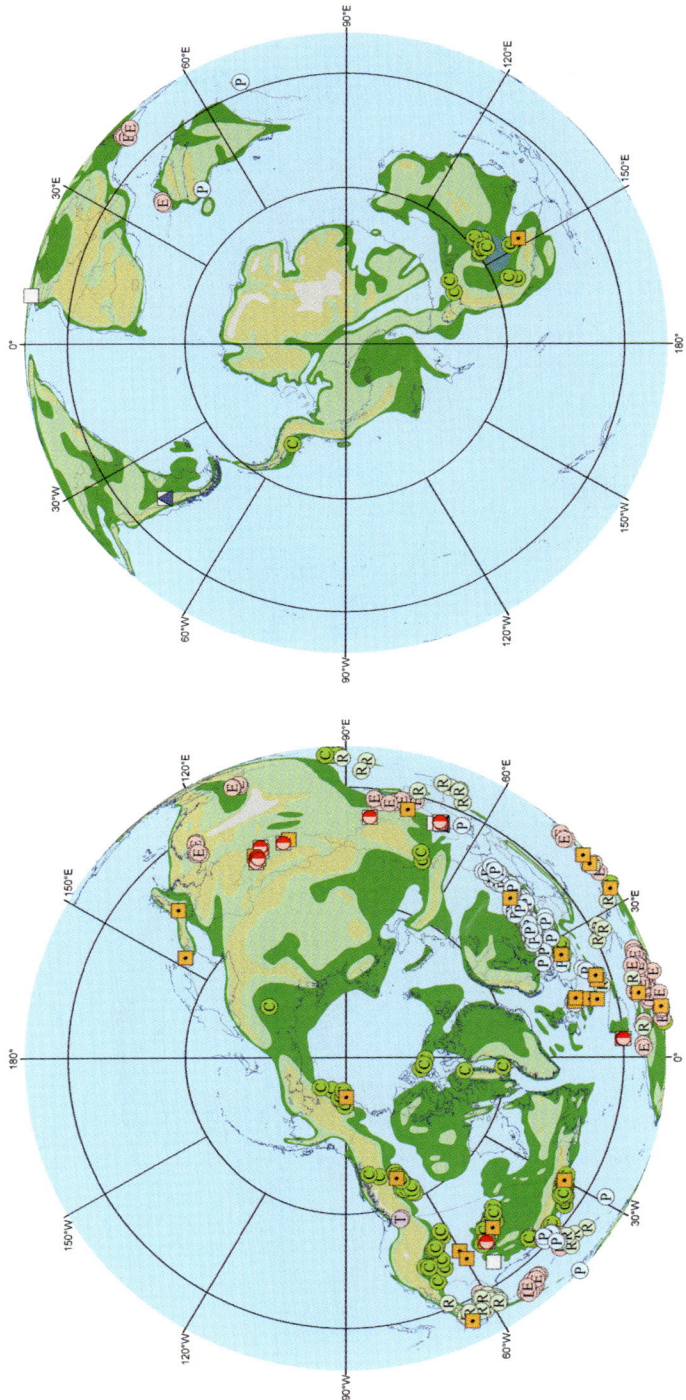

Fig. 31. Palaeogeographic map of the Cenomanian. See Figure 28 for legend. Left, northern hemisphere; right, southern hemisphere.

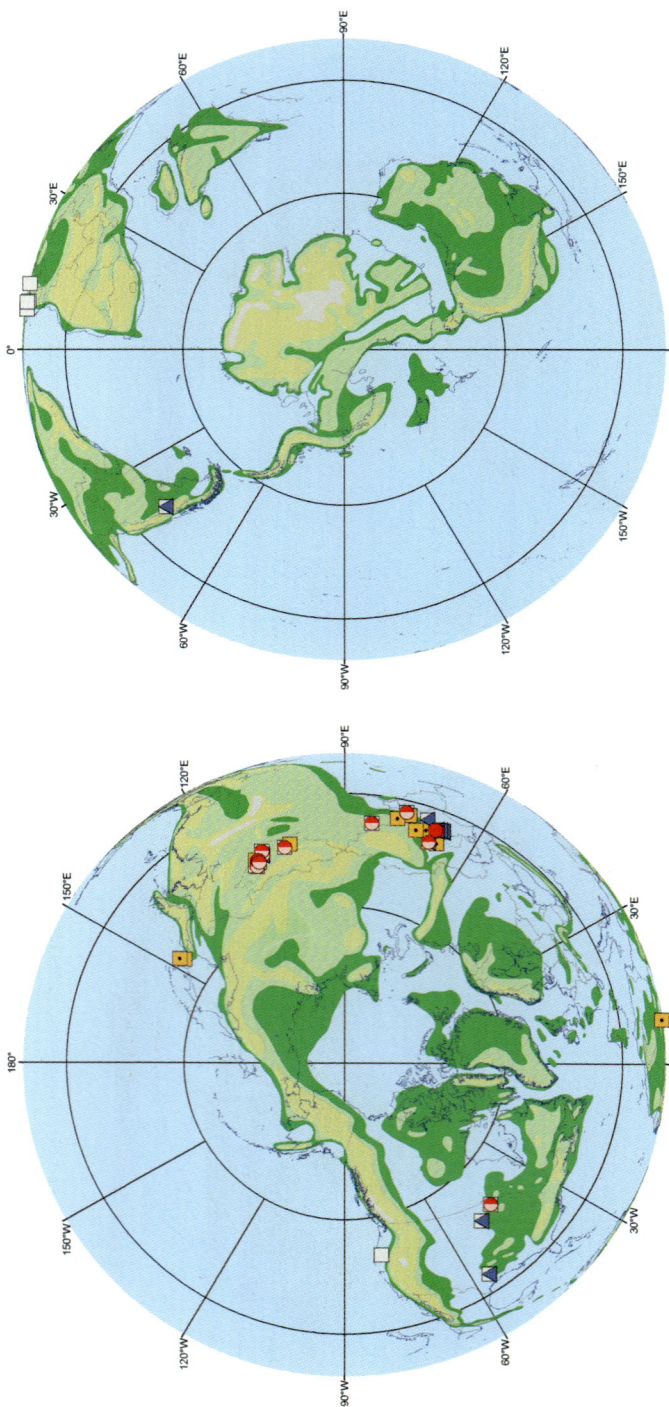

Fig. 32. Palaeogeographic map of the Turonian. See Figure 28 for legend. Left, northern hemisphere; right, southern hemisphere.

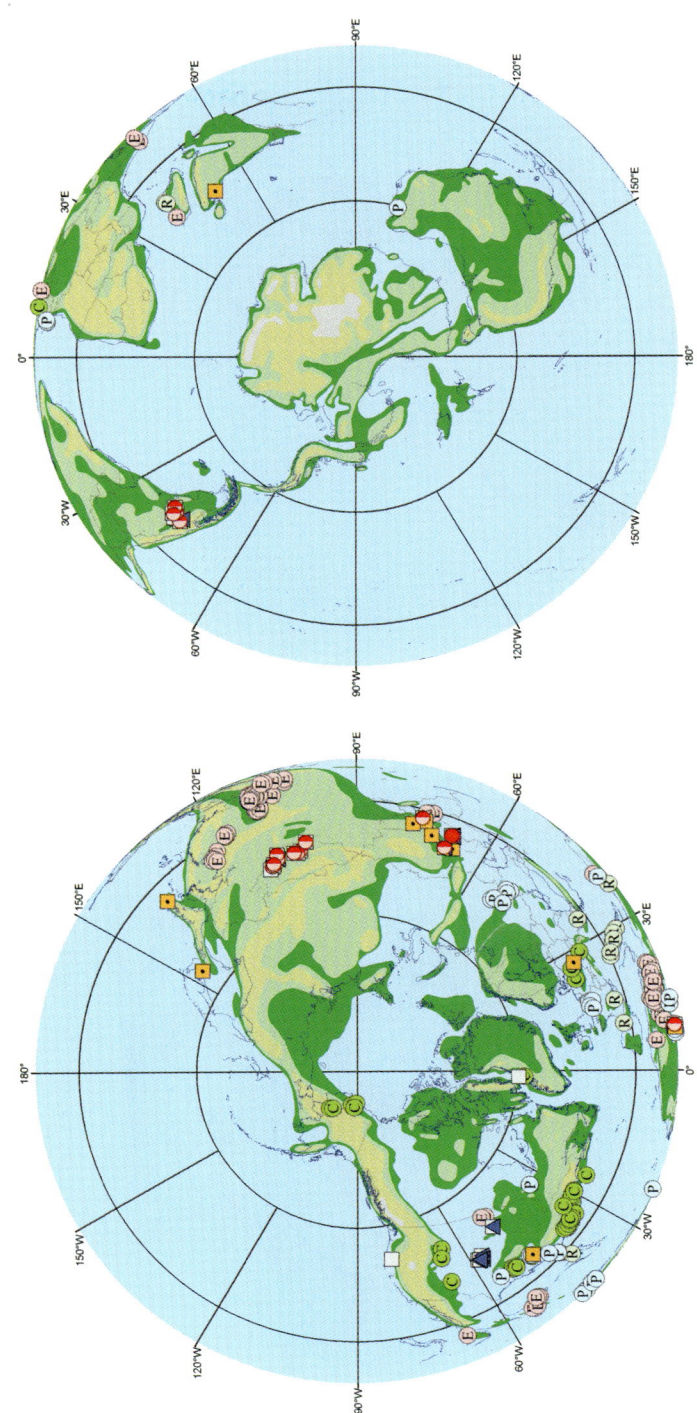

Fig. 33. Palaeogeographic map of the Coniacian. See Figure 28 for legend. Left, northern hemisphere; right, southern hemisphere.

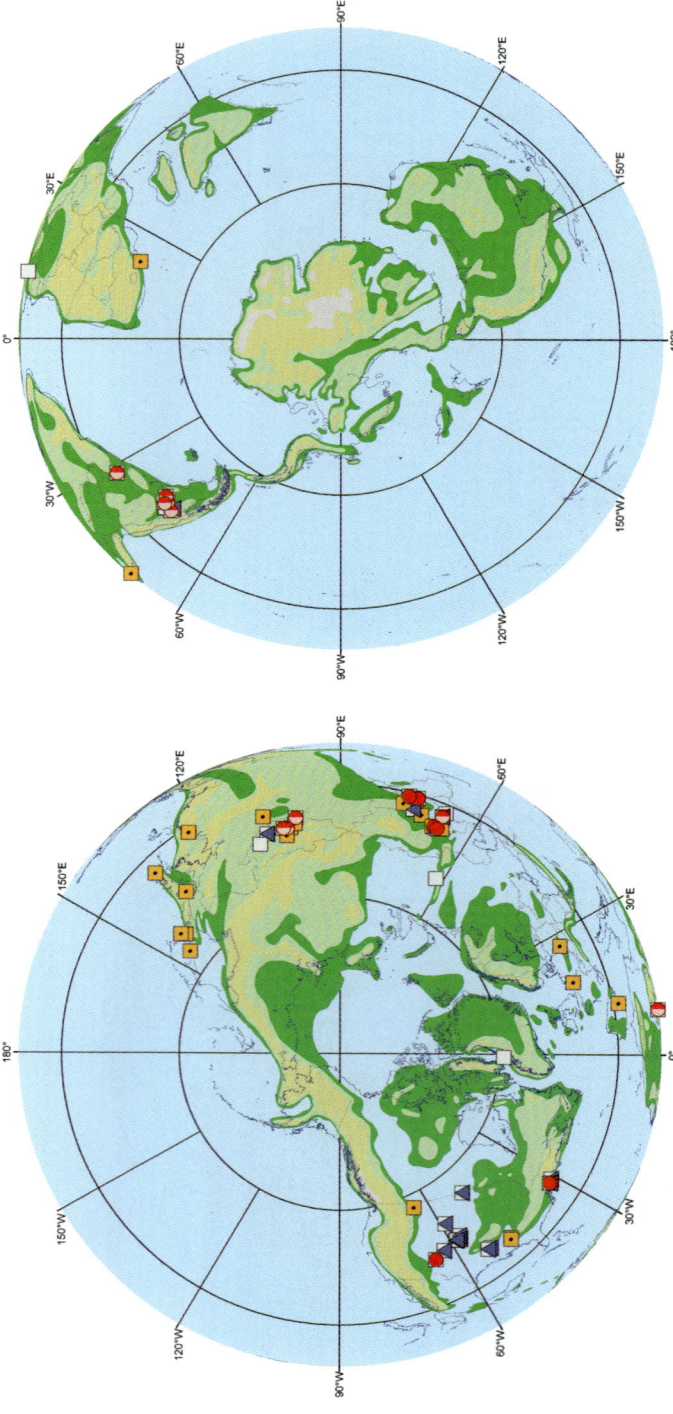

Fig. 34. Palaeogeographic map of the Santonian. See Figure 28 for legend. Left, northern hemisphere; right, southern hemisphere.

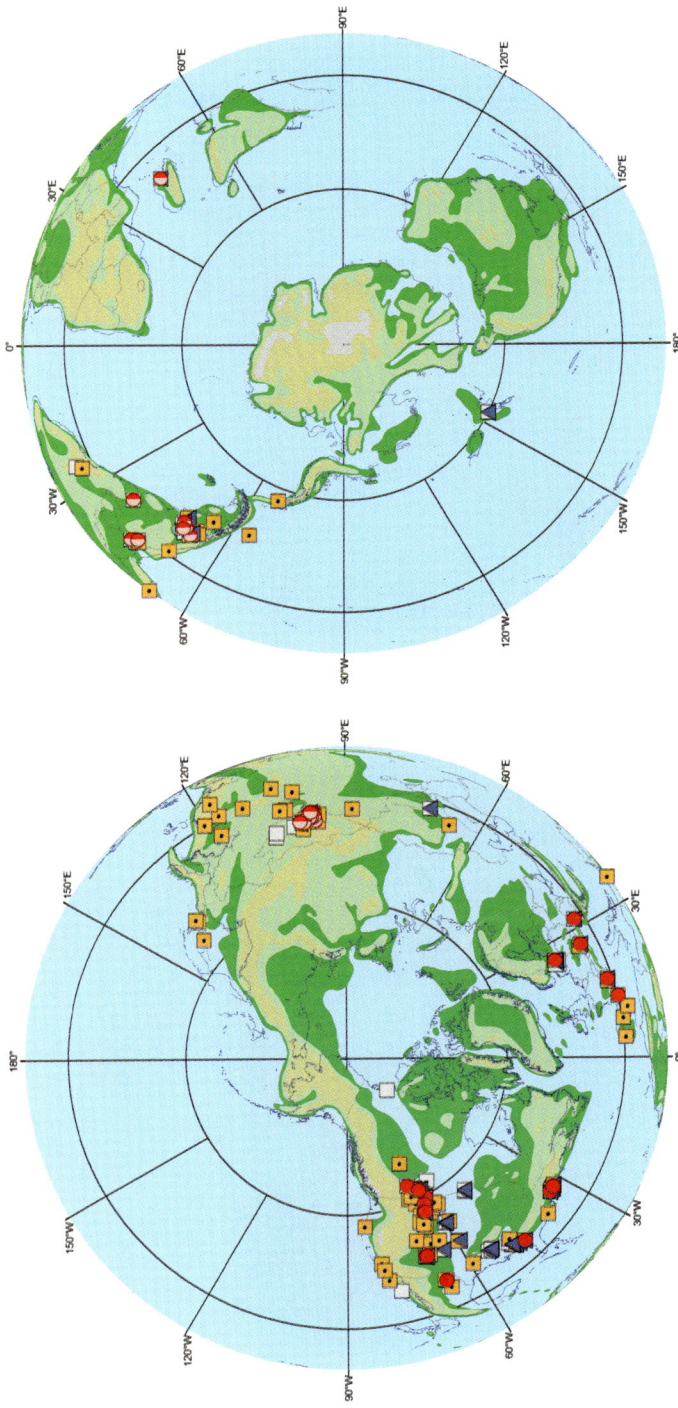

Fig. 35. Palaeogeographic map of the Campanian. See Figure 28 for legend. Left, northern hemisphere; right, southern hemisphere.

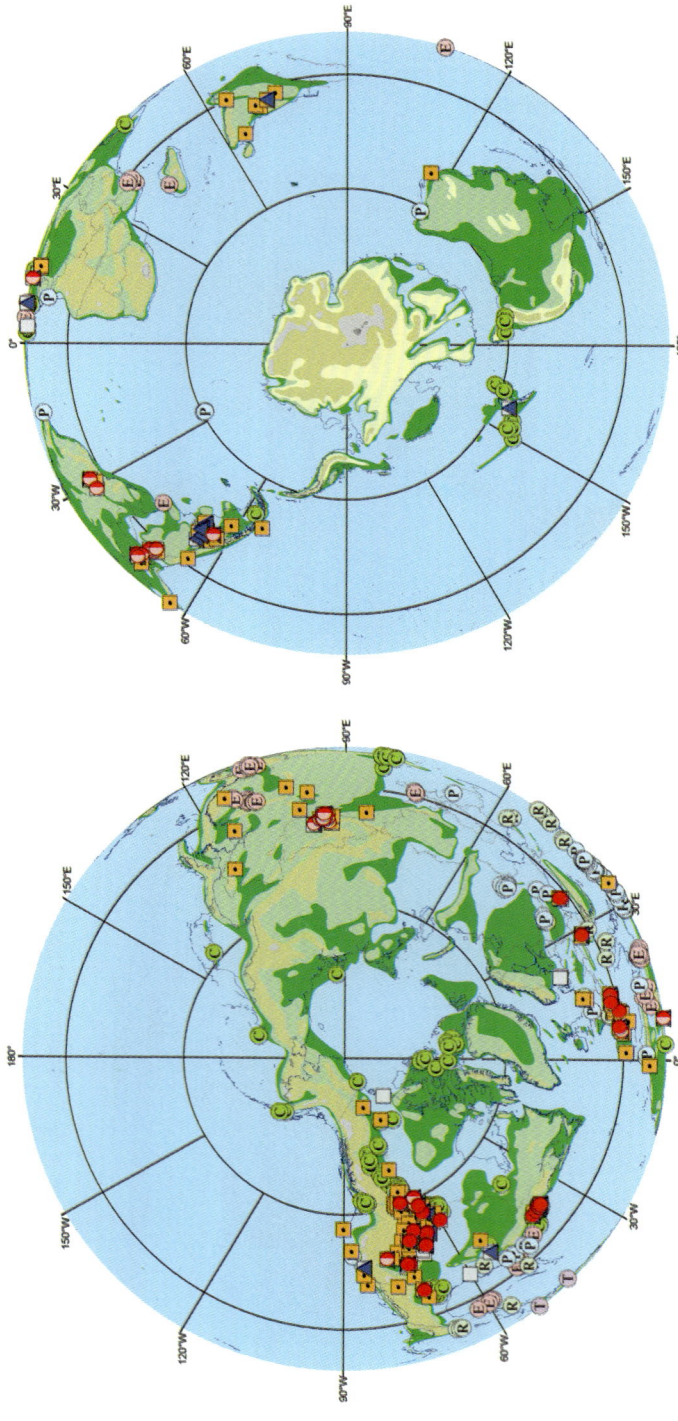

Fig. 36. Palaeogeographic map of the Maastrichtian. See Figure 28 for legend. Left, northern hemisphere; right, southern hemisphere.

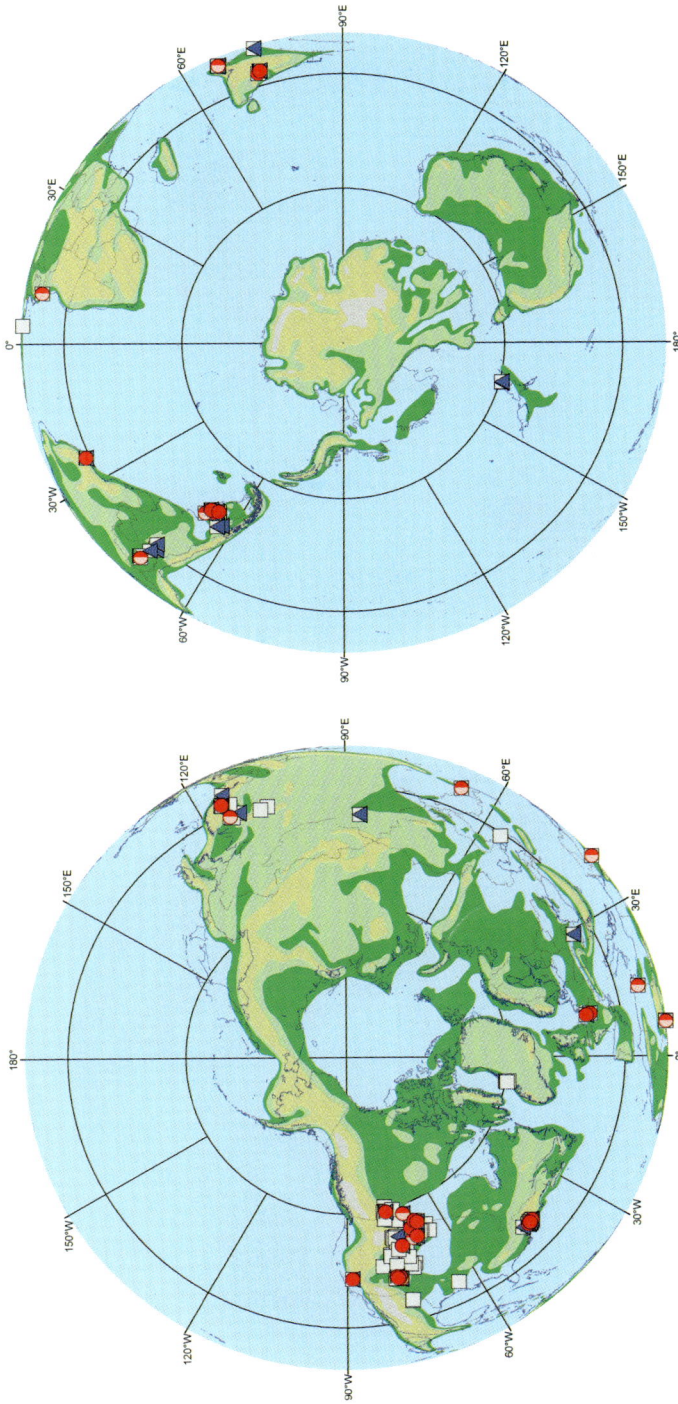

Fig. 37. Palaeogeographic map of the Early Palaeocene. See Figure 28 for legend. Left, northern hemisphere; right, southern hemisphere.

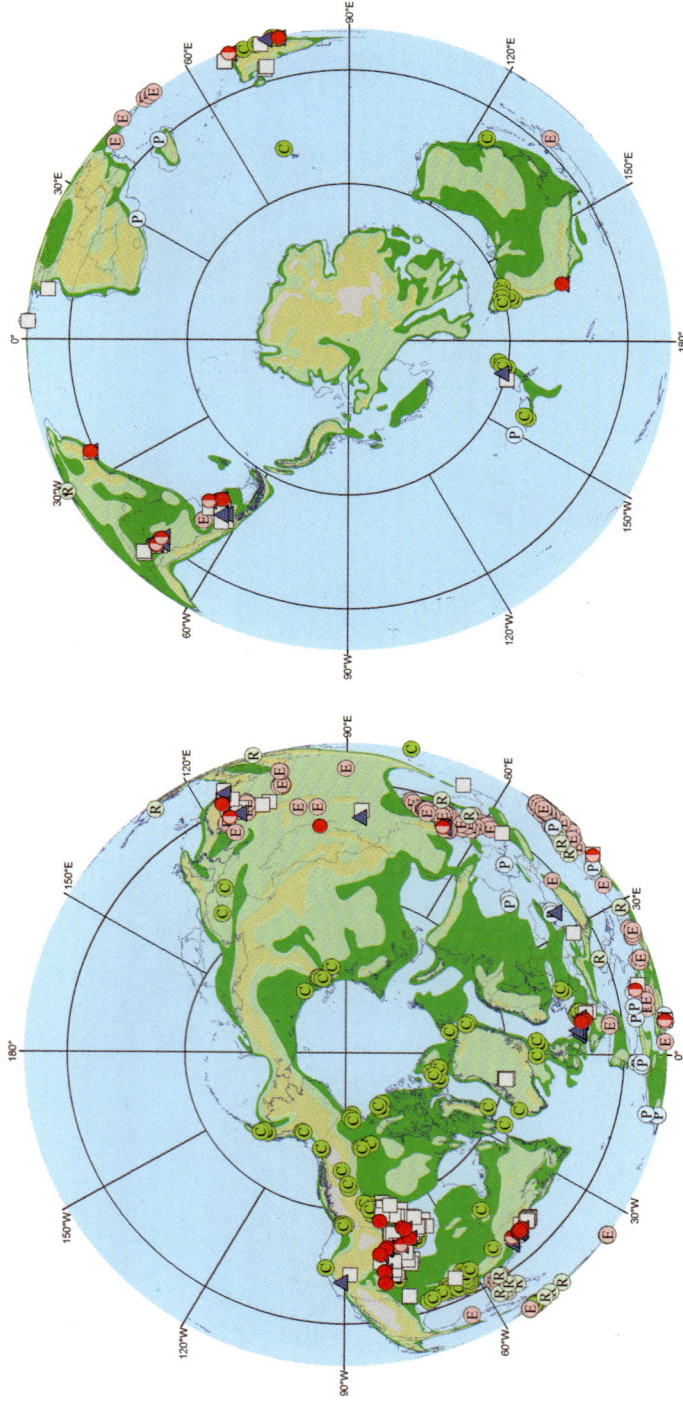

Fig. 38. Palaeogeographic map of the Late Palaeocene. See Figure 28 for legend. Left, northern hemisphere; right, southern hemisphere.

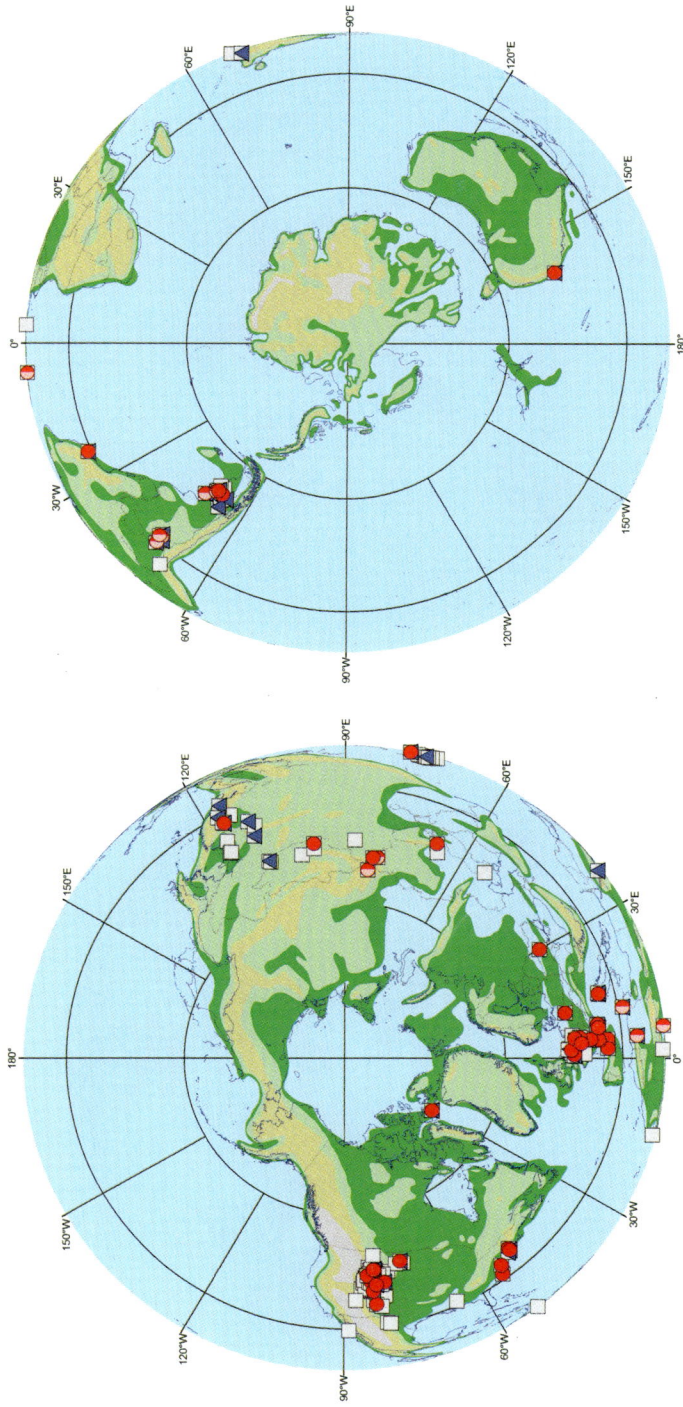

Fig. 39, Palaeogeographic map of the Early Eocene. See Figure 28 for legend. Left, northern hemisphere; right, southern hemisphere.

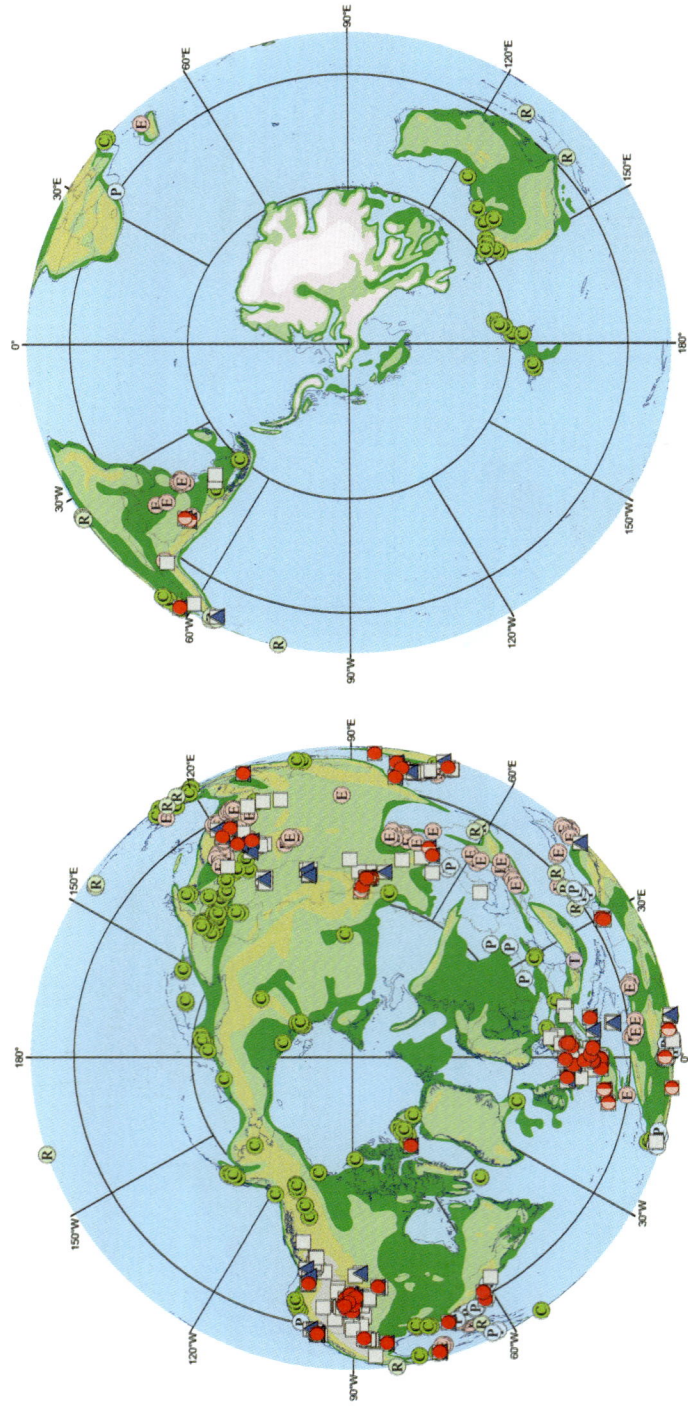

Fig. 40. Palaeogeographic map of the Middle Eocene. See Figure 28 for legend. Left, northern hemisphere; right, southern hemisphere.

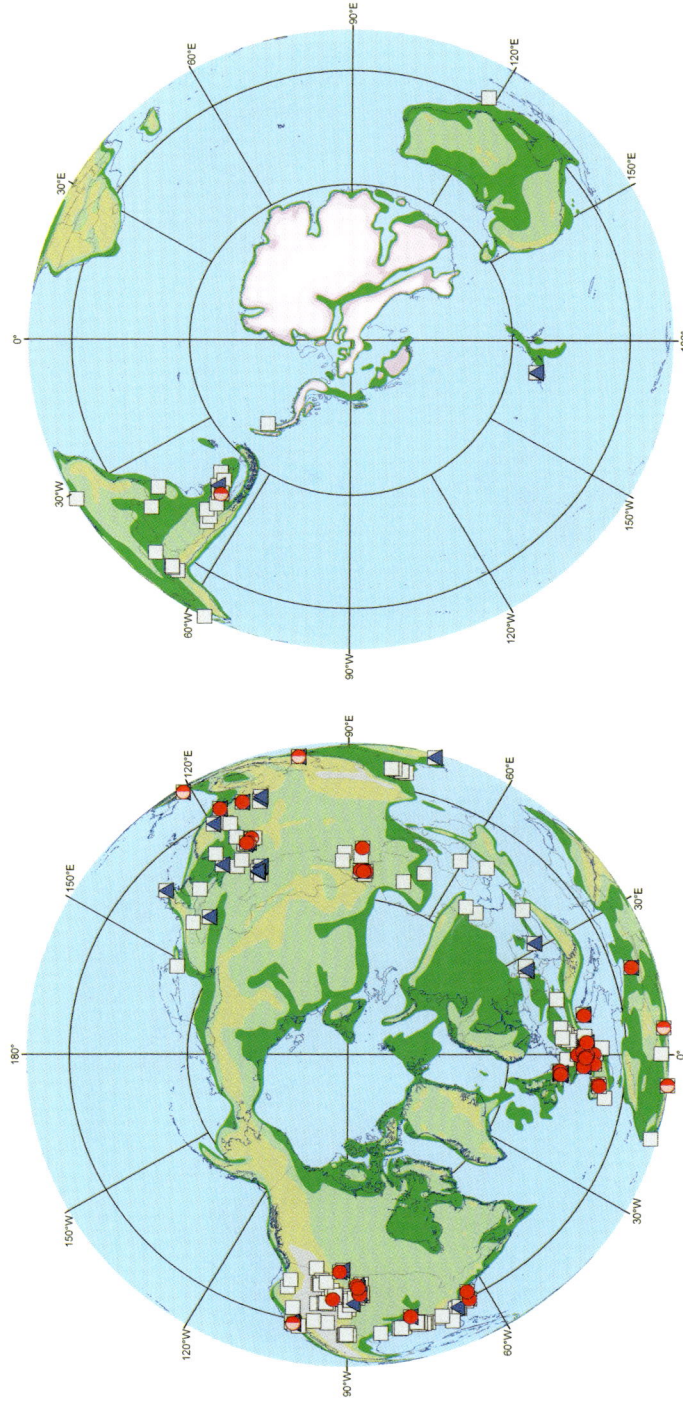

Fig. 41. Palaeogeographic map of the Late Eocene. See Figure 28 for legend. Left, northern hemisphere; right, southern hemisphere.

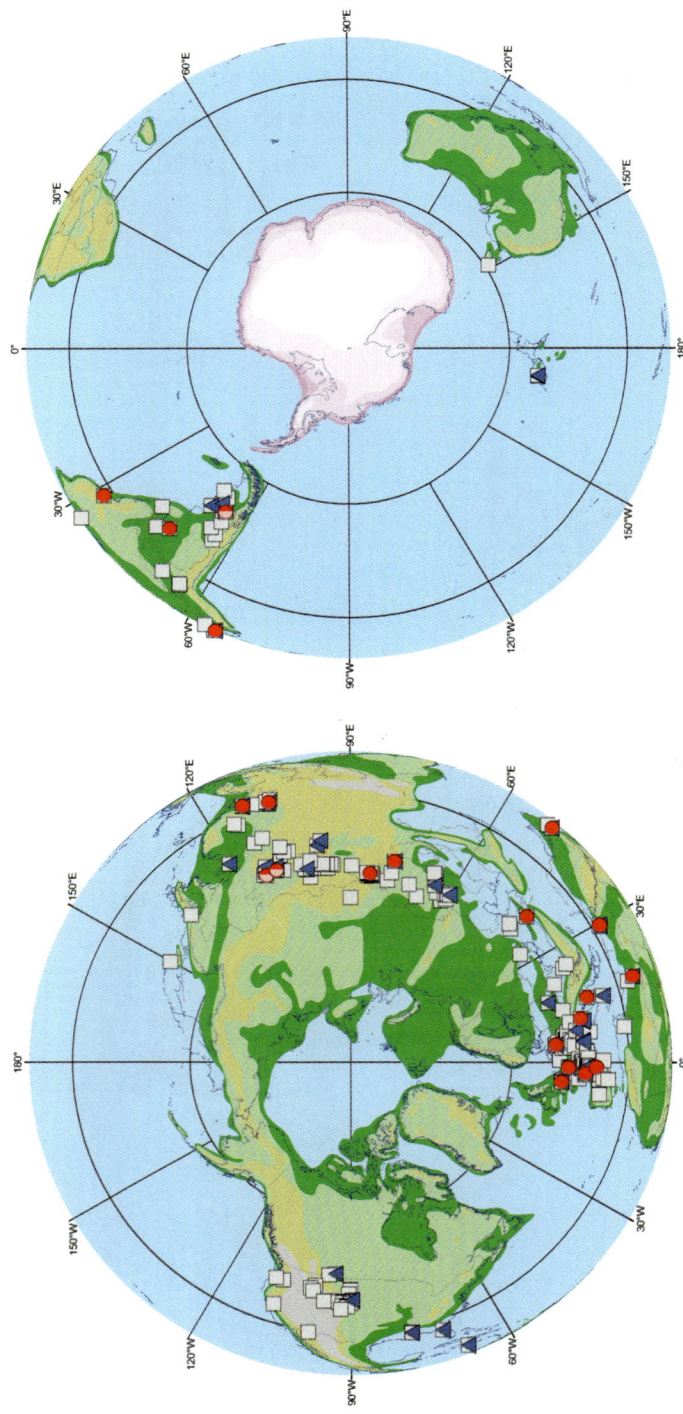

Fig. 42. Palaeogeographic map of the Early Oligocene. See Figure 28 for legend. Left, northern hemisphere; right, southern hemisphere.

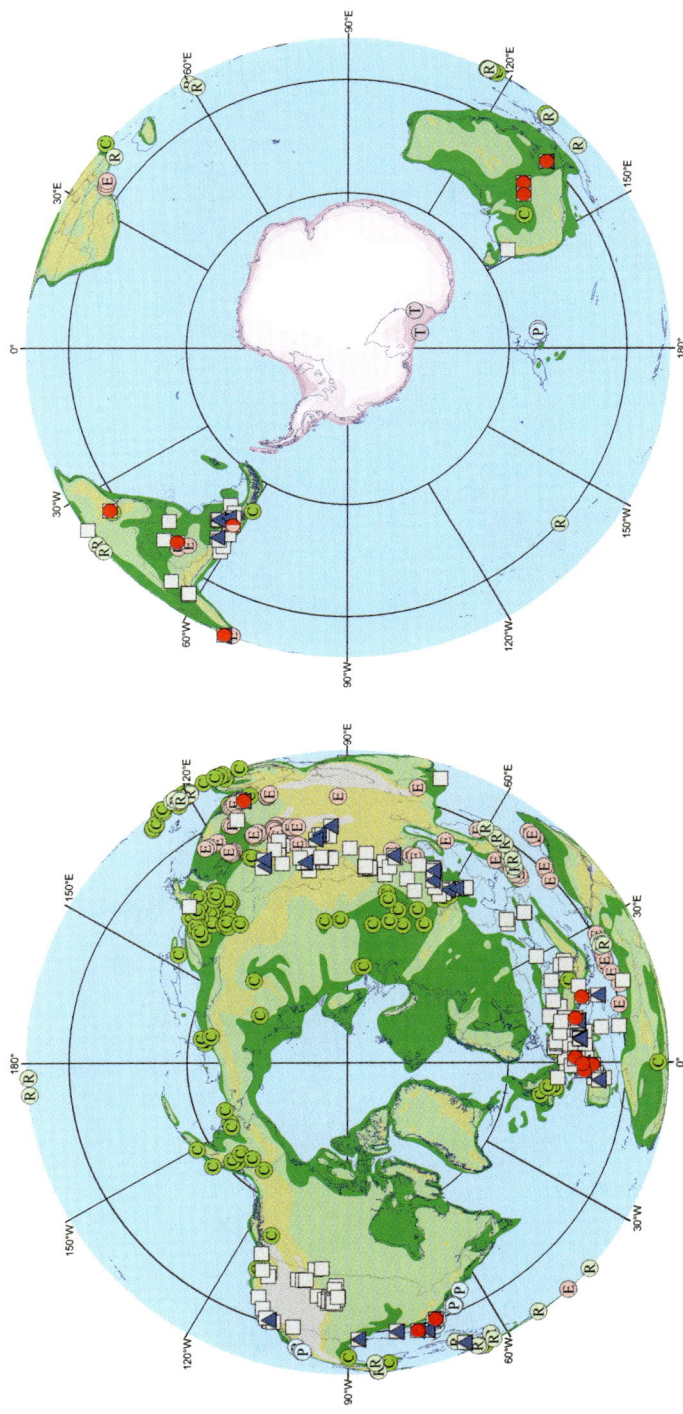

Fig. 43. Palaeogeographic map of the Late Oligocene. See Figure 28 for legend. Left, northern hemisphere; right, southern hemisphere.

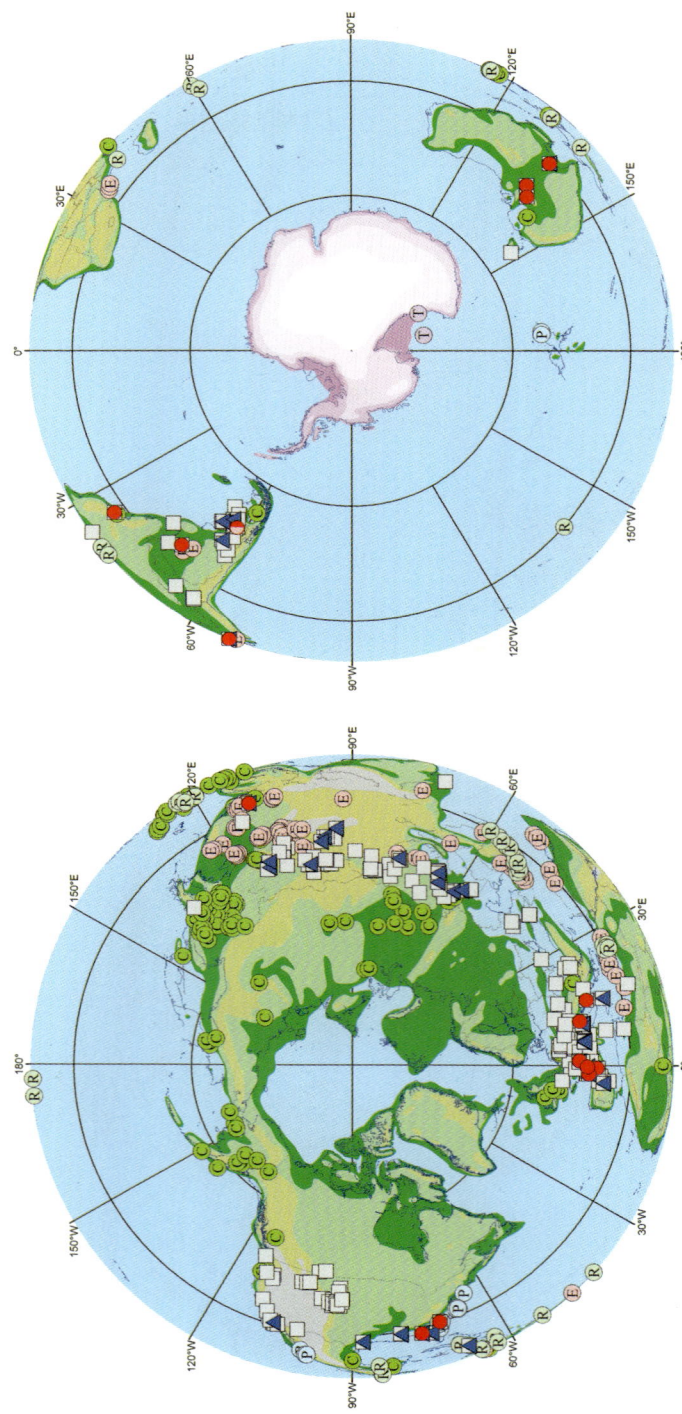

Fig. 44. Palaeogeographic map of the Early Miocene. See Figure 28 for legend. Left, northern hemisphere; right, southern hemisphere.

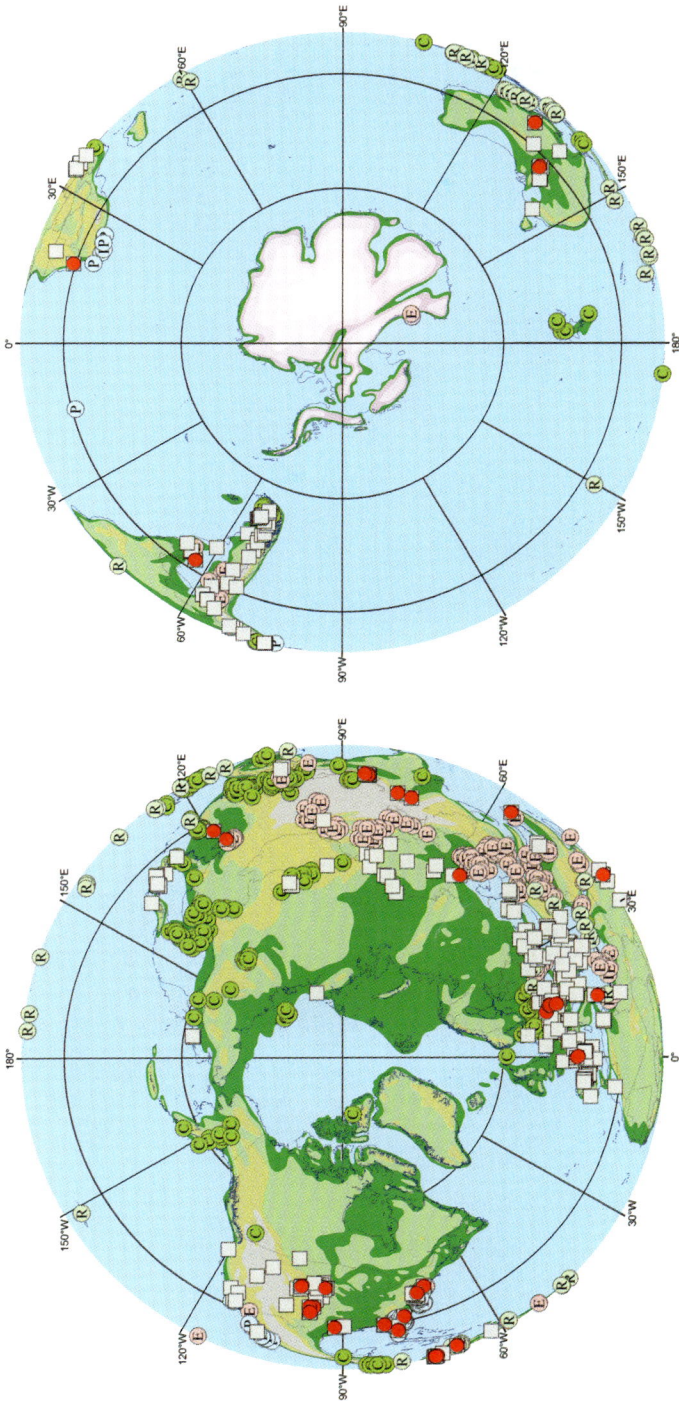

Fig. 45. Palaeogeographic map of the Middle Miocene. See Figure 28 for legend. Left, northern hemisphere; right, southern hemisphere.

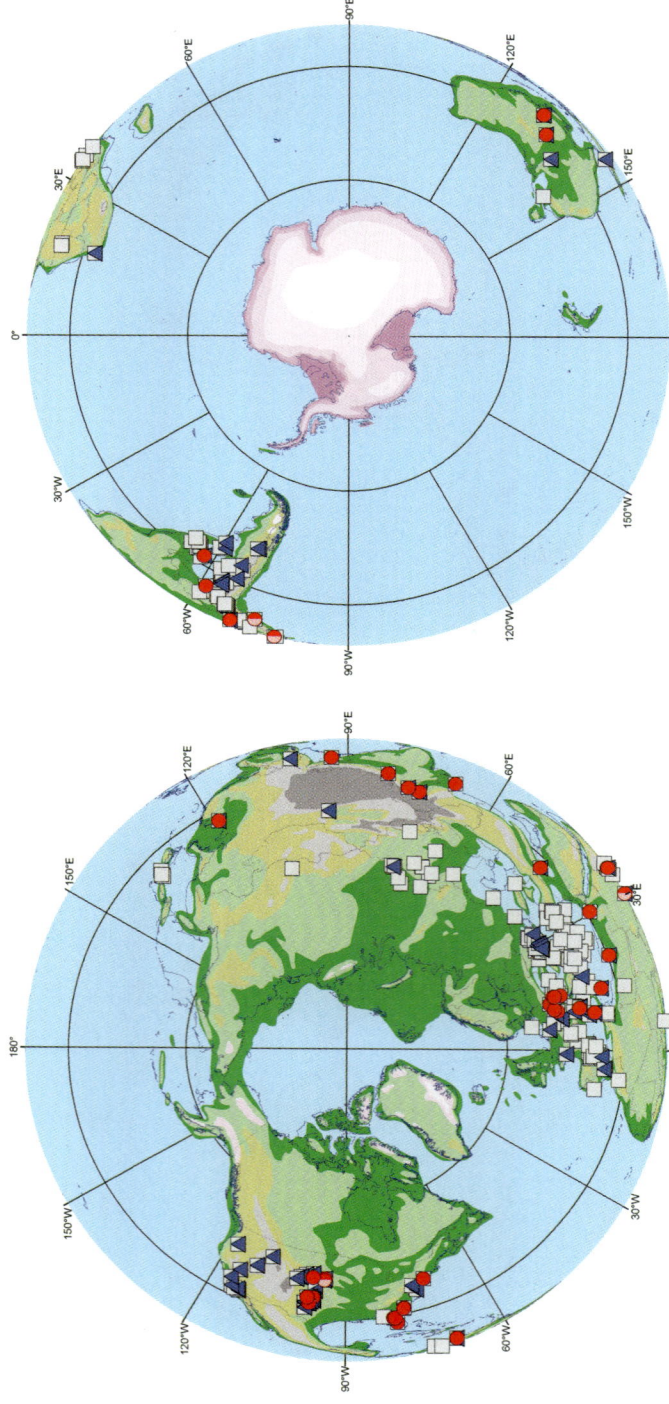

Fig. 46. Palaeogeographic map of the Late Miocene. See Figure 28 for legend. Left, northern hemisphere; right, southern hemisphere.

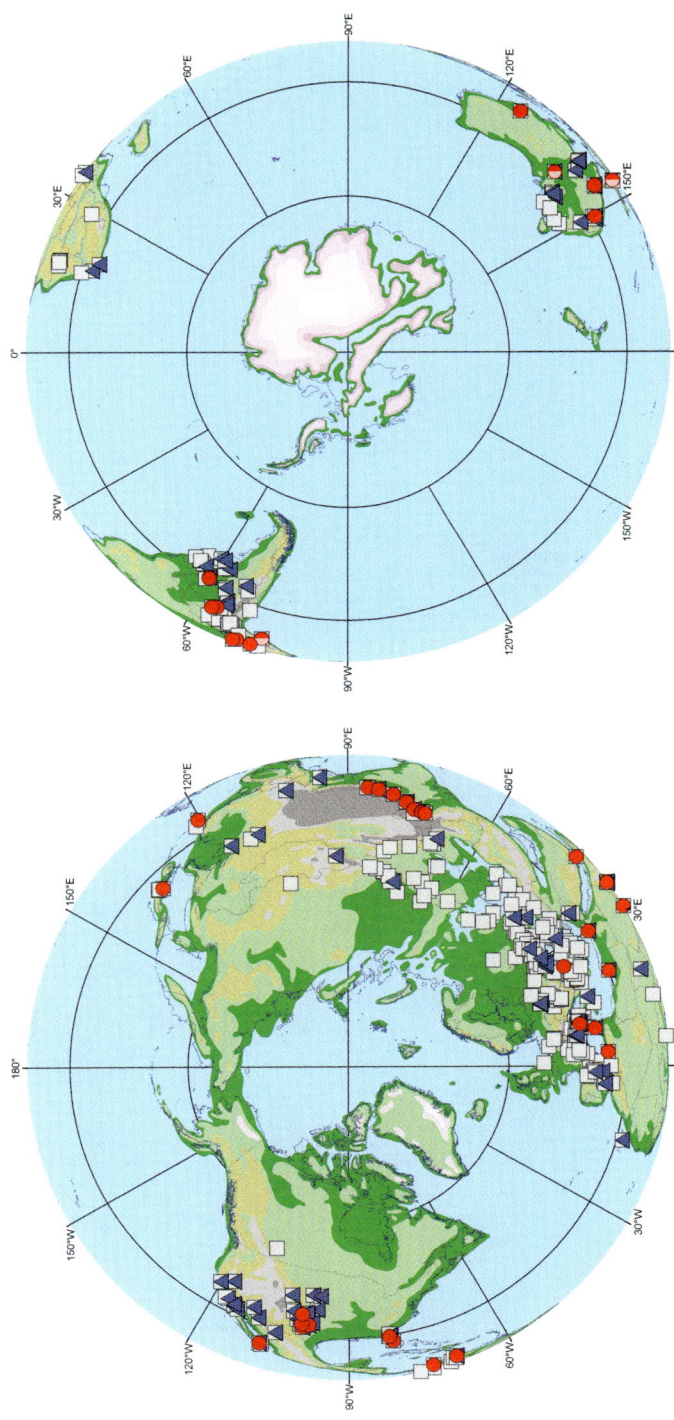

Fig. 47. Palaeogeographic map of the Pliocene. The ice extent shown for Antarctica is the probable minimum suggested by Webb and others (Webb *et al.* 1984; Andersen 1990) if their hypothesis on the age and origin of diatoms in the Sirius Group is correct. See Figure 28 for legend. Left, northern hemisphere; right, southern hemisphere.

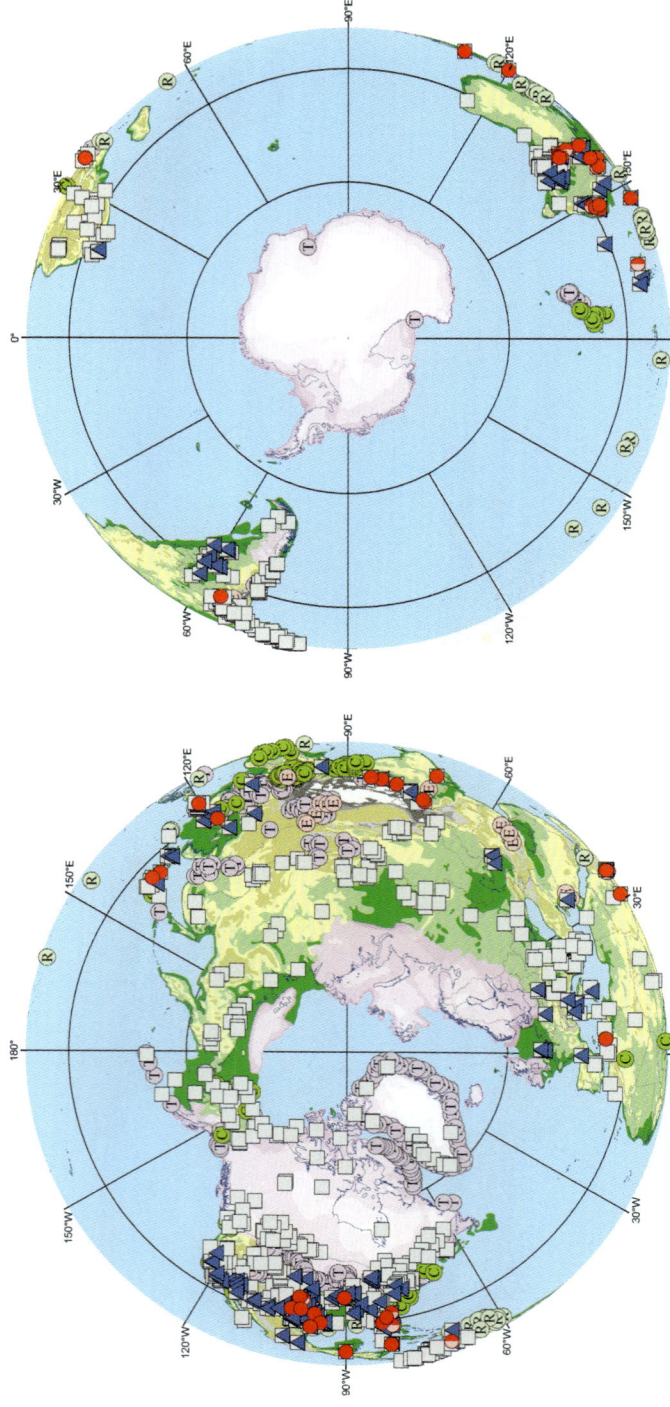

Fig. 48. Palaeogeographic map of the Pleistocene (geography shows ice extent at Last Glacial Maximum from Peltier (1994). Data from the entire Pleistocene. See Figure 28 for legend. Left, northern hemisphere; right, southern hemisphere.

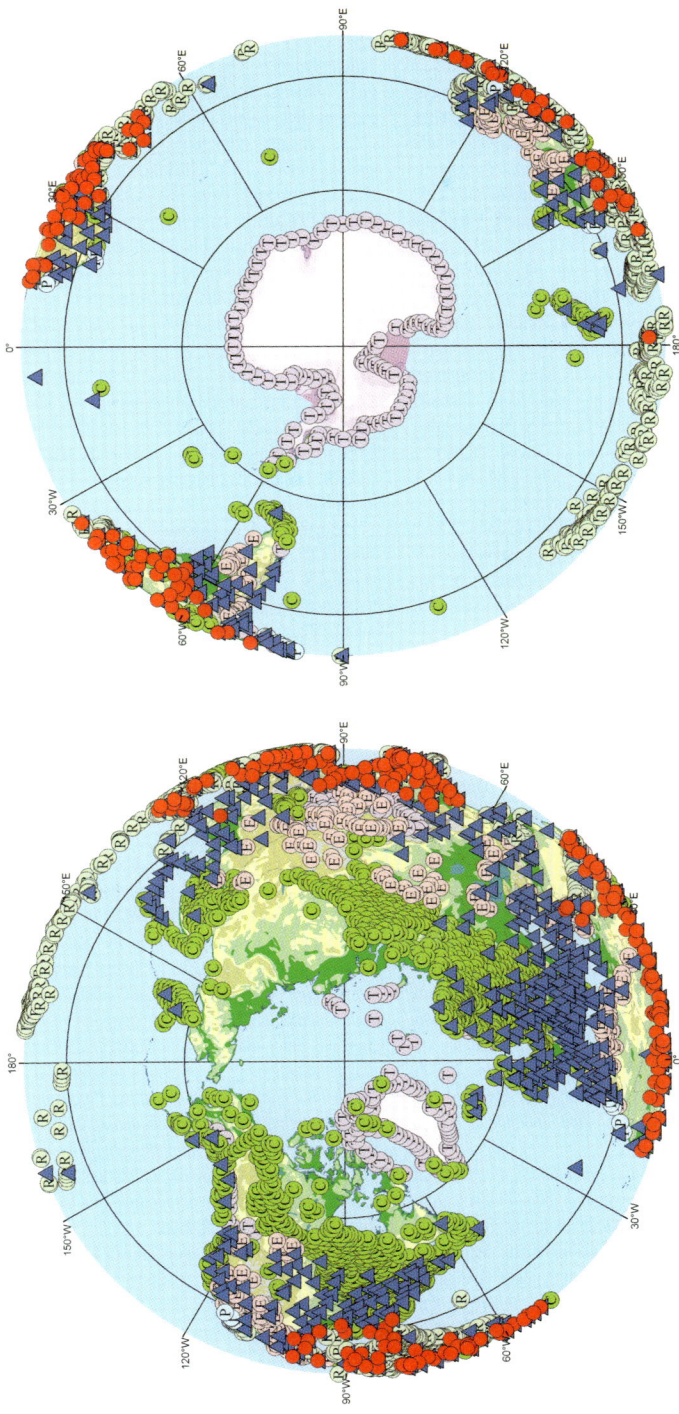

Fig. 49, Geographic map of the Recent. See Figure 28 for legend. Left, northern hemisphere; right, southern hemisphere.

Models

The climate and ocean experiments presented in this study were run using the HadCM3 coupled atmosphere–ocean model developed at the Hadley Centre of the Meteorological Office (Gordon *et al.* 2000; Pope *et al.* 2000). HadCM3 has no flux adjustment, which makes it ideal for palaeoclimate studies. It is a grid-point model, with 19 atmospheric levels with a horizontal resolution of 2.5° (latitude) × 3.75° (longitude). The land surface scheme includes surface runoff and soil drainage, and is unique amongst the current generation of coupled atmosphere–ocean models in that it explicitly requires outfall points for terrestrial drainage to be defined (Cox *et al.* 1999). This in turn necessitates an understanding of the hinterland geography and palaeodrainage, but also means that it can be used to directly model freshwater and sediment fluxes to precise points within a palaeogeography (Markwick *et al.* 2002; Markwick & Valdes 2004). HadCM3 includes an interactive vegetation model called 'Triffid' (Cox 2000, 2001). The radiation scheme is that of Edwards & Slingo (1996). The ocean component comprises 20 levels with a horizontal resolution of 1.25° × 1.25°.

The modern-day experiment used here is for pre-Industrial conditions (i.e. prior to 1780), which is designed to try to eliminate the major anthropogenic affects of the last two centuries and represent the 'natural' climate state, although this is not without caveats (specifically, the role of human activity on the natural environment and thereby any climate signal). However, a major consequence of this is that comparisons with climate station observations (which for robust records are invariably 20th century), would be expected to result in some mismatch, which must be taken into consideration. For the Maastrichtian experiments, atmospheric CO_2 is set at three different levels (2×, 3× and 4× pre-Industrial levels) consistent with the range of published estimates of Late Cretaceous CO_2 (Freeman & Hayes 1992; Pearson & Palmer 2000; Berner & Kothavala 2001) and variations during the Maastrichtian stage (see Table 3).

The Maastrichtian palaeogeography is that of Markwick & Valdes (2004), which has been compiled in ArcGIS as a paleoDEM in order to facilitate grid-based analyses. The Maastrichtian drainage systems are based on an investigation of the subsequent evolution of each drainage, using the modern as a tie-point (drainage analysis), and a survey of the published literature. This is discussed in detail in Markwick & Valdes (2004). The modern drainage systems are based on those given in the USGS Hydro 1K dataset (Verdin 2001).

Methods

ESRI's ArcGIS software has been used throughout this study and provides the GIS platform for data compilation, data-model comparisons and analysis, and visualization of results. Statistical analysis, especially multivariate techniques, is external to GIS, using several readily available software packages: Excel, Statistica and Statview. GIS has the advantage that diverse formats of spatial and temporal data can be quantitatively queried and analyzed within a spatial context (maps), which is ideal for data-model comparisons, with model results stored and manipulated as grids (rasters), and geological climate proxies and other observational data stored as points (unique occurrences), polygons (range data), and also grids (satellite data, interpolated observational data such as the CRU datasets (New *et al.* 1999), and climate proxy masks, which represent the potential distribution of each climate proxy based on its defined climate limits).

Logistically, model results were imported as NetCDF formatted files, which were converted to ArcGIS point shapefiles (points downloaded in this way represent the centre of each model grid box). These were interpolated into grids in whatever map projection is required (equal area projections are necessary for area-dependent analyses, such as hydrological studies of drainage basins). Different interpolation methods can impart analytic grid-boundary issues that must be taken into account. I have used various methods to generate grids, including simple tension splines, and a bilinear interpolation method that preserves area averaging. An analysis of the results of these various techniques has allowed the consequences of interpolation for data-model comparisons to be examined, although this is not discussed further here. A further operational consideration, when using the ArcGIS Spatial Analyst extension is that only square grid cells can be dealt with. Since model grid cells are usually rectangular for grid-point models (2.5 × 3.75 degrees for the HadCM3 model), the original grid geometry cannot be exactly replicated. In general, this is not a problem, but in certain palaeogeographic settings, such as narrow seaways, interpolations can lead to spurious results (usually anomalously extreme values, which are readily identifiable).

Two methods have been applied for comparing models and data: direct comparison and proxy models.

Direct comparison

The most direct method of data-model comparison intersects each individual observation (as a point,

Table 3. *Model experiments mentioned in this study (model results courtesy of Professor Paul Valdes, University of Bristol)*

Model name	Model type	Total run (years)	Averaging period (years)	CO_2 (ppmv)	CH_4 (ppbv)	N_2O (ppbv)	Notes
Present day (pre-Industrial)							
xakxa	Coupled Ocean-atmosphere	106	86	279	703	276	The use of a pre-Industrial model experiment would be expected to lead to some mismatches between model and 20th century observational data, but this should not be as great as observed in this study
Early Eocene							
xaokt	Atmosphere only	16	10	560	703	276	Palaeogeography is the same as that shown in Figure 39 of this study. SST gradient used: $33\cos$ (latitude)
Maastrichtian							
xayda	Coupled Ocean-atmosphere	101	10	558	703	276	$2\times$ Pre-Industrial CO_2; Palaeogeographic boundary condition is that of Markwick and Valdes (2004)
xaydb	Coupled Ocean-atmosphere	101	10	838	703	276	$3\times$ Pre-Industrial CO_2; Palaeogeographic boundary condition is that of Markwick and Valdes (2004)
xaydc	Coupled Ocean-atmosphere	101	10	1117	703	276	$4\times$ Pre-Industrial CO_2; Palaeogeographic boundary condition is that of Markwick and Valdes (2004)
xanir	Coupled Ocean-atmosphere	40	10	1116	790	284	$4\times$ Pre-Industrial CO_2; Palaeogeographic boundary condition is that of Markwick and Valdes (2004)

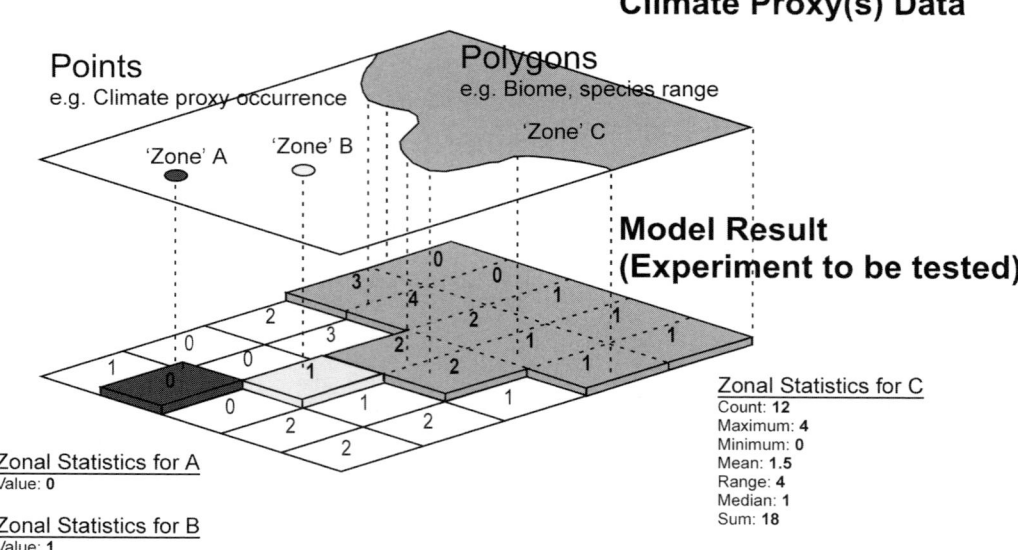

Fig. 50. Schematic representation illustrating the 'zonal statistics' functionality in ArcGIS.

polygon or grid) with the spatially coincident model result (usually gridded data). In ArcGIS, this can be done in three main ways: spatial joins; zonal statistics; grid algebra (Fig. 50). 'Spatial joins' link the attribute (database) tables of spatially coincident features (in this case, observations and model results), although this can only be applied between point, line and polygon data types. The 'zonal statistics' functionality, on the other hand, is used to assign grid values (or summary statistics if more than one grid cell is involved) to points, lines, polygons or integer grids that intersect the grid cells, based on the 'zone' (category) they represent. This 'zone' can be designated by a locality number, if the intersection is with individual observations, or a category representing associated features, such as species range for a biological climate proxy, or a biome (Fig. 50). 'Grid algebra' can be used in several ways: to calculate difference grids between observations and model results; to generate a grid of only those model results that correspond with the presence of a climate proxy (by multiplying a presence–absence grid for a proxy by the model result grid). Of these three methods, 'zonal statistics' is the most efficient for comparisons between palaeoclimate model results and climate proxies, and discussed further here.

Model values extracted in this way can be 'joined' (or 'linked') to the observation record as an additional attribute, or item, in its data table, and then exported into external spreadsheet programs or statistical packages, where the results can be analysed statistically to quantitatively assess similarity. The advantage of this method over proxy models is that results can be compared directly within climate space, and provide information on not only whether a model agrees or disagrees with observations, but the nature and magnitude of any differences (see Fig. 51 below).

Proxy models

The second method is to generate a model or 'mask' (a conditional grid) that shows the spatial distribution of a proxy's climate space derived from the model results being evaluated (Price *et al.* 1997). Model success or failure is then based on the degree to which the distribution of the proxy occurrences match the climate space distribution indicated by the model experiment.

Data-model comparisons using climate proxies in the modern

The modern day provides the only opportunity to directly compare model results with well-constrained observations, whilst also providing palaeoclimatologists with the means to examine

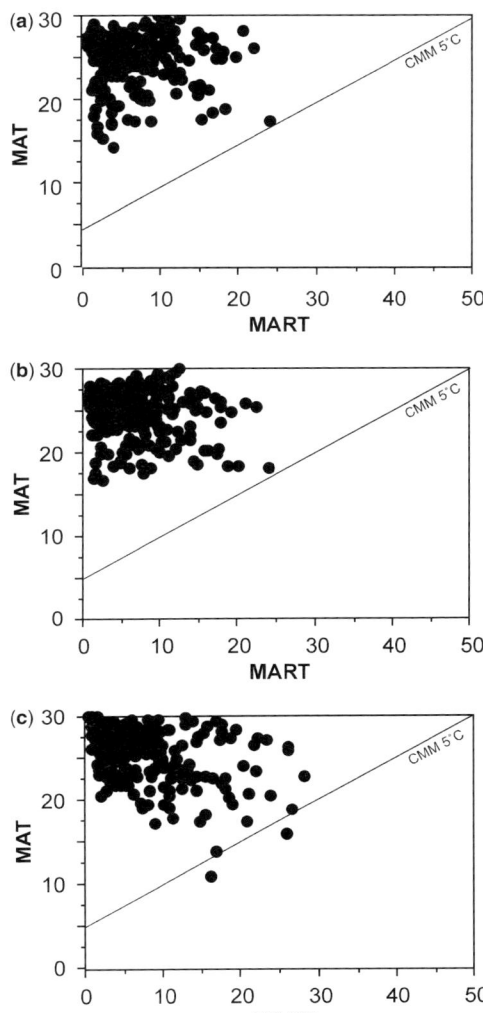

Fig. 51. Crocodilian climate space as indicated by intersection of the modern geographic range with (**a**) Climate Stations (Müller 1982); (**b**) CRU 1961–1990 climatology (New *et al.* 1999); (**c**) modelled Pre-Industrial Present Day experiment (see Table 3). The CMM value of 5 °C marks the modern thermal limit for crocodilians (Markwick 1998).

the behaviour of proxies and climate models simultaneously. Here, I investigate the differences between the climate space defined for climate proxies using station-based observations, and those from model simulations of the modern. Discrepancies between these results are systematically examined in order to identify the baseline of potential errors, which

must be considered when interpreting data-model comparisons for palaeoclimate experiments. The modern also provides the means of investigating sampling effects on data-model comparisons (Markwick & Lupia 2001).

Figure 51 shows the climate space represented by climate station locations that fall within the modern range of crocodilians, for the weather station values themselves (see also Figs 2 and 3), the CRU 1961–1990 interpolated gridded climatology, which is itself based on climate stations (New *et al.* 1999), and the results of a modern (pre-Industrial) climate model simulation (experiment 'xakxa' in Table 3). In this example, the grid values have been extracted from the model results using the 'zonal statistics' function described above.

The results show that this particular model experiment indicates a slightly colder and slightly more seasonal climate than the climate space defined from observations, although only two localities actually fall outside of the 5 °C CMM 'limit', which might be considered model 'failures' or 'mismatches' (i.e. >95% of localities fall within crocodilian climate space and would, if this had been a simulation of a geological time-slice, be considered as successful matches).

The question is how to explain this result and assess what it means for data-model evaluations of palaeoclimate experiments. At face value, any discrepancies could be interpreted as evidence of model failure, but this would be naive. As discussed above, the definition of climate proxies themselves comes with caveats, but in this case the discrepancy is between modelled and climate station observations, which happen to also fall within the modern geographic range of crocodilians, and so the two should be similar (given the caveat that the model is a pre-Industrial 'present day' simulation). The working assumption has to be that the station data are relatively sacrosanct at the point of observation, but this raises the prospect that model resolution is the cause of discrepancies, since a climate station observation is a function of local factors, especially for hydrological variables, which have short de-correlation distances. The stations used by Muller (1982) are generally city- or town-based, chosen because they have relatively long, consistent records, but which therefore also inherently include the thermal effects of conurbations; for major cities, this can be up to a 1–2 °C warming (Barry & Chorley 1987). The model resolution here is 2.5 × 3.75 degrees. If this is the case, it should be apparent in how the model replicates topography over a grid cell compared with the elevation of the observation itself. Figure 52 shows regression plots for the points shown in

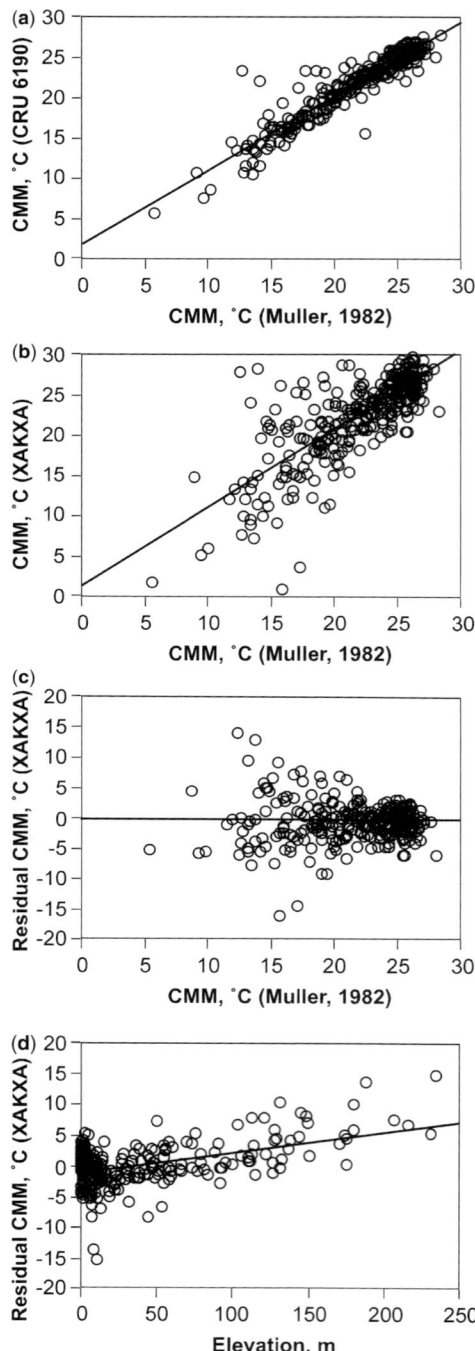

Fig. 52. Correlation for between-station and gridded observations of CMM values for crocodilian localities (**a**), and stations and modelled CMM (**b**). The residuals (**c**) show the strongest correlation with the elevation of each location (**d**).

Figure 51. The differences between stations and CRU data are shown to be minor, but the disparity with modelled thermal results is clearly seen. An analysis of the residuals of this regression shows the strongest correlation with elevation, which would appear to support the hypothesis that model resolution is at least partly responsible for the disparity found, and therefore something that must be considered in palaeoclimate simulations.

Spatial resolution is clearly an important factor. For biodiversity studies, Markwick (2001) used a 50 km radius 'buffer' around each climate station, which was considered the area over which the climate of the station could be considered applicable (except in areas of rapid relief changes). A similar approach is used here to emulate the area over which a climate proxy might be considered to represent the same climate. Figure 53 shows the results when buffered stations are intersected with the model results, to see if this has an effect on data-model discrepancies.

Temporal climate heterogeneity is also a potential issue, with the observations representing a short-term signal and the model the longer-term general state of the atmosphere (in this case, also the model is for pre-Industrial conditions so should be expected to be slightly different from 20th century observations, but not in the direction or magnitude found). The modern climate values used here represent the period 1961–1990 (New et al. 1999) and approximately 1951–1980 (Müller 1982), which is relatively short in relation to the stability of the ecosystems occupied by biological climate proxies. We have examined this by comparing the results for crocodilians using the

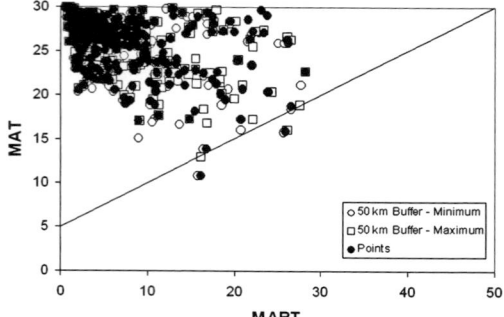

Fig. 53. Crocodilian climate space using the xakxa climate model results, for point extraction (black points), and the minimum and maximum values for 50 km radius areas around each locality. The differences between these two approaches do not explain the points that fall out of crocodilian climate space using the 'xakxa' result (see Table 3).

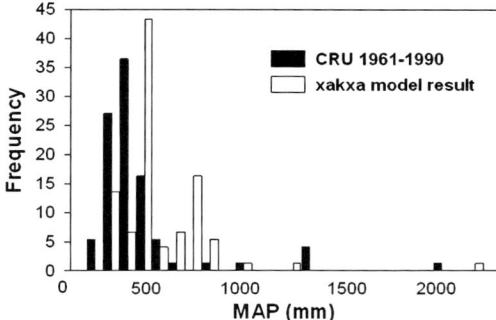

Fig. 54. Frequency distribution of modern evaporite localities (Gyllenhaal 1991) with respect to mean annual precipitation (MAP), using precipitation values extracted for each location using the CRU 1961–1990 climatology (New *et al.* 1999), and model xakxa (see Table 3). The difference between them has been examined using the Mann-Whitney Rank Test and found to be statistically significant ($p < 0.0001$).

CRU 1901–1930, CRU 1931–1960 and CRU 1961–1990 climatologies and found no significant differences between them, which would suggest that for thermal parameters this is not the major cause of the discrepancies between the station and modelled results shown in Figure 51, unless longer-wavelength fluctuations are the issue (greater than century-scale).

These two issues do not negate the possibility of model error, especially due to missing, or misrepresenting, a process, but would suggest that the major cause of the discrepancies in proxy climate space here are intrinsic resolution issues, which therefore must comprise the operational error associated with model results.

The differences between modelled and observed thermal values are important, but as already discussed hydrological variables would be expected to have larger potential errors, given shorter de-correlation distances, and more uncertainties in dynamical processes. Given that hydrology plays such a crucial role in surface runoff, water availability and flood risk, it is important to understand the degree of uncertainty in such variables. Figure 54 shows the precipitation climate space for evaporites based on observations and model results (again using experiment xakxa). At low precipitation values, observations and model are statistically significantly different, with the model being wetter. This is less apparent at higher precipitation values, and for peats and coals there is no statistical difference between modelled and observed precipitation values, although this also reflects the broad range of values over which peats can occur. Although the causes for such

model behaviour are unclear, except to say that annual evaporation may be being underestimated (see evaporites section above), the implications for data-model comparisons in the geological record are that the extent of evaporite climate space may be underestimated.

Data-model comparisons using climate proxies in the past

The modern data-model comparisons provide an indication of potential model behaviour that could be true of simulations of geological time-slices (especially for a model such as the HadCM3 that has no flux correction and should therefore be largely independent of modern-day tuning). The question is: do we see the same trends? The same methods used for the modern can be applied to the palaeo-examples.

Figure 55 shows the CCM result for three Maastrichtian sensitivity experiments, with varying concentrations of atmospheric CO_2 (Table 3). The results show that for all three experiments, several crocodilian localities fall out of the defined climate space, indicating model 'failure' in those areas, but the general pattern is very reminiscent of the results for the modern, as can be seen when these results are plotted in climate space (Fig. 56).

The vectors in Figure 56 show how different localities respond to changing boundary conditions. Fossil crocodilian sites from mid-latitude continental interior locations, tend to have higher MART values, even at lower CO_2 concentrations (bottom right in Fig. 56), and respond to increases in CO_2 by major increases in WMM, such that the trajectories are almost parallel with CMM isotherms in this diagram. In contrast, maritime or low-latitude sites tend to become warmer (increasing MAT), but with little or no change in seasonality. If we examine the three fossil crocodilian localities from Figures 55 and 56 that remain outside of crocodilian climate space, even at $4\times$ CO_2, we can again see the influence of topography and how it is represented by the model (Fig. 57).

The same pattern of results is seen for other time-slices, which further supports the idea that at least part of the cause of these particular data-model discrepancies may be intrinsic factors to the model (resolution), rather than time-slice specific issues. Figure 58 shows the modelled climate space for Early Eocene crocodilians (Markwick *et al.* 2000*b*). In this case, the outliers to crocodilian climate space lie principally in the continental interiors of western North America and central Asia, as well as the enigmatic Arctic locality of Ellesmere Island (Fig. 59). The issue of model failure in continental interiors has been raised by

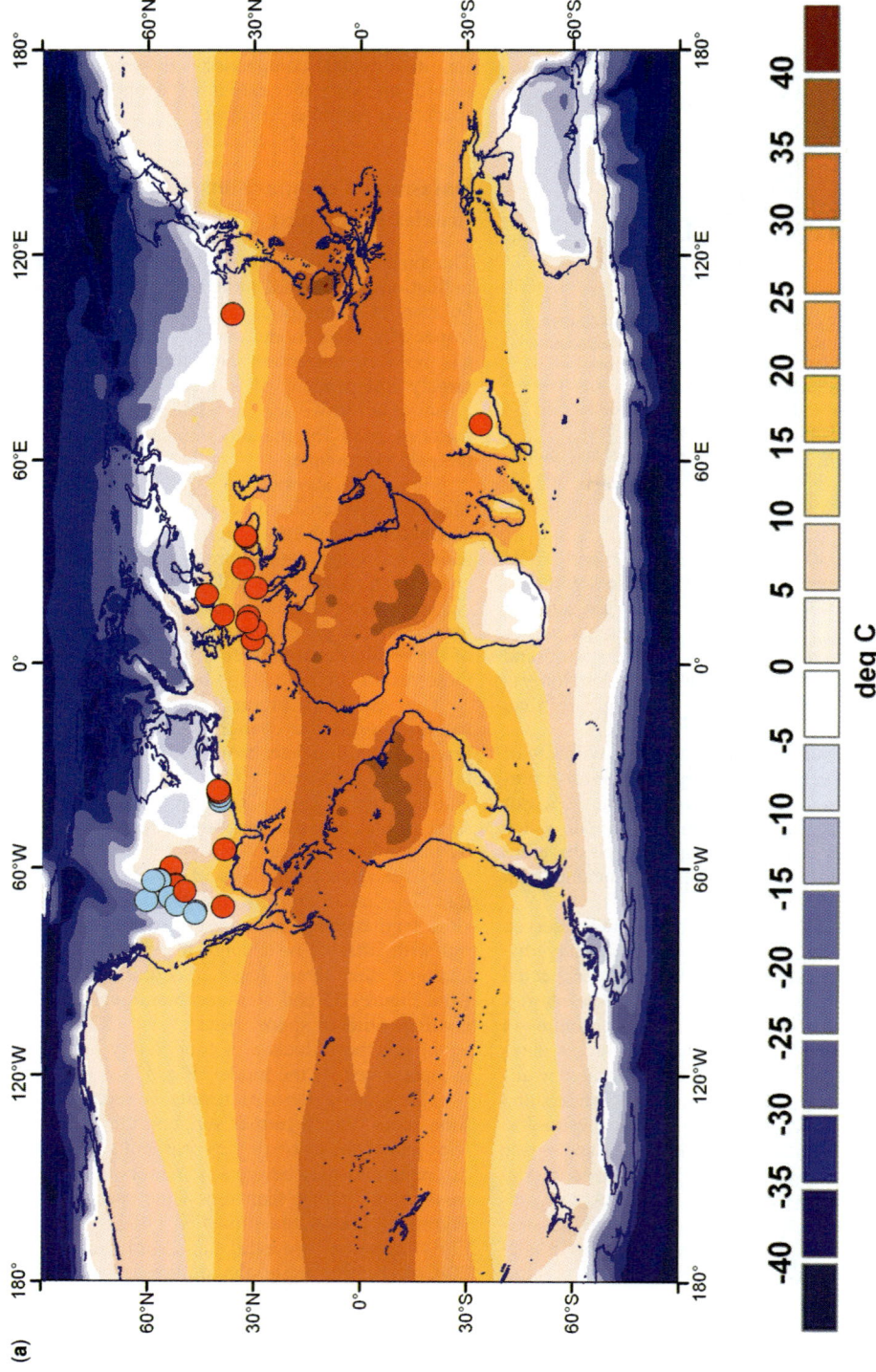

Fig. 55. Maastrichtian model results for CCM for 2× (**a**), 3× (**b**) and 4× (**c**) pre-Industrial atmospheric CO_2 concentrations (Table 3). Blue circles are crocodilian localities that fall out of crocodilian climate space in each experiment, red are those with CMM > 5 °C. The position of these results in climate space is shown in Figure 56.

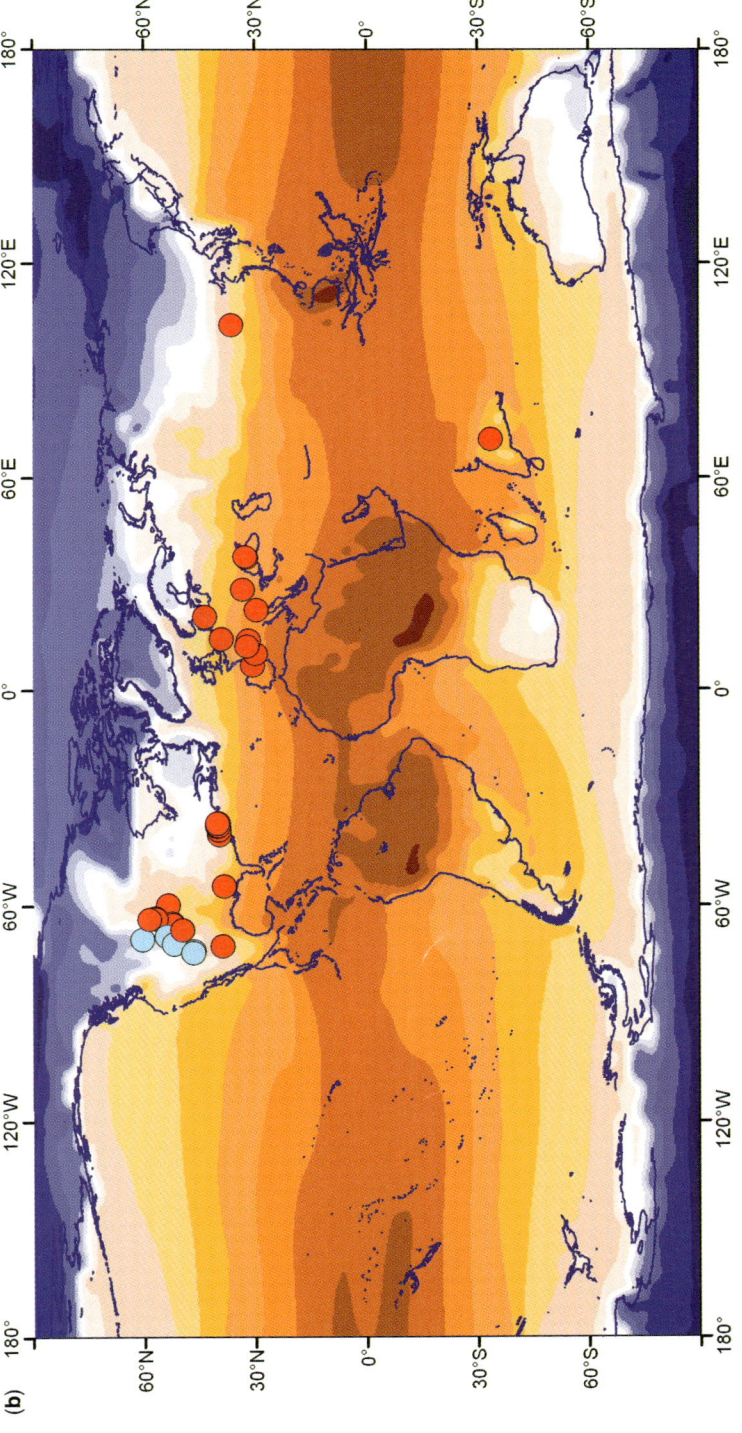

Fig. 55. (*Continued*).

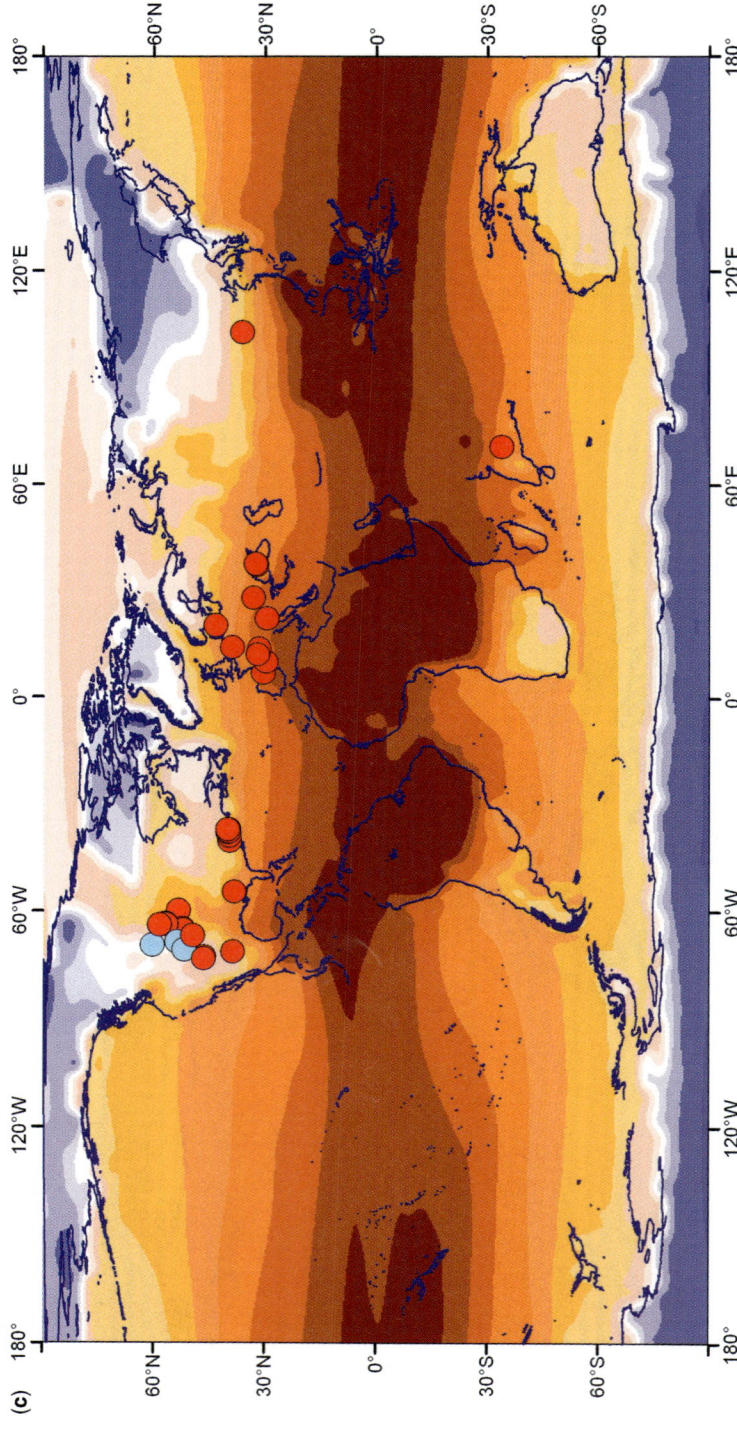

Fig. 55. (*Continued*).

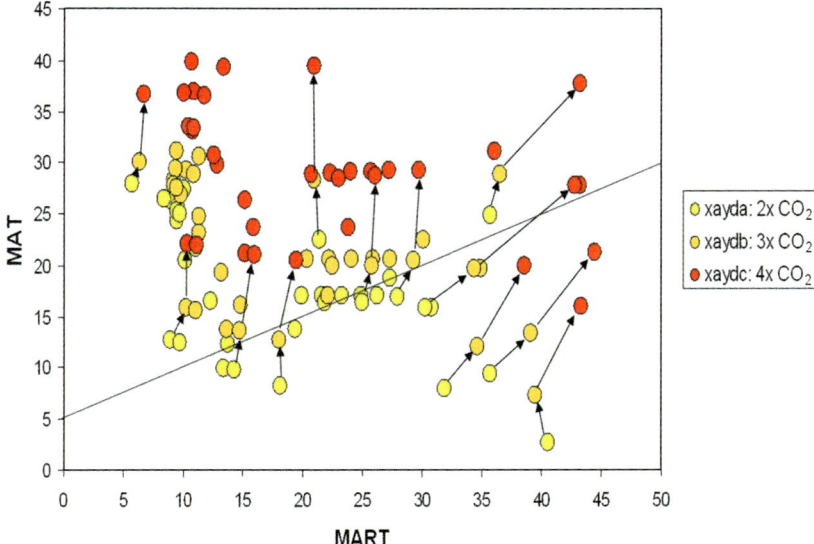

Fig. 56. Crocodilian climate space defined using the three Maastrichtian sensitivity experiments (Table 3), showing the general warming as CO_2 increases, which moves the crocodilian localities into crocodilian climate space. Trajectories illustrate the behaviour of individual localities and how the response to changing CO_2 varies. Localities below the CMM threshold are those shown in blue in Figure 55.

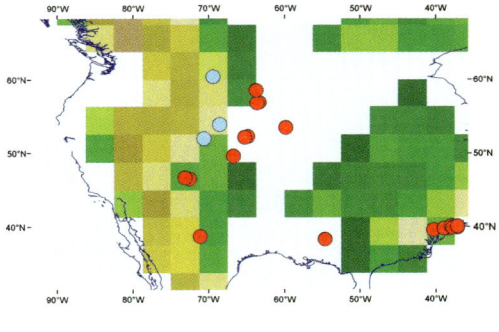

numerous authors, and may also reflect geographic resolution issues; Sloan (1994) showed how the inclusion of lakes in the Eocene boundary conditions could warm modelled local climates. It is also possible that higher CO_2 experiments, and/or increases in CH_4, may generate enough warming (Sloan *et al.* 1992; Beerling & Valdes 2003) to move many of these localities into crocodilian climate space, as they do for the Maastrichtian

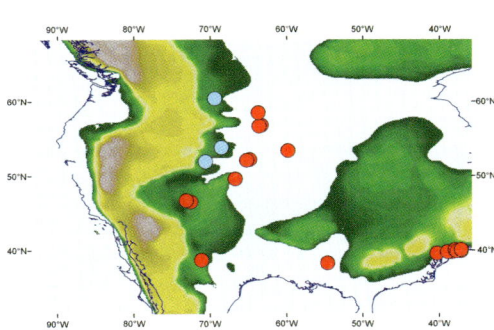

Fig. 57. Outliers to crocodilian climate space for $4\times CO_2$ from Figure 56 (blue), showing how in each case the topography used by the model is higher than the reconstructed elevation (right) using the DEM of Markwick & Valdes (2004).

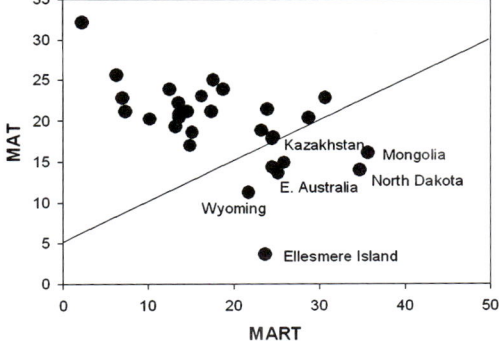

Fig. 58. Crocodilian climate space defined using the Early Eocene model experiment (xaokt, see Table 3). Outliers to crocodilian climate space are dominated by continental interior locations and Ellesmere Island.

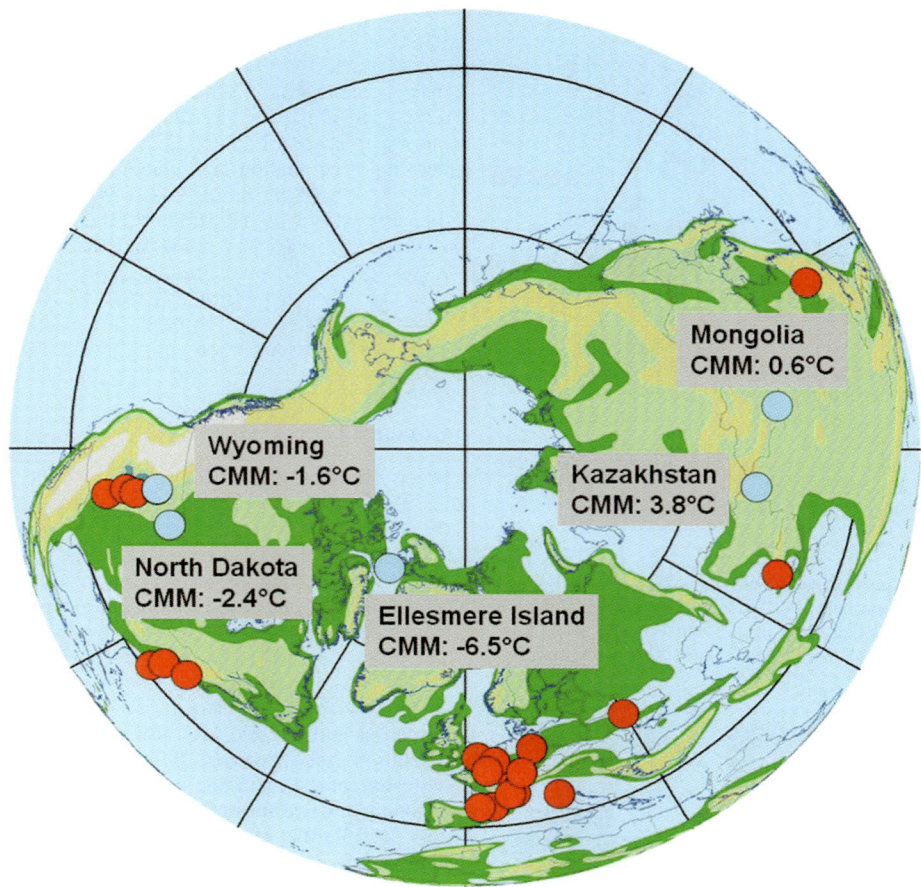

Fig. 59. Distribution of those Early Eocene crocodilian localities that fall out of crocodilian climate space (blue circle) when using the xaokt model experiment (CMM values from xaokt are given for each outlier). This may reflect the consequences of elevational resolution, but also the effects of transient rapid climate events, such as the PETM.

experiments (Fig. 56). However, the Early Eocene results, and especially the warmth indicated by the Ellesmere Island fauna and flora (West *et al.* 1977; Estes & Hutchison 1980; McKenna 1980; Francis 1988), raise again the issue of climate heterogeneity within time-slices. The Early Eocene represents the warmest stage of the Tertiary (Zachos *et al.* 2001) but the maximum poleward extent of many warm climate proxies may in fact be the result of short-term (10,000–20,000 years) rapid warming events, of which the PETM is the most well documented (Corfield & Norris 1998; Bolle *et al.* 2000; Hooker & Dashzeveg 2003; Bloch *et al.* 2004; Shellito & Sloan 2004; Wing *et al.* 2004).

The importance of such major, transient events to data-model comparisons is that they are currently beyond the resolution of global-scale terrestrial

biological proxy datasets, because of the limitations of global correlation. Only in a few well-constrained terrestrial localities, such as the Bighorn Basin, can such events be recognized (Bloch *et al.* 2004; Wing *et al.* 2004). This places greater emphasis on the use of sensitivity experiments to examine the range of possible climates through an interval by changing boundary conditions. This requires the careful use of model-data comparisons, because, as discussed throughout this paper and elsewhere (Markwick & Lupia 2001; Markwick & Valdes 2004), the proxy data are highly heterogeneous.

Such heterogeneities also apply to hydrological climate proxies. Transient events, such as the PETM, would be expected to result in periods of enhanced evaporation and precipitation as temperatures increase (Blecha & Gardner 2004; Bowen

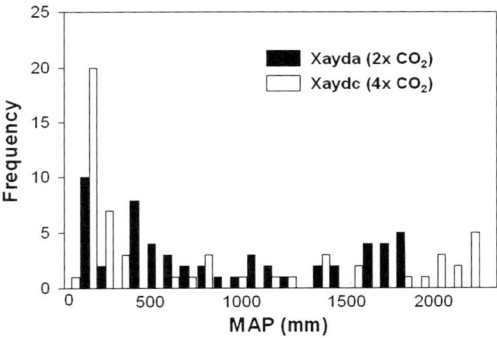

Fig. 60. Distribution of Maastrichtian gypsums defined by the results of the 2× and 4× Maastrichtian sensitivity experiments. The principal difference from the modern (Fig. 54) is that results spill into much wetter conditions. But as CO_2 is increased, the distribution becomes polarized, reflecting the overall precipitation results with intensification of the wet and dry belts.

et al. 2004; Nicolo 2004). This is the same result as found for the Maastrichtian due to increased atmospheric concentrations of CO_2 (Markwick *et al.* 2002; Valdes & Markwick 2006). Evaluating hydrological model results is problematic, with floras giving the most precise estimates of precipitation though not without caveats (Wolfe 1993; Wilf *et al.* 1998). Vertebrates and sediments are more problematic still. Figure 60 shows the frequency distribution of Maastrichtian gypsums for comparison with the modern climate space shown in Figure 54.

Comparison between precipitation and gypsums is inconclusive, in that the results are broadly distributed across the range of precipitation values. This may be due to a number of problems including

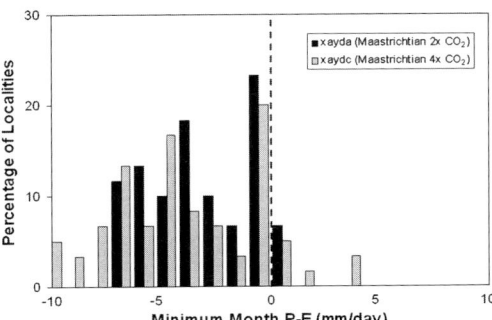

Fig. 61. Distribution of Maastrichtian gypsums defined by the precipitation and evaporation of the 2× and 4× Maastrichtian sensitivity experiments. The majority of results lie in the net evaporation side of the graph as has to be the case for evaporites to form.

temporal variations within the data and modelling issues. It may also reflect the problem of using precipitation in the modern as the basis for defining hydrological climate proxies, especially when the area applied to demonstrably more hydrological intense intervals. The use of P–E (the difference between precipitation and evaporation) or relative humidity may be more appropriate, except for their inconsistent reporting in modern climate stations. Figure 61 shows the position of the same Maastrichtian gypsums in terms of the minimum monthly value of P–E, which must be negative if evaporites are to form. The results are more convincing than for precipitation.

Conclusion

Models are powerful tools for reconstructing and understanding the dynamics of palaeoclimate, but they are experiments that must be constantly tested against observations, which for the geological past comprise climate proxies. Here, I have used a selection of terrestrial climate proxies, especially vertebrates, to show how climate proxies must be defined, their palaeogeographic distribution during the Cretaceous and Cenozoic determined, the logistics of using them in GIS to evaluate climate model results, and how data-model comparisons can tell us about the behaviour of models and data alike.

Models are numerical and therefore the observations (climate proxies) used to evaluate them must also be quantitative. This in turn requires the compilation and analysis of large, diverse, global datasets, compiled in well-designed databases that record information on the nature of the data (reliability, temporal and spatial heterogeneity) as well as its presence and climatic significance. GIS provides a powerful platform for this, as well as the logistics of data-model evaluation itself, especially through direct intersection of results and observations, or through the construction of proxy models. My preference is for direct comparison since this provides the opportunity for more-detailed analysis of results, such as examining the trajectories of locations as boundary conditions are changed (e.g. the Maastrichtian CO_2 sensitivity experiments reported here).

The results shown here indicate the effects of the coarse elevational resolution used by the models, and how this leads to cooler-than-observed temperatures, especially in mountainous areas or continental interiors (which results in discrepancies between modelled temperatures and climate proxies such as crocodilians, tortoises and palms). These apply to both modern and geological experiments and highlight the importance of a thorough understanding of palaeogeography and correctly representing this in the model. Hydrological variables are more problematic still, and geography is again important in how it partitions precipitation, through palaeodrainage systems that can then affect sediment distributions. Comparison between evaporites and modelled precipitation in the modern finds that the model is wetter than observations, especially in arid areas (<500mm annual precipitation). But comparisons between Maastrichtian gypsums, and model results also suggest that precipitation may not be the best variable for defining hydrological proxies, especially for 'hot-house' time intervals with a greatly enhanced hydrological system (higher precipitation and evaporation rates). Comparisons between evaporites and P−E give more consistent results, but with the caveat that evaporation records in the modern are generally inconsistently reported.

These findings demonstrate the value of data-model comparisons, but also illustrate the care that must be taken. Neither models nor data come without caveats. As Conybeare (1829) noted almost 200 years ago, geologists do not rely on any one line of evidence in reconstructing the climate of the past.

I would like to thank Mark Williams and Alan Haywood for the original invitation to contribute to this volume, and John Gregory and Dave Martill for their very useful reviews of earlier versions of the manuscript. The databases used in this study were initially compiled whilst I was a member of Fred Ziegler's Palaeogeographic Atlas Project at the University of Chicago. My thanks go to Fred and all of the current and former members of this group, especially Dave Rowley, Mike Hulver and Ann Lottes.

References

ANDERSEN, B. G. 1990. Cenozoic glacier fluctuations in polar regions – terrestrial records from Antarctica and the North Atlantic sector of the Arctic. *In*: BLEIL, U. & THIEDE, J. (eds) *Geological History of the Polar Oceans: Arctic Versus Antarctic*. Kluwer Academic Publishers, 245–254.

ARNOLD, E. N. & BURTON, J. A. 1978. *A Field Guide to the Reptiles and Amphibians of Britain and Europe*. Collins, London, 272pp.

BÁRDOSSY, G. 1982. *Karst Bauxites: Bauxites on Carbonate rocks*. Elsevier, Amsterdam, 441pp.

BÁRDOSSY, G. & ALEVA, G. J. J. 1990. *Lateritic Bauxites*. Developments in economic geology, **27**, Elsevier, Amsterdam, 624pp.

BARRICK, R. E. & SHOWERS, W. J. 1999. Thermophysiology and biology of *Gigantosaurus*: comparison with *Tyrannosaurus*. *Palaeontologia Electronica*, 2: http://www-odp.tamu.edu/paleo/1999_2/gigan/main.htm

BARRON, E. J. 1981. Paleogeography as a climatic forcing factor. *Geologische Rundschau*, **70**, 737–747.

BARRON, E. J. 1985. Numerical climate modeling, a frontier in Petroleum Source Rock prediction: results based on Cretaceous simulations. *The American Association of Petroleum Geologists Bulletin*, **69**, 448–459.

BARRON, E. J. & MOORE, G. T. 1994. *Climate Model Application in Paleoenvironmental Analysis*. SEPM Short Course, **33**. SEPM, Tulsa, 338pp.

BARRON, E. J. & WASHINGTON, W. M. 1984. The role of geographic variables in explaining paleoclimates: results from Cretaceous climate model sensitivity studies. *Journal of Geophysical Research*, **89**, 1267–1279.

BARRON, E. J., FAWCETT, P. J., POLLARD, D. & THOMPSON, S. 1993. Model simulations of Cretaceous climates: the role of geography and carbon dioxide. *Philosophical Transactions of the Royal Society of London*, **B341**, 307–316.

BARRY, R. G. & CHORLEY, R. J. 1987. *Atmosphere, Weather and Climate*. Routledge, London, 460pp.

BEERLING, D. J. 2002. Low atmosphere CO_2 levels during the Permo-Carboniferous glaciation inferred from fossil lycopsids. *Proceedings of the National Academy of Sciences*, **99**, 12567–12571.

BEERLING, D. J. & ROYER, D. L. 2002. Reading a CO_2 signal from fossil stomata. *New Phytologist*, **153**, 387–397.

BEERLING, D. J. & VALDES, P. J. 2003. Global warming in the early Eocene: was it driven by carbon dioxide? *EOS Transactions, AGU, Fall Meeting Supplement*, **84**, Abstract PP22B-04.

BEERLING, D. J., OSBORNE, C. P. & CHALONER, W. G. 2001. Evolution of leaf-form in land plants linked to atmospheric CO_2 decline in the Late Palaeozoic era. *Nature*, **410**, 352–354.

BEHRENSMEYER, A. K. & CHAPMAN, R. E. 1993. Models and simulations of time-averaging in terrestrial vertebrate accumulations. *In*: KIDWELL, S. M. & BEHRENSMEYER, A. K. (eds) *Taphonomic Approaches to Time Resolution in Fossil Assemblages*. Short courses in paleontology No. 6. The University of Tennessee, Knoxville, 125–149.

BENTON, M. J. & CLARK, J. M. 1988. Archosaur phylogeny and the relationships of the Crocodylia. *In*: BENTON, M. J. (ed.) *The Phylogeny and Classification of the Tetrapods, Vol. 1: Amphibians, Reptiles, Birds*. Systematics Association Special Volume. Clarendon Press, Oxford, 295–338.

BERNER, R. A. & KOTHAVALA, Z. 2001. GEOCARB III: a revised model of atmospheric CO_2 over Phanerozoic time. *American Journal of Science*, **301**, 182–204.

BICE, K. L., BARRON, E. J. & PETERSON, W. H. 1998. Reconstruction of realistic Early Eocene paleobathymetry and ocean GCM sensivity to specified basin configuration. *In*: CROWLEY, T. J. & BURKE, K. C. (eds) *Tectonic Boundary Conditions for Climate Reconstructions*. Oxford University Press, New York, 227–247.

BLECHA, A. M. & GARDNER, M. H. 2004. Does rapid climate change affect sedimentation in a fluvial system? Documenting the PETM using pedogenic carbonate in the Wasatch Formation of Western

Colorado. *EOS Transactions, AGU, Fall Meeting Supplement*, **85**, Abstract PP11B-0569.

BLOCH, J. I., BOYER, D. M., STRAIT, S. G. & WING, S. L. 2004. New Sections and Fossils from the Southern Bighorn Basin, Wyoming Document Faunal Turnover During the PETM. *EOS Transactions, AGU, Fall Meeting Supplement*, **85**.

BÖHME, M., ILG, A., OSSIG, A. & KÜCHENHOFF, H. 2006. New method to estimate paleoprecipitation using fossil amphibians and reptiles and the middle and late Miocene precipitation gradients in Europe. *Geology*, **34**, 425–428.

BOLLE, M. P., PARDO, A., ADATTE, T., VON SALIS, K & BURNS, S. 2000. Climatic evolution on the southeastern margin of the Tethys (Negev, Israel) from the Palaeocene to the early Eocene: focus on the late Palaeocene thermal maximum. *Journal of the Geological Society, London*, **157**, 929–941.

BOTTJER, D. J. & JABLONSKI, D. 1988. Paleoenvironmental patterns in the evolution of post-Paleozoic benthic marine invertebrates. *Palaios*, **3**, 540–560.

BOUCHARDY, C. & MOUTOU, F. 1989. *Observing British and European Mammals*. British Museum (Natural History), London.

BOWEN, G. J., BEERLING, D. J., KOCH, P. L., ZACHOS, J. C. & QUATTLEBAUM, T. 2004. A humid climate state during the Palaeocene/Eocene thermal maximum. *Nature*, **432**, 495–499.

BRANCH, B. 1988. *Field Guide to the Snakes and Other Reptiles of Southern Africa*. New Holland, London, 328pp.

BRATTSTROM, B. H. 1961. Some new fossil tortoises from western North America with remarks on the zoogeography and paleoecology of tortoises. *Journal of Paleontology*, **35**, 543–560.

BRATTSTROM, B. H. 1965. Body temperatures of reptiles. *The American Midland Naturalist*, **73**, 376–422.

BRISBIN, I. L., STANDORA, E. A. & VARGO, M. J. 1982. Body temperatures and behavior of American alligators during cold winter weather. *The American Midland Naturalist*, **107**, 209–218.

BROIN, F. D. 1984. Rôle des tortues comme indicateurs de climat. *Studia Geologica Salmanticensia*, **1**, 99–103.

BSC 1964. *A World Survey of Phosphate Deposits*. The British Sulphur Corporation Ltd., London, 206pp.

BURGGRAF, D. R., HARRIS, J. P., SUTER, J. *ET AL.* 2006. Coupled ocean-atmosphere global paleo-climate modeling for source rock prediction in frontier basins, *AAPG Annual Convention*, April 9–12, AAPG, Houston, TX.

BURNETT, W. C. & RIGGS, S. R. 1990. *Phosphate Deposits of the World. Vol. 3. Neogene to Modern Phosphorites*. International Geological Correlation Programme Project 156, Phosphorites. Cambridge University Press, Cambridge.

CIHLAR, J., ST.-LAURENT, L. & DYER, J. A. 1991. Relation between the Normalized Difference Vegetation Index and ecological variables. *Remote Sensing of Environment*, **35**, 279–298.

COGGER, H. G. 1992. *Reptiles and Amphibians of Australia*. Comstock, Cornell, 775pp.

COLBERT, E. H. 1964a. The relevance of palaeontological data concerning evidence of aridity and hot climates in past geologic ages. *In*: NAIRN, A. E. M. (ed.) *Problems in Palaeoclimatology*. Interscience Publishers, London, 378–381.

COLBERT, E. H. 1964b. Climatic zonation and terrestrial faunas. *In*: NAIRN, A. E. M. (ed.) *Problems in Palaeoclimatology*. Interscience Publishers, London, 617–638.

COLBERT, E. H., COWLES, R. B. & BOGERT, C. M. 1946. Temperature tolerances in the American alligator and their bearing on the habits, evolution, and extinction of the Dinosaurs. *Bulletin of the American Museum of Natural History*, **86**, 327–374.

COLLINSON, M. E. & HOOKER, J. J. 2003. Palaeogene vegetation of Eurasia: framework for mammalian faunas. *In*: REUMER, J. W. F. & WESSELS, W. (eds) *Distribution and Migration of Tertiary Mammals in Eurasia. A Volume in Honour of Hans de Bruijn*. DEINSEA, 41–83.

CONANT, R. & COLLINS, J. T. 1991. *A Field Guide to Reptiles and Amphibians. Eastern/central North America*. Houghton Mifflin Co., Boston, 450pp.

CONYBEARE, W. D. 1829. Answer to Dr Fleming's view of the evidence from the animal kingdom, as to the former temperature of the northern regions. *The Edinburgh New Philosophical Journal*, **7**, 142–152.

COOK, F. R. 1984. *Introduction to Canadian Amphibians and Reptiles*. National Museums of Canada, Ottawa, 200pp.

CORFIELD, R. M. & NORRIS, R. D. 1998. The oxygen and carbon isotopic context of the Paleocene/Eocene epoch boundary. *In*: AUBRY, M.-P., LUCAS, S. G. & BERGGREN, W. A. (eds) *Late Paleocene – Early Eocene Climatic and Biotic Events in the Marine and Terrestrial Records*. Columbia University Press, New York, 124–137.

COWLES, R. B. & BOGERT, C. M. 1944. A preliminary study of the thermal requirements of desert reptiles. *Bulletin of the American Museum of Natural History*, **83**, 261–296.

COX, P. M. 2000. Incorporating vegetation as a dynamic element in the Hadley Centre GCM, *11th Symposium on Global Change Studies*.

COX, P. M. 2001. Description of the 'TRIFFID' dynamic global vegetation model. *Hadley Centre technical note 24*, Hadley Centre, Meteorological Office, Bracknell, Berkshire.

COX, P., BETTS, R., BUNTON, C., ESSERY, R., ROWNTREE, P. R. & SMITH, J. 1999. The impact of new land surface physics on the GCM simulation of climate and climate sensivity. *Climate Dynamics*, **15**, 183–203.

CRANE, P. R. & LIDGARD, S. 1989. Angiosperm diversification and paleolatitudinal gradients in Cretaceous floristic diversity. *Science*, **246**, 675–678.

CURRIE, D. J. 1991. Energy and large-scale patterns of animal- and plant-species richness. *The American Naturalist*, **137**, 27–49.

CURRIE, D. J. & FRITZ, J. T. 1993. Global patterns of animal abundance and species energy use. *OIKOS*, **67**, 56–68.

DUBIEL, R. F. & SMOOT, J. P. 1994. Criteria for interpreting paleoclimate from red beds – a tool for Pangean reconstructions. *In*: EMBRY, A. F., BEAUCHAMP, B. & GLASS, D. J. (eds) *Pangea: Global Environments and Resources.* Canadian Society of Petroleum Geologists, Calgary, 295–310.

EDWARDS, J. M. & SLINGO, A. 1996. Studies of a flexible new radiation code. I: Choosing a configuration for a large scale model. *Quarterly Journal of the Royal Meteorological Society of London*, **122**, 689–719.

EISENBERG, J. F. 1989. *Mammals of the Neotropics. The northern Neotropics. Vol. 1. Panama, Colombia, Venezuela, Guyana, Suriname, French Guiana.* The University of Chicago Press, Chicago, 449pp.

ESTES, R. & HUTCHISON, J. H. 1980. Eocene lower vertebrates from Ellesmere Island, Canadian Arctic Archipelago. *Palaeogeography, Palaeoclimatology, Palaeoecology*, **30**, 325–347.

FLEMING, R. J. 1829.On the value of the evidence from the animal kingdom, tending to prove that the Arctic regions formerly enjoyed a milder climate than at present. *The Edinburgh New Philosophical Journal*, **6**, 277–286.

FLEMING, R. J. 1830. Additional remarks on the climate of the Arctic regions, in answer to Mr Conybeare. *The Edinburgh New Philosophical Journal*, **8**, 65–74.

FRANCIS, J. E. 1988. A 50-million-year-old fossil forest from Strathcona Fiord, Ellesmere Island, Arctic Canada: evidence for a warm polar climate. *Arctic*, **41**, 314–318.

FRANCKO, D. A. 2003. *Palms Won't Grow Here and Other Myths. Warm-Climate Plants for Cooler Areas.* Timber Press, Inc., Portland, Oregon, 267pp.

FRANCKO, D. A. & WILSON, K. G. 2001. The Miami University hardy palm project, Rhapidophyllum, 12–15.

FREEMAN, K. H. & HAYES, J. M. 1992. Fractionation of carbon isotopes by phytoplankton and estimates of ancient CO^2 levels. *Global Biogeochemical Cycles*, **6**, 185–198.

GANS, C. 1982. *Biology of the Reptilia.* Academic Press, London.

GANS, C. 1989. Crocodilians in perspective. *American Zoologist*, **29**, 1051–1054.

GLENNIE, K. W. 1970. *Desert Sedimentary Environments.* Elsevier Publishing Co., Amsterdam, 222pp.

GORDON, C., COOPER, C., SENIOR, C. A. ET AL. 2000. The simulation of SST, sea ice extents and ocean heat transports in a version of the Hadley Centre coupled model without flux adjustments. *Climate Dynamics*, **16**, 147–168.

GOWARD, S. N. & DYE, D. G. 1987. Evaluating North American net primary productivity with satellite observations. *Advances in Space Research*, **7**, 165–174.

GOWARD, S. N., TUCKER, C. J. & DYE, D. G. 1985. North American vegetation patterns observed with the NOAA-7 advanced very high resolution radiometer. *Vegetatio*, **64**, 3–14.

GRADSTEIN, F. M., OGG, J. & SMITH, A. G. 2004. *A Geologic Time Scale 2004.* Cambridge University Press, Cambridge, 384pp.

GREENWOOD, D. R. & WING, S. L. 1995. Eocene continental climates and latitudinal temperature gradients. *Geology*, **23**, 1044–1048.

GRENARD, S. 1991. *Handbook of Alligators and Crocodiles.* Krierger Publishing Co., Malabar, Florida, 210pp.

GYLLENHAAL, E. D. 1991. How accurately can paleoprecipitation and paleoclimatic change be interpreted from subaerial disconformities? PhD Thesis, University of Chicago, Chicago, 529pp.

HALL, E. R. & KELSON, K. R. 1959. *The Mammals of North America.* Ronald Press Co., New York, 1083pp.

HARRISON, S. P. 2004. The Role of Global Palaeodata Syntheses in Benchmarking Climate Model Simulations. *EOS Transactions, AGU, Fall Meeting Supplement*, **85**.

HAYWOOD, A. M., VALDES, P. J. & MARKWICK, P. J. 2004. Cretaceous (Wealden) climates: a modelling perspective. *Cretaceous Research*, **25**, 303–311.

HEAD, J. J. 2005. Snakes of the Siwalik Group (Miocene of Pakistan): systematics and relationship to environmental change. *Palaeontologia Electronica*, **8**, p18A:33p.

HIBBARD, C. W. 1960. An interpretation of Pliocene and Pleistocene climates in North America. *Annual Report of the Michigan Academy of Science, Arts, and Letters*, **62**, 5–30.

HOOKER, J. J. & DASHZEVEG, D. 2003. Evidence for direct mammalian faunal interchange between Europe and Asia near the Paleocene–Eocene boundary. *In*: WING, S. L., GINGERICH, P. D., SCHMITZ, B. & THOMAS, E. (eds) *Causes and Consequences of Globally Warm Climates in the Early Paleogene.* Geological Society of America Special Paper. Geological Society of America, Boulder, CO, 479–500.

HOPSON, J. A. 1976. Review: Hot-, cold-, or lukewarm-blooded dinosaurs? *The Hot-Blooded Dinosaurs: a Revolution in Paleontology* DESMOND, Adrian J. *Paleobiology*, **2**, 271–275.

HORRELL, M. A. 1991. Phytogeography and paleoclimatic interpretation of the Mastrichtian. *Palaeogeography, Palaeoclimatology, Palaeoecology*, **86**, 87–138.

HOUGHTON, J. T., DING, Y., GRIGGS, D. J. ET AL. (eds) 2001. Climate Change 2001: *The Scientific Basis. Contribution of Working Group I to the third Assessment Report of the Intergovernmental Panel on Climate Change.* Cambridge University Press, Cambridge, 881pp.

HUC, A.-Y. 1990. *Deposition of Organic Facies*, **30**, AAPG, Tulsa.

HULVER, M. L., MARKWICK, P. J., ZIEGLER, A. M. & ROWLEY, D. B. 1993. Paleogeographic Atlas Project. *Houston Geological Society Bulletin*, **36**, 50–52.

HUMBOLDT, A.V. 1828. On the principal causes of the difference of temperature on the globe. *The Edinburgh New Philosophical Journal*, 329–346.

HUTCHISON, V. H., VINEGAR, A. & KOSH, R. J. 1966. Critical thermal maxima in turtles. *Herpetologica*, **22**, 32–41.

IVERSON, J. B. 1992a. *A Revised Checklist with Distribution Maps of the Turtles of the World.* Privately published, Richmond, Indiana, 363pp.

IVERSON, J. B. 1992b. Global correlates of species richness in turtles. *Herpetological Journal*, **2**, 77–81.

JANIS, C. M. 1989. A climatic explanation for patterns of evolutionary diversity in ungulate mammals. *Palaeontology*, **32**, 463–481.

JANIS, C. M. 1993. Tertiary mammal evolution in the context of changing climates, vegetation, and tectonic events. *Annual Review of Ecology and Systematics*, **24**, 467–500.

KIESSLING, W. 2001. Paleoclimatic significance of Phanerozoic reefs. *Geology*, **29**, 751–754.

KINGDON, J. 1990. *Arabian Mammals. A Natural History*. Academic Press, London, 279pp.

KÖPPEN, W. 1931. *Grundriss der Klimakunde*. Walter de Gruyter Co., Berlin, 388pp.

KRUIJS, E. & BARRON, E. J. 1990. Climate model prediction of paleoproductivity and potential source-rock distribution. *In*: HUC, A. Y. (ed.) *Deposition of Organic Facies*. AAPG Studies in Geology. AAPG, Tulsa, 195–216.

KUTZBACH, J. E. & ZIEGLER, A. M. 1993. Simulation of Late Permian climate and biomes with an atmosphere–ocean model: comparisons with observations. *Philosophical Transactions of the Royal Society of London. Series B.*, **341**, 327–340.

LARCHER, W. 1980. *Physiological Plant Ecology*. Springer-Verlag, Berlin, 303pp.

LEGENDRE, S. 1989. Les communautés de mammifères du Palèogène (Eocène supérieur et Oligocène) d'Europe occidentale: structures, milieux et évolution. *Münchner Geowiss. Abh.*, **16**, 1–110.

LEGENDRE, S. & HARTENBERGER, J.-L. 1992. Evolution of mammalian faunas in Europe during the Eocene and Oligocene. *In*: PROTHERO, D. R. & BERGGREN, W. A. (eds) *Eocene–Oligocene Climatic and Biotic Evolution*. Princeton University Press, Princeton, 399–420.

LITTLE, E. L. 1977. *Atlas of United States Trees*. Miscellaneous Publication. United States Department of Agriculture Forest Service. United States Government Printing Office, Washing, D.C.

LOTTES, A. L. & ZIEGLER, A. M. 1994. World peat occurrence and the seasonality of climate and vegetation. *Palaeogeography, Palaeoclimatology, Palaeoecology*, **106**, 23–37.

LUPIA, R. 1997. Palynological record of the Cretaceous angiosperm radiation: diversity, abundance and morphological patterns. PhD Thesis, University of Chicago, Chicago, 625pp.

LUPIA, R. & MARKWICK, P. J. 1998. *Dealing with Paleontological Uncertainty: Data Collection and the Remote Past*. Penrose.

LYELL, C. 1830. *Principles of Geology, Being an Attempt to Explain the Former Changes of the Earth's Surface, by Reference to Causes Now in Operation, 1*. John Murray, London, 511pp.

MACLEOD, K. G. & BERGEN, J. A. 2004. Apparent cooling in the tropical Pacific during the Maastrichtian and diagenetic artefacts in Late Cretaceous stable isotopic trends in bulk carbonate from Ontong Java Plateau. *In*: FITTON, J. G., MAHONEY, J. J., WALLACE, P. J. & SAUNDERS, A. D. (eds) *Proceedings of the Ocean Drilling Project, Scientific Results*, **192**, 1–15.

MARKWICK, P. J. 1994. 'Equability', continentality and Tertiary 'climate': the crocodilian perspective. *Geology*, **22**, 613–616.

MARKWICK, P. J. 1996. Late Cretaceous to Pleistocene climates: nature of the transition from a 'hot-house'

to an 'ice-house' world. PhD Thesis, The University of Chicago, Chicago, 1197pp.

MARKWICK, P. J. 1998. Fossil crocodilians as indicators of Late Cretaceous and Cenozoic climates: implications for using palaeontological data for global change. *Palaeogeography, Palaeoclimatology, Palaeoecology*, **137**, 205–271.

MARKWICK, P. J. 2000. Biodiversity gradients as a palaeoclimate tool? *Geoscience 2000*, 17–20 April. Geological Society, London, Manchester.

MARKWICK, P. J. 2001. Integrating the present and past records of climate, biodiversity and biogeography: implications for palaeoecology and palaeoclimatology. *In*: CRAME, J. A. & OWEN, A. W. (eds) *Palaeobiogeography and Biodiversity Change: a Comparison of the Ordovician and Mesozoic-Cenozoic Radiations*. Geological Society, London, Special Publications, 179–200.

MARKWICK, P. J. & LUPIA, R. 2001. Palaeontological databases for palaeobiogeography, palaeoecology and biodiversity: a question of scale. *In*: CRAME, J. A. & OWEN, A. W. (eds) *Palaeobiogeography and Biodiversity Change: a Comparison of the Ordovician and Mesozoic–Cenozoic Radiations*. Geological Society, London, Special Publications, 169–174.

MARKWICK, P. J. & ROWLEY, D. B. 1998. The geologic evidence for Triassic to Pleistocene glaciations: Implications for Eustacy. *In*: PINDELL, J. L. & DRAKE, C. L. (eds) *Paleogeographic Evolution and Non-Glacial Eustasy, Northern South America*. SEPM Special Publication, **58**, SEPM Special Publication, SEPM, Tulsa, 17–43.

MARKWICK, P. J. & VALDES, P. J. 2003. Modeling the Maastrichtian (Late Cretaceous) Landscape Using a Coupled Ocean-Atmosphere GCM and Detailed Global Paleo-DEM, *AAPG Annual Meeting*, Salt Lake City, USA.

MARKWICK, P. J. & VALDES, P. J. 2004. Palaeo-digital elevation models for use as boundary conditions in coupled ocean-atmosphere GCM experiments: a Maastrichtian (late Cretaceous) example. *Palaeogeography, Palaeoclimatology, Palaeoecology*, **213**, 37–63.

MARKWICK, P. J., CROSSLEY, R. & VALDES, P. J. 2002. A comparison of "Ice-House" (Modern) and "Hot-House" (Maastrichtian) drainage systems: the implications of large-scale changes in the surface hydrological scheme, *AGU Fall Meeting*, San Francisco.

MARKWICK, P. J., ROWLEY, D. B., ZIEGLER, A. M., HULVER, M., VALDES, P. J. & B. W., S. 2000a. Late Cretaceous and Cenozoic global palaeogeographies: mapping the transition from a "hot-house" world to an "ice-house" world. *GFF*, **122**, 103.

MARKWICK, P. J., VALDES, P. J., SELLWOOD, B. & PIERREHUMBERT, R. T. 2000b. 'Equability' in an unequal world: the early Eocene revisited. *GFF*, **122**, 101–102.

MARKWICK, P. J., VALDES, P. J. & SELLWOOD, B. W. 2000c. Maastrichtian dinosaur paleobiogeography and climate. *Geoscience 2000*, 17–20 April. Geological Society, London, Manchester.

MARKWICK, P. J., VALDES, P. J. & CROSSLEY, R. 2005. A quantitative, multi-proxy assessment of hydrological results from coupled-ocean models: Modern and Maastrichtian examples. *Geophysical Research Abstracts*, **7**.

McKENNA, M. C. 1980. Eocene paleolatitude, climate, and mammals of Ellesmere Island. *Palaeogeography, Palaeoclimatology, Palaeoecology*, **30**, 349–362.

McKENNA, M. C. 1983. Cenozoic paleogeography of North Atlantic land bridges. *In*: BOTT, M. H. P., SAXOV, S., TALWANI, M. & THIEDE, J. (eds) *Structure and Development of the Greenland–Scotland Ridge*. NATO Advanced Research Institute. Plenum Press, New York, 351–399.

MOORE, P. D. & BELLAMY, D. J. 1974. *Peatlands*. Elek Science, London, 184pp.

MÜLLER, M. J. 1982. *Selected Climatic Data for a Global Set of Standard Stations for Vegetation Science*. Tasks for Vegetation Science. Dr W. Junk Publishers, The Hague, 306pp.

NEW, M., HULME, M. & JONES, P. D. 1999. Representing twentieth century space-time climate variability. Part 1: development of a 1961–90 mean monthly terrestrial climatology. *Journal of Climate*, **12**, 829–856.

NICOLO, M. J. 2004. The composition and flux of Terrigenous material from the late Palaeocene to the early Eocene in the Indian Ocean. *EOS Transactions, AGU, Fall Meeting Supplement*, **85**, Abstract PP11B-0566.

NOAA-EPA Global Ecosystems Database Project 1992. *Global Ecosystems Database Version 1.0. User's Guide, Documentation, Reprints, and Digital Data on CD-ROM*. USDOC/NOAA National Geophysical Data Center, Boulder, CO.

NOBLICK, L. R. 1998. Predicting hardiness in palms. World Wide Web Address: http://www.bg-map.com/noblick.html.

NOTHOLT, A. J. G., SHELDON, R. P. & DAWSON, D. F. (eds) 1989. *Phosphate Deposits of the World. Vol. 2. Phosphate Rock Resources*. International Geological Correlation Programme Project 156: Phosphorites. Cambridge University Press, Cambridge, 566pp.

ORRIS, G. J. & CHERNOFF, C. B. 2002*a*. Data set of world phosphate mines, deposits, and occurrences – Part B. Location and mineral economic data. *USGS Open-File Report*, **02-156-B**, 1–327.

ORRIS, G. J. & CHERNOFF, C. B. 2002*b*. Data set of world phosphate mines, deposits, and occurrences – Part A. Geologic data. *USGS Open-File Report*, **02-156-A**, 1–352.

OSTROM, J. H. 1970. *Terrestrial vertebrates as indicators of Mesozoic climates*, Proceedings of the North American Paleontological Convention. Allen Press, Inc., Chicago, 347–376.

OTTO-BLIESNER, B. L., BRADY, E. C. & KOTHAVALA, Z. 2004. Climate Sensitivity of the Last Glacial Maximum from Paleoclimate Simulations and Observations. *EOS Transactions, AGU, Fall Meeting Supplement*, **85**.

OWEN, R. M. 1850*a*. On the fossil crocodilia of England. *The Edinburgh New Philosophical Journal*, **49**, 248–250.

OWEN, R. M. 1850*b*. *Monograph on the Fossil Reptilia of the London Clay, and of the Bracklesham and other Tertiary Beds. Part II. Crocodilia (Crocodilus, &c.)*. The Palaeontographical Society, London, 50pp.

OWEN, R. M. 1851. *Monograph on the Fossil Reptilia of the Cretaceous Formations*. The Palaeontological Society, London, 118pp.

PARRISH, J. T., ZIEGLER, A. M. & SCOTESE, C. R. 1982. Rainfall patterns and the distribution of coals and evaporites in the Mesozoic and Cenozoic. *Palaeogeography, Palaeoclimatology, Palaeoecology*, **40**, 67–101.

PARRISH, J. T. 2001. *Interpreting Pre-Quaternary Climate from the Geological Record*. Critical moments and perspectives in Earth History and Paleobiology Series. Columbia University Press, New York, 354pp.

PAUL, G. S. 1992. Physiology and migration of North Slope dinosaurs, *International Conference on Arctic Margins*. OCS Study, 405–408.

PEARSON, P. N. & PALMER, M. R. 2000. Declining atmospheric carbon dioxide in the last sixty million years. *Nature*, **406**, 695–699.

PEARSON, P. N., DITCHFIELD, P. W., SINGANO, J. *ET AL.* 2001. Warm tropical sea surface temperatures in the Late Cretaceous and Eocene epochs. *Science*, **413**, 481–487.

PELTIER, W. R. 1994. Ice age paleotopography. *Science*, **265**, 195–201.

PEUQUET, D. J. 1988. Issues involved in selecting appropriate data models for global databases. *In*: MOUNSEY, H. (ed.) *Building Databases for Global Science*. Taylor and Francis, London, 66–78.

PIANKA, E. R. 1977. Reptilian species diversity. *In*: GANS, C. (ed.) *Biology of the Reptilia*. Academic Press, London, 1–34.

POPE, V. D., GALLANI, M. L., ROWNTREE, P. R. & STRATTON, R. A. 2000. The impact of new physical parametrizations in the Hadley Centre climate model – HadCM3. *Climate Dynamics*, **16**, 123–146.

PORTER, W. P. & GATES, D. M. 1969. Thermodynamic equilibria of animals with environment. *Ecological Monographs*, **39**, 227–244.

POULSEN, C. J., SEIDOV, D., BARRON, E. J. & PETERSON, W. H. 1998. The impact of paleogeographic evolution on the surface oceanic circulation and the marine environment within the mid-Cretaceous Tethys. *Paleoceanography*, **13**, 546–559.

PRICE, G. D., SELLWOOD, B. W. & VALDES, P. J. 1995. Sedimentological evaluation of general circulation model simulations for the "greenhouse" Earth: Cretaceous and Jurassic case studies. *Sedimentary Geology*, **100**, 159–180.

PRICE, G. D. & SELLWOOD, B. W. 1997. "Warm" palaeotemperatures from high Late Jurassic palaeolatitudes (Falkland Plateau): ecological, environmental or diagenetic controls? *Palaeogeography, Palaeoclimatology, Palaeoecology*, **129**, 315–327.

PRICE, G. D., VALDES, P. J. & SELLWOOD, B. W. 1997. Prediction of modern bauxite occurrence: implications for climate reconstruction. *Palaeogeography, Palaeoclimatology, Palaeoecology*, **131**, 1–13.

RAUP, D. M. 1972. Taxonomic diversity during the Phanerozoic. *Science*, **177**, 1065–1071.

REDFORD, K. H. & EISENBERG, J. F. 1992. *Mammals of the Neotropics. The Southern Cone. Vol. 2. Chile, Argentina, Uruguay, Paraguay*. The University of Chicago Press, Chicago, 430pp.

RETALLACK, G. J. 1990. *Soils of the Past. An Introduction to Paleopedology.* Unwin Hyman, Boston, 520pp.

ROSENZWEIG, M. L. 1995. *Species Diversity in Space and Time.* Cambridge University Press, Cambridge, 436pp.

ROYER, D. L., BERNER, R. A. & BEERLING, D. J. 2001. Phanerozoic atmospheric CO_2 change: evaluating geochemical and paleobiological approaches. *Earth-Science Reviews*, **54**, 349–392.

SCHALL, J. J. & PIANKA, E. R. 1977. Species densities of reptiles and amphibians on the Iberian Peninsula. *Doñana Acta Vertebrata*, **4**, 27–34.

SCHALL, J. J. & PIANKA, E. R. 1978. Geographical trends in numbers of species. *Science*, **201**, 679–686.

SELLWOOD, B. W. & PRICE, G. D. 1994. Sedimentary facies as indicators of Mesozoic palaeoclimate. *In*: ALLEN, J. R. L., HOSKINS, B. J., SELLWOOD, B. W., SPICER, R. A. & VALDES, P. J. (eds) *Palaeoclimates and their Modeling with Special Reference to the Mesozoic Era.* Chapman & Hall, London, 17–25.

SEPKOSKI, J. J., JR., 1982. A compendium of fossil marine families. *Milwaukee Public Museum Contributions in Biology and Geology*, **51**.

SHELLITO, C. J. & SLOAN, L. C. 2004. Using a Dynamic Global Vegetation Model to Simulate the Response of Vegetation to Warming at the Paleocene–Eocene Boundary. *EOS Transactions, AGU, Fall Meeting Supplement*, **85**, Abstract PP11B-0568.

SINKA, K. J. & ATKINSON, T. C. 1999. A mutual climatic range method for reconstructing palaeoclimate from plant remains. *Journal of the Geological Society, London*, **156**, 381–396.

SLOAN, L. C. 1994. Equable climates during the early Eocene: significance of regional paleogeography for North American climate. *Geology*, **22**, 881–884.

SLOAN, L. C. & BARRON, E. J. 1992. A comparison of Eocene climate model results to quantified paleoclimatic interpretations. *Palaeogeography, Palaeoclimatology, Palaeoecology*, **93**, 183–202.

SLOAN, L. C., WALKER, J. C. G., MOORE JR., T. C. M., REA, D. K. & ZACHOS, J. C. 1992. Possible methane-induced polar warming in the early Eocene. *Nature*, **357**, 320–322.

SMITH, E. N. 1975. Thermoregulation of the American alligator, *Alligator mississippiensis. Physiological Zoölogy*, **48**, 177–194.

SMITH, E. N. 1979. Behavioral and physiological thermoregulation of crocodilians. *American Zoologist*, **19**, 239–247.

SONNENFELD, P. 1984. *Brines and Evaporites.* Academic Press, Inc., Orlando, 613pp.

STACH, E. 1982. *Stach's Textbook of Coal Petrology.* Gebrüder Borntraeger, Berlin, 535pp.

STEBBINS, R. C. 1985. *A Field Guide to Western Reptiles and Amphibians.* Houghton Mifflin Co., Boston, 336pp.

STOREY, K. B. 1990. Life in a frozen state: adaptive strategies for natural freeze tolerance in amphibians and reptiles. *American Journal of Physiology*, **258**, R559–R568.

STOREY, K. B. & STOREY, J. M. 1988. Freeze tolerance in Animals. *Physiological Reviews*, **68**, 27–84.

STRAHAN, R. 1992. *Encylopoedia of Australian Animals. Mammals.* Angus & Robertson, Pymble, NSW, 184pp.

THORNWAITE, C. W. 1948. An approach toward a rational classification of climate. *Geographical Review*, **38**, 55–89.

UHL, N. W. & DRANSFIELD, J. 1987. *Genera Palmarum. A Classification of Palms Based on the Work of Harold E. Moore, Jr.* Allen Press, Lawrence, Kansas, 610pp.

VAKHRAMEEV, V. A. 1981. *Classopolis* pollen indicator of Jurassic and Cretaceous climate. *International Geological Review*, **24**, 1190–1196.

VALDES, P. J. & MARKWICK, P. J. 2006. The sensitivity of modelled ocean upwelling and terrestrial runoff to changing boundary conditions: implications for predictions of source, reservoir and seal facies. *AAPG Annual Convention*, April 9–12, AAPG, Houston, TX.

VAN DER BRINK, F. H. 1967. *A Field Guide to the Mammals of Britain and Europe.* Collins, London, 221pp.

VERDIN, K. 2001. HYDRO1k documentation. USGS.

WALTER, H. 1985. *Vegetation of the Earth and Ecological Systems of the Geo-Biosphere.* Springer-Verlag, Berlin, 318pp.

WALTER, H. & LIETH, H. 1967. *Klimadiagramm – Weltatlas.* Gustav Fischer Verlag, Jena.

WALTERS, L. A. 1998. Minimum temperatures for some cold-hardy palms. World Wide Web Address: http://www.hardiestpalms.com.

WEBB, P. N., HARWOOD, D. M., MCKELVEY, B. C., MERCER, J. H. & STOTT, L. D. 1984. Cenozoic marine sedimentation and ice-volume variation on the East Antarctic craton. *Geology*, **12**, 287–291.

WENG, Q., LU, D. & SCHUBRING, J. 2004. Estimation of land surface temperature – vegetation abundance relationship for urban heat island studies. *Remote Sensing of Environment*, **89**, 467–483.

WEST, R. M., DAWSON, M. R. & HUTCHISON, J. H. 1977. Fossils from the Paleogene Eureka Sound Formation, N.W.T., Canada: occurrence, climatic and paleogeographic implications. *In*: WEST, R. M. (ed.) *Paleontology and Plate Tectonics with Special Reference to the History of the Atlantic Ocean.* Milwaukee Public Museum Special Publications in Biology and Geology. Milwaukee Public Museum, Milwaukee, 77–93.

WILF, P., WING, S. L., GREENWOOD, D. R. & GREENWOOD, C. L. 1998. Using fossil leaves as paleoprecipitation indicators: an Eocene example. *Geology*, **26**, 203–206.

WILLIS, P. M. A. & STILWELL, J. D. 2000. A probable piscivorous crocodile from Eocene deposits of McMurdo Sound, East Antarctica. *In*: STILWELL, J. D. & FELDMANN, R. M. (eds) *Paleobiology and Paleoenvironments of Eocene Rocks, McMurdo Sound, East Antarctica.* Antarctic Research Series. American Geophysical Union, Washington, D.C., 355–358.

WING, S. L. & GREENWOOD, D. R. 1993. Fossils and fossil climate: the case for equable continental interiors in the Eocene. *Philosophical Transactions of the Royal Society of London. Series B.*, **341**, 243–252.

WING, S. L., HARRINGTON, G. J., BLOCH, J. I., BOYER, D. M. & SMITH, F. 2004. Major Transient Floral Change during the Paleocene-Eocene Thermal Maximum. *EOS Transactions, AGU, Fall Meeting Supplement*, **85**.

WOLFE, J. A. 1971. Tertiary climatic fluctuations and methods of analysis of Tertiary floras. *Palaeogeography, Palaeoclimatology, Palaeoecology*, **9**, 27–57.

WOLFE, J. A. 1978. A paleobotanical interpretation of Tertiary climates in the northern hemisphere. *American Scientist*, **66**, 694–703.

WOLFE, J. A. 1993. A method of obtaining climatic parameters from leaf assemblages. *U.S. Geological Survey Bulletin*, 2040, 1–71.

WOLFE, J. A. 1994. Alaskan palaeogene climates as inferred from the CLAMP database. *In*: BOULTER, M. C. & FISHER, H. C. *Cenozoic Plants and Climates of the Arctic*. NATO ASI Series. Springer-Verlag, Berlin, 223–237.

WU, C. & ZUE, S. 1997. *Sedimentology of Petroliferous Basins in China*. Petroleum Industry Press, Beijing, 504pp.

ZACHOS, J., PAGANI, M., SLOAN, L., THOMAS, E. & BILLUPS, K. 2001. Trends, rhythms, and aberrations in global climate 65 Ma to Present. *Science*, **292**, 686–693.

ZIEGLER, A. M., SCOTESE, C. R. & BARRETT, S. F. 1983. Mesozoic and Cenozoic paleogeographic maps. *In*: BROSCHE/SÜNDERMANN, (ed.) *Tidal Friction and the Earth's Rotation II*. Springer-Verlag, Berlin, 240–252.

ZIEGLER, A. M., HULVER, M. L., LOTTES, A. L. & SCHMACHTENBERG, W. F. 1984. Uniformitarianism and paleoclimates: inferences from the distribution of carbonate rocks. *In*: BRENCHLY, P. (ed.) *Fossil and Climate*. John Wiley & Sons Ltd., Chichester, 3–25.

ZIEGLER, A. M., ROWLEY, D. B., LOTTES, A. L., SAHAGIAN, D. L., HULVER, M. L. & GIERLOWSKI, T. C. 1985. Paleogeographic interpretation: with an example from the Mid-Cretaceous. *Annual Review of Earth and Planetary Sciences*, **13**, 385–425.

ZIEGLER, A. M., RAYMOND, A. L., GIERLOWSKI, T. C., HORRELL, M. A., ROWLEY, D. B. & LOTTES, A. L. 1987. Coal, climate and terrestrial productivity: the Present and Early Cretaceous compared. *In*: SCOTT, A. C. (ed.) *Coal and Coal-Bearing Strata: Recent Advances*. Geological Society Special Publication, **32**. Geological Society, London, 25–49.

ZIEGLER, A. M. 1990. Phytogeographic patterns and continental configurations during the Permian Period. *In*: MCKERROW, W. S. & SCOTESE, C. R. (eds) *Palaeogeography and Biogeography*. Geological Society Memoir. The Geological Society of London, London, 363–377.

ZIEGLER, A. M., ESHEL, G., REES, P. M., ROTHFUS, T. A., ROWLEY, D. B. & SUNDERLIN, D. 2003. Tracing the tropics across land and sea: Permian to present. *Lethaia*, **36**, 227–254.

Mg/Ca palaeothermometry: a new window into Cenozoic climate change

C. H. LEAR

School of Earth, Ocean and Planetary Sciences, Cardiff University, Main Building, Park Place, Cardiff, CF10 3YE, UK (e-mail: carrie@earth.cf.ac.uk)

Abstract: Mg/Ca palaeothermometry provides a means of constructing quantitative records of Cenozoic climate change. Its application is not limited to past temperature estimates, for when this proxy is coupled with oxygen isotope palaeothermometry, past variations in the oxygen isotopic composition of seawater may also be estimated. In this way, the Cenozoic history of continental ice-sheet growth has been outlined. Further insights into the dynamics of the Cenozoic climate system have arisen from regional seawater salinity reconstructions, which shed light on changes in ocean circulation patterns. Finally, Mg/Ca palaeothermometry provides a means of testing the accuracy of climate models simulating past warm climates.

Many proxy records have been developed to allow us to look back into Earth's climate history and define the timing and magnitude of the overall cooling trend through the Cenozoic. Arguably, the most established quantitative proxy of Cenozoic climate change is the oxygen isotopic composition ($\delta^{18}O$) of calcium carbonate, in particular that of well-preserved foraminifera (e.g. Shackleton & Kennett 1975). Foraminifera are single-celled animals that construct calcite tests usually less than 1 mm in diameter. These tests accumulate in marine sediments, forming archives of past oceanic conditions. The oxygen isotopic composition of the calcite tests reflects that of the seawater in which they lived, which in turn varies with the global ice budget, because the hydrologic cycle concentrates the light isotope of oxygen (oxygen-16) in continental ice-sheets, leaving oceans more enriched in the heavy isotope of oxygen (oxygen-18) as glaciations advance. However, the foraminiferal oxygen isotope proxy is complicated by several additional factors: (1) foraminifera fractionate the seawater oxygen isotope ratio during calcification of test calcite, and the magnitude of this fractionation is temperature-dependent; (2) the oxygen isotopic composition of seawater varies within the oceans, dependent on circulation patterns and the local balance of evaporation to precipitation; (3) most fossil foraminifera that are buried within thick piles of carbonate-rich sediments are affected by post-depositional diagenetic calcite, which offsets absolute values and attenuates shifts within down-core oxygen isotope records to some degree (e.g. Killingley 1983; Norris & Wilson 1998; Pearson et al. 2001; Wilson & Norris 2001).

Mg/Ca palaeothermometry is a relatively new proxy where the amount of magnesium incorporated into biogenic calcite (usually foraminifera, but also for example ostracodes) is used to obtain estimates of seawater temperatures (Dwyer et al. 1995; Nürnberg et al. 1996; Rosenthal et al. 1997). Because these temperature estimates are salinity-independent, they turn some of the complexities of the oxygen isotope proxy into distinct advantages, as the seawater oxygen isotope ratio may be then calculated from co-existing $\delta^{18}O$ records and interpreted in terms of changes in continental ice volume and/or salinity (e.g. equation 1, Shackleton 1974):

$$\text{temperature} = 16.9 - 4.0$$
$$\times (\delta^{18}O_{FORAM} - \delta^{18}O_{SEAWATER}). \quad (1)$$

A temperature control on calcite Mg/Ca

Laboratory experiments show that the partition coefficient of Mg^{2+} into inorganic calcite increases with increasing temperature (e.g. Katz 1973; Burton & Walter 1991; Hartley & Mucci 1996), so it is perhaps not too surprising that a positive Mg/Ca-temperature relationship in biogenic calcite has been found in both empirical (core-top/sediment trap) and culture experiments (Dwyer et al. 1995; Nürnberg et al. 1996; Rosenthal et al. 1997; Lea et al. 1999; Elderfield & Ganssen 2000; Anand et al. 2003). The magnitude of the foraminiferal Mg/Ca-temperature relationship is about twice that observed in inorganic calcite laboratory experiments, suggesting that the relationship is not purely thermodynamic (Rosenthal et al. 1997). Mg/Ca offsets between species living at the same temperature also suggest a biological influence on this calcification process. Nevertheless, even without a complete understanding of the biochemical

From: WILLIAMS, M., HAYWOOD, A. M., GREGORY, F. J. & SCHMIDT, D. N. (eds) *Deep-Time Perspectives on Climate Change: Marrying the Signal from Computer Models and Biological Proxies.* The Micropalaeontological Society, Special Publications. The Geological Society, London, 313–322.
1747-602X/07/$15.00 © The Micropalaeontological Society 2007.

mechanisms that control the uptake of Mg^{2+} into biogenic calcite, the consistency of the Mg/Ca-temperature sensitivity of both benthic and planktonic foraminifera from a wide variety of oceanic and artificial laboratory settings gives us confidence in applying this proxy to the fossil record (Fig. 1; Eqn 2, where the exponent A varies between 0.05 and 0.12 depending on species). The exponent in Equation 2 dictates the 'sensitivity' of a given foraminiferal species Mg/Ca to temperature; the higher the exponent, the greater the absolute change in Mg/Ca per degree change in temperature. Recent core-top calibration data suggest that an exponential fit might not be appropriate for all species (Marchitto et al. unpublished), further underlining the observation that temperature records constructed using species-specific calibrations are more consistent than those constructed using a generic calibration (Lear et al. 2003):

$$Mg/Ca_{FORAM} = B \exp(A \times temperature). \quad (2)$$

The choice of species-specific calibration decreases as we delve back further into the Cenozoic, and this problem is particularly enhanced for planktonic foraminifera, which tend to evolve faster than benthic foraminifera. Nevertheless, a recent multi-species planktonic foraminiferal Mg/Ca calibration study based on modern sediment trap material provides a single temperature calibration equation for 10 species of planktonic foraminifera, which has an exponent of 0.09 ± 0.003 (Anand et al. 2003). Care must be taken not to apply this calibration to inappropriate foraminifera such as *O. universa* (which has distinctly higher and more variable Mg/Ca relative to other species: Anand et al. 2003). However, this calibration has been successfully applied to planktonic foraminiferal Mg/Ca records of the early Cenozoic (e.g. Tripati et al. 2003).

A small saturation state effect on benthic foraminiferal Mg/Ca

Although temperature is the dominant control on benthic foraminiferal Mg/Ca, the degree of carbonate saturation appears to exert a secondary control on Mg/Ca ratios (Martin et al. 2002). Although small, this potential saturation effect must be taken into account when interpreting down-core and site-to-site Mg/Ca records where a significant change or difference in seawater saturation state is suspected. As discussed below, Mg/Ca records across a dramatic deepening of the ocean's calcite compensation depth display a larger signal at the deepest site, supporting the suggestion that this secondary control is most apparent at very low levels of saturation (Lear et al. 2004).

Temporal variations in seawater Mg/Ca

Magnesium substitutes for calcium in the calcite ($CaCO_3$) lattice. The degree of substitution is dependent on the partition coefficient, D_{CATION}, specific to the calcifying organism, and also the concentration of that cation relative to calcium in the fluid used by the organism (Eqn 3). As we have just seen, D_{Mg} is temperature-dependent:

$$D_{CATION} = \frac{(cation/calcium)_{CALCITE}}{(cation/calcium)_{FLUID}} \quad (3)$$

The amount of Mg^{2+} relative to Ca^{2+} is more or less invariant within the ocean today, yet Equation 3 demonstrates that for Cenozoic studies the Mg/Ca temperature calibration should include past variations in seawater Mg/Ca (Eqn 4):

$$Mg/Ca_{FORAM} = \frac{(Mg/Ca)_{SW-T}}{(Mg/Ca)_{SW-0}}$$
$$\times B \exp(A \times temperature) \quad (4)$$

While there are no direct records of ancient seawater Mg/Ca ratios, indirect evidence as to how seawater Mg/Ca ratios have varied through time

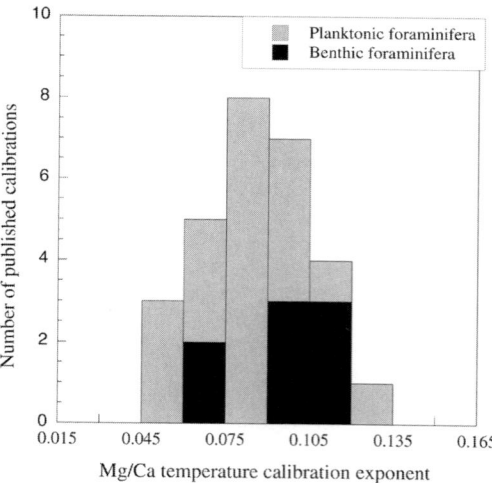

Fig. 1. The exponent in the general Mg/Ca-temperature equation (Eqn 2) describes the sensitivity of a particular foraminiferal species Mg/Ca to temperature. These values represent the calibrations determined via core-top, sediment trap and culture experiments that are compiled in Lear et al. (2002) and Anand et al. (2003).

allow us to put reasonable constraints on our uncertainty envelopes. In addition, the long residence times of Mg^{2+} and Ca^{2+} in seawater (c. one million years and ten million years respectively: Broecker & Peng 1982) means that any change in the seawater Mg/Ca ratio must have occurred very slowly. This means that when we look at snapshots of time within the Cenozoic, and when we compare records from different locations but of the same age, the seawater Mg/Ca problem essentially falls away. In other words, even in the worst-case scenario (having no idea how seawater Mg/Ca changed throughout the Cenozoic), we can still estimate relative temperature changes with a high degree of certainty, even though we may have some uncertainty regarding the absolute temperature estimates. As discussed below, one of the most important ways to test climate models is to reconstruct spatial maps of relative ocean temperatures for given snapshots in geological time.

Fortunately, we are not in the worst-case scenario, as three lines of evidence help us to estimate how seawater Mg/Ca has varied throughout the Cenozoic. The first line of evidence is the composition of fluid inclusions caught up in evaporite minerals. This method is not without its uncertainties and assumptions, but it suggests that at around 50 Ma, seawater Mg/Ca was probably between 2.5 and 3.5 mol/mol (e.g. Lowenstein et al. 2001). The second line of evidence comes from biogenic carbonate Mg/Ca ratios, such as the Mg/Ca of foraminifera that lived in the greenhouse world of the early Cenozoic. With zero (or near-zero) continental ice volume, we can use benthic foraminiferal oxygen isotope ratios to calculate bottom-water temperatures for these times. We can then use Equation 4 to calculate seawater Mg/Ca, which suggests that during the early Cenozoic, seawater Mg/Ca was between 3 and 5 mol/mol (the window reflects current estimated calibration uncertainties) (Lear et al. 2002). Mg/Ca of fossil echinoderm calcite produce a lower estimate of seawater Mg/Ca for 50 Ma, at around 2 mol/mol (Dickson 2002). The final (and least constrained) line of evidence is through modelling the processes that may change the concentrations of Mg^{2+} and Ca^{2+} in the ocean. Essentially, these processes are hydrothermal alteration of basaltic crust, calcification of marine organisms, dolomite formation, and the flux of Mg^{2+} and Ca^{2+} ions brought to the ocean by rivers. Predicted seawater ratios vary between models, from around 1.6 mol/mol (Demicco et al. 2005) to around 3.9 mol/mol (Wilkinson & Algeo 1989) at c. 50 Ma (note that the latter model predicts more or less constant seawater Mg/Ca when run with a constant residence time of Mg passing through basaltic crust). The large range in predicted

seawater Mg/Ca reflects different input parameters, and whether the model attempts to constrain temporal variations in dolomite formation. Estimates of seawater Mg/Ca using the different techniques converge towards the present-day known value of 5.2 mol/mol, so that uncertainties in calculating absolute temperatures and seawater $\delta^{18}O$ are reduced in younger sediments (Fig. 2). Future work on the isotopic systems of calcium and magnesium may provide additional constraints on secular variations in seawater composition (e.g. De La Rocha & DePaolo 2000).

Foraminiferal sample preparation and analysis

Most palaeoceanographic laboratories analysing foraminiferal Mg/Ca prepare the foraminifera by using a variant of the cleaning method described by Boyle & Keigwin (1985/86), which is comprised of a number of 'steps'. These steps have been found to be effective at removing specific contaminant phases, although some of the corrosive steps may also introduce an artificial dissolution bias on planktonic and benthic foraminiferal Mg/Ca (Hastings et al. 1998; Martin & Lea 2002; Barker et al. 2003; Lear et al. 2003; Rosenthal et al. 2004). The removal of silicate contaminants is the most crucial step of the entire procedure, and it has therefore been recommended that levels of iron, titanium or aluminium are also analysed to provide a basis for removing or correcting contaminated samples from downcore records (Barker et al. 2003; Rosenthal et al. 2004; Lea et al. 2005). Barker et al. (2003) use Fe/Mg ratios greater than 0.1 mol/mol as a rejection criterion where clays are believed to be the prime contaminant.

However, the variable composition of contaminant phases precludes universal 'threshold criteria' (e.g. Lea et al. 2005). Foraminiferal tests may be contaminated by Fe–Mn oxides, which hinder the analysis of trace metals such as cadmium, but should not be a significant problem for the analysis of magnesium (Barker et al. 2003). Laboratories analysing metals such as cadmium in addition to magnesium therefore include a reductive step in the cleaning method, whereas laboratories solely interested in magnesium or strontium generally do not include this time-consuming stage in the sample preparation. However, it has been shown that inclusion of this step tends to decrease foraminiferal Mg/Ca by 10–15% (Martin & Lea 2002; Barker et al. 2003; Rosenthal et al. 2004). There are two likely explanations for this decrease: (i) the metal oxides contain bound contaminants that do contain significant amounts of magnesium

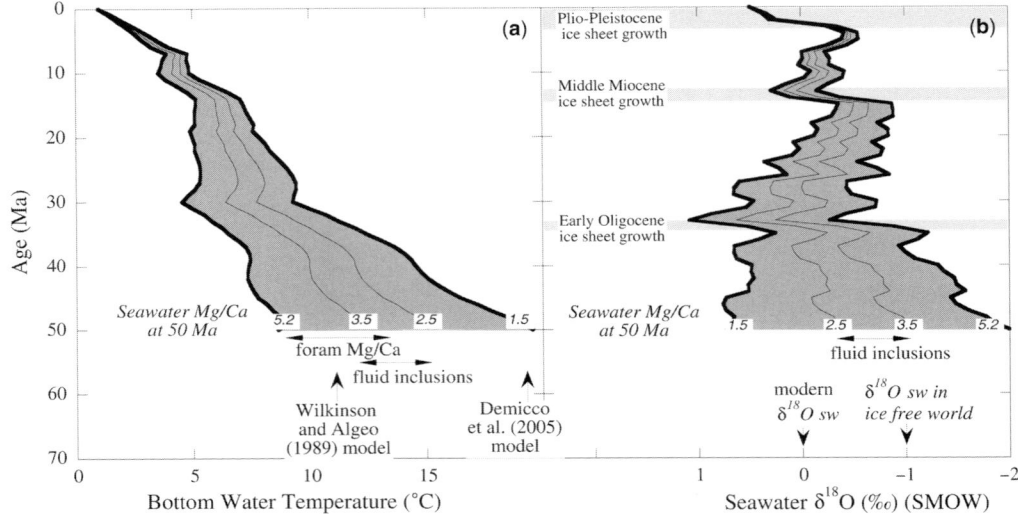

Fig. 2. (a) Cenozoic bottom-water temperatures calculated from a low-resolution Mg/Ca compilation (Lear *et al.* 2000), using the *O. umbonatus* calibration from Lear *et al.* (2002), and assuming different scenarios for seawater Mg/Ca. All scenarios assume a linear increase in seawater Mg/Ca since 50 Ma and converge on the modern-day value of 5.2 mol/mol. Sources of seawater Mg/Ca estimates are indicated by arrows. This figure does not take into account potential effects of seawater saturation state with respect to calcium carbonate on foraminiferal Mg/Ca. (b) Cenozoic seawater $\delta^{18}O$ calculated from temperatures in Figure 2A and a composite benthic foraminiferal $\delta^{18}O$ record (Miller *et al.* 1987). Seawater $\delta^{18}O$ corresponding to the modern ocean and an ice-free world (assumed to be the case at 50 Ma) are indicated by arrows.

(e.g. organic matter or clays), or (ii) the reductive step is so corrosive to foraminiferal calcite that the observed Mg/Ca decrease is a dissolution artefact (Barker *et al.* 2003). Supporting the latter mechanism is the observation that the reductive step did not significantly decrease Mg/Ca of planktonic foraminifera that had been pretreated with several acid leaches (Barker *et al.* 2003). This is an interesting observation because it implicitly suggests that the decrease in Mg/Ca with increasing dissolution eventually reaches some sort of limit, or plateau, hence potentially providing more 'reproducible' data.

A study of several European and American laboratories found that intralaboratory analytical precision was excellent (generally <0.5%) indicating that the majority of the palaeoceanographic laboratories in the study are producing precise records of downcore Mg/Ca-temperature variations (Rosenthal *et al.* 2004). However, the interlaboratory results were quite variable, with an RSD of *c.* 8%. About 2% of this 'scatter' was attributed to inaccurate standards at some of the laboratories. The remainder of the scatter was attributed to varying cleaning techniques (e.g. with or without reductive step), and instrumental matrix effects especially associated with Inductively Coupled Plasma (ICP) instruments (e.g. de Villiers *et al.*

2002; Lear *et al.* 2002). At first sight, this interlaboratory scatter raises some concerns over the accuracy of palaeotemperature reconstructions, although in reality, the biases imparted by using various cleaning protocols are effectively corrected for by using appropriate core top calibrations obtained using the same cleaning technique (Rosenthal *et al.* 2004).

Preservation of foraminiferal calcite

Several post-mortem diagenetic processes may impact the preservation of geochemical signals in foraminiferal calcite. Increased levels of impurities such as magnesium contained within a calcite crystal lattice increases the solubility of calcite. Therefore, as post-mortem dissolution progresses and the test mass decreases, so does the average Mg/Ca of the remaining test calcite (e.g. Savin & Douglas 1973; Brown & Elderfield 1996; Rosenthal *et al.* 2000). The relative constancy of a benthic foraminifer's habitat means that this selective dissolution process has greater potential to alter the bulk Mg/Ca of planktonic test calcite than benthic test calcite (Rosenthal *et al.* 2000). For Pleistocene cores, it has been shown that the extent of dissolution and thus Mg/Ca depletion may be estimated and corrected for using shell weights, although the assumption of constant initial shell weight through time may not be valid (Barker & Elderfield 2002; Rosenthal & Lohmann 2002). On longer timescales, the dissolution issue is typically circumvented by using samples from relatively shallow-water depths, assuming minimal dissolution above the lysocline (e.g. Tripati *et al.* 2003). However, the observed decrease in core-top planktonic foraminiferal Mg/Ca above the calcite saturation horizon points to the complexity of this issue, with the extent of dissolution itself a function of the Mg content in the foraminiferal calcite, and levels of pore water carbonate saturation state being dependent on sediment composition as well as the overlying seawater chemistry (Emerson & Bender 1981; Brown & Elderfield 1996).

Excess surface free energy of small intricately structured carbonate microfossils, and pressure-solution at grain-to-grain contacts during sediment burial leads to recrystallization of foraminiferal calcite, with a concomitant alteration in the bulk test geochemistry (e.g. Baker *et al.* 1982; Killingley 1983; Schrag *et al.* 1995; Norris & Wilson 1998; Pearson *et al.* 2001; Wilson & Norris 2001). Foraminiferal recrystallization processes are enhanced in carbonate-rich sediments; foraminifera collected from moderately deeply buried (>100 m) clay-rich sediments appear to retain their original microstructure and geochemistry (e.g. Norris & Wilson 1998; Pearson *et al.* 2001). Recrystallization processes have a greater effect on downcore records when there is a larger geochemical contrast between test calcite and the recrystallized calcite (Schrag *et al.* 1995). The effect of recrystallization on foraminiferal stable isotope records is fairly well understood (e.g. Schrag *et al.* 1995; Norris & Wilson 1998; Pearson *et al.* 2001; Wilson & Norris 2001), but its effect on foraminiferal trace metal contents is poorly constrained. Laboratory determinations of inorganic partition coefficients of Mg^{2+} into calcite suggest that diagenetic calcite should be characterized by significantly higher Mg/Ca ratios than pristine foraminiferal calcite (Katz 1973). This situation would make foraminiferal Mg/Ca more sensitive to recrystallization processes, with the flip side of the coin being reduced ambiguity in identifying the presence of diagenetic calcite.

However, Mg/Ca ratios of deeply buried bulk carbonate sediments do not trend towards the high values predicted using laboratory-derived partition coefficients (Baker *et al.* 1982), implying that foraminiferal Mg/Ca ratios are relatively robust to the effects of early foraminiferal recrystallization processes. A relatively small contrast between Mg/Ca of test calcite and recrystallized calcite is also supported by a recent study comparing bulk Mg/Ca of 'frosty' versus 'glassy' Eocene planktonic foraminifera (Sexton *et al.* in press). A significant influence to foraminiferal Mg/Ca records from such diagenetic calcite is assumed where an inverse correlation is observed between paired Mg/Ca and Sr/Ca ratios, as diagenetic process tends to alter these ratios in opposite directions (Baker *et al.* 1982).

Finally, even weakly dolomitized samples should be avoided for Mg/Ca palaeothermometry, as dolomite has a Mg/Ca ratio about 500 times that of typical foraminifera. An example where weak dolomitization has prevented palaeoceanographic interpretation of foraminiferal Mg/Ca records is the basal part of the sedimentary package at ODP Site 1218, where much of the Eocene sediments are close to the basalt basement (Lear *et al.* 2004).

Cenozoic deep-water temperatures

Cenozoic cooling and glaciation

Benthic foraminifera living in the deep oceans are insulated from large seasonal, latitudinal and geographical variations in temperature and salinity. For this reason, they are used in preference to planktonic foraminifera to depict global temperature trends and to calculate changes in average seawater salinity. Low-resolution benthic foraminiferal Mg/Ca records depict the general cooling trend throughout the Cenozoic (Lear *et al.* 2000; Billups & Schrag 2002; Billups & Schrag 2003). The most significant cooling phases appear to have occurred during the Eocene and Plio-Pleistocene (Fig. 2a). These Mg/Ca temperature estimates have been used in conjunction with the benthic oxygen isotope record to estimate the oxygen isotopic composition of seawater (Lear *et al.* 2000; Fig. 2b). The results suggest that there were three major permanent or semi-permanent increases in the volume of continental ice-sheets through the Cenozoic; in the earliest Oligocene, the middle Miocene and Plio-Pleistocene. As one might expect, the middle Miocene and Plio-Pleistocene climate transitions appear to be associated with an overall cooling of deep-sea and polar surface temperatures (Lear *et al.* 2000; Billups & Schrag 2002; Lear *et al.* 2003; Shevenell *et al.* 2004).

However, to date there is no evidence for an overall cooling associated with the establishment of the Antarctic ice-sheet in the early Oligocene ('Oi-1', following the terminology of Miller *et al.* (1991), 'cycle $84_{Eo-C13n}$' following the terminology of Wade & Pälike (2004)) (Lear *et al.* 2000; Billups & Schrag 2003; Lear *et al.* 2004) (Table 1). Three of the four sites display no

C. H. LEAR

Table 1. *Summary of benthic Mg/Ca records across the glaciation event in the earliest Oligocene (Oi-1)*

Site	Ocean basin	Approximate resolution (10^3 yr)	Approx palaeo-water depth (m)	Mg/Ca signal	Source*
DSDP 522	Atlantic	25	3000	No net change	(a)
ODP 689	Southern	150	1500	Transient peak	(b)
ODP 757	Indian	250	1500	No net change	(b)
ODP 1218	Pacific	30	3700	Increase	(c)

*Sources: (a) Lear *et al.* 2000; (b) Billups & Schrag 2003; (c) Lear *et al.* 2004.

overall change in Mg/Ca, while the deepest site displays a small increase in Mg/Ca across Oi-1 (Table 1). Taken at face value, this could indicate a different temperature history in the Pacific Ocean compared to the other oceans. However, it has been suggested that the earliest Oligocene increase in Mg/Ca observed at the deep Pacific site primarily reflects a change in seawater chemistry that occurred across Oi-1 (Lear *et al.* 2004). The increase in the benthic foraminiferal oxygen isotope record occurs in two *c.* 40,000-year steps, each of which is coincident with a dramatic increase in seawater carbonate saturation state, evidenced by increasing carbonate content of deep-sea sediments and a 1.2 km deepening of the calcite compensation depth (Coxall *et al.* 2005; Rea & Lyle 2005). As noted above, core-top benthic foraminifera from sites very close to saturation with respect to calcite have lower-than-expected Mg/Ca (Martin *et al.* 2002). Therefore, some portion of the overall increase in Mg/Ca across Oi-1 at ODP Site 1218 may be attributed to the stepwise increase in saturation levels. The three shallower sites do not display an overall increase in benthic foraminiferal Mg/Ca across Oi-1 (Table 1), suggesting that benthic foraminiferal Mg/Ca is less sensitive to changes in saturation state in more saturated waters. At present, the temperature history across Oi-1 remains poorly constrained, although the case for some component of cooling is strengthened by the observation that the oxygen isotope increase seen in the Pacific record is too large to be attributed to ice growth on Antarctica alone (Coxall *et al.* 2005).

The Cenozoic also witnessed transient increases in continental ice volume, most notably the 'Mi-1 event' in the early Miocene (Zachos *et al.* 2001*a*). The onset of the glaciation, as recorded by benthic foraminiferal $\delta^{18}O$, appears to lag a *c.* 2 °C cooling of bottom-water (and hence polar surface) temperatures. However, each pulse of glaciation is followed by a *c.* 2 °C warming. It has been suggested that the warming reflects a negative feedback process associated with the ice-sheet expansion, perhaps linked to chemical weathering rates (Lear *et al.* 2004).

Mg/Ca as a proxy for ocean circulation changes

Ocean circulation patterns affect global climate via the distribution of heat, moisture and CO_2. A commonly cited example is the evaporative cooling of North Atlantic Deep Water (NADW), which supplies heat and moisture to high northern latitudes, thereby affecting the stability of Northern Hemisphere ice-sheets (Broecker & Denton 1990). Conversely, climate affects ocean circulation patterns via the distribution of heat and moisture, an extreme example being the postulated presence of warm saline deep water in the greenhouse world of the early Cenozoic, resulting from the high rates of evaporation in relatively shallow seaways (e.g. Woodruff & Savin 1989). Changing palaeogeography has presumably also had a major effect on circulation patterns, through opening or closing of ocean gateways and other physical barriers (e.g. Kennett & Shackleton 1976; Woodruff & Savin 1989; Wright & Miller 1996). The relative importance of ocean circulation versus greenhouse gas concentrations on the long-term transition from the greenhouse world of the early Cenozoic to today's icehouse world remains unresolved.

To date, most reconstructions of past deep-water circulation rely on tracing a characteristic chemical signal of a given water mass, such as its carbon isotopic or neodymium isotopic ratio (Wright *et al.* 1992; Frank 2002). While such tracers may point to source regions and circulation paths, they are independent of the physical properties of the water mass. Combining Mg/Ca-palaeothermometry with traditional circulation tracers therefore adds a new dimension of information, as the key to understanding how deep waters formed and hence their effect on climate lies in reconstructing their relative densities, a function of temperature and salinity. In addition, there are periods within the Cenozoic when the chemical differences between different water masses are negligible, and so we must turn to proxies for their physical properties to reconstruct circulation patterns in these intervals.

The simplest way to reconstruct the history of a particular water mass is to compare parallel records of temperature, salinity and chemistry inside and outside the water mass. The long residence times of Mg and Ca in seawater mean that site-to-site contrasts in foraminiferal Mg/Ca will be independent of any secular changes in seawater Mg/Ca. Despite this advantage, at the time of writing there is only one published study that uses Mg/Ca palaeothermometry to help reconstruct Cenozoic circulation patterns (Lear *et al.* 2003). This study compared benthic foraminiferal Mg/Ca, $\delta^{13}C$ and $\delta^{18}O$ from a site within modern-day NADW to a deep-water site in the equatorial Pacific. Intervals of increased interbasin $\delta^{13}C$ gradient within the Miocene have previously been interpreted in terms of intensification of NADW (Wright *et al.* 1991). As one might expect, these intervals also coincide with increased temperature contrast between the two ocean basins. However, a more intriguing aspect of this study is that prior to *c.* 5 Ma, apparent intensification of NADW is associated with a cooling of the deep Atlantic. The simplest explanation is that NADW has not always been the warm, salty bottom water mass we are familiar with today. It appears to have taken on these 'modern' characteristic properties around the same time that the Panama Gateway closed, which perhaps reflects the increased residence time of surface waters at low latitudes prior to traveling northwards to act as a source for NADW (Lear *et al.* 2003).

Cenozoic sea-surface temperatures

Planktonic foraminiferal Mg/Ca palaeothermometry is not without its complications. The lifestyle of a planktonic foraminifer is complex, with some species living at different depths in the water column at different stages in their life cycle. Temperatures calculated from bulk Mg/Ca values therefore reflect an 'average' temperature of the habitat range. Just prior to reproduction, many species precipitate a layer of gametogenic calcite, which may be enriched in magnesium relative to the rest of the calcite test (Nürnberg *et al.* 1996). Another complication is selective post-mortem dissolution of test calcite as outlined above. Nevertheless, provided the caveats outlined here are borne in mind, planktonic foraminiferal Mg/Ca palaeothermometry has a great deal still to add to our understanding of Cenozoic climate. Planktonic foraminiferal Mg/Ca records may be used to generate spatial maps of sea-surface temperature, thus providing a means to test the validity of coupled ocean-atmosphere general circulation models. Planktonic foraminiferal Mg/Ca

records used to test model predictions regarding the Palaeocene–Eocene Thermal Maximum (PETM) are described below. An additional case study, where a planktonic foraminiferal Mg/Ca record has shed light on the triggering mechanism of the middle Miocene climate transition, is also discussed.

Planktonic Mg/Ca records of a hydrocarbon-fuelled global warming

The Palaeocene–Eocene Thermal Maximum (PETM) has been the subject of much research in recent years, and for good reason: it is widely believed to represent the closest analogue to anthropogenic CO_2 emissions in the fossil record. Coupled ocean-atmosphere models predict that the release of the greenhouse gases inferred from the PETM carbon isotope excursion would have caused warming at all latitudes, with the most extreme warming occurring at higher latitudes. Warming around Antarctica has been estimated to be on the order of 8–10 °C, which can be simulated by climate models that assume a pCO_2 increase of 1500 ppmv on top of a baseline of 560 ppmv (Kennett & Stott 1991; Shellito *et al.* 2003; Zachos *et al.* 2003). Attempts to test these climate models using the $\delta^{18}O$ of low-latitude planktonic foraminifera have proved inconclusive, in part a result of the competing effects of temperature and sea-surface salinity. However, a recent study of both the isotopic and the trace metal composition of planktonic foraminifera from Ocean Drilling Program Site 1209 in the Pacific Ocean has allowed the first robust estimates of tropical sea-surface temperature change across the PETM (Zachos *et al.* 2003). At this low-latitude site, the oxygen isotopic composition of planktonic foraminifera display a decrease equivalent to a warming of only 2–3 °C. The planktonic foraminifera Sr/Ca record does not change across the PETM at Site 1209, and this is neatly used as evidence that the concurrent Mg/Ca record is not biased by seawater chemistry changes or diagenetic alteration (Zachos *et al.* 2003).

How does the corresponding Mg/Ca temperature record compare with the $\delta^{18}O$ record and the model predictions? The planktonic foraminiferal Mg/Ca displays a substantial increase across the PETM at Site 1209, equivalent to a 4–5 °C warming. This indicates that at this low-latitude site, the surface $\delta^{18}O$ record reflects not only increasing temperature but also increasing salinity. Significantly, the 4–5 °C low-latitude warming inferred from the Mg/Ca record is in good agreement with that predicted by the models that use a

1500 ppmv increase in pCO_2 to produce the high-latitude warming of 8–10 °C. Although the nature and extent of the forcing used by the models to simulate the PETM may be inaccurate, the study by Zachos *et al.* (2003) demonstrates that coupled ocean-atmosphere models can successfully build up a general picture of the climatic response to a perturbation in atmospheric greenhouse gas concentrations.

Planktonic Mg/Ca records across the middle Miocene climate transition

The middle Miocene climate transition is marked by one of the three large 'steps' in the Cenozoic benthic foraminiferal oxygen isotope record (Shackleton & Kennett 1975; Miller *et al.* 1987; Zachos *et al.* 2001*b*). Benthic foraminiferal Mg/Ca palaeothermometry suggests that roughly two-thirds of the $\delta^{18}O$ increase can be attributed to an increase in continental ice volume, with the rest representing a 2–3 °C cooling of deep-sea temperatures (Lear *et al.* 2000; Billups & Schrag 2002). Changes in ocean circulation and the global carbon cycle, perhaps astronomically paced, have both been called upon as triggers for this fundamental shift in Earth's climate regime. A high-resolution planktonic foraminiferal Mg/Ca record from the Southern Ocean sheds some light on this interval of Earth's climate history. The record depicts a *c.* 7 °C cooling of high-latitude sea-surface temperatures, a picture which is not apparent from the planktonic $\delta^{18}O$ records (Shevenell *et al.* 2004). A large part of the difference between the paired planktonic $\delta^{18}O$ and Mg/Ca records is attributed to a freshening of Southern Ocean surface waters, consistent with intensification of the Antarctic Circumpolar Current. Rather than episodic drawdown of carbon dioxide as an ultimate trigger for the climate transition (Flower & Kennett 1993), it appears that orbitally paced circulation changes altered meridional heat/vapour transport, which ultimately led to the permanent establishment of the Antarctic ice-sheet (Shevenell *et al.* 2004).

Conclusions

Ongoing research is aimed at achieving a complete understanding of the Mg/Ca palaeothermometer. Despite current uncertainties, which include calibration issues, Cenozoic seawater Mg/Ca variations, and foraminiferal geochemical preservation, this relatively new proxy has improved our quantitative records of past climate change, and prompted re-evaluation of long-held paradigms in Cenozoic climate dynamics.

The author would like to thank Yair Rosenthal for many enjoyable and fruitful discussions, and the Leverhulme Trust for financial support. This manuscript benefited from the thoughtful comments of two anonymous reviewers.

References

ANAND, P., ELDERFIELD, H. & CONTE, M. H. 2003. Calibration of Mg/Ca thermometry in planktonic foraminifera from a sediment trap time series. *Paleoceanography*, **18**, 1050, doi:10.1029/2002PA000846.

BAKER, P. A., GIESKES, J. M. & ELDERFIELD, H. 1982. Diagenesis of carbonates in deep-sea sediments – evidence from Sr/Ca ratios and interstitial dissolved Sr^{2+} data. *Journal of Sedimentary Petrology*, **52**, 71–82.

BARKER, S. & ELDERFIELD, H. 2002. Foraminiferal calcification response to glacial-interglacial changes in atmospheric CO_2. *Science*, **297**, 833–836.

BARKER, S., GREAVES, M. & ELDERFIELD, H. 2003. A study of cleaning procedures for foraminiferal Mg/Ca palaeothermometry. *Geochemistry, Geophysics, Geosystems*, **4**, 8407, doi:10.1029/2003GC000559.

BILLUPS, K. & SCHRAG, D. P. 2002. Paleotemperatures and ice volume of the past 27 Myr revisited with paired Mg/Ca and $^{18}O/^{16}O$ measurements on benthic foraminifera. *Paleoceanography*, **17**, 3-1–3-11.

BILLUPS, K. & SCHRAG, D. P. 2003. Application of benthic foraminiferal Mg/Ca ratios to questions of Cenozoic climate change. *Earth and Planetary Science Letters*, **209**, 181–195.

BOYLE, E. A. & KEIGWIN, L. D. 1985/86. Comparison of Atlantic and Pacific paleochemical records for the last 215,000 years: changes in deep ocean circulation and chemical inventories. *Earth and Planetary Science Letters*, **76**, 135–150.

BROECKER, W. S. & DENTON, G. H. 1990. The role of ocean-atmosphere reorganization in glacial cycles. *Quaternary Science Reviews*, **9**, 305–341.

BROECKER, W. S. & PENG, T.-H. 1982. Tracers in the Sea: Palisades, Eldigio Press, New York.

BROWN, S. J. & ELDERFIELD, H. 1996. Variations in Mg/Ca and Sr/Ca ratios of planktonic foraminifera caused by postdepositional dissolution: Evidence of shallow Mg-dependent dissolution. *Paleoceanography*, **11**, 543–551.

BURTON, E. A. & WALTER, L. M. 1991. The effects of pCO_2 and temperature on magnesium incorporation in calcite in seawater and $MgCl_2$-$CaCl_2$ solutions. *Geochimica et Cosmochimica Acta*, **55**, 775–785.

COXALL, H. K., WILSON, P. A., PÄLIKE, H., LEAR, C. H. & BACKMAN, J. 2005. Rapid stepwise onset of Antarctic glaciation and deeper calcite compensation in the Pacific Ocean. *Nature*, **433**, 53–57.

DE LA ROCHA, C. L. & DEPAOLO, D. J. 2000. Isotopic Evidence for Variations in the Marine Calcium Cycle Over the Cenozoic. *Science*, **289**, 1176–1178.

DEMICCO, R. V., LOWENSTEIN, T. K., HARDIE, L. A. & SPENCER, R. J. 2005. Model of seawater composition for the Phanerozoic. *Geology*, **33**, 877–880.

DE VILLIERS, S., GREAVES, M. & ELDERFIELD, H. 2002. An intensity ratio calibration method for the accurate determination of Mg/Ca and Sr/Ca of marine carbonates by ICP-AES. *Geochemistry, Geophysics, Geosystems*, **3**, doi:10.1029/2001GC000169.

DICKSON, J. A. D. 2002. Fossil echinoderms as monitor of the Mg/Ca ratio of Phanerozoic Oceans. *Science*, **298**, 1222–1224.

DWYER, G. S., CRONIN, T. M., BAKER, P. A., RAYMO, M. E., BUZAS, J. S. & CORREGE, T. 1995. North Atlantic deepwater temperature change during late Pliocene and late Quaternary climatic cycles. *Science*, **270**, 1347–1351.

ELDERFIELD, H. & GANSSEN, G. 2000. Past temperature and $\delta^{18}O$ of surface ocean waters inferred from foraminiferal Mg/Ca ratios. *Nature*, **405**, 442–445.

EMERSON, S. & BENDER, M. 1981. Carbon fluxes at the sediment–water interface of the deep-sea: Calcium carbonate preservation. *Journal of Marine Research*, **39**, 139–162.

FLOWER, B. P. & KENNETT, J. P. 1993. Relations between Monterey Formation deposition and middle Miocene global cooling: Naples Beach section, California. *Geology*, **21**, 877–880.

FRANK, M. 2002. Radiogenic isotopes: Tracers of past ocean circulation and erosional input. *Reviews of Geophysics*, **40**, 1001, doi:10.1029/2000RG000094.

HARTLEY, G. & MUCCI, A. 1996. The influence of pCO$_2$ on the partitioning of magnesium in calcite overgrowths precipitated from artificial seawater at 25° and 1 atm total pressure. *Geochimica et Cosmochimica Acta*, **60**, 315–324.

HASTINGS, D. W., RUSSELL, A. D. & EMERSON, S. R. 1998. Foraminiferal magnesium in *Globigerinoides sacculifer* as a paleotemperature proxy. *Paleoceanography*, **13**, 161–169.

KATZ, A. 1973. The interaction of magnesium with calcite during crystal growth at 25–90 °C and one atmosphere. *Geochimica et Cosmochimica Acta*, **37**, 1563–1586.

KENNETT, J. P. & SHACKLETON, N. J. 1976. Oxygen isotopic evidence for the development of the psychrosphere 38 Myr ago. *Nature*, **260**, 513–515.

KENNETT, J. P. & SCOTT, L. D. 1991. Abrupt deep-sea warming, palaeoceanographic changes and benthic extinctions at the end of the Palaeocene. *Nature*, **353**, 225–229.

KILLINGLEY, J. S. 1983. Effects of diagenetic recrystallization on $^{18}O/^{16}O$ values of deep sea sediments. *Nature*, **301**, 594–597.

LEA, D. W., MASHIOTTA, T. A. & SPERO, H. 1999. Controls on magnesium and strontium uptake in planktonic foraminifera determined by live culturing. *Geochimica et Cosmochimica Acta*, **63**, 2369–2379.

LEA, D. W., PAK, D. K. & PARADIS, G. 2005. Influence of volcanic shards on foraminiferal Mg/Ca in a core from the Galápagos region. *Geochemistry, Geophysics, Geosystems*, **6**, Q11P04, doi:10.1029/2005GC000970.

LEAR, C. H., ELDERFIELD, H. & WILSON, P. A. 2000. Cenozoic deep-sea temperatures and global ice volumes from Mg/Ca in benthic foraminiferal calcite. *Science*, **287**, 269–272.

LEAR, C. H., ROSENTHAL, Y. & SLOWEY, N. 2002. Benthic foraminiferal Mg/Ca-paleothermometry: A revised core–top calibration. *Geochimica et Cosmochimica Acta*, **66**, 3375–3387.

LEAR, C. H., ROSENTHAL, Y. & WRIGHT, J. D. 2003. The closing of a seaway: Ocean water masses and global climate change. *Earth and Planetary Science Letters*, **210**, 425–436.

LEAR, C. H., ROSENTHAL, Y., COXALL, H. K. & WILSON, P. A. 2004. Late Eocene to early Miocene ice-sheet dynamics and the global carbon cycle. *Paleoceanography*, **19**, PA4015, 10.1029/2004PA001039.

LOWENSTEIN, T. K., TIMOFEEFF, M. N., BRENNAN, S. T., HARDIE, L. A. & DEMICCO, R. V. 2001. Oscillations in Phanerozoic seawater chemistry: Evidence from fluid inclusions. *Science*, **294**, 1086–1088.

MARTIN, P. A. & LEA, D. W. 2002. A simple evaluation of cleaning procedures on fossil benthic foraminiferal Mg/Ca. *Geochemistry, Geophysics, Geosystems*, **3**, 8401, doi:10.1029/2001GC000280.

MARTIN, P. A., LEA, D. W., ROSENTHAL, Y., SHACKLETON, N. J., SARNTHEIN, M. & PAPENFUSS, T. 2002. Quaternary deep sea temperature histories derived from benthic foraminiferal Mg/Ca. *Earth and Planetary Science Letters*, **198**, 193–209.

MILLER, K. G., FAIRBANKS, R. G. & MOUNTAIN, G. S. 1987. Tertiary oxygen isotope synthesis, sea level history, and continental margin erosion. *Paleoceanography*, **2**, 1–9.

MILLER, K. G., WRIGHT, J. D. & FAIRBANKS, R. G. 1991. Unlocking the icehouse: Oligocene–Miocene oxygen isotopes, eustasy, and margin erosion. *Journal of Geophysical Research*, **96**, 6829–6848.

NORRIS, R. D. & WILSON, P. A. 1998. Low-latitude sea-surface temperatures for the mid-Cretaceous and the evolution of planktic foraminifera. *Geology*, **26**, 823–826.

NÜRNBERG, D., BIJMA, J. & HEMLEBEN, C. 1996. Assessing the reliability of magnesium in foraminiferal calcite as a proxy for water mass temperature. *Geochimica et Cosmochimica Acta*, **60**, 803–814.

PEARSON, P. N., DITCHFIELD, P. W., SINGANO, J. ET AL. 2001. Warm tropical sea surface temperatures in the Late Cretaceous and Eocene epochs. *Nature*, **413**, 481–487.

REA, D. K. & LYLE, M. 2005. Paleogene calcite compensation depth in the eastern subtropical Pacific: Answers and questions. *Paleoceanography*, **20**, PA1012, doi:10.1029/2004PA001064.

ROSENTHAL, Y. & LOHMANN, G. P. 2002. Accurate estimation of sea surface temperatures using dissolution-corrected calibrations for Mg/Ca paleothermometry. *Paleoceanography*, **17**, 1044, doi:10.1029/2001PA000749.

ROSENTHAL, Y., BOYLE, E. A. & SLOWEY, N. 1997. Environmental controls on the incorporation of Mg, Sr, F and Cd into benthic foraminiferal shells from Little Bahama Bank: Prospects for thermocline paleoceanography. *Geochimica et Cosmochimica Acta*, **61**, 3633–3643.

ROSENTHAL, Y., LOHMANN, G. P., LOHMANN, K. C. & SHERRELL, R. M. 2000. Incorporation and preservation of Mg in *Globigerinoides sacculifer*: Implications for reconstructing the temperature and $^{18}O/^{16}O$ of seawater. *Paleoceanography*, **15**, 135–145.

ROSENTHAL, Y., PERRON-CASHMAN, S., LEAR, C. H.
ET AL. 2004. Interlaboratory comparison study of
Mg/Ca and Sr/Ca measurements in planktonic
foraminifera for paleoceanographic research. Geochemistry, Geophysics, Geosystems, 5, Q04D09,
doi:10.1029/2003GC000650.

SAVIN, S. M. & DOUGLAS, R. G. 1973. Stable isotope and
magnesium geochemistry of recent planktonic foraminifera from the South Pacific, Geological Society of
America Bulletin, 84, 2327–2342.

SCHRAG, D. P., DePAOLO, D. J. & RICHTER, F. M. 1995.
Reconstructing past sea surface temperatures: Correcting for diagenesis of bulk marine carbonate. Geochimica et Cosmochimica Acta, 59, 2265–2278.

SEXTON, P. F., WILSON, P. A. & PEARSON, P. N. 2006.
Microstructural and geochemical perspectives on
planktic foraminiferal preservation – 'glassy' versus
'frosty'. Geochemistry, Geophysics, Geosystems, 7,
Q12P19, doi:10.1029/2006GC001291.

SHACKLETON, N. J. 1974. Attainment of isotopic equilibrium between ocean water and the benthonic foraminifera genus Uvigerina: Isotopic changes in the ocean
during the last glacial. Colloques Internationaux du
C. N. R. S., 219.

SHACKLETON, N. J. & KENNETT, J. P. 1975. Paleotemperature history of the Cenozoic and the initiation of
Antarctic glaciation: Oxygen and carbon isotope
analyses in DSDP Sites 277, 279, and 281. Initial
Reports of the Deep Sea Drilling Project, 29, 743–755.

SHELLITO, C. J., SLOAN, L. C. & HUBER, M. 2003.
Climate model sensitivity to atmospheric CO_2 levels
in the Early-Middle Paleogene, Palaeogeography,
Palaeoclimatology, Palaeoecology, 193, 113–123.

SHEVENELL, A., KENNETT, J. P. & LEA, D. W. 2004.
Middle Miocene Southern Ocean cooling and Antarctic cryosphere expansion. Science, 305, 1766–1770.

TRIPATI, A. K., DELANEY, M. L., ZACHOS, J. C.,
ANDERSON, L. D., KELLY, D. C. & ELDERFIELD,
H. 2003. Tropical sea-surface temperature

reconstruction for the early Paleogene using Mg/Ca
ratios of planktonic foraminifera. Paleoceanography,
18, 1101, doi:10. 1029/2003PA000937.

WADE, B. S. & PÄLIKE, H. 2004. Oligocene climate
dynamics. Paleoceanography, 19, PA4019, doi:
10.1029/2004PA001042.

WILKINSON, B. H. & ALGEO, T. J. 1989. Sedimentary
carbonate record of calcium-magnesium cycling.
American Journal of Science, 289, 1158–1194.

WILSON, P. A. & NORRIS, R. D. 2001. Warm tropical
ocean surface and global anoxia during the mid-
Cretaceous period. Nature, 412, 425–429.

WOODRUFF, F. & SAVIN, S. M. 1989. Miocene
deepwater oceanography. Paleoceanography, 4,
87–140.

WRIGHT, J. D. & MILLER, K. G. 1996. Control of
North Atlantic deep water circulation by the
Greenland–Scotland Ridge. Paleoceanography, 11,
157–170.

WRIGHT, J. D., MILLER, K. G. & FAIRBANKS, R. G.
1991. Evolution of modern deepwater circulation:
evidence from the late Miocene Southern Ocean.
Paleoceanography, 6, 275–290.

WRIGHT, J. D., MILLER, K. G. & FAIRBANKS, R. G.
1992. Early and Middle Miocene stable isotopes:
Implications for deep-water circulation and climate.
Paleoceanography, 7, 357–389.

ZACHOS, J. C., SHACKLETON, N. J., REVENAUGH, J. S.,
PALIKE, H. & FLOWER, B. P. 2001a. Climate
response to orbital forcing across the Oligocene–
Miocene boundary. Science, 292, 274–278.

ZACHOS, J. C., PAGANI, M., SLOAN, L., THOMAS, E. &
BILLUPS, K. 2001b. Trends, rhythms, and aberrations
in global climate 65 Ma to Present. Science, 292,
686–693.

ZACHOS, J. C., WARA, M. W., BOHATY, S. ET AL. 2003.
A transient rise in tropical sea surface temperature
during the Paleocene–Eocene Thermal Maximum.
Science, 302, 1551–1554.

The Palaeocene–Eocene Thermal Maximum super greenhouse: biotic and geochemical signatures, age models and mechanisms of global change

A. SLUIJS[1], G. J. BOWEN[2], H. BRINKHUIS[1], L. J. LOURENS[3] & E. THOMAS[4]

[1]*Palaeoecology, Institute of Environmental Biology, Utrecht University, Laboratory of Palaeobotany and Palynology, Budapestlaan 4, 3584 CD Utrecht, The Netherlands (e-mail: A.Sluijs@uu.nl)*

[2]*Earth and Atmospheric Sciences, Purdue University, 550 Stadium Mall Drive, West Lafayette, IN 47907, USA*

[3]*Faculty of Geosciences, Department of Earth Sciences, Utrecht University, Budapestlaan 4, 3584 CD Utrecht, The Netherlands*

[4]*Center for the Study of Global Change, Department of Geology and Geophysics, Yale University, New Haven CT 06520-8109, USA; also at Department of Earth & Environmental Sciences, Wesleyan University, Middletown, CT, USA*

Abstract: The Palaeocene–Eocene Thermal Maximum (PETM), a geologically brief episode of global warming associated with the Palaeocene–Eocene boundary, has been studied extensively since its discovery in 1991. The PETM is characterized by a globally quasi-uniform 5–8 °C warming and large changes in ocean chemistry and biotic response. The warming is associated with a negative carbon isotope excursion (CIE), reflecting geologically rapid input of large amounts of isotopically light CO_2 and/or CH_4 into the exogenic (ocean–atmosphere) carbon pool. The biotic response on land and in the oceans was heterogeneous in nature and severity, including radiations, extinctions and migrations. Recently, several events that appear similar to the PETM in nature, but of smaller magnitude, were identified to have occurred in the late Palaeocene through early Eocene, with their timing possibly modulated by orbital forcing. Although debate continues on the carbon source, the mechanisms that caused the input, the mechanisms of carbon sequestration, and the duration and pacing of the event, the research carried out over the last 15 years has provided new constraints and spawned new research directions that will lead to improved understanding of PETM carbon cycle and climate change.

A distinct period of extreme global warmth was initiated close to the boundary between the Palaeocene and Eocene epochs, approximately 55.5 Ma ago (Gradstein *et al.* 2004). This event, termed the Palaeocene–Eocene Thermal Maximum (PETM), occurred during a time of generally warm, 'greenhouse' climate conditions, but stands out against the background warmth as an abrupt and short-lived spike in global temperatures. Evidence for global warming is preserved by the $TEX_{86}{'}$ temperature proxy (Sluijs *et al.* 2006; Zachos *et al.* 2006), oxygen isotope ($\delta^{18}O$) excursions in marine foraminiferal calcite (Fig. 1) (Kennett & Stott 1991; Thomas *et al.* 2002) and terrestrial carbonates (Koch *et al.* 1995), increased Mg/Ca values in planktonic and benthic foraminifera (Zachos *et al.* 2003; Tripati & Elderfield 2005), poleward migrations of (sub)tropical marine plankton (Kelly *et al.* 1996; Crouch *et al.* 2001) and terrestrial

plant species (Wing *et al.* 2005), and mammal migrations across high northern latitudes (Bowen *et al.* 2002, 2006; Smith *et al.* 2006). Associated with the warming is a negative 2.5–6‰ carbon isotope ($\delta^{13}C$) excursion (CIE) (Kennett & Stott 1991; Koch *et al.* 1992; Thomas *et al.* 2002; Pagani *et al.* 2006), generally accepted to reflect the geologically rapid injection of ^{13}C-depleted carbon, in the form of CO_2 and/or CH_4, into the global exogenic carbon pool (Fig. 1).

The apparent conjunction between carbon input and warming has fuelled the hypothesis that increased greenhouse gas concentrations resulted in greenhouse warming during the PETM. The total amount of carbon input during the PETM, which is known to within an order of magnitude (Dickens *et al.* 1997; Zachos *et al.* 2005; Pagani *et al.* 2006), was about 4–8 times the anthropogenic carbon release from the start of the industrial era up

From: WILLIAMS, M., HAYWOOD, A. M., GREGORY, F. J. & SCHMIDT, D. N. (eds) *Deep-Time Perspectives on Climate Change: Marrying the Signal from Computer Models and Biological Proxies.* The Micropalaeontological Society, Special Publications. The Geological Society, London, 323–349.
1747-602X/07/$15.00 © The Micropalaeontological Society 2007.

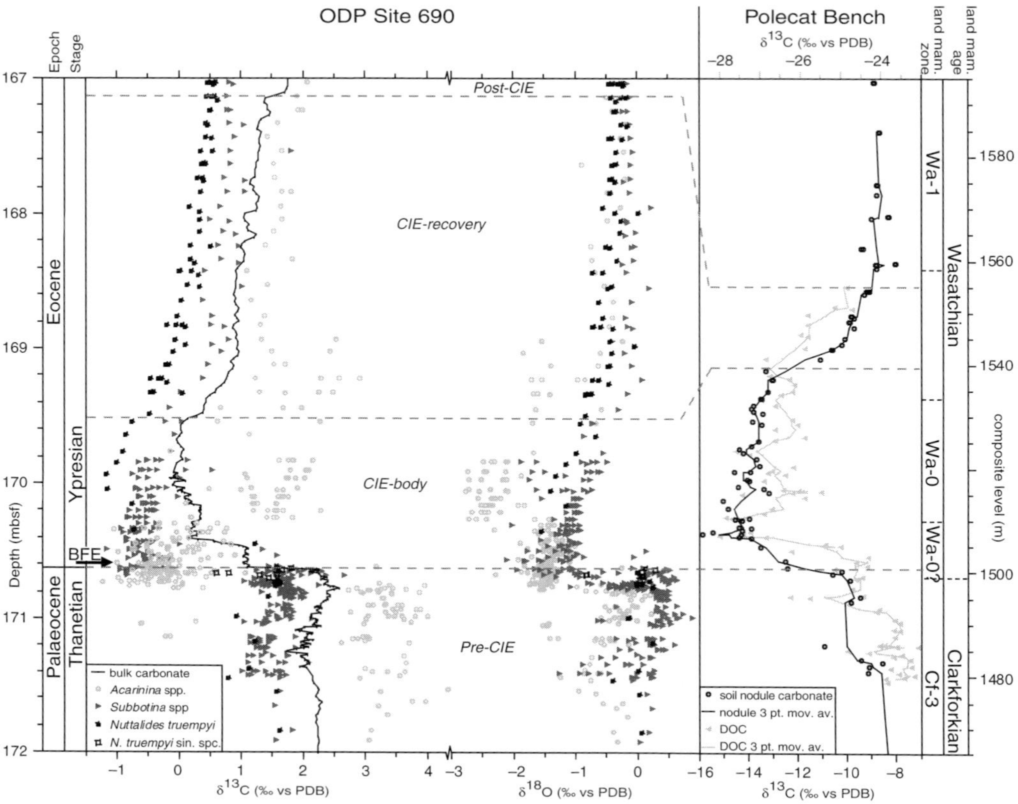

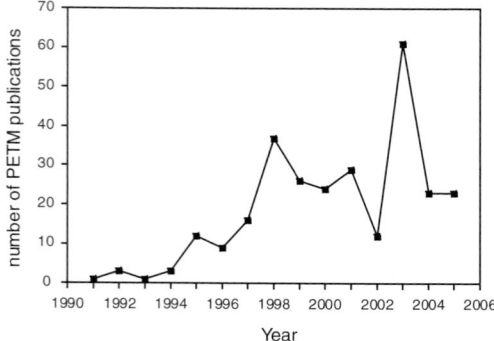

Fig. 1. Compilation of δ^{13}C and δ^{18}O values of planktonic foraminifera (surface-dweller *Acarinina* and thermocline-dweller *Subbotina* spp.; mostly single specimen), benthic foraminifera (*Nuttallides truempyi*) and bulk carbonate from ODP Site 690 in the Weddell Sea (data from (Kennett & Stott 1991; Bains *et al.* 1999; Thomas *et al.* 2002; Kelly *et al.* 2005), soil carbonate nodule (Bowen *et al.* 2001) and dispersed organic carbon (DOC) (Magioncalda *et al.* 2004) δ^{13}C records from the Polecat Bench section in the Bighorn Basin, Wyoming, USA. BFE refers to the main phase of benthic foraminifer extinction according to Thomas (2003). Mbsf, metres below sea-floor.

to today (Marland *et al.* 2005), and comparable to that expected from gross anthropogenic emissions through the end of the 21st century (Intergovernmental Panel on Climate Change 2001). In association with carbon cycle and climatic change, the PETM also stands out as a time of major biotic restructuring. Given the probable ties between releases of near-modern levels of carbon-based greenhouse gases and PETM climatic and biotic change, the PETM has developed as a provocative geological case study in global change, and many of the event's characteristics and mechanisms are under intensive study. A large volume of research on the PETM has appeared over the past decade (Fig. 2), and in this paper we aim to review and synthesize this material, including the duration and magnitude of carbon cycle perturbation, magnitude of warming, changes in ocean chemistry and marine and terrestrial biotic response.

Fig. 2. Number of studies focused published per year on the PETM since the first publication on its CIE and warming in 1991. Numbers are based on a Web-of-Science search using the keywords Palaeocene, Palaeocene (both American and English spellings), Eocene, PETM, IETM and LPTM.

The age of the PETM

Initially, the PETM was placed within the latest Palaeocene (Berggren *et al.* 1995) and as such termed the Late Palaeocene Thermal Maximum (LPTM) (Zachos *et al.* 1993; see papers in Knox *et al.* 1996 and in Aubry *et al.* 1998). In 2000, the Palaeocene–Eocene (P/E) boundary global stratotype section and point (GSSP) was formally defined at the base of the clay layer coinciding with the steepest slope of the negative CIE (Aubry & Ouda 2003; Gradstein *et al.* 2004) in the Gabal Dababiya section (Egypt). As a consequence, the base of the CIE has been used as global reference to pinpoint the P/E boundary. In a few publications, the term Initial Eocene Thermal Maximum (IETM) has been used because the maximum absolute temperatures occurred after the Palaeocene–Eocene boundary (see Wing *et al.* 2003; Zachos *et al.* 2003; Bowen *et al.* 2004). We will use the term PETM throughout this paper.

In the marine realm, the PETM is approximately coeval with the short planktonic foraminiferal zone E1, of which the base coincides with the Palaeocene–Eocene boundary (Pearson & Berggren 2005). The PETM is located within calcareous nannoplankton zone NP9 (Martini 1971) and CP8 (Okada & Bukry 1980), and its base occurs close to the benthic foraminiferal extinction (BFE) event (Fig. 1) (Thomas & Shackleton 1996). In the North Sea, the CIE and its recovery cover the dinoflagellate cyst (dinocyst) zone *Apectodinium augustum* (Powell *et al.* 1996; Sluijs 2006) and coincide with the interval without foraminifera (planktonic and benthic, including agglutinated forms), with large pyritized diatoms (see e.g. Gradstein *et al.* 1994; van Eetvelde & Dupuis 2004). In the terrestrial realm, the base of the CIE coincides with the Clarkforkian–Wasatchian North American Land Mammal Age (NALMA) zone boundary, and is correlative or nearly so (within 10^1 to 10^5 years) with the Gashatan–Bumbanian land mammal age boundary in Asia and the Cernaysian–Neustrian European land mammal age boundary (Hooker 1998). The CIE occurs within radiolaria zone RP7 in the low latitudes, but marks the RP6-7 boundary in the South Pacific (Gradstein *et al.* 2004). Considering the uncertainties in radiometric dating and orbital tuning, the PETM occurred between 55.8 and 55.0 Ma ago (Berggren *et al.* 1992; Norris & Röhl 1999; Röhl *et al.* 2003; Gradstein *et al.* 2004; Lourens *et al.* 2005; Westerhold *et al.* 2007).

The carbon isotope excursion

The negative CIE has been shown to be a distinctive, globally identifiable geochemical marker for the PETM in marine and continental sedimentary rocks (Fig. 1) (e.g. Kennett & Stott 1991; Koch *et al.* 1992; Stott *et al.* 1996). The CIE is considered to reflect the injection and subsequent removal of huge amounts of ^{13}C-depleted carbon into the ocean–atmosphere system (Dickens *et al.* 1995; Dickens 2001*a*).

Shape

Details of the rate and timing of $\delta^{13}C$ change in the oceans, atmosphere and continental reservoirs, which we will refer to as characterizing the 'shape' of the CIE, are needed for $\delta^{13}C$ curves to serve as a tool for correlation of PETM sections and to provide information on the dynamics of carbon release and sequestration during the event (Dickens *et al.* 1997; Dickens 2001*a*). Defining the detailed shape of the CIE is complicated, however, in that the temporal evolution of $\delta^{13}C$ values through the PETM may be different for various substrates and environments and may also be obscured by short-term changes in sedimentation rates that are difficult to recognize (Bowen *et al.* 2006) (see discussion below).

Following attempts to account for accumulation rate changes in deep marine sections affected by carbonate dissolution, a fairly consistent picture of the general shape of the CIE has emerged from terrestrial and marine environments (Bowen *et al.* 2006) (Fig. 1). The initiation of the CIE is marked by an abrupt, negative shift in $\delta^{13}C$ values, occurring within several hundreds or thousands of years. This is followed by a phase of relatively stable, low values, which has been termed the 'body' of the CIE, and a subsequent recovery to higher $\delta^{13}C$ values that follows an exponential trend. The general shape of the CIE has been interpreted to reflect a potentially pulsed, geologically rapid input of ^{13}C-depleted carbon into the system, followed by gradual sequestration of the excess carbon through burial in rocks (e.g. Dickens *et al.* 1997; Zachos *et al.* 2005). The existence of a $\delta^{13}C$ plateau during the body of the CIE may reflect a lag between carbon release and the onset of net sequestration (Bowen *et al.* 2006), but the exact duration of this lag and mechanisms causing it remain unclear.

For both marine and continental records, several uncertainties affect estimation of the pace and duration of carbon release, which is relevant to constraining the source of carbon and modelling the effect of carbon input on ocean carbonate chemistry (Dickens *et al.* 1997; Schmidt & Schindell 2003; Cramer & Kent 2005). The $\delta^{13}C$ records based on different foraminifer species and/or size fractions of the intensively studied Ocean Drilling Program (ODP) Site 690 (Maud Rise, Weddell Sea) indicate several cm of stratigraphic offset for the onset of the CIE (Fig. 1; see also Thomas 2003). In the $\delta^{13}C$ records from continental sections in the Bighorn Basin (Wyoming, USA), the CIE onset as marked by soil nodule carbonate and dispersed organic carbon (DOC) is offset by several metres. Several hypotheses have been proposed to explain stratigraphic offsets in marine carbonate records, including lags in the propagation of injected carbon from the atmosphere to the surface and deep oceans, local changes in productivity, and post-depositional mechanisms (such as diagenesis and differential bioturbation; (e.g. Thomas *et al.* 2002; Thomas 2003; Stoll 2005). Lags in the Bighorn Basin organic carbon $\delta^{13}C$ records have been hypothesized to reflect enhanced downward diffusion of atmospheric CO_2 with subsurface diagenesis in an environment of good drainage and elevated temperature and pCO_2 (Magioncalda *et al.* 2004) or the detrital origin of DOC in these sections (Bowen, unpublished).

Substrate-specific differences also exist in the shape of the body of the CIE. In the Bighorn Basin soil nodule $\delta^{13}C$ record and the *Subbotina* and *Acarinina* records of ODP Site 690, the body is characterized by relatively stable $\delta^{13}C$ values with a gentle trend towards higher values. Minimum $\delta^{13}C$ values in these records are thus located at the base of the body of the CIE. However, minimum bulk $\delta^{13}C$ values occur *c.* 50 cm higher.

One aspect of the CIE shape that has been invoked as evidence of PETM carbon cycle dynamics is a multi-stepped drop in $\delta^{13}C$ values during the onset of the CIE in several relatively complete bulk carbonate records (Bains *et al.* 1999; Zachos *et al.* 2005) (Fig. 1), which has been interpreted as evidence for multiple injections of carbon (Bains *et al.* 1999, 2003). The soil nodule record from the Bighorn Basin also appears to record a short plateau during the onset of the CIE, perhaps representing a few thousand years of relative $\delta^{13}C$ stasis (Fig. 1) (Bowen *et al.* 2001; Bowen *et al.* 2006). While the plateau also appeared in the 3–5 μm size fraction – which is dominated by the calcareous nannofossil species *Toweius* (Stoll 2005) – at Site 690, it has not been reproduced in single foraminifer $\delta^{13}C$ records (Fig. 1) (e.g. Thomas *et al.* 2002), which consistently lack intermediate values. Resolving the pattern – whether pulsed or unique – of carbon release at the onset of the PETM remains a central challenge to understanding carbon cycle perturbation during the PETM.

Above differences must reflect factors other then the $\delta^{13}C$ evolution of the exogenic carbon pool, such as dissolution, local productivity and vital effects, and complicate attempts to correlate

PETM sections based on the details of carbon isotope curves (e.g. Bains *et al.* 1999). As additional data is gathered, these discrepancies may reveal more detailed information on PETM carbon cycle dynamics within the study systems.

During the recovery phase, $\delta^{13}C$ values rise from values characteristic of the CIE body, but remain lower than pre-CIE values. This shift from pre-PETM to post-PETM values is in pace with a longer-term decline in exogenic $\delta^{13}C$ values during the Palaeocene–early Eocene (Zachos *et al.* 2001), suggesting a remarkable decoupling of the mechanisms underlying the transient carbon cycle changes of the PETM and the longer-term evolution of the early Palaeogene carbon cycle. The stratigraphic thickness of the recovery phase relative to the thickness of the body of the CIE is larger in at least some of the marine records than in the terrestrial realm. This can be explained, at least in part, by strong variations in deep marine sedimentation rates due to the fluctuations in the depths of the lysocline and CCD (Zachos *et al.* 2005), and possible changes in production rate of calcite (Kelly *et al.* 2005). As improved high-resolution age models and estimates of accumulation rate changes are developed, a consensus should emerge regarding the pace of carbon input and sequestration during the recovery phase and more accurate correlation based on the shape of the CIE should be possible.

Magnitude

The maximum magnitude of the CIE varies greatly between various records, indicating that the manifestation of the CIE at any single site and substrate reflects both changes in the global exogenic $\delta^{13}C$ budget and local or substrate-specific effects. Separation of local and global effects is therefore necessary in order to understand the 'true' global magnitude of the CIE, and could additionally allow $\delta^{13}C$ records to be interpreted in terms of local environmental or biotic change. Because the CIE magnitude remains a primary form of evidence for the size of the PETM carbon release, and a 1‰ difference in the global CIE magnitude could alter release estimates by 1000 Gt or more, this issue is critical to accurate understanding of PETM carbon cycle forcing.

Planktonic foraminifera show a 2.5–>4‰ excursion with a high level of variability among sites and taxa; e.g. at Site 690 up to 4‰ in the mixed-layer dweller *Acarinina*, 2–3‰ in the mixed-layer dweller *Morozovella* and 2‰ in thermocline dweller *Subbotina* (e.g. Thomas & Shackleton 1996; Thomas *et al.* 2002; Zachos *et al.* 2003; Tripati & Elderfield 2004; Cramer & Kent 2005). In contrast, only a *c.* 2‰ CIE is recorded in isolated calcareous nannofossils at Site 690 (Stoll 2005). A large $\delta^{13}C$ offset between planktonic data has also been reported from the Kerguelen Plateau (Southern Indian Ocean) (Lu & Keller 1993), whereas a lesser offset was noted in the Bay of Biscay (Pardo *et al.* 1997) although this may be associated with the relatively low resolution of this study. The average magnitude of the CIE based on benthic foraminifera is *c.* 2.5‰ (e.g. Kennett & Stott 1991; Thomas & Shackleton 1996; Zachos *et al.* 2001; Nunes & Norris 2006; Fig. 1), although considerable variation is observed between different benthic foraminiferal species and locations. The CIE in soil carbonate nodules is 5–6‰ (Koch *et al.* 1992; Bowen *et al.* 2001, 2002, 2004) and it is 4–5‰ in terrestrial higher plant n-alkanes (Pagani *et al.* 2006). The magnitude recorded in total organic carbon $\delta^{13}C$ records generated in terrestrial (Magioncalda *et al.* 2004) and marine locations (e.g. Dupuis *et al.* 2003; Steurbaut *et al.* 2003; Sluijs 2006) is *c.* 5‰.

Bowen *et al.* (2004) analysed a number of local effects that might account for differences in the magnitude of the CIE in marine and mid-latitude terrestrial records. Changes in marine fractionation processes, such as the tendency for foraminiferal calcite to become ^{13}C-enriched with lower pH and $[CO_3^{2-}]$ (Spero *et al.* 1997), were argued to be relatively minor, explaining up to *c.* 0.5‰ damping of the CIE in foraminiferal records (Bowen *et al.* 2004). The study attributed the larger terrestrial excursion primarily to increased carbon isotope fractionation by plants and changes in soil carbon cycling rates due to increased relative humidity and soil moisture during the body of the PETM. The CIE magnitude in biomarkers (C29 n-alkanes) records from high northern latitudes are consistent with the Bowen *et al.* (2004) analysis, but it has been argued that water-cycle induced changes in plant and soil carbon balance would be minimal given the relatively wet, cool climate of the early Palaeogene Arctic (Pagani *et al.* 2006). In addition, new evidence from Bighorn Basin fossil floras may indicate drier, rather than wetter, mid-latitude PETM climate (Wing *et al.* 2005). The implications of magnitude differences among various CIE records and the magnitude of the 'global' excursion therefore remain unclear. Among the terrestrial records, both the biomarker and palaeosol records may be affected by changes in the local composition of vegetation during the PETM. On the other hand, changes in growth rate and cell size and geometry of nannoplankton which are likely to have occurred with the environmental change recorded at the PETM, might have affected the CIE magnitude in some bulk carbonate and nannofossil-specific isotope records (e.g. Popp *et al.* 1998; Bralower *et al.* 2004).

Nunes & Norris (2006) suggested that variations in CIE magnitude among benthic $\delta^{13}C$ records from multiple sites reflect changes in deep oceanic circulation. However, the benthic isotope records comprise mostly multi-specimen records which could be influenced by bioturbation (evidenced by single specimen planktonic foraminifer isotope data within the PETM of for example Site 690, Fig. 1; Thomas *et al.* 2002) causing mixing between pre-CIE and CIE values. Moreover, some sites have suffered severe dissolution, while at other sites only very small benthic foraminifera are present within the body of the CIE, implying that these records are incomplete (see Thomas *et al.* 2000 for a record from Site 690 where the species *Nuttallides truempyi* is absent during the peak PETM). In addition, deep-water $\delta^{13}C$ gradients were very small during the early Eocene warm period, making reconstructions of deep-water circulation difficult (Sexton *et al.* 2006).

Duration and age models

Age models for the PETM and/or CIE generally agree that the whole event took 100 000 to 250 000 years, although none of the various approaches revealed consistent estimates for the total duration as well as the duration of its different phases (Fig. 1). These differences may be partly explained by uncertainties in the detailed 'shape' of the CIE itself, as discussed above. Here, we will discuss and update previously published age models.

An estimate for the duration of the PETM in the terrestrial realm was derived from the Polecat Bench section in the Bighorn Basin, based on the relative stratigraphic thickness of the CIE compared to that of Chron C24r, and the 2.557 million-year estimate for the duration of this chron by Cande & Kent (1995). Sedimentation rates in the Bighorn Basin depend largely on the accommodation space resulting from the apparently constant subsidence (Wing *et al.* 2000). Bowen *et al.* (2001) found a *c.* 40 m stratigraphic thickness for the body of the CIE and *c.* 15 m for the recovery based on the soil nodule $\delta^{13}C$ record (Fig. 1), implying a *c.* 55 m thickness for the total CIE. Applying the average sedimentation rates of 47.5 cm per thousand years for Chron C24r (Gingerich 2000), based on the magnetostratigraphic results of Butler *et al.* (1981), results in a *c.* 84 000-year duration for the body of the CIE (Bowen *et al.* 2001) (Fig. 1). More recently, Koch *et al.* (2003) updated the magnetostratigraphy, and showed that *c.* 1030 m of sediment accumulated during C24r, which would result in average sedimentation rates of 40.2 cm kyr^{-1}, implying a *c.* 71 kyr duration for the body of the CIE.

Recent studies on astronomically derived cycles in lowermost Palaeogene successions from ODP Leg 208 on the Walvis Ridge (Lourens *et al.* 2005) have confirmed interpretations from previous studies (e.g. Röhl *et al.* 2003) that much more time is represented in the interval between the CIE and the C24r/24n reversal than estimated by Cande & Kent (1995), and that the duration of Chron 24r was in the order of 3.118 million years (Westerhold *et al.* 2007). This implies that average sedimentation rates in the Bighorn Basin during C24r were approximately 33.0 cm kyr^{-1} (i.e. 1030 m/ 3.118 million years), significantly lower than previous estimates, resulting in an estimate of the duration of the body of the CIE at Polecat Bench of *c.* 120,000 years (i.e. 40 m/33.0 cm kyr^{-1}) and at least *c.* 170 kyr (i.e. 55 m/33.0 cm kyr^{-1}) for the entire PETM.

Two commonly used age models were derived from the relatively expanded marine PETM section at ODP Site 690. Both age models agree across the main body of the CIE, but differ significantly for the duration of the recovery phase. The first of these (Röhl *et al.* 2000) was based on core-scan X-Ray Fluorescence (XRF) Fe and Ca measurements through the CIE, and recognition of precession cycles (duration *c.* 20 kyr) in these records. The authors counted four cycles within the body of the CIE, and another seven within the recovery phase, recognized by the location of inflection points of the bulk carbonate δ^{13}C record, thus arriving at an estimate for the entire CIE of 210 to 220 kyr. The second of these models is based on the extraterrestrial He (^{3}He$_{ET}$) concentrations of the Site 690 PETM sediments and the assumption that the flux of this isotope to the Earth remained constant during the PETM (Farley & Eltgroth 2003). The background absolute flux of ^{3}He$_{ET}$ to the sea-floor during the PETM was estimated from the ^{3}He$_{ET}$ concentration in 13 samples taken from C24r and C25n above and below the PETM and average sedimentation rates obtained from Aubry *et al.* (1996). These average sedimentation rates were based on a relatively low resolution magnetostratigraphy (Spiess 1990) – which has since been revised by Ali *et al.* (2000) – and durations for C24r and C25n of 2.557 and 1.650 Ma (Cande & Kent 1995), respectively. Accordingly, Farley & Eltgorth (2003) arrived at an estimate of *c.* 90 kyr for the duration of the body of the CIE and *c.* 120 kyr for the entire PETM.

The ^{3}He$_{ET}$–based age model is subject to several important uncertainties. First, the depth level of the reversal between C24r and C24n has not been clearly identified at Site 690, being adjacent to an unconformity. The carbon isotope excursion associated with Eocene Thermal Maximum 2 (ETM2), which is *c.* 150 kyr older

than this reversal (Cramer *et al.* 2003; Lourens *et al.* 2005; Westerhold *et al.* 2007), is present at Site 690 (Cramer *et al.* 2003; Lourens *et al.* 2005) and could thus be used as an alternate calibration point. Secondly, Chron C24r was likely 561 kyr longer than assumed (Westerhold *et al.* 2007). Thirdly, ODP Site 690 recovered the upper Palaeocene and lower Eocene in a single hole only, but we know from multiple-hole drilling that sediment cores expand when they are released from the overlying load, so that part of the core is lost. The average expansion factor for Sites 1262–1267 at Walvis Ridge, for instance, varied between 110 and 118% (Zachos *et al.* 2004), which implies recovery gaps of 1 to 1.8 m between each core, similar to the value of 1.2–1.8 m found by Florindo & Roberts (2005) for the double-cored, younger part of Site 690. This core loss due to expansion has not been accounted for in Site 690 studies, but the whole CIE is within one core at Site 690, so that cycle counts limited to this interval are not affected (Röhl *et al.* 2000). Fourthly, the 13 samples used to calculate the background ^{3}He$_{ET}$ flux may be affected by temporal (possibly orbitally forced) variations in sedimentation rates, probably as covered by the 'minimum' and 'maximum' estimates based on the standard errors in the background flux values, which vary between 0.38 and 0.97 pcc cm^{-2} kyr^{-1} (1 pcc = 10^{-12} cm^3 of He at STP). Fifthly, ^{3}He$_{ET}$ fluxes may not have been constant during C24r and C25n (Mukhopadhyay *et al.* 2001; Kent *et al.* 2003), and are known to vary by an order of magnitude over millions of years (Farley 2001). Background flux values appear significantly higher during C24r than during C25 (Table 1c).

In order to re-evaluate the He-based age model and test its sensitivity to the above uncertainties, we calculated several age models for the PETM using different assumptions and corrections for the above issues (Table 1, Fig. 3). We adopt durations of C25n (504 kyr) and the interval between the onset of C24r and the ETM2 (2940 kyr) from Westerhold *et al.* (2007) and apply different sediment expansion factors. Because background ^{3}He$_{ET}$ content/g sediment is quite sensitive to small changes in these assumptions is large, the resulting estimates for the total duration of the PETM vary between 90 ± 10 and 140 ± 30 kyr for the most likely scenarios (Fig. 3). Regardless of the assumptions, however, the helium model requires a very large increase in sedimentation rate towards the end of the PETM, thus giving a rapid recovery period relative to the body of the PETM (Fig. 3) (Farley & Eltgroth 2003).

A rapid recovery may be supported by the theory that calcite production in the photic zone increased during the recovery phase (Kelly *et al.* 2005) or

Table 1. Calculation of sedimentation rates and $^3He_{ET}$ fluxes (in $pcc\ cm^{-2}\ kyr^{-1}$) through the upper Palaeocene–lower Eocene section of ODP Hole 690B. Average values and standard deviations for the background $^3He_{ET}$ content/g sediment (from Farley & Eltgroth 2003) in Table 1c are used to calculate absolute background $^3He_{ET}$ fluxes through the following intervals: onset – termination C25n, onset C24r – ETM2 and onset C25n – ETM2. For these calculations, we adopt the durations of these intervals from (Westerhold et al. in press) and exclude (Table 1a) and include (Table 1b) 11% core loss due to sediment expansion (see text)

(a)

Background ET3He flux and sedimentation rate model for Hole 690B. Expansion not included

model (Fig. 5)	interval	duration (kyr)	thickness (cm)	sed. rates (cm/kyr)		
I	C25n	504	937	1.86		cm/kyr
				dens/(kyr/cm) =	2.48	g/kyr
				av. flux	0.63	pcc/cm²/kyr
				minus stand. dev.	0.47	pcc/cm²/kyr
				plus stand. dev.	0.79	pcc/cm²/kyr
II	base 24r – ETM 2	2940	5181	1.76		cm/kyr
				dens/(kyr/cm) =	2.35	g/kyr
				av. flux	0.79	pcc/cm²/kyr
				minus stand. dev.	0.71	pcc/cm²/kyr
				plus stand. dev.	0.87	pcc/cm²/kyr
III	C25n – ETM 2	3444	6118	1.78		cm/kyr
				dens/(kyr/cm) =	2.37	g/kyr
				av. flux	0.71	pcc/cm²/kyr
				minus stand. dev.	0.56	pcc/cm²/kyr
				plus stand. dev.	0.86	pcc/cm²/kyr

(b)

Background ET3He flux and sedimentation rate model for Hole 690B. Expansion (11%) included

	interval	duration (kyr)	thickness (cm)	sed rates (cm/kyr)		
IV	C25n	504	1003	1.99		cm/kyr
				dens/(kyr/cm) =	2.66	g/kyr
				av. flux	0.68	pcc/cm²/kyr
				minus stand. dev.	0.51	pcc/cm²/kyr
				plus stand. dev.	0.85	pcc/cm²/kyr
V	base 24r – ETM 2	3150	5809	1.84		cm/kyr
				dens/(kyr/cm) =	2.46	g/kyr
				av. flux	0.82	pcc/cm²/kyr
				minus stand. dev.	0.74	pcc/cm²/kyr
				plus stand. dev.	0.91	pcc/cm²/kyr
VI	C25n – ETM 2	3654	6812	1.86		cm/kyr
				dens/(kyr/cm) =	2.49	g/kyr
				av. flux	0.74	pcc/cm²/kyr
				minus stand. dev.	0.58	pcc/cm²/kyr
				plus stand. dev.	0.90	pcc/cm²/kyr

(c)

Background flux measurements

Chron 24r		Chron 25n+r	
ET3He pcc/g		ET3He pcc/g	
0.28		0.30	
0.35		0.23	
0.35		0.19	
0.31		0.36	
0.39		0.21	
0.33		0.24	
0.33			
av	0.33	av	0.25
stdev	0.03	stdev	0.06

Chron 25 – ETM2 (all samples)

ET3He pcc/g	
av	0.30
stdev	0.06

DB Density	1.34

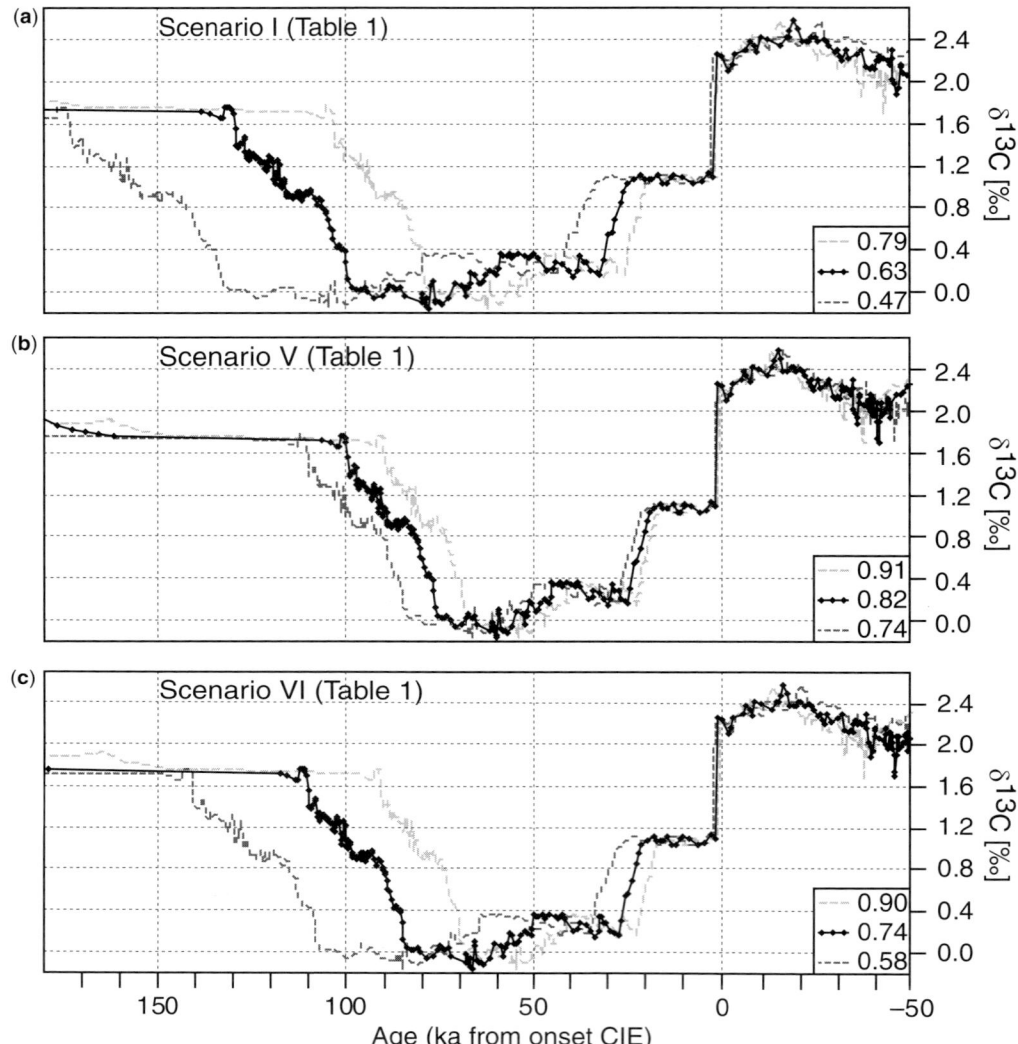

Fig. 3. Shape and duration of the CIE (data from Bains *et al.* 1999) assuming the various options and uncertainties in sedimentation rates and background ³He$_{ET}$ fluxes calculated in Table 1. Options (**a**, **b** and **c**) represent scenarios I, V and VI from Table 1: (a) ³He$_{ET}$ fluxes calculated from the measurements of C25 only (Table 1c) and sediment expansion not included. (b) ³He$_{ET}$ fluxes calculated from the measurements of C24r and sediment expansion included. (c) ³He$_{ET}$ fluxes calculated from the measurements of C25n through C24r and sediment expansion included. Dotted lines represent standard deviations of background flux measurements.

calcite preservation increased due to the 'overshoot' of the lysocline following silicate weathering (Dickens *et al.* 1997; Zachos *et al.* 2005). The ³He$_{ET}$ and cyclostratigraphic age models each provide estimates of the magnitude of sedimentation rate increases due to these mechanisms. Our recalculated He-based estimates suggest duration of 30 to 35 ka for the recovery phase. Röhl *et al.* (2000) counted seven precession cycles through

the recovery interval at ODP Site 690, implying a relatively prolonged, *c.* 140 kyr, recovery and negligible changes in sediment accumulation rates. Clear identification of cycles in this interval of the 690 cores is complicated, however, by the high carbonate content of the sediments. A cyclostratigraphic study on an Italian PETM section (Giusberti *et al.* 2007) counted only five precession cycles (i.e. *c.* 100 kyr) within the recovery interval.

Direct site-to-site correlation of the termination of the CIE is complicated, however, due to the lack of a clear $\delta^{13}C$ inflection point, which makes the definition of this level somewhat subjective. These differences demonstrate that significant uncertainty remains regarding the pace of carbon cycle recovery from the PETM. Studies in progress, including cycle and $^3He_{ET}$ studies, will probably provide better constraints regarding the durations of the various parts of the CIE.

The temperature anomaly

Deep sea benthic foraminiferal calcite consistently shows a $>1‰$ negative excursion in $\delta^{18}O$ (e.g. Kennett & Stott 1991; Bralower et al. 1995; Thomas & Shackleton 1996; Zachos et al. 2001) during the PETM (see above for $\delta^{13}C$). Application of empirical temperature – $\delta^{18}O$ relations (e.g. Shackleton 1967; Erez & Luz 1983) indicates a deep-water temperature rise of c. 4–5 °C, which is corroborated by benthic foraminifer Mg/Ca values (Tripati & Elderfield 2005). This warming was first interpreted to reflect a shift in deep-water formation from high to low latitudes (Kennett & Stott 1991). Evidence accumulating since then indicates that the dominant source of intermediate and deep-water formation probably remained at high latitudes, although it may have switched from southern to northern high latitudes (Pak & Miller 1992; Bice & Marotzke 2001; Thomas 2004; Tripati & Elderfield 2005; Nunes & Norris 2006). Although some component of deep-water warming could reflect different temperatures at Northern and Southern Hemisphere sites of deep-water formation, it is likely that most of the c. 5 °C warming was the result of surface-water warming in sub-polar regions (Tripati & Elderfield 2005).

Planktonic foraminiferal $\delta^{18}O$ and Mg/Ca excursions generally point towards c. 5 °C surface-water warming over a large range of latitudes (Kennett & Stott 1991; Thomas & Shackleton 1996; Charisi & Schmitz 1998; Thomas et al. 2002; Zachos et al. 2003; Tripati & Elderfield 2004). At Site 690, the maximum warming appears to be as much as 6–8 °C (Kennett & Stott 1991; Thomas et al. 2002). The $\delta^{18}O$ records for the mixed-layer dweller Acarinina and thermocline-dweller Subbotina follow somewhat different trajectories through the event, with Subbotina recording its lowest (warmest) $\delta^{18}O$ values at the base of the CIE-body and Acarinina reaching its $\delta^{18}O$ minimum halfway through the CIE-body (c. 171 m below sea-floor, mbsf). The apparent decoupling between thermocline and surface-water temperatures at Site 690 seems also to be reflected

in Mg/Ca-based temperature reconstructions from subtropical Pacific Site 1209 (Zachos et al. 2003). At Site 690, and perhaps to a lesser degree at Site 1209, it is accompanied by a sudden increase in $\delta^{13}C$ values of Acarinina, suggesting the possible influence of surface-water stratification and productivity on temperature and carbon isotope values.

A salinity-independent organic palaeothermometer TEX_{86}' record across a PETM succession in the Arctic Ocean revealed a warming of c. 5 °C close to the North Pole (Sluijs et al. 2006), implying that the magnitude of temperature change in tropical and (north-) polar surface waters were equal and excluding polar amplification scenario for the PETM at least for the Northern Hemisphere (Tripati & Elderfield 2005; Sluijs et al. 2006). The absence of sea ice, as implied by the high TEX_{86}'-derived average absolute temperatures (Sluijs et al. 2006), was probably a critical factor determining the lack of polar temperature change amplification.

Records from nearshore marine sediments provide evidence for greater temperature-change heterogeneity in coastal settings. A benthic foraminiferal $\delta^{18}O$ record from the neritic sediments recovered at Bass River, New Jersey (USA), exhibits a relatively large excursion that, assuming no change in salinity, suggests 8° of PETM warming (Cramer et al. 1999). This finding has recently been supported by the salinity-independent TEX_{86} record from the nearby Wilson Lake borehole (New Jersey) (Zachos et al. 2006). In contrast, a benthic foraminiferal $\delta^{18}O$ record from the Egyptian shelf shows a c. 1‰ negative excursion (Schmitz et al. 1996), which might imply c. 4 °C of warming but hardly stands out from the background scatter. Application of salinity-independent temperature proxies at this site may help constrain the interpretation of the $\delta^{18}O$ data, but the general suggestion is that the magnitude of coastal warming may have varied by at least a factor of 2.

Warming estimates from the Bighorn Basin are also in the range of 5 °C. Koch et al. (2003) and Bowen et al. (2001) calculate a 3–7 °C warming based on carbonate soil nodule $\delta^{18}O$. Fricke and colleagues (Fricke et al. 1998; Fricke & Wing 2004) infer that PETM mean annual temperature was 4–6 °C warmer than before and after the PETM, based on $\delta^{18}O$ of biogenic phosphate. Estimates from leaf margin analysis of fossil leaves are consistent with these other data, implying a c. 5 °C rise in mean annual temperature during the PETM (Wing et al. 2005).

Acidification of the ocean

According to theory, and as observed and expected in the present and future ocean (Caldeira & Wickett

2003; Feely *et al.* 2004; Delille *et al.* 2005; Orr *et al.* 2005; Kleypas *et al.* 2006), the instantaneous induction of large amounts of CO_2 or CH_4 (which would rapidly be oxidized to CO_2 in the atmosphere) into the ocean–atmosphere system at the PETM should have increased the carbonic acid (H_2CO_3) concentration in the oceans. This acidification would reduce the depth of the calcite compensation depth (CCD) and lysocline in the oceans, leading to sea-floor carbonate dissolution and reduced carbonate burial, which would restore the pH balance in deep ocean waters over 10^3 to 10^4-year time-scales (Dickens *et al.* 1997; Dickens 2000). The extent of globally averaged CCD shoaling is related to the amount of CO_2 injected into the ocean–atmosphere system, and thus may be used to constrain the amount of carbon (Dickens *et al.* 1997).

Sea-floor carbonate dissolution during the PETM has been documented in sediments across a *c.* 2 km depth transect (palaeodepths *c.* 1500–3600 m) at Walvis Ridge in the southeast Atlantic Ocean (Zachos *et al.* 2005). These data allow estimation of PETM carbon release as at least *c.* 4500 Gt if it is assumed that the magnitude of shoaling at Walvis Ridge was characteristic of the global ocean, that the geometry of the ocean basins and seawater alkalinity were equal to those of the modern ocean, and that the release rates of carbon compounds were comparable to projected rates of anthropogenic emissions. It is unlikely that any of these assumptions is strictly correct, and therefore the estimated carbon release can only be taken as a rough estimate that is broadly consistent with, but perhaps suggests a somewhat larger release than, previous estimates (e.g. Dickens *et al.* 1995). In particular, it is not clear that the dramatic shoaling of the CCD at Walvis Ridge was characteristic of the world oceans (e.g. Thomas 1998). Carbonate content at Site 690 (palaeodepth *c.* 1900 m), only decreased from *c.* 85% to *c.* 60% (e.g. Bralower *et al.* 2004). At central Pacific Site 1209 (Shatsky Rise), which was deeper than the shallowest Walvis Ridge site, carbonate content decreased by only *c.* 10% (Colosimo *et al.* 2005), and decreased carbonate content at the Mead Stream section in New Zealand (continental slope) has been interpreted to reflect dilution by increased terrigenous influx rather than dissolution (Hollis *et al.* 2005*a*). These data seem to mimic the general pattern of more severe dissolution in the Atlantic relative to the Pacific than has been predicted by a modelling study that assumes ocean circulation patterns in the late Palaeocene were analogous to those of the present day (Dickens *et al.* 1997), but the magnitude of lysocline and CCD shoaling at Walvis Ridge was much larger than predicted in that study. Severe dissolution has also been observed in many marginal

basins, such as the North Sea region (e.g. Gradstein *et al.* 1994), and the marginal Tethys (e.g. Ortiz 1995; Speijer *et al.* 1997; Speijer & Wagner 2002; Dupuis *et al.* 2003; Schmitz *et al.* 2004; Ernst *et al.* 2006). Understanding the controls on dissolution at these sites, including local factors such as carbonate and organic matter production rates, clastic influx and post-burial dissolution, as well as large-scale factors such as ocean circulation, will be critical to extrapolating to derive a global estimate of ocean acidification and robust estimates of the magnitude of carbon release during the PETM.

The recovery of the oceanic carbonate system may have occurred through increased rates of marine or terrestrial organic carbon burial (Bains *et al.* 2000; Beerling 2000) and through a feedback involving increased marine carbonate carbon burial driven by silicate weathering (Dickens *et al.* 1997; Kelly *et al.* 2005). At Walvis Ridge, the lysocline deepened below pre-PETM levels during the recovery phase, suggesting elevated deep-water alkalinity and supporting the hypothesis of feedback through silicate weathering (Zachos *et al.* 2005). Marine organic-rich PETM successions have been recorded from shallow marine sections from the Tethyan margins (e.g. Bolle *et al.* 2000; Speijer & Wagner 2002; Gavrilov *et al.* 2003), the North Sea (e.g. Bujak & Brinkhuis 1998) and the Arctic Ocean (Sluijs *et al.* 2006). Despite these organic-rich deposits, and although abundant Palaeocene and early Eocene peat and coal deposits exist on the continents (e.g. Collinson *et al.* 2003; Kurtz *et al.* 2003), it is not clear to which extent (if at all) the rate of organic carbon deposition and burial increased during the PETM (e.g. Bowen *et al.* 2004).

Biotic response

Patterns of benthic turnover

The PETM stands out in the geological record as one of the largest extinctions in deep marine calcareous benthic foraminifera, when 35–50% of deep-sea species rapidly became extinct (Tjalsma & Lohmann 1983; Thomas 1989; Pak & Miller 1992; Thomas & Shackleton 1996; Thomas 1998, 2003). Benthic foraminifer extinction (BFE) events of this magnitude are rare in the geological record and deep sea turnover usually takes millions of years (Thomas 2007). Rapid extinction in the deep sea is expected to be rare: many species are cosmopolitan, and are rapid colonizers. Extinction of a large fraction of deep-sea species thus requires a rapid change affecting the whole deep ocean, the largest habitat on Earth. Discussion of the cause of the extinction has concentrated on food availability, acidification, bottom-water oxygen depletion and temperature, based on the palaeoecological interpretation of post-BFE assemblages, which are unfortunately not straightforward (Thomas 1998, 2003).

Food availability in the present oceans depends on surface ocean production and export production of organic matter. Deep sea benthic foraminiferal assemblages do not indicate a global increase or decrease in the food supply during the PETM (e.g. Thomas 1998). In the central Pacific (ODP Site 865) and Southern Ocean (Site 690), benthic foraminifer assemblages point to an increase in food supply, whereas the opposite is found at Atlantic and Indian Ocean sites (Thomas 2003). Nannofossil assemblages (see below) suggest a decrease in surface productivity at Sites 690 and 865, whereas benthic foraminiferal assemblages suggest an increased food supply to the sea-floor. To explain this discrepancy, Thomas (2003) speculated that during the PETM there may have been a chemosynthetic food source at the ocean floor or that export production was more efficient, possibly because lower oxygen levels caused lower organic matter decomposition. If the latter is true, decreased oxygen concentrations of the deep ocean due to higher temperatures and possibly methane oxidation (see below) may have contributed to the BFE (Thomas 1998, 2003; Dickens 2000). Nevertheless, there is no geochemical or sedimentological evidence for severe low oxygen conditions in open ocean (Thomas 2007). Small thin-walled benthic foraminifera (as well as ostracodes) in the interval just above the BFE, are associated with increased calcite corrosiveness of deep waters, which would have been detrimental to calcifiers and could have provided a selective force driving extinction (Steineck & Thomas 1996; Thomas 1998; Orr et al. 2005; Zachos et al. 2005). Extinctions, however, occurred also among non-calcifying deep marine agglutinated foraminifera (Kaminski et al. 1996; Galeotti et al. 2005). There is thus no evidence that deep-water oxygen depletion, carbonate dissolution or increased or decreased food supply to the deep sea were global. Therefore, the global deep ocean temperature increase, which would have raised metabolic rates and lead to ecosystem restructuring, may have been the most important cause of the BFE (Thomas 2007).

Benthic foraminifer studies on neritic and upper bathyal assemblages are largely restricted to the Tethyan basin and Atlantic margins. Extinction and temporal changes in composition in these settings were less severe than in the deep sea (Gibson et al. 1993; Speijer & Schmitz 1998; Thomas 1998; Cramer et al. 1999; Speijer & Wagner 2002). Speijer and colleagues argue that late Palaeocene through early Eocene assemblages

generally indicate relatively oligotrophic conditions along the southern Tethyan margin, but show a transient increase in food supply during the PETM (Speijer & Schmitz 1998; Speijer & Wagner 2002; Scheibner et al. 2005), consistent with multiproxy evidence from neritic realms around the world (see sections on dinocysts and nannofossils).

Larger foraminifera show a large turnover in the Tethyan realm, which has long been recognized but only recently correlated to the PETM (Pujalte et al. 2003). This turnover is characterized by a decrease in generic diversity, but the few genera that became dominant during the PETM show rapid diversification on the species level, possibly due to migration from a previously isolated basin or fast specialization as a result of changing environments (Pujalte et al. 2003).

Unlike benthic foraminifera, the PETM does not stand out as a major extinction event in the deep sea ostracode fossil record (Boomer & Whatley 1995), but it has not been well studied at high resolution. The only reasonable-resolution record from the deep sea is by Steineck & Thomas (1996) on assemblages from Site 689, close to Site 690 but c. 900 m shallower. Their results indicate that ostracodes were smaller and more thin-walled during the PETM, suggesting that within-lineage changes in ostracode morphology may reflect calcite corrosiveness.

Tethyan neritic ostracode assemblages, on the other hand, show a turnover at the PETM (Speijer & Morsi 2002, and references therein), when long-ranging Palaeocene taxa were out-competed by a species that is thought to thrive in upwelling areas. Hence, the dominance of this species is interpreted as a response to enhanced food supply and decreased bottom-water oxygenation (Speijer & Morsi 2002). These Tethyan assemblages suggest a sea-level rise at the PETM, an interpretation consistent with information from other shallow marine successions (Powell et al. 1996; Cramer et al. 1999; Oreshkina & Oberhänsli 2003; Crouch & Brinkhuis 2005; Sluijs et al. 2006).

Migration and radiation patterns in the planktonic realm

The most dramatic planktonic microfossil signature at the PETM is recorded in organic-walled dinoflagellate cysts (dinocysts). Most organic cyst-forming dinoflagellates have life strategies that require a restricted water depth and are, hence, specialized in neritic settings (Fensome et al. 1996). They are very sensitive to changes in the physiochemical characteristics of the surface waters, including salinity, temperature, nutrient availability and stratification (Sluijs et al. 2005). The taxon Apectodinium originated close to the Danian–Selandian boundary

(Brinkhuis *et al.* 1994; Guasti *et al.* 2005), and abundant occurrences remained largely restricted to low latitudes throughout the Palaeocene (Bujak & Brinkhuis 1998). In contrast, every studied dinocyst-bearing succession across the PETM has yielded abundant *Apectodinium*, usually >40% of the dinocyst assemblage (Heilmann-Clausen 1985; Bujak & Brinkhuis 1998; Heilmann-Clausen & Egger 2000; Crouch *et al.* 2001; Sluijs *et al.* 2005, 2006) (Fig. 4). Such a global, synchronous acme is unique in the dinocyst fossil record, which spans the late Triassic to the Recent. During the PETM temperate to polar, sea-surface temperatures increased to allow poleward migration of *Apectodinium* (Bujak & Brinkhuis 1998; Crouch *et al.* 2001; Sluijs *et al.* 2006). The *Apectodinium* acme appears to be associated not only with high temperatures, but also with a strong increase in nutrient availability in marginal marine settings (Powell *et al.* 1996; Crouch *et al.* 2001, 2003a; Crouch & Brinkhuis 2005): the motile dinoflagellates that formed *Apectodinium* cysts were probably heterotrophic and fed on organic detritus or other plankton (Bujak & Brinkhuis 1998). Increased nutrient input by rivers is consistent with results from fully coupled general circulation models predicting an

intensified hydrological cycle at elevated greenhouse gas concentrations (Pierrehumbert 2002; Huber *et al.* 2003; Caballero & Langen 2005). Other microfossil, geochemical and lithological evidence at least supports locally intensified runoff during the PETM, although several clay mineral records suggest that this started prior to the CIE (Robert & Kennett 1994; Gibson *et al.* 2000; Ravizza *et al.* 2001; Speijer & Wagner 2002; Egger *et al.* 2003; Gavrilov *et al.* 2003; Gibbs *et al.* 2006; Pagani *et al.* 2006). *Apectodinium* was probably euryhaline, i.e. tolerant to a wide range of salinities, as the acme has been recorded from the relatively fresh (Pagani *et al.* 2006; Sluijs *et al.* 2006) Arctic Ocean to the probably more salty subtropical regions and even open ocean settings (Egger *et al.* 2000). Because of this complex palaeoecology, it is difficult to elucidate which environmental parameter(s) was the driving force behind the global *Apectodinium* acme during the PETM.

Marine dinoflagellate assemblages usually show a strong proximal–distal signal, so that dinocyst assemblages can be used to reconstruct the influence of inshore waters in a more offshore locality (Brinkhuis 1994; Pross & Brinkhuis 2005; Sluijs *et al.* 2005). Dinocyst assemblages show a trend

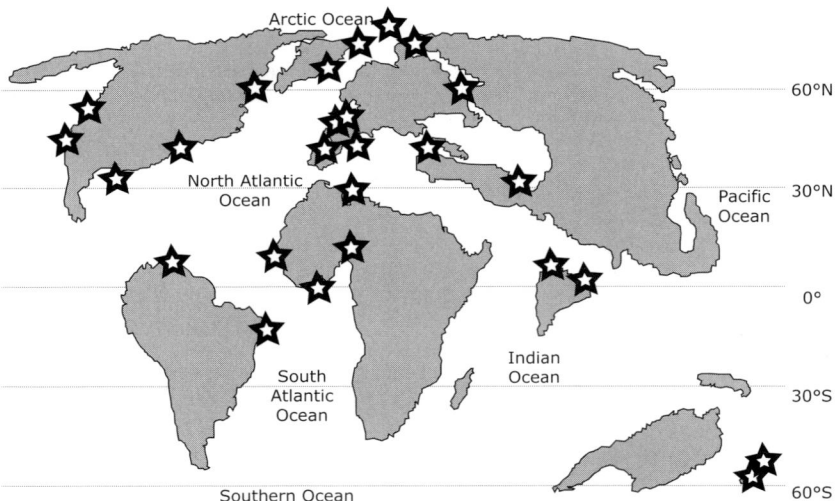

Fig. 4. Palaeogeographic reconstruction of the continental constellation during PETM times (modified from Scotese & Golanka 1992) with the distribution of the acme of dinocyst *Apectodinium*. All studied PETM sections that bear dinocysts yield abundant *Apectodinium*. Records are from the North Sea (Bujak & Brinkhuis 1998 and references therein; Heilmann-Clausen 1985; Steurbaut *et al.* 2003; Sluijs 2006), Greenland, Spitsbergen (e.g. Boulter & Manum 1989; Nohr-Hansen 2003), the Tethyan Ocean (N. Africa, Austria, Tunisia, Uzbekistan, Pakistan, India, Kazakhstan) (e.g. Köthe *et al.* 1988; Bujak & Brinkhuis 1998; Iakovleva *et al.* 2001; Crouch *et al.* 2003a; Egger *et al.* 2003), equatorial Africa (JanDuChêne & Adediran 1984), the eastern (e.g. Edwards 1989; Sluijs 2006) and western USA. (J. Lucas-Clark, pers. comm., 2003; Brinkhuis, Sluijs, unpublished), Barents Sea, South America (Brinkhuis, pers. obs.) and New Zealand (Crouch *et al.* 2001; Crouch *et al.* 2003b; Crouch & Brinkhuis 2005) and the Arctic Ocean (Sluijs *et al.* 2006).

towards more offshore surface-water conditions at many locations during the PETM (Crouch & Brinkhuis 2005; Sluijs *et al.* 2005) implying that eustatic transgression took place (Sluijs 2006). Although the magnitude of this transgression is unclear, qualitatively it may be expected during a time of significant global warming because of thermal expansion and potential cryosphere melting, which together potentially comprised in the order of *c.* 10–20 m across the PETM (Miller *et al.* 2005). Furthermore, it is consistent with data on shallow marine benthic foraminifer, ostracode and grain-size information (Gibson & Bybell 1994; Cramer *et al.* 1999; Speijer & Morsi 2002; Speijer & Wagner 2002; Sluijs 2006).

Apectodinium shows a large variation of morphotypes through this event. As for planktonic foraminifera (see below), it is hard to assess which of these dinocyst types represent true biological species, but intermediate forms occur, implying that they represent ecophenotypes. *Apectodinium* is a member of the family Wetzeliellaceae, which underwent major radiation during or close to the PETM. There are few high-resolution studies of upper Palaeocene material (Heilmann-Clausen 1985; Bujak & Brinkhuis 1998; Iakovleva *et al.* 2001; Guasti *et al.* 2005), but associated genera, such as *Wilsonidium*, *Dracodinium* and *Rhombodinium* originated close to or at the PETM. Afterwards, new genera and species within the Wetzeliellaceae, including the genus *Wetzeliella*, developed, their evolution potentially related to other early Eocene global warming events such as ETM2 (Lourens *et al.* 2005) and ETM3 (Röhl *et al.* 2005).

Compared to dinocysts, planktonic foraminifera show a relatively minor response to the PETM. Poleward migrations include the only occurrence of the low-latitude genus *Morozovella* in the Weddell Sea (Thomas & Shackleton 1996) just prior to and during the lower part of the CIE. Extinctions and radiations are largely absent but evidence of faunal turnover has been recorded at many locations (e.g. Lu & Keller 1993, 1995; Pardo *et al.* 1997). The genera *Morozovella* and *Acarinina* developed extreme morphotypes during the PETM in tropical regions (Kelly *et al.* 1996, 1998). The dominance of these newly developed taxa within the assemblages has been interpreted as indicative of relatively oligotrophic conditions in the open ocean due to changes in the thermal structure of the water column (Kelly *et al.* 1996). These PETM morphotypes might represent true evolutionary transitions or ecophenotypes reflecting unusual environmental conditions, such as stratification and nutrient depletion (Kelly *et al.* 1998).

Only few high-resolution calcareous nannofossil studies focused on palaeoecology through the

PETM, although assemblage changes are extensively described (e.g. Backman 1986; Aubry *et al.* 1996; Raffi *et al.* 2005 and references therein; see also fig. 17 in Raffi *et al.* 2005 for geographic distribution maps of recorded changes in the nannofossil assemblage). Neritic assemblages from the New Jersey Shelf were interpreted to reflect increased productivity during the PETM (Gibbs *et al.* 2006). Bralower (2002) argued that nannofossil assemblages at open ocean Site 690 reflect a change from abundant *r*-mode species (comprising opportunistic species, indicating eutrophic conditions with a well-mixed upper water column and a shallow thermocline) to abundant *K*-mode species (specialized species, indicating oligotrophic conditions with a stratified water column and a deep thermocline) at the onset of the CIE. This interpretation is consistent with that of nannofossil assemblage studies of open marine sites from the Indian Ocean (Tremolada & Bralower 2004), the Pacific Ocean (Gibbs *et al.* 2006) and Tethys (Monechi *et al.* 2000), and supported by model studies (e.g. Boersma *et al.* 1998; Huber *et al.* 2003). However, it appears directly contradicted for Site 690 by observations on accumulation rates of Ba (Bains *et al.* 2000) and Sr/Ca in coccoliths (Stoll & Bains 2003): both suggest increased productivity. Both arguments are in debate: the latter by Bralower *et al.* (2004), the former by (Dickens *et al.* 2003), who suggested that dissociation of methane hydrate (see below) resulted in Ba^{2+} release into the ocean, thus elevating dissolved Ba^{2+} concentrations in the deep sea and causing improved preservation of sedimentary barite. In addition, changes in bottom-water CO_3^- (described above) may have had an effect on barite preservation (Schenau *et al.* 2001).

Not many upper Palaeocene to lower Eocene sections contain well-preserved siliceous microfossil assemblages so the response of diatoms and radiolarians to the PETM is not well constrained. Sediment sequences that bear radiolaria commonly do not have a good record through the PETM (Sanfilippo & Nigrini 1998), and the well-preserved PETM assemblage from low-latitude Blake Nose is unique (Sanfilippo & Blome 2001). At that site, the composition of the fauna shows no net, overall change during the PETM, although there is a minor increase in the number of first and last occurrences. In contrast, the influx of several low-latitude taxa at the PETM is observed in New Zealand, suggesting poleward migrations (Hollis *et al.* 2005*a, b*; Hollis 2006).

Diatom records through the PETM are also rare. ODP Site 752 in the eastern Indian Ocean recovered a diatom-bearing PETM succession, but only biostratigraphical and taxonomical work has been done on this material (Fourtanier 1991). A considerable turnover of the diatom flora at generic and species levels has been observed in sections from the epicontinental sea in Russia, including the appearance of short-lived taxa (Radionova *et al.* 2001; Oreshkina & Oberhänsli 2003). It has been suggested that these turnovers are associated with a more vigorous exchange of surface waters with the Tethys resulting from transgression (Oreshkina & Oberhänsli 2003).

Terrestrial mammals

The PETM stands out as a time of significant changes in terrestrial biota (Fig. 5). Perhaps the most dramatic and best-known of these is the abrupt introduction of four major taxonomic groups to terrestrial mammalian assemblages on the Northern Hemisphere continents at or near the P–E boundary (Gervais 1877; McKenna 1983; Gingerich 1989; Krause & Maas 1990; Smith & Smith 1995; Hooker 1998). The first appearance of the ordinal-level ancestors of modern hoofed mammals (orders Artiodactyla and Perissodactyla), Euprimates (those bearing the complete set of anatomical characteristics uniting modern primates), and a now-extinct family of carnivores (Hyaenodontidae) has long been held by palaeontologists to represent the base of the Eocene in western North America and Europe (Gingerich & Clyde 2001; Gingerich 2006). Recent high-resolution stratigraphic studies demonstrated that they occur within metres of the CIE base at *c.* six sites across the Holarctic continents (e.g. Koch *et al.* 1992, 1995; Cojan *et al.* 2000; Bowen *et al.* 2002; Steurbaut *et al.* 2003; Ting *et al.* 2003).

These first appearances are associated with long- and short-term (transient) changes in terrestrial mammal faunas and initiated a profound modernization of terrestrial mammal communities that were still dominated by archaic forms despite prolific diversification following the Cretaceous–Palaeogene boundary (Alroy *et al.* 2000). The new groups rapidly impacted the communities that they entered, producing significant increases in species richness, average species size, and proportional representation of herbivorous and frugiverous taxa, characteristics of community structure which persisted in early Eocene assemblages from the Bighorn Basin (Clyde & Gingerich 1998) (Fig. 5). The first appearances also provided a seed for longer-term changes in community composition and structure: in the context of longer-term changes in species diversity, the first appearances are overshadowed by rapid post-PETM diversification within the new clades. Other impacts of the PETM on land mammals were transient, including a reduction in average individual body size among both new groups and lineages that ranged through

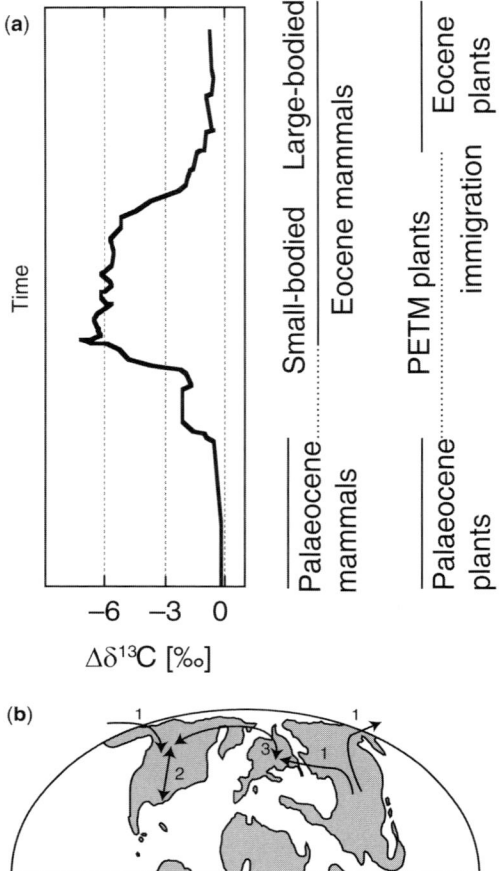

Fig. 5. Patterns of terrestrial biotic change through the PETM. (**a**) Temporal sequence of changes in mammal and plant assemblages, shown relative to the palaeosol carbonate $\delta^{13}C$ curve for Polecat Bench, Wyoming, USA (carbon isotope values are shown here as anomalies ($\Delta\delta^{13}C$) relative to the average latest Palaeocene values; after Bowen *et al.* (2006). (**b**) Spatial pathways of migration thought to have been used by PETM intra- and intercontinental migrants. 1, Directional exchange of mammals and turtles from Asia to North America and/or Europe. 2, Northward range expansion of thermophilic plants from southern North America. 3, Exchange of plant and mammal taxa between North America and Europe, including immigration of plant taxon to North America and homogenization of mammal faunas, including new PETM immigrants.

the Palaeocene–Eocene boundary in the Bighorn Basin (Gingerich 1989; Clyde & Gingerich 1998) (Fig. 5). A significant taxonomic turnover between two groups of faunas assigned to the Bumbanian Asian land mammal age may reflect a shift from transient PETM fauna to a more stable early

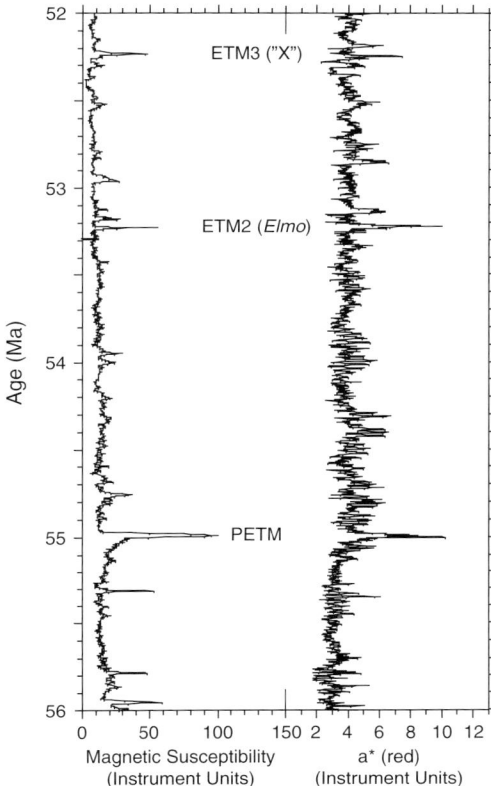

Fig. 6. Magnetic susceptibility and a* colour reflectance records through the uppermost Palaeocene and lower Eocene at Site 1267, Walvis Ridge (from Zachos *et al.* 2004). The peaks are associated with the PETM and the early Eocene hyperthermal events ETM2 and ETM3. The age model used follows that of Zachos *et al.* (2004), which has recently been updated (Lourens *et al.* 2005; Westerhold *et al.* 2007).

Warming during the PETM may have allowed mammals previously restricted to lower latitudes to access these intercontinental corridors, resulting in widespread dispersal of the new groups and homogenization of the Holarctic fauna (McKenna 1983; Krause & Maas 1990; Peters & Sloan 2000). A concomitant dispersal event affecting the Holarctic turtle fauna may reflect a similar climate-mediated mechanism (Holroyd *et al.* 2001).

This mechanism provides a compelling link between climate and PETM mammal turnover, but does not address the questions of where, when, or why the new immigrant groups originated. Two models for the origination of these groups have been proposed. Beard and colleagues have argued that the similarity between the early representatives of the new 'PETM' groups and outgroup taxa from the Palaeocene of Asia suggests an Asian origin, and that some primitive Asian representatives of the 'new' clades may be of Palaeocene age (Beard 1998; Beard & Dawson 1999). In contrast, Gingerich has argued that rapid origination of these clades at unspecified locations in response to environmental perturbations associated with the PETM is both possible and plausible (Gingerich 2006).

This debate centres on the issue where and when the missing links between the new PETM groups and their ancestors occurred, and thus it has been difficult to test the competing ideas. The hypothesis for Asian origination, however, presents the testable prediction that the 'new' mammal groups were present in the Palaeocene of Asia. A combination of chemo-, magneto-, and biostratigraphic data shows that Hyaenodontidae were indeed present in Asian Palaeocene faunas (Bowen *et al.* 2002; Ting *et al.* 2003; Meng *et al.* 2004; Bowen *et al.* 2005). Smith *et al.* (2006) argued for a slightly earlier appearance (by *c.* 10–25 ka) of primates in Asia. However, the failure to find ubiquitous support for the hypothesis of Asian origination during the Palaeocene, and recent evidence against other potential loci of Palaeocene origination such as the Indian subcontinent (Clyde *et al.* 2003), has led to renewed interest in the idea that environmental change during the PETM may have spurred the evolution of several important orders (Gingerich 2006).

Terrestrial plants

Palynological and macrofloral remains from the uppermost Palaeocene and lowermost Eocene have revealed no evidence for net long-term taxonomic turnover or long-lasting major changes in community structure associated with the PETM (Jaramillo & Dilcher 2000; Harrington & Kemp 2001; Wing & Harrington 2001; Collinson *et al.* 2003; Crouch *et al.* 2003*b*; Crouch & Visscher 2003;

Eocene fauna, and be somewhat analogous to North American faunal changes at the end of the PETM (Ting *et al.* 2003), although the data are at much lower time resolution than those in the Bighorn Basin. No clear equivalent has been proposed for European faunas.

There is near-universal consensus that the abrupt first appearances in early Eocene mammal faunas represent synchronous dispersal of new taxa across Holarctica. Fossil biogeography (e.g. McKenna 1983) and recent studies documenting Arctic Ocean palaeosalinity (Brinkhuis *et al.* 2006; Pagani *et al.* 2006; Sluijs *et al.* 2006) suggest that the Northern Hemisphere continents must have been linked at least intermittently by land bridges during the early Palaeogene, providing high-latitude corridors for faunal exchange.

Harrington 2003; Wing *et al.* 2003; Harrington *et al.* 2004). These studies documented modest floral change across the P–E boundary, including the introduction of a small number of immigrant taxa (e.g. introduction of some European taxa to North America) and increases in the diversity of floras from the late Palaeocene to early Eocene. Changes in terrestrial floras across the P–E boundary may have in some cases been diachronous and do not stand out as highly anomalous relative to background spatial and temporal taxonomic variation (Harrington *et al.* 2004).

Two recently discovered macrofloral collections of distinctive composition, however, show that major transient changes in the taxonomic composition of floras occurred during the PETM in the Bighorn Basin (Wing *et al.* 2005). These floras document the immigration of thermophillic taxa previously known from the southern United States and adjacent basins of the western United States, and the first appearance of the European immigrant palynospecies *Platycarya platycaryoides* later within the PETM. Many early Palaeogene plant taxa underwent major geographic range shifts during the PETM, both within and between continents. These shifts are comparable to floral range shifts at the end of the Pleistocene: in both cases, rapid and plastic reorganization of plant communities occurred in response to climatic and environmental change (Overpeck *et al.* 1992; Jackson & Overpeck 2000; Wing *et al.* 2005).

Summary of biotic response

The response of marine organisms to the PETM was heterogeneous. Benthic foraminifera comprise the only group that underwent significant extinction, potentially related to the increased temperature, although lower oxygen levels, changes in primary productivity or the biological carbon pump, and the corrosiveness of carbonate, may have played a role. Deep-sea ostracode faunas may also reflect this increased corrosivity, but the geographical extent is unclear. In the surface ocean, a global acme of the (sub)tropical dinocyst *Apectodinium* occurred along the continental margins, and has been interpreted as a combination of warming and an increase in trophic level of marginal seas, as supported by neritic lithological, ostracode, benthic foraminifer and nannofossil information. Trophic changes in the open ocean are still debated: planktonic foraminiferal and nannofossil assemblages suggest that relatively oligotrophic conditions existed during the PETM, whereas increased barite accumulation rates and Sr/Ca of nannoplankton (at some sites) have been interpreted as indicative of elevated surface productivity. Benthic foraminifera suggest an increase in food supply to the

sea-floor at some (but not all) open ocean locations, but not necessarily higher surface productivity. Both dinocysts and neritic ostracodes indicate that eustatic sea level rose during the PETM. Finally, extreme morphotypes of several planktonic protist microfossil groups are restricted to the PETM and probably represent ecophenotypes.

The terrestrial biotic record of the PETM demonstrates that climate change induces changes in the geographic distribution of terrestrial organisms. Migration appears to be the dominant mechanism of change within PETM terrestrial ecosystems, not only within the mammal and plant records, but also among early Palaeogene turtle faunas. This shifting of geographic distributions introduced new and unique taxonomic assemblages to PETM terrestrial ecosystems, mainly through addition or substitution of taxa without significant loss or modification of existing groups (the example of transient body-size reduction in North American mammals being an important exception). One of the striking aspects of terrestrial biotic change through the PETM is the lack of evidence for a PETM extinction event within any of the groups studied. This suggests a surprising level of adaptability within terrestrial ecosystems, although many details of the conditions and timing of PETM terrestrial environmental and biotic change remain to be resolved.

Hypotheses on causes and effects of the PETM

We have summarized the broad range of biotic, geochemical and climatic changes associated with the massive release of ^{13}C-depleted carbon into the ocean–atmosphere system during the PETM. A major, ongoing research focus is to develop an integrated understanding of the cause and effect scenarios that brought about these changes, beginning with the carbon release itself. Several hypotheses describing the trigger and source of PETM carbon release have been proposed, and these must be critically evaluated in terms of our knowledge of the event's timing, of PETM carbon cycle changes, and of the biotic and climatic responses to carbon cycle perturbation. As described above, changes in the δ^{13}C values of the exogenic carbon pool and widespread dissolution of marine carbonates provide constraints on the magnitude of carbon released (Kennett & Stott 1991; Koch *et al.* 1992; Thomas *et al.* 2002; Zachos *et al.* 2005; Pagani *et al.* 2006). Orbital cycle- and helium-based age models give an order-of-magnitude constraint on the rates of release, which imply large, catastrophic or chronic fluxes of carbon occurred over *c.* 10^4 years. Global warming of 5–6 °C implies large, sustained increases in atmospheric greenhouse gas levels (Dickens *et al.* 1995; Dickens 2001*a*). Critically, the existence of similar events in the early Eocene, including the ETM2 (Lourens *et al.* 2005) and ETM3 (Röhl *et al.* 2005), demonstrates that the PETM was not a unique event, and suggests ties between the initiation of these events and orbital cyclicity (Lourens *et al.* 2005).

Many proposed hypotheses cannot satisfy the latter constraint (Lourens *et al.* 2005). In theory, climatic extremes on a long-term gradual climate trend are expected to occur during eccentricity maxima, because at such times seasonal contrasts on both hemispheres are maximized, so that critical climate thresholds are likely to be surpassed. The climate of the late Palaeocene through early Eocene followed a long-term warming trend, as evidenced by benthic foraminifer $\delta^{18}O$ (Zachos *et al.* 2001) and many biotic proxies (e.g. Haq & Lohmann 1976; Hallock *et al.* 1991; Bujak & Brinkhuis 1998). This warming may have been caused by increasing atmospheric CO_2 levels through high volcanic activity in the North Atlantic Igneous Province (Eldholm & Thomas 1993; Schmitz *et al.* 2004; Thomas & Bralower 2005; Maclennan & Jones 2006) and/or along Indian Ocean spreading zones (Cogné & Humler 2006). The late Palaeocene is also characterized by a long-term decrease in benthic foraminiferal and bulk carbonate (and likely global exogenic) $\delta^{13}C$ after the major positive event in the mid-Palaeocene (Corfield & Norris 1998; Zachos *et al.* 2001). The eccentricity maxima superimposed on these trends could have comprised thresholds for transient events and resulting climate change.

Spectral characteristics of magnetic susceptibility and colour reflectance records from Walvis Ridge show that the PETM, ETM2 (Lourens *et al.* 2005) and ETM3 (Röhl *et al.* 2005), all transient global greenhouse warming events, started during short (100 kyr) and long-term (405 kyr) eccentricity maxima, suggesting an insolation-driven forcing mechanism (Lourens *et al.* 2005). Westerhold *et al.* (2007), however, using similar statistical treatments of high-resolution Fe and red/green (a*) records of the same Walvis Ridge sites, argued that there were two short-term eccentricity cycles less between the PETM and ETM2 than indicated in (Lourens *et al.* 2005), with the implication that the PETM and ETM2 do not correspond exactly to a maximum in the long-term eccentricity cycle, but to one of the short-term (100 kyr) eccentricity extremes superimposed on the long-term cycles. A definite clue to a possible orbital-based forcing mechanism is complicated, however, due to large uncertainties in available astronomical computations for the late Palaeocene to early Eocene (see Lourens *et al.* 2005). A more complete insight into

possible forcing mechanisms may become available when a new generation of astronomical solutions is launched (Laskar, pers. comm., 2006).

An important consequence of the hypothesis that the forcing of the PETM and similar (but less severe events) was orbitally based is that triggering mechanisms involving unique tectonic, volcanic or cosmic events can be excluded as the cause of the PETM. Proposed singular events include a comet impact (Kent *et al.* 2003; Cramer & Kent 2005), explosive volcanism (Bralower *et al.* 1997; Schmitz *et al.* 2004), intrusion-forced injection of thermogenic methane (Svensen *et al.* 2004), tectonic uplift-forced methane hydrate release (Maclennan & Jones 2006), sub-lithospheric gas explosions (Phipps Morgan *et al.* 2004), and tectonically forced desiccation of epicontinental seas (Higgins & Schrag 2006). Regardless of the specific orbital parameter association of the PETM, ETM2 and ETM3 and possible additional hyperthermal events, the simple fact that there were multiple events within the late Palaeocene and early Eocene implies a non-singular trigger as their cause.

What type of carbon-release scenarios could be triggered by orbital fluctuations in a greenhouse world and lead to rapid global warming and biotic change? In the present-day situation, non-anthropogenic carbon reservoirs that can inject the required amount of ^{13}C-depleted carbon needed to generate the CIE in the atmosphere and ocean on the required short timescale are scarce (Dickens *et al.* 1995, 1997). One potential reservoir is formed by methane hydrates, although the size of the present hydrate reservoir is the subject of discussion (Milkov 2004). The methane incorporated in the hydrates is produced by anoxic bacterial decomposition (average $\delta^{13}C$ of *c.* -60‰) or thermogenic breakdown (average $\delta^{13}C$ of *c.* -30‰) of organic matter (Kvenvolden 1988, 1993). In the present ocean, these hydrates are stable along continental slopes at relatively high pressure and low temperatures (at high latitudes at the surface, at lower latitudes below a few hundreds of metres depth) and they have been argued to dissociate when pressure falls or temperature rises (MacDonald 1990; Xu *et al.* 2001). During the much warmer latest Palaeocene, hydrates were stable at deeper-water depths (e.g. below *c.* 900 m) (Dickens *et al.* 1995), suggesting that the reservoir was smaller than at present (Buffett & Archer 2004). However, methane hydrates might have been present at greater depths than today (Dickens 2001*b*) if more organic matter was present further away from the continents than nowadays, perhaps in conjunction with lower oxygen content of the bottom waters resulting from higher temperatures, as suggested by benthic foraminiferal assemblages (Thomas & Zachos 2000; Thomas 2007). In addition to concerns about the standing volume of methane hydrate at the time of the PETM, the rate of replenishment of these reservoirs is problematic if the CIEs related to ETM2 and ETM3 are also to be attributed to hydrate dissociation (Fehn *et al.* 2000). This might imply that a significant fraction of the hydrate reservoir was maintained through the PETM and ETM2, but would compound discrepancies between the implied and predicted hydrate inventories.

Dissociation of methane clathrates along continental slopes has been invoked to explain the CIE and part of the climatic warming (Dickens *et al.* 1995; Matsumoto 1995). In the present atmosphere, CH_4 is oxidized to CO_2 within a decade. The atmospheric residence time of methane could increase at high atmospheric concentrations of this gas, but even this longer residence time has been argued to be only about one to two centuries (Schmidt & Schindell 2003). The relatively short residence time of CH_4 implies that the long-term (10^4 years and more) climate effects of methane release would result from CO_2 forcing after the oxidation of methane in the atmosphere and/or oceans. As already pointed out by Dickens *et al.* (1997), the radiative forcing of the CO_2 resulting from the injection of *c.* 2000 Gt of biogenic methane is not enough to explain the magnitude of PETM climate warming (Dickens *et al.* 1997; Schmidt & Schindell 2003; Renssen *et al.* 2004; Archer 2007).

In their study to assess terrestrial and marine carbon burial rates, Kurtz *et al.* (2003) use the coupling between the carbon and sulphur cycles to quantify the very high rates of continental organic carbon burial through the middle Palaeocene qualitatively indicated by the global abundance of Palaeocene coals. These authors suggest that rapid oxidation (burning) of terrestrial organic carbon, in their words '*global conflagration*', could have at least contributed to the CIE and climate warming. Terrestrial organic matter (*c.* $-30‰$) is much less ^{13}C-depleted than biogenic methane hydrates, implying that much more carbon would have entered the atmosphere and ocean and provided a greater potential radiative forcing than that resulting from the release of methane hydrates (Kurtz *et al.* 2003; Higgins & Schrag 2006). The greatest challenge to the peat-burning hypothesis is the lack of direct evidence for a massive loss of continental organic carbon at the PETM. High concentrations of macroscopic charcoal have been recorded locally at the PETM (Collinson *et al.* 2003), but these remains do not support a scenario of peat burning (Collinson *et al.* 2006). Huge deposits of upper Palaeocene peat deposits are still found today (e.g. Collinson *et al.* 2003), raising questions as to how these deposits survived the '*global conflagration*' and whether it is plausible that the late

Palaeocene continental organic carbon reservoir contained sufficient, unseen organic-rich deposits to account for the PETM.

Problems thus still exist for the hypotheses of clathrate dissociation and continental organic carbon burning, although both are potentially climatically induced and thus could be associated with orbital forcing. An orbitally forced mechanism to trigger dissociation of methane hydrates might be found in temperatures of intermediate waters, which increased due to warming of their high-latitude source regions or due to orbitally forced changes in circulation of deep waters. An orbitally forced trigger for massive burning of peat could be found in the occurrence of droughts in regions with large peat deposits. To invoke such mechanisms as a cause of the PETM with confidence, however, requires better documentation of the character, trends and dynamics of late Palaeocene climate, including its response to orbital cycles.

Thus, integration of available information on carbon cycle, climate change, biotic response as well as the timescale involved in the PETM still leaves us short of a clearly supported, all-encompassing cause-and-effect scenario for the PETM. Data on atmospheric CO_2 and CH_4 (although no proxy has been developed yet for the latter) concentrations through the event could help in evaluating the form and magnitude of injected carbon, which in combination with an improved understanding of the CIE magnitude would clearly distinguish between the hydrate and peat-burning hypotheses. As the causes and mechanisms of carbon release become clearer, focused analysis of the cascade of effects of this perturbation should eventually lead to the identification of mechanisms and feedbacks through which climate, environmental and biotic change occurred (Bowen *et al.* 2004, 2006; Sluijs 2006). More-detailed studies of ETM2 and ETM3 are required to confirm that these events are similar in nature to the PETM and the degree to which these events are different from 'background climate' trends. Furthermore, comparative study of these events should help to resolve the importance of different climatic boundary conditions on the nature and magnitude of abrupt, transient, carbon-induced greenhouse events, improving our understanding of the dynamics of past hyperthermal events and their relevance to understanding climate change today.

References

ALI, J. R., KENT, D. V. & HAILWOOD, E. A. 2000. Magnetostratigraphic reinvestigation of the Palaeocene/ Eocene boundary interval in Hole 690B, Maud Rise, Antarctica. *Geophysical Journal International*, **141**, 639–646.

ALROY, J., KOCH, P. L. & ZACHOS, J. C. 2000. Global climate change and North American mammalian evolution. *Paleobiology*, **26**, 259–288.

ARCHER, D. 2007 Methane hydrates and anthropogenic climate change. *Biogeosciences*, **4**, 521–544.

AUBRY, M. P. & OUDA, K. 2003. Introduction. *In*: OUDA, K. & AUBRY, M. P. (eds) *The Upper Paleocene–Lower Eocene of the Upper Nile Valley, Part 1, Stratigraphy*. Micropaleontology Press, New York, ii–iv.

AUBRY, M.-P., BERGGREN, W. A., STOTT, L. & SINHA, A. 1996. The upper Paleocene–lower Eocene stratigraphic record and the Paleocene–Eocene boundary carbon isotope excursion: implications for geochronology. *In*: KNOX, R. W. O. B., CORFIELD, R. M. & DUNAY, R. E. (eds) *Correlation of the Early Paleogene in Northwest Europe*. Geological Society of London Special Publication, **101**, 353–380.

AUBRY, M. P., LUCAS, S. G. & BERGGREN, W. A. (eds) 1998. *Late Paleocene–Early Eocene Climatic and Biotic Events in the Marine and Terrestrial Records*. Columbia University Press, New York.

BACKMAN, J. 1986. Late Paleocene to middle Eocene calcareous nannofossil biochronology from Shatsky Rise, Walvis Ridge and Italy. *Palaeogeography, Palaeoclimatology, Palaeoecology*, **57**, 43–59.

BAINS, S., CORFIELD, R. M. & NORRIS, G. 1999. Mechanisms of climate warming at the end of the Paleocene. *Science*, **285**, 724–727.

BAINS, S., NORRIS, R. D., CORFIELD, R. M. & FAUL, K. L. 2000. Termination of global warmth at the Paleocene/Eocene boundary through productivity feedbacks. *Nature*, **407**, 171–174.

BAINS, S., NORRIS, R. D., CORFIELD, R. M., BOWEN, G. J., GINGERICH, P. D. & KOCH, P. L. 2003. Marine-terrestrial linkages at the Paleocene–Eocene boundary. *In*: WING, S. L., GINGERICH, P. D., SCHMITZ, B. & THOMAS, E. (eds) *Causes and Consequences of Globally Warm Climates in the Early Paleogene*. Geological Society of America Special Paper, **369**. Geological Society of America, Boulder, CO, 1–9.

BEARD, K. C. 1998. East of Eden; Asia as an important center of taxonomic origination in mammalian evolution. *In*: BEARD, K. C. & DAWSON, M. R. (eds) *Dawn of the Age of Mammals in Asia*. Bulletin of the Carnegie Museum of Natural History, **34**, 5–39.

BEARD, K. C. & DAWSON, M. R. 1999. Intercontinental dispersal of Holarctic land mammals near the Paleocene/Eocene boundary; paleogeographic, paleoclimatic and biostratigraphic implications. *Bulletin de la Société Géologique de France*, **170**, 697–706.

BEERLING, D. J. 2000. Increased terrestrial carbon storage across the Palaeocene–Eocene boundary. *Palaeogeography, Palaeoclimatology, Palaeoecology*, **161**, 395–405.

BERGGREN, W. A., KENT, D. V., OBRADOVICH, J. D. & SWISHER, III, C. C. 1992. Towards a revised Paleogene geochronology. *In*: PROTHERO, D. R. & BERGGREN, W. A. (eds) *Eocene–Oligocene Climatic and Biotic Evolution*. Princeton University Press, Princeton, 29–45.

BERGGREN, W. A., KENT, D. V., SWISHER, III, C. C. & AUBRY, M.-P. 1995. A revised Cenozoic geochronology and chronostratigraphy. *In*: BERGGREN, W. A.,

KENT, D. V. & HARDENBOL, J. (eds) *Geochronology, Time Scales and Global Stratigraphic Correlation*. SEPM (Society for Sedimentary Geology) Special Publication **54**, Tulsa, Oklahoma, 129–212.

BICE, K. L. & MAROTZKE, J. 2001. Numerical evidence against reversed thermohaline circulation in the warm Paleocene/Eocene ocean. *Journal of Geophysical Research*, **106**, 11,529–11,542.

BOERSMA, A., PREMOLI SILVA, I. & HALLOCK, P. 1998. Trophic models for the well-mixed and poorly mixed warm oceans across the Paleocene/Eocene epoch boundary. *In*: AUBRY, M.-P., LUCAS, S. G. & BERGGREN, W. A. (eds) *Late Paleocene – Early Eocene Climatic and Biotic Events in the Marine and Terrestrial Records*. Columbia University Press, New York, 204–213.

BOLLE, M.-P., PARDO, A., HINRICHS, K.-U. *ET AL*. 2000. The Paleocene–Eocene transition in the marginal northeastern Tethys (Kazakhstan and Uzbekistan). *International Journal of Earth Sciences*, **89**, 390–414.

BOOMER, I. & WHATLEY, R. 1995. Cenozoic Ostracoda from guyots in the western Pacific: Holes 865B and 866B (Leg 143). *In*: WINTERER, E. L., SAGER, W. W., FIRTH, J. V. & SINTON, J. M. (eds) *Proceedings of the Ocean Drilling Program, Scientific Results*, **143**, Ocean Drilling Program, College Station, TX, 75–86.

BOULTER, M. C. & MANUM, S. B. 1989. The Brito-Arctic igneous province flora around the Paleocene/Eocene boundary. *In*: ELDHOLM, O., THIEDE, J. & TAYLOR, E. (eds) *Proceedings of the Ocean Drilling Program, Scientific Results*, **104**, Ocean Drilling Program, College Station, TX, 663–680.

BOWEN, G. J., KOCH, P. L., GINGERICH, P. D., NORRIS, R. D., BAINS, S. & CORFIELD, R. M. 2001. Refined isotope stratigraphy across the continental Paleocene–Eocene boundary on Polecat Bench in the Northern Bighorn Basin. *In*: GINGERICH, P. D. (ed.) *Paleocene–Eocene Stratigraphy and Biotic Change in the Bighorn and Clarks Fork Basins, Wyoming*. University of Michigan Papers on Paleontology **33**, 73–88.

BOWEN, G. J., CLYDE, W. C., KOCH, P. L. *ET AL*. 2002. Mammalian dispersal at the Paleocene/Eocene boundary. *Science*, **295**, 2062–2065.

BOWEN, G. J., BEERLING, D. J., KOCH, P. L., ZACHOS, J. C. & QUATTLEBAUM, T. 2004. A humid climate state during the Palaeocene/Eocene thermal maximum. *Nature*, **432**, 495–499.

BOWEN, G. J., KOCH, P. L., MENG, J., YE, J. & TING, S. 2005. Age and correlation of fossiliferous Late Paleocene–Early Eocene strata of the Erlian Basin, Inner Mongolia, China. *American Museum Novitates*, **3474**, 1–26.

BOWEN, G. J., BRALOWER, T. J., DELANEY, M. L. *ET AL*. 2006. Eocene Hyperthermal Event Offers Insight Into Greenhouse Warming. *EOS Transactions, American Geophysical Union*, **87**, 165, 169.

BRALOWER, T. J. 2002. Evidence of surface water oligotrophy during the Paleocene–Eocene thermal maximum: nannofossil assemblage data from Ocean Drilling Program Site 690, Maud Rise, Weddell Sea. *Paleoceanography*, **17**, doi:10.1029/2001PA000662.

BRALOWER, T. J., ZACHOS, J. C., THOMAS, E. *ET AL*. 1995. Late Paleocene to Eocene paleoceanography of the equatorial Pacific Ocean: stable isotopes recorded at Ocean Drilling Program Site 865, Allison Guyot. *Paleoceanography*, **10**, 841–865.

BRALOWER, T. J., THOMAS, D. J., ZACHOS, J. C. *ET AL.* 1997. High-resolution records of the late Paleocene thermal maximum and circum-Caribbean volcanism: Is there a causal link? *Geology*, **25**, 963–966.

BRALOWER, T. J., KELLY, D. C. & THOMAS, E. 2004. Comment on "Coccolith Sr/Ca records of productivity during the Paleocene–Eocene thermal maximum from the Weddell Sea" by Heather M. Stoll and Santo Bains. *Paleoceanography*, **17**, doi:10.1029/2003PA000953.

BRINKHUIS, H. 1994. Late Eocene to Early Oligocene dinoflagellate cysts from the Priabonian type-area (Northeast Italy); biostratigraphy and palaeoenvironmental interpretation. *Palaeogeography, Palaeoclimatology, Palaeoecology*, **107**, 121–163.

BRINKHUIS, H., ROMEIN, A. J. T., SMIT, J. & ZACHARIASSE, W. J. 1994. Danian–Selandian dinoflagellate cysts from lower latitudes with special reference to the El Kef section, NW Tunesia. *Geologiska Föreningens i Stockholm Förhandlingar (GFF; Transactions of the Geological Society in Stockholm)*, **116**, 46–48.

BRINKHUIS, H., SCHOUTEN, S., COLLINSON, M. E. *ET AL.* 2006. Episodic fresh surface waters in the Eocene Arctic Ocean. *Nature*, **441**, 606–609.

BUFFETT, B. & ARCHER, D. 2004. Global inventory of methane clathrate: sensitivity to changes in the deep ocean. *Earth and Planetary Science Letters*, **227**, 185–199.

BUJAK, J. P. & BRINKHUIS, H. 1998. Global warming and dinocyst changes across the Paleocene/Eocene Epoch boundary. *In*: AUBRY, M.-P., LUCAS, S. G. & BERGGREN, W. A. (eds) *Late Paleocene–Early Eocene Biotic and Climatic Events in the Marine and Terrestrial Records*. Columbia University Press, New York, 277–295.

BUTLER, R. F., GINGERICH, P. D. & LINDSAY, E. H. 1981. Magnetic polarity stratigraphy and biostratigraphy of Paleocene and lower Eocene continental deposits, Clarks Fork Basin, Wyoming. *Journal of Geology*, **89**, 299–316.

CABALLERO, R. & LANGEN, P. L. 2005. The dynamic range of poleward energy transport in an atmospheric general circulation model. *Geophysical Research Letters*, **32**, doi:10.1029/2004GL021581.

CALDEIRA, K. & WICKETT, M. E. 2003. Anthropogenic carbon and ocean pH. *Nature*, **425**, 365–365.

CANDE, S. & KENT, D. V. 1995. Revised calibration of the geomagnetic polarity timescale for the Late Cretaceous and Cenozoic. *Journal of Geophysical Research*, **111**, 6093–6095.

CHARISI, S. D. & SCHMITZ, B. 1998. Paleocene to early Eocene paleoceanography of the Middle East: the $\delta^{13}C$ and $\delta^{18}O$ isotopes from foraminiferal calcite. *Paleoceanography*, **13**, 106–118.

CLYDE, W. C. & GINGERICH, P. D. 1998. Mammalian community response to the latest Paleocene thermal maximum: An isotaphonomic study in the northern Bighorn Basin, Wyoming. *Geology*, **26**, 1011–1014.

CLYDE, W. C., KHAN, I. H. & GINGERICH, P. D. 2003. Stratigraphic response and mammalian dispersal during initial India–Asia collision: Evidence from the Ghazij Formation, Balochistan, Pakistan. *Geology*, **31**, 1097–1100.

COGNÉ, J.-P. & HUMLER, E. 2006. Trends and rhythms in global seafloor generation rate. *Geochemistry, Geophysics, Geosystems*, **7**, doi:10.1029/2005GC001148.

COJAN, I., MOREAU, M. G. & STOTT, L. E. 2000. Stable carbon isotope stratigraphy of the Paleogene pedogenic series of southern France as a basis for continental–marine correlation. *Geology*, **28**, 259–262.

COLLINSON, M. E., HOOKER, J. J. & GRÖCKE, D. R. 2003. Cobham lignite bed and penecontemporaneous macrofloras of southern England: A record of vegetation and fire across the Paleocene–Eocene Thermal Maximum. *In* WING, S. L., GINGERICH, P. D., SCHMITZ, B. & THOMAS, E. (eds) *Causes and Consequences of Globally Warm Climates in the Early Paleogene*. Geological Society of America Special Paper, **369**, Geological Society of America, Boulder, CO, 333–349.

COLLINSON, M. E., STEART, D., HANDLEY, L. *ET AL.* 2006. Fire regimes and palaeoenvironments across the onset of the Palaeocene/Eocene thermal maximum, S. England, *Climate and Biota of the Early Paleogene, Bilbao, Volume of Abstracts*, 28.

COLOSIMO, A. B., BRALOWER, T. J. & ZACHOS, J. C. 2005. Evidence for lysocline shoaling at the Paleocene/Eocene Thermal Maximum on Shatsky rise, northwest Pacific. *In*: BRALOWER, T. J., PREMOLI-SILVA, I. & MALONE, M. J. (eds) *Proceedings of the Ocean Drilling Program 198*, World Wide Web Address: http://www-odp.tamu.edu/publications/198_SR/VOLUME/CHAPTERS/112.PDF, 1–36.

CORFIELD, R. M. & NORRIS, R. D. 1998. The Oxygen and Carbon Isotopic Context of the Paleocene/Eocene Epoch Boundary. *In*: AUBRY, M.-P., LUCAS, S. G. & BERGGREN, W. A. (eds) *Late Paleocene–Early Eocene Climatic and Biotic Events in the Marine and Terrestrial Records*. Columbia University Press, New York, 124–137.

CRAMER, B. S. & KENT, D. V. 2005. Bolide summer: The Paleocene/Eocene thermal maximum as a response to an extraterrestrial trigger. *Palaeogeography, Palaeoclimatology, Palaeoecology*, **224**, 144–166.

CRAMER, B. S., AUBRY, M.-P., MILLER, K. G., OLSSON, R. K., WRIGHT, J. D. & KENT, D. V. 1999. An exceptional chronologic, isotopic, and clay mineralogic record of the latest Paleocene thermal maximum, Bass River, NJ, ODP 174AX. *Bulletin de la Société Géologique de France*, **170**, 883–897.

CRAMER, B. S., WRIGHT, J. D., KENT, D. V. & AUBRY, M.-P. 2003. Orbital climate forcing of $\delta^{13}C$ excursions in the late Paleocene–early Eocene (chrons C24n–C25n). *Paleoceanography*, **18**, doi:10.1029/2003PA000909.

CROUCH, E. M. & BRINKHUIS, H. 2005. Environmental change across the Paleocene–Eocene transition from eastern New Zealand: A marine palynological approach. *Marine Micropaleontology*, **56**, 138–160.

CROUCH, E. M. & VISSCHER, H. 2003. Terrestrial vegetation record across the initial Eocene thermal maximum at the Tawanui marine section, New Zealand. *In*: WING, S. L., GINGERICH, P. D., SCHMITZ, B. & THOMAS, E. (eds) *Causes and Consequences of Globally Warm Climates in the Early Paleogene*. Geological Society of America Special

Paper 369, Ecological Society of America, Boulder, CO, 351–363.

CROUCH, E. M., HEILMANN-CLAUSEN, C., BRINKHUIS, H., MORGANS, H. E. G., ROGERS, K. M., EGGER, H. & SCHMITZ, B. 2001. Global dinoflagellate event associated with the late Paleocene thermal maximum. *Geology*, 29, 315–318.

CROUCH, E. M., BRINKHUIS, H., VISSCHER, H., ADATTE, T. & BOLLE, M.-P. 2003a. Late Paleocene–early Eocene dinoflagellate cyst records from the Tethys: Further observations on the global distribution of *Apectodinium*. *In*: WING, S. L., GINGERICH, P. D., SCHMITZ, B. & THOMAS, E. (eds) *Causes and Consequences of Globally Warm Climates in the Early Paleogene*. Geological Society of America Special Paper 369. Geological Society of America, Boulder, CO, 113–131.

CROUCH, E. M., DICKENS, G. R., BRINKHUIS, H., AUBRY, M.-P., HOLLIS, C. J., ROGERS, K. M. & VISSCHER, H. 2003b. The *Apectodinium* acme and terrestrial discharge during the Paleocene–Eocene thermal maximum: new palynological, geochemical and calcareous nannoplankton observations at Tawanui, New Zealand. *Palaeogeography, Palaeoclimatology, Palaeoecology*, 194, 387–403.

DELILLE, B., HARLAY, J., ZONDERVAN, I. *ET AL.* 2005. Response of primary production and calcification to changes of pCO2 during experimental blooms of the coccolithophorid *Emiliania huxleyi*. *Global Biogeochemical Cycles*, 19, doi:10.1029/2004 GB002318.

DICKENS, G. R. 2000. Methane oxidation during the late Paleocene Thermal Maximum. *Bulletin de la Société Géologique de France*, 171, 37–49.

DICKENS, G. R. 2001a. Carbon addition and removal during the Late Palaeocene Thermal Maximum: Basic theory with a preliminary treatment of the isotope record at Ocean Drilling Program Site 1051, Blake Nose. *In*: KROON, D., NORRIS, R. D. & KLAUS, A. (eds) *Western North Atlantic Paleogene and Cretaceous Paleoceanography*, Geological Society of London, Special Publication, 183.

DICKENS, G. R. 2001b. The potential volume of oceanic methane hydrates with variable external conditions. *Organic Geochemistry*, 32, 1179–1193.

DICKENS, G. R., O'NEIL, J. R., REA, D. K. & OWEN, R. M. 1995. Dissociation of oceanic methane hydrate as a cause of the carbon isotope excursion at the end of the Paleocene. *Paleoceanography*, 10, 965–971.

DICKENS, G. R., CASTILLO, M. M. & WALKER, J. C. G. 1997. A blast of gas in the latest Paleocene: Simulating first-order effects of massive dissociation of oceanic methane hydrate. *Geology*, 25, 259–262.

DICKENS, G. R., FEWLESS, T., THOMAS, E. & BRALOWER, T. J. 2003. Excess barite accumulation during the Paleocene/Eocene thermal maximum: Massive input of dissolved barium from seafloor gas hydrate reservoirs. *In*: WING, S. L., GINGERICH, P. D., SCHMITZ, B. & THOMAS, E. (eds) *Causes and Consequences of Globally Warm Climates in the Early Paleogene*. Geological Society of America Special Publication, 369. Geological Society of America, Boulder, CO, 11–23.

DUPUIS, C., AUBRY, M.-P., STEURBAUT, E. *ET AL.* 2003. The Dababiya Quarry Section: Lithostratigraphy, clay mineralogy, geochemistry and paleontology. *Micropaleontology*, 49, 41–59.

EDWARDS, L. E. 1989. Dinoflagellate Cysts from the Lower Tertiary Formations, Haynesville Cores, Richmond County, Virginia *Geology and Paleontology of the Haynesville Cores – Northeastern Virginia Coastal Plain*. US Geological Survey professional paper, 1489-C. United States Government Printing Office, Washington, 23.

EGGER, H., HEILMANN-CLAUSEN, C. & SCHMITZ, B. 2000. The Paleocene–Eocene boundary interval of a Tethyan deep-sea section and its correlation with the North Sea basin. *Bulletin de la Société Géologique de France*, 171, 207–216.

EGGER, H., FENNER, J., HEILMANN-CLAUSEN, C., RÖGL, F., SACHSENHOFER, R. & SCHMITZ, B. 2003. Paleoproductivity of the northwestern Tethyan margin (Anthering section, Austria) across the Paleocene–Eocene transition. *In*: WING, S. L., GINGERICH, P. D., SCHMITZ, B. & THOMAS, E. (eds) *Causes and Consequences of Globally Warm Climates in the Early Paleogene*. Geological Society of America Special Paper, 369. Geological Society of America, Boulder, CO 133–146.

ELDHOLM, O. & THOMAS, E. 1993. Environmental impact of volcanic margin formation. *Earth and Planetary Science Letters*, 117, 319–329.

EREZ, J. & LUZ, B. 1983. Experimental paleotemperature equation for planktonic foraminifera. *Geochimica et Cosmochimica Acta*, 47, 1025–1031.

ERNST, S. R., GUASTI, E., DUPUIS, C. & SPEIJER, R. P. 2006. Environmental perturbation in the southern Tethys across the Paleocene/Eocene boundary (Dababiya, Egypt): Foraminiferal and clay mineral records. *Marine Micropaleontology*, 60, 89–111.

FARLEY, K. A. 2001. Extraterrestrial helium in seafloor sediments: identification, characteristics, and accretion rate over geologic time. *In*: PEUCKER-EHRINBRINK, B. & SCHMITZ, B. (eds) *Accretion of Extraterrestrial Matter Throughout Earth's History*. Kluwer, New York, 179–204.

FARLEY, K. A. & ELTGROTH, S. F. 2003. An alternative age model for the Paleocene–Eocene thermal maximum using extraterrestrial ³He. *Earth and Planetary Science Letters*, 208, 135–148.

FEELY, R. A., SABINE, C. L., LEE, K., BERELSON, W., KLEYPAS, J., FABRY, V. J. & MILLERO, F. J. 2004. Impact of Anthropogenic CO2 on the CaCO3 System in the Oceans. *Science*, 305, 362–366.

FEHN, U., SNYDER, G. & EGEBERG, P. K. 2000. Dating of Pore Waters with 129I: Relevance for the Origin of Marine Gas Hydrates. *Science*, 289, 2332–2335.

FENSOME, R. A., GOCHT, H. & WILLIAMS, G. L. 1996. The Eisenack Catalog of Fossil Dinoflagellates. New Series, 4. E. Schweizerbart'sche Verlagsbuchhandlung, Stuttgart, Germany, 2009–2548.

FLORINDO, F. & ROBERTS, A. P. 2005. Eocene–Oligocene magnetobiochronology of ODP Sites 689 and 690, Maud Rise, Weddell Sea, Antarctica. *Geological Society of America Bulletin*, 117, 46–66.

FOURTANIER, E. 1991. Paleocene and Eocene diatom stratigraphy and taxonomy of eastern Indian Ocean Site 752. *In*: WEISSEL, J., PEIRCE, J., TAYLOR, E. *ET AL.* (eds) *Proceedings of the Ocean Drilling*

Program, Scientific Results, **121**. Texas A&M University, College Station, TX, 171–187.

FRICKE, H. C. & WING, S. L. 2004. Oxygen isotope and paleobotanical estimates of temperature and δ^{18}O-latitude gradients over North America during the early Eocene. *American Journal of Science*, **304**, 612–635.

FRICKE, H. C., CLYDE, W. C., O'NEIL, J. R. & GINGERICH, P. D. 1998. Evidence for rapid climate change in North America during the latest Paleocene thermal maximum: oxygen isotope compositions of biogenic phosphate from the Bighorn Basin (Wyoming). *Earth and Planetary Science Letters*, **160**, 193–208.

GALEOTTI, S., KAMINSKI, M. A., COCCIONI, R. & SPEIJER, R. P. 2005. High resolution deep water agglutinated foraminiferal records across the Paleocene/Eocene transition in the Contessa Road Section (Italy). *In*: BUBIK, M. & KAMINSKI, M. A. (eds) *Proceedings of the Sixth International Workshop on Agglutinated Foraminifera*. Grzybowski Foundation Special Publication, **8**, 83–103.

GAVRILOV, Y., SHCHERBININA, E. A. & OBERHÄNSLI, H. 2003. Paleocene–Eocene boundary events in the northeastern Peri-Tethys. *In*: WING, S. L., GINGERICH, P. D., SCHMITZ, B. & THOMAS, E. (eds) *Causes and Consequences of Globally Warm Climates in the Early Paleogene*. Geological Society of America Special Paper, **369**. Geological Society of America, Boulder, CO, 147–168.

GERVAIS, P. 1877. Enumeration de quelques ossements d'animaux vertebres recueillis aux environs de Reims par M. Lemoine. *Journal de Zoologie*, **6**, 74–79.

GIBBS, S. J., BRALOWER, T. J., BOWN, P. R., ZACHOS, J. C. & BYBELL, L. M. 2006. Shelf and open-ocean calcareous phytoplankton assemblages across the Paleocene–Eocene Thermal Maximum: Implications for global productivity gradients. *Geology*, **34**, 233–236.

GIBSON, T. G. & BYBELL, L. M. 1994. Sedimentary Patterns across the Paleocene–Eocene boundary in the Atlantic and Gulf coastal plains of the United States. *Bulletin de la Société Belge de Géologie*, **103**, 237–265.

GIBSON, T. G., BYBELL, L. M. & OWENS, J. P. 1993. Latest Paleocene lithologic and biotic events in neritic deposits of Southwestern New Jersey. *Paleoceanography*, **8**, 495–514.

GIBSON, T. G., BYBELL, L. M. & MASON, D. B. 2000. Stratigraphic and climatic implications of clay mineral changes around the Paleocene/Eocene boundary of the northeastern US margin. *Sedimentary Geology*, **134**, 65–92.

GINGERICH, P. D. 1989. New earliest Wasatchian mammalian fauna from the Eocene of Northwestern Wyoming: Composition and diversity in a rarely sampled high-floodplain assemblage. *University of Michigan Papers on Paleontology*, **28**, 1–97.

GINGERICH, P. D. 2000. Paleocene–Eocene boundary and continental vertebrate faunas of Europe and North America. *In*: SCHMITZ, B., SUNDQUIST, B. & ANDREASSON, F. P. (eds) *Early Paleogene Warm Climates and Biosphere Dynamics*. GFF (Geologiska Föreningens Förhandlingar), **122**, Geological Society of Sweden, Uppsala, 57–59.

GINGERICH, P. D. 2006. Environment and evolution through the Paleocene–Eocene thermal maximum. *Trends in Ecology & Evolution*, **21**, 246–253.

GINGERICH, P. D. & CLYDE, W. C. 2001. Overview of mammalian biostratigraphy of the Paleocene–Eocene Fort Union and Willwood formations of the Bighorn and Clarks Fork basins. *In*: GINGERICH, P. D. (eds) *Paleocene–Eocene Stratigraphy and Biotic Change in the Bighorn and Clarks Fork Basins, Wyoming*. University of Michigan Papers on Paleontology, **33**, 1–14.

GIUSBERTI, L., RIO, D., AGNINI, C., BACKMAN, J., FORNACIARI, E., TATEO, F. & ODDONE, M. 2007. An expanded marine PETM section in the Venetian Pre-Alps, Italy. *Geological Society of America Bulletin*, **119**, 391–412.

GRADSTEIN, F. M., KAMINSKI, M. A., BERGGREN, W. A., KRISTIANSEN, I. L. & D'IORO, M. A. 1994. Cenozoic biostratigraphy of the North Sea and Labrador Shelf. *Micropaleontology*, **40**, 152.

GRADSTEIN, F. M., OGG, J. G. & SMITH, A. G. 2004. *A Geologic Time Scale 2004*. Cambridge University Press, Cambridge.

GUASTI, E., KOUWENHOVEN, T. J., BRINKHUIS, H. & SPEIJER, R. P. 2005. Paleocene sea-level and productivity changes at the southern Tethyan margin (El Kef, Tunisia). *Marine Micropaleontology*, **55**, 1–17.

HALLOCK, P., PREMOLI SILVA, I. & BOERSMA, A. 1991. Similarities between planktonic and larger foraminiferal evolutionary trends through Paleogene paleoceanographic changes. *Palaeogeography, Palaeoclimatology, Palaeoecology*, **83**, 49–64.

HAQ, B. U. & LOHMANN, G. P. 1976. Early Cenozoic calcareous nannoplankton biogeography of the Atlantic Ocean. *Marine Micropaleontology*, **1**, 119–194.

HARRINGTON, G. J. 2003. Geographic patterns in the floral response to Paleocene–Eocene warming. *In*: WING, S. L., GINGERICH, P. D., SCHMITZ, B. & THOMAS, E. (eds) *Causes and Consequences of Globally Warm Climates in the Early Paleogene*. Geological Society of America Special Paper, **369**, Boulder, 381–393.

HARRINGTON, G. J. & KEMP, S. J. 2001. US Gulf Coast vegetation dynamics during the latest Palaeocene. *Palaeogeography Palaeoclimatology Palaeoecology*, **167**, 1–21.

HARRINGTON, G. J., KEMP, S. J. & KOCH, P. L. 2004. Palaeocene–Eocene paratropical floral change in North America: Responses to climate change and plant immigration. *Journal of the Geological Society*, **161**, 173–184.

HEILMANN-CLAUSEN, C. 1985. Dinoflagellate stratigraphy of the Uppermost Danian to Ypresian in the Viborg 1 borehole, Central Jylland, Denmark. *DGU*, **A7**, 1–69.

HEILMANN-CLAUSEN, C. & EGGER, H. 2000. The Anthering outcrop (Austria), a key-section for correlation between Tethys and Northwestern Europe near the Paleocene/Eocene boundary. *In*: SCHMITZ, B., SUNDQUIST, B. & ANDREASSON, F. P. (eds) *Early Paleogene Warm Climates and Biosphere Dynamics*. GFF (Geologiska Föreningens Förhandlingar), **122**, Geological Society of Sweden. Uppsala, 69.

HIGGINS, J. A. & SCHRAG, D. P. 2006. Beyond methane: Towards a theory for the Paleocene–Eocene Thermal Maximum. *Earth and Planetary Science Letters*, **245**, 523–537.

HOLLIS, C. J. 2006. Radiolarian faunal turnover through the Paleocene–Eocene transition, Mead Stream, New Zealand. *Eclogae Geologicae Helvetiae*, **99**, Supp. 1, 579–599.

HOLLIS, C. J., DICKENS, G. R., FIELD, B. D., JONES, C. M. & PERCY STRONG, C. 2005a. The Paleocene–Eocene transition at Mead Stream, New Zealand: a southern Pacific record of early Cenozoic global change. *Palaeogeography, Palaeoclimatology, Palaeoecology*, **215**, 313–343.

HOLLIS, C. J., FIELD, B. D., JONES, C. M., STRONG, C. P., WILSON, G. J. & DICKENS, G. R. 2005b. Biostratigraphy and carbon isotope stratigraphy of uppermost Cretaceous–lower Cenozoic Muzzle Group in middle Clarence valley, New Zealand. *Journal of the Royal Society of New Zealand*, **35**, 345–383.

HOLROYD, P. A., HUTCHISON, J. H. & STRAIT, S. G. 2001. Turtle diversity and abundance through the lower Eocene Willwood Formation of the southern Bighorn Basin. *University of Michigan Papers on Paleontology*, **33**, 97–107.

HOOKER, J. J. 1998. Mammalian faunal change across the Paleocene–Eocene transition in Europe. *In*: AUBRY, M. P., LUCAS, S. & BERGGREN, W. A. (eds) *Late Paleocene–Early Eocene Climatic and Biotic Events in the Marine and Terrestrial Records*. Columbia University Press, New York, 428–450.

HUBER, M., SLOAN, L. C. & SHELLITO, C. J. 2003. Early Paleogene oceans and climate: A fully coupled modeling approach using the NCAR CCSM. *In*: WING, S. L., GINGERICH, P. D., SCHMITZ, B. & THOMAS, E. (eds) *Causes and Consequences of Globally Warm Climates in the Early Palaeogene*. Geological Society of America Special Paper, **369**. Geological Society of America, Boulder, 25–47.

IAKOVLEVA, A. I., BRINKHUIS, H. & CAVAGNETTO, C. 2001. Late Palaeocene–Early Eocene dinoflagellate cysts from the Turgay Strait, Kazakhstan; correlations across ancient seaways. *Palaeogeography, Palaeoclimatology, Palaeoecology*, **172**, 243–268.

Intergovernmental Panel on Climate Change 2001. Climate Change 2001, The Scientific Basis. http://www.grida.no/climate/ipcc_tar/wg1/index.htm.

JACKSON, S. T. & OVERPECK, J. T. 2000. Responses of plant populations and communities to environmental changes of the late Quaternary. *Paleobiology*, **26** 194–220.

JANDUCHÈNE, R. E. & ADEDIRAN, S. A. 1984. Late Paleocene to Early Eocene Dinoflagellates from Nigeria. *Cahiers de Micropaléontologie*, **3**, 88.

JARAMILLO, C. A. & DILCHER, D. L. 2000. Microfloral diversity patterns of the late Paleocene–Eocene interval in Colombia, northern South America. *Geology*, **28**, 815–818.

KAMINSKI, M. A., KUHNT, W. A. & RADLEY, J. D. 1996. Palaeocene–Eocene deep water agglutinated foraminifera from the Namibian Flysch (Rif, Northern Morocco), their significance for the palaeoceanography of the Gibraltar Gateway. *Journal of Micropalaeontology*, **15**, 1–19.

KELLY, D. C., BRALOWER, T. J., ZACHOS, J. C., PREMOLI SILVA, I. & THOMAS, E. 1996. Rapid diversification of planktonic foraminifera in the tropical Pacific (ODP Site 865) during the late Paleocene thermal maximum. *Geology*, **24**, 423–426.

KELLY, D. C., BRALOWER, T. J. & ZACHOS, J. C. 1998. Evolutionary consequences of the latest Paleocene thermal maximum for tropical planktonic foraminifera. *Palaeogeography, Palaeoclimatology, Palaeoecology*, **141**, 139–161.

KELLY, D. C., ZACHOS, J. C., BRALOWER, T. J. & SCHELLENBERG, S. A. 2005. Enhanced terrestrial weathering/runoff and surface ocean carbonate production during the recovery stages of the Paleocene–Eocene thermal maximum. *Paleoceanography*, **20**, doi:10.1029/2005PA001163.

KENNETT, J. P. & STOTT, L. D. 1991. Abrupt deep-sea warming, palaeoceanographic changes and benthic extinctions at the end of the Palaeocene. *Nature*, **353**, 225–229.

KENT, D. V., CRAMER, B. S., LANCI, L., WANG, D., WRIGHT, J. D. & VOO, R. V. d. 2003. A case for a comet impact trigger for the Paleocene/Eocene thermal maximum and carbon isotope excursion. *Earth and Planetary Science Letters*, **211**, 13–26.

KLEYPAS, J. A., FEELY, R. A., FABRY, V. J., LANGDON, C., SABINE, C. L. & ROBBINS, L. L. 2006. Impacts of Ocean Acidification on Coral Reefs and Other Marine Calcifiers, A Guide for Future Research. Available at: http://www.isse.ucar.edu/florida/. Institute for the Study of Society and Environment, National Center for Atmospheric Research, 89.

KNOX, R. W. O. B., CORFIELD, R. M. & DUNAY, R. S. (eds) 1996. Correlation of the Early Paleogene in Northwest Europe. *Geological Society of London Special Publication*, **101**, Geological Society, London, UK.

KOCH, P. L., ZACHOS, J. C. & GINGERICH, P. D. 1992. Correlation between isotope records in marine and continental carbon reservoirs near the Palaeocene/Eocene boundary. *Nature*, **358**, 319–322.

KOCH, P. L., ZACHOS, J. C. & DETTMAN, D. L. 1995. Stable isotope stratigraphy and paleoclimatology of the Paleogene Bighorn Basin (Wyoming, USA). *Palaeogeography, Palaeoclimatology, Palaeoecology*, **115**, 61–89.

KOCH, P. L., CLYDE, W. C., HEPPLE, R. P., FOGEL, M. L., WING, S. L. & ZACHOS, J. C. 2003. Carbon and oxygen isotope records from paleosols spanning the Paleocene–Eocene boundary, Bighorn Basin, Wyoming. *In*: WING, S. L., GINGERICH, P. D., SCHMITZ, B. & THOMAS, E. (eds) *Causes and Consequences of Globally Warm Climates in the Early Paleogene*. Geological Society of America Special Paper, **369**, Boulder, CO, 49–64.

KÖTHE, A., KHAN, A. M. & ASHRAF, M. 1988. Biostratigraphy of the Surghar Range, Salt Range, Sulaiman Range and the Kohat area, Pakistan, according to Jurassic through Paleogene Calcareous Nannofossils and Paleogene Dinoflagellates. Geologisches Jahrbuch, Reihe B, **71**. *Bundesanstalt für Geowissenschaften und Rohnstoffe und den Geologischen Landesämtern in der Bundesrepublik Deutschland*, Hannover, 87.

KRAUSE, D. W. & MAAS, M. C. 1990. The biogeographic origins of late Paleocene–early Eocene mammalian immigrants to the western interior of North America. *In*: BOWN, T. M. & ROSE, K. D. (eds) *Dawn of the Age of Mammals in the Northern Part of the Rocky Mountain Interior, North America*. Geological Society of America Special Paper **243**, Boulder, CO, 71–105.

KURTZ, A., KUMP, L. R., ARTHUR, M. A., ZACHOS, J. C. & PAYTAN, A. 2003. Early Cenozoic decoupling of the global carbon and sulphur cycles. *Paleoceanography*, **18**, doi:10.1029/2003PA000908.

KVENVOLDEN, K. A. 1988. Methane hydrate – a major reservoir of carbon in the shallow geosphere? *Chemical Geology*, **71**, 41–51.

KVENVOLDEN, K. A. 1993. Gas hydrates: Geological perspective and global change. *Review of Geophysics*, **31**, 173–187.

LOURENS, L. J., SLUIJS, A., KROON, D. *ET AL.* 2005. Astronomical pacing of late Palaeocene to early Eocene global warming events. *Nature*, **435**, 1083–1087.

LU, G. & KELLER, G. 1993. The Paleocene–Eocene transition in the Antarctic Indian Ocean: Inference from planktonic foraminifera. *Marine Micropaleontology*, **21**, 101–142.

LU, G. & KELLER, G. 1995. Ecological stasis and saltation: species richness change in planktonic foraminifera during the late Paleocene to early Eocene, DSDP Site 577. *Palaeogeography, Palaeoclimatology, Palaeoecology*, **117**, 211–227.

MACDONALD, G. J. 1990. Role of methane clathrates in past and future climates. *Climatic Change*, **16**, 247–281.

MACLENNAN, J. & JONES, S. M. 2006. Regional uplift, gas hydrate dissociation and the origins of the Paleocene–Eocene Thermal Maximum. *Earth and Planetary Science Letters*, **245**, 65–80.

MAGIONCALDA, R., DUPUIS, C., SMITH, T., STEURBAUT, E. & GINGERICH, P. D. 2004. Paleocene–Eocene carbon isotope excursion in organic carbon and pedogenic carbonate: Direct comparison in a continental stratigraphic section. *Geology*, **32**, 553–556.

MARLAND, G., BODEN, T. A. & ANDRES, R. J. 2005. Global, Regional, and National CO_2 Emissions. *Trends: A Compendium of Data on Global Change*. Carbon Dioxide Information Analysis Center, Oak Ridge National Laboratory, US Department of Energy, Oak Ridge, USA.

MARTINI, E. 1971. Standard Tertiary and Quaternary calcareous nannoplankton zonation. *In*: A. FARINACCI (ed.) *Proceedings of the II Planktonic Conference, Roma 1970. Vol. 2*. Edizioni Tecnoscienza, Rome, 739–785.

MATSUMOTO, R. 1995. Causes of the $\delta^{13}C$ anomalies of carbonates and a new paradigm "gas-hydrate hypothesis". *Journal of the Geological Society of Japan*, **101**, 902–904.

MCKENNA, M. C. 1983. Holarctic landmass rearrangement, cosmic events, and Cenozoic terrestrial organisms. *Annals of the Missouri Botanical Garden*, **70**, 459–489.

MENG, J., BOWEN, G. J., YE, J., KOCH, P. L., TING, S., LI, Q. & JIN, X. 2004. *Gomphos elkema* (Glires, Mammalia) from the Erlian Basin: Evidence for the Early Tertiary Bumbanian Land Mammal Age in Nei-Mongol, China. *American Museum Novitates*, **3425**, 1–25.

MILKOV, A. V. 2004. Global estimates of hydrate-bound gas in marine sediments: how much is really out there? *Earth-Science Reviews*, **66**, 183–197.

MILLER, K. G., WRIGHT, J. D. & BROWNING, J. V. 2005. Visions of ice sheets in a greenhouse world. *Marine Geology*, **217**, 215–231.

MONECHI, S., ANGORI, E. & VON-SALIS, K. 2000. Calcareous nannofossil turnover around the Paleocene/Eocene transition at Alamedilla (southern Spain). *Bulletin de la Société Géologique de France*, **171**, 477–489.

MUKHOPADHYAY, S., FARLEY, K. A. & MONTANARI, A. 2001. A 35 Myr record of helium in pelagic limestones from Italy: implications for interplanetary dust accretion from the early Maastrichtian to the middle Eocene. *Geochimica et Cosmochimica Acta*, **65**, 653–669.

NOHR-HANSEN, H. 2003. Dinoflagellate cyst stratigraphy of the Palaeogene strata from the Hellefisk-1, Ikermiut-1, Kangamiut-1, Nukik-1, Nukik-2 and Qulleq-1 wells, offshore West Greenland. *Marine and Petroleum Geology*, **20**, 987–1016.

NORRIS, R. D. & RÖHL, U. 1999. Carbon cycling and chronology of climate warming during the Palaeocene/Eocene transition. *Nature*, **401**, 775–778.

NUNES, F. & NORRIS, R. D. 2006. Abrupt reversal in ocean overturning during the Palaeocene/Eocene warm period. *Nature*, **439**, 60–63.

OKADA, H. & BUKRY, D. 1980. Supplementary modification and introduction of code numbers to the low-latitude coccolith biostratigraphic zonation (Bukry, 1973; 1975). *Marine Micropaleontology*, **5**, 321–325.

ORESHKINA, T. V. & OBERHÄNSLI, H. 2003. Diatom turnover in the early Paleogene diatomite of the Sengiley section, middle Povolzhie, Russia: A response to the initial Eocene thermal maximum? *In*: WING, S. L., GINGERICH, P. D., SCHMITZ, B. & THOMAS, E. (eds) *Causes and Consequences of Globally Warm Climates in the Early Paleogene*. Geological Society of America Special Paper, **369**, Boulder, CO, 169–179.

ORR, J. C., FABRY, V. J., AUMONT, O. *ET AL.* 2005. Anthropogenic ocean acidification over the twenty-first century and its impact on calcifying organisms. *Nature*, **437**, 681–686.

ORTIZ, N. 1995. Differential patterns of benthic foraminiferal extinctions near the Paleocene/Eocene boundary in the North Atlantic and western Tethys. *Marine Micropaleontology*, **26**, 341–359.

OVERPECK, J. T., WEBB, R. S. & WEBB III, T. 1992. Mapping eastern North American vegetation change of the past 18 ka: No-analogs and the future. *Geology*, **20**, 1071–1074.

PAGANI, M., PEDENTCHOUK, N., HUBER, M. *ET AL.* 2006. Arctic hydrology during global warming at the Palaeocene–Eocene thermal maximum. *Nature*, **442**, 671–675.

PAK, D. K. & MILLER, K. G. 1992. Paleocene to Eocene benthic foraminiferal isotopes and assemblages: Implications for deep water circulation. *Paleoceanography*, **7**, 405–422.

PARDO, A., KELLER, G., MOLINA, E. & CANUDO, J. 1997. Planktonic foraminiferal turnover across the Paleocene–Eocene transition at DSDP Site 401, Bay of Biscay, North Atlantic. *Marine Micropaleontology*, **29**, 129–158.

PEARSON, P. N. & BERGGREN, W. A. 2005. Revised Tropical to Subtropical Paleogene Planktonic Foraminiferal Zonation. *Journal of Foraminiferal Research* **35**, 279–298.

PETERS, R. B. & SLOAN, L. C. 2000. High concentrations of greenhouse gases and polar stratospheric clouds: A possible solution to high-latitude faunal migration at the latest Paleocene thermal maximum. *Geology*, **28**, 979–982.

PHIPPS MORGAN, J., RESTON, T. J. & RANERO, C. R. 2004. Contemporaneous mass extinctions, continental flood basalts, and 'impact signals': are mantle plume-induced lithospheric gas explosions the causal link? *Earth and Planetary Science Letters*, **217**, 263–284.

PIERREHUMBERT, R. T. 2002. The hydrologic cycle in deep-time climate problems. *Nature*, **419**, 191–198.

POPP, B. N., LAWS, E. A., BIDIGARE, R. R., DORE, J. E., HANSON, K. L. & WAKEHAM, S. G. 1998. Effect of Phytoplankton Cell Geometry on Carbon Isotopic Fractionation. *Geochimica et Cosmochimica Acta*, **62**, 69–77.

POWELL, A. J., BRINKHUIS, H. & BUJAK, J. P. 1996. Upper Paleocene–Lower Eocene dinoflagellate cyst sequence biostratigraphy of southeast England. *In*: KNOX, R. W. O. B., CORFIELD, R. M. & DUNAY, R. S. (eds) *Correlation of the Early Paleogene in Northwest Europe*. Geological Society Special Publication, **101**, 145–183.

PROSS, J. & BRINKHUIS, H. 2005. Organic-walled dinoflagellate cysts as paleoenvironmental indicators in the Paleogene; a synopsis of concepts. *Paläontologische Zeitschrift*, **79**, 53–59.

PUJALTE, V., ORUE-ETXEBARRIA, X., SCHMITZ, B. ET AL. 2003. Basal Ilerdian (earliest Eocene) turnover of larger Foraminifera: Age constraints based on calcareous plankton and $\delta^{13}C$ isotopic profiles from new southern Pyrenean sections (Spain). *In*: WING, S. L., GINGERICH, P. D., SCHMITZ, B. & THOMAS, E. (eds) *Causes and Consequences of Globally Warm Climates in the Early Paleogene*. Geological Society of America Special Paper, **369**, Boulder, CO, USA, 205–221.

RADIONOVA, E. P., KHOKHLOVA, I. E., BANIAMOVSKII, V. N., SHCHERBININA, E. A., IAKOVLEVA, A. I. & SADCHIKOVA, T. A. 2001. The Paleocene/Eocene transition in the northeastern Peri-Tethys area: Sokolovskii key section of the Turgay Passage (Kazakhstan). *Bulletin de la Société Géologique de France*, **172**, 245–256.

RAFFI, I., BACKMAN, J. & PÄLIKE, H. 2005. Changes in calcareous nannofossil assemblages across the Paleocene/Eocene transition from the paleo-equatorial Pacific Ocean. *Palaeogeography, Palaeoclimatology, Palaeoecology*, **226**, 93–126.

RAVIZZA, G., NORRIS, R. N. & BLUSZTAJN, J. 2001. An osmium isotope excursion associated with the late Paleocene thermal maximum: Evidence of intensified chemical weathering. *Paleoceanography*, **16**, 155–163.

RENSSEN, H., BEETS, C. J., FICHEFET, T., GOOSSE, H. & KROON, D. 2004. Modeling the climate response to a massive methane release from gas hydrates. *Paleoceanography*, **18** (PA2010), doi:10.1029/2003PA000968.

ROBERT, C. & KENNETT, J. P. 1994. Antarctic subtropical humid episode at the Paleocene–Eocene boundary: clay mineral evidence. *Geology*, **22**, 211–214.

RÖHL, U., BRALOWER, T. J., NORRIS, G. & WEFER, G. 2000. A new chronology for the late Paleocene thermal maximum and its environmental implications. *Geology*, **28**, 927–930.

RÖHL, U., NORRIS, R. D. & OGG, J. G. 2003. Cyclostratigraphy of upper Paleocene and lower Eocene sediments at Blake Nose Site 1051 (western North Atlantic). *In*: WING, S. L., GINGERICH, P. D., SCHMITZ, B. & THOMAS, E. (eds) *Causes and Consequences of Globally Warm Climates in the Early Paleogene*. Geological Society of America Special Paper, **369**. Geological Society of America, Boulder, CO, 567–589.

RÖHL, U., WESTERHOLD, T., MONECHI, S., THOMAS, E., ZACHOS, J. C. & DONNER, B. 2005. The third and final early Eocene thermal maximum: characteristics, timing, and mechanisms of the "X" event. *Geological Society of America Annual Meeting. Abstr.* **37**, 264.

SANFILIPPO, A. & NIGRINI, C. 1998. Upper Palaeocene–Lower Eocene deep-sea radiolarian stratigraphy and the Palaeocene–Eocene series boundary. *In*: AUBRY, M.-P., LUCAS, S. G. & BERGGREN, W. A. (eds) *Late Palaeocene–Early Eocene Climatic and Biotic Events in the Marine and Terrestrial Records*. Columbia University Press, New York, 244–276.

SANFILIPPO, A. & BLOME, C. D. 2001. Biostratigraphic implications of mid-latitude Palaeocene–Eocene radiolarian faunas from Hole 1051A, ODP Leg 171B, Blake Nose, western North Atlantic. *In*: KROON, D., NORRIS, R. D. & KLAUS, A. (eds) *Western North Atlantic Palaeogene and Cretaceous Palaeoceanography*. Geological Society, London, Special Publications, **183**, 185–224.

SCHEIBNER, C., SPEIJER, R. P. & MARZOUK, A. M. 2005. Turnover of larger foraminifera during the Paleocene–Eocene Thermal Maximum and paleoclimatic control on the evolution of platform ecosystems. *Geology*, **33**, 493–496.

SCHENAU, S. J., PRINS, M., DE LANGE, G. J. & MONNIN, C. 2001. Barium accumulation in the Arabian Sea: Controls on barite preservation in marine sediments. *Geochimica et Cosmochimica Acta*, **65**, 1545–1556.

SCHMIDT, G. A. & SCHINDELL, D. T. 2003. Atmospheric composition, radiative forcing, and climate change as a consequence of a massive methane release from gas hydrates. *Paleoceanography*, **18**, doi:1010. 1029/2002PA000757.

SCHMITZ, B., SPEIJER, R. P. & AUBRY, M.-P. 1996. Latest Paleocene benthic extinction event on the southern Tethyan shelf (Egypt): Foraminiferal stable isotopic ($\delta^{13}C$ and $\delta^{18}O$) records. *Geology*, **24**, 347–350.

SCHMITZ, B., PEUCKER-EHRENBRINK, B., HEILMANN-CLAUSEN, C., ÅBERG, G., ASARO, F. & LEE, C.-T. A. 2004. Basaltic explosive volcanism, but no comet impact, at the Paleocene–Eocene boundary: high-resolution chemical and isotopic records from Egypt, Spain and Denmark. *Earth and Planetary Science Letters*, **225**, 1–17.

SCOTESE, C. R. 2002. http://www.scotese.com, (PALEOMAP website).

SEXTON, P. F., WILSON, P. A. & NORRIS, R. D. 2006. Testing the Cenozoic multisite composite $\delta^{18}O$ and $\delta^{13}C$ curves: New monospecific Eocene records from a single locality, Demerara Rise (Ocean Drilling Program Leg 207). *Paleoceanography*, **21**, doi:10.1029/2005PA001253.

SHACKLETON, N. J. 1967. Oxygen isotope analyses and Pleistocene temperatures reassessed. *Nature*, **215**, 15–17.

SLUIJS, A. 2006. Global Change during the Paleocene–Eocene thermal maximum. PhD thesis, Laboratory of Palaeobotany and Palynology Foundation, **21**, Utrecht, The Netherlands, 228.

SLUIJS, A., PROSS, J. & BRINKHUIS, H. 2005. From greenhouse to icehouse; organic-walled dinoflagellate cysts as paleoenvironmental indicators in the Paleogene. *Earth-Science Reviews*, **68**, 281–315.

SLUIJS, A., SCHOUTEN, S., PAGANI, M. *ET AL.* 2006. Subtropical Arctic Ocean temperatures during the Palaeocene/Eocene thermal maximum. *Nature*, **441**, 610–613.

SMITH, T. & SMITH, R. 1995. Synthese des donnees actuelles sur les vertebres de la transition Paleocene–Eocene de Dormaal (Belgique). *Bulletin de la Société belge de Géologie*, **104**, 119–131.

SMITH, T., ROSE, K. D. & GINGERICH, P. D. 2006. Rapid Asia–Europe–North America geographic dispersal of earliest Eocene primate *Teilhardina* during the Paleocene–Eocene Thermal Maximum. *Proceedings of the National Academy of Sciences of the United States of America*, **103**, 11223–11227.

SPEIJER, R. P. & SCHMITZ, B. 1998. A benthic foraminiferal record of Paleocene sea level and trophic/redox conditions at Gebel Aweina, Egypt. *Palaeogeography, Palaeoclimatology, Palaeoecology*, **137**, 79–101.

SPEIJER, R. P. & MORSI, A.-M. M. 2002. Ostracode turnover and sea-level changes associated with the Paleocene–Eocene thermal maximum. *Geology*, **30**, 23–26.

SPEIJER, R. P. & WAGNER, T. 2002. Sea-level changes and black shales associated with the late Paleocene thermal maximum: Organic-geochemical and micropaleontologic evidence from the southern Tethyan margin (Egypt-Israel). *Geological Society of America Special Paper*, **356**, 533–549.

SPEIJER, R. P., SCHMITZ, B. & VAN DER ZWAAN, G. J. 1997. Benthic foraminiferal extinction and repopulation in response to latest Paleocene Tethyan anoxia. *Geology*, **27**, 683–686.

SPERO, H. J., BIJMA, J., LEA, D. W. & BEMIS, B. E. 1997. Effect of seawater carbonate concentration on foraminiferal carbon and oxygen isotopes. *Nature*, **390**, 497–500.

SPIESS, V. 1990. Cenozoic magnetostratigraphy of Leg 113 drill sites, Maud Rise, Weddell Sea, Antartcica. *In*: BARKER, P. F., KENNETT, J. P., *ET AL.* (eds) *Proceedings of the Ocean Drilling Project, Initial Reports*, **113**, 261–315.

STEINECK, P. L. & THOMAS, E. 1996. The latest Paleocene crisis in the deep sea: Ostracode succession at Maud Rise, Southern Ocean. *Geology*, **24**, 583–586.

STEURBAUT, E., MAGIONCALDA, R., DUPUIS, C., VAN SIMAEYS, S., ROCHE, E. & ROCHE, M. 2003. Palynology, paleoenvironments, and organic carbon isotope evolution in lagoonal Paleocene–Eocene boundary settings in North Belgium. *In*: WING, S. L., GINGERICH, P., SCHMITZ, B. & THOMAS, E. (eds) *Causes and Consequences of Globally Warm Climates in the Early Paleogene*. Geological Society of America Special Paper, **369**. Geological Society of America, Boulder, CO, 291–317.

STOLL, H. M. 2005. Limited range of interspecific vital effects in coccolith stable isotopic records during the Paleocene–Eocene thermal maximum. *Paleoceanography*, **20**, doi:10.1029/2004PA001046.

STOLL, H. M. & BAINS, S. 2003. Coccolith Sr/Ca records of productivity during the Paleocene–Eocene thermal maximum from the Weddell Sea. *Paleoceanography*, **18**, doi:10.1029/2002PA000875.

STOTT, L. D., SINHA, A., THIRY, M., AUBRY, M.-P. & BERGGREN, W. A. 1996. Global $\delta^{13}C$ changes across the Paleocene–Eocene boundary; criteria for terrestrial-marine correlations. *In*: KNOX, R. W. O. B., CORFIELD, R. M. & DUNAY, R. E. (eds) *Correlation of the Early Paleogene in Northwestern Europe*. Geological Society London Special Publication, **101**, Geological Society of London, UK.

SVENSEN, H., PLANKE, S., MALTHE-SØRENSEN, A., JAMTVEIT, B., MYKLEBUST, R., EIDEM, T. R. & REY, S. S. 2004. Release of methane from a volcanic basin as a mechanism for initial Eocene global warming. *Nature*, **429**, 542–545.

THOMAS, D. J. 2004. Evidence for deep-water production in the North Pacific Ocean during the early Cenozoic warm interval. *Nature*, **430**, 65–68.

THOMAS, D. J. & BRALOWER, T. J. 2005. Sedimentary trace element constraints on the role of North Atlantic Igneous Province volcanism in late Paleocene–early Eocene environmental change. *Marine Geology*, **217**, 233–254.

THOMAS, D. J., ZACHOS, J. C., BRALOWER, T. J., THOMAS, E. & BOHATY, S. 2002. Warming the fuel for the fire: Evidence for the thermal dissociation of methane hydrate during the Paleocene–Eocene thermal maximum. *Geology*, **30**, 1067–1070.

THOMAS, E. 1989. Development of Cenozoic deep-sea benthic foraminiferal faunas in Antarctic waters. *Geological Society London Special Publication*, **47**, 283–296.

THOMAS, E. 1998. Biogeography of the late Paleocene benthic foraminiferal extinction. *In*: AUBRY, M.-P., LUCAS, S. G. & BERGGREN, W. A. (eds) *Late Paleocene–Early Eocene Climatic and Biotic Events in the Marine and Terrestrial Records*. Columbia University Press, New York, 214–243.

THOMAS, E. 2003. Extinction and food at the seafloor: A high-resolution benthic foraminiferal record across the Initial Eocene Thermal Maximum, Southern Ocean Site 690. *In*: WING, S. L., GINGERICH, P. D., SCHMITZ, B. & THOMAS, E. (eds) *Causes and Consequences of Globally Warm Climates in the Early Paleogene*. Geological Society of America Special

Paper **369**. Geological Society of America, Boulder, CO, 319–332.

THOMAS, E. 2007. Cenozoic mass extinctions in the deep sea: what perturbs the largest habitat on earth? *Geological Society of America Special Paper*, **424**. Geological Society of America, Bolder, Colorado, 1–24.

THOMAS, E. & SHACKLETON, N. J. 1996. The Palaeocene–Eocene benthic foraminiferal extinction and stable isotope anomalies. *In*: KNOX, R. W. O. B., CORFIELD, R. M. & DUNAY, R. E. (eds) *Correlation of the Early Paleogene in Northwestern Europe*. Geological Society London Special Publication, **101**. Geological Society of London, UK, 401–441.

THOMAS, E. & ZACHOS, J. C. 2000. Was the late Paleocene thermal maximum a unique event? *Geologiska Föreningens i Stockholm Förhandlingar (GFF; Transactions of the Geological Society in Stockholm)*, **122**, 169–170.

THOMAS, E., ZACHOS, J. C. & BRALOWER, T. J. 2000. Deep-sea environments on a warm Earth: Latest Paleocene-early Eocene. *In*: HUBER, B. T., MACLEOD, K. & WING, S. L. (eds) *Warm Climates in Earth History*. Cambridge University Press, 132–160.

TING, S., BOWEN, G. J., KOCH, P. L. *ET AL.* 2003. Biostratigraphic, chemostratigraphic, and magnetostratigraphic study across the Paleocene/Eocene boundary in the Hengyang Basin, Hunan, China. *In*: WING, S. L., GINGERICH, P. D., SCHMITZ, B. & THOMAS, E. (eds) *Causes and Consequences of Globally Warm Climates in the Early Paleogene*, Geological Society of America Special Paper, **369**, Boulder, CO, 521–535.

TJALSMA, R. C. & LOHMANN, G. P. 1983. Paleocene–Eocene bathyal and abyssal benthic foraminifera from the Atlantic Ocean, *Micropaleontology, Special Publication*, **4**, 1–90.

TREMOLADA, F. & BRALOWER, T. J. 2004. Nannofossil assemblage fluctuations during the Paleocene–Eocene Thermal Maximum at Sites 213 (Indian Ocean) and 401 (North Atlantic Ocean): palaeoceanographic implications. *Marine Micropaleontology*, **52**, 107–116.

TRIPATI, A. K. & ELDERFIELD, H. 2004. Abrupt hydrographic changes in the equatorial Pacific and subtropical Atlantic from foraminiferal Mg/Ca indicate greenhouse origin for the thermal maximum at the Paleocene–Eocene Boundary. *Geochemistry, Geophysics, Geosystems*, **5**, doi:10.1029/2003GC000631.

TRIPATI, A. & ELDERFIELD, H. 2005. Deep-Sea Temperature and Circulation Changes at the Paleocene–Eocene Thermal Maximum. *Science*, **308**, 1894–1898.

VAN EETVELDE, Y. & DUPUIS, C. 2004. Upper Palaeocene and lower Eocene interval in the Dieppe–Hampshire Basin; biostratigraphic analysis based on pyritized diatoms. *In*: BEAUDOIN, A. B. & HEAD, M. J. (eds) *The Palynology and Micropaleontology of Boundaries*. Geological Society Special Publications, **230**. Geological Society of London, UK, 275–291.

WESTERHOLD, T., RÖHL, U., LASKAR, J., RAFFI, I., BOWLES, J., LOURENS, L. J. & ZACHOS, J. C. 2007. On the duration of Magnetochrons C24r and C25n, and the timing of early Eocene global warming events: Implications from the ODP Leg 208 Walvis Ridge depth transect. *Paleoceanography*, **22**, doi:10.1029/2006PA001322.

WING, S. L. & HARRINGTON, G. J. 2001. Floral response to rapid warming in the earliest Eocene and implications for concurrent faunal change. *Paleobiology*, **27**, 539–563.

WING, S. L., BAO, H. & KOCH, P. L. 2000. An early Eocene cool period? Evidence for continental cooling during the warmest part of the Cenozoic. *In*: HUBER, B. T., MACLEOD, K. G. & WING, S. L. (eds) *Warm climates in earth history*. Cambridge University Press, Cambridge, UK.

WING, S. L., HARRINGTON, G. J., BOWEN, G. J. & KOCH, P. L. 2003. Floral change during the Initial Eocene Thermal Maximum in the Powder River Basin, Wyoming. *In*: WING, S. L., GINGERICH, P. D., SCHMITZ, B. & THOMAS, E. (eds) *Causes and Consequences of Globally Warm Climates in the Early Paleogene*. Geological Society of America Special Paper, **369**, Boulder, CO, 425–440.

WING, S. L., HARRINGTON, G. J., SMITH, F. A., BLOCH, J. I., BOYER, D. M. & FREEMAN, K. H. 2005. Transient Floral Change and Rapid Global Warming at the Paleocene–Eocene Boundary. *Science*, **310**, 993–996.

XU, W., LOWELL, R. P. & PELTZER, E. T. 2001. Effect of seafloor temperature and pressure variations on methane flux from a gas hydrate layer: Comparison between current and late Paleocene climate conditions. *Journal of Geophysical Research*, **106**, 26,413–26,423.

ZACHOS, J. C., LOHMANN, K. C., WALKER, J. C. G. & WISE, S. W. 1993. Abrupt climate change and transient climates during the Palaeogene: A marine perspective. *Journal of Geology*, **101**, 191–213.

ZACHOS, J., PAGANI, M., SLOAN, L., THOMAS, E. & BILLUPS, K. 2001. Trends, rhythms, and aberrations in global climate 65 Ma to present. *Science*, **292**, 686–693.

ZACHOS, J. C., WARA, M. W., BOHATY, S. *ET AL.* 2003. A transient rise in tropical sea surface temperature during the Paleocene–Eocene thermal maximum. *Science*, **302**, 1151–1154.

ZACHOS, J. C., KROON, D., BLUM, P. *ET AL.* 2004. *Proceedings of the Ocean Drilling Program, Initial Reports*, 208. World Wide Web Address: http://www-odp.tamu.edu/publications/208_IR/208ir.htm.

ZACHOS, J. C., RÖHL, U., SCHELLENBERG, S. A. *ET AL.* 2005. Rapid Acidification of the Ocean during the Paleocene–Eocene Thermal Maximum. *Science*, **308**, 1611–1615.

ZACHOS, J. C., SCHOUTEN, S., BOHATY, S. *ET AL.* 2006. Extreme warming of mid-latitude coastal ocean during the Paleocene–Eocene Thermal Maximum: Inferences from TEX$_{86}$ and Isotope Data. *Geology*, **34**, 737–740.

The Eocene–Oligocene Transition

H. K. COXALL & P. N. PEARSON

School of Earth, Ocean and Planetary Sciences, Cardiff University, Main Building, Park Place, Cardiff, CF10 3AT, UK (e-mail: Helen.Coxall@earth.cf.ac.uk)

Abstract: A diverse array of fossil, geochemical and sedimentary data shows patterns of major change at or near the Eocene–Oligocene boundary, indicating a period of fundamental climatic and biotic reorganization on Earth. Multiple lines of evidence support the hypothesis that these changes signal major global cooling, especially at high latitudes, and rapid growth of semi-permanent ice-sheet on Antarctica in the early Oligocene. The quality and temporal resolution of Eocene–Oligocene fossil and sediment records and the diversity of climatic proxy tools have increased enormously in recent years, bringing a new level of detail to the study of this transition. The higher-resolution records have revealed that the climatic shifts across the Eocene–Oligocene boundary occurred in multiple stages that can be calibrated on orbital time-scales. Coupled with increasingly sophisticated computer models, these records have led to significant advances in our understanding of possible causal mechanisms and feedbacks driving this global shift. Here we review the wealth of evidence for Eocene–Oligocene climate change, summarize the current state of understanding, and highlight the key areas still requiring work that might guide the direction of future research. Obtaining a thorough understanding of this critical climatic transition is important for highlighting the mechanisms and sensitivities of Cenozoic climate, and addressing topical questions relating to the dynamics of global change during greenhouse–icehouse climate switching.

The transition from the Eocene to the Oligocene was a period of global change lasting about 500,000 years that marks a major step towards the development of the modern glaciated climate. Several decades of research (reviewed in this paper), which builds on work stretching back to the early 20th century, have revealed that this interval is associated with extinctions and evolutionary turnover, on land and in the oceans, and major shifts in geochemical and sedimentological proxies. These records provide strong evidence for a phase of oceanic reorganization, global cooling and the growth of the first semi-permanent continental-scale ice-sheets on Antarctica. The causative mechanisms of Eocene–Oligocene (E–O) climate change are widely debated, with much of the discussion centring on the relative roles of declining greenhouse gases and the opening of Southern Hemisphere oceanic gateways in permitting substantial ice build-up on Antarctica. Following Quaternary and Neogene models, attention is now being focused on defining the significance of orbital configurations, which affect the distribution of solar radiation received by the Earth, as well as the role of many possible feedbacks within the Earth System that may have interacted to give the record of E–O change that we see.

Terrestrial and shallow marine records of E–O climate change from around the world reveal significant biotic turnover in plants and animals across a range of latitudes that are linked to climatic cooling, widespread regression and changes in the hydrological cycle. These records are sporadic, however, and the most complete archives are found in the deep-sea realm where sedimentation is often more continuous. Here, E–O environmental changes have left their mark as major shifts in microfossil communities, microfossil geochemistry, sedimentation and mineralogy. Much of the recent progress in understanding the E–O transition has come from palaeoclimate proxies derived from these marine records that are accessed through deep-sea coring.

A major problem has been that most Deep-Sea Drilling Project (DSDP) and Ocean Drilling Program (ODP) sites spanning the E–O boundary are condensed and/or interrupted by hiatuses, which have been attributed to an increase in ocean circulation vigour and glacioeustatic sea-level fall associated with climate change (Kennett & Shackleton 1976; Aubry 1991; Miller *et al.* 1991; Zachos *et al.* 1996). There has, therefore, been a shortage of sequences appropriate for continuous palaeoclimate analysis. Recent advances in drilling technology and methods of stratigraphic correlation have lead to improved recovery of deep-sea Palaeogene sequences, and the quality of E–O sediment archives and derived proxy records has increased significantly. These newly available records, combined with continuously more sophisticated computer models, have led to advances in our understanding of the timing and mechanisms of

From: WILLIAMS, M., HAYWOOD, A. M., GREGORY, F. J. & SCHMIDT, D. N. (eds) *Deep-Time Perspectives on Climate Change: Marrying the Signal from Computer Models and Biological Proxies.* The Micropalaeontological Society, Special Publications. The Geological Society, London, 351–387.
1747-602X/07/$15.00 © The Micropalaeontological Society 2007.

E–O climate change, and helped constrain theories on the feedbacks involved in the inferred cryosphere and biosphere responses.

Here we review the current state of understanding of the E–O transition. The account will begin with a discussion of the systematics of E–O boundary stratigraphic terminology. The core of the paper will review evidence for climatic change from (i) fossils and (ii) palaeoclimatic proxies. Finally, we provide a short synthesis summarizing progress to date and our vision of where future research is and should be directed.

Terminology, correlation and calibration

The E–O boundary is formally defined at the Global Stratotype Section and Point (GSSP) at Massignano, Italy, and corresponds to the extinction of the planktonic foraminiferal Family Hantkeninidae (see Coccioni et al. 1988; Nocchi et al. 1988; Premoli Silva & Jenkins 1993; Berggren et al. 1995). The boundary is a tie-point in the timescale of Cande & Kent (1995) where it is fixed at 33.7 Ma, although this figure is likely to be refined by astronomical tuning (Coxall et al. 2005; Gale et al. 2006; Jovane et al. 2006). We use the term 'Eocene–Oligocene Transition' (EOT) to encompass a phase of accelerated climatic and biotic change lasting 500 kyr that began before and ended after the boundary. This transition interval is most clearly recognized using deep-sea benthic foraminifera stable isotope data. In most records, it begins with a phase of relatively negative carbon isotopes and ends with a peak in both oxygen and carbon isotope ratios. Our definition of the transition interval excludes longer-term events that have sometimes been included in discussion of 'terminal Eocene events'. We start with a clarification and revision of the nomenclature.

The $\delta^{18}O$ peak at the end of the transition interval coincides with the base of an Oligocene isotope Zone defined by Miller et al. (1991) as 'Oi-1', and is widely regarded as signalling a peak in glaciation on Antarctica (Fig. 1). There is some confusion in the literature regarding the meaning of Oi-1, despite the fact that it was clearly defined by Miller et al. (1991). Most workers use Oi-1 to identify the climax of the early Oligocene $\delta^{18}O$ excursion that lasted c. 400 kyr (Zachos et al. 1996; Coxall et al. 2005). As an isotope zone, however, Oi-1 extends between two peaks in the oxygen isotope record, and in fact spans much of the lower Oligocene, equivalent to several million years of time (Miller et al. 1991, Fig. 6, p. 6839). The base of the Oi-1 zone is defined in DSDP Site 522 by the maximum $\delta^{18}O$ value in benthic foraminifera Stilostomella spp. (Oberhansli et al. 1984).

This occurs at 133.13 mbsf, which equates to 33.494 Ma on the geomagnetic polarity timescale of Cande & Kent (1995) (19.38% through C13n) (Miller et al. 1991; see summary by Vergnaud-Grazzini & Oberhansli 1986). This level is slightly above the level of the abrupt shift to high $\delta^{18}O$ recorded in Cibicidoides spp. from the same site (133.59 m, c.33.588 Ma), which represents the base of 'Oi-1' as identified by Zachos et al. (1996). Oi-1 does not correspond to or include an isotopic shift, as implied in some of the literature (e.g. Zachos et al. 2001, Fig. 2, p. 688; Gale et al. 2006, p. 412; Van Mourik & Brinkhuis 2005, p. 13).

The isotopic shift and the $\delta^{18}O$ maximum are clearly both important events that represent different climatic and environmental phases. Here we use the term 'Early Oligocene Glacial Maximum' (EOGM), advocated by Liu et al. (2004) and Tuo et al. (2006) (after Zachos et al. 1996) to identify the phase of maximum $\delta^{18}O$ (Oi-1 of Zachos et al. 1996), and to differentiate from Miller et al.'s (1991) isotope zone. The phase of rapidly increasing $\delta^{18}O$ that precedes the EOGM we refer to as the E-O 'Shift'. In the type section (DSDP Site 522) and elsewhere (Zachos et al. 1996; Coxall et al. 2005), the Shift occurs over a period of several hundred thousand years and encompasses the E-O boundary. In some isotope records (see below), it is resolved as two or more steps (Fig. 1). The 'E-O Transition' in its strict sense, therefore, is equivalent to the 'E-O Shift'. Under an alternative astronomical naming scheme (based on the 400 kyr cycle of Earth's eccentricity) that has recently been proposed for the Oligocene glaciations (Wade & Palike 2005) the EOGM is referred to as event '84Eo C13n'.

Fossils

The fossil record across the E–O boundary has been periodically reviewed as knowledge has increased (Pomerol & Premoli Silva 1986; Prothero & Berggren 1992; Prothero 1994; Ivany et al. 2003; Prothero et al. 2003) and new data continue to be published every year. While the E–O transition interval was not one of the 'Big Five' mass extinctions' (Sepkoski 1986), it was nevertheless a time of substantial extinction and ecological reorganization in many biological groups. Some of these groups are vital for biostratigraphic correlation, so the disappearance or appearance of key taxa in the fossil record is often tied up with attempts to recognize the E–O boundary itself in different facies. Because of fundamental problems in precisely correlating stratigraphic sections across environments, oceans, continents and latitudes, the detailed sequence of biotic events is

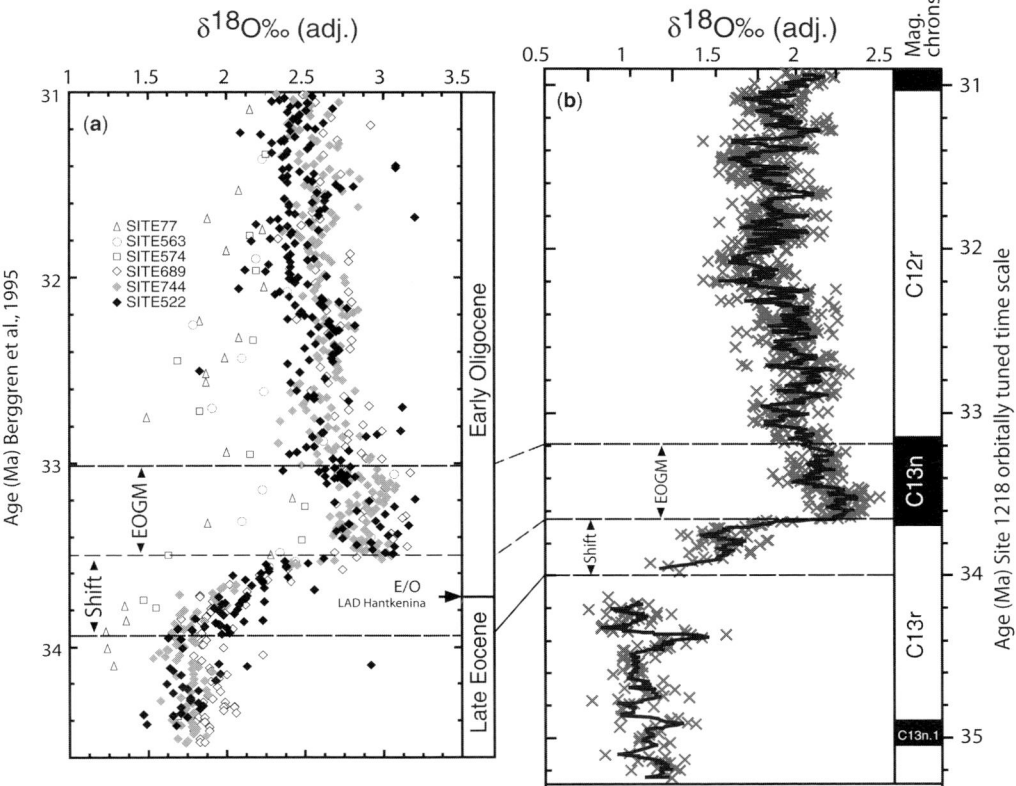

Fig. 1. Terminology, calibration and δ¹⁸O signature of the Eocene–Oligocene Transition. (**a**) Benthic foraminifera $\delta^{18}O$ compilation (after Zachos *et al.* 2001) from mid- to high southern latitudes (diamonds) and the equatorial Pacific (other symbols) plotted on a common time scale (Berggren *et al.* 1995) showing the globally recognizable *c.* $\delta^{18}O$ 1.5‰ shift and positive $\delta^{18}O$ anomaly. The extinction of *Hantkenina* spp. has been identified in South Atlantic DSDP Site 522 (136.7 mbsf; Poore 1984) and is here calibrated (arrow) against the corresponding benthic $\delta^{18}O$ record (black diamonds). (**b**) *Cibicidoides* spp. $\delta^{18}O$ (+0.64) from ODP Site 1218 (3800 m water depth; Coxall *et al.* 2005) (5-point moving average trend line) plotted on an orbitally tuned timescale. We distinguish the interval of $\delta^{18}O$ 'Shift' (which is time equivalent to the Eocene-Oligocene Transition), from the early Oligocene $\delta^{18}O$ maximum, here referred to as the Early Oligocene Glacial Maximum (EOGM) (after Zachos *et al.* 1996; Liu *et al.* 2004). These features are also recognizable in the compilation. The site 1218 data gap corresponds with a zone of severe dissolution through which no calcareous benthic forams were available (see below). Lower $\delta^{18}O$ values at Site 1218 and other tropical sites suggest bottom-water temperatures *c.* 2 °C warmer than high southern latitudes. Species-specific adjustments have been applied (+0.64 or +0.4) to account for vital effects (after Zachos *et al.* 2001).

still not clear at high resolution (Ivany *et al.* 2003). However, it is likely that some of the extinctions are associated with the phases of rapid climatic change and sea-level fall across the E–O boundary and others may be associated with the maximum glacial conditions of the early Oligocene (EOGM, Fig. 1; corresponding to the base of the 'Oi-1' isotope zone of Miller *et al.* 1991). The radiation of more cold-adapted forms in many biotic groups probably began during this glacial period, and continued thereafter into the Oligocene. Ongoing efforts directed at relating patterns

of biotic turnover in different environments to the global isotope and palaeomagnetic record is likely to lead to greater clarity as work progresses.

Planktonic foraminifera

In the Global Stratotype Section at Massignano in Italy, the E–O boundary marker (the 'Golden Spike') is placed at the last occurrence of the planktonic foraminiferal Family Hantkeninidae (Nocchi *et al.* 1988; Premoli Silva & Jenkins 1993). This Family, although seldom dominant in planktonic

Fig. 2. SEM micrographs of species of the planktonic foraminiferal family Hantkeninidae that went extinct at the E–O boundary (see Coxall & Pearson 2006). (**1**) *Hantkenina alabamensis*; (**2**) *H. primitiva*; (**3**) *H. compressa*; (**4**) *H. nanggulanensis*; and (**5**) *Cribrohantkenina inflata*. Specimens are from various deep-sea sites. Scale bar = 100 μm. The extinction of these species denotes the Eocene–Oligocene boundary worldwide.

assemblages, is a very distinctive component of middle and upper Eocene pelagic carbonates world-wide (Fig. 2). Coccioni *et al.* (1988) recognized five species and two genera (*Hantkenina* and *Cribro-hantkenina*) of Hantkeninidae at Massignano, although in their analysis they do not all persist to the boundary, and a similar set of species can be recognized in the Spanish sections such as that at Fuente Caldera (Molina 1986; Molina *et al.* 2006). The earlier suggestion by Blow (1979) that the extinction of *Cribrohantkenina* preceded *Hantkenina* (which was based on limited sampling) can now be discounted. Our own work (Coxall & Pearson 2006) suggests that the extinction of *Hantkenina* and *Cribrohantkenina* was essentially simultaneous and involved all five species. More-over, it exactly coincides with the local extinction of another common species, *Pseudohastigerina micra* (Molina *et al.* 2006), or its dwarfing (Nocchi *et al.* 1986), and is close to the first appearance of a typically Oligocene form, *Globoquadrina tapur-iensis* (Blow & Banner 1962; Coccioni *et al.* 1988; see also Keller 1983 and Keller *et al.* 1992). This means that the E–O boundary is one of the best-defined biostratigraphic levels of the Cenozoic for planktonic foraminifera and in the latest tropical subtropical biozonation it is used as the top of the topmost Eocene zone, E16 (Berggren & Pearson 2005). Although good stratigraphic sections with acceptable carbonate preservation are rare, there is as yet no evidence that the extinction of the Hantkeninidae was locally controlled or diachro-nous (as claimed by Van Mourick & Brinkhuis 2005), except in the high polar latitudes where the group only occasionally occurs during climatically favourable episodes.

Another major extinction in the planktonic fora-miniferal records can be found close to the last occurrence of the Hantkeninidae, namely the extinction of the *Turborotalia cerroazulensis* group, which in most taxonomies consist of three separate species (see Pearson *et al.* 2006 for a recent review). In the Massignano stratotype, this occurs just 60 cm below the Golden Spike, which equates to about 65 kyr (Coccioni *et al.* 1988; Berggren & Pearson 2005). It also narrowly pre-dates the *Hantkenina* extinction at Fuente Caldera in Spain (Molina *et al.* 2006) and several deep-sea drilling sites in the North and South Atlantic and Indian Oceans (e.g. Poore 1984; Pearson & Chaisson 1997). The *T. cerroazulensis* group of species was very abundant and widespread. Both the Hantkeninidae and the *T. cerroazulensis* group existed for many millions of years before their eventual demise, which hints that their closely spaced extinctions were very likely more than coincidental and related to a prolonged phase of environmental disruption. Taken together, they rep-resent one of the most obvious extinctions of the Cenozoic among planktonic foraminifera.

Boersma & Premoli Silva (1986) and Keller *et al.* (1992) noted that the long-term trend of planktonic foraminifera evolutionary turnover from the middle Eocene into the Oligocene largely involves extinction of warm-water, tropical, surface-dwelling species. The extinctions at and near the E–O boundary might therefore be due to rapid environmental change and cooling, and may also have been influenced by changing water mass stratification and patterns of biological productivity.

Unfortunately, it still not possible to confidently (and precisely) locate the planktonic foraminifer extinctions with reference to the stable isotope events (see below). Probably the best published record is DSDP Site 522 on the Walvis Ridge, South Atlantic, where the hantkeninid extinction (Poore 1984) occurs within the isotope shift that precedes the basal EOGM (see Fig. 1), but in Site 522 the planktonic foraminifera are somewhat dissolved and fragmentary. Another relevant section is ODP Site 925 in the tropical North Atlantic. Here the preservation is even worse (in hard limestone), but the hantkeninid extinction (Pearson & Chaisson 1997) occurs above a zone of dissolution and below the most prominent carbon isotopic shift that leads into the EOGM (Diester-Haass & Zachos 2003).

Nannofossils

The other calcareous group that is widely used for deep-sea biostratigraphy is the nannoplankton. Perch-Nielsen (1986) and Aubry (1992) documented a long-term decline in diversity from the middle Eocene into the Oligocene but little change is associated with the boundary interval itself. The only significant extinctions that occur near the E–O boundary interval are the stepped disappearance of rosette-shaped discoasters (Aubry 1992). The disappearance of *Discoaster saipanensis* at approximately 34.2 Ma (i.e. about 500 kyr before the boundary: Berggren *et al.* 1995) is preceded slightly by the extinction of another species, *D. barbadiensis*. The former of these extinctions marks the boundary between nannofossil Zones NP20 and NP21. A more minor event, namely the extinction of *Pemma papillatum*, may be more closely associated with the E–O boundary itself (Varol 1998) but requires further study. Beyond that, no major changes to nannofossil assemblages have been reported, but there are significant biogeographic changes associated with the basal Oligocene glaciation and the onset of a relatively unstratified ocean around Antarctica, where nannoplankton locally disappear from the record (Hay *et al.* 2005).

Radiolaria

Although it has long been thought that radiolaria were largely unaffected by the E–O transition (e.g. Riedel & Sanfillipo 1986), a major turnover of tropical species has recently been described by Funakawa *et al.* (2006) from sites drilled in the Pacific. This turnover includes the extinction of several taxa and a corresponding sudden drop in diversity and radiolarian accumulation rates, and is combined with the expansion of cool-water cosmopolitan taxa. It appears to be closely associated with the stable isotope shifts that precede the

EOGM and awaits study in other areas, especially high latitudes.

Dinoflagellate cysts

Brinkhuis (1992), Brinkhuis & Biffi (1993) and more recently Van Mourik & Brinkhuis (2005) have reviewed the evolution of dinoflagellate cysts (dinocysts) across the Eocene–Oligocene boundary interval in Italy. Van Mourik & Brinkhuis (2005) present new data from the GSSP and a neighbouring sediment core (the 'Massicore'). Across the E–O boundary interval in Italy, two successive influxes of cool-water high-latitude species occur (Brinkhuis & Biffi 1993), the first of which correlates directly with the E–O boundary *sensu stricto* and the second with the onset of a more severe cold episode and inferred sea-level lowstand. Although these effects are local migrations, they may relate to the stepwise isotopic shifts that precede the early Oligocene glaciation and the maximum glaciation itself. A later event, identified in the Massicore, is the extinction of *Areosphaeridium diktyoplokum*, which occurs near the top of Chron C13n. This is several hundred thousand years younger than the $\delta^{18}O$ maximum.

Elsewhere, the record of dinocysts across the E–O boundary interval is patchy with the record from local basins probably affected by sea-level fall and changing local environments (e.g. Gedl 2004). Sluijs *et al.* (2003) have described the transition interval in the Southern Ocean where typical early Palaeogene assemblages are sequentially replaced with assemblages dominated by *Brigantedinium*, which is interpreted as related to the onset of upwelling conditions in the water column. Records from the Weddell Sea and off Dronning Maud Land (Antarctic margin) show decreasing dinocyst diversity through the Eocene and at the E–O boundary, falling to only two species by the end of the early Oligocene (Mohr 1990). This pattern is thought to indicate development of cold surface waters.

Diatoms

Baldauf (1992) brought together deep-sea drilling data available at that time and identified a significant turnover in marine diatoms as occurring at the E–O transition, with a notable increase in diversity in the tropics. He interpreted this as being related to a decrease in the vertical stratification of the water column and an increase in the latitudinal thermal gradient. The most profound changes seem to have occurred in the higher latitudes, associated with the changes in high-latitude water masses across the E–O transition. For example, Suto (2006) noted a rapid diversification of *Chaetoceros* resting spores in the Norwegian Sea, which they associate with a change from a stable water column with a constant nutrient supply in the Eocene to an unstable and more vertically mixed water column in the Oligocene. Similarly in the southern high latitudes, Olney *et al.* (2005) record an increase in cold-adapted species in the Oligocene of Antarctica, and Whitehead (2005) describes an increase in the diversity of benthic diatoms in the Oligocene associated with the increasing influence of the Antarctic polar current.

Benthic foraminifera

Benthic foraminifera occur in a range of environments and habitats, and the pattern of evolution differs markedly between these habitats. In abyssal and bathyal environments, there is little major change associated with the boundary interval itself (Thomas 1992; Coccioni & Galeotti 2003). Nevertheless, Diester-Haass & Zahn (2001) and Diester-Haass & Zachos (2003) record a sudden increase in the rate of accumulation of benthic foraminifera in deep-sea environments in various parts of the world, which they correlate to the rapid isotopic shifts and attribute to an increase in productivity associated with the more vigorous overturning of the earliest Oligocene ocean (see also Boersma 1986, for a similar suggestion). Thomas and Gooday (1996) also observe declining deep-sea benthic diversity and an increase in dominance of opportunistic phytodetritus-exploiting species (e.g. *Epistominella exigua* and *Alabaminella weddellensis)* that suggest a switch to a more unpredictable and seasonally fluctuating food supply, especially at high latitudes during the transition. Kaminski (2005) has recorded an acme of deep-water agglutinated species in the North Atlantic and Western Tethys during the E–O transition interval, one of several such events in the Palaeogene.

The pattern of turnover in smaller benthic foraminifera from shallower environments seems to contrast with the deep-sea record in that significant changes are more obvious. McGowran & Beecroft (1986) and McGowran *et al.* (1992) noted an abrupt turnover of neritic benthic foraminifera in Australia, which they linked to the sudden cooling and sea-level drop associated with the early Oligocene glaciation. The pattern is similar in the US Gulf Coast, although, as in Australia, it is complicated by facies changes across the boundary (Fluegeman 2003). A slightly delayed radiation of Oligocene shallow-water smaller benthic foraminifera in the US Gulf Coast was noted by Fluegeman (2003).

By far the most dramatic evidence for extinction and turnover occurs in larger foraminifera. These organisms were important carbonate producers throughout the Eocene, where they were abundant and widespread. According to Kiessling *et al.* (2003) and Nebelsick *et al.* (2005), Eocene carbonate platforms, which are dominated by coralline algae and larger foraminifera, declined rapidly in the Late Eocene and across the E–O boundary interval, reaching a 'post-Cambrian low' in the earliest Oligocene. Unfortunately, continuous sections in carbonate facies are rare, possibly because of the sea-level drops associated with the glacial maximum. The most complete section available seems to be the Melinau limestone of Sarawak, which was studied in detail by Adams (1965). Adams *et al.* (1986) reviewed the Indo-Pacific records and noted a mass extinction of some important long-ranging genera and species, notably *Asterocyclina, Discocyclina* and some species of *Nummulites*. These events occur in the Melinau limestone associated with a brief switch to algal-dominated facies. The mass extinction has not been reliably correlated to isotope or magnetostratigraphy, nor has it been demonstrated how abrupt or gradual the extinctions were, but Adams *et al.* (1986) suggest a relation to the climate deterioration and sea-level fall across the E–O transition. The Americas were a different faunal province, but from limited evidence, the pattern may be similar; for example, Robinson (2003) has noted a rapid decline in larger foraminifera in the Caribbean that may be associated with the boundary interval.

Ostracodes

Benson (1975) suggested that the E–O transition to be the most significant in the Cenozoic history of ostracodes, but it is not clear how precisely this turnover correlates to the boundary events and to what extent different habitats were affected. Schellenberg (1998) found significant drops in diversity of ostracodes across the E–O boundary interval, and noted that the extinction seems to have been most severe among deposit-feeding species. In contrast, Dall'Antonia *et al.* (2003) found little variation in assemblages across the E–O boundary in the Massignano stratotype section, and found little change in either the shallow and deep records from the US Gulf Coast and Barbados respectively, except for some minor extinctions in both. Like benthic foraminifera, ostracodes lived in a wide variety of habitats and the Eocene turnover pattern probably varies between them.

Shallow marine invertebrates

The fossil record of shallow marine invertebrates across the E–O boundary worldwide is quite patchy, with the best records coming from North America and Europe. Dockery (1986) and

Dockery & Lozouet (2003) have described a major turnover of molluscs in the early Oligocene of the US Gulf Coast and Paris Basin, including the disappearance of many long-ranging forms, which they attribute to sea-level fall and global cooling. Dockery & Lozouet (2003) also document a major influx of European taxa into North America in the earliest Oligocene that may be related to changing water mass circulation in the North Atlantic. Squires (2003), Hickman (2003) and Nesbitt (2003) described a major and relatively abrupt turnover of marine molluscs in the earliest Oligocene in the western United States, in which over half of pre-existing genera disappeared. During these events, warm-water taxa suffered disproportionately, while cold-water forms radiated and expanded their ranges. The higher-latitude record from Alaska and Kamchatka (Oleinik & Marincovich 2003) is similar to the record further south. More recently, minor extinction and size reduction in veneroid bivalves has been described from the eastern United States (Lockwood 2005). Other events, both before and after the E–O boundary interval, are also in the record and some groups, while others such as the echinoids of the United States (Carter 2003) do not show an unusual extinction pattern.

The fundamental problem involved with shallow-water marine invertebrates (as with benthic foraminifera) is distinguishing global and local effects. The problem of sea-level fall associated with the transition is bound to have caused many local changes; far better geographic sampling would be necessary to distinguish local from global effects.

Terrestrial vegetation

There is clear evidence for major changes in terrestrial vegetation worldwide across the E–O transition, but the pattern of change varied from continent to continent and across the latitudes. The change in North American floras was recognized by Wolfe (1978) as the 'Terminal Eocene Event' and has been studied in more detail by Wolfe (1992, 1994) and recently by Liv et al. (2007). The dominant pattern in North America is for widespread replacement of the subtropical broad-leaved Evergreen vegetation of the Eocene with cooler, deciduous forms in the Oligocene, with regional extinctions, especially at higher latitudes. Similarly in the south, Patagonian floral records show the disappearance of tropical vegetation and the rise to dominance of subtropical and cool-temperate species from the middle Eocene of Early Oligocene (Palazzessi & Barreda 2007). In other places in North America, there was a diversification of desert species as aridity

increased in the continental interior (Yancey et al. 2003; Moore & Jansen 2005). In some areas such as the US Gulf Coast, there is less evidence for dramatic change (Oboh-Ikuenobe & Jaramillo 2003). Collinson (1992) has reviewed the botanical data from Europe, where there is a similar change to a more seasonal temperate flora accompanied by loss of tropical and subtropical taxa. These records suggest that the main phase of change shortly pre-dates the mammalian extinctions (discussed below) (see also Hooker et al. 2004).

The evolution of extensive North American grassland habitats, and associated mammalian radiation, has been linked with the Eocene–Oligocene Transition but the timing is controversial (Stucky 1992; Strömberg 2005 and references therein). A recent study suggests that the spread of this habitat, and the mammalian grazing ecology, occurred between the late Oligocene and Early Miocene, driven by post-EOGM climate changes that led to increased seasonal aridity (Strömberg 2005, 2006).

With regard to the tropics, Jaramillo et al. (2006) have recently produced a detailed composite floral record from Columbia and Venezuela, based on pollen and spore occurrences that produce an unprecedented view of long-term floral diversity. The most rapid phase of change in this record occurs around the E–O boundary interval, which Jaramillo et al. (2006) link to the environmental changes associated with the onset of Antarctic glaciation. The resolution, however, is too low to correlate floral changes with the events of the transition itself, as seen in detailed deep-sea records (see below).

Perhaps the most dramatic changes would be expected in the high southern latitudes close to the supposed expansion of Antarctic ice. Macro-floral and palynological evidence from Antarctica has been reviewed by Francis (1999) and Francis & Poole (2002). The E–O transition interval coincides with a change from evergreen forest to sparse tundra on the Antarctic Peninsula and Seymour Island, and Francis (1999) goes further to suggest that vegetation–climate feedbacks may have played a role in the rapid environmental changes at that time. Similar patterns are seen in the high northern latitudes, where there was widespread displacement of subtropical broad-leaved vegetation, especially at altitude (Wolfe 1994; Myers 2003).

Evidence from beyond America and Europe is sparser. Kemp (1978) reviewed the Australian palaeobotanical record and found a major change near the E–O boundary, with a trend towards lower diversity, higher seasonality and a spread of cool-temperate plants. Leopold et al. (1992) document significant changes in Asian floras, with diversity reduction across the boundary interval. Ramussen *et al.* (1992) found little variation in North Africa. More recently, however, Pan *et al.* (2006) have suggested that the E–O transition was the most important step in the extinction of African palms (Arecaceae), although they continued to flourish on other continents.

Mammals

A mass extinction of mammalian faunas in Europe (principally known from records in the Paris and Hampshire basins) at approximately the E–O boundary was noted almost a century ago by Stehlin (1909) as the 'Grand Coupure' (great break). It represents a major turnover in both perissodactyls and artiodactyls (hoofed herbivores). This has recently been dated as approximately coinciding with the EOGM (Hooker *et al.* 2004) and the associated sea-level regression. Approximately 60% of taxa seem to have disappeared at this time and it has been argued that the extinctions were caused either by climatic deterioration or competition with immigrant taxa from Asia (Savage & Russell 1983). A major turnover of mammals in Mongolia at this time has also been linked to climatic change (Meng & McKenna 1998).

In North America, the mammalian record has long been confused because of a misalignment of the land mammal stages with the global stratigraphy (see Prothero & Swisher 1992 for discussion). It is now clear, however, that there was a more minor extinction, coinciding with the diversification of hypsodont mammals close to the E–O boundary (Stucky 1992), and there are similarities with the record in South America (Marshall & Cifelli 1989). There may be a similar pattern in Asia but the issue is further confused by uncertain chronology (see comments by Berggren *et al.* 1992). The Oligocene in general marks the time when representatives of modern land mammals became the dominant vertebrate life, including the appearance of the first primates and apes, but precise timing is problematic.

In the marine realm, it has long been recognized that the extinction of the archeocete whales such as the giant *Basilosaurus* occurred at or near the E–O transition (Fordyce 1992, 2003; Manning 2003) although there is no evidence for a substantial mass extinction (Fordyce 1992). The cold high-latitude water masses of the early Oligocene seem to have been a spur to the evolution of crown-group whales, and the first baleen whales are known from the early Oligocene (Fordyce 1992, 2003), most likely in parallel with increased Southern Ocean upwelling and increases in plankton abundance. In addition, it has been demonstrated that toothed whales (suborder Odontoceti), which include dolphins and porpoises, underwent a significant increase in brain size with respect to body size near the E–O boundary (Marino *et al.* 2004).

Others

The amphibians and freshwater turtle records from North America seem to suggest declining diversity (Hutchinson 1992), possibly related to increasing aridity around the E–O boundary. Corsini *et al.* (2006) also identified changes in turtle carapace size that may be a response to E–O climatic change. A similar picture emerges from Europe (Rage 1986) where tropical taxa disappear in large numbers near the end of the Eocene. The first diversification of cat-like carnivores (mostly sabre-toothed) and other predators also occurs in the early Oligocene and is thought to coincide with the development of grassland habitats in central North America (Bryant 1996). Evidence for other tetrapods tends to be patchy and difficult to interpret (e.g. see Prothero & Emry 1996), although it may be noteworthy that there seems to have been a radiation of passerine birds in the early Oligocene (Mayr 2005).

Summary of palaeontological evidence

There is widespread evidence for enhanced extinction in many groups across the E–O boundary interval, followed either immediately, or after a slight delay, by renewed radiation in the Oligocene. Precise correlations to the isotope shifts, glacial maxima and magnetostratigraphy are still problematic in many groups. In several groups where there is sufficient age-discrimination, the extinctions seem to have been stepwise over a period of change that may equate to several hundred thousand years, starting before the stage boundary as formally defined and ending with the maximum glaciation that corresponds to the base of the Oi-1 isotope zone. It is interesting that in groups as diverse as foraminifera, primates, whales and birds, the Oligocene saw the diversification of recognizably modern taxa, whereas most of the Eocene forms are from 'prehistoric', now-extinct groups that lie outside of the modern lineages. This pattern almost certainly relates to the fact that the E–O boundary interval was one of the most profound periods of change that led to the modern climatic regime, the change that is widely described as the greenhouse–icehouse transition (Miller *et al.* 1991).

The close coincidence of widespread extinction and stepwise climatic changes seems to imply causal linkage, but the causes of extinction must be different in many groups. Possibilities include temperature change, sea-level fall and the associated exposure of the continental shelf, change in biological production on land and in the oceans, water mass properties (especially at high latitudes but also in equatorial and upwelling areas), atmospheric and ocean chemistry, etc. The whole Earth System seems to have entered a period of prolonged change, a factor that makes the E–O boundary extinctions rather different from the more sudden mass extinctions at the end of the Cretaceous and probably at the end of the Permian as well. In the following sections, we will review the geochemical and other evidence for prolonged and interlinked changes in the Earth System at this time.

Proxies

Other than the occurrence of glaciomarine sediments on and around Antarctica (e.g. Barrett *et al.* 1989; Zachos *et al.* 1999), there is little direct evidence of E–O climate change. Therefore, climate-sensitive proxies are used to trace responses to climatic parameters indirectly. These proxies provide information on global temperature, ice volume, ocean productivity, water mass structure, circulation, carbon cycling, atmospheric carbon dioxide concentration and continental weathering. Many of the processes being recorded are related to the global carbon cycle, either as cause or effect, and have the potential to influence and be influenced by atmospheric CO_2 concentration. The search for cause and effect relationship is therefore similar to the lesser (but still-unresolved) problem of understanding glacial–interglacial climatic switching.

Other important techniques that are integral to modern E–O palaeoclimatic study are climate modelling on various scales and complexities, and time series analysis, the latter having the power to identify patterns of external orbital forcing on Milankovitch timescales. Tectonic evolution is also fundamental to the system and feeds into all questions related to E–O climate change. Descriptions and applications of the various proxies and modelling techniques that have been utilized in the study of the E–O transition are discussed below under headings that represent the principal process or parameter they are tracing. As with the discussion of fossil evidence, there is no way it can be completed in this discussion, but we can identify a snapshot of important and promising areas.

Ice, temperature and sea level

Benthic foraminifera oxygen stable isotopes – Undoubtedly, the most significant proxy for documenting E–O climatic change is benthic foraminifera $\delta^{18}O$. As classically demonstrated, the principal controls on benthic $\delta^{18}O$ are: (i) seawater temperature and, (ii) $\delta^{18}O$ of seawater (δ_w), which

changes during glaciations. The benthic $\delta^{18}O$ proxy, therefore, helps constrain deep-water temperatures and global ice volumes. Since the Southern Ocean was probably the dominant deep-water source from at least the Late Eocene (e.g. Wright & Miller 1993), benthic $\delta^{18}O$ also provides an estimate of high-latitude surface water temperatures.

As discussed above, studies of numerous deep-sea cores demonstrates the existence of an abrupt positive *c.* 1.2–1.5‰ shift in benthic $\delta^{18}O$ associated with the E–O boundary that builds to a sustained maximum at the base of the Oligocene (Fig. 1). This interval of maximum $\delta^{18}O$, which we refer to as the EOGM, is now widely believed to represent the maximum extent of the first semi-permanent ice sheet on Antarctica (e.g. Kennett & Shackleton 1976; Kennett 1977; Miller *et al.* 1987; Zachos *et al.* 1996, 1999, 2001; Lear *et al.* 2001; Coxall *et al.* 2005), 50% (or more) of the present Antarctic ice-sheet volume (Barker *et al.* 1999; DeConto & Pollard 2003*a*). Following termination of the EOGM, $\delta^{18}O$ recovered to a new equilibrium value that was on average 1‰ higher than the Late Eocene, suggesting subsequent deglaciation and stabilization of Oligocene ice volumes.

High resolution (2–10 kyr) $\delta^{18}O$ records have improved constraints on the timing, magnitude and structure of the E–O isotopic shift and $\delta^{18}O$ maximum and identified additional structure within it. These records demonstrate that the EOGM can be correlated fairly precisely to the base of magnetic Chron C13n with an estimated duration of 400 kyr (e.g. Oberhansli *et al.* 1984; Miller 1985; Zachos *et al.* 1996; Coxall *et al.* 2005). The *c.* 1.5‰ $\delta^{18}O$ shift into the EOGM lasted about 500 kyr and occurred in two phases (Zachos *et al.* 1996; Coxall *et al.* 2005). Orbitally tuned records from equatorial Pacific ODP Site 1218 suggest that this involved two 40 kyr steps, separated by a well-defined 200 kyr-long plateau (Coxall *et al.* 2005) (Fig. 1b), although this is not a unique interpretation of the orbital evidence. The first step, which accounts for less than half of the total shift, and the plateau occur in reversed interval C13r and the second step was completed at approximately the base of C13n. This relationship to the magnetic reversal record holds up in DSDP Site 522 and ODP site 744 (Zachos *et al.* 1996). The step-form is remarkably similar to the pattern of non-linear ice growth simulated by a coupled GCM-ice sheet models (DeConto & Pollard 2003*a*, *b*), although it is of significantly greater magnitude than predicted by the model. At ODP Site 744 (Southern Indian Ocean), the EOGM comprises two pronounced peaks lasting approximately 100 and 150 kyr respectively. These have been interpreted as two distinct glacial maxima, termed Oi-1a and Oi-1b (Zachos *et al.* 1996). These features are less well defined elsewhere. The termination of the EOGM is more difficult to define and correlate between sites than its initiation because it consists of a series of small stepped decreases in $\delta^{18}O$, totalling *c.* 0.5–0.6‰, interspersed with minor positive excursions spread over a few tens of thousands of years (Zachos *et al.* 1996; Coxall *et al.* 2005). Most of this decrease, however, occurs in the top of magnetochron C13n (which is where the upper boundary of the EOGM is placed), coincident with a decrease in glaciomarine sediments close to Antarctica (e.g. Wise *et al.* 1991).

Quantifying the relative contribution of temperature and ice growth to the *c.* 1.5 ‰ $\delta^{18}O$ shift is a problem, but separating the two effects is not a trivial issue. One method is to obtain independent palaeotemperature estimates using Mg/Ca ratios in order to extract the δ_w component. This method is summarized here but discussed fully by Lear (2007). Other methods that have been used include looking for covariance in deep-sea benthic and low-latitude planktonic foraminifera $\delta^{18}O$ records (e.g. Miller *et al.* 1991) and modelling of Antarctic ice-sheet sensitivity to climate change (Oerlemans 2004, 2005).

Magnesium calcium ratios and ice volume – Benthic foraminiferal Mg/Ca palaeothermometry has been useful for providing independent temperature estimates and, therefore, resolving temperature and ice volume contributions to $\delta^{18}O$ records in the Quaternary and early Cenozoic (Mashiotta *et al.* 1999; Lear *et al.* 2001, 2004; Martin *et al.* 2002; Billups & Schrag 2003). Surprisingly, existing E–O Mg/Ca records show no evidence of deepwater cooling. In fact, if anything they suggest a 2 °C warming (Lear *et al.* 2001, 2004; Billups & Schrag 2003) (Fig. 3a). Taken at face value, this suggests that the entire E–O $\delta^{18}O$ shift is attributable to ice growth. This implied ice volumes seem unrealistic for Antarctica alone (Coxall *et al.* 2005), and raises several possibilities: (i) that changes in the hydrological system (moisture supply) rather than cool temperatures were important for ice-sheet expansion (Lear *et al.* 2001) (cf the 'snow gun hypothesis', Prentice & Matthews 1991), (ii) there was more ice elsewhere; i.e. possible contemporaneous northern hemisphere glaciation, consistent with sedimentary and modelling (DeCouto & Pollard 2007) evidence, (Davies *et al.* 2001; Via & Thomas 2006; Eldrett *et al.* 2007), and/or (iii) that some additional factor acts to mask the Mg/Ca cooling signal, such as seawater pH and/or carbonate ion concentration (Billups & Schrag 2003; Lear *et al.* 2004).

The lack of a cooling signal in deep water proxy records is puzzling because, although cooling might not have been the trigger for E–O Antarctic glaciation, it seems logical that widespread Antarctic glaciation would lead to local cooling in the vicinity of Antarctica in the early Oligocene (Zachos *et al.* 1996; Oerlemans 2005). A likely explanation for the lack of Mg/Ca evidence for cooling is that the extreme increase in deep ocean carbonate ion concentration/alkalinity associated with a 1 km deepening of the calcite compensation depth (see below), synchronous with the shift into the EOGM, masked the cooling (Lear *et al.* 2004; Coxall *et al.* 2005). Until these effects are better understood, the Mg/Ca record across the E–O should be interpreted cautiously.

Covariance of $\delta^{18}O$ in global benthic and nonupwelling tropical planktonic foraminifera (i.e. from regions with the most thermally stable surface waters) has been used as an indicator of global ice volume fluctuations in the Quaternary (e.g. Shackleton & Opdyke 1973). However, there is a general shortage of suitable low-latitude and/or high-resolution E–O planktonic records because of poor preservation of dissolution-susceptible planktonic shells in notoriously carbonate-poor deep-sea sequences of this age. Existing planktonic $\delta^{18}O$ records are mainly restricted to the mid- to high latitudes, i.e. DSDP Site 522 (Oberhansli *et al.* 1984), DSDP Sites 592 and 593 (Murphy & Kennett 1986) and ODP Site 744 (Stott *et al.* 1990; Barrera & Huber 1991). These show highly variable increases in planktonic $\delta^{18}O$, ranging from 0.5 to 1.5 times benthic $\delta^{18}O$, that reflect complex contributions of regional sea-surface temperatures, salinity, depth habitat and post-burial diagenesis effects, in addition to sea-level shift related to ice growth. Prentice & Matthews (1988) compared compilations of low-latitude planktonic and deep-sea benthic $\delta^{18}O$ during the Cenozoic. They concluded that benthic $\delta^{18}O$ primarily reflects deep-ocean temperature variation whereas tropical planktonic $\delta^{18}O$ show responses to Cenozoic ice volume and suggested the presence of a significant ice budget for the past 40 myr. The Prentice & Matthews (1988) record, however, lacks the detail necessary to resolve the pattern of change during the E–O transition. In any case, assumptions about the stability of low-latitude temperatures during major climatic transitions may be ill founded (Miller *et al.* 1991). Acquisition of high-quality planktonic isotope records from low and high latitudes, of the kind that have been produced from Eocene and Cretaceous 'glassy forams' (Pearson *et al.* 2001; Wilson & Norris 2001), therefore, are a priority for better constraining global sea-surface temperatures and E–O latitudinal thermal gradients.

Other temperature proxies – Much of the study of E–O palaeoclimates is limited by our ability to accurately determine temperature variability. As

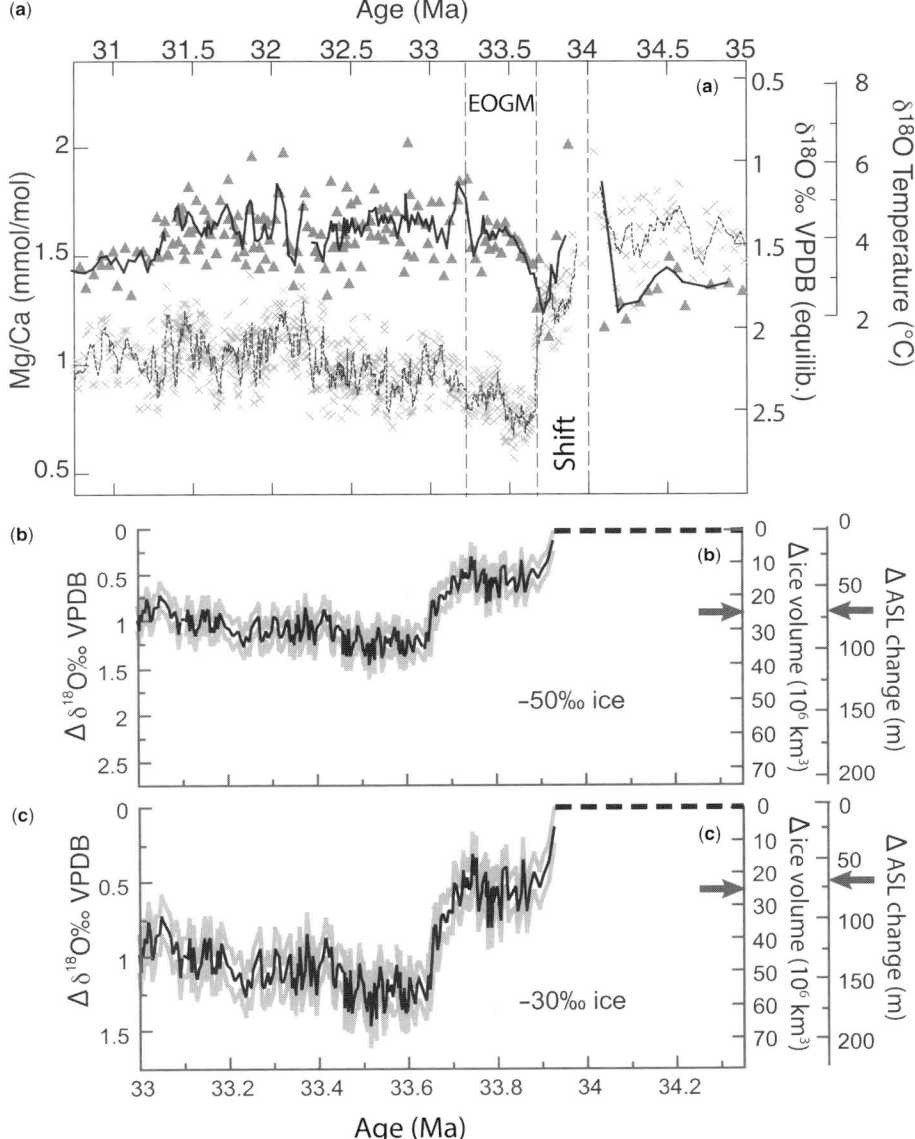

Fig. 3. Eocene–Oligocene benthic foraminifera palaeoclimate proxy records from ODP Site 1218 examining possible changes in sea level, ice volume and temperature (after Coxall *et al.* 2005). Data are plotted on the ODP Site 1218 orbitally tuned timescale. (**a**) $\delta^{18}O$ (crosses, 5-point moving average trend line) versus Mg/Ca (triangles, 3-point moving average trend line), in principle an independent palaeothermometer (*c.* 35 to 31 Ma) from ODP Site 1218 (Mg/Ca data from Lear *et al.* 2004). The records show no decrease in Mg/Ca across the E–O transition and into the EOGM. In fact, the Mg/Ca data show an increase from latest Eocene to earliest Oligocene suggesting either bottom-water warming or that Mg partitioning into benthic foraminiferal calcite is sensitive to factors other than temperature (e.g. increasing pH with CCD deepening, see Fig. 7). $\delta^{18}O$ temperatures shown apply to a world free of continental-scale ice-sheets ($\delta^{18}O_w = -1\permil$ Standard Mean Ocean Water (Kennett & Shackleton 1976). Equilibrium calcite values $= \delta^{18}O_c + 0.64\permil$) (Kennett & Shackleton 1976). (**b**) & (**c**) Estimated global ice budgets and glacioeustatic sea-level fall associated with onset of Antarctic glaciation for ice with oxygen isotope values of $-50\permil$ and $-30\permil$, assuming that all of the $\delta^{18}O$ increase associated with Oi-1 is attributable to increased ice volume. Arrows indicate modern Antarctic ice volume (*c.* 25.4×10^6 km³) and apparent sea-level fall (ASL, defined as eustacy plus the effects of water loading on the crust) (70 m) estimated for the Eocene–Oligocene Transition by sequence stratigraphy (Pekar *et al.* 2002).

discussed above, the deep-sea climate proxy records used routinely across the E–O interval that are temperature sensitive are also affected by other factors such as changing global ice volume and possibly alkalinity related to CCD shift and or changes in atmospheric pCO_2. The range of additional independent palaeotemperature proxies is limited for this time interval because of preservation issues, but there are several that provide information about E–O changes especially in continental regions.

Palaeobotanical assemblage associations (see above) and leaf margin analysis document significant cooling in the Late Eocene to Early Oligocene of Saxony, Germany (Roth-Nebelsick et al. 2004). $\delta^{18}O$ variations in low latitude shallow-marine molluscs and fish otoliths also indicate E-O cooling, and increased seasonality (i.e. cooler winters) (Kobashi et al. 2001; Ivany et al. 2000). Elsewhere in Northern Europe, summer palaeotemperature estimates for continental freshwater derived from rodent tooth enamel, molluscs and fish otoliths, suggest a fluctuating mesothermal climate during Eocene Oligocene time (Grimes et al. 2005) but no clear evidence for climatic cooling. Cooling in this record, however may be masked by the large error bars associated with the mean fresh water $\delta^{18}O$ estimates (5–6‰.) because other continental records derived from fossil bone and tooth enamel indicate a large drop in mean annual temperatures of c. 8°C over 400 kyrs (Zanazzi et al. 2007). In the latter study, continental cooling appears to have been delayed in time with respect to the marine changes by c. 400 kyrs. Sedimentary records from the Tibetan Plateau provide additional evidence for cooling and aridification in continental Asia precisely at the time of the Eocene-Oligocene Transition (Dupont-Nivet et al. 2007).

In the future, organic biomarkers, such asTEX$_{86}$ (TetraEther indeX of tetraethers with 86 carbon atoms), a proxy for determining sea-surface temperature from the fossilized membranes of marine archaea (or Archaebacteria) (Schouten et al. 2002), may prove useful for obtaining accurate temperature reconstructions. Recent recovery of well-preserved organic biomarkers of early Palaeogene age from Tanzania (van Dongen et al. 2006) suggests that such temperature estimates are not far off (Pearson et al. 2007).

Sea level – Widespread glaciation of Antarctica at the E–O boundary, and possibly elsewhere, would have resulted in a significant drop in global sea level proportional to the magnitude of ice growth. Identification and quantification of E–O eustatic change, therefore, is important for understanding this climatic event. Although there are doubts concerning the relative contributions of temperature and ice volume to the $\delta^{18}O$ record, it is generally considered that Antarctic ice volume

achieved at least 50% of its present development at the E–O boundary (Barker et al. 1999; DeConto & Pollard 2003a). Associated sea-level fall has been estimated at 30–90 m by Miller et al. (1991) and modelled more conservatively at 40–50 m by DeConto & Pollard (2003a). The classic Cenozoic sea-level curves, based on sequence stratigraphy, failed to recognize sea-level fall close to the E–O boundary and place the largest change in the middle Oligocene (e.g. Vail & Hardenbol 1979; Haq et al. 1987). However, there are large uncertainties in dating and correlation in these early reconstructions and recent sequence stratigraphic studies on the New Jersey continental margin have identified evidence for a prominent eustatic lowering, indicating as much as 70 m absolute sea-level fall, coincident with the EOGM (Miller & Mountain 1996; Pekar & Miller 1996; Pekar et al. 2002). Eustatic change is further supported by covariance between benthic and tropical planktonic $\delta^{18}O$ records (Miller et al. 1991). Evidence of significant E–O sea-level fall has been also been recorded on the West African margin (Séranne 1999), in the North Sea region (Vandenberghe et al. 2003) and Southern Australia (McGowran et al. 1992), although these studies do not provide estimates of the magnitude of change.

Surprisingly, E–O sea-level fall does not appear to be ubiquitous, and in other regions there is little or no evidence for large-scale change, even in quasimarine continental facies that should be very sensitive to change. For example, estuarine sedimentary records from the Hampshire Basin, UK, suggest maximum relative sea-level fall of 15 m (Gale et al. 2006) and in Mississippi, USA, the stratigraphy has been interpreted as indicating sea-level highstand at the E–O boundary with evidence for increasing water depth across the transition (Echols et al. 2003). These studies question the extent of E–O sea-level change and/or suggest that it coincided with widespread tectonic activity that masked the full magnitude of eustatic response. The lack of consensus and quantitative agreement of benthic proxies is one of the current areas of disagreement regarding the E–O transition and there is still sckepticism among some workers over whether there was any global change in sea level, and thus, ice volume, at all (Hay et al. 2005).

Carbon cycling and productivity

Marine carbonate carbon stable isotopes – Deep-sea benthic foraminifera and bulk carbonate sediment $\delta^{13}C$ reflect regional and local changes in surface to deep organic carbon cycling as well as global changes in organic and inorganic carbon burial as carbon moves between the lithosphere, oceanic and atmospheric reservoirs. The commencement of Oligocene glaciation was accompanied by an oceanwide positive $\delta^{13}C$ anomaly of up to 1.0‰ (e.g. Zachos *et al.* 1996, 2001; Diester-Haass & Zachos 2003; Coxall *et al.* 2005) (Fig. 4). The shift into the $\delta^{13}C$ anomaly was also rapid (several hundred kyr) and stepwise (Coxall *et al.* 2005), with peak values recorded at the base of the EOGM. ODP Site 1218 records show a distinctive $\delta^{13}C$ 'overshoot' of typical early Oligocene values followed by a longer recovery phase

(400–600 kyr) to near pre-excursion values (Fig. 4b). The long recovery is probably a response to the rapid deepening of the oceanic carbonate compensation depth (see below) (Coxall *et al.* 2005; Zachos & Kump 2005). In the high-resolution record from ODP Site 1218, the $\delta^{18}O$ and $\delta^{13}C$ step-shifts are similar in form, but $\delta^{13}C$ lags by *c.* 10 kyr (Coxall *et al.* 2005). A lag between $\delta^{13}C$ and $\delta^{18}O$ was also recognized in Site 522 records (Zachos *et al.* 1996). This pattern had reversed by the Oligocene because spectral analysis shows a phase lag of $\delta^{18}O$ to $\delta^{13}C$ of *c.* 8 kyr with respect to

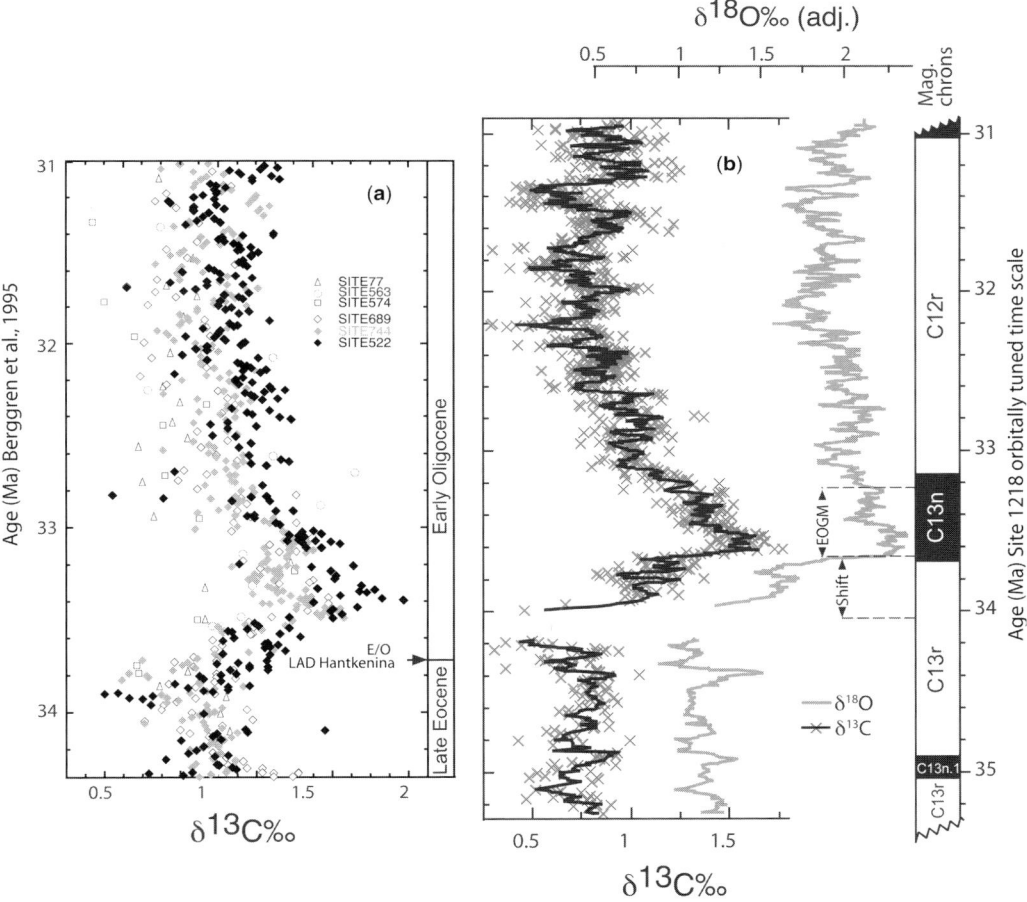

Fig. 4. Benthic foraminifer $\delta^{13}C$ from deep-sea drill sites across the Eocene–Oligocene Transition. (**a**) Compilation (after Zachos *et al.* 2001) from mid- to high southern latitudes (diamonds) and the equatorial Pacific (other symbols) plotted on a common timescale (Berggren *et al.* 1995) showing the globally recognizable *c.* 1‰ shift and positive excursion. Note that the two figures are plotted on different timescales. The E–O boundary, as identified by Poore (1984), is shown in relation to the Site 522 data (see Fig. 2). (**b**) $\delta^{13}C$ (crosses) from ODP Site 1218 (Coxall *et al.* 2005) plotted on an orbitally tuned timescale and compared to $\delta^{18}O$ (5-point running mean from Fig. 2), the isotope 'Shift' and the EOGM. A stepped pattern of change, including an intermediate plateau, is clear in the Site 1218 $\delta^{13}C$ record. There is an also an indication of step features in the compilation, especially Site 522.

the 40 kyr band, suggesting that the response of the global carbon cycle may have helped force changes in early Oligocene climate (Coxall *et al.* 2005).

One hypothesis regarding the carbon isotope anomaly relates it to a temporary increase in the ratio of organic carbon/CaCO$_3$ burial rates because of enhanced marine export production, brought about by climate-induced intensification of wind stress, upwelling and oceanic turnover (e.g. Shackleton & Kennett 1975; Diester-Haass 1992, 1995; Zachos *et al.* 1996 and references therein – see below). In addition, the supply of production-limiting nutrients, especially iron, is likely to have increased in response to sea-level fall and enhanced continental weathering, as has been suggested during Quaternary glacial/interglacials (e.g. Martin 1990). Observations are also compatible with recent modelling studies that suggest the δ^{13}C shift reflects reorganization of the carbon cycle in response to a rapid drawdown of carbon dioxide related to increased biological production and CCD deepening (Zachos & Kump 2005). Qualitative and quantitative proxy evidence for these parameters and processes is presented in more detail below.

Marine productivity – Variations in palaeomarine productivity and export production provide important information about global carbon cycling and the causal/feedback roles they play in regulating climate. Significant effort has been focused on obtaining proxy records of E–O productivity changes. Commonly used methods include benthic foraminifera and other marine micro- and macrofossil accumulation rates, opal/carbonate accumulation, carbonate dissolution indices, carbonate δ^{13}C and marine barite accumulation (e.g. Salamy & Zachos 1999; Diester-Haass & Zachos 2003; Averyt *et al.* 2005). Fossil accumulation rate proxies are thought to vary in direct response to standing biomass in the water column and phytodetritus availability at the sea floor but can be biased by age model problems (Diester-Haass & Zachos 2003). Carbonate dissolution indices (based on foraminiferal shell fragmentation) provide insight into primary production because dissolution is affected by the rain ratio of organic carbon/CaCO$_3$ (Diester-Haass & Zachos 2003).

Palaeoproductivity proxy data point towards a sharp increase in export production associated with the E–O transition, especially in the southern high latitudes. Benthic and planktonic foraminifera, radiolaria, fish debris, echinoderms and ostracodes all show significant increases in accumulation rate in parallel with the isotopic shifts into the EOGM, in both Southern Ocean (Ehrmann & Mackensen 1992; Diester-Haass 1995, 1996; Diester-Haass & Zahn 1996; Salamy & Zachos 1998) and mid- and low-latitude regions (Diester-Haass & Zahn 2001; Diester-Haass & Zachos 2003; Vanden Berg &

Jarrard 2004). Other reports, however, find no evidence for increased productivity in equatorial regions (Nilsen *et al.* 2003; Schumacher & Lazarus 2004) suggesting that there was considerable regional variation in circulation and productivity responses to the climate change outside of the Southern Ocean region. The highest-resolution records suggest that the increase in productivity was initiated during the isotopic shift, i.e. during the E–O transition, and was sustained for at least the duration of the EOGM (Salamy & Zachos 1999; Diester-Haass & Zachos 2003).

Records from the Southern Ocean suggest that net productivity increased by several-fold in response to the EOGM (e.g. Baldauf & Barron 1990; Salamy & Zachos 1999) with evidence for increased seasonality in the delivery of organic matter to the sea floor (Thomas & Gooday 1996; Schumacher & Lazarus 2004) that may have facilitated more rapid burial of organic carbon. Calcareous primary producers, dominant in the Late Eocene, were partially replaced by opaline organisms suggesting a trend toward seasonally greater surface divergence and upwelling. A corresponding increase in organic carbon content has not been widely recorded, but organic content is generally very low in Palaeogene-age deep-sea sediments because of rapid post-burial consumption, and the signal may not be preserved (Curry *et al.* 1995). Increases in carbonate accumulation proxies is complicated by the major deepening of the CCD synchronous with the onset of the E–O shift, thus, integration of multiple independent proxies is important for differentiating productivity signals from secondary processes, such as dissolution related to lysocline shift.

Other productivity proxies such as barium, reactive phosphorus and nitrogen isotopes may provide complementary evidence for changes in export production that are independent of carbonate dissolution issues. For example, Nilsen *et al.* (2003) concluded that although biogenic silica production/preservation increased during the E–O transition in equatorial Atlantic ODP Site 925, barium and reactive phosphorus suggested no change in productivity or nutrient burial, implying that the increased opal production was relative and occurred at the expense of carbonate production.

The cause of worldwide increased palaeoproductivity from the Late Eocene to the Oligocene is attributed to ocean mixing/circulation changes that increased the availability of nutrients in surface waters. Whether this was driven by increased latitudinal thermal gradients stimulating stronger winds, upwelling and mixing, or reorganization of global circulation related to the opening and closing of tectonic ocean gateways is still a matter of much debate (Diester-Haass 1996; Diester-Haass & Zahn 1996;

Salamy & Zachos 1999; Diester-Haass & Zahn 2001; Diester-Haass & Zachos 2003; Hay *et al.* 2005). Either way, the result of the widespread palaeoproductivity increase, and a phase of increased organic carbon burial, might have led to a lowering of oceanic and atmospheric CO_2 that could have played a key role in E–O climate change by: (1) directly forcing the extreme transient cooling and glaciation, and/or (2) providing a positive feedback that further contributed to global cooling once the transition was initiated (Diester-Haass & Zachos 2003; Zachos & Kump 2005).

Atmospheric carbon dioxide

Changes in the partial pressure of atmospheric carbon dioxide (pCO_2) is thought to be a primary driver of climate over Phanerozoic time (Raymo 1991; Berner 1992; Crowley 2000) and especially during transitions into and out of glaciations (e.g. Barnola *et al.* 1987; Petit *et al.* 1999; Royer *et al.* 2004). General circulation models of Palaeogene climate have been used to show that continuous depletion of $pCO2$, amplified by Milankovitch forcing and ice-albedo feedbacks, could cause significant temperature reduction resulting in permanent continental ice-sheets in high latitudes (e.g. DeConto & Pollard 2003*a*, *b*; Pollard & DeConto 2003*a*; Zachos & Kump 2005). Reconstructing E–O pCO_2, therefore, is a primary goal for constraining the links between climate and radiative forcing of the Earth's surface temperatures through the E–O transition.

A variety of proxies are used to estimate pCO_2 during the early Cenozoic including; isotope ratios of marine carbonates or soil carbonates; stomatal density/index in fossil leaves; alkenone proxy in organic biomarkers; boron isotopic ratios in fossil surface-dwelling foraminifera (see Royer *et al.* 2001; Boucot & Gray 2001 for reviews). These are discussed further below. In addition, models have been used to simulate CO_2 variations over Phanerozoic time. Such models are based on changes in carbon burial on land (due to the evolution of land plants) and weathering changes (caused by organic acids in soils) (e.g. Berner 1990) and suggest a general decline in the early Cenozoic. Although there is a good first-order agreement between model results and proxy CO_2 estimates on long timescales (Berner 1990, 1997; Crowley 2000; Berner & Kothavala 2001; Royer *et al.* 2007), the records mostly lack the detail and accuracy to fully resolve the magnitude and timing of decline during the E–O transition and other key climatic events.

Fossil stomata proxies – The palaeobotanical approach for determining past CO_2 concentration is based on the inverse relationship between leaf stomatal density and pCO_2 in some types of plants (e.g. Beerling *et al.* 1993). Application of this method using early Cenozoic plant fossils has yielded conflicting results. Stomatal counts on the gymnosperms *Ginkgo biloba* and *Metasequoia glyptostroboides* suggest that pCO_2 was stable at between 300 and 450 parts per million (ppm) by volume during the Eocene and Neogene (Royer 2001; Royer *et al.* 2001), whereas results of a later study on *Ginkgo* suggest a decrease in pCO_2 at the E–O boundary (Retallack 2002). Stomatal densities in fossil leaves from three dicotyledon angiosperm taxa (*Eotrigonobalanus furcinervis*, *Laurophyllum pseudoprinceps* and *Laurophyllum acutimontanum*) suggest that pCO_2 was higher during the Late Eocene than during the Early Oligocene (Roth-Nebelsick *et al.* 2004). These differing results indicate large uncertainties in the method and probably different stomatal responses to pCO_2 in different species. This is not surprising because, although stomatal density may decrease somewhat systematically in laboratory experiments with increasing pCO_2, the response on evolutionary timescales is unknown and the calibration curve used to calculate pCO_2 with the stomatal indices are subject to wide uncertainties especially at higher than pre-industrial values (Roth-Nebelsick *et al.* 2004). Consequently, results using stomatal proxy methods should be interpreted cautiously.

Boron isotopes – The boron-isotope ratios of planktonic foraminifera have been used to estimate the pH of surface-layer seawater, which is in turn used to estimate pCO_2 (e.g. Pearson & Palmer 1999, 2000; Hšnisch & Hemming 2004). Results suggest an erratic decline in pCO_2 from 3000–4000 ppm between 55 and 40 Myr ago to 1000 ppm for the Late Eocene but are based on various assumptions (see e.g. Pagani *et al.* 2005*a*). Data for the E–O transition are lacking but values are suggested to have decreased by at least a factor of two by the Miocene (Pearson & Palmer 2000).

Alkenones – pCO_2 estimates using the carbon stable isotope composition of marine alkenones indicate a similar pattern of pCO_2 decline through the Palaeogene and Neogene to the boron isotope method, but there are also significant differences (Pagani 2002; Pagani *et al.* 2005*b*). Results suggest that pCO_2 ranged between 1000 to 1500 ppm in the middle to Late Eocene then decreased in several steps during the Oligocene, reaching 'modern' levels by the latest Oligocene. In detail, the alkenone record shows that pCO_2 remained high across the E–O boundary and into the lower Oligocene and began falling rapidly at *c.* 32 Ma (Pagani *et al.* 2005*b*).

Palaeosol estimates – pCO_2 estimates based on palaeosols isotopic data support a pattern of higher concentrations during the Eocene and Oligocene compared to the Neogene (e.g. Cerling 1991, 1999; Ekart *et al.* 1999; Lowenstein & Demicco 2006).

In summary, despite some discrepancies, independent proxies largely support a pattern of higher than modern (2 to 5 times) atmospheric pCO_2 in the Late Eocene, falling through the Cenozoic and reaching near-modern values in the lower Miocene. Possible causes are reduced sea-floor spreading (Berner 1990), increased silicate weathering (Raymo & Ruddiman 1992; Ruddiman *et al.* 1997) and increased organic carbon burial (Diester-Haass & Zachos 2003). Although there is a suggestion of a trend towards decreasing pCO_2 through the Eocene (Pearson & Palmer 2000; Pagani *et al.* 2005*b*), the alkenone data suggest that pCO_2 remained high (1000–1500 ppm) until after the E–O climatic transition, possibly following termination of the EOGM (Pagani *et al.* 2005*b*). This questions the proposed relation between pCO_2 and the initiation of glaciation on Antarctica (DeConto & Pollard 2003*a*). However, none of the proxies or age correlations are perfect (e.g. Pagani *et al.* 2005*a*; Roth-Nebelsick *et al.* 2005) and despite the issues of timing it is still possible that the pCO_2 decrease was a critical factor that allowed expansion of ice-sheets on Antarctica.

Carbonate and the CCD

The carbonate compensation depth (CCD) is the depth in the ocean at which the supply of carbonate produced in surface waters is balanced by dissolution. The CCD is sensitive to ocean fertility, acidity and pCO_2 and, thus, is closely linked to global climate. Cenozoic palaeodepth reconstructions showing the distribution of carbonate sediments across the ocean basins reveal a general increase in deep-sea carbonate content in the Oligocene compared to the Eocene associated with a major global deepening of the CCD (Van Andel 1975; Thunell & Corliss 1986) (Fig. 5a). The deepening was most extreme in the eastern equatorial Pacific (estimated at 1200 m, Rea & Lyle 2005) but has also been recorded in the Atlantic Ocean (*c.* 1 km; Hsü *et al.* 1984; Moore *et al.* 1984) and the Indian Ocean (max. 700 m; Peterson & Backman 1990), indicating that it was a global phenomenon reflecting large changes in ocean palaeochemistry. The effect of the CCD deepening was to more than double the area of sea-floor available for $CaCO_3$ deposition (Rea & Lyle 2005).

Recent CCD proxy records from equatorial Pacific ODP Site 1218 have provided improved constraints on the timing of the CCD shift and its relation to the EOGM (Coxall *et al.* 2005). These records show a very rapid increase in $CaCO_3$ content and $CaCO_3$ mass accumulation rate (MAR) close to the E–O boundary (Fig. 5b). The shift occurred in parallel with the $\delta^{18}O$ and $\delta^{13}C$ increase, and shows the same two-step structure, indicating that deepening CCD shift was synchronous with the climatic shifts that preceded the EOGM. Data and model simulations suggest that under Pleistocene conditions, the increase in deep-sea carbonate ion concentration associated with a 1-km deepening of the global lysocline yields a drawdown of atmospheric CO_2 of less than 25 ppm (Sigman & Boyle 2000; Zeebe & Westbroek 2003). Thus, E–O CCD deepening is unlikely on its own to have caused glaciation through the reduction of pCO_2. More likely, the glaciation triggered the CCD shift. The CCD probably deepened to compensate for a reduction in the global ratio of $CaCO_3$ to organic carbon burial, as suggested by the calcite $\delta^{13}C$ increase. The observed increase in the carbonate saturation state of the deep ocean could have been achieved in various ways, which may have acted in combination: (1) a shift in the locus of carbonate sedimentation from the shelf to the deep sea (Berger & Winterer 1974; Opdyke & Wilkinson 1989; Coxall *et al.* 2005); (2) a rapid increase in the amount of calcium entering the oceans (Rea & Lyle 2005); (3) an increase in global siliceous (at the expense of calcareous) plankton export production (e.g. Thunell & Corliss 1986; Harrison 2000). The biomineral shift hypothesis is supported by microfossil evidence from the Southern Ocean and elsewhere that demonstrates increased opal accumulation in the early Oligocene (e.g. Diester-Haass & Zahn 1996; Salamy & Zachos 1999; Diester-Haass & Zachos 2003). The relative importance of these hypotheses can be tested using biogeochemical models such as developed in Zachos & Kump (2005).

Also of note is an interval of severe carbonate dissolution in the top of magnetic Chron C13r that pre-dates the CCD shift (Fig. 5b). At ODP Site 1218, $CaCO_3$ content of the sediment and $CaCO_3$ MAR fall to zero for an interval of *c.* 0.6 m (~200 kyr) close to the base of the shift into the EOGM. In fact, the precise timing of the base of the $\delta^{18}O$ shift is lost in this record because there are no benthic calcitic foraminifera for analysis. A similar interval of dissolution, correlative with upper Chron C13r, occurs at DSDP Site 522 (Zachos *et al.* 1996), Equatorial Atlantic ODP Site 925 (Diester-Haass & Zachos 2003) and ODP Site 1265 (Liu *et al.* 2004; Tuo *et al.* 2006). This suggests a temporary shallowing of the CCD immediately before the build-up to the EOGM, the significance of which has yet to be established.

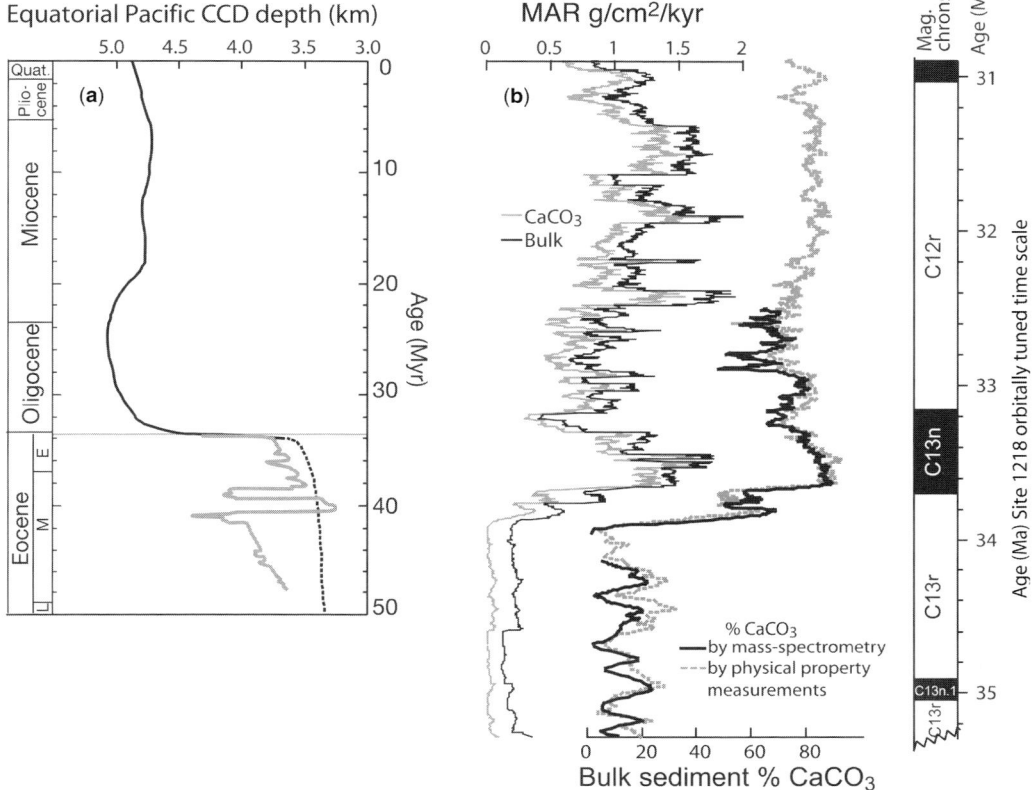

Fig. 5. Deep-sea records showing changes in the Eocene–Oligocene equatorial Pacific Calcite Compensation Depth (CCD). (**a**) Long-term Cenozoic record of CCD depth from 50 Ma to Present sediments based on carbonate mass accumulation rate (MAR) (after Tripati *et al.* 2005). Black line is the CCD depth reconstructed from classic low resolution Deep Sea Drilling Project records (modified from Van Andel 1975). Grey line is the revised E–O CCD history based on ODP Leg 199 sediments. (**b**) High-resolution %CaCO₃ and CaCO₃ MAR (5-point running means) across the E–O transition from ODP Site 1218 (palaeodepth *c.* 3800 m) (Coxall *et al.* 2005) on an orbitally tuned timescale. The records show that CCD deepening (increase in CaCO₃) occurred (i) faster than previously documented, (ii) in two 40 kyr steps. The timing of CCD deepening is synchronous with the stepwise δ¹⁸O shift into the EOGM (Coxall *et al.* 2005).

Weathering and sediments

Changes in the flux, nature and chemistry of deep-sea marine sediments indicate the style and intensity of contemporaneous weathering and the possible sources of incoming sedimentary material, all of which vary with climate.

Glaciomarine sediments – The most direct evidence of E–O Antarctic glaciation is the occurrence of glaciomarine sediments, i.e. water-lain glacial tills (including ice-rafted debris), sands and diamictites (coarsely sorted boulder- to sand-sized grains), on or close to Antarctica and elsewhere. These sediments represent material that was scraped off and entrained by moving ice and subsequently

deposited at the front of retreating glaciers or transported offshore by icebergs. The pattern of early Cenozoic glaciomarine sedimentation as recovered in deep-sea cores indicates several episodes of expansive continental glaciation during the Oligocene and several short-lived episodes in the Late Eocene in east Antarctica (e.g. Barrett *et al.* 1989; Wise *et al.* 1991; Breza & Wise 1992; Zachos *et al.* 1992, 1999). There is now also glaciomarine evidence for a regionally extensive West Antarctica ice-sheet of E–O age (Birkenmajer 1996; Gilbert *et al.* 2003; Birkenmajer *et al.* 2005; Ivany *et al.* 2006). These records suggest that the Oligocene glaciation was not restricted to East Antarctica and, thus, imply an extreme climatic response to the forcing factors that facilitated high-latitude ice expansion in the earliest Oligocene (Ivany *et al.* 2006).

Information for the Northern Hemisphere is limited. Results from recent drilling on Lomonosov Ridge, central Arctic Ocean, recovered the first evidence of ice-rafted debris from the middle Eocene, *c.* 35 Myr earlier than previously thought (Moran *et al.* 2006). These sediments are interpreted as indicating Arctic cooling and significant ice coverage coincident with Antarctic glaciation and are used in support of arguments for bipolar symmetry in early Cenozoic climate change (e.g. Tripati *et al.* 2005). Understanding of the extent of Arctic glaciation and pattern of climatic change, however, remains limited because carbonate sediments, including calcitic micofossils that provide important geochemical climate proxies, are missing from Lomonosov cores, and the Late Eocene and Oligocene fall in a hiatus (Moran *et al.* 2006). Further drilling, therefore, is necessary to corroborate the initial findings from Lomonosov Ridge and obtain appropriate stratigraphic coverage of the E–O Transition. Recent reports of stratigraphically extensive ice rafted material in the Late Eocene to early Oligocene from the Norwegian-Greenland Sea add support to the idea of contemporaneous Northern Hemisphere glaciation (Eldrett *et al.* 2007). However, these records cannot determine the extent of glaciation, i.e. whether the material was deposited by icebergs calving off small ephemeral glaciers or larger ice-sheets.

Strontium isotope ratios – Variations in marine carbonate $^{87}Sr/^{86}Sr$ during the Cenozoic represent changes in the balance of Sr inputs from hydrothermal flux and continental weathering (e.g. Palmer & Edmond 1992; Richter *et al.* 1992; Reilly *et al.* 2002). Therefore, as well as providing a useful stratigraphic tool (e.g. Burke *et al.* 1982), $^{87}Sr/^{86}Sr$ has the potential to record information on various processes relevant to climate change, especially concerning weathering shifts related to longer-term variations in atmospheric CO_2. The pattern of post-Eocene $^{87}Sr/^{86}Sr$ shows a general increase in values with a series of stepped increases in the rate of Sr isotopic variation superimposed. The first of these steps occurs around the Late Eocene to early Oligocene (Hess *et al.* 1986; Miller *et al.* 1991; Mead & Hodell 1995; Zachos *et al.* 1999; Reilly *et al.* 2002). Although changes in hydrothermal flux and carbonate weathering are thought to have contributed to the general $^{87}Sr/^{86}Sr$ increase over the last 100 myr, increased riverine inputs due to enhanced continental weathering are believed to have been responsible for the larger-scale isotopic variation (e.g. Reilly *et al.* 2002; Lear *et al.* 2003).

The cause of the inferred enhanced weathering is controversial. This is in part because of differences in identifying the timing of the Sr shift (see Reilly *et al.* 2002) and has been attributed to: tectonic uplift of the Himalayan Plateau, which increased elevation and altered exposure of rock lithologies (e.g. Raymo & Ruddiman 1992); initiation of Antarctic glaciation, whereby increased ice-sheet growth resulted in increased mechanical and chemical weathering of continental rocks (Miller *et al.* 1991; Zachos *et al.* 1999; Zachos & Kump 2005); or a combination of these processes (Mead & Hodell 1995; Reilly *et al.* 2002).

Clay mineralogy – Variations in clay mineralogy indicate changes in depositional environment and provide insight into erosional processes related to climate. Cenozoic clay mineral records show a general trend towards illite–chlorite-dominated assemblages and a reduction in smectite and kaolinite. This is thought to reflect an increase in the rock-derived supply and a decrease in soil-derived supply of clay minerals to the ocean (Chamley 1986), although the pattern is also affected by local conditions. The shift begins in the Late Eocene and has been recorded in marine sediments globally (Chamley 1986; Ehrman & Mackensen 1992; Diester-Haass *et al.* 1993; Robert *et al.* 2002; Gale *et al.* 2006). Close to Antarctica, the illite–chlorite shift has been interpreted as recording a switch from a climate typified by alternating wet and dry seasons resulting in predominantly chemical erosion, to intensified physical (mechanical) weathering by glaciers on eastern Antarctica under a cooler climate (Ehrman & Mackensen 1992; Diester-Haass *et al.* 1993; Zachos *et al.* 1999; Robert *et al.* 2002). Antarctic environmental magnetic records support this hypothesis. For example, Ross Sea sediments show a decrease in detrital magnetite during the early Oligocene, which is interpreted as indicating a shift to a cooler, drier climate leading to reduced chemical weathering of igneous rocks on Antarctica (Sagnotti *et al.* 1998; Wilson *et al.* 1998). The environmental magnetic records and clay mineral behaviour is thus consistent with other evidence for widespread cooling and glaciation of Antarctica in the earliest Oligocene. These records also constrain the timing of glacial activity on Antarctica in the Late Eocene and suggest intensification of glaciation in the basal Oligocene.

Osmium isotopes – Os isotope variation in seawater ($^{187}Os/^{188}Os$), as recorded by marine sediments, is believed to reflect the contribution of Os to the oceans from continents (rivers) and submarine alteration, and the influx of extraterrestrial material (cosmic dust). The long-term Cenozoic record of $^{187}Os/^{188}Os$ variation shows a pattern of increasing values that was probably caused by some combination of changes in weathering or increased influx of cosmic dust (Peucker-Erenbrink *et al.* 1995; Pegram & Turekian 1999; Peucker-Erenbrink & Ravizza 2000). The increasing numbers of detailed Os isotope records across the E–O transition provide good evidence for the global influence of glaciation on the supply of Os to the ocean (Ravizza & Peucker-Erenbrink 2003; Dalai *et al.* 2006).

Bulk sediment $^{187}Os/^{188}Os$ records from the Atlantic and Pacific reveal a *c.* 30% stepwise shift towards higher values in the early Oligocene after a minimum in the Late Eocene (Pegram & Turekian 1999; Ravizza & Peucker-Erenbrink 2003; Dalai *et al.* 2006). The stepwise shift is interpreted as coinciding with the growth and decay of major ice-sheets. Comparison of E–O $^{187}Os/^{188}Os$ records with high-resolution benthic foraminiferal $\delta^{18}O$ from Site 1218 (Coxall *et al.* 2005) suggests that Os flux to the oceans decreased during cooling and ice growth, whereas subsequent decay of ice-sheets and deglacial weathering drove seawater $^{187}Os/^{188}Os$ to higher values, i.e. higher $^{187}Os/^{188}Os$ values occur following termination of the EOGM. The post-EOGM $^{187}Os/^{188}Os$ shift is interpreted to represent an increase in radiogenic Os to the oceans derived from the weathering of easily erodible glacial moraines following termination of the glaciation (Ravizza & Peucker-Ehrenbrink 2003; Dalai *et al.* 2006). It is suggested that $^{187}Os/^{188}Os$ remained high through the Oligocene because subsequent cyclic variations in ice volume regularly replenished sediment supply (Ravizza & Peucker-Ehrenbrink 2003).

Dust – Aeolian grains in pelagic clays, i.e. wind-blown dust particles, show an increase in mean grain size beginning in the Late Eocene close to the E–O boundary. This has been attributed to more vigorous atmospheric circulation associated with the climatic reorganization (Rea *et al.* 1985; Ravizza Peucker-Ehrenbrink 2003; Vanden Berg & Jarrad 2004). A more controversial idea is that cosmic dust flux in the Late Eocene may have contributed to global cooling and the EOGM by supplying bio-essential trace elements to the oceans and thereby resulting in higher ocean productivity, enhanced burial of organic carbon and drawdown of atmospheric CO_2 (Dalai *et al.* 2006).

The sedimentary and geochemical evidence for changes in weathering associated with the E–O transition is consistent with other proxies that indicate widespread glaciation. Weathering-CO_2 climate feedbacks undoubtedly play a role in forcing or moderating climate but their importance is widely debated (Volk 1987; Kump *et al.* 2000; Lear *et al.* 2004).

Late Eocene impacts

The discovery of three bolide impact craters (Chesapeake Bay, Toms Canyon and Popigai; e.g. Poag 1995; Bottomley *et al.* 1997) and associated breccia and ejecta deposits in the Late Eocene has provoked speculation that the biotic turnover and climatic cooling associated with the E–O boundary was triggered by the impact of an extraterrestrial body or bodies (e.g. Keller 1986; Vonhof *et al.* 2000). Recent age correlation of the impact sites shows that they all occurred within magnetic Chron C16n.2n (Poag *et al.* 2003), which precedes the E–O boundary by about 2 million years, and may have formed part of a 'comet shower' (Poag *et al.* 2003). Minor extinction episodes and assemblage shifts among planktonic foraminifera (e.g. Keller 1986) and dinocysts (Vonhof *et al.* 2000) and a negative $\delta^{13}C$ excursion associated with the impact horizons at several sites (Poag *et al.* 2003) has been taken as evidence for global-scale long-term enviornmental disturbance related to the impacts. However, it seems unlikely that the impacts had a major influence on the Earth's biosphere at the E–O boundary and they are discounted here as a possible cause of the climatic transition. It is possible, however, that climatic disruption caused by atmospheric dust loading associated with the Late Eocene impacts, coupled with an ice-albedo feedback mechanism that amplified impact-induced climatic cooling, may have contributed to global cooling in the run-up to the E–O transition (Vonhof *et al.* 2000).

Cyclostratigraphy and Milankovitch cyclicity

Orbital dynamics that affect the amount and distribution of incoming solar radiation received by the Earth are thought to have been the principal 'pacemaker' of Quaternary climate cycles, forcing the repeated growth and decay of continental glaciers (Imbrie *et al.* 1984). The effects of orbital fluctuations in the early Cenozoic, however, when climate boundary conditions differed and global ice volumes may have been significantly less than in the Quaternary, have been difficult to resolve. One problem has been uncertainties in tracing astronomical solutions that predict the position of individual cycles back into the Palaeogene (Laskar 1999). However, much progress has been made recently, and astronomical models have been extended back to the early Oligocene and beyond (Laskar *et al.* 2004; Pälike *et al.* 2004), allowing orbital tuning of E–O records (Coxall *et al.* 2005; Gale *et al.* 2006; Jovane *et al.* 2006).

Improved deep-sea core recovery and automated quasi-continuous core and downhole logging analysis have resulted in a number of E–O climate proxy records, including magnetic susceptibility, percent carbonate and benthic $\delta^{18}O$ and $\delta^{13}C$, of sufficient resolution for resolving patterns of cyclic variation on Milankovitch timescales (e.g. Mead et al. 1986; Hartl et al. 1995; Zachos et al. 1996; Coxall et al. 2005; Jovane et al. 2006; Gale et al. 2006). Changes in surface productivity and bottom-water redox conditions in response to climate are thought to be responsible for the observed periodicity in sediment physical property records, whereas fluctuating ice volumes and behaviour of the global carbon cycle are thought to be the major climatic control on $\delta^{18}O$ and $\delta^{13}C$ respectively (Zachos et al. 1996).

Spectral analysis of ODP Site 1218 $\delta^{18}O$ and $\delta^{13}C$ records, which are the highest-resolution E–O time series available, show that all Milankovitch periods are encoded (i.e. c. 20, c.40, c.100 and c. 400). Power is concentrated at the obliquity (40 kyr) frequency for $\delta^{18}O$ and at the 400-kyr eccentricity frequency, with a weaker 40 kyr component for $\delta^{13}C$ (Coxall et al. 2005). A similar c. 40 kyr spectral peak was recognized in DSDP Site 522 $\delta^{18}O$ records (Zachos et al. 1996). In the Oligocene, a phase lag of $\delta^{18}O$ with respect to $\delta^{13}C$ of c. 8 kyr in the 40 kyr band suggests that the response of the global carbon cycle helped to force climate changes in early Oligocene climate rather than reacting to it (Coxall et al. 2005). As in Neogene records, the clear c. 40 kyr spectral peak in $\delta^{18}O$ is probably associated with changes in Earth's tilt axis and is consistent with a strong high-latitude (ice volume) influence on global climate (e.g. Imbrie et al. 1984). A new 13-million-years-long record from ODP Site 1218 for the entire Oligocene, which includes the E–O dataset, confirms this pattern and identifies 405, 127 and 96 kyr Earth's eccentricity and 1.2-million-year obliquity cycles pacing periodically re-occurring glacial and carbon cycle events throughout the Oligocene (Pälike et al. 2006).

Time-series analysis of other datasets shows similar patterns. Spectral analysis of astronomically tuned terrestrial illitic clay data from the Hampshire Basin, UK, identifies strong c. 400 kyr eccentricity and 41 kyr obliquity frequencies, with a smaller peak at around 100 ka (Gale et al. 2006). Illite is interpreted to have formed in palaeosols through repeated wetting and drying in response to high seasonality. The obliquity frequency cyclic variation probably reflects changes in sea level driven by minor fluctuations in the volume of Antarctic ice. Longer-term environmental magnetic records from the E–O stratotype in Massignano, Italy, reveal cycles of c. 405 kyr (Jovane et al. 2004, 2006) as

a robust feature through the Late Eocene and into the Oligocene. External forcing by orbital parameters at eccentricity frequencies has also been detected in E–O Southern Ocean (Diester-Haass & Zahn 1996) and low-latitude productivity proxies (Nilsen et al. 2003), suggesting that orbital forcing caused a sedimentary response to global climate change from at least Late Eocene.

The dominance of the long Milankovitch periods in these records is unexpected because the climate system response predicted by insolation calculations based on the traditional Milankovitch summer insolation hypothesis (Loutre et al. 2004) is for the dominance of climatic precession and obliquity, with only a very minor contribution by Earth's eccentricity periods. Modelling experiments using a carbon cycle box model coupled to models with orbital variations suggest that the explanation for this phenomenon at least for $\delta^{13}C$ records is amplification of the 405 kyr cycle due to the long residence time of carbon in the oceans (c. 0.1 million years) (Pälike et al. 2006).

In summary, time-series analyses performed on early Oligocene datasets have revealed that 40 and 400 kyr Milankovitch cycles are a pervasive feature of isotopic and other climate proxy records, as in much of the Neogene (e.g. Zachos et al. 1997; Paul et al. 2000), and further support the presence of ice-sheets responding to orbital fluctuations during this time period.

Deep-water circulation

The history of deep-water formation and circulation provides important information about global thermal gradients and circulation regimes. Today, and for much of the later Cenozoic, the near-freezing conditions at high northerly and southerly latitudes produce the most dense in situ water masses that sink to fill the deep ocean basins. During the early Cenozoic, however, when the high latitudes were warmer, the Southern Ocean and even mid-latitudes are thought to have played greater roles in deep-water formation (e.g. Miller & Fairbanks 1985; Kennett & Stott 1990; Wright & Miller 1993). Considerable effort has been put into trying to reconstruct E–O deep-water circulation history using a variety of different techniques. Early reconstructions that utilized $\delta^{18}O$ (temperature), $\delta^{13}C$ (water mass age) and sediment distribution suggest that the Southern Ocean was the dominant source of deep water for the Late Eocene, with pulses of bipolar deep-water formation, i.e. an additional Northern Hemisphere-sourced deep-water component, from the earliest Oligocene (e.g. Wright & Miller 1993 and references therein). This is supported by sedimentary and seismic evidence that indicates the occurrence of early Oligocene-age sediment drift deposits reflecting a southern-flowing deep-water mass in the North Atlantic (Kidd & Hill 1987; Davies *et al.* 2001) and North Pacific (Scholl *et al.* 2003). Deep-sea benthic foraminifera also provide indications of bottom-water conditions. The gradual faunal changes in lower bathyal benthic foraminiferal assemblages observed in the Southern Ocean and elsewhere have been interpreted as suggesting that the cold, high-latitude-sourced deep waters became established gradually as a result of progressive cooling of surface waters at high latitudes (Thomas 1992; Takata *et al.* 2007).

Neodymium isotopes – Additional evidence for northern-sourced deep water is provided by neodymium isotopes. Nd isotopic variations in seawater (ε_{Nd}) reflect local Nd supply to the ocean through weathering of continental rocks. Deep-water masses sinking in a particular region therefore carry a ε_{Nd} signature characteristic of the proximal Nd source (e.g. Goldstein & Jacobsen 1988; Via & Thomas 2006 and references therein). This signature is preserved in the teeth and bones of fossil fish (e.g. Wright *et al.* 1984) and can be extracted to provide a useful tracer of deep-water influences on geological timescales. Nd isotope data from E–O sediments indicate that the initial transition to a bipolar mode of deep-water circulation, involving the onset of Northern Component deep-water formation, occurred in the early Oligocene (Scher & Martin 2004; Via & Thomas 2006). Such data seem to imply substantial cooling and/or ice growth in the Northern Hemisphere at the climax of the E–O transition.

Mechanisms and modelling

The cause of the E–O climate transition is still widely debated and will doubtless remain so, because such complex shifts in the Earth System probably do not have a single identifiable cause. The most widely held explanations for the fundamental climatic shift and initiation of the EOGM are that it was caused by (i) cooling of Antarctica due to changes in ocean circulation controlling poleward heat transport, (ii) a gradually forced threshold response to declining atmospheric carbon dioxide levels. New drill sites and high-fidelity proxy records are important for exploring these hypotheses by documenting patterns of change across the planet. Computer models, however, which are becoming ever more sophisticated in their ability to couple ocean, atmosphere, cryosphere and biosphere climatic parameters, have a valuable role in hypothesis testing and allow exploration of climate behaviour that includes

a range of complex forcing and feedback mechanisms operating on global scales. Here the high-quality proxy records provide essential boundary conditions for constraining the models. The following section summarizes evidence for and against the possible mechanisms of E–O climate change including the results of modelling studies, but we emphasize that there is still more that we 'don't know' than that which we 'know'.

Ocean gateways – One classical hypothesis suggests that E–O glaciation was achieved through thermal isolation of Antarctica following the tectonic opening of ocean gateways between Antarctica and Australia (Tasmanian Passage) and Antarctica and South America (Drake Passage), which led to the organization of the Antarctic Circumpolar Current (ACC) (Kennett *et al.* 1975; Kennett 1977; Murphy & Kennett 1986). This idea draws on foraminiferal isotopic evidence from deep-sea drill cores, principally from the Tasman Sea, that suggests a shift from warm to cold currents close to the E–O boundary and a shift from calcareous to biosiliceous microfossils in the early Oligocene. The theory has been extended recently following a new campaign of deep-sea drilling in the Tasman Sea region (Shipboard Scientific Party 2001; Exon *et al.* 2004). Renewed support for the hypothesis that thermal isolation of Antarctica was the principal cause of E–O climate change is also provided by the results of (GCM) simulations that suggest opening of Drake Passage and the organization of an ACC would have reduced southward oceanic heat transport and facilitated cooling of Southern Ocean sea-surface temperatures (Toggweiler & Bjornsson 2000; Nong *et al.* 2000).

This view, however, remains controversial. An alternative interpretation of siliceous microfossil distribution from the region, principally dinocysts and diatoms that are abundant in productive polar regions, suggests that the Tasman Gateway opened two million years before permanent glaciation at about 35.5 Ma and that, although possibly a factor contributing to the changes, this tectonic event probably was not the cause of the E–O climatic shift (Stickley *et al.* 2004). Moreover, an alternate coupled ocean–atmosphere GCM challenges the idea that the opening of the Tasmanian Passage contributed to Antarctic thermal isolation by interrupting the southward flow of warm tropical waters, because model simulations suggest that warm waters did not penetrate to high latitudes in the first place (Huber *et al.* 2004). This interpretation is supported by another set of GCM simulations which suggest that the opening of Southern Ocean gateways plays a secondary role in this transition, relative to pCO$_2$ concentration (DeConto & Pollard 2003*a*, *b*). Similarly, in an assessment of the potential influence of the ACC on Antarctic glaciation, it was concluded that the presence of biosiliceous biofacies in the early Oligocene cannot be used as evidence for the existence of a continuous deep-reaching front, in the form of an ACC, and that colder SSTs were an effect of glaciation rather than the cause (Barker & Thomas 2004). According to this study, Eocene–Oligocene deepening of the Tasmanian gateway and timing of glaciation are coincidental.

Precise timing of the opening of these tectonic gateways is necessary for assessing their importance as possible mechanisms of E–O climate change. However, there are differences and uncertainties in the age estimates of events, especially for the opening of Drake Passage. Using tectonic plate reconstructions, Lawver & Gahagan (2003) suggested that a deep-water passage opened in the Tasman Seaway close to the E–O boundary (c. 33 Ma), and that the Drake Passage may have been open by 31 Ma. The authors recognize a major change in circulation at this time, with increased production of Antarctic Bottom Water, but they are cautious about linking this change directly to the early Oligocene glaciation because there is not a great deal of evidence to support vigorous ACC until the Miocene (Rack 1993).

Using geophysical-modelling techniques, Livermore et al. (2005) and Eagles et al. (2006) estimate Drake Passage opened near to the E–O boundary (34–30 Ma) and suggest that this tectonic event was the trigger for rapid onset of early Oligocene glaciation. Geochemical proxy data (Latimer & Fillipelli 2002) and opal accumulation records (Diekmann et al. 2004) from the Agulhas Basin have also been interpreted as evidence for Drake Passage opening to deep ACC flow close to the E–O boundary at around 32–33 Ma. The Livermore et al. (2005) kinematic model predicts that initial opening of the Drake Passage occurred in the middle Eocene, which is considerably earlier than previous estimates, and suggest that flow of deep water through a series of restricted basins may have contributed to gradual cooling through the middle Eocene.

Nd isotopes also provide constraints on the timing of Drake Passage opening. Scher and Martin (2006) interpret Nd records from Agulhas Ridge as indicating an influx of shallow Pacific water in the late middle Eocene at c. 41 Ma, suggesting that a connection through Drake Passage was already open and probably facilitating complete circum-Antarctic circulation by the Late Eocene, i.e. before the opening of the Tasman Seaway. Based on this timing, the authors conclude that Drake Passage opening could not have initiated the EOGM glaciation and propose circulation/productivity linkages as a mechanism for declining

atmospheric carbon dioxide that may have been the primary cause.

Despite recent and ongoing research into these questions, opinions on the role of circulation changes in the initiation of E–O climate change remain divided. There appears to be a variety of evidence to suggest that Southern Ocean gateways had opened by the E–O boundary and that some form of proto-ACC, which might have had a direct impact on Antarctic climate, had been established by the earliest Oligocene. Alternatively, new interpretations of microfossil records and alternative GCM simulations suggest that circulation changes leading to thermal isolation of Antarctica did not play a primary role in initiating glaciation. Considering these issues, Barker and Thomas (2004) outline ideas for future work. They suggest that future efforts should be focused on obtaining: (i) better definition of the properties of oceanic fronts, (ii) Cenozoic atmospheric pCO_2 proxy records, (iii) further developments in models of Antarctic glaciation, (iv) marine geology studies, including grain-size, geochemical and mineralogical as a way of examining bottom-water circulation patterns, and (vi) reappraisal of palaeontological records previously interpreted as evidence for an ACC.

Declining carbon dioxide and orbital geometry – The second widely held hypothesis is that E–O Antarctic glaciation was triggered by a threshold response to declining atmospheric pCO_2, consistent with proxy records (see above), in combination with a planetary orbital configuration that favoured accumulation of annual snow. Modelling experiments have bolstered support for this hypothesis. In one set of simulations, sudden glaciation was forced by gradually declining pCO_2 (DeConto & Pollard 2003a). The results suggest that a threshold may be reached whereby ice-sheet height/mass-balance feedbacks cause isolated ice-caps to expand rapidly, eventually coalescing into a continental-scale ice-sheet (Fig. 6). Comparison of ODP Site 1218 proxy records with an orbital template reveals that the beginning of the shift into the EOGM coincides with a rare orbital configuration comprising a phase of low eccentricity and low-amplitude change in obliquity, favouring cool austral summers and permitting accumulation of summer snow that probably tipped the balance towards glaciation (Coxall et al. 2005) (Fig. 7). Further model sensitivity experiments suggest that the range of atmospheric CO_2 variability needed to induce such a rapid transition in the presence of orbital forcing is 2 to 4 times pre-industrial levels (Pollard & DeConto 2005). The pattern of stepchange in geochemical proxy records substantiates the model prediction of a non-steady mode of ice-sheet growth but the magnitude of the $\delta^{18}O$

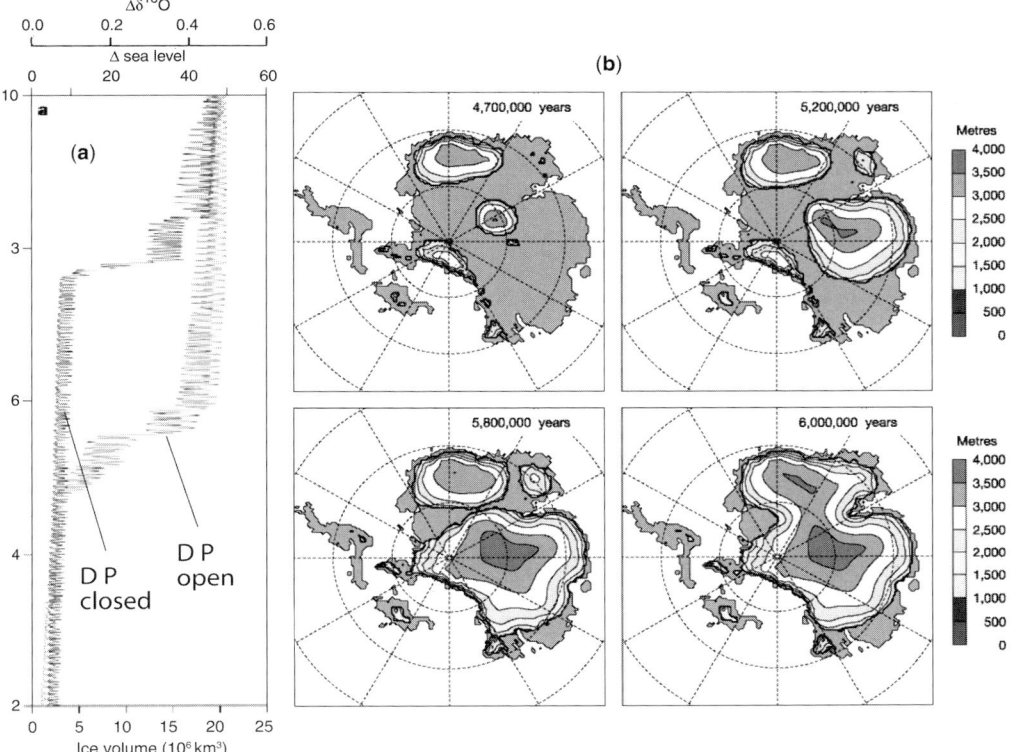

Fig. 6. A General Circulation Model simulation of the glacial inception and early growth of the East Antarctic Ice Sheet (reprinted from DeConto & Pollard (2003*a*) with permission from authors and Nature Publishing Group). (**a**) Model simulation of the transient climate-cryosphere response to a prescribed decline in CO_2 from 4 times to 2 times the pre-industrial atmospheric level over a 10-million year period. Ice volume, equivalent changes in sea level (Δ sea level) and the mean isotopic composition of the ocean ($\Delta\delta^{18}O$) are shown for two pairs of simulations: (i) Drake Passage (DP) open; (ii) DP closed (global climate model ocean heat transport coefficient increased in the Southern Hemisphere). Equivalent sea-level changes were calculated according to the global ocean-area fraction in the 34-million-year palaeogeography (0.731). $\Delta\delta^{18}O$ was calculated assuming a 0.0091 change in $\delta^{18}O$ per 1 m change of sea level. (**b**) Modelled ice-surface elevations for four time slices during the E–O transition in a 10-million-year simulation (DeConto & Pollard 2003*a*). The simulation predicts that as CO_2 declined through the Palaeogene, the gradual lowering of the Antarctic snowline began intersecting areas of high topography, first producing small, isolated ice-caps (4.7 and 5.2 million years). As CO_2 continued to fall, height/mass-balance feedbacks were initiated suddenly, producing larger dynamic ice-sheets that alternately coalesced and separated in response to orbital forcing (5.8 million years). With further decline of CO_2, a single, large EAIS became a more permanent feature, almost insensitive to orbital forcing, with very little summer melting, and accumulation zones reaching sea level around most of the continent (6.0 million years).

response is underestimated by the model (Coxall *et al.* 2005).

A biogeochemical modelling study that was used to investigate the role of the carbon cycle in the glaciation process also supports a drawdown of CO_2 as a likely initial forcing mechanism (Zachos & Kump 2005). The models were able to reproduce the observed positive excursions in the mean $\delta^{13}C$ of inorganic marine carbon and biogenic sediment accumulation rates by simple CO_2 balancing feedbacks in the climate system associated

with ice-sheet coverage, silicate weathering rates, and increased global carbon burial. However, the post-EOGM timing of the stepped reduction in $p CO_2$, as suggested by recent proxy records, questions this proposed relation between $p CO_2$ and the initiation of Antarctic glaciation (Pagani *et al.* 2005*b*).

Additional forcing and feedback mechanisms – A range of additional climate models is providing further insight into the importance of other forcing and feedback mechanisms on E–O climate

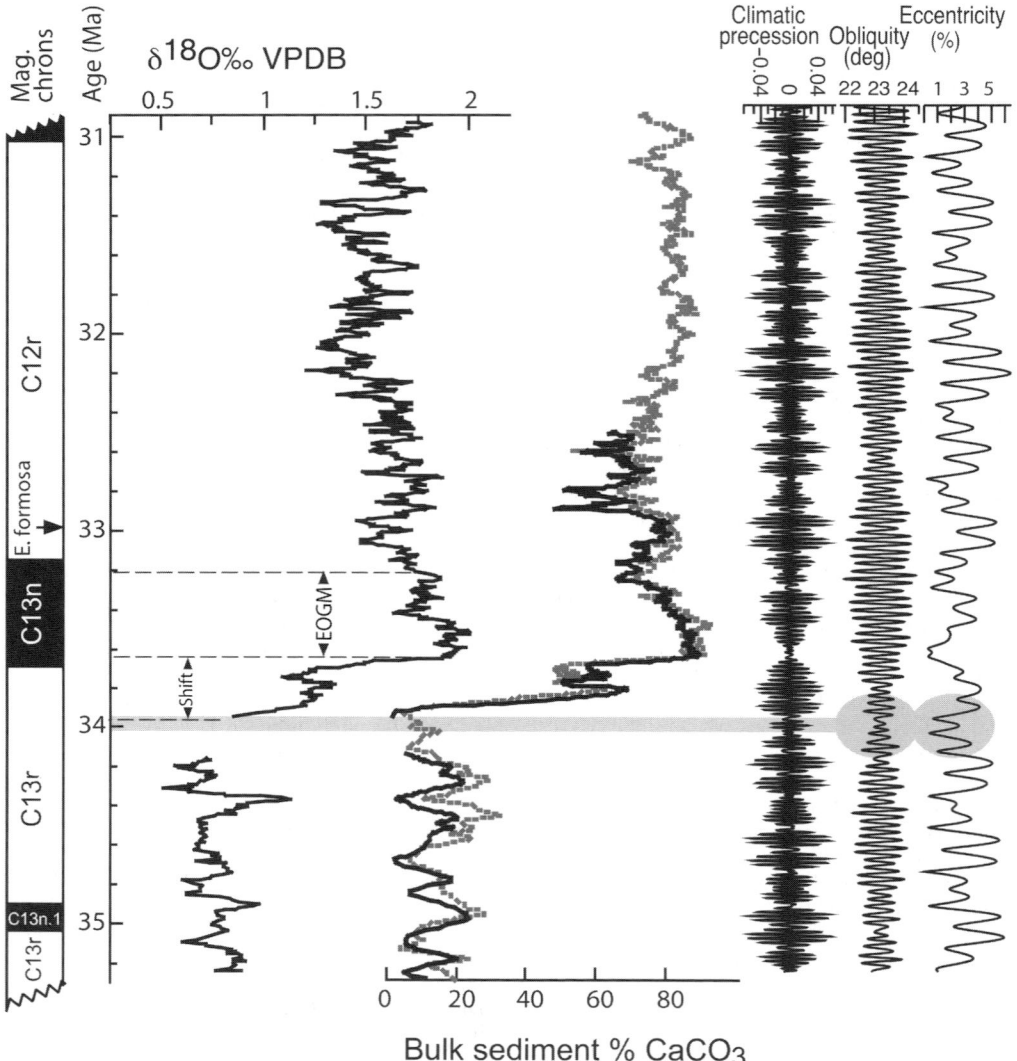

Fig. 7. ODP Site 1218 glacial and CCD proxy records during the Late Eocene to Early Oligocene showing the orbital pacing of Eocene–Oligocene climate change. Comparison of the records indicates that stepwise growth of Antarctic ice-sheets (benthic foraminifera $\delta^{18}O$ increase), was synchronous with CCD deepening in the equatorial Pacific and occurred during an eccentricity minimum and interval of low-amplitude obliquity change (grey shading) favouring cool summers (Coxall *et al.* 2005).

change. For example, modelling has suggested that changes in Antarctic vegetation (i.e. needle leaf forest to tundra) could have caused positive feedbacks through changes in albedo that would have played a significant role in the rapid glaciation (Thorn & DeConto 2006). Other experiments have shown that geothermal heat flux could strongly influence aspects of Cenozoic Antarctic evolution such as ice-sheet basal hydrology, sediment deformation and discharge that appear to increase

orbital variability of ice-sheet behaviour (DeConto & Pollard 2003*a,b*; Pollard *et al.* 2005). Models have also suggested that non-linear Antarctic ice-sheet transitions involving hysteresis have played important roles in many of the observed patterns of $\delta^{18}O$ change (Pollard & DeConto 2005). The hysteresis effect in this context relates to the sudden transitions in ice-sheet size between multiple stable states as a geometric consequence of the intersection of ice-sheet surfaces with typical spatial patterns, such as topography and coastline geography (Pollard *et al.* 2005). The results of these simulations suggest that the E–O global climate shift may have been sensitive to the particular Antarctic conditions operating at the time. Even so, it would be wise to consider Northern Hemisphere changes as well (e.g. Moran *et al.* 2006; Eldrett *et al.* 2007; DeConto & Pollard 2007).

Change in ocean structure and circulation – An alternate view of E–O climate change is presented by Hay *et al.* (2005). These authors suggest that the transition from an Earth without perennial polar ice to one with quasi-permanent polar ice could have taken place rapidly only if both polar regions became extensively ice-covered in a short time. Since there is currently only limited evidence for rapid E–O Northern Hemisphere glaciation, Hay *et al.* (2005) suggest the global transition to an ice-dominated climate system must have been gradual. Their explanation relates the observed changes in climate proxy data to changes in ocean structure and deep-water circulation, specifically attributing the E–O $\delta^{18}O$ shift to three factors: (1) a major component of the $\delta^{18}O$ shift is related to temperature, representing the filling of the deep ocean with cold water and development of a 'psychrosphere' (Kennett & Shackleton 1976); (2) a secondary part is attributed to the formation of permanent sea-ice in the Arctic and perhaps around Antarctica, with only minor growth of an East Antarctic ice-sheet; (3) a large part of the $\delta^{18}O$ shift change may be related to changes in alkalinity and decrease in the pH of seawater as a result of uptake of CO_2 by the ocean from the atmosphere. This hypothesis rests largely on the lack of contemporaneous E–O sea-level change interpreted from earlier sea-level reconstructions but loses power in light of reconstructions that link sequence boundaries implying sea-level fall with global $\delta^{18}O$ increases (e.g. Pekar *et al.* 2002).

Synthesis

The weight of fossil and climate proxy evidence strongly indicates the Late Eocene to Early Oligocene as a time of major global climate change, with many implications for marine and terrestrial ecosystems. Unlike the Cretaceous–Palaeogene or Permian–Triassic boundaries, there is no indication of a sudden catastrophic event; rather, fossil records document a pattern of enhanced yet gradual turnover signalling adjustment to changes in food and nutrient availability, habitat and climate regime.

Terrestrial biotas suggest significant and varied environmental change across the latitudes, including cooling, increased aridity and increased seasonality. These changes are consistent with a scenario of reduced thermal gradients, strengthening of zonal winds and changes in precipitation patterns that might result from cooling of one or both poles. Most floral, and other terrestrial, fossil records, however, lack the time resolution to correlate them precisely with the EOGM as recorded by marine records. Evolutionary turnover is recorded in many terrestrial organisms, with highly varied ecologies, which indicates a variety of external-forcing mechanisms. The clear signal from the terrestrial realm is that there was no 'Terminal Eocene Event'. Instead, the transition was marked by a series of gradual extinctions and speciations, before and after the E–O boundary, that suggest interplay of environmental selection pressures at variable intensities.

Patterns of evolutionary turnover in the pelagic realm suggest changes to ocean thermal structure, food availability and ocean temperatures. Phytoplankton show patterns of gradual turnover and replacement by cosmopolitan, cold-tolerant species. Tropical micro-zooplankton experienced more extinction than other groups, although not on a wide scale, and there was significant reduction in tropical plankton diversity, especially among the radiolaria. These records support the hypothesis of increased oceanic turnover, a reduction in tropical pelagic niches and more intense high-latitude, equatorial and coastal upwelling associated with the climatic shift into the EOGM. These records can be tightly correlated with the proxy records of Antarctic glaciation and show extinctions within the isotopic shift that precedes the EOGM (i.e. planktonic foraminifera and shallow-water benthics), as well as major changes in plankton communities during the EOGM. Deep-sea benthic communities appear to have survived relatively unchanged. Some even benefited from increased food availability because of increased ocean mixing and higher surface ocean productivity. Diversification of large plankton-feeding whales, in parallel with increased opal accumulation, suggests that other elements of the pelagic ecosystem were responding to an expanding nutritional opportunity that arose during the EOGM.

Shallow marine environments appear to have suffered severely during the E–O transition, which brings support to the hypothesis that there was significant sea-level fall associated with the E–O climatic transition. Collection and correlation of fossil records from these types of environments may help further constrain the problem of identifying and quantifying evidence for sea-level change associated with the EOGM.

Multiple proxies support the idea that the biotic changes were associated with major climatic and oceanographic reorganization brought about by widespread glaciation on Antarctica and possibly in the Northern Hemisphere. The causal mechanism for this fundamental shift is controversial. The most popular explanations are that glaciation was caused by (i) cooling of Antarctica due to plate tectonic reorganization and related changes in ocean circulation controlling poleward heat transport and (ii) a threshold response to declining atmospheric $p\mathrm{CO}_2$. Models and proxy records are being interpreted very differently to bring support to both theories. In our view, the balance is tipped in favour of the idea that the Eocene–Oligocene climate transition was forced by a gradual change in a key parameter (i.e. $p\mathrm{CO}_2$) past a switching point, rather than rapid forcing, which caused a shift towards conditions favourable for ice accumulation and leading to the geologically sudden build-up of the Antarctic ice-cap. This mechanism seems to provide the most parsimonious explanation linking up proxy data observations, orbital configurations and complex carbon system and biosphere feedbacks. The effects were not instantaneous, however, and were spread over a timeframe of several hundred thousand years. On the other hand, it is difficult to dismiss the tectonic separation of Antarctica leading to geographic isolation as pure coincidence. Most likely, the E–O transition, which represents a very significant shift in the Earth System, was caused by some complex combination of forcing factors rather than a single identifiable cause.

Incoming datasets have a tendency to raise new, more focused questions that direct future research. For example, $\delta^{18}\mathrm{O}$ and $\mathrm{Mg/Ca}$ proxies and the problems associated with separating temperature and ice volume effects have peaked interest in the possibility of contemporaneous Northern Hemisphere glaciation. This highlights the necessity for future deep-sea drilling to target the Northern Hemisphere, particularly Arctic, sediments that will resolve this question. In addition, future drilling should aim to sample sediments close to Antarctica that might provide further constraints on the timing and extent of Antarctic glaciation. Recognition of the synchrony of CCD deepening and the $\delta^{18}\mathrm{O}$ shift into the EOGM has raised questions about the importance of deep-ocean alkalinity changes as a primary cause or effect of the climatic transition. Attention to these issues has also prompted studies into the role of deep-water alkalinity and carbonate ion concentration on the $\mathrm{Mg/Ca}$ palaeotemperature proxy (see Lear 2007).

In the future, production of high-resolution records will be vital. First, these are essential for resolving leads and lags between ice fluctuations and carbon cycling. This has already been achieved in one site (e.g. Coxall et al. 2005) but needs to be duplicated elsewhere. Secondly, high-resolution records on 1000-yr timescales are necessary to assess the importance of orbital

pacing of climate, and identify changes in the influence of different forcing periods during the transition from a (relatively) ice-free to a glaciated climate state. The Site 1218 record, which combines various proxies, represents an important step towards this goal and should provide a model for future high-resolution studies.

In summary, the improved resolution of incoming proxy records together with the variety of additional proxies that provide independent checks on key climatic parameters are providing better constraints for modelling E–O climate change scenarios. These data allow us to test hypotheses on climate change mechanisms that incorporate many factors. Model outputs force us to cross-check data interpretations, seek new ways of constraining key processes and, in some cases even re-evaluate the principles underpinning proxies. In particular, new SST estimates from 'glassy forams' and the TEX_{86} method will be important in this respect and recent data for other time periods (Pearson et al. unpublished) are already much closer to the modelled view of warmer poles and tropics under greenhouse conditions. Only then can the influence of global extinction patterns on climate itself start to be understood.

Still requiring further study is the question of the relative importance of changing ocean circulation patterns (as a consequence of the opening of Southern Ocean tectonic gateways) and threshold-global cooling related to gradually declining atmospheric CO_2. Resolving this issue requires improved proxy records of atmospheric CO_2 at sufficient resolution to resolve leads and lags that reveal the primary forcing mechanism. Future efforts should be concentrated on refining the calibrations and reducing assumptions to produce more accurate $p\mathrm{CO}_2$ estimates and obtaining higher-resolution time series that can be accurately correlated with glaciation and carbon cycle proxies, not forgetting the power of multiple proxies to bring conciliation. In addition, there is a great need to expand floral and faunal datasets that document climatic and biotic changes in the terrestrial realm, and to more accurately correlate them with marine records. A fundamental hole in our understanding is the extent of climate change and ice growth in the Arctic region during the E–O transition. Efforts could be focused on identifying and accessing records on the Arctic margins and locating suitable deep-sea drilling targets in the Arctic Ocean basin where mid-Cenozoic sediments are preserved.

At the time of writing (August 2006), there is a need for refinement of $p\mathrm{CO}_2$ proxies, expansion of datasets and improved understanding of high northern latitude and continental E–O climate change. Future application of new and existing proxies, recovery of additional high-quality sediment archives from the oceans and continents and more-complex models that integrate chemical and biological processes across the Eocene–Oligocene Transition will no doubt yield exciting results and should rapidly advance the field of palaeoclimatology as a whole.

References

ADAMS, C. G. 1965. The Foraminifera and stratigraphy of the Melinau Limestone, Sarawak, and its importance in Tertiary correlation. *Quarterly Journal of the Geological Society of London*, **121**, 283–338.

ADAMS, C. G., BUTTERLIN, J. & SAMANTA, B. K. 1986. Larger foraminifera and events at the Eocene–Oligocene boundary in the Indo-West Pacific. *In*: POMEROL, C. & PREMOLI SILVA, I. (eds) *Terminal Eocene Events*, Amsterdam, Elsevier, 237–252.

AUBRY, M.-P. 1991. Sequence stratigraphy: Eustacy or tectonic imprint? *Journal of Geophysical Research*, **96**, 6641–6679.

AUBRY, M.-P. 1992. Late Paleogene nannoplankton evolution: a tale of climatic deterioration. *In*: PROTHERO, D. R. & BERGGREN, W. A. (eds) *Eocene–Oligocene Climatic and Biotic Evolution*, Princeton, NJ, Princeton University Press.

AVERY, K. B., CALHOUN, M., SCHMALZ, L. S. & PAYTAN, A. 2005. Data report: Carbonate and barite trends across the Eocene–Oligocene Boundary at Shatsky Rise, ODP Leg 198. *In*: BRALOWER, T. J., PREMOLI SILVA, I. & MALONE, M. J. (eds) *Proceedings of the Ocean Drilling Program, Scientific Results*, College Station,TX, Ocean Drilling Program [Online]. Available from http://www-odp.tamu.edu/publications/198_SR VOLUME/CHAPTERS/106. PDF, 1–16.

BALDAUF, J. G. 1992. Middle Eocene through early Miocene diatom floral turnover. *In*: PROTHERO, D. R. & BERGGREN, W. A. (eds) *Eocene–Oligocene Climatic and Biotic Evolution*, Princeton, NJ, Princeton University Press.

BALDAUF, J. G. & BARRON, J. A. 1990. Evolution of biosiliceous sedimentation patterns – Eocene through Quaternary: paleoceanographic response to polar cooling. *In*: BLEIL, U. & THIEDE, R. (eds) *Geological History of the Polar Oceans: Arctic versus Antarctic*, Norwell, MA, Kluwer Academic, 575–607.

BARKER, P. F. & THOMAS, E. 2004. Origin, signature and palaeoclimatic influence of the antarctic circumpolar current. *Earth Science Reviews*, **66**, 143–162.

BARKER, P. F., BARRETT, P. J., COOPER, A. K. & HUYBRECHTS, P. 1999. Antarctic glacial history from numerical models and continental margin sediments. *Palaeogeography, Palaeoclimatology, Palaeoecology*, **150**, 247–267.

BARNOLA, J.-M., RAYNAUD, D., KOROTKEVICH, Y. S. & LORIUS, C. 1987. Vostok ice core provides 160,000–year record of atmospheric CO_2. *Nature*, **329**, 408–414.

BARRERA, E. & HUBER, B. T. 1991. Paleogene and early Neogene oceanography of the southern Indian Ocean: Leg 119 foramincfer stable isotope results.

Scientific Results of the Ocean Drilling Program, **119**, 693–717.

BARRETT, P. J., HAMBREY, M. J., HARWOOD, D. M., PYNE, A. R. & WEBB, P. N. 1989. Synthesis. *In*: BARRETT, P. J. (ed.) *Antarctic Cenozoic History from CIROS–1 Drill Hole, McMurdo Sound Antarctica*, Wellington, New Zealand, DSIR Publishing, Department of Scientific and Industrial Research Bulletin, **245**, 241–251.

BEERLING, D. J., CHALONER, W. G., HUNTLEY, B., PEARSON, J. A. & TOOLEY, M. J. 1993. Stomatal density responds to the glacial cycle of environmental change. *Proceedings of the Royal Society of London Series B*, **251**, 133–138.

BENSON, R. 1975. The origin of the psychrosphere as recorded in changes of deep-sea ostracode assemblages. *Lethaia*, **8**, 69–83.

BERGER, W. H. & WINTERER, E. L. 1974. Plate stratigraphy and the fluctuating carbonate line. *In*: HSÜ, K. J. & JENKYNS, H. C. (eds) *Pelagic Sediments on Land and Under the Sea, Special Publication of the International Association of Sedimentologists*, 11–48.

BERGGREN, W. A. & PEARSON, P. 2005. A revised tropical to subtropical Paleogene planktonic foraminiferal zonation. *Journal of Foraminiferal Research*, **35**, 279–298.

BERGGREN, W. A., KENT, D. V., OBRADOVICH, J. D. & SWISHER, C. C. III. 1992. Toward a revised Paleogene geochronology. *In*: PROTHERO, D. R. & BERGGREN, W. A. (eds) *Eocene–Oligocene Climatic and Biotic Evolution*, Princeton, NJ, Princeton University Press, 29–45.

BERGGREN, W. A., KENT, D. V., SWISHER, I. C. C. & AUBRY, M. P. 1995. A Revised Cenozoic geochronology and Chronostratigraphy. *In*: BERGGREN, W. A., KENT, D. V. & HARDENBOL, J. (eds) *Geochronology, Time Scales and Global Stratigraphic Correlation: A Unified Temporal Framework for an Historical Geology*, Society for Sedimentary Geology, Special Publications, 129–212.

BERNER, R. A. 1990. Atmospheric carbon dioxide levels over Phanerozoic time. *Science*, **249**, 1382–1386.

BERNER, R. A. 1992. Palaeo-CO_2 and climate; discussion. *Nature*, **358**, 114.

BERNER, R. A. 1997. The carbon cycle and CO_2 over Phanerozoic time: the role of land plants. *Philosophical Transactions of the Royal Society Series B*, **353**, 75–82.

BERNER, R. A. & KOTHAVALA, Z. 2001. GEOCARB III. A revised model of atmospheric CO_2 over Phanerozoic time. *American Journal of Science*, **291**, 182–204.

BILLUPS, K. & SCHRAG, D. P. 2003. Application of benthic foraminiferal Mg/Ca ratios to questions of Cenozoic climate change. *Earth and Planetary Science Letters*, **209**, 181–195.

BIRKENMAJER, K. 1996. Tertiary glacial/interglacial palaeoenvironments and sea-level changes, King George Island, West Antarctica. An overview. *Bulletin of the Polish Academy of Sciences, Earth Sciences*, **44**, 157–181.

BIRKENMAJER, K., GAZDZICKI, A., KRAJEWSKI, K. P., PRZYBYCIN, A., SOLECKI, A. & TATUR, YOON, H. II. 2005. First Cenozoic glaciers in West Antarctica. *Polish Polar Research*, **26**, 3–12.

BLOW, W. H. 1979. The Cainozoic globigerinida. A study of the morphology, taxonomy, evolutionary relationships and the stratigraphical distribution of some of the Globigerinida (mainly Globigerinacea). *Leiden, E. J. Brill*, **3**, 1413pp.

BLOW, W. H. & BANNER, F. T. 1962. The mid-Tertiary (upper Eocene to Aquitanian) Globigerinacae. *In*: EAMES, F. T. *ET AL*. (eds). *Fundamentals of mid-Tertiary Stratigraphical Correlation*, London, Cambridge University Press, 61–151.

BOERSMA, A. 1986. Eocene–Oligocene Atlantic paleo–oceanography, using benthic foraminifera. *In*: POMEROL, C. & PREMOLI SILVA, I. (eds) *Terminal Eocene Events*, Elsevier, 225–236.

BOERSMA, A. & PREMOLI SILVA, I., 1986. Terminal Eocene events – planktonic foraminifera and isotopic evidence. *In*: POMEROL, C. & PREMOLI SILVA, I. (eds) *Terminal Eocene Events*, Elsevier, 213–224.

BOTTOMLEY, R., GRIEVE, R., YORK, D. & MASALTIS, V. 1997. The age of the Popigai impact event and its relation to events at the Eocene–Oligocene boundary. *Nature*, **388**, 365–368.

BOUCOT, A. J. & GRAY, J. 2001. A critique of Phanerozoic climatic models involving changes in the CO_2 content of the atmosphere. *Earth–Science Reviews*, **56**, 1–159.

BREZA, J. R. & WISE, S. W. JR. 1992. Lower Oligocene ice–rafted debris on the Kerguelen Plateau: evidence for East Antarctic continental glaciation. *In*: WISE, S. W. & SCHLICH, R. *ET AL*. (eds) *Proceedings of the Ocean Drilling Program Scientific Result*, vol. **120**, College Station, TX, Ocean Drilling Program, 161–178.

BRYANT, H. N. 1996. Nimravidae. *In*: PROTHERO, D. R. & EMRY, R. J. (eds) *The Terrestrial Eocene–Oligocene Transition in North America*, Cambridge University Press, 453–475.

BRINKHUIS, H. 1992. Late Paleogene Dinoflagellate Cysts with special reference to the Eocene–Oligocene boundary. *In*: PROTHERO, D. R. & BERGGREN, W. A. (eds) *Eocene–Oligocene Climatic and Biotic Evolution*, Princeton, NJ, Princeton University Press, 327–340.

BRINKHUIS, H. & BIFFI, U. 1993. Dinoflagellate cysts stratigraphy of the Eocene–Oligocene transition in central Italy. *Marine Micropalaeontology*, **22**, 131–183.

BURKE, W. H., DENISON, R. E., HEATHERINGTON, E. A., KOEPNICK, R. B., NELSON, H. F. & OTTO, J. B. 1982. Variation of seawater $^{87}Sr/^{86}Sr$ throughout Phanerozoic time. *Geology*, **10**, 516–519.

CANDE, S. C. & KENT, D. V. 1995. Revised calibration of the geomagnetic polarity timescale for the Late Cretaceous and Cenozoic. *Journal of Geophysical Research*, **100**, 6093–6095.

CARTER, B. D. 2003. Diversity patterns in Eocene and Oligocene echinoids of the Southeastern United States. *In*: PROTHERO, D. R., IVANY, L. & NESBITT, E. A. (eds) *From Greenhouse to Icehouse*, New York, Columbia University Press, 354–365.

CERLING, T. E. 1991. Carbon dioxide in the atmosphere: evidence from Cenozoic and Mesozoic paleosols. *American Journal of Science*, **291**, 377–400.

CERLING, T. E. 1999. Stable carbon isotopes in paleosol carbonates: evidence from Cenozoic and Mesozoic paleosols. *Special Publication of the International Association of Sedimentologists*, **27**, 43–60.

CHAMLEY, H. 1986. Clay mineralogy at the Eocene–Oligocene boundary. *In*: POMEROL, C. & PREMOLI SILVA, I. (eds) *Terminal Eocene Events*, Elsevier, 381–386.

COCCIONI, R. & GALEOTTI, S. 2003. Deep water benthic foraminiferal events from the Massignano Eocene–Oligocene boundary stratotype, Central Italy. *In*: PROTHERO, D. R., IVANY, L. & NESBITT, E. A. (eds) *From Greenhouse to Icehouse*, Columbia University Press.

COCCIONI, R., MONACO, P., MONECHI, S., NOCCHI, M. & PARISI, G. 1988. Biostratigraphy of the Eocene–Oligocene Boundary at Massignano (Ancona, Italy). International Subcommission on Paleogene Stratigraphy, E–O Meeting, 59–75.

COLLINSON, M. E. 1992. Vegetation and floristic changes around the Eocene–Oligocene transition in western and central Europe. *In*: PROTHERO, D. R. & BERGGREN, W. A. (eds) *Eocene–Oligocene Climatic and Biotic Evolution*, Princeton, NJ, Princeton University Press.

CORSINI, J., SMITH, T. & LEITE, M. 2006. Palaeoenvironmental implications of size, carapace position, and incidence of non-shell elements in White River turtles. *Palaeogeography, Palaeoclimatology, Palaeoecology*, **234**, 287–303.

COXALL, H. K. & PEARSON, P. N. 2006. Taxonomy, biostratigraphy and phylogeny of Hantkeninidae (*Clavigerinella, Hantkenina* and *Cribrohantkenina*). *In*: PEARSON, P. N., OLSSON, R. K., HEMLEBEN, C., HUBER, B. T. & BERGGREN, W. A. (eds) *Atlas of Eocene Planktonic Foraminifera*, Cushman Foundation of Foraminiferal Research, Special Publication No. 41, Lawrence, Kansas, Allen Press, 213–252.

COXALL, H. K., WILSON, P. A., PÄLIKE, H., LEAR, C. H. & BACKMAN, J. 2005. Rapid stepwise onset of Antarctic glaciation and deeper calcite compensation in the Pacific Ocean. *Nature*, **433**, 53–57.

CROWLEY, T. J. 2000. Carbon dioxide and Phanerozoic climate: An overview. *In*: HUBER, B. T., MACLEOD, K. G. & WING, S. L. (eds) *Warm Climates in Earth History*, Cambridge University Press, 425–444.

CURRY, W. B., SHACKLETON, N. J., RICHTER, C. *ET AL*. 1995. *Proceedings of the Ocean Drilling Program, Initial Reports, 154*: College Station,TX, Ocean Drilling Program, 1111pp.

DALL'ANTONIA, B., BOSSIO, A. & GUERNET, C. 2003. The Eocene–Oligocene boundary and the psychrospheric event in the Tethys as recorded by deep-sea ostracodes from the Massignano Global Boundary Stratotype Section and Point, Central Italy. *Marine Micropaleontology*, **48**, 91–106.

DALAI, T. K., RAVIZZA, G. E. & PEUCKER–EHRENBRINK, B. 2006. The Late Eocene $^{187}Os/^{188}Os$ excursion: Chemostratigraphy, cosmic dust flux and the Early Oligocene glaciation. *Earth and Planetary Science Letters*, **241**, 477–492.

DAVIES, R., CARTWRIGHT, J., PIKE, J. & LINE, C. 2001. Early Oligocene initiation of North Atlantic deep water formation. *Nature*, **410**, 917–920.

DeConto, R. M. & Pollard, D. 2003a. Rapid Cenozoic glaciation of Antarctica induced by declining atmospheric CO$_2$. *Nature*, **421**, 245–249.

DeConto, R. M. & Pollard, D. 2003b. A coupled climate-ice sheet modeling approach to the Early Cenozoic history of the Antarctic ice sheet. *Palaeogeography, Palaeoclimatology, Palaeoecology*, **198**, 39–52.

DeConto, R. & Pollard, D. 2007. Rethinking Cenozoic glacial history: a model-data perspective. *Geophysical Research Abstracts*, **9**, 1607-7962/gra/EGU2007-A-09083.

Diekmann, B., Kuhn, G., Gersonde, R. & Mackensen, A. 2004. Middle Eocene to early Miocene environmental changes in the sub-Antarctic Southern Ocean: evidence from biogenic and terrigenous depositional patterns at ODP Site 1090. *Global and Planetary Change*, **40**, 295–313.

Diester-Haass, L. 1992. Late Eocene–Oligocene sedimentation in the Antarctic Ocean, Atlantic sector (Maud Rise, ODP Leg 113, Site 689): development of surface and bottom water circulation. *Antarctic Research Series*, **56**, 185–202.

Diester-Haass, L. 1995. Middle Eocene to early Oligocene palaeoceanography of the Antarctic Ocean (Maud Rise, ODP Leg 113. Site 689): change from a low productivity to a high productivity ocean. *Palaeogeography, Palaeoclimatology, Palaeoecology*, **113**, 311–334.

Diester-Haass, L. 1996. Late Eocene–Oligocene paleoceanography in the southern Indian Ocean (ODP Site 744). *Marine Geology*, **130**, 99–119.

Diester-Haass, L. & Zachos, J. C. 2003. The Eocene–Oligocene transition in the Equatorial Atlantic (ODP Site 925): Paleoproductivity increase and positive δ^{13}C excursion. *In*: Prothero, D. R., Ivany, L. & Nesbitt, E. A. (eds) *From Greenhouse to Icehouse*, New York, Columbia University Press, 397–418.

Diester-Haass, L. & Zahn, R. 1996. Eocene–Oligocene transition in the Southern Ocean; history of water mass circulation and biological productivity. *Geology*, **24**, 163–166.

Diester-Haass, L. & Zahn, R. 2001. Palaeoproductivity increase at the Eocene–Oligocene climatic transition; ODP/DSDP sites 763 and 592. *Palaeogeography, Palaeoclimatology, Palaeoecology*, **172**, 153–170.

Diester-Haass, L., Robert, C. & Chamley, H. 1993. Paleoceanographic and paleoclimatic evolution in the Weddell Sea (Antarctica) during the middle Eocene–late Oligocene, from a coarse sediment fraction and clay mineral data (ODP Site 689). *Marine Geology*, **114**, 233–250.

Dockery, D. T. III. 1986. Punctuated succession of marine molluscs in the northern Gulf Coastal Plain. *Palaios*, **1**, 582–589.

Dockery, D. T. III. & Lozouet, P. 2003. Biotic patterns in Eocene–Oligocene Molluscs of the Atlantic Coastal Plain, USA. *In*: Prothero, D. R., Ivany, L. & Nesbitt, E. A. (eds) *From Greenhouse to Icehouse*, New York, Columbia University Press, 303–340.

Dupont-Nivet, G., Krijgsman, W., Langereis, C. G., Abels, H. A. & Fang, X. 2007. Tibetan plateau aridification linked to global cooling at the Eocene-Oligocene transition. *Nature*, **445**, 635–638.

Eagles, G., Livermore, R. & Morris, P. 2006. Small basins in the Scotia Sea: The Eocene Drake Passage gateway. *Earth and Planetary Science Letters*, **242**, 343–353.

Echols, R., Armentrout, J. M., Root, S. A., Fearn, L. B., Cooke, J. C., Rodgers, B. K. & Thompson, P. R. 2003. Sequence stratigraphy of the Eocene–Oligocene Boundary Interval: Southeastern Mississippi. *In*: Prothero, D. R., Ivany, L. & Nesbitt, E. A. (eds) *From Greenhouse to Icehouse*, New York, Columbia University Press, 189–222.

Ehrmann, W. U. & Mackensen, A. 1992. Sedimentological evidence for the formation of an East Antarctic ice sheet in Eocene–Oligocene time. *Palaeogeography, Palaeoclimatology, Palaeoececology*, **93**, 85–112.

Ekart, D. D., Cerling, T. E., Montañez, I. P. & Tabor, N. J. 1999. A 400 million year carbon isotope record of pedogenic carbonate: implications for palaeoatmospheric carbon dioxide. *American Journal of Science*, **299**, 805–827.

Eldrett, J. S., Harding, I. C., Wilson, P. A., Butler, E. & Roberts, A. P. 2007. Continental ice in Greenland during the Eocene and Oligocene. *Nature*, doi:10.1038/nature05591

Exon, N. F., Kennett, J. P. & Malone, M. J. 2004. Leg 189 Synthesis: Cretaceous–Holocene history of the Tasmanian gateway. *In*: Exon, N. F., Kennett, J. P. & Malone, M. J. et al. (eds) *Proceedings of the Ocean Drilling Program: Scientific Results, 189*. [Online]. Available from http://www-odp//www.odp.tamu. edu/publications/189_SR/synth/synth. htm.

Fluegeman, R. H. 2003. Late Eocene–early Oligocene in the Gulf Coastal Plain: regional vs. global influences. *In*: Prothero, D. R., Ivany, L. & Nesbitt, E. A. (eds) *From Greenhouse to Icehouse*, New York, Columbia University Press, 283–293.

Francis, J. E. 1999. Evidence from fossil plants for Antarctica palaeoclimates over the past 100 million years, *Terra Antarctica*, **3**, 43–52.

Francis, J. E. & Poole, I. 2002. Cretaceous and Tertiary climates of Antarctica: evidence from fossil wood. *Palaeogeography, Palaeoclimatology, Palaeoecology*, **182**, 47–64.

Fordyce, R. E. 1992. Cetacean evolution and Eocene–Oligocene environments. *In*: Prothero, D. R. & Berggren, W. A. (eds) *Eocene–Oligocene Climatic and Biotic Evolution*, Princeton, Princeton University Press, 368–382.

Fordyce, R. E. 2003. Cetacean evolution and Eocene–Oligocene oceans revisited. *In*: Prothero, D. R., Ivany, L. & Nesbitt, E. A. (eds) *From Greenhouse to Icehouse*, New York, Columbia University Press, 368–382.

Funakawa, S., Nishi, H., Moore, T. C. & Nigrini, C. A. 2006. Radiolarian faunal turnover and palaeoceanographic change around the Eocene–Oligocene boundary in the Central Equatorial Pacific, ODP Leg 199, Holes 1218A, 1219A and 1220A. *Palaeogeography, Palaeoclimatology, Palaeoecology*, **230**, 183–203.

Gale, A. S., Huggett, J. M., Pälike, H., Laurie, E., Haliwood, E. A. & Hardebol, J. 2006. Correlation of Eocene–Oligocene marine and continental records: orbital cyclicity, magnetostratigraphy and sequence stratigraphy of the Solent Group, Isle of Wight, UK. *Journal of the Geological Society of London*, **163**, 401–415.

GEDL, P. 2004. Dinoflagellate cyst record of the Eocene–Oligocene boundary succession in flysch deposits at Leluchów, Carpathian Mountains, Poland. *Geological Society Special Publication*, **230**, 309–324.

GILBERT, R., DOMACK, E. W. & CAMERLENGHI, A. 2003. Deglacial history of the Greenpeace Trough: Icesheet to ice shelf transition in the north-western Weddell Sea. *Antarctic Research Series*, **79**, 195–204.

GOLDSTEIN, S. L. & JACOBSEN, S. B. 1988. Nd and Sr isotope systematics of river water suspended material: Implications for crustal evolution. *Earth and Planetary Science Letters*, **87**, 249–265.

GRIMES, S. T., HOOKER, J. J., COLLINSON, M. E. & MATTEY, D. P. 2005. Summer temperatures of late Eocene to early Oligocene freshwaters. *Geology*, **33**, 189–192.

HAQ, B. U., HARDENBOL, J. & VEIL, P. U. 1987. Chronology of fluctuating sea levels since the Triassic. *Science*, **235**, 1156–1166.

HARRISON, K. G. 2000. Role of increased marine silica input on paleo-pCO$_2$ levels. *Paleoceanography*, **13**, 292–298.

HARTL, P., TAUXE, L. & HERBERT, T. 1995. Earliest Oligocene increase in South Atlantic productivity as interpreted from "rock magnetics" at Deep-sea Drilling Project Site 522. *Paleoceanography*, **10**, 311–326.

HAY, W., FLÖGEL, S. & SÖDING, E. 2005. Is the initiation of glaciation on Antarctica related to a change in the structure of the ocean? *Global and Planetary Change*, **45**, 23–33.

HESS, J., BENDER, M. L. & SCHILLING, J.-G. 1986. Evolution of the ratio of strontium-87 to strontium-86 in seawater from Cretaceous to Present. *Science*, **231**, 979–984.

HICKMAN, C. S. 2003. Evidence for Abrupt Eocene–Oligocene molluscan faunal change in the Pacific Northwest. *In*: PROTHERO, D. R., IVANY, L. & NESBITT, E. A. (eds) *From Greenhouse to Icehouse*, New York, Columbia University Press, 71–87.

HOOKER, J. J., COLLINSON, M. E. & SILLE, N. P. 2004. Eocene–Oligocene mammalian faunal turnover in the Hampshire Basin, UK; calibration to the global time scale and the major cooling event. *Journal of the Geological Society of London*, **161**, 161–172.

HÖNISCH, B. & HEMMING, N. G. 2004. Ground-truthing the boron isotope-paleo-pH proxy in planktonic foraminifera shells: Partial dissolution and shell size effects. *Paleoceanography*, **19**, doi:10.1029/2004PA001026.

HSÜ, J. K., LABREQUE, J. J., CARMAN, M. F. *ET AL.* 1984. *Initial Reports of the Deep-Sea Drilling Project*, **73**: Washington DC, US Government Printing Office, 798pp.

HUBER, M., BRINKHUIS, H., STICKLEY, C. E. *ET AL.* 2004. Eocene circulation of the Southern Ocean: Was Antarctica kept warm by subtropical waters? *Paleoceanography*, **19**, doi:10.1029/2004PA001014.

HUTCHINSON, J. H. 1992. Western North American reptile and amphibian record across the Eocene–Oligocene boundary and its climatic implications. *In*: PROTHERO, D. R. & BERGGREN, W. A. (eds) *Eocene–Oligocene Climatic and Biotic Evolution*, Princeton, NJ, Princeton University Press, 451–463.

IMBRIE, J., HAYS, J. D., MCINTYRE, A. *ET AL.* 1984. The orbital theory of Pleistocene climate: Support from a revised chronology of the marine δ^{18}O record. *In*: BERGER, A., IMBRIE, J., HAYS, J., KUKLA, G. J. & SALTZMAN, E. (eds) *Milankovitch and Climate*, Boston, Reidel, 269–305.

IVANY, L. C., PATTERSON, W. P. & LOHMANN, K. C. 2000. Cooler winters as a possible cause of mass extinctions at the Eocene–Oligocene boundary. *Nature*, **407**, 887–890.

IVANY, L., NESBITT, E. A. & PROTHERO, D. R. 2003. The Marine Eocene–Oligocene transition – A Synthesis. *In*: PROTHERO, D. R., IVANY, L. & NESBITT, E. A. (eds) *From Greenhouse to Icehouse*, New York, Columbia University Press, 522–534.

IVANY, L. C., VAN SIMAEYS, S., DOMACK, E. W. & SAMSON, S. D. 2006. Evidence for an earliest Oligocene ice sheet on the Antarctic Peninsula. *Geology*, **34**, 377–380.

JARAMILLO, C., RUEDA, M. J. & MORA, G. 2006. Cenozoic plant diversity in the Neotropics. *Science*, **311**, 1893–1896.

JOVANE, L., FLORINDO, F. & DINARÈS-TURELL, J. 2004. Environmental magnetic record of paleoclimate change from the Eocene–Oligocene stratotype section, Massignano, Italy. *Geophysical Research Letters*, **31**, doi:10.1029/2004GL020554.

JOVANE, L., FLORINDO, F., SPROVIERI, M. & PÄLIKE, H. 2006. Astronomic calibration of the late Eocene/early Oligocene Massignano section (central Italy). *Geochemistry, Geophysics, Geosystems*, **7**, doi:10.1029/2005GC001195.

KAMINSKI, M. A. 2005. The utility of Deep-Water Agglutinated Foraminiferal acmes for correlating Eocene to Oligocene abyssal sediments in the North Atlantic and Western Tethys. *Studia Geologica Polonica*. **124**, 325–339.

KELLER, G. 1983. Biochronology and paleoclimatic implications of middle Eocene to Oligocene planktonic foraminiferal faunas. *Micropalaeontolgy*, **7**, 463–486.

KELLER, G. 1986. Stepwise mass extinctions and impact events: Late Eocene to early Oligocene. *Marine Micorpaleontology*, **10**.

KELLER, G., MACLEOD, N. & BARRERA, E. 1992. Eocene–Oligocene faunal turnover in planktonic foraminifera, and Antarctic glaciation. *In*: PROTHERO, D. R. & BERGGREN, W. A. (eds) *Eocene and Oligocene Climatic and Biotic Evolution*, Princeton, Princeton University Press, 218–244.

KEMP, E. M. 1978. Tertiary climatic evolution and vegetation history in the southeast Indian Ocean Region. *Palaeogeography, Palaeoclimatology, Palaeoecology*, **24**, 169–208.

KENNETT, J. P. 1977. Cenozoic evolution of Antarctic glaciation, the circum-Antarctic Ocean, and their impact on global paleoceanogrpahy. *Journal of Geophysical Research*, **82**, 384–386.

KENNETT, J. P. & SHACKLETON, N. J. 1976. Oxygen isotopic evidence for the development of the psychrosphere 38 Myr ago. *Nature*, **260**, 513–515.

KENNETT, J. P. & STOTT, L. D. 1990. Proteus and Proto-Oceanus: ancestral Paleogene oceans as revealed from stable isotopic results: ODP Leg 113. *In*: BARKER, P. F. & KENNETT, J. P. *ET AL*. (eds) *Proceedings of the Ocean Drilling Program, Scientific Results*, College Station, TX, Ocean Drilling Program, 937–960.

KENNETT, J. P., HOUTZ, R. E., ANDREWS, P. B. & EDWARDS, A. R. 1975. Development of the circum-Antarctic current. *Science*, **186**, 144–147.

KIDD, R. B. & HILL, P. R. 1987. Sedimentation on Feni and Gardar sediment drifts. *Initial Reports of the Deep-Sea Drilling Project*, **94**, 1217–1244.

KIESSLING, W., FLÜGEL, E. & GOLONKA, J. 2003. Patterns of Phanerozoic carbonate platform sedimentation. *Lethaia*, **36**, 195–225.

KOBASHI, T., GROSSMAN, E. L., YANCEY, T. E. & DOCKERY, D. T. III. 2001. Reevaluation of conflicting Eocene tropical temperature estimates: Molluscan oxygen isotope evidence for warm low latitudes. *Geology*, **29**, 983–986.

KUMP, L. R., BRANTLEY, S. L. & ARTHUR, M. A. 2000. Chemical weathering, atmospheric CO_2, and climate. *Annual Review of Earth and Planetary Sciences*, **28**, 611–667.

LASKAR, J. 1999. The limits of Earth orbital calculations for geological time-scale use. *Royal Society Philosophical Transactions: Mathematical, Physical and Engineering Sciences (Series A)*, **193**, 1735–1759.

LASKAR, J., ROBUTEL, P., JOUTEL, F., GASTINEAU, M., CORREIA, A. & LEVRARD, B. 2004. A long term numerical solution for the insolation quantities of the Earth. *Astronomy and Astrophysics*, **428**, 261–285.

LATIMER, J. & FILIPPELLI, G. 2002. Eocene to Miocene terrigenous inputs and export production: geochemical evidence from ODP Leg 177, Site 1090. *Palaeogeography, Palaeoclimatology, Palaeoecology*, **182**, 151–164.

LAWVER, L. A. & GAHAGAN, L. M. 2003. Evolution of Cenozoic seaways in the circum-Antarctic region. *Palaeogeography, Palaeoclimatology, Palaeoecology*, **198**, 11–37.

LEAR, C. H. 2007. Mg/Ca Palaeothermometry: A New Window into Cenozoic Climate Change. *In*: WILLIAMS, M., HAYWOOD, A. M., GREGORY, F. J. & SCHMIDT, D. N. (eds) *Deep-time Perspectives on Climate Change: Marrying the Signal from Computer Models and Biological Proxies*. The Micropalaeontological Society, Special Publications. The Geological Society, London, 313–322.

LEAR, C. H., ELDERFIELD, H. & WILSON, P. A. 2001. Cenozoic deep-sea temperatures and global ice volumes from Mg/Ca in benthic foraminiferal calcite. *Science*, **287**, 269–272.

LEAR, C. H., ELDERFIELD, H. & WILSON, P. A. 2003. A Cenozoic sea water Sr/Ca record from benthic foraminiferal calcite and its application in determining global weathering fluxes. *Earth and Planetary Science Letters*, **208**, 69–84.

LEAR, C. H., ROSENTHAL, Y., COXALL, H. K. & WILSON, P. A. 2004. Late Eocene to early Miocene ice-sheet dynamics and the global carbon cycle. *Paleoceanography*, **19**, PA4015, doi:10.1029/2004PA001039.

LEOPOLD, E. B., LIU, G. & CLAY–POOLE, S. 1992. Low–biomass vegetation in the Oligocene? *In*: PROTHERO, D. R. & BERGGREN, W. A. (eds) *Eocene and Oligocene Climatic and Biotic Evolution*, Princeton, Princeton University Press, 399–420.

LIU, Y. -J., ARENS, N. C. & LI, C.-S. 2007. Range change in Metasequoia: Relationship to palaeoclimate. *Botanical Journal of the Linnean Society*, **154**, 115–127.

LIU, Z., TOU, S., ZHAO, Q., CHENG, X. & HUANG, W. 2004. Deep-water Earliest Oligocene Glacial Maximum (EOGM) in the South Atlantic. *Chinese Science Bulletin*, **49**, 2190–219.

LIVERMORE, R., NANKIVELL, A., EAGLES, G. & MORRIS, P. 2005. Paleogene opening of Drake Passage. *Earth and Planetary Science Letters*, **236**, 459–470.

LOCKWOOD, R. 2005. Body size, extinction events, and the early Cenozoic record of veneroid bivalves: A new role for recoveries? *Paleobiology*, **31**, 578–590.

LOUTRE, M. F., PAILLARD, D., VIMEUX, F. & CORTIJO, E. 2004. Does mean annual insolation have the potential to change the climate? *Earth and Planetary Science Letters*, **221**, 1–14.

LOWENSTEIN, T. K. & DEMICCO, R. V. 2006. Elevated Eocene atmospheric CO_2 and its subsequent decline. *Science*, **313**, 1928.

MANNING, E. M. 2003. The Eocene–Oligocene transition in marine vertebrates of the Gulf Coastal Plain. *In*: PROTHERO, D. R., IVANY, L. & NESBITT, E. A. (eds) *From Greenhouse to Icehouse*, New York, Columbia University Press, 386–398.

MARINO, L., MCSHEA, D. W. & UHEN, M. D. 2004. Origin and evolution of large brains in toothed whales. *Anatomical Record – Part A Discoveries in Molecular, Cellular, and Evolutionary Biology*, **281**, 1247–1255.

MARSHALL, L. G. & CIFELLI, R. L. 1989. Analysis of changing diversity patterns in Cenozoic land mammal age faunas, South America. *Palaeovertebrata*, **19**, 169–210.

MARTIN, J. H. 1990. Glacial–interglacial CO_2 change: the iron hypothesis. *Paleoceanography*, **5**, 1–13.

MARTIN, P. A., LEA, D. W., ROSENTHAL, Y., SHACKLETON, N. J., SARNTHEIN, M. & PAPENFUSS, T. 2002. Quaternary deep-sea temperature histories derived from benthic foraminiferal Mg/Ca. *Earth and Planetary Science Letters*, **189**, 193–209.

MASHIOTTA, T. A., LEA, D. W. & SPERO, H. J. 1999. Glacial–interglacial changes in Subantarctic surface temperature and δ^{18}O-water using foraminiferal Mg. *Earth and Planetary Science Letters*, **170**, 417–432.

MAYR, G. 2005. The Paleogene fossil record of birds in Europe. *Biological Reviews of the Cambridge Philosophical Society*, **80**, 515–542.

MCGOWRAN, B. & BEECROFT, A. 1986. Neritic, southern extratropical foraminifera and the terminal Eocene event. *Palaeogeography, Palaeoclimatology, Palaeoecology*, **55**, 23–34.

MCGOWRAN, B., MOSS, G. & BEECROFT, A. 1992. Late Eocene and early Oligocene in Southern Australia: Local neritic signals of global oceanic changes. *In*: PROTHERO, D. R. & BERGGREN,

W. A. (eds) *Eocene–Oligocene Climatic and Biotic Evolution*, Princeton, NJ, Princeton University Press, 178–201.

MEAD, G. A. & HODELL, D. A. 1995. Controls on the $^{87}Sr/^{86}Sr$ composition of seawater from the middle Eocene to Oligocene: Hole 689B, Maud Rise, Antarctica. *Paleoceanography*, **10**, 327–346.

MEAD, G. A., TAUXE, L. & LABREQUE, J. J. 1986. Oligocene paleoceanography of the South Atlantic: paleoclimatic implications of sediment accumulation rates and magnetic susceptibility measurements. *Paleoceanography*, **1**, 273–284.

MENG, J. & MCKENNA, M. C. 1998. Faunal turnovers of Paleogene mammals from the Mongolian Plateau. *Nature*, **394**, 364–367.

MILLER, K. G. & FAIRBANKS, R. G. 1985. Oligocene to Miocene carbon isotope cycles and abyssal circulation changes. *In*: SUNQUIST, E. T. & BROECKER, W. S. (eds) *The Carbon Cycle and Atmospheric CO_2; Natural Variations Archean to Present*, American Geophysical Union, Geophysics Monograph **32**, 469–486.

MILLER, K. G. & MOUNTAIN, G. S. 1996. Drilling and dating New Jersey Oligocene–Miocene sequences: Ice volume, global sea level, and Exxon records. *Science*, **271**, 1092–1095.

MILLER, K. G., FAIRBANKS, R. & MOUNTAIN, G. S. 1987. Tertiary Oxygen isotope synthesis, sea level history and continental margin erosion. *Paleoceanography*, **1**, 1–19.

MILLER, K. G., WRIGHT, J. & FAIRBANKS, R. 1991. Unlocking the icehouse: Oligocene–Micoene oxygen isotopes, eustacy and margin erosion. *Journal of Geophysical Research*, **96**, 6829–6848.

MOHR, B. 1990. Eocene and Oligocene sporomorphs and dinoflagellate cysts from Leg 113 drill sites, Weddell Sea, Antarctica. *In*: BARKER, P. F. & KENNETT, J. P. *ET AL.* (eds) *Proceedings of the Ocean Drilling Program, Scientific Results*, College Station, TX, Ocean Drilling Program, 595–612.

MOLINA, E. 1986. Description and biostratigraphy of the main reference section of the Eocene–Oligocene boundary in Spain: Fuente Caldera section. *In*: POMEROL, C. & PREMOLI SILVA, I. (eds) *Terminal Eocene Events*, Elsevier, 53–64.

MOLINA, E., GONZALVO, C., ORTIZ, S. & CRUZ, L. E. 2006. Foraminiferal turnover across the Eocene–Oligocene transition at Fuente Caldera, southern Spain: No cause-effect relationship between meteorite impacts and extinctions. *Marine Micropaleontology*, **58**, 270–286.

MOORE, M. J. & JANSEN, R. K. 2005. Molecular evidence for the age, origin, and evolutionary history of the American desert plant genus Tiquilia (Boraginaceae). *Molecular Phylogenetics and Evolution*, **39**, 668–687.

MOORE, T. C. J., RABINOWITZ, P. D. *ET AL.* 1984. *Initial Reports of the Deep-Sea Drilling Project*, **74**, Washington, DC, US Government Printing Office.

MORAN, K., BACKMAN, J. & BRINKHUIS, H. *ET AL.* 2006. The Cenozoic palaeoenvironment of the Arctic Ocean. *Nature*, **441**, 601–605.

MURPHY, M. G. & KENNETT, J. P. 1986. Development of latitudinal thermal gradients during the Oligocene: Oxygen isotope evidence from the southwest Pacific.

Initial Reports of the Deep-Sea Drilling Project, Washington DC, US Government Printing Office, 1347–1360.

MYERS, J. A. 2003. Terrestrial Eocene–Oligocene vegetation and climate in the Pacific Northwest. *In*: PROTHERO, D. R., IVANY, L. & NESBITT, E. A. (eds) *From Greenhouse to Icehouse*, New York, Columbia University Press, 171–188.

NEBELSICK, J. H., RASSER, M. W. & BASSI, D. 2005. Facies dynamics in Eocene to Oligocene circumalpine carbonates. *Facies*, **51**, 197–216.

NESBITT, E. A. 2003. Changes in shallow marine faunas from the Northeastern Pacific margin across the Eocene–Oligocene boundary. *In*: PROTHERO, D. R., IVANY, L. & NESBITT, E. A. (eds) *From Greenhouse to Icehouse*, New York, Columbia University Press, 57–70.

NILSEN, E. B., ANDERSON, J. B. & DELANEY, M. L. 2003. Paleoproductivity, nutrient burial, climate change and the carbon cycle in the western equatorial Atlantic across the Eocene–Oligocene boundary. *Paleoceanography*, **18**, doi:10.1029/2002PA000804.

NOCCHI, M., PARISI, G., MONACO, P. *ET AL.* 1986. The Eocene–Oligocene boundary event in the Umbrian pelagic sequence, Italy. *In*: POMEROL, C. & PREMOLI SILVA, I. (eds) *Terminal Eocene Events*, Amsterdam, Elsevier, 25–40.

NOCCHI, M., MONECHI, S., COCCIONI, R. *ET AL.* 1988. The extinction of Hantkeninidae as a marker for defining the Eocene–Oligocene boundary: A proposal. *In*: PREMOLI SILVA, I., COCCIONI, R. & MONTANARI, A. (eds) *International Union of Geological Sciences Commission on Stratigraphy, The International Subcommission of Paleogene Stratigraphy: The Eocene–Oligocene Boundary in the Marche–Umbria Basin (Italy)*, Ancona, Monte Cònero, 249–252.

NONG, G. T., NAJJAR, R. G., SEIDOV, D. & PETERSON, W. 2000. Simulation of ocean temperature change due to the opening of Drake Passage. *Geophysical Research Letters*, **27**, 2689–2692.

OBERHANSLI, H., MCKENZIE, J., TOUMARKINE, M. & WEISSERT, H. 1984. A paleoclimatic and paleoceanographic record of the Paleogene in the central South Atlantic (Leg 73, Sites 522, 523, and 524). *In*: HSÜ, K. J. & LA BREQUE, J. L. *ET AL.* (eds) *Initial Reports DSDP, Leg 73*, Washington DC, US Government Printing Office, 737–747.

OBOH-IKUENOBE, F. & JARAMILLO, C. 2003. Palynological patterns in uppermost Eocene to lower Oligocene sedimentary rocks in the US Gulf Coast. *In*: PROTHERO, D. R., IVANY, L. & NESBITT, E. A. (eds) *From Greenhouse to Icehouse*. New York, Columbia University Press, 269–282.

OERLEMANS, J. 2004. Correcting the Cenozoic $\delta^{18}O$ deep-sea temperature record for Antarctic ice volume. *Palaeogeography, Palaeoclimatology, Palaeocecology*, **208**, 195–205.

OERLEMANS, J. 2005. Antarctic ice volume and deep-sea temperature during the last 50 Myr: A model study. *Annals of Glaciology*, **39**, 13–19.

OLEINIK, A. E. & MARINCOVICH, L. J. 2003. Biotic response to the Eocene-Oligocene transition: gastropod assemblages in the high-latitude North Pacific. *In*: PROTHERO, D. R., IVANY, L. & NESBITT, E. A.

(eds) *From Greenhouse to Icehouse*, New York, Columbia University Press, 36–56.

OLNEY, M. P., SCHERER, R. P., BOHATY, S. M. & HARWOOD, D. M. 2005. Eocene–Oligocene paleoecology and the diatom genus *Kisseleviella Sheshukova-Poretskaya* from the Victoria Land Basin, Antarctica. *Marine Micropaleontology*, **58**, 56–72.

OPDYKE, B. N. & WILKINSON, B. H. 1989. Surface area control of shallow cratonic to deep marine carbonate accumulation. *Paleoceanography*, **3**, 685–703.

PALAZZESI, L. & BARREDA, V. 2007. Major vegetation trends in the Tertiary of Patagonia (Argentina): A qualitative paleoclimatic approach based on palynological evidence. *Flora: Morphology, Distribution, Functional Ecology of Plants*, **202**, 328–337.

PAGANI, M. 2002. The alkenone-CO_2 proxy and ancient atmospheric carbon dioxide. *Philosophical Transactions of the Royal Society of London*, **360**, 609–632.

PAGANI, M., LEMARCHAND, D., SPIVACK, A. & GAILLARDET, J. 2005a. A critical evaluation of the boron isotope-pH proxy: The accuracy of ancient ocean pH estimates. *Geochimica et Cosmochimica Acta*, **69**, 953–961.

PAGANI, M., ZACHOS, J. C., FREEMAN, K. H., TIPPLE, B. & BOHATY, S. 2005b. Marked decline in atmospheric carbon dioxide concentrations during the Paleogene. *Science*, **309**, 600–603.

PÄLIKE, H., LASKAR, J. & SHACKLETON, N. J. 2004. Geologic constraints on the chaotic diffusion of the solar system. *Geology*, **32**, 929–932.

PÄLIKE, H., NORRIS, R. N., HERRLE, J. ET AL. 2006. The heartbeat of the Oligocene climate system. *Science*, **314**, 1894–1898.

PALMER, M. R. & EDMOND, J. M. 1992. Controls over the strontium isotope composition of river water. *Geochimica et Cosmochimica Acta*, **56**, 2099–2111.

PAN, A. D., JACOBS, B. F., DRANSFIELD, J. & BAKER, W. J. 2006. The fossil history of palms (Arecaceae) in Africa and new records from the Late Oligocene (28–27 Ma) of north-western Ethiopia. *Botanical Journal of the Linnean Society*, **151**, 69–81.

PAUL, H. A., ZACHOS, J. C., FLOWER, B. P. & TRIPATI, A. 2000. Orbitally induced climate and geochemical variability across the Oligocene/Miocene boundary. *Paleoceanography*, **15**, 471–485.

PEARSON, P. N. & CHAISSON, P. 1997. Late Paleocene to middle Miocene planktonic foraminifer biostratigraphy of the Ceara Rise. *In*: SHACKLETON, N. J., CURRY, W. B., RICHTER, C. & BRALOWER, T. J. (eds) *Proceedings of the Ocean Drilling Program Scientific Results*, **154**, College Station, TX (Ocean Drilling Program).

PEARSON, P. N. & PALMER, M. R. 1999. Middle Eocene seawater pH and atmospheric carbon dioxide concentrations. *Science*, **284**, 1824–1826.

PEARSON, P. N. & PALMER, M. R. 2000. Atmospheric carbon dioxide concentrations over the past 60 million years. *Nature*, **406**, 695–699.

PEARSON, P. N., DITCHFIELD, P. W., SINGANO, J., HARCOURT-BROWN, K. G., NICHOLAS, C. J., SHACKLETON, N. J. & HALL, M. A. 2001. Warm tropical sea surface temperatures in the Late Cretaceous and Eocene epochs. *Nature*, **413**, 481–487.

PEARSON, P. N., OLSSON, R. K., HEMLEBEN, C., HUBER, B. T. & BERGGREN, W. A. 2006. *Atlas of Eocene planktonic foraminifera*. Cushman Foundation of Foraminiferal Research, Special Publication: Lawrence, Kansas, Allen Press.

PEARSON, P. N., VAN DONGEN, B. E., NICHOLAS, C. J., PANCOST, R. D., SCHOUTEN, S., SINGANO, J. & WADE, B. S. 2007. Stable warm tropical climate through the Eocene epoch. *Geology*, **35**, 211–214.

PEGRAM, W. J. & TUREKIAN, K. K. 1999. The osmium isotopic composition change of Cenozoic sea water as inferred from a deep-sea core corrected for meteoritic contributions. *Geochimica et Cosmochimica Acta*, **63**, 4053–4058.

PEKAR, S. & MILLER, K. G. 1996. New Jersey Oligocene 'icehouse' sequences (ODP Leg 150X) correlated with global $\delta^{18}O$ and Exxon eustatic records. *Geology*, **24**, 567–570.

PEKAR, S. F., CHRISTIE-BLICK, N., KOMINZ, M. A. & MILLER, K. M. 2002. Calibration between eustatic estimates from backstripping and oxygen isotopic records for the Oligocene. *Geology*, **30**, 903–906.

PERCH-NIELSEN, K. 1986. Calcareous nannofossil events at the Eocene–Oligocene boundary. *In*: POMEROL, C. & PREMOLI SILVA, I. (eds) *Terminal Eocene Events*, Elsevier, 275–282.

PETIT, J. R., JOUZEL, J., RAYNAUD, D. ET AL. 1999. Climate and atmospheric history of the past 420,000 years from the Vostok ice core, Antarctica. *Nature*, **399**, 429–436.

PETERSON, L. C. & BACKMAN, J. 1990. Late Cenozoic carbonate accumulation and the history of the carbonate compensation depth in the western equatorial Indian Ocean. *In*: BACKMAN, J. & PETERSON, L. C. ET AL. (eds) *Proceedings of the Ocean Drilling Program, Scientific Results*, **115**, College Station, TX, Ocean Drilling Program, 467–489.

PEUCKER-EHRENBRINK, B. & RAVIZZA, G. 2000. The marine osmium isotope record. *Terra Nova*, **12**, 205–219.

PEUCKER-EHRENBRINK, B., RAVIZZA, G. & HOFMANN, A. W. 1995. The marine $^{187}Os/^{186}Os$ record of the past 80 millions years. *Earth and Planetary Science Letters*, **130**, 155–167.

POAG, C. W. 1995. Upper Eocene impactites of the U.S. East Coast: Depositional origins, biostratigraphic fram work, and correlation. *Palaios*, **10**, 16–43.

POAG, C. W., MANKINEN, E. & NORRIS, R. 2003. Late Eocene Impacts: geological record, correlation, and paleoenvironmental consequences. *In*: PROTHERO, D. R., IVANY, L. & NESBITT, E. A. (eds) *From Greenhouse to Icehouse: The Marine Eocene–Oligocene Transition*, New York, Columbia University Press, 495–510.

POLLARD, D. & DECONTO, R. M. 2003. Antarctic ice and sediment flux in the Oligocene simulated by a climate-ice sheet-sediment model. *Palaeogeography, Palaeoclimatology, Palaeoecology*, **198**, 53–67.

POLLARD, D. & DECONTO, R. M. 2005. Hysteresis in Cenozoic Antarctic ice-sheet variations. *Global and Planetary Change*, **45**, 9–21.

POLLARD, D., DECONTO, R. M. & NYBLADE, A. A. 2005. Sensitivity of Cenozoic Antarctic ice sheet variations to geothermal heat flux. *Global and Planetary Change*, **49**, 63–74.

POMEROL, C. & PREMOLI SILVA, I. 1986. *Terminal Eocene Events*: Amsterdam, Elsevier, 412pp.

POORE, R. Z. 1984. Middle Eocene through Quaternary planktonic foraminifers from the southern Angola Basin: Deep-sea Drilling Project Leg 73. *In*: HSÜ, K. J. & LA BREQUE, J. L. (eds) *Initial Reports of the Deep-Sea Drilling Project*, Washington DC, US Government Printing Office, 429–448.

PREMOLI SILVA, I. & JENKINS, D. G. 1993. Decision on the Eocene–Oligocene boundary stratotype. *Episodes*, **16**, 379–381.

PRENTICE, M. L. & MATTHEWS, R. K. 1988. Cenozoic ice-volume history: development of a composite oxygen isotope record. *Geology*, **16**, 963–966.

PRENTICE, M. L. & MATTHEWS, R. K. 1991. Tertiary ice sheet dynamics: The snow gun hypothesis. *Journal of Geophysical Research*, **96**, 6811–6827.

PROTHERO, D. R. 1994. *The Eocene–Oligocene Transition: Paradise Lost*. New York, Columbia University Press, 291pp.

PROTHERO, D. R. & BERGGREN, W. A. 1992. *Eocene–Oligocene Climatic and Biotic Evolution*, Princeton University Press, 568.

PROTHERO, D. & EMRY, R. J. 1996. *The Terrestrial Eocene–Oligocene Transition in North America*, Cambridge University Press, 688.

PROTHERO, D. R. & SWISHER, C. C. I. 1992. Magnetostratigraphy and geochronology of the Terrestrial Eocene–Oligocene transition in North America. *In*: PROTHERO, D. R. & BERGGREN, W. A. (eds) *Eocene–Oligocene Climatic and Biotic Evolution*, Princeton, Princeton University Press, 46–73.

PROTHERO, D. R., IVANY, L. C. I. & NESBITT, E. A. 2003. *From Greenhouse to Icehouse: The Marine Eocene–Oligocene Transition*, Columbia University Press, 560.

RACK, F. R. 1993. A geologic perspective on the Miocene evolution of the Antarctic Circumpolar Current system. *Tectonophysics*, **222**, 397–415.

RAGE, J. C. 1986. The amphibians and reptiles at the Eocene–Oligocene boundary in western Europe: an outline of the faunal alterations. *In*: POMEROL, C. & PREMOLI SILVA, I. (eds) *Terminal Eocene Events*, Elsevier, 309–310.

RAMUSSEN, D. T., BOWN, T. M. & SIMONS, E. L. 1992. The Eocene–Oligocene transition in continental Africa. *In*: PROTHERO, D. R. & BERGGREN, W. A. (eds) *Eocene–Oligocene Climatic and Biotic Evolution*, Princeton, Princeton University Press, 548–567.

RAVIZZA, G. & PEUCKER-EHRENBRINK, B. 2003. The marine $^{187}Os/^{188}Os$ record of the Eocene–Oligocene transition: The interplay of weathering and glaciation. *Earth and Planetary Science Letters*, **210**, 151–165.

RAYMO, M. E. 1991. Geochemical evidence supporting T.C. Chamberlin's theory of glaciation. *Geology*, **19**, 344–347.

RAYMO, M. E. & RUDDIMAN, W. F. 1992. Tectonic forcing of late Cenozoic climate. *Nature*, **359**, 117–122.

REA, D. K. & LYLE, M. W. 2005. Paleogene calcite compensation depth in the eastern subtropical Pacific: Answers and questions. *Paleoceanography*, **20**, doi:10.1029/2004PA001064.

REA, D. K., LEINEN, M. & JANECEK, T. R. 1985. Geologic approach to the long-term history of atmospheric circulation. *Science*, **227**, 721–725.

REILLY, T. J., MILLER, K. G. & FEIGENSON, M. D. 2002. Latest Eocene–earliest Miocene Sr isotopic reference section, Site 522, eastern South Atlantic. *Paleoceanography*, **17**, doi:10.1029/2001PA000745.

RETALLACK, G. J. 2002. Carbon dioxide and climate over the past 300 Myr. *Philosophical Transactions of the Royal Society of London. Series A*, **360**, 659–673.

RICHTER, F. M., ROWLEY, D. B. & DEPAOLO, D. J. 1992. Sr isotope evolution of seawater: The role of tectonics. *Earth and Planetary Science Letters*, **109**, 11–23.

RIEDEL, W. R. & SANFILIPPO, A. 1986. Radiolarian events and the Eocene–Oligocene boundary. *In*: POMEROL, C. & PREMOLI SILVA, I. (eds) *Terminal Eocene Events*, Elsevier, 253–258.

ROBERT, C., DIESTER-HAASS, L. & CHAMLEY, H. 2002. Late Eocene–Oligocene oceanographic development at southern high latitudes, from terrigenous and biogenic particles: A comparison of Kerguelen Plateau and Maud Rise, ODP Sites 744 and 689. *Marine Geology*, **191**, 37–54.

ROBINSON, E. 2003. Upper Paleogene larger foraminiferal succession on a tropical carbonate bank, Nicaragua Rise, Caribbean region. *In*: PROTHERO, D. R., IVANY, L. & NESBITT, E. A. (eds) *From Greenhouse to Icehouse*, New York, Columbia University Press, 294–302.

ROTH-NEBELSICK, A., UTESCHER, T., MOSBRUGGER, V., DIESTER-HAASS, L. & WALTHER, H. 2004. Changes in atmospheric CO_2 concentrations and climate from the Late Eocene to Early Miocene: Palaeobotanical reconstruction based on fossil floras from Saxony, Germany. *Palaeogeography, Palaeoclimatology, Palaeoecology*, **205**, 43–67.

ROYER, D. L. 2001. Stomatal density and stomatal index as indicators of palaeoatmospheric CO_2 concentration. *Review of Palaeobotany and Palynology*, **114**, 1–28.

ROYER, D. L., WING, S. L., BEERLING, D. J., JOLLEY, D. W., KOCH, P. L., HICKEY, L. J. & BERNER, R. A. 2001. Paleobotanical evidence for near present-day levels of atmospheric CO_2 during part of the Tertiary. *Science*, **292**, 2310–2313.

ROYER, D. L., BERNER, R. A., MONTAÑEZ, I. P., TABOR, N. J. & BEERLING, D. J. 2004. CO_2 as a primary driver of Phanerozoic climate. *GSA Today*, **14**, 4–10.

ROYER, D. L., BERNER, R. A. & PARK, J. 2007. Climate sensitivity constrained by CO_2 concentrations over the past 420 million years. *Nature*, **446**, 530–532.

RUDDIMAN, W. F., KUTZBACH, J. E. & PRENTICE, I. C. 1997. Testing the climatic effects of orography and CO_2 with general circulation and biome models. *In*: RUDDIMAN, W. F. (ed.) *Tectonic Uplift and Climate Change*, New York, Plenum 204–235.

SAGNOTTI, L., FLORINDO, F., VEROSUB, K. L., WILSON, G. S. & ROBERTS, A. P. 1998. Environmental magnetic record of Antarctic palaeoclimate from Eocene–Oligocene glaciomarine sediments, Victoria Land Basin. *Geophysical Journal International*, **134**, 653–662.

SALAMY, K. A. & ZACHOS, J. C. 1999. Latest Eocene–Early Oligocene climate change and Southern Ocean fertility: inferences from sediment accumulation and stable isotope data. *Palaeoceanography, Palaeoclimatology, Palaeoecology*, **145**, 61–67.

SAVAGE, D. E. & RUSSELL, D. E. 1983. *Mammalian Paleofaunas of the World*: Reading, MA, Addison-Wesley, 432pp.

SCHELLENBERG, S. A. 1998. Ecological and evolutionary response of deep-sea ostracodes to the Eocene–Oligocene transition: High-resolution data from ODP Site 744, Kerguelen Plateau. *GSA Annual Meeting Abstracts with Programs*, A-286.

SCHER, H. D. & MARTIN, E. E. 2004. Circulation in the Southern Ocean during the Paleogene inferred from neodymium isotopes. *Earth and Planetary Science Letters*, **228**, 391–405.

SCHER, H. D. & MARTIN, E. E. 2006. Timing and climatic consequences of the opening of Drake Passage. *Science*, **312**, 428–430.

SCHOLL, D. W., STEVENSON, A. J., NOBLE, M. A. & REA, D. K. 2003. The Meiji Drift Body and Late Paleogene–Neogene Paleoceanography of the North Pacific–Bering Sea region. *In*: PROTHERO, D. R., IVANY, L. & NESBITT, E. A. (eds) *From Greenhouse to Icehouse*, New York, Columbia University Press, 119–153.

SCHOUTEN, S., HOPMANS, E. C., SCHEFU, E. & SINNINGHE DAMSTÉ, J. S. 2002. Distributional variations in marine crenarchaeotal membrane lipids: A new tool for reconstructing ancient sea water temperatures? *Earth and Planetary Science Letters*, **204**, 265–274.

SCHUMACHER, S. & LAZARUS, D. 2004. Regional differences in pelagic productivity in the late Eocene to early Oligocene-a comparison of southern high latitudes and lower latitudes. *In*: KIESSLING, W. & LAZARUS, D. (eds) *Mesozoic–Cenozoic Bioevents, Palaeogeography, Palaeoclimatology, Palaeoecology*, 243–263.

SÉRANNE, M. 1999. Early Oligocene stratigraphic turnover on the west Africa continental margin: a signature of the Tertiary greenhouse-to-icehouse transition? *Terra Nova*, **11**, 135–140.

SEPKOSKI, J. J. J. 1986. Phanerozoic overview of mass extinction. *In*: RAUP, D. M. & JABLONSKI, D. (eds) *Patterns and Processes in the History of Life*, Berlin, Springer Verlag, 277–295.

SHACKLETON, N. J. & KENNETT, P. 1975. Paleotemperature history of the Cenozoic and the initiation of Antarctic glaciation: oxygen and carbon isotope analysis in DSDP Sites 277, 279 and 281. *In*: KENNETT, J. P. & HOUTZ, R. E. *ET AL*. (eds) *Initial Reports of the Deep-Sea Drilling Project 29*, Washington DC, US Government Printing Office, 881–884.

SHACKLETON, N. J. & OPDYKE, N. D. 1973. Oxygen isotope and paleomagnetic stratigraphy of equatorial Pacific core V28–238: Oxygen isotope temperatures and ice volumes on a 10^5 and 10^6 year scale. *Quaternary Research*, **3**, 39–55.

SHIPBOARD SCIENTIFIC PARTY, *ET AL*. 2001. Leg 189 summary. *In*: EXON, N. F., KENNETT, J. P., MALONE, M. J. (eds) *Proceedings of the Ocean Drilling Program Initial Reports, 189*, College Station, TX,

Ocean Drilling Program, [Online]. Available from: http://www-odp.tamu.edu/publications/189_IR/chap_01/chap_01. htm

SIGMAN, D. & BOYLE, E. 2000. Glacial/Interglacial variations in atmospheric carbon dioxide. *Nature*, **407**, 859–869.

STICKLEY, C. E., BRINKHUIS, H., SCHELLENBERG, S. A. *ET AL*. 2004. Timing and nature of the deepening of the Tasmanian Gateway. *Palaeoceanography*, **19**, doi:10.1029/2004PA001022.

SLUIJS, A., BRINKHUIS, H., STICKLEY, C. E., WARNAAR, J. & WILLIAMS, G. L. 2003. Dinoflagellate cysts from the Eocene–Oligocene transition in the Southern Ocean; results from ODP Leg 189. *In*: EXON, N. F., KENNETT, J. P. & MALONE, M. J. (eds) *Proceedings of the Ocean Drilling Program Scientific Results ODP Leg 189*, College Station, TX, Ocean Drilling Program, 1–42 [Online]. Available from: http://www://www.odp.tamu.edu/publications/189_SR/VOLUME/CHAPTERS/104. PDF.

SQUIRES, R. L. 2003. Turnovers in marine gastropod assemblages in the high-latitude North Pacific. *In*: PROTHERO, D. R., IVANY, L. & NESBITT, E. A. (eds) *From Greenhouse to Icehouse*, New York, Columbia University Press, 14–35.

STEHLIN, H. G. 1909. Remarques sur les faunules de Mammifères des couches Eocènes et Oligocènes du Bassin de Paris. *Bulletin de la Société Géologique de France*, **29**, 488–520.

STOTT, L. D., KENNETT, J. P., SHACKLETON, N. J. & CORFIELD, R. M. 1990. The evolution of Antarctic surface waters during the Paleogene: Inferences from the stable isotopic composition of planktonic foraminifers, ODP Leg 113. *In*: BARKER, P. F., KENNETT, J. P. *ET AL*. (eds) *Proceedings of the Ocean Drilling Program, Scientific Results*, College Station, TX, Ocean Drilling Program, 849–863.

STRÖMBERG, C. A. E. 2005. Decoupled taxonomic radiation and ecological expansion of open-habitat grasses in the Cenozoic of North America. *Geology*, **102**, 11980–11984.

STRÖMBERG, C. A. E. 2006. Evolution of hypsodonty in equids: testing a hypothesis of adaptation. *Palaeobiology*, **32**, 236–258.

STUCKY, R. K. 1992. Mammalian faunas in North America of Bridgerian to early Arikareean 'ages' (Eocene and Oligocene). *In*: PROTHERO, D. R. & BERGGREN, W. A. (eds) *Eocene–Oligocene Climatic and Biotic Evolution*, 464–493.

SUTO, I. 2006. The explosive diversification of the diatom genus *Chaetoceros* across the Eocene–Oligocene and Oligocene–Miocene boundaries in the Norwegian Sea. *Marine Geology*, **58**, 259–269.

TAKATA, H., NOMURA, R. & SETO, K. 2007. Review of paleoceanographic transition during the Oligocene in the eastern equatorial Pacific Ocean. *Fossils*, **81**, 15–14.

THOMAS, E. 1992. Middle Eocene–late Oligocene bathyal benthic foraminifera (Weddell Sea): faunal changes and implications for ocean circulation. *In*: PROTHERO, D. R. & BERGGREN, W. A. (eds) *Eocene–Oligocene Climatic and Biotic Evolution*, Princeton, NY, Princeton University Press, 245–271.

THOMAS, E. & GOODAY, A. J. 1996. Cenozoic deep-sea benthic foraminifers: tracers for changes in oceanic productivity? *Geology*, **24**, 355–358.

THORN, V. C. & DeCONTO, R. 2006. Antarctic climate at the Eocene–Oligocene boundary – Climate model sensitivity to high latitude vegetation type and comparisons with the palaeobotanical record. *Palaeogeography, Palaeoclimatology, Palaeoecology*, **231**.

THUNELL, R. & CORLISS, B. H. 1986. Late Eocene–early Oligocene carbonate sedimentation in the deep-sea. *In*: POMEROL, C. & PREMOLI SILVA, I. (eds) *Terminal Eocene Events*, Elsevier, 363–380.

TOGGWEILER, J. R. & BJORNSSON, H. 2000. Drake Passage and paleoclimate. *Journal of Quaternary Science*, **15**, 319–328.

TRIPATI, A., BACKMAN, J., ELDERFIELD, H. & FERRETTI, P. 2005. Eocene bipolar glaciation associated with global carbon cycle changes. *Nature*, **436**, 341–346.

TUO, S.-T., LIU, Z.-F., ZHAO, Q.-H. & CHENG, X.-R. 2006. Earliest Oligocene glacial maximum: records from ODP Site 1265, South Atlantic. *Diqiu Kexue – Zhongguo Dizhi Daxue Xuebao/Earth Science – Journal of China University of Geosciences*, **31**, 151–158.

VAIL, P. R. & HARDENBOL, J. 1979. Sea level changes during the Tertiary. *Oceanus*, **22**, 171–179.

VAN ANDEL, T. H. 1975. Mesozoic/Cenozoic calcite compensation depth and the global distribution of calcareous sediments. *Earth and Planetary Science Letters*, **26**, 187–194.

VAN DEN BOLD, W. A. 1986. Ostracodes at the Eocene–Oligocene Boundary in the Aquitaine basin. Stratigraphy, phylogeny, palaeoenvironments. *In*: POMEROL, C. & PREMOLI SILVA, I. (eds) *Terminal Eocene Events*, Elsevier, 259–264.

VAN DONGEN, B. E., TALBOT, H. M., SCHOUTEN, S., PEARSON, P. N. & PANCOST, R. D. 2006. Well preserved Palaeogene and Cretaceous biomarkers from the Kilwa area, Tanzania. *Organic Geochemistry*, **37**, 539–557.

VAN MOURIK, C. A. & BRINKHUIS, H. 2005. The Massignano Eocene–Oligocene golden spike section revisited. *Stratigraphy*, **2**, 13–29.

VANDEN BERG, M. D. & JARRARD, R. D. 2004. Cenozoic mass accumulation rates in the equatorial Pacific based on high-resolution mineralogy of Ocean Drilling Program Leg 199. *Paleoceanography*, **19**, doi:10.1029/2003PA000928.

VANDENBERGHE, N., BRINKHUIS, H. & STEURBAUT, E. 2003. The Eocene–Oligocene boundary in the North Sea Area: a sequence stratigraphic approach. *In*: PROTHERO, D. R., IVANY, L. C. & NESBITT, E. A. (eds) *From Greenhouse to Icehouse*, New York, Columbia University Press, 419–438.

VAROL, O. 1998. Paleogene. *In*: BOWN, P. R. (ed). *Calcareous Nannofossil Biostratigraphy*, Dordrecht, Kluwer Academic Publishing, 201–224.

VERGNAUD-GRAZZINI, C. & OBERHANSLI, H. 1986. Isotopic events at the Eocene Oligocene boundary – A review. *In*: POMEROL, C. & PREMOLI SILVA, I. (eds) *Terminal Eocene Events*, Elsevier, 311–330.

VIA, R. K. & THOMAS, D. J. 2006. Evolution of Atlantic thermohaline circulation: Early Oligocene onset of deep-water production in the North Atlantic. *Geology*, **34**, 441–444.

VOLK, T. 1987. Feedbacks between weathering and atmospheric CO_2 over the last 100 million years. *American Journal of Science*, **287**, 763–779.

VONHOF, H. B., SMIT, J., BRINKHUIS, H., MONTANARI, A. & NEDERBRAGT, A. J. 2000. Global cooling accelerated by early late Eocene impacts? *Geology*, **28**, 267–293.

WADE, B. S. & PÄLIKE, H. 2004. Oligocene climate dynamics. *Paleoceanography*, **19**, PA4019, doi: 10.1029/2004PA001042.

WHITEHEAD, J. M. 2005. Cenozoic history of Antarctic benthic diatoms. *Alcheringa*, **29**, 151–169.

WILSON, G. S., ROBERTS, A. P., VEROSUB, K. L., FLORINDO, F. & SAGNOTTI, L. 1998. Magnetobiostratigraphic chronology of the Eocene–Oligocene transition in the CIROS-1 core, Victoria Land margin, Antarctica: Implications for Antarctic glacial history. *Bulletin of the Geological Society of America*, **110**, 35–47.

WILSON, P. A. & NORRIS, R. D. 2001. Warm tropical ocean surface and global anoxia during the mid-Cretaceous period. *Nature*, **412**, 425–429.

WISE, S. W. JR., BREZA, J. R., HARWOOD, D. M. & WEI, W. 1991. Paleogene glacial history of Antarctica. *In*: MUELLER, D. W., McKENZIE, J. A. & WEISSERT, H. (eds) *Controversies in Modern Geology; Evolution of Geological Theories in Sedimentology, Earth History and Tectonics*, London, Academic Press, 133–171.

WOLFE, J. A. 1978. A paleobotanical interpretation of Tertiary climates in the northern hemisphere. *American Scientist*, **66**, 694–703.

WOLFE, J. A. 1992. Climatic, floristic, and vegetational changes near the Eocene–Oligocene boundary in North America. *In*: PROTHERO, D. R. & BERGGREN, W. A. (eds) *Eocene–Oligocene Climatic and Biotic Evolution*, Princeton, NJ, Princeton University Press, 421–436.

WOLFE, J. A. 1994. Tertiary climatic changes at middle latitudes of western North America. *Palaeogeography, Palaeoclimatology, Palaeoecology*, **108**, 195–205.

WRIGHT, J., SEYMOUR, R. S. & SHAW, H. 1984. REE and Nd isotopes in conodont apatite: Variations with geological age and depositional environment. *In*: CLARK, D. L. (ed.) *Conodont Biofacies and Provincialism: Geological Society of America Special Paper 196*, 325–340.

WRIGHT, A. & MILLER, K. G. 1993. Southern ocean influences on late Eocene to Miocene deepwater circulation. *The Antarctic paleoenvironment: A perspective on global change Antarctic research series*, 1–25.

YANCEY, T. E., ELSIK, W. C. & SANCY, R. H. 2003. The Palynological record of the Late Eocene climate change, Northwest Gulf of Mexico. *In*: PROTHERO, D. R., IVANY, L. & NESBITT, E. A. (eds) *From Greenhouse to Icehouse*, New York, Columbia University Press, 252–268.

ZACHOS, J. C. & KUMP, L. R. 2005. Carbon cycle feedbacks and the initiation of Antarctic glaciation in the earliest Oligocene. *Global and Planetary Change*, **47**, 51–66.

ZACHOS, J. C., BREZA, J. R. & WISE, S. W. 1992. Early Oligocene ice sheet expansion on Antarctica. *Geology*, **569–573**, 839–854.

ZACHOS, J. C., QUINN, T. M. & SALAMY, K. A. 1996. High-resolution (10^4 years) deep-sea foraminiferal stable isotope records of the Eocene–Oligocene climate transition. *Paleoceanography*, **11**, 251–266.

ZACHOS, J. C., FLOWER, B. P. & PAUL, H. 1997. Orbitally paced climate oscillations across the Oligocene/Miocene boundary. *Nature*, **388**, 567–570.

ZACHOS, J. C., OPDYKE, B. N., QUINN, T. M., JONES, C. E. & HALLIDAY, A. N. 1999. Early Cenozoic glaciation, Antarctic weathering, and seawater $^{87}Sr/^{86}Sr$; is there a link? *Chemical Geology*, **161**, 165–180.

ZACHOS, J. C., PAGANI, M., SLOAN, L. C., THOMAS, E. & BILLUPS, K. 2001. Trends, rhythms, and aberrations in global climate 65 Ma to present. *Science*, **292**, 686–693.

ZANAZZI, A., KOHN, M. J., MACFADDEN, B. J. & TERRY, D. O. 2007. Large temperature drop across the Eocene–Oligocene transition in central North America. *Nature*, **445**, 639–642.

ZEEBE, R. E. & WESTBROEK, P. 2003. A simple model for the $CaCO_3$ saturation state of the ocean: The 'Strangelove', the 'Neritan', and the 'Cretan' ocean. *Geochemistry, Geophysics, Geosystems*, **4**, doi: 10.1029/2003GC000538.

The Oligocene–Miocene boundary – cause and consequence from a Southern Ocean perspective

H. A. PFUHL[1] & I. N. McCAVE[2]

[1]*Department of Earth & Environmental Sciences, Ludwig Maximilians University, Theresienstrasse 41, 80333 Munich, Germany (e-mail: pfuhl@lmu.de)*

[2]*Godwin Laboratory for Palaeoclimate Research, Department of Earth Sciences, Cambridge University, Downing Street, Cambridge CB2 3EQ, UK*

Abstract: Understanding of Earth's transition from a warm, ice-free Cretaceous to today's bipolar glaciation is hotly debated. The Oligocene–Miocene boundary is marked by a brief glacial event followed by an interval of colder temperatures. Changes are small compared to the major Antarctic ice build-up at the Eocene–Oligocene boundary and establishment of a permanent Antarctic ice-sheet in the mid-Miocene. However, fossil evidence from low latitudes, including the faunal turnover which originally defined the Oligocene–Miocene boundary, indicates a reversal in trans-Atlantic flow, i.e. from westward to eastward, at this time. Modelling results suggest that a combined narrowing of the Tethys Seaway and deep opening of Drake Passage, and hence inception of Antarctic circumpolar circulation, drove reorganization of the patterns of ocean circulation. Despite evidence for a shallow Drake Passage opening in earliest Eocene time and subsequent deepening, a comparison of Southern Ocean isotopic records suggests that circumpolar circulation did not exist prior to *c.* 26 Ma. In fact, sedimentary records of a grain-size current-speed indicator from the Tasman Gateway reveal a singular, marked increase immediately preceding the initial Miocene event. The likely driver of this increase is inception of the full Antarctic Circumpolar Current. Among possible causes of early Cenozoic climate deterioration, the opening of seaways appears to play the major role. Extreme orbital configurations and pCO$_2$-drawdown may act as reinforcing factors.

The early Cenozoic, which encompasses the Oligocene–Miocene boundary (OMB), is significant within the context of Earth's transition from the warm, probably ice-free, climate of the Cretaceous to the bipolar glaciation of modern times. Over the last two to three years, the geological community has gained exciting new insights from the study of medium- to high-resolution marine sedimentary records recovered during recent deep-sea drilling cruises. Together with new modelling results, the debate on when and how much ice existed in what location is being reopened. Until these new results emerged, the consensus was that first widespread expansion of the East Antarctic Ice-Sheet (EAIS) coincided with the Eocene–Oligocene boundary (EOB). Permanent ice-coverage including the West Antarctic Ice-Sheet (WAIS) was a consequence of mid-Miocene cooling, and Northern Hemisphere glaciation did not happen until the Pliocene. However, benthic foraminiferal δ18O values, as an indicator of temperature and ice-volume, already experience a continuous decline following the Early Eocene Climate Optimum (EECO). Modelling by DeConto & Pollard (2005) suggests that the observed increases in Cenozoic benthic δ18O cannot be explained by Southern Hemisphere ice build-up alone and recent data indicate that sporadic major glaciation commenced in the Eocene (Tripati *et al.* 2005).

The role of the OMB within the context of the climate change debate is often neglected as the focus has been on the more prominent, rapid and lasting shifts to cooler temperatures and ice build-up at, for example, the EOB and in the mid-Miocene. In the following, we reflect on earlier and recent interpretations of cause and consequence of OMB climate change. The most recent conclusions are then discussed within a Cenozoic context.

Events associated with the OMB include the observed fossil turnover, aspects of which initially served to identify the boundary, and the negative climatic shift known as Mi-1. We cite evidence suggesting a reversal in Panamanian throughflow and a narrowing of the Tethys Seaway as causes for the Caribbean and Mediterranean fossil turnovers. We also show sedimentary evidence from the Southern Ocean which suggests that the onset of the Antarctic Circumpolar Current (ACC) immediately preceded the Mi-1 cooling event at the OMB.

In comparing records from the EOB, the OMB and the mid-Miocene cooling, we find that although individual factors like pCO$_2$ drawdown and orbital

From: WILLIAMS, M., HAYWOOD, A. M., GREGORY, F. J. & SCHMIDT, D. N. (eds) *Deep-Time Perspectives on Climate Change: Marrying the Signal from Computer Models and Biological Proxies*. The Micropalaeontological Society, Special Publications. The Geological Society, London, 389–407.

configurations may have acted to reinforce each other, (deep) opening/closure of seaways may be the only event immediately preceding all three climatic incidents.

Timing and age

The Oligocene–Miocene boundary – relative age

The earliest approaches to dating sediments used criteria of relative age, i.e. superposition, with sequences originally identified on the basis of their fossil content (first and last occurrence – FO and LO). The most prominent changes were used to define boundaries between geological time-intervals.

Ever greater detail in fossil resolution, especially from marine microfossil records, has increased available time-resolution over the past 40 years (Harland et al. 1964; Gradstein et al. 2004). However, at the same time these records revealed geographical restrictions, e.g. different species dwell at low, mid- and high latitudes, and depending on oceanographic conditions (depth, carbonate and silica undersaturation) their preservation may be limited. This severely limits identification of global events and integration of existing zonal schemes. Identification of the global record of Earth's magnetic polarity reversals, however, permits the creation of detailed stratigraphic records across geographic boundaries (in combination with biostratigraphy).

The Oligocene–Miocene boundary (OMB) was first defined in order to separate observed distinctions in molluscan faunas (% of living species in the unit) of these intervals (Lyell 1830: actually, Eocene/Miocene as the Oligocene was not carved out until 1854 by von Beyrich). Later identification of additional fossil transitions allowed description of the Oligocene and Miocene intervals with their subdivisions.

The recent compilations by Luterbacher et al. (2004) and Lourens et al. (2004) provide an up-to-date review of the stratigraphic markers defining the OMB (see their papers for details). The boundary (Fig. 1) is placed on the basis of the following:

- magnetic polarity record: at the reversal from C6Cn.2r to C6Cn.2n;
- planktic foraminiferal record: above the *Globigerinoides* Acme and just below the FO of *Globorotalia kugleri* s.s.;
- calcareous nannoplankton record: between the LO of *Sphenolithus ciperoensis* and the FO of *Discoaster druggi*; at low latitudes, it is also coincident with the short ranges of *S. delphix* and *S. capricornutus*;
- northwest European dinoflagellate record: coincident with the LO of *Membranophoridium aspinatum* and FO of *Invertocysta tabulata*;
- radiolarian record: just below the FO of *Cyrtocapsella tetrapera*.

With ever more detail in sedimentary and isotopic chemistry becoming available, it is possible to identify the OMB on the basis of non-fossil indicators. Independent of geographic location is the glacial event identified as Mi-1 (Miller et al.

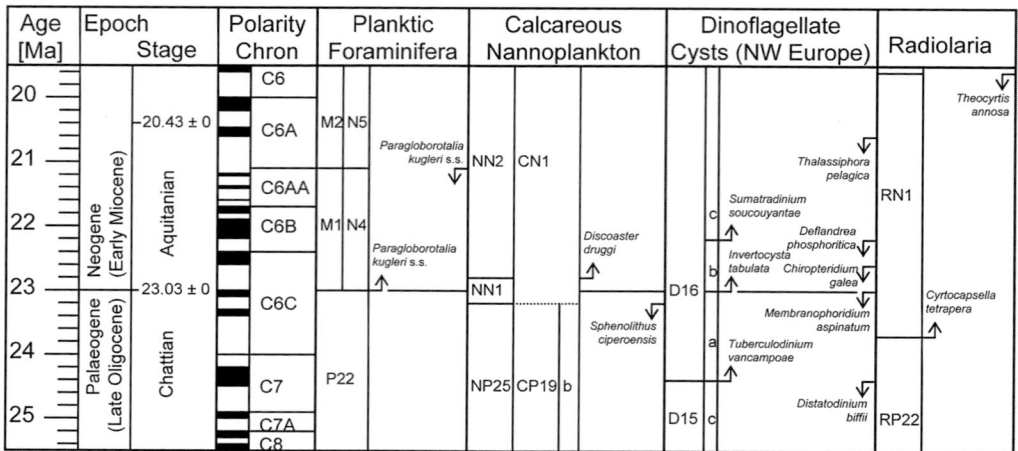

Fig. 1. Stratigraphic summary focusing on marine micropalaeontology, magnetostratigraphy and chronostratigraphy. The dinocyst record is representative of NW Europe. Ages are calculated using the latest astronomical solution La2004 (Laskar et al. 2004). See Luterbacher et al. (2004) and Lourens et al. (2004) for details.

1991). This short-duration (*c.* 200 kyr) increase in foraminiferal $\delta^{18}O$ is evident at all latitudes and readily recognizable in an early Cenozoic $\delta^{18}O$ record even at limited resolution (Fig. 2a–b). At Site 1170 in the southwest Pacific sector of the Southern Ocean, the Mi-1 event is followed by an interval of heavier $\delta^{18}O$ values (lasting *c.* 1 Ma, hashed background) and shift to lighter values with a relatively weak Mi-1a event (Fig. 2b).

The Oligocene–Miocene boundary – absolute age

Dating of rocks by radioactive decay methods allowed a first assignment of absolute ages to stratigraphic events. This method prevails in older sediments where ages are assigned by linear interpolation between magnetic polarity reversals. This method has been replaced in younger sediments by the higher precision of astronomical dating (Hilgen *et al.* 1999).

Analyses of sedimentological and physical properties of marine records with high resolution detail have documented the sensitivity of the ocean to astronomical forcing of Earth's climate by solar insolation. Periodic perturbations in Earth's orbit and the tilt of its inclination axis are caused by gravitational interactions between the large bodies of our Solar System, including the Moon (Berger 1977). However, astronomical timescales are only as good as their target curves, which are based on the most recent astronomical solution (Laskar 1999). At present, little retuning is expected for the Neogene and in fact calculations have been improved to date back as far as the base of the late Eocene (La2004, also referred to as La2003: Laskar *et al.* 2004).

At present, the astronomical age of the OMB is given at a retuned age of 23.03 Ma (Shackleton *et al.* 2000; Laskar *et al.* 2004; Lourens *et al.* 2004). The error of incorporating uncertainties from changes in the dynamical ellipticity of the Earth and/or tidal dissipation by the Moon amounts to a maximum of *c.* 125 kyr at the OMB (Lourens *et al.* 2004). For a more detailed discussion of the history of absolute ages attributed to the OMB, see Lourens *et al.* (2004).

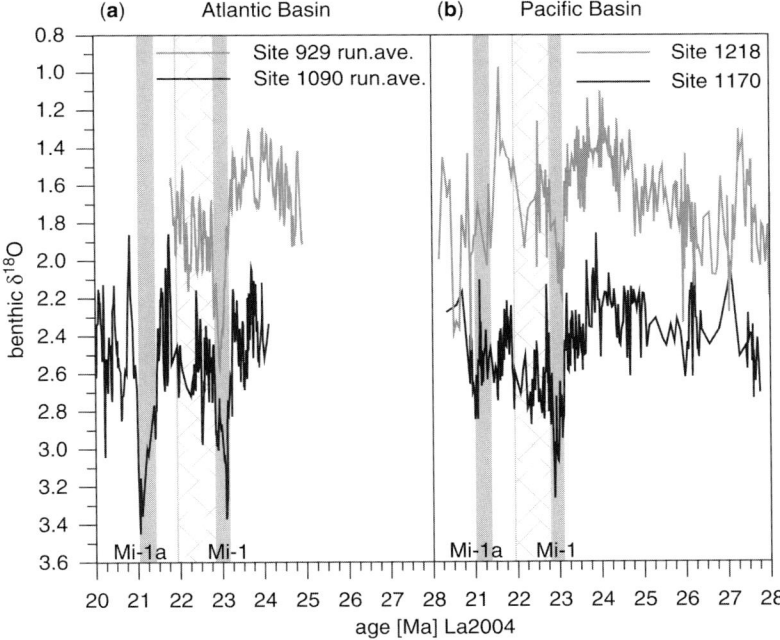

Fig. 2. Comparison of benthic $\delta^{18}O$ records from the equatorial and Southern Ocean regions of the Atlantic and Pacific (Flower *et al.* 1997; Billups *et al.* 2004; Lear *et al.* 2004; Pfuhl & McCave 2005). All records show a strong Mi-1 event lasting about 200 kyr and, if data are of high-enough resolution, the Mi-1a event, which is weakest at Site 1170 (grey background). This contrasts with evidence for an interval of heavier $\delta^{18}O$ values at Site 1170 on South Tasman Rise, which is more pronounced than at Site 1090 on Agulhas Ridge and much less evident at the equatorial sites (hashed lines).

In search of cause and impact

The Oligocene–Miocene boundary (OMB) was originally identified on the basis of a molluscan faunal turnover in the fossil record. The boundary is important, separating Palaeogene from Neogene. Large-scale faunal turnover may conveniently be explained by a profound change in living conditions affecting the organisms directly or indirectly through perturbation of the food chain. The likely driving force was climate change, but what was the cause of this change?

Climate responds to changes in atmospheric and oceanic conditions. Rapid responses are seen in El Niño–Southern Oscillation (ENSO), North Atlantic Oscillation (NAO) and ocean mixing times of c. 1000 years. Climate on Earth also responds to the 19 and 23 kyr cycles of precession, the 41 kyr cycle of obliquity, as well as the c. 100 and 400 kyr cycles of eccentricity (see Berger 1977; Berger & Tricot 1986). The impact of longer periods of obliquity and eccentricity has equally been observed.

Despite the potential for reinforcing positive climate feedbacks associated with extreme planetary alignments, the question remains as to whether these alone are sufficient to drive climate across a critical threshold or whether major changes are a consequence of plate tectonics and orogenic uplift, and/or oceanographic reorganization. Orogenic uplift takes millions of years to accomplish, while large-scale oceanographic

reorganization due to the opening or closing of gateways may culminate in the rapid (on geological timescales) breach of a critical threshold (e.g. Smith & Pickering 2003).

In the following, we will address (1) major tectonic movements affecting the OMB, and (2) other forcing factors cited as part of the climate discussion.

Orogenic uplift and plate tectonics across the OMB

Over the years, three major tectonic events have been linked with OMB climate change (compare Fig. 3):

In the Northern Hemisphere:
1. uplift of the Tibetan Plateau.

In the Southern Hemisphere:
2. the rise of the Andes (formation of the Altiplano–Puna Plateau) and
3. the deep opening of the Drake Passage.

These tectonic events are of two types affecting either the atmosphere or ocean. There is no doubt about the existing feedback between climate change and the mean elevation of mountain belts, i.e. the first two events, by causing changes in atmospheric circulation (Ruddiman *et al.* 1997). The third event is different in that it drives changes in ocean circulation and only indirectly affects climate.

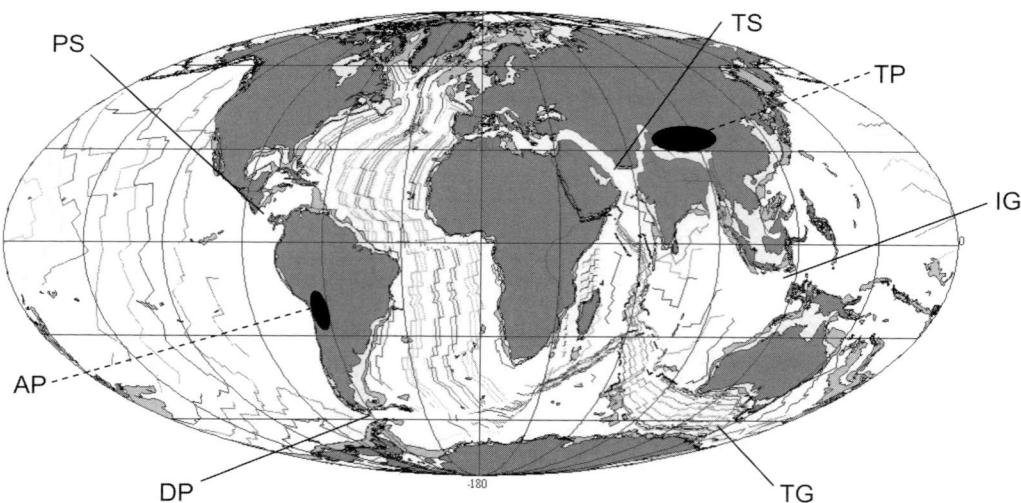

Fig. 3. Global plate reconstruction near the Oligocene–Miocene boundary (OMB) at 23 Ma. Note an open Drake Passage (DP) and Panamanian Seaway (PS), wider Tasman Gateway (TG), narrow Tethys Seaway (TS) and Indonesian Gateway (IG). Also shown are later locations of the Altiplano–Puna Plateau (AP) and Tibetan Plateau (TP) with major uplift post-dating the OMB. Courtesy of Dr A. G. Smith, Cambridge Palaeomap Services Ltd.

Uplift of the Tibetan Plateau

The Tibetan Plateau is an imposing topographic feature bounded on the south by the Himalayan mountain range with several of the highest mountains on Earth. At a mean elevation of $c.$ 5000 m and covering an area well over $c.$ 5×10^6 km^2, it is the main cause of the Asian monsoon system. Moist winds from the Indian Ocean are forced to rise, producing large amounts of seasonal rain that causes flooding and feeds eight of Earth's largest rivers. Furthermore, Tibetan uplift may well have altered patterns of atmospheric circulation in the Northern Hemisphere on a larger scale (e.g. Ruddiman & Kutzbach 1989).

The sequence of events is important to this discussion with the increasing impact of uplift on atmospheric circulation causing cooling of the Northern Hemisphere, higher levels of precipitation and river runoff, as well as glaciation at high altitudes. Weathering feedbacks led to removal of CO_2, resulting in further cooling, increased precipitation, and erosion. Modelling has shown that the climatic impact of the Tibetan uplift together with the increase in surface area of rock available for weathering (relative to a flat plain) have significantly raised the rate of CO_2-removal from the atmosphere, with the oceans acting as the ultimate sink. In fact, it has been suggested that only an increasingly cold Cenozoic climate has prevented a substantial removal of CO_2 from the atmosphere. As increased erosion results in mountain mass decrease, isostatic uplift and continued plate tectonic movement would cause further rises in elevation, counteracting the forces of erosion which might otherwise create a negative feedback eventually reducing the impact of the Tibetan Plateau on Earth's climate. Glaciation at high altitudes may therefore provide the essential negative feedback to prevent extreme CO_2-drawdown. For details, see the book edited by Ruddiman (1997) and its synthesis-chapter (Ruddiman *et al.* 1997, pp. 494–503).

Initially, the timing of Tibetan uplift was poorly constrained and it was only known to have commenced some time in the Eocene. This allowed discussion of Tibetan uplift as a possible cause of the dramatic drop in temperature at the EOB (Raymo & Ruddiman 1992). Near coincident with the evolution of the chemical erosion-driven drop in pCO$_2$-hypotheses, publications began to place Tibetan, Himalayan and Tian Shan uplift in latest Oligocene to earliest Miocene time (e.g. Copeland *et al.* 1987; Harrison *et al.* 1992; Hendrix *et al.* 1994). However, recent studies and improved dating place the episode of rapid and therefore most significant part of the uplift after 13 Ma (see Clark *et al.* 2005 for more details). The implication is that

Tibetan uplift cannot have been the cause of early to mid-Cenozoic climate change.

The rise of the Andes

The Altiplano–Puna plateau of the Central Andes is second only to Tibet in terms of height (3600 m) and extent ($c.$ 675×10^3 km^2). In the Southern Hemisphere, it forms the only barrier to atmospheric circulation. The main Andean thrust with contemporaneous development of the Altiplano Plateau and sub-Andean basins commenced in the late Oligocene lasting $c.$ eight million years (Sempere *et al.* 1990). Additional uplift in the Puna region commenced around 20–15 Ma (Allmendinger *et al.* 1997). Palaeobotanical data suggest that at $c.$ 20 Ma the plateau had attained no more than a third of its modern elevation and no more than half by $c.$ 10 Ma (see Gregory-Wodzicki 2000 for more details). These findings preclude Andean uplift from playing a major role in early to mid-Cenozoic climate deterioration; indeed Lamb & Davis (2003) argue it was the other way around: climate-forced erosion and tectonic change leading to uplift.

Deep opening of the Drake Passage

Much debate on Antarctica's transition from Cretaceous greenhouse to modern-day icehouse focuses on Antarctica's climatic isolation driven by inception of the Antarctic Circumpolar Current (ACC), which contains strong frontal systems keeping warmer equatorial waters at bay. This contrasts with the Gulf Stream in the Northern Hemisphere which transports warm equatorial waters from the Caribbean to northwestern Europe. Both currents are surface wind-driven currents; however, the ACC in parts reaches all the way to the ocean floor (Rintoul *et al.* 2001).

Crucial in establishing circum-Antarctic flow was deep opening of the last two remaining land connections in the Cenozoic Southern Ocean – the Tasman Gateway and the Drake Passage:

1. Ocean Drilling Program Leg 189 to the Tasman Gateway provided the necessary marine records for reconstruction of the history of separation between Antarctica and Tasmania/Australia. The process from shallow to deep took only a few million years with the transition's completion almost coinciding with the EOB (35.5–30.2 Ma, Stickley *et al.* 2004). The widely observed Marshall Paraconformity (33–30 Ma, Fulthorpe *et al.* 1996) from the southwest Pacific region is a likely consequence of established through flow (Carter *et al.* 1996).

ACC-inception cannot pre-date Tasman Gateway opening and must therefore be a latest Eocene to Miocene event.

2. The history, timing and magnitude of Drake Passage opening is still disputed, as direct evidence from deep ocean-drilling in the Scotia Sea region at present does not exist. Most authors addressing Drake Passage opening agree on a history of progressive deepening, which may have taken several million years to accomplish. Yet, as their interpretations are based on the movement of several crustal fragments and opening or closing of small basins (Eagles *et al.* 2006), it is not surprising that two time-frames for 'deep' opening have emerged. (By 'deep', we mean at least 1000 m which requires a passage at least 600 km wide to carry something like the present ACC of >130 10^6 m^3 s^{-1} at a mean

speed of 0.2 m s^{-1}). New work links early shallow Drake Passage opening to the middle Eocene climate optimum (Eagles *et al.* 2006; Scher & Martin 2006), while deep opening centres on the EOB (*c.* 31–28.5 Ma) (also Lawver & Gahagan 2003; Livermore *et al.* 2005). Earlier work favoured the OMB (*c.* 22–17 Ma) for deep opening (e.g. Barker 2001).

ACC flow is seen as directly linked to Drake Passage depth passing a critical threshold (>1000 m palaeodepth, Sijp & England 2004), and evidence for ACC-inception is therefore interpreted in terms of deep opening of Drake Passage. Sijp & England (2004) modelled that an increase in Drake Passage depth from 690 m to >1000 m caused an increase in throughflow from 64 Sv (1 Sv = 10^6 m^3 s^{-1}) with some sea-ice growth to the present *c.* 130 Sv and major ice growth.

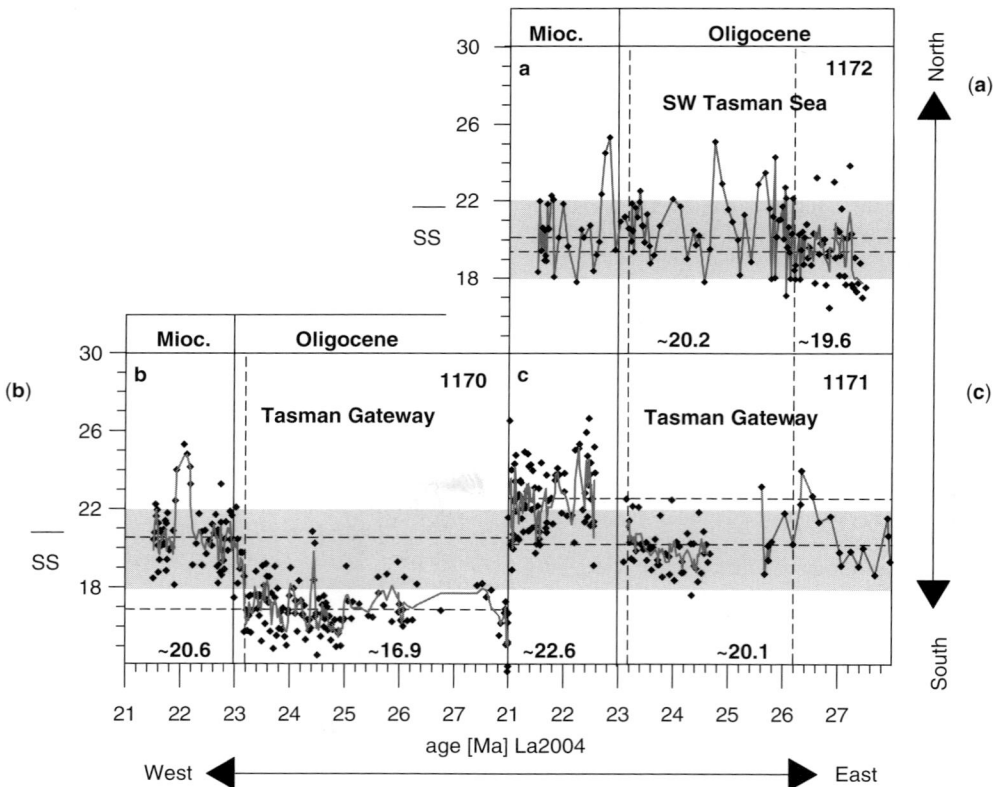

Fig. 4. Records of sortable silt mean diameter (10–63 μm, labelled $\overline{SS}$ on vertical axes) at Ocean Drilling Program Sites 1170–1172 (smoothed records, except in low-resolution intervals). Sites 1170–1171 in the path of the Antarctic Circumpolar Current (ACC) reveal a strong increase just prior to the Oligocene–Miocene boundary (**b–c**), while at the same time no obvious change is apparent at Site 1172 to the north of the ACC (**a**). Enlarged numbers at the bottom of the graphs indicate average values [μm] of the $\overline{SS}$ records over a period of time. At Site 1170, the change represents a remarkable increase in mean diameter of *c.* 22%. From Pfuhl & McCave (2005).

Sedimentary evidence for ACC inception is cited in support of the earlier EOB (Wright & Miller 1993), as well as the later OMB time-frame (Moore *et al.* 2004). Records from Site 1090 on Agulhas Ridge alone are interpreted in support of ages for ACC inception ranging from 32.8 Ma (Latimer & Filippelli 2002), 27.5–23 Ma (Diekmann *et al.* 2004), to possibly less than 20 Ma (Anderson & Delaney 2005). For a detailed review of the various publications on ACC inception, see the work by Barker & Thomas (2004).

The, as yet, most convincing support for a latest Oligocene inception of the full ACC was presented by Pfuhl & McCave (2005). Their records of sortable silt (an indicator of current speed: McCave *et al.* 1995; McCave & Hall 2006) from two sites in the Tasman Gateway, i.e. directly within the path of ACC type circulation (Sites 1170–1171), and a third site from the Tasman Sea, i.e. well north of any ACC flow (Site 1172), provide evidence for a singular strong increase in current speed at 23.2–23.0 Ma at the two southern sites (Fig. 4a–c). The time-interval investigated, 28–21 Ma, covers nearly the entire Oligocene period following the completion of Tasman Gateway opening.

Pfuhl & McCave (2005) base their interpretations on the fact that at the OMB the Tasman Gateway was still sufficiently narrow (300–900 km, Shipboard Scientific Party 2001) to have enhanced flow speeds resulting from increases in the current's volume (Shipboard Scientific Party 2001) (see tectonic reconstructions in Figs 5–6). Citing information from other Southern Ocean and global studies, Pfuhl & McCave (2005) argue

that this increase in current speed can only be interpreted as evidence for full ACC inception. Based on additional data obtained from these three sites, as well as Site 1168 on the southwestern margin of Tasmania, the authors propose a change in regional flow-patterns depicted in Figure 5a–b. Pfuhl & McCave (2005) then quote further support from Southern Ocean sediments which suggest OMB changes in the hydrographic regime and most importantly onsets of major hiatuses on Falkland Plateau (SW Atlantic Ocean) and northern Kerguelen Plateau (S Indian Ocean) (compare Fig. 6).

Due to the complex Drake Passage – and Scotia Sea – evolution, one would expect deep and shallow pathways through the region to have varied over time, which would have led to variations in through-flow and vertical mixing. Pfuhl & McCave (2005) allow that partially circumpolar circulation with a small volume flux in the Drake Passage region probably existed prior to 23 Ma; however, other than a crucial increase in Drake Passage sill depth, there are no other obvious tectonic explanations for the Southern Ocean imprint of current increase and hiatus formation at the OMB.

Climate change across the OMB

Until recently, the debate on Antarctica's transition from Cretaceous 'greenhouse' (as seen in fossilized birch tree leaves and pollen) to present-day 'icehouse' (terms originally coined by Fischer 1984) in mid-Miocene time (following a climate optimum at *c.* 15 Ma) focused on the establishment of permanent ice-sheets on East and West

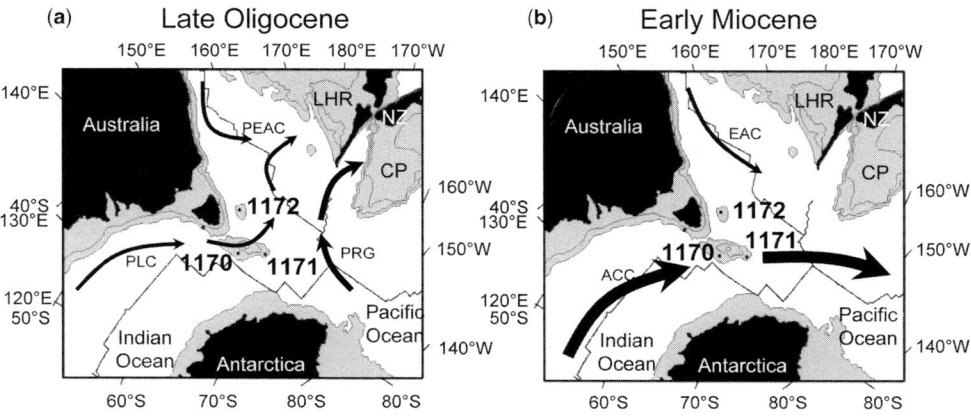

Fig. 5. Detailed map of the Tasman Gateway region at *c.* 26 Ma, giving an impression of ocean topography with Sites 1170 and 1171 situated on the South Tasman Rise and Site 1172 on the East Tasman Plateau in the Tasman Sea. Arrows indicate current flow in **(a)** the late Oligocene and **(b)** early Miocene. Abbreviations: CP, Campbell Plateau; LHR, Lord Howe Rise; NZ, New Zealand for topographic features; ACC, Antarctic Circumpolar Current; PEAC, Proto East Australia Current; PLC, Proto Leeuwin Current; PRG, Proto Ross Gyre for ocean currents. From Pfuhl & McCave (2005).

Antarctica at 15–13 and *c.* 6 Ma. The East Antarctic Ice-Sheet (EAIS) is a land-based ice-sheet separated from the marine-based, and therefore less stable, West Antarctic Ice-Sheet (WAIS) by the Transantarctic Mountains. Their combined present ice masses (90% EAIS, 10% WAIS) represent a volume equivalent to about 55–60 m of sea level (Denton *et al.* 1971).

Sedimentary evidence suggests that the Oligocene interval is crucial to major build-up of the East Antarctic Ice-Sheet, in line with the original 'icehouse' definition which covered Oligocene to modern time (Fischer 1984). The WAIS build-up is assumed to have followed eventually, though its initial steps are poorly constrained. This period of EAIS and WAIS build-up has been referred to as 'doubt house' (Miller *et al.* 1991) and the debate surrounding its process – along with its causes – appears far from over. This interval is thus the focus of recent, current and new drilling efforts onshore and offshore of Antarctica.

A reflection of the existing uncertainties may be found in recent publications that seem to reopen a debate believed by most workers to be over. On one side, Miller *et al.* (2005) argue that the paradigm of an ice-free greenhouse world in the Cretaceous to Eocene interval needs to be re-evaluated. They and Tripati *et al.* (2005) maintain that some isotopic records in combination with evidence for sea-level change suggest the existence of intermittent land-based ice-sheets in pre-Oligocene time.

On the other side, even the onset of Northern Hemisphere glaciation, which might reasonably have been presumed to be settled at 3.0–2.7 Ma (Shackleton *et al.* 1984; Haug *et al.* 2005), has been questioned and assigned a much earlier date. Indeed, Hay *et al.* (2005) conclude on the basis of microfossil distributions that Northern Hemisphere glaciation occurred at the same time as the establishment of Southern Hemisphere glaciation, i.e. at the Eocene–Oligocene boundary, and may even have produced the larger ice-sheet. DeConto & Pollard (2005) argue that on the basis of GCM (General Circulation Model) ice-sheet models the EAIS may have been too small to explain differences in benthic $\delta^{18}O$ and Mg/Ca records (respectively indicators of temperature plus

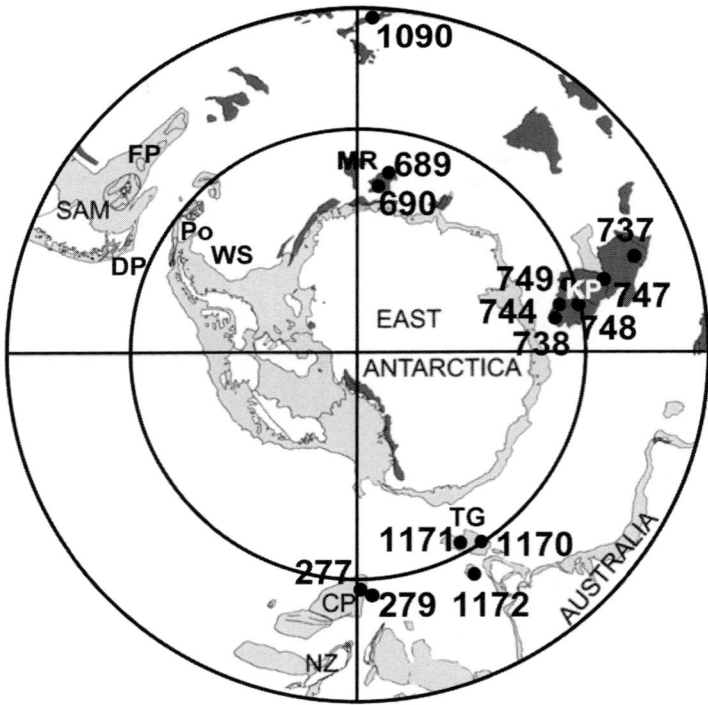

Fig. 6. View of the Southern Ocean at *c.* 25 Ma after Lawver & Gahagan (2003), who suggest a fully open Drake Passage at this time. Indicated are Deep Sea Drilling Program and Ocean Drilling Program sites from the area. Abbreviations: CP, Campbell Plateau; DP, Drake Passage; KP, Kerguelen Plateau; FP, Falkland Plateau; MR, Maud Rise; NZ, New Zealand; Po, Powell Basin; SAM, South America; TG, Tasman Gateway; WS, Weddell Sea.

ice-volume and temperature only), and that the onset of Antarctic glaciation may have preconditioned the climate system for episodic bipolar glaciation. In a similar fashion, Coxall *et al.* (2005) argue that the Eocene–Oligocene boundary increase in benthic $\delta^{18}O$ is too large to be explained by Antarctic ice build-up alone and therefore must be accompanied by global cooling and/or Northern Hemisphere glaciation.

In the following, we focus on the study of benthic foraminiferal climatic records from the late Eocene to early Miocene Southern Ocean, as at present most evidence points towards the Southern Ocean as the crucial location for events revolving around early to mid-Cenozoic climate deterioration. To better understand events defining the OMB, it is necessary to include the whole late Eocene to mid-Miocene interval in the discussion.

Unfortunately, interpretation is hampered by incompleteness of the records, with those closer to Antarctica ending in mid-Oligocene time, and those at greater distance from Antarctica commencing only in late Oligocene time with little to no overlap. The medium-resolution dataset from Ocean Drilling Program Leg 189 (Sites 1170–1171) is particularly important for two reasons: (a) it provides data from the southwest Pacific across the OMB, and (b) together with a few very old datapoints from the 1970s (Shackleton & Kennett 1975; Kennett & Shackleton 1976, retuned) it allows us to compare the records from all three ocean basins across the entire interval. Figure 7 shows that (a) bottom water conditions at the various sites in the Tasman Gateway region were near indistinguishable and (b) that the offset between southwest and equatorial Pacific bottom waters remains constant until 26 Ma, after which it increases suggesting southwest Pacific cooling.

Within a Southern Ocean context, we observe the following (compare Fig. 8):

1. Eocene–Oligocene boundary (EOB):
   - identical increase in benthic $\delta^{18}O$ (known as Oi-1 event) at Maud Rise (Atlantic) and southern Kerguelen Plateau (Indian), contrasting with the southwest Pacific;
   - southwest Pacific late Eocene and early Oligocene $\delta^{18}O$ values are lighter than in the other two basins;
   - southwest Pacific Oi-1 event shows a sharp increase in $\delta^{18}O$, but not only are peak values lighter by 0.7‰ relative to the other two basins, the increase is also less pronounced by 0.2‰.

2. Early Oligocene:
   - after *c.* 29 Ma, southernmost Atlantic $\delta^{18}O$ values become heavier than at southern Kerguelen Plateau; while

following *c.* 26 Ma we observe a trend to rapidly decreasing values at southern Kerguelen Plateau, where the $\delta^{18}O$ values converge upon values at central Kerguelen Plateau;
   - at the same time, $\delta^{18}O$ values in the southwest Pacific continue to rise following the Oi-1 event.

3. Late Oligocene:
   - there is a good match between the $\delta^{18}O$ records from all three basins;
   - the trend to lighter $\delta^{18}O$ values following 26 Ma at southern Kerguelen Plateau is not matched at the other sites; however, the trend allows speculation that eventually values might have converged onto those from proximal central Kerguelen Plateau (compare Fig. 6).

4. Oligocene–Miocene boundary (OMB):
   - good match in $\delta^{18}O$ at all three basins; however, as no data are available from Maud Rise situated in the Weddell Gyre south of the Antarctic Circumpolar Current (ACC), we can only assume that these remained heavier than within the ACC path.

5. Entire interval:
   - the benthic $\delta^{13}C$ values suggest no basin-to-basin differentiation – increases at the EOB and OMB are separated via a trough in the early Oligocene.

What are the implications? The stable isotopic data imply that a common Southern Ocean water mass did not exist before 26 Ma. This supports interpretation of the sortable silt records as only one increase is observed over the interval 28–21 Ma, i.e. well after the date for an earliest common Southern Ocean water mass (which should be a result of ACC circulation).

In addition to the timing of ACC inception (i.e. mainly Drake Passage deepening), two other factors form important foci for discussions of the driving forces behind climate change over the Eocene to Miocene interval:

- a decline/drop in $p\mathrm{CO}_2$ independent of tectonic uplift theories;
- orbital configurations supporting cooling and ice build-up.

Decline/drop in $p\mathrm{CO}_2$

A decline in atmospheric $p\mathrm{CO}_2$ has been proposed as a driving force behind Cenozoic climate deterioration since the 1970s (refer Crowley & North 1991, pp. 166–169), but as explained earlier orogenic uplift does not appear to cause it prior to mid- to late Miocene time. DeConto & Pollard (2003*a*, *b*,

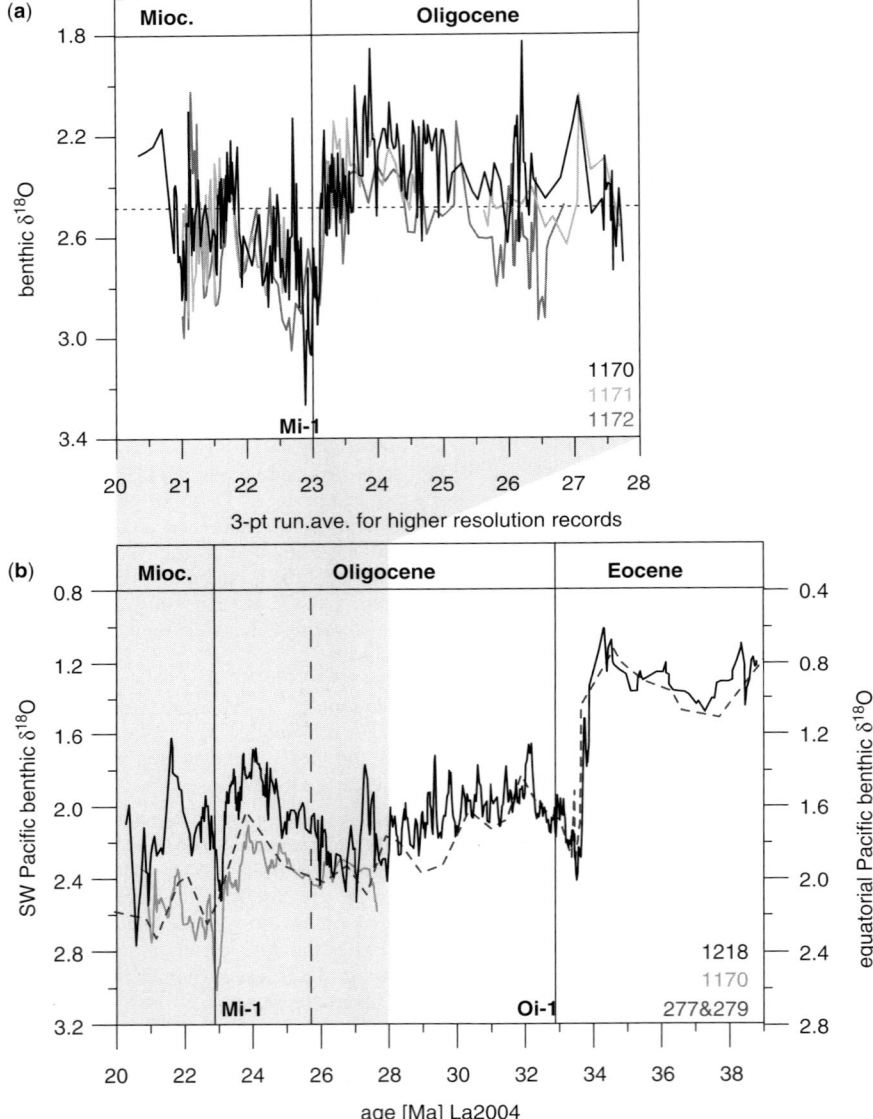

Fig. 7. All three Leg 189 benthic $\delta^{18}O$ records from Sites 1170–1171 on South Tasman Rise and Site 1172 on East Tasman Plateau are near-indistinguishable across the investigated interval (**a**) (Pfuhl & McCave 2005). (**b**) A comparison between the benthic $\delta^{18}O$ records from Sites 1170, 277 and 279 (Pfuhl & McCave 2005) and 1218 in the equatorial Pacific (Lear *et al.* 2004) reveals a constant offset until 26 Ma after which values in the southwest Pacific become relatively heavier.

2005) simulate a long-term linear decline in atmospheric pCO_2 (values reflecting those from deep-sea core data, Pagani *et al.* 1999; Pearson & Palmer 2000) with imposed orbital forcing. Their model results show a relatively sudden non-linear transition from very small ice amounts to a near-continental expanse with higher-frequency orbital variations bearing a strong resemblance to the observed two-step EOB increase in the $\delta^{18}O$ records (Coxall *et al.* 2005). The authors argue that the opening of the Drake and Tasman Passages and changes in ocean circulation probably only played a minor role relative to the impact of atmospheric pCO_2-drawdown.

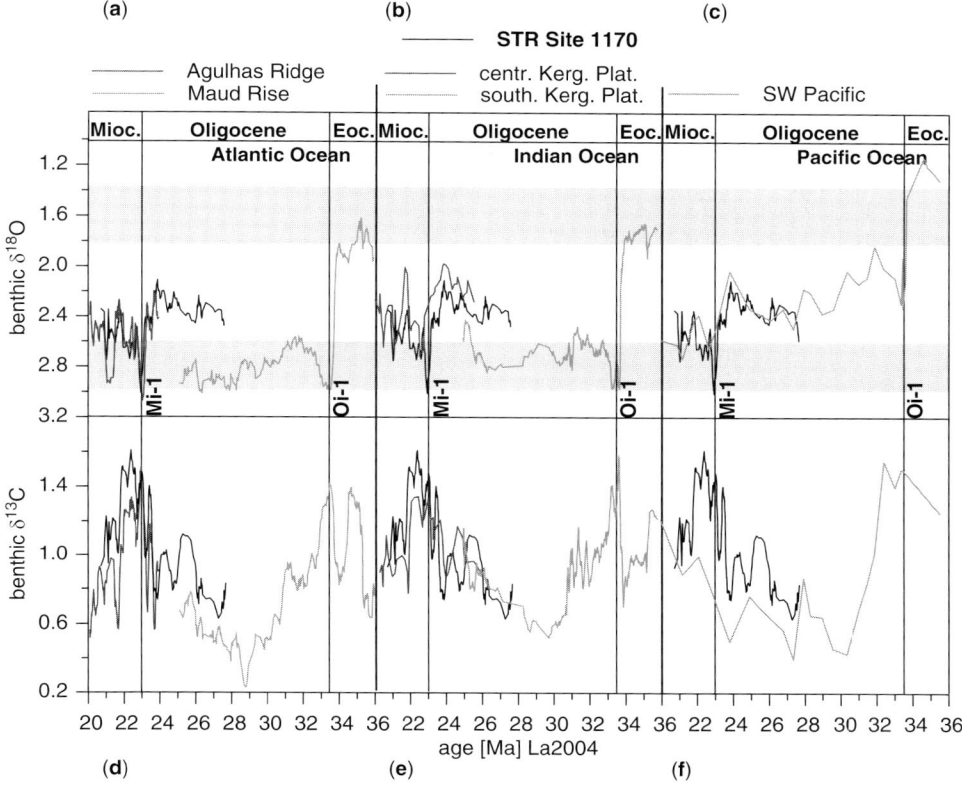

(a)   **(b)**   — STR Site 1170   **(c)**

— Agulhas Ridge   — centr. Kerg. Plat.
···· Maud Rise   ···· south. Kerg. Plat.   — SW Pacific

Fig. 8. A Southern Ocean comparison of benthic $\delta^{18}O$ records from 36–20 Ma, separated by regional trends and ocean basins, suggests a North–South temperature gradient for the Oligocene–Miocene southern Atlantic and Indian Oceans (**a–b**). Pacific late Eocene and early Oligocene values are lighter than in the other two basins, while the Oi-1 event increase is less than in the other two basins (**c**). For better comparison, the record from Site 1170 is superimposed on the records for all three basins. At the same time, we observe no basin-to-basin differentiation in the $\delta^{13}C$ records (**d–f**).

Although no agreement exists on the most significant controls on $p\mathrm{CO}_2$ over long timescales, different hypotheses have been voiced. Increases due to volcanic and hydrothermal outgassing (Owen & Rea 1985; Berner *et al.* 1993), and metamorphic decarbonation reactions (Kerrick & Caldeira 1998), stand against decreases caused by weathering of silicate minerals and limestone formation (e.g. Brady 1991; Raymo & Ruddiman 1992) or organic carbon burial (e.g. Berger & Vincent 1986; McGowran 1989). As ice build-up and sea-level fall leads to exposure of additional land to erosion, which increases $p\mathrm{CO}_2$-drawdown, this last 'control' can only serve as a reinforcing factor, but cannot be responsible for driving the 'climate-changing' $p\mathrm{CO}_2$-drawdown.

It has hitherto been difficult to sufficiently test the role of $p\mathrm{CO}_2$-drawdown in climate change. An independent proxy (preferably more than one) for palaeo-$p\mathrm{CO}_2$ estimates is essential. The first major

reconstruction used the boron isotopic composition of foraminiferal carbonate to establish ocean pH (an inverse proxy for $p\mathrm{CO}_2$, Pearson & Palmer 2000, but see Pagani *et al.* 2005*a*, for a critical evaluation of the method). This reconstruction was constrained by the limited availability of high-quality foraminiferal tests to fill the important gap across the EOB. Nonetheless, their data, supported by other $p\mathrm{CO}_2$ reconstructions for the Miocene interval (Pagani *et al.* 1999; Royer *et al.* 2001) suggest rather low values with little variance despite inferred periods of global cooling (OMB and mid-Miocene) and global warming (mid-Miocene climatic optimum) for this interval. Recently, Pagani *et al.* (2005*b*) used their method of palaeo-$p\mathrm{CO}_2$ reconstruction based on the $\delta^{13}C$ values of the organic molecules of alkenones (a product of phytoplankton) to close the Oligocene gap. Experiments have shown that alkenones reflect strongly growth conditions of their algal producers (see Pagani *et al.* 2005*b* for

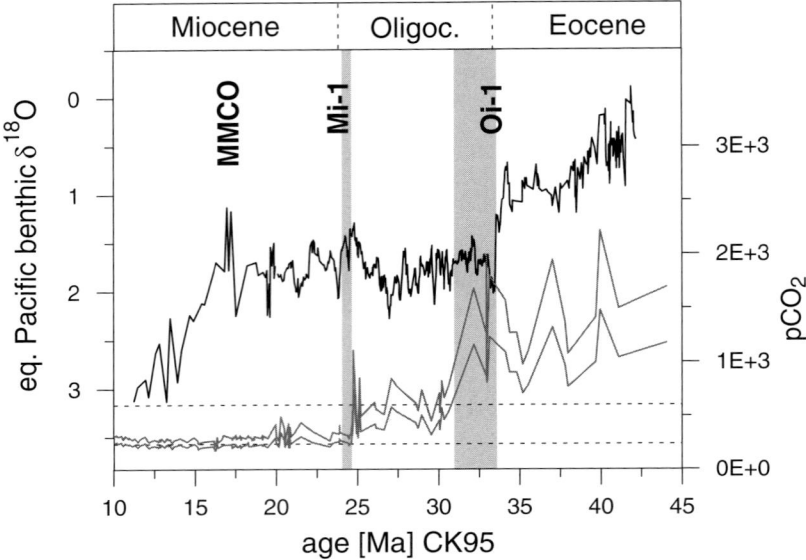

Fig. 9. Alkenone-based reconstruction of pCO_2 values (min & max) over the Eocene to Miocene interval by Pagani *et al.* (2005*b*, for details) show a strong decline after the Oi-1 and before the Mii-1 event.

discussion). By reconstructing a composite record for the period 45–5 Ma, the authors present data with error-bars, which nevertheless reveal a strong decline in pCO_2 in the earliest, and attainment of modern pCO_2-levels in the latest Oligocene (Fig. 9). Both incidents occur either after or before the Oi-1 and Mi-1 climate shifts and therefore confound the proposed causal link between pCO_2-drawdown and global cooling incidents.

Orbital configurations and gateway opening/closure

Eocene–Oligocene boundary (EOB) – Oi-1 event
Based on older records, the Oi-1 event (*c.* 33.7–33 Ma) was seen as a $\delta^{18}O$ shift of 1.2‰ interpreted as a 3 °C cooling and a drop in sea level of 30–40 m. Kennett and Shackleton, who were the first to observe this shift in sediments from the southwest Pacific, attributed this event to the development of more vigorous, colder deepwater circulation (Shackleton & Kennett 1975; Kennett & Shackleton 1976). This is certainly in accordance with the detailed reconstruction of Tasman Gateway opening, i.e. beginning around 35.5 Ma and completed by *c.* 30.2 Ma (Stickley *et al.* 2004), and timing of the Marshall Paraconformity of the Australasian region from *c.* 33–30 Ma (Fulthorpe *et al.* 1996).

Support for cooling at this time is drawn from faunal and floral fossil records (Premoli Silva *et al.* 1988; Vonhof *et al.* 2000), e.g. the major turnover of the deepwater ostracod fauna (Benson 1975). Evidence for ice-coverage is found in the appearance of ice-rafted debris in the Southern Ocean and glaciological evidence on land (e.g. Hayes & Frakes 1975; Leckie & Webb 1983; Barrett *et al.* 1987).

However, the discrepancy between the $\delta^{18}O$ shift in the Pacific and the other two ocean basins implies that cooling in the Southern Ocean and southern Atlantic and Indian Ocean basins was more intense at this time. This imbalance is consistent with evidence from terrestrial fossil carbonates at low latitudes, which suggests a barely noticeable drop in summer temperatures (Grimes *et al.* 2005).

Coxall *et al.* (2005) presented data at a benthic $\delta^{18}O$ resolution of up to 2 kyr, and thus were able to show that the climate shift at the EOB is in fact a stepwise increase matching a similar stepwise deepening in the carbonate compensation depth (CCD) and increase in the $\delta^{13}C$ record. Most change took place in two 40 kyr steps separated by a 200 kyr plateau. The observed changes occur during an interval of low eccentricity forcing and a low-amplitude change in obliquity, i.e. damped seasonality (absence of warm summers inhibiting summer snow melt, not the occurrence of cool winters favouring accumulation). A phase lag in the $\delta^{18}O$ response to obliquity lags that of the $\delta^{13}C$ record by *c.* 8 kyr, which is interpreted as a response to the carbon cycle, i.e. a drop in pCO_2

as proposed by DeConto & Pollard (2003*b*), helping to force climate change. However, this interpretation fails to explain the 10 kyr lag in the $\delta^{13}C$ increase relative to the $\delta^{18}O$ increase at the Oi-1 event, as well as the lag in pCO_2-drawdown as mentioned earlier (Pagani *et al.* 2005*b*) (Fig. 9). Coxall *et al.* (2005) suggest that orbital configuration was the ultimate trigger for the Oi-1 event and pacemaker for ice-sheet growth, but they also point out that there is a lack of evidence to suggest that the low-eccentricity obliquity 'node' conditions at 34 Ma were more extreme than those occurring every 2.4 million years and 1.2 million years (eccentricity and obliquity minima respectively).

Oligocene–Miocene boundary (OMB) – Mi-1 event Sedimentary evidence from Cape Roberts on the edge of the West Antarctic Ice-Sheet (WAIS) suggests a shift from periodic glacially influenced sediments in the Oligocene to permanent glacial dominance in the early Miocene (Naish *et al.* 2001; Florindo *et al.* 2005). This would imply that at least during the Mi-1 event and following cold interval an initial WAIS must have existed, which is supported by data from the South Shetland Islands (Troedson & Riding 2002).

Evidence from McMurdo Sound provides support for a major expansion of the East Antarctic Ice-Sheet (EAIS) during the Mi-1 event (Roberts *et al.* 2003), while seismic reflection data suggest that the EAIS at the time may even have been as large as today (Bartek *et al.* 1996). Once the ice-sheets reached the coast, their growth was limited, at which point they began to be an influence on, not just a response to, global climate change (Miller *et al.* 2005). One consequence is an imprint of the 41 kyr cycle of obliquity on the $\delta^{18}O$ record caused by ice-volume change (Zachos *et al.* 1997) which replaces the Eocene and earlier dominance of eccentricity (DeConto & Pollard 2003*b*; Miller *et al.* 2005). However, records of ice-rafted debris from the early Miocene at Prydz Bay indicate eccentricity-paced ice-advance (Williams & Handwerger 2005), which contrasts with the strong obliquity signal in the $\delta^{18}O$ isotopic record and the aforementioned explanations.

It appears, as an additional conundrum, that despite the accepted evidence for large-scale expansion of the ice-sheets, there is fossil evidence of *Nothofagus* (southern beech) pollen and leaves from Antarctica, which suggests the persistence of coastal forests across the Oligocene–Miocene transition and thus, at least in places, mean summer temperatures across the Oligocene–Miocene boundary of above 5 °C (Mildenhall 1989; Hill *et al.* 1996; Roberts *et al.* 2003).

Similar to speculations concerning the Eocene–Oligocene boundary, the factors most frequently cited to explain Oligocene–Miocene boundary

climate change are a response to changing orbital parameters and/or feedback internal to the climate system, i.e. threshold ice-sheet size or transient carbon reservoirs (Flower *et al.* 1997; Zachos *et al.* 1997; Paul *et al.* 2000). The most recent explanation is a proposed orbital amplification, i.e. low-amplitude variance in obliquity and a minimum in eccentricity resulting in low seasonality which favours ice-sheet expansion (Zachos *et al.* 2001), similar to the constellation at the Eocene–Oligocene boundary.

However, just as there appears to be a link between Tasman Gateway opening and the Oi-1 event, Pfuhl & McCave (2005) have shown that the rapid increase in mean diameter of the sortable silt range precedes the Mi-1 event at the OMB by *c.* 60 kyr (Fig. 10). This suggests that the implied ACC-inception – and hence inferred deep opening of Drake Passage driven by plate tectonics – was a (if not the most) significant driving force behind climate change at the OMB.

Mid-Miocene cooling The mid-Miocene increase in $\delta^{18}O$ from 15–10 Ma was initially interpreted as the time of major Antarctic ice-sheet growth (Shackleton & Kennett 1975), but has also been described as deepwater cooling of 4–5 °C leading up to increases in ice-volume and a drop

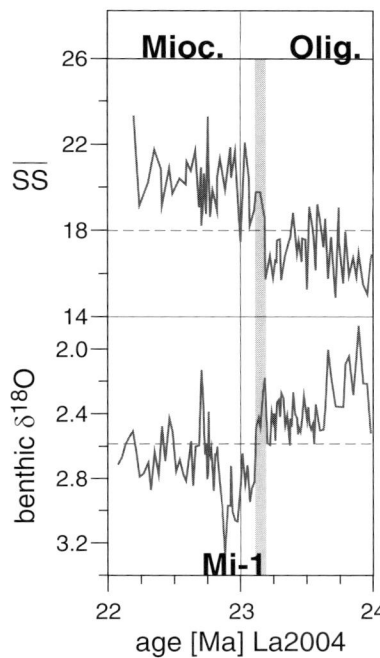

Fig. 10. Details of the $\overline{SS}$ and benthic $\delta^{18}O$ records at Site 1170 showing that the increase in current speed at *c.* 23.2 Ma precedes the onset of the Mi-1 event at *c.* 23.15 Ma. From Pfuhl & McCave (2005).

in sea level at 11–10 Ma (Miller *et al.* 1987; Moore *et al.* 1987). Escutia *et al.* (2005) suggest that sedimentary evidence from Wilkes Land provides support for a transition from a dynamic but intermittently established ice-sheet regime to a permanent one with oscillatory movements. Permanent establishment of the marine-based, and therefore less stable, West Antarctic Ice-Sheet (WAIS) probably occurred at this time.

Greater variance in all $\delta^{18}O$ records accompanied by decreases at low latitudes following this interval are consistent with increased ice-volume fluctuations and the development of a zonal temperature gradient from the cool poles to the warm equator, inferred strengthened surface water circulation and eventual establishment of a Northern Hemisphere ice-cap (see also earlier discussion on bipolar ice-sheets in the Oligocene) (Moore *et al.* 1981; Savin *et al.* 1985; Thunell *et al.* 1992; Flower & Kennett 1993; Haug & Tiedemann 1998).

Shevenell *et al.* (2004) present a Mg/Ca sea-surface temperature (SST) reconstruction for the South Tasman Rise across the mid-Miocene cooling, which shows a stepwise drop over 300 kyr preceding the increase in planktic $\delta^{18}O$ (lasting 100 kyr) and also, more importantly, the rather abrupt increase in benthic $\delta^{18}O$. A clear imprint of eccentricity-paced orbital frequencies on the climate record, however, does not coincide with a similar orbital anomaly as calculated for the OMB. The authors also show that the drop in temperature precedes a positive peak in foraminiferal $\delta^{13}C$ (minimum in pCO_2), which implies that changes in the carbon cycle, often cited at other climate transitions, was not the key driving or amplifying force. Shevenell *et al.* (2004) prefer a strengthening of the ACC due to further closure of the eastern Tethys (e.g. Woodruff & Savin 1989; Flower & Kennett 1993), as the likely cause that drove climate across a threshold. Figure 3 suggests that the Tethys was still closing at 23 Ma (i.e. the Oligocene–Miocene boundary). It was actually shut in the Tethys closure or Terminal Tethys Event (TTE) which occurred between 18 and 12 Ma (Rögl & Steininger 1983; Bellwood & Wainwright 2002; Golonka 2004; Vallejo 2005). It is therefore likely to have had a significant impact on mid-Miocene events. We note that closure of the Indonesian Gateway at *c.* 15 Ma to major throughflow (Hall 2002) may have served as another factor in causing mid-Miocene cooling.

on the basis of molluscan faunal turnover. As its cause, we proposed climatic change as the driving force behind a change in the molluscan palaeoecological conditions. We have shown that the OMB nearly coincides with a climatic cooling and ice-growth shift known as the Mi-1 event, which appears to have been followed by an interval of global or at least high latitude cooling.

When discussed within the context of early to middle Cenozoic climatic shifts to a cooler climate, the OMB cold interval appears less significant than those of the Eocene–Oligocene boundary (EOB), i.e. the Oi-1 event, and the mid-Miocene. However, sedimentary and fossil evidence suggests that all three incidents are accompanied by major ice-sheet expansion. Of the many proposed causes for Cenozoic climate deterioration, only one emerges as a major contender in all three cases, although, of course, a positive feedback between two or more of the proposed causes at any one time cannot be excluded:

1. pCO_2-drawdown – We have shown that rapid orogenic uplift was confined to the middle to late Miocene and that palaeo-pCO_2 reconstructions show a decrease following the Oi-1 event and the establishment of modern levels predating the Mi-1 event, with no obvious change at the mid-Miocene glaciation; in addition, high-resolution records reveal a lag in $\delta^{13}C$ increase relative to $\delta^{18}O$. Thus, pCO_2-drawdown does not come top of the list.

2. Orbital configurations – The patterns of orbital forcing differ on all three occasions (in pattern and/or strength of its components), and neither do they appear unique on long timescales, which suggests that they only had a supporting role.

3. Gateway opening/closure – Ocean circulation may be largely wind-driven; however, its pattern is forced by physical boundaries. Opening or closure of gateways leads to a reorganization of water mass flow which for example affects the meridional heat transport in surface ocean waters; all three cooling events post-date evidence for either gateway opening (Tasman Gateway – Oi-1 event), gateway deepening (Drake Passage – Mi-1 event) or gateway restriction (Tethys Seaway – Oligocene–Miocene boundary, Indonesian Gateway – mid-Miocene cooling event). A change in ocean circulation forces atmospheric and hence climate change.

Evaluation

As mentioned in our introduction, the Oligocene–Miocene boundary (OMB) was originally defined

Far-field consequences

How can high-latitude deepening of Drake Passage and ACC-inception have impacted molluscan turnover at low latitudes in the Mediterranean? The observed gastropod extinction event in Europe did not affect America, and distribution of gastropod provinces within the Tethys region suggests a restriction of the eastern basin to throughflow in the early Miocene (with distinct faunas in the Burdigalian) (Harzhauser *et al.* 2002). In addition, studies have shown that, accompanying these events, trans-Atlantic migration switched from westward in the Eocene and Oligocene to eastward in the early Miocene (Squires & Advocate 1986).

Similar to the gastropod extinction event in the Tethys region, a coral extinction event in the Caribbean focusing on the Oligocene–Miocene boundary (OMB) is observed. Out of the Oligocene genera, those that survived best were cold-tolerant and able to survive in turbid waters (Edinger & Risk 1994, 1995). This event was gradual (10 genera became extinct at the OMB, 5 in early Miocene) and coincided with a disruption of trans-Atlantic dispersal (4 new 'American' genera appeared in early Miocene) (Budd 2000). Evidence from coral studies supports the gastropod evidence that before the disruption, dispersal in the Eocene to Oligocene was from the Tethys region westwards (Frost 1977).

While restriction of the eastern Tethys Seaway, as suggested by the gastropod evidence, may have been one cause for the disruption and reversal of the trans-Atlantic dispersal, it is unlikely to have been the sole driving force. Out of the central American gateways, the Isthmus of Panama remained open until the Pliocene, while closure of the other two passages (between Nicaragua and Costa Rica at *c.* 28 Ma; Guatemala–Honduras area at any time between *c.* 32–20 Ma) was probably of little relevance in changing the influence of throughflow (dates from Hay *et al.* 1999).

Von der Heydt & Dijkstra (2005) simulated throughflow in the Panama Seaway with various 'gateway settings'. Their results suggest that in combination with enhanced flow through Drake Passage and a closed Tethys Seaway the net transport through the Panama Seaway reverses from westward (bringing warm, high-salinity Atlantic waters to the Caribbean) to eastward (bringing cold, low-salinity Pacific waters to the Caribbean). These modelling results are consistent with the fossil evidence cited above and sedimentological evidence (Nesbitt & Young 1997). The Pacific waters entering the Caribbean Sea in the Miocene were primarily derived from upwelling regions, and it was their reduced temperature together with increased sediment and nutrient load which may have raised the turbidity levels to values damaging to some coral genera (Edinger & Risk 1994).

We suggest that, while the crucial increase in Drake Passage throughflow occurred over a geologically short period (Pfuhl & McCave 2005), a protracted process of Tethys Seaway narrowing may have caused the defining faunal extinction events at the OMB to be more gradual and regional than originally perceived. But most importantly, together with the modelling results, fossil evidence and sedimentological data, it is possible to link high and low latitude geological records and establish that Drake Passage deepening across a crucial threshold allowed inception of the Antarctic Circumpolar Current, which played an important role in changing the patterns of trans-Atlantic circulation and therefore drove European and Caribbean faunal extinction events across the OMB.

Outlook

The evidence cited here for Drake Passage deepening across a critical threshold causing ACC-inception is at best circumstantial. Direct evidence from the Drake Passage region, i.e. drilling through the Eocene, is essential in guiding interpretations of the marine and terrestrial records and furthering our understanding of Earth's climatic past. A drilling-leg to the Scotia Sea region near Drake Passage is proposed under the new Integrated Ocean Drilling Project and may provide the final answers to a contentious question.

NERC supported our post-cruise work on ODP Leg 189 samples, for which we are most grateful. We would also like to thank two anonymous reviewers for their comments and Dr. Alan G. Smith from the Cambridge Palaeomap Services Ltd for providing the global plate tectonic reconstruction presented in Figure 3.

Abbreviations

ACC	Antarctic Circumpolar Current
EAIS	East Antarctic Ice-Sheet
EOB	Eocene–Oligocene boundary
FO	First Occurrence
LO	Last Occurrence
OMB	Oligocene–Miocene boundary
WAIS	West Antarctic Ice-Sheet

References

ALLMENDINGER, R. W., JORDAN, T. E., KAY, S. M. ET AL. 1997. The evolution of the Altiplano–Puna Plateau of the central Andes. *Annual Reviews of Earth and Planetary Science*, **25**, 139–174.

ANDERSON, L. D. & DELANEY, M. L. 2005. Use of multiproxy records on the Agulhas Ridge, Southern Ocean (Ocean Drilling Project Leg 177, Site 1090) to investigate sub-Antarctic hydrography from the Oligocene to the early Miocene. *Paleoceanography*, **20/1**, 1–16, doi:10.1029/2004PA001082

BARKER, P. F. 2001. Scotia Sea regional tectonic evolution: implications for mantle flow and palaeocirculation. *Earth Science Reviews*, **55**, 1–39.

BARKER, P. F. & THOMAS, E. 2004. Origin, signature and palaeoclimatic influence of the Antarctic Circumpolar Current. *Earth Science Reviews*, **66**, 143–162.

BARRETT, P. J., ELSTON, D. P., HARWOOD, D. M. *ET AL.* 1987. Mid-Cenozoic record of glaciation and sea-level change on the margin of the Victoria Land basin, Antarctica. *Geology*, **15**, 634–637.

BARTEK, L. R., HENRYS, S. A., ANDERSON, J. B. *ET AL.* 1996. Seismic stratigraphy of McMurdo Sound, Antarctica: implications for glacially influenced early Cenozoic eustatic change? *Marine Geology*, **130**, 79–98.

BELLWOOD, D. R. & WAINWRIGHT, P. C. 2002. The history and biogeography of fishes on coral reefs. *In*: SALE, P. F. (ed.) *Coral Reef Fishes: Dynamics and Diversity in a Complex Ecosystem*. Academic Press, San Diego, 5–32.

BENSON, R. H. 1975. The origin of the psychrosphere as recorded in changes of deep–sea ostracode assemblages. *Lethaia*, **8**, 69–83.

BERGER, A. L. 1977. Long-term variations of the Earth's orbital elements. *Celestial Mechanics*, **15**, 53–74.

BERGER, A. L. & TRICOT, C. 1986. Global climatic change and the theory of paleoclimates. *In*: CAZENAVE, A. (ed.) *Earth Rotation: Solved and Unsolved Problems*. Reidel Publishing Co., Dordrecht, 111–129.

BERGER, W. H. & VINCENT, E. 1986. Deep-sea carbonates: reading the carbon-isotope signal. *Geologische Rundschau*, **75**, 249–269.

BERNER, R. A., LASAGA, A. C. & GARRELS, R. M. 1993. The carbonate–silicate geochemical cycle and its effect on atmospheric carbon dioxide over the past 100 million years. *American Journal of Science*, **283**, 641–683.

BILLUPS, K., PÄLIKE, H., CHANNELL, J. E. T. *ET AL.* 2004. Astronomic calibration of the late Oligocene through early Miocene geomagnetic polarity time scale. *Earth and Planetary Science Letters*, **224**, 33–44, doi:10.1016/j.epsl.2004.1005.1004

BRADY, P. V. 1991. The effect of silicate weathering on global temperature and atmospheric CO_2. *Journal of Geophysical Research*, **96**, 18101–18106.

BUDD, A. F. 2000. Diversity and extinction in the Cenozoic history of Caribbean reefs. *Coral Reefs*, **19**, 25–35.

CARTER, R. M., CARTER, L. & McCAVE, I. N. 1996. Current controlled sediment deposition from shelf to the deep ocean: the Cenozoic evolution of circulation through the SW Pacific gateway. *Geologische Rundschau*, **85**, 438–451.

CLARK, M. K., HOUSE, M. A., ROYDEN, L. H. *ET AL.* 2005. Late Cenozoic uplift of southeastern Tibet. *Geology*, **33/6**, 525–528, doi:510.1130/G21265. 21261

COPELAND, P., HARRISON, T. M., KIDD, W. S. F. *ET AL.* 1987. Rapid early Miocene acceleration of uplift in the Gangdese Belt, Xizang (southern Tibet), and its bearing on accomodation mechanisms of the India–Asia collision. *Earth and Planetary Science Letters*, **86**, 240–252.

COXALL, H. K., WILSON, P. A., PÄLIKE, H. *ET AL.* 2005. Rapid stepwise onset of Antarctic glaciation and deeper calcite compensation in the Pacific Ocean. *Nature*, **433**, 53–57.

CROWLEY, T. J. & NORTH, G. R. 1991. *Paleoclimatology*. Oxford University Press, Oxford, 349pp.

DECONTO, R. M. & POLLARD, D. 2003a. A coupled climate-ice sheet modeling approach to the early Cenozoic history of the Antarctic ice sheet. *Palaeogeography, Palaeoclimatology, Palaeoecology*, **198**, 39–52.

DECONTO, R. M. & POLLARD, D. 2003b. Rapid Cenozoic glaciation of Antarctica induced by declining atmospheric CO_2. *Nature*, **421**, 245–249.

DECONTO, R. M. & POLLARD, D. 2005. Rethinking the Cenozoic record of ice volume: a modeling perspective on the relative contributions of Southern and Northern Hemispheres, Fall Meet. Suppl., Abstract PP52B-01. *EOS Transactions of the AGU*, **86/52**.

DENTON, G. H., ARMSTRONG, R. L. & STUIVER, M. 1971. The late Cenozoic glacial history of Antarctica. *In*: TUREKIAN, K. K. (ed.) *The Late Cenozoic Glacial Ages*. Yale Univ. Press, New Haven, CT, 267–306.

DIEKMANN, B., KUHN, G., GERSONDE, R. *ET AL.* 2004. Middle Eocene to early Miocene environmental changes in the sub-Antarctic Southern Ocean: evidence from biogenic and terrigenous patterns at ODP Site 1090. *Global and Planetary Change*, **40**, 295–313.

EAGLES, G., LIVERMORE, R. & MORRIS, P. 2006. Small basins in the Scotia Sea: the Eocene Drake Passage gateway. *Earth and Planetary Science Letters*, **242**, 343–353.

EDINGER, E. N. & RISK, M. J. 1994. Oligocene–Miocene extinction and geographic restriction of Caribbean corals: roles of turbidity, temperature, and nutrients. *Palaios*, **9**, 576–598.

EDINGER, E. N. & RISK, M. J. 1995. Preferential survivorship of brooding corals in a regional extinction. *Paleobiology*, **21**, 200–219.

ESCUTIA, C., DE SANTIS, L., DONDA, F. *ET AL.* 2005. Cenozoic ice sheet history from East Antarctic Wilkes Land continental margin sediments. *Global and Planetary Change*, **45**, 51–81.

FISCHER, A. G. 1984. Two Phanerozoic supercycles. *In*: BERGGREN, W. A. & VAN COUVERING, J. A. (eds) *Catastrophes and Earth History*. Princeton University Press, Princeton, NJ, 129–150.

FLORINDO, F., WILSON, G. S., ROBERTS, A. P. *ET AL.* 2005. Magnetostratigraphic chronology of a late Eocene to early Miocene glacimarine succession from the Victoria Land Basin, Ross Sea, Antarctica. *Global and Planetary Change*, **45**, 207–236, doi: 210.1016/j.gloplacha.2004.1009.1009

FLOWER, B. P. & KENNETT, J. P. 1993. Middle Miocene ocean-climate transition: high-resolution oxygen and carbon isotopic records from deep sea drilling project Site 588A, southwest Pacific. *Paleoceanography*, **8**, 811–843.

FLOWER, B. P., ZACHOS, J. C. & PAUL, H. 1997. Milankovitch-scale variability recorded near the Oligocene/Miocene boundary. *In*: SHACKLETON, N. J. *ET AL.* (eds) *Proceedings of the Ocean Drilling Program, Scientific Results*, **154**, 433–439.

FROST, S. H. 1977. Oligocene reef coral biogeography, Caribbean and western Tethys. *Mémoires du Bureau Recherches Géologique et Minières*, **89**, 342–352.

FULTHORPE, C. S., CARTER, R. M., MILLER, K. G. ET AL. 1996. Marshall Paraconformity: a mid-Oligocene record of inception of the Antarctic Circumpolar Current and coeval glacioeustatic lowstand? *Marine and Petroleum Geology*, **13**, 61–77.

GOLONKA, J. 2004. Plate tectonic evolution of the southern margin of Eurasia in the Mesozoic and Cenozoic. *Tectonophysics*, **381**, 235–273.

GRADSTEIN, F. M., OGG, J. G. & SMITH, A. G. 2004. *A Geologic Timescale 2004*. Cambridge University Press, Cambridge, UK, 589pp.

GREGORY-WODZICKI, K. M. 2000. Uplift history of the Central and Northern Andes: a review. *Geological Society of America Bulletin*, **112/7**, 1091–1105.

GRIMES, S. T., HOOKER, J. J., COLLINSON, M. E. ET AL. 2005. Summer temperatures of late Eocene to early Oligocene freshwaters. *Geology*, **33/3**, 189–192, doi:110.1130/G21019.21011

HALL, R. 2002. Cenozoic geological and plate tectonic evolution of SE Asia and the SW Pacific: computer-based reconstructions, model and animations. *Journal of Asian Earth Sciences*, **20**, 353–431.

HARLAND, W. B., SMITH, A. G. & WILCOCK, B. 1964. The Phanerozoic time-scale. *Quarterly Journal of the Geological Society of London*, Supplement volume, **120S**, 1–458.

HARRISON, T. M., COPELAND, P., KIDD, W. ET AL. 1992. Raising Tibet. *Science*, **255**, 1663–1670.

HARZHAUSER, M., PILLER, W. E. & STEININGER, F. F. 2002. Circum-Mediterranean Oligo-Miocene biogeographic evolution – the gastropods' point of view. *Palaeogeography, Palaeoclimatology, Palaeoecology*, **183**, 103–133.

HAUG, G. H. & TIEDEMANN, R. 1998. Effect of the formation of the Isthmus of Panama on Atlantic Ocean thermohaline circulation. *Nature*, **393**, 673–676.

HAUG, G. H., GANOPOLSKI, A., SIGMAN, D. M. ET AL. 2005. North Pacific seasonality and the glaciation of North America 2.7 million years ago. *Nature*, **433**, 821–825, doi:810.1038/nature03332

HAY, W. W., DE CONTO, R. M., WOLD, C. N. ET AL. 1999. Alternative global cretaceous paleogeography. In: BARRERA, E. & JOHNSON, C. (eds) *The Evolution of Cretaceous Ocean/Climate Systems*. Geological Society of America, Boulder, CO, **332**, 1–47.

HAY, W. W., FLÖGEL, S. & SÖDING, E. 2005. Is the initiation of glaciation on Antarctica related to a change in the structure of the ocean? *Global and Planetary Change*, **45**, 23–33, doi:10.1016/j.gloplacha.2004.1009.1005

HAYES, D. E. & FRAKES, L. A. 1975. General synthesis, Deep Sea Drilling Project Leg 28. In: HAYES, D. E., FRAKES, L. A. ET AL. (eds) *Initial Reports of the Deep Sea Drilling Project*, **28**, 919–942.

HENDRIX, M. S., DUMITRU, T. A. & GRAHAM, S. A. 1994. Late Oligocene–early Miocene unroofing in the Chinese Tian Shan: an early effect of the India–Asia collision. *Geology*, **22**, 487–490.

HILGEN, F. J., KRIJGSMAN, W., LANGEREIS, C. G. ET AL. 1999. Present status of the astronomical (polarity) time scale for the Mediterranean late Neogene. *Phil. Trans. R. Soc. Lond. A*, **357**, 1907–1929.

HILL, R. S., HARWOOD, D. M. & WEBB, P.-N. 1996. *Nothofagus beardmorensis* (Nothofagaceae), a new species based on leaves from the Pliocene Sirius Group, Transarctic Mountains, Antarctica. *Reviews in Palaeobotany and Palynology*, **94**, 11–24.

KENNETT, J. P. & SHACKLETON, N. J. 1976. Oxygen isotopic evidence for the development of the psychrosphere 38 Myr ago. *Nature*, **260**, 513–315.

KERRICK, D. M. & CALDEIRA, K. 1998. Metamorphic CO_2 degassing from orogenic belts. *Chemical Geology*, **145**, 213–232.

LAMB, S. & DAVIS, P. 2003. Cenozoic climate change as a possible cause for the rise of the Andes. *Nature*, **425**, 792–797.

LASKAR, J. 1999. The limits of Earth orbital calculations for geological time-scale use. *Phil. Trans. R. Soc. Lond. A*, **357**, 1735–1759.

LASKAR, J., ROBUTEL, P., JOUTEL, F. ET AL. 2004. A numerical solution for the insolation quantities for the Earth. *Astronomy and Astrophysics*, **428**, 261–285.

LATIMER, J. C. & FILIPPELLI, G. M. 2002. Eocene to Miocene terrigenous inputs and export production: geochemical evidence from ODP Leg 177, Site 1090. *Palaeogeography, Palaeoclimatology, Palaeoecology*, **182**, 151–164.

LAWVER, L. A. & GAHAGAN, L. M. 2003. Evolution of Cenozoic seaways in the circum-Antarctic region. *Palaeogeography, Palaeoclimatology, Palaeoecology*, **198**, 11–37.

LEAR, C. H., ELDERFIELD, H. & WILSON, P. A. 2000. Cenozoic deep-sea temperatures and global ice volumes from Mg/Ca in benthic foraminiferal calcite. *Science*, **287**, 269–272.

LEAR, C. H., ROSENTHAL, Y., COXALL, H. K. ET AL. 2004. Late Eocene to early Miocene ice sheet dynamics and the global carbon cycle. *Paleoceanography*, **19**, 1–11, PA4015, doi:4010.1029/2004PA001039

LECKIE, R. M. & WEBB, P.-N. 1983. Late Oligocene–early Miocene glacial record of the Ross Sea, Antarctica: evidence from DSDP Site 270. *Geology*, **11**, 578–582.

LIVERMORE, R. A., NANKIVELL, A., EAGLES, G. ET AL. 2005. Paleogene opening of Drake Passage. *Earth and Planetary Science Letters*, **236**, 459–470, doi:410.1016/j.epsl.2005.1003.1027

LOURENS, L. J., HILGEN, F. J., SHACKLETON, N. J. ET AL. 2004. The Neogene Period. In: GRADSTEIN, F. M., OGG, J. G. & SMITH, A. G. (eds) *A Geologic Timescale 2004*. Cambridge University Press, Cambridge, UK, 409–440.

LUTERBACHER, H. P., ALI, J. R., BRINKHUIS, H. ET AL. 2004. The Paleogene period. In: GRADSTEIN, F. M., OGG, J. G. & SMITH, A. G. (eds) *A Geologic Timescale 2004*. Cambridge University Press, Cambridge, UK, 384–408.

LYELL, C. 1830. *Principles of Geology*. **Vol. 1**, John Murray, London.

MATTHEWS, R. K. & POORE, R. Z. 1980. Tertiary ^{18}O record and glacio-eustatic sea-level fluctuations. *Geology*, **8**, 501–504.

MCCAVE, I. N. & HALL, I. R. 2006. Size sorting in marine muds: processes, pitfalls and prospects for palaeoflow-speed proxies. *Geochemistry, Geophysics, Geosystems*, **7**, 1–36, doi:10.1029/2006GC001284.

McCAVE, I. N., MANIGHETTI, B. & ROBINSON, S. G. 1995. Sortable silt and fine sediment size/composition slicing: Parameters for palaeocurrent speed and palaeoceanography. *Paleoceanography*, **10/3**, 593–610.

McGOWRAN, B. 1989. Silica burp in the Eocene ocean. *Geology*, **17**, 857–860.

MILDENHALL, D. C. 1989. Terrestrial palynology. *In*: BARRETT, P. J. (ed.) *Antarctic Cenozoic history from the CIROS-1 drillhole, McMurdo Sound*, DSIR Bulletin, **245**, 119–127.

MILLER, K. G., FAIRBANKS, R. G. & MOUNTAIN, G. S. 1987. Tertiary oxygen isotope synthesis, sea level history, and continental margin erosion. *Paleoceanography*, **2**, 1–19.

MILLER, K. G., WRIGHT, J. D. & FAIRBANKS, R. G. 1991. Unlocking the ice house: Oligocene–Miocene oxygen isotopes, eustasy, and margin erosion. *Journal of Geophysical Research*, **96/B4**, 6829–6848.

MILLER, K. G., WRIGHT, J. D. & BROWNING, J. V. 2005. Visions of ice sheets in a greenhouse world. *Marine Geology*, **217**, 215–231, doi:210.1016/j.margeo.2005.1002.1007

MOORE, T. C., LOUTIT, T. S. & GRENLEE, S. M. 1987. Estimating short-term changes in eustatic sea level. *Paleoceanography*, **2**, 625–637.

MOORE, T. C., HUTSON, W. H., KIPP, N. G. *ET AL.* 1981. Ocean basin and depth variability of oxygen isotopes in Cenozoic benthic foraminifera. *Marine Micropaleontology*, **6**, 465–481.

MOORE, T. C., BACKMAN, J., RAFFI, I. *ET AL.* 2004. Paleogene tropical Pacific: Clues to circulation, productivity, and plate motion. *Paleoceanography*, **19/3**, 1–16, doi:10.1029/2003PA000998

NAISH, T. R., WOOLFE, K. J., BARRETT, P. J. *ET AL.* 2001. Orbitally induced oscillations in the East Antarctic ice sheet at the Oligocene/Miocene boundary. *Nature*, **413**, 719–722.

NESBITT, H. W. & YOUNG, G. M. 1997. Sedimentation in the Venezuelan Basin, circulation in the Caribbean Sea, and onset of Northern Hemisphere glaciation. *Journal of Geology*, **105**, 531–544.

OWEN, R. M. & REA, D. K. 1985. Sea floor hydrothermal activity links climate to tectonics – the Eocene carbon dioxide greenhouse. *Science*, **227**, 166–169.

PAGANI, M., ARTHUR, M. A. & FREEMAN, K. H. 1999. Miocene evolution of atmospheric carbon dioxide. *Paleoceanography*, **14**, 273–292.

PAGANI, M., LEMARCHAND, D., SPIVACK, A. *ET AL.* 2005*a*. A critical evaluation of the boron isotope-pH proxy: the accuracy of ancient pH estimates. *Geochimica et Cosmochimica Acta*, **69/4**, 953–961, doi: 910.1016/j.gca.2004.1007.1029

PAGANI, M., ZACHOS, J. C., FREEMAN, K. H. *ET AL.* 2005*b*. Marked decline in atmospheric carbon dioxide concentrations during the Paleogene. *Science*, **309**, 600–603.

PAUL, H. A., ZACHOS, J. C., FLOWER, B. P. *ET AL.* 2000. Orbitally induced climate and geochemical variability across the Oligocene/Miocene boundary. *Paleoceanography*, **15/5**, 471–485.

PEARSON, P. N. & PALMER, M. R. 2000. Atmospheric carbon dioxide concentrations over the past 60 million years. *Nature*, **406**, 695–699.

PFUHL, H. A. & McCAVE, I. N. 2005. Evidence for late Oligocene establishment of the Antarctic Circumpolar Current. *Earth and Planetary Science Letters*, **235**, 715–728, doi:710.1016/j.epsl.2005.1004.1025

POLLARD, D. & DeCONTO, R. M. 2005. Hysteresis in Cenozoic Antarctic ice-sheet variations. *Global and Planetary Change*, **45**, 9–21, doi:10.1016/j.gloplacha.2004.1009.1011

PREMOLI SILVA, I., COCCIONI, R. & MONTANARI, A. 1988. The Eocene–Oligocene boundary in the Marche–Umbria Basin (Italy). *Proceedings of the Eocene-Oligocene Boundary Meeting*, Ancona: Annibali.

RAYMO, M. E. & RUDDIMAN, W. F. 1992. Tectonic forcing of late Cenozoic climate. *Nature*, **359**, 119–122.

RINTOUL, S. R., HUGHES, C. W. & OLBERS, D. 2001. The Antarctic Circumpolar Current system. *In*: SIEDLER, G., CHURCH, J. & GOULD, J. (eds) *Ocean circulation and climate*. Academic Press, 271–302.

ROBERTS, A. P., WILSON, G. S., HARWOOD, D. M. *ET AL.* 2003. Glaciation across the Oligocene–Miocene boundary in southern McMurdo Sound, Antarctica: new chronology from the CIROS-1 drill hole. *Palaeogeography, Palaeoclimatology, Palaeoecology*, **198**, 113–130, doi:110.1016/S0031-0182 (1003)00399–00397

RÖGL, F. & STEININGER, F. 1983. Neogene Paratethys, Mediterranean and Indo-Pacific seaways: implications for the paleobiogeography of marine and terrestrial biotas. *In*: BRENCHLEY, P. J. (ed.) *Fossils and Climate*, Geological Journal Special Issue, **11**, 171–200.

ROYER, D. L., WING, S. L., BEERLING, D. J. *ET AL.* 2001. Paleobotanical evidence for near present-day levels of atmospheric CO_2 during part of the Tertiary. *Science*, **292**, 2310–2313.

RUDDIMAN, W. F. (ed.) 1997. *Tectonic Uplift and Climate Change*. Plenum Press, New York, 535pp.

RUDDIMAN, W. F. & KUTZBACH, J. E. 1989. Forcing of late Cenozoic northern hemisphere climate by plateau uplift in southeast Asia and the American southwest. *Journal of Geophysical Research*, **94**, 18409–18427.

RUDDIMAN, W. F., RAYMO, M. E., PRELL, W. L. *ET AL.* 1997. The Uplift-Climate Connection: A Synthesis. *In*: RUDDIMAN, W. F. (ed.) *Tectonic Uplift and Climate Change*. Plenum Press, New York, 471–515.

SAVIN, S. M., ABEL, L., BARRERA, E. *ET AL.* 1985. The evolution of Miocene surface and near-surface marine temperatures: Oxygen isotopic evidence. *In*: KENNETT, J. P. (ed.) *The Miocene ocean: Paleoceanography and biogeography*. Geological Society of America, Memoir, Boulder, CO, **163**, 49–81.

SCHER, H. D. & MARTIN, E. E. 2006. Timing and climate consequences of the opening of Drake Passage. *Science*, **312**, 428–430, doi:410.1126/science.1120044

SEMPERE, T., HÉRAIL, G., OLLER, J. *ET AL.* 1990. Late Oligocene–early Miocene major tectonic crisis and related basins in Bolivia. *Geology*, **18**, 946–949.

SHACKLETON, N. J. 1967. Oxygen isotope analyses and Pleistocene temperatures re-assessed. *Nature*, **215**, 15–17.

SHACKLETON, N. J. & KENNETT, J. P. 1975. Paleotemperature history of the Cenozoic and the initiation of Antarctic glaciation: Oxygen and carbon isotope analyses in DSDP Sites 277, 279, and 281. *Deep Sea Drilling Project Initial Reports*, **29**, 743–755.

SHACKLETON, N. J., HALL, M. A., RAFFI, I. *ET AL.* 2000. Astronomical calibration age for the Oligocene–Miocene boundary. *Geology*, **28/5**, 447–450.

SHACKLETON, N. J., BLACKMAN, J., ZIMMERMAN, H. *ET AL.* 1984. Oxygen isotope calibration of the onset of ice-rafting and history of glaciation in the North Atlantic region. *Nature*, **307**, 620–623.

SHEVENELL, A. E., KENNETT, J. P. & LEA, D. W. 2004. Middle Miocene Southern Ocean cooling and Antarctic cryosphere expansion. *Science*, **305**, 1766–1770.

Shipboard Scientific Party 2001. Leg 189 The Tasmanian Gateway: Cenozoic Climatic and Oceanographic Development. *In*: EXON, N. F. *ET AL.* (eds) *Proceedings of the Ocean Drilling Program, Initial Reports*, **189**, online.

SIJP, W. P. & ENGLAND, M. H. 2004. Effect of the Drake Passage throughflow on global climate. *Journal of Physical Oceanography*, **34/5**, 1254–1266.

SKINNER, L. C. & ELDERFIELD, H. 2005. Constraining ecological and biological bias in planktonic foraminiferal Mg/Ca and $d^{18}O_{cc}$: A multispecies approach to proxy calibration testing. *Paleoceanography*, **20**, 1–15, doi:10.1029/2004PA001058

SMITH, A. G. & PICKERING, K. T. 2003. Oceanic gateways as a critical factor to initiate icehouse Earth. *Journal of the Geological Society, London*, **160**, 337–340.

SQUIRES, R. L. & ADVOCATE, D. M. 1986. New Early Eocene molluscs from the Orocopia Mountains, Southern California. *Journal of Paleontology*, **60**, 851–864.

STICKLEY, C. E., BRINKHUIS, H., SCHELLENBERG, S. A. *ET AL.* 2004. Timing and nature of the deepening of the Tasmanian Gateway. *Paleoceanography*, **19**, 1–18, doi:10.1029/2004PA001022

THUNELL, R. C., QINGMIN, M., CALVERT, S. E. *ET AL.* 1992. Glacial-Holocene biogenic sedimentation patterns in the South China Sea: productivity variations and surface water pCO_2. *Paleoceanography*, **7/2**, 143–162.

TRIPATI, A., BACKMAN, J., ELDERFIELD, H. *ET AL.* 2005. Eocene bipolar glaciation associated with global carbon cycle changes. *Nature*, **436**, 341–346, doi:310.1038/nature03874

TROEDSON, A. L. & RIDING, J. B. 2002. Upper Oligocene to lowermost Miocene strata of King George Island, South Shetland Islands, Antarctica: stratigraphy, facies analysis, and implications for the glacial history of the Antarctic Peninsula. *Journal of Sedimentary Research*, **72/4**, 510–523.

VALLEJO, B. 2005. Inferring the mode of speciation in Indo-West Pacific *Conus* (Gastropoda: Conidae). *Journal of Biogeography*, **32**, 1429–1439.

VON BEYRICH, H. E. 1854. Über die Stellung der Hessischen Tertiärbildungen. *Berichte Verhandlungen der Preussischen Akademie der Wissenschaften, Berlin*, 640–666.

VON DER HEYDT, A. & DIJKSTRA, H. A. 2005. Flow reorganizations in the Panama Seaway: a cause for the demise of Miocene corals? *Geophysical Research Letters*, **32**, 1–4, doi:10.1029/2004GL020990

VONHOF, H. B., SMIT, J., BRINKHUIS, H. *ET AL.* 2000. Global cooling accelerated by early late Eocene impacts? *Geology*, **28**, 687–690.

WILLIAMS, T. & HANDWERGER, D. A. 2005. A high-resolution record of early Miocene Antarctic glacial history from ODP Site 1165, Prydz Bay. *Paleoceanography*, **20/2**, 1–17, doi:10.1029/2004PA001067

WOODRUFF, F. & SAVIN, S. M. 1989. Miocene deepwater oceanography. *Paleoceanography*, **4/1**, 87–140.

WRIGHT, J. D. & MILLER, K. G. 1993. Southern Ocean influences on Eocene to Miocene deepwater circulation. *Antarctic Research Series*, **60**, 1–25.

ZACHOS, J. C., FLOWER, B. P. & PAUL, H. A. 1997. Orbitally paced climate oscillations across the Oligocene/Miocene boundary. *Nature*, **388**, 567–570.

ZACHOS, J. C., SHACKLETON, N. J., REVENAUGH, J. S. *ET AL.* 2001. Climate response to orbital forcing across the Oligocene–Miocene boundary. *Science*, **292**, 274–278.

The origin of modern oceanic foraminiferal faunas and Neogene climate change

M. KUCERA[1] & J. SCHÖNFELD[2]

[1]Institut für Geowissenschaften, Eberhard Karls Universität Tübingen, Sigwartstraße 10, 72076 Tübingen, Germany (e-mail: michal.kucera@uni-tuebingen.de)

[2]Leibniz-Institute of Marine Sciences, IFM-GEOMAR. Wischhofstrasse 1-3, 24148 Kiel, Germany (e-mail: jschoenfeld@ifm-geomar.de)

Abstract: Planktonic and benthic foraminifera are the most significant providers of information on the state of surface and deep oceans in the past. Many foraminiferal proxies rely on the knowledge of ecological preferences of individual species and the assumption that these remained similar through time. Consequently, the applicability of such proxies is limited in time by the extent of the modern fauna. By analysing extensive datasets of species occurrences, we show that the modern oceanic foraminifer fauna originated during the Neogene. This occurred during two distinct diversification pulses: one in the Middle Miocene (17–14 Ma) and the second at the Miocene/Pliocene transition (7–4 Ma). The first diversification coincides with the time of a major change in the frequency of the dominant climate cycles during the Miocene Climatic Optimum. The environmental driver of the second diversification could be related to an increased provincialism induced by the closure of the Panama Seaway, but the exact link is not clear, particularly for the plankton. Surprisingly, major changes of ocean circulation due to the growth of Antarctic ice-sheet and closure of low-latitude seaways appear to have caused mainly extinctions. Given the age of the latest diversification and extinction pulses that shaped the modern foraminiferal fauna, we conclude that calibrated proxies based on assemblage properties should not be interpreted quantitatively in sediments older than the late Pliocene.

Much of what we know about the Cenozoic oceans comes from the study of fossil foraminifera. These microscopic protozoans occur in nearly all marine environments – from the bottom of the deepest trenches (Todo *et al.* 2005) to brine channels in Antarctic sea-ice (Dieckmann *et al.* 1991) – and their ornate shells are readily preserved in sediments. Foraminiferal species are sensitive to a range of environmental parameters and the chemical and isotopic composition of their calcite skeleton records the properties of ambient seawater and the nature of metabolic and kinetic processes that take place during calcification. This remarkable capacity to reflect the state of their habitat, coupled with the abundance of their fossils in marine cores, made foraminifera the main tool for reconstructions of past oceans and climate.

Benthic foraminifera convey information about conditions prevailing on the ocean floor. Early research on benthic foraminifera was qualitative, aimed at assessing the faunal inventory, and obtaining a general understanding of past environmental conditions. Recent studies focused on the development of quantitative tools, in particular for the reconstruction of organic matter flux to the sea floor (Herguera & Berger 1991; de Rijk *et al.* 2000), oxygen content of near-bottom waters (Kaiho 1994), and intensity of bottom currents

(Schönfeld 2002). Planktonic foraminiferal assemblages have long been used to estimate sea-surface temperature (Imbrie & Kipp 1971; Kucera *et al.* 2005), upwelling intensity (Thiede 1975; Conan *et al.* 2002) and productivity (Ivanova *et al.* 2003), and the modification of their assemblages on the sea floor has been used to reconstruct past bottom-water carbonate ion concentration (Conan *et al.* 2002). The elemental and isotopic chemistry of calcareous tests from both benthic and planktonic foraminifera forms the basis of an ever-increasing battery of proxies used to reveal the age, origin, chemical and physical properties of ocean waters (Fischer & Wefer 1999; Henderson 2002).

The application of foraminifera as palaeoceanographic proxies is based on the recognition of ecological relationships in the modern ocean and their translation into the fossil realm. This method relies on a number of assumptions (e.g. Birks 1995), the most pertinent one being the stationarity through time of the individual ecological relationships. Some foraminiferal proxies are based on determination of broad functional and ecological types (benthic vs. planktonic) or preservational state (fragmentation) and their temporal applicability is therefore less of an issue. However, most quantitative proxies rely on species-specific calibrations and the interpretation of almost all chemical

From: WILLIAMS, M., HAYWOOD, A. M., GREGORY, F. J. & SCHMIDT, D. N. (eds) *Deep-Time Perspectives on Climate Change: Marrying the Signal from Computer Models and Biological Proxies.* The Micropalaeontological Society, Special Publications. The Geological Society, London, 409–425.
1747-602X/07/$15.00 © The Micropalaeontological Society 2007.

signals in foraminifera requires knowledge of the habitat and phenology of the analysed species.

Because of circular evidence, the stability of the ecological preferences of taxa through time is often difficult to assess. Therefore, morphological similarity has been used as a first approximation of ecology and the fossil range of a given species has been taken to represent the duration of its specific, present-day ecological behaviour. This approximation is inevitably crude, as it is known that species' ecology can change with little or no shift in morphology (Norris *et al.* 1996; Kucera & Kennett 2002), but it provides a consistent and objective means to estimate the maximum range of modern environmental calibrations.

The purpose of this study is to review data on the age of extant species of planktonic and benthic foraminifera, determine the diversification history of the groups and assess its links with oceanographic events in the Cenozoic. A further aim is to produce objective guidelines for the use of foraminiferal proxies based on modern environmental calibrations in 'deep time'. This review will only deal with oceanic foraminifera; shelf seas, brackish facies, marshes and other marginal environments will not be considered.

Origin of the modern foraminiferal fauna

The number of living benthic foraminiferal species is estimated at approximately 4000 (Thies 1991), and about 45 living planktonic species are commonly recognized (Hemleben *et al.* 1989). The number of all benthic and planktonic species since the Cambrian is estimated at about 10,000 (Vickermann 1992). This estimate would suggest

that foraminiferal faunas are dominated by conservative, long-ranging taxa and that the modern fauna should have its roots in deep time. A direct assessment of the antiquity of extant foraminiferal species would require a thorough analysis of species occurrences on a global scale. However, information on the ranges of living benthic foraminifera is rather scattered. Biostratigraphic compendia typically cover a limited number of stages or a subsystem, have a regional bias and concentrate on stratigraphically important taxa. Only a few studies in the framework of the Deep Sea Drilling Project and Ocean Drilling Program go beyond that scope (see Berggren & Miller 1989; Thomas 1992; and Hayward 2001 for further discussion). Jones (1994) gives a list of stratigraphic ranges for 305 recent, mostly deep-sea benthic foraminifera and 26 planktonic species figured by Brady (1884). Less than 10% of the presumed living benthic species and only a half of the living planktonic foraminifera are covered by this compilation. Nonetheless, the dataset is methodologically and taxonomically consistent and provides the first approximation of the distribution of ages of modern oceanic foraminifera (Fig. 1).

As many as 13 (4.2%) of the living benthic species listed by Jones (1994) have originated in the Cretaceous; the oldest was reported as being of Valanginian age. The number of survivors from the Palaeogene is with 58 (19.0%) still low. Clearly, the Miocene was the crucial time for the diversification of the modern benthic fauna. More than 50% of the recent benthic species included in Jones' (1994) compilation have originated during this stage. Remarkably, almost a quarter of the modern species appear to have originated during the last 5 Ma in the Pliocene or Pleistocene.

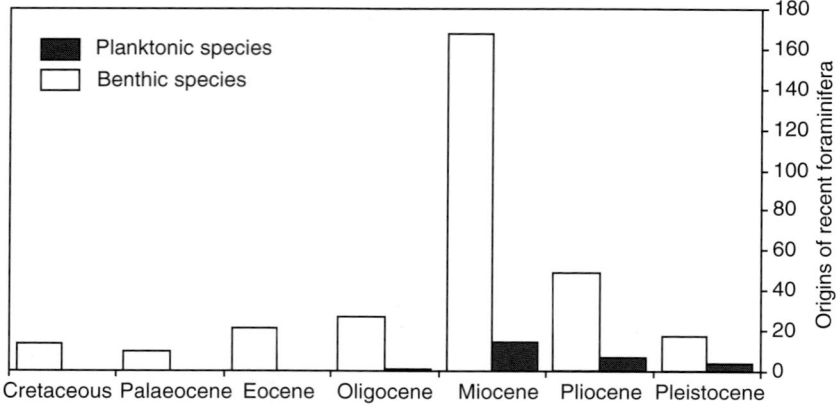

Fig. 1. Distribution of the first appearance ages of modern benthic and planktonic foraminifera included in the compilation by Jones (1994).

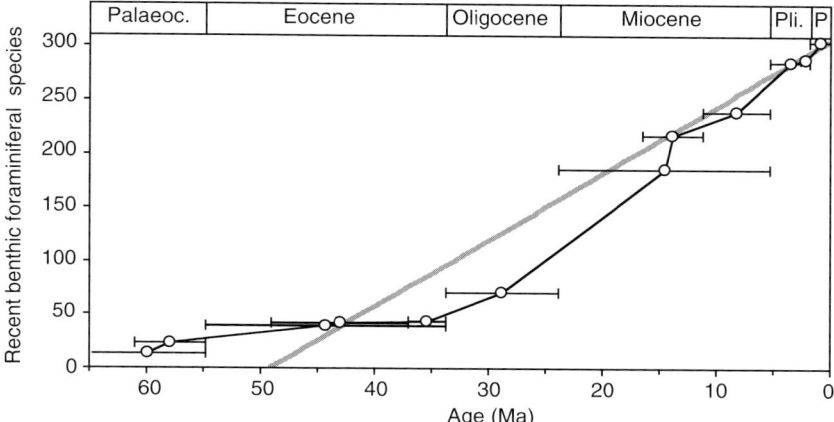

Fig. 2. Cumulative number of recent benthic foraminiferal species versus age of their origin during the Cenozoic (Jones 1994). The data are centred on the time interval depicted by the error bars. The grey line indicates a background turnover rate of 2% per Ma as estimated for deep-sea foraminifera (McKinney 1987). Ages are after Berggren *et al.* (1995).

Assuming a constant background turnover rate of 2% per million years for all deep-sea benthic foraminifera as proposed by McKinney (1987), none of the 305 extant species listed by Jones (1994) should be older than 49 Ma. The observed distribution shows a rather different pattern with a disproportionately large number of species originating in the Miocene and too many species being older than the mid-Eocene (Fig. 2). The latter would imply either that the deep-sea benthos includes a significant number of extremely long-ranging species, or that the taxonomic concept is not adequate for distinguishing modern species from their morphologically similar ancient relatives. The elevated diversification rate during the Miocene echoes the pattern seen in Figure 1: the modern benthic fauna is not the result of constant addition of new species through time. It is only since the mid-Miocene, when over 70% of all recent benthic species were already present, that the Jones' (1994) data match the inferred constant background turnover rate.

Only one of the 26 modern planktonic foraminiferal species listed by Jones (1994) originated in the Oligocene. As with the benthic foraminifera, the main diversification of planktonic species occurred during the Miocene (Fig. 1). The number of newly appearing planktonic foraminifera decreased successively through the Pliocene and Pleistocene, but still as much as 40% of the modern species appear to have originated during the last 5 Ma. The comparatively younger age of the planktonic foraminiferal fauna is not surprising. Cenozoic planktonic foraminifera evolved in two iterative radiations separated by a marked reduction in diversity and disparity at the Eocene/Oligocene

boundary. The modern fauna is largely the result of adaptive radiation among the Oligocene survivors. Norris (1991) showed that the mean longevity among all Neogene planktonic foraminifera varied between 14 Ma for species without a keel and 5.6 Ma for keeled species. Similarly short longevities were calculated by Norris (1991) for all Cretaceous and Cenozoic species, indicating a much faster turnover rate among planktonic foraminifera in comparison with benthic foraminifera.

Estimating the age of extant foraminiferal species

Although the database of Jones (1994) gives useful insight into the evolution of modern foraminifera, its scope and particularly its temporal resolution is too limited for an assessment of diversification rates during the Miocene and their links with palaeoceanographic events at that time. Therefore, in order to resolve the Neogene foraminiferal evolution in greater detail, we have analysed the range chart data of 84 benthic species (total fauna >63 µm) from five DSDP and ODP Sites from the North Atlantic and Equatorial Pacific (Thomas 1986), and the stratigraphic ranges of 83 species (index fossils only) from different drill sites and land sections in the Gulf of Mexico area (van Morkhoven *et al.* 1986). The ranges of benthic foraminifers were reported relative to planktonic foraminiferal and nannoplankton standard zones, making it possible to calibrate them to a common chronostratigraphic framework (Berggren *et al.* 1995; Lyle *et al.* 2002).

The first appearances of extant planktonic species that are stratigraphically significant are well known, but such species represent less than one-quarter of the modern fauna (Table 1). The latest comprehensive compilation of Neogene planktonic foraminifera is that by Kennett & Srinivasan (1983). The ranges of species presented there form the basis of all subsequent studies of evolutionary patterns in Neogene planktonic foraminifera (Wei & Kennett 1983, 1986; Stanley et al. 1988; Norris 1991). Yet, during the last 20 years, a large amount of new data on species occurrences of planktonic foraminifera have been collected from DSDP and ODP cores and a more precise Neogene chronology has been developed (e.g. Berggren et al. 1995). In order to benefit from this extensive dataset, we have extracted all dated occurrences of 32 species of extant planktonic foraminifera (Table 1) from the CHRONOS database (http://chronos.org; search generated by Michal Kucera using CHRONOS XML searches of the Janus database on 11 June 2006). Species prone to taphonomic bias such as the monolamellar hastigerinids and pink-pigmented *Globigerinoides ruber* were not considered, as well as species that are exceedingly rare (e.g. *Globigerinella adamsi*) or smaller than the commonly investigated size fraction (some microperforate species and all bi- and triserial forms). Obvious outliers and records indicating taxonomic uncertainty (*s.l.*, subspecies or synonyms of unclear significance) have been manually removed. This database is an extension of the Neptune compilation which has been used previously in the assessment of biochronology of Neogene planktonic foraminiferal events (Spencer-Cervato et al. 1994).

Evolution of Neogene benthic foraminifera

The analysis of the range data by Thomas (1986) and van Morkhoven et al. (1986) reveals that average innovation and disappearance rates during the last 20 Ma among the investigated deep-sea benthic foraminifera ranged from 0.96 to −3.46 species per Ma (Fig. 3). These figures agree well with estimated background turnover rate of 2% per Ma (McKinney 1987), which for 84 species corresponds to 1.7 species per Ma. The Neogene disappearance rates of bathyal foraminifera were on average higher than their innovation rates, implying a net loss of diversity at that time. A slight, long-term decline in diversity was indeed recognized in DSDP Site 573 and ODP Site 608 (Thomas 1986; her Fig. 1). In contrast, a habitat diversity analysis of neritic benthic foraminifera from temperate and tropical latitudes indicates a large diversity increase

during the Neogene. The average Fisher alpha index increased at mid-latitudes 1.4 times but 2.2 times in the tropics (Buzas et al. 2002).

The innovation and disappearance rates show strong fluctuations and distinct maxima during certain time intervals. The earliest maximum is centred in the Middle Miocene between 16 and 13 Ma. It is characterized by higher disappearance than innovation rates and shows an asymmetric pattern with higher disappearance rates towards the end of this time interval. The younger part of the Middle Miocene and most of the Late Miocene exhibit comparatively low innovation and disappearance rates. Elevated disappearance rates can also be observed between 11.5 and 10 Ma, but only in the Gulf of Mexico data (Fig. 3). The latest Miocene shows a distinct, sharp peak in turnover between 7 and 5.5 Ma. Again, the disappearance rates were higher than the innovation rates during this time interval. The highest innovation and disappearance rates in the Atlantic and Pacific data occur between 3.6 and 1.0 Ma during the late Pliocene and early Pleistocene. The early part is characterized by higher innovation rates while the later part depicts a strong disappearance rate maximum. Clearly, the Neogene benthic foraminiferal evolution is rather discontinuous. Two of the periods of enhanced evolutionary activity have been previously recognized and named the 'Mid-Miocene turnover' (Thomas 1992) and the early to mid-Pleistocene '*Stilostomella* Extinction' (Schönfeld 1996; Hayward 2001). The events at the end of the Late Miocene and in the Late Pliocene have not been described to date.

The Mid-Miocene turnover

The Mid-Miocene is characterized by a pronounced climatic optimum (17 to 14 Ma) associated with massive deposition of siliceous sediments in the Pacific (Monterrey Formation) and a positive $\delta^{13}C$ excursion (Holbourn et al. 2004). This 'Monterey Event' (Vincent & Berger 1985) was the last period in Earth's history when a complete oceanic basin may have become nearly anoxic (Smart & Ramsay 1995). It was immediately followed by a period of rapid expansion of the East Antarctic Ice-Sheet (Flower & Kennett 1993, 1995; Holbourn et al. 2005), marking the final step in the transition towards the current 'icehouse' climate (Zachos et al. 2001).

The benthic faunal change occurred between 16 and 13 Ma (Berggren & Miller 1989; Flower 1999). The faunal turnover started well before the oxygen isotopic increase, during the early Middle Miocene Climate Optimum (e.g. Thomas 1986; Thomas & Vincent 1987, 1988; Nomura et al. 1992; Holbourn

Table 1. *Ages of first occurrence of extant planktonic foraminifer species with adequate fossil record*

Species	No. of records (1)	Nr/myr	Age of first occurrence (2)	Age of first occurrence at two sites	FAD 99th percentile	Age of first consistent occurrences (3)	Reported age (4)	Note
Tropical:								
Sphaeroidinella dehiscens	800	133	5.9	5.8	5.4	6.0	5.54	
Pulleniatina obliquiloculata	544	91	6.0	5.4	5.3	6.0		
Globorotalia menardii	1755	130	13.5	13.4	12.0	13.5		
Globorotalia tumida	795	122	6.5	6.3	5.9	6.5	5.82	
Globigerinoides sacculifer	1897	84	23.6	23.2	18.6	22.5		
Globigerinoides trilobus	2154	88	24.7	24.4	22.6	24.5	23.4	
Globorotalia crassaformis	1313	202	6.5	6.4	4.7	6.5	4.31	(5)
Globigerinella siphonifera	1925	143	15.1	14.7	12.1	13.5		(5)
Globigerinoides ruber (white)	1353	85	18.2	18.2	15.1	16.0		(6)
Globoturborotalita tenella	72	29	2.5	2.2	2.2	2.5		
Globigerinoides conglobatus	1257	157	12.9	8.6	7.5	8.0		
Globoquadrina conglomerata	337	21	15.7	13.2	15.4	16.0		(6)
Globorotaloides hexagonus	931	55	20.6	16.8	16.5	17.0		(6)
Dentigloborotalia anfracta	66	10	7.1	6.5	6.5	6.5		(6)
Candeina nitida	441	42	10.3	10.2	9.9	10.5		
Subtropical:								
Neogloboquadrina dutertrei	662	88	10.5	9.4	7.4	7.5		
Globigerinella calida	203	51	5.0	4.6	4.6	4.0		(5)
Beella digitata	93	21	5.9	4.1	4.1	4.5		
Globorotalia truncatulinoides	555	185	3.0	2.8	2.8	3.0	1.92	
Globoturborotalita rubescens	343	62	12.4	8.8	11.8	5.5		(6)
Globigerina falconensis	1083	62	23.8	19.5	15.8	17.5		
Orbulina universa	2795	186	15.1	15.0	14.2	15.0	15.1	
Globorotalia hirsuta	225	50	6.9	4.7	6.8	4.5		(5)
Turborotalita humilis	337	26	13.1	12.8	12.9	13.0		(6)
Temperate:								
Globorotalia inflata	973	216	4.9	4.7	4.2	4.5		
Globorotalia scitula	2218	139	17.3	17.1	15.3	16.0		
Globigerinita glutinata	3289	113	30.3	29.0	26.9	29.0		(6)
Globigerina bulloides	2929	108	34.2	30.2	23.5	27.0		
Subpolar:								
Neogloboquadrina incompta	1864	155	12.5	11.8	10.1	12.0		
Turborotalita quinqueloba	1387	68	24.6	23.8	20.1	20.5		
Globigerinita uvula	1052	41	28.9	28.7	25.6	25.5		
Polar:								
Neogloboquadrina pachyderma	844	106	11.8	11.7	9.6	8.0		

(1) Excluding synonyms, s.l. and dubious subspecies.
(2) Excluding obvious outliers.
(3) Defined as the bottom age of the oldest of all consecutive 0.5 myr intervals with >2 records/myr.
(4) For FADs defining zonal boundaries, ages from Lyle *et al.* (2002).
(5) Chronospecies, early phylogeny not fully resolved.
(6) Early phylogeny and taxonomy not resolved.

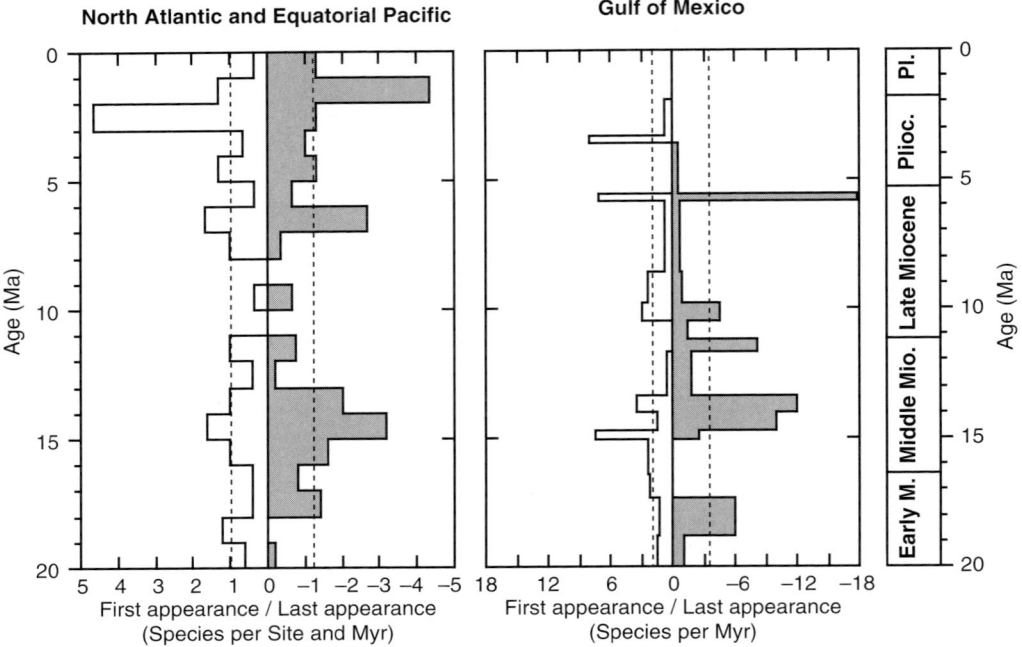

Fig. 3. Innovation rates (white, positive values) and disappearance rates (grey, negative values) of benthic foraminifera from the Atlantic and Pacific (left panel, Thomas 1986) and from the Caribbean (right panel, van Morkhoven 1986) throughout the Neogene. Dashed lines: mean values.

et al. 2005). It was initiated by a decrease in relative abundance of uniserial species such as nodosariids and pleurostomellids that had been frequent throughout the Palaeogene and Early Miocene. Miliolid species subsequently increased in abundance. The faunal change affected about 20% of all deep-water foraminifers (Boltovskoy & Boltovskoy 1988; Thomas 1986). The common extant species *Cibicidoides wuellerstorfi* and *Pyrgo murrhina* evolved at the beginning of the faunal turnover. The evolution of *C. wuellerstorfi* took place in the equatorial Pacific between 15.6 and 15.3 Ma (Thomas 1985; Woodruff & Savin 1991; Holbourn *et al.* 2004) and the species then spread into other oceans; for instance, it first appears in the equatorial Atlantic as late as around 14.3 Ma (Guasti & Iaccarino 2002). The exact significance of this pattern is difficult to assess because intergrades between the ancestor *Cibicidoides renzi* and *C. wuellerstorfi* are common where the stratigraphic ranges of the two species overlap (Holbourn & Henderson 2002). Orbitally paced fluctuations between faunal groups have been recognized throughout the period of faunal turnover (Holbourn *et al.* 2004) indicating that the deep-sea fauna was tightly connected with climatic processes affecting the surface ocean.

The highest innovation and disappearance rates occur between 14 and 15 Ma, immediately before the expansion of the East Antarctic Ice Sheet and coincide with the centre of the Monterey Event (Vincent & Berger 1985). This period was characterized by elevated benthic $\delta^{13}C$ values reflecting enhanced burial of organic carbon in deep-sea sediments (Woodruff & Savin 1985; Thomas & Vincent 1987; Flower & Kennett 1993) that may have been sufficient to draw down CO_2 from the atmosphere (Pagani *et al.* 1999; Flower 1999). The organic carbon burial was estimated to 50 to 60 atmospheric carbon masses (Berger & Vincent 1986), and the entire Pacific Ocean was as productive as all high-productivity areas in this ocean are today (Woodruff & Savin 1985).

It has been suggested that the Mid-Miocene faunal turnover was induced by changes in productivity and flux of particulate organic matter to the ocean floor (Miller & Katz 1987; Thomas & Vincent 1987). Alternatively, Woodruff & Savin (1989) linked the event to changing patterns of deep-water circulation. The deep-water exchange between the Pacific and Indian Oceans has diminished at that time, the production of the Tethyan Indian Saline Water terminated, the production of Southern Component Water increased, and the

Antarctic Circumpolar Current was established in its present configuration (Flower & Kennett 1995; Holbourn et al. 2004; Anderson & Delaney 2005; Pfuhl & McCave 2006). After a short period of extensive high productivity due to the advection of nutrient-enriched Subantarctic Mode Waters from a deep thermocline in the eastern Atlantic and western Indian Ocean between 18.9 and 17.2 Ma, upwelling concentrated at its modern locations (Sarmiento et al. 2004; Smart & Thomas 2006). In the South China Sea, the early phase of the Monterey Excursion between 16.4 and 15.5 Ma still saw a dominance of endobenthic, low-oxygen-tolerant taxa, while the time of the faunal turnover (15–14 Ma) was characterized by a mixed endo-benthic and epibenthic assemblage (Holbourn et al. 2004).

At present, there is no clear and unequivocal correlation between faunal change and oxygen or carbon isotopic events, or changes in productivity and deep-ocean ventilation during the Middle Miocene. However, spectral analysis of high-resolution isotopic records revealed that the Mid-Miocene faunal turnover coincides with a switch from low-latitude eccentricity to high-latitude obliquity-forcing of the ocean–climate system (Holbourn et al. 2004, 2005). The ensuing repetitive and comparatively severe environmental changes could have exerted a high level of stress and demanded a high migratory and adaptive ability from deep-sea foraminifera which could not be afforded by many 'old' species. The evolution of the modern species was, however, relatively slow. This conclusion is corroborated by the persist-ently higher disappearance rates and innovation rates and their asymmetric distribution along the Mid-Miocene turnover (Fig. 3).

The Terminal Miocene Event

Late Miocene bathyal foraminiferal faunas are characterized by relatively abundant lagenids, pla-nulids, cibicidoids and uvigerinids of which at least 13 species disappeared at the end of the Miocene, predominantly in the Caribbean and Atlantic. Late Miocene assemblages from the Gulf Coast and Caribbean contain a suite of locally dis-tinctive species which were not observed elsewhere. In contrast, the early Pliocene faunas are very similar in their principal composition to those seen in present-day environments, and most species have a widespread distribution (Berggren & Miller 1989).

Provincialism in the Gulf Coast and Caribbean region could have been facilitated by shoaling of the Panama Isthmus to a sill depth of about 1000 m after 12 to 13 Ma. This threshold inter-rupted the exchange of intermediate waters

between the Caribbean and Pacific. The disconnec-tion is contemporaneous with the extinctions between 12 and 11 Ma in the Caribbean (van Mor-khoven et al. 1986). The sill depth of the Panama-nian Seaway rose to 200 m and the influence of Californian benthic faunal elements diminished in the latest Miocene at 6 Ma (Duque-Caro 1990; Coates et al. 1992). The Kazakhstan Seaway closed and the Mediterranean became isolated and finally dried out during the same time (Hsü et al. 1973). The Atlantic Ocean thus completely lost deep-water connections with the Indo-Pacific realm at mid-latitudes at that time. This temporary isolation and ensuing palaeoceanographic changes could have caused extinctions of bathyal benthic foraminifera in the Caribbean and Atlantic during the Late Miocene.

The Late Pliocene Event

During the late Pliocene between 3 and 2 Ma, the bathyal benthic foraminiferal assemblages saw immigration or significant increase in abundance of Lusitanian and Boreal faunal elements such as *Uvigerina mediterranea*, *Hyalinea balthica*, *Pyrgo elongata*, *Eponides repandus* and *Cassidulina teretis*. This faunal change is associated with high innovation rates, predominantly in the North Atlan-tic (Thomas 1985; Seidenkrantz 1995). It is also visible but less constrained in the Caribbean and Mediterranean (Thomas 1985; Berggren & Miller 1989; Seidenkrantz 1995). The disappearance rate at this time is low (Fig. 3), even though the older faunas were highly diverse (ODP Site 984, M. Weinelt, Kiel, oral communication 2006). The faunal change coincides with or slightly lags the onset of significant ice rafting in the North Atlantic at 2.7 Ma (Shackleton & Hall 1984). The faunal turnover is superimposed over strong fluctuations in the abundances of the dominant species with an orbitally driven periodicity of 20 ka (Schnitker 1984). Conspicuous is the massive occurrence of *Cassidulina teretis*, which is assumed to indicate pulsed food flux (Smart et al. 1994). Such climati-cally driven fluctuations and the dominance of *Cas-sidulina teretis* point to an increased seasonality with plankton blooms and a pulsed deposition of particulate organic matter as the governing environ-mental factors behind the Late Pliocene Event.

The Mid-Pleistocene Stilostomella Extinction

This last, profound change in deep-sea benthic for-aminiferal assemblages is depicted by the highest disappearance rates for the entire Neogene, occur-ring between 1 and 2 Ma (Fig. 3). The extinction affected 98 species and thus about 20% of the total benthic foraminiferal diversity at bathyal depths at that time (Gupta 1993; Schönfeld 1996; Hayward 2001). The extinction mainly concerned foraminifera with elongate, cylindrical, uniserial or multiserial tests with small rounded, dentate, cribrate and lunate apertures, such as *Orthomorphina, Pleurostomella, Siphonodosaria* and *Stilostomella* (Hayward 2002). These species constituted 3–20% of the mid- to lower bathyal benthic foraminiferal faunas during the late Pliocene. The extinction rate was approximately 10 species per 0.1 million years, thus 50 times greater than the background turnover rate of bathyal benthic foraminifera (Kawagata *et al.* 2005).

The extinctions were preceded by a pulsed decline in abundances of the dieback groups with major declines during the onset of cold intervals and partial recoveries during warm intervals. The local disappearances of individual species were mostly diachronous but the last occurrence of any member of the *Stilostomella* extinction group was markedly consistent worldwide between 0.7 and 0.58 Ma. This time interval matches the beginning of the Milankovitch Epoch at 0.62 Ma which is characterized by eccentricity-dominated variations

in the Earth's orbit and ice-ages reoccurring every 100,000 years. It has been suggested that the *Stilostomella* group was unable to keep pace with the marked and rapid environmental changes during modern deglaciations and therefore died out (Schönfeld 1996). A correlation of *Stilostomella* abundances with low-oxygen or high organic input species (e.g. *Uvigerina, Bulimina* and *Bolivina*) suggests a specific adaptation of the former taxa to lower oxygen levels and high food supply. As the declines of stilostomellids coincide with high benthic $\delta^{13}C$ levels during glacial intervals, their final extinction may be related to unprecedented high oxygen content of glacial deep waters (Kawagata *et al.* 2005).

Neogene diversification of planktonic foraminifera

Constraining first appearance dates

An analysis of the CHRONOS/Janus database reveals that the oldest records of extant species are often associated with a significant amount of noise. This is not surprising given the fact that the

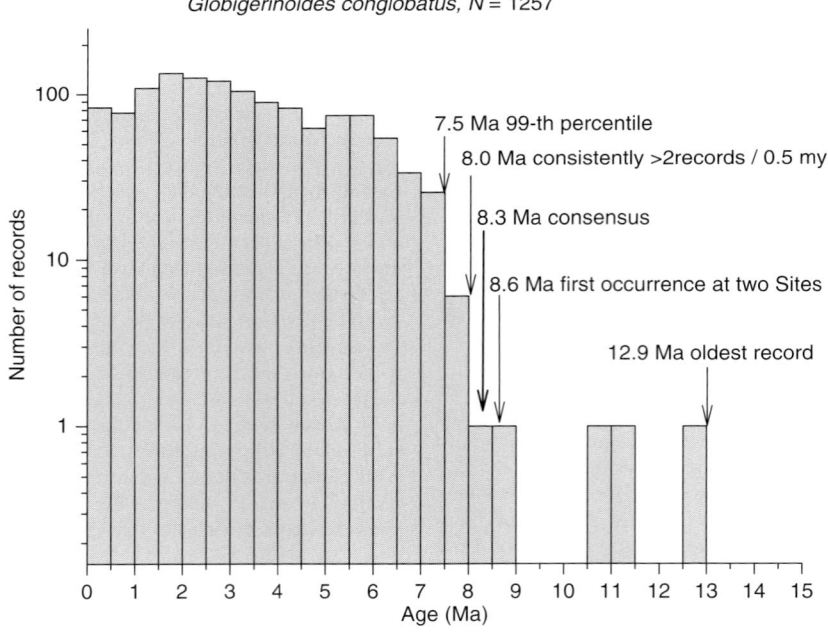

Fig. 4. Determination of the most likely age of the first occurrence of the planktonic foraminifer species *Globigerinoides conglobatus* on the basis of 1257 dated occurrences of the species in the CHRONOS Janus dataset. Consensus age is defined as the average between the onset of the consistent occurrence of the species and its first occurrence at multiple locations. The oldest record of the species is not considered reliable and the age of the 99th percentile of the occurrence appears too conservative.

data have been produced by several generations of micropalaeontologists and that they integrate over a range of sites with varying degree of recovery, sampling effort and sedimentological disturbance. A combination of taxonomic inconsistency, mis-identification and dating errors leads to a pattern where the absolute first occurrence of every species is manifested by a small number of scattered records which significantly pre-date its generally accepted first occurrence (Fig. 4). In order to circumvent this problem, we have adopted the following strategy: the occurrences of each species were sorted by age and besides the absolute first occurrence, the oldest occurrence from a second Site was recorded as well as the bottom age of the oldest 0.5 million years interval marking the onset of the consistent (> two records per 0.5 million years) occurrence of the species (Fig. 4). The rationale for this procedure is that taxonomic and dating errors are rare and unlikely to be repeated in different studies. Both the 'multiple occurrence' and the 'consistent occurrence' algorithms yielded first

appearance data highly consistent with each other (Table 1), and their average was used as a 'consensus' estimate for the FAD of each species. An alternative, objective approach defining the FAD of a species as the age of the 99th percentile sample (Fig. 4) yielded substantially younger ages (Table 1), which we felt were too conservative.

An inspection of the divergences among the ages of each species as suggested by the various algorithms (Table 1, Fig. 5) reveals a pattern with a generally high level of consistency in the recognition of the first appearance of most species accompanied by large uncertainties in the first appearance ages of a few taxa. The largest discrepancy is associated with the FAD of *Globoturborotalita rubescens*. This species is difficult to differentiate morphologically and its relationship with the ancestral form, *Globoturborotalita decoraperta*, is not well understood. The first appearance of *Neogloboquadrina pachyderma* (*sensu* Darling *et al.* 2006) is associated with the second-largest uncertainty. At present, *N. pachyderma* and *N. incompta* can be

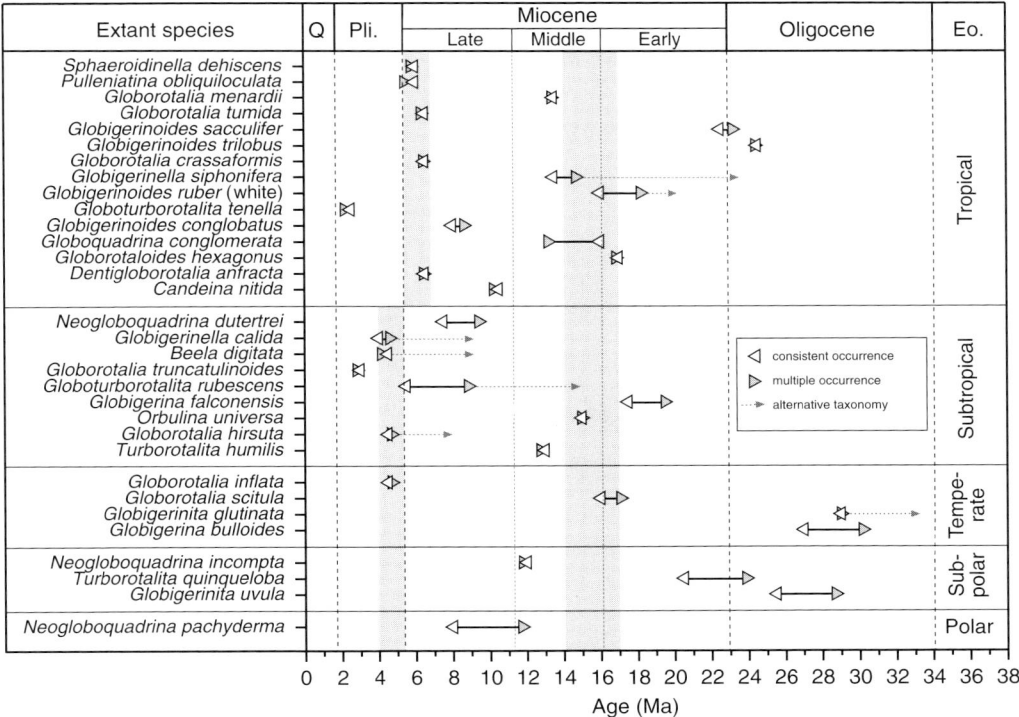

Fig. 5. Graphic representation of FAD estimates for 32 extant planktonic foraminifer species (Table 1) extracted from the CHRONOS Janus compilation using the procedure depicted in Figure 4. FAD ages of selected lineages using broader taxonomic concepts are shown by dashed lines. Species are assigned to faunal provinces on the basis of the position of their relative abundance maxima in the MARGO North Atlantic dataset (Kucera *et al.* 2005). Pacific endemics, rare and small species were added according to their accepted biogeography (Hemleben *et al.* 1989). Grey bars show periods of major diversification.

best differentiated by their opposite coiling directions, but this character has not been systematically
recorded in the past and it is not known how far
back the species can indeed be discriminated on
the basis of coiling. Therefore, the FAD of *N.
pachyderma* can only be constrained as being some-
where between the FAD of *N. incompta* and the
onset of the consistent occurrence of sinistrally
coiled populations in the lineage. Several other
species show some divergence between the FAD
estimates, pointing at taxonomic problems in the
identification of the oldest representatives of the
lineage and/or their comparative rarity.

It is appropriate to point out at this place that
many of the modern planktonic species are
defined as segments of anagenetic lineages, and
their first occurrences do not necessarily reflect cladogenetic divergence. The FADs of such species
reflect gradual (*Globorotalia inflata*: Malmgren &
Kennett 1981; *G. crassaformis*: Arnold 1983) or
abrupt (*G. tumida*: Malmgren *et al.* 1983) shifts in
the lineage morphology. The taxonomy of such
lineages is prone to subjective treatment and the
FADs may in part reflect different taxonomic
praxis, particularly in those cases where a thorough
morphometrical treatment of the lineage is not
available. This is exemplified in Figure 5 by
showing range extension of selected lineages
when an alternative taxonomic treatment of the
chronospecies within the lineage is used. Nonetheless, in many cases the FADs of the extant species
reflect well-constrained cladogenetic events
(*G. truncatulinoides*: Lazarus *et al.* 1995; *Orbulina*

universa: Pearson *et al.* 1997) or are associated with
the appearance of morphological innovations
(*Sphaeroidinella dehiscens*: Malmgren *et al.*
1996). On the whole, we believe that the FAD
data are a useful approximation of the duration of
stable morphology within species and, by a leap
of faith, their ecological preferences.

Neogene diversification rates

Although the number of planktonic species considered is substantially smaller than that of the
benthic species (Fig. 3), making it more difficult to
detect significant patterns, it is clear that the rate
of appearance of extant species was not constant
through time. Almost 50% of the modern fauna originated during two relatively short periods: nine
species originated between 4 and 7 Ma and five
species originated between 14 and 17 Ma (Figs 5 and 6). The
turnover rate during these events was 2–3 times
higher than the background rate of one species per
million years. Interestingly, the mean net turnover
rate based on the survivorship curve of the extant
species is 3%, which is only slightly higher than
the observed (Figs 2 and 3) mean turnover rates
for deep-sea benthic foraminifera.

The two peaks of diversification among
Neogene planktonic foraminifera have been documented in earlier analyses (Wei & Kennett 1986),
but their significance for the origin of the modern
fauna has not been recognized. The Middle
Miocene peak corresponds exactly to the time
of the Miocene Climatic Optimum and to the

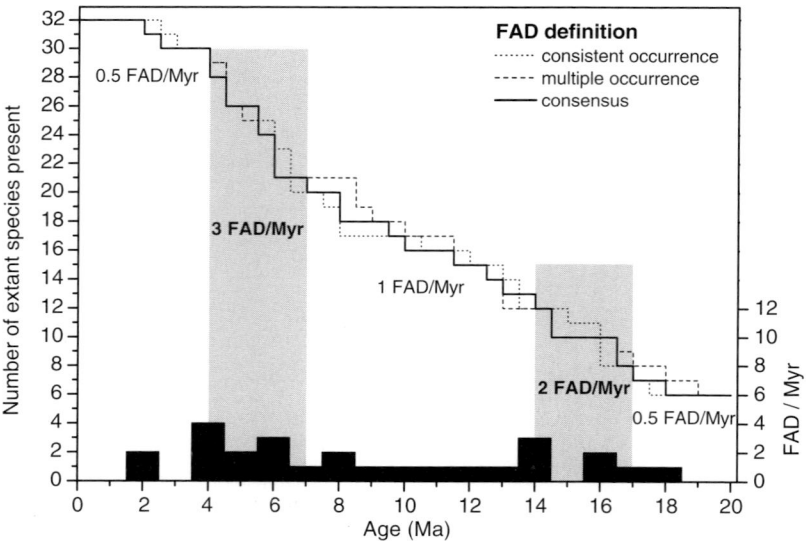

Fig. 6. Cumulative survivorship curve and rate of diversification for extant planktonic foraminifer species based
on data from Table 1. Grey bars highlight periods of elevated diversification rates (bold figures).

Mid-Miocene turnover event in benthic foraminifera (Fig. 3). The exact structure of this event remains unclear as many of the first appearances recorded during this interval are associated with considerable uncertainty (Fig. 5). Remarkably, both of the Pacific endemics considered here, *Globoquadrina conglomerata* and *Globorotaloides hexagonus*, trace their origin to this time. This may indicate that like with benthic foraminifera, plankton evolution in the Middle Miocene responded to the unique productivity regime at that time (Vincent & Berger 1985), which seems to have been particularly significant for the Pacific Ocean. Interestingly, the Middle Miocene expansion of the Antarctic ice-sheet and the presumably ensuing strengthening of thermal gradients in the oceans did not have any discernible effect on the diversification of the modern fauna. Neither do the modern species reflect the peak of diversification recorded by Wei & Kennett (1986) between 10 and 12 Ma. The majority of species that originated during this turnover peak were short-lived keeled globorotaliids (Stanley *et al.* 1988), which apparently did not contribute to the modern fauna.

Almost one-third of the modern species originated between 4 and 7 Ma (Fig. 5). This event corresponds to the highest turnover rate for the entire Neogene and, unlike the Middle Miocene event, involves species whose ages are well constrained. This event shows a clear structure with tropical species originating in the latest Miocene and subtropical and temperate species originating in the early Pliocene. This latitudinal diachrony is not seen in earlier compilations (Wei & Kennett 1986), most likely as a result of insufficient temporal resolution. The causes of this major turnover are not obvious and the reason why tropical species diversified earlier even less so. The newly evolved tropical forms are mostly chronospecies, segments of evolving anagenetic lineages (*Globorotalia tumida*: Malmgren *et al.* 1983; *G. crassaformis*: Arnold 1983; *Sphaeroidinella dehiscens*: Malmgren *et al.* 1996), and the event could thus be interpreted as a concerted development of morphological innovations within the existing fauna in response to some form of external or internal forcing. In terms of environmental forcing, the event clearly precedes the final closure of the Panama Isthmus and the subsequent onset of large-scale glacial–interglacial cycles (Haug & Tiedemann 1998). The effects of the oceanic reorganization at that time on planktonic foraminifera were not negligible, but appear to have resulted in elevated extinction rates rather than diversification (Wei & Kennett 1983, 1986). The progressive closure of the Panamanian Seaway led to some diversification and increased faunal provincialism, but these effects were temporary and the newly evolved species were short-lived (Norris 2000; Chaisson 2003). Only two extant species of planktonic foraminifera are younger than 4 Ma.

Age of extant planktonic species

Using the consensus FAD estimates (Fig. 4), the average age of the 32 analysed extant planktonic foraminifer species is 12.6 Ma. This is significantly at odds with the calculated mean longevity of planktonic foraminiferal species of 5–10 Ma, which appears to have been relatively constant during the Cenozoic and Cretaceous (Wei & Kennett 1983; Levinton & Ginzburg 1984; Norris 1991). This is even more remarkable if we realize that the extant species have not reached their actual longevity yet and that the observed mean age is thus merely an estimate of their minimum potential longevity. The most likely explanation of this pattern is that the extant species represent resilient lineages that survived the Pliocene extinction pulse, which led to the disappearance of many short-lived taxa (see also Levinton & Ginzburg 1984). The modern fauna is characterized not only by disproportionately high species ages but also by the lowest diversity since the Middle Miocene (Wei & Kennett 1986; Norris 1991) and the largest shell sizes since the late Cretaceous (Schmidt *et al.* 2004). Clearly, the modern planktonic fauna is unusual in comparison to its earlier evolutionary history. This possibly reflects the fact that the extant species inhabit a rather unique oceanic habitat characterized by lack of low-latitude seaways and dominated by large-magnitude orbitally paced glacial cycles. Such configuration has never previously occurred during the evolutionary history of the group.

Another remarkable pattern seen in the distribution of species ages (Table 1, Fig. 5) is that tropical and subtropical species are significantly ($p < 0.01$) younger (10.6 Ma; $N = 24$) than temperate to polar species (18.7 Ma; $N = 8$). This is contrary to the observation by Wei and Kennett (1983), who found no difference in the longevity of warm and cool Neogene species. Significant differences in longevity have been observed, however, between the globigerinid and globorotaliid clades (Stanley *et al.* 1988). Norris (1991) concluded that keeled globorotaliid species are on average half as long-ranging as unkeeled species and that this pattern reflects the lack of truly long-ranging taxa among the keeled forms. However, three of the eight modern cool-water species are globorotaliid and the one showing the typical globorotaliid compressed shell (*G. scitula*) is the oldest extant globorotaliid species with such morphology. Conversely, many of the young warm-water species are spinose globigerinids (*Globigerinella calida, Beela digitata, Globigerinoides conglobatus*).

Cryptic genetic divergences

The utility of species duration data for the assessment of the applicability of calibrated proxies relies on the assumption that morphology is an adequate predictor of diversity. In the last decade, this notion has been challenged, as many planktonic foraminiferal species were found to consist of several distinct genetic types, which could often be linked to different ecologies and specific geographical regions (Huber *et al.* 1997; de Vargas *et al.* 1999; Darling *et al.* 2004; Kucera & Darling 2002). Estimating the age of such cryptic genetic types is not possible. First, if the genetic difference between such types is not manifested in their mineralized skeleton, they cannot be traced in the fossil record. Secondly, the number of extinct genetic types within a lineage is unknown and molecular clock estimates of divergences between extant genetic types thus do not necessarily represent the

time of establishment of the extant form. However, molecular clock estimates of cryptic divergences can be used to indicate maximum ages of genetic types.

Because of substantial differences in fixation rates (de Vargas *et al.* 1997), only a small number of studies present molecular clock data for planktonic foraminifer species (de Vargas *et al.* 1999, 2001, 2002; Darling *et al.* 1999, 2003, 2004). All of the calibrations are inevitably associated with a large uncertainty, but the dates thus obtained are useful and informative, even if 50% error is assigned to each of the published divergences (Fig. 7). Clearly, many of the cryptic divergences are of Quaternary age, particularly among the cool-water taxa. In fact, the pre-Quaternary divergences within *Globigerina bulloides* and *Turborotalita quinqueloba* represent splits between warm-water and cold-water lineages within these species (Darling *et al.* 2003). On the other hand, three of the four analysed warm-water lineages show relatively ancient cryptic divergences (Fig. 7). This pattern is exactly opposite to the distribution of species durations between the provinces

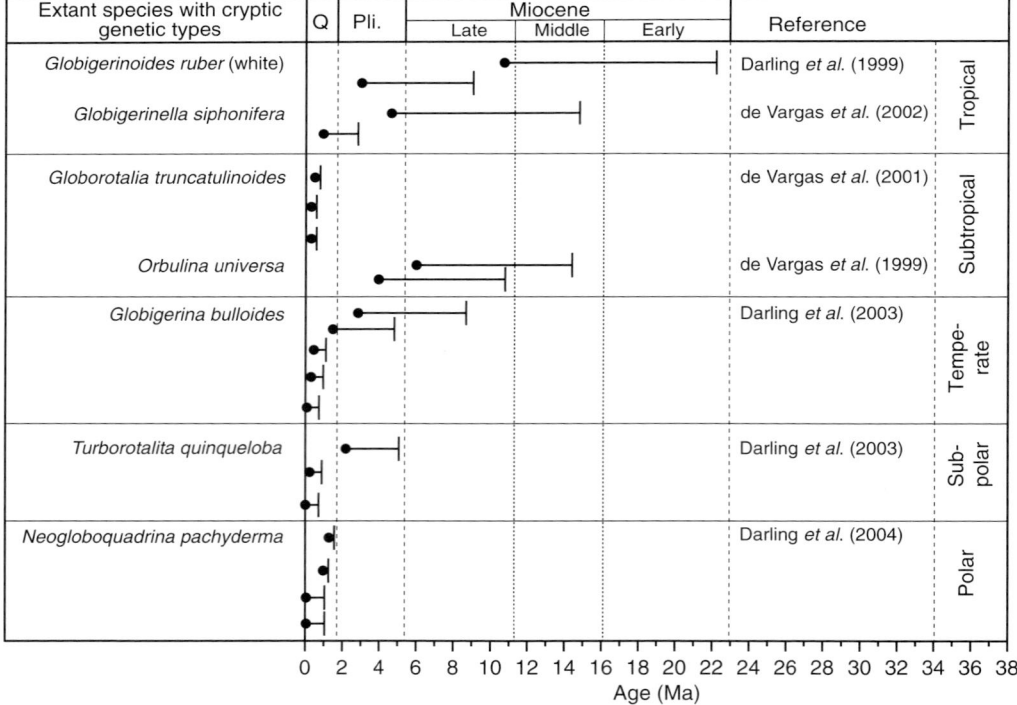

Fig. 7. Compilation of molecular-clock estimates of ages of cryptic genetic divergences among selected extant planktonic foraminifer species. Note that the data shown represent ages of nodes on the phylogenetic trees, which represent maximum ages of the respective genetic types. Lines depict the intervals of 0.5 to 1.5 times the published age of each divergence, truncated by the present day and the FAD of the species.

(Fig. 5). Although the dataset is still too small and the dating is imperfect, if this pattern proves to be of general validity it would indicate that the reason for the disproportionate longevity of cool-water taxa is the inadequacy of the species concept. Morphologically stable, long-ranging cold-water taxa appear to hide a substantial level of Quaternary diversification. Interestingly, elevated turnover rates among Quaternary high-latitude planktonic foraminifera are also indicated by data from the northeastern Pacific (Kucera & Kennett 2000) and there is good morphological evidence that the present-day polar affinity of *N. pachyderma* evolved in the Pleistocene (Huber *et al.* 2000; Kucera & Kennett 2002).

Consequences for the applicability of foraminiferal palaeoproxies

Although the modern deep-sea benthic and planktonic foraminiferal faunas are mainly of Miocene origin, the most significant origination pulse in both groups took place between 7 and 4 Ma (Figs 3 and 5) and the latest extinction pulse in both groups took place in the Late Pliocene and Pleistocene (Wei & Kennett 1986; Schönfeld 1996; Hayward 2001, 2002). Thus, whilst the use of individual species as proxies may be justified throughout their respective stratigraphic range, the use of calibrated proxies based on faunal assemblages cannot. Calculating percentages of low-oxygen indicators in benthic assemblages or polar species in planktonic assemblages in sediments older than 3 Ma is only meaningful as a qualitative indicator, and often even that may not be justified. This is particularly the case for planktonic foraminifera where the Quaternary diversification of cryptic genetic types limits the range of ecological calibrations to the (Late) Pleistocene; and attempts to quantify past ocean properties by applying assemblage-based transfer functions calibrated in the modern ocean (Dowsett 1991; Andersson 1997) must be viewed with utmost caution. Cryptic genetic diversification has been observed in shallow-water benthic foraminifera as well (e.g. Tsuchiya *et al.* 2003) and it is not unlikely that the deep-sea benthos is also more diverse than currently thought, although a case study of '*Uvigerina peregrina*' by Schweizer *et al.* (2005) revealed no consistent genetic differences among morphologically diverse forms. Notwithstanding the presence of cryptic genetic types in deep-sea benthos, the use of calibrated proxies, including those based on functional groups like BFAR (Herguera & Berger 1991) or BFOI (Kaiho 1994), is only justified in the Pleistocene and Pliocene.

Similar limitations apply for proxies that are based on the chemical composition of foraminiferal calcite. Geochemical proxies that require species-specific calibrations can only be applied with confidence as far back in time as the relationship between species ecology and biology, and the target chemical process, remain stationary. Even if the metabolic or biochemical pathway responsible for incorporation of the chemical signal remained the same, any change in calcification depth or production season will alter the meaning of the reconstructed record. Therefore, 'deep-time' geochemical proxy applications require elaborate matching and cross-calibration of the ecology of fossil species. Lastly, as discussed above, the modern planktonic foraminiferal fauna is in many respects unique, and ecological and biogeochemical processes involving modern planktonic foraminifera thus should not be uncritically extrapolated to the past.

Conclusions

An analysis of literature data and species distribution datasets reveals that the modern oceanic foraminiferal fauna originated in the Miocene; more than 70% of the extant benthic and planktonic species evolved by that time. The rate of species addition was not constant through time and the modern fauna originated largely through a series of distinct diversification pulses punctuating periods of low, constant turnover rate.

The two main turnover pulses occurred during the Middle Miocene (17–14 Ma) and the Miocene/Pliocene transition (7–4 Ma). The first diversification coincides with the Miocene Climatic Optimum, the warmest interval of the last 30 million years. This interval saw a major shift in the frequency of climatic fluctuations and we speculate that the environmental stress may have facilitated diversification of new species in both the benthos and the plankton. The environmental driver of the Miocene/Pliocene diversification is not clear. Major changes in oceanic thermal gradients associated with ice-sheet growth in the Middle Miocene, Pliocene and Pleistocene, as well as the closure of the Tethyan Seaway in the Late Miocene, seem to have caused mainly extinctions or appearance of short-lived species and as such did not contribute to the diversity ,of the extant faunas.

Modern planktonic foraminiferal species are unusually old and there is a significant difference in ages of warm- and cool-water taxa. The pattern of cool-water taxa being characterized by older species is reversed when cryptic genetic divergences within morphological species are considered, indicating that either shell morphology is not adequate to distinguish cold-water species or that these faunas are unique, characterized by extensive incipient speciation linked to Quaternary climate cyclicity.

Given the age of the latest diversification and extinction pulses that shaped the modern fauna, calibrated proxies based on assemblage properties should not be interpreted quantitatively in sediments older than the late Pliocene onset of large ice-age cycles.

We wish to express our thanks to the CHRONOS consortium and the ODP for providing the micropalaeontological community with access to an unprecedented compilation of taxa occurrence data whose potential is only beginning to be realized. The manuscript has benefited from constructive comments by Gerd Schmiedl and John Murray.

References

ANDERSSON, C. 1997. Transfer function vs. modern analog technique for estimating Pliocene sea-surface temperatures based on planktic foraminiferal data, western Equatorial Pacific Ocean. *Journal of Foraminiferal Research*, **27**, 123–132.

ANDERSON, L. D. & DELANEY, M. L. 2005. Use of multiproxy records on the Agulhas Ridge, Southern Ocean (Ocean Drilling Project Leg 177, Site 1090) to investigate sub-Antarctic hydrography from the Oligocene to the early Miocene, *Paleoceanography*, **20**, PA3011, doi:10.1029/2004PA001082.

ARNOLD, A. 1983. Phyletic evolution in the *Globorotalia crassaformis* (Galloway and Wissler) lineage: a preliminary report. *Paleobiology*, **9**, 390–397.

BERGER, W. H. & VINCENT, E. 1986. Deep-sea carbonates: Reading the carbon isotope signal. *Geologische Rundschau*, **75**, 249–269.

BERGGREN, W. A. & MILLER, K. G. 1989. Cenozoic bathyal and abyssal calcareous benthic foraminiferal zonation, *Micropaleontology*, **35**, 308–320.

BERGGREN, W. A., KENT, D. V., SWISHER, C. C., III AUBRY, M. 1995. A revised Cenozoic geochronology and chronostratigraphy. *In*: BERGGREN, W. A. *ET AL.* (eds) *Geochronology, Time Scales and Global Stratigraphic Correlation*. SEPM Special Publication, **54**, 129–212.

BIRKS, H. J. B. 1995. Quantitative palaeoenvironmental reconstructions. *In*: MADDY, D. & BREW, J. S. (eds) *Statistical Modelling of Quaternary Science Data*. Technical Guide 5, Quaternary Research Association, Cambridge, 161–254.

BOLTOVSKOY, E. & BOLTOVSKOY, D. 1988. Cenozoic deep-sea benthic foraminifera: Faunal turnovers and paleobiographic differences, *Revue de Micropaléontologie*, **31**, 67–84.

BRADY, H. B. 1884. Report on the foraminifera dredged by H.M.S. Challenger during the years 1873–1876. Report of the scientific results of the voyage of H.M.S. Challenger, 1873–1876, *Zoology*, **9**, 1–814.

BUZAS, M. A., COLLINS, L. S. & CULVER, S. J. 2002. Latitudinal difference in biodiversity caused by higher tropical rate of increase. *PNAS*, **99**, 7841–7843.

CHAISSON, W. P. 2003. Vicarious living: Pliocene menardellids between an isthmus and an ice sheet. *Geology*, **31**, 1085–1088.

COATES, A. G., JACKSON, J. B. C., COLLINS, L. S. *ET AL.* 1992. Closure of the Isthmus of Panama: the near-shore marine record of Costa Rica and Panama. *Geological Society of America Bulletin*, **104**, 814–828.

CONAN, S. M.-H., IVANOVA, E. M. & BRUMMER, G.-J. A. 2002. Quantifying carbonate dissolution and calibration of foraminiferal dissolution indices in the Somali Basin. *Marine Geology*, **182**, 325–349.

DARLING, K. F., WADE, C. M., KROON, D., LEIGH BROWN, A. J. & BIJMA, J. 1999. The diversity and distribution of modern planktonic foraminiferal small subunit ribosomal RNA genotypes and their potential as tracers of present and past ocean circulations. *Paleoceanography*, **14**, 3–12.

DARLING, K. F., KUCERA, M., WADE, C. M., VON LANGEN, P. & PAK, D. 2003. Seasonal occurrence of genetic types of planktonic foraminiferal morphospecies in the Santa Barbara Channel. *Paleoceanography*, **18**, 1032, doi:10.1029/2001PA000723.

DARLING, K. F., KUCERA, M., PUDSEY, C. J. & WADE, C. M. 2004. Molecular evidence links cryptic diversification in polar plankton to Quaternary climate dynamics. *PNAS*, **101**, 7657–7662.

DARLING, K. F., KUCERA, M., KROON, D. & WADE, C. M. 2006. A resolution for the coiling direction paradox in *Neogloboquadrina pachyderma*. *Paleoceanography*, **21**, PA2011, doi:10.1029/2005PA001189.

DE RIJK, S., JORISSEN, F. J., ROHLING, E. J. & TROELSTRA, S. R. 2000. Organic flux control on bathymetric zonation of Mediterranean benthic foraminifera. *Marine Micropaleontology*, **40**, 151–166.

DE VARGAS, C., ZANINETTI, L. HILBRECHT, H. & PAWLOWSKI, J. 1997. Phylogeny and rates of molecular evolution of planktonic foraminifera: SSU rDNA sequences compared to the fossil record. *Journal of Molecular Evolution*, **45**, 285–294.

DE VARGAS, C., NORRIS, R., ZANINETTI, L., GIBB, S. W. & PAWLOVSKI, J. 1999. Molecular evidence of cryptic speciation in planktonic foraminifers and their relation to oceanic provinces. *PNAS*, **96**, 2864–2868.

DE VARGAS, C., RENAUD, S., HILBRECHT, H. & PAWLOVSKI, J. 2001. Pleistocene adaptive radiation in *Globorotalia truncatulinoides*: genetic, morphologic, and environmental evidence. *Paleobiology*, **27**, 104–125.

DE VARGAS, C., BONZON, M., REES, N., PAWLOWSKI, J. & ZANINETTI, L. 2002. A molecular approach to biodiversity and ecology in the planktonic foraminifera *Globigerinella siphonifera* (d'Orbigny). *Marine Micropalaeontology*, **45**, 101–116.

DIECKMANN, G. S., SPINDLER, M., LANGE, M. A., ACKLEY, S. F. & EICKEN, H. 1991. Antarctic sea ice: A habitat for the foraminifer *Neogloboquadrina pachyderma*. *Journal of Foraminiferal Research*, **21**, 182–189.

DOWSETT, H. J. 1991. The development of a long-range foraminifer transfer function and application to Late Pleistocene North Atlantic climatic extremes. *Paleoceanography*, **6**, 259–273.

DUQUE-CARO, H. 1990. Neogene stratigraphy, palaeoceanography, and palaeobiology in northwest South America and the evolution of the Panama seaway. *Palaeogeography, Palaeoclimatology, Palaeoecology*, **777**, 203–234.

FISCHER, G. & WEFER, G. 1999. Use of Proxies in Paleoceanography (pp. 1–68). Berlin, Heidelberg: Springer.

FLOWER, B. P. 1999. Warming without high CO_2? *Nature*, **399**, 313–314.

FLOWER, B. P. & KENNETT, J. P. 1993. Middle Miocene ocean/climate transition: high-resolution oxygen and carbon isotopic records from DSDP Site 588A, Southwest Pacific. *Paleoceanography*, **8**, 811–843.

FLOWER, B. P. & KENNETT, J. P. 1995. Middle Miocene deepwater paleoceanography in the southwest Pacific: relations with East Antarctic Ice Sheet development. *Paleoceanography*, **10**, 1095–1112.

GUASTI, E. & IACCARINO, S. 2002. Middle Miocene paleoceanography of Equatorial Atlantic Ocean (Leg 154, Site 926): benthic foraminifera evidences. Abstract, International Society of Environmental Meiobenthology Microbiology Micropaleontology, 'The micropaleontology applied to the environmental and paleoenvironmental research', Urbino, Italy, 4–6 June 2002.

GUPTA, A. K. 1993. Biostratigraphic vs. palaeoceanographic importance of *Stilostomella lepidula* (Schwager) in the Indian Ocean. *Micropaleontology*, **39**, 47–51.

HAUG, G. & TIEDEMANN, R. 1998. Effect of the formation of the Isthmus of Panama on Atlantic Ocean thermohaline circulation. *Nature*, **393**, 673–676.

HAYWARD, B. W. 2001. Global deep-sea extinctions during the Pleistocene ice ages. *Geology*, **29**, 599–602.

HAYWARD, B. W. 2002. Late Pliocene to middle Pleistocene extinctions of deep-sea benthic foraminifera (*Stilostomella* Extinction) in the Southwest Pacific. *Journal of Foraminiferal Research*, **32**, 274–307.

HEMLEBEN, C., SPINDLER, M. & ANDERSON, O. 1989. Modern planktonic foraminifera. 363pp., Springer-Verlag, Heidelberg, Tokyo, New York.

HENDERSON, G. M. 2002. New oceanic proxies for paleoclimate. *Earth and Planetary Science Letters*, **203**, 1–13.

HERGUERA, J. C. & BERGER, W. H. 1991. Paleoproductivity from benthic foraminifera abundance: glacial to post-glacial change in west-equatorial Pacific. *Geology*, **19**, 1173–1176.

HOLBOURN, A. & HENDERSON, A. S. 2002. Re-illustration and Revised Taxonomy for Selected Deep-sea Benthic Foraminifers. *Palaeontologia Electronica*, **4**: 34pp. (http://palaeo-electronica.org/paleo/2001_2/foram/issue2_01.htm)

HOLBOURN, A., KUHNT, W. & SCHULZ, M. 2004. Orbitally paced climate variability during the middle miocene: high resolution benthic foraminiferal stable-isotope records from the Tropical Western Pacific. *Geophysical Monograph Series*, **149**, 321–337.

HOLBOURN, A., KUHNT, W., SCHULZ, M. & ERLENKEUSER, H. 2005. Impacts of orbital forcing and atmospheric carbon dioxide on Miocene ice-sheet expansion. *Nature*, **438**, 483–487.

HSÜ, K. J., RYAN, W. B. F. & CITA, M. B. 1973. Late Miocene desiccation of the Mediterranean Sea. *Nature*, **242**, 240–244.

HUBER, B. T., BIJMA, J. & DARLING, K. 1997. Cryptic speciation in the living planktonic foraminifer *Globigerinella siphoniphera* (d'Orbigny). *Paleobiology*, **23**, 33–62.

HUBER, R., MEGGERS, H., BAUMANN, K.-H., RAYMO, M. E. & HENRICH, R. 2000. Shell size variation of the planktonic foraminifer *Neogloboquadrina pachyderma* sin. in the Norwegian–Greenland Sea during the last 1.3 Myrs: implications for palaeoceanographic reconstructions. *Palaeogeography, Palaeoclimatology, Palaeoecology*, **160**, 193–212.

IMBRIE, J. & KIPP, N. G. 1971. A new micropaleontological method for quantitative paleoclimatology: application to a Late Pleistocene core. *In*: TUREKIAN, K. (ed.) *The Late Cenozoic Glacial Ages* (pp. 71–181). New Haven, tConnecticut: Yale University Press.

IVANOVA, E., SCHIEBEL, R., SINGH, A. D., SCHMIEDL, G., NIEBLER, H.-S. & HEMLEBEN, C. 2003. Primary production in the Arabian Sea during the last 135 000 years. *Palaeogeography, Palaeoclimatology, Palaeoecology*, **197**, 61–82.

JONES, R. W. 1994. The Challenger Foraminifera. Oxford University Press, Oxford, 1–149.

KAIHO, K. 1994. Benthic foraminiferal dissolved-oxygen levels in the modern ocean. *Geology*, **22**, 719–722.

KAWAGATA, S., HAYWARD, B. H., GRENFELL, H. R. & SABAA, A. 2005. Mid-Pleistocene extinction of deep-sea foraminifera in the North Atlantic Gateway (ODP sites 980 and 982). *Palaeogeography, Palaeoclimatology, Palaeoecology*, **221**, 267–291.

KENNETT, J. P. & SRINIVASAN, M. S. 1983. Neogene planktonic foraminifera. Hutchinson Ross: Stroudsburg, Pennsylvania.

KUCERA, M. & DARLING, K. F. 2002. Cryptic species of planktonic foraminifera: their effect on palaeoceanographic reconstructions. *Philosophical Transactions of the Royal Society of London*, **A360**, 695–718.

KUCERA, M. & KENNETT, J. P. 2000. Biochronology and evolutionary implications of Late Neogene California margin planktonic foraminiferal events. *Marine Micropaleontology*, **40**, 67–81.

KUCERA, M. & KENNETT, J. P. 2002. Causes and consequences of a middle Pleistocene origin of the modern planktonic foraminifer *Neogloboquadrina pachyderma* sinistral. *Geology*, **30**, 539–542.

KUCERA, M., WEINELT, MARA, KIEFER, T. ET AL. 2005. Reconstruction of sea-surface temperatures from assemblages of planktonic foraminifera: Multitechnique approach based on geographically constrained calibration datasets and its application to glacial Atlantic and Pacific Oceans. *Quaternary Science Reviews*, **24**, 951–998.

LAZARUS, D., HILBRECHT, H., SPENCER-CERVATO, C. & THIERSTEIN, H. 1995. Sympatric speciation and phyletic change in *Globorotalia truncatulinoides*. *Paleobiology*, **21**, 28–51.

LEVINTON, J. S. & GINZBURG, L. 1984. Repeatability of taxon longevity in successive foraminifera radiations and a theory of random appearance and extinction. *PNAS*, **81**, 5478–5481.

LYLE, M. W., WILSON, P. A., JANECEK, T. R. & SHIPBOARD SCIENTIFIC PARTY 2002. Explanatory Notes (Ch. 2). *Proceedings of the Ocean Drilling Programme, Initial Reports*, **199**, 1–87.

MALMGREN, B. A. & KENNETT, J. P. 1981. Phyletic gradualism in a Late Cenozoic planktonic foraminiferal lineage; DSDP Site 284, southwest Pacific. *Paleobiology*, **7**, 230–240.

MALMGREN, B. A., BERGGREN, W. A. & LOHMAN, G. P. 1983. Evidence for punctuated gradualism in the Late Neogene *Globorotalia tumida* lineage of planktonic foraminifera. *Paleobiology*, **9**, 377–389.

MALMGREN, B., KUCERA, M. & EKMAN, G. 1996. Evolutionary changes in supplementary apertural characteristics of the Late Neogene *Sphaeroidinella dehiscens* lineage (Planktonic Foraminifera). *Palaios*, **11**, 192–206.

MCKINNEY, M. L. 1987. Taxonomic selectivity and continuous variation in mass and background extinctions of marine taxa. *Nature*, **325**, 143–145.

MILLER, K. G. & KATZ, M. 1987. Oligocene to Miocene benthic foraminiferal and abyssal circulation changes in the North Atlantic. *Micropaleontology*, **33**, 97–149.

NOMURA, R., SETO, K. & NIITSUMA, N. 1992. Late Cenozoic Deep-sea Benthic Foraminiferal Changes and Isotopic Records in the Eastern Indian Ocean. *In*: TAKAYANAGI, Y. & SAITO, T. (eds.) *Studies in Benthic Foraminifera*, 227–234, Tokai University Press.

NORRIS, R. D. 1991. Biased extinctions and evolutionary trends. *Paleobiology*, **17**, 388–399.

NORRIS, R. D. 2000. Pelagic species diversity, biogeography, and evolution. *Paleobiology*, **26**, 236–258.

NORRIS, R. D., CORFIELD, R. M. & CARTLIDGE, J. 1996. What is gradualism? Cryptic speciation in globorotalid foraminifera. *Paleobiology*, **22**, 386–405.

PAGANI, M., ARTHUR, M. A. & FREEMAN, K. H. 1999. Miocene evolution of atmospheric carbon dioxide. *Paleoceanography*, **14**, 273–292.

PEARSON, P. N., SHACKLETON, N. J. & HALL, M. A. 1997. Stable isotopic evidence for the sympatric divergence of *Globigerinoides trilobus* and *Orbulina universa* (planktonic foraminifera). *Journal of the Geological Society, London*, **154**, 295–302.

PFUHL, H. A. & MCCAVE, I. N. 2006. Investigating the link between Antarctic Circumpolar Current inception and events at the Oligocene–Miocene transition. *Geophysical Research Abstracts*, 8, 06047. SRef-ID: 1607-7962/gra/EGU06-A-06047.

SARMIENTO, J. L., GRUBER, N., BRZEZINSKI, M. & DUNNE, J. P. 2004. High-latitude controls of thermocline nutrients and low latitude biological productivity. *Nature*, **427**, 56–60.

SCHMIDT, D. N., THIERSTEIN, H. R., BOLLMANN, J. & SCHIEBEL, R. 2004. Abiotic forcing of plankton evolution in the Cenozoic. *Science*, **303**, 207–210.

SCHNITKER, D. 1984. High resolution records of benthic foraminifers in the Late Neogene of the northeastern Atlantic. *Initial Reports of the Deep Sea Drilling Project*, **81**, 611–622.

SCHÖNFELD, J. 1996. The '*Stilostomella* Extinction'. Structure and dynamics of the last turnover in deep-sea benthic foraminiferal assemblages. *In*: MOGULIEVSKY, E. A. & WHATLEY, R. (eds) *Microfossils and Oceanic Environments*. Aberystwyth Press, University of Wales, 27–37.

SCHÖNFELD, J. 2002. A new benthic foraminiferal proxy for near-bottom current velocities in the Gulf of Cadiz, northeastern Atlantic Ocean. *Deep-Sea Research I*, **49**, 1853–1875.

SCHWEIZER, M., PAWLOWSKI, J., DUIJNSTEE, I. A. P., KOUWENHOVEN, T. J. & VAN DER ZWAAN, G. J. 2005. Molecular phylogeny of the foraminiferal genus *Uvigerina* based on ribosomal DNA sequences. *Marine Micropaleontology*, **57**, 51–67.

SEIDENKRANTZ, M.-S. 1995. *Cassidulina teretis* and *Cassidulina neoteretis* new species (Foraminifera): stratigraphic markers for deep sea and outer shelf areas. *Journal of Micropalaeontology*, **14**, 145–157.

SHACKLETON, N. J. & HALL, M. A. 1984. Oxygen and carbon isotope stratigraphy of Deep Sea Drilling Project Hole 552A: Plio-Pleistocene glacial history. *In*: ROBERTS, D. G., & SCHNITKER, D. ET AL. (eds) *Initial Reports of the Deep Sea Drilling Project*, **81**. Washington (US Govt. Printing Office), 599–609.

SMART, C. W. & RAMSAY, A. T. S. 1995. Benthic foraminiferal evidence for the existence of an early Miocene oxygen-depleted oceanic water mass? *Journal of the Geological Society, London*, **152**, 735–738.

SMART, C. W. & THOMAS, E. 2006. The enigma of early Miocene biserial planktic foraminifera. *Geology*, **34**, 1041–1044.

SMART, C. W., KING, S. C., GOODAY, A. J., MURRAY, J. W. & THOMAS, E. 1994. A benthic foraminiferal proxy of pulsed organic matter paleofluxes. *Marine Micropaleontology*, **23**, 89–99.

SPENCER-CERVATO, C., THIERSTEIN, H. R., LAZARUS, D. B. & BECKMANN, J.-P. 1994. How synchronous are Neogene marine plankton events? *Paleoceanography*, **9**, 739–763.

STANLEY, S. M., WETMORE, K. L. & KENNETT, J. P. 1988. Macroevolutionary differences between the two major clades of Neogene planktonic foraminifera. *Paleobiology*, **14**, 235–249.

THIEDE, J. 1975. Distribution of foraminifera in coastal waters of an upwelling area. *Nature*, **253**, 712–714.

THIES, A. 1991. Die Benthos-Foraminiferen im Europäischen Nordmeer. *Berichte des Sonderforschungsbereiches 313 University of Kiel*, **31**, 1–97.

THOMAS, E. 1985. Late Eocene to Recent deep-sea benthic foraminifers from the central equatorial Pacific Ocean. *Initial Reports of the Deep Sea Drilling Project*, **85**, 655–679.

THOMAS, E. 1986. Changes in composition of Neogene benthic foraminiferal faunas in equatorial Pacific and North Atlantic. *Palaeogeography, Palaeoclimatology, Palaeoecology*, **53**, 47–61.

THOMAS, E. 1992. Cainozoic deep-sea circulation: evidence from deep-sea benthic foraminifera. *In*: KENNETT, J. P. & WARNKKE, D. A. (eds.) *The Antarctic Paleoenvironment: a Perspective on*

Global Change. Antarctic Research Series, **56**, pp. 141–165. Washington, DC.

THOMAS, E. & VINCENT, E. 1987. Major changes in benthic foraminifera in the equatorial Pacific before the middle Miocene polar cooling. Geology, **15**, 1035–1039.

THOMAS, E. & VINCENT, E. 1988. Early to middle Miocene deep-sea benthic foraminifera in the equatorial Pacific. Rev. Paleobiol. Spec., **2**, 583–588.

TODO, Y., KITAZATO, H., HASHIMOTO, J. & GOODAY, A. J. 2005. Simple foraminifera flourish at the ocean's deepest point. Science, **307**, 689.

TSUCHIYA, M., KITAZATO, H. & PAWLOWSKI, J. 2003. Analysis of internal transcribed spacer of ribosomal DNA reveals cryptic speciation in Planoglabratella opercularis. Journal of Foraminiferal Research, **334**, 285–293.

VAN MORKHOVEN, F. P. C. M., BERGGREN, W. A. & EDWARDS, A. S. 1986. Cenozoic Cosmopolitan Deep-Water Benthic Foraminifera. Bull. Cent. Rech. Explor. —Prod. Elf-Aquitaine, **11**.

VICKERMAN, K. 1992. The diversity and ecological significance of protozoa. Biodiversity Conservation, **1**, 334–341.

VINCENT, E. & BERGER, W. H. 1985. Carbon dioxide and polar cooling in the Miocene: the Monterey hypothesis. In: SUNDQUIST, E. T. & BROECKER, W. S. (eds.) The Carbon Cycle and Atmospheric CO_2: Natural Variations Archean to Present. Geophysical Monograph, **32**, 455–468.

WEI, K.-Y. & KENNET, J. P. 1983. Nonconstant extinction rates of Neogene planktonic foraminifera. Nature, **305**, 218–220.

WEI, K.-Y. & KENNET, J. P. 1986. Taxonomic evolution of Neogene planktonic foraminifera and palaeoceanographic relations. Paleoceanography, **1**, 67–84.

WOODRUFF, F. & SAVIN, S. M. 1985. $\delta^{13}C$ values of Miocene Pacific benthic foraminifera: Correlations with sea level and productivity. Geology, **13**, 119–122.

WOODRUFF, F. & SAVIN, S. M. 1989. Miocene deep water oceanography, Paleoceanography, **4**, 87–140.

WOODRUFF, F. & SAVIN, S. M. 1991. Mid-Miocene isotope stratigraphy in the deep sea: High resolution correlations, paleoclimatic cycles, and sediment preservation. Paleoceanography, **6**, 755–806.

ZACHOS, J. C., PAGANI, M., SLOAN, L., THOMAS, E. & BILLUPS, K. 2001. Trends, rhythms, and aberrations in global climate 65 Ma to present. Science, **292**, 686–693.

The closure history of the Central American seaway: evidence from isotopes and fossils to models and molecules

D. N. SCHMIDT

Department of Earth Sciences, University of Bristol, Wills Memorial Building, Bristol, BS8 1RJ
(e-mail: d.schmidt@bristol.ac.uk)

Abstract: The rise of the Panama Isthmus was the last step in the closure of the circumtropical seaways. The closure of the Panama Isthmus had fundamental consequences for global ocean circulation, evolution of the tropical ecosystems and potentially influenced the switch to the modern 'cold house' climate mode. The Atlantic and Pacific marine ecosystems became gradually separated whereas terrestrial organisms suddenly had the means to migrate between North and South America. Combining high-resolution geochemical proxies for the closure history with data on fossil distributions and genetic data provides independent evidence on the closure history. These datasets provide new boundary conditions for Earth System models to simulate the effects of palaeoceanographic change on global climate and allow exploration of hypotheses for the Northern Hemisphere glaciation.

The Miocene was a critical time of palaeoceanographic reorganization during which the oceanic circulation became more similar to that of today. The profound changes in climate and ocean circulation have been linked in part to the final closure of the Tethyan Ocean. There is considerable debate about the respective influence of Central American (Keigwin 1982b) and Indonesian seaways (Cane & Molnar 2001) on Neogene climate change. The persistence of entirely different views of the effect and relative importance of a closure of the circumtropical seaways on global circulation reflects in part the considerable uncertainty surrounding the geological evolution of the Central American Seaway (CAS) and the Indonesian Seaway.

This review focuses on the CAS. Its closure and the emergence of the Panama Isthmus in the Miocene to Pliocene stopped the exchange of tropical Atlantic and Pacific water (Fig. 1). This caused the evolution of two different tropical ecosystems. Furthermore, high salinity warm waters of the Caribbean were transported via the Loop Current, Florida Current and Gulf Stream northwards and potentially triggered the switch to the modern 'cold house' climate mode. As an additional consequence, the eastern equatorial Pacific became the locus of upwelling of cold, high CO_2, low-alkalinity, high-nutrient water (Fig. 2). The resulting high productivity and subsequent decay of organic material in the water column caused higher oxygen utilization (Fig. 3) on the Pacific side of the Isthmus in contrast to the Caribbean. Because water vapour is transported from the Caribbean to the Pacific, the Caribbean has an increased surface salinity of around 1‰ relative to the Pacific (Figs 2 and 3). The closure split a homogeneous plankton community into two vastly different ecosystems (e.g. Jackson *et al.* 1993; Collins *et al.* 1996b; Kameo & Sato 2000) while connecting the North and South American terrestrial organisms (Whitmore & Stewart 1965; Marshall *et al.* 1982).

Both palaeontological and geochemical data describing the progressive closure history are available, though not from the same samples or sites, which makes it difficult to constrain the chronology. This review will discuss the current knowledge of the chronology of events and the consequences of the progressive closure for climate and marine/terrestrial ecosystems.

Chronology of events

This review is based on publications over the course of the last 25 years. During this time, the absolute ages for the biostratigraphic framework of these studies have changed, in parts of the record by up to 1 Ma. To facilitate comparisons between records, all biostratigraphic ages were amended to the ATS2004 timescale (Lourens *et al.* 2005).

Today, the Panama Isthmus blocks the flow of Pacific waters into the Atlantic (Fig. 1). During the Miocene, Central America consisted of a complex island-arc archipelago/peninsula (Fig. 4) with several marine corridors/basins: the San Carlos Basin (Northern Costa Rica–southern Nicaragua), the Panama Canal Basin, and the Atrato Basin (northwestern Colombia) (Duque-Caro 1990). There is significant debate about the palaeogeographic shape of the emerging Isthmus, hypothesizing of either a continuous land bridge

From: WILLIAMS, M., HAYWOOD, A. M., GREGORY, F. J. & SCHMIDT, D. N. (eds) *Deep-Time Perspectives on Climate Change: Marrying the Signal from Computer Models and Biological Proxies.* The Micropalaeontological Society, Special Publications. The Geological Society, London, 427–442.
1747-602X/07/$15.00 © The Micropalaeontological Society 2007.

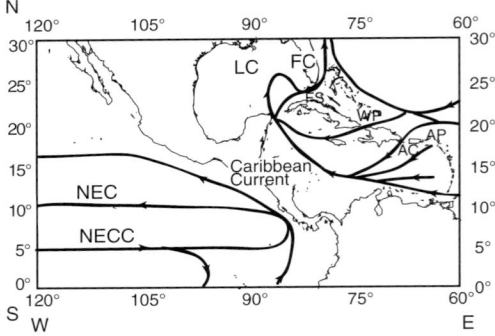

Fig. 1. Modern current systems in the Caribbean and eastern equatorial Pacific. Modified after Tomczak (1994). LC, Loop Current; FC, Florida Current; NEC, North Equatorial Current; NECC, North Equatorial Counter Current; FS, Florida Strait; WP, Windward Passage; AP, Anegada Passage; AC, Antilles Current.

with North America or an extensive island-arc system (Fig. 4). Evidence from marine sediments and volcanic records suggests an active volcanic arc during the Miocene consisting of small islands separated by shallow straits (Coates & Obando 1996). In contrast, palaeontological data (Kirby & MacFadden 2005) point towards gene flow with the North American continent and hence support the peninsula hypothesis (see Terrestrial ecosystems, Fig. 4b).

The history of the closure is not unidirectional, since there is evidence for closing and re-establishment of exchange between the Caribbean and the Pacific (Duque-Caro 1990), but stepwise and sequential at different water depths. The deep-sea record points towards gradual shoaling beginning between 15.4 and 14.7 Ma (Keller & Barron 1983) with the final closure around 3 Ma (Keigwin 1982*b*). Sedimentological evidence (Coates *et al.* 2004) indicates a shallowing of the Central American seaway from bathyal to inner neritic depth from 12.8 to 7.4 Ma. The reliability of the dates for the beginning and end of this process is hampered by two regional unconformities creating sediment gaps from about 15 to 13.5 Ma and 8.3 to 7.4 Ma (Coates *et al.* 2004). At the same time, the neodymium isotopic record in Atlantic and Pacific manganese crust starts to diverge, indicating a gradual establishment of modern circulation patterns (Burton *et al.* 1997). During the middle Miocene, the initial uplift of the Panama sill changed bottom-water circulation and sedimentation in the coastal areas of Central America (Duque-Caro 1990). The earliest time for uplift of the sill to upper middle bathyal depths (1000– 500 m) and subsequent blocking of the deep-water

flow through the seaway is suggested for 13.45 to 13.0 Ma for the Atrato Basin (Duque-Caro 1990).

Marine carbon isotope records (δ^{13}C) of deepsea benthic foraminifers from Atlantic and Pacific sediments (Fig. 5) have been interpreted to indicate proto-North Atlantic Deep Water (NADW) production starting around 13 Ma (Wright *et al.* 1991), subtly visible in Figure 5. The beginning of NADW production has been linked to a first phase of deep-water blockage in the CAS (Wright *et al.* 1991) but alternatively also to subsidence of the Greenland–Scotland Ridge (Wright & Miller 1996). Intermittent closure of the shallow-water connection was suggested for the 10.71 to 9.36 Ma interval (Roth *et al.* 2000) based on coccolithophorid assemblages (Kameo & Sato 2000). Increasing clay supply from the Amazon to the Atlantic indicates uplift phases in the Andes from 9 to 8 Ma (Dobson *et al.* 1997) and tectonic change in northern South America. At the same time (8.2 to 7.8 Ma), intensive volcanic activity is indicated by ash in deep-sea cores (Ledbetter 1982) and the first significant δ^{13}C gradients evolved around 8.3 to 8.5 Ma (Zachos *et al.* 2001). The earliest terrestrial interchange of 'waif immigrants' from Northern to South America is dated between 10.1 and 9.1 Ma. The dispersal is assumed to have happened along an island arc system before the establishment of a land bridge (Webb 1985).

Equally, neodymium isotopes in manganese crust show a gradual reduction of exchange between Pacific and northwestern Atlantic waters between 8 and 5 Ma with the largest change between 8 and 7.5 Ma (Frank *et al.* 1999). These data are corroborated by an increase of northern component waters in the Atlantic part of the Southern Ocean between 6.6 and 6 Ma (Billups 2002), indicating invigorated circulation transporting NADW far south. This dataset on the other hand also points to very low export of NADW to the South Atlantic between 8 and 6.6 Ma, which suggested a re-establishment of water exchange between the Caribbean and the Pacific in the latest Miocene. These data cannot be corroborated by faunal data, since most sediments around the seaway are eroded in the Neogene hiatus NH6 ranging from 7.8 to 6.9 Ma (Keller & Barron 1983).

During the last 4 Ma, the amount of clay transported via the Amazon increased a second time (Dobson *et al.* 1997) parallel to the uplift in the Colombian Andes at 4 Ma (Gregory-Wodzicki 2000). A marked reorganization of the surface circulation (Haug & Tiedemann 1998) and ecosystems (Keller *et al.* 1989; Chaisson & Ravelo 2000) started at 4.7 Ma. Comparisons of oxygen isotope records (δ^{18}O) of planktic, surfacedwelling foraminifers, a proxy for temperature, ice volume and salinity changes, show a

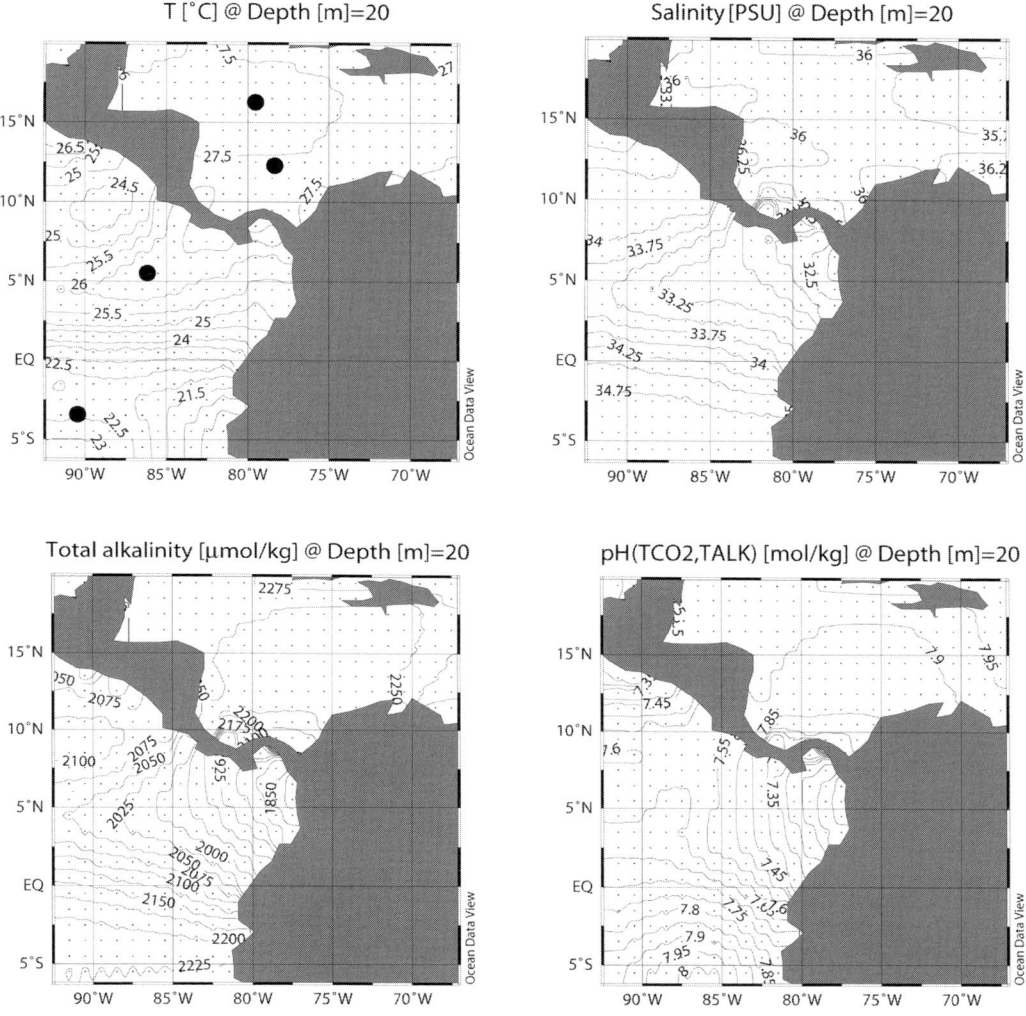

Fig. 2. Temperature, salinity and pH for the Caribbean and eastern equatorial Pacific (from Ocean Data View, Schlitzer (2006)). The modern Caribbean is warmer and more saline than the Pacific. The upwelling of cold and CO_2-rich water is reducing the pH in the Pacific. The black dots indicate the positions of the most important Ocean Drilling Program Sites (from north to south): Site 1000, Site 999, Site 1241, Site 846.

divergence of surface water isotope values starting at 4.7 Ma (Haug *et al.* 2001) indicating a shoaling of the CAS to less than 100 m. A significant sea-level lowstand period starting at 4.6 and lasting until 3.1 Ma (Haq *et al.* 1987) enhanced the shallowing of the Isthmus. Brief reversals of the isotope differential are explained by either short breaching of the Pacific waters into the Caribbean across the still-submerged sill (Haug *et al.* 2001) or by short-lasting re-openings at 3.8 and 3.4–3.3 Ma (Fig. 5) as indicated by reduced ventilation and hence preservation (Haug & Tiedemann

1998). The closure was almost complete at 3.8–3.6 Ma (Coates *et al.* 1992) though a shallow-water connection continued beyond 3.0 Ma (Coates & Obando 1996) most likely until 2.5 Ma though some records even point towards 1.9 Ma as the date for the last breach of the Isthmus (Keller *et al.* 1989). The Great Interchange between North American and South American land mammals indicates the finalization of the land bridge at 2.7 Ma (Webb 1997) when modern elevations in the Colombian Andes were reached (Gregory-Wodzicki 2000).

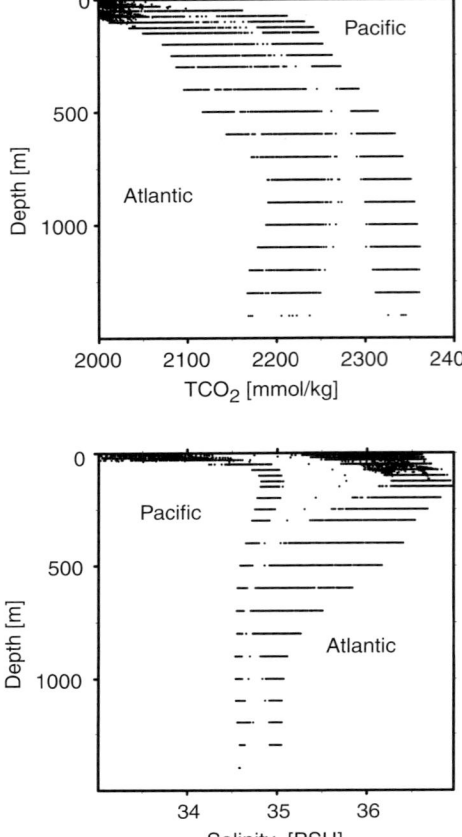

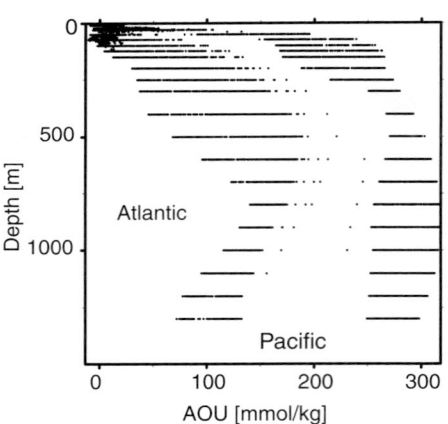

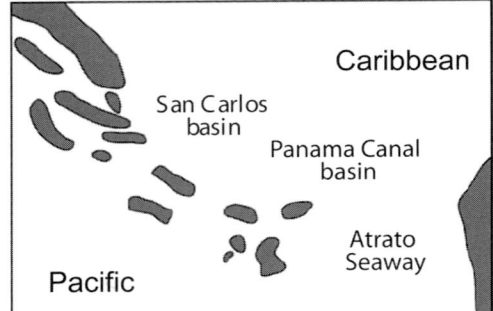

Fig. 4. Two possible palaeographic interpretations showing the distribution of land (in grey) in the Central American Seaway during the Middle Miocene: top, the Island model; bottom, the Peninsula model (modified after Kirby & MacFadden 2005).

Consequences of closure

The closure of the CAS changed the boundary conditions of the oceans and created a new state of the oceanic and atmospheric system. The Isthmus blocked the exchange of tropical water masses between the Atlantic and Pacific. The closure of the circumtropical seaways is assumed to have triggered and/or strengthened the North Atlantic Deep Water production, initiated the Caribbean Current, strengthened the Gulf Stream, and, therefore, changed the global distribution of deep-water masses, heat and salinity (Haug & Tiedemann 1998). The intensification of the circulation caused the build-up of sediment drifts in the Caribbean (Anselmetti *et al.* 2000) and later in the North Atlantic (Wold 1994).

Sedimentology

Oxygen-rich, nutrient-poor, northern-sourced deep-water masses (such as NADW) preferentially preserve carbonates, whereas southern-sourced, nutrient-rich, oxygen-poor corrosive water masses (such as the modern Antarctic Intermediate Water,

Fig. 3. Water column property profiles from both sides of the Isthmus showing the difference in total CO_2, salinity and apparent oxygen utilization [AOU] between the modern Atlantic and Pacific throughout the water column (from Ocean Data View, Schlitzer (2006)).

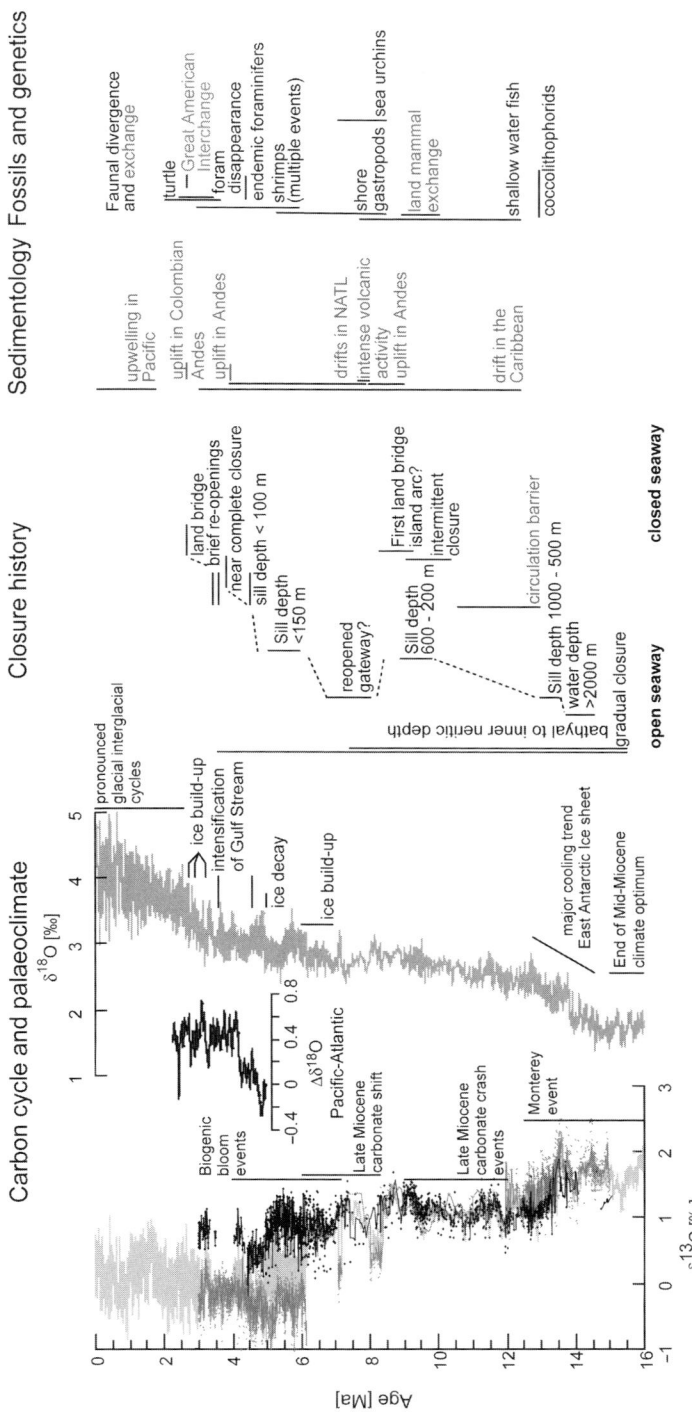

Fig. 5. Overview of the closure history of the Isthmus. Global climate change is exemplified in carbonate [δ¹³C] and oxygen isotope [δ¹⁸O] records of benthic foraminiferal carbonate (Zachos et al. 2001): major events highlighted. Oxygen isotopes represent changes in ice volume and temperature, whereas carbon isotopes represent changes in water masses and carbonate/carbon deposition history. The development of the salinity contrast between Atlantic and Pacific (Haug et al. 2001) is shown as oxygen isotope differentials between planktic foraminifers from both sides of the Isthmus [Δδ¹⁸O$_{Pacific-Atlantic}$]. The sill depth history is synthesized from the references in the text, as well as sedimentological evidence for events in relationship to the closure history, such as the build-up of drift bodies and tectonic events. The last column represents biological evidence for separation of species on both sides of the Isthmus (terrestrial animals in grey type, marine in black). Gradual changes are indicated with vertical lines, abrupt changes, e.g. separation of ecosystems or sill depth estimates, as horizontal lines. Additional information and references in the text.

AAIW) cause dissolution (Roth *et al.* 2000). The exchange of the corrosive southern water mass against the NADW in the Caribbean related to the re-establishment and intensification of the NADW production (Roth *et al.* 2000) led to better ventilation and carbonate preservation in the Caribbean (Tiedemann & Franz 1997; Haug *et al.* 2001).

The intensification of deep-water flow eroded the Bahamas carbonate platforms (Reijmer *et al.* 2002). The same process lead to the erosion of sediments from the continental margins. Their subsequent redeposition led to an increase in the growth rates of sediment drifts, for example the Santaren Drift in the Caribbean from 12.5 Ma onwards (Anselmetti *et al.* 2000) or in the North Atlantic (Feni and Gandar Drift) and Labrador Sea (Eirik Drift) from 7.9 and 3.1 Ma (Wold 1994). The drift deposition at the Santaren Drift is synchronous with the first shallowing of the seaway and the consequent increase in loop current in the Gulf of Mexico which intensified the Gulf Stream and NADW production (Anselmetti *et al.* 2000) whereas the North Atlantic Drift's deposition corresponds to the second closing phase. Three distinct reflectors in seismic profiles of the Bahamas platform at 12.2, 12.6 and 10.8 Ma corroborate the importance of erosion as a result of the current intensification (Anselmetti *et al.* 2000). A prominent erosional surface in the seismic data at 4.2 to 3.3 Ma (Anselmetti *et al.* 2000) indicates a second increase of loop current transport associated with the final closure steps of the CAS.

The exchange a Southern Ocean water mass (the AAIW is an example for a water mass) as the Caribbean deep water with less-corrosive northern component (NADW) water mass at 4.7 Ma led to an increase in deep-water ventilation and carbonate preservation in the Caribbean (Haug & Tiedemann 1998). Similarly, the increased thermohaline circulation amplified the global interbasin carbonate fractionation, resulting in enhanced preservation in the Atlantic Ocean whereas dissolution prevailed in the Pacific (Haug & Tiedemann 1998).

NADW production

Prior to the late Mid-Miocene, the carbon isotope chemistry of the deep basins was homogeneous because of the lack of low-latitude barriers to deep communication and weak to non-existent NADW production (Fig. 5). The first divergence, though very small, of $\delta^{13}C$ of deep-sea benthic foraminifers from Atlantic and Pacific sediments can be recognized in the middle Miocene around 13 Ma (Fig. 5). The first significant isotopic differential was established in the late Miocene around 8.3 to 8.5 Ma and the largest step around 6.1 Ma (Delaney 1990; Wright & Miller 1996; King *et al.*

1997). This most significant step in the deep-water divergence is synchronous with the rise of the sill to a water depth of 500–200 m (Fig. 5). This circulation-related basin-to-basin fractionation results in a more than 1.0‰ difference between the deep Pacific and Atlantic (Zachos *et al.* 2001). An increase in NADW production after 4.7 Ma and 4.2 Ma has been documented in the $\delta^{13}C$ record of benthic foraminifers (Haug *et al.* 2001) and is interpreted to reflect ventilation changes in the North Atlantic (Tiedemann & Franz 1997; Haug & Tiedemann 1998). The similarity of the stepwise changes in deep-water chemistry and planktic foraminifer isotope records suggests a link between the increased salinity contrast and deep-water circulation (Haug *et al.* 2001). During the early to middle Pliocene (4.2 to 3.7 Ma), the production of NADW is considered to have been stronger than modern values (Haug & Tiedemann 1998; Billups 2002).

The necessity of a closure of the CAS for the production of NADW is heavily debated (see section entitled 'Computer models assessing the impact of the closure of the CAS'). Though the onset of NADW production is often related to the closure of the Central American Seaway, some data suggest significant NADW production during the mid-Miocene (Keller & Barron 1983; Delaney 1990; Wright *et al.* 1992).

Surface-water properties

Surface-water properties of the ocean were influenced by the closure of the CAS in two ways: first, the water mass exchange was physically blocked; and secondly, the physicochemical properties of the surface water changed subsequently. The surface-water circulation was altered due to the initiation of the Caribbean Current (Fig. 1) and the strengthening of the Loop Current, Florida Current and Gulf Stream. The global distribution of heat and salinity changed (Haug & Tiedemann 1998), influencing the physical properties of water transported to the Nordic seas and ultimately of NADW (Lear *et al.* 2003).

The restricted exchange of surface-water masses led to the establishment of the modern Atlantic/Pacific salinity contrast. Due to intense evaporation in the Caribbean-freshwater is transported from the tropical Atlantic and Caribbean into the equatorial east Pacific (Broecker & Denton 1989). The high evaporation increases salinity in the Caribbean (Haug *et al.* 2001) and the land bridge or shallow-water strait prohibits the flow of comparatively less-saline waters back into the Caribbean. The divergence of $\delta^{18}O$ records of planktic foraminifers from both sides of the Isthmus starts at 4.7 Ma (Fig. 5) and is interpreted to reflect the establishment of the modern Pacific–Caribbean salinity contrast and formation of the Caribbean Warm Pool (Keigwin 1982a; Haug *et al.* 2001). The establishment of the modern salt asymmetry between Atlantic and Pacific is the key driver of the thermohaline circulation pumping heat northwards, which resulted in the Pliocene warm period (Haug *et al.* 2001).

The northward transport of warm waters increased zonal temperature and pressure gradients, changed the trade winds (Billups *et al.* 1999) and shifted the intertropical convergence zone (ICTZ) southwards to its modern position around 4.36 Ma (Cannariato & Ravelo 1997; Chaisson & Ravelo 1997; Billups *et al.* 1999). The southward shift of the ITCZ would have led to an intensification of the North Brazil Current which transports warm and saline water along the northern coast of Brazil towards the Caribbean and further increased the advection of warmer and more saline Caribbean surface waters (Billups *et al.* 1999).

The change in the surface-water structure is visible in the differential $\delta^{18}O$ ($\Delta\delta^{18}O$) values between shallow- and intermediate-dwelling planktic foraminifers from three tropical sites (Chaisson & Ravelo 1997, 2000). The record from the eastern equatorial Pacific shows the beginning of upwelling of cold water around 4–5 Ma whereas the western Pacific did not change at all. In the tropical Atlantic, there is also evidence for changes in surface hydrography between 5 and 4 Ma (Chaisson & Ravelo 1997; Billups *et al.* 1999). This east–west hydrographic gradient in the Pacific and its associated faunal provinciality developed between 4.4 and 4.0 Ma (Chaisson & Ravelo 2000). The upwelling of cold nutrient-rich waters in the eastern tropical Pacific changed the depth of the thermocline and shifted the locus of maximum opal accumulation, leading to a decrease in carbonate and opal accumulation around 4.5 Ma (Farrell *et al.* 1995).

The divergence between Pacific and Atlantic oxygen isotope values was not unidirectional, but shows reduced gradients around 3.3 and 2.8 Ma (Fig. 5). The reversals at 3.8 and 3.4–3.3 (Fig. 5) can be best explained by short breaching of the Pacific waters into the Caribbean across the still-submerged sill due to either sea-level change or short-lasting re-openings (Haug *et al.* 2001) They are also visible in the carbonate preservation changes indicating reduced ventilation in the Caribbean (Haug & Tiedemann 1998).

Northern Hemisphere glaciation

The ice build-up in the Northern Hemisphere is the most profound change in the Earth's climate in the Neogene (Shackleton *et al.* 1984). The link between

the closure of the CAS and Northern Hemisphere glaciation (NHG), the build-up of the Laurentide and Scandinavian ice-sheets, is a matter of debate. Hypotheses about the influence of the Panama Isthmus on NHG range from it being the cause for the onset (Berggren & Hollister 1974), a delaying factor due to changes in heat transport (Berger & Wefer 1996), or setting the preconditions of the Northern Hemisphere glaciation (Driscoll & Haug 1998; Haug & Tiedemann 1998).

The earliest evidence for glaciation of the Northern Hemisphere has recently been pushed back to the Eocene (Moran *et al.* 2006). Larger pulses of ice-rafted debris are first recognized in the Mid-Miocene, around 14.4 Ma (Thiede *et al.* 1998). Increasing evidence for ice-sheets, based on $\delta^{18}O$ records and ice-rafted debris (Lear *et al.* 2003), is linked to the initiation of cold proto-NADW between 7 and 6 Ma during the start of the second closing phase of the CAS. These ice-sheets decayed in the early Pliocene (Lear *et al.* 2003), which has been related to the increased warm-water transport to the north potentially due to the closure of the shallow-water throughflow. Between 3.2 and 2.7 Ma, a gradual change towards the current icehouse conditions is indicated by progressively heavier $\delta^{18}O$ values during the glacials (Shackleton *et al.* 1984; Tiedemann *et al.* 1994). The first major glaciation happened during MIS 110 at 2.73 Ma (Lisiecki & Raymo 2005). This large increase of ice build-up on the continents must have resulted in a significant sea-level drop and may have led to the final closure of the seaway.

The closure of the CAS has often been linked to the NHG via increased thermohaline circulation (Haug *et al.* 2001). Next to heat, the advection of warm and saline waters to northern high latitudes led to increased moisture supply, which may have increased Siberian runoff and consequently changed the freshwater balance in the Arctic Ocean (Driscoll & Haug 1998; Haug & Tiedemann 1998). The less-saline water could freeze more easily, form sea-ice, increase the Earth's albedo, and isolate the heat of the ocean from the atmosphere (Driscoll & Haug 1998).

Though this scenario is plausible, the increase in heat transport to the North Atlantic would first delay the onset of glaciation in the Northern Hemisphere instead of facilitating it (Berger & Wefer 1996). The *c.* 1.5 Ma gap between the closure of the seaway and the onset of the glaciation indicates that the closure may have been a pre-condition but not the cause. A second tectonic mechanism related to the NHG is the restriction of the Indonesian Seaway between 4 and 3 Ma and a subsequent reduction in atmospheric heat transport from the tropics to higher northern latitudes in the

Pacific region (Cane & Molnar 2001). The timing of the restriction of the Indonesia throughflow is by far not as certain as the CAS (Hall 2001), making this relationship less convincing.

Regional differences in the timing of cooling imply that global cooling was a gradual process (Ravelo *et al.* 2004) rather than the response to a single threshold or episodic event. Alternative hypotheses for the NHG are plentiful. One scenario relates the climate change to a weathering-induced reduction in the greenhouse gas CO_2 (Crowley & Berner 2001) as a result of Tibetan uplift (Raymo *et al.* 1988). However, this hypothesis is not corroborated by CO_2 proxy data (Kürschner *et al.* 1996; Pearson & Palmer 2000).

Alternatively, long-term reorganization of tropical surface-water conditions has been suggested to have influenced global climate and initiated the change from Pliocene warmth to increased build-up of ice. Wara *et al.* (2005) suggest that the temperature structure of the equatorial Pacific during the warm early to middle Pliocene has been more zonally uniform than today – similar to the modern Pacific in an El Niño state. Modern El Niño events result in a weakening of the easterly trade winds over the equatorial central Pacific (Cane & Molnar 2001), reduced upwelling in the eastern Pacific, a reduction of temperature gradients between the eastern and western Pacific and a reduced low-latitude latent heat transport (Wara *et al.* 2005). The development of a strong Walker circulation and a weakening of the Hadley circulation at *c.* 1.7 Ma coincides with the development of colder interglacials and an increase in the sensitivity of climate to solar forcing (Ravelo *et al.* 2004). In stark contrast, Rickaby & Halloran (2005) give evidence for enhanced thermocline tilt and cold upwelling in the equatorial Pacific which would suggest a prevalence of a La Niña-like state, rather than the proposed persistent warm El Niño-like conditions. Independent evidence is therefore necessary to solve this contradiction.

Another suggestion for the cause of the NHG intensification is changes in Milankovitch cyclicity, such as an increase in obliquity minima (Maslin *et al.* 1998). In contrast, Mudelsee & Raymo (2005) suggest that Milankovitch variability may have influenced Pliocene to Pleistocene climate evolution but was not the trigger.

The ice build-up may have initiated positive-feedback mechanisms such as increased vertical stratification in high latitudes at 2.7 Ma, which in turn would limit CO_2 exchange (Haug *et al.* 1999; Sigman *et al.* 2004). Haug *et al.* (2005) found evidence for increased late summer sea-surface temperatures in the high-latitude Pacific as a response to this increase in stratification while, at the same time, winter sea-surface temperatures cooled. The

warmer summer temperatures are suggested to provide increased moisture transport to northern North America (Haug *et al.* 2005). This scenario would amplify the effects of increased moisture transport in the Atlantic, such as the increase in Siberian runoff and a change to the freshwater balance of the Arctic Ocean suggested by the same authors (Driscoll & Haug 1998; Haug & Tiedemann 1998). Causes for and consequences of NHG are discussed in more detail in the paper by Ravelo *et al.* (2007).

Ecosystems and palaeobiogeography

The formation of the Isthmus of Panama allowed terrestrial faunas of North and South America to mix via the emerging land connection (Webb 1976) while, at the same time, separating the tropical marine ecosystem into separate and ecologically distinct Pacific and Caribbean communities (Vermeij & Petuch 1986; Jackson *et al.* 1993). Combining the fossil record with modern molecular data allows dating of some of the splitting events caused by the vicarious living and provides new fascinating evidence for stepwise separation of the marine ecosystems. The phylogenetic relations of species in cosmopolitan genera can point towards barriers for gene flow. Using the molecular clock approach, these divergence times can be dated and provide additional information for the closure history of the CAS.

The palaeogeographic shape of this land connection – either a continuous land bridge with North America or an extensive island-arc system (Fig. 4) – is heavily debated. Evidence from marine sediments and volcanic records suggests an arc during the Miocene consisting of small islands separated by shallow straits (Coates & Obando 1996). Kirby & MacFadden (2005) used body size to determine the palaeogeography of the connection between the Americas. Their work is based on the assumption that the separated population, without exchange and immigration from other populations and the absence of large predators, has effects on body size: the so-called island effect. The island effect leads to gigantism (in small mammals) and dwarfism (in large mammals). Kirby & MacFadden (2005) determined that the body size of land mammals from Central America did not significantly differ from their North America counterparts, which points towards gene flow with the North American continent and hence is evidence for the peninsula hypothesis (Kirby & MacFadden 2005).

Terrestrial ecosystems

The closure of the CAS and the emergence of the Isthmus allowed exchange of terrestrial species between North and South America in several immigration events. The mammalian fossil record suggests a continuous connection of Panama with North America in the Miocene from 19–16 Ma onwards (MacFadden 2005). The earliest terrestrial interchange of 'waif immigrants' (racoons) from North to South America is dated between 10.1 and 9.1 Ma. The dispersal is assumed to have happened along an island arc system before the establishment of a land bridge (Webb 1985). A similar migration path has been suggested for the New World rats and ground sloths arriving in South America in the early Pliocene (Marshall *et al.* 1982) at a time where the sill depth was less than 100 m.

The biogeographic distribution and the genetic divergence in a group of freshwater and marsh fishes (synbranchid eels) in South and Central America indicated that the South American clade was present in Central America before the final closure of the Isthmus between 7.7 and 12.4 Ma (Perdices *et al.* 2005). This is time-equivalent with the intermittent closure or rather shallow-water connections. These fish are very salinity-tolerant and hence were present before primary freshwater fish could colonize Central America (Perdices *et al.* 2005).

The most significant terrestrial migration, the Great American Interchange around 2.8–2.6 Ma (Lundelius 1987), resulted in exchange of mammals in both directions (Marshall *et al.* 1982) indicating a land bridge connecting North and South America. During the Pleistocene, the exchange became asymmetrical with more North American mammals occurring in South America than *vice versa* (Marshall 1988), which is assumed to be a result of increasing climatic variability rather than a tectonic feature (Webb 1991). The total number of mammalian genera in South America increased markedly as a result of North American immigrants.

Despite the great dispersal potential of seabirds, the Isthmus of Panama is also an effective barrier to gene flow in several pantropical species such as terns (Avise *et al.* 2000) and boobies (Schreiber *et al.* 2002). These seabirds avoid flying over land (Steeves *et al.* 2003) and hence the 35 km-wide Isthmus, with an interior dominated by steep mountains, provides a barrier which these birds cannot pass. Interestingly, the divergence in masked booby (*Sula dactylatra*) populations is much younger than the closure with a divergence time of *c.* 640 kyr (Steeves *et al.* 2005) and might be more related to the uplift history of Central America than with the formation of the land bridge.

The history of vegetation shows that the contact between Central America and South America decreased the extent of endemism in both areas (Burnham 1999). Palynoflora lost their distinct differences between North and South America at

c. 4 Ma, though the age is not very well constrained (Graham 1992).

Marine ecosystem

The closure of the circumtropical seaways caused a tectonic constellation which is a novelty in the evolutionary history of all plankton groups and significantly influenced their environment and dispersal possibilities. Speciation in the ocean is hard to determine, mainly because of the dispersal abilities of marine organisms, particularly those species with planktic larvae. Consequently, this distinct geographical event provided classic textbook examples for allopatric speciation (speciation through geographical separation). Different species responded to these changes in a different manner, and thus become separated at different times.

The two ecosystems on both sides of the Isthmus changed profoundly. The closure thus split a homogeneous plankton community into two vastly different ecosystems. Studying the faunal and floral evolution provides a unique opportunity to investigate the effects of changes in temperature, stratification and ocean acidification on ecosystems. Though the evolutionary consequences of the closure of the seaway are well studied, the ecological importance of this unrivalled local case study for climatically induced restructuring of the planktic ecosystem has not yet been exploited.

The closure increased thermohaline circulation (Haug & Tiedemann 1998) and in turn enhanced stratification (Billups *et al.* 1999). This in consequence has changed the density of the surface water, changed the vertical water mass structure and has caused an elevation of the thermocline into the photic zone (Billups *et al.* 1999). Consequently, a distinct planktic foraminiferal faunal difference developed from 7–6.5 Ma between the eastern and western equatorial Pacific (Chaisson & Ravelo 2000). In the eastern Pacific, upwelling of deep water cooled the subsurface waters, increased the temperature gradients in the upper photic zone and hence led to elevated opal deposition (Cortese *et al.* 2004). The higher stratification in the tropics increased the number of available niches in the upper water column (Schmidt *et al.* 2004*a*). By 4.2 Ma, the closure of the Panamanian Isthmus appears to have facilitated the evolution of unprecedentedly large sizes in planktic foraminifer assemblages (Schmidt *et al.* 2004*c*) with the largest size in the western tropical Pacific and Atlantic (Schmidt *et al.* 2004*b*).

The first steps towards closure

Coccolithophore assemblages in deep-sea cores on both sides of the Isthmus were identical from

16.21 to 13.57 Ma (Kameo & Sato 2000). The divergence, starting at 13.57 Ma, led to an increasingly different assemblage with the next step between 10.71 and 9.36 Ma, suggesting an intermittent first closure of this Isthmus. A gradual decrease of the ratio of planktic to benthic foraminifers points towards a shallowing of the seaway by 14.7 to 13.5 Ma (Duque-Caro 1990). The faunal composition of benthic foraminifers and their poor preservation indicates nutrient-rich, oxygen-poor water masses. The benthic fauna in the Atrato Basin indicated a water depth between 500 and 1000 m around 13.45 Ma. By 13.0 to 11.64 Ma, the sill is assumed to have deflected the intermediate water currents and caused a circulation barrier (Duque-Caro 1990). Another shallowing step is indicated by increased abundances of the *Uvigerina–Valvulineria* assemblages, pointing towards an upper bathyal depth between 10.4 and 9.9 Ma. Clear differences between Pacific and Caribbean foraminiferal fauna at this time are interpreted as evidence for an interoceanic biogeographic barrier (Duque-Caro 1990) due to intensification of the California Current. Planktic foraminiferal assemblages change towards cool low-diversity faunas (Keller & Barron 1983).

Caribbean reef corals started to diversify rapidly between *c*. 12 Ma until 8 Ma (Collins *et al.* 1996a). Tropical rocky shore gastropods from both sides of the Isthmus show a high amount of divergence between 8.5 and 5.3 Ma, pointing towards isolation (Collins *et al.* 1996c) which can also be seen in the genetic divergence of sea-urchins between 8.5 and 6.1 Ma (Lessios *et al.* 2001). The similarity between the Pacific and Caribbean benthic foraminiferal assemblages indicates a cessation of the oceanic barrier and a brief re-establishment of the water exchange in the Late Miocene. The shallow-water benthic foraminiferal community in the Caribbean changed from species preferring siliciclastic habitats to a carbonate-associated assemblage around 7.1 Ma exploiting the new reef habitats. The largest change was between 5.3 and 3.5 Ma (Collins *et al.* 1996a). In the latest Miocene and Pliocene, seagrass-associated bryozoans underwent dramatic extinction (Cheetham & Jackson 1996). The diversification of reefs points towards oligotrophic conditions and warm-water temperatures which favour corals and produce carbonate sediments, the new habitats which the benthic foraminifers became adapted to.

The final closure

An increasing abundance of species, today adapted to higher-salinity conditions, in the Caribbean foraminiferal fauna by 4.5 Ma indicates an increase in salinity (Keller *et al.* 1989). The first appearance of endemic species of planktic foraminifers (*Menardella exilis*) at 4.45 Ma indicates the different evolutionary histories in both ocean basins. Though there is an increase in endemism, diversity starts to decrease around 4.5 Ma (Chaisson & Ravelo 2000). Some species were briefly absent or very rare in one ocean basin but not in the other. For example, *Menardella miocenia*, first appearing at 3.48 Ma, rarely migrated through the Gulf of Panama (Kennett & Srinivasan 1983). *Globorotalia tumida* was absent between 3.7 and 2.1 Ma at Ceara Rise in the western equatorial Atlantic (Chaisson & Pearson 1997) whereas in the western Pacific *G. tumida* showed a marked increase in relative abundance in the Pliocene (Chaisson & Leckie 1993). During the same time interval (3.48 to 2.28 Ma), the pulleniatinids disappeared in the Atlantic (Bolli 1971) whereas they were continuously present in the Pacific (Chaisson & Leckie 1993). By 3 Ma, finely perforate menardellids were dominant in the Atlantic and rare to absent in the Pacific (Kennett & Srinivasan 1983; Chaisson 2003). The occurrences of all these species indicate that the necessary water depth for migration from the Pacific to the Atlantic was not available anymore around 3.7 Ma and definitely by 3.4 Ma. Divergence events in sea-urchins (benthic organisms with planktic larvae) from both sides of the Isthmus, for example, *Diadema mexicanum* from *D. antillarum*, at 3.1 Ma point towards the restriction of larval exchange (Lessios *et al.* 2001).

By 2.5 Ma, permanent divergence of faunal provinces between the Pacific and the Caribbean are interpreted as the closure of the gateway. The further abundance increase of *G. ruber* points towards even higher salinity in the Caribbean (Keller *et al.* 1989) in agreement with changes in nannofossils (Gartner *et al.* 1987) while a reduction in surface dwellers in the Pacific indicates increased upwelling. A decrease in the diversity of Caribbean reef corals starting at 4 Ma (Collins *et al.* 1996a) might be related to the high salinity. The biogeographic migration pattern of planktic species indicates a blocked exchange between the Atlantic and Pacific which forced *Globorotalia truncatulinoides*, for example, to colonize the oceans from the Pacific via the Indic and the Agulhas retroflection into the Atlantic (Spencer-Cervato & Thierstein 1997). New data (Schmidt submitted) points to an earlier first appearance in the South Atlantic (2.1 Ma) than in the tropical Atlantic and hence corroborates a migration around South Africa.

Caribbean gastropod (*Strombina* group) diversity peaked in the early Pliocene and declined thereafter from 23 to 3 species, whereas eastern Pacific diversity is highest today with 33 species (Jackson *et al.* 1993) indicating the profound ecological changes in the shallow-water habitats. Shallow-water molluscs, living on the inner shelves, indicate a complete closure by 3.7 Ma (Coates *et al.* 1992). The occurrence of similar pairs of late Pliocene gastropods (2.6 to 1.8 Ma) on both sides of the Isthmus, however, suggests some interchange may still have been possible. The increased separation of the habitats led to disparate histories in gastropod evolution with the Caribbean extinction (1.8–1.6 Ma) being a time of explosive Pacific first appearances (Jackson *et al.* 1993). The extinction did not just affect molluscs but also reef corals throughout the Caribbean. The diversity of Caribbean reef corals decreased dramatically in the Pleistocene (Collins *et al.* 1996*a*). Marked alteration in the trophic structures in shelf ecosystems since the late Pliocene (2.6–1.8 Ma) led to an increase in reef dwellers and a decrease in suspension feeders in the Caribbean. These food-web changes strongly support the hypothesis that declining nutrient supply had an increasing impact on regional macroecology, culminating in a faunal turnover (Todd *et al.* 2002).

Closure history of nekton

The Panama Isthmus is an important barrier for marine nekton such as turtles and fish (Bowen *et al.* 1997; Lessios *et al.* 2001, 2003). The closing of the seaway led to allopatric separation and consequent divergence. Many sister taxa in the Caribbean and eastern Pacific became isolated during the initial shoaling rather than in final closure (Knowlton *et al.* 1993; Cronin & Dowsett 1996), e.g. the reef fish *Ophioblennius* of about 7 million years ago (Muss *et al.* 2001). The reorganization of Atlantic circulation involved an initial period of high energy in major current systems (Haug & Tiedemann 1998) which some species exploited, e.g. the *Ophioblennius atlanticus* larvae were able to traverse the mid-Atlantic barrier via the Equatorial Undercurrent. Once regularized circulation patterns became established, larvae of *O. atlanticus* were not able to bridge the Amazon–Orinoco or mid-Atlantic barriers anymore (Muss *et al.* 2001).

The genetic divergence between Kemp's ridley sea turtle *Lepidochelys kempi* and the olive ridley turtle *L. olivacea* indicates a restriction of the seaway to a water depth too shallow for these near-shore and inshore organisms to cross between 2.5 and 3.5 Ma. The snapping shrimp genus *Alpheus* shows a staggered separation rather than one event (Knowlton *et al.* 1993). Though some divergences within *Alpheus* are assumed to be synchronous with the final steps of the closure around 3 Ma, several events are significantly earlier (6.1 to 4.4 and 6.3 to 4 Ma). The oldest split in the genus *Alpheus* is dated for the time interval 9.1 to 6.8 Ma in a species with the adaptation to the relatively deepest habitat in the intertidal zone. This is puzzling though, since the depth of the Atrato Strait (Fig. 2) has been suggested to be at least 150 m (Duque-Caro 1990) which would have allowed larvae to cross from one ocean basin to the other as indicated by the gene flow between several taxa. The smallest trans-isthmian difference in this group is measured in those species that live in mangroves, an environment that was likely to be the last habitat to be separated by the Isthmus (Knowlton & Weight 1998).

Evidence for breaching of the Isthmus

Interestingly, several species point towards gene flow even in the Pleistocene which has been interpreted as evidence for breaching of the Panama Isthmus. Evidence from planktic foraminifers (Cronin & Dowsett 1996) points towards two breaching events during high sea-level stands, the most recent occurring at 2 Ma. Keller *et al.* (1989) also suggested the existence of seawater connections between the Caribbean and eastern Pacific that lasted until 1.8 Ma, based on foraminiferal data. A short-lived breach of the Isthmus would be expected to affect a small number of species, which (because of their ecological attributes or due to pure chance) either re-established genetic connections with their geminates, or replaced the resident population.

Computer models assessing the impact of the closure of the CAS

Computer models were used to identify consequences of the closure of the Panama Isthmus, specifically its influence on the production of North Atlantic Deep Water (NADW) and the glaciation of the Northern Hemisphere. Using the Hamburg Ocean general circulation model (GCM), Maier-Reimer *et al.* (1990) suggested that the closure was the prerequisite for development of the modern thermohaline circulation. Their model suggests that an open seaway would inhibit NADW production. Similarly, Mikolajewicz & Crowley's (1997) model results in severely reduced NADW production when the CAS was partially open due to a flow of relatively fresh Pacific water into the North Atlantic (Mikolajewicz *et al.* 1993). Nisancioglu *et al.* (2003) agree that the shoaling of the CAS had a significant impact on global circulation, but they deduced from their model that significant amounts of NADW formed even though the CAS was open. They calculated that, if the sill depth is larger than 1000 m, most

of the NADW passes as a westward jet through the
CAS into the Pacific, greatly reducing NADW
transport to the South Atlantic. When the sill
depth is shallower than the depth of the NADW
outflow, the NADW flow to the Pacific Ocean
would be blocked and transport to the South
Atlantic increased. Several independent datasets
strongly suggest that NADW was formed before
the closure of the seaway (e.g. Keller & Barron
1983; Burton *et al.* 1997) and increased in the
Miocene (e.g. Delaney 1990) to modern values in
the Pliocene (Haug & Tiedemann 1998) and therefore this model is corroborated. Interestingly,
Nisancioglu *et al.* (2003) suggest that the similarity
of δ^{13}C and Cd/Ca data, traditionally interpreted as
an indication for the absence of NADW (Delaney
1990; Wright *et al.* 1992), should be interpreted as
the presence of the same water mass – young
NADW – in the Atlantic and western Pacific. The
divergence of the records in the late Miocene indicates the aging of the NADW on its way to the
Pacific as a result of the closure of the Isthmus
(Nisancioglu *et al.* 2003).

Klocker *et al.* (2005) assessed the importance of
the closure of the CAS for the NHG with an
intermediate-complexity coupled atmosphere–
ocean model. They found that the closure of the
Isthmus lead to an increase in sea-surface temperature and salinity in the northern North Atlantic
similar to proxy results but not to the predicted
increase in snow accumulation in northern latitudes
and the initiation of the Northern Hemisphere glaciations (Klocker *et al.* 2005). In contrast, the
increased heat transport results in higher temperatures and a retreat of perennial snow cover in line
with the arguments of Berger & Wefer (1996).
Using the same model, Prange & Schulz (2004)
suggested that the closure of the CAS caused an
intensification of upwelling of southwest Africa,
similar to the sedimentary record (Marlow *et al.*
2000), and an intensified equatorial upwelling
system, while the northwest African upwelling
around the Canary Islands would be reduced. This
change is caused by altered sea-surface temperature
patterns influencing atmospheric dynamics and
hence wind stress (Prange & Schulz 2004).

Summary and outlook

The Isthmus is the last step in the closure of the circumtropical seaways, resulting in the separate evolution of the Caribbean/Atlantic and the Pacific
Ocean. The closure of the Panama Isthmus had fundamental consequences for global ocean circulation, evolution of the tropical ecosystems and
potentially the switch to the modern 'cold house'
climate mode. It influenced global ocean circulation

and climate. Although the connection between
NHG and closure of the CAS is plausible, there
are ongoing arguments regarding the causal
relationship between Panamanian tectonic and climatic and ecological changes (Schmittner *et al.*
2004). Available studies have generated single
proxy records, focusing mostly on the last part of
the closure history, some of them at large distances
from the Isthmus.

The shoaling of the Isthmus started around
14.7 Ma with the first consequences for deep-water
flow between 13.5 and 13 Ma when the sill depth
was around 1000 to 500 m. This changed bottom-
water circulation led to the formation of drift
deposits in the Atlantic. Although intermittent
closure has been suggested for the time interval
from 10.7 to 9.4 Ma, the data can also be explained
by an oceanic barrier such as the intensification of
the California Current, separating the Pacific from
the Caribbean bioprovince. Uplift, as indicated by
proxy records of weathering products, combined
with active volcanism may have led to an island
arc chain which allowed the first terrestrial organisms to migrate between the Americas from 9 to
8 Ma. This increased restriction is also evident
from the divergence of bottom water characteristics
between the Pacific and Atlantic. The continued shallowing led to genetic isolation of several groups of
species and consequently to vicarious speciation.
Uncertainty is related to the time between 8 and
6.6 Ma, where several datasets point towards a
reopening of the Isthmus whereas others point
towards continued restriction. By 6 Ma, the sill
depth had decreased to 150 m.

The limitation of surface-water exchange starting at 4.7 Ma resulted in an increase in salinity in
the Caribbean, changes in the faunal composition,
the origin of endemic planktic species and a profound crisis in Caribbean reef communities. This
closing process had two brief reversals at 3.8 and
3.4 Ma. The emergence of a land bridge by 2.8 to
2.6 Ma enabled the Great American Interchange
of terrestrial species. The shallow-water connections finally closed at 2.5 Ma, blocking the
exchange of shore organisms and influencing the
reef communities, resulting in profoundly different
faunal provinces. Evidence for potential breaching
of the Isthmus at sea-level highstands continues
into the early Pleistocene.

I would like to thank Michael Sarnthein and the partici-
pants of the Gateways Workshop on the closure of the
Panama Isthmus in Kiel for stimulating discussions,
Gerald Haug and an anonymous reviewer for their excel-
lent reviews, Derek Vance, Tim Elliott and Samantha
Gibbs for comments on an earlier version and Sandra
Jasinoski for English corrections. This review is based
on samples and data provided by the Ocean Drilling

Program, which is sponsored by the US National Science Foundation and participating countries under the management of Joint Oceanographic Institutions. Funding came through the grants of the Deutsche Forschungsgemeinschaft (Schm 1668/1-2) and NERC (NE/B500874/1).

References

ANSELMETTI, F. S., EBERLI, G. P. & DING, Z.-D. 2000. From the Great Bahama Bank into the Straits of Florida: A margin architecture controlled by sea-level fluctuations and ocean currents. *Geological Society of America Bulletin*, **112**, 829–844.

AVISE, J. C., NELSON, W. S., BOWEN, B. W. & WALKER, D. 2000. Phylogeography of colonially nesting seabirds, with special reference to global matrilineal patterns in the sooty tern (*Sterna fuscata*). *Molecular Ecology*, **9**, 1783–1792.

BERGER, W. H. & WEFER, G. 1996. Expeditions into the Past: Paleoceanographic studies in the South Atlantic. *In*: WEFER, G., BERGER, W. H., SIEDLER, G. & WEBB, D. J. (eds) *The South Atlantic: Present and Past Circulation*, Springer-Verlag, Berlin, pp. 363–410.

BERGGREN, W. A. & HOLLISTER, C. D. 1974. Paleogeography, paleobiogeography, and the history of circulation of the Atlantic Ocean. *In*: HAY, W. W. (ed.) *Studies in Paleoceanography*, Special Publication, vol. **20**, SEPM, pp. 126–186.

BILLUPS, K. 2002. Late Miocene through early Pliocene deep water circulation and climate change viewed from the sub-Antarctic South Atlantic. *Palaeogeography, Palaeoclimatology, Palaeoecology*, **185**, 287–307.

BILLUPS, K., RAVELO, A. C., ZACHOS, J. C. & NORRIS, R. N. 1999. Link between oceanic heat transport, thermohaline circulation, and the Intertropical Convergence Zone in the early Pliocene Atlantic. *Geology*, **27**, 319–322.

BOLLI, H. M. 1971. The direction of coiling in planktonic foraminifera. *In*: FUNNELL, B. M. & RIEDEL, W. R. (eds) *The Micropaleontology of the Oceans*, Cambridge University Press, Cambridge, pp. 639–648.

BOWEN, B. W., CLARK, A. M., ABREU-GROBOIS, F. A., CHAVES, A., REICHART, H. A. & FERL, R. J. 1997. Global phylogeography of the ridley sea turtles (*Lepidochelys* spp.) as inferred from mitochondrial DNA sequences. *Genetica*, **101**, 179–189.

BROECKER, W. S. & DENTON, G. H. 1989. The role of ocean-atmosphere reorganizations in glacial cycles. *Geochimica et Cosmochimica Acta*, **53**, 2465–2501.

BURNHAM, R. 1999. The history of neotropical vegetation: New developments and status. *Annals of the Missouri Botanical Garden*, **86**, 546–589.

BURTON, K. W., LING, H.-F. & O'NIONS, R. K. 1997. Closure of the Central American Isthmus and its effect on deep-water formation in the North Atlantic, **386**, 382–385.

CANE, M. A. & MOLNAR, P. 2001. Closing of the Indonesian seaway as a precursor to east African aridification around 3–4 million years ago. *Nature*, **411**, 157–162.

CANNARIATO, K. G. & RAVELO, A. C. 1997. Plio-Pleistocene evolution of eastern tropical Pacific surface water ciruclation and thermocline depth. *Paleoceanography*, **12**, 805–820.

CHAISSON, W. P. 2003. Vicarious living: Pliocene menardellids between an isthmus and an ice sheet. *Geology*, **31**, 1085–1088.

CHAISSON, W. P. & LECKIE, R. M. 1993. High-resolution planktonic foraminifer biostratigraphy of Site 806, Ontong Java Plateau (Western Equatorial Pacific). *In*: BERGER, W. H., KROENKE, L. W. & MAYER, L. A. (eds) *Proceedings of the Ocean Drilling Program, Scientific Results*, **130**, Ocean Drilling Project, College Station, TX, pp. 137–178.

CHAISSON, W. P. & PEARSON, P. N. 1997. Planktonic foraminifer biostratigraphy at Site 925: Middle Miocene–Pleistocene. *In*: SHACKLETON, N. J., CURRY, W. B., RICHTER, C. & BRALOWER, T. J. (eds) *Proceedings of the Ocean Drilling Program, Scientific Results*, **154**, Ocean Drilling Program, College Station, TX, pp. 3–31.

CHAISSON, W. P. & RAVELO, A. C. 1997. Changes in upper water-column structure at Site 925, Late Miocene–Pleistocene: Planktonic foraminifer assemblage and isotope evidence. *In*: SHACKLETON, N. J., CURRY, W. B., RICHTER, C. & BRALOWER, T. J. (eds) *Proceedings of the Ocean Drilling Program, Scientific Results*, **154**, Ocean Drilling Program, College Station, TX, pp. 255–268.

CHAISSON, W. P. & RAVELO, A. C. 2000. Pliocene development of the east–west hydrographic gradient in the equatorial Pacific. *Paleoceanography*, **15**, 497–505.

CHEETHAM, A. H. & JACKSON, J. B. C. 1996. Speciation, extinction, and the decline of arborescent grwoth in Neogene and Quaternary cheilostome Bryozoa of tropical America. *In*: JACKSON, J. B. C., BUDD, A. F. & COATES, A. G. (eds) *Evolution and Environment in Tropical America*, University of Chicago Press, Chicago, pp. 205–233.

COATES, A. G. & OBANDO, J. A. 1996. The geologic evolution of the central American isthmus. *In*: JACKSON, J. B. C., BUDD, A. F. & COATES, A. G. (eds) *Evolution and Environment in Tropical America*, University of Chicago Press, Chicago, pp. 21–56.

COATES, A. G., COLLINS, L. S., AUBRY, M. P. & BERGGREN, W. A. 2004. The Geology of the Darien, Panama, and the late Miocene–Pliocene collision of the Panama arc with northwestern South America. *GSA Bulletin*, **116**, 1327–1344.

COATES, A. G., JACKSON, J. B. C., COLLINS, L. S. ET AL., 1992. Closure of the Isthmus of Panama: the near-Shore Marine Record of Costa-Rica and Western Panama. *Geological Society of America Bulletin*, **104**, 814–828.

COLLINS, L. S., BUDD, A. F. & COATES, A. G. 1996a. Earliest evolution associated with closure of the Tropical American Seaway. *Proceedings of the National Academy of Sciences of the United States of America*, **93**, 6069–6072.

COLLINS, L. S., COATES, A. G., BERGGREN, W. A., AUBRY, M.-P. & ZHANG, J. 1996b. The late Miocene Panama isthmian strait. *Geology*, **24**, 687–690.

COLLINS, T. M., FRAZER, K., PALMER, A. R., VERMEIJ, G. J. & BROWN, W. M. 1996c. Evolutionary history of northern hemisphere *Nucella* (Gastropoda, Muricidae): molecules, ecology, and fossils. *Evolution*, **50**, 2287–2304.

CORTESE, G., GERSONDE, R., HILLENBRAND, C.-D. & KUHN, G. 2004. Opal sedimentation shifts in the World Ocean over the last 15 Myr. *Earth and Planetary Science Letters*, **224**, 509–527.

CRONIN, T. M. & DOWSETT, H. J. 1996. Biotic and oceanographic response to the Pliocene closing of the Central American isthmus. *In*: JACKSON, J. B. C., BUDD, A. F. & COATES, A. G. (eds) *Evolution and Environment in Tropical America*, University of Chicago Press, Chicago, pp. 76–104.

CROWLEY, T. J. & BERNER, R. A. 2001. CO_2 and climate change. *Science*, **292**, 870–872.

DELANEY, M. L. 1990. Miocene benthic foraminiferal Cd/Ca records: South Atlantic and western equatorial Pacific. *Paleoceanography*, **5**, 743–760.

DOBSON, M. D., DICKENS, G. R. & REA, D. K. 1997. Terrigenous sedimentation of Ceara Rise. *In*: CURRY, R., SHACKLETON, N. J. & RICHTER, C. (eds) *Proceedings of the Ocean Drilling Program, Scientific Results*, **154**, Ocean Drilling Program, College Station, TX.

DRISCOLL, N. W. & HAUG, G. H. 1998. A short circuit in thermohaline circulation: a cause for Northern Hemisphere glaciation? *Science*, **282**, 436–438.

DUQUE-CARO, H. 1990. Neogene stratigraphy, paleoceanography and paleobiography in northwest South America and the evolution of the Panama Seaway. *Palaeogeography, Palaeoclimatology, Palaeoecology*, **77**, 203–234.

FARRELL, J. W. ET AL., 1995. Late Neogene sedimentation patterns in the eastern equatorial Pacific Ocean. *In*: PISIAS, N., MAYER, L. & JANECEK, T. R. (eds) *Proceedings of the Ocean Drilling Program, Scientific Results*, **138**, Ocean Drilling Program, College Station, TX, pp. 717–756.

FRANK, M., REYNOLDS, B. C. & O'NIONS, K. 1999. Nd & Pb isotopes in Atlantic and Pacific water masses before and after the closure of the Panama gateway. *Geology*, **27**, 1147–1150.

GARTNER, S., CHOW, J. & STANTON, R. J. 1987. Late Neogene paleoceanography of the eastern Caribbean, the Gulf of Mexico, and the eastern Equatorial Pacific. *Marine Micropaleontology*, **12**, 255–304.

GRAHAM, A. 1992. Utilization of the isthmian land bridge during the Cenozoic – paleobotanical evidence for timing, and the selective influence of altitudes and climate. *Review of Palaeobotany and Palynology*, **72**, 119–128.

GREGORY-WODZICKI, K. M. 2000. Andean paleoelevation estimates: A review and critique. *Geological Society of America Bulletin*, **112**, 1091–1105.

HALL, R. 2001. Cenozoic reconstructions of SE Asia and the SW Pacific: changing patterns of land and sea. *In*: METCALFE, I., SMITH, J. M. B., MORWOOD, M. & DAVIDSON, I. D. (eds) *Faunal and Floral Migrations and Evolution in SE Asia–Australasia*, A.A. Balkema, Lisse, 35–56.

HAQ, B. U., HARDENBOL, J. & VAIL, P. R. 1987. The new chronostratigraphic basis of Cenozoic and Mesozoic sea level cycles. *Special Publication of the Cushman Foundation of Foraminiferal Research*, **24**, 7–13.

HAUG, G. & TIEDEMANN, R. 1998. Effect of the formation of the Isthmus of Panama on Atlantic Ocean thermohaline circulation. *Nature*, **393**, 673–676.

HAUG, G. H., SIGMAN, D. M., TIEDEMANN, R., PEDERSEN, T. & SARNTHEIM, M. 1999. Onset of permanent stratification in the subarctic Pacific Ocean. *Nature*, **401**, 779–782.

HAUG, G., TIEDEMANN, R., ZAHN, R. & RAVELO, A. C. 2001. Role of Panama uplift on oceanic freshwater balance. *Geology*, **29**, 207–210.

HAUG, G. H. ET AL. 2005. North Pacific seasonality and the glaciation of North America 2.7 million years ago. *Nature*, **433**, 821–825.

JACKSON, J. B. C., JUNG, P., COATES, A. G. & COLLINS, L. S. 1993. Diversity and extinction of tropical American mollusks and emergence of the Isthmus of Panama. *Science*, **260**, 1624–1626.

KAMEO, K. & SATO, T. 2000. Biogeography of Neogene calcareous nannofossils in the Caribbean and the eastern equatorial Pacific – floral response to the emergence of the Isthmus of Panama. *Marine Micropaleontology*, **39**, 201–218.

KEIGWIN, L. 1982a. Isotopic paleoceanography of the Caribbean and East Pacific: role of Panama uplift in late Neogene time. *Science*, **217**, 350–353.

KEIGWIN, L. D. 1982b. Isotopic paleoceanography of the Caribbean and east Pacific: role of Panama uplift in late Neogene time. *Science*, **217**, 350–353.

KELLER, G. & BARRON, J. A. 1983. Paleoceanographic implications of the Miocene deep-sea hiatuses. *Geological Society of America Bulletin*, **94**, 590–613.

KELLER, G., ZENKER, C. E. & STONE, S. M. 1989. Late Neogene history of the Pacific-Caribbean gateway. *Journal of South American Earth Science*, **2**, 73–108.

KENNETT, J. P. & SRINIVASAN, M. S. 1983. *Neogene Planktonic Foraminifera: A Phylogenetic Atlas*. Hutchinson Ross, Stroudsburg, PA.

KING, T. A., ELLIS, W. G., MURRAY, D. W., SHACKLETON, N. J. & HARRIS, S. 1997. Miocene evolution of carbonate sedimentation at the Ceara Rise: a multivariate data/proxy approach. *In*: SHACKLETON, N. J., CURRY, W. B., RICHTER, C. & BRALOWER, T. J. (eds) *Proceedings of the Ocean Drilling Program, Scientific Results*, **154**, Ocean Drilling Program, College Station, TX, pp. 349–365.

KIRBY, M. X. & MACFADDEN, B. 2005. Was southern Central America an archipelago or a peninsula in the middle Miocene? A test using land-mammal body size. *Palaeogeography, Palaeoclimatology, Palaeoecology*, **228**, 193–202.

KLOCKER, A., PRANGE, M. & SCHULZ, M. 2005. Testing the influence of the Central American Seaway on orbitally forced northern hemisphere glaciation. *Geophysical Research Letters*, **32**.

KNOWLTON, C. & WEIGHT, L. A. 1998. New dates and new rates for divergence across the Isthmus of Panama. *Proceedings of the Royal Society B: Biological Sciences*, **265**, 2257–2263.

KNOWLTON, N., WEIGHT, L. A., SOLORZANO, L. A., MILLS, D. K. & BERMINGHAM, E. 1993. Divergence in proteins, mitochondrial DNA, and reproductive compatibility across the Isthmus of Panama. *Science*, **260**, 1629–1631.

KÜRSCHNER, W. M., VAN DER BURGH, J., VISSCHER, H. & DILCHER, D. L. 1996. Oak leaves as biosensors of late Neogene and early Pleistocene paleoatmospheric CO_2 concentrations. *Marine Micropaleontology*, **27**, 299–312.

LEAR, C. H., ROSENTHAL, Y. & WRIGHT, J. D. 2003. The closing of a seaway; ocean water masses and global climate change. *Earth and Planetary Science Letters*, **210**, 425–436.

LEDBETTER, M. T. 1982. Tephrachronology at Sites 502 and 503. *In*: PRELL, W. L. & GARDNER, J. V. (eds) *Initial Reports of the Deep Sea Drilling Project*, **68**, Governmental Printing Office, Washington, pp. 403–408.

LESSIOS, H. A., KESSING, B. D. & PEARSE, J. S. 2001. Population structure and speciation in tropical seas: Global phylogeography of the sea urchin *Diadema*. *Evolution*, **55**, 955–975.

LESSIOS, H. A., KANE, J. & ROBERTSON, D. R. 2003. Phylogeography of the pantropical sea urchin *Tripneustes*: constraining patterns of population structure between oceans. *Evolution*, **57**, 2026–2036.

LISIECKI, L. E. & RAYMO, M. E. 2005. A Pliocene-Pleistocene stack of 57 globally distributed benthic $\delta^{18}O$ records. *Paleoceanography*, **20**, PA1003, doi:10.1029/2004PA001071.

LOURENS, L. J., HILGEN, F. J., LASKAR, J., SHACKLETON, N. J. & WILSON, D. 2005. The Neogene Period. *In*: GRADSTEIN, F. M., OGG, J. & SMITH, A. G. (eds) *A Geologic Time Scale*, Cambridge University Press, Cambridge.

LUNDELIUS, E. L. 1987. The North American quaternary sequence. *In*: WOODBURNE, M. O. (ed.), *Cenozoic Mammals of North America*, University of California Press, pp. 211–235.

MACFADDEN, B. J. 2005. Middle Miocene land mammals from the Cucaracha Formation (Hemingfordian-Barstovian) of Panama. *Journal of Vertebrate Paleontology*.

MAIER-REIMER, E., MIKOLAJEWICZ, U. & CROWLEY, T. 1990. Ocean general circulation model sensitivity experiment with an open Central American Isthmus. *Paleoceanography*, **5**, 349–366.

MARLOW, J. R., LANGE, C. B., WEFER, G. & ROSELL-MELÉ, A. 2000. Upwelling intensification as part of the Pliocene-Pleistocene climate transition. *Science*, **290**, 2288–2291.

MARSHALL, L. G. 1988. Land mammals and the Great American interchange. *American Scientist*, **76**, 380–386.

MARSHALL, L. G., WEBB, S. D., SEPKOSKI, J. & RAUP, D. 1982. Mammalian Evolution and the Great American Interchange. *Science*, **215**, 1351–1357.

MASLIN, M. A., LI, X. S., LOUTRE, M.-F. & BERGER, A. 1998. The contribution of orbital forcing to the progressive intensification of Northern Hemisphere glaciation. *Quaternary Science Reviews*, **17**, 411–426.

MIKOLAJEWICZ, U. & CROWLEY, J. L. 1997. Response of a coupled ocean/energy balance model to restricted flow through the Central American Isthmus. *Paleoceanography*, **12**, 429–442.

MIKOLAJEWICZ, U., MAIER-REIMER, E., CROWLEY, J. L. & KIM, K.-Y. 1993. Effect of Drake and Panamanian gateways on the circulation of an ocean model. *Paleoceanography*, **8**, 409–426.

MORAN, K. *ET AL.* 2006. The Cenozoic palaeoenvironment of the Arctic Ocean. **441**, 601–605.

MUDELSEE, M. & RAYMO, M. E. 2005. Slow dynamics of the Northern Hemisphere glaciation. *Paleoceanography*, **20**, PA4022, http://dx.doi.org/10.1029/2005PA001153.

MUSS, A., ROBERTSON, D. R., STEPIEN, C. A., WIRTZ, P. & BOWEN, B. W. 2001. Phylogeography of *Ophioblennius*: The role of ocean currents and geography in reef fish evolution. *Evolution*, **55**, 561–572.

NISANCIOGLU, K. H., RAYMO, M. E. & STONE, P. H. 2003. Reorganisation of Miocene deep water circulation in response to the shoaling of the Central American Seaway. *Paleoceanography*, **18**, dio:10.1029/PA2002PA000767.

PEARSON, P. N. & PALMER, M. R. 2000. Atmospheric carbon dioxide concentrations over the past 60 million years. *Nature*, **406**, 695–699.

PERDICES, A., DOADRIO, I. & BERMINGHAM, E. 2005. Evolutionary history of the synbranchid eels (Teleostei: Synbranchidae) in Central America and the Caribbean islands inferred from their molecular phylogeny. *Molecular Phylogenetics and Evolution*, **37**, 460–473.

PRANGE, M. & SCHULZ, M. 2004. A coastal upwelling seesaw in the Atlantic Ocean as a result of the closure of the Central American Seaway. *Geophysical Research Letters*, **31**, L17207, doi:10.1029/2004GL020073.

RAVELO, A. C., ANDREASEN, D. H., LYLE, M., LYLE, A. O. & WARA, M. W. 2004. Regional climate shifts caused by gradual global cooling in the Pliocene epoch. *Nature*, **429**, 263–267.

RAVELO, A. C., BILLUPS, K., DEKENS, P. S., HERBERT, T. D & LAWRENCE, K. T 2007. Onto the ice ages: proxy evidence for the onset of Northern Hemisphere glaciation. *In*: WILLIAMS, M., HAYWOOD, A. M., GREGORY, F. J. & SCHMIDT, D. N. (eds) *Dept-Time Perspectives on Climate Change: Marrying the Signal from Computer Models and Biological Proxies*. The Micropalaeontological Society, Special Publications. The Geological Society, London, pp. 000–000.

RAYMO, M. E., RUDDIMAN, W. F. & FROELICH, P. N. 1988. Influence of late Cenozoic mountain building on ocean geochemical cycles. *Geology*, **16**, 649–653.

REIJMER, J. J. G., BETZLER, C., KROON, D., TIEDEMANN, R. & EBERLI, G. P. 2002. Bahamian carbonate platform development in response to sea-level changes and the closure of the Isthmus of Panama. *International Journal of Earth Sciences*, **91**, 482–489.

RICKABY, R. E. M. & HALLORAN, P. 2005. Cool La Nina During the Warmth of the Pliocene? *Science*, **307**, 1948–1952.

ROTH, J. M., DROXLER, A. W. & KAMEO, K. 2000. The Caribbean carbonate crash at the Middle to Late Miocene transition: Linkage to the establishment of the modern gobal ocean conveyor. *In*: LECKIE, R. M., SIGURDSSON, H., ACTON, G. D. & DRAPER, G. (eds) *Proceedings of the Ocean Drilling*

Program, Scientific Results, **165**, Ocean Drilling Program, College Station, TX, pp. 249–273.

SCHLITZER, R. 2006. Ocean Data View - 3.1.0, http://odv.awi.de://odv.awi.de.

SCHMIDT, D. N. submitted. Neogene foraminiferal biostratigraphy in the Mid-latitude Atlantic. *In*: KROON, D. (ed.), Leg 208, **208**, *Ocean Drilling Program, Scientific Reports*, College Station, TX.

SCHMIDT, D. N., RENAUD, S., BOLLMANN, J., SCHIEBEL, R. & THIERSTEIN, H. R. 2004*a*. Size distribution of Holocene planktic foraminifer assemblages: biogeography, ecology and adaptation. *Marine Micropaleontology*, **50**, 319–338.

SCHMIDT, D. N., THIERSTEIN, H. R. & BOLLMANN, J. 2004*b*. The evolutionary history of size variation of planktic foraminiferal assemblages in the Cenozoic. *Palaeogeography, Palaeoclimatology, Palaeoecology*, **212**, 159–180.

SCHMIDT, D. N., THIERSTEIN, H. R., BOLLMANN, J. & SCHIEBEL, R. 2004*c*. Abiotic Forcing of Plankton Evolution in the Cenozoic. *Science*, **303**, 207–210.

SCHMITTNER, A. *ET AL*. 2004. Global Impact of the Panamanian Seaway Closure. *Eos, Transactions, American Geophysical Union*, **85**, 526–528.

SCHREIBER, E. A., FEARE, C. J., HARRINGTON, B. A., MURRAY, B. G., ROBERSTON, W. B., ROBERSTON, M. J. & WOOLFENDEN, G. E. 2002. Sooty tern (*Sterna fuscata*). *In*: POOLE, A. & GILL, F. (eds) *The Birds of North America*, **665**, The Academy of Natural Sciences, Philadelphia, pp. 1–31.

SHACKLETON, N. J. *ET AL*. 1984. Oxygen isotope calibration of the onset of ice-rafting and history of glaciation in the North Atlantic region. *Nature*, **207**, 620–623.

SIGMAN, D. M., JACCARD, S. L. & HAUG, G. H. 2004. Polar ocean stratification in a cold climate. *Nature*, **428**, 59–63.

SPENCER-CERVATO, C. & THIERSTEIN, H. R. 1997. First appearance of *Globorotalia truncatulinoides*; cladogenesis and immigration. *Marine Micropaleontology*, **30**, 267–291.

STEEVES, T. E., ANDERSON, D. J., MCNALLY, H., KIM, M. J. & FRIESEN, V. L. 2003. Phylogeography of Sula: the role of physical barriers to gene flow in the diversification of tropical seabirds. *Journal of Avian Biology*, **34**, 217–223.

STEEVES, T. E., ANDERSON, D. J. & FRIESEN, V. L. 2005. The Isthmus of Panama: A major physical barrier to gene flow in a highly mobile pantropical seabird. *Journal of Evolutionary Biology*, **18**, 1000–1008.

THIEDE, J., WINKLER, A., WOLF-WELLING, T. *ET AL*., 1998. Late Cenozoic history of the Polar North Atlantic: results from ocean drilling. *Quaternary Science Reviews*, **17**, 185–208.

TIEDEMANN, R. & FRANZ, S. O. 1997. Deep-water circulation, chemistry, and terrigenous sediment supply in the Equatorial Atlantic during the Pliocene, 3.3–2.6 Ma and 5–4.5 Ma. *In*: SHACKLETON, N. J.,

CURRY, W. B., RICHTER, C. & BRALOWER, T. J. (eds) *Proceedings of the Ocean Drilling Program, Scientific Results*, **154**, Ocean Drilling Program, College Station, TX, 299–318.

TIEDEMANN, R., SARNTHEIN, M. & SHACKLETON, N. J. 1994. Astronomic timescale for the Pliocene Atlantic $\delta^{18}O$ and dust flux records of Ocean Drilling Program site 659. *Paleoceanography*, **9**, 619–638.

TODD, J. A., JACKSON, J. B. C., JOHNSON, K. G., FORTUNATO, H. M., HEITZ, A., ALVAREZ, M. & JUNG, P. 2002. The ecology of extinction: molluscan feeding and faunal turnover in the Caribbean Neogene. *Proceedings of the Royal Society of London B*, **269**, 571–577.

TOMCZAK, M. & GODFREY, J. S. 1994. *Regional Oceanography: An Introduction*. Pergamon, Elsevier Science Ltd., Oxford, 422pp.

VERMEIJ, G. J. & PETUCH, E. J. 1986. Differential extinction in tropical American molluscs: endemism, architecture, and the Panama land bridge. *Malacologia*, **27**, 29–41.

WARA, M. W., RAVELO, A. C. & DELANEY, M. L. 2005. Permanent El Nino-Like Conditions During the Pliocene Warm Period. *Science*, **309**, 758–761.

WEBB, S. D. 1976. Mammalian faunal dynamics of the great American interchange. *Paleobiology*, **2**, 220–234.

WEBB, S. D. 1985. Late Cenozoic mammal dispersals between the Americas. *In*: STEHLI, F. G. & WEBB, S. D. (eds) *The Great American Biotic Interchange*, Plenum Press, New York, 357–388.

WEBB, S. D. 1991. Ecogeography and the Great American Interchange. *Paleobiology*, **17**, 266–280.

WEBB, S. D. 1997. The great American faunal interchange. *In*: COATES, A. G. (ed.) *Central America: A Natural and Cultural History*, Yale University Press, New Haven, CT, 97–122.

WHITMORE, F. C. & STEWART, R. H. 1965. Miocene mammals and Central American seaways. *Science*, **148**, 180–185.

WOLD, C. N. 1994. Cenozoic sediment accumulation on drifts in the northern North Atlantic. *Paleoceanography*, **9**, 917–942.

WRIGHT, J. D. & MILLER, K. G. 1996. Control of the North Atlantic Deep Water circulation by the Greenland-Scotland Ridge. *Paleoceanography*, **11**, 157–170.

WRIGHT, J. D., MILLER, K. G. & FAIRBANKS, R. G. 1991. Evolution of modern deepwater circulation: evidence from the Late Miocene Southern Ocean. *Paleoceanography*, **6**, 275–290.

WRIGHT, J. D., MILLER, K. G. & FAIRBANKS, R. G. 1992. Early and Middle Miocene stable Isotopes: implication for deepwater circulation and climate. *Paleoceanography*, **7**, 357–389.

ZACHOS, J. C., PAGANI, M., SLOAN, L., THOMAS, E. & BILLUPS, K. 2001. Trends, rhythms, and aberrations in global climate 65 Ma to present. *Science*, **292**, 686–693.

The mid-Pliocene warm period: A test-bed for integrating data and models

A. M. HAYWOOD[1,*], P. J. VALDES[2], D. J. HILL[1] & M. WILLIAMS[1,†]

[1]*Geological Sciences Division, British Antarctic Survey, High Cross, Madingley Road, Cambridge, CB3 0ET, UK (e-mail: a.haywood@su.leeds.ac.uk); [*]Present address: School of Earth & Environment, Environment Building, University of Leeds, LS2 9JT, UK; [†]Present address: University of Leicester, Department of Geology, University Road, Leicester LE1 7RH, UK*

[2]*School of Geographical Sciences, The University of Bristol, University Road, Bristol, BS8 1SS, UK*

Abstract: In this paper, we present a summary of palaeoclimate modelling activities carried out for the mid-Pliocene, and demonstrate how a combined data and modelling approach has led to significant advances in our understanding and 'retrodiction' of the last great warm period of the Cenozoic Era. The development and refinement of mid-Pliocene palaeoenvironmental datasets is discussed, as are the steps that have been taken to increase the sophistication of the modelling effort. Initially, the discussion focuses on the use of Atmosphere-only General Circulation Models (AGCMs) and summarizes the results produced by such models. We then consider the importance of incorporating slab-ocean, dynamic ocean, mechanistic and dynamic vegetation models with AGCMs. Recent developments are considered and a strategy for further investigation is presented.

The period of mid-Pliocene warmth is defined by the United States Geological Survey's PRISM Group (Pliocene Research Interpretations and Synoptic Mapping) as the interval between 3.29 and 2.97 Ma (according to the geomagnetic polarity timescale of Berggren *et al.* 1995), lying between the transition of oxygen isotope stages M2/M1 and G19/G18 (Shackleton *et al.* 1995), in the middle part of the Gauss Normal Polarity Chron (Dowsett *et al.* 1999). The 'Time Slab' represents a climatically distinct period during the Pliocene when Earth's climate was, on the whole, warmer than present (Dowsett *et al.* 1999; Dowsett this volume; see Fig. 1).

The mid-Pliocene has been the subject of intense study for the last two decades. There are many reasons for this, but the most important driver has been our desire to understand the dynamics of past warm climates as a potential guide to understanding climate change in the future. The mid-Pliocene is well suited to this task. The climatic signal (change from modern) is sufficiently large to be differentiated from the noise generated by the uncertainties and limitations inherent in the techniques used for palaeoclimatic/palaeoenvironmental research. The interval was the last time in Earth history when global temperatures were significantly warmer than modern, over a period longer than any Quaternary interglacial. It is unique in that continental configurations were relatively unchanged from today, and geological proxies are superior to preceding warm periods due to improved geographic coverage, more reliable biota-environment correlations and higher resolution stratigraphy.

If a palaeoclimate modeller were asked why so much attention has been paid to the mid-Pliocene, the honest answer would be 'because it is the only pre-Quaternary period for which a series of palaeoenvironmental datasets have been developed, in a synthesized and internally consistent manner, that are suitable for use as boundary conditions within climate models'. In reality, the reason why modellers have paid such attention to the mid-Pliocene is not simply because it is scientifically interesting, but because it also represents a pragmatic choice; geologists have worked for years providing modellers with the tools needed for their trade (see Dowsett this volume). The lack of similar datasets for other pre-Quaternary time intervals means that there is often a considerable degree of freedom regarding the boundary conditions used within the models. This limits the usefulness of deep-palaeo studies in our quest to evaluate the ability of GCMs to simulate large-scale climate changes of the past because, when data/model inconsistencies are recognized, it is not clear where the problem lies (i.e. with the

From: WILLIAMS, M., HAYWOOD, A. M., GREGORY, F. J. & SCHMIDT, D. N. (eds) *Deep-Time Perspectives on Climate Change: Marrying the Signal from Computer Models and Biological Proxies.* The Micropalaeontological Society, Special Publications. The Geological Society, London, 443–457.

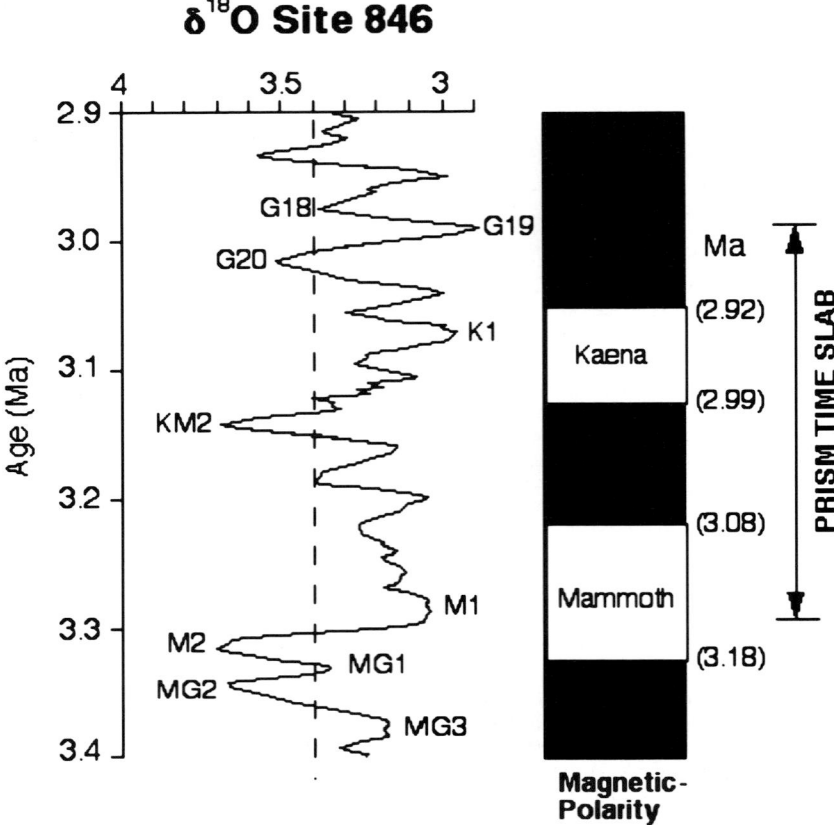

Fig. 1. The period of mid-Pliocene warmth and the PRISM2 time slab correlated to the geomagnetic polarity timescale of Berggren *et al.* (1995) and the benthic oxygen isotope records from ODP Site 846 (Shackleton *et al.* 1995). Numbers in parentheses adjacent to magnetic polarity indicate ages of subchron boundaries according to Berggren *et al.* (1985). The modern isotopic value from Shackleton *et al.* (1995) is shown as a vertical dashed line (from Dowsett *et al.* 1999).

model or with the boundary conditions used within the model).

This paper presents a history of palaeoclimate modelling for the mid-Pliocene. It will be shown how the sophistication of the modelling has grown in step with the palaeoenvironmental datasets, and how an iterative approach between data and models has led to significant advances in our understanding of mid-Pliocene climates and environments.

Forcings and feedbacks on mid-Pliocene climate

After two decades, the answer to the question 'why was the mid-Pliocene warm' remains difficult to answer with any degree of confidence. The one thing we can say is that the answer is complex

and involves a number of, perhaps contemporaneous, forcing mechanisms. No climate driver can unlock all of the climate secrets of the mid-Pliocene. Indeed, the scientific usefulness of the question 'what was the cause of mid-Pliocene warmth' is arguable. This is due to (a) the multiple interactions of different components of the Earth's climate system, (b) equifinality (different climate-forcing mechanisms leading to the production of a similar response in climate) and (c) climatic memory effects (the existence of a global climate regime at any one time depends not only on forcing mechanisms in operation during the said regime but also on those which occurred previously). A more scientifically useful question to pose might be 'what were the processes that helped promote, sustain and ultimately terminate the climates of the mid-Pliocene'. Such an approach has the advantage of being process-driven, making the investigation of past warm periods more relevant to future climate change.

The drivers of mid-Pliocene warmth have been widely discussed (e.g. Raymo *et al.* 1996) and can be crudely summarized as relating to: (a) palaeogeographic change, e.g. altered elevations of major mountain chains such as the western cordillera of North and South America (Rind & Chandler 1991); (b) altered atmospheric trace gas concentrations and water vapour content (e.g. Van der Burgh *et al.* 1993; Raymo & Rau 1992); (c) changes to ocean circulation (Ravelo & Andreasen 2000; Cane & Molnar 2001), ocean heat transport (e.g. Dowsett *et al.* 1992; Haug & Tiedemann 1998; Kim & Crowley 2000), or the thermal structure of the oceans (e.g. Wara *et al.* 2005; Philander & Fedorov 2003); and (d) feedbacks generated through reduced land ice-cover, surface albedo, cloud cover and temperature (Haywood & Valdes 2004).

The most common forcing mechanisms invoked to explain mid-Pliocene warmth are elevated concentrations in atmospheric CO_2 and/or altered ocean heat transports. Estimates of atmospheric CO_2 levels have been derived from analyses of the stomatal density of fossil leaves (Kürschner *et al.* 1996), from analyses of $\delta^{13}C$ ratios of marine organic carbon (Raymo & Rau 1992; Raymo *et al.* 1996) and through measurement of the differences between the carbon isotope composition of surface and deep waters (Shackleton *et al.* 1992). Use of all of these proxy methods provides a consensus and suggests that absolute CO_2 levels during the epoch ranged from 360 to 400 ppmv, compared to mid-19th century levels of approximately 280 ppmv and modern concentrations of 380 ppmv. The cause for this potentially higher (than pre-industrial) mid-Pliocene concentration is not well understood and may be attributable to the long-term trend of CO_2 change throughout the Tertiary (Berner 1990). Alternatively, feedbacks between climate and the carbon cycle may also be important.

More recent papers have questioned the idea of elevated atmospheric CO_2 levels at this time. Based on boron isotope data and stable carbon isotopic compositions of sedimentary organic molecules, Pearson & Palmer (2000) and Pagani *et al.* (2005) suggest that the concentration of atmospheric CO_2 declined to approximately pre-industrial levels by the end of the Oligocene. If so, this indicates a decoupling between atmospheric CO_2 concentration and global climate change during the Miocene and Pliocene.

A likely additional factor is a change in ocean heat transport, in particular increased meridional ocean heat transport. This could be achieved through an increase in thermohaline or surface gyre circulation (Dowsett *et al.* 1992; Ravelo & Andreasen 2000; Haywood *et al.* 2000*a*). This possibility was highlighted in early modelling studies (e.g. Rind & Chandler 1991), but many difficulties and unexplained issues remain. Paradoxically, a reduced latitudinal SST gradient implies weaker atmospheric forcing of surface oceanic circulation, and hence weaker oceanic heat transport from equator to higher latitudes (Crowley 1996). This is a common problem associated with past warm climate dynamics. Problems exist in any explanation of Pliocene warmth which is based solely upon strengthening of the thermohaline circulation as it is (a) difficult to ascribe a thermohaline coupling argument to all ocean basins, and (b) difficult to generate the correct reconstructed hemispheric temperature distribution (Crowley 1996).

Palaeoclimate studies using atmosphere-only GCMs

Initial palaeoclimate modelling studies for the mid-Pliocene utilized atmosphere-only GCMs (AGCMs) with prescribed SSTs derived from the PRISM project. The principal advantage of this approach was that it facilitated the identification of the processes which influenced the simulations. Furthermore, by examining the energy balance of the models, it was possible to deduce if the models were consistent with the SSTs or whether increased concentrations of greenhouse gases, for example, were needed to bring the models into equilibrium (Valdes 1993).

The first AGCM study was published by Chandler *et al.* (1994), and used the initial PRISM dataset (PRISM0) and the GISS (Goddard Institute for Space Studies) GCM ($8° \times 10°$ model resolution). The PRISM0 dataset provided information on mid-Pliocene SSTs, sea ice, the terrestrial vegetation cover, land-sea cover, and land ice cover on an $8° \times 10°$ grid (see PRISM Project Members 1994 and Dowsett *et al.* 1994). PRISM0 only incorporated mid-Pliocene information for the Northern Hemisphere because, at that stage of the PRISM Project, insufficient information had been collected to produce a global mid-Pliocene dataset. However, the state of the cryosphere was of first order importance to the PRISM Group in understanding the changes that they depicted in the Northern Hemisphere. The volume of ice contained on Antarctica has major implications for global sea level as well as influencing planetary albedo and the production of ocean bottom water (Dowsett *et al.* 1994). Therefore, PRISM supplied mid-Pliocene seasonal reconstructions of sea ice and mean land ice configurations for both hemispheres. In PRISM0, planktonic foraminifer census data from 24 ODP and DSDP sites were used to reconstruct Northern Hemisphere SSTs. The geographical coverage of these data was skewed toward the North Atlantic. Fifty terrestrial localities were used to reconstruct mid-Pliocene Northern Hemisphere vegetation. The geographical locations of these data points were skewed towards Western Europe and North America.

Using this dataset, the GISS GCM predicted a global mean annual surface temperature 1.4 °C higher and Northern Hemisphere precipitation 5.1% greater than the present day. Warming was greatest at high latitudes, and consequently, the equator-to-pole temperature gradient was decreased by 11.5 °C. Surface air temperature warming was largest in the winter hemisphere, as decreased snow and sea ice triggered a positive albedo feedback effect. At low latitudes, temperatures were almost unchanged, except for a 3 °C cooling over eastern Africa which was caused by the increase in elevation of the East African Rift Zone specified in the PRISM0 dataset (Chandler *et al.* 1994).

Sloan *et al.* (1996) described the second AGCM investigation in which a 2° × 2° version of the original PRISM dataset (PRISM1) was used to prescribe the boundary conditions for the NCAR (National Center for Atmospheric Research) GENESIS GCM (R15 model, approximately equivalent to a 4.5° in latitude × 7.5° in longitude model resolution). In this version of the dataset (see Poore & Sloan 1996 and references therein), as well as an enhanced resolution, the geographical distribution of data points was greatly improved. Sixty-four marine sites (Dowsett *et al.* 1996) and 74 terrestrial sites (Thompson & Fleming 1996) were incorporated. Crucially, PRISM1 included more data sites in the North Pacific and Southern Hemisphere which enabled a global dataset of boundary conditions to be developed. However, no marine data for the Indian Ocean, Mediterranean Sea, or South Pacific Ocean were included. The data in PRISM1 represent annual vegetation and land ice, monthly SST and sea ice fields, sea level and topography. PRISM1 used a sea level +35 m higher than modern. This is in contrast with PRISM0 that used a more conservative estimate of +25 m. This change in PRISM1 represents a 33% reduction in Antarctic ice cover compared to modern (Thompson & Fleming 1996). In PRISM0, Antarctic ice cover was reduced by 25%. Greenland ice cover in both PRISM0 and PRISM1 remained the same and represented a 50% reduction compared to modern conditions.

Sloan *et al.* (1996) reported that the main influence of the specified warm SSTs upon the climatology predicted by the NCAR GENESIS GCM was the occurrence of warmer and more humid high latitudes than observed in the current

climate, especially in the winter season. The global mean surface temperature was 3.6 °C warmer than in a present-day simulation, and global mean precipitation increased by 5%, with most excess precipitation occurring over the African continent and the oceans. Zonal winds weakened slightly but Hadley cell extent and jet stream locations remained unchanged.

In 1999, the next iteration of the PRISM dataset (PRISM2) was released by the USGS PRISM Group (Dowsett *et al.* 1999). PRISM2 incorporated the following important differences from PRISM1:

- One hundred and fifty-one sites were included within PRISM2. Seventy-seven sites were used in the marine reconstruction. Added marine data were incorporated from the Mediterranean Sea, Indian Ocean, and SW Pacific Ocean for the first time. The terrestrial coverage of locations remained the same as PRISM1 (see Thompson & Fleming 1996).

- All Pliocene SST estimates were recalculated based on a new core-top calibration to the Reynolds & Smith (1995) Adjusted Optimum Interpolation (AOI) SST dataset (see Dowsett this volume for further details).

- PRISM2 incorporated a sea level equivalent to a +25 m increase for the mid-Pliocene. In contrast, PRISM1 used a +35 m sea level increase. This change to a more conservative level was in keeping with new data that had become available (e.g. Kennett & Hodell 1993).

- PRISM2 used model results from Michael Prentice (pers. comm. to Harry Dowsett, cited in Dowsett *et al.* 1999) to guide the areal topographic distribution of Antarctic ice. The Antarctic ice sheet in PRISM1 was represented by blocks of ice that had steep sides and resulted in an unrealistic ice sheet which caused significant problems in modelling with GCMs. The changes made in PRISM2 resulted in a more appropriate configuration of ice (compared to PRISM1) in tune with a +25 m sea level increase relative to modern.

The PRISM2 dataset was used to provide the boundary conditions for a new series of mid-Pliocene experiments using the UK Meteorological Office Atmospheric GCM (HadAM3; Haywood *et al.* 2000*b*). HadAM3 benefited from having a much higher resolution than models used previously (2.5° of latitude by 3.75° of longitude) as well as updated physical parameterizations. By comparison with present day, the HadAM3 predicted a 1.9 °C annual mean warming over the globe. Warming was at its greatest in high latitudes during the winter (Fig. 2) season which resulted in a reduced equator-to-pole temperature gradient of 6 °C and a reduction in the general circulation of the

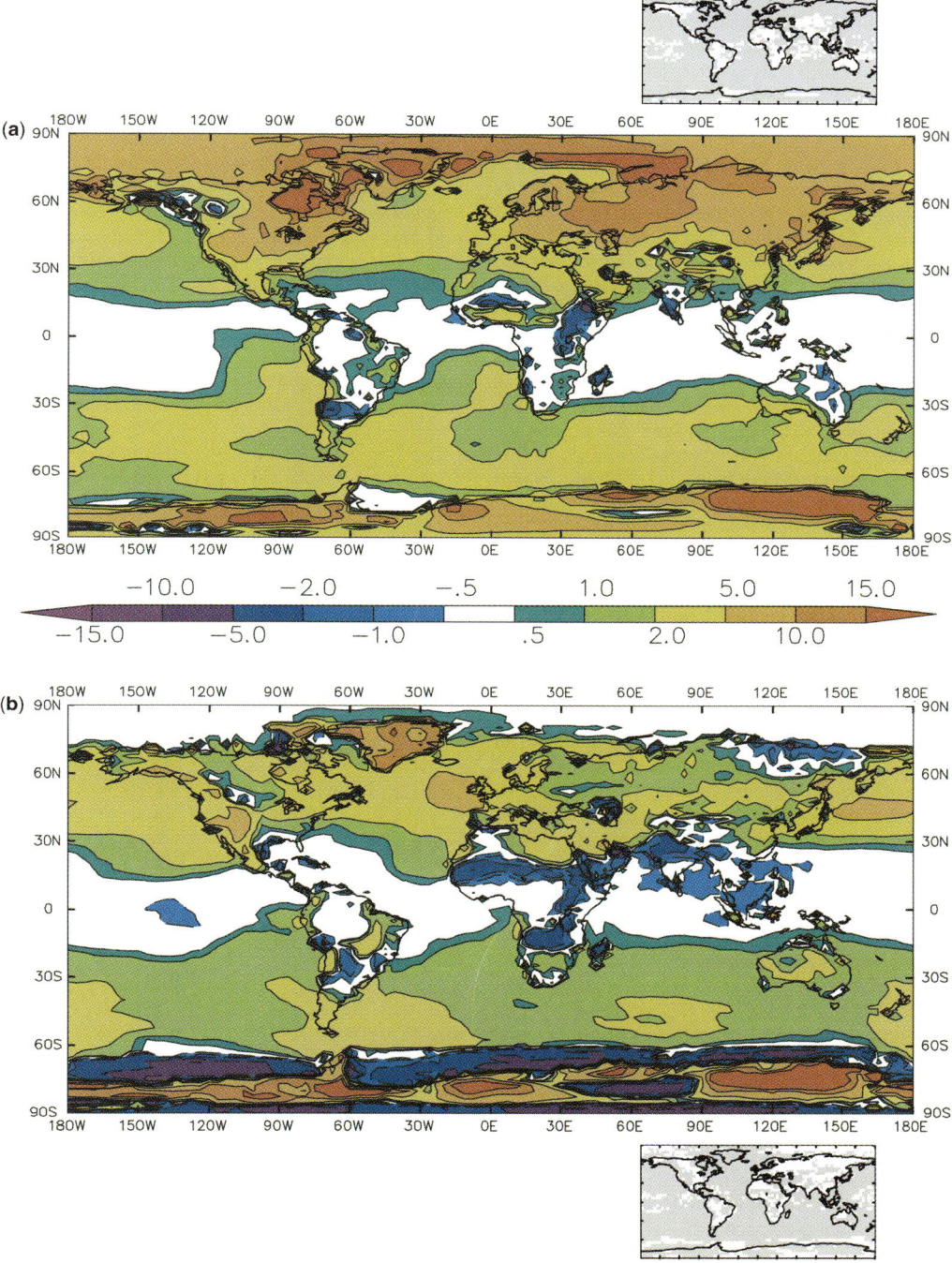

Fig. 2. HadAM3 prediction: difference in mean surface temperature between the mid-Pliocene and present-day experiments (°C) during (**a**) December, January & February (DJF) and (**b**) June, July & August (JJA). Note that the contour interval is not equally spaced and regions where temperature changes are less than 0.5 °C are unshaded. Grey shaded areas in accompanying mini-boxes show the geographical areas in which the difference in surface temperature is significant to a 95% significance level using a simple student t-test (redrawn from Haywood *et al.* 2000*b*; Haywood 2001).

atmosphere. Annual mean values for total precipitation (mm/day) increased by 6%. In low latitude and equatorial regions, temperatures decreased by a maximum of 8 °C (e.g. over East Africa). This change was generated by a precipitation increase in this region, associated with a broadening of the Hadley cell and also due to the increase in the specified elevation of the East African Rift Zone.

The results from all of these modelling studies are closely linked, and dependent on, the specified boundary conditions provided by the PRISM datasets. It is clear that alterations in SSTs and sea and land ice volumes within the PRISM datasets were primarily responsible for the difference in the mid-Pliocene climate relative to present-day simulations. The variations in climate were caused by the alteration of processes, including the release of sensible and latent heat (SST-driven) and also driven by variations in heat exchange between the ocean and atmosphere caused by differences in sea ice.

Modelling studies utilizing the PRISM2 dataset have not just considered the global climate. Output from GCMs has been used to examine particular regions in detail. For example, Haywood *et al.* (2000*a*) presented modelling results, supported by geological data, which suggested that in the European and Mediterranean region during the mid-Pliocene the climate was warmer (by 5 °C), wetter (by 400–1000 mm yr^{-1}) and less seasonal than present. Modelling results predicted an intensification of the Icelandic low-pressure system and the Azores high-pressure system as a direct result of higher annual SSTs and reduced ice cover in the Northern Hemisphere (Fig. 3). This change increased the surface pressure gradient over the region and strengthened annual westerly wind velocity by 4 m s^{-1}. Haywood *et al.* (2000*a*) postulated that associated increases in wind stress (by 20 N m^{-2}) over the North Atlantic Ocean may have enhanced the flow of surface currents such as the Gulf Stream and North Atlantic Current, which in turn would have sustained the higher SSTs. These changes to the regional climate system could have driven greater atmospheric and oceanic transport of heat from equatorial regions to the North Atlantic Ocean, particularly during winter. This mechanism provides a way to explain how enhanced meridional ocean heat transport could be achieved even when the equator-to-pole temperature gradient is reduced, and emphasized the potential for regional-scale processes to override global scale trends.

Combining vegetation models with AGCMs

The need to consider vegetation within palaeoclimate modelling experiments, particularly the interaction between vegetation and climate, is well documented (e.g. Cosgrove *et al.* 2002). A growing number of studies have demonstrated the importance of climate–vegetation interaction in understanding climate sensitivity and climate change (e.g. Bonan *et al.* 1992; Foley *et al.* 1994; Gallimore & Kutzbach 1996; Dutton & Barron 1996, 1997; Otto-Bleisner & Upchurch 1997; DeConto *et al.* 1999; Cosgrove *et al.* 2002). Initial vegetation modelling efforts concentrated on coupling the outputs from global climate models to empirical models of vegetation (e.g. Henderson-Sellers 1993). However, the complexity of vegetation models quickly increased to allow for the representation of numerous physiological processes such as photosynthesis, respiration, transpiration and soil water uptake (e.g. Prentice *et al.* 1992; Neilson 1995; Woodward *et al.* 1995; Haxeltine & Prentice 1996; Foley *et al.* 1998). Whereas initially, outputs from global climate models were simply used offline to the vegetation model itself, many of the vegetation models are now asynchronously or dynamically coupled to global climate models (VEMAP 1995; de Noblet *et al.* 1996; Betts *et al.* 1997; Texier *et al.* 1997; Claussen *et al.* 1998; Ganopolski *et al.* 1998; Doherty *et al.* 2000; Levis *et al.* 1999*a–c*, 2000; Cosgrove *et al.* 2002).

The rapid development of vegetation models has greatly expanded the number of applications for which they can be used, and therefore the usefulness of the models themselves (Cosgrove *et al.* 2002). These applications now include investigating the effects of changing carbon dioxide levels on primary productivity and competition (Culotta 1995; Jolly & Haxeltine 1997; Levis *et al.* 1999*b*, 2000) and the exploration of transient vegetation dynamics (Foley *et al.* 1996, 1998; Beerling *et al.* 1997; Friend *et al.* 1997).

Haywood *et al.* (2002*a–c*) used a mechanistically based biome model (BIOME 4) offline to the HadAM3 GCM to predict global mid-Pliocene biome distributions. These studies were useful in that they helped to quantify Pliocene climate–vegetation feedbacks and aid in the identification of mid-Pliocene vegetation patterns that are in equilibrium with different ice sheets and insolation scenarios. However, to better understand mid-Pliocene climate–vegetation feedbacks, the land cover had to be treated as an interactive element (i.e. actively growing vegetation) by incorporating a dynamic global vegetation model (DGVM).

Haywood & Valdes (2006) employed the TRIFFID Dynamic Global Vegetation Model (DGVM) and the HadAM3 GCM to investigate vegetation distributions and climate-vegetation feedbacks during the mid-Pliocene, and examine the implications of these results for the origins of hominid bipedalism. The model was initialized

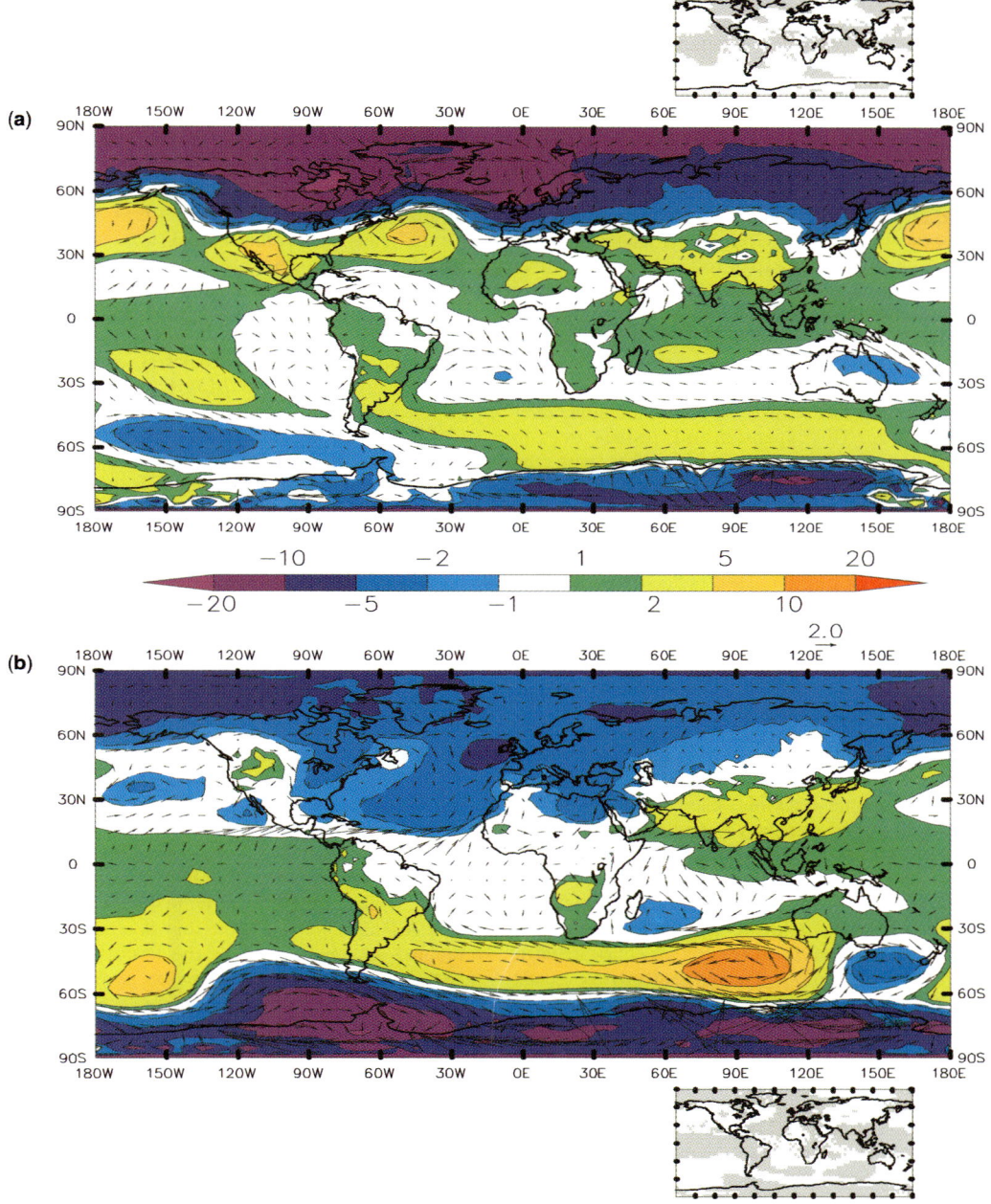

Fig. 3. HadAM3 predicted difference in surface pressure (hPa) between the mid-Pliocene and present day for (**a**) DJF and (**b**) JJA. Vector surface mean winds are also shown. The grey shading in accompanying mini-boxes shows the geographical areas in which the difference in surface pressure is significant to a 95% statistical significance level (redrawn from Haywood *et al.* 2000*b*; Haywood 2001).

using a Pliocene vegetation state translated from the PRISM2 vegetation scheme (Haywood & Valdes 2006). TRIFFID model outputs broadly supported the palaeoenvironmental reconstructions for the mid-Pliocene provided by the PRISM Group. TRIFFID simulated a significant increase in forest cover, composed of needleleaf trees in the higher latitudes of the Northern Hemisphere and Broadleaf trees in other regions. Needleleaf trees extended from the Arctic coast into the northern mid-latitudes. The fractional coverage of bare soil declined in North Africa, the Arabian Peninsula, Australia and southern South America, a pattern consistent with PRISM's assertion of reduced arid deserts. A significant increase in the fractional coverage of broadleaf trees in both Africa and South America was predicted but was found not to be indicative of a major expansion in tropical rainforests. Rather, it represented an expansion of general forest and woodland-type habitats in these regions. The principal impact of using a DGVM on the GCM-predicted climatology was to reduce minimum and maximum temperature extremes, thus reducing the seasonality of temperature over wide regions. The expansion in broadleaf trees in Africa was found to be incompatible with the 'savannah hypothesis' for the evolution of hominid bipedalism. Rather, the results lent credence to an alternative hypothesis which suggested that bipedalism evolved in wooded to forested ecosystems and was, for several million years, linked to arborealism (e.g. Pickford *et al.* 2004).

Palaeoclimate studies using coupled atmosphere and slab ocean models

Although AGCMs running with fixed SSTs have been widely applied to the mid-Pliocene and many other time intervals, they are only suitable for specific types of study. For example, it is inappropriate to use fixed SSTs in modelling experiments that aim to analyse the effect of varying concentrations of greenhouse gases or orbital parameters (Milankovitch forcing) on climate. This is because increasing or decreasing greenhouse gases acts to warm or cool oceans respectively, whist the orbital configuration will affect the geographical distribution of insolation over ocean areas. If an atmosphere-only simulation with fixed SSTs were used, the model would be unable to calculate the true effect that changing the greenhouse gas concentration had on ocean temperatures. If prescribed SSTs are used in modelling experiments that alter orbital parameters and insolation values, the effect will be to artificially dampen the resulting climate change. Such problems can, in part, be overcome by using a slab-ocean model that is capable of simulating SSTs, but the prescription of ocean heat transports is still required.

The UK Meteorological Office GCM has been used to address questions regarding the magnitude of climatic variability during the PRISM time slab. Mean climatic conditions during the

PRISM2 time slab are distinct when compared to immediately surrounding intervals. Benthic foraminiferal oxygen isotope values during this interval were either equal to, or isotopically lighter than (except for the KM2 (*c.* 3.12 Ma) and G20 (3.01 Ma) isotope excursions), those of today (Shackleton *et al.* 1995). Nevertheless, a degree of variability is clearly present within the time slab (e.g. Dowsett & Poore 1991; Barron 1992; Hodell & Venz 1992; Shackleton *et al.* 1995), especially over obliquity and precessional timescales (e.g. Tiedemann *et al.* 1994). Haywood *et al.* (2002*a*) presented a suite of palaeoclimate modelling experiments designed to examine the magnitude of climate variability during the PRISM time slab. In this study, the HadAM3 model was coupled to a Q-flux ocean model (HadSM3). Experiments were run for 3.29, 3.12 and 2.97 Ma before present (BP) Astronomical solutions for the periods were derived from the Berger and Loutre BL2 astronomical solution. Boundary conditions, excluding SSTs and sea-ice coverage, which were predicted by the slab-ocean model, were provided from the USGS PRISM2 dataset.

The model results indicated that little annual variation ($>0.5\,°C$) in SSTs, relative to the control experiment, occurred in response to the altered orbital configurations. Annual surface air temperatures also displayed little variation. Seasonally, surface air temperatures displayed a trend of cooler temperatures during December, January and February, and warmer temperatures during June, July and August. This pattern is consistent with altered seasonality resulting from the prescribed orbital configurations. Precipitation changes followed the seasonal trend observed for surface air temperature. Relative to present day, surface wind strength and wind stress over the North Atlantic, North Pacific and Southern Ocean remained greater in each of the Pliocene experiments, suggesting that wind-driven gyral circulation may have been consistently greater during the mid-Pliocene. However, it was unclear if the model correctly represented the magnitude of climate variability through the time slab. This was principally due to a lack of detailed time series data concerning changes to terrestrial ice cover and greenhouse gas concentrations for the mid-Pliocene which, if available, could have had a significant impact on model results.

Fully coupled Ocean Atmosphere GCMs

All of the palaeoclimate modelling studies for the mid-Pliocene discussed thus far have utilized atmospheric GCMs with prescribed SSTs and/or relatively simple slab-ocean models. Although slab-ocean models are capable of simulating part of the feedbacks of the oceans on climate, they are incapable of simulating changes to horizontal ocean heat transports, and related changes to ocean currents and thermohaline circulation. Yet these mechanisms have been recognized as being potentially critical drivers of mid-Pliocene climate. Haywood & Valdes (2004) remedied this situation through applying a fully coupled ocean-atmosphere model (HadCM3) to the mid-Pliocene for the first time. The aim of their study was to examine the relative roles of the atmosphere and oceans in generating and/or maintaining mid-Pliocene warmth. Since HadCM3 was capable of predicting SSTs and sea-ice coverage/depth, it was not necessary to force the model with those conditions from the PRISM2 dataset. However, due to concerns regarding multiple equilibria in coupled models, and their desire to put the coupled model into a mid-Pliocene state, Haywood & Valdes (2004) initialized the surface ocean with monthly SSTs derived from the PRISM2 dataset. For the first 30 years of the model integration, a simple Haney restoring term fixed the model-predicted SSTs to the PRISM2 SST reconstruction and sea-surface salinity (SSS) to the present-day value. However, after 30 simulated years, the Haney function was gradually reduced to zero, after which the model was completely free to simulate changes in SST and SSS.

The simulation resulted in a global surface temperature warming of 3 °C compared to present day. In contrast to earlier modelling experiments for the mid-Pliocene, surface temperatures warmed in most areas including the tropics (1 to 5 °C; Fig. 4). Compared with present day, the model predicted a general pattern of ocean warming (1 to 5 °C) in both hemispheres to a depth of 2000 m, below which no significant differences were noted. Sea-ice coverage was massively reduced (up to 90%). The velocity of the Gulf Stream/North Atlantic Drift was up to 100 mm s^{-1} greater in the mid-Pliocene. Analysis of the model-predicted meridional streamfunction suggested a global pattern of reduced outflow of Antarctic bottom water (AABW; up to 5 Sverdrups), a shallower depth for North Atlantic Deep Water formation and weaker thermohaline circulation (3 Sv). The decrease in AABW occurs mainly in the Pacific rather than in the Atlantic Ocean. Model diagnostics for heat transports indicated that neither the oceans nor the atmosphere transported significantly more heat in the mid-Pliocene case. Rather, the results indicated that the major contributing mechanism to mid-Pliocene warmth was the reduced extent of high latitude terrestrial ice sheets (50% reduction on Greenland, 33% reduction on Antarctica) and sea-ice cover resulting in a strong ice-albedo feedback (Haywood & Valdes 2004).

Significant differences were noted between absolute PRISM SSTs and those predicted by HadCM3. Specifically, the model appeared to underestimate SST increases in the North Atlantic and North Pacific and overestimate SST increases in the tropics and low latitudes (Fig. 4). These differences either relate to errors in the PRISM SST and/or the model itself. In an effort to assess this, Haywood et al. (2005) carried out a comparison between HadCM3 predicted SSTs and those predicted by an independent geochemical proxy, specifically alkenone palaeothermometry carried out on three Pacific Ocean ODP sites. These estimates, combined with published alkenone SST data from two Atlantic sites (Herbert et al. 1998; Marlow et al. 2000), were located in tropical and subtropical regions where the distribution of PRISM SST estimates is sparse. Significant differences were noted between absolute PRISM SSTs interpolated to the core locations and alkenone-derived SSTs. These differences may relate to errors in the PRISM and alkenone paleothermometry estimates, or to regional differences between localities where data were collected. However, the available alkenone and model-based SST estimates were consistent in the sign of temperature change they predicted, and provided the first indication of warmer SSTs during the mid-Pliocene in the tropics and sub-tropics. SST estimates derived using Mg/Ca palaeothermometry in the eastern equatorial Pacific support the results obtained by alkenone palaeothermometry and the fully coupled ocean-atmosphere GCM (Wara et al. 2005). These results contrast with PRISM's estimates of unchanged or slightly cooler SSTs for the same geographical regions.

The pattern of SSTs, produced by alkenone estimates and HadCM3, was not characteristic of that produced by enhanced meridional ocean heat transport or thermohaline circulation. Instead, the pattern is similar to that which might be expected as a result of higher concentrations of atmospheric CO_2, which would act to warm the oceans at the tropics and other latitudes. Haywood et al. (2005) concluded that further work aimed at validating these conclusions should concentrate on expanding the mid-Pliocene alkenone SST dataset, particularly at low latitudes and in areas other than ocean upwelling regions.

Recent developments and the way forward

The PRISM 3D project

A recent development has been the launch of the PRISM 3D collaborative data analysis and climate modelling effort. The primary goal of this iteration

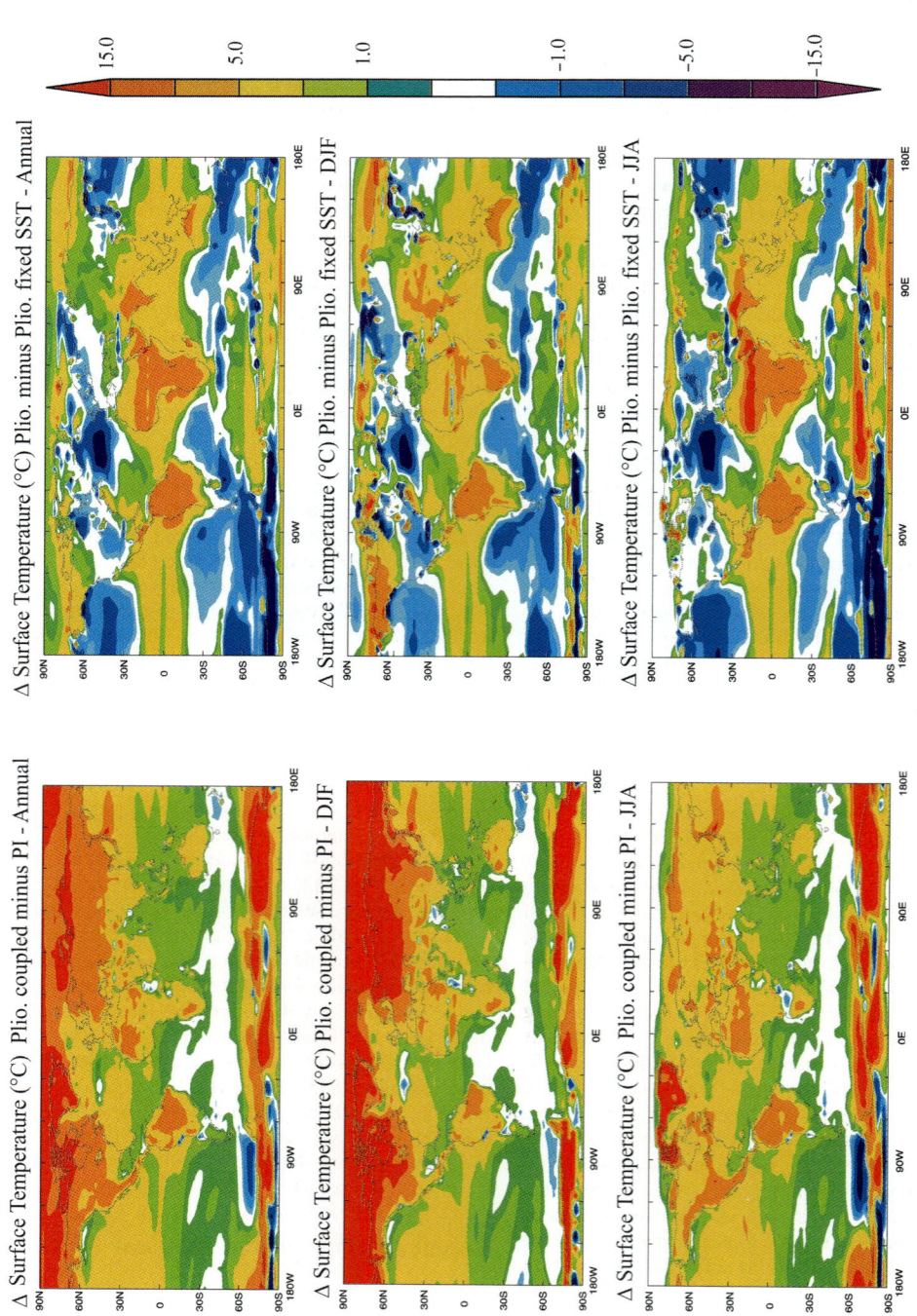

Fig. 4. Model-predicted sea-surface temperatures (SST) and surface temperatures over land (°C). From top left to bottom right: the difference between the mid-Pliocene coupled and pre-industrial coupled experiment (PI) for DJF; the difference between the mid-Pliocene coupled experiment and a mid-Pliocene experiment using fixed PRISM2 SSTs for DJF; the difference between mid-Pliocene coupled and pre-industrial coupled experiments for JJA and the difference between the mid-Pliocene coupled experiment and a mid-Pliocene experiment using fixed PRISM2 SSTs for JJA (redrawn from Haywood *et al.* 2005).

of the PRISM project is to create three-dimensional global datasets of mid-Pliocene ocean temperature and salinity, which will form the most comprehensive global reconstruction for any warm period prior to the recent past. The datasets will then be used to drive coupled ocean-atmosphere GCM simulations designed to explore the impact of climate forcings and feedbacks during the mid-Pliocene (see http://geology.er.usgs.gov/eespteam/ prism/prism3main.html). Deep-water temperatures and salinities are to be included to reproduce a three-dimensional reconstruction of mid-Pliocene ocean temperatures. Although the PRISM 3D project is relatively new, significant progress has already been made. Specifically, two new SST reconstructions have been constructed which are designed to provide a climatological error bar for warm peak phases of the Pliocene and to document the spatial distribution and magnitude of SST variability within the mid-Pliocene warm period. These data suggest long-term stability of low-latitude SSTs and document greater variability in regions of maximum warming (Dowsett 2004; Dowsett et al. 2005).

Magnesium : calcium (Mg/Ca) ratios in shells of the mid-Pliocene deep-sea ostracode *Krithe* from deep-sea drilling sites in the North and South Atlantic have also been examined in order to estimate bottom-water temperatures (BWT). Results from DSDP and ODP Sites 552A, 610A, 607, 658A, 659A, 661A and 704 for the period reveal both depth and latitudinal gradients of mean Mg/Ca values. North Atlantic deep-water temperature change during late Pliocene and late Quaternary climatic cycles suggests that mean mid-Pliocene bottom-water temperatures at the study sites in the deep Atlantic were about the same as modern temperatures. However, brief pulses of elevated BWT occurred several times between 3.29 and 2.97 Ma in both the North and South Atlantic Oceans, suggesting short-term changes in deep ocean circulation (Cronin et al. 2005).

Chandler et al. (2006) noted that initializing Pliocene-coupled ocean-atmosphere simulations using these deep ocean temperatures led to a 15% increase in North Atlantic stream function in the GISS model compared to Pliocene experiments that used modern ocean values to initiate the runs (Levitus & Boyer 1994). Such increases are not large compared to modern values, because the GISS Pliocene runs had previously shown a reduction in NADW production compared to the modern. It is interesting however that inclusion of deep ocean temperatures in the set-up of the model run had the effect of increasing ocean overturning which, qualitatively at least, agrees with interpretations derived from proxy data (Chandler et al. 2006).

Multi-proxy reconstructions of mid-Pliocene ocean temperatures

Sufficient palaeoceanographic data and model outputs exist for the mid-Pliocene to facilitate combined multi-proxy reconstructions of mid-Pliocene ocean temperatures. Such an approach is useful for testing the robustness of model outputs, but also for assessing the relative efficacy of different proxies. Williams et al. (2005) reconstructed Atlantic SSTs using the $\delta^{18}O$ composition of fossil planktonic foraminifer calcite for the Kaena Subchron (3.12 to 3.05 Ma BP). These estimates were compared to the SST estimates derived from PRISM and outputs from a fully coupled ocean-atmosphere GCM (Haywood & Valdes 2004). It was found that most SST estimates derived from the $\delta^{18}O$ data indicated cooler, by as much as 5 °C, than modern SSTs. Difficulties in interpreting the ecology of fossil foraminifer assemblages and inaccurate estimate of the mid-Pliocene seawater $\delta^{18}O$ composition could only partly explain the reconstructed differences. Instead, Williams et al. (2005) ascribe the difference as being a product of (a) calcite formed at a level deep within or below the ocean mixed-layer during the life cycle of the foraminifera, or (b) secondary calcite with higher $\delta^{18}O$ formed in the planktonic foraminifer tests in sea bottom pore waters. It was concluded that reconstructing accurate mid-Pliocene SSTs with much of the published oxygen isotope data requires a detailed re-assessment of taphonomy.

Modelling mid-Pliocene ice sheets

Coupled ocean-atmosphere GCM studies (Haywood & Valdes 2004; Haywood et al. 2005) have indicated that the largest contribution to mid-Pliocene warmth may have been delivered by the cryosphere through strong ice-albedo feedbacks. Through the efforts of the PRISM Group, our knowledge of mid-Pliocene sea-surface temperatures, vegetation cover and topography is the most complete of any pre-Quaternary time slab. However, PRISM's reconstructions of the mid-Pliocene ice sheets are based solely upon past sea-level estimates, derived from well-dated palaeo-shorelines (Dowsett & Cronin 1990; Wardlaw & Quinn 1991) that may have significant uncertainties associated with the estimates. Because of these uncertainties, estimates of the mid-Pliocene Antarctic ice-sheet volume range from near present-day values to >33% reduction compared to modern. This uncertainty severely hampers modelling efforts designed to investigate the cause of mid-Pliocene warmth and restricts the value of this period as a test bed for numerical climate models because we do not even know if the current

PRISM Antarctic ice-sheet reconstruction is consistent with other palaeoenvironmental data (e.g. global SST distributions, palaeobiological evidence).

Over the past two decades, there has been great controversy over the extent of the East Antarctic Ice Sheet (EAIS) during the Pliocene. One group of researchers claims that the Pliocene saw huge changes in East Antarctic ice volume (e.g. Webb *et al.* 1984), while the conventional view is of a stable EAIS since the Miocene (e.g. Sugden 1996). The geological record on Antarctica is sparse, but data apparently supporting and refuting both viewpoints have been obtained. As the geological evidence is currently inconclusive and the debate shows no sign of resolution, it is important that the extent of the EAIS during the Pliocene is tested through modelling. In the future, outputs from prescribed ice sheet GCM modelling experiments should be used to drive three-dimensional, thermomechanical ice-sheet models. This will allow the validity of the PRISM ice sheets to be tested (Hill *et al.* this volume).

The need for model/model intercomparisons

Finally, the need to perform model/model intercomparisons has been clearly demonstrated by the PMIP and PMIPII projects (Palaeoclimate Modelling Intercomparison Project; Joussaume & Taylor 2000). The aim of the current phase of PMIP (PMIPII) is to study the role of climate feedbacks arising for the different climate subsystems (atmosphere, ocean, land surface, sea ice and land ice) and evaluate the capability of state-of-the-art climate models to reproduce climate states that are radically different from those of today (i.e. Last Glacial Maximum and the mid-Holocene). Results from both coupled ocean-atmosphere models and ocean-atmosphere-vegetation models are being considered in this second phase. PMIPII is continuing to stimulate development and improvement of palaeoenvironmental datasets (for more information, see: http://www.lsce.cea.fr/pmip2/). Although there have been numerous modelling experiments performed for the mid-Pliocene, the question of how model-dependent the results are has yet to be considered. This is an essential next step if the importance of modelling mid-Pliocene warmth is to be fully realized, and a vital ingredient in ensuring that palaeoenvironmental datasets for the interval continue to be developed and refined.

Conclusions

In this paper, we have examined the use of the mid-Pliocene as a test bed for the integration between data and numerical climate models. The mid-Pliocene is seen to be a climatically distinct interval between 2.97 and 3.29 Ma BP when Earth's climate appears to have been generally warmer than present day. The interval has been the subject of intense study by geologists and palaeoclimatologists. The period is ideal for examining dynamics of past warm climates and for testing the retrodictions produced by the same climate models that are used for future climate change prediction. A comprehensive review of the research of the US Geological Survey's PRISM (Pliocene Research Interpretations & Synoptic Mapping) Group, along with all of the modelling studies which have been derived from this primary work, illustrates the importance and usefulness of a combined data/modelling strategy, and it is apparent that both the models and the proxies used for reconstructing past climates need further testing. Strategies for future research must include the use of multiproxy studies, coupled atmosphere-ocean-vegetation-ice sheet models, and involve model/model intercomparisons.

The Natural Environment Research Council is acknowledged for supporting the work of AH and PJV through the provision of High Performance Computing. This paper is published as part of the British Antarctic Survey's GEACEP programme that seeks to investigate climate change over geological timescales. We also thank Prof. Bruce Sellwood for help and advice regarding Pliocene palaeoclimate modelling and Harry Dowsett and the PRISM Group for use of their materials. Mark Chandler and John Whittaker are thanked for their constructive reviews.

References

BARRON, J. A. 1992. Pliocene paleoclimatic interpretation of DSDP Site 580 (NW Pacific) using diatoms. *Marine Micropaleontology*, **20**, 23–44.

BEERLING, D. J., WOODWARD, F. I., LOMAS, M. & JENKINS, A. J. 1997. Testing the responses of a dynamic global vegetation model to environmental change: a comparison of observations and predictions. *Global Ecology & Biogeography*, **6**, 439–450.

BERGGREN, W. A., KENT, D. V. & VAN COUVERING, J. A. 1985. Neogene geochronology and chronostratigraphy. *In*: SNELLING, N. J. (ed.) The Chronology of the Geological Record. *Geological Society of London Memoir*, **10**, 211–260.

BERGGREN, W. A., KENT, D. V., SWISHER, C. C. & AUBRY, M. P. 1995. A revised Cenozoic geochronology and chronostratigraphy. *In*: BERGGREN, W. A., KENT AUBRY, M. P. & HARDENBOL, J. (eds) Geochronology, time scales and global stratigraphic correlation. *Tulsa, Society for Sedimentary Geology Special Publication*, **54**, 129–212.

BERNER, R. A. 1990. Atmospheric carbon dioxide over Phanerozoic time. *Science*, **249**, 1382–1386.

BETTS, R. A., COX, P. M., LEE, S. E. & WOODWARD, F. I. 1997. Contrasting physiological and structural vegetation feedbacks in climate change simulations. *Nature*, **387**, 796–799.

BONAN, G. B., POLLARD, D. & THOMPSON, S. L. 1992. Effects of boreal forest vegetation on global climate. *Nature*, **359**, 716–718.

CANE, M. A. & MOLNAR, P. 2001. Closing of the Indonesian seaway as a precursor to East African aridification around 3–4 million years ago. *Nature*, **411**, 157–162.

CHANDLER, M., RIND, D. & THOMPSON, R. 1994. Joint investigations of the middle Pliocene climate II: GISS GCM Northern Hemisphere results. *Global and Planetary Change*, **9**, 197–219.

CHANDLER, M., DOWSETT, H., DWYER, G., CRONIN, T. & ROBINSON, M. 2006. The last great global warming: a modelling perspective into mid-Pliocene climate. *Annual Meeting of the American Association for the Advancement of Science*, St Louis Missouri,

CLAUSSEN, M., BROVKIN, V., GANOPOLSKI, A., KUBATZKI, C. & PETOUKHOV, V. 1998. Modelling global terrestrial vegetation-climate interactions. *Philosophical Transactions of the Royal Society of London, Series B*, **353**, 53–63.

COSGROVE, B. A., BARRON, E. J. & POLLARD, D. 2002. A simple interactive vegetation model coupled to the GENESIS GCM. *Global & Planetary Change*, **32**, 253–278.

CRONIN, T. M., DOWSETT, H. J., DWYER, G. S., BAKER, P. A. & CHANDLER, M. A. 2005. Mid Pliocene deep-sea bottom water temperatures based on ostracode Mg/Ca ratios. *Marine Micropaleontology*, **54**, 249–261.

CROWLEY, T. J. 1996. Pliocene climates: the nature of the problem. *Marine Micropaleontology*, **27**, 3–12.

CULOTTA, E. 1995. Will plants profit from high CO_2? *Science*, **268**, 654–656.

DECONTO, R., BRADY, E., BERGENGREN, J., THOMPSON, S., POLLARD, D. & HAY, W. 1999. Late Cretaceous climate, vegetation, and ocean interactions. *In*: HUBER, B. *ET AL*. (eds) *Warm Climates in Earth History*, 275–297.

DE NOBLET, N. I., PRENTICE, I. C., JOUSSAUME, S., TEXIER, D., BOTTA, A. & HAXELTINE, A. 1996. Possible role of atmosphere–biosphere interactions in triggering the last glaciation. *Geophysical Research Letters*, **23**, 3191–3194.

DOHERTY, R., KUTZBACH, J., FOLEY, J. & POLLARD, D. 2000. Fully coupled climate/dynamical vegetation model simulations over northern Africa during the mid-Holocene. *Climate Dynamics*, **16**, 561–573.

DOWSETT, H. J. 2004. Bracketing mid Pliocene sea surface temperature. Maximum and minimum possible warming. US Geological Survey Data Series, DS114 (http://pubs.usgs.gov/ds/2004/114/).

DOWSETT, H. J. & CRONIN, T. M. 1990. High eustatic sea level during the Middle Pliocene: Evidence from the southeastern U.S. Atlantic Coastal Plain. *Geology*, **18**, 435–438.

DOWSETT, H. J. & POORE, R. Z. 1991. Pliocene sea surface temperatures of the North Atlantic Ocean at 3.0 Ma. *Quaternary Science Reviews*, **10**, 189–204.

DOWSETT, H. J., CRONIN, T. M., POORE, P. Z., THOMPSON, R. S., WHATLEY, R. C. & WOOD, A. M. 1992. Micropaleontological evidence for increased

meridional heat-transport in the North Atlantic Ocean during the Pliocene. *Science*, **258**, 1133–1135.

DOWSETT, H. J., THOMPSON, R. S., BARRON, J. A. *ET AL*. 1994. Joint investigations of the middle Pliocene climate I: PRISM paleoenvironmental reconstructions. *Global and Planetary Change*, **9**, 169–195.

DOWSETT, H., BARRON, J. & POORE, R. 1996. Middle Pliocene sea surface temperatures: a global reconstruction. *Marine Micropaleontology*, **27**, 13–26.

DOWSETT, H. J., BARRON, J. A., POORE, R. Z., THOMPSON, R. S., CRONIN, T. M., ISHMAN, S. E. & WILLARD, D. A. 1999. Middle Pliocene palaeoenvironmental reconstruction: PRISM2. *USGS Open File Report 99–535* (http://pubs.usgs.gov/of/1999/of99-535/).

DOWSETT, H. J., CHANDLER, M. A., CRONIN, T. M. & DWYER, G. S. 2005. Middle Pliocene sea surface temperature variability. *Paleoceanography*, **20**, PA2014.

DOWSETT, H. J. (2007). Pliocene Sea-Surface Temperature and the PRISM Paleoclimate Reconstructions. *In*: WILLIAMS, M., HAYWOOD, A. M., GREGORY, F. J. & SCHMIDT, D. N. (eds) *Deep-time Perspectives on Climate Change: marrying the signal from computer models and biological proxies*. The Micropalaeontological Society, Special Publications. The Geological Society, London, pp. 000–000

DUTTON, J. F. & BARRON, E. J. 1996. GENESIS sensitivity to changes in past vegetation. *Palaeoclimates: Data & Modelling*, **1**, 325–354.

DUTTON, J. F. & BARRON, E. J. 1997. Miocene to present vegetation changes: a possible piece of the Cenozoic puzzle. *Geology*, **25**, 39–41.

FOLEY, J. A., KUTZBACH, J. E., COE, M. T. & LEVIS, S. 1994. Feedbacks between climate and boreal forests during the Holocene epoch. *Nature*, **371**, 52–54.

FOLEY, J. A., PRENTICE, I. C., RAMANKUTTY, N., LEVIS, S., POLLARD, D., SITCH, S. & HAXELTINE, A. 1996. An integrated biosphere model of land surface processes, terrestrial carbon balance, and vegetation dynamics. *Global Biogeochemical Cycles*, **10**, 603–628.

FOLEY, J. A., LEVIS, S., PRENTICE, I. C., POLLARD, D. & THOMPSON, S. L. 1998. Coupling dynamic models of climate and vegetation. *Global Change Biology*, **4**, 561–579.

FRIEND, A. D., STEVENS, A. K., KNOX, R. G. & CANNELL, M. G. R. 1997. A process-based terrestrial biosphere model of ecosystem dynamics (Hybrid v 3.0). *Ecological Modelling*, **95**, 249–287.

GALLIMORE, R. G. & KUTZBACH, J. E. 1996. Role of orbitally induced changes in tundra area in the onset of glaciation. *Nature*, **381**, 503–505.

GANOPOLSKI, A., KUBATZKI, C., CLAUSSEN, M., BROVKIN, V. & PETOUKHOV, V. 1998. The influence of vegetation–atmosphere–ocean interaction on climate during the mid-Holocene. *Science*, **280**, 1916–1919.

HAUG, G. H. & TIEDEMANN, R. 1998. Effect of the formation of the Isthmus of Panama on Atlantic Ocean thermohaline circulation. *Nature*, **393**, 673–676.

HAYWOOD, A. M. 2001. Evaluating General Circulation Climate Model Reliability Against the Pliocene

Geological Record. Unpublished PhD Thesis, The University of Reading, UK, 295pp.

HAYWOOD, A. M., SELLWOOD, B. W. & VALDES, P. J. 2000a. Regional warming: Pliocene (c. 3 Ma) paleoclimate of Europe and the Mediterranean. *Geology*, **28**, 1063–1066.

HAYWOOD, A. M., VALDES, P. J. & SELLWOOD, B. W. 2000b. Global scale palaeoclimate reconstruction of the middle Pliocene climate using the UKMO GCM: initial results. *Global & Planetary Change*, **25**, 239–256.

HAYWOOD, A. M., VALDES, P. J., FRANCIS, J. E. & SELLWOOD, B. W. 2002a. Global middle Pliocene biome reconstruction: A data/model synthesis. *Geochemistry, Geophysics, Geosystems*, **3**, 1072, doi:10.1029/2002GC000358.

HAYWOOD, A. M., VALDES, P. J. & SELLWOOD, B. W. 2002b. Magnitude of middle Pliocene climate variability: A palaeoclimate modelling study. *Palaeogeography, Palaeoclimatology, Palaeoecology*, **188**, 1–24.

HAYWOOD, A. M., VALDES, P. J., SELLWOOD, B. W. & KAPLAN, J. O. 2002c. Antarctic climate during the middle Pliocene: model sensitivity to ice sheet variation. *Palaeogeography, Palaeoclimatology, Palaeoecology*, **182**, 93–115.

HAYWOOD, A. M. & VALDES, P. J. 2004. Modelling Pliocene warmth: contributions of atmosphere, oceans and cryosphere. *Earth and Planetary Science Letters*, **218**, 363–377.

HAYWOOD, A. M., DEKENS, P., RAVELO, A. C. & WILLIAMS, M. 2005. Warmer tropics during the mid Pliocene: evidence from alkenone paleothermometry and a fully coupled ocean-atmosphere GCM. *Geochemistry, Geophysics, Geosystems*, **6**, doi: 10.1029/2004GC000799.

HAYWOOD, A. M. & VALDES, P. J. 2006. Vegetation Cover in a Warmer World Simulated using a Dynamic Global Vegetation Model for the mid Pliocene: Implications for the Origin of Hominid Bipedalism. *Palaeogeography, Palaeoclimatology, Palaeoecology*, **237**, 412–427.

HAXELTINE, A. & PRENTICE, I. C. 1996. BIOME3: an equilibrium terrestrial biosphere model based on ecophysiological constraints, resource availability, and competition among plant functional types. *Global Biogeochemical Cycles*, **10**, 693–709.

HENDERSON-SELLERS, A. 1993. Continental vegetation as a dynamic component of a global climate model: A preliminary assessment. *Climatic Change*, **23**, 337–377.

HERBERT, T. D., SCHUFFERT, J. D., THOMAS, D., LANGE, K., WEINHEIMER, A. & HERGUERA, J. C. 1998. Depth and seasonality of alkenone production along the California margin inferred from a core-top transect. *Paleoceanography*, **13**, 263–271.

HILL, D., HAYWOOD, A. M., HINDMARSH, R. C. & VALDES, P. J. (2007). Characterizing ice sheets during the Pliocene: evidence from data and models. *In*: WILLIAMS, M., HAYWOOD, A. M., GREGORY, F. J. & SCHMIDT, D. N. (eds) *Deep-Time Perspectives on Climate Change: Marrying the Signal from Computer Models and Biological Proxies*. The Micropalaeontological Society, Special Publications. The Geological Society, London, pp. 000–000.

HODELL, D. A. & VENZ, K. 1992. Toward a high-resolution stable isotopic record of the Southern Ocean during the Pliocene-Pleistocene (4.8 to 0.8 Ma). *Antarctic Research Series*, **56**, 265–310.

JOLLY, D. & HAXELTINE, A. 1997. Effect of low glacial atmospheric CO_2 on tropical African montane vegetation. *Science*, **276**, 786–787.

JOUSSAUME, S. & TAYLOR, K. E. 2000. Modelling extreme climates of the past: what we have learned from PMIP and related experiments. *PAGES Newsletter/CLIVAR exchanges*, **8/5**, 18–19.

KENNETT, J. P. & HODELL, D. A. 1993. Evidence for relative climatic stability of Antarctica during the early Pliocene: A marine perspective. *Geografiska Annaler*, **75**, 205–220.

KIM, S. J. & CROWLEY, T. J. 2000. Increased Pliocene North Atlantic Deep Water: Cause or consequence of Pliocene warming. *Paleoceanography*, **15**, 451–455.

KÜRSCHNER, W. M., VAN DER BURGH, J., VISSCHER, H. & DILCHER, D. L. 1996. Oak leaves as biosensors of late Neogene and early Pleistocene paleoatmospheric CO_2 concentrations. *Marine Micropaleontology*, **27**, 299–31.

LEVIS, S., FOLEY, J. A. & POLLARD, D. 1999a. Potential high-latitude vegetation feedbacks on CO_2-induced climate change. *Geophysical Research Letters*, **26**, 747–750.

LEVIS, S., FOLEY, J. A. & POLLARD, D. 1999b. CO_2, climate, and vegetation feedbacks at the Last Glacial Maximum. *Journal of Geophysical Research*, **104** (D24), 31191–31198.

LEVIS, S., FOLEY, J. A., BROVKIN, V. & POLLARD, D. 1999c. On the stability of the high latitude climate–vegetation system in a coupled atmosphere–biosphere model. *Global Ecology & Biogeography*, **8**, 489–500.

LEVIS, S., FOLEY, J. A. & POLLARD, D. 2000. Large-scale vegetation feedbacks on a doubled CO_2 climate. *Journal of Climate*, **13**, 1313–1325.

LEVITUS, S. & BOYER, T. P. 1994. World Ocean Atlas 1994 (Vol. 4): Temperature. *NOAA Atlas NESDIS* 4.

MARLOW, J. R., LANGE, C. B., WEFER, G. & ROSELL-MELÉ, A. 2000. Upwelling intensification as part of the Plio-Pleistocene climate transition. *Science*, **290**, 2288–2294.

NEILSON, R. P. 1995. A model for predicting continental-scale vegetation distribution and water balance. *Ecological Applications*, **5**, 362–385.

OTTO-BLEISNER, B. & UPCHURCH, G. 1997. Vegetation induced warming of high-latitude regions during the Late Cretaceous period. *Nature*, **385**, 804–807.

PAGANI, M., ZACHOS, J. C., FREEMAN, K. H., TIPPLE, B. & BOHATY, S. 2005. Marked decline in atmospheric carbon dioxide concentrations during the Paleogene. *Science*, **309**, 600–603.

PEARSON, P. N. & PALMER, M. R. 2000. Atmospheric carbon dioxide over the past 60 million years. *Nature*, **406**, 695–699.

PHILANDER, S. G. & FEDOROV, A. V. 2003. Role of tropics in changing the response to Milankovitch forcing some three million years ago. *Paleoceanography*, **18**, 1045, doi:10.1029/2002PA000837.

PICKFORD, M., SENUT, B. & MOURER-CHAUVIRÉ, C. 2004. Early Pliocene Tragulidae and peafowls in the

Rift Valley, Kenya: evidence for rainforest in East Africa. *Comptes Rendus Palevol*, **3**, 179–189.

POORE, R. Z. & SLOAN, L. C. (eds) 1996. Climates and Climate Variability of the Pliocene, *Marine Micropaleontology*, **27**.

PRENTICE, I. C., CRAMER, W., HARRISON, S. P., LEEMANS, R., MONSERUD, R. A. & SOLOMON, A. M. 1992. A global biome model based on plant physiology and dominance, soil properties and climate. *Journal of Biogeography*, **19**, 117–134.

PRISM Project Members 1994. PRISM $8° \times 10°$ Northern Hemisphere paleoclimate reconstruction: Digital data. *US Geological Survey Open File Report*, **94–281**, 23pp.

RAVELO, A. C. & ANDREASEN, D. H. 2000. Enhanced circulation during a warm period. *Geophysical Research Letters*, **27**, 1001–1004.

RAYMO, M. E. & RAU, G. H. 1992. Plio-Pleistocene atmospheric CO_2 levels inferred from POM $\delta^{13}C$ at DSDP Site 607. *EOS, Transactions of the American Geophysical Union, Spring Meeting*, **73** (Suppl.).

RAYMO, M. E., GRANT, B., HOROWITZ, M. & RAU, G. H. 1996. Mid-Pliocene warmth: stronger greenhouse and stronger conveyor. *Marine Micropaleontology*, **27**, 313–326.

REYNOLDS, R. W. & SMITH, T. M. 1995. A high-resolution global sea-surface temperature climatology. *Journal of Climatology*, **7**, 1571–1583.

RIND, D. & CHANDLER, M. 1991. Increased ocean heat transports and warmer climate. *Journal of Geophysical Research*, **96**, 7437–7461.

SHACKLETON, N. J., LE, J., MIX, A. & HALL, M. A. 1992. Carbon isotope records from Pacific surface waters and atmospheric carbon dioxide. *Quaternary Science Reviews*, **11**, 387–400.

SHACKLETON, N. J., HALL, M. A. & PETE, D. 1995. Pliocene stable isotope stratigraphy of Site 846. *In*: PISIAS, N. G., MAYER, L. A., JANECSEK, T. R., PALMER-JULSON, A. & VAN ANDEL, T. H. (eds) *Proceedings of the Ocean Drilling Program, Scientific Results*, **138**, 337–353.

SLOAN, L. C., CROWLEY, T. J. & POLLARD, D. 1996. Modeling of middle Pliocene climate with the NCAR GENESIS general circulation model. *Marine Micropaleontology*, **27**, 51–61.

SUGDEN, D. E. 1996. The East Antarctic Ice Sheet: unstable ice or unstable ideas? *Transactions of the Institute of British Geographers*, **21**, 443–454.

TEXIER, D., DE NOBLET, N., HARRISON, S. P. *ET AL.* 1997. Quantifying the role of biosphere–atmosphere feedbacks in climate change: coupled model simulations for 6000 years BP and comparison with paleodata for northern Eurasia and Africa. *Climate Dynamics*, **13**, 865–882.

THOMPSON, R. S. & FLEMING, R. F. 1996. Middle Pliocene vegetation: reconstructions, paleoclimatic inferences, and boundary conditions for climatic modeling. *Marine Micropaleontology*, **27**, 13–26.

TIEDEMANN, R., SARNTHEIN, M. & SHACKLETON, N. J. 1994. Astronomic timescale for the Pliocene $\delta^{18}O$ and dust records of Ocean Drilling Program Site 659. *Paleoceanography*, **9**, 619–638.

VALDES, P. J. 1993. Atmospheric general circulation models for the Jurassic. *Philosophical Transactions of the Royal Society of London*, **B341**, 317–326.

VAN DER BURGH, J., VISSCHER, H., DILCHER, D. L. & KÜRSCHNER, W. M. 1993. Paleoatmospheric signatures in Neogene fossil leaves. *Science*, **260**, 1788–1790.

VEMAP Members 1995. VEMAP: a comparison of biogeography and biogeochemistry models in the context of global climate change. *Global Biogeochemical Cycles*, **9**, 407–437.

WARA, M., RAVELO, A. C. & DELANEY, M. L. 2005. Permanent El Niño conditions during the Pliocene warm period. *Science*, **309**, 758–761.

WARDLAW, B. R. & QUINN, T. M. 1991. The record of Pliocene sea-level changes at Enewetak Atoll. *Quaternary Science Reviews*, **10**, 247–258.

WEBB, P. N., HARWOOD, D. M., MCKELVEY, B. C., MERCER, J. H. & SCOTT, L. D. 1984. Cenozoic marine sedimentation and ice-volume variation on the East Antarctic craton. *Geology*, **12**, 287–291.

WILLIAMS, M., HAYWOOD, A. M., HILLENBRAND, C. D. & WILKINSON, I. P. 2005. Efficacy of $\delta^{18}O$ data from Pliocene planktonic foraminifer calcite for spatial sea surface temperature reconstruction: comparison with a fully coupled ocean-atmosphere GCM and fossil assemblage data for the mid Pliocene. *Geological Magazine*, **142**, 399–417.

WOODWARD, F. I., SMITH, T. M. & EMANUEL, W. R. 1995. A global land primary productivity and phytogeography model. *Global Biogeochemical Cycles*, **9**, 471–490.

The PRISM palaeoclimate reconstruction and Pliocene sea-surface temperature

H. J. DOWSETT

US Geological Survey, 926A National Center, Reston VA 20192, USA
(e-mail: hdowsett@usgs.gov)

Abstract: In this paper, I present a summary of the Pliocene Research, Interpretation and Synoptic Mapping (PRISM) palaeoenvironmental reconstruction, with emphasis on its historical development and range of boundary condition datasets. Sea-surface temperature (SST), sea level, sea ice, land cover (vegetation and ice) and topography are discussed as well as many of the assumptions required to create an integrated global-scale reconstruction. New multiproxy research shows good general agreement on the magnitude of mid-Pliocene SST warming. Future directions, including maximum and minimum SST analyses and deep ocean temperature estimates aimed at a full three-dimensional reconstruction, are presented.

Anthropogenic greenhouse gas emissions and modification of land surfaces are expected to cause the Earth's climate to warm (Intergovernmental Panel on Climate Change 1995). However, the amount and details of the warming are still highly uncertain. Identifying and predicting human-related changes must take into account natural climate variability and the complex interactions of the different components of the Earth's climate system (National Research Council 2002). As part of the US Geological Survey climate change research effort, the PRISM (Pliocene Research, Interpretation and Synoptic Mapping) Project has documented the characteristics of mid-Pliocene climate on a global scale. The mid-Pliocene was selected for detailed study because it spans the transition from relatively warm global climates of the early Pliocene epoch when glaciers were greatly reduced in the Northern Hemisphere to the generally cooler climates of the Pleistocene with expanded Northern Hemisphere ice-sheets and prominent glacial–interglacial cycles.

Originally, the PRISM Project had two primary goals. The first was to identify and characterize the nature and variability of climate during this time of past global warming to indicate how the Earth might respond to future warming. The second goal was to develop a series of integrated global-scale, quantitative datasets that could be used in experiments modelling climate and environmental conditions during the mid-Pliocene epoch. Today, the PRISM reconstruction is being used to test the ability of climate models to simulate past warmer conditions on Earth and to provide insights into the causes, mechanisms and effects of global warming (Dowsett *et al.* 1992; Chandler *et al.* 1994; Sloan *et al.* 1996; Haywood *et al.* 2002; Haywood & Valdes 2004; Dowsett *et al.* 2005; Jiang *et al.* 2005; Haywood *et al.* 2007).

The PRISM reconstruction currently consists of a series of 39 global-scale datasets (Table 1) covering sea- and land-surface conditions on a 2° latitude by 2° longitude grid. As such, it is the most comprehensive and detailed global reconstruction of climate and environmental conditions older than the last glacial interval (*c.* 20 ka). The purpose of this chapter is to review the PRISM mid-Pliocene reconstruction with special emphasis placed on planktic foraminifer-based sea-surface temperature (SST) estimates.

If one has the mindset that pre-Pleistocene palaeoclimate reconstruction is too uncertain to be useful, there is no need to read further. However, it is worth adding a word of caution. There are no arguments against the methods applied to Pliocene sequences that cannot be used, to some degree, to undermine reconstructions of time periods of any age, including those as young as the Holocene. Faunal components of older sequences have less in common with modern assemblages than younger sequences due to the evolutionary process. Since this commonality forms the basis of analogous reconstruction, it is cause for concern. Even those taxa common to both ancient deposits and the modern sea are suspect since ecological tolerances most likely drift with time.

The mid-Pliocene was chosen for reconstruction for several reasons. It was clear two decades ago that climate models needed to be able to successfully predict past conditions significantly warmer than today if they were ever to be used to project future climate scenarios. Continued work on intervals like marine isotope stage 5 or 11 seemed unwise since by all accounts they were not significantly different from the present (Dowsett 1991; Dowsett & Poore 1991; Dowsett *et al.* 1992). The Early Pliocene was generally accepted to be a time of warm equable climate (Zubakov & Borzenkova 1988, 1990);

From: WILLIAMS, M., HAYWOOD, A. M., GREGORY, F. J. & SCHMIDT, D. N. (eds) *Deep-Time Perspectives on Climate Change: Marrying the Signal from Computer Models and Biological Proxies.* The Micropalaeontological Society, Special Publications. The Geological Society, London, 459–480.

Table 1. *PRISM data design*

Type	Datasets	Description
Sea level	1	Land-sea distribution on a $2° \times 2°$ grid (serves as base map for the reconstruction)
Topography	1	Elevations above Pliocene sea level
Land cover	1	Distribution of grassland, desert, deciduous forest, rainforest, coniferous forest, tundra and ice on Pliocene land areas.
SST (PRISM2)	12	Distribution of sea-surface temperatures for ocean areas (monthly). Includes sea-ice distribution.
SST (PRISM3)	24	Distribution of maximum and minimum possible sea-surface temperatures for ocean areas (12 months each). Includes sea-ice distribution.

however, early Pliocene faunal assemblages are taxo-nomically distant from present-day faunas. Thus, the mid-Pliocene was chosen as a compromise – the most recent interval of time clearly and significantly warmer than modern. Mid-Pliocene faunal (and floral) biogeography shows distribution patterns similar to those of today, and the majority of the assemblage (in a morphological sense) is still extant.

Definition of the PRISM interval and time-slab concept

The PRISM reconstruction is a global synthesis of a period of warm and stable (relative to high-amplitude Pleistocene glacial–interglacial cycles) climate lying between the transition of oxygen isotope stages M2/M1 and G19/G18 (Shackleton *et al.* 1995) in the middle part of the Gauss Normal Polarity Chron. The reconstruction spans the interval of 3.29 Ma to 2.97 Ma (geomagnetic polarity timescale of Berggren *et al.* 1995; astro-nomically tuned timescale of Lourens *et al.* 1996) (Fig. 1). It ranges from near the bottom of C2An1 (just above Kaena reversed polarity) to within C2An2r (Mammoth reversed polarity). This interval correlates in part to planktonic foraminif-eral zones PL3 (*Globorotalia margaritae– Sphaeroidinellopsis seminulina* Interval Zone), PL4 (*Sphaeroidinellopsis seminulina–Dentoglobi-gerina altispira* Interval Zone) and PL5 (*Dentoglo-bigerina altispira–Globorotalia miocenica* Interval Zone) of Berggren *et al.* (1995). It falls within calcareous nannofossil zone NN16 of Martini (1971) or CN12a of Bukry (1973, 1975). This interval occurs prior to the first strong oxygen isotope excursions which represent a change toward modern conditions (Northern Hemisphere ice volume increased, polar fronts were strength-ened and glacial–interglacial variation intensified) (Sancetta & Silvestri 1986; Raymo *et al.* 1989; Hodell & Ciesielski 1991; Dowsett *et al.* 1994).

While the interval of time between 3.29 and 2.97 Ma (PRISM time-slab) is easily distinguished from the intervals immediately surrounding it, there is a

high degree of variability within the time-slab (Dowsett & Poore 1991; Barron 1992a; Hodell & Venz 1992; Shackleton *et al.* 1995; Draut *et al.* 2003; Lisiecki & Raymo 2005; Dowsett *et al.* 2005) (Fig. 1). The 41 ka period of Earth's obliquity dominates the Pliocene climate record (Tiedemann *et al.* 1994). Other than glacial stages KM2 (*c.* 3.12 Ma) and G20 (*c.* 3.01 Ma), benthic foraminiferal oxygen isotope values were either equal to or isoto-pically lighter than those of today (Shackleton *et al.* 1995; Lisiecki & Raymo 2005).

The PRISM interval is long enough to be reliably identified and correlated between marine sequences independent of climatic characteristics because of its proximity to a number of biostrati-graphic and magnetostratigraphic events (Berggren *et al.* 1985, 1995; Dowsett 1989a, b; Dowsett & Robinson 2006). Deep sea records and, to varying degrees, ocean margin records, can be correlated with some confidence to this interval. Many Plio-cene terrestrial records come from short sequences that rely on limited radiometric dates and magnetos-tratigraphy for chronology. The sparseness of long terrestrial time-series with multiple age control points makes identification of high-frequency varia-bility and integration of our terrestrial palaeocli-mate estimates into our time-slab interval less certain than our marine estimates. This is a problem with most pre-Quaternary terrestrial palaeoclimate reconstructions. In the remainder of this chapter, '3 Ma' and 'mid-Pliocene epoch' indi-cate the PRISM time-slab interval as defined above.

The PRISM reconstruction is presented as a series of matrices (Table 1), each containing 90 rows and 180 columns (16,200 cells), representing the Earth at a resolution of $2°$ latitude by $2°$ longi-tude. Each cell in this grid is designated either land or water, based upon which element comprises the majority (>50%) of the cell (Fig. 2). If water, a cell has either a sea-surface temperature or is covered by sea ice (-1.8 °C). If a cell is desig-nated land, it contains one of seven land cover categories (desert, tundra, grassland, deciduous forest, coniferous forest, rainforest or ice). In addition, land cells are given a topographic

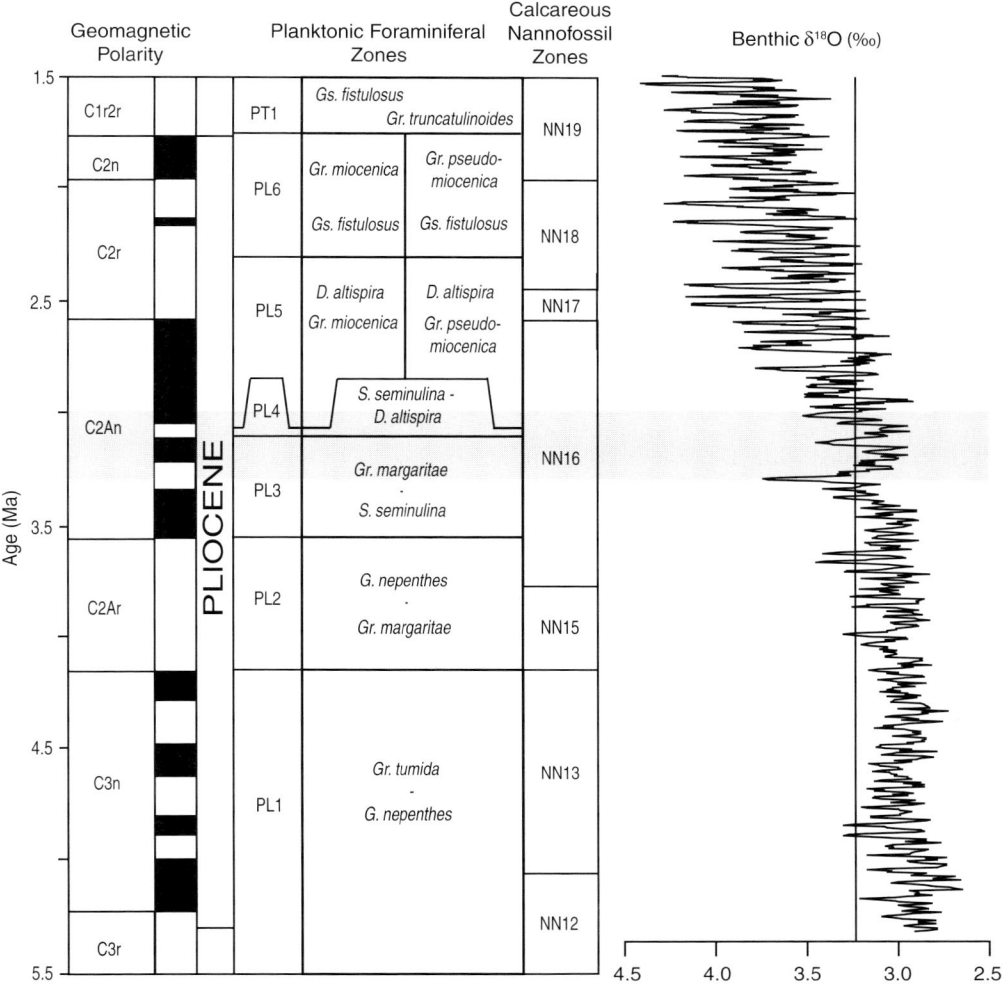

Fig. 1. Pliocene magnetobiostratigraphic framework after Berggren *et al.* (1995). Grey band approximates the PRISM time-slab. Benthic δ¹⁸O record from Lisiecki & Raymo (2005).

elevation. PRISM datasets are available by contacting the author.

History of the PRISM palaeoclimate reconstructions

There have been four generations of PRISM palaeoclimate reconstructions (PRISM0–PRISM3) that evolved from a series of studies summarizing conditions at a large number of marine and terrestrial sites and regions (e.g. Cronin & Dowsett 1991; Poore & Sloan 1996). Prior to PRISM0, Dowsett & Poore (1991) analysed a series of North Atlantic core sites and compared Pliocene SST to last interglacial conditions (Table 2). Their study concluded that although the modern North Atlantic differed little from the Last Interglacial, it was warmer during the mid-Pliocene epoch. Using additional data, Dowsett *et al.* (1992) determined the Pliocene anomaly (Pliocene SST minus modern conditions) for a transect in the North Atlantic region and showed that middle- and high-latitude SST was warmer during the Pliocene epoch than during either the last interglacial or present day. Conversely, low-latitude SST showed no change during climate extremes (Last Glacial Maximum, Last Interglacial and mid-Pliocene). These results favoured increased meridional ocean heat transport over increased CO_2 as the primary forcing behind mid-Pliocene warming.

(a)

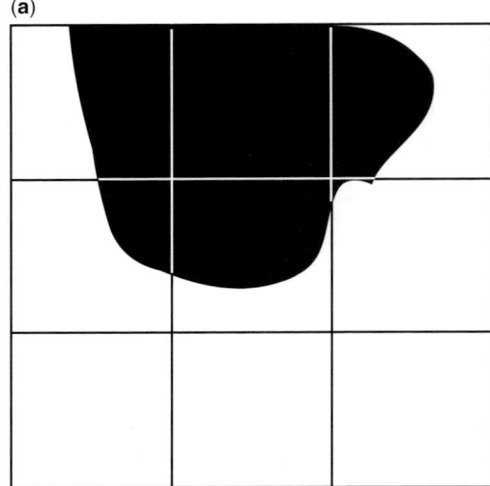

(b)

84m grassland	165m tundra	-1.8°C
9.2°C	410m ice	7.4°C
15.2°C	18.1°C	14.4°C

Fig. 2. Hypothetical grid showing data model used by PRISM. (**a**) Actual land area (black) superimposed on $2° \times 2°$ grid system. (**b**) Cells covered by $\geq 50\%$ land are designated as land for the reconstruction (shaded green) and given elevation in metres and a predominant land cover. Cells designated as water are given SST estimates. In this example, cell with cross-hatching is covered by sea ice which is designated by SST of -1.8 °C.

Barron (1992*a*, *b*), Ikeya & Cronin (1993) and Cronin *et al.* (1993) provided new data from the Pacific and circum-Arctic regions that was used to produce the first Northern Hemisphere SST analysis (PRISM0) for the mid-Pliocene epoch (Dowsett *et al.* 1994). One of the chief findings of that reconstruction was the documentation of warming in the Pacific at essentially the same time as the North Atlantic warming, suggesting that the mid-Pliocene warming was more than a regional event.

By 1996, a number of Southern Hemisphere sites had been analysed for foraminifers, diatoms and ostracodes, and the first mid-Pliocene global reconstruction of SST was produced (Dowsett *et al.* 1996). That first global reconstruction of mid-Pliocene climate (PRISM1) was based upon 64 marine and 74 terrestrial sites, and included datasets

representing annual vegetation, land ice, monthly SST and sea ice, sea level and topography (Dowsett *et al.* 1996; Thompson & Fleming 1996).

PRISM2 (Dowsett *et al.* 1999) was a revision of PRISM1, incorporating additional marine sites to improve geographic coverage. For example, sites from the Mediterranean Sea and Indian Ocean were included for the first time. All SST estimates were recalculated using a new core-top calibration based upon the Reynolds & Smith (1995) adjusted optimum interpolation (AOI) data set. The Pliocene sea level was reset from $+35$ m to $+25$ m in keeping with re-evaluation of glaciological, isotopic, shoreline and other sea-level data (Haq *et al.* 1987*a*, *b*; Dowsett & Cronin 1990; Wardlaw & Quinn 1991; Krantz 1991; Wilson 1993; Shackleton *et al.* 1995). PRISM2 used model results to guide the

Table 2. *History of PRISM SST reconstructions*

Reconstruction	Region	Resolution (lat × lon)	Calibration	Reference
NATL	North Atlantic	–	USNO[*]	Dowsett & Poore (1991)
PRISM0	Northern Hemisphere	8 × 10	USNO[*]	Dowsett *et al.* (1994)
PRISM1	Global	2 × 2	USNO[*,**]	Dowsett *et al.* (1996)
PRISM2	Global	2 × 2	REYNLD[§]	Dowsett *et al.* (1999)
PRISM3	Global	2 × 2	REYNLD[§]	Dowsett *et al.* (2005)
PRISM3D	Global 3D	4 × 5	LEVITUS[¶]	Dowsett *et al.* (2006)

[*] US Naval Oceanographic Office (1958, 1967).
[**] Diatoms and Ostracodes calibrated to other best available SST data.
[§] Reynolds & Smith (1995).
[¶] Levitus & Boyer (1994).

areal and topographic distribution of Antarctic ice, which resulted in a more realistic Antarctic ice configuration in keeping with the 25 m sealevel rise.

PRISM3, still in development, has three primary components. It incorporates a multiproxy re-examination of SST, maximum and minimum probable SST reconstructions which provide a climatological error bar for model experiments, and for the first time, a deep-ocean reconstruction of bottom water temperature based upon Mg/Ca

palaeothermometry on the ostracode genus *Krithe* (Cronin *et al.* 2005; Dowsett *et al.* 2005, 2006).

Localities used in the PRISM reconstruction

The geographic distribution of the marine and terrestrial sites from which fossil data were analysed is shown in Figure 3. Marine localities are listed longitudinally in Table 3, along with relevant

(a)

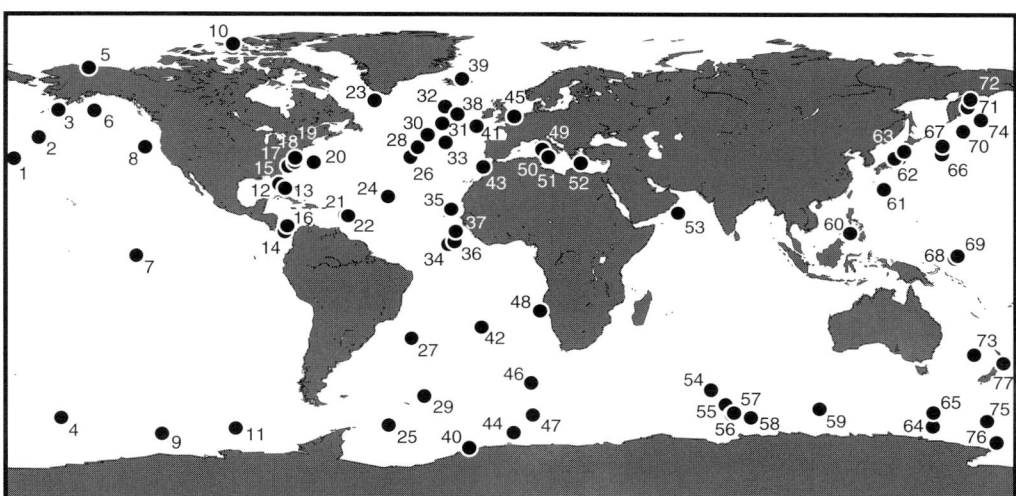

(b)

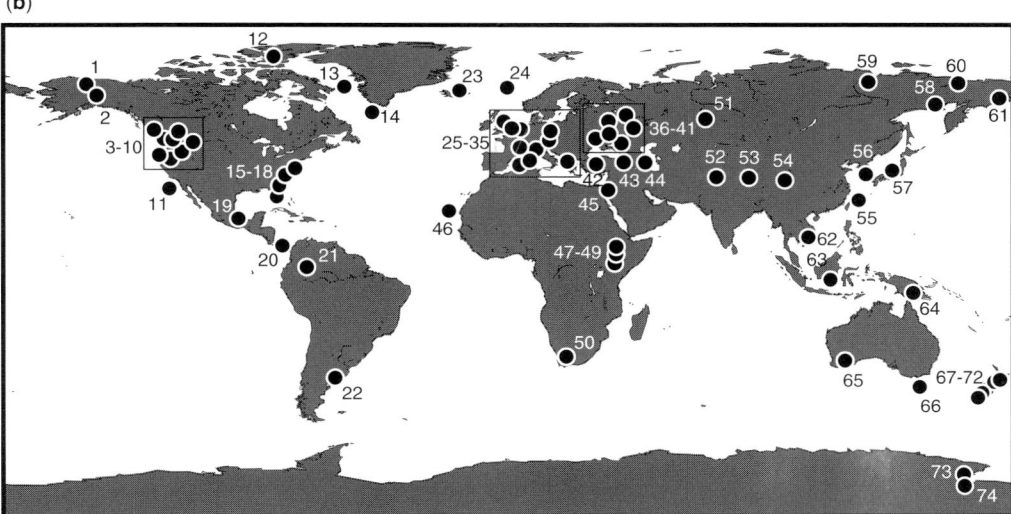

Fig. 3. Location of marine (**a**) and terrestrial (**b**) locations used in the PRISM2 reconstructions. For detailed information on localities, see Dowsett *et al.* (1994, 1996, 1999) and Thompson & Fleming (1996). Marine localities are identified in Table 3.

Table 3. *PRISM Marine localities and corresponding SST (°C) estimates*

	Locality	Latitude	Longitude	Modern SST[1]			PRISM2 Pliocene SST		PRISM3[2] Pliocene (MAX) SST		Pliocene (MIN) SST	
				Feb	Dec–Mar	Aug	Feb	Aug	Feb	Aug	Feb	Aug
1	DSDP 310	36.87	−176.9	14.1	–	24.2	17.1	26.2	18.6	27.2	16.3	25.7
2	ODP 886C	44.69	−168.24	7.3	–	14.9	4.7	18.0	3.4	19.5	5.3	17.2
3	DSDP 183	54.58	−161.21	2.9	–	11.3	1.7	13.6	1.1	14.8	2.0	13.0
4	E14-8	−59.67	−160.29	3.7	3.1	1.2	>4.6	–	5.2	–	4.4	–
5	Colvillian	70.29	−150.42	−1.8	–	−1.8	0.2	2.2	1.2	4.2	−0.3	1.2
6	ODP 887A	54.37	−148.45	4.1	–	12.3	6.1	13.3	7.8	13.8	6.0	13.0
7	DSDP 573	0.49	−133.3	25.6	–	24.9	24.6	24.7	24.9	24.8	24.1	24.6
8	DSDP 36	40.98	−130.12	11.7	–	17.9	13.7	19.9	14.7	20.9	13.2	19.4
9	E13-17	−65.68	−124.11	1.1	−0.4	−1.5	<5.5	–	7.7	–	4.4	–
10	Meighen Island	79	−99	−1.8	–	−1.8	0.2	3.2	1.2	5.7	−0.3	2.0
11	DSDP 323	−63.68	−97.99	3.2	2.7	0.4	>4.4	–	5.0	–	4.1	–
12	Sarasota	27.25	−82.66	20.5	–	29.4	19.9	27.0	19.6	25.8	20.0	27.6
13	Pinecrest Beds	27.35	−82.43	20.5	–	29.4	20.5	29.4	20.5	29.4	20.5	29.4
14	Cayo Aqua	9.15	−82.05	28.0	–	28.4	28.0	29.5	28.0	30.1	28.0	29.2
15	SEFlor(G-182)	25.78	−80.28	23.3	–	29.6	21.3	30.0	20.3	31.8	21.8	28.0
16	DSDP 502	11.49	−79.38	26.7	–	28.1	27.8	27.6	28.4	27.3	27.5	27.7
17	Duplin	34	−79	20.4	–	28.6	19.7	28.2	19.4	28.0	19.9	28.3
18	Lee Creek	35.38	−76.75	14.5	–	27.0	16.8	30.1	17.9	31.7	16.2	29.4
19	Yorktown	37	−76.5	10.1	–	25.9	15.3	27.7	17.9	28.6	14.0	27.3
20	DSDP 603	35.49	−70.03	19.9	–	27.1	25.2	27.5	27.9	27.8	23.9	27.4
21	DSDP 541	15.52	−58.72	26.0	–	28.1	27.8	27.6	28.8	27.3	27.4	27.7
22	ODP 672A	15.5	−58.5	26.0	–	28.1	27.6	26.9	28.4	26.3	27.2	27.2
23	ODP 646B	58.25	−48.33	2.2	–	7.4	3.6	9.5	4.4	10.6	3.3	9.0
24	DSDP 396A	22.9	−43.5	23.5	–	26.4	24.8	27.5	25.4	28.1	24.4	27.2
25	ODP 695A	−62.39	−43.45	0.1	−0.7	−1.8	<4.7	–	7.0	–	3.6	–
26	DSDP 606	37.34	−35.5	16.5	–	24.4	17.6	25.6	18.1	26.3	17.3	25.3
27	DSDP 516	−30.27	−35.28	24.0	–	18.5	26.0	20.5	27.0	21.5	25.5	20.0
28	DSDP 607	41	−32.96	15.3	–	22.9	15.3	22.6	15.3	22.4	15.3	22.6
29	ODP 699A	−51.54	−30.68	3.8	3.3	1.0	5.8	–	6.8	–	5.3	–
30	DSDP 410	45.51	−29.48	13.3	–	19.2	16.1	25.0	17.5	27.8	15.4	23.5
31	DSDP 609B	49.88	−24.24	11.5	–	16.6	20.5	28.6	25.0	34.6	18.2	25.6
32	DSDP 552	56.04	−23.23	9.1	–	13.3	12.4	21.1	14.0	25.0	11.5	19.2
33	DSDP 608	42.84	−23.09	13.7	–	20.7	17.8	25.6	19.8	28.1	16.7	24.4
34	ODP 667A	4.55	−21.9	27.3	–	26.8	27.3	26.8	27.3	26.8	27.3	26.8

#	Site												
35	ODP 659A	18.08	−21.03	20.9	—	24.5	22.3	27.7	22.9	29.3	21.9	26.9
36	DSDP 366A	5.68	−19.85	27.3	—	26.8	27.8	27.2	28.1	27.4	27.7	27.1
37	ODP 661A	9.45	−19.39	24.8	—	26.9	27.1	26.7	28.3	26.6	26.5	26.8
38	DSDP 610A	53.22	−18.89	10.5	—	15.0	14.2	24.3	16.0	29.0	13.2	22.0
39	Tjornes	66.16	−17.25	1.4	—	7.4	5.4	13.6	7.4	16.7	4.4	12.0
40	ODP 693A	−70.83	−14.57	−0.9	—	−1.8	<4.3	—	6.6	—	3.0	—
41	DSDP 548	48.85	−12	11.3	—	17.2	18.3	27.6	21.8	32.8	16.6	25.0
42	DSDP 521	−26.07	−10.27	24.7	—	19.5	26.7	21.5	27.7	22.5	26.2	21.0
43	DSDP 546	33.8	−9.6	16.7	—	22.3	19.1	24.2	20.3	25.1	18.5	23.7
44	ODP 690B	−65.16	1.21	0.9	−0.1	−1.8	<5.8	—	6.9	—	3.9	—
45	North Sea	52.5	1.5	5.7	—	16.3	10.4	16.7	12.8	16.9	9.2	16.6
46	ODP 704B	−46.88	7.42	7.1	—	4.8	9.6	7.3	10.8	8.6	8.9	6.7
47	PS1448	−58.64	7.92	1.1	0.2	−1.8	<2.0	—	5.6	—	3.3	—
48	DSDP 532	−19.74	10.52	20.9	—	15.8	22.7	17.6	23.6	18.5	22.2	17.1
49	DSDP 132	40.25	11.43	13.8	—	25.0	15.4	26.6	16.2	27.4	15.0	26.2
50	Punta di Maiata	37.33	13.5	14.6	—	25.8	17.5	21.3	18.9	19.0	16.8	22.4
51	Punta Piccola	37.33	13.58	14.6	—	25.8	18.8	23.9	21.0	22.9	17.8	24.3
52	Finikia	35.25	25.17	15.4	—	24.3	16.1	19.9	16.5	17.7	16.0	21.0
53	ODP 722	16.62	59.8	25.0	—	24.8	26.1	27.0	26.6	28.2	25.8	26.5
54	ODP 736C	−49.4	71.66	4.5	4.0	1.8	<7.5	—	9.0	—	6.7	—
55	ODP 747	−54.81	76.79	2.4	1.9	0.1	>4.5	—	5.6	—	4.0	—
56	ODP 748	−58.44	78.98	1.4	0.8	−1.7	<4.6	—	6.2	—	3.8	—
57	ODP 751	−57.73	79.81	1.7	1.1	−1.3	<4.6	—	6.0	—	3.9	—
58	ODP 745B	−59.6	85.86	1.8	1.2	−1.6	<4.7	—	6.2	—	4.0	—
59	DSDP 266	−56.4	110.11	3.5	2.9	0.7	>4.7	—	5.3	—	4.2	—
60	ODP 769	8.78	121.29	26.9	—	28.7	26.9	29.2	26.9	29.5	26.9	29.1
61	DSDP 445	25.52	133.2	21.0	—	28.5	23.0	29.5	24.0	30.0	22.5	29.2
62	Yabuta	37	137	9.9	—	25.0	9.9	25.0	9.9	25.0	9.9	25.0
63	Sasaoka	39.5	140.5	7.0	—	22.1	8.5	24.1	9.3	25.1	8.1	23.6
64	E50-28	−62.9	150.68	1.9	0.9	−1.7	<5.0	—	6.4	—	4.2	—
65	E36-33	−57.75	150.88	3.7	3.0	0.8	>4.7	—	5.2	—	4.4	—
66	DSDP 579	38.63	153.84	12.0	—	23.3	17.5	27.7	20.3	29.9	16.2	26.6
67	DSDP 580	41.63	153.98	6.2	—	19.8	12.4	23.1	15.5	24.7	10.8	22.3
68	DSDP 586	−0.5	158.5	29.3	—	29.0	29.3	29.0	29.3	29.0	29.3	29.0
69	ODP 806	0.31	159.36	29.2	—	29.0	29.0	29.0	28.9	29.0	29.0	29.0
70	ODP 881C	47.1	161.49	2.4	—	12.1	4.3	15.9	5.3	17.8	3.9	15.0
71	E. Kamchatka	56	163	0.1	—	11.4	4.1	10.9	6.1	10.6	3.1	11.0
72	Karaginsky	58.85	164.04	−0.9	—	11.3	3.1	13.3	5.1	14.3	2.1	12.8
73	DSDP 592	−36.47	165.44	20.3	—	15.1	22.3	17.1	23.3	18.1	21.8	16.6
74	ODP 883C	51.2	167.77	2.7	—	10.4	3.2	14.9	3.5	17.1	3.1	13.8
75	E50-33	−61.09	170.06	4.0	3.3	0.2	>4.5	—	4.8	—	4.4	—
76	DSDP 274	−68.99	173.43	−0.5	—	−1.8	<4.5	—	5.5	—	3.3	—
77	Rangitikei R.	−39.5	175.87	18.0	—	13.2	20.0	15.2	21.0	16.2	19.5	14.7

[1] Modern SST from Reynolds and Smith (1995). [2] Dowsett et al. (2005).

environmental estimates. Discussion of terrestrial localities used to reconstruct land cover and vegetation can be found in Dowsett *et al.* (1994, 1999) and Thompson & Fleming (1996).

Calibration of sea-surface temperature estimates

A major change between the PRISM1 and PRISM2 reconstruction was the recalibration of all modern marine data to the modern SST of Reynolds & Smith (1995). Comparison of the CLIMAP modern dataset (see Prell 1985) and the modern SST data used by the Goddard Institute for Space Science (GISS) model showed significant differences, sometimes exceeding the magnitude of estimated Pliocene temperature change. In addition, at some shallow marine sites we utilized the 'best available temperature data' that was not always equivalent to the temperatures being used at nearby deep-sea core sites. Because of inconsistencies between modern climatologies and interfossil-group calibration problems, we recalibrated all modern samples and all Pliocene localities for all fossil groups, using the $1° \times 1°$ SST of Reynolds & Smith (1995). Modern 'core-top' faunal census data for planktic foraminifers comes from Prell *et al.* (1999), Dowsett & Poore (1999) and Dowsett *et al.* (2003).

Methods of planktic foraminifer-based SST estimation

Middle Pliocene planktic foraminifers were identified to species level and counted to produce census data (see Dowsett *et al.* 1999 and references therein). In many cases, magnetobiostratigraphy was sufficient to create age models but in some instances new biochronologic analyses were employed to help scale the sequences in the time domain (Dowsett & Robinson 2006). Great care was taken to employ a consistent taxonomy across this dataset. More than 477,000 specimens were identified to species level by two investigators and cross-checked to assure taxonomic consistency.

Sea-surface temperatures were estimated from these census data using factor analytic transfer functions, modern analogue technique (MAT), or semi-quantitative comparison to modern faunas (Dowsett & Poore 1990, 1991; Dowsett 1991; Dowsett *et al.* 1996; Dowsett & Robinson 1998; Dowsett & Poore 1999). In order to compare the mid-Pliocene assemblage to the core-top thanatocoenosis, taxonomic grouping schemes were employed to simplify the taxonomic structure of both the present and mid-Pliocene assemblages. The assumption is that

these taxonomic categories had the same environmental preferences during the mid-Pliocene as they do today. These categories were designed so that any potential errors in assumption produce overly conservative (too cool) SST.

Transfer function technique

In 1971, Imbrie and Kipp developed a transfer function involving factor analysis and multiple regression to create equations relating microfossil abundance data in modern (core-top) samples to physical parameters such as SST and salinity. The equations can be applied to downcore microfossil abundance data to estimate past sea-surface conditions. The transfer function approach has been widely used in palaeoclimate studies of the late Pleistocene based upon planktic foraminifers, diatoms, radiolarians and ostracodes, and was the primary technique used in reconstruction of global SST for the last glacial maximum and last interglacial (CLIMAP 1981, 1984). Use of the technique on older sequences was limited by concern over changing ecological tolerances and preferences of species as well as addition and elimination of species from fossil assemblages due to evolutionary changes and extinction.

Early attempts to use the transfer function on sequences older than the Last Interglacial were carried out by Briskin & Berggren (1975), Thunell (1979*a*, *b*), and Ruddiman *et al.* (1986). Keigwin (1976), Thunell (1979*a*, *b*) and Poore (1981) used factor analysis on Pliocene- to Miocene-age sequences of planktic foraminifers by assuming that taxa in downcore samples, now extinct, were ecologically equivalent to closely related forms in modern samples.

PRISM built upon the pioneering work of these earlier studies to create a set of 18 counting categories for both modern and Pliocene taxa that would allow a transfer function approach to be used on sequences extending through the mid-Pliocene epoch (Dowsett & Poore 1990; Dowsett 1991). Factor analysis of modern core-top data using the 18 categories yielded five planktic foraminifer assemblages that account for *c.* 95% of the variance in the faunal data. Standard multiple regression techniques were applied to the five assemblages (varimax factors) to write palaeoecological equations of the form:

$$Y_{est} = B_{ct}^2 K + k_0 \qquad (1)$$

where Y_{est} = the palaeoecological estimates, B_{ct}^2 = the cross-product matrix, and K = a vector of regression coefficients corresponding to the columns of B_{ct}^2 and k_0 is the intercept of the equation. The resulting function, GSF18, exhibits standard error of estimate of 1.47 °C and 1.36 °C for cold and warm seasons respectively (Dowsett 1991). Sensitivity tests have been performed on this equation to determine the effects of unstable taxonomy and varying ecological tolerance (e.g. Dowsett & Poore 1990). These tests show that cold season SST estimates are less sensitive to partitioning of taxa within our categories and cold season estimates are probably biased toward cooler temperatures. GSF18 has been applied to Atlantic basin sequences to generate SST time-series (Dowsett & Poore 1990, 1991; Dowsett & Loubere 1992; Dowsett et al. 1994, 1996).

One of the advantages of the factor analytic transfer function is the ability to extrapolate beyond the modern calibration data. While estimates based on extrapolation outside of the calibration envelope should be viewed with caution, qualitative change rather than absolute magnitude of change is fairly robust. For example, Early Pliocene temperature estimates, while unreliable due to significant differences in the assemblage, clearly point to conditions warmer than the mid-Pliocene or present.

Modern analogue technique

MAT quantifies faunal changes within deep-sea cores in terms of modern oceanographic conditions (Hutson 1980) using a measure of faunal dissimilarity to compare downcore samples to each reference sample in a modern oceanographic database. Working with late Pleistocene planktonic foraminifers, Hutson (1980) originally used cosine-theta distance:

$$\cos \theta_{ij} = \frac{\sum_{k=1}^{m} p_{ik} p_{jk}}{\sqrt{\sum_{k=1}^{m} p_{ik}^2 \sum_{k=1}^{m} p_{jk}^2}} \qquad (2)$$

where $\cos \theta_{ij}$ is the distance between two multivariate samples i and j, m is the number of species, and p_{ik} is the proportion of species k in sample i, to match modern Indian Ocean samples to late Pleistocene-age core samples and then employed a weighted average of sea-surface temperature and salinity associated with the closest analogues of each core sample to derive downcore environmental estimates. Overpeck et al. (1985) investigated the responsiveness of eight dissimilarity coefficients to palynological changes caused by differences in modern vegetation and applied the MAT technique to late Quaternary pollen assemblages from eastern North America. Their analyses suggested that, whereas all coefficients give roughly similar results, signal-to-noise measures performed better than unweighted or equal-weight dissimilarity coefficients (Overpeck et al. 1985).

After extensive experimentation using Pliocene data with different dissimilarity coefficients (Dowsett & Robinson 1998; Dowsett & Poore 1999), PRISM chose the squared chord distance (SCD) signal-to-noise measure:

$$d_{ij} = \sum k(p_{ik}^{1/2} - p_{jk}^{1/2})^2 \qquad (3)$$

where d_{ij} is the squared chord distance between two multivariate samples i and j, and p_{ik} is the proportion of species k in sample i. Squared chord distance values can range from 0.0 to 2.0, with 0.0 indicating identical proportions of species within the samples being compared. Values above 0.15 are considered non-analogues for the purpose of temperature estimation. The SIMMAX method of analogues (Pflaumann et al. 1996) offers many additional enhancements but has not been used because of the geographic bias inherent in the methodology (Telford et al. 2004). The MAT has been applied to Pacific basin sequences by Dowsett & Robinson (1998).

Diatom and ostracode SST estimation

Mid-Pliocene diatoms were analysed by Barron (1995, 1996a, b) from 24 Southern Ocean and six North Pacific deep-sea cores. Age models for these cores were based on existing magnetostratigraphy and both published and refined diatom biostratigraphy (Barron 1996a, b). Diatom-based SST estimates for the Southern Ocean were determined by estimating the position of the Antarctic Polar Front (APF) relative to the various sites (Dowsett et al. 1996). In general, the Polar Front Zone (PFZ) expanded during the middle Pliocene with the Sub-Antarctic Front (SAF) essentially at its modern position; mid-Pliocene APF was shifted toward the south. North Pacific SST was estimated using equations generated by Barron (1995) based upon the relative ratios of key taxa. See Barron (1995, 1996a, b) and Dowsett et al. (1996) for expanded discussion.

Mid-Pliocene ostracodes were counted from 12 sites in the Northern Hemisphere including Central America, the eastern United States Coastal Plain, Tjörnes (Iceland), Meighen Island, Alaskan Arctic Coastal Plain, North Sea, and Japan. In each region, age models were constructed using magnetobiochronology and shallow-sea bottom temperatures were quantitatively estimated by transfer function, MAT, or environmental preference matching (Cronin & Dowsett 1990; Cronin 1991a, b; Wood et al. 1994; Dowsett et al. 1996).

Selection of SST estimates representative of PRISM time-slab

Seventy-seven marine localities/sections were used for the PRISM2 SST reconstructions (Fig. 3). The uneven distribution of control points used for the reconstruction primarily reflects the availability of suitable material for study. When available, temperature estimates published by other workers (e.g. Herbert & Schuffert 1998; Pflaumann *et al.* 1998; Whitehead & Bohaty 2003; Cooper & O'Brien 2004; Li *et al.* 2004; Bartoli *et al.* 2005; Haywood *et al.* 2005) were used to augment and cross-check the estimates derived by the PRISM2 study.

For marine data points that were generated from time-series studies, we adopted a strategy to develop an estimate of mean 'interglacial' conditions within the 300 ka time-slab. This minimized problems associated with point-to-point correlation of peaks and troughs in temperature time-series separated by large geographic distances. The late Pleistocene parallel would be to select a single SST value representing average peak-interglacial conditions (e.g. average SST of marine isotope stages 5, 7 and 9).

This peak-averaging method (Dowsett & Poore 1991; Dowsett *et al.* 2005; Dowsett & Robinson 2006) examines an SST time-series for warm peaks. Warm peaks are defined as temperature estimates warmer than those immediately above and below in a stratigraphic sense. Those identified estimates that are considered valid (see below) are summed and averaged to provide a 'warm peak average' (Fig. 4). These warm-peak averages at all sites are then contoured to produce SST reconstructions. By convention, the cold season SST time-series is analysed and the corresponding warm season estimates are also tabulated.

The 'validity test' for factor analytic transfer function estimates is a communality cutoff. The communality h^2 is given by the formula:

$$h_k^2 = \sum_{i=1}^{k} s_k^2 \qquad (4)$$

where the square of the correlation of variable k with factor i gives the part of the variance accounted for by that factor or the proportion of information explained by the chosen factor model. Therefore, communality ranges from 0 to 1. For example, a factor model that explains 80% of a multivariate sample's variance has a sample communality of 0.640. The PRISM reconstructions routinely use a communality threshold of 0.70 indicating the factor model explains a minimum of approximately 84% of the variance in the data.

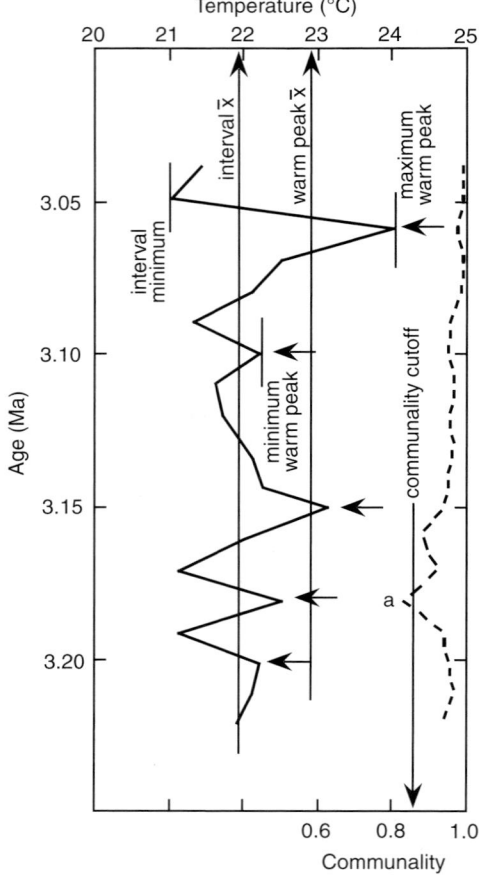

Fig. 4. Hypothetical SST record illustrating warm peak averaging (PRISM2) and maximum and minimum warming (PRISM3) as described in Dowsett *et al.* 2005 and Dowsett & Robinson 2006. Solid jagged line represents SST estimates from factor analytic transfer function (Dowsett & Poore 1990; Dowsett 1991). Dashed line shows downcore communality estimates. Vertical line pointing to the communality axis indicates communality of 0.85 (92% of variance explained by factor model). Horizontal arrows indicate warm peaks. Note that one warm peak (a) near 3.175 Ma has a communality less than the chosen threshold of 0.85 and is therefore not used to develop the warm peak average (22.8 °C, indicated by vertical line pointing toward the SST axis) for the interval. Maximum warming, minimum warming, and interval average also shown.

SST time-series developed with MAT use 0.15 as a cutoff; samples with nearest analogues >0.15 are considered non-analogues for the purpose of our temperature estimation and are excluded (Dowsett & Robinson 1998; Dowsett & Poore 1999).

In the example shown in Figure 4, the four warm peaks passing the validity test (in this case set to a communality $= 0.85$) have a mean value of 22.8 °C. The PRISM methodology would use this value to represent how warm it gets during warm peak phases of the interval being investigated. The warmest peak on Figure 4 is $c.$ 24 °C and the 'coolest' warm peak still passing the validity test is just over 22 °C. The minimum within the PRISM interval is $c.$ 21 °C.

Several different averaging techniques have been developed, and these have been tested to determine how best to portray mid-Pliocene SST at each locality by a single value. The benefits and problems associated with each method are discussed in Dowsett & Robinson (2006).

Following the methodology outlined above, all time-series were processed to obtain warm peak average temperatures (PRISM2) and minimum and maximum warm peak temperatures (PRISM3) (Dowsett et al. 2005).

Mid-Pliocene SST analysis

Contouring the global SST field from 77 irregularly spaced data points is a problematic task. Early attempts at automated contouring of the raw data, using algorithms from Middleton (2000), resulted in unrealistic oceanographic solutions. In order to facilitate model–model comparisons, PRISM elected to have researchers most familiar with the data and its strengths and weaknesses produce subjectively contoured datasets using the methodology outlined below. Much of the design of these datasets (e.g. monthly SST and sea ice) clearly goes beyond the resolution of the raw data, but was dictated necessary by the modelling community.

At each locality, modern SST was subtracted from the PRISM time-slab value in order to produce a mid-Pliocene SST anomaly for cold and warm seasons. To create a global dataset, these SST anomalies were plotted as individual points on a $2° \times 2°$ grid representing the Earth. Modern SST contours served as a rough guide to draw anomaly contours around the control points, because it was assumed that the general pattern of modern oceanic surface current systems was present in the mid-Pliocene. Boundaries between anomaly bands were smoothed to make even temperature gradients. Finally, this smoothed, contoured anomaly field was added to the modern SST of Reynolds & Smith (1995) to create mid-Pliocene February and August SST maps. The remaining 10 months of the year were derived by fitting a sine curve to the February and August SST estimates (Dowsett et al. 1996). The resulting PRISM2 (average), PRISM3 (maximum and minimum) SST analyses as well as the Reynolds & Smith (1995) modern climatology are shown in Figure 5 (Dowsett et al. 1999, 2005).

Other PRISM datasets

Sea level and global ice volume

The initial PRISM sea-level estimate was 35 ± 18 m above present sea level based upon the altitude (corrected for uplift) of the Orangeburg Scarp, a mid-Pliocene palaeoshoreline from the southeastern US Atlantic Coastal Plain (ACP) (Dowsett & Cronin 1990). The stratigraphic distribution of disconformities at Enewetak Atoll (equatorial Pacific) indicate a mid-Pliocene sea level 36 m higher than present (Wardlaw & Quinn 1991). Krantz (1991) correlated ACP sediments with $\delta^{18}O$ isotopic records from deep-sea cores and estimated sea level to have been 30–35 m higher at 4.0–3.2 Ma and 25 m higher at 3.0 Ma.

Estimates of potential sea-level rise from melting of the present day Greenland and West Antarctic Ice-Sheets are 6.55 m and 8.06 m respectively (Poore et al. 2000). The present-day East Antarctic Ice-Sheet contains the equivalent of $c.$ 65 m of sea-level rise, thus it is impractical to discuss Neogene sea-level apart from Antarctic ice-volume history. The question of long-term stability of Antarctic ice-sheets and concomitant surface temperatures (see Harwood & Webb 1998; Miller & Mabin 1998; Stroeven et al. 1998) is controversial and both stabilists and dynamicists have presented excellent data supporting both views (Dowsett & Cronin 1990; Kennett & Hodell 1993; Dwyer et al. 1995; Shackleton et al. 1995; Sugden et al. 1995; Marchant et al. 1996; Marchant & Denton 1996; Webb et al. 1996; Barrett et al. 1997; Stroeven & Prentice 1997; Jonkers & Kelley 1998; Van der Wateren et al. 1999; Harwood et al. 2000; Quilty et al. 2000; Marchant et al. 2002; Ashworth & Kuschel 2003; Ashworth & Cantrill 2004). New geological data from Alpine glacier deposits in the McMurdo Dry Valleys, previously used to bolster the stability hypothesis, now suggest the probability of warmer surface conditions and fluctuating ice volume in the early to mid-Pliocene epoch (Prentice & Krusic 2005). PRISM uses the evidence for reduced mid-Pliocene sea ice, geological evidence for higher than present sea level, and these new data from Antarctica suggesting a dynamic ice-sheet and warmer surface temperatures, to drive the land ice and land vs. ocean (sea level) distributions shown in Figure 6. The ice volume and areal distribution used for PRISM2 are compatible with a $+25$ m sea level (relative to present). A detailed discussion of the rationale behind the PRISM land ice distribution and explanation of the methodology used is found in Dowsett et al. (1994, 1999).

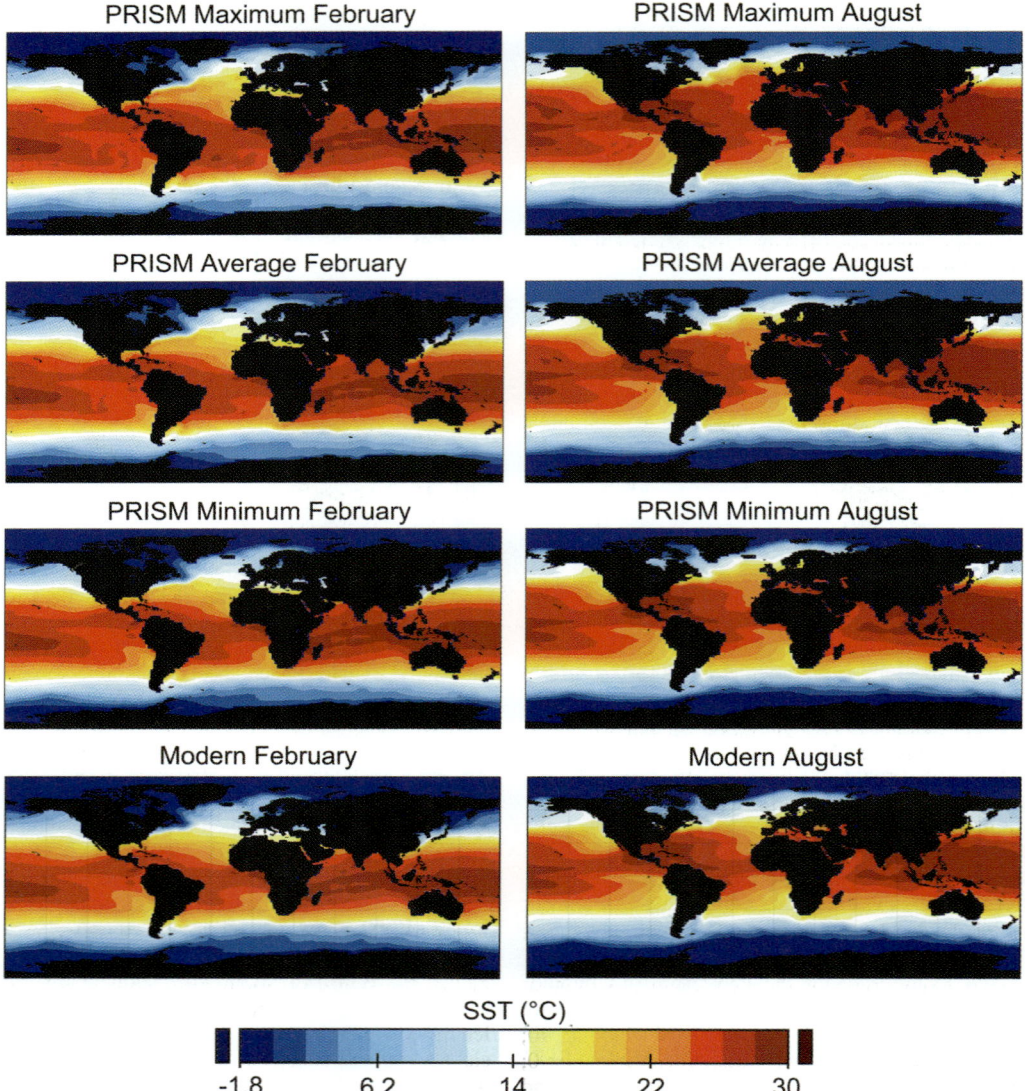

Fig. 5. Maximum (Dowsett *et al.* 2005), Average (Dowsett *et al.* 1999), Minimum (Dowsett *et al.* 2005) and Modern (Reynolds & Smith 1995) sea-surface temperature. Comparison of position of 14 °C isotherm (white) shows poleward displacement of contours from modern through maximum possible warming scenarios.

Sea-ice

Northern Hemisphere sea-ice distribution for the mid-Pliocene epoch is based primarily upon marine submerged and emerged (outcrops) deposits along the margins of the Arctic Basin. Shallow marine ostracodes from Pliocene transgressive sequences on the North Slope of Alaska and northern Greenland suggest sea ice was greatly reduced

during these intervals (Brouwers *et al.* 1991; Brigham-Grette & Carter 1992; Cronin *et al.* 1993; Brouwers 1994). Pollen, macrofossils and insects from the Alaskan North Slope, Canadian Arctic, Russian Arctic and Iceland likewise suggest a warmer terrestrial climate with boreal forests growing at the edge of the modern Arctic during some intervals of the Pliocene epoch (Nelson & Carter 1985; Matthews & Ovenden

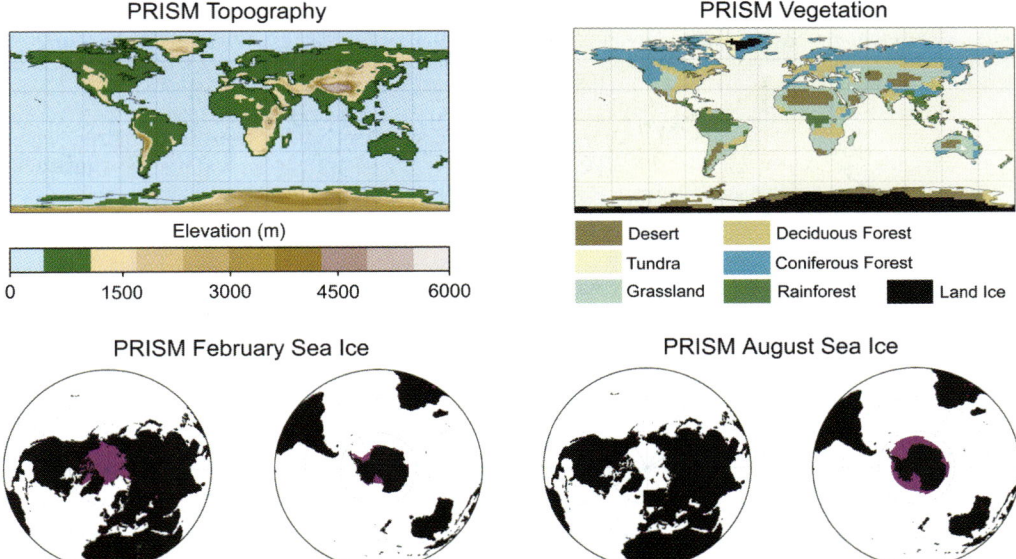

Fig. 6. PRISM2 topography, vegetation and sea-ice data. Topography and vegetation show data shown at $2° \times 2°$ resolution. Present-day coastlines shown by thin black line. Sea ice shown by purple areas superimposed on modern geography. See text for details.

1990; Fradkina 1991; Fyles *et al.* 1991; White *et al.* 1997; Elias & Matthews 2002; Buchardt & Simonarson 2003; Tedford & Harrington 2003). North Atlantic and North Pacific SST estimates indicate that relatively warm waters entered the Arctic Ocean during periods within the Pliocene epoch (Dowsett *et al.* 1999, 2005).

Southern Ocean sea-ice extent is primarily based upon the distribution of sea-ice-related diatoms and the presence of diatomaceous oozes that do not occur today south of the northernmost extent of Southern Hemisphere sea ice (Harwood 1986; Burckle & Cirilli 1987; Abelmann *et al.* 1990; Gersonde & Buckle 1990; Kennett & Barker 1990; Barron 1996a, b; Dowsett *et al.* 1996; Whitehead *et al.* 2005.) Other evidence for reduced sea ice comes from silicoflagellates (Whitehead & Bohaty 2003) and opal sedimentation (Hillenbrand & Fütterer 2001).

Taken *in toto*, these observations from both hemispheres have been interpreted to suggest greatly reduced sea ice during the mid-Pliocene epoch (Fig. 6). PRISM uses a seasonally ice-free Northern Hemisphere summer and sets winter sea ice equivalent to modern summer extent (Dowsett *et al.* 1994, 1996, 1999, 2005). Southern Hemisphere winter sea ice is set between 4° and 6° latitude further south than present day. Summer sea ice is minimal, and a functional relationship between sea ice cover and SST, derived from

modern seasonal meltback and ice formation, is used to infer intermediate states for spring and autumn (Barron 1996a, b; Dowsett *et al.* 1996, 1999, 2005).

Land-surface topography

As discussed in Thompson & Fleming (1996), Pliocene elevations were apparently lower than present in the western Cordillera of western North America and in the Andes of South America (Fig. 6). The reduction in the size of the Greenland and Antarctic Ice Sheets also resulted in a net reduction in elevation in those areas. Conversely, data discussed in Thompson & Fleming (1996) suggest that the East African rift zone was higher than at present. PRISM Pliocene elevations in parts of North and South America are set to half of their modern values, and the elevation of portions of the Greenland Ice Sheet were reduced. The elevations of Antarctica were taken directly from M. Prentice (pers. comm.). The elevations of grid points in the East African rift zone were set at 500 m above those on the modern grid.

Vegetation distribution

Information on mid-Pliocene vegetation was compiled from fossil pollen and plant macrofossil data from over 75 terrestrial sites (Fig. 3) covering all continents. PRISM2 vegetation (Fig. 6) is essentially identical to PRISM1 (see Thompson & Fleming 1996). Whenever practical, pollen was recovered from marine sequences to establish a true continental–marine correlation (Willard et al. 1993; Willard 1994, 1997; Dowsett & Willard 1996; Fleming & Barron 1996). PRISM uses seven land cover categories (desert, tundra, grassland, deciduous forest, coniferous forest, rainforest, and land ice) that are a simplification of the 22 land cover types of Matthews (1985).

Discussion and summary

The PRISM2 reconstruction is an internally consistent global summary of many important components of the Earth's climate and surface conditions during the mid-Pliocene epoch and thus is the most complete climate reconstruction older than the last interglacial. As such, it provides an opportunity to test and refine our ability to decipher past environments and model climates that are different from the present. Important features of the present PRISM2 mid-Pliocene reconstruction compared to modern are:

1. Increased SST in mid-to-high latitudes and unchanged SST in low latitudes (outside of upwelling regions). Warming is most pronounced in the northeastern North Atlantic sector. Mid-Pliocene warming occurs in the Atlantic and Pacific and Northern and Southern Hemispheres (Fig. 5).

2. Greatly reduced sea ice in both hemispheres with the Arctic being seasonally ice-free (Figs 5 & 6).

3. Greatly reduced continental ice volume with a small ice cap on Greenland being the only continental ice in the Northern Hemisphere (Fig. 6).

4. Global sea level c. 25 m higher than present which implies an Antarctic Ice-Sheet smaller than today (Fig. 6).

5. Expansion of evergreen forests to the margins of the Arctic Ocean, a reduction of desert area in equatorial Africa and essential elimination of polar desert and tundra regions in the Northern Hemisphere. A small amount of deciduous vegetation occurred at the edge of the Antarctic continent (Fig. 6).

PRISM3 multiproxy analysis: low latitude changes?

A multiproxy (faunal; Mg/Ca; alkenones) comparison of PRISM SST estimates, initiated in the North Atlantic but now expanding to other regions, shows overall agreement among proxies. At sites away from upwelling, both Mg/Ca and alkenone palaeothermometry provide annual average SST

estimates that fall intermediate between independently derived planktic foraminiferal-based faunal estimates of winter and summer SST (Fig. 7). In general, all three proxies at DSDP Site 552A show concordance in magnitude of mid-Pliocene warming. Fine-scale comparison of the three proxies shows differences that may be attributable to (1) problems with any and/or all of the techniques; (2) differences owing to deviation in sampling strategies (when same samples are not used for different proxy analyses); and (3) complexities arising from what each proxy is actually monitoring (e.g. depth or season of estimate) and how these compare in time-averaged samples (Dowsett & Robinson 2006).

In other areas of the North Atlantic, particularly in upwelling cells, the agreement is not as good and the PRISM conclusion that low latitude areas were not warmer in the mid-Pliocene epoch relative to present day has been contested (Haywood et al. 2005). Unfortunately, none of the relevant localities have multiple proxies, temporal resolution is low, and variability is high, making comparisons tenuous. It is entirely possible and even probable, based upon the available data, that upwelling decreased and SST rose above present cool levels at times during the PRISM time-slab interval.

In low latitude areas away from upwelling, no conclusive evidence of warmer than modern surface temperatures exists. This is particularly evident in the Western Equatorial Pacific (WEP) warm pool. Faunal-based estimates of SST at DSDP Site 586 and ODP Site 806 (Dowsett et al. 1999, 2005) are essentially the same as Mg/Ca-based estimates from Site 806 (Wara et al. 2005) and modern conditions.

While MAT cannot be used to estimate temperatures warmer than those found in the calibration dataset, PRISM uses changes to planktic foraminifer assemblage structure to search for temperature values in the equatorial Pacific over 30 °C. The rationale is that in the present-day ocean, multiple elements of the planktic assemblage change as temperatures transition from c. 25° to c. 30 °C. The working hypothesis is that during the mid-Pliocene, SSTs in the WEP were similar to modern values. Values exceeding 30 °C would have resulted in changes to the structure of the faunal assemblage. To date, no such changes have been observed in mid-Pliocene samples from the WEP. An alternative (albeit highly complicated) hypothesis is that warming did occur but ecological tolerances of individual taxa changed (the species adapted) and therefore the structure of the assemblage did not change. This second hypothesis is less appealing since it requires multiple taxa to change preferences simultaneously. Furthermore, comparison of the WEP assemblage structure

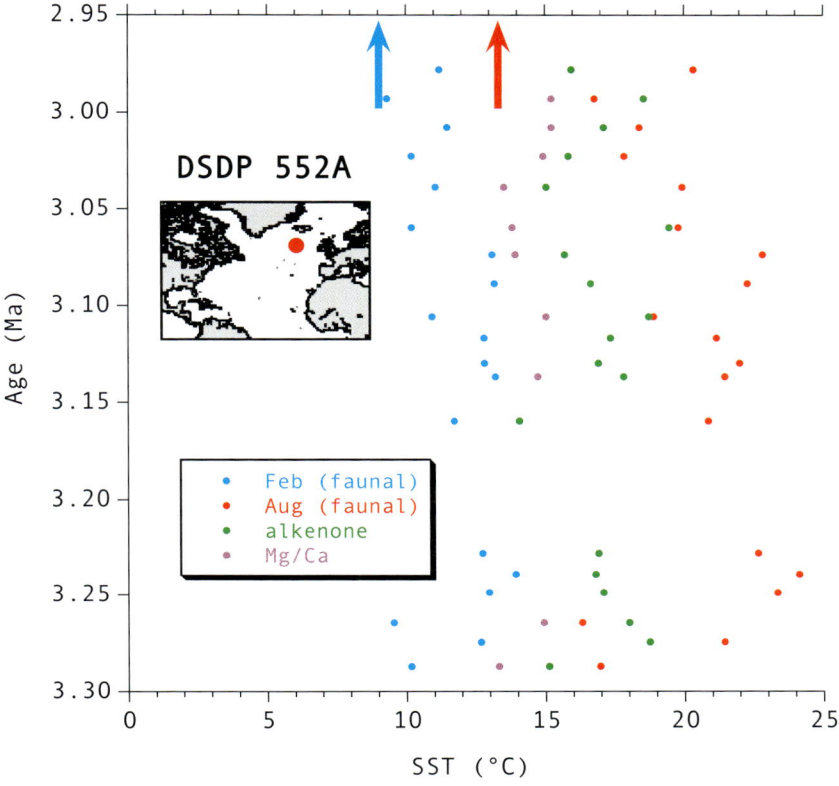

Fig. 7. Multiproxy comparison of SST at DSDP 552A in the North Atlantic. Note Mg/Ca (*G. bulloides*) and alkenone temperature estimates are considered annual averages (at shallow subsurface and surface, respectively) and fall between annual surface extremes determined by transfer function technique.

between repeated late Pleistocene glacial–interglacial extremes and the mid-Pliocene shows no distinguishable differences. Therefore, if ecological strategies changed, they must have changed back-and-forth repeatedly, in multiple taxa, which is highly unlikely.

PRISM geographic coverage (marine) in the equatorial Pacific has been poor, and is lacking entirely in the Eastern Equatorial Pacific (EEP). Preliminary analysis of core material from new PRISM sites in the EEP (ODP 847 and 852) suggests conditions 3–4 °C warmer than present. These results are in agreement with Mg/Ca (Wara *et al.* 2005) and alkenone (Ravelo *et al.* 2006) palaeothermometry estimates from the same region (ODP Site 847). These latter estimates have been used to suggest a permanent El Niño state or lack of a strong east–west surface temperature contrast during the mid-Pliocene epoch (Ravelo *et al.* 2006). New PRISM faunal data from Sites 807, 847 and 852, when taken with the existing geochemical and faunal data, suggest a

relaxation of the EEP cool tongue and corroborate the lack of a strong WEP–EEP contrast during the PRISM time-slab.

Deep ocean temperature

The role of the deep ocean in climate change is becoming better known through the application of the newest coupled ocean–atmosphere general circulation models (Haywood & Valdes 2004). PRISM3 is centred around the development of a deep ocean reconstruction based primarily upon Mg/Ca palaeothermometry on the ostracode genus *Krithe* (Dwyer *et al.* 1995; Cronin *et al.* 1996, 2005). New Atlantic Basin data suggest cyclic bottom-water temperature (BWT) changes throughout the mid-Pliocene epoch, increased North Atlantic Deep Water (NADW) production and decreased Antarctic Bottom Water (AABW) production (Dowsett *et al.* 2006). Initial experiments using the GISS coupled ocean–atmosphere model and PRISM2 surface boundary conditions produced a

North Atlantic overturning that was lower than modern and clearly not in agreement with mid-Pliocene interpretations of a greater deep-water production. The addition of Pliocene deep ocean temperature data (PRISM3D; Dowsett *et al.* 2006) to the coupled model initialization led to a strengthening of the North Atlantic deep-water production to near-modern levels (M. Chandler, pers. comm.). It remains to be seen whether coupled model experiments initialized with new extra-Atlantic deep-water reconstructions of PRISM3D and a new tropical Pacific SST reconstruction with reduced EEP upwelling, will result in surface conditions similar to the PRISM2 SST reconstruction.

Taxonomic Uniformitarianism

Dowsett & Robinson (1998) discussed the problem of Taxonomic Uniformitarianism and an updated version of that discussion is worth including here. The assumption of ecological stability (basic to most paleontological studies) is subject to criticism. The question is whether in the course of three million years of evolution, planktonic foraminifers (or other organisms) adapt in a manner that significantly changes their environmental preferences from those of their ancestors. Since it is nearly impossible to answer this question with certainty, regardless of timescale, the grouping of ancestors with modern descendants and/or fossil taxa with modern taxa having similar morphologies or biogeographic distributions is inherently open to criticism. Nevertheless, various techniques are often employed to determine whether a particular taxon occupies the same position with respect to stable isotopes ($\delta^{18}O$, $\delta^{13}C$) over time, or whether an ancestor and descendant occupy the same ecological space. Interestingly, these geochemical tests also rely on an equally challenging and untestable assumption: that organismic vital effects remain constant over time. In truth, it is not possible to ascertain whether the assumptions that must be made to apply transfer functions to Pliocene-age sequences can ever be unequivocally proven correct. Any attempt to infer environmental states from geochemical observations is flawed from the outset by the assumption that a biological system is known in sufficient detail today and that the system did not interact differently with the physical environment in the past.

Irrespective of these considerations, we hold some confidence that the assumptions made, after careful experimentation, are valid, not because the results obtained are those anticipated, but because the same results were obtained with different types of organisms and different quantitative methods. For example, the Pliocene of the northeastern Atlantic Ocean has been intensively studied over the past 15 years. Willard (1994) suggests that during the middle Pliocene areas of Iceland now covered by tundra vegetation were covered by deciduous forest, indicating temperatures 3° to 5 °C warmer than today. Ostracods were used by Cronin (1991) to show that shallow sea-bottom temperatures (the same zone sampled by planktonic foraminifers except closer to the shoreline) off Iceland were warmer than today by about 5 °C (both winter and summer) during the warm intervals of the middle Pliocene. Dowsett & Poore (1990) and Dowsett *et al.* (1992) estimated sea-surface temperatures to be 3.6 °C and 7.9 °C (winter and summer) warmer than today during mid Pliocene 'interglacials' in the same region. These three fossil groups (plants, animals and protists) all give more-or-less the same results using three independent techniques. More recent analysis of Mg/Ca ratios in planktic foraminifers and alkenone unsaturation indices from bulk sediment (Fig. 7) from the same region show annual average temperature, at surface and shallow depth, in close agreement with the previous palaeontological proxies. While assumptions about environmental preferences are inherent in all five proxies, it is highly unlikely that these investigations would all suggest the same general answer unless they were sensing a strong and consistent signal.

Therefore, I offer no answers for the critics of palaeontologically based deep-time environmental reconstruction, but also hold that through careful integrated analysis using different biological and geochemical systems one can gain improved confidence in results.

Future needs and conclusion

The PRISM marine SST dataset is the cornerstone of the PRISM reconstruction. It is based upon multiple fossil groups, overlapping wherever possible. The assumptions and rationale for all analyses are published. Over the years, new data have become available, warranting changes to the existing datasets. In general, the overall signal of increased warming at higher latitudes has stood the test of time. As indicated above, changes to the low-latitude Pacific reconstruction are now warranted. In the not-too-distant future, multiple proxies will be understood in sufficient detail, and enough suitable material may be available, for a true synchronous (in a deep time sense) reconstruction tied possibly to a single obliquity or precession cycle.

Recent analyses of terrestrial material derived from the Meyer Desert Formation in Antarctica alone suggest that the PRISM land cover dataset needs thorough revision (Ashworth & Cantrill 2004). Other new data from areas not well represented in the current reconstruction (Africa, South America) warrant inclusion in the next PRISM vegetation/land cover dataset (e.g. Franz-Odendaal *et al.* 2002; Hartley & Chong 2002; Bonnefille *et al.* 2004; Feakins *et al.* 2005).

The presence and distribution of land ice is a fundamentally important element of the climate system. Despite years of research, we have at best a poor understanding of the distribution of ice during the mid-Pliocene epoch and even that is, in part, highly controversial. Warmer than modern SSTs in both hemispheres, evidence for diminished sea ice in the Arctic and Antarctic, physical and isotopic evidence for elevated sea level, and geological and biotic evidence for reduced ice from Antarctica, suggest a diminished global ice volume relative to present. Accurate models of stable ice volumes and distributions that would be in tune with the plethora of information suggesting warmer conditions are needed for both Antarctica and Greenland (see Hill *et al.* 2007).

PRISM is as an evolving work (call it the adhesive) marrying the signal between palaeoenvironmental data and climate models. In an iterative fashion, data and model specialists listen to each other to improve both the palaeoclimate analyses and model experiments utilizing the PRISM boundary conditions. With humankind at the threshold of historically unprecedented environmental changes, both natural and anthropogenic, it is vital that climate scientists accurately read the deep-time palaeoclimate record and make best use of the lessons it contains.

I thank the editors for inviting this submission. Tom Cronin, Deb Willard and two anonymous reviewers provided many useful suggestions. Marci Robinson, as always, was invaluable throughout the preparation of this chapter and the PRISM Project. This chapter would not have been possible without the collaboration and support of key people. Dick Poore facilitated the original PRISM effort and deserves much credit for its success. Tom Cronin and John Barron provided key information on many aspects of the Pliocene marine environment which forms the core of our understanding of this complex time period. PRISM is a reality because of their dedicated involvement. Deb Willard and Bob Thompson are responsible for a tremendous accumulation of carefully vetted vegetation data as well as research that led to true marine–terrestrial correlations in key regions. Elizabeth Dowsett is a constant source of encouragement and support. This paper is a product of the US Geological Survey Earth Surface Dynamics Program.

References

ABELMANN, A., GERSONDE, R. & SPIESS, V. 1990. Pliocene-Pleistocene paleoceanography in the Weddell Sea: siliceous microfossil evidence. *In*: BLEIL, U. & THIEDE, J. (eds) *Geologic History of the Polar Oceans: Arctic Versus Antarctic:* NATO ASI Series C, **308**, 729–759.

ASHWORTH, A. C. & CANTRILL, G. 2004. Neogene vegetation of the Meyer Desert Formation (Sirius Group) Transantarctic Mountains, Antarctica. *Palaeogeography, Palaeoclimatology Palaeoecology*, **213**, 65–82.

ASHWORTH, A. C. & KUSCHEL, G. 2003. Fossil weevils (Coleoptera: Curculionidae) from latitude 85°S Antarctica. *Palaeogeography, Palaeoclimatology Palaeoecology*, **191**, 191–202.

BARRETT, P. J., BLEAKLEY, N. L., DICKINSON, W. W., HANNAH, M. J. & HARPER, M. A. 1997. Distribution of siliceous microfossils on Mount Feather, Antarctica, and the age of the Sirius Group. *In*: RICCI-CARLO, A. (ed.) *The Antarctic Region: Geological Evolution and Processes*, Terra Antarctica, **4**, 763–770.

BARRON, J. A. 1992*a*. Pliocene paleoclimatic interpretation of DSDP Site 580 (NW Pacific) using diatoms. *Marine Micropaleontology*, **20**, 23–44.

BARRON, J. A. 1992*b*. Paleoceanographic and tectonic controls on the Pliocene diatom record of California. *In*: TSUCHI, R. & INGLE, J. C. JR. (eds) *Pacific Neogene: Environment, Evolution, and Events*, University of Tokyo Press, Tokyo, 25–41.

BARRON, J. A. 1995. High resolution diatom paleoclimatology of the middle part of the Pliocene of the northwest Pacific. *In*: REA, D. K., BASOV, I. A., SCHOLL, D. W. & ALLAN, J. F. (eds) *Proceedings of the Ocean Drilling Program, Scientific Results*, **145**, 43–53, College Station, TX.

BARRON, J. A. 1996*a*. Diatom constraints on the position of the Antarctic Polar Front in the middle part of the Pliocene. *Marine Micropaleontology*, **27**, 195–213.

BARRON, J. A. 1996*b*. Diatom constraints on sea surface temperatures and sea ice distribution during the middle part of the Pliocene. *U.S. Geological Survey Open-File Report*, **96–713**, 1–45.

BARTOLI, G., SARNTHEIN, M., WEINELT, M., ERLENKEUSER, H., GARBE-SCHÖNBERG, D. & LEA, D. W. 2005. Final closure of Panama and the onset of northern hemisphere glaciation. *Earth and Planetary Science Letters*, **237**, 33–44.

BERGGREN, W. A., KENT, D. V. & VAN COUVERING, J. A. 1985. Neogene geochronology and chronostratigraphy. *In*: SNELLING, N. J. (ed.) *The Chronology of the Geological Record*. London, Geological Society of London *Memoir*, **10**, 211–260.

BERGGREN, W. A., KENT, D. V, SWISHER, C. C. & AUBRY, M. P. 1995. A revised Cenozoic geochronology and chronostratigraphy. *In*: BERGGREN, W. A., KENT, D. V., AUBRY, M.-P. & HARDENBOL, J. (eds) Geochronology, time scales and global stratigraphic correlation. Tulsa, *Society for Sedimentary Geology Special Publication*, **54**, 129–212.

BONNEFILLE, R., POTTS, R., CHALIE, F., JOLLY, D. & PEYRON, O. 2004. High resolution vegetation and climate change associated with Pliocene *Australopithecus afarensis*. *PNAS*, **101**, 12125–12129.

BRIGHAM-GRETTE, J. & CARTER, L. D. 1992. Pliocene marine transgressions of Northern Alaska: Circumarctic Correlations and Paleoclimatic Interpretations. *Arctic*, **45**, 74–89.

BRISKIN, M. & BERGGREN, W. A. 1975. Pleistocene stratigraphy and quantitative paleo-oceanography of tropical North Atlantic core V216-205. In: SAITO, T. & BURCKLE, L. H. (eds) Late Epoch Boundaries. Micropaleontology, Special Publication, 1, 167–198.

BROUWERS, E. M. 1994. Late Pliocene paleoecologic reconstructions based on ostracode assemblages from the Sagavanirktok and Gubik Formations, Alaskan North Slope. Arctic, 47, 16–33.

BROUWERS, E. M., JORGENSEN, N. O. & CRONIN, T. M. 1991. Climatic significance of the ostracode fauna from the Pliocene Kap Kobenhavn Formation, north Greenland. Micropaleontology, 37, 245–276.

BUCHARDT, B. & SIMONARSON, L. A. 2003. Isotope paleotemperatures from the Tjornes beds in Iceland: evidence of Pliocene cooling. Palaeogeography, Palaeoclimatology, Palaeoecology, 189, 71–95.

BUKRY, D. 1973. Low-latitude coccolith biostratigraphic zonation. Initial Reports of the Deep Sea Drilling Project, 15, 685–703.

BUKRY, D. 1975. Coccolith and silicoflagellate stratigraphy, northwestern Pacific Ocean, Deep Sea Drilling Project Leg 32, Initial Reports of the Deep Sea Drilling Project, 32, 677–701.

BURCKLE, L. H. & CIRILLI, J. 1987. Origin of diatom ooze belt in the Southern Ocean: Implications for Late Quaternary paleoceanography. Micropaleontology, 33, 82–86.

CHANDLER, M., RIND, D. & THOMPSON, R. 1994. Joint investigations of the middle Pliocene climate II: GISS GCM Northern Hemisphere results. Global and Planetary Change, 9, 197–219.

CLIMAP Project Members. 1981. Seasonal reconstructions of the Earth's surface at the last glacial maximum. Geological Society of America Map and Chart Series, MC-36.

CLIMAP Project Members. 1984. The last interglacial ocean. Quaternary Research, 21, 123–224.

COOPER, A. K. & O'BRIEN, P. E. 2004. Leg 188 synthesis: transitions in the glacial history of the Prydz Bay region, East Antarctica, from ODP drilling. Proceedings of the Ocean Drilling Program, Scientific Results, 188, 1–42.

CRONIN, T. M., DWYER, G. S., BAKER, P. A., RODRIGUEZ-LAZARO, J. & BRIGGS, Jr., W. M. 1996. Deep-sea ostracode shell chemistry (Mg : Ca ratios) and late Quaternary Arctic Ocean History. In: ANDREWS, J. T., AUSTIN, W. E. N., BERGSTEN, H. & JENNINGS, A. E. (eds) Late Quaternary Paleoceanography of North Atlantic Margins, Special Publications, 111, Geological Society, London, 117–134.

CRONIN, T. M. 1991a. Late Neogene marine ostracoda from Tjörnes, Iceland. Journal of Paleontology, 65, 767–794.

CRONIN, T. M. 1991b. Pliocene shallow water paleoceanography of the North Atlantic Ocean based on marine ostracodes. Quaternary Science Reviews, 10, 175–188.

CRONIN, T. M. & DOWSETT, H. J. 1990. A quantitative micropaleontologic method for shallow marine paleoclimatology: Application to Pliocene deposits of the western North Atlantic Ocean. Marine Micropaleontology, 16, 117–148.

CRONIN, T. M. & DOWSETT, H. J. (eds) 1991. Preface, Pliocene Climates, Quaternary Science Reviews, 10, v–vi.

CRONIN, T. M., WHATLEY, R. C., WOOD, A., TSUKAGOSHI, A., IKEYA, N., BROUWERS, E. M. & BRIGGS, W. M. 1993. Microfaunal evidence for elevated mid-Pliocene temperatures in the Arctic Ocean. Paleoceanography, 8, 161–173.

CRONIN, T. M., DOWSETT, H. J., DWYER, G. S., BAKER, P. A. & CHANDLER, M. A. 2005. Mid-Pliocene deep-sea bottom water temperatures based on ostracode Mg/Ca ratios, Marine Micropaleontology, 54, 249–261.

DOWSETT, H. J. 1989a. Application of the graphic correlation method to Pliocene marine sequences. Marine Micropaleontology, 14, 3–32.

DOWSETT, H. J. 1989b. Improved dating of the Pliocene of the eastern South Atlantic using graphic correlation: Implications for paleobiogeography and paleoceanography. Micropaleontology, 35, 279–292.

DOWSETT, H. J. 1991. The development of a long-range foraminifer transfer function and application to Late Pleistocene North Atlantic climatic extremes. Paleoceanography, 6, 259–273.

DOWSETT, H. J. & CRONIN, T. M. 1990. High eustatic sea level during the Middle Pliocene: Evidence from the southeastern U.S. Atlantic Coastal Plain. Geology, 18, 435–438.

DOWSETT, H. J. & LOUBERE, P. 1992. High resolution Late Pliocene sea-surface temperature record from the Northeast Atlantic Ocean. Marine Micropaleontology, 20, 91–105.

DOWSETT, H. J. & POORE, R. Z. 1990. A new planktic foraminifer transfer function for estimating Pliocene through Holocene sea surface temperatures. Marine Micropaleontology, 16, 1–23.

DOWSETT, H. J. & POORE, R. Z. 1991. Pliocene sea surface temperatures of the North Atlantic Ocean at 3.0 Ma. Quaternary Science Reviews, 10, 189–204.

DOWSETT, H. J. & POORE, R. Z. 1999. Last interglacial sea-surface temperature estimates from the California margin: Improvements to the modern analog technique. U.S. Geological Survey Bulletin, 2171. http://pubs.usgs.gov/bul/b2171/

DOWSETT, H. & ROBINSON, M. 1998. Application of the modern analog technique (MAT) of sea surface temperature estimation to middle Pliocene North Pacific planktic foraminifer assemblages. Paleontologia Electronica, 1. http://www-odp.tamu.edu/paleo/1998_1/dowsett/issue1.htm

DOWSETT, H. J. & ROBINSON, M. M. 2006. Stratigraphic framework for Pliocene paleoclimate reconstruction: the correlation conundrum. Stratigraphy, 3, 53–64.

DOWSETT, H. J. & WILLARD, D. A. 1996. Southeast Atlantic marine and terrestrial response to middle Pliocene climate change. Marine Micropaleontology, 27, 181–93.

DOWSETT, H. J., CRONIN, T. M., POORE, R. Z., THOMPSON, R. S., WHATLEY, R. C. & WOOD, A. M. 1992. Micropaleontological evidence for increased meridional heat transport in the North Atlantic Ocean during the Pliocene. Science, 258, 1133–1135.

DOWSETT, H. J., BARRON, J. & POORE, R. 1996. Middle Pliocene sea surface temperatures: a global reconstruction. Marine Micropaleontology, 27, 13–26.

DOWSETT, H. J., BARRON, J. A., POORE, R. Z., THOMPSON, R. S., CRONIN, T. M., ISHMAN, S. E. & WILLARD, D. A. 1999. Pliocene paleoenvironmental reconstruction: PRISM2, *U.S. Geological Survey Open File Report*, **99–535**. (http://pubs.usgs.gov/of/of99-535/)

DOWSETT, H. J., BRUNNER, C. A., VERARDO, S. & POORE, R. Z. 2003. Gulf of Mexico Planktic Foraminifer Core-top Calibration Data Set: Raw Data. *U.S. Geological Survey Open File Report*: **03–008**. http://pubs.usgs.gov/of/2003/of03-008/text.htm

DOWSETT, H. J., THOMPSON, R. S., BARRON, J. A. *ET AL.* 1994. Joint investigations of the middle Pliocene climate I: PRISM paleoenvironmental reconstructions. *Global and Planetary Change*, **9**, 169–195.

DOWSETT, H. J., CHANDLER, M. A., CRONIN, T. M. & DWYER, G. S. 2005. Middle Pliocene sea surface temperature variability. *Paleoceanography*, **20**, PA2014, doi:10.1029/2005PA001133.

DOWSETT, H., ROBINSON, M., DWYER, G., CHANDLER, M. & CRONIN, T. 2006. PRISM3 DOT1 Atlantic Basin Reconstruction. *U.S. Geological Survey Data Series*, **189**. http://pubs.usgs.gov/ds/2006/189/

DRAUT, A. E., RAYMO, M. E., MCMANUS, J. F. & OPPO, D. W. 2003. Climate stability during the Pliocene warm period. *Paleoceanography*, **18**, 1078, doi:10.1029/2003PA000889.

DWYER, G. S., CRONIN, T. M., BAKER, P. A., RAYMO, M. E., BUZAS, J. S. & CORREGE, T. 1995. North Atlantic deepwater temperature change during late Pliocene and late Quaternary climatic cycles. *Science*, **270**, 1347–1351.

ELIAS, S. A. & MATTHEWS, J. V. 2002. Arctic North American seasonal temperatures from the latest Miocene to the Early Pleistocene, based upon mutual climatic range analysis of fossil beetle assemblages. *Canadian Journal of Earth Sciences*, **39**, 911–920.

FEAKINS, S., DEMENOCAL, P. & EGLINTON, T. 2005. Biomarker records of late Neogene changes in northeast African vegetation. *Geology*, **33**, 977–980.

FLEMING, R. F. & BARRON, J. A. 1996. Evidence of Pliocene *Nothofagus* in Antarctica from Pliocene marine sedimentary deposits (DSDP Site 274). *Marine Micropalaeontology*, **27**, 227–236.

FRADKINA, A. F. 1991. Pliocene climatic fluctuations in the Far North-East of the USSR. Pliocene Climates of the Northern Hemisphere: abstracts of the Joint US/USSR Workshop on Pliocene Paleoclimates, Moscow, USSR, April, 1990. *U.S. Geological Survey Open-File Report*, **91–447**, 22.

FRANZ-ODENDAAL, T., LEE-THORP, J. & CHINSAMY, A. 2002. New evidence for the lack of C4 grassland expansion during the early Pliocene at Langebaanweg, South Africa. *Paleobiology*, **28**, 378–388.

FYLES, J. G., MARINCOVICH, L., MATTHEWS, J. V. & BARENDREGT, R. 1991. Unique mollusc find in the Beaufort Formation (Pliocene) on Meighen Island, Arctic Canada. Geologic Survey of Canada, *Current Research*, Part B **91–1B**, 105–112.

GERSONDE, R. & BURCKLE, L. H. 1990. Neogene diatom biostratigraphy of ODP Leg 113, Weddell Sea (Antarctic Ocean). *In*: BARKER, P. F., KENNETT, J. P. *ET AL.* (eds) *Proceedings of the Ocean Drilling Program, Scientific Results*, **113**, 761–789.

HAQ, B. H., HARDENBOL, J. & VAIL, P. R. 1987*a*. Chronology of fluctuating sea levels since the Triassic. *Science*, **235**, 1156–1167.

HAQ, B. U., HARDENBOL, J. & VAIL, P. R. 1987*b*. The new chronostratigraphic basis of Cenozoic and Mesozoic sea level cycles. Cushman Foundation for Foraminiferal Research, *Special Publication*, **24**, 7–13.

HARTLEY, A. J. & CHONG, G. 2002. A late Pliocene age for the Atacama Desert: Implications for the desertification of western South America. *Geology*, **30**, 43–46.

HARWOOD, D. M. 1986. Recycled siliceous microfossils from the Sirius Formation. *Antarctic Journal of the United States*, **21**, 101–103.

HARWOOD, D. M. & WEBB, P. 1998. Glacial transport of diatoms in the Antarctic Sirius Group: Pliocene refrigerator. *GSA Today*, **8**, 1–8.

HARWOOD, D. M., MCMINN, A. & QUILTY, P. G. 2000. Diatom biostratigraphy and age of the Pliocene Sorsdal Formation, Vestfold Hills, East Antarctica. *Antarctic Science*, **12**, 443–462.

HAYWOOD, A. M. & VALDES, P. J. 2004. Modeling Pliocene warmth: contributions of atmosphere, oceans and cryosphere. *Earth and Planetary Science Letters*, **218**, 363–377.

HAYWOOD, A. M., VALDES, P. J., FRANCIS, J. E. & SELLWOOD, B. W. 2002. Global middle Pliocene biome reconstruction: A data/model synthesis. *Geochemistry, Geophysics, Geosystems*, **3**, 1072, doi:10.1029/2002GC000358.

HAYWOOD, A. M., DEKENS, P., RAVELO, A. C. & WILLIAMS, M. 2005. Warmer tropics during the mid Pliocene: evidence from alkenone paleothermometry and a fully coupled ocean-atmosphere GCM. *Geochemistry, Geophysics, Geosystems*, **6**, doi: 10.1029/2004GC000799.

HAYWOOD, A. M., VALDES, P. J., HILL, D. & WILLIAMS, M. 2007. The mid-Pliocene warm period: a test-bed for integrating data & models. *In*: WILLIAMS, M., HAYWOOD, A. M., GREGORY, F. J. & SCHMIDT, D. N. (eds) *Deep-Time Perspectives on Climate Change: Marrying the Signal from Computer Models and Biological Proxies*. The Micropalaeontological Society, Special Publications. Geological Society, London, 000–000.

HERBERT, T. D. & SCHUFFERT, J. D. 1998. Alkenone unsaturation estimates of Late Miocene through Late Pliocene sea-surface temperatures at Site 958. *In*: FIRTH, J. V. (ed.) *Proceedings of the Ocean Drilling Program, Scientific Results*, **159T**, 17–22.

HILL, D., HAYWOOD, A. M., HINDMARSH, R. C. A. & VALDES, P. J. 2007. Characterising Pliocene ice-sheets: evidence from data and models. *In*: WILLIAMS, M., HAYWOOD, A. M., GREGORY, F. J. & SCHMIDT, D. N. (eds) *Deep-Time Perspectives on Climate Change: Marrying the Signal from Computer Models and Biological Proxies*. The Micropalaeontological Society, Special Publications. Geological Society, London, 485–506.

HILLENBRAND, C.-D. & FÜTTERER, D. K. 2001. Neogene to Quaternary deposition of opal on the continental rise west of the Antarctic Peninsula, ODP Leg 178, Sites 1095, 1096, and 1101. *Proceedings of the Ocean Drilling Program, Scientific Results*, **178**.

http://www.odp.tamu.edu/publications/178_SR/chap_23/chap_23.htm

HODELL, D. A. & CIESIELSKI, P. F. 1991. Stable isotopic and carbonate stratigraphy of the Plio-Pleistocene of Ocean Drilling Program (ODP) Hole 704A: Eastern subantarctic South Atlantic. *Proceedings of the Ocean Drilling Program, Scientific Results*, **114**, 409–436.

HODELL, D. A. & VENZ, K. 1992. Toward a high-resolution stable isotopic record of the Southern Ocean during the Pliocene-Pleistocene (4.8 to 0.8 Ma). *Antarctic Research Series*, **56**, 265–310.

HUTSON, W. H. 1980. The Agulhas Current during the Late Pleistocene: Analysis of modern faunal analogs. *Science*, **207**, 64–66.

IKEYA, N. & CRONIN, T. 1993. Quantitative analysis of ostracoda and water masses around Japan: application to Pliocene and Pleistocene paleoceanography. *Micropaleontology*, **39**, 263–281.

IMBRIE, J. & KIPP, N. G. 1971. A new micropaleontological method for quantitative paleoclimatology: Application to a Late Pleistocene Caribbean core. *In*: TUREKIAN, K. K. (ed.) *The Late Cenozoic Glacial Ages*. Yale University Press, New Haven, 71–181.

JIANG, D., WANG, H., DING, Z., LANG, X. & DRANGE, H. 2005. Modeling the middle Pliocene climate with a global atmospheric general circulation model. *Journal of Geophysical Research*, **110**, D14107, doi:10.1029/2004JD005639.

JONKERS, H. A. & KELLEY, S. P. 1998. A reassessment of the Cockburn Island Formation, northern Antarctic Peninsula, and its palaeoclimatic implications. *Journal of the Geological Society of London*, **155**, 737–740.

KEIGWIN, L. D. 1976. Late Cenozoic planktonic foraminiferal biostratigraphy and paleoceanography of the Panama Basin. *Micropaleontology*, **22**, 419–442.

KENNETT, J. P. & BARKER, P. F. 1990. Latest Cretaceous to Cenozoic climate and oceanographic developments in the Weddell Sea, Antarctica: An ocean drilling perspective. *Proceedings of the Ocean Drilling Program, Scientific Results*, **113**, 937–960.

KENNETT, J. P. & HODELL, D. A. 1993. Evidence for relative climatic stability of Antarctica during the early Pliocene: a marine perspective. *Geografiska Annaler*, **75A**, 205–220.

KRANTZ, D. E. 1991. A chronology of Pliocene sea level fluctuations: The U.S. middle Atlantic Coastal Plain record. *Quaternary Science Reviews*, **10**, 163–174.

LEVITUS, S. & BOYER, T. P. 1994. World Ocean Atlas 1994 (Vol. 4): Temperature. *NOAA Atlas NESDIS*, **4**.

LI, B., WANG, J., HUANG, B. *ET AL*. 2004. South China Sea surface water evolution over the last 12 Myr: A south-north comparison from Ocean Drilling Program Sites 1143 and 1146. *Paleoceanography*, **19**, PA1009, doi:10.1029/2003PA000906.

LISIECKI, L. E. & RAYMO, M. E. 2005. A Pliocene-Pleistocene stack of 57 globally distributed benthic $\delta^{18}O$ records. *Paleoceanography*, **20**, PA1003, doi:10.1029/2004PA001071.

LOURENS, L. J., ANTONARAKOU, A., HILGEN, F. J., VAN HOOF, A. A. M., VERGNAUD-GRAZZINI, C. & ZACHARIASSE, W. J. 1996. Evaluation of the Plio-Pleistocene astronomical timescale. *Paleoceanography*, **11**, 391–413.

MARCHANT, D. R. & DENTON, G. H. 1996. Miocene and Pliocene paleoclimate of the Dry Valleys region, Southern Victoria land: a geomorphological approach. *Marine Micropaleontology*, **27**, 253–271.

MARCHANT, D., DENTON, G. H., SWISHER, C. & POTTER, N. 1996. Late Cenozoic Antarctic paleoclimate reconstructed from volcanic ashes in the Dry Valleys region of southern Victoria Land. *Geological Society of America Bulletin*, **108**, 181–194.

MARCHANT, D. R., LEWIS, A. R., PHILLIPS, W. M. *ET AL*. 2002. Formation of patterned ground and sublimation till over Miocene glacier ice in Beacon Valley, southern Victoria Land, Antarctica. *Geological Society of America Bulletin*, **114**, 718–730.

MARTINI, E. 1971. Standard Tertiary and Quaternary calcareous nannoplankton zonation. *In*: FERINACCI, A. (ed.) *Proc. II Int. Plankt. Conf., Roma, 1970*. Tecnoscienza, **2**, 739–785.

MATTHEWS, E. 1985. Prescription of land-surface boundary conditions in GISS GCM II: a simple method based on high-resolution vegetation databases. *NASA Report*, **TM 86096**, 20pp.

MATTHEWS, J. V. JR & OVENDEN, L. E. 1990. Late Tertiary plant macrofossils from localities in Arctic/Subarctic North America: a review of the data. *Arctic*, **43**, 364–392.

MIDDLETON, G. V. 2000. *Data analysis in the Earth Sciences using MATLAB*. Prentice Hall, Upper Saddle River, New Jersey, 260pp.

MILLER, M. & MABIN, M. 1998. Antarctic Neogene landscapes: In the refrigerator or in the deep freeze? *GSA Today*, **8**, 1–3.

National Research Council Committee on Abrupt Climate Change 2002. *Abrupt Climate Change: Inevitable Surprises*. National Academies Press, Washington, DC. 244pp.

NELSON, R. E. & CARTER, L. D. 1985. Pollen analysis of a late Pliocene and early Pleistocene section from the Gubik Formation of arctic Alaska. *Quaternary Research*, **24**, 295–306.

OVERPECK, J. T., WEBB, T. III & PRENTICE, I. C. 1985. Quantitative interpretation of fossil pollen spectra: dissimilarity coefficients and the method of modern analogs. *Quaternary Research*, **23**, 87–108.

PFLAUMANN, U., DUPRAT, J., PUJOL, C. & LABEYRIE, L. D. 1996. SIMMAX: A modern analog technique to deduce Atlantic sea surface temperatures from planktonic foraminifera in deep-sea sediments. *Paleoceanography*, **11**, 15–35.

PFLAUMANN, U., SARNTHEIN, M., FICKEN, K., GROTHMANN, A. & WINKLER, A. 1998. Variations in eolian and carbonate sedimentation, sea-surface temperature, and productivity over the last 3 M.Y. at Site 958 off Northwest Africa. *In*: FIRTH, J. V. (ed.) *Proceedings of the Ocean Drilling Program, Scientific Results* **159T**, 3–16.

POORE, R. Z. 1981. Temporal and spatial distribution of ice-rafted mineral grains in Pliocene sediments of the North Atlantic: Implications for Late Cenozoic climatic history. *In*: WARME, J. E. A. (ed.) The Deep Sea Drilling Project: A Decade of Progress. *Society* of Economic Paleontologists and Mineralogists Special Publication, 505–515.

POORE, R. Z. & SLOAN, L. C. 1996. Climates and Climate Variability of the Pliocene. Marine Micropaleontology, 27, 326pp.

POORE, R. Z., WILLIAMS, R. S. JR. & TRACEY, C. 2000. Sea level and climate. U.S. Geological Survey Fact Sheet FS-002-00, 2pp.

PRELL, W. L. 1985. The stability of low-latitude sea-surface temperatures: An evaluation of the CLIMAP reconstruction with emphasis on the positive SST anomalies. Washington, DC, US Department of Energy.

PRELL, W., MARTIN, A., CULLEN, J. & TREND, M. 1999. The Brown University Foraminiferal Data Base, IGBP PAGES/World Data Center-A for Paleoclimatology, Data Contribution Series #1999-027. NOAA/NGDC Paleoclimatology Program, Boulder, USA.

PRENTICE, M. L. & KRUSIC, A. G. 2005. Early Pliocene Alpine glaciation in Antarctica: terrestrial versus tide-water glaciers in Wright Valley. Geografiska Annaler, 87A, 87–109.

QUILTY, P. G., LIRIO, J. M. & JILLETT, D. 2000. Stratigraphy of the Pleistocene Sorsdal Formation, Marine Plain, Vestfold Hills, East Antarctica. Antarctic Science, 12, 205–216.

RAVELO, A. C., DEKENS, P. & MCCARTHY, M. 2006. Evidence for El Nino-like conditions during the Pliocene. GSA Today, 16, 4–11.

RAYMO, M. E., RUDDIMAN, W. F., BACKMAN, J., CLEMENT, B. M. & MARTINSON, D. G. 1989. Late Pliocene variation in Northern Hemisphere ice sheets and North Atlantic deep water circulation. Paleoceanography, 4, 413–446.

REYNOLDS, R. W. & SMITH, T. M. 1995. A high resolution global sea surface temperature climatology. Journal of Climatology, 8, 1571–1583.

RUDDIMAN, W. F., MCINTYRE, A. & RAYMO, M. 1986. Paleoenvironmental results from North Atlantic Sites 607 and 609. Initial Reports of the Deep Sea Drilling Project, 94, 855–878.

SANCETTA, C. & SILVESTRI, S. 1986. High-resolution biostratigraphic and oceanographic events in the late Pliocene and Pleistocene North Pacific Ocean. Paleoceanography, 1, 163–180.

SHACKLETON, N. J., HALL, M. A. & PATE, D. 1995. Pliocene stable isotope stratigraphy of Site 846. In: PISIAS, N. G., MAYER, L. A., JANECSEK, T. R. ET AL., Proceedings of the Ocean Drilling Program, Scientific Results, 138, 337–355.

SLOAN, L. C., CROWLEY, T. J. & POLLARD, D. 1996. Modelling of middle Pliocene climate with the NCAR GENESIS general circulation model. Marine Micropaleontology, 27, 51–61.

STROEVEN, A. P. & PRENTICE, M. L. 1997. A case for the Sirius Group alpine glaciation at Mount Fleming, South Victoria Land, Antarctica: a case against Pliocene East Antarctic Ice Sheet reduction. Geological Society of America Bulletin, 109, 825–840.

STROEVEN, A. P., BURCKLE, L., KLEMAN, J. & PRENTICE, M. 1998. Atmospheric transport of diatoms in the Antarctic Sirius Group: Pliocene Deep Freeze. GSA Today, 8, 1–5.

SUGDEN, D. E., MARCHANT, D. R., POTTER, N., SOUCHEZ, R. A., DENTON, G. H., SWISHER, C. C. & TISON, J.-L. 1995. Preservation of Miocene ice in East Antarctica. Nature, 376, 412–414.

TEDFORD, R. H. & HARRINGTON, C. R. 2003. An Arctic mammal fauna from the Early Pliocene of North America. Nature, 425, 388–390.

TELFORD, R. J., ANDERSSON, C., BIRKS, H. J. B. & JUGGINS, S. 2004. Biases in the estimation of transfer function prediction errors. Paleoceanography, 19, PA4014, doi:10.1029/2004PA001072.

THOMPSON, R. S. & FLEMING, R. F. 1996. Middle Pliocene vegetation: reconstructions, paleoclimatic inferences, and boundary conditions for climatic modeling. Marine Micropaleontology, 27, 13–26.

THUNELL, R. C. 1979a. Climatic evolution of the Mediterranean Sea during the last 5.0 million years. Sedimentary Geology, 23, 67–79.

THUNELL, R. C. 1979b. Pliocene-Pleistocene paleotemperature and paleosalinity history of the Mediterranean Sea: Results from DSDP Sites 125 and 132. Marine Micropaleontology, 4, 173–187.

TIEDEMANN, R., SARNTHEIN, M. & SHACKLETON, N. J. 1994. Astronomic timescale for the Pliocene δ18O and dust records of Ocean Drilling Program Site 659. Paleoceanography, 9, 619–638.

US NAVAL OCEANOGRAPHIC OFFICE 1958. Oceanographic Atlas of the Polar Seas, Part II, Arctic. US Naval Printing Office, Washington, DC, 700, 1–300.

US NAVAL OCEANOGRAPHIC OFFICE 1967. Oceanographic Atlas of the North Atlantic Ocean, Section II, Physical Properties (Pub. 700). US Naval Printing Office, Washington, DC, 300pp.

VAN DER WATEREN, F. M., DUNAI, T. J., VAN BALEN, R. T., WERNER, K., VERBERS, A. L. L. M., PASSCHIER, S. & HERPERS, U. 1999. Contrasting Neogene denudation histories of different structural regions in the Transantarctic Mountains rift flank constrained by cosmogenic isotope measurements. Global and Planetary Change, 23, 145–172.

WARA, M. W., RAVELO, A. C. & DELANEY, M. L. 2005. Permanent El Nino-like conditions during the Pliocene warm period. Science, 309, 758–761.

WARDLAW, B. R. & QUINN, T. M. 1991. The record of Pliocene sea-level change at Enewetak Atoll. Quaternary Science Reviews, 10, 247–258.

WEBB, P.-N., HARWOOD, D. M., MABIN, M. G. C. & MCKELVEY, B. C. 1996. A marine and terrestrial Sirius Group succession, middle Beardmore Glacier-Queen Alexandra Range, Transantarctic Mountains, Antarctica. Marine Micropaleontology, 27, 273–297.

WHITE, J. M., AGER, T. A., ADAM, D. P., LEOPOLD, E. B., LIU, G., JETTE, H. & SCHWEGER, C. E. 1997. An 18 million year record of vegetation and climate change in northwestern Canada and Alaska: tectonic and global climate correlates. Palaeogeography, Palaeoclimatology, Palaeoecology, 130, 293–306.

WHITEHEAD, J. M. & BOHATY, S. M. 2003. Pliocene summer sea surface temperature reconstruction using silicoflagellates from Southern Ocean ODP Site 1165. Paleoceanography, 18, 1075, doi:10.1029/2002PA000829.

WHITEHEAD, J. M., WOTHERSPOON, S. & BOHATY, S. M.
2005. Minimal Antarctic sea ice during the Pliocene.
Geology, **33**, 137–140.

WILLARD, D. A. 1994. Palynological record from the
North Atlantic region at 3 Ma: vegetational distri-
bution during a period of global warmth: *Review of
Palaeobotany and Palynology*, **83**, 275–297.

WILLARD, D. A. 1997. Plio-Pleistocene pollen assem-
blages from the Yermak Plateau, Arctic Ocean: ODP
Sites 910 and 911. *Proceedings of the Ocean Drilling
Program, Scientific Results*, **151**, 297–305.

WILLARD, D. A., CRONIN, T. M., ISHMAN, S. E. &
LITWIN, R. J. 1993. Terrestrial and marine records of
climate and environmental change during the
Pliocene in subtropical Florida. *Geology*, **21**, 679–682.

WILSON, G. S. 1993. *Ice induced sea level change in the
late Neogene*. Victoria University, Wellington.

WOOD, A. M., WHATLEY, R. C., CRONIN, T. M. &
HOLTZ, T. 1994. Pliocene palaeotemperature
reconstruction for the southern North Sea based on
Ostracoda: a review. *Quaternary Science Reviews* **12**,
747–767.

ZUBAKOV, V. A. & BORZENKOVA, I. I. 1988.
Pliocene palaeoclimates: past climates as possible
analogues of mid-twenty-first century climate. *Palaeogeography, Palaeoclimatology, Palaeoecology*. **65**,
35–49.

ZUBAKOV, V. A. & BORZENKOVA, I. I. 1990. *Global
Palaeoclimate of the Cenozoic*. Elsevier, Amsterdam,
456pp.

Latitudinal climatic gradients in the Western European and Mediterranean regions from the Mid-Miocene (*c*. 15 Ma) to the Mid-Pliocene (*c*. 3.5 Ma) as quantified from pollen data

S. FAUQUETTE[1], J.-P. SUC[2], G. JIMÉNEZ-MORENO[3], A. MICHEELS[4], A. JOST[5],
E. FAVRE[2], N. BACHIRI-TAOUFIQ[6], A. BERTINI[7], M. CLET-PELLERIN[8], F. DINIZ[9],
G. FARJANEL[10], N. FEDDI[11] & Z. ZHENG[12]

[1]*Institut des Sciences de l'Evolution de Montpellier, UMR CNRS 5554, case courrier 061,
Université de Montpellier II, Place Eugène Bataillon, 34095 Montpellier cedex 5, France
(e-mail: fauquet@isem.univ-montp2.fr)*

[2]*Laboratoire Paléoenvironnements et Paléobiosphère, UMR CNRS 5125, Université Claude
Bernard - Lyon1, Boulevard du 11 Novembre, 69622 Villeurbanne cedex, France*

[3]*Departamento de Estratigrafía y Paleontología, Universidad de Granada, Avda. Fuente Nueva
S/N, 18002 Granada, Spain; Department of Earth and Planetary Sciences, Northrop Hall,
University of New Mexico, Albuquerque, New Mexico 87131 and Center for Environmental
Sciences & Education, Box 5694, Northern Arizona University, Flagstaff, AZ 86011, USA*

[4]*Senckenberg Forschungsinstitut und Naturmuseum, Senckenberganlage 25, 60325
Frankfurt am Main, Germany*

[5]*UMR CNRS 7619* Sisyphe, *Université de Paris VI, Case 105, 4 Place Jussieu,
75252 Paris Cedex 05, France*

[6]*Département de Géologie, Faculté des Sciences de Ben M'Sik, Université
Hassan II – Mohammedia, BP 7955 Sidi Othmane, Casablanca, Morocco*

[7]*Università degli Studi di Firenze, Dipartimento di Scienze della Terra, Via G. La Pira 4,
50121 Firenze, Italy*

[8]*Morphodynamique continentale et côtière, UMR CNRS 6143, Université de Caen,
24 rue des Tilleuls, 14000 Caen, France*

[9]*Departamento de Geologia, Universidade de Lisboa, 1294 Lisbon codex, Portugal*

[10]*Rue du Faubourg Bonnefoy, 31500 Toulouse, France*

[11]*Département des Sciences de la Terre, Faculté des Sciences, Université Caddi Ayyad,
Avenue Prince Moulay Abdellah, BP S15, Marrakech, Morocco*

[12]*Department of Earth Sciences, Zhongshan University, 510275 Guangzhou, China*

Abstract: In Europe and the Mediterranean region, the vegetation and climate of the Neogene is well understood, due to the abundance of pollen data, allowing the climate evolution at a time of global cooling to be described. This paper presents a climatic reconstruction of four key time-slices of the Neogene: the Mid-Miocene (*c*. 14 Ma), the Late Miocene (*c*. 10 Ma), the Early Pliocene (*c*. 5–5.3 Ma) and the Mid-Pliocene (*c*. 3.6 Ma). The results show that Neogene climate was warmer than today and that the transition from a weak latitudinal thermic gradient (around 0.48 °C/degree in latitude) to a gradient similar to that of today (0.6 °C/degree in latitude) took place at the end of the Miocene. The latitudinal precipitation gradient was more accentuated than today from the Mid-Miocene to the Mid-Pliocene, with higher precipitation than today in northwestern Europe and the northwestern Mediterranean but with conditions that were drier than or equivalent to today in the southwestern Mediterranean region.

From: WILLIAMS, M., HAYWOOD, A. M., GREGORY, F. J. & SCHMIDT, D. N. (eds) *Deep-Time Perspectives on Climate Change: Marrying the Signal from Computer Models and Biological Proxies.* The Micropalaeontological Society, Special Publications. The Geological Society, London, 481–502.
1747-602X/07/$15.00 © The Micropalaeontological Society 2007.

The Neogene is a period of intense climatic changes, from the 'greenhouse' climate of the Early to Middle Cenozoic to the 'icehouse' climate of the Late Cenozoic (Zagwijn 1960; Shackleton *et al.* 1995; Suc *et al.* 1999), and many factors, such as atmospheric CO_2, orbital parameters, ocean heat transport and palaeogeographical modifications, may have played a role in these changes.

The continental configuration of the world during the Miocene was similar to the present. However, plate tectonics led to intense palaeogeographical changes around the world during the Miocene, especially the Early and Middle Miocene. These changes contributed to fluctuations in the Neogene climate, in particular the opening of some ocean gateways (Drake Passage, Bering Strait) and the closure of others (the Atlantic-Pacific passage across Panama, the passage between the Indian Ocean and the Tethys) (Pagani *et al.* 2000; Hall *et al.* 2003). Changes in oceanic circulation at that time led to the establishment of the modern ocean circulation pattern (e.g. the Antarctic Circumpolar Current) that in turn affected the global climate. Ocean general circulation model simulations have shown the influence of ocean on global climate through changes in oceanic heat transport (e.g. Nisancioglu *et al.* 2003; Mikolajewicz *et al.* 1993). In addition, many atmospheric general circulation model (GCM) simulations have shown the influence of the uplift of mountain ranges and plateaus (Rocky Mountains, Andes, Himalayas, Alps, Tibetan Plateau) on global climate through changes in the atmospheric circulation (e.g. Ruddiman & Kutzbach 1989; Kutzbach *et al.* 1993; Ramstein *et al.* 1997; Fluteau *et al.* 1999; Kutzbach & Behling 2004).

Other authors have demonstrated that the Miocene climate variability was driven by fluctuations in the amplitude of obliquity and eccentricity (Westerhold *et al.* 2005). DeConto & Pollard (2003*a*, *b*) argue for a combination of atmospheric CO_2, orbital forcing and ice-climate feedbacks as the primary causes of climate transitions. Recently, Moran *et al.* (2006) have shown the dominance of greenhouse gases on climate control over tectonic forcing. The vegetation also had a significant influence on the Neogene climate. Climate model experiments demonstrate that the presence of high-latitude forests caused a warming in polar regions in the Miocene and, therefore, contributed to a weaker-than-present equator-to-pole temperature gradient (Dutton and Barron 1997; Micheels *et al.* unpublished data). Palaeovegetation changes, such as the evolution of grasslands during the Neogene (Retallack 2001), have an influence on the climate and must be considered when attempting to explain climatic fluctuations.

For Western European and Mediterranean regions, the Neogene vegetation history is well known as many pollen sequences have been studied during the last few decades. At present, more than 120 pollen records from the Early Miocene to the Early Pleistocene are available in this area (Zagwijn 1960; Suc 1980; Diniz 1984*a*, *b*; Bessedik 1985; Zheng 1986; Bertini 1992; 1994; 2001, 2003; Clet-Pellerin 1996; Bertini & Roiron, 1997; Bachiri-Taoufiq 2000; 2003; Jiménez-Moreno 2005), providing a reliable and accurate view of latitudinal and altitudinal vegetation change (Suc 1989; Suc *et al.* 1995*a*, *b*; Jiménez-Moreno & Suc 2007). The pollen-based descriptions of the palaeovegetation are supported by a number of macrofossils studies (Kovar-Eder *et al.* 2006).

In this paper, we reconstruct the evolution of climatic gradients in Europe and the Mediterranean region during the Neogene based on pollen data for four periods: the Middle Miocene, around 14 million years (Ma); the Late Miocene around 10 Ma (Tortonian); the Early Pliocene, around 5–5.3 Ma; and the Middle Pliocene, around 3.5 Ma. For each time-slice, the vegetation is briefly described based on the pollen records.

Methodology for climate reconstruction from pollen data

In order to produce comparable and homogenous results, the same transfer function was applied to all selected pollen sequences. The climate was estimated using the 'Climatic Amplitude Method' developed by Fauquette *et al.* (1998*a*, *b*) to quantify the climate of periods for which no modern analogues of the pollen spectra exist. The Neogene spectra contain a mixture of temperate, warm-temperate and subtropical plants (even tropical plants during the Miocene) that today live in different parts of the world. The past climate is estimated by transposing the climatic requirements of the maximum number of modern taxa to the fossil data. This method may be applied to the Neogene period as the pollen flora of the region has been defined following botanical nomenclature for many years now (Zagwijn 1960; Pons 1964; Elhai 1969; Suc 1976; Diniz 1984*a*, *b*; Bessedik 1985).

In contrast to other methods such as the best analogue method (Guiot 1990), this approach does not rely on the analysis of entire pollen assemblages, but on the relationship between the relative pollen abundance of each individual taxa and the climate. Presence/absence limits, as well as abundance thresholds, have been defined for 60 taxa from modern pollen spectra and the literature. This method takes into account not only the presence/absence criterion but also pollen percentages to provide more reliable reconstruction. Low abundances of some tropical and subtropical taxa

taxa exceeding
their threshold

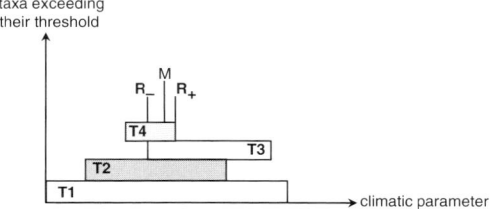

Fig. 1. Principle of the 'Climatic Amplitude Method' (Fauquette *et al.* 1998*a*, *b*). The most probable climate for a set of taxa exceeding their presence/absence and/ or abundance thresholds in a pollen spectrum corresponds to the smallest climatic interval suitable for a maximum number of taxa [R−; R+]. A 'most likely value' (M) is then calculated (see text).

(e.g. *Microtropis fallax*, *Avicennia*) are meaningful and should be taken into account as these plants produce relatively small numbers of pollen grains. Conversely, low abundances of wind-pollinated taxa (e.g. *Quercus*, *Alnus*, *Corylus*) may reflect long-distance transport of these high pollen producers by air and water. In this case, very low pollen percentages are not significant.

With this method, the most probable climate for a fossil pollen assemblage is estimated as the climatic interval in which the highest number of taxa can exist (Fig. 1). The climatic estimate is presented as an interval [R−; R+] and as a 'most likely value' (M), which corresponds to a mean that is weighted according to the size of the climatic intervals of all taxa exceeding their presence/absence and/or abundance thresholds. As the precision of the information obtained from a taxon's climatic interval is inversely related to the breadth of this interval, the weights are greater for taxa with smaller intervals.

In this paper, we present reconstructions of two climatic parameters estimated from the pollen data: the mean annual temperature (Ta) and the mean annual precipitation (Pa).

High latitude/altitude taxa were excluded from the reconstruction process. The identification and exclusion of high latitude/altitude plants is based on numerous palynological studies (e.g. Suc *et al.* 1995*a*, *b*, 1999; Jiménez-Moreno, 2005) that show the Neogene vegetation zonation to follow a similar latitudinal and altitudinal zonation to that observed in present-day south-eastern China (Wang 1961), where most of the taxa that had disappeared from Europe by the late Neogene may be found. The estimates obtained, therefore, correspond to the climate at low to middle–low altitude (Fauquette *et al.* 1998*a*).

Pinus and non-identified Pinaceae (due to poor preservation of these disaccate pollen grains) have been excluded from the pollen sum of the fossil spectra (Fauquette *et al.* 1998*a*, 1999). The pollen grains of these taxa are often over-represented in the sediments due to their high production and over-abundance in air and water (fluvial and marine) transport (Heusser 1988; Cambon *et al.* 1997).

The climatic latitudinal gradient during the Middle Miocene (*c.* 14–15 Ma)

A number of new pollen samples covering this period have recently been published (Jiménez-Moreno 2005). The samples are located along a latitudinal range in western Europe, from 47° to 36° N. The Mid-Miocene palaeogeography, which is now well established (Rögl 1998; Meulenkamp & Sissingh 2003; Goncharova *et al.* 2004; Ilyna *et al.* 2004; Paramonova *et al.* 2004) has shown that these sites were (a) separated by around 12° in latitude at that time (instead of 11° today) and (b) situated a few degrees further south than today (Rögl 1998).

The study sites are (from north to south): Le Locle outcrop (western Switzerland), Les Mées borehole (southern France), Bayanne outcrop (southern France), Farinole outcrop (Corsica, France), La Rierussa outcrop (northeastern Spain), Gor outcrop (southern Spain), Alborán A-1 borehole (southern Spain), Andalucía G-1 borehole (southern Spain) (Fig. 2). Le Locle locality is dated from the upper mammal unit MN16 (Kälin *et al.* 2001). The other sites are marine deposits and are generally dated by micropalaeontology Gor: calcareous nannofossil Zone CN-3 (Martín-Pérez & Viseras 1994); the time-interval taken from the Andalucia G1 borehole has been dated by planktonic foraminifera and ranges from zones N10 to N16 (Calandra in ELF 1984); the time-interval taken from the Alboran A1 borehole has been dated by planktonic foraminifera and ranges from zones N10 to N14 (Bailey *et al.* in CHEVRON 1986); La Rierussa: planktonic foraminifera (zone N4: Magné 1978) and calcareous nannofossil (zone NN4: C. Müller in Bessedik 1985); Farinole: planktonic foraminifera (zone N10) and calcareous nannofossils (zone NN6) (Ferrandini *et al.* 1998); Bayanne: planktonic foraminifera (zones N7 and N8: Besson *et al.* 2005). The Les Mées 1 borehole samples contain no micropalaeontological information. These samples have been allocated to the late Burdigalian as they correspond to the transgressive maximum according to the regional stratigraphy (Dubois & Curnelle 1978). Complete information is available in Jiménez-Moreno (2005).

Pollen taxa have been grouped following Suc (1989) and are detailed in Table 1.

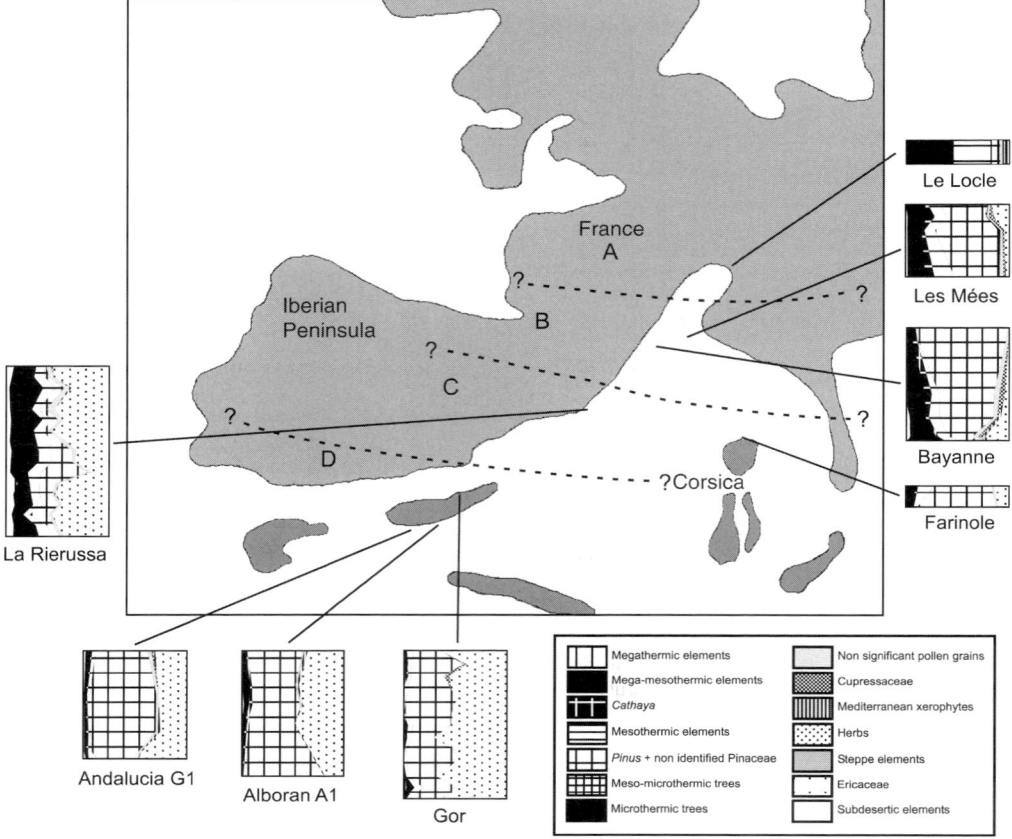

Fig. 2. Location of the studied sites covering the Middle Miocene and their synthetic pollen diagrams
in the palaeogeographical framework of the early Serravallian (from Rögl 1998). Pollen localities
(Jiménez-Moreno 2005): Le Locle-Combe Girard, Les Mées, Bayanne, La Rierussa, Farinole, Gor, Alborán A1,
Andalucía G1. The four vegetational regions reconstructed from pollen data are indicated (A, B, C, D, see text for
explanation).

The studied area can be subdivided into four
vegetational domains from north to south on the
basis of the pollen floras (Fig. 2) (Jiménez-Moreno
2005; Jiménez-Moreno & Suc 2007):

- In Western Switzerland (Fig. 2, zone A), the
pollen flora is characterized by a high abundance
of mega-mesothermic (= subtropical) elements,
in particular *Taxodium* type and *Engelhardia*,
and by high percentages of mesothermic
(=warm-temperate) elements (mainly *Quercus*
deciduous type). Some megathermic (= tropical)
elements are present at low abundances. Percentages of herbs are very low.
- Southern France (Fig. 2, zone B) is characterized by the dominance of mega-mesothermic
elements as well as temperate elements. *Avicennia*, a mangrove plant, is regularly present.

Herbaceous percentages are higher, but do not
dominate the pollen spectra. Semiarid taxa
such as *Acacia*, Caesalpiniaceae or *Prosopis*
are recorded at very low values.
- In northeastern Spain and Corsica (Fig. 2, zone C),
pollen spectra are rich in herbs and shrubs. Caesalpiniaceae and *Acacia* occur at very low percentages in the pollen spectra. Mega-mesothermic
and mesothermic taxa are highly abundant. Megathermic elements are abundant in all the samples.
Avicennia plays an important role in this area,
indicating the presence of an impoverished mangrove along the coast. Meso-microthermic and
microthermic taxa (i.e. inhabiting middle and
high altitudes, respectively), including *Cathaya*,
a conifer living today in the subtropical mid-
altitude forests of southern China (Wang 1961),
occur infrequently in these samples.

Table 1. *Taxa groups*

Megathermic elements	Rutaceae, *Mussaenda* type, Acanthaceae, *Acacia, Sindora, Croton, Alchornea, Bombax, Buxus bahamensis* type, *Mappianthus*, Rubiaceae, Euphorbiaceae, *Avicennia, Phyllanthus* type, Melastomataceae, Simarubaceae
Mega-mesothermic elements	*Symplocos, Engelhardia*, Sapotaceae, *Platycarya, Distylium, Rhoiptelea*, Taxodiaceae, *Taxodium* type, Hamamelidaceae, *Rhodoleia*, Loranthaceae, *Microtropis fallax, Embolanthera, Corylopsis, Mallotus*, Celastraceae, *Parthenocissus, Leea, Myrica*, Menispermaceae, Theaceae, *Aesculus*
Mesothermic elements	*Quercus* deciduous type, *Fagus, Ostrya, Carpinus, Carya, Pterocarya, Juglans, Juglans cathayensis* type, *Parrotia* cf *persica, Liquidambar, Tilia, Castanea-Castanopsis* type, *Parrotiopsis jacquemontiana*, Restionaceae, *Buxus sempervirens, Ilex, Eucommia, Ligustrum, Populus, Ulmus, Zelkova, Celtis, Elaeagnus*
Meso-microthermic elements	*Cedrus, Tsuga, Sciadopitys*
Microthermic elements	*Picea, Abies, Keteleeria*
Mediterranean xerophytes	*Quercus ilex-coccifera* type, *Olea, Phillyrea, Ceratonia*, Cistaceae, *Pistacia, Nerium*
Herbs	Poaceae, Asteraceae Asteroideae, Asteraceae Cichorioideae, Centaurea, Convolvulaceae, *Plantago*, Ericaceae, Brasicaceae, *Helianthemum*, Geraniaceae, *Erodium*, Caryophyllaceae …
Steppe elements	*Artemisia, Ephedra*
Subdesertic elements	*Lygeum, Neurada, Calligonum, Nitraria, Prosopis*, Agavaceae

- Southern Spain (Fig. 2, zone D) is characterized by the dominance of herbs and shrubs in the pollen spectra; Poaceae and halophytes are repeatedly found at high values. Further, subdesertic elements, such as *Nitraria, Lygeum, Prosopis, Neurada* and *Calligonum* are very abundant. Significant amounts of megathermic elements, including *Avicennia*, occur in all samples. Mega-mesothermic and mesothermic are regularly present. Meso-microthermic and microthermic elements appear sporadically at very low values. These pollen spectra with high percentages of herbs are typical of an open environment.

The presence, all along this transect, of plants characterized by high thermic requirements such as *Engelhardia, Myrica, Taxodium*-type, *Mussaenda*-type and *Avicennia*, indicates that the latitudinal temperature gradient was lower than today. This is consistent with the presence in other pollen data, covering the Mid-Miocene of Central Europe, of thermophilous taxa at high latitudes (Jiménez-Moreno 2005; Jiménez-Moreno & Suc 2007). The occurrence of thermophilous plants at higher latitudes has also been observed in North America by Liu & Leopold (1994). These authors estimated a thermic gradient of 0.3 °C per degree of latitude for North America (between 35° N–65° N) during the Mid-Miocene.

There are, however, important changes in the vegetation from north to south in Western Europe, occurring gradually between Switzerland and southern Spain (between 36° N and 47° N). The vegetation becomes more and more open from north to south with the presence of subdesertic taxa in southern Spain, reflecting a latitudinal gradient in precipitation.

The results of the climatic quantification (Fig. 3) show, from north to south, increasing annual temperatures but decreasing annual precipitation. The reconstructed most likely values show higher mean annual temperatures than today all along the gradient (*c.* 2 to 8 °C higher) and higher mean annual precipitation than today in Southern France, Corsica and northeastern Spain (between 400 mm and 700 mm higher). In southwestern Europe, the mean annual precipitation is almost equivalent to modern values (maximum 200 mm higher). This is also the case at Le Locle, where little change in mean annual precipitation values is shown. The thermic gradient is weaker than the modern one as the differences between the Miocene and the modern temperatures are between *c.* 2 °C in southern Spain, *c.* 4 °C in northern Spain, *c.* 5/6 °C in southern France and *c.* 8 °C in western Switzerland. On the basis of the most likely value reconstructed from pollen data, the thermic gradient in Western Europe was around 0.48 °C per degree in latitude whereas it is around 0.6 °C today (Ozenda 1989). This result is in agreement with the estimations obtained by Bruch *et al.* (2004) from fossil floras of Europe where they find a lower latitudinal temperature gradient than today.

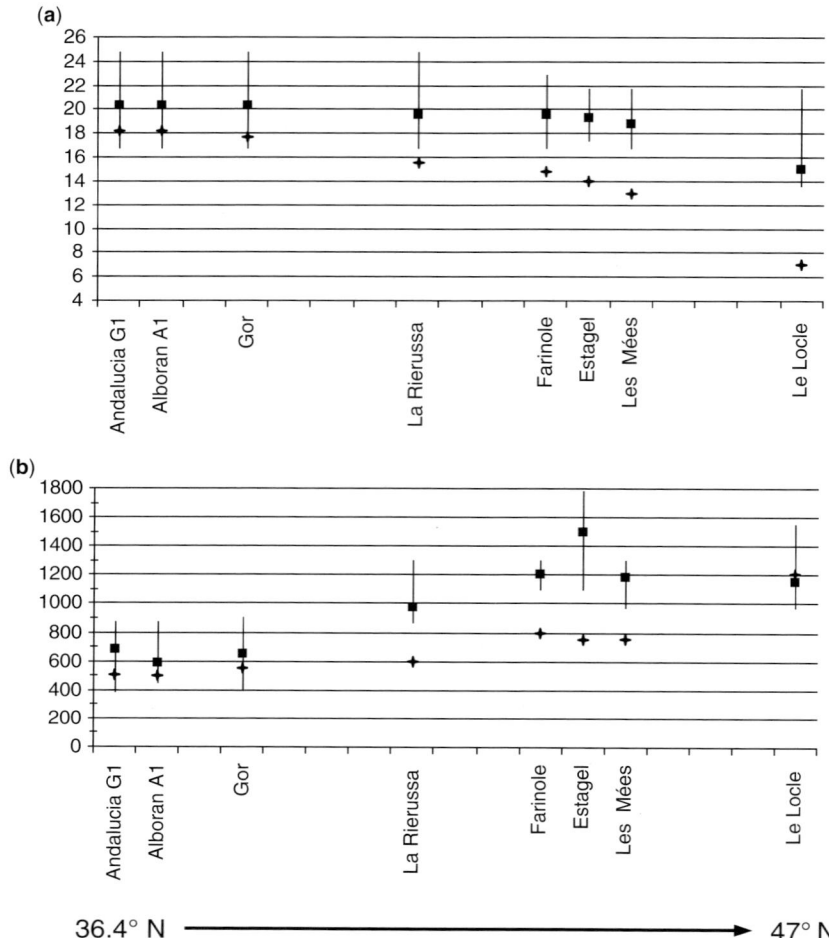

Fig. 3. Climatic reconstruction from pollen data (climatic interval and most likely value) in Western Europe and Mediterranean region showing (**a**) the gradient of temperature (mean annual temperature in °C) and (**b**) precipitation (mean annual precipitation in mm) for the Middle Miocene. Modern values are indicated by a cross to show the differences between modern and Miocene latitudinal gradients. Pollen data of Le Locle-Combe Girard, Les Mées, Bayanne, La Rierussa, Farinole, Gor, Alborán A-1, Andalucía G-1 (Jiménez-Moreno 2005) have been used.

The latitudinal climatic gradient at the end of the Miocene (Tortonian, *c.* 10 Ma)

For this period, pollen spectra from the sites of Ambérieu (Farjanel & Mein 1984), Mirabel (Naud & Suc 1975), Sanabastre and Sampsor in Cerdanya (Bessedik 1985), Zaratan (Rivas-Carballo *et al.* 1994), Capodarso in Sicily (Suc *et al.* 1995*c*) and MSD 1 borehole in Morocco (Bachiri-Taoufiq 2000) have been used to estimate the climatic gradient (Fig. 4). The palaeogeography has been established by Paramonova *et al.* (2004). The Cerdanya

sites (Agusti & Roca 1987), Zaratan (Rivas-Carballo *et al.* 1994) and Ambérieu (Farjanel & Mein 1984) were assigned to the Tortonian on the basis of mammal biochronology. The site of Mirabel belongs to the volcanic Coirons area and has a radiometric age (Naud & Suc 1975). Planktonic foraminifera are available for the MSD1 borehole for zones N16 and N17 (Barhoun 2000). The Capodarso section covers the late Tortonian to early Messinian according to planktonic foraminifera and calcareous nannofossils (Cita *et al.* 1973; Suc *et al.* 1995*c*). Only the lower part of this section is considered here.

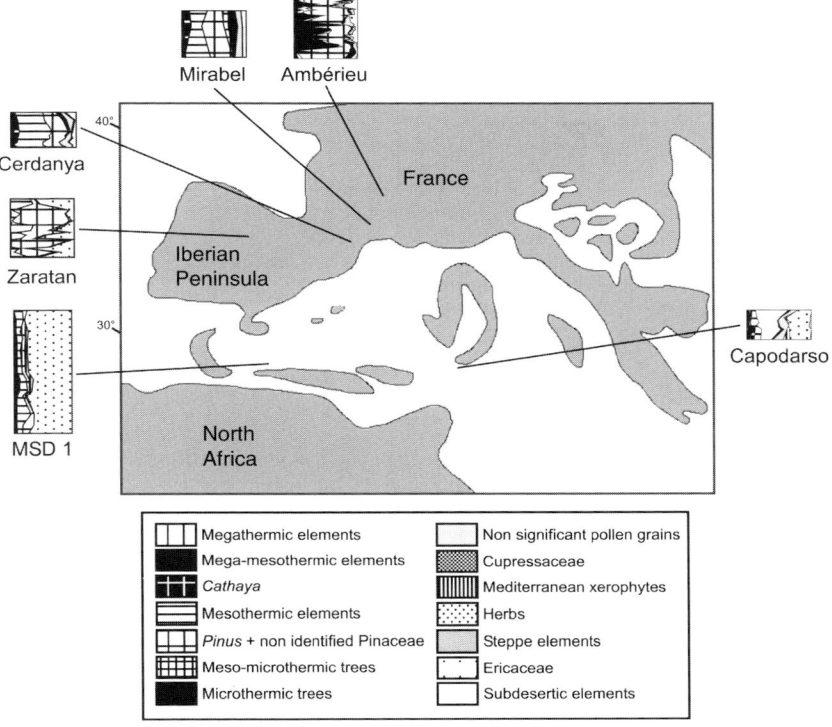

Fig. 4. Location of the studied sites covering the Tortonian around 10 Ma in Western Europe and Western Mediterranean region and their synthetic pollen diagrams in the palaeogeographical framework of the Late Tortonian (Paramonova *et al.* 2004). Pollen localities: Ambérieu (Farjanel and Mein 1984), Mirabel (Naud & Suc 1975), Sanabastre and Sampsor sites in Cerdanya (Bessedik 1985), Zaratan (Rivas-Carballo *et al.* 1994), Capodarso in Sicily (Suc *et al.* 1995c) and MSD 1 borehole in Morocco (Bachiri-Taoufiq 2000).

Although less pollen data exist for this period, a clear latitudinal gradient is observed for both temperature and precipitation. In northwestern Europe (Ambérieu, Mirabel sites), pollen data indicate forested environments, characterized by taxa growing under a wet climate (Taxodiaceae, *Engelhardia, Symplocos, Platycarya . . .*). Forested environments are also indicated in the northwestern Mediterranean region (Sanabastre/Sampsor sites in Cerdanya, Bessedik 1985), where arboreal pollen dominates with low values of herbaceous taxa. The microflora is characterized by the dominance of *Quercus, Fagus, Alnus* and conifers (*Cathaya, Pinus,* Taxodiaceae), reflecting the presence of mixed deciduous forests. Pollen grains of *Abies* are also recorded. Only few megathermic plants are present at low values. However, the presence, in the Cerdanya Basin, of plants such as evergreen *Quercus* (in the microflora) or even *Cassia, Mahonia, Cinnamomum, Banksia,* Combretaceae (in the macroflora, Menendez Amor 1955) indicates a warmer climate than today. At Zaratan, the pollen

assemblages are similar to those found today in the southwestern Mediterranean region with sclerophyllous woods of *Quercus* and pines associated with species characteristic of open vegetation as Cistaceae, Cupressaceae, Ericaceae, Geraniaceae and *Plantago.* The presence of deciduous taxa indicates a warm-temperate climate in this region (Rivas-Carballo *et al.* 1994). Finally, in the southwestern Mediterranean region (Capodarso in Sicily and MSD 1 borehole in Morocco), the pollen spectra are largely dominated by herbaceous taxa, indicating dry open environments with the presence of subdesertic herbs such as *Lygeum.* However, pollen data also indicate the presence of forests on the surrounding uplands. The record of *Avicennia* in the MSD 1 borehole (Bachiri-Taoufiq 2000) indicates an impoverished mangrove along the south Mediterranean shoreline during the Tortonian.

The climatic reconstruction based on these pollen sequences shows that temperatures were higher than today during the Tortonian, in particular in the northwestern Mediterranean area. The climate was warm and humid in Western Europe (most likely values 4 to 9 °C and annual precipitation rainfall 100 to 600 mm higher than today), and warm and dry in the south Mediterranean region (most likely values of 3 to 4 °C higher and less than 200 mm higher than today). The difference between the Tortonian and the modern annual temperature is larger for the site of Sampsor. Indeed, this site is currently situated at about 1000 m above sea level but was at lower altitude during the Tortonian (Mauffret *et al.* 2001).

The climatic estimates show that the north–south climatic gradient that existed during the Tortonian was similar to today, with increasing temperature and decreasing precipitation, but with

higher temperatures (Fig. 5). On the basis of the most likely values reconstructed from pollen data of MSD 1 borehole and Ambérieu section, the thermic gradient is around 0.6 °C per degree in latitude. This does not agree with the estimated reduction of about 50% in the thermic gradient calculated by Bruch *et al.* (2006) from fossil floras of Central and Eastern Europe.

The climatic latitudinal gradient for the Early Pliocene (*c.* 5.0–5.3 Ma)

During this period, the palaeogeography was similar to today, with the exception of the existence

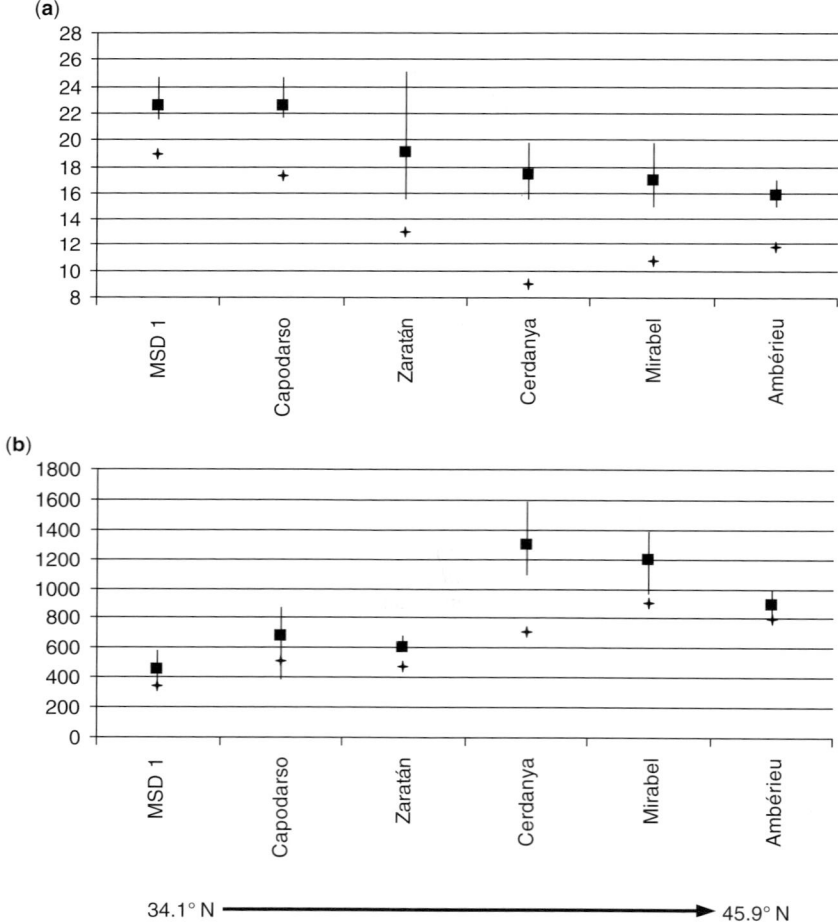

Fig. 5. Climatic reconstruction (climatic interval and most likely value) from pollen data covering the Tortonian in Western Europe and the Mediterranean region showing (**a**) the temperature gradient (mean annual temperature in °C) and (**b**) the precipitation gradient (mean annual precipitation in mm). Modern values are indicated by a cross.

of rias (corresponding to the excision of deep canyons by rivers during the desiccation of the Mediterranean Sea margins at the time of the Messinian Salinity Crisis), which penetrated lands (Clauzon *et al.* 1995). During the earliest part of the Pliocene (5.33 Ma), the reflooding of the Mediterranean basin by Atlantic waters (to about 70–80 m above the present-day sea level, Haq *et al.* 1987) resulted in the accumulation of terrigenous sediments in a large number of areas. As a result, this period is rich in pollen sites. Here, we have used pollen spectra from the sites of Susteren 752/72 (Zagwijn 1960), Stirone (Bertini 1994, 2001), Saint-Martin du Var (Zheng 1986), Cap d'Agde 1 (Suc 1989), Le Boulou (Suc *et al.* 1999), Garraf 1 (Suc & Cravatte 1982), Tarragone E2 (Bessais & Cravatte 1988), Rio Maior F16 (Diniz 1984*a*, *b*), Capo Rossello (Suc *et al.* 1995*c*), Andalucia G1 (Suc *et al.* 1995*a*), Oued Tellil (Suc *et al.* 1999), Nador 1 (Fauquette *et al.* 1999), Habibas 1 (Suc *et al.* 1999), to estimate the climatic gradient during the Early Pliocene. All sites (except Susteren and Rio Maior F16) are marine deposits belonging to the Mediterranean earliest Pliocene and are well-dated by both planktonic foraminifera (zones MPl1-3) and calcareous nannofossils (zones NN12-13) (Suc *et al.* 1995*b*). In addition, the lowermost layers of these sequences directly overly the Messinian erosional surface, providing a synchronous chronological marker at 5.33 Ma (Clauzon *et al.* 1996). The lower part of the Susteren borehole belongs to the Brunssumian climatic phase (Zagwijn 1960). The lower part of the Rio Maior F16 borehole has been correlated to the Brunssumian climatic phase on the basis of a similar evolution of the vegetation to the Early Pliocene changes recorded in the Garraf 1 borehole (Suc & Zagwijn 1983; Suc *et al.* 1995*b*).

Three main vegetation domains in Western Europe and the Mediterranean region (Fig. 6) have been described by Suc (1989) and Suc *et al.* (1995a), during the Early Pliocene, with a clear latitudinal zonation of vegetation. Sites on the Atlantic coast of Western Europe (Fig. 6, zone A, Susteren and Rio Maior sites) show forested vegetation dominated by Taxodiaceae, Ericaceae and mesothermic deciduous trees *(Quercus, Carya, Pterocarya, Acer, Carpinus, Fagus, Liquidambar, Parrotia persica)*. In the north Mediterranean region (Fig. 6, zone B), the forests were dominated by Taxodiaceae (*Taxodium*/*Glyptostrobus* or *Sequoia* dependent on local environment conditions, respectively swamps and slopes), accompanied by mega-mesothermic plants such as *Engelhardia, Symplocos* and *Platycarya*. These latter taxa were reduced later, in the Mid-Pliocene. The South Mediterranean region (Fig. 6, zone C) was characterized by Mediterranean xerophytic

ecosystems ('matorral' composed by *Olea, Phillyrea, Pistacia, Ceratonia*, evergreen *Quercus, Nerium, Cistus*) and, to the south, by open environments dominated by subdesertic plants like *Lygeum, Neurada, Nitraria, Calligonum*, Geraniaceae and Agavaceae.

The climatic reconstruction shows that temperatures at the beginning of the Pliocene at around 5.0–5.33 Ma were higher than today, particularly in the northwestern Mediterranean area. The average climate was warm and humid in Europe and the north Mediterranean region (most likely values 1 to 4 °C and precipitation 400 to 700 mm higher than today), and warm and dry in the south Mediterranean region (most likely values equal to or 5 °C higher and drier than or equal to today) (Fauquette *et al.* 1998*b*, 1999; Fauquette & Bertini 2003; this study). A north–south climatic gradient existed at the beginning of the Pliocene, with, as today, increasing temperatures and decreasing precipitation (Fig. 7). The thermic gradient calculated on the 'most likely values' of mean annual temperatures of Susteren and Habibas sites is around 0.65 °C per degree in latitude.

The climatic gradient reconstructed for the West European Pliocene, both for temperatures and precipitation, seems to be very similar to that observed today in northwestern North America and in particular in western California and Lower California. This gradient may be summarized as follows. The climate of northern California is particularly humid, with annual precipitation from *c.* 1000 to more than 2000 mm, especially in the Coast Ranges to the north of San Francisco (summer is the drier season). Mean annual temperatures range from 9 °C to 14 °C. To the south, in central California, the climate is less humid with precipitation ranging from *c.* 600 to 1500 mm/year. This region is characterized by decreasing humidity and an increasing summer drought from north to south. Mean annual temperatures are between 10 and 18 °C. In southern California, mean annual temperatures are between 14 and 24 °C and mean annual precipitation between 400 and 800 mm. Finally, in Lower California, the climate is arid with annual precipitation from *c.* 100 to 500 mm (Walter 1979; Thompson *et al.* 1999). Mean annual temperatures are comprised between 17 and 30 °C.

The vegetation zonation imposed by the latitudinal/altitudinal climatic gradients in this region is also similar to that of the European and Mediterranean Pliocene (Fig. 8). Humidity, as either rainfall or fog, allows the installation of dense *Sequoia* forests in Northern California in the littoral plain as well as in the Coast Ranges (Quézel & Barbero 1989; Thompson *et al.* 1999). These forests occur at up to 900 m a.s.l. on the

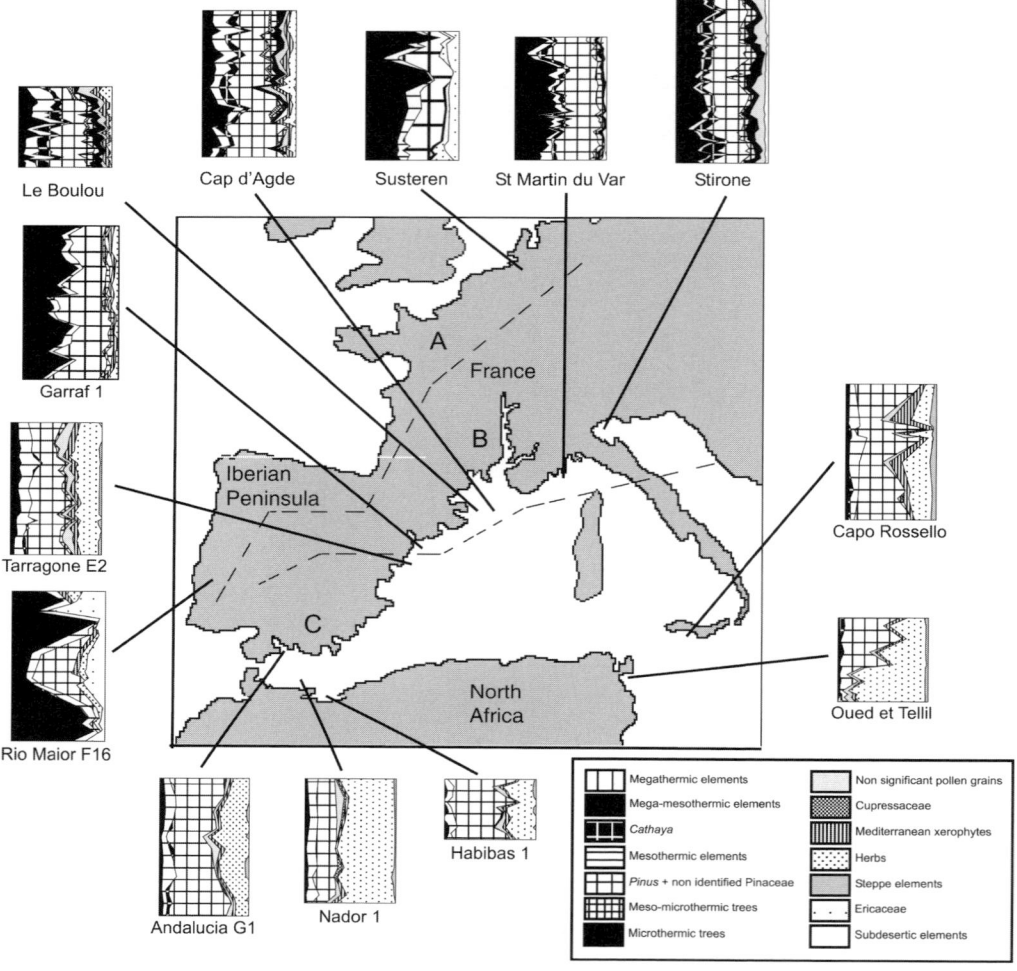

Fig. 6. Location of the studied sites covering the Early Pliocene, at around 5–5.3 Ma, and their synthetic pollen diagrams in the palaeogeographical framework of the Early Pliocene (Clauzon *et al.* 1995; Clauzon 1996; Jolivet *et al.* 2006). Pollen localities: Susteren 752/72 (Zagwijn 1960), Saint-Martin du Var (Zheng 1985), Stirone (Bertini 1994, 2001), Cap d'Agde 1 (Suc 1989), Rio Maior F16 (Diniz 1984), Andalucia G1 (Suc *et al.* 1995*a*), Le Boulou (Suc *et al.* 1999), Nador 1 (Fauquette *et al.* 1999), Habibas 1 (Suc *et al.* 1999), Tarragone E2 (Bessais & Cravatte 1988), Garraf 1 (Suc & Cravatte 1982), Oued et Tellil (Suc *et al.* 1999), Capo Rossello (Suc *et al.* 1995*c*).

Coast Ranges to the north of San Francisco. To the south, the occurrence of sequoias in the uplands becomes sparse (Quézel & Barbero 1989). This vegetation zone closely resembles the Europe and northwestern Mediterranean area dominated by Taxodiaceae and other taxa growing under a wet climate during the Early Pliocene.

Central and southern California are dominated by the chaparral vegetation type (Walter 1979), which resembles the Mediterranean 'matorral' *s.s.* defined by Quézel & Barbero (1989). This vegetation type developed in Europe during the Pliocene

with the appearance of the summer drought (Axelrod 1973; Thompson 1991; Thompson & Fleming 1996). The very low precipitation in Lower California prevents the installation of forests, and only a subtropical desert vegetation type may develop in this arid zone (Walter 1979). This zone may be compared to the subdesertic vegetation reconstructed for the Pliocene in North Africa (Suc *et al.* 1995*b*; Fauquette *et al.* 1999).

In California, therefore, the vegetation zonation imposed by the latitudinal climatic gradient along the Pacific coast shows very strong similarities to

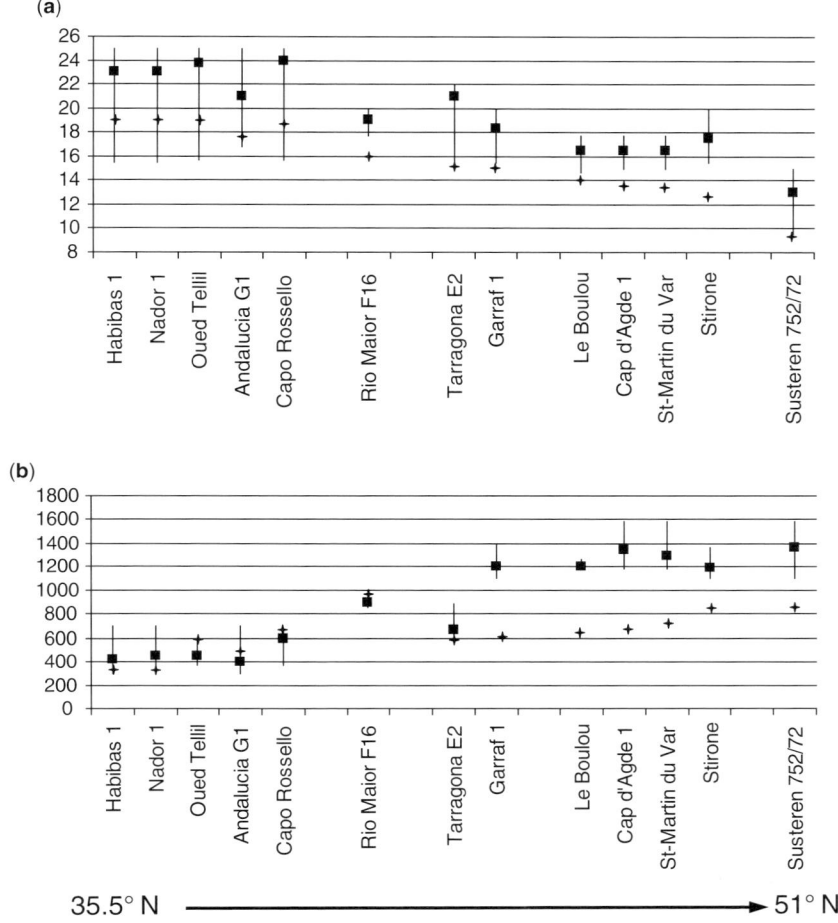

Fig. 7. Climatic reconstruction (climatic interval and most likely value) from pollen data covering the Early Pliocene (*c.* 5–5.3 Ma) in Western Europe and the Mediterranean region showing (**a**) the temperature gradient (mean annual temperature in °C) and (**b**) the precipitation gradient (mean annual precipitation in mm). Modern values are indicated by a cross.

the Pliocene in European and the Mediterranean region, with a development from dense humid forests in the north to subdesertic/desertic vegetation in the south. Changes in vegetation types to the north and south of the region considered also support this comparison. Today, the desertic zone is replaced to the south by intertropical forest in Central America (Walter 1979); a pollen sequence obtained from a borehole in the Guinea Gulf shows a similar change during the Pliocene in Central Africa, to the south of our study area (Suc *et al.* 1995*b*). In northern California, *Sequoia* forests are replaced to the north by forests

composed of *Tsuga* in association with other conifers and some deciduous trees such as *Alnus* (Walter 1979); during the Pliocene in Europe, *Tsuga* forests replaced *Sequoia* forests at higher altitudes and latitudes.

The climatic latitudinal gradient for the Middle Pliocene (*c.* 3.5 Ma)

The climatic gradient in Western Europe at around 3.5 Ma has been estimated using pollen spectra from the sites of Susteren (Zagwijn 1960),

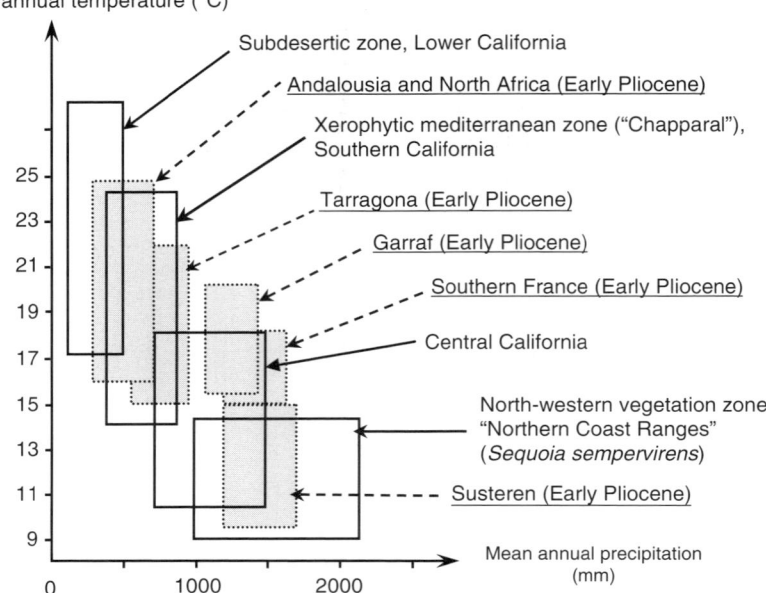

Mean annual temperature (°C)

Subdesertic zone, Lower California

Andalousia and North Africa (Early Pliocene)

Xerophytic mediterranean zone ("Chapparal"),
Southern California

Tarragona (Early Pliocene)

Garraf (Early Pliocene)

Southern France (Early Pliocene)

Central California

North-western vegetation zone:
"Northern Coast Ranges"
(*Sequoia sempervirens*)

Susteren (Early Pliocene)

Mean annual precipitation (mm)

Fig. 8. Comparison between modern climate/vegetation structures of California and those of the Mediterranean Pliocene defined from pollen data and climate reconstruction.

La Londe (Clet & Huault 1987), Saint-Isidore (Zheng 1986), Garraf 1 (Suc & Cravatte 1982), Tarragona E2 (Bessais & Cravatte 1988), Andalousia G1 (Suc *et al.* 1995a), Habibas 1 (Suc *et al.* 1999) and Oued Galaa (Suc 1989) (Fig. 9). All sites, except Susteren, La Londe and Rio Maior F16, are marine deposits dated using planktonic foraminifera (zone MPl4): Saint-Isidore (Zheng & Cravatte 1986), Garraf 1 (Suc & Cravatte 1982), Tarragone E2 (Bessais & Cravatte 1988), Andalucia G1 (Suc *et al.* 1995b), Habibas 1 (J. Cravatte, unpublished information), Oued Galaa (J. Cravatte, unpublished information). The section of the Susteren borehole sequence used here belongs to the Reuverian climatic phase (Zagwijn 1960) and has been climatostratigraphically correlated to the upper part of the Rio Maior F16 pollen diagram (Suc *et al.* 1995b). The same method has been used to attribute uppermost part of the La Londe pollen diagram to the Praetiglian climatic phase (Clet & Huault 1987). The lower part of this section therefore belongs to the Reuverian climatic phase, correlated to the Mediterranean Piacenzian Stage (Suc & Zagwijn 1983; Suc *et al.* 1995b).

These pollen data show a clear climatic latitudinal gradient for both temperature and precipitation. In northwestern Europe, along the Atlantic coast (Fig. 9, zone A), the vegetation was characterized by taxa growing under a wet climate, as during

the Early Pliocene, but with a reduction in mega-mesothermic taxa (Suc *et al.* 1995a). In the northwestern Mediterranean region (Fig. 9, zone B), mega-mesothermic taxa are still well represented but there is a general increase in deciduous mesothermic taxa, especially at Saint-Isidore (southern France) where microthermic plants, which developed at higher altitudes, also increase. In the southwestern Mediterranean region (Fig. 9, zone C), pollen spectra have the same composition as those of the Early Pliocene, characterizing open steppe-like vegetation and dry and warm environments.

As during the Early Pliocene, the climate reconstructed from these pollen data show higher mean annual temperatures than today along the entire gradient (most likely values 3 to 6 °C higher than modern values) and mean annual precipitation that is higher than today in northwestern Europe (most likely values 400 to 700 mm higher), but equivalent to modern values in southwestern Europe (maximum 200 mm higher) (Fig. 10).

The difference in the climate reconstructions between the Early and Middle Pliocene is negligible despite notable variations in pollen assemblages between these two periods (e.g. decrease in pollen percentages of some mega-mesothermic trees but increase in pollen percentages of some mesothermic trees at Garraf in Catalonia, at Saint-Martin du Var and Saint-Isidore in southern France). These

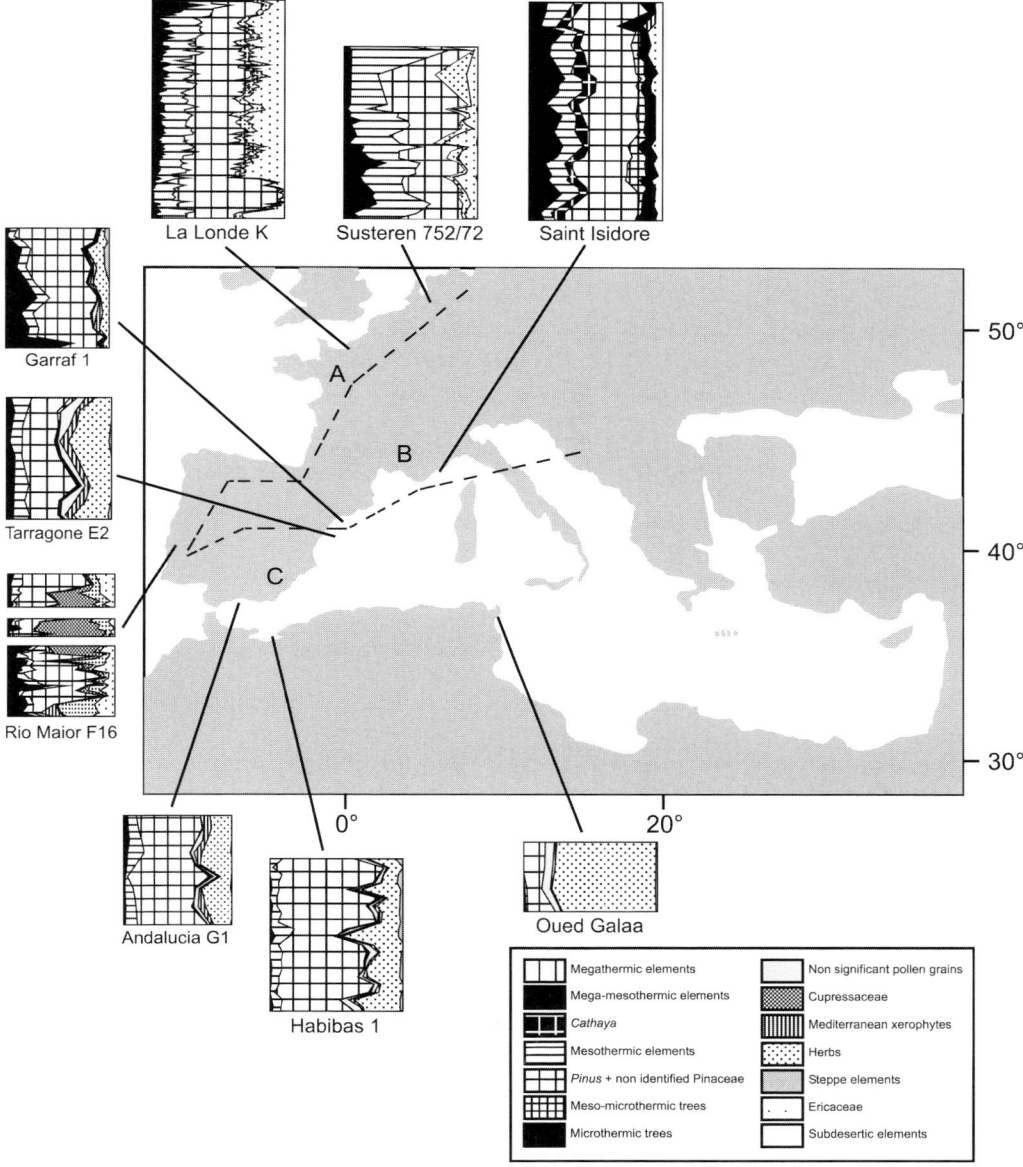

Fig. 9. Location of the studied sites covering the Middle Pliocene, at around 3.5–3 Ma in Western Europe and the Western Mediterranean region and their synthetic pollen diagrams in the palaeogeographical framework of the Middle Pliocene (Khondkarian *et al.* 2004). Pollen localities: Susteren 752/72 (Zagwijn 1960), La Londe (Clet & Huault 1987), Saint-Isidore du Var (Zheng 1985), Garraf 1 (Suc & Cravatte 1982), Tarragona E2 (Bessais & Cravatte 1988), Andalucia G1 (Suc *et al.* 1995*a*), Oued Galaa (Suc unpublished) and Rio Maior F16 (Diniz 1984).

variations in pollen records are often too slight to result in changes in the reconstructed climate. This result indicates the sensitivity limit of the 'Climatic Amplitude Method' and of all the methods based on the principle of co-existence intervals.

The West European climatic gradients during the Middle Pliocene, in particular the thermic gradient, appear to be very similar to that observed today, i.e. around 0.6 °C per degree in latitude (Fig. 10).

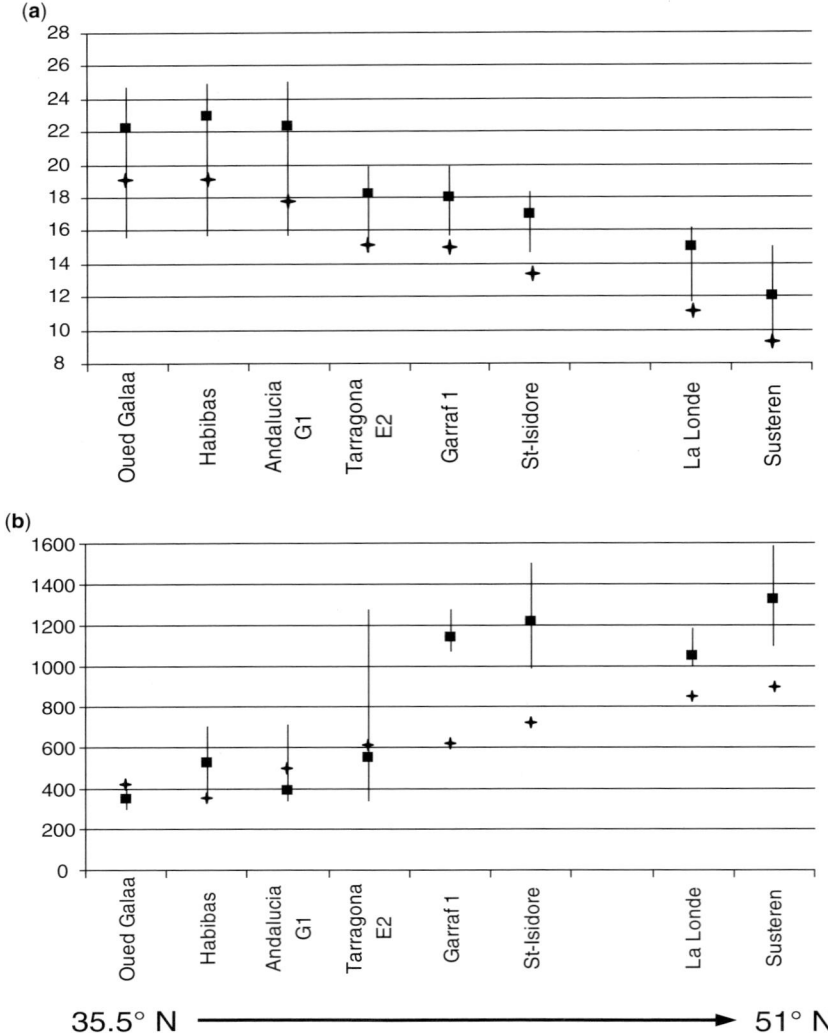

Fig. 10. Climatic reconstruction (climatic interval and most likely value) from pollen data covering the Middle Pliocene in Western Europe and Mediterranean region showing (**a**) the temperature gradient (mean annual temperature in °C) and (**b**) the precipitation gradient (mean annual precipitation in mm). Modern values are indicated by a cross.

Discussion

During the Middle Miocene, the reconstructed thermic gradient was approximately 0.48 °C per degree of latitude. Whilst this estimated value is certainly not the exact value as it is calculated using solely the most likely values, the weakening of the gradient is supported by changes in the entire climatic intervals. A weaker gradient during this period has also been suggested by other

studies based on macro- and microfloras of the circumalpine region and Central Europe (Bruch *et al.* 2004) and also in North America (Liu & Leopold 1994).

From the end of the Miocene (Tortonian) to the mid-Pliocene, the vegetation and climatic latitudinal gradients (in particular the thermic one) differ from those of the Middle Miocene. Temperatures were higher than today, as during the Mid-Miocene, but the difference between the north and the south was greater than during the Mid-Miocene, resulting in a similar gradient in temperature and vegetation to today.

Our results place the transition, for mid-latitude regions, from the weak thermic gradient of the Mid-Miocene to the modern-like gradient of the Pliocene during the Middle–Late Miocene, before or during the Tortonian.

Simulations with the AGCM ECHAM4 coupled to a slab ocean model have been made in order to study the climate response during the Tortonian to a generally low palaeo-orography, a weaker-than-present palaeo-oceanic heat transport and a changed palaeovegetation (Steppuhn *et al.* 2006; Micheels *et al.* unpublished data). Climate trends in the Tortonian model simulations show an overall reduction of the meridional temperature gradient (Steppuhn *et al.* 2006; Micheels *et al.* unpublished data). For the Mediterranean region, the simulated climate is slightly warmer and drier than today (Micheels *et al.* unpublished data), which agrees with our data from southwestern Europe (Sicily, Central Spain). In northern Africa, the Tortonian model simulates warmer and less-arid conditions than today (Micheels *et al.* unpublished data). Our evidence for subdesertic herbs and forest elements in Morocco supports the climate modelling results. However, it should be noted that these two groups of plants certainly developed at different altitudes that may not be adequately resolved by the climatic model. The pollen flora of the Ambérieu and Mirabel sites indicate forest

environments that developed under a warm and wet climate. The Tortonian simulation indicates an increased precipitation over Central Europe. Although this simulation tends to be too cool at higher latitudes (Micheels *et al.* unpublished data), our data from Europe largely agree with the model simulation.

On the basis of fossil floras of Central and Eastern Europe, Bruch *et al.* (2006) indicate a much weaker thermic gradient than today during the Tortonian, with a reduction of approximately 50%. Whilst the ECHAM model is unable to reproduce this weak temperature gradient, our results show higher mean annual temperatures than the simulated zonal average temperature for land surfaces but agree with the thermic gradient simulated (Fig. 11).

Simulations have been made of the Mid-Pliocene climate by Haywood *et al.* (2000*a*) using the HadAM3 version of the UK Meteorological Office's (UKMO) general circulation model and by Jost (2005) using the LMDz (Laboratoire de Météorologie Dynamique, zoom, Institut Pierre Simon Laplace) atmospheric general circulation model.

Boundary conditions of the models are those established by the Pliocene Research, Interpretation and Synoptic Mapping group of the US Geological Survey (PRISM 2, see Cronin & Dowsett 1990; Dowsett *et al.* 1994, 1999). Jost (2005) confirms the increase in global temperatures compared to the present (Fig. 12a), previously shown by Haywood *et al.* (2000*a*). This increase is greater at mid- to high latitudes than at the equator.

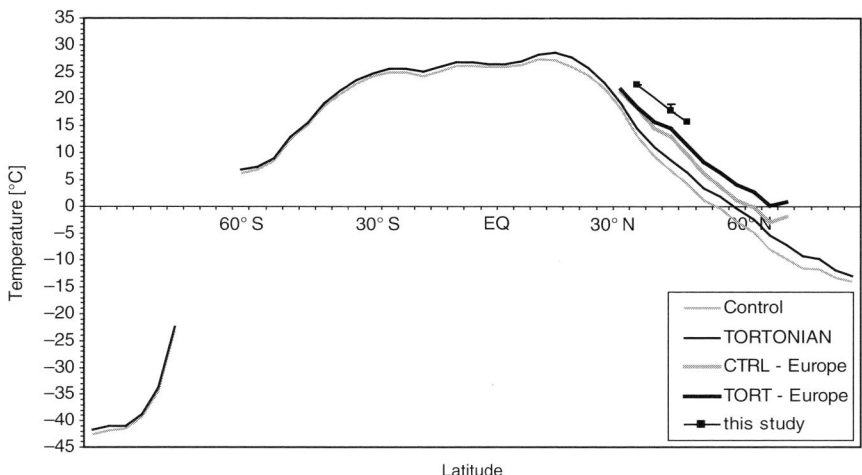

Fig. 11. The zonal average temperature (in °C) for land surfaces of the Tortonian simulation (black line), the present-day control simulation (grey line) and terrestrial proxy-data (squares and reconstructed interval). In order to obtain the zonal averages of proxy-data, the data locations are transformed into the grid point resolution (3.75°) of the model ECHAM4. For Europe, the zonal average temperatures of the model runs are shown separately.

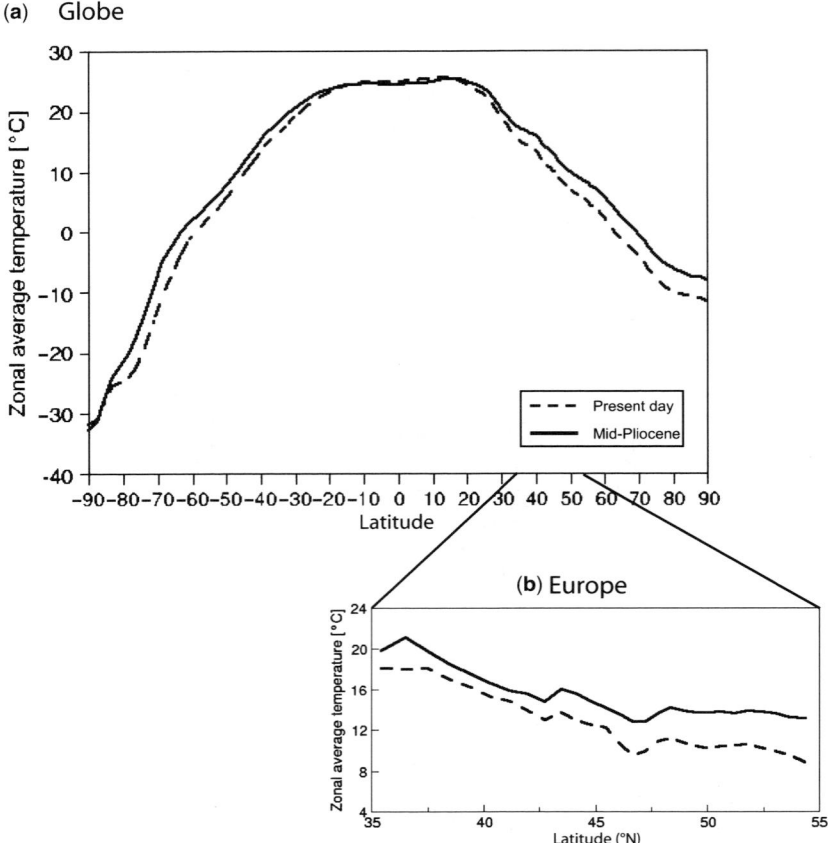

Fig. 12. The zonal average temperatures (in °C) for land surfaces of the Middle Pliocene simulation (bold line) and the present-day control simulation (dotted line), (**a**) for the globe, (**b**) for Europe (from −10 to 20° East and 35 to 55° West), from the LMDz AGCM (modified from Jost 2005).

However, despite this weakened pole-to-equator thermic gradient, the simulation shows similar changes of temperature at mid-latitudes of the Northern Hemisphere between the Middle Pliocene and today (Fig. 12b), suggesting that, for this region, the latitudinal thermic gradient was close to the modern one (i.e. around 0.6 °C/degree in latitude). This pattern is completely consistent with our reconstructed climate and vegetation distribution. The mean annual temperatures of the simulations cited above are in good agreement with the reconstructed climate (Fig. 13a) with temperatures that are clearly higher than today in the study region.

There are, however, some important differences between the reconstructed annual precipitation and changes simulated by the AGCM LMDz (Fig. 13b). Pollen data indicate in western Europe and in the western Mediterranean higher annual precipitation than the LMDz model.

On the contrary, the pollen-based precipitation estimates are in better agreement with the simulations by Haywood *et al.* (2000*a*, *b*). As shown on Figures 7 and 10, the precipitation gradient, whilst still decreasing from north to south, was more accentuated, with a larger difference between the Pliocene and today in the north than in the south. Haywood *et al.*'s simulations (2000*b*) show annual zonal average precipitation similar to modern values between around 30 and 42° N and higher than today between 42 and 51° N. This pattern is particularly true in Western Europe and western Mediterranean and has been explained by the authors by an increased arrival of southwestern air masses. During the Mid-Pliocene, the enhancement of the Icelandic low- and Azores high-pressure systems and the stronger pressure gradient in the North Atlantic caused an intensification of annual westerly wind strength. Combined

(a) Mean annual temperatures (°C)

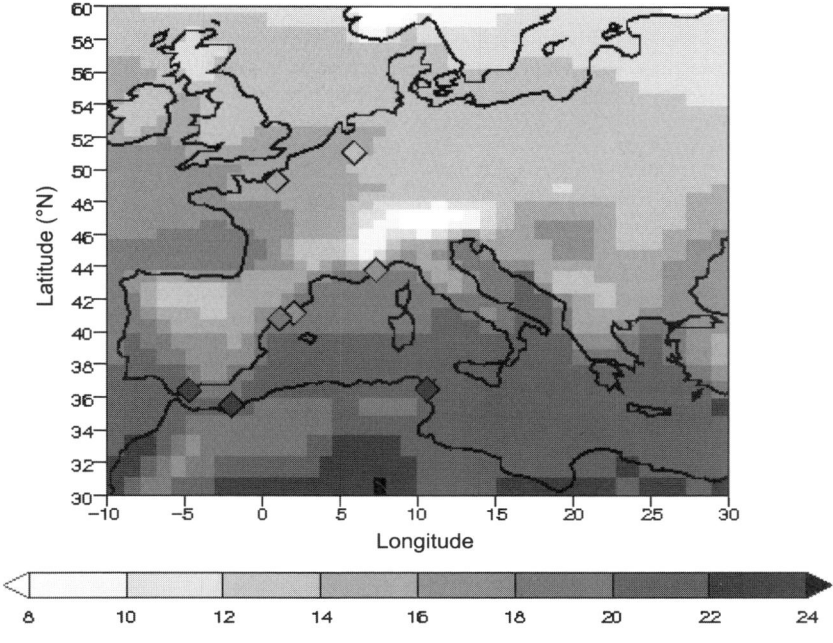

(b) Mean annual precipitation (mm)

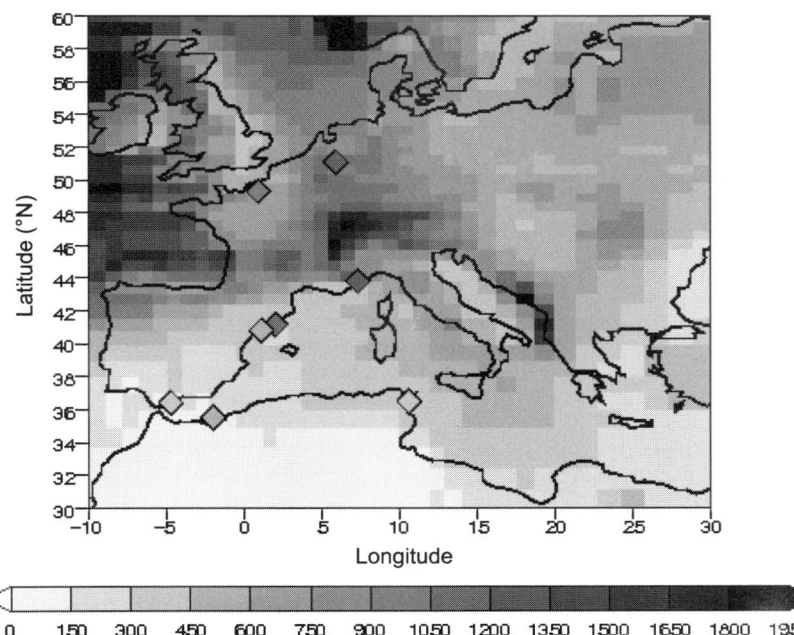

Fig. 13. Comparison for the Mid-Pliocene (3.5–3 Ma) between (**a**) mean annual temperatures and (**b**) annual precipitation simulated by the LMDz atmospheric general circulation model (modified from Jost 2005) and the values obtained from pollen data (diamonds).

with warmer sea-surface temperatures, the atmospheric transport of heat and moisture over Europe and the Mediterranean region was increased (Haywood *et al.* 2000*a*). This pattern is in agreement with the development of Ericaceae moors along the European Atlantic coast that suggest an enhanced westerly wind pattern (Suc *et al.* 1995*a, b*).

New simulations realized by Haywood *et al.* (2002) with the HadAM3 AGCM coupled with the BIOME 4 vegetation model (Kaplan 2001) show that, due to higher temperatures and higher precipitation, Europe and the Mediterranean region were dominated by forest biomes during the Mid-Pliocene. This agrees with the pollen data, especially in higher latitudes where forests extended into what is now tundra. This confirms the good agreement between our pollen data and the HadAM3 AGCM.

Conclusion

A thorough review of existing pollen data has allowed us to present a regional climate synthesis for the Neogene, from the Middle Miocene (*c*. 14–15 Ma) to the Middle Pliocene (*c*. 3.5 Ma).

In Western Europe and the Mediterranean region, from the Middle Miocene to the Middle Pliocene, the vegetation cover reflects a clear latitudinal gradient. In all of the considered periods, the quantitatively reconstructed climate shows, in comparison to today, higher mean annual temperatures along the gradient and increases in mean annual precipitation in northwestern Europe and the northwestern Mediterranean, but similar precipitation to today in the southwestern Mediterranean region. The results also show a clear latitudinal gradient of temperature and precipitation, increasing for temperatures but decreasing for precipitation from north to south.

The results show that the thermic gradient has evolved in time. During the Middle Miocene, the presence of mega-mesothermic taxa in pollen spectra at all sites, even in Switzerland, shows that the thermic latitudinal gradient was weaker than the modern one. Based on the climate reconstructed from pollen data, the thermic gradient was around 0.48 °C per degree of latitude whereas it is today around 0.6 °C degree in Western Europe. This result is in agreement with estimations obtained from fossil floras of Europe (Bruch *et al.* 2004) and America (Liu & Leopold 1994). During the Tortonian and the Pliocene, the vegetation distribution and the climate reconstruction show the thermic latitudinal gradient to have been close to the modern one. The transition from the weak thermic latitudinal gradient of the Mid-Miocene to the modern-like gradient of the Pliocene appears to take place during the Middle–Late Miocene, before or during the Tortonian.

The precipitation gradient was more accentuated than today from the Mid-Miocene to the Mid-Pliocene: the precipitation anomalies between the Neogene and today are larger in the north than in the south. The transition from this pattern to the modern latitudinal gradient took place after the Mid-Pliocene, at time of the first glacial–interglacial cycles.

W.H. Zagjwin and The Netherlands Geological Survey are acknowledged for the pollen data of Susteren. G. Jiménez-Moreno was funded by a PhD grant ('Junta de Andalucía', Spain) and a co-supervised grant from the French Ministry of Universities. This paper is a contribution to the Project 'La diversité végétale du domaine méditerranéen: son évolution depuis 6 millions d'années' of the French Programme 'Environnement, Vie et Sociétés' (Institut Français de la Biodiversité). Financial support was also partly provided by the EEDEN ('Environments and Ecosystems Dynamic of the Eurasian Neogene') program of the European Science Foundation and by the CNRS (ECLIPSE program: 'Quantification de l'impact des forçages climatiques/ anthropiques passés et futurs sur les circulations dans le bassin de Paris'). We are grateful to the two referees for their helpful suggestions and corrections on the manuscript. Simon Brewer (CEREGE, University of Aix-Marseille, France) is thanked for help with the linguistic editing of this paper. This paper is ISEM contribution.

References

AGUSTI, J. & ROCA, E. 1987. Sintesis biostratigrafica de la fossa de la cerdanya (Pirineos orientales). *Estudios Geologicos*, **43**, 521–529.

AXELROD, D. I. 1973. History of the Mediterranean ecosystem in California. *In*: DI CASTRI, F. & MOONEY, H. A. (eds) *Mediterranean Type Ecosystems – Origin and Structure.* Springer-Verlag, 225–277.

BACHIRI-TAOUFIQ, N. 2000. *Les environnements marins et continentaux du corridor rifain au Miocène supérieur d'après la palynologie.* Habilitation thesis, University Hassan II – Mohammedia, Casablanca (Morocco).

BARHOUN, N. 2000. *Biostratigraphie et paléoenvironnement du Miocène supérieur et du Pliocène inférieur du maroc septentrional. Apport des foraminifères planctoniques.* PhD thesis, University of Casablanca, 206pp.

BERTINI, A. 1992. *Palinologia ed aspetti ambientali del versante Adriatico dell'Appennino centro-settentrionale durante il Messiniano e lo Zancleano.* PhD thesis, University of Florence (Italy).

BERTINI, A. 1994. Palynological investigations on Upper Neogene and Lower Pleistocene sections in Central and Northern Italy. *Memorie della Società Geologica Italiana*, **48**, 431–443.

BERTINI, A. 2001. Pliocene climatic cycles and altitudinal forest development from 2.7 Ma in the Northern Apennines (Italy): evidence from the pollen record of the

Stirone section (*c.* 5.1 to *c.* 2.2 Ma). *Géobios*, **34**, 253–265.

BERTINI, A. 2003. Early to Middle Pleistocene changes of the Italian flora and vegetation in the light of a chronostratigraphic framework. *Il Quaternario*, **16**, 19–36.

BERTINI, A. & ROIRON, P. 1997. Evolution de la végétation et du climat pendant le Pliocène moyen en Italie centrale: apport de la palynologie et des macroflores à l'étude du bassin du Valdarno Supérieur, (coupe de Santa Barbara). *Compte Rendus de l'Académie des Sciences, Série IIa*, **324**, 763–771.

BESSAIS, E. & CRAVATTE, J. 1988. Les écosystèmes végétaux pliocènes de Catalogne méridionale. Variations latitudinales dans le domaine nord-ouest méditerranéen. *Geobios*, **21**, 49–63.

BESSEDIK, M. 1985. *Reconstitution des environnements Miocènes des régions Nord-ouest méditerranéennes à partir de la Palynologie.* Habilitation thesis, University of Montpellier II (France).

BESSON, D., PARIZE, O., RUBINO, J.-L. ET AL. 2005. Un réseau fluviatile d'âge burdigalien terminal dans le sud-est de la France: remplissage, extension, âge, implications. *Comptes-Rendus Geoscience*, Stratigraphie, Géomorphologie, **337**, 1045–1054.

BRUCH, A. A., UTESCHER, T., OLIVARES, C. A., DOLAKOVA, N., IVANOV, D. & MOSBRUGGER, V. 2004. Middle and Late Miocene spatial temperature patterns and gradients in Europe – preliminary results based on palaeobotanical climate reconstructions. *Courier Forschungsinstitut Senckenberg*, **249**, 15–27.

BRUCH, A. A., UTESCHER, T., MOSBRUGGER, V., GABRIELYAN, I. & IVANOV, D. A. 2006. Late Miocene climate in the circum-Alpine realm – a quantitative analysis of terrestrial palaeofloras. *Palaeogeography, Palaeoclimatology, Palaeoecology*, **238**, 270–280.

CAMBON, G., SUC, J.-P., ALOISI, J.-C. ET AL. 1997. Modern pollen deposition in the Rhône delta area (lagoonal and marine sediments). France, *Grana*, **36**, 105–113.

CHEVRON, 1986. *Informe final sondeo Alborán A-1.* Internal Report.

CITA, M. B., STRADNER, H. & CIARANFI, N. 1973. Biostratigraphical investigations on the Messinian stratotype and on the overlying "trubi" formation. *Rivista Italiana di Paleontologia*, **79**, 393–446.

CLAUZON, G., RUBINO, J.-L. & SAVOYE, B. 1995. Marine Pliocene Gilbert-type fan deltas along the French Mediterranean coast. Publication *Association de Sédimentologie Française*, **23**, 145–222.

CLAUZON, G. 1996. Limites de séquences et évolution géodynamique. *Géomorphologie*, **1**, 3–22.

CLAUZON, G., SUC, J.-P., GAUTIER, F., BERGER, A. & LOUTRE, M.-F. 1996. Alternate interpretation of the Messinian salinity crisis: Controversy resolved? *Geology*, **24**, 363–366.

CLET-PELLERIN, M. 1996. *Palynologie, paléoenvironnements et cycles glaciaire-interglaciaire: Applications au Plio-Quaternaire de Normandie et de la vallée du Saint-Laurent.* Habilitation thesis, University of Caen (France).

CLET, M. & HUAULT, M.-F. 1987. Les dépôts lagunaires du Reuvérien dans les argiles de la Londe (Normandie, France). *Bulletin de l'association française pour l'étude du Quaternaire*, **4**, 195–202.

CRONIN, T. M. & DOWSETT, H. J. 1990. A quantitative micropalaeontologic method for shallow marine paleoclimatology: application to Pliocene deposits of the western North Atlantic Ocean. *Marine Micropaleontology*, **16**, 117–148.

DECONTO, R. M. & POLLARD, D. 2003a. Rapid Cenozoic glaciation of Antarctica induced by declining atmospheric CO_2. *Nature*, **421**, 245–249.

DECONTO, R. M. & POLLARD, D. 2003b. A coupled climate-ice sheet modelling approach to the Early Cenozoic history of the Antarctic ice sheet. *Palaeogeography, Palaeoclimatology, Palaeoecology*, **198**, 39–52.

DINIZ, F. 1984a. *Apports de la palynologie à la connaissance du Pliocène portugais. Rio Maior: un bassin de référence pour l'histoire de la flore, de la végétation et du cimat de la façade atlantique de l'Europe méridionale.* Habilitation thesis, University of Montpellier 2 (France).

DINIZ, F. 1984b. Etude palynologique du bassin pliocène de Rio Maior. *Paléobiologie Continentale*, **14**, 259–267.

DOWSETT, H. J., THOMPSON, R. S., BARRON, J. A. ET AL. 1994. Joint investigations of the middle Pliocene climate I: PRISM paleoenvironmental reconstructions. *Global and Planetary Change*, **9**, 169–195.

DOWSETT, H. J., BARRON, J. A., POORE, R. Z., THOMPSON, R. S., CRONIN, T. M., ISHMAN, S. E. & WILLARD, D. A. 1999. *Middle Pliocene Paleoenvironmental Reconstruction: PRISM2.* US Geological Survey open file report 99–535.

DUBOIS, P. & CURNELLE, R. 1978. Résultats apportés par le forage Les Mées n°1 sur le plateau de Valensole (Alpes-de-Haute-Provence). *Comptes Rendus sommaires de la Société géologique de France*, **4**, 181–184.

DUTTON, J. F. & BARRON, E. J. 1997. Miocene to present vegetation changes: A possible piece of the Cenozoic puzzle. *Geology*, **25**, 39–41.

ELF 1984. *Informe final sondeo Andalucía G-1.* Internal report.

ELHAI, H. 1969. La flore sporo-pollinique du gisement Villafranchien de Senèze (Massif Central, France). *Pollen et Spores*, **11**, 127–139.

FARJANEL, G. & MEIN, P. 1984. Une association de mammifères et de pollens dans la formation continentale des 'Marnes de Bresse' d'âge Miocène supérieur, à Ambérieu (Ain). *Géologie de la France*, **1–2**, 131–148.

FAUQUETTE, S. & BERTINI, A. 2003. Quantification of the northern Italy Pliocene climate from pollen data – evidence for a very peculiar climate pattern. *Boreas*, **32**, 361–369.

FAUQUETTE, S., GUIOT, J. & SUC, J.-P. 1998a. A method for climatic reconstruction of the Mediterranean Pliocene using pollen data. *Palaeogeography, Palaeoclimatology, Palaeoecology*, **144**, 183–201.

FAUQUETTE, S., QUÉZEL, P., GUIOT, J. & SUC, J.-P. 1998b. Signification bioclimatique de taxons – guides du Pliocene Méditerranéen. *Geobios*, **31**, 151–169.

FAUQUETTE, S., SUC, J.-P., GUIOT, J. *ET AL.* 1999. Climate and biomes in the West Mediterranean area during the Pliocene. *Palaeogeogeography, Palaeoclimatology, Palaeoecology,* **152,** 15–36.

FERRANDINI, M., FERRANDINI, J., LOYE-PILOT, M.-D., BUTTERLIN, J., CRAVATTE, J. & JANIN, M.-C. 1998. Le Miocène du bassin de Saint-Florent (Corse): modalités de la transgression du Burdigalien supérieur et mise en évidence du Serravallien. *Geobios,* **31,** 1, 125–137.

FLUTEAU, F., RAMSTEIN, G. & BESSE, J. 1999. Simulating the evolution of the Asian and African monsoons during the past 30 Myr using an atmospheric general circulation model. *Journal of Geophysical Research,* **104,** 11995–12018.

GONCHAROVA, I. G., SHCHERBA, I. G., KHONDKARIAN, S. O. *ET AL.* 2004. Map 5: Early middle Miocene (16–15 Ma). *In:* POPOV, S. V., RÖGL, F., ROZANOV, A. Y., STEININGER, FRITZ, F., SHCHERBA, I. G. & KOVAC, M. (eds) *Lithological-Paleogeographic maps of Paratethys, 10 maps Late Eocene to Pliocene. Courier Forschungsinstitut Senckenberg,* **250,** 19–21.

GUIOT, J. 1990. Methodology of paleoclimatic reconstruction from pollen in France. *Palaeogeography, Palaeoclimatology, Palaeoecology,* **80,** 49–69.

HALL, I. R., MCCAVE, I. N., ZAHN, R., CARTER, L., KNUTZ, P. C. & WEEDON, G. P. 2003. Paleocurrent reconstruction of the deep Pacific inflow during the middle Miocene: Reflections of East Antarctic Ice Sheet growth. *Paleoceanography,* **18,** 1040.

HAQ, B. U., HARDENBOL, J. & VAIL, P. R. 1987. Chronology of fluctuating sea levels since the Triassic (250 million years ago to present), *Science,* **235,** 1156–1167.

HAYWOOD, A. M., SELLWOOD, B. W. & VALDES, P. J. 2000*a.* Regional warming: Pliocene (3 Ma) paleoclimate of Europe and the Mediterranean. *Geology,* **28,** 1063–1066.

HAYWOOD, A. M., VALDES, P. J. & SELLWOOD, B. W. 2000*b.* Global scale palaeoclimate reconstruction of the middle Pliocene climate using the UKMO CGM: initial results. *Global and Planetary Change,* **25,** 239–256.

HAYWOOD, A. M., VALDES, P. J., FRANCIS, J. E. & SELLWOOD, B. W. 2002. Global middle Pliocene biome reconstruction: a data/model synthesis. *Geochemistry, Geophysics, Geosystems,* **3,** 1072.

HEUSSER, L. 1988. Pollen distribution in marine sediments on the continental margin of Northern California. *Marine Geology,* **80,** 131–147.

ILYNA, L. B., SHCHERBA, I. G., KHONDKARIAN, S. O. & GONCHAROVA, I. A. 2004. Map 6: Mid-Middle Miocene (14–13 Ma). *In:* POPOV, S. V., RÖGL, F., ROZANOV, A. Y., STEININGER, FRITZ, F., SHCHERBA, I. G. & KOVAC, M. (eds) *Lithological-Paleogeographic maps of Paratethys, 10 maps Late Eocene to Pliocene. Courier Forschungsinstitut Senckenberg,* **250,** 23–25.

JIMÉNEZ-MORENO, G. 2005. *Utilizacion del analisis polinico para la reconstruccion de la vegetacion, clima y estimacion de paleoaltitudes a lo largo de arco alpino europeo durante el Mioceno (21–8 Ma).* PhD thesis, University Claude Bernard Lyon 1 (France) and University of Granada (Spain).

JIMÉNEZ-MORENO, G. & SUC, J.-P. 2007. Middle Miocene latitudinal climatic gradient in western Europe: evidence from pollen records. *Palaeogeography, Palaeoclimatology, Palaeoecology,* in press.

JIMÉNEZ-MORENO, G., RODRÍGUEZ-TOVAR, F. J., PARDO-IGÚZQUIZA, E., FAUQUETTE, S., SUC, J.-P. & MULLER, P. 2005. High-resolution palynological analysis in late Early-Middle Miocene Tengelic-2 core from the Pannonian Basin, Hungary: Climatic changes, astronomical forcing and eustatic fluctuations in the Central Paratethys. *Palaeogeography, Palaeoclimatology, Palaeoecology,* **216,** 73–97.

JOLIVET, L., AUGIER, R., ROBIN, C., SUC, J.-P. & ROUCHY, J.-M. 2006. Lithospheric-scale geodynamic context of the Messinian salinity crisis. *Sedimentary Geology,* **188/189,** 9–33.

JOST, A. 2005. *Caractérisation des forçages climatiques et géomorphologiques des cinq derniers millions d'années et modélisation de leurs conséquences sur un système aquifère complexe: le bassin de Paris.* PhD thesis, University of Paris VI (France).

KÄLIN, D., WEIDMANN, M., ENGESSER, B. & BERGER, J.-P. 2001. Paleontologie et âge de la Molasse d'eau douce supérieure (OSM) du Jura neuchâtelois. *Mémoires suisses de Paléontologie,* **121,** 66–99.

KAPLAN, J. O. 2001. *Geophysical applications of vegetation modelling.* PhD thesis, Lund University (Sweden).

KHONDKARIAN, S. O., PARAMONOVA, N. P., SCHIERBA, I. G. *ET AL.* 2004. Map 10: Middle-Late Pliocene (3.4–1.8 Ma). *In:* POPOV, S. V., RÖGL, F., ROZANOV, A. Y., STEININGER, FRITZ, F., SHCHERBA, I. G. & KOVAC, M. (eds) *Lithological-Paleogeographic maps of Paratethys, 10 maps Late Eocene to Pliocene. Courier Forschungsinstitut Senckenberg,* **250,** 39–41.

KOVAR-EDER, J., KVACEK, Z., MARTINETTO, E. & ROIRON, P. 2006. Late Miocene to Early Pliocene vegetation of southern Europe (7–4 Ma) as reflected in the mega fossil plant record. *Palaeogeography, Palaeoclimatology, Palaeoecology,* **238,** 321–339.

KUTZBACH, J. E., PRELL, W. L. & RUDDIMAN, W. F. 1993. Sensitivity of Eurasian climate to surface uplift of Tibetan plateau. *The Journal of Geology,* **101,** 177–190.

KUTZBACH, J. E. & BEHLING, P. 2004. Comparison of simulated changes of climate in Asia for two scenarios: Early Miocene to present, and present to future enhanced greenhouse. *Global and Planetary Change,* **41,** 157–165.

LIU, G. & LEOPOLD, E. B. 1994. Climatic comparison of Miocene pollen floras from northern-east China and south-central Alaska, USA. *Palaeogeography, Palaeoclimatology, Palaeoecology,* **108,** 217–228.

MAGNÉ, J. 1978. *Etudes microstratigrafiques sur le Néogène de la Méditerranée nord-occidentale. Les Bassins néogènes catalans.* Editions CNRS, Paris, 259pp.

MARTÍN PÉREZ, J. A. & VISERAS, C. 1994. Sobre la posición estratigráfica de las "Margas de Gor", Sierra de Baza, Cordillera Bética. *Geogaceta,* **15,** 63–66.

MAUFFRET, A., DURAND DE GROSSOUVRE, B., DOS REIS, A. T., GORINI, G. & NERCESSIAN, A. 2001. Structural geometry in the eastern Pyrenees and western Gulf of Lion (Western Mediterranean). *Journal of Structural Geology*, **23**, 1701–1726.

MENENDEZ AMOR, J. 1955. *La depresion ceretana espanola y sus vegetales fosiles. Caracteristica fitopaleontologica del Neogeno de la Cerdana espanola*, Memorias de la Real Academia de Ciencias exactas, fisicas y naturales de Madrid. Serie de Ciencias Naturales, **XVIII**, 232pp.

MEULENKAMP, J. E. & SISSINGH, W. 2003. Tertiary palaeogeography and tectonostratigraphic evolution of the Northern and Southern Peri-Tethys platforms and the intermediate domains of the African-Eurasian convergent plate boundary zone. *Palaeogeography, Palaeoclimatology, Palaeoecology*, **196**, 209–228.

MIKOLAJEWICZ, U., MAIER-REIMER, E., CROWLEY, T. J. & KIM, K. Y. 1993. Effect of Drake and Panamanian gateways on the circulation of an ocean model. *Paleoceanography*, **8**, 409–426.

MORAN, K., BACKMAN, J., BRINKHUIS, H. *ET AL.* 2006. The Cenozoic palaeoenvironment of the Arctic Ocean. *Nature*, **441**, 601–605.

NAUD, G. & SUC, J.-P. 1975. Contribution à l'étude paléofloristique des Coirons (Ardèche): premières analyses polliniques dans les alluvions sous-basaltiques et interbasaltiques de Mirabel (Miocène supérieur). *Bulletin de la Société Géologique de France*, ser. 7, **17, 5**, 820–827.

NISANCIOGLU, K. H., RAYMO, M. E & STONE, P. H. 2003. Reorganization of Miocene deep water circulation in response to the shoaling of the Central American Seaway. *Paleoceanography*, **18**, 1006.

OZENDA, P. 1989. Le déplacement vertical des étages de végétation en fonction de la latitude: un modèle simple et ses limites. *Bulletin de la Société Géologique de France*, **8**, t. V, n°3, 535–540.

PAGANI, M., ARTHUR, M. A. & FREEMAN, K. H. 2000. Variations in Miocene phytoplankton growth rates in the southwest Atlantic: evidence for change in ocean circulation. *Paleoceanography*, **15**, 486–496.

PARAMONOVA, N. P., SHCHERBA, I. G., KHONDKARIAN, S. O. *ET AL.* 2004. Map 7: Late Middle Miocene (12–11 Ma). *In*: POPOV, S. V., RÖGL, F., ROZANOV, A. Y., STEININGER, FRITZ, F., SHCHERBA, I. G. & KOVAC, M. (eds) *Lithological-Paleogeographic maps of Paratethys, 10 maps Late Eocene to Pliocene. Courier Forschungsinstitut Senckenberg*, **250**, 27–29.

PONS, A. 1964. Contribution palynologique à l'étude de la flore et de la végétation pliocènes de la région rhodanienne. *Annales de Sciences Naturelles, Botanique*, **série 12, 5**, 499–722.

QUÉZEL, P. & BARBERO, M. 1989. Zonation altitudinale des structures forestières de végétation en Californie méditerranéenne. Leur interprétation en fonction des méthodes utilisées sur le pourtour méditerranéen. *Annales des Sciences forestières*, **46**, 233–250.

RAMSTEIN, G., FLUTEAU, F., BESSE, J. & JOUSSEAUME, S. 1997. Effect of orogeny, plate motion and land-sea distribution on Eurasian climate change over the past 30 million years. *Nature*, **386**, 788–795.

RETALLACK, G. 2001. Cenozoic expansion of grasslands and climatic cooling. *Journal of Geology*, **109**, 407–426.

RIVAS-CARBALLO, M. R., ALONSO-GAVILAN, G., VALLE, M. F. & CIVIS, J. 1994. Miocene palynology of the central sector of the Duero Basin (Spain) in relation to palaeogeography and palaeoenvironment. *Review of Palaeobotany and Palynology*, **82**, 251–264.

RÖGL, V. F. 1998. Palaeogeographic considerations for Mediterranean and Paratethys seaways (Oligocene to Miocene). *Annalen der Naturhistorisches Museum Wien*, **99A**, 279–310.

RUDDIMAN, W. F. & KUTZBACH, J. E. 1989. Forcing of the late Cenozoic uplift northern hemisphere climate by plateau uplift in the southern Asia and American West. *Journal of Geophysical Research*, **94**, 18409–18427.

SHACKLETON, N. J., HALL, M. A. & PATE, D. 1995. Pliocene stable isotope stratigraphy of Site 846. *Proceedings of the Ocean Drilling Program, Scientific Results*, **138**, 337–355.

STEPPUHN, A., MICHEELS, A., GEIGER, G. & MOSBRUGGER, V. 2006. Reconstructing the Late Miocene climate and oceanic heat flux using the AGCM ECHAM4 coupled to a mixed-layer ocean model with adjusted flux correction. *Palaeogeography, Palaeoclimatology, Palaeoecology*, **238**, 399–423.

SUC, J.-P. 1976. Apports de la palynologie à la connaissance du Pliocène du Roussillon (Sud de la France). *Geobios*, **9**, 741–771.

SUC, J.-P. 1980. *Contribution à la connaissance du Pliocène et du Pléistocène inférieur des régions méditerranéennes d'Europe occidentale par l'analyse palynologique des dépôts du Languedoc-Roussillon (sud de la France) et de la Catalogne (nord-est de l'Espagne)*. Habilitation thesis, University of Montpellier 2 (France).

SUC, J.-P. 1989. Distribution latitudinale et étagement des associations végétales au Cénozoïque supérieur dans l'aire ouest-méditerranéenne. *Bulletin de la Société Géologique de France*, **8**, t. V, n°3, 541–550.

SUC, J.-P. & CRAVATTE, J. 1982. Etude palynologique du Pliocène de Catalogne (Nord-est de l'Espagne). *Paléobiologie Continentale*, **13**, 1–31.

SUC, J.-P., BERTINI, A., COMBOURIEU-NEBOUT, N., DINIZ, F. *ET AL.* 1995a. Structure of West Mediterranean and climate since 5,3 Ma. *Acta zoologica cracovia*, **38**, 3–16.

SUC, J.-P., DINIZ, F., LEROY, S. *ET AL.* 1995b. Zanclean (~Brunssumian) to early Piacenzian (~early-middle Reuverian) climate from 4° to 54° north latitude (West Africa, West Europe and West Mediterranean areas). *Mededelingen Rijks Geologische Dienst*, **52**, 43–56.

SUC, J.-P., VIOLANTI, D., LONDEIX, L. *ET AL.* 1995c. Evolution of the Messinian Mediterranean environments: the Tripoli Formation at Capodarso (Sicily, Italy). *Review of Palaeobotany and Palynology*, **87**, 51–79.

SUC, J.-P., FAUQUETTE, S., BESSEDIK, M. *ET AL.* 1999. Neogene vegetation changes in West European and West circum-Mediterranean areas. *In*: AGUSTI, J., ROOK, L. & ANDREWS, P. (eds) *Hominid Evolution and Climate in Europe, 1, Climatic and Environmental Change in the Neogene of Europe*, Cambridge University Press, 370–385.

SUC, J.-P. & ZAGWIJN, W. H. 1983. Plio-Pleistocene correlations between the northwestern Mediterranean region and northwestern Europe according to recent biostratigraphic and paleoclimatic data. *Boreas*, **12**, 153–166.

THOMPSON, R. S. 1991. Pliocene environments and climates in the Western United States. *Quaternary Science Reviews*, **10**, 115–132.

THOMPSON, R. S. & FLEMING, R. F. 1996. Middle Pliocene vegetation: reconstructions, paleoclimatic inferences, and boundary conditions for climate modelling. *Marine Micropaleontology*, **27**, 27–49.

THOMPSON, R. S., ANDERSON, K. H. & BARTLEIN, P. J. 1999. *Atlas of relations between climatic parameters and distributions of important trees and shrubs in North America*. US Geological Survey professional paper, 1650 A to 1650 C.

WALTER, H. 1979. *Vegetation of the Earth and ecological systems of the Geo-biosphere*. Springer-Verlag, Heidelberg Science Library, Berlin, 274pp.

WANG, C. W. 1961. *The Forest of China (with a survey of grassland and desert vegetation)*. Maria Moors Cabot Foundation. Harvard University, 313pp.

WESTERHOLD, T., BICKERT, T. & RÖHL, U. 2005. Middle to late Miocene oxygen isotope stratigraphy of ODP site 1085 (SE Atlantic): new constraints on Miocene climate variability and sea-level fluctuations. *Palaeogeography, Palaeoclimatology, Palaeoecology*, **217**, 205–222.

ZAGWIJN, W. H. 1960. Aspects of the Pliocene and early Pleistocene vegetation in the Netherlands. *Mededelingen van de Geologische Stichting*, **3**, 1–78.

ZHENG, Z. 1986. *Contribution palynologique à la connaissance du néogène du Sud-Est français et de Ligurie*. PhD thesis, University of Montpellier 2 (France).

ZHENG, Z. & CRAVATTE, J. 1986. Etude palynologique du Pliocène de la Côte d'Azur (France) et du littoral ligure (Italie). *Geobios*, **19**, 815–823.

Neogene flora, vegetation and climate dynamics in southeastern Europe and the northeastern Mediterranean

G. JIMENEZ-MORENO[1,2,3], S.-M. POPESCU[1], D. IVANOV[4] & J.-P. SUC[1]

[1]*Laboratoire PaléoEnvironnement et PaléobioSphère (UMR CNRS 5125), Université Claude Bernard - Lyon 1, 27-43 boulevard du 11 Novembre, 69622 Villeurbanne, France (e-mail: gonzaloj@ugr.es; popescu@univ-lyon1.fr; jean-pierre.suc@univ-lyon1.fr)*

[2]*Departamento de Estratigrafía y Paleontología, Universidad de Granada, Avda. Fuente Nueva S/N 18002, Granada, Spain*

[3]*Center for Environmental Sciences & Education, Box 5694, Northern Arizona University, Flagstaff, AZ 86011USA. (present address) (e-mail: Gonzalo.Jimenez-Moreno@NAU.EDU)*

[4]*Institute of Botany, Bulgarian Academy of Sciences, Acad. G. Bonchev Str., 23, 1113 Sofia, Bulgaria (e-mail: dimiter@bio.bas.bg)*

Abstract: Pollen analysis of Miocene and Pliocene sediments from southeastern Europe and the northeastern Mediterranean is represented in pollen synthetic diagrams based on ecological criteria in order to clearly visualize changes in the composition and structure of the vegetation through time. New pollen data, together with abundant existing palynological information from this area, show a progressive reduction in plant diversity caused by a decrease in the most thermophilous and high-water requirement plants and, on the contrary, an increase in warm-temperate (mesothermic) and seasonal-adapted taxa during the Middle–Late Miocene and Pliocene. At the same time, an increase in high-elevation trees and herbs has been recorded, with a strong augmentation in *Artemisia*, first in the eastern Mediterranean and later on in the western Mediterranean area. This has been interpreted as a response of the vegetation to global and regional processes, including climate cooling related to the development of the East Antarctic Ice Sheet (EAIS), uplift of regional mountains during Alpine orogenesis and progressive movement of Eurasia towards northern latitudes as a result of the northwards collision of Africa.

Pollen analyses dealing with Miocene–Pliocene sediments from the Paratethys are rare. Studies have focused on the Miocene and Pliocene palynology of the Central Paratethys (Petrescu *et al.* 1989*a*, *b*; Planderová 1990; Nagy 1991, 1992, 1999; Petrescu & Malan 1992) and Turkey (Benda 1971; Benda *et al.* 1975; Akgün & Akyol 1999), but the lack of any quantitative information render these analyses limited. However, palynological data, with reliable botanical identification, are available for the Miocene (Ivanov 1995; Ivanov & Koleva-Rekalova 1999; Palamarev & Ivanov 2001; Ivanov *et al.* 2002; Jiménez-Moreno *et al.* 2005) and the Pliocene (Drivaliari 1993; Drivaliari *et al.* 1999; Popescu 2001, 2002, 2006; Popescu *et al.* 2006*a*, *b*) of the same region. In these studies, pollen was not used for biostratigraphy but for climatic information, as independent biostratigraphic dating was available (see below). Pollen counts and a statistical treatment of the data were made to obtain reliable information about floral diversity, organization of the vegetation and to better visualize vegetation and climate change.

The geographical position of the studied area, between Africa and Eurasia and between a Mediterranean and temperate climate, makes this region of great interest for palaeobotanic studies. Today, the southeastern part of the area is mainly occupied by steppe vegetation rich in *Artemisia* (i.e. the central Anatolian steppes), that is the main refuge area of thermophilous plants (mostly along the Turkish coastlines: Zohary 1973; Quézel & Médail 2003). Alpine tectonics were active during the Neogene, producing uplift of the Carpathians, Dinarides, Balkan, Rhodope and Taurides mountains. Then, important palaeogeographical changes occurred (see below; Rögl 1998; Meulenkamp & Sissingh 2003; Popov *et al.* 2004) that may have contributed to the pattern of vegetation distribution seen today.

In this paper, we present a synthesis of palynological data, interpreted vegetation and climate dynamics based on Miocene and Pliocene deposits from Eastern Europe. New sections of Middle and Late Miocene age from this area have been analysed, adding new information to the already

From: WILLIAMS, M., HAYWOOD, A. M., GREGORY, F. J. & SCHMIDT, D. N. (eds) *Deep-Time Perspectives on Climate Change: Marrying the Signal from Computer Models and Biological Proxies*. The Micropalaeontological Society, Special Publications. The Geological Society, London, 503–516.

published data. Changes in vegetation have been
observed from the Langhian to the early Pliocene
(16.3–3 Ma). These are mainly related to global
climatic changes, in temperature and precipitation,
that are linked to atmospheric and palaeogeographic
changes that were of significant importance during
the Neogene.

Regional setting

The studied area comprises Neogene basins formed
within the Central–Eastern Paratethys Sea. They
were generated during the Neogene, like the rest
of the basins belonging to the Paratethys, as a
product of the collision of the African plate and
Eurasia. These basins are delimited by the
Carpathians, Balkan, Dinarides and Taurides, occupying parts of Hungary, Romania, Bulgaria, Serbia,

Greece and Turkey (Fig. 1; Kojumdgieva & Popov
1989; Rögl 1998; Meulenkamp & Sissingh 2003;
Goncharova *et al.* 2004; Ilyina *et al.* 2004; Paramonova *et al.* 2004; Khondkarian *et al.* 2004*a, b*).
During the Neogene, the Paratethys displayed a
long-term trend of decreasing marine influence
and a correlative reduction in size with regard to
the marine depositional domains. Marine deposition
lasted throughout the Early and Middle Miocene up
to approximately 12 Ma, when uplift caused the
sea to retreat from the Pannonian basin complex
where a brackish lake formed instead (Rögl 1998).
However, during the Early and Middle Miocene,
connection between the Mediterranean Sea and
the Paratethys existed that allowed for a free
marine faunal exchange (Harzhauser *et al.* 2003).
The first impairment of marine connections is
evident in the Late Badenian (Early Serravallian)

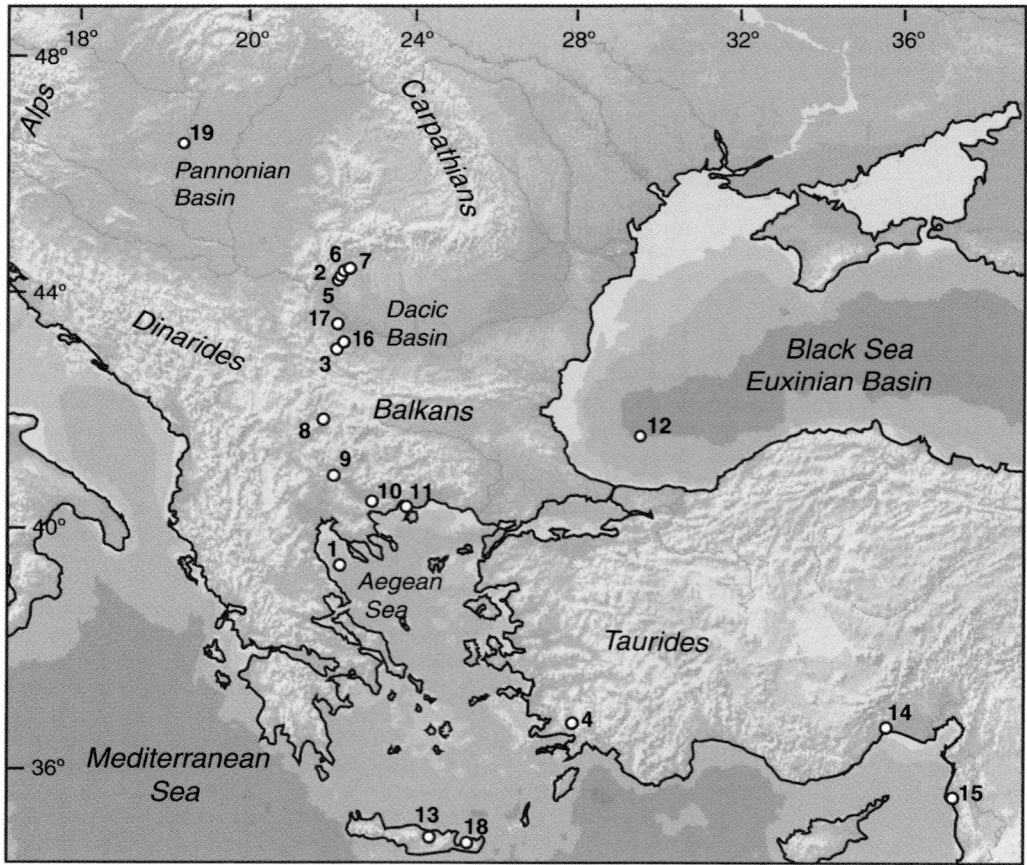

Fig. 1. Geographic map of the studied area and location of the sites. 1 Nireas-1; 2 Valea Morilor; 3 Ruzhintsi; 4
Catakbagyaka; 5 Hinova; Husnicioara and Valea Visenilor; 6 Lupoaia; 7 Ticleni-1; 8 Ravno Pole and Lozenec; 9
Sandanski; 10 Lion of Amphipoli; 11 Nestos-2; 12 Site 380 A; 13 Aghios Vlassios; 14 Avadan; 15 Lataquie; 16
Drenovets C-1; 17 Deleina C-12; 18 Makrilia; 19 Tengelic-2.

when dysaerobic bottom conditions and a stratified water column characterized the Paratethyan realm (Kovac *et al.* 2004). With the onset of the Sarmatian, marine connection to the Mediterranean almost completely ceased, and was reflected by the development of a highly endemic molluscan fauna (Harzhauser & Piller 2004). Finally, at the Sarmatian/Pannonian boundary (Serravallian/Tortonian boundary), the Central Paratethys became entirely restricted and the brackish Lake Pannon was established. Sporadic brief connections occurred during the Late Miocene and Pliocene between the Eastern Paratethys (Dacic and Euxinian basins) and the Mediterranean Sea as documented by nannoplankton influxes (Mărunţeanu & Papaianopol 1998; Semenenko & Olejnik 1995). One of these short connections also concerned the southeastern Pannonian Basin during the so-called Portaferrian regional Stage (Pontian). Some of these connections occurred just before and just after the Messinian salinity crisis (Clauzon *et al.* 2005; Snel *et al.* 2006), resulting in the same responses (i.e. an intense erosion, then the construction of Gilbert-type fan deltas) to the Messinian desiccation and Zanclean flooding as in the Mediterranean Basin itself (Clauzon *et al.* 2005). However, during the late Neogene, most of the Paratethyan basins were disconnected and evolved as isolated lakes, some of them being temporarily connected with the Mediterranean Sea (Mărunţeanu & Papaianopol 1995).

The independent evolution of the different subdomains of the Paratethys led to the construction of several regional stratigraphies, constituted by stages based on diverse groups of organisms, mainly bivalves and ostracods, and benthic and planktonic foraminifera etc. (Marinescu 1978; Papaianopol & Motas 1978; Papaianopol & Marinescu 1995; Rögl 1998; Fig. 2). Reliable correlations are established between the Eastern Paratethys regional stratigraphy and the Mediterranean standard stratigraphy using nannoplankton (Papaianopol & Mărunţeanu 1993; Mărunţeanu & Papaianopol 1995, 1998; Drivaliari *et al.* 1999; Clauzon *et al.* 2005; Snel *et al.* 2006).

Chronological background

A total of 19 sections and a total of 680 samples have been studied for pollen. Of those 19 sections, 12 (or a part) belong to the Miocene and 14 (or a part) to the Pliocene (Fig. 2). As far as possible, an independent age control has been obtained; it is indicated in Table 1 with the authors of the pollen analyses. The timescale of Gradstein *et al.* (2004) has been used.

Methods

Identification was performed comparing the Neogene pollen grains with those of the living relative plants using databanks of modern pollen grains and modern and past pollen grain photographs. Based on the results of the pollen spectra, standard synthetic diagrams (Suc 1984) with *Pinus* and Pinaceae have been constructed. In these pollen diagrams, taxa have been arranged into 12 different groups based on ecological criteria in order to obtain some visualization of the vegetation (see below) and more easily compare with reference oxygen isotopic curves. This method has been proven to be a very efficient tool for high-resolution climatic studies characterizing warm–cold alternations related to Milankovitch cycles for both the Miocene (Jiménez-Moreno *et al.* 2005) and the Pliocene (Popescu 2001, 2006; Popescu *et al.* 2006a, b).

Pollen data will be available, after publication, on the web from the 'Cenozoic pollen and climatic values' database (CPC) (http://medias.obs-mip.fr/cpc).

Results

Plant diversity and vegetation

Even if some parts of the studied region are characterized today by a very diverse flora and are main refuge areas of thermophilous plants (i.e. the Ponto-Euxinian area) (Quézel & Médail 2003), a richer and more diverse flora has been identified for the Mio-Pliocene that consisted of elements found presently in different geographic areas:

(1) Tropical and subtropical Africa, America and Asia (*Avicennia*, *Bombax*, Caesalpiniaceae, *Engelhardia*, *Platycarya*, Taxodiaceae, Hamamelidaceae, *Myrica*, Sapotaceae, etc.).

(2) Warm-temperate latitudes of the Northern Hemisphere (*Acer*, *Alnus*, *Betula*, Cupressaceae, *Fagus*, *Populus*, deciduous *Quercus*, *Salix*, etc.).

(3) Mediterranean region (*Olea*, *Phillyrea*, *Ceratonia*, evergreen *Quercus*, etc.).

All of these taxa grew in the Eastern European area during the Miocene.

We use the Chinese flora as a present-day comparison for the southeastern Europe and Middle East flora during the Neogene as it is the closest living example of this floral inventory (Suc 1984). Flora of the broad-leaved evergreen forest was represented by 45 typical tropical and subtropical taxa (i.e. megathermic and mega-mesothermic elements, respectively) in the studied region during the middle Miocene's warmest phase; only 21 of them persisted until the early Pliocene and have presently disappeared from the area. Flora of the evergreen and deciduous mixed forest was represented by 21 subtropical and warm-temperate taxa (i.e. mega-mesothermic and mesothermic

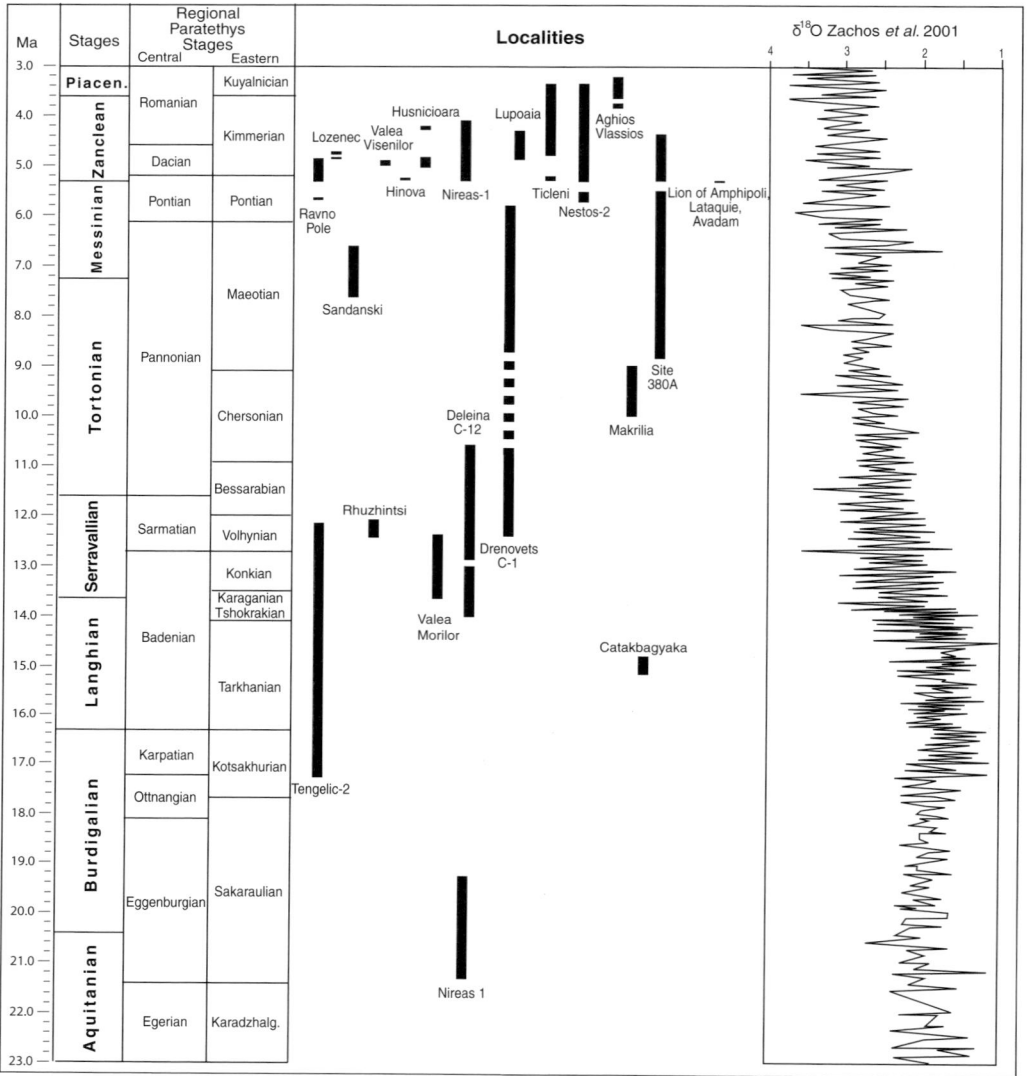

Fig. 2. Miocene and Pliocene chronostratigraphy and temporal situation of the studied sites. Correlations between standard stages and Paratethys stages by Harzhauser & Piller (in press) after data of Steininger (1999), Sprovieri (1992), Sprovieri *et al.* (2002), Fornaciari & Rio (1996) and Fornaciari *et al.* (1996). Oxygen isotope curve after Zachos *et al.* (2001); all stages recalibrated according to Gradstein & Ogg (2004), Gradstein *et al.* (2004) and Lourens *et al.* (2004).

elements, respectively) in the studied region during the Middle Miocene's warmest phase; they persisted here during the Early Pliocene and 17 among them are still living in the area.

The vegetation was characterized by a complex mosaic due to its dependency on several factors (water availability, characteristics of the soils, orientation of relief slopes, etc.) which superimposed its latitudinal–altitudinal organization. The most important factor, similar to

present day, would be altitude, controlling both temperature and precipitation. Therefore, the vegetation would be organized in altitude belts, which have been compared with those found today in subtropical to temperate southeastern China, the most reliable model. The following have been distinguished:

(1) a coastal marine environment characterized by the presence of an impoverished mangrove composed of *Avicennia* which is mainly

Table 1. *Age control of the 19 considered pollen localities indicating the authors of the pollen analyses*

#	Section	Location	Datation	Age	Pollen analysis by
1	Nireas 1	Greece	* Drivaliari 1993	Aquitanian-Burdigalian Zanclean	Drivaliari, A.
2	Valea Morilor	Romania	*) Papaianopol et al. 1995	Badenian-Sarmatian	Jiménez-Moreno, G.
3	Ruzhintsi	Bulgaria	) Kojumdjieva 1976; Palmarev & Ivanov 2001	Sarmatian	Jiménez-Moreno, G.
4	Catakbagyaka	Turkey	¡ Sickenberg et al. 1975; Heissig 1976	Langhian	Jiménez-Moreno, G.
5	Hinova	Romania	) Marinescu 1978 * Clauzon et al. 2005 * § ♣ Popescu et al. 2006a	Early Zanclean Bosphorian	Popescu, S.-M.
5	Husnicioara	Romania	¶ Ticleanu & Diaconita 1997 § ♣ Popescu et al. 2006b	Early Zanclean Dacian	Popescu, S.-M.
5	Valea Visenilor	Romania	¶ Ticleanu & Diaconita 1997 ♣ Popescu et al. 2006b	Early Zanclean Dacian	Popescu, S.-M.
6	Lupoaia	Romania	¡ Radulescu et al. 1997; Apostol & Enache 1979 § Radan & Radan 1998; Van Vugt 2001 ¶ Ticleanu & Diaconita 1997 ♣ Popescu S.-M. 2002, Popescu et al. 2006b	Zanclean Dacian-Romanian	Popescu, S.-M.
7	Ticleni 1	Romania	) ♣ Drivaliari et al. 1999	Zanclean Dacian-Romanin	Drivaliari, A. 1993
8	Ravno Pole	Bulgaria	) Drivaliari 1993	Pontian-Dacian	Drivaliari, A.
8	Lozenec	Bulgaria	¡ Gromolard & Guerin 1980; Thomas et al. 1986	Dacian	Drivaliari, A.
9	Sandanski	Bulgaria	¡ Kojumdgieva et al. 1982; Spassov N. (personal information)	Maeotian	Ivanov, D.
10	Lion of Amphipoli	Greece	* Melinte, M.C. (personal information)	Early Zanclean	Suc, J.-P.
11	Nestos 2	Greece	+ Drivaliari 1993	Messinian-Zanclean	Drivaliari, A.
12	Site 380 A	Black Sea	¶ Hsü 1978; Hsü & Giovanoli 1979; Letouzey et al. 1978 ♣ Popescu 2006	Late Miocene-Early Pliocene	Popescu, S.-M.
13	Aghios Vlassios	Greece	+ Spaak 1983; Drivaliari 1993	Early Pliocene	Drivaliari, A. 1993
14	Avadam	Turkey	¶ Robertson A.H.F. (personal information)	Early Zanclean	Suc, J.-P.
15	Lataquie	Syria	¶ Rubino J.-L. (personal information)	Early Zanclean	Suc J.-P.
16	Drenovets C-1	Bulgaria	+) ∧ Kojumdgieva et al. 1989	Sarmatian to Pontian	Ivanov, D.
17	Deleina C-12	Bulgaria	+) ∧ Kojumdgieva & Popov 1989	Badenian to Pannonian	Ivanov, D.
18	Makrilia	Greece	* Sachse et al. 1999	Tortonian	Sachse, M.
19	Tengelic-2	Hungary	*+) Nagyamarosi 1982; Bohn-Havas 1982; Korecz-Laky 1982	Burdigalian Ottnangian to Sarmatian	Jiménez-Moreno, G.

+ Foraminifera ∧ Ostracods ¡ Mammals) Bivalves § Palaeomagnetism ¶ Lithostratigraphy ♣ Climatostratigraphy * Nannoplankton accompanied by halophytes (Amaranthaceae–Chenopodiaceae, *Armeria*, etc.);

(2) a broad-leaved evergreen forest, from sea level to around 700 m altitude characterized by *Taxodium* or *Glyptostrobus, Myrica, Rhus,* Theaceae, Cyrillaceae–Clethraceae, *Bombax,* Euphorbiaceae, *Distylium, Castanopsis,* Sapotaceae, Rutaceae, *Mussaenda, Ilex, Hedera, Ligustrum, Jasminum,* Hamamelidaceae, *Engelhardia, Rhoiptelea,* etc.;

(3) an evergreen and deciduous mixed forest, above 700 m altitude; characterized by deciduous *Quercus, Engelhardia, Platycarya, Carya, Pterocarya, Fagus, Liquidambar, Parrotia, Carpinus, Celtis, Acer,* etc. Within this vegetation belt, riparian vegetation has been identified, composed of *Salix, Alnus, Carya, Carpinus, Zelkova, Ulmus, Liquidambar,* etc. The shrub level was dominated by Ericaceae, *Ilex,* Caprifoliaceae, etc.;

(4) above 1000 m, a mid-altitude deciduous and coniferous mixed forest with *Betula, Fagus, Cathaya, Cedrus, Tsuga.*

(5) above 1800 m altitude, a coniferous forest with *Abies* and *Picea.*

Vegetation dynamics

The following description of the Miocene and Pliocene vegetation dynamics in the southern Forecarpathian Basin and Greece–Turkey is a brief summary of the pollen analysis of Drivaliari (1993), Ivanov (1995), Drivaliari *et al.* (1999), Popescu (2001, 2006), Popescu *et al.* (2006*a, b*), Jiménez-Moreno *et al.* (2005) and Jiménez-Moreno (2005).

Burdigalian-Langhian (20.4–13.6 Ma). The regular occurrence and abundance of thermophilous species typical of the lowest altitudinal belts described above and the relative scarcity of altitudinal elements (Fig. 3) are characteristic for vegetation of this time. The coastal marine environment was then occupied by an impoverished *Avicennia* mangrove and several halophytes (Nagy & Kókay 1991; Nagy 1999; Plaziat *et al.* 2001; Jiménez-Moreno 2005). In the hinterland, lowlands were populated by a broad-leaved evergreen forest, characterized by *Alchornea,* Passifloraceae, *Pandanus, Rhus,* Theaceae, Cyrillaceae–Clethraceae, *Bombax,* Rubiaceae, Chloranthaceae, *Reevesia,* Euphorbiaceae, *Distylium, Castanopsis,* Sapotaceae, Rutaceae, *Mussaenda, Ilex, Hedera, Itea, Alangium,* cf. Mastixiaceae, *Ligustrum, Jasminum,* Hamamelidaceae, *Engelhardia, Rhoiptelea,* Schizaeaceae, Gleicheniaceae, etc. Within this vegetation belt, swamp forests were also well developed during this time period. Its components, such as *Taxodium* or *Glyptostrobus, Nyssa, Myrica, Planera,* show

comparatively high values in the pollen spectra. Probably the low elevation palaeogeography and very humid conditions at that time in the studied area favoured the wide distribution of swamp forests and of ecologically related riparian forests with *Platanus, Liquidambar, Zelkova, Carya, Pterocarya* and *Salix.*

An evergreen and deciduous mixed forest mainly composed of mesothermic elements such as *Quercus, Carya, Pterocarya, Fagus,* Ericaceae, *Ilex,* Caprifoliaceae, *Liquidambar, Parrotia, Carpinus, Celtis, Acer,* but also *Engelhardia, Platycarya,* etc., characterized areas of higher altitude. Within this vegetation belt, riparian vegetation has been identified, composed of *Salix, Alnus, Carya, Carpinus, Zelkova, Ulmus,* etc.

It should also be mentioned that conifer pollen, mainly *Pinus* and indeterminate Pinaceae, can be particularly abundant, presumably because of the capacity of saccate pollen for long-distance transport (Heusser 1988; Suc & Drivaliari 1991; Cambon *et al.* 1997; Beaudouin 2003): during the Badenian, the basin developed its largest extension so that the studied sections had the maximum distance from the coastline (Fig. 3). Mid- and high-altitude elements (*Tsuga, Cedrus, Abies* and *Picea*) and *Cathaya* seem not to vary significantly in sections of this age (Fig. 3).

Serravallian–Tortonian–Messinian (13.6–5.3 Ma). During this time-interval, important changes in the vegetation are observed: *Avicennia,* which populated the coastal areas in previous times, is not found commonly and several megathermic elements (*Buxus bahamensis* group, *Alchornea, Bombax,* Iacacinaceae, *Croton,* Melastomataceae, etc.), typical from the broad-leaved evergreen forest, became rare and most of them disappeared (Fig. 3). The evergreen–deciduous mixed forest suffered a great transformation due to the loss and decrease in the abundance of several mega-mesothermic evergreen plants. This kind of vegetation was progressively enriched by deciduous mesothermic plants, such as deciduous *Quercus,* and *Fagus, Alnus, Acer, Eucommia, Betula, Alnus, Carpinus, Ulmus, Zelkova, Tilia,* etc. Thus, the vegetation shows a tendency towards increasing proportions of mesothermic deciduous elements coming from higher altitudes.

Even if the thermophilous elements decreased during this period, the swamp forest continued to be well developed. At the same time, the vegetation from mid- (*Cathaya, Tsuga* and *Cedrus*) and high-altitude (*Picea* and *Abies*) belts clearly strengthened. For instance, *Tsuga* (mid-altitude indicator) is absent in the Badenian (Langhian and Early Serravallian) or very rare, it is still rare at the base of the Volhynian (approx. 12.7 Ma), but reaches

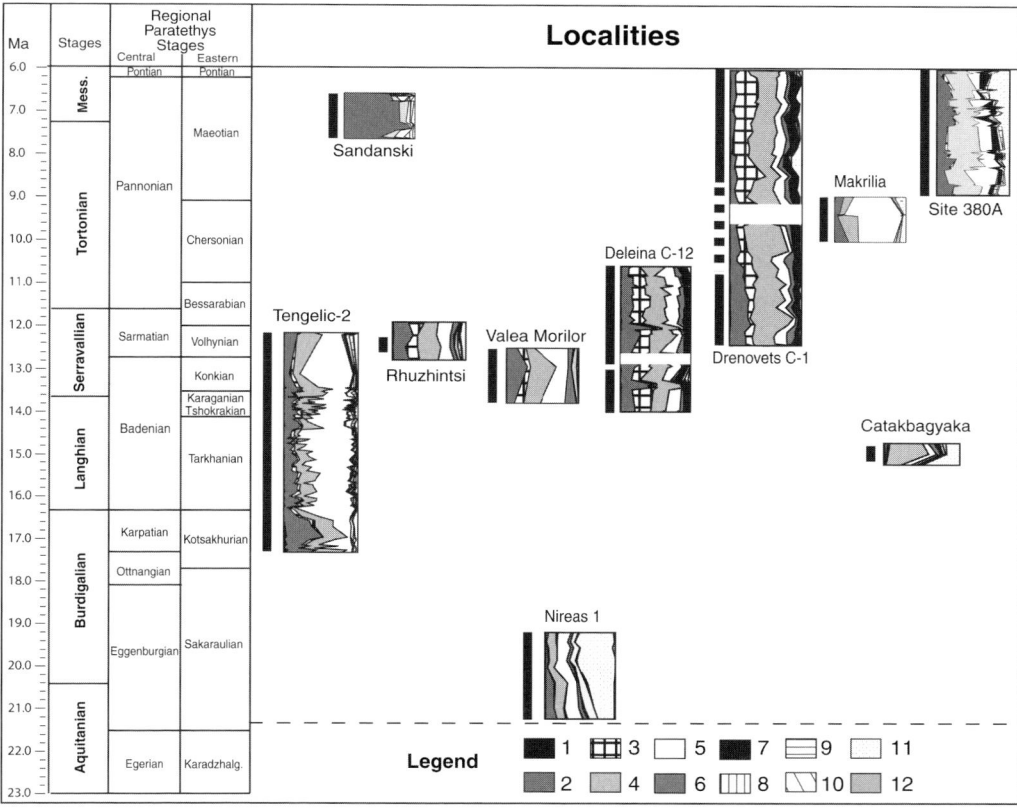

Fig. 3. Synthetic pollen diagrams of the sections spanning the Miocene until 6 Ma. Taxa have been grouped according to their ecological significance as follows: 1 Megathermic (= tropical) elements (*Avicennia, Amanoa, Alchornea, Fothergilla, Exbucklandia,* Euphorbiaceae, Sapindaceae, Loranthaceae, Arecaceae, Acanthaceae, *Canthium* type, Passifloraceae, etc.). 2 Mega-mesothermic (= subtropical) elements (Taxodiaceae, *Engelhardia, Platycarya, Myrica,* Sapotaceae, *Microtropis fallax, Symplocos, Rhoiptelea, Distylium* cf. *sinensis, Embolanthera, Hamamelis,* Cyrillaceae–Clethraceae, Araliaceae, *Nyssa, Liriodendron,* etc.). 3 *Cathaya,* an altitudinal conifer living today in Southern China. 4 Mesothermic (= warm-temperate) elements (deciduous *Quercus, Carya, Pterocarya, Carpinus, Juglans, Celtis, Zekkova, Ulmus, Tilia, Acer, Parrotia* cf. *persica, Liquidambar, Alnus, Salix, Populus, Fraxinus, Buxus sempervirens* type, *Betula, Fagus, Ostrya, Parthenocissus* cf. *henryana, Hedera, Lonicera, Elaeagnus, Ilex, Tilia,* etc.). 5 *Pinus* and poorly preserved Pinaceae pollen grains. 6 Meso-microthermic (= mid-altitude) trees (*Tsuga, Cedrus*). 7 Microthermic (= high-altitude) trees (*Abies, Picea*). 8 Non-significant pollen grains (undetermined ones, poorly preserved pollen grains, some cosmopolitan or widely distributed elements such as Rosaceae and Ranunculaceae). 9 Cupressaceae. 10 Mediterranean xerophytes (*Quercus ilex* type, *Carpinus* cf. *orientalis, Olea, Phillyrea, Ligustrum, Pistacia, Ziziphus, Cistus,* etc.). 11 Herbs (Poaceae, *Erodium, Geranium, Convolvulus,* Asteraceae Asteroideae, Asteraceae Cichorioideae, Lamiaceae, *Plantago, Euphorbia,* Brassicaceae, Apiaceae, *Knautia, Helianthemum, Rumex, Polygonum, Asphodelus,* Campanulaceae, Ericaceae, Amaranthaceae–Chenopodiaceae, Caryophyllaceae, Plumbaginaceae, Cyperaceae, *Potamogeton, Sparganium, Typha,* Nympheaceae, etc.) including some subdesertic elements (*Lygeum, Neurada, Nitraria, Calligonum*). 12 Steppe elements (*Artemisia, Ephedra*).

up to 10% in the middle and upper part of the Volhynian (Fig. 3). This palaeofloristic change occurs slowly and gradually without major fluctuations. A similar vegetation change is observed during the same time-interval in other areas of Europe (e.g. Spain, southern France, Switzerland and Austria: Bessedik 1985; Jiménez-Moreno 2005).

The herbs (mainly Poaceae, Amaranthaceae–Chenopodiaceae, *Artemisia,* Caryophyllaceae, Polygalaceae, Lamiaceae, Asteraceae Asteroideae and Asteraceae Cichorioideae) also became more abundant (Fig. 3). This may be due to a somewhat drier climate during that time as is also indicated by macrofloras of the same area (Palamarev 1991; Palamarev &

Ivanov 2004) and confirmed by sedimentological data (Koleva-Rekalova 1994; Ivanov & Koleva-Rekalova 1999): in Bessarabian to Chersonian sediments (12–9.1 Ma) of northeast Bulgaria, aragonite sediments occur which are assumed to have been formed under a seasonally dry climate. This trend continued during the Late Miocene. Presumably, open landscapes covered by more xerophytic herbaceous communities existed during that time.

Pliocene (5.3–c. 3.2 Ma). The vegetation was then characterized by a mosaic of different plant associations inherited from the Miocene. The same vegetation dynamics marked by disappearance of thermophilous plants and increase in mesothermic and micro-mesothermic plants continued. Some of the coastal areas of this region were still inhabited by *Avicennia* mangrove (*Avicennia* pollen at Site 380A at 781.63 m) and several megathermic elements typical from the broad-leaved evergreen forest occupying the lowlands, such as *Amanoa*, *Pachysandra*, *Entada*, Meliaceae, Mimosaceae, Sapindaceae, Tiliaceae, Euphorbiaceae, Acanthaceae and *Fothergilla*, are sporadically present. They disappeared during the early Pliocene (between 4–3.5 Ma) (Popescu 2001). The mega-mesothermic plants, belonging to these plant associations, such as *Engelhardia*, *Microtropis*, *Distylium*, *Parthenocissus*, Sapotaceae, Arecaceae, etc., are still abundant and persisted through the Pliocene (Fig. 4). Swampy (mainly *Taxodium* or *Glyptostrobus*, *Nyssa*, *Myrica*) and marshy (Cyperaceae, Poaceae, Cyrillaceae–Clethraceae, *Myrica*) elements, populating deltaic areas, were very abundant. Trees from the family Taxodiaceae did not disappear from this area until the middle Pleistocene (Mamatsashvili 1975).

The mixed deciduous forest (mainly made up of conifers like *Pinus*, and several deciduous tress such as *Quercus*, *Acer*, *Carpinus*, *Parrotia*, *Carya*, *Pterocarya*, *Liquidambar*, *Platanus*, *Tilia*, *Ulmus*, *Zelkova*, etc.), situated at higher altitude, as well as the trees belonging to the highest altitudinal belts, become more abundant during this period (*Cathaya*, *Cedrus*, *Tsuga*, *Picea* and *Abies*) (increasing percentages of these elements are compared on Fig 3 and 4).

Another important fact that makes a difference between the Pliocene and the Miocene is the strong development of the steppe with *Artemisia* in the Ponto-Euxinian region since the early Pliocene (Site 380A, Fig. 4).

Climatic evolution: regional *vs.* global climatic change

The high presence of mega- and mega-mesothermic elements during the Early and early Mid-Miocene suggests the existence of a warm, subtropical climate and a tendency towards slightly cooler conditions in the late Mid-Miocene. Climate was also quite humid, to support the development of such a large association of thermic elements (of present-day 'Asiatic' affiliation and climate) which require very humid conditions all year (Wang 1961). The major change is the impoverishment in plant diversity produced by the disappearance of the most thermophilous plants and the consequent enrichment in mesothermic plants (mainly deciduous *Quercus*, *Alnus*, etc.) and high-elevation conifers, from the Serravallian to the Pliocene.

The floral assemblages during the Early and early Mid-Miocene clearly reflect the Miocene Climatic Optimum (MCO: Zachos *et al.* 2001; Shevenell *et al.* 2004) well-recorded at Tengelic-2 (Jiménez-Moreno *et al.* 2005). The major change registered in plant diversity is related to a gradual decrease in temperature and precipitation after the MCO (Ivanov *et al.* 2002; Jiménez-Moreno *et al.* 2005). This fact is well documented on a worldwide scale and has been correlated with the general decrease in temperature observed by several authors as a gradual increase in the isotopic $\delta^{18}O$ values of foraminifera from deep-sea sediments (DSDP Sites 608: Miller *et al.* (1991) and 588: Zachos *et al.* (2001)) during this timespan and related to an increase in the size of the EAIS (East Antarctic Ice Sheet) (Zachos *et al.* 2001) (Fig. 2). The isotopic values also indicate that this cooling continued during the Late Miocene and Pliocene (Zachos *et al.* 2001) (Fig. 2).

High-elevation conifers seem not to vary along the sections of early and early Mid-Miocene; however, these elements are abundant in the samples and indicate that the surrounding mountains were already significantly uplifted. Mid- (including *Cathaya*) and high-elevation conifers clearly increase during the late Mid-Miocene and Late Miocene. This can be observed in the boreholes Deleina C-12 and Drenovets C-1 (Fig. 3).

In addition, an augmentation in herbs, mainly *Artemisia*, Amaranthaceae–Chenopodiaceae, Poaceae, Asteraceae, etc., during the Late Miocene and Pliocene, indicates more open vegetation, and drier conditions. Supporting this interpretation is the substitution of thermophilous elements with high humidity requirements all year (Asiatic-like vegetation) by mesothermic (mainly deciduous) elements which can survive under seasonal climate with respect to the precipitation (Popescu 2001; Ivanov *et al.* 2002; Jiménez-Moreno *et al.* 2005).

The noticeable increase in mesothermic plants and high-elevation conifers can be interpreted as a result of climate cooling, or by uplift of surrounding mountains (Kuhlemann & Kempf 2002). In both situations, altitudinal elements would increase.

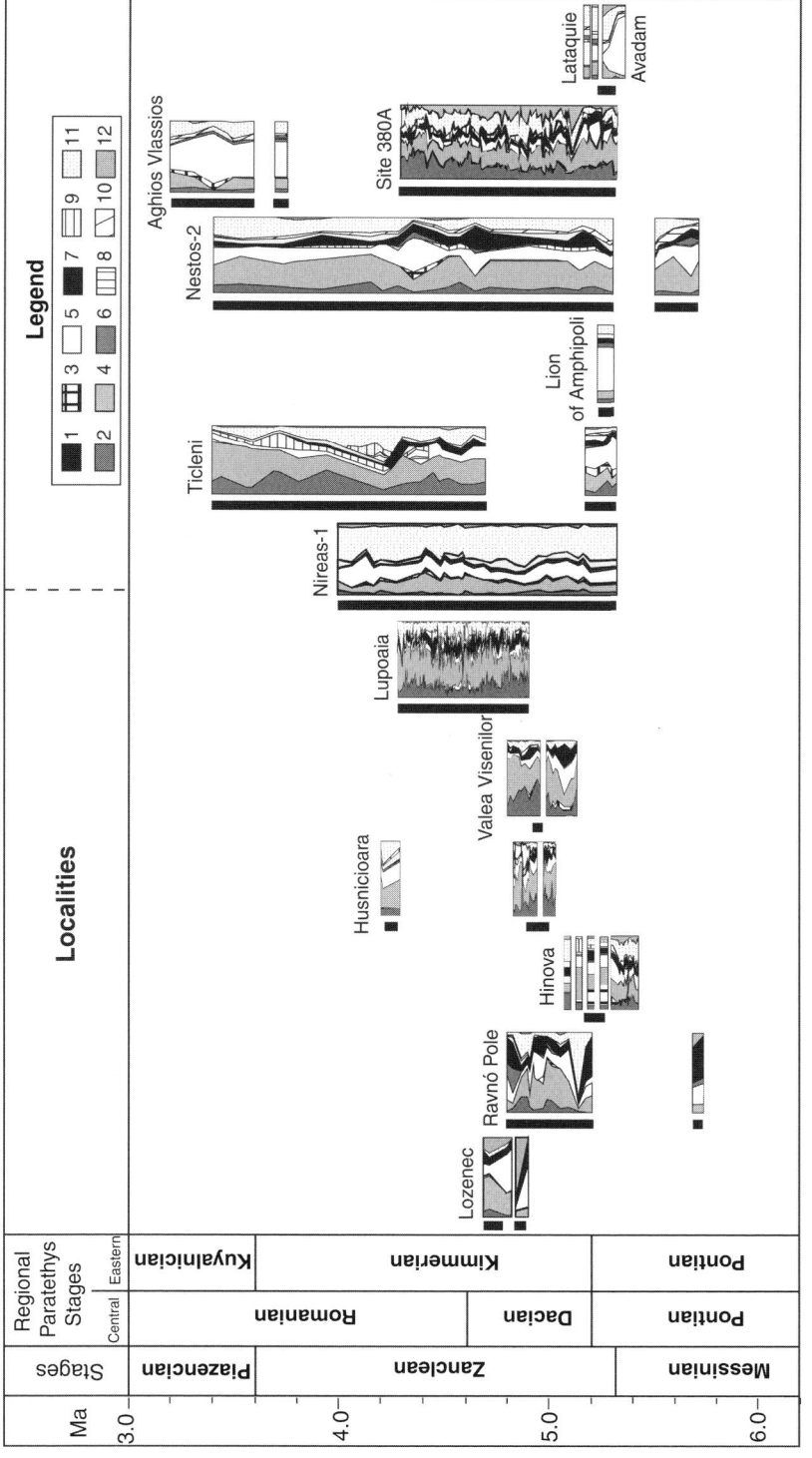

Fig. 4. Synthetic pollen diagrams of the studied sections spanning the Late Miocene (from 6 Ma) and Pliocene. For legend of plant groups, see Figure 3.

It is quite difficult to separate one process from another (global climatic forcing vs. the regional one), due to the tectonic situation of the studied area and the fact that they may have interfered. However, the vanishing of several thermophilous plants, which lived at low elevations and thus were not affected by the regional uplift, and the climate reconstructions using mainly taxa growing at low to middle–low altitude confirm a decrease in mean annual temperatures (Ivanov *et al.* 2002; Jiménez-Moreno *et al.* 2005; Mosbrugger *et al.* 2005). Then, it is clear that even if the uplift of the surrounding mountains may have influenced the regional climate, the evolution of the vegetation during both the Miocene and Pliocene was very dependent on the global climatic signal as shown in previous studies (Popescu 2001, 2002; Ivanov *et al.* 2002; Jiménez-Moreno *et al.* 2005). Hence, according also to the rapid nature of the recorded change in vegetation, we consider that global cooling was the most efficient forcing.

The origin of the steppe with Artemisia

Open herbaceous formations in the southern Mediterranean area are known since the Burdigalian (Suc *et al.* 1995*a, b*; Bachiri Taoufiq *et al.* 2001; Jiménez-Moreno 2005; Jiménez-Moreno & Suc in press). They were already well-developed during the Zanclean in other regions of the Mediterranean area (Suc *et al.* 1999) but were relatively poor in *Artemisia*. It is at the end of the Pliocene, as the climate got cooler and glacial–interglacial cycles appeared in the Northern Hemisphere, when the steppes with *Artemisia* became of significant importance (Suc *et al.* 1995*b*) during the glacial periods (Suc & Cravatte 1982; Combourieu-Nebout & Vergnaud Grazzini 1991; Beaudouin 2003) and even during interglacials (Subally *et al.* 1999) because of the ambivalent significance of *Artemisia* from the temperature viewpoint (cold *vs.* warm species: Subally & Quézel 2002).

The presence of steppe vegetation with *Artemisia* in the Ponto-Euxinian region (i.e. in Anatolia according to Site 380A pollen record; Popescu 2001, 2006) in the Late Miocene and their significant strengthening in the Early Pliocene is very informative. Their early presence and development in this region, contrary to the extreme scarcity of *Artemisia* in the Moroccan steppes in the Late Miocene and Early to Middle Pliocene (Bachiri Taoufiq 2000; Suc *et al.* 1999), indicates that Anatolia and neighbouring areas could have been the source area of this kind of vegetation for the rest of the Mediterranean region, a style of vegetation that became very abundant during the cold periods of the Quaternary (Popescu 2001; Suc & Popescu

2005). The early settlement and then development of *Artemisia* steppe vegetation in Anatolia may have resulted from migration from the east of this genus as a consequence of uplift of the Tibetan Plateau (where *Artemisia* species are still abundant today) and the succeeding reinforced Asiatic monsoon (Zhisheng *et al.* 2001).

Conclusions

Pollen data show a progressive reduction in the most thermophilous and high-water requirement plants typical of a broad-leaved evergreen forest and, in contrast, an increase in seasonal-adapted plants coming from higher altitude belts, including mesothermic (mainly deciduous) elements, altitudinal trees and herbs, during the Middle–Late Miocene and Pliocene. This has been interpreted as the response of the vegetation to global climate cooling, accentuated by the regional uplift of the surrounding mountains during Alpine tectonics. This process may also have been favoured by progressive movement of Eurasia towards northern latitudes.

The appearance of steppe vegetation with *Artemisia* on the Anatolian Plateau since the Late Miocene and its development in the Early Pliocene, significantly earlier than in the rest of Southern Europe, is informative. This suggests that the Anatolian *Artemisia*-rich steppes could have been the source area of this kind of open vegetal formation for the rest of the Mediterranean area during the Quaternary.

This paper is a contribution to the French Programme 'Environnement, Vie et Sociétés' (Institut Français de la Biodiversité). The authors thank J. Agustí and M. Harzhauser for their helpful reviews, and the EEDEN Programme (ESF) for invitations to participate in international workshops about the subject. Nurdan Yavuz-Isik is thanked for providing the Miocene samples from Turkey.

References

AKGÜN, F. & AKYOL, E. 1999. Palynostratigraphy of the coal-bearing Neogene deposits graben in Büyük Menderes Western Anatolia. *Géobios*, **32**, 367–383.

APOSTOL, L. & ENACHE, C. 1979. Etude de l'espèce *Dicerorhinus megarhinus* (de Christol) du bassin carbonifère de Motru. *Travaux du Musée d'Histoire Naturelle "Grigore Antipa"*, Bucarest, **20**, 533–540.

BACHIRI TAOUFIQ, N. 2000. Les environnements marins et continentaux du corridor rifain au Miocène supérieur d'après la palynologie. PhD thesis, Université Hassan II – Mohammedia, Casablanca (Morocco).

BACHIRI TAOUFIQ, N., BARHOUN, N., SUC, J.-P., MEON, H., ELAOUAD, Z. & BENBOUZIANE, A.

2001. Environment, végétation et climat du Messinien au Maroc. *Paleontologia i Evolució,* **32–33,** 127–138.

BEAUDOUIN, C. 2003. Effets du dernier cycle climatique sur la vegetation de la basse vallée du Rhône et sur la sédimentation de la plate-forme du golfe du Lion d'après la palynologie. PhD thesis, Université Claude Bernard Lyon-1, France, 403pp.

BENDA, L. 1971. Grundzüge einer pollenanalytischen Gliederung des türkischen Jungtertiärs. *Beihefte zum Geologischen Jahrbuch,* **113,** 46pp.

BENDA, L., HEISSIG, K. & STEFFENS, P. 1975. Die Stellung der vertebraten-faunengruppen der Türkei innerhalb der chronostratigraphischen systeme von Tethys und Paratethys. *Geologische Jahrbuch, B,* **15,** 109–116.

BESSEDIK, M. 1985. Reconstitution des énvironnéments Miocènes de régions nord-ouest méditerranéennes à partir de la palynologie. PhD thesis, Université de Montpellier, France, 162pp.

BOHN-HAVAS, M. 1982. Mollusca fauna of Badenian and Sarmatian stage from the borehole Tengelic 2. *In:* NAGY, E., BODOR, E., HAGYAMAROSI, A. *ET AL.* (eds) *Palaeontological examination of the geological log of the borehole Tengelic 2.* Annales Instituti Geologici Publici Hungarici, **65,** 200–203.

CAMBON, G., SUC, J.-P., ALOISI, J.-C. *ET AL.* 1997. Modern pollen deposition in the Rhône delta area (lagoonal and marine sediments) France. *Grana,* **36,** 105–113.

CLAUZON, G., SUC, J.-P., POPESCU, S.-M., MARUNTEANU, M., RUBINO, J.-L., MARINESCU, F. & MELINTE, M. C. 2005. Influence of the Mediterranean sea-level changes over the Dacic Basin (Eastern Paratethys) in the Late Neogene. *Basin Research,* **17,** 437–462.

COMBOURIEU-NEBOUT, N. & VERGNAUD GRAZZINI, C. 1991. Late Pliocene Northern hemisphere glaciation: the continental and marine responses in Central Mediterranean. *Quaternary Science Reviews,* **10,** 319–334.

DRIVALIARI, A. 1993. Images polliniques et paléoenvironnement au Néogène supérieur en Méditerranée orientale. Aspects climatiques et paléogéographiques d'un transect latitudinal (de la Roumanie au Delta du Nil). PhD thesis, Université Montpellier-2, France, 333pp.

DRIVALIARI, A., ŢICLEANU, N., MARINESCU, F., MĂRUNŢEANU, M. & SUC, J.-P. 1999. A Pliocene climatic record at Ticleni (Southwestern Romania). *In:* WRENN, J. H., SUC, J.-P. & LEROY, S. A. G. (eds) *The Pliocene: Time of Change.* American Association of Stratigraphic Palynologists Foundation, Dallas, 103–108.

FORNACIARI, E. & RIO, D. 1996. Latest Oligocene to early Middle Miocene quantitative calcareous nanno-fossil biostratigraphy in the Mediterranean region. *Micropaleontology,* **42,** 1–36.

FORNACIARI, E., DI STEFANO, A., RIO, D. & NEGRI, A. 1996. Middle Miocene quantitative calcareous nanno-fossil biostratigraphy in the Mediterranean region. *Micropaleontology,* **42,** 37–63.

GONCHAROVA, I. A., SHCHERBA, I. G., KHONDKARIAN, S. O. *ET AL.* 2004. Lithological-Paleogeographic

maps of Paratethys. Map 5: Early Middle Miocene. *Courier Forschungsinstitut Senckenberg,* **250,** 19–21.

GRADSTEIN, F. M. & OGG, J. G. 2004. Geologic time scale 2004 – why, how, and where next! *Lethaia,* **37,** 175–181.

GRADSTEIN, F. M., OGG, J. G., SMITH, A. G., BLEEKER, W. & LOURENS, L. J. 2004. A new geologic time scale with special reference to Precambrian and Neogene. *Episodes,* **27,** 83–100.

GROMOLARD, C. & GUERIN, C. 1980. Mise au point sur *Parabos cordieri* (de Crystol), un Bovidé (Mammalia, Artiodactyla) du Pliocène d'Europe Occidentale. *Géobios,* **13,** 741–755.

HARZHAUSER, M., MANDIC, O. & ZUSCHIN, M. 2003. Changes in Parathethyan marine molluscs at the Early/Middle Miocene transition: diversity, palaeogeography and palaeoclimate. *Acta Geologica Polonica,* **53,** 323–339.

HARZHAUSER, M. & PILLER, W. E. 2004. The Early Sarmatian – hidden seesaw changes. *Courier Forschungsinstitut Senckenberg,* **246,** 89–112.

HARZHAUSER, M. & PILLER, W. E. in press. Benchmark data of a changing sea – palaeogeography, palaeobiography and events in the Central Paratethys during the Miocene. *Palaeogeography, Palaeoclimatology, Palaeoecology*

HEISSIG, K. 1976. Rhinocerotidae (Mammalia) aus der Anchitherium-Fauna Anatoliens. *Geologisches Jahrbuch Reihe B,* **19,** 121pp.

HEUSSER, L. 1988. Pollen distribution in marine sediments on the continental margin of Northern California. *Marine Geology,* **80,** 131–147.

HSÜ, K. J. 1978. Correlation of Black Sea sequences. *In:* ROSS, D. A., NEPROCHNOV, Y. P. *ET AL.* (eds) *Initial Reports of the Deep Sea Drilling Project,* US Government Printing Office, **42,** 489–497.

HSÜ, K. J. & GIOVANOLI, F. 1979. Messinian event in the Black Sea. *Palaeogeography, Palaeoclimatology, Palaeoecology,* **29,** 75–94.

ILYINA, L. B., SHCHERBA, I. G., KHONDKARIAN, S. O. *ET AL.* 2004. Lithological-Paleogeographic maps of Paratethys. Map 6: Mid Middle Miocene. *Courier Forschungsinstitut Senckenberg,* **250,** 23–25.

IVANOV, D. 1995. Palynological investigations of Miocene sediments from North-West Bulgaria (in Bulgarian, English abstract). PhD thesis, Institute of Botany BAS, Sofia, 45pp.

IVANOV, D. A. & KOLEVA-REKALOVA, E. 1999. Palynological and sedimentological data about Lake Sarmatian palaeoclimatic changes in the Forecarpathian and Euxinian basins (Northern Bulgaria). *Acta Paleobotanica, Supplement,* **2,** 307–313.

IVANOV, D., ASHRAF, A. R., MOSBRUGGER, V. & PALAMAREV, E. 2002. Palynological evidence for Miocene climate change in the Forecarpathian Basin (Central Paratethys, NW Bulgaria). *Palaeogeography, Palaeoclimatology, Palaeoecology,* **178,** 19–37.

JIMÉNEZ-MORENO, G. 2005. Utilización del análisis polínico para la reconstrucción de la vegetación, clima y estimación de paleoaltitudes a lo largo de arco alpino europeo durante el Mioceno (21-8 m.a.). PhD thesis, Univ. Granada and Univ. C. Bernard – Lyon 1, 318pp.

JIMÉNEZ-MORENO, G., RODRÍGUEZ-TOVAR, F. J., PARDO-IGÚZQUIZA, E., FAUQUETTE, S., SUC, J.-P.

& MÜLLER, P. 2005. High resolution palynological analysis in the late early-middle Miocene core from the Pannonian Basin, Hungary: climatic changes, astronomical forcing and eustatic fluctuations in the Central Paratethys. *Palaeogeography, Palaeoclimatology, Palaeoecology*, **216**, 73–97.

JIMÉNEZ-MORENO, G. & SUC, J.-P. in press. Middle Miocene Latitudinal Climatic Gradient in Western Europe: Evidence from Pollen Records. *Palaeogeography, Palaeoclimatology, Palaeoecology*.

KOJUMDGIEVA, E. 1976. Paléoecologie des communautés des mollusques du Miocène en Bulgarie du Nord-Ouest. III. Communautés des mollusques du Volhynien (Sarmatien inférieur). *Geologica Balcanica*, **6**, 53–63.

KOJUMDGIEVA, E. & POPOV, N. 1989. Paléogéographie et évolution géodynamique de la Bulgarie Septentrionale au Néogene. *Geologica Balcanica*, **19**, 73–92.

KOJUMDGIEVA, E., NIKOLOV, I., NEDJALKOV, P. & BUSEV, A. 1982. Stratigraphy of the Neogene in the Sandanski Graben. *Geologica Balcanica*, **12**, 69–81.

KOJUMDJIEVA, E., POPOV, N., STANCHEVA, M. & DARAKCHIEVA, S. 1989. Correlation of the biostratigraphic subdivision of the Neogene in Bulgaria after molluscs, foraminifers and ostracods. *Geologica Balcanica*, **19**, 9–22.

KOLEVA-REKALOVA, E. 1994. Sarmatian aragonite sediments in North-eastern Bulgaria – origin and diagenesis. *Geologica Balcanica*, **25**, 47–64.

KORECZ-LAKY, I. 1982. Miocene foraminifera fauna from the borehole Tengelic 2. *In:* NAGY, E., BODOR, E., HAGYAMAROSI, A. *ET AL.* (eds) *Palaeontological examination of the geological log of the borehole Tengelic 2*, Annales Instituti Geologici Publici Hungarici, **65**, 186–187.

KOVAC, M., BARATH, I., HARZHAUSER, M., HLAVATY, I. & HUDACKOVA, N. 2004. Miocene depositional systems and sequence stratigraphy of the Vienna Basin. *Courier Forschungsinstitut Senckenberg*, **246**, 187–212.

KUHLEMANN, J. & KEMPF, O. 2002. Post-Eocene evolution of the North Alpine Foreland Basin and its response to Alpine tectonics. *Sedimentary Geology*, **152**, 45–78.

LETOUZEY, J., GONNARD, R., MONTADERT, L., KRISTCHEV, K. & DORKEL, A. 1978. Black Sea: Geological setting and recent deposits distribution from seismic reflection data. *In:* ROSS, D. A., NEPROCHNOV, Y. P. *ET AL.* (eds) *Initial Reports of the Deep Sea Drilling Project*, US Government Printing Office, **42**, 1077–1084.

LOURENS, L., HILGEN, F., SHACKLETON, N. J., LASKAR, J. & WILSON, D. 2004. The Neogene Period. *In:* GRADSTEIN, F., OGG, J. & SMITH, A. (eds) *Geologic Time Scale 2004*. Cambridge University Press.

MAMATSASHVILI, N. S. 1975. The palynological characteristics of the Kolkhida Quaternary continental deposits (The Georgian SSR). *Metsniereba*, Tbilisi, 114pp.

MARINESCU, F. 1978. Stratigrafia Neogenului superior din sectorul vestic al Bazinului Dacic. Editura Academiei Republicii Socialista România (in Romanian), 155pp.

MĂRUNŢEANU, M. & PAPAIANOPOL, I. 1995. L'association de nannoplancton dans les dépôts romaniens situés entre les vallées de Cosmina et de Cricovu Dulce (Munténie, bassin dacique, Roumanie). *Roman nian Journal of Paleontology*, **76**, 169–170.

MĂRUNŢEANU, M. & PAPAIANOPOL, I. 1998. Mediterranean calcareous nannoplankton in the Dacic Basin. *Romanian Journal of Stratigraphy*, **78**, 115–121.

MEULENKAMP, J. E. & SISSINGH, W. 2003. Tertiary palaeogeography and tectonostratigraphic evolution of the Northern and Southern Peri-Tethys platforms and the intermediate domains of the African-Eurasian convergent plate boundary zone. *Palaeogeography, Palaeoclimatology, Palaeoecology*, **196**, 209–228.

MILLER, K. G., FEIGENSON, M., WRIGHT, J. D. & CLEMENT, B. 1991. Miocene isotope reference section, Deep Sea Drilling Project Site 608: an evaluation of isotope and biostratigraphic resolution. *Palaeoceanography*, **6**, 33–52.

MOSBRUGGER, V., UTESCHER, T. & DILCHER, D. L. 2005. Cenozoic continental climatic evolution of Central Europe. *PNAS*, **102**, 14964–14969.

NAGY, E. 1991. Climatic changes in the Hungarian Neogene. *Review of Palaeobotany and Palynology*, **65**, 71–74.

NAGY, E. 1992. Magyarorszag Neogen sporomorphainak ertekelese. *Geologica Hungarica*, **53**, 1–379.

NAGY, E. 1999. *Palynological correlation of the Neogene of the Central Paratethys*. Geological Institute of Hungary, Budapest, 149pp.

NAGY, E. & KÓKAY, J. 1991. Middle Miocene mangrove vegetation in Hungary. *Acta geologica Hungarica*, **34**, 45–52.

NAGYMAROSI, A. 1982. Badenian-Sarmatian nannoflora from the borehole Tengelic 2. *In:* NAGY, E., BODOR, E., HAGYAMAROSI, A. *ET AL.* (eds) *Palaeontological examination of the geological log of the borehole Tengelic 2*, Annales Instituti Geologici Publici Hungarici, **65**, 145–149.

PALAMAREV, E. 1991. Composition, structure and main stages in the evolution of Miocene paleoflora in Bulgaria. Dsc thesis, (in Bulgarian). BAS, Sofia, 60pp.

PALAMAREV, E. & IVANOV, D. 2001. Charakterzüge der vegetation des Sarmatien (Mittel- bis Obermiozän im südlichen Teil des Dazischen Beckens (Südost Europa). *Palaeontographica*, **B259**, 209–220.

PALAMAREV, E. & IVANOV, D. 2004. Badenian vegetation of Bulgaria: biodiversity, palaeoecology and palaeoclimate. *Courier Forschungsinstitut Senckenberg*, **249**, 63–69.

PAPAIANOPOL, I. & MĂRUNŢEANU, M. 1993. Biostratigrapy (molluscs and calcareous nannoplankton) of the Sarmatian and Meotian in eastern Muntenia (dacic basin-Rumania), *Zemni plyn a nafta*, **38**, 9–15.

PAPAIANOPOL, I. & MARINESCU, F. 1995. Lithostratigraphy and age of Neogene deposits on the Moesian Platform, between Olt and Danube Rivers. *Romanian Jornal of Stratigraphy*, **76**, 67–70.

PAPAIANOPOL, I. & MOTAS, I. C. 1978. Marqueurs biostratigraphiques pour dépôt post-chersoniens du Bassin Dacique. *Dari de Seama ale Institutului de Geologie si Geofizica*, Stratigrafie, **64**, 283–294.

PAPAIANOPOL, I., JIPA, D., MARINESCU, F., ŢICLEANU, N. & MACALET, R. 1995. Upper Neogene from the

Dacic Basin – Guide to excursion B2 (post-congress) X congress RCMNS, Bucuresti. *Romanian Journal of Stratigraphy*, **76**, 1–43.

PARAMONOVA, N. P., SHCHERBA, I. G. & KHONDAKARIAN, S. O. 2004. Lithological-Paleogeographic maps of Paratethys. Map 7: Late Middle Miocene (Late Serravallian, Sarmatian s.s., Middle Sarmatian s.l.). *Courier Forschungsinstitut Senckenberg*, **250**, 27–31.

PETRESCU, I. & MALAN, L. 1992. *Contributions to the knowledge of Upper Neogene microflora East of Turnu-Severin (Summary)*. Univ. Babes-Bolyai, Cluj-Napoca Gradina Botanica, Contributii Botanice 1991–1992: 135–143.

PETRESCU, I., CERNITA, P., MEILESCU, C. ET AL. 1989*a*. Preliminary approaches to the palynology of the Lower Pliocene (Dacian) deposits in the Husnicioara area (Mehedinti county, SW Romania). *Studia Universitatis Babes-Bolyai Geologia-Geografia*, **34**, 67–74.

PETRESCU, I., NICA, T., FILIPESCU, S. ET AL. 1989*b*. Paleoclimatical significance of the palynlogical approach to the Pliocene deposits of Lupoaia (Gorj county). *Studia Univeritatis Babes-Bolyai Geologia-Geografia*, **34**, 75–81.

PLANDEROVÁ, E. 1990. *Miocene microflora of slovak Central Paratethys and its biostratigraphical significance*. Dionyz Stur Institute of Geology, Bratislava (Slovakia), 143pp.

PLAZIAT, J.-C., CAVAGNETTO, C., KOENIGUER, J.-C. & BALTZER, F. 2001. History and biogeography of the mangrove ecosystem, based on a critical reassessment of the paleontological record. *Wetlands Ecology and Management*, **9**, 161–179.

POPESCU, S.-M. 2001. Végétation, climat et cyclostratigraphie en Paratéthys centrale au Miocène supérieur et au Pliocène inférieur d'après la palynologie. PhD thesis. Université Claude Bernard Lyon-1, Lyon, France.

POPESCU, S.-M. 2002. Repetitive changes in Early Pliocene vegetation revealed by high-resolution pollen analysis: revised cyclostratigraphy of southwestern Romania. *Review of Palaeobotany and Palynology*, **120**, 181–202.

POPESCU, S.-M. 2006. Upper Miocene and Lower Pliocene environments in the southwestern Black Sea region from high-resolution palynology of DSDP site 380A (Leg 42B). *Palaeogeography, Palaeoclimatology, Palaeoecology*, **238**, 64–77.

POPESCU, S.-M., KRIJGSMAN, W., SUC, J.-P., CLAUZON, G., MARUNTEANU, M. & NICA, T. 2006*a*. Pollen record and integrated high-resolution chronology of the early-Pliocene Dacic Basin (Southwestern Romania). *Palaeogeography, Palaeoclimatology, Palaeoecology*, **238**, 78–90.

POPESCU, S.-M., SUC, J.-P. & LOUTRE, M.-F. 2006*b*. Early Pliocene vegetation changes forced by eccentricity-precession. Example from Southwestern Romania. *Palaeogeography, Palaeoclimatology, Palaeoecology*, **238**, 340–348.

POPOV, S. V., RÖGL, F., ROZANOV, A. Y., STEININGER, F. F., SHCHERBA, I. G. & KOVAC, M. (eds) 2004. Lithological-Paleogeographic maps of Paratethys. 10 maps Late Eocene to Pliocene. *Courier Forschungsinstitut Senckenberg*, **250**, 1–46.

QUÉZEL, P. & MÉDAIL, F. 2003. Ecologie et biogéographie des forêts du bassin méditerranéen, Elsevier France, 571pp.

RADAN, S. C. & RADAN, M. 1998. Study of the geomagnetic field structure in the Tertiary in the context of magnetostratigraphic scale elaboration. I – The Pliocene. *An. Inst. Geol. Rom.*, **70**, 215–231.

RADULESCU, C., SAMSON, P.-M., SEN, S., STIUCA, E. & HOROI, V. 1997. Les micromammifères pliocènes de Deanic (bassin Dacique, Roumanie). *In*: AGUILAR, J.-P., LEGENDRE, S. & MICHAUX, J. (eds) *BiochroM'97*, Mémoires Travaux E.P.H.E., Institute Montpellier, **21**, 635–647.

RÖGL, V. F. 1998. Palaeogeographic considerations for Mediterranean and Paratethys seaways (Oligocene to Miocene), *Annalen des Naturhistorischen Museums in Wien*, **99A**, 279–310.

SACHSE, M., MOHR, B. & SUC, J.-P. 1999. The Makrilaflora (Crete, Greece) – a contribution to the Neogene history of the climate and vegetation of the Eastern Mediterranean. *Acta palaeobotanica*, **Supplement 2**, 365–372.

SEMENENKO, V. N. & OLEJNIK, E. S. 1995. Stratigraphic correlation of the Eastern Paratethys Kimmerian and Dacian stages by molluscs, dinocyst and nannoplankton data. *Rom. J. Stratigraphy*, **76**, 113–114.

SHEVENELL, A. E., KENNETT, J. P. & LEA, D. W. 2004. Middle Miocene southern cooling and Antarctic cryosphere expansion. *Science*, **305**, 1766–1770.

SICKENBERG, O., BECKER-PLATEN, J. D., BENDA, L. ET AL. 1975. Die Gliederung des höheren Jungtertiärs und Altquartärs in der Türkei nach Vertebraten und ihre Bedeutung für die inernationale Neogen-Stratigraphie. *Geologisches Jahrbuch Reihe B*, **15**, 167pp.

SNEL, E., MĂRUNŢEANU, M., MACALET, R., MEULENKAMP, J. E. & VAN VUGT, N. 2006. Late Miocene to Early Pliocene chronostratigraphic framework for the Dacic Basin, Romania. *Palaeogeography, Palaeoclimatology, Palaeoecology*, **238**, 107–124.

SPAAK, P. 1983. Accuracy in correlation and ecological aspects of the planktonic foraminiferal zonation of the Mediteranean Pliocene. *Utrecht Micropalaeontological Bulletin*, **28**, 160pp.

SPROVIERI, R. 1992. Mediterranean Pliocene biochronology: a high resolution record based on quantitative planktonic foraminifera distribution. *Rivista Italiana di Paleontologia e Stratigrafia*, **98**, 61–100.

SPROVIERI, R., BONOMO, S., CARUSO, A. ET AL. 2002. An Integrated calcareous plankton biostratigraphic scheme and biochronology of the Mediterranean Middle Miocene. *Rivista Italiana di Paleontologia e Stratigrafia*, **108**, 337–353.

STEININGER, F. F. 1999. Chronostratigraphy, Geochronology and Biochronology of the "European Land Mammal Mega-Zones" (ELMMZ) and the Miocene "Mammal-Zones" (MN-Zones). *In*: RÖSSNER, G. E. & HEISSIG, K. (eds) *The Miocene Land Mammals of Europe*. Dr. Friedrich Pfeil, München, Germany, 9–24.

SUBALLY, D., BILLODEAU, G., TAMRAT, E., FERRY, S., DEBARD, E. & HILLAIRE-MARCEL, C. 1999. Cyclic climatic records during the Olduvai subchron (uppermost Pliocene) on Zakynthos Island (Ionian Sea). *Geobios*, **32**, 793–803.

SUBALLY, D. & QUÉZEL, P. 2002. Glacial or interglacial: *Artemisia* a plant indicator with dual responses. *Review of Palaeobotany and Palynology*, **120**, 123–130.

SUC, J.-P. 1984. Origin and evolution of the Mediterranean vegetation and climate in Europe. *Nature*, **307**, 429–432.

SUC, J.-P. & CRAVATTE, J. 1982. Etude palynologique du Pliocène de Catalogne (nord-est de l'Espagne). *Paléobiologie Continentale*, **13**, 1–31.

SUC, J.-P. & DRIVALIARI, A. 1991. Transport of bisaccate coniferous fossil pollen grains to coastal sediments: an example from the earliest Pliocene Orb Ria (Languedoc, Southern France). *Review of Palaeobotany and Palynology*, **70**, 247–253.

SUC, J.-P. & POPESCU, S.-M. 2005. Pollen records and climatic cycles in the North Mediterranean region since 2.7 Ma. *In*: HEAD, M. J. & GIBBARD, P. L. (eds) *Early–Middle Pleistocene Transitions: The Land-Ocean Evidence*. Geological Society of London, Special Publication, **247**, 147–158.

SUC, J.-P., DINIZ, F., LEROY, S. *ET AL.* 1995*a*. Zanclean (∼ Brunssumian) to early Piacenzian (∼ early-middle Reuverian) climate from 4° to 54° north latitude (West Africa, West Europe and West Mediterranean areas). *Mededelingen Rijks Geologische Dienst*, **52**, 43–56.

SUC, J.-P., BERTINI, A., COMBORIEU-NEBOUT, N. *ET AL.* 1995*b*. Structure of West Mediterranean vegetation and climate since 5.3 Ma. *Acta zoologica Cracoviense*, **38**, 3–16.

SUC, J.-P., FAUQUETTE, S., BESSEDIK, M. *ET AL.* 1999. Neogene vegetation changes in West European and West circum-Mediterranean areas. *In*: AGUSTÍ, J.,

ROOK, L. & ANDREWS, P. (eds) *The Evolution of Neogene Terrestial Ecosystems in Europe*. Cambridge University Press, Cambridge, 378–388.

THOMAS, H., SPASSOV, N., KODJUMJIEVA, E. *ET AL.* 1986. Résultats préliminaires de la première mission paléontologique franco-bulgare à Dorkovo (arrondissement de Pazardjik, Bulgarie). *Comptes Rendus de lÁcademie des Sciences, Paris*, **302**, 1037–1042.

TICLEANU, N. & DIACONITA, D. 1997. The main coal facies and lithotypes of the Pliocene coal basin, Oltenia, Romania. *In*: GAYER, R. & PESEK, J. (eds) *European Coal Geology and Technology*, Geological Society, London, Special Publication, **125**, 131–139.

VAN VUGT, N., LANGEREIS, C. G. & HILGEN, F. J. 2001. Orbital forcing in Pliocene-Pleistocene Mediterranean lacustrine deposits: dominant expression of eccentricity versus precession. *Palaeogeography, Palaeoclimatology, Palaeoecology*, **172**, 193–205.

WANG, C. W. 1961. The forests of China with a survey of grassland and desert vegetation. Maria Moors Cabot Foundation, **5**, Harvard University Cambridge, Massachusetts, 313pp.

ZACHOS, J., PAGANI, M., SLOAN, L. & BILLUPS, K., 2001. Trends, rhythms, and aberrations in global climate 65 Ma to present. *Science*, **292**, 686–693.

ZHISHENG, A., KUTZBACH, J., PRELL, W. L. & PORTER, S. C. 2001. Evolution of Asian monsoons and phased uplift of the Himalaya-Tibetan plateau since Late Miocene times. *Nature*, **411**, 62–66.

ZOHARY, M. 1973. Geobotanical foundations of the Middle East. Fischer ed., Stuttgart, **2 vol.**, 739pp.

Characterizing ice sheets during the Pliocene: evidence from data and models

D. J. HILL[1], A. M. HAYWOOD[1,*], R. C. A. HINDMARSH[2] & P. J. VALDES[3]

[1]*Geological Sciences Division, British Antarctic Survey, High Cross, Madingley Road, Cambridge, CB3 0ET, UK (e-mail: dahi@bas.ac.uk)*

[2]*Physical Sciences Division, British Antarctic Survey, High Cross, Madingley Road, Cambridge, CB3 0ET, UK*

[3]*School of Geographical Science, University of Bristol, University Road, Bristol, BS8 1SS, UK*

[*]*Now at: School of Earth and Environment, Environment Building, University of Leeds, Leeds, LS2 9JT, UK*

Abstract: The Pliocene (*c.* 5.3–1.8 Myr BP) was the last epoch of geological time in which global temperatures were generally higher than modern. It is important if we are to understand the dynamics of warm climates. This is particularly true of the interaction of climate and cryosphere, where the Pliocene may represent the first epoch in which ice sheets, at least on Antarctica, were a permanent feature. In this paper, we review the available evidence for the state of ice sheets during the Pliocene as well as previous attempts to model them. We then present new models and sensitivity studies of the mid-Pliocene East Antarctic Ice Sheet (EAIS) and consider the implications for the debate on ice-sheet stability during the Pliocene. These new reconstructions suggest that the mid-Pliocene EAIS was significantly smaller than modern, but the modelled average mid-Pliocene climate is not sufficient to cause the widespread deglaciation suggested by Sirius Group diatom evidence.

Understanding the role of the cryosphere within the climate system is important if the mechanisms of climate are to be fully understood and future climate predictions well-constrained. Ice sheets play an important role in the modern climate, having direct and indirect influences on atmospheric temperatures, wind patterns, surface albedo, vegetation, continental water balance, ocean temperatures, sea-ice formation and ocean circulation patterns (Clark *et al.* 1999). However, discussion continues over their significance in deep-time palaeoclimates. Whether ice sheets are one of the primary drivers of Cenozoic climate change or a largely passive responder to other changes in the climate system remains an open question.

After the initial transition into a world with significant glaciation at *c.* 34 Ma, the waxing and waning of both Northern Hemisphere and Antarctic ice sheets has been used to explain the variation of a large variety of Cenozoic palaeoclimate proxies, particularly fluctuations in marine $\delta^{18}O$ oxygen isotope ratios (Kennett 1977; Matthews 1986; Zachos 2001). However, there are no direct proxies for ice volume, since, for example, $\delta^{18}O$ values also depend on bottom-water temperatures and ocean salinity (Rostek *et al.* 1993). Even when assumptions are made which allow a global ice volume to be inferred, there is typically no method to identify the volume or location of individual ice sheets. This ambiguity is further complicated by the non-linearity of climate, ice-sheet dynamics and the interactions between the two. In a system as complex as the ice sheet–climate system, it is not always immediately obvious what the observed values and fluctuations of climate-dependent variables represent. In order to reconstruct the ice sheets from deep-time periods of the Earth's history, and thus investigate glaciations in warm climates, it is important that direct evidence of ice-sheet extent is collected in order to constrain models. These models can in turn interpret the consequences of this evidence and reconstruct the ice sheet as a whole.

The Pliocene epoch is important, as it is the last time when the Earth was warmer for a period longer than any Quaternary interglacial, providing a unique window into the workings of the ice sheet-climate system. Deep-time cryospheric data are limited within the geological record; however, even limited data coverage can constrain models and provide important boundary conditions. In this paper, we review existing ice-sheet data and models, showing that much can already be inferred regarding Pliocene ice sheets, whilst debate continues over some of these interpretations. We then

From: WILLIAMS, M., HAYWOOD, A. M., GREGORY, F. J. & SCHMIDT, D. N. (eds) *Deep-Time Perspectives on Climate Change: Marrying the Signal from Computer Models and Biological Proxies.* The Micropalaeontological Society, Special Publications. The Geological Society, London, 517–538.
1747-602X/07/$15.00 © The Micropalaeontological Society 2007.

present new model reconstructions of the East Antarctic Ice Sheet (EAIS) and consider the implications of the results.

Importance of the Pliocene

If we are to understand the future of our climate, we need to understand the past Earth system and its potential modes of operation. Greater understanding is required of how the climate will respond to increased greenhouse gas concentration and how transitions, from one climate state to another, occur. These required areas of understanding necessitate the study of palaeoclimates. Of particular interest to those looking at future climate change are periods with increased temperatures, greater concentrations of greenhouse gases in the atmosphere and periods in which large climate transitions occurred.

Cenozoic records show that climate has cooled from the hot, greenhouse worlds of the Palaeocene and Eocene, through a number of gradual and stepwise transitions to the bipolar glaciations of the Pleistocene (Miller *et al.* 1987; Lear *et al.* 2000; Zachos *et al.* 2001). The Pliocene is crucial to our understanding of the response of ice sheets to these changes in climate. It is believed that large-scale, permanent Antarctic glaciation was established in the middle Miocene (Flower & Kennett 1994; Zachos *et al.* 2001), whilst the first indication of limited Greenland glaciation occurs in the late Miocene (Larsen *et al.* 1994). This means that the late Miocene and Pliocene are a unique Cenozoic testing ground for the interaction of large-scale Antarctic and limited Northern Hemisphere ice sheets with a warmer-than-modern climate. If the nature and extent of the Pliocene cryosphere can be determined, then this could help us to understand the way the ice sheets, and therefore climate, may respond to future global warming.

The Pliocene epoch also has great significance in a wider understanding of climate change, not least because it is 'geologically recent', providing large quantities of palaeoenvironmental data. The reliability of these data, compared with previous warm periods, is increased by improved geographical distribution, biota–environment correlations and stratigraphy (Dowsett *et al.* 1992). The mid-Pliocene is believed to be the last time in Earth history when global temperatures were sustained at levels higher than today. A number of geological (Dowsett *et al.* 1999) and modelling studies (Sloan *et al.* 1996; Haywood *et al.* 2000; Haywood & Valdes 2004) indicate that mean temperatures during this interval were between 1.4 and 3.6 °C higher than present. After this warmth, the climate cooled and underwent a transition that saw the onset of widespread Northern Hemisphere

glaciation, beginning approximately 2.7 Myr BP in the late Pliocene. Thus, the Pliocene is the last interval at which temperatures were sustained at levels equating to those predicted for the coming century (Fig. 1). Unfortunately, the concentration of CO_2 in the Pliocene atmosphere is only partially constrained. Some proxies indicate that levels were similar to, or slightly higher than, those measured today (Van der Burgh *et al.* 1993; Raymo *et al.* 1996), whilst others suggest values closer to the Last Glacial Maximum (Pearson & Palmer 2000). These indirect proxies predict similar values for the Pleistocene and Eocene, but have very different results for the Pliocene. This suggests that the calibration of one or more of these may not reflect the true sensitivity of the proxy to atmospheric carbon.

Although there is a consensus that the global climate of the Pliocene was significantly warmer than modern, the mechanisms that caused the enhanced temperatures and the pattern of warming have yet to be definitively determined. The polar amplification of temperatures, combined with little, if any, tropical warming in mid-Pliocene sea-surface temperature reconstructions, suggests that enhanced oceanic poleward heat transport could be the major contributing factor (Dowsett *et al.* 1992; Kwiek & Ravelo 1999). However, coupled ocean-atmosphere General Circulation Models (GCMs) fail to reproduce the proposed increase in meridional heat transport and suggest significant tropical warming (Haywood & Valdes 2004), more indicative of CO_2-induced heating. However, it remains unclear whether the discrepancy between the data-based reconstructions and the models arises from insufficient data coverage, lack of proxy sensitivity, errors in the GCM boundary conditions or inaccuracies in the physics of ocean circulation changes. Recent data using alkenone palaeothermometry techniques support the model results, showing enhanced temperatures in the equatorial Pacific and Atlantic (Haywood *et al.* 2005; Williams *et al.* 2005).

A modelling study aimed at establishing the relative contributions of the various mechanisms for producing enhanced global temperatures found that the most significant forcing was provided by the smaller prescribed ice sheets (Haywood & Valdes 2004). The PRISM (Pliocene Research, Interpretation and Synoptic Mapping) mid-Pliocene environmental reconstructions incorporate a 50% reduction in Greenland ice volume and a 33% reduction in Antarctic ice volume (Dowsett *et al.* 1999). These ice-sheet volume reconstructions were based on sea level and marine oxygen isotope data that have large error bars associated with them (Lietz & Schmincke 1975; Dowsett & Cronin 1990; Krantz 1991; Wardlaw & Quinn 1991). These error bars encompass vastly different global

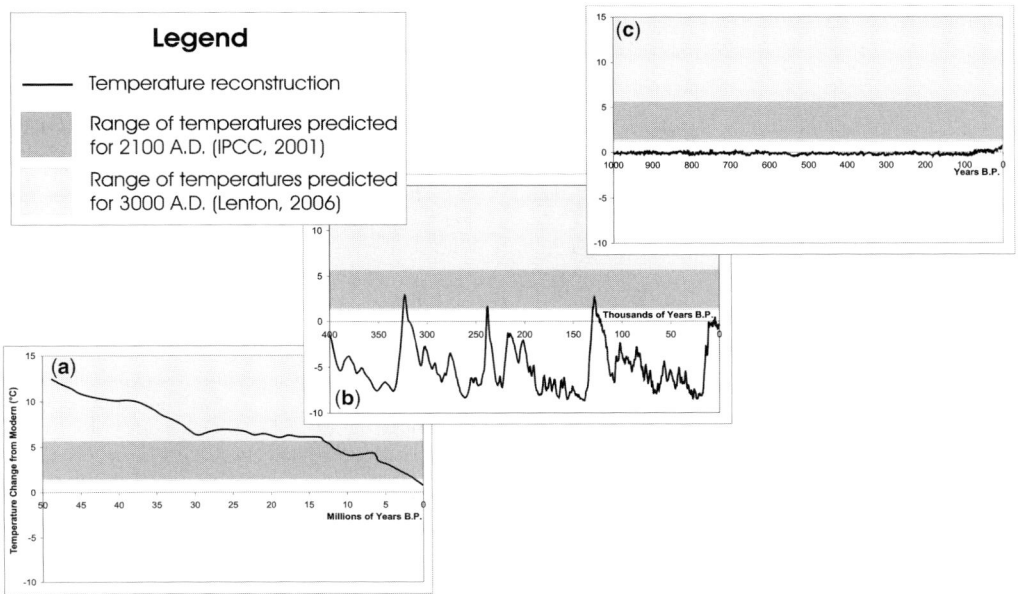

Fig. 1. Comparison of temperature datasets for the (**a**) Cenozoic (Lear *et al.* 2000), (**b**) last four glacial cycles (Petit *et al.* 1999) and (**c**) last 1000 years (Mann *et al.* 1999) with predictions of future global temperatures.

ice-sheet scenarios, from significant deglaciation of East Antarctica, to all the sea-level increase originating from Greenland and West Antarctica.

Pliocene Northern Hemisphere glaciation

The Plio–Pleistocene climate transition, *c.* 2.7 Myr BP, principally records the development of large ice sheets over Fennoscandinavia and North America, which probably had not previously existed during the Cenozoic. However, the Greenland Ice Sheet (GrIS) is more resilient to warmer temperatures and a large portion probably survived the warmer-than-modern last interglacial (Otto-Bleisner *et al.* 2006; Overpeck *et al.* 2006). Small amounts of ice-rafted debris (IRD) have been found in Greenland continental shelf records for the last seven million years (Larsen *et al.* 1994), but this is not necessarily indicative of large ice sheets in Greenland. Mountain glaciers can produce IRD when they reach down to the coast, requiring only small total ice volumes. Indeed, the first significant volume of IRD on the Greenland continental margin is observed at approximately 3 Ma, which may represent the first major increase in Greenland ice volume (Jansen *et al.* 2000).

Although the Neogene geological record on Greenland is sparse, there is some evidence that the GrIS was significantly smaller than at present during the Pliocene epoch. The shallow marine deposits from Ile de France in northeast Greenland indicate summer temperatures more than 6 °C warmer than today during the mid-Pliocene warm period (Bennike *et al.* 2002). Furthermore, evidence for forestation in the northernmost reaches of Greenland, as late as 2.4 Myr BP is found at the Kap Kopenhavn formation (Funder *et al.* 2001) and willow and pine fragments in the basal ice of the NGRIP (North Greenland Ice core Project) ice core suggests forestation in central Greenland at the point of glacial inception (Dahl-Jensen 2006).

Pliocene deposits across the Canadian Arctic islands, northern Alaska and northern Russia show that the Arctic was generally warmer in the Pliocene and that the treeline had migrated well into the Arctic Circle (Fig. 2). Particularly telling is the evidence for evergreen forests on Meighen (Matthews 1987) and Ellesmere Islands (Fyles 1989), today a polar desert. The terrestrial biota found there and around the Arctic suggests that large-scale glaciation in the Pliocene Northern Hemisphere could only have existed on Greenland, with possible localized glaciation on Ellesmere Island (de Vernal & Mudie 1989) and in Alaska (Lagoe & Zellers 1996). The available data from Greenland are indicative of a significant reduction in the extent of the ice sheet during the Pliocene.

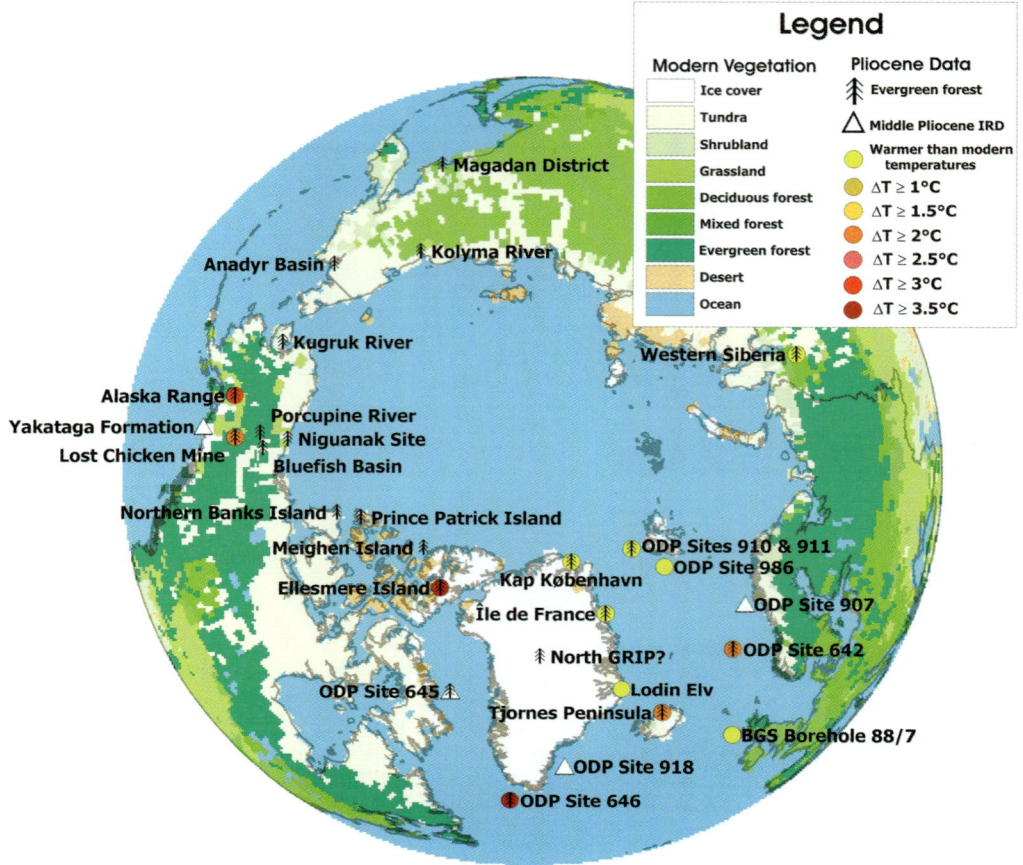

Fig. 2. Pliocene palaeoenvironmental data for the Arctic (Table 1), superimposed on the $1° × 1°$ modern vegetation distribution (Matthews 1983).

Pliocene Antarctic Peninsula Ice Sheet

The Antarctic Peninsula Ice Sheet (APIS) is the smallest and most northerly component of Antarctic ice volume. It is also the area of the Antarctic that has shown most response to recent climate changes (Vaughan *et al.* 2003; Cook *et al.* 2005). A number of ice shelves have collapsed, probably due to the formation of surface-melt features, as warmer summer temperatures propagate further south (Vaughan & Doake 1996; Scambos *et al.* 2000). The removal of this buffer has been shown to cause the retreat of grounding lines and acceleration of the glaciers that once fed the ice shelf (De Angelis & Skvarca 2003; Rignot *et al.* 2004). Owing to this high degree of climate sensitivity, many researchers have assumed that the peninsula would generally be ice free in the warmer

pre-Quaternary Cenozoic palaeoclimates (Abreu & Anderson 1998; Dowsett *et al.* 1999).

However, recent evidence has shown that significant ice must have existed on the Antarctic Peninsula for large portions of the post-Eocene Cenozoic (Dingle & Lavelle 1998; Ivany *et al.* 2006). Marine geological and geophysical studies of the Antarctic Peninsula Pacific margin, at Ocean Drilling Program (ODP) Sites 1095, 1096, 1097 and 1101 (Fig. 3), seem to show that ice must have been present during the Neogene. Seismic reflectors, which have been interpreted as glacial unconformities, give evidence for a large APIS as long ago as the middle Miocene. Correlation with biostratigraphic dating of ODP Leg 178 cores allowed a date of between 5.12 and 7.94 Ma to be assigned to a series of unconformities, suggesting that the APIS advanced out onto

Table 1. *Pliocene Arctic palaeoenvironmental data set*

Site	Location	Evidence	Paper
Kap København	Greenland	Macrofossils, ostracoda, foraminifera	Funder *et al.* 2001
Île de France	Greenland	Sedimentology, macrofossils, foraminifera	Bennike *et al.* 2002
Lodin Elv	Greenland	Ostracoda	Penney 1993
North GRIP	Greenland	Macrofossils (in basal ice)	Dahl-Jensen 2006
Tjornes Peninsula	Iceland	Palynology	Willard 1994
ODP Sites 910 & 911	Arctic Ocean	Palynology	Willard 1996
ODP Site 986	Arctic Ocean	Dinoflagellates	Smelror 1999
ODP Site 907	Norwegian Sea	IRD	Jansen *et al.* 2000
ODP Site 642	Norwegian Sea	Palynology	Willard 1994
BGS Borehole 88/7	Northern Atlantic	Sedimentology, nanofossils	Stoker *et al.* 1994
ODP Site 918	Northern Atlantic	IRD	St. John & Krissek 2002
ODP Site 646	Northern Atlantic	Palynology	Willard 1994
ODP Site 645	Baffin Bay	Palynology, IRD	de Vernal & Mudie 1989
Ellesmere Island	Canadian Arctic	Palynology, macrofossils	Csank 2006
Meighen Island	Canadian Arctic	Macrofossils	Matthews 1987
Prince Patrick Island	Canadian Arctic	Macrofossils	Fyles 1990
Northern Banks Island	Canadian Arctic	Macrofossils	Fyles *et al.* 1994
Bluefish Basin	Yukon, Canada	Palynology, macrofossils	Matthews & Ovenden 1990
Niguanak Site	Alaska	Macrofossils	Matthews & Ovenden 1990
Porcupine River	Alaska	Macrofossils	Matthews & Ovenden 1990
Lost Chicken Mine	Alaska	Palynology, macrofossils	Matthews *et al.* 2003
Yakataga Formation	Alaska	IRD	Lagoe & Zellers 1996
Alaska Range	Alaska	Palynology	Ager 1994
Kugruk River	Alaska	Palynology	Matthews & Ovenden 1990
Anadyr Basin	Arctic Russia	Palynology, macrofossils	Fradkina 1991
Kolyma River	Arctic Russia	Palynology, macrofossils	Fradkina 1991
Magadan District	Arctic Russia	Palynology, macrofossils	Fradkina 1991

the continental shelf at least 12 times during the late Miocene (Bart *et al.* 2005). The number of unconformities seems to increase in the Pliocene, with 30 recorded by Bart & Anderson (2000), but this may represent preservation rather than intrinsic rate of occurrence. The Antarctic Peninsula frequency was compared to similar sites in the eastern and western Ross Sea, which was speculated to show that the APIS was more dynamic during the late Neogene than either the West Antarctic Ice Sheet (WAIS) or EAIS, although preservation issues and accurate dating of the seismic sequences remain unresolved. The advance of the APIS over the continental shelf is substantiated by the occurrence of Pliocene sediments in large drift deposits located on the adjacent continental rise. They indicate repeated advance and retreat of grounded ice masses across the shelf and are associated with IRD of Antarctic Peninsula provenance (Cowan 2002; Hillenbrand & Ehrmann 2002, 2005; Pudsey 2002). Physical and geochemical records and X-ray images, derived from ODP Leg 178 Site 1095, show variations attributed to glacial–interglacial cycles during the early Pliocene (Hepp *et al.* 2006).

The limited terrestrial record seems to support the picture of a fluctuating Pliocene APIS. Deposits, interbedded with igneous rocks, at Cockburn Island (Dingle *et al.* 1997) and James Ross Island (Kristjansson *et al.* 2005; Hambrey & Smellie 2006; Smellie *et al.* 2006), are believed to represent at least periodic interglacial conditions, with faunal assemblages indicative of warmer-than-modern marine conditions (Zinsmeister & Webb 1982) and little evidence of IRD (Jonkers *et al.* 2002). Other Pliocene volcanism on the Antarctic Peninsula gives further evidence of a complex cryospheric history with both englacial and subaerial eruptions at Seal Nunataks (Smellie & Hole 1997) and Hornpipe Heights (Smellie 1999) respectively.

Pliocene West Antarctic Ice Sheet

The WAIS is currently the only marine ice sheet (one that is largely grounded on bedrock below sea level). It is particularly difficult to define its history prior to the Last Glacial Maximum, as erosion of the surrounding continental shelf has removed all surface evidence of previous

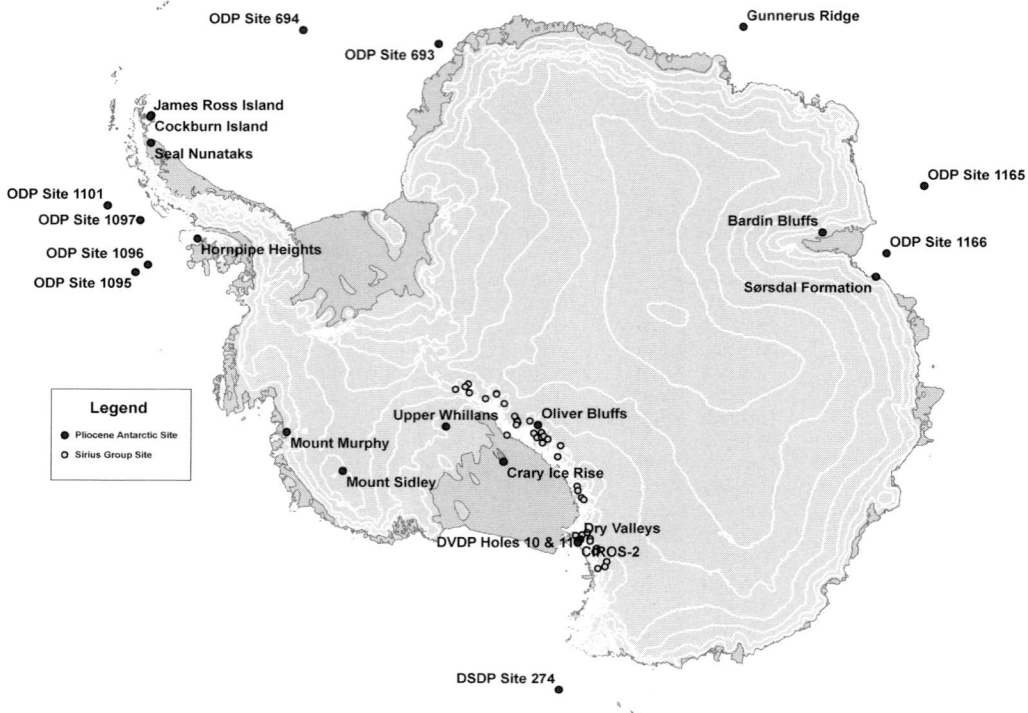

Fig. 3. Antarctic Pliocene sites (Table 2) and Sirius Group deposits (Stroeven 1997).

glaciations on the seabed. Few rock outcrops occur in the interior of the ice sheet, making any evidence of pre-Quaternary glaciations very difficult to obtain. However, a limited record of the state of the WAIS during the Pliocene does exist (Fig. 3).

Mount Sidley in the Executive Committee Range, Marie Byrd Land, is the highest volcano in Antarctica. The outcrops are dominated by subaerially erupted volcanic rocks, hence the most recent eruptions must have occurred when there was little ice cover in the region. These rocks have been dated to between 5.7 and 4.2 Ma, suggesting the early Pliocene WAIS was reduced in size compared to today (Panter *et al.* 1994). Further Pliocene dates are found on another West Antarctic volcano, Mount Murphy, which is located on the coast, beside the Crosson Ice Shelf. Mount Murphy has an unusual association of volcanic rocks, glacially derived sediments and recycled microfossils as well as a varied ice sheet history (LeMasurier *et al.* 1994; Smellie 2000). A similar scenario has been proposed for these sediments as for the Sirius Group deposits in the Transantarctic Mountains (see later), meaning the diatoms found there could represent periods of marine deposition in the interior of West Antarctica. These sediments

are of particular note in this discussion due to the presence of diatoms restricted to the Pliocene in the Southern Ocean, suggesting that the Byrd Subglacial Basin, to the south and upstream of Mount Murphy, may have been ice free during some of the Pliocene (LeMasurier *et al.* 1994).

Attempts to extract microfossils from underneath the currently grounded WAIS have been undertaken in the Siple Coast region. During a campaign to drill through the Crary Ice Rise, at the mouth of Whillans Ice Stream (formerly Ice Stream B), sediment was removed from under the ice dome. The diatoms recovered were from the late Miocene, with no diagnostically post-Miocene species (Scherer *et al.* 1988). This does not necessarily mean that marine deposition stopped at the end of the Miocene, as the ice rise is believed to have only been grounded for around 1100 years (Bindschadler *et al.* 1989), but the sediments above the Miocene layers may have been subsequently eroded by the WAIS. Microfossils were also found in the sediments underneath the upstream section of the Whillans Ice Stream. Diatoms from the middle Miocene to the present were found, with the temporal ranges of the species potentially showing a continuous

Table 2. *Pliocene Antarctic sites*

Site	Location	Evidence	Paper
Cockburn Island	Northern Antarctic Peninsula	Micropalaeontology	Zinsmeister & Webb 1988
James Ross Island	Northern Antarctic Peninsula	Micropalaeontology, macrofossils	Jonkers *et al.* 2002
Seal Nunataks	Northern Antarctic Peninsula	Volcanology	Smellie & Hole 1997
Hornpipe Heights	Alexander Island	Volcanology	Smellie 1999
ODP Site 1095	Western Antarctic Peninsula	Sedimentology, geochemistry	Hepp *et al.* 2006
ODP Site 1096	Western Antarctic Peninsula	Sedimentology	Hillenbrand & Ehrmann 2005
ODP Site 1097	Western Antarctic Peninsula	Sedimentology	Hillenbrand & Ehrmann 2005
ODP Site 1101	Western Antarctic Peninsula	Ice-rafted debris	Cowan 2002
Mount Sidley	West Antarctic Ice Sheet	Volcanology	Panter *et al.* 1994
Mount Murphy	West Antarctic Ice Sheet	Diatoms	LeMasurier *et al.* 1994
Crary Ice Rise	Ross Sea Ice Streams	Diatoms	Scherer *et al.* 1988
Upper Whillans	Ross Sea Ice Streams	Diatoms	Scherer 1991
ODP Site 693	Weddell Sea	Sedimentology	Barker & Kennett 1988
ODP Site 694	Weddell Sea	Sedimentology	Barker & Kennett 1988
Bardin Bluffs	Amery Basin	Diatoms	Whitehead *et al.* 2004
Sørsdal Formation	Amery Basin	Diatoms	Whitehead *et al.* 2001
Oliver Bluffs	Transantarctic Mountains	Palaeontology, sedimentology	Francis & Hill 1996
Dry Valleys	Transantarctic Mountains	Geomorphology	Sugden *et al.* 1995
DVDP Holes 10 & 11	Transantarctic Mountains	Micropalaeontology	Ishman & Reick 1992
ODP Site 1165	Prydz Bay	Diatoms	Whitehead *et al.* 2005
ODP Site 1166	Prydz Bay	Diatoms	Whitehead *et al.* 2005
Gunnerus Ridge	Enderby Land	Sedimentology	Hillenbrand & Ehrmann 2003
DSDP Site 274	Transantarctic Mountains	Palynology	Fleming & Barron 1996

deposition record. However, none of the species is diagnostic of a Pliocene age, whereas some of the diatoms are limited to the Miocene and Pleistocene, suggesting both these periods saw deglaciation of central West Antarctica (Scherer 1991). It is hard to draw specific conclusions about the Pliocene, but if the WAIS grounding line can retreat beyond the upstream sections of the Ross Sea sector during both the late Miocene and Pleistocene, then this may also be expected in the Pliocene.

Marine evidence for the state of the WAIS is available from the Ross Sea and Weddell Sea regions. Neogene glacial unconformities have been observed in seismic stratigraphic studies of the eastern and western continental shelf of the Ross Sea. These are thought to represent periods of sufficient ice-sheet expansion to cause the grounding line to advance and predicate subglacial erosion. Both regions were found to have eight unconformities during the Neogene, which suggests at least eight grounding events, representing expansions of the EAIS and WAIS respectively (Bart & Anderson 2000). This would seem to show that, at least during the Neogene, both the EAIS and WAIS experienced fluctuations and were prone to large-scale changes in volume.

Beyond the Weddell Sea shelf margin, continuous deposition occurred throughout the late Miocene, Pliocene and Pleistocene (Barker & Kennett 1988). In fact, the Pliocene saw faster sedimentation than the Pleistocene, under conditions that, looking at other evidence from marine sediments, should have seen a highly variable sediment supply, as the ice sheet advanced and retreated across the Weddell Sea continental margin. It could be argued that during this period the Weddell Sea locality must have been a stable, heavily glaciated environment, in order to continuously supply the ocean basin with sediment. However, a further scenario has been proposed, where sediment is supplied by local glaciers transporting material through deep canyons, bypassing the continental shelf and negating the need for large-scale glaciation in the region (Barker 1992).

Pliocene East Antarctic Ice Sheet

The behaviour of the EAIS is the most contentious aspect of the Pliocene ice-sheet stability debate. The two scenarios that have been proposed are (i) a dynamic ice sheet, experiencing large-scale deglaciations throughout the late Miocene and Pliocene

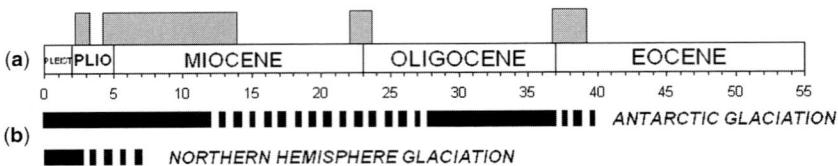

Fig. 4. Differing views of Cenozoic glaciations (**a**) periods of proposed Antarctic deglaciation, dated from Sirius Group diatoms (Harwood 1983); (**b**) glaciations (solid black represents permanent and dashed black transient) inferred from Cenozoic $\delta^{18}O$ values (Zachos *et al.* 2001).

and (ii) a continent-wide cold-based EAIS, since the inception of a permanent Antarctic ice sheet in the middle Miocene (Fig. 4). The two hypotheses and their supporters, daubed 'dynamicists' and 'stabilists', have been at loggerheads for more than 20 years and the questions they raise still lack completely satisfactory answers.

The controversy began with the discovery of Sirius Group marine diatoms high in the Transantarctic Mountains. The mechanism originally proposed for the emplacement of these diatoms is glacial transport in times of an expanded EAIS from their marine deposition sites in the modern day Wilkes-Pensacola Subglacial Basins. The location of the tills in which diatoms are found would require the almost complete deglaciation of the basins, which could correspond to a rapid ice-volume reduction of up to 60%. The chronology of these events has been approximated by dating the diatoms found in the Sirius Group sediments (Fig. 3) and forms a discrete set of marine incursions in the Miocene and Pliocene (Fig. 4).

Evidence of grounding-line retreat from the Lambert Glacier region supports these interpretations. Miocene and Pliocene marine deposits, subsequently uplifted above the modern ice surface in the Prince Charles Mountains, hundreds of kilometres inland from the modern grounding line, seem to correlate well with the deposition periods of the Sirius Group diatoms. The deposition of the Pliocene member of the group, the Bardin Bluffs sediments, has been biostratigraphically dated as occurring between 3.4 and 2.6 Ma (Whitehead *et al.* 2004). The presence of three planktonic taxa in reasonable abundance points to the existence of ice-free marine surface waters when the Bardin Bluffs sediments were deposited (McKelvey *et al.* 2001). Micropalaeontological studies of ODP Sites 1165 and 1166, in Prydz Bay, show that the sea ice in the region was reduced by up to 78% during the Pliocene (Whitehead *et al.* 2005). There is also evidence of significantly increased early Pliocene temperatures (summer sea-surface temperatures $>3\ °C$) and open water conditions at the nearby Sørsdal Formation (Whitehead *et al.* 2001).

The Beardmore Glacier region also provides some possible evidence of a very different East Antarctica during the Pliocene. The Oliver Bluffs deposits, suggested as mid-Pliocene in age by the diatom assemblage, contain exceptionally preserved examples of *Nothofagus* (Southern Beech) leaf mats, twigs and wood (Francis & Hill 1996). Ecological interpretation of these and other Oliver Bluffs finds, which include palaeosols, palynomorphs, mosses, liverworts, conifers, angiosperms and two species of weevil (Askin & Markgraf 1987; Retallack *et al.* 2001; Ashworth & Kuschel 2003; Ashworth & Cantrill 2004), suggests a warmer-than-modern polar environment, with mean annual temperatures of approximately $-8\ °C$. The use of the diatoms to date these sediments is controversial. However, provenance studies of the Sirius Group deposits suggests the Oliver Bluffs may be one of the youngest members, as they seem to represent erosion of the local glacial troughs (Passchier 2004). This finding, along with marine evidence of *Nothofagus* in Pliocene sediments in the outer Ross Sea (Fleming & Barron 1996), supports the existence of localized dwarf *Nothofagus* forests on Antarctica until relatively recently.

The assertion of a dynamic EAIS is disputed by the 'stabilists', who maintain that there has been little change in East Antarctica since the middle Miocene, when large-scale glaciation was in place. This hypothesis was established with the compilation of Cenozoic $\delta^{18}O$ records, which showed a clear shift towards heavier values (colder climate and/or greater global ice volume) in the middle of the Miocene (Kennett 1977; Zachos *et al.* 2001). The terrestrial evidence for this view comes mainly from the remarkable landscape stability of the Dry Valleys region of the Transantarctic Mountains. Geomorphological studies (Sugden *et al.* 1995) and ashfall-dating techniques (Marchant *et al.* 1996) have been used to infer that the region has been essentially unchanged climatically (Sugden *et al.* 1999; Summerfield *et al.* 1999), tectonically (Fitzgerald & Stump 1997; Ackert & Kurz 2004) and glaciologically (Stroeven & Prentice 1997; Goff *et al.* 2002) since the Miocene. Some researchers have suggested that the Dry Valleys region is glaciologically unique, and could have seen little change during periods of major deglaciation elsewhere in East Antarctica (Kerr & Huybrechts 1999). However, the level of landscape stability reported is difficult to reconcile with a scenario of EAIS dynamism and significant Antarctic warming. This gives credence to alternative suggestions, such as atmospheric transport (Burckle & Potter 1996; Kellogg & Kellogg 1996; Stroeven *et al.* 1996) and meteorite impact fallout (Gersonde *et al.* 1997), for the emplacement of the Sirius Group diatoms.

The Pliocene marine record around East Antarctica is limited and, in many cases, poorly dated. However, there are a few coring expeditions that have obtained Pliocene material. At opposite sides of the continent, at Ferrar Fjord (CIROS-2 drill site), Taylor palaeofjord (Dry Valleys Drilling Program (DVDP) holes 10 and 11), Prydz Bay (ODP Site 1165) and Gunnerus Ridge (Polarstern cores PS1811-8 and PS1812-6), sediments appear to alternate between glacial- and interglacial-type deposition (Barrett *et al.* 1992; Ishman & Rieck 1992; Hillenbrand & Ehrmann 2003; Warnke *et al.* 2004). Combined with evidence of erosional unconformities in the Ross Sea sediments (Bart & Anderson 2000), this suggests that the EAIS fluctuated, at least at the margins, during the Pliocene.

Clearly, the history of the EAIS is a complex one and it seems increasingly likely that the ice sheet does not generally behave uniformly. The individual drainage basins (Vaughan *et al.* 1999) and tectonic provinces (Fitzgerald *et al.* 1986; Fitzgerald 1994; Hindmarsh *et al.* 1998) could have responded very differently to climatic changes. Establishing exactly what this response has been to the various known climate events is severely hampered by the lack of available terrestrial evidence. It is in response to this scarcity of evidence and the differing interpretations of the existing geological record that modelling of the EAIS during warm periods of the past, such as the Pliocene, must test the different paradigms of ice-sheet behaviour.

Modelling ice sheets in warm climates

There is a reasonable amount of information on the state of the ice sheets during the Pliocene, but this has not been sufficient to finally settle many of the debates regarding the cryosphere. In such a situation, modelling should allow the gaps in the data to be filled in. However, accurate ice-sheet modelling requires high-resolution boundary conditions for any particular time period. Ice-sheet models (ISMs) are generally driven by temperature,

accumulation and melting, as well as being heavily dependent on the bedrock on which the model is run. Clearly, extracting such parameters from the geological record, on a grid of high-enough resolution to be useful within these models (e.g. at 50 km intervals), for time intervals before satellite observations are available, is extremely difficult.

There are a number of potential ways to overcome this problem. One of the ways utilized in previous studies is to use modern observations or a modern parameterization as a basis from which to increase temperatures and examine the effect these increases have on the ice sheets. Huybrechts (1993) used a 3-D ISM of Antarctica to test the effect of a constant increase in surface temperature, up to 20 °C above modern. Temperature increases of less than 5 °C were found to increase East Antarctic ice volumes, owing to the dominance of a parameterized increase in precipitation over melting, modelled by the positive degree-day (pdd) method. Warming of greater than 15 °C was required to deglaciate the Wilkes-Pensacola Subglacial Basins, as hypothesized by Harwood & Webb (1986; Webb & Harwood 1991; Harwood 1983). It was considered that 15 °C of temperature increase was probably more than would be expected in late Tertiary warm climates (Huybrechts 1993).

More sophisticated modelling has allowed the potential errors caused by insufficient knowledge of the boundary conditions to be examined. Kerr & Huybrechts (1999) undertook similar experiments to those of earlier work (Huybrechts 1993), but examined the effect that changes in the elevation of individual tectonic blocks within the Transantarctic Mountains had on ice-surface elevations. The tectonic history of the Transantarctic Mountains is poorly understood and contributes significantly to the uncertainty in reconstructing Antarctic palaeogeography. Evidence exists that some areas have been lower in the recent past (Behrendt & Cooper 1991; McKelvey *et al.* 1991; Webb *et al.* 1996; Hambrey *et al.* 2003). However, the region has experienced a complex history, with the Transantarctic Mountains split into a number of tectonic blocks (Fitzgerald *et al.* 1986; van der Wateron *et al.* 1999) that may have undertaken very different movements (Fitzgerald 1994; Hindmarsh *et al.* 1998). The study of Kerr & Huybrechts (1999) reveals that tectonic changes can have a large effect locally, but do not seem to cause significant long-range changes in ice-sheet elevation. This result could hamper attempts to compare ISM reconstructions with available data, which often tends to relate to single localities. This is particularly true in the Dry Valleys region, which is key in the arguments between the 'dynamicists' and 'stabilists'. In this region, a modelled temperature increase of 10 °C caused the EAIS margin to retreat significantly, but a local ice cap remained on the Dry Valley mountaintops. This suggests that the exceptional stability observed could reflect the local glaciological and climatic conditions, with inferences and conclusions drawn from the region not reflecting the wider EAIS behaviour.

Modelling mid-Pliocene ice sheets

Although there is much to be learnt from previous sensitivity studies, their applicability to specific ice sheets within the geological record is uncertain. This is because of the assumptions implicit within these models. Ice sheets experience altitude-temperature, ice-albedo, sea ice and ocean circulation feedbacks as well as internal non-linearities, and the results are therefore highly dependent on the boundary and initial conditions, particularly the ice-surface height that the models are initiated with. None of the studies described above has explored the effect that initial conditions would have on the conclusions. A further assumption implemented in the application of changes in temperature is that Antarctic temperatures increase smoothly, with little regional variation. However, the climate is highly non-linear and changes to the circulation and temperature patterns could be significant in warmer climates of the past (Rial *et al.* 2004). It is therefore important that more sophisticated climate modelling of Antarctica in warm periods of the geological past is applied to the boundary condition of ISMs.

The two existing models utilized in this study are the British Antarctic Survey ISM (BASISM) (Hindmarsh 1999, 2001) and the United Kingdom Met. Office (UKMO) GCM, specifically the HadAM3 atmospheric GCM. The atmospheric component of the UKMO GCM consists of 19 vertical layers, with a horizontal resolution of 2.5° in latitude and 3.75° in longitude. A more complete technical summary of the UKMO GCM is described in Pope *et al.* (2000) in which the impacts of improvements to the model, compared to the previous Hadley Centre model, are discussed. BASISM is a thermomechanically coupled, three-dimensional ice-sheet model, similar to those of Huybrechts (1990) and Ritz *et al.* (1997). The model uses the shallow ice approximation to close the mass, momentum and thermal balance equations that govern ice flow, and describe the grounded ice component. This numerical scheme has yet to be extended to include the more complex flow regimes of ice streams and ice shelves, as the understanding of the physics in these regimes is incomplete.

The ISM was run on a 50 km × 50 km polar stereographic grid, in a domain covering the modern grounded EAIS. The GCM is run on a 2.5° lat × 3.75° long grid and hence output climate fields require downscaling onto the ISM grid. The relevant fields from the GCM model were downscaled using a bilateral interpolation technique and simple lapse rate and albedo corrections. The pdd method was then employed to convert these climate fields into an accumulation/melt rate (Reeh 1991; Braithwaite 1995). This technique assumes that the melting of the ice-sheet surface can be fully described by three physical constants and the temperature record, which, although many other factors could contribute, has been shown to have some physical justification (Ohmura 2001).

The mid-Pliocene EAIS can thus be modelled by taking the output of extant mid-Pliocene GCMs and driving BASISM with the climatological output fields. Rather than previous modelling, which essentially tested the sensitivity of the modern ice sheet to uniform temperature changes, a realistic mid-Pliocene climate can now be reproduced and the ice sheet, during this particular period of time, reconstructed from the modelled climatology (Fig. 5) and BEDMAP topographic reconstruction (Fig. 6). Using an ISM in conjunction with the GCM allows us to attribute the ice-sheet reconstruction to this particular period of the geological record and hence test the results against the geological record. The employed modelling strategy also allows the effects of ice-sheet hysteresis to be explored, rather than assuming the ice sheet deglaciated from the modern configuration. The 'standard' mid-Pliocene climate reconstruction is a HadAM3 experiment (Haywood *et al.* 2000), whose boundary conditions are based on the environmental reconstructions of the PRISM group (Dowsett *et al.* 1999). These are designed to represent the values averaged over the mid-Pliocene time slice (3.29–2.97 Ma).

The models depicted in Figure 7 are the end members of a suite of EAIS reconstructions, under the climate forcing of the 'standard' mid-Pliocene GCM experiment. The difference between the reconstructions is due to the feedback-induced hysteresis in the ice sheet–climate system. The coupling between the GCM and ISM is offline, meaning the GCM experiment is not rerun with the BASISM-predicted ice sheets. This eliminates a number of important feedbacks; however, altitude–temperature and ice–albedo feedbacks are modelled and account for the differences between ISM results.

The models show that there were significant differences between the modern and mid-Pliocene ice sheets on Antarctica. Modern Antarctic temperatures rarely climb above 0 °C, especially in East Antarctica (Comiso 2000), therefore very little ice sheet runoff is observed or predicted

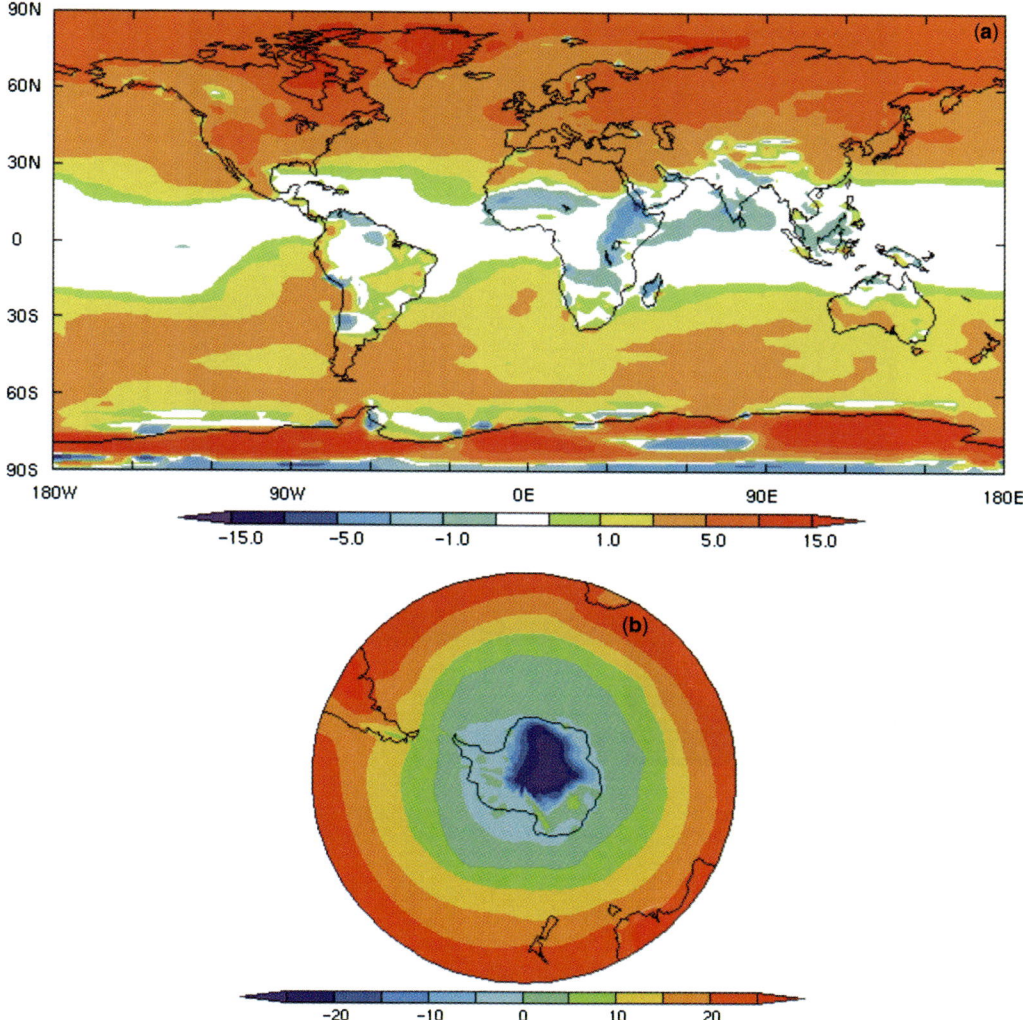

Fig. 5. Mid-Pliocene temperatures, utilized as boundary conditions within ice-sheet models, as simulated by HadAM3 GCM (Haywood *et al.* 2000). (**a**) Difference between mid-Pliocene and modern annual temperatures (°C). (**b**) Mid-Pliocene Antarctic January temperatures (°C).

(Jacobs *et al.* 1992). This is not the case in the mid-Pliocene climate simulations, where average January temperatures are significantly above the melting point of ice across some parts of East Antarctica (Fig. 5b). Hence, surface melting can become a significant factor in the equations of ice-sheet mass balance and drive deglaciation of the coastal regions of Antarctica (Fig. 7c). This type of ice sheet–climate regime is today restricted to Greenland, where surface melting constitutes at least 50% of all ice loss (Hanna *et al.* 2005).

In all cases, the extent of the ice sheet is reduced in the Wilkes and Aurora subglacial basins;

however, the average mid-Pliocene climate, modelled with the GCMs, seems insufficient to cause the large-scale deglaciation suggested by the diatom evidence. Although, in the lower ice volume reconstructions, the northern reaches of the Wilkes Subglacial Basin are deglaciated, this does not extend southwards to some of the drainage basins from which the proposed transport of marine diatoms into the central Transantarctic Mountains occurred (Harwood & Webb 1986).

Clearly, there are uncertainties in the boundary conditions that are utilized in these models. In order to test the effect of two of the most important

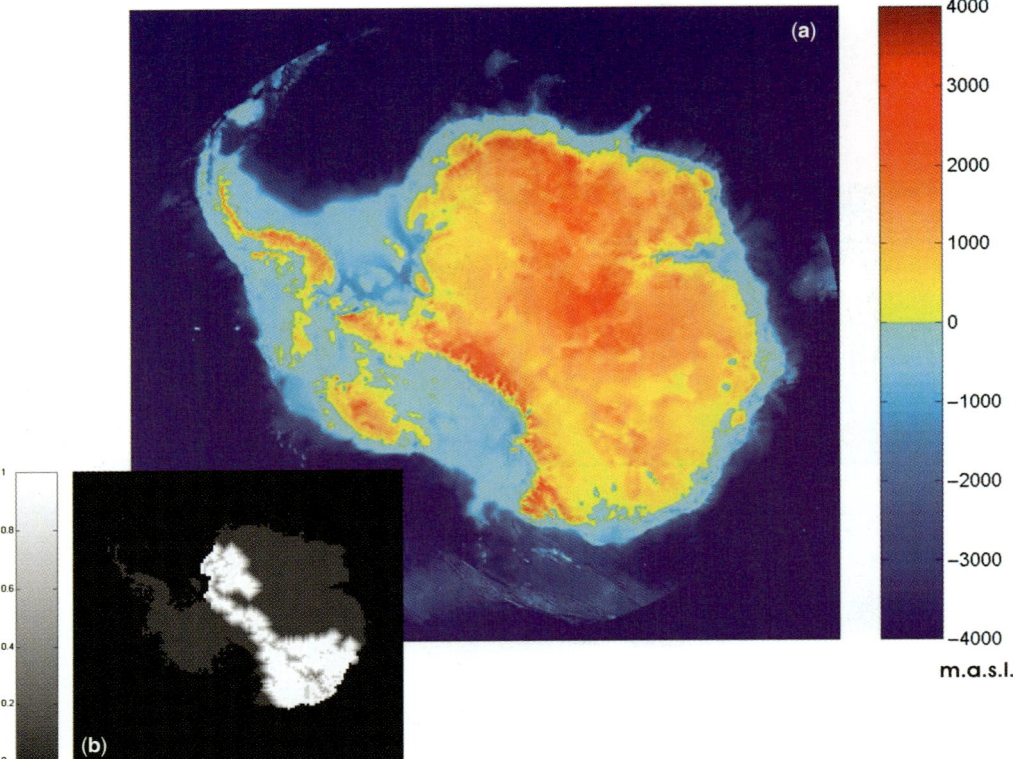

Fig. 6. (**a**) Isostatically rebounded Antarctic bed topography (Lythe *et al.* 2001). (**b**) Basin mask applied to Antarctic bed topography in order to change the depth of subglacial basins. Quoted change is multiplied by this field, enabling smooth transition zones.

boundary conditions – surface temperature and the bedrock topography of the subglacial basins – on the conclusions drawn from the equilibrium ice-sheet reconstructions, sensitivity experiments were undertaken. Antarctic temperatures are notoriously difficult to reproduce, either within a GCM modelling context (Connolley & Cattle 1994) or from observational datasets (Comiso 2000). The potential errors in modelling mid-Pliocene Antarctic temperatures are only increased by the lack of palaeoenvironmental data and could introduce significant uncertainty to the Pliocene ice-sheet model reconstructions. Thus, in order to test the sensitivity to the modelled temperature, the 'standard' mid-Pliocene GCM temperatures were incremented in discrete steps up to 10 °C. The large magnitude of this temperature anomaly should also encompass the potentially large, orbitally driven, temperature changes that occur within the PRISM-defined mid-Pliocene time slab (Haywood *et al.* 2002).

Some areas of the Antarctic have been explored extensively in recent years and many now have good coverage of bedrock topography data. However, large parts of the continent, particularly in East Antarctica, are unexplored and have little or no data on the subglacial environment (Lythe *et al.* 2001, p. 11338, Fig. 2b). Although a limited geological history of Antarctica can be reconstructed from the outcrops at the ice-sheet margin, the tectonic history of the Antarctic interior since the Pliocene is largely unknown. There are three main mountain ranges in East Antarctica. The geological history of the subglacial Gamburtsev Mountains is essentially unknown (Dalziel 1992), although it has been suggested they may have formed during the Carboniferous (Veevers 1994). The mountains of Dronning Maud Land are a series of ancient mountain belts, believed to have formed during the middle to late Mesozoic (Näslund 2004). Indeed, evidence from cosmogenic dating of glacial tills in the Sør Rondane Mountains

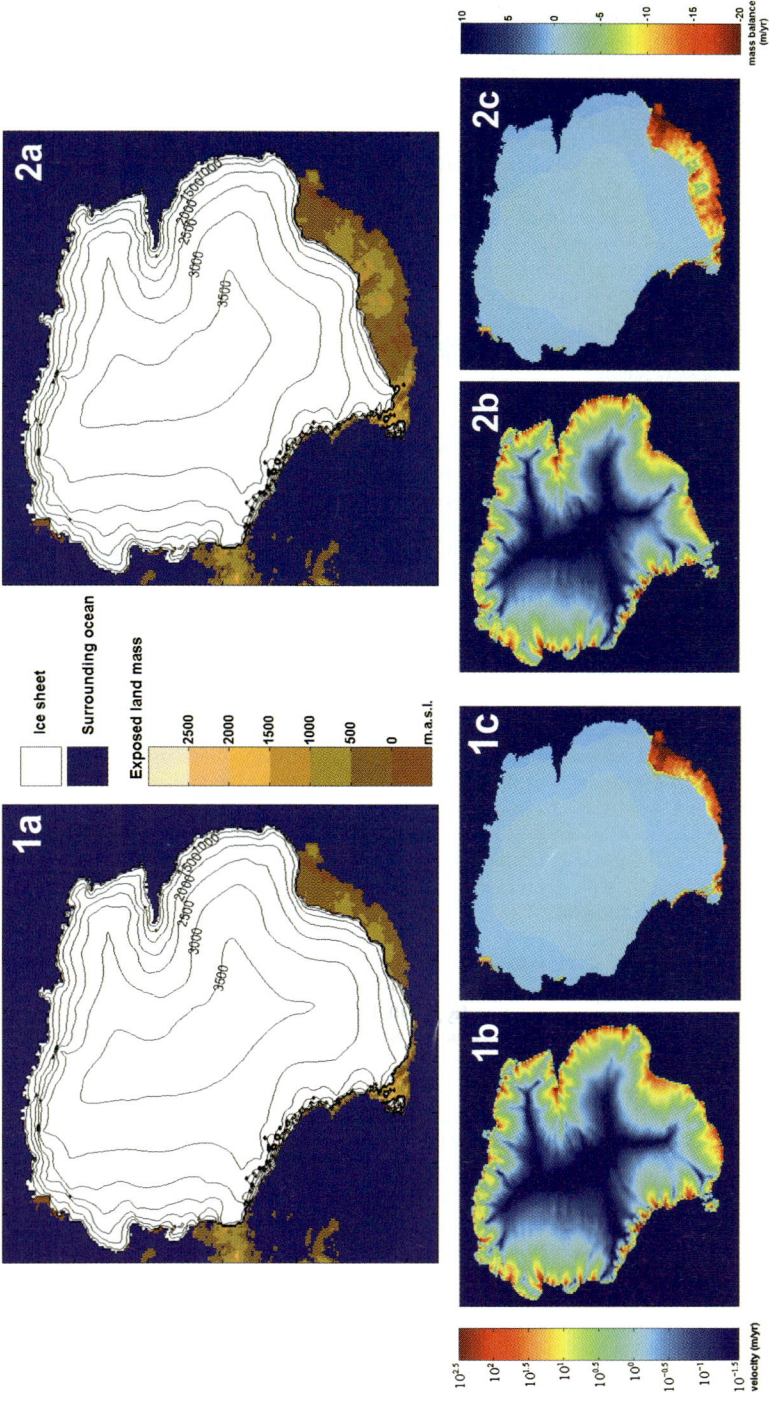

Fig. 7. Largest (**1**) and smallest (**2**) BASISM reconstructions of the mid-Pliocene EAIS under the modelled Hadley Centre GCM climate. (**a**) Surface height (m); (**b**) ice velocities (m/yr); and (**c**) mass balance of the modelled ice sheets (m/yr).

shows Pliocene ages for tills >300 m above the present ice surface (Moriwaki *et al.* 1992), suggesting this region had sufficient topography to benefit from the increased precipitation in the warmer Pliocene, without significant melting. The Transantarctic Mountains are the most well studied of East Antarctica's mountains, but provide an altogether more complex picture. Their Tertiary and Quaternary history is not well established and differing lines of evidence produce very different conclusions (Behrendt & Cooper 1991; Brook *et al.* 1995; Sugden *et al.* 1999; van der Wateren *et al.* 1996). This would seem to add a large uncertainty in the boundary conditions for our ice-sheet models. However, previous modelling studies seem to indicate that changes to the Transantarctic Mountain altitudes have only local consequences, with little impact across the ice sheet as a whole (Kerr & Huybrechts 1999). Altogether, this evidence suggests that the East Antarctic mountain topography does not add significant uncertainty to Pliocene ice-sheet modelling; however, the same may not be true of the subglacial basin topography.

The altitude of the Wilkes and Pensacola subglacial basins could have a number of potential impacts on the plausibility of the Pliocene deglaciation scenarios. The greater ice depth provides an increased potential altitude–temperature feedback. It also reduces the required rapidity of the proposed deglaciation, as a deeper basin would take longer to isostatically rebound to an altitude above sea level. If it is assumed that the dynamicist hypothesis is correct, then there should be a significant depth of sediment in the basins that would not have existed during the deglaciation phase of the mid-Pliocene. Hence in our sensitivity studies, the altitude of the subglacial basins (Fig. 6b) was decreased by up to 1000 m, simulating the removal of possible sediment packages.

Even in the most extreme scenarios modelled here, the southernmost areas of the subglacial basins remain glaciated (Fig. 8), meaning that marine deposition could not have occurred here. It appears from this study that the average mid-Pliocene climate is not sufficient to explain possible Pliocene ice-sheet collapse, enabling marine deposition in the southernmost Wilkes–Pensacola Subglacial Basin, even when tested errors in Antarctic temperature reconstructions and basin topography are taken into account. However, there remain a number of possible explanations of Pliocene marine deposition in the southern Wilkes Basin.

By analogy with the rapid deposition of sediments at the mouths of fast-flowing, polythermal glaciers in East Greenland (Dowdeswell *et al.* 1994), it was estimated that the mid-Pliocene sediments at Bardin Bluffs could represent as little as hundreds to thousands of years of marine deposition

(Hambrey & McKelvey 2000). If this proposed scenario occurred in the Lambert Glacier region, then it is plausible that a similarly short collapse could have occurred in the Wilkes Basin during the warmest period of the mid-Pliocene. The climatic conditions for such a collapse would not be simulated in the existing reconstructions of the mid-Pliocene climate, which are averaged over a 300,000-year interval (Dowsett *et al.* 1999).

One of the regions that recent measurements suggest is changing most rapidly is the Pine Island region, which has the greatest observed ice-thinning rate in Antarctica (Thomas *et al.* 2004; Davis *et al.* 2005). The oceans have been implicated in this rapid thinning, possibly a result of warm water flowing onto the continental shelf towards the ice stream grounding line (Jenkins *et al.* 2004; Payne *et al.* 2004; Shepherd *et al.* 2004; Bindschadler 2006). Such an ice sheet–ocean interaction could potentially lead to rapid and extensive ice-sheet retreat. However, the oceanographic perturbation would probably be beyond the resolution of the current generation of GCMs and such dynamical interaction is not possible with the current ISMs.

Another factor that may have led to an overestimation of the mid-Pliocene EAIS stability is the lack of a self-consistent representation of ice streams within existing ice-sheet models. Ice streams are the most dynamic of Antarctic Ice Sheet features (Rose 1979; Jacobel *et al.* 1996; Shepherd *et al.* 2002) and are responsible for over 90% of modern ice and sediment discharge (Bamber *et al.* 2000). They may also play an important role by propagating changes from the margins deep into the ice sheet's interior (Bamber *et al.* 2000; Payne *et al.* 2004). If a palaeo-ice stream existed in the Wilkes Basin, this could have a major implication on the ease with which ice could be removed from this region.

Our study only couples the GCM and the ISM in an offline scheme. This misses some of the major feedbacks in the ice sheet–climate system and as such may introduce uncertainties in the results. These potentially important feedbacks include sea ice, precipitation, atmospheric circulation, melting, iceberg-calving rate, ocean temperature and circulation feedbacks. Currently, there are no ice-sheet models that effectively reproduce the dynamics of the sea-ice boundary. A full appreciation of the physics of grounding-line migration is only now being gained (Schoof 2007). Whereas records of ice-sheet calving and oceanic boundary melting rates are spatially limited and only extend back a few decades (Losev 1963; Jacobs *et al.* 1992; Rignot & Thomas 2002), meaning that the dependence of rates on climatic and topographic parameters is poorly understood. In this modelling study, the problems presented by grounding-line

Increase in temperature above mid-Pliocene 'standard' HadAM3 model (°C)

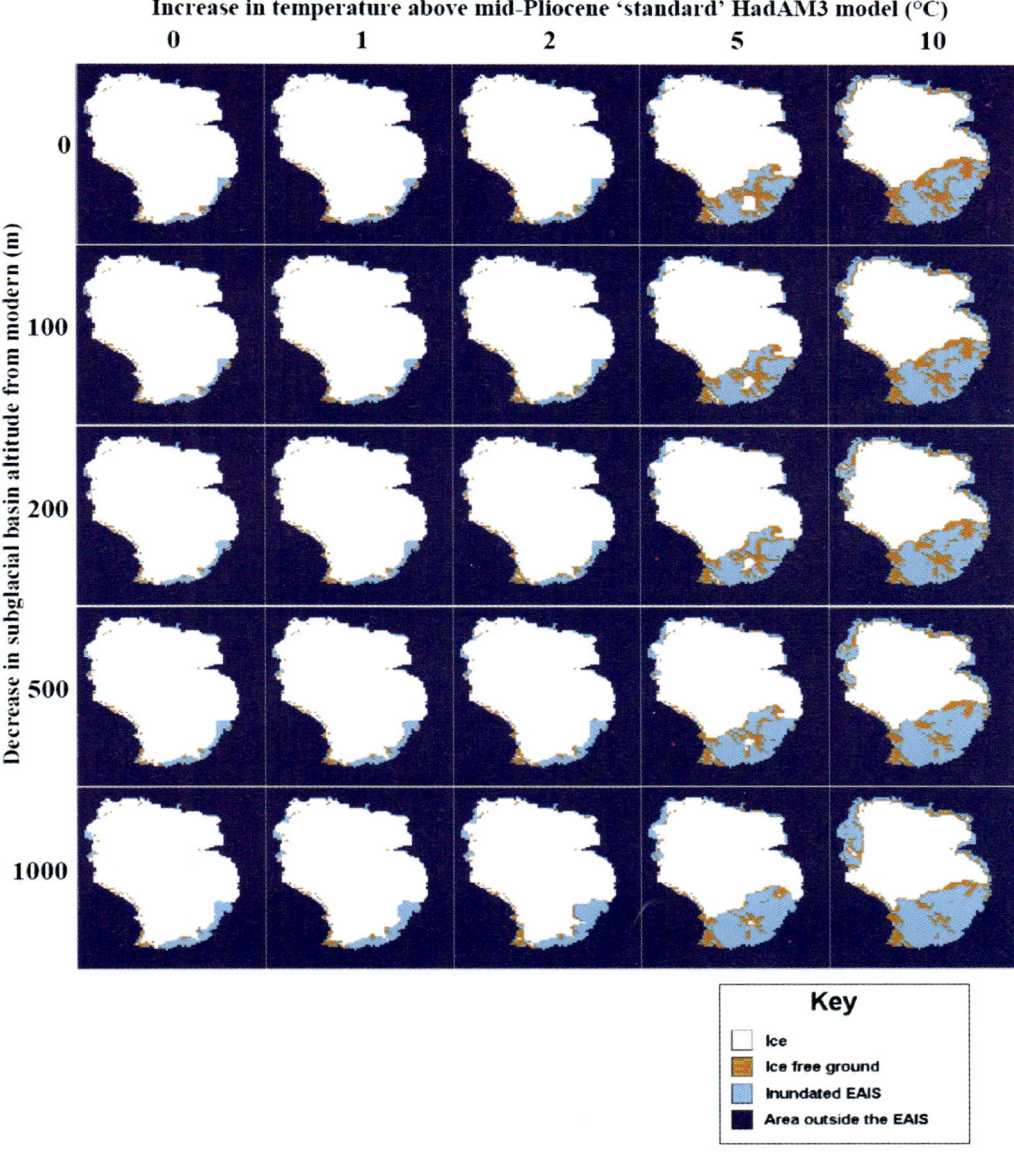

Fig. 8. Results of a sensitivity study of modelled temperatures and basin altitude for the EAIS. For each scenario, the depicted ice sheet represents the maximum marine incursion into the subglacial basins and minimum ice-sheet extent. These reconstructions are not representative of the equilibrium state, but also capture any transient ice-sheet responses.

migration are overcome by fixing the model grounding line at the observed, modern grounding line (Lythe *et al.* 2001). Sensitivity experiments (not presented here) on calving rates (in which only a water depth dependence was included) showed that even for large rates the results presented here are not conceptually different. Although calving can cause significant drawdown of the ice sheet, the reduction in extent that this can produce is limited by the topography and isostatic response rather than the magnitude of the calving rate.

Clearly, there is much work to do before a complete understanding of Pliocene ice sheets can be achieved. Terrestrial data from Antarctic exposures will always be sparse and important new revelations from this source, although possible, are probably unlikely. There is some scope for important discoveries, such as the nature of the Wilkes Basin bed, from geophysical studies of the Antarctic continent. Continental shelf marine records also have much to offer, with a far greater possibility of new Pliocene sediments coming to light. Drilling around Antarctica offers the possibility of being able to constrain the advances of the EAIS and the WAIS onto the continental shelf. However, this record will never give us the complete picture, as sediments will have been eroded away by subsequent ice advances and evidence of the extent of retreats will, naturally, be limited.

Modelling remains an important tool if we are to understand Antarctic palaeoclimates. There are many improvements that remain to be implemented if we are to be able to accurately reconstruct past ice sheets. However, advancements both in our understanding of the physics of the ice sheet–climate system and in modelling techniques continue to be made. In the near future, more-sophisticated ISMs will be fully integrated into climate models and careful reconstruction of the boundary conditions will allow more accurate constraint of palaeo-ice sheets.

Conclusion

The Pliocene is an important epoch for palaeo-glaciology. It marks the transition from warmer climates with smaller, possibly more dynamic (Rebesco et al. 2006) ice sheets into the bipolar, cold, icehouse world of the Pleistocene. Evidence and modelling of the Pliocene ice sheets suggests that they existed in a generally smaller form compared to present, but fluctuated significantly. The major question that remains, particularly so for the EAIS, is what were the magnitude and limits of these fluctuations? Modelling of the climate and ice sheets can help constrain these values. However, these techniques are only now being turned to the Pliocene and there remains room for further development.

Sensitivity modelling of the EAIS suggests that the climate, averaged over the c. 300,000 years (3.29–2.97 Ma) of the mid-Pliocene warm period, is not sufficient to cause the deglaciation scenarios proposed by Harwood and Webb (1986) based on Sirius Group diatom deposition. Conversely, even the largest reconstructed extents of the EAIS are smaller than modern, suggesting that the inferences made from Dry Valleys landscape stability cannot be extended to the entire continent. The Pliocene seems to have experienced significant changes in cryosphere volume and extent, with temperature increases within the range predicted for the coming century (IPCC 2001). Ice-sheet volumes and extents do not evolve linearly with temperature. Therefore hysteresis in the ice sheet–climate system may mean a similar future temperature increase produces less cryospheric change than seen in the Pliocene. However, there are many examples of potentially climate-induced changes in modern ice sheets, and palaeoclimate studies of warmer climates such as the Pliocene do not seem to rule out significant changes to the cryosphere.

The Natural Environment Research Council is acknowledged for funding the PhD of DJH and supporting the work of AH and PJV through the provision of High Performance Computing. This paper is published as part of the British Antarctic Survey's GEACEP programme that seeks to investigate climate change over geological timescales. We also thank Harry Dowsett, the PRISM Group and the BEDMAP consortium for use of their materials, John Smellie for his helpful comments on this manuscript and Bob Oglesby and an anonymous reviewer for their encouraging remarks and useful suggestions.

References

ABREU, V. S. & ANDERSON, J. B. 1998. Glacial eustacy during the Cenozoic; sequence stratigraphic implications. American Association of Petroleum Geologists Bulletin, 82, 1385–1400.

ACKERT, R. P. & KURZ, M. D. 2004. Age and uplift rates of Sirius Group sediments in the Dominion Range, Antarctica, from surface exposure dating and geomorphology. Global and Planetary Change, 42, 207–225.

AGER, T. A. 1994. Terrestrial palynology and paleobotanical records of Pliocene age from Alaska and Yukon Territory. In: THOMPSON, R. S. (ed.) Pliocene terrestrial environments and data/model comparisons. US Geological Survey Open-File Report, 94-23, 2–3.

ASHWORTH, A. C. & KUSCHEL, G. 2003. Fossil weevils (Coleoptera: Curculionidae) from latitude 85° S Antarctica. Palaeogeography, Palaeoclimatology, Palaeoecology, 191, 191–202.

ASHWORTH, A. C. & CANTRILL, D. J. 2004. Neogene vegetation of the Meyer Desert Formation (Sirius Group), Transantarctic Mountains, Antarctica. Palaeogeography, Palaeoclimatology, Palaeoecology, 213, 65–82.

ASKIN, R. A. & MARKGRAF, V. 1987. Palynomorphs from the Sirius Fromation, Dominion Range, Antarctica. Antarctic Journal of the United States, 21, 34–35.

BAMBER, J. L., VAUGHAN, D. G. & JOUGHIN, I. 2000. Widespread complex flow in the interior of the Antarctic Ice Sheet. Science, 287, 1248–1250.

BARKER, P. F. & KENNETT, J. P. (eds) 1988. Proceedings of the Ocean Drilling Program, Scientific Results, 113. College Station, TX, (Ocean Drilling Program).

BARKER, P. F. 1992. The sedimentary record of Antarctic climate change. Philosophical Transactions of the Royal Society, London, 338, 259–267.

BARRETT, P. J., ADAMS, C. J., MCINTOSH, W. C., SWISHER, III, C. C. & WILSON, G. S. 1992.

Geochronological evidence supporting Antarctic deglaciation three million years ago. *Nature*, **359**, 816–818.

BART, P. J. & ANDERSON, J. B. 2000. Relative temporal stability of the Antarctic Ice Sheets during the late Neogene based on the maximum frequency of outer shelf grounding events. *Earth and Planetary Science Letters*, **182**, 259–272.

BART, P. J., EGAN, D. & WARNY, S. A. 2005. Direct constraints on Antarctic Peninsula Ice Sheet grounding events between 5.12 and 7.94 Ma. *Journal of Geophysical Research – Earth Surface*, **110**, Art. No. F04008.

BEHRENDT, J. C. & COOPER, A. 1991. Evidence of rapid Cenozoic uplift of the shoulder escarpment of the Cenozoic West Antarctic rift system and a speculation on possible climate forcing. *Geology*, **19**, 315–319.

BENNIKE, O., ABRAHAMSEN, N., BAK, M., ISRAELSON, C., KONRADI, P., MATTHIESSEN, J. & WITKOWSKI, A. 2002. A multi-proxy study of Pliocene sediments from Ile de France, North-East Greenland. *Palaeogeography, Palaeoclimatology, Palaeoecology*, **186**, 1–23.

BINDSCHADLER, R. A. 2006. Hitting the ice sheets where it hurts. *Science*, **311**, 1720–1721.

BINDSCHADLER, R. A., ROBERTS, E. P. & MACAYEAL, D. R. 1989. Distribution of net mass balance in the vicinity of Crary Ice Rise, Antarctica. *Journal of Glaciology*, **35**, 370–377.

BRAITHWAITE, R. J. 1995. Positive degree-day factors for ablation on the Greenland ice sheet studied by energy-balance modelling. *Journal of Glaciology*, **41**, 153–160.

BROOK, E. J., BROWN, E. T., KURZ, M. D., ACKERT, R. P., RAISBECK, G. M. & YIOU, F. 1995. Constraints on age, erosion and uplift of Neogene glacial deposits in the Transantarctic Mountains determined from *in situ* cosmogenic [10]Be and [26]Al. *Geology*, **23**, 1063–1066.

BURCKLE, L. H. & POTTER, N. 1996. Pliocene–Pleistocene diatoms in Paleozoic and Mesozoic sedimentary and igneous rocks from Antarctica: A Sirius problem solved. *Geology*, **24**, 235–238.

CLARK, P. U., ALLEY, R. B. & POLLARD, D. 1999. Northern Hemisphere ice-sheet influences on global climate change. *Science*, **286**, 1104–1111.

COMISO, J. C. 2000. Variability and trends in Antarctic surface temperatures from *in situ* and satellite infrared measurements. *Journal of Climate*, **13**, 1674–1696.

CONNOLLEY, W. M. & CATTLE, H. 1994. The Antarctic climate of the UKMO Unified Model. *Antarctic Science*, **6**, 115–122.

COOK, A. J., FOX, A. J., VAUGHAN, D. G. & FERRIGNO, J. G. 2005. Retreating glacier fronts on the Antarctic Peninsula over the past half-century. *Science*, **308**, 541–545.

COWAN, E. A. 2002. Identification of the glacial signal from the Antarctic Peninsula since 3.0 Ma at Site 1101 in a continental rise sediment drift. *In*: BARKER, P. F., CAMERLENGHI, A., ACTON, G. D. & RAMSEY, A. T. S. (eds) *Proceedings of the Ocean Drilling Program*, Scientific Results, **178**, 1–22.

CSANK, A. Z. 2006. *Pliocene climate change on Ellesmere Island, Canada: annual variability determined from stable isotopes of fossil wood*. MSc. Thesis University of Saskatchewan.

DAHL-JENSEN, D. 2006. NGRIP ice core reveals detailed climatic history 123 kyrs back in time. *PAGES News*, **14**, 15–16.

DALZIEL, I. W. D. 1992. Antarctica: a tale of two super-continents? *Annual Review of Earth and Planetary Sciences*, **20**, 501–526.

DAVIS, C. H., LI, Y., MCCONNELL, J. R., FREY, M. M. & HANNA, E. 2005. Snowfall-driven growth in East Antarctic Ice Sheet mitigates recent sea-level rise. *Science*, **308**, 1898–1901.

DE ANGELIS, H. & SKARCA, P. 2003. Glacier surge after ice shelf collapse. *Science*, **299**, 1560–1562.

DE VERNAL, A. & MUDIE, P. J. 1989. Late Pliocene to Holocene palynostratigraphy at ODP Site 645, Baffin Bay. *In*: SRIVASTAVA, S. P., ARTHUR, M., CLEMENT, B. *ET AL.* (eds) *Proceedings of the Ocean Drilling Program, Scientific Results*, **105**, 401–422.

DINGLE, R. V., MCARTHUR, J. M. & VROON, P. 1997. Oligocene and Pliocene interglacial events in the Antarctic Peninsula dated using strontium isotope stratigraphy. *Journal of the Geological Society*, **154**, 257–264.

DINGLE, R. V. & LAVELLE, M. 1998. Antarctic Peninsula cryosphere: Early Oligocene (*c.* 30 Ma) initiation and a revised glacial chronology. *Journal of the Geological Society, London*, **15**, 433–437.

DOWDESWELL, J. A., WITTINGTON, R. J. & MARIENFELD, P. 1994. The origin of massive diamicton facies by iceberg rafting and scouring, Scoresby Sund, East Greenland. *Sedimentology*, **41**, 21–35.

DOWSETT, H. J. & CRONIN, T. M. 1990. High eustatic sea level during the middle Pliocene: Evidence from the southeastern U.S. Atlantic Coastal Plain. *Geology*, **18**, 435–438.

DOWSETT, H. J., CRONIN, T. M., POORE, R. Z., THOMPSON, R. S., WHATLEY, R. C. & WOOD, A. M. 1992. Micropaleontological evidence for increased meridonal heat-transport in the North-Atlantic Ocean during the Pliocene. *Science*, **258**, 1133–1135.

DOWSETT, H. J., BARRON, J. A., POORE, R., THOMPSON, R. S., CRONIN, T. M., ISHMAN, S. E. & WILLARD, D. 1999. Middle Pliocene paleoenvironmental reconstruction: PRISM2, *US Geological Survey Open File Report* **99–535**, http://pubs.usgs.gov/openfile/of99-535.

FITZGERALD, P. G. 1994. Thermochronologic constraints on post-Paleozoic tectonic evolution of the central Transantarctic Mountains, Antarctica. *Tectonics*, **13**, 818–836.

FITZGERALD, P. G., SANDIFORD, M., BARRETT, P. J. & GLEADOW, A. J. W. 1986. Asymmetric extension associated with uplift and subsidence in the Transantarctic Mountains and Ross Embayment. *Earth and Planetary Science Letters*, **81**, 67–78.

FITZGERALD, P. G. & STUMP, E. 1997. Cretaceous and Cenozoic episodic denudation of the Transantarctic Mountains, Antarctica: new constraints from apatite fission track thermochronology in the Scott Glacier region. *Journal of Geophysical Research*, **102**, 7747–7765.

FLEMING, R. P. & BARRON, J. A. 1996. Evidence of Pliocene *Nothofagus* in Antarctica from Pliocene marine sedimentary deposits (DSDP Site 274). *Marine Micropaleontology*, **27**, 227–236.

FLOWER, B. P. & KENNETT, J. P. 1994. The middle Miocene climatic transition: East Antarctic ice sheet development, deep ocean circulation and global carbon cycling. *Palaeogeography, Palaeoclimatology, Palaeoecology,* **108**, 537–555.

FRADKINA, A. F. 1991. Pliocene climate fluctuations in the far north-east of the USSR. Pliocene climates of the Northern Hemisphere: abstracts of the Joint US/USSR Workshop on Pliocene Paleoclimates, Moscow, USSR, April, 1990. *US Geological Survey Open-File Report* **91-447**, 22–23.

FRANCIS, J. E. & HILL, R. S. 1996. Fossil plants from the Pliocene Sirius Group, Transantarctic Mountains: Evidence for climate from growth rings and fossil leaves. *Palaios,* **11**, 389–396.

FUNDER, S., BENNIKE, O., BOCHER, J., ISRAELSON, C., PETERSEN, K. S. & SIMONARSON, L. A. 2001. Late Pliocene Greenland – The Kap København Formation in North Greenland. *Bulletin of the Geological Society of Denmark,* **48**, 117–134.

FYLES, J. G. 1989. High terrace sediments, probably of Neogene age, west-central Ellesmere Island, Northwest Territories. *Current Research, Geological Survey of Canada.* Paper **89-1D**, 101–104.

FYLES, J. G. 1990. Beaufort Formation (Late Tertiary) as seen from Prince Patrick Island, Arctic Canada. *Arctic,* **43**, 393–403.

FYLES, J. G., HILLS, L. V., MATTHEWS, J. V., BARENDREGT, R., BAKER, J., IRVING, E. & JETTÉ, H. 1994. Ballast Brook and Beaufort Formations (Late Tertiary) on northern Banks Island, Arctic Canada. *Quaternary International,* **22/23**, 141–171.

GERSONDE, R., KYTE, F. T., BLEIL, U. *ET AL.* 1997. Geological record and reconstruction of the late Pliocene impact of the Eltanin asteroid in the Southern Ocean. *Nature,* **390**, 357–363.

GOFF, J. R., JENNINGS, I. W. & DICKINSON, W. W. 2002. Depositional environment of Sirius Group sediments, Table Mountains, Dry Valleys area, Antarctica. *Geografiska Annaler,* **84A**, 11–24.

HAMBREY, M. J. & MCKELVEY, B. C. 2000. Neogene fjordal sedimentation on the western margin of the Lambert Graben, East Antarctica. *Sedimentology,* **47**, 577–607.

HAMBREY, M. J., WEBB, P.-N., HARWOOD, D. M. & KRISSEK, L. A. 2003. Neogene glacial record from the Sirius Group. *Geological Society of America Bulletin,* **115**, 994–1015.

HAMBREY, M. J. & SMELLIE, J. L. 2006. Distribution, lithofacies and environmental context of Neogene glacial sequences on James Ross and Vega Islands, Antarctic Peninsula. *In:* FRANCIS, J. E., PIRRIE, D. & CRAME, J. A. (eds) *Cretaceous–Tertiary High-Latitude Palaeoenvironments.* Geological Society, London, Special Publications, **258**, 187–200.

HANNA, E., HUYBRECHTS, P., JANSSENS, I., CAPPELEN, J., STEFFEN, K. & STEPHENS, A. 2005. Runoff and mass balance of the Greenland ice sheet: 1958–2003. *Journal of Geophysical Research,* **110**, D13108, doi:10.1029/2004JD005641.

HARWOOD, D. M. 1983. Diatoms from the Sirius Formation, Transantarctic Mountains. *Antarctic Journal of the United States,* **18**, 98–100.

HARWOOD, D. M. & WEBB, P.-N. 1986. Recycled marine microfossils from basal debris-ice in ice-free valleys of southern Victoria Land. *Antarctic Journal of the United States,* **21**, 87–88.

HAYWOOD, A. M., VALDES, P. J. & SELLWOOD, B. W. 2000. Global scale palaeoclimate reconstruction of the middle Pliocene climate using the UKMO GCM: initial results. *Global and Planetary Change,* **25**, 239–256.

HAYWOOD, A. M., VALDES, P. J. & SELLWOOD, B. W. 2002. Magnitude of climate variability during middle Pliocene warmth: a palaeoclimate modelling study. *Palaeogeography, Palaeoclimatology, Palaeoecology,* **188**, 1–24.

HAYWOOD, A. M. & VALDES, P. J. 2004. Modelling Pliocene warmth: contribution of atmosphere, oceans and cryosphere. *Earth and Planetary Science Letters,* **218**, 363–377.

HAYWOOD, A. M., DEKENS, P., RAVELO, A. C. & WILLIAMS, M. 2005. Warmer tropics in the mid-Pliocene? Evidence from alkenone palaeothermometry and a fully coupled ocean-atmosphere GCM. *Geochemistry, Geophysics, Geosystems,* **6**, art. no. Q03010.

HEPP, D. A., MORZ, T. & GRUTZNER, J. 2006. Pliocene glacial cyclicity in a deep-sea sediment drift (Antarctic Peninsula Pacific Margin). *Palaeogeography, Palaeoclimatology, Palaeoecology,* **231**, 181–198.

HILLENBRAND, C.-D. & EHRMANN, W. 2002. Distribution of clay minerals in drift sediments on the continental rise west of the Antarctic Peninsula, ODP Leg 178, Sites 1095 and 1096. *In:* BARKER, P. F., CAMERLENGHI, A., ACTON, G. D. & RAMSEY, A. T. S. (eds) *Proceedings of the Ocean Drilling Program, Scientific Results,* **178**, 1–29.

HILLENBRAND, C.-D. & EHRMANN, W. 2003. Palaeo-environmental implications of Tertiary sediments from Kainan Maru Seamount and northern Gunnerus Ridge. *Antarctic Science,* **15**, 522–536.

HILLENBRAND, C.-D. & EHRMANN, W. 2005. Late Neogene to Quaternary environmental changes in the Antarctic Peninsula region: evidence from drift sediments. *Global and Planetary Change,* **45**, 165–191.

HINDMARSH, R. C. A. 1999. On the numerical computation of temperature in an ice sheet. *Journal of Glaciology,* **45**, 568–574.

HINDMARSH, R. C. A. 2001. Influence of channelling on heating in ice-sheet flows. *Geophysical Research Letters,* **28**, 3681–3684.

HINDMARSH, R. C. A., VAN DER WATEREN, F. M. & VERBERS, A. L. L. M. 1998. Sublimation of ice through sediment in Beacon Valley. *Geographiska Annaler,* **80A**, 209–219.

HUYBRECHTS, P. 1990. A 3-D model for the Antarctic ice sheet: a sensitivity study on the glacial–interglacial contrast. *Climate Dynamics,* **5**, 79–92.

HUYBRECHTS, P. 1993. Glaciological modelling of the late Cenozoic east Antarctic ice sheet: Stability or dynamism? *Geografiska Annaler,* **75A**, 221–238.

IPCC 2001. Intergovernmental Panel on Climate Change Third Assessment Report: Climate Change 2001: The Scientific Basis. Cambridge University Press.

ISHMAN, S. E. & RIECK, H. J. 1992. A late Neogene Antarctic glacio-eustatic record Victoria Land Basin margin, Antarctica. *In*: KENNETT, J. P. & WARNKE, D. A. (eds) The Antarctic Palaeoenvironment: A Perspective on Global Change, *Antarctic Research Series*, **56**, 327–347.

IVANY, L. C., VAN SIMAEYS, S., DOMACK, E. W. & SAMSON, S. D. 2006. Evidence for an earliest Oligocene ice sheet on the Antarctic Peninsula. *Geology*, **34**, 377–380.

JACOBEL, R. W., SCAMBOS, C. F., RAYMOND, C. F. & GADES, A. M. 1996. Changes in the configuration of ice stream flow from the West Antarctic Ice Sheet. *Journal of Geophysical Research*, **101**, 5499–5504.

JACOBS, S. S., HELMER, H. H., DOAKE, C. S. M., JENKINS, A. & FROLICH, R. M. 1992. Melting ice shelves and the mass balance of Antarctica. *Journal of Glaciology*, **38**, 375–387.

JANSEN, E., FRONVAL, T., RACK, F. & CHANNELL, J. E. T. 2000. Pliocene–Pleistocene ice rafting history and cyclicity in the Nordic Seas during the last 3.5 Myr. *Paleoceanography*, **15**, 709–721.

JENKINS, A., HAYES, D., BRANDON, M., POZZI-WALKER, Z., HARDY, S. & BANKS, C. 2004. Oceanographic observations at the Amundsen Sea shelf break. *Forum for Research into Ice Shelf Processes (FRISP) Report*, **No. 15**.

JONKERS, H. A., LIRIO, J. M., DEL VALLE, R. A. & KELLEY, S. P. 2002. Age and environment of Miocene – Pliocene glaciomarine deposits, James Ross Island, Antarctica. *Geological Magazine*, **139**, 577–594.

KELLOGG, D. E. & KELLOGG, T. B. 1996. Diatoms in South Pole ice: Implications for eolian contaminations of Sirius Group deposits. *Geology*, **24**, 115–118.

KENNETT, J. P. 1977. Cenozoic evolution of Antarctic glaciation, the circum-Antarctic Ocean and their impact on global paleoceanography. *Journal of Geophysical Research*, **82**, 3843–3860.

KERR, A. & HUYBRECHTS, P. 1999. The response of the East Antarctic ice-sheet to the evolving tectonic configuration of the Transantarctic Mountains. *Global and Planetary Change*, **23**, 213–229.

KRANTZ, D. E. 1991. A chronology of Pliocene sea-level fluctuations: the U.S. middle Atlantic Coastal Plain record. *Quaternary Science Review*, **10**, 163–174.

KRISTJANSSON, L., GUDMUNDSSON, M. T., SMELLIE, J. L., MCINTOSH, W. C. & ESSER, R. 2005. Palaeomagnetic and stratigraphical mapping of Miocene–Pliocene basalts in the Brandy Bay area, James Ross Island, Antarctica. *Antarctic Science*, **17**, 409–417.

KWIEK, P. B. & RAVELO, A. C. 1999. Pacific Ocean intermediate and deep water circulation during the Pliocene. *Palaeogeography, Palaeoclimatology, Palaeoecology*, **154**, 191–217.

LAGOE, M. B. & ZELLERS, S. D. 1996. Depositional and microfaunal response to Pliocene climate change and tectonics in the eastern Gulf of Alaska. *Marine Micropaleontology*, **27**, 121–140.

LARSEN, H. C., SAUNDERS, A. D., CLIFT, P. D. *ET AL.* 1994. 7-million years of glaciation in Greenland. *Science*, **264**, 952–955.

LEAR, C. H., ELDERFIELD, H. & WILSON, P. A. 2000. Cenozoic deep-sea temperatures and global ice volumes from Mg/Ca in benthic foraminiferal calcite. *Science*, **287**, 269–273.

LEMASURIER, W. E., HARWOOD, D. M. & REX, D. C. 1994. Geology of Mount Murphy Volcano: An 8-m.y. history of interaction between a rift volcano and the West Antarctic Ice Sheet. *Geological Society of America Bulletin*, **106**, 265–280.

LENTON, T. M. 2006. Climate change to the end of the millennium. *Climatic Change*, **76**, 7–29.

LIETZ, J. & SCHMINCKE, H. U. 1975. Miocene–Pliocene sea-level changes and volcanic phases on Gran Canaria (Canary Islands) in light of new K–Ar ages. *Palaeogeography, Palaeoclimatology, Palaeoecology*, **18**, 213–239.

LOSEV, K. S. 1963. Computations of the mass balance of the Antarctic ice cap. *Soviet Antarctic Expedition Information Bulletin*, **5**, 60–63.

LYTHE, M. B. & VAUGHAN, D. G. the BEDMAP Consortium 2001. BEDMAP: a new thickness and subglacial topographic model of Antarctica. *Journal of Geophysical Research*, **106**, 11335–11351.

MANN, M. E., BRADLEY, R. S. & HUGHES, M. K. 1999. Northern Hemisphere Temperatures During the Past Millennium: Inferences, Uncertainties, and Limitations, *Geophysical Research Letters*, **26**, 759–762.

MARCHANT, D. R., DENTON, G. H., SWISHER, C. C. & POTTER, N. 1996. Late Cenozoic Antarctic paleoclimate reconstructed from volcanic ashes in the Dry Valley region of southern Victoria Land. *Geological Society of America Bulletin*, **108**, 181–194.

MATTHEWS, E. 1983. Global vegetation and land use: new high resolution data bases for climate studies, *Journal of Climate and Applied Meteorology*, **22**, 474–487.

MATTHEWS, J. V. 1987. Plant macrofossils from the Neogene Beaufort Formation on Banks and Meighen Islands, District of Franklin. *Current Research, Geological Survey of Canada*, Paper **87-1A**, 73–87.

MATTHEWS, J. V. & OVENDEN, L. E. 1990. Late Tertiary plant macrofossils from localities in Arctic/subarctic North America: a review of the data. *Arctic*, **43**, 364–392.

MATTHEWS, J. V., WESTGATE, J. A., OVENDEN, L. E., CARTER, L. D. & FOUCH, T. 2003. Stratigraphy, fossils and age of sediments at the upper pit of the Lost Chicken gold mine: new information on the late Pliocene environment of east central Alaska. *Quaternary Research*, **60**, 9–18.

MATTHEWS, R. K. 1986. The $\delta^{18}O$ signal of deep-sea planktonic foraminifera at low latitudes as an ice-volume indicator. *South African Journal of Science*, **82**, 521–522.

MCKELVEY, B. C., WEBB, P.-N., HARWOOD, D. M. & MABIN, M. G. C. 1991. The Dominion Range Sirius Group – A record of the late Pliocene – early Pleistocene Beardmore glacier. *In*: THOMSON, M.R.A. (eds) *Geological Evolution of Antarctica*, 675–682.

MCKELVEY, B. C., HAMBREY, M. J., HARWOOD, D. M., MABIN, M. C. G., WEBB, P.-N. & WHITEHEAD, J. M. 2001. The Pagodroma Group – a Cenozoic record of the East Antarctic Ice Sheet in the northern Prince Charles Mountains. *Antarctic Science*, **13**, 455–468.

MILLER, K. G., FAIRBANKS, R. G. & MOUNTAIN, G. S. 1987. Tertiary oxygen isotope synthesis, sea level history and continental margin erosion. *Paleoceanography*, **2**, 1–19.

MORIWAKI, K., HIRAKAWA, M., HAYASHI, M. & IWATA, S. 1992. Late Cenozoic glacial history of the Sør Rondane Mountains, East Antarctica. *In*: YOSHIDA, Y., KAMINUMA, K. & SHIRAISHI, K. (eds) *Recent Progress in Antarctic Earth Science*, 661–667.

NÄSLUND, J. O. 2004. Landscape development in western and central Dronning Maud Land, East Antarctica. *Antarctic Science*, **13**, 302–331.

OHMURA, A. 2001. Physical basis for the temperature-based melt-index method. *Journal of Applied Meteorology*, **40**, 753–761.

OTTO-BLIESNER, B. L., MARSHALL, S. J., OVERPECK, J. T., MILLER, G. H. & HU, A. CAPE Interglacial Project members 2006. Simulating Arctic climate warmth and icefield retreat in the Last Interglacial. *Science*, **311**, 1751–1753.

OVERPECK, J. T., OTTO-BLIESNER, B. L., MILLER, G. H., MUHS, D. R., ALLEY, R. B. & KIEHL, J. T. 2006. Paleoclimatic evidence for future ice-sheet instability and rapid sea-level rise. *Science*, **311**, 1747–1750.

PANTER, K. S., MCINTOSH, W. C. & SMELLIE, J. L. 1994. Volcanic history of Mount Sidley, a major alkaline volcano in Marie Byrd Land, Antarctica. *Bulletin of Volcanology*, **56**, 361–376.

PASSCHIER, S. 2004. Variability in geochemical provenance and weathering history of the Sirius Group strata, Transantarctic Mountains: Implications for Antarctic glacial history. *Journal of Sedimentary Research*, **74**, 607–619.

PAYNE, A. J., VIELI, A., SHEPHERD, A. P., WINGHAM, D. J. & RIGNOT, E. 2004. Recent dramatic thinning of largest West Antarctic ice stream triggered by oceans. *Geophysical Research Letters*, **31**, doi:10.1029/2004GL021284.

PEARSON, P. N. & PALMER, M. R. 2000. Atmospheric carbon dioxide concentrations over the past 60 million years. *Nature*, **406**, 695–699.

PENNEY, D. N. 1993. Late Pliocene to Early Pleistocene ostracod stratigraphy and palaeoclimate of the Lodin Elv and Kap København formations, East Greenland. *Palaeogeography, Palaeoclimatology, Palaeoecology*, **101**, 49–66.

PETIT, J. R., JOUZEL, J., RAYNAUD, D. *ET AL.* 1999. Climate and atmospheric history of the past 420,000 years from the Vostok ice core, Antarctica. *Nature*, **399**, 429–436.

POPE, V. D., GALLANI, M. L., ROWNTREE, P. R. & STRATTON, R. A. 2000. The impact of new physical parameterisations in the Hadley Centre model – HadAM3. *Climate Dynamics*, **16**, 123–146.

PUDSEY, C. J. 2002. Neogene record of Antarctic Peninsula glaciation in continental rise sediment: ODP Leg 178, Site 1095. *In*: BARKER, P. F., CAMERLENGHI, A., ACTON, G. D. & RAMSEY, A. T. S. (eds) *Proceedings of the Ocean Drilling Program, Scientific Results*, **178**, 1–25.

RAYMO, M. E., GRANT, M., HOROWITZ, M. & RAU, G. H. 1996. Mid-Pliocene warmth: Stronger greenhouse and stronger conveyor. *Marine Micropaleontology*, **27**, 313–326.

REBESCO, M., CAMERLENGHI, A., GELETTI, R. & CANALS, M. 2006. Margin architecture reveals the transition to the modern Antarctic ice sheet ca. 3 Ma. *Geology*, **34**, 301–304.

REEH, N. 1991. Parameterisation of melt rate and surface temperature on the Greenland ice sheet. *Polarforschung*, **59**, 113–128.

RETALLACK, G. J., KRULL, E. S. & BOCKHEIM, J. G. 2001. New grounds for reassessing palaeoclimate of the Sirius Group, Antarctica. *Journal of the Geological Society, London*, **158**, 925–935.

RIAL, J. A., PIELKE, S. R., BENISTON, M. *ET AL.* 2004. Nonlinearities, feedbacks and critical thresholds within the Earth's climate system. *Climatic Change*, **65**, 11–38.

RIGNOT, E. & THOMAS, R. H. 2002. Mass balance of polar ice sheets. *Science*, **297**, 1502–1506.

RIGNOT, E., CASASSA, G., GOGINENI, P., KRABILL, W., RIVERA, A. & THOMAS, R. 2004. Accelerated ice discharge from the Antarctic Peninsula following the collapse of Larsen B ice shelf. *Geophysical Research Letters*, **31**, doi:10.1029/2004GL020697.

RITZ, C., FABRE, A. & LETRÉGUILLY, A. 1997. Sensitivity of a Greenland ice sheet model to ice flow and ablation parameters: consequences for the evolution through the last glacial cycle. *Climate Dynamics*, **13**, 11–24.

ROSE, K. E. 1979. Characteristics of ice flow in Marie-Byrd Land, Antarctica. *Journal of Glaciology*, **24**, 63–75.

ROSTEK, F., RUHLAND, G., BASSINOT, F. C., MULLER, P. J., LABEYRIE, L. D., LANCELOT, Y. & BARD, E. 1993. Reconstructing sea-surface temperature and salinity using delta-O-18 and alkenone records. *Nature*, **364**, 319–321.

SCAMBOS, T. A., HULBE, C., FAHNESTOCK, M. & BOHLANDER, J. 2000. The link between climate warming and break-up of ice shelves in the Antarctic Peninsula. *Journal of Glaciology*, **46**, 516–530.

SCHERER, R. P., HARWOOD, D. M., ISHMAN, R. & WEBB, P.-N. 1988. Micropaleontological analyses of sediments from Crary Ice Rise. *Antarctic Journal of the United States*, **24**, 65–67.

SCHERER, R. P. 1991. Quaternary and Tertiary microfossils from beneath Ice Stream B: Evidence for dynamic West Antarctic Ice Sheet history. *Palaeogeography, Palaeoclimatology, Palaeoecology*, **90**, 395–412.

SCHOOF, C. 2007. Marine ice sheet dynamics. Part 1. The case of rapid sliding. *Journal of Fluid Mechanics*, **573**, 27–55.

SHEPHERD, A., WINGHAM, D. J. & MANSLEY, J. A. D. 2002. Inland thinning of the Amundsen Sea sector, Antarctica. *Geophysical Research Letters*, **29**, doi:10.1029/2001GL014183.

SHEPHERD, A., WINGHAM, D. J. & RIGNOT, E. 2004. Warm ocean is eroding West Antarctic Ice Sheet. *Geophysical Research Letters*, **31**, doi:10.1029/2004GL021106.

SLOAN, L. C., CROWLEY, T. J. & POLLARD, D. 1996. Modeling of middle Pliocene climate with the NCAR GENESIS general circulation model. *Marine Micropaleontology*, **27**, 51–61.

SMELLIE, J. L. 1999. Lithostratigraphy of Miocene–recent, alkaline volcanic fields in the Antarctic Peninsula and eastern Ellsworth Land. *Antarctic Science*, **11**, 362–378.

SMELLIE, J. L. 2000. Subglacial eruptions. *In*: SIGURDSSON, H. (ed.) *Encyclopaedia of Volcanoes*. Academic Press, 403–418.

SMELLIE, J. L. & HOLE, M. J. 1997. Products and processes in Pliocene – recent, subaqueous to emergent volcanism in the Antarctic Peninsula: examples of englacial Surtseyan volcano construction. *Bulletin of Volcanology*, **58**, 628–646.

SMELLIE, J. L., MCARTHUR, J. M., MCINTOSH, W. C. & ESSER, R. 2006. Late Neogene interglacial events in the James Ross Island region, northern Antarctic Peninsula, dated by Ar/Ar and Sr-isotope stratigraphy. *Palaeogeography, Palaeoclimatology, Palaeoecology*, **242**, 169–187.

SMELROR, M. 1999. Pliocene–Pleistocene and redeposited dinoflagellate cysts from the western Svalbard margin (site 986): Biostratigraphy, paleoenvironments and sediment provenance. *In*: RAYMO, M. E., JANSEN, E., BLUM, P. & HERBERT, T. D. (eds) *Proceedings of the Ocean Drilling Program, Scientific Results*, **162**, 83–97.

ST. JOHN, K. E. K. & KRISSEK, L. A. 2002. The late Miocene to Pleistocene ice-rafting history of southeast Greenland. *Boreas*, **31**, 28–35.

STOKER, M. S., LESLIE, A. B., SCOTT, W. D. *ET AL.* 1994. A record of late Cenozoic stratigraphy, sedimentation and climate change from the Hebrides Slope, NE Atlantic Ocean. *Journal of the Geological Society, London*, **151**, 235–249.

STROEVEN, A. P. 1997. The Sirius Group of Antarctica: age and environments. *In*: RICCI, C. A. *The Antarctic Region: Geological Evolution and Processes*, 747–761.

STROEVEN, A. P., PRENTICE, M. L. & KLEMEN, J. 1996. On marine microfossil transport and pathways in Antarctica during the late Neogene: Evidence from the Sirius Group at Mount Fleming. *Geology*, **24**, 727–730.

STROEVEN, A. P. & PRENTICE, M. L. 1997. A case for Sirius Group alpine glaciation at Mount Fleming, South Victoria Land, Antarctica: A case against Pliocene East Antarctic Ice Sheet reduction. *GSA Bulletin*, **109**, 825–840.

SUGDEN, D. E., DENTON, G. H. & MARCHANT, D. R. 1995. Landscape evolution of the Dry Valleys, Transantarctic Mountains: Tectonic implications. *Journal of Geophysical Research*, **100**, 9949–9967.

SUGDEN, D. E., SUMMERFIELD, M. A., DENTON, G. H., WILCH, T. I., MCINTOSH, W. C., MARCHANT, D. R. & RUTFORD, R. H. 1999. Landscape development in the Royal Society Range, southern Victoria Land, Antarctica: stability since the mid-Miocene, *Geomorphology*, **28**, 181–200.

SUMMERFIELD, M. A., STUART, F. M., COCKBURN, H. A.P., SUGDEN, D. E., DENTON, G. H., DUNAI, T. & MARCHANT, D. R. 1999. Long-term rates of denudation in the Dry Valleys, Transantarctic Mountains, southern Victoria Land, Antarctica based on in-situ-produced cosmogenic ^{21}Ne. *Geomorphology*, **27**, 113–129.

THOMAS, R., RIGNOT, E., CASASSA, G. *ET AL.* 2004. Accelerated sea-level rise from West Antarctica, *Science*, **306**, 255–258.

VAN DER BURGH, J., VISSCHER, H., DILCHER, D. L. & KÜRSCHNER, W. M. 1993. Paleoatmospheric signatures in Neogene fossil leaves. *Science*, **260**, 1788–1790.

VAN DER WATEREN, F. M., DUNAI, T. J., VAL BALEN, R. T., KLAS, W., VERBERS, A. L. L. M., PASSCHIER, S. & HERPERS, U. 1999. Contrasting Neogene denudation histories of different structural regions in the Transantarctic Mountains rift flank constrained by cosmogenic isotope measurements. *Global and Planetary Change*, **23**, 145–172.

VAUGHAN, D. G. & DOAKE, S. M. 1996. Recent atmospheric warming and retreat of ice shelves on the Antarctic Peninsula. *Nature*, **379**, 328–331.

VAUGHAN, D. G., BAMBER, J. L., GIOVINETTO, M., RUSSELL, J. & COOPER, A. P. R. 1999. Reassessment of net surface mass balance in Antarctica. *Journal of Climate*, **12**, 933–946.

VAUGHAN, D. G., MARSHALL, G. J., CONNELLY, W. M. *ET AL.* 2003. Recent rapid warming on the Antarctic Peninsula, *Climate Change*, **60**, 243–274.

VEEVERS, J. J. 1994. Case for the Gamburtsev subglacial mountains of East Antarctica originating by mid-Carboniferous shortening of an intracratonic basin. *Geology*, **22**, 593–596.

WARDLAW, B. R. & QUINN, T. M. 1991. The record of Pliocene sea-level change at Enewetak Atoll. *Quaternary Science Reviews*, **10**, 247–258.

WARNKE, D. A., RICHTER, C., FLORINDO, F. *ET AL.* 2004. Data report: HiRISC (High-Resolution Integrated Stratigraphy Committee) Pliocene–Pleistocene interval, 0-50mbsf at ODP Leg 188 Site 1165, Prydz Bay, Antarctica. *In*: COOPER, A. K., O'BRIEN, P. E. & RICHTER, C. (eds) *Proceedings of the Ocean Drilling Program, Scientific Results*, **188**, 1–38.

WEBB, P.-N. & HARWOOD, D. M. 1991. Late Cenozoic glacial history of the Ross Embayment, Antarctica. *Quaternary Science Reviews*, **10**, 215–223.

WEBB, P.-N., HARWOOD, D. M., MABIN, M. G. C. & MCKELVEY, B. C. 1996. A marine and terrestrial Sirius Group succession, middle Beardmore Glacier–Queen Alexandra Range, Transantarctic Mountains, Antarctica. *Marine Micropaleontology*, **27**, 273–297.

WHITEHEAD, J. M., QUILTY, P. G., HARWOOD, D. M. & MCMINN, A. 2001. Early Pliocene paleoenvironment of the Sørsdal Formation, Vestfold Hills, based on diatom data. *Marine Micropaleontology*, **41**, 125–152.

WHITEHEAD, J. M., HARWOOD, D. M., MCKELVEY, B. C., HAMBREY, M. J. & MCMINN, A. 2004. Diatom biostratigraphy of the Cenozoic glaciomarine Pagodroma Group, northern Prince Charles Mountains. *Australian Journal of Earth Science*, **51**, 521–547.

WHITEHEAD, J. M., WOTHERSPOON, S. & BOHATY, S. M. 2005. Minimal Antarctic sea ice during the Pliocene. *Geology*, **33**, 137–140.

WILLARD, D. A. 1994. Palynological record from the North Atlantic reion at 3 Ma: vegetational distribution during a period of global warmth. *Review of Palaeobotany and Palynology*, **83**, 275–297.

WILLARD, D. A. 1996. Pliocene–Pleistocene pollen assemblages from the Yermak Plateau, Arctic Ocean: Sites 910 and 911. *In*: MYHRE, T. J., FIRTH, A. M., JOHNSON, J. V. & RUDDIMAN, W. F. (eds) *Proceedings of the Ocean Drilling Program, Scientific Results*, **151**, 297–307.

WILLIAMS, M., HAYWOOD, A. M., HILLENBRAND, C.-D. & WILKINSON, I. P. 2005. Efficacy of $\delta^{18}O$ data from Pliocene foraminifer calcite for spatial sea surface temperature reconstruction: comparison with a fully coupled ocean-atmosphere GCM and fossil assemblage data for the mid-Pliocene. *Geological Magazine*, **142**, 399–417.

ZACHOS, J., PAGANI, M., SLOAN, L., THOMAS, E. & BILLUPS, K. 2001. Trends, rhythms and aberrations in global climate change 65 Ma to present. *Science*, **292**, 686–693.

ZINSMEISTER, W. J. & WEBB, P. N. 1982. Cretaceous–Tertiary geology and paleontology of Cockburn Island. *Antarctic Journal of the United States*, **17**, 41–42.

The application of the alkenone organic proxy to the study of Plio-Pleistocene climate

K. T. LAWRENCE[1], T. D. HERBERT[2], P. S. DEKENS[3] & A. C. RAVELO[3]

[1]*Lafayette College, Department of Geology and Environmental Geosciences,
102 Van Wickle Hall, Easton, PA 18042, USA (e-mail: lawrenck@lafayette.edu)*

[2]*Brown University, Department of Geological Sciences, Box 1846,
Providence, RI 02912, USA*

[3]*University of California Santa Cruz, Department of Ocean Sciences,
1156 High Street Santa Cruz, CA 95064, USA*

Abstract: The last major transition in Earth's history, from a world with unipolar to bipolar ice, occurred between the Pliocene and Pleistocene epochs. Yet a variety of challenges associated with most previous methods of determining Earth's surface temperatures results in a paucity of estimates constraining the evolution of this critical climatic variable through this transition. Here, we review the alkenone organic proxy, which allows for the rapid, independent, characterization of past sea-surface temperature, as well as the construction of qualitative records of past ocean productivity. We discuss the development, calibration and implementation of the alkenone method for the study of palaeoclimate problems and identify the limitations as well as some of the important considerations for its application. We specifically explore the use of these proxies to study Plio-Pleistocene climate, briefly summarizing the insights that the application of the alkenone method has provided about the Plio-Pleistocene transition. We conclude with an optimistic view of the potential use of these proxies to greatly augment our understanding of past climates.

Benthic oxygen isotope ($\delta^{18}O$) records form the foundation of our knowledge of Cenozoic climate, recording both the warming and cooling of the Earth's deep oceans and the waxing and waning of polar ice caps (Zachos *et al.* 2001). Our fundamental understanding of the Plio-Pleistocene transition as a change from the warmer, more ice-free, less variable conditions of the Pliocene, to the colder, more variable Pleistocene in which large ice-sheets expanded across the continents of the Northern Hemisphere is largely derived from these records. Yet little is known about the Plio-Pleistocene evolution of the Earth's surface temperature field, the most widely used variable for characterizing modern climates. Past efforts to monitor the evolution of surface temperatures have been hampered by time-intensive analysis techniques (e.g. faunal analysis) (Dowsett *et al.* 1992, 1996; Nikolaev *et al.* 1998; Dowsett *et al.* 2005), discontinuous records of sedimentation (e.g. terrestrial records, Thompson & Fleming 1996), or by the use of proxies with multiple climate sensitivities (e.g. planktonic $\delta^{18}O$, e.g. Shackleton & Opdyke 1973; Chappell & Shackleton 1986). Recently, three new palaeothermometers, the alkenone-based $U^{K'}_{37}$ index (Brassell *et al.* 1986; Prahl & Wakeham 1987; Prahl *et al.* 1988), Mg/Ca ratios from foraminifera (Nürnberg 1995; Nürnberg *et al.* 1996; Rosenthal *et al.* 1997; Lea *et al.* 1999; Rosenthal *et al.* 2000), and the TEX$_{86}$ index (Schouten *et al.* 2002, 2003), each of which allows for the independent characterization of past near-surface ocean temperatures, have begun to offer new insight into the evolution of past climates. Here, we focus on the alkenone organic proxy, which not only allows for the rapid generation of independent sea-surface temperature (SST) estimates, but also offers qualitative estimates of past surface productivity.

This review is directed to readers who have little or no prior knowledge of the use of alkenones for constraining past sea-surface conditions. We provide a broad overview of the alkenone organic proxy, discussing its development and application to palaeoclimate questions as well as some important considerations and limitations of applying this method. In light of the focus of this volume on deep time, we specifically address the application of the alkenone organic proxy to timescales longer than one Million years. For a more comprehensive discussion of alkenone palaeothermometry, we direct readers to several previous reviews of this proxy (Volkman 2000; Herbert 2001, 2004).

From: WILLIAMS, M., HAYWOOD, A. M., GREGORY, F. J. & SCHMIDT, D. N. (eds) *Deep-Time Perspectives on Climate Change: Marrying the Signal from Computer Models and Biological Proxies.* The Micropalaeontological Society, Special Publications. The Geological Society, London, 539–562.
1747-602X/07/$15.00 © The Micropalaeontological Society 2007.

Occurrence of alkenones in ocean sediments

Alkenones are a set of long-chained organic compounds uniquely synthesized by a few species of haptophyte algae (Volkman *et al.* 1980; Marlowe *et al.* 1984*b*; Conte *et al.* 1994; Volkman *et al.* 1995; Versteegh *et al.* 2001). These lipid compounds are distinguished from one another by their chain length (i.e. their number of carbons – C_{37}, C_{38}, C_{39}), their degree of unsaturation (i.e. the number of double carbon bonds they have: di- (2), tri- (3) and tetra-(4) unsaturated), and the structure of the terminal ketone group (methyl or ethyl)

(Volkman *et al.* 1980; Marlowe *et al.* 1984*b*, 1990) (Fig. 1). The C_{37} alkenones typically used in alkenone palaeothermometery are di- ($C_{37:2}$) and tri- ($C_{37:3}$) methyl unsaturated ketones. C_{38} alkenones have both di- and tri-unsaturated methyl ($C_{38:2\ me}$; $C_{38:3me}$) as well as ethyl ($C_{38:2et}$; $C_{38:3et}$) ketone forms. The sum of the concentrations of all four C_{38} ketones typically is approximately equal to that of the C_{37} ketones (Volkman *et al.* 1995; Müller *et al.* 1997; Herbert 2001; Prahl *et al.* 2001). C_{39} alkenones have di- and tri-unsaturated ethyl ketone forms, $C_{39:2et}$ and $C_{39:3et}$, respectively. However, the concentration of C_{39} ketones is only 10–20% of C_{37} ketones (Prahl *et al.* 2001).

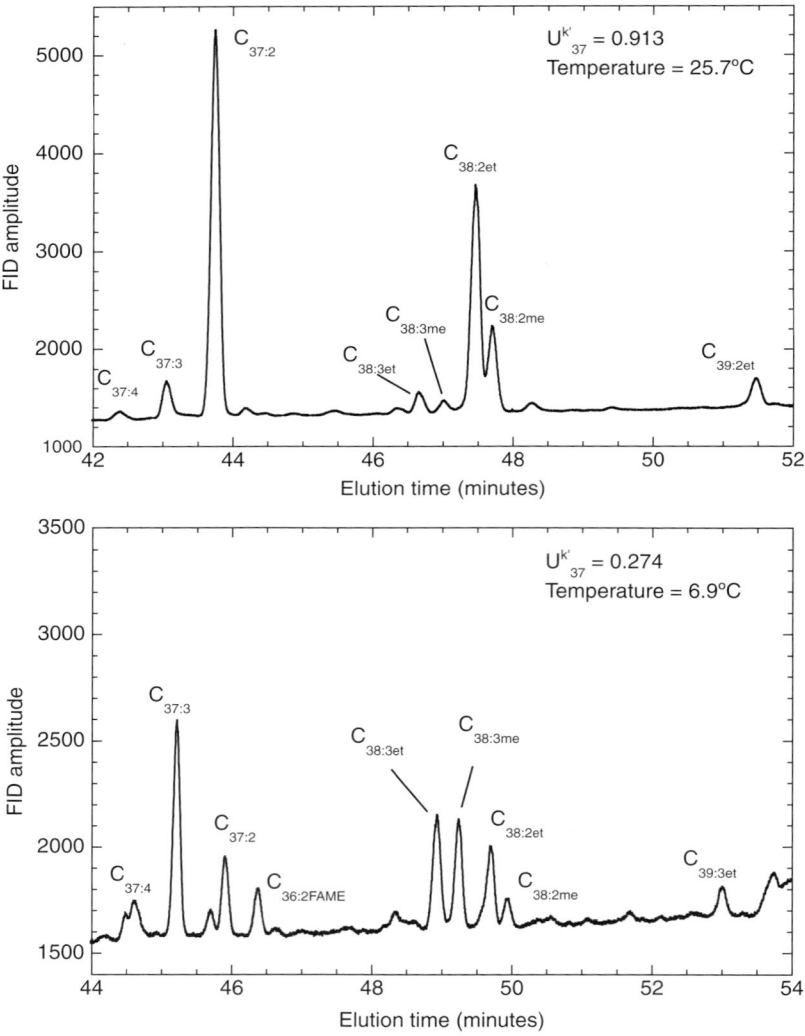

Fig. 1. GC-FID chromatograms of the organic extracts from ocean sediments showing the typical suite of alkenones.

The production of C_{38} and C_{39} alkenones is also temperature sensitive (e.g. Conte *et al.* 1995*a*, *b*; Volkman *et al.* 1995). However, the analytical difficulty of separating the four C_{38} ketones from each other and the typically lower concentrations of C_{39} ketones means these compounds are not widely used in alkenone palaeothermometry.

Alkenones were first found in Miocene through Pleistocene age sediments from the Walvis Ridge off of southwest Africa (Boon *et al.* 1978). Shortly thereafter, De Leeuw *et al.* (1980) identified the structure of these compounds by examining organic compounds recovered from Walvis Ridge and Black Sea sediments. The same year, Volkman *et al.* (1980) showed that the living haptophyte algal species *Emiliani huxleyi* synthesizes alkenones. Subsequent work characterized the chemical structure of alkenones (Rechka & Maxwell 1988) and established that *E. huxleyi* and *Gephyrocapsa oceanica*, two geographically widespread species of surface-dwelling, coccolith-producing, haptophyte algae of the order Isochrysidales, are the primary producers of alkenones in the modern ocean (Conte *et al.* 1994; Volkman *et al.* 1995). Alkenones are also produced by a few species in two other non-calcifying genera of haptophytes from the order Isochrysidales – *Isochrysis* and *Chrysotila* (Marlowe *et al.* 1984*a*; Volkman *et al.* 1989; Conte *et al.* 1994; Versteegh *et al.* 2001; Rontani *et al.* 2004). However, these genera, which dwell primarily in coastal areas, comprise only a very small fraction of the modern marine ecosystem (Volkman 2000). No other marine coccolithophorid (calcifying) or non-calcifying haptophytes are known to produce alkenones.

The cosmopolitan species *E. huxleyi* is abundant and geographically widespread in the modern ocean, tolerating a wide range of temperatures, salinities, nutrient availability and light levels and sometimes composing 60–90% of modern coccolithophorid assemblages (Okada & Honjo 1973; Okada & McIntyre 1979; Brand 1994; Winter *et al.* 1994). *G. oceanica* is limited to the tropics and subtropics in waters warmer than *c.* 12 °C (Okada & Honjo 1973; Okada & McIntyre 1979), but is tolerant of a considerable range of salinities (25‰–45‰) (Brand 1984). Given the wide geographic range of modern alkenone producers and the recalcitrance of these compounds to degradation at depth in the sediment column, it is not surprising that alkenones are common in ocean sediments of Pleistocene through Eocene age (Marlowe *et al.* 1990) (De Leeuw *et al.* 1980; Marlowe *et al.* 1984*a*, 1990; Lichtfouse *et al.* 1992; Brassell 1993; Rinna *et al.* 2002) and that they have been found in sediments as old as Cretaceous age (Farrimond *et al.* 1986; Brassell & Dumitrescu 2004).

$U_{37}^{K'}$ index and its calibration

Shortly after the identification of alkenone-producing species, laboratory culturing studies demonstrated that the degree of alkenone unsaturation (i.e. the number of double carbon bonds) varied with organism growth temperature (Marlowe *et al.* 1984*a*). Observing that the relative abundances of $C_{37:2}$ and $C_{37:3}$ alkenones collected in previous studies of ocean sediments varied with latitude, Brassell *et al.* (1986) proposed that the relative abundances of these alkenones in ocean sediments could be an indirect measure of sea-surface temperature (SST). To test this hypothesis, they compared late Pleistocene records of alkenone unsaturation and oxygen isotope ratios of planktonic foraminifera, an accepted indicator of past SSTs, from a sediment core in the eastern equatorial Atlantic. They found a strong positive covariance of these variables and noted that like the $\delta^{18}O$ of planktonic foraminifera, alkenone unsaturation fluctuates to the beat of glacial–interglacial cycles. These results suggested that the degree of alkenone unsaturation in ocean sediments could be used as an index of past SSTs. To quantitatively express the degree of alkenone unsaturation, Brassell *et al.* (1986) developed the alkenone unsaturation index:

$$U_{37}^{K} = \frac{[C_{37:2} - C_{37:4}]}{[C_{37:2} + C_{37:3} + C_{37:4}]}$$

Increases in the U_{37}^{K} index imply a decrease in the degree of alkenone unsaturation (i.e. a decrease in the number of double bonds) and thus warmer SSTs. The absence of tetra-unsaturated C_{37} alkenones from most marine sediments prompted a simplification of the U_{37}^{K} index to the now widely used $U_{37}^{K'}$ alkenone unsaturation index:

$$U_{37}^{K'} = \frac{[C_{37:2}]}{[C_{37:2} + C_{37:3}]}$$

Prahl & Wakeham (1987) initially calibrated the covariance of the alkenone unsaturation index with organism growth temperature using laboratory cultures of *E. huxleyi* at a range of different temperatures. Applying the revised $U_{37}^{K'}$ index to their culturing data, Prahl & Wakeham (1987) obtained a linear calibration curve ($U_{37}^{K'} = 0.033T + 0.043$; $r^2 = 0.997$). By comparing the relationship between $U_{37}^{K'}$ measurements from alkenones collected from organisms in the marine water column, from the North Pacific, and their associated ocean surface temperatures with the laboratory-derived calibration, they demonstrated that the laboratory-based calibration curve reliably predicted absolute ocean surface-water temperatures. This initial calibration was later revised based on more extensive culturing study and ocean sediment data (Prahl *et al.* 1988). The Prahl *et al.* (1988) calibration ($U_{37}^{K'} = 0.034T + 0.039$) is presently the most extensively used calibration for alkenone SST estimates.

The use of the $U_{37}^{K'}$ index as a measure of past SST fundamentally rests on the consistent results found from a number of extensive core-top sediment calibration studies, which compare SST estimates at a given locality with $U_{37}^{K'}$ temperature estimates derived from the underlying ocean sediments. Large regional calibrations from the Atlantic Ocean (Rosell-Melé *et al.* 1995), the Indian Ocean (Sonzogni *et al.* 1997) and the California Margin (Herbert *et al.* 1998) as well as extensive (>350 sites and >600 sites respectively) global calibrations by Müller *et al.* (1998) and Conte *et al.* (2006), find robust linear correlations between $U_{37}^{K'}$ and mean annual SST. Both, the Müller *et al.* (1998) and more recent Conte *et al.* (2006) core-top calibrations include sites that cover the entire range of temperatures and geographic distribution of alkenone-producing species. Remarkably, these global calibration curves yield SST predictions that are essentially the same as those from the original linear calibration of Prahl *et al.* (1988) with differences between the original and the global calibration curves of at most 0.7 °C (Fig. 2). Mean annual SST estimates from these

calibrations are accurate to within ± 2.6 °C at the 95% confidence level.

C_{37} Total index

Palaeoclimate investigations have used the total abundance of C_{37} alkenones preserved in sediments as an index of past productivity (Brassell 1993; Villanueva *et al.* 1997; Villanueva *et al.* 1998, 2001; Sicre *et al.* 2001; Seki *et al.* 2004; Sachs & Anderson 2005). The index is defined as the sum of the concentrations of the $C_{37:2}$ and $C_{37:3}$ ketones per unit dry weight of sediment. While these compounds represent only a fraction of the organic matter produced in the surface ocean, they comprise a significant portion of total cellular carbon (15–20%) (Epstein *et al.* 1998; Prahl *et al.* 2003) of alkenone-producing species, which dominate phytoplankton communities in many regions of the modern ocean (Okada & Honjo 1973; Okada & McIntyre 1979; Winter *et al.* 1994). Like most other commonly employed indices of past production (e.g. total organic carbon (TOC), opal, % calcium carbonate), the C_{37} Total index is not calibrated to variations in surface production. Calibration of these indices is inhibited by the potential for differential preservation of different sedimentary components to overprint or alter the relative concentration of these components in the sediment. Thus, these indices, including C_{37} Total, provide only qualitative estimates of past productivity. Previous studies, which have employed the C_{37} Total index in conjunction with other indicators of past productivity, demonstrate strong agreement between C_{37} Total and these other indices (Rostek *et al.* 1997; Villanueva *et al.* 1997; Schubert *et al.* 1998; Hinrichs *et al.* 1999; Budziak *et al.* 2000; Sicre *et al.* 2001; Moreno *et al.* 2004; Seki *et al.* 2004). However, because only a small fraction of alkenones synthesized in the surface ocean end up buried in ocean sediments (Prahl *et al.* 1989; Prahl *et al.* 1993*a*; Müller & Fischer 2001), degradation of alkenones is a substantial concern for the application of the C_{37} Total proxy. These concerns are discussed more extensively below.

Quantification of alkenones

Prior to analysis, ocean sediment samples are freeze-dried to remove any pre-existing water. Organic compounds, including alkenones, are extracted from ocean sediments using organic solvents, typically 100% dichloromethane or a combination of dichloromethane and methanol, via either Soxhlet apparatus, through repeated sonication at room temperature, or by use of a Dionex 200 Accelerated Solvent Extractor (ASE). A recent

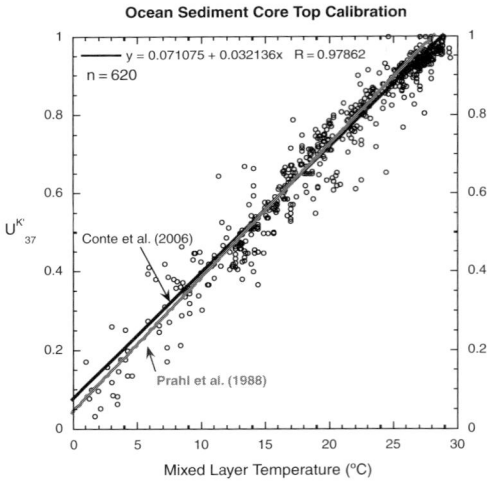

Fig. 2. Global compilation of core-top sediment $U_{37}^{K'}$ data plotted against mean annual SST. All data are from Conte *et al.* (2006). The linear regression line (black line) produces estimates that are no more than 0.7 °C different from those produced by the original regression line of Prahl *et al.* (1988) (thick grey line).

study indicates that alkenones can also be extracted from sediments through a microwave-assisted extraction procedure (Kornilova & Rosell-Melé 2003). Typically, 0.1–5 g of sediment (dry weight) is sufficient to yield measurable quantities of alkenones. If the sample has a high organic content or there are compounds present that co-elute (i.e. have the same elution time) with alkenone compounds, after extraction silica gel chromatography or other sample clean-up procedures may be necessary prior to analysis. After extraction and any clean-up, a small volume of sample (0.2–5 μl) is then injected into a gas chromatograph (GC), which enables the separation of alkenone compounds and the quantification of their abundances. A long (50–60 m), nonpolar column and a slow-temperature program separates organic compounds primarily based on their boiling points, but also through interaction with the column film. A flame ionization detector (FID) is the most commonly used detector for alkenone analysis allowing for simple, reliable and precise quantification of organic compounds. Using this detector, alkenones are identified based on their elution times and by reference to external standards that are added to the injection solvent. Because the FID does not give information about the structure of the compounds detected, accuracy of estimates produced by GC-FID analysis strongly depends on whether or not co-eluting compounds are present. Fortunately, compounds with elution times similar to those of alkenones are rare in most ocean sediments. The general absence of co-eluting compounds enables the accurate determination of alkenone abundances from most marine sediment samples and illustrates the hardiness of alkenones relative to other high molecular weight compounds during diagenesis.

Rapid ($c.$ 75–100 samples/week) and precise (reproducibility ± 0.005 $U_{37}^{K'}$ units equivalent to a temperature uncertainty of $\pm 0.2\,°C$) analysis allows for the efficient simultaneous generation of long, high-resolution time-series of both SST and past productivity. Consistent evaluation and maintenance of GC performance is important to assuring the long-term reproducibility of $U_{37}^{K'}$ estimates. While there is no community-wide standard for $U_{37}^{K'}$ analysis, internal laboratory standards are typically analysed at the beginning and end of every GC sequence (approximately every 10 to 20 samples) to ensure intralaboratory consistency in estimates. A recent interlaboratory comparison study (Rosell-Melé et al. 2001) compared the results of the analysis of a set of homogenized unidentified sample splits from alkenone laboratories worldwide. This study showed that the maximum difference in $U_{37}^{K'}$ SST estimates between any two labs at the 95% confidence level was 2.1°C. Excluding one sample

as an outlier, the average difference between any two labs for a particular sample was 0.034 $U_{37}^{K'}$ units at the 95% confidence level, equivalent to a modest average interlaboratory difference of $\pm 1\,°C$. Results for alkenone concentrations were not as encouraging. Interlaboratory reproducibility was estimated to be a meagre 32%, which was attributed to the diversity of procedures used to quantify alkenones (Rosell-Melé et al. 2001).

Limitations to application

The $U_{37}^{K'}$ index enables the accurate estimation of SST. However, the shallow habitation depths of marine, alkenone-synthesizing algae prevent characterization of the thermal properties of the ocean below the mixed layer, such as the thermal structure of the thermocline or deep ocean, using this method. Additionally, the application of the alkenone organic proxy is restricted to regions of the ocean where alkenones are preserved in ocean sediments. To our knowledge, $c.$ 20 years after the first application of the $U_{37}^{K'}$ index, only sediments from some oligotrophic gyre regions and the high latitudes of the central North Pacific and Southern Ocean have failed to yield alkenones in measurable quantities. In general, the overall low biological productivity of gyre sites yields characteristically low sedimentation rates and biomarker contents, making these regions more challenging for application of the alkenone organic proxy. Finally, using the Prahl et al. (1988) equation, the alkenone palaeothermometer saturates at $c.$ 28 °C (i.e. the $U_{37}^{K'}$ ratio reaches a value of 1), which means that the palaeothermometer does not have temperature sensitivity above this value. Furthermore, it is hard to accurately quantify the less-abundant compound as the limit of detection is approached at either end member of the temperature range (0 °C and $c.$ 28 °C, respectively) (Grimalt et al. 2001; Pelejero & Calvo 2003). Because of the higher absorption of $C_{37:3}$ to gas chromatographic columns, error in palaeotemperature estimation is more common at the low end of the temperature range (Grimalt et al. 2001). In our experience, $U_{37}^{K'}$ estimates that yield temperatures in the range of 27.5 °C and 5 °C via the Prahl et al. (1988) equation are reliably reproducible while those outside this range are subject to much greater errors. While this temperature range covers most modern ocean surface temperatures, some regions (e.g. the warm pools of the tropics or the very high latitudes in both hemispheres) have temperatures outside of this range. In these regions, application of other palaeothermometers is required to accurately characterize SST.

Thus far, the alkenone organic proxy has been primarily used to estimate SST in marine environments. However, recent studies have quantified alkenone abundances in lake sediments (e.g. Sun *et al.* 2004; Chu *et al.* 2005; D'Andrea & Huang 2005), illustrating the potential to use these estimates as a relative measure of past temperatures in terrestrial environments. Lacustrine, estuarine and some marginal marine environments have much higher abundances of $C_{37:4}$ alkenones, which are rarely observed in open ocean marine sediments. A number of studies that measured alkenones in particulate organic matter and/or surface sediments found that the percentage of $C_{37:4}$ alkenones relative to other C_{37} alkenones varies as a function of salinity (e.g. Rosell-Melé 1998; Sicre *et al.* 2002; Harada *et al.* 2003; Bendle *et al.* 2005; Blanz *et al.* 2005; Seki *et al.* 2005). Use of this proxy to explore Plio-Pleistocene climates is in its preliminary stages, but this index may provide a new tool for estimating past salinity in regions where $C_{37:4}$ alkenones are abundant. Additionally, work is underway to isolate alkenone producers in lake environments, to identify the essential controlling factors of alkenone unsaturation in lacustrine settings, and to determine an appropriate temperature calibration (D'Andrea, personal communication). In the next several years, it is likely that the use of alkenone palaeothermometry will no longer be restricted to open-ocean marine environments.

Important considerations to the application of the alkenone organic proxy

An important set of factors involving the genetics, ecology and physiology of alkenone producers as well as the deposition and preservation of the compounds they synthesize warrant careful consideration when applying the alkenone organic proxy. Study of alkenone producers both in laboratory-culturing experiments under controlled conditions and in marine samples taken directly from the water column have been pivotal in shaping our present understanding of these organisms and help us to characterize the important outstanding questions related to the application of alkenone proxies.

Culture and water column studies of alkenone-producing species confirm the primary dependence of $U_{37}^{K'}$ on temperature (Conte *et al.* 1995*b*, 1998; Volkman *et al.* 1995; Sawada *et al.* 1996). However, these studies do not yield the consistent linear calibration derived from core-top sediment studies. Culture studies document large variations in the $U_{37}^{K'}$ temperature calibration for different strains of alkenone-producing species. In contrast to the widely used calibration of Prahl *et al.*

(1988), many of these culturing studies suggest that the $U_{37}^{K'}$-temperature relationship should be non-linear (Volkman *et al.* 1995; Conte *et al.* 1998). In addition, the results of culturing studies suggest a second-order dependence of $U_{37}^{K'}$ on the physiological factors of growth phase, light availability and nutrient levels (Conte *et al.* 1998; Epstein *et al.* 1998; Epstein *et al.* 2001; Prahl *et al.* 2003).

Water column studies yield more universally consistent calibrations than culture studies and show more agreement with the initial Prahl *et al.* (1988) calibration (Brassell 1993; Conte & Eglinton 1993; Sikes *et al.* 1997; Ternois *et al.* 1997; Goni *et al.* 2001; Conte *et al.* 2006). However, many of them also suggest that the relationship between $U_{37}^{K'}$ and temperature may be non-linear (e.g. Sikes & Volkman 1993; Goni *et al.* 2001; Bendle & Rosell-Melé 2004; Conte *et al.* 2006). In a new synthesis study of all existing surface water alkenone unsaturation measurements, Conte *et al.* (2006) conclude that a single third-order polynomial equation accurately describes the relationship between $U_{37}^{K'}$ and organism growth temperature across the entire range of open ocean environments and alkenone-producing organisms. Non-linearity of the calibration curve stems from a flattening of the $U_{37}^{K'}$ temperature relationship at both the high and low ends of the temperature range (Fig. 3). Inconsistencies in calibration at the extreme ends of the index are not surprising. Calibration in

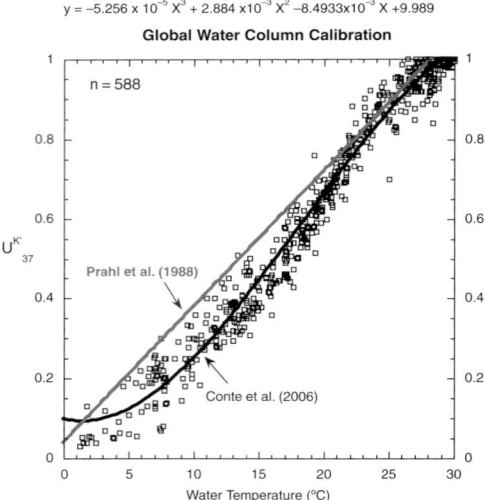

$y = -5.256 \times 10^{-5} X^3 + 2.884 \times 10^{-3} X^2 - 8.4933 \times 10^{-3} X + 9.989$

Global Water Column Calibration

Fig. 3. Global compilation of water column $U_{37}^{K'}$ values *vs.* the measured temperature for the surface mixed layer (0–30 m) (Conte *et al.* 2006). Data are fit with a third order polynomial after Conte *et al.* (2006) (black line) and the original linear regression of Prahl *et al.* (1988) (grey line).

these end-member regions is analytically challenging as the $C_{37:2}$ and $C_{37:3}$ ketones become minor peaks at the low and high ends of the index respectively (Grimalt *et al.* 2001).

Although culture and water column studies both have their merits, they also have their limitations when it comes to informing our application of the $U_{37}^{K'}$ index to palaeoclimate questions. Culturing studies allow for the isolation of variables that may influence the relative abundances of C_{37} ketones. However, they can only serve as models and not actual representations of natural conditions. Water column studies, while capturing the production of alkenones in the natural environment, provide only a limited picture of that environment as they sample a highly temporally and spatially variable parameter over a limited time interval at a finite number of sites. The temporal and spatial averaging effects of slow deep ocean sediment accumulation probably serve to average out much of the variability introduced by genetic, ecological and physiological factors. The non-linear calibration implied by the water column data stands in contrast to the results from global datasets of surface ocean sediments, which indicate that despite scatter at the high and low end of the temperature range, these $U_{37}^{K'}$ – SST datasets are best fit by linear calibrations (Müller *et al.* 1998; Conte *et al.* 2006), which yield predicted temperatures that are very nearly the same as those from the original Prahl *et al.* (1988) calibration (Fig. 2). However, Conte *et al.* (2006) report that the $U_{37}^{K'}$ from sediments is systematically higher than that predicted by their water column calibration. They show that this offset, which is greater at colder temperatures, can be explained by seasonality of production or by the combination of seasonality and a very small differential degradation rate between the $C_{37:3}$ and $C_{37:2}$ ketones. Additional work investigating these calibration inconsistencies is required.

Ecology

Seasonal variations in the production of alkenones and the depth ranges in which coccolithophorids dwell are observed in the modern ocean (e.g. Okada & Honjo 1973; Okada & McIntyre 1979; Prahl *et al.* 1993*a*; Broerse *et al.* 2000*c*; Cortes *et al.* 2001; Haidar & Thierstein 2001). These variations raise questions about which season and what portion of the water column $U_{37}^{K'}$-derived temperature estimates reflect. Results from studies that examined the modern depth habitat of alkenone producers suggest that alkenone synthesizers dwell primarily in the mixed layer in the mid- to upper euphotic zone at depths <100 m (Bentaleb

et al. 1999; Conte *et al.* 2001; Cortes *et al.* 2001; Haidar & Thierstein 2001). The global core-top calibration of Müller *et al.* (1998) corroborates these findings. Müller *et al.* (1998) compared the $U_{37}^{K'}$ index and temperatures at various depths. They found that a strong positive correlation exists at all depths down to 100 m, indicating that alkenone production could occur throughout the mixed layer and upper thermocline. Yet, by far the best correlation exists with mean annual temperatures for water depths of 0 to 10 m, suggesting that the $U_{37}^{K'}$ index is most accurately measuring near-surface temperatures rather than deep mixed layer or thermocline temperatures. However, a few studies suggest that some regions of the ocean, particularly oceanic gyres, may not conform to this simple model (e.g. Prahl *et al.* 1993*b*; Ternois *et al.* 1997; Ohkouchi *et al.* 1999; Goni *et al.* 2001). Additional work is required to determine the effects of seasonal and regional differences in mixed layer and nutricline depth on alkenone producers.

Alkenone production in the ocean varies with the seasonal cycle. Water column sediment trap data indicate that $U_{37}^{K'}$ responds rapidly to changes in ocean temperature, accurately reflecting seasonal SST variations (Goni *et al.* 2001). The maximum flux of alkenones and the maximum abundance of *E. huxleyi* and *G. oceanica* typically occur synchronously with the maximum in export flux to the sediments. Following the seasonal progression of maximum production with increasing latitude, peak production occurs in the spring in most subtropical and mid-latitude locations (e.g. Prahl *et al.* 1993*a*; Broerse *et al.* 2000*a*; Broerse *et al.* 2000*b*; Sprengel *et al.* 2000; Cortes *et al.* 2001; Haidar & Thierstein 2001), whereas at high latitudes alkenone production usually occurs during the summer with little or no production during the winter season (e.g. Ternois *et al.* 1998; Broerse *et al.* 2000*c*; Prahl *et al.* 2001). The Müller *et al.* (1998) calibration study shows that while $U_{37}^{K'}$ is highly correlated with the SST of all major seasons, the best correlation is with mean annual SST, suggesting that the seasonal variations in surface production do not leave a large imprint in the underlying sediments. The $U_{37}^{K'}$ signal derived from surface sediment is an integrated record of longer timescale (100s to 1000s of years) variations. Müller *et al.* (1998) argue that interannual temperature variations, which can be large, may overshadow the effect of seasonal variations on the surface sediments $U_{37}^{K'}$ signal suppressing any dramatic seasonal temperature signal.

Because the seasonal cycle in high latitude regions is much larger than in mid- to low latitudes, it is much more likely that a seasonal production bias will cause estimated $U_{37}^{K'}$ temperatures to deviate from actual mean annual temperature in these regions (e.g. Rosell-Melé *et al.* 1995; Sikes *et al.* 1997). Model results from Conte *et al.* (2006) lend some support to this concern. They found that seasonality in production yields smaller offsets relative to mean annual SST in a mid latitude scenario (*c.* 1 °C) with a dominant springtime production bloom and a smaller autumn production maximum, in comparison with a high latitude case (*c.* 2.5 °C) with a strong summer maximum in production. While more work exploring the effect of seasonal variations in production is warranted, these studies suggest that $U_{37}^{K'}$ is a reasonable measure of near-surface mean annual temperature throughout the low and mid-latitudes, but at high latitudes where seasonal variations in temperature are greater and production is skewed toward summer months $U_{37}^{K'}$ values may be somewhat biased toward summer temperatures.

Physiology

The observed variation in alkenone ratios as a function of temperature may be related to the physiologic function that alkenones serve in haptophyte cells. Early studies suggested that alkenones were membrane lipids and that unsaturation changes potentially regulated membrane fluidity in response to changing temperatures (Marlowe *et al.* 1984*a*; Brassell *et al.* 1986). This notion arose from culturing studies, which demonstrated that some algae adjust their membrane fluidity in response to environmental stress by changing either their lipid length or degree of unsaturation (Harwood & Russell 1984). This hypothesized function logically explains the temperature dependence of alkenone unsaturation state because increases in unsaturation at lower temperatures decrease the lipid melting point, promoting greater membrane fluidity.

However, more recent results from culturing experiments of alkenone producers indicate that alkenones serve as metabolic storage devices, responding to both nutrient and light limitation (Epstein *et al.* 1998; Epstein *et al.* 2001; Prahl *et al.* 2003; Eltgroth *et al.* 2005). These studies show that the concentration of alkenones increases under nutrient stress and decreases in the absence of light. Alkenone-producing species synthesize only small quantities of triacylglycerols, which are common metabolic storage molecules for many marine microalgae, suggesting that alkenones may act as surrogates for these molecules in the species that synthesize them (Epstein *et al.* 2001; Prahl *et al.* 2003; Eltgroth *et al.* 2005). Eltgroth *et al.* (2005) found that alkenones as well as other polyunsaturated long-chain molecules are primarily concentrated in lipid bodies found within the cell and within cellular chloroplasts, their potential site of synthesis, rather than in cell membranes. Thus, rather than serving as a means of maintaining membrane fluidity, the temperature dependence of alkenone unsaturation may be a consequence of differences in melting point, density or the enzymatic optima of biochemical pathways (Epstein *et al.* 2001). Additional work is required to ascertain the specific role that temperature dependency of alkenone unsaturation plays in alkenone synthesizers.

Genetics

While different species of haptophytes and different strains within those species produce alkenones, existing molecular evidence indicates that there is strong genetic similarity among all marine alkenone producers (e.g. Edvardsen *et al.* 2000; Fujiwara *et al.* 2001). However, *E. huxleyi*, the dominant producer of alkenones in the modern ocean, evolved in geologically recent times. This species first appeared in the geologic record 280 ka ago and only rose to dominance after *c.* 80 ka (Thierstein *et al.* 1977). *E. huxleyi* is believed to have evolved from the genus *Gephyrocapsa* (McIntyre 1970). Micropalaeontological data indicate that *G. oceanica*, the other main producer of alkenones in the modern ocean and a member of the genus *Gephyrocapsa*, was globally dominant for some time periods during the Pleistocene (e.g. Thierstein *et al.* 1977; Bollmann *et al.* 1998). The genus *Gephyrocapsa* as a whole extends back to at least the middle Miocene, and coccoliths from Gephyrocapsaceae (Noelaerhabdaceae in the current taxonomic scheme of Jordan & Green 1994), the family of the genus *Gephyrocapsa*, have been found in sediments as old as the Eocene (Marlowe *et al.* 1990). A synthesis study comparing coccolithophorid assemblage data with the occurrence of alkenones in sediments found that Gephyrocapsaceae was the only family of nannofossil producers to occur in all sediments that contained alkenones (Marlowe *et al.* 1990). These micropalaeontological data suggest that alkenone production occurs and probably has occurred only in closely related species.

In applying the $U_{37}^{K'}$ index to longer timescales ($>10^5$ yr), the robustness of the calibration through geologic time becomes an important consideration. Are $U_{37}^{K'}$ variations still recording meaningful climate information before the evolution of the modern producing species? This concern is not new to palaeoclimatic studies. All biologically based proxies are potentially affected by a change in the response of the palaeoclimate index over evolutionary time. However, this is a particularly important concern in light of the relatively recent evolution (0.28 Ma) of *E. huxleyi* (Thierstein *et al.* 1977). If the alkenone unsaturation indices of predecessor species to *E. huxleyi* have different temperature sensitivities, palaeo-SST estimates derived from the $U_{37}^{K'}$ index may be biased for the time period preceding the evolution of *E. huxleyi*. A number of studies have explored this possibility by pairing $U_{37}^{K'}$ analysis with micropalaeontological estimates of the abundance of coccolithophorids in Quaternary sediments. These studies found no indication that changes in the relative proportions of alkenone-producing species with time affected $U_{37}^{K'}$ SST reconstructions (e.g. Müller *et al.* 1997; Doose-Rolinski *et al.* 2001; Villanueva *et al.* 2002; McClymont *et al.* 2005).

Examination of other empirical alkenone indices further supports the conclusion that the modern $U_{37}^{K'}$ SST calibration can be applied to longer timescales. The ratio of C_{37}:C_{38} alkenones in time series derived from Pleistocene and Pliocene sediments show a limited range of values (Rostek *et al.* 1993; Müller *et al.* 1997; McClymont *et al.* 2005), which fall within the observed modern range of these values measured from water column and sediment samples (Prahl *et al.* 1988; Conte *et al.* 1998; Conte *et al.* 2001). These results suggest that alkenone producers throughout the Plio-Pleistocene had biochemical properties similar to modern alkenone producers and thus also had similar temperature sensitivities. Several studies examined the variation of alkenones and coccolithophorid abundances in late Quaternary ocean sediments and compared their $U_{37}^{K'}$-derived temperature estimates to planktonic foraminifera $\delta^{18}O$ records of ocean SST (Müller *et al.* 1997; Villanueva *et al.* 2002; McClymont *et al.* 2005). All of these studies found a strong correlation between $U_{37}^{K'}$ and $\delta^{18}O$ throughout the interval studied and no indication that changes in $U_{37}^{K'}$ were associated with changes in the coccolithophorid assemblage. All of these studies supported the use of the modern $U_{37}^{K'}$ SST calibration for past SST reconstructions during intervals preceding the evolution of the modern coccolithophorid assemblage.

Preservation

Because the $U_{37}^{K'}$ palaeothermometer involves measuring the ratio of two C_{37} alkenones, good overall preservation of these compounds is not required to yield accurate estimates of past SST. However, application of the $U_{37}^{K'}$ SST proxy does require that the relative abundances of di- and tri-unsaturated alkenones be unchanged during deposition and diagenesis. Use of the C_{37} Total index has the more stringent requirement that the degree of degradation of C_{37} alkenones in sediments does not change appreciably over time. In other words, the C_{37} Total index assumes that synthesis of alkenones in the surface ocean is the primary control on the concentration of alkenones preserved in ocean sediments.

In the interest of assessing the effects of postdepositional diagenesis on alkenones, several studies have examined the effect of dramatically different depositional environments on alkenone abundances. These studies assess changes in the alkenone unsaturation patterns and alkenone concentrations under contrasting environmental conditions both in ocean sediments (Prahl *et al.* 1989; McCaffrey *et al.* 1990; Hoefs *et al.* 1998; Gong & Hollander 1999) and under controlled laboratory conditions (Rontani *et al.* 1997; Teece *et al.* 1998). All studies found extensive degradation of C_{37} alkenones over time. In some cases, >85% of C_{37} alkenones were lost under oxic conditions (Prahl *et al.* 1989; Teece *et al.* 1998). Despite the marked decline in overall abundance of alkenones as a consequence of degradation, most studies show that the $U_{37}^{K'}$ index remains remarkably unchanged by diagenesis (Prahl *et al.* 1989; McCaffrey *et al.* 1990; Rontani *et al.* 1997; Teece *et al.* 1998). In contrast to $\delta^{18}O$ and Mg/Ca proxies, which are strongly susceptible to diagenetic effects (e.g. Brown & Elderfield 1996; Schrag 1999; Pearson *et al.* 2001; Rosenthal & Lohmann 2002; Lawrence & Herbert 2005), the $U_{37}^{K'}$ index does not require good overall preservation of alkenones to yield accurate estimates of past SST, but rather proportional preservation of the two compounds. However, these alkenone preservation studies illustrate that degradation does affect the total amount of alkenones preserved, which has significant implications for the application of the C_{37} Total productivity proxy.

Most organic degradation occurs in a biologically active zone at the sediment–water interface and in the upper few centimetres of the sediment column. The loss of alkenones to degradation in this zone is not a surprising result when compared to the similar fate of most other organic compounds. Only a few percent or less of the TOC reaching the deep ocean is typically preserved in sediments. The preservation efficiency of alkenones is even lower. Typically 1% or less of alkenones reaching the deep ocean are preserved in sediments (Prahl *et al.* 1989; Prahl *et al.* 1993a; Müller & Fischer 2001).

Despite loss of a significant portion of the total flux of organic compounds to the sediments, a few observations suggest that variations in concentration of C_{37} alkenones are still primarily driven by changes in surface productivity (e.g. Müller & Fischer 2001; Blanz *et al.* 2005). At many sites, the concentration of alkenones increases at greater depth in the sediment column (i.e. greater age) (e.g. Rostek *et al.* 1997; Budziak *et al.* 2000; Liu & Herbert 2004; Lawrence *et al.* 2006). This trend is opposite that of the monotonic loss of ketones with time one would expect if degradation were the primary control on C_{37} Total. Where samples are measured at sufficient resolution, changes in C_{37} Total are periodic in nature and vary systematically on glacial–interglacial timescales (e.g. Rostek *et al.* 1997; Villanueva *et al.* 1998; Liu & Herbert 2004; Lawrence *et al.* 2006). These variations in C_{37} Total are typically one to two orders of magnitude in amplitude, far too large to be explained by dilution, dissolution or preservation of the major sediment components.

Variations in deep-sea oxygenation, which control deep ocean degradation rates, could also potentially account for observed variations in C_{37} Total (Prahl *et al.* 1989). Although past fluctuations in deep-sea oxygenation are poorly constrained, unless the depositional environment is highly sensitive to oxygenation, variations in this parameter are not the likely explanation to the several orders of magnitude changes in C_{37} Total observed. However, given the absence of data constraining the nature and extent of this effect, C_{37} Total records should not be interpreted in isolation. Other indices of past production, including those that are not similarly susceptible to changes in deep water oxygenation should be employed to confirm results from C_{37} Total records. Additionally, variations smaller than a 10% change in C_{37} Total concentration, the average error in concentration associated with replicate samples, should not be interpreted as real variations in production.

Sedimentation

One additional concern related to the application of the alkenone organic proxy comes from recent work using ^{14}C to radiometrically date different sediment fractions at a number of sites with very high deposition rates (Ohkouchi *et al.* 2002; Mollenhauer *et al.* 2003, 2005). These studies show that lighter sediment constituents (i.e. alkenones, other organic material and fine carbonates) at drift sites and in drift-like settings on continental margins can be significantly older (0–7 kyr) than the larger and denser foraminiferal fraction. Greatest age offsets (>5 kyr) occur at drift sites, where a significant portion of the accumulating sediments are advected from other localities by strong currents (Ohkouchi *et al.* 2002). Along continental margins, sites with high organic matter content, which may enable the formation of easily resuspendable organic aggregates, are associated with larger age offsets (1–4.5 kyr), whereas sites with

low sedimentary organic matter have little or no age offset between the fine and coarse fractions (Mollenhauer *et al.* 2003, 2005).

The relative age of different sediment fractions has not yet been studied at deep-water open ocean sites where the sediments for most longer-term palaeoclimate studies are collected. The association of sediment fraction age offsets with known drift site or continental margin localities with high amounts of organic matter means these types of offsets are less likely a concern for deep water pelagic sites, which typically have very low organic matter contents and are not strongly influenced by the long-range lateral transport of sediment or tidal currents. These sediment fraction age offsets do not invalidate the use of multi-proxy approaches to palaeoclimate problems, but rather imply that when applying the alkenone organic proxy great care must be taken in study site selection, making use of published seismic profiles and avoiding known drift sites and localities that have strong bottom currents.

Assessment of the $U^{K'}_{37}$ SST proxy

Even though some uncertainty remains about aspects of the ecology, physiology and modern genetic variation of alkenone producers, these concerns have only a minimal effect on $U^{K'}_{37}$ estimates. These factors can account for at most the variance not explained by the $U^{K'}_{37}$ temperature core-top calibration curve, which is ± 2.6 °C at the 95% confidence level (Fig. 2). Given the potential influence of other sources of error, the error attributable to these factors is probably considerably smaller than ± 2.6 °C. Since the core-top calibration incorporates data from a number of laboratories, some of this error is probably due to interlaboratory differences. Additionally, errors in the modern SST estimates themselves or potential age mismatches between modern surface ocean temperatures and the time represented by the surface sediments may contribute to scatter about the core-top calibration curve.

The widespread occurrence of alkenones in marine sediments, the global applicability of the $U^{K'}_{37}$ temperature calibration, and a negligible degradation effect on the $U^{K'}_{37}$ index make it much more universally applicable than most other commonly used methods of estimating past SST. The application of transfer function methods and the modern analogue technique (MAT) are inhibited by the significant deviation of some past assemblages from modern faunas (i.e. no-analogue faunas) resulting in either an inability to fit these assemblages into a modern regression model or modern analogues that are highly dissimilar in comparison to the assemblage (e.g. CLIMAP 1981; Molfino *et al.* 1982; Prell 1985; Pflaumann *et al.* 1996; Trend-Staid & Prell 2002). Furthermore, both transfer functions and the MAT are labour-intensive methodologies. SST estimates derived from the $\delta^{18}O$ of planktonic foraminifera are confounded by the dual dependence of foraminiferal $\delta^{18}O$ on water temperature as well as the $\delta^{18}O$ of seawater. Additionally, SST estimates based on the $\delta^{18}O$ of planktonic foraminifera are potentially subject to error caused by the post-depositional recrystallization of foraminiferal calcite (e.g. Schrag *et al.* 1996; Pearson *et al.* 2001). Mg/Ca palaeothermometry lacks a universal calibration necessitating depth and species corrections to make comparisons between studies and regions (e.g. Lea *et al.* 1999; Elderfield & Ganssen 2000; Dekens *et al.* 2002; Anand *et al.* 2003). Both faunal techniques and Mg/Ca palaeothermometry are strongly affected by dissolution of foraminiferal shells (e.g. Thompson 1976; Brown & Elderfield 1996), which has been shown to play a significant role in a number of marine environments (Schrag 1999; Rosenthal & Lohmann 2002; Lawrence & Herbert 2005).

Application to long timescales

The $U_{37}^{K'}$ index has been widely applied to Holocene and late Pleistocene palaeoclimate questions. It has been used to study El Niño Southern Oscillation cycles in the late Holocene (e.g. Farrington *et al.* 1988; McCaffrey *et al.* 1990; Herbert *et al.* 1998; Zhao *et al.* 2000), to examine the temperature stability of the Holocene epoch (e.g. Bard *et al.* 1997; Rühlemann *et al.* 1999; Cacho *et al.* 2001; Calvo *et al.* 2001; Doose-Rolinski *et al.* 2001; Steinke *et al.* 2001), to characterize the intensity and timing of the last glacial maximum (e.g. Zhao *et al.* 1993; Sikes & Keigwin 1994; Bard *et al.* 1997; Steinke *et al.* 2001; Rosell-Melé *et al.* 2004) to investigate late Quaternary millennial-scale events at localities worldwide (e.g. Rühlemann *et al.* 1999; Bard *et al.* 2000; Zhao *et al.* 2000; Kienast *et al.* 2001; Cacho *et al.* 2002; Seki *et al.* 2002), and to constrain the response of the surface ocean temperature field to the ice-age cycles of the late Pleistocene (e.g. Lyle *et al.* 1992; Villanueva & Grimalt 1996; Kirst *et al.* 1999; Pelejero *et al.* 1999; Herbert *et al.* 2001). Some of these studies compare results from alkenone palaeothermometry to other indices of past sea-surface temperature (e.g. Nürnberg *et al.* 2000; Calvo *et al.* 2001; Steinke *et al.* 2001; Seki *et al.* 2002). Results of absolute temperature comparisons are mixed, ranging from negligible differences between proxies to discrepancies on the

order of 4 °C. However, discrepancies between the absolute values of $U_{37}^{K'}$, Mg/Ca, planktonic $\delta^{18}O$, and fauna assemblage SST estimates should not evoke great alarm, as they may stem from differences in the seasonality of production, habitation depth of the primary producers, or calibration uncertainties. In fact, if the physical or ecological factors accounting for offsets between proxies can be reasonably constrained, more subtle and complex palaeoclimatic stories can potentially be elucidated (Haug *et al.* 2005). Studies that compare late Quaternary time series of alkenone SST estimates to those from different indices show strong agreement in the timing and structure of variations in these records (e.g. Müller *et al.* 1997; Villanueva *et al.* 1998; Kirst *et al.* 1999; Pelejero *et al.* 1999; Calvo *et al.* 2001), suggesting that while the differences mentioned above may cause discrepancies in absolute temperature estimates the alkenone-derived estimates are reliably reproducing temperature variations at these sites.

Even if the uncertainty of absolute estimates of SST increases on longer timescales, the $U_{37}^{K'}$ index should still yield useful estimates of the temporal and spatial variability of SST over these time intervals. Comparisons between existing climate data and results from alkenone studies demonstrate that alkenone records produce the now familiar patterns of Plio-Pleistocene climatic change. Pleistocene records of $U_{37}^{K'}$-derived SST mimic those derived from benthic oxygen isotope records in structure and periodicity. A saw-toothed, 100 kyr cycle dominates alkenone records from the late Pleistocene (e.g. Müller *et al.* 1997; Rostek *et al.* 1997; Villanueva *et al.* 1998; Calvo *et al.* 2001; Herbert *et al.* 2001) and the few existing orbitally resolved records spanning the mid-Pleistocene transition show the same shift from 41-kyr to 100-kyr cycles that is recorded by oxygen isotope records (Liu & Herbert 2004; McClymont & Rosell-Melé 2005; McClymont *et al.* 2005). Until very recently, the alkenone organic proxy has been only sparsely applied to time intervals preceding the late Pleistocene. While alkenones have been found in ocean sediments from as early as the Cretaceous (Farrimond *et al.* 1986; Brassell & Dumitrescu 2004), to our knowledge time series of $U_{37}^{K'}$ ratios have not yet been generated for time periods earlier than the late Miocene. Two low-resolution (40–50 kyr) studies using $U_{37}^{K'}$ document the evolution of SSTs in the Atlantic from *c.* 6 Ma to the present (Herbert & Schuffert 1998; Marlow *et al.* 2000). Both records closely mirror the $\delta^{18}O$ records at these sites without the suggestion of any major interruptions or excursions in the $U_{37}^{K'}$ SST record across evolutionary events in the alkenone-producing lineage.

While we may never be able to answer absolutely the question of whether or not the modern calibrations are applicable over long timescales, comparisons between independent methodologies for constraining past sea-surface conditions can help confirm or refute our notion that these indices are recording meaningful climatic information as we move back in time. Initial results from studies applying the $U^{K'}_{37}$ index on long timescales lend credence to the long-term applicability of the alkenone palaeothermometer. A recent study from the Plio-Pleistocene eastern equatorial Pacific (EEP) illustrates the similarity in overall trend between a $U^{K'}_{37}$ SST record and a Mg/Ca-derived SST record (Ravelo *et al.* 2006) (Fig. 4). While the Mg/Ca record suggests temperatures *c.* 1 °C warmer than those derived from alkenone palaeothermometry, both records illustrate a similar evolution of long-term cooling of the EEP during the Plio-Pleistocene. The discrepancy between absolute temperature estimates from the proxies may stem from real differences in the ecology of the synthesizing organism for each of these methods (i.e. haptophytes versus *G. sacculifer*), which could potentially cause slight differences in the seasonality of production or in the depth habitat of these organisms. Alternatively, it could be due a difference in the temperature calibration of these two methodologies. A number of recent studies applying Mg/Ca palaeothermometry to longer timescales illustrate the uncertainty in the calibration of this palaeothermometer, each using different calibrations to generate their SST estimates (de Garidel-Thoron *et al.* 2005; Medina-Elizalde & Lea 2005; Wara *et al.* 2005; Steph *et al.* 2006). A new high-resolution study

employing alkenone palaeothermometry also focuses on the Plio-Pleistocene evolution of the EEP (Lawrence *et al.* 2006). This study demonstrates the first order similarity in overall trend in both long-term and orbital-scale structure between $U^{K'}_{37}$ SST estimates and benthic $\delta^{18}O$, an index of both ice volume and deep ocean temperature change (Fig. 5). This work suggests that the $U^{K'}_{37}$ index faithfully captures the global cooling trend and the glacial–interglacial fluctuations in climate that occurred during the Plio-Pleistocene.

At present, there is only one locality where long-timescale, high-resolution records of $U^{K'}_{37}$ SST can be directly compared at orbital resolution to another index of past SST (i.e. planktonic $\delta^{18}O$, Mg/Ca SST, or fauna assemblage SST estimates) (McClymont *et al.* 2005). McClymont *et al.* (2005) found strong similarities in the timing and structure of variations between alkenone SST estimates and variations in planktonic $\delta^{18}O$ (Fig. 6). The limited number of long-timescale, inter-proxy comparison studies is largely a consequence of the fact that very few high-resolution records of SST spanning timescales longer than a million years presently exist. Practitioners of each of these methodologies recognize the need for inter-proxy comparisons, and several more investigations employing different proxies for past SST at high resolution at the same locality are currently underway.

In the absence of same-site comparisons, comparisons of high-resolution alkenone SST records with SST proxies from nearby sites or in similar latitudinal belts offer the best means of evaluating the reliability of the alkenone SST proxy for high-resolution study of deep palaeoclimate.

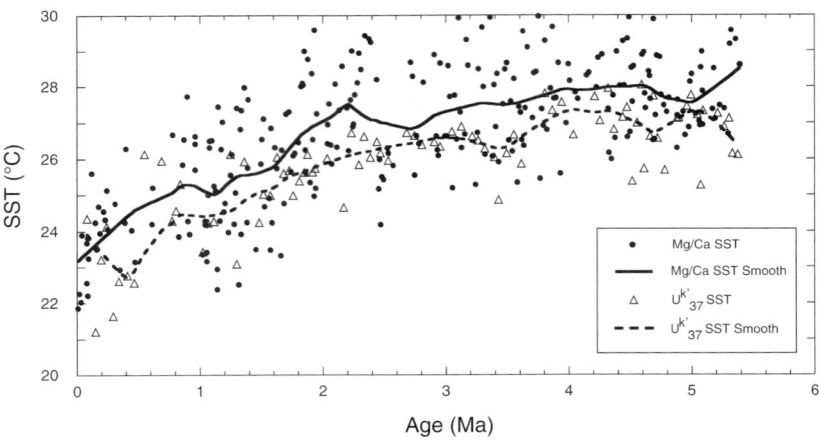

Fig. 4. Comparison of alkenone-derived (triangles and dashed black line) and Mg/Ca-derived (dots and solid black line) estimates for SST from ODP Site 847 in the eastern equatorial Pacific over the past *c.* 5 Myr (Ravelo *et al.* 2006). The curves were fit using the locally weighted least-squared error method.

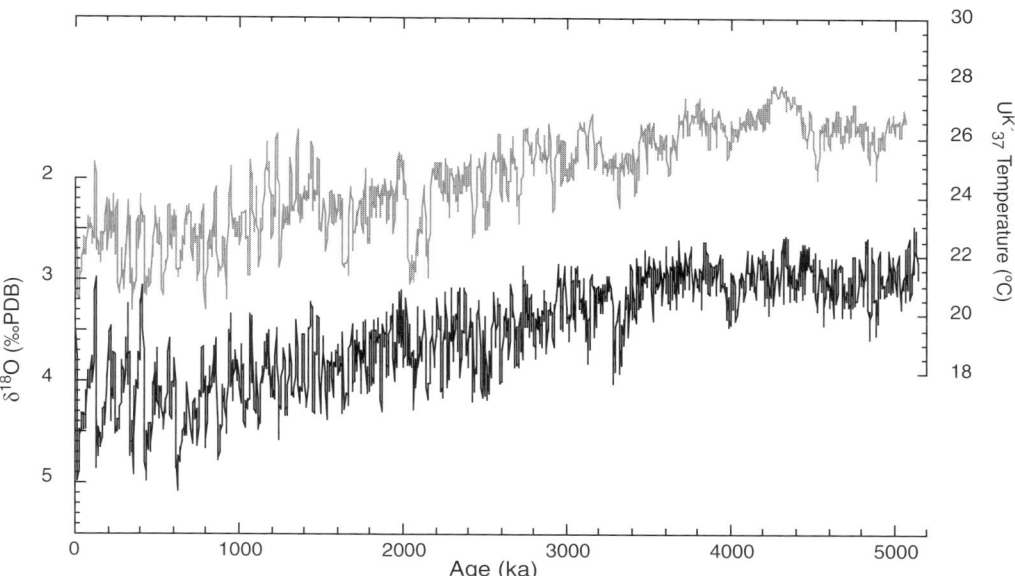

Fig. 5. Comparison of the evolution of high-resolution alkenone-derived records of SST (grey line) and benthic δ¹⁸O (black line) (Mix *et al.* 1995; Shackleton *et al.* 1995) from ODP Site 846 in the eastern equatorial Pacific over the past 5 Myr.

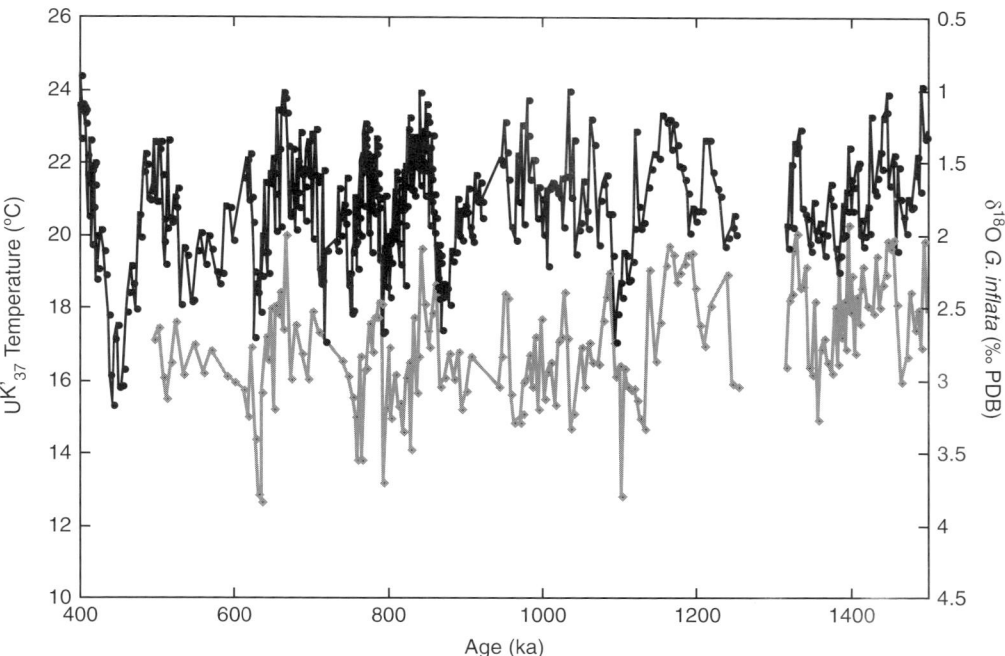

Fig. 6. Comparison of the Pleistocene evolution of alkenone-derived estimates of SST (grey line with diamonds) and planktonic δ¹⁸O (*G. inflata*) (black line with dots) from ODP 1087 (31° S, 15° E) (McClymont *et al.* 2005).

A comparison of Pliocene variations in planktonic $\delta^{18}O$ from Site 851 (Cannariato & Ravelo 1997) to $U^{K'}_{37}$ SST estimates from Site 846 (Lawrence *et al.* 2006), both sites located in the eastern equatorial Pacific, shows a marked similarity in glacial–interglacial structure (Fig. 7). Difference in amplitudes between the two records may stem from actual differences in the sea-surface temperature between the two sites. Site 851 is located west of Site 846, which is in the heart of the EEP upwelling zone. As discussed above, differences in the ecology of the organisms, which synthesize the target material of these analyses, or the confounding effects of changes in moisture balance and ice volume on $\delta^{18}O$ may also account for the discrepancies. Two orbital-scale Mg/Ca SST records from the western equatorial Pacific (de Garidel-Thoron *et al.* 2005; Medina-Elizalde & Lea 2005), spanning 1.4 Myr and 1.75 Myr respectively, offer comparisons with the high-resolution alkenone-derived record of SST from the EEP (Liu & Herbert 2004). Despite the difference in proxies (i.e. Mg/Ca versus $U^{K'}_{37}$) and the location of these sites on opposite sides of the equatorial Pacific Ocean, the western Pacific SST records show remarkable similarity in structure and phasing to the glacial–interglacial variations in

the eastern Pacific (Fig. 8). These similarities provide strong support for the robustness of both palaeothermometers.

While numerous studies have utilized C_{37} Total as an indicator of past productivity for late Quaternary climate studies (Brassell 1993; Villanueva *et al.* 1997; Villanueva *et al.* 1998; Sicre *et al.* 2001; Villanueva *et al.* 2001; Seki *et al.* 2004), the C_{37} Total proxy is in its fledgling stages of being applied to longer timescales. The 5 Myr-long C_{37} Total record from ODP Site 846 in the EEP shows broad similarity to low-resolution records of other proxies of past productivity (Lawrence *et al.* 2006) (Fig. 9). Additionally, the long-term evolution of C_{37} Total at Site 846 is very similar to that of the $CaCO_3$ mass accumulation rate at Site 1014 along the California Margin (Ravelo *et al.* 2004), both of which indicate an interval of markedly higher productivity from *c.* 2.9 to *c.* 1.6 Ma. Furthermore, the Site 846 C_{37} Total record shows strong orbital-scale variability at characteristic Milankovitch periods and strong coherence with SST records at this site with higher alkenone concentrations occurring during cold intervals (Lawrence *et al.* 2006). This predictable periodic behaviour lends support to the C_{37} Total proxy as a reliable recorder of past productivity variations.

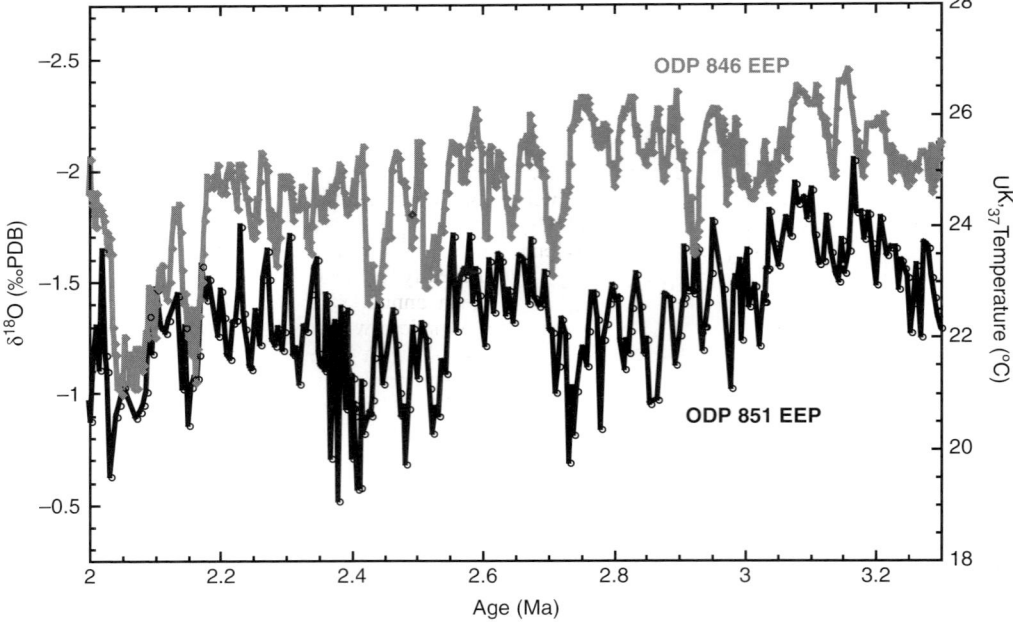

Fig. 7. Comparison of Pliocene orbital-scale, alkenone-derived estimates from ODP 846 (3° S, 91° W) (grey line with diamonds) (Lawrence *et al.* 2006) and a planktonic $\delta^{18}O$ record from nearby site ODP Site 851 (3° N, 111° W) (black line with circles) (Cannariato & Ravelo 1997).

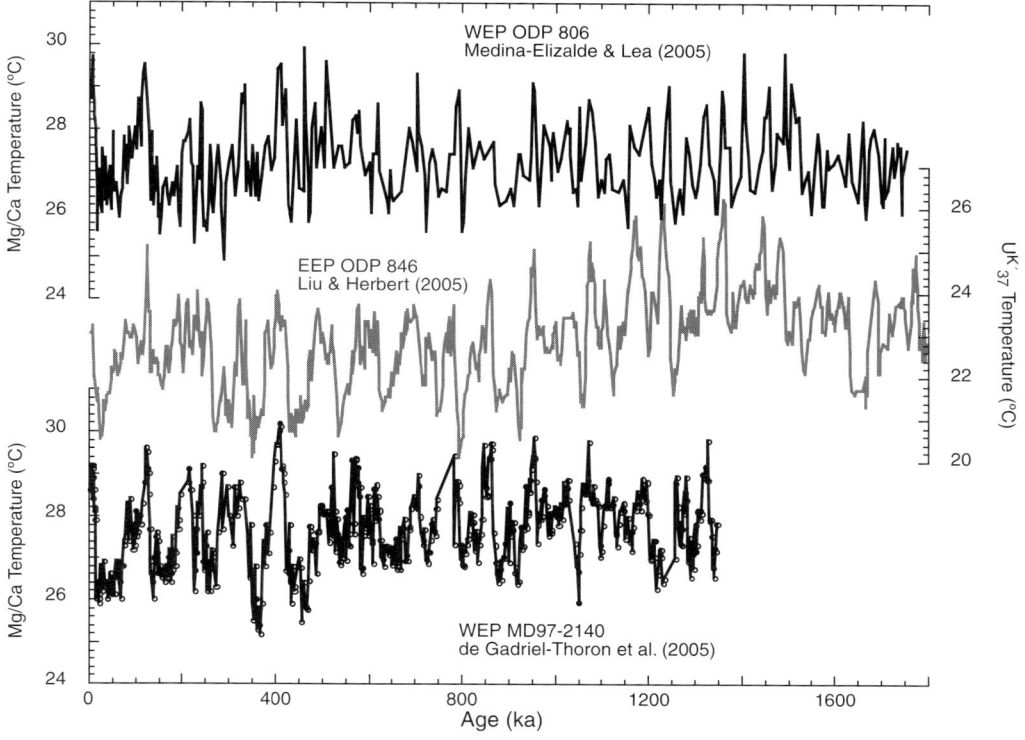

Fig. 8. Comparison of orbital-resolution Pleistocene alkenone-derived SST estimates from eastern equatorial Pacific Site ODP 846 (3° S, 91° W) (light grey line) (Liu & Herbert 2004) with Mg/Ca-derived SST estimates from western equatorial Pacific sites ODP 806 (0°, 159° E) (black line) (Medina-Elizalde & Lea 2005) and MD97-2140 (lower dark grey line with dots) (2° N, 142° E) (de Garidel-Thoron *et al.* 2005).

In the near future, a number of long-timescale, high-resolution studies comparing C_{37} Total and other indices of past productivity (e.g. total organic carbon, biogenic opal, percent nitrogen) will enable a more comprehensive evaluation of C_{37} Total as a proxy for past productivity on long timescales (Fig. 9).

New perspectives on Plio-Pleistocene climate

Alkenone-based studies of Plio-Pleistocene climates have lead to some new insights into this important climatic transition. The first $U^{K'}_{37}$ SST record to span the Plio-Pleistocene transition indicated that a pronounced cooling (*c.* 10 °C) occurred between 3.2 and 1 Ma in the Benguela upwelling system off of southwest Africa (Marlow *et al.* 2000). Subsequently, additional studies in other subtropical and tropical upwelling zones have shown that rapid cooling in upwelling regions during the Plio-Pleistocene transition was

a widespread phenomenon occurring along the California and Peruvian margins (Z. Liu, personal communication; Dekens *et al.* 2007), as well as in the EEP (Ravelo *et al.* 2006). The initial Marlow *et al.* (2000) study also showed that temperatures in the Benguela Current upwelling system during the early Pliocene (*c.* 26 °C) were similar to the mean annual SSTs for the oligotrophic water of the modern western South Atlantic. Similarly, two records from the EEP demonstrate that early Pliocene temperatures in the eastern tropical Pacific (*c.* 28 °C) approached those found in the modern western Pacific warm pool (Lawrence *et al.* 2006; Ravelo *et al.* 2006) (Figs 4 and 5).

These new SST records, which show substantial early Pliocene warmth relative to modern conditions in tropical and subtropical regions, have several important implications. First, a number of studies have recently suggested that the early Pliocene resembled a permanent El Niño state (Cane & Molnar 2001; Molnar & Cane 2002; Philander & Fedorov 2003). The new $U^{K'}_{37}$-derived SST datasets from the EEP along with a study based on

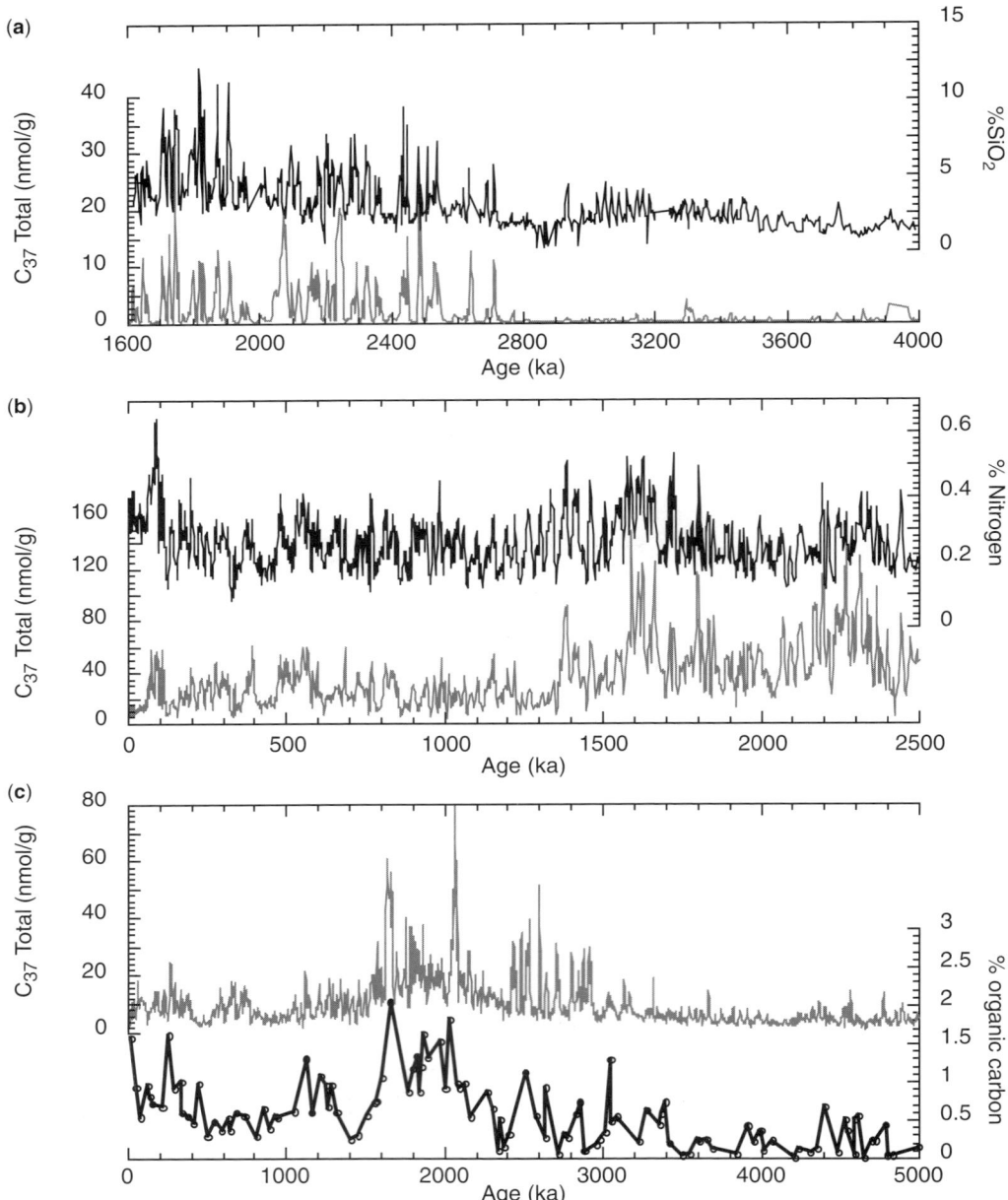

Fig. 9. Comparisons between C_{37} Total and other indices of past productivity. (**a**) Orbital resolution comparison from 4 to 1.6 Ma between SiO_2 variations (Ruddiman & Janecek 1989) (black line) and alkenone-derived C_{37} Total productivity (grey line) (Lawrence, unpublished data) from eastern equatorial Atlantic site ODP 662 ($1°$ S, $12°$ W). (**b**) Orbital-resolution comparison from 2.5 to 0 Ma between percent nitrogen variations (black line) and alkenone-derived C_{37} Total productivity (grey line) at California margin ODP Site 1012 ($32°$ N, $118°$ W) (Liu, unpublished data). (**c**) Plio-Pleistocene orbital-resolution alkenone-derived C_{37} Total productivity (grey line) (Lawrence *et al.* 2006) and a lower-resolution record of percent organic carbon (black line with circles) (Shipboard Scientific Party 1992) from ODP Site 846 in the eastern equatorial Pacific ($3°$ S, $91°$ W).

Mg/Ca palaeothermometery (Wara *et al.* 2005) provide strong evidence in support of the assertion that conditions in the tropical Pacific were indeed El Niño-like during the early Pliocene. Secondly, a recent comparison between the few existing sites with $U^{K'}_{37}$ SST estimates for the mid Pliocene PRISM (Pliocene Research Interpretations & Synoptic Mapping) timeslice (2.97–3.29 Ma) and the results of a full-coupled ocean–atmosphere GCM found strong agreement between observational and modelling results (Haywood *et al.* 2005). Like the models and in contrast to results from faunally based SST reconstructions for this timespan (Dowsett *et al.* 1999), $U^{K'}_{37}$ SST estimates suggest that the subtropics and tropics were considerably warmer than today (Haywood *et al.* 2005). These results are pivotal to understanding possible mechanisms for early Pliocene warmth because different mechanisms should produce different spatial patterns of surface temperature. The global pattern of SST change relative to the modern derived from PRISM timeslice faunal analysis studies was one of significant high latitude Pliocene warmth and slightly cooler or unchanged temperatures in low latitudes (Dowsett *et al.* 1996). This pattern was attributed to enhanced meridional ocean heat transport driven by more vigorous surface ocean and thermohaline circulation (Dowsett *et al.* 1992; Haywood *et al.* 2000). In contrast, the $U^{K'}_{37}$ SST implied pattern of significant Pliocene warmth in the tropics and subtropics as well as high latitudes is suggestive of higher atmospheric CO_2 concentration being a key driver of early Pliocene warmth. If $U^{K'}_{37}$ SST estimates from additional sites support the pattern of global warmth implied by existing data, this new picture will force a re-examination of the role that changing atmospheric CO_2 has played in the Plio-Pleistocene transition.

The one published orbital-resolution study spanning the Plio-Pleistocene transition indicates that Plio-Pleistocene cooling of the EEP preceded the onset of the Northern Hemisphere glaciation (NHG), indicating that glaciation did not force or initiate tropical cooling and implying that NHG was the culmination of more gradual changes in Earth's climate (Lawrence *et al.* 2006). Using both alkenone SST and C_{37} Total records, this study demonstrates that a linkage between the EEP and high latitude climates existed in the obliquity band for at least the past 5 Myr (Lawrence *et al.* 2006), which suggests that the plumbing of the EEP upwelling system has remained unchanged over the past 5 Myr. The early cooling of the EEP relative to NHG, the decoupling of C_{37} Total changes from SST changes on long timescales and the importance of Southern Hemisphere mode waters in supplying nutrients to the EEP led the

authors to suggest that the Southern Hemisphere may have played a more important role in modulating the climate of the Plio-Pleistocene than previously thought (Lawrence *et al.* 2006).

Conclusion

By providing independent estimates of past ocean surface temperatures and productivity, the alkenone organic proxy provides two important constraints on past sea-surface conditions. The high throughput and precision of this technique, as well as the wide temporal and spatial distribution of alkenones in ocean sediments, enables the rapid and reliable characterization of both of these parameters for most regions in the modern and past ocean. Important questions regarding the ecology, physiology and genetic variability of alkenone producers remain to be addressed. However, core-top calibrations indicate that these factors play a secondary role to the effect of near-surface mean annual temperature on the $U^{K'}_{37}$ index. Comparisons between existing alkenone-derived SST records and those from other proxies provide strong support for the application of this index over long timescales.

The poor overall preservation of alkenones in ocean sediments is of significant concern to the application of the C_{37} Total productivity proxy. However, existing comparisons with other indices of past production imply that C_{37} Total is as reliable as other existing measures of past productivity. While this index provides only a qualitative measure of past production, when applied with other palaeoclimate indicators, these records can help constrain potential mechanisms for climatic change and provide another perspective on past changes in the global carbon cycle. Although some important questions still remain, the alkenone organic proxy is an extremely valuable tool for the study of past climate on long timescales.

Comments and suggestions made by L. C. Cleaveland and C. A. Riihimaki were helpful in writing this review. We thank C. E. Lawrence for statistical advice, M. H. Conte for providing core-top and water column data, and E. L. McClymont for providing data from ODP Site 1087.

References

ANAND, P., ELDERFIELD, H. & CONTE, M. H. 2003. Calibration of Mg/Ca thermometry in planktonic foraminifera from a sediment trap time series. *Paleoceanography*, **18**, 1050, doi:10.1029/2002PA000846.

BARD, E., ROSTEK, F. & SONZOGNI, C. 1997. Interhemispheric synchrony of the last deglaciation inferred from alkenone palaeothermometry. *Nature*, **385**, 707–710.

BARD, E., ROSTEK, F., TURON, J. L. & GENDREAU, S. 2000. Hydrological impact of Heinrich events in the subtropical northeast Atlantic. *Science*, **289**, 1321–1324.

BENDLE, J. & ROSELL-MELÉ, A. 2004. Distributions of U_{37}^K and $U_{37}^{K'}$ in the surface waters and sediments of the Nordic Seas: Implications for paleoceanography. *Geochemistry, Geophysics, Geosystems*, **5**, doi:10.1029/2004GC000741.

BENDLE, J., ROSELL-MELÉ, A. & ZIVERI, P. 2005. Variability of unusual distributions of alkenones in the surface waters of the Nordic seas. *Paleoceanography*, **20**, doi:10.1029/2004PA001025.

BENTALEB, I., GRIMALT, J. O., VIDUSSI, F. *ET AL.* 1999. The C-37 alkenone record of seawater temperature during seasonal thermocline stratification. *Marine Chemistry*, **64**, 301–313.

BLANZ, T., EMEIS, K. C. & SIEGEL, H. 2005. Controls on alkenone unsaturation ratios along the salinity gradient between the open ocean and the Baltic Sea. *Geochimica et Cosmochimica Acta*, **69**, 3589–3600.

BOLLMANN, J., BAUMANN, K. H. & THIERSTEIN, H. R. 1998. Global dominance of *Gephyrocapsa* coccoliths in the late Pleistocene: Selective dissolution, evolution, or global environmental change? *Paleoceanography*, **13**, 517–529.

BOON, J. J., VAN DER MEER, F. W., SCHUYL, J. W. *ET AL.* 1978. Organic Geochemical Analysis of Core Samples from Site 362. *Initial Reports of the Deep Sea Drilling Project*, **40**, 627–637.

BRAND, L. E. 1984. The salinity tolerance of forty-six marine phytoplankton isolates. *Estuarine Coastal Shelf Science*, **18**, 543–556.

BRAND, L. E. 1994. Physiological ecology of marine coccolithophores. *In*: WINTER, A. & SIESSER, W. G. (eds) *Coccolithophores*. Cambridge, UK, Cambridge University Press, 39–49.

BRASSELL, S. C., EGLINTON, G., MARLOWE, I. T. *ET AL.* 1986. Molecular stratigraphy: a new tool for climatic assessment. *Nature*, **320**, 129–133.

BRASSELL, S. C. 1993. Applications of biomarkers for delineating marine paleoclimate fluctuations during the Quaternary. *In*: ENGEL, M. H. & MACKO, S. A. (eds) *Organic Geochemistry*, 669–738.

BRASSELL, S. C. & DUMITRESCU, M. 2004. Recognition of alkenones in a lower Aptian porcellanite from the west–central Pacific. *Organic Geochemistry*, **35**, 181–188.

BROERSE, A. T. C., BRUMMER, G. J. A. & VAN HINTE, J. E. 2000*a*. Coccolithophore export production in response to monsoonal upwelling off Somalia (northwestern Indian Ocean). *Deep-Sea Research Part Ii-Topical Studies in Oceanography*, **47**, 2179–2205.

BROERSE, A. T. C., ZIVERI, P., VAN HINTE, J. E. & HONJO, S. 2000*b*. Coccolithophore export production, species composition, and coccolith-CaCO₃ fluxes in the NE Atlantic (34 degrees N 21 degrees W and 48 degrees N 21 degrees W). *Deep-Sea Research Part II-Topical Studies in Oceanography*, **47**, 1877–1905.

BROERSE, A. T. C., ZIVERI, P. & HONJO, S. 2000*c*. Coccolithophore (-CaCO₃) flux in the Sea of Okhotsk: seasonality, settling and alteration processes. *Marine Micropaleontology*, **39**, 179–200.

BROWN, S. J. & ELDERFIELD, H. 1996. Variations in Mg/Ca and Sr/Ca ratios of planktonic foraminifera caused by postdepositional dissolution: Evidence of shallow Mg-dependent dissolution. *Paleoceanography*, **11**, 543–551.

BUDZIAK, D., SCHNEIDER, R. R., ROSTEK, F. *ET AL.* 2000. Late Quaternary insolation forcing on total organic carbon and C-37 alkenone variations in the Arabian Sea. *Paleoceanography*, **15**, 307–321.

CACHO, I., GRIMALT, J. O., CANALS, M. *ET AL.* 2001. Variability of the Western Mediterranean Sea surface temperature during the last 25,000 years and its connection with the Northern Hemisphere climatic changes. *Paleoceanography*, **16**, 40–52.

CACHO, I., GRIMALT, J. O. & CANALS, M. 2002. Response of the Western Mediterranean Sea to rapid climatic variability during the last 50,000 years: a molecular biomarker approach. *Journal of Marine Systems*, **33**, 253–272.

CALVO, E., PELEJERO, C., HERGUERA, J. C. *ET AL.* 2001. Insolation dependence of the southeastern Subtropical Pacific sea surface temperature over the last 400 kyrs. *Geophysical Research Letters*, **28**, 2481–2484.

CANE, M. A. & MOLNAR, P. 2001. Late Cenozoic Closing of the Indonesian Seaway as the Missing Link between the Pacific and East African Aridification. *Nature*, **411**, 157–162.

CANNARIATO, K. G. & RAVELO, A. C. 1997. Pliocene-Pleistocene evolution of eastern tropical Pacific surface water circulation and thermocline depth. *Paleoceanography*, **12**, 805–820.

CHAPPELL, J. & SHACKLETON, N. J. 1986. Oxygen isotopes and sea level. *Nature*, **324**, 137–140.

CHU, G. Q., SUN, Q., LI, S. Q. *ET AL.* 2005. Long-chain alkenone distributions and temperature dependence in lacustrine surface sediments from China. *Geochimica et Cosmochimica Acta*, **69**, 4985–5003.

CLIMAP 1981. Seasonal Reconstructions of the Earth's Surface at the Last Glacial Maximum, *Geological Society of America Chart Series 36*, Boulder, Colorado, Geological Society of America.

CONTE, M. H. & EGLINTON, G. 1993. Alkenone and alkenoate distributions within the euphotic zone of the eastern North Atlantic–Correlation with Production Temperature. *Deep-Sea Research Part I-Oceanographic Research Papers*, **40**, 1935–1961.

CONTE, M. H., VOLKMAN, J. K. & EGLINTON, G. 1994. Lipid biomarkers of the Haptophyta. *In*: GREEN, J. C. & LEADBEATER, B. S. C. (eds) *The Haptophyte Algae*. Oxford, Clarendon Press, 351–377.

CONTE, M. H., EGLINTON, G. & MADUREIRA, L. A. S. 1995*a*. Origin and fate of organic biomarker compounds in the water column and sediments of the eastern North-Atlantic. *Philosophical Transactions of the Royal Society of London Series B-Biological Sciences*, **348**, 169–177.

CONTE, M. H., THOMPSON, A., EGLINTON, G. & GREEN, J. C. 1995*b*. Lipid biomarker diversity in the coccolithophrid *Emiliania huxleyi* (Prymnesiophyceae) and the related species *Gephyrocapsa oceanica*. *Journal of Phycology*, **31**, 272–282.

CONTE, M. H., THOMPSON, A., LESLEY, D. & HARRIS, R. P. 1998. Genetic and physiological influences on the alkenone/alkenoate versus growth temperature relationship in *Emiliania huxleyi* and *Gephyrocapsa oceanica*. *Geochimica et Cosmochimica Acta*, **62**, 51–68.

CONTE, M. H., WEBER, J. C., KING, L. L. & WAKEHAM, S. G. 2001. The alkenone temperature signal in western North Atlantic surface waters. *Geochimica et Cosmochimica Acta*, **65**, 4275–4287.

CONTE, M. H., SICRE, M. A., RÜHLEMANN, C. ET AL. 2006. Global temperature calibration of the alkenone unsaturation index ($U_{37}^{K'}$) in surface waters and comparison with surface sediments. *Geochemistry, Geophysics, Geosystems*, **7**, doi:10.129/2005GC001054.

CORTES, M. Y., BOLLMANN, J. & THIERSTEIN, H. R. 2001. Coccolithophore ecology at the HOT station ALOHA, Hawaii. *Deep-Sea Research Part II-Topical Studies in Oceanography*, **48**, 1957–1981.

D'ANDREA, W. J. & HUANG, Y. 2005. Long chain alkenones in Greenland lake sediments: Low $\delta^{13}C$ values and exceptional abundance. *Organic Geochemistry*, **36**, 1234–1241.

DE GARIDEL-THORON, T., ROSENTHAL, Y., BASSINOT, F. C. & BEAUFORT, L. 2005. Stable sea surface temperatures in the western Pacific warm pool over the past 1.75 million years. *Nature*, **433**, 294–298.

DE LEEUW, J. W., VAN DER MEER, F. W., RIJPSTRA, I. C. & SCHENCK, P. A. 1980. On the occurrence and structural identification of long chain unsaturated ketones and hydrocarbons in sediments. *In*: DOUGLAS, A. G. & MAXWELL, J. R. (eds) *Advances in Organic Geochemistry*: Tarrytown, Pergamon, 211–217.

DEKENS, P. S., LEA, D. W., PAK, D. K. & SPERO, H. J. 2002. Core top calibration of Mg/Ca in tropical foraminifera: Refining paleotemperature estimation. *Geochemistry, Geophysics, Geosystems*, **3**, 1022, doi:10.1029/2001GC000200.

DEKENS, P. S., RAVELO, A. C. & MC CARTHY, M. D. 2007. Warm upwelling regions in the Pliocene warm period. *Paleoceanography*, **22**, doi:10.1029/2006PA001394.

DOOSE-ROLINSKI, H., ROGALLA, U., SCHEEDER, G. ET AL. 2001. High-resolution temperature and evaporation changes during the late Holocene in the northeastern Arabian Sea. *Paleoceanography*, **16**, 358–367.

DOWSETT, H. J., CRONIN, T. M., POORE, R. Z. ET AL. 1992. Micropaleontological evidence for increased meridional heat transport in the North Atlantic Ocean during the Pliocene. *Science*, **258**, 1133–1135.

DOWSETT, H. J., BARRON, J. & POORE, R. 1996. Middle Pliocene sea surface temperatures: a global reconstruction. *Marine Micropaleontology*, **27**, 13–25.

DOWSETT, H. J., BARRON, J. A., POORE, R. Z. ET AL. 1999. *Middle Pliocene paleoenvironmental reconstruction: PRISM2*, US Geological Survey Open File Report, 99–535.

DOWSETT, H. J., CHANDLER, M. A., CRONIN, T. M. & DWYER, G. S. 2005. Middle Pliocene sea surface temperature variability. *Paleoceanography*, **20**, PA2014, doi:10.1029/2005PA001133.

EDVARDSEN, B., EIKREM, W., GREEN, J. C. ET AL. 2000. Phylogenetic reconstructions of the Haptophyta inferred from 18S ribosomal DNA sequences and available morphological data. *Phycologia*, **39**, 19–35.

ELDERFIELD, H. & GANSSEN, G. M. 2000. Past temperature and $\delta^{18}O$ of surface ocean waters inferred from foraminiferal Mg/Ca ratios. *Nature*, **405**, 442–445.

ELTGROTH, M. L., WATWOOD, R. L. & WOLFE, G. V. 2005. Production and cellular localization of neutral long-chain lipids in the haptophyte algae *Isochrysis galbana* and *Emiliania huxleyi*. *Journal of Phycology*, **41**, 1000–1009.

EPSTEIN, B., D'HONDT, S., QUINN, J. ET AL. 1998. An effect of dissolved nutrient concentrations on alkenone-based temperature estimates. *Paleoceanography*, **13**, 122–126.

EPSTEIN, B. L., D'HONDT, S. & HARGRAVES, P. E. 2001. The possible metabolic role of C_{37} alkenones in *Emiliania huxleyi*. *Organic Geochemistry*, **32**, 867–875.

FARRIMOND, P., EGLINTON, G. & BRASSELL, S. C. 1986. Alkenones in Cretaceous Black Shales, Blake-Bahama Basin, Western North Atlantic. *Organic Geochemistry*, **10**, 897–903.

FARRINGTON, J. W., DAVIS, A. C., SULANOWSKI, J. ET AL. 1988. Biogeochemistry of Lipids in Surface Sediments of the Peru Upwelling Area at 15-Degrees-S. *Organic Geochemistry*, **13**, 607–617.

FUJIWARA, S., TSUZUKI, M., KAWACHI, M. ET AL. 2001. Molecular phylogeny of the Haptophyta based on the rbcL gene and sequence variation in the spacer region of the RUBISCO operon. *Journal of Phycology*, **37**, 121–129.

GONG, C. & HOLLANDER, D. J. 1999. Evidence for differential degradation of alkenones under contrasting bottom water oxygen conditions: Implication for paleotemperature reconstruction. *Geochimica et Cosmochimica Acta*, **63**, 405–411.

GONI, M. A., HARTZ, D. M., THUNELL, R. C. & TAPPA, E. 2001. Oceanographic considerations for the application of the alkenone-based paleotemperature U-37(K') index in the Gulf of California. *Geochimica et Cosmochimica Acta*, **65**, 545–557.

GRIMALT, J. O., CALVO, E. & PELEJERO, C. 2001. Sea surface paleotemperature errors in $U_{37}^{K'}$ estimation due to alkenone measurements near the limit of detection. *Paleoceanography*, **16**, 226–232.

HAIDAR, A. T. & THIERSTEIN, H. R. 2001. Coccolithophore dynamics off Bermuda (N. Atlantic). *Deep-Sea Research Part II-Topical Studies in Oceanography*, **48**, 1925–1956.

HARADA, N., SHIN, K. H., MURATA, A. ET AL. 2003. Characteristics of alkenones synthesized by a bloom of *Emiliania huxleyi* in the Bering Sea. *Geochimica et Cosmochimica Acta*, **67**, 1507–1519.

HARWOOD, J. R. & RUSSELL, N. J. 1984. *Lipids in Plants and Microbes,* G. Allen & Unwin, p. 162.

HAUG, G., GANOPOLSKI, A., SIGMAN, D. M. ET AL. 2005. North Pacific seasonality and the glaciation of North America 2.7 million years ago. *Nature*, **433**, 821–825.

HAYWOOD, A. M., VALDES, P. J. & SELLWOOD, B. W. 2000. Global scale palaeoclimate reconstruction of the middle Pliocene climate using the UKMO GCM: initial results. *Global and Planetary Change*, **25**, 239–256.

HAYWOOD, A. M., DEKENS, P., RAVELO, A. C. & WILLIAMS, M. 2005. Warmer tropics during the mid-Pliocene? Evidence from alkenone paleothermometry and a fully coupled ocean-atmosphere GCM. *Geochemistry, Geophysics, Geosystems*, **6**, doi:10.1029/2004/GC000799.

HERBERT, T. D. 2001. Review of alkenone calibrations (culture, water column, and sediments). *Geochemistry, Geophysics, Geosystems*, **2**, doi:10.1029/2000GC 000055.

HERBERT, T. D. 2004. Alkenone Paleotemperature Determinations. *In*: ELDERFIELD, H. (ed.) *Treatise on Geochemistry: The Oceans and Marine Geochemisty*, **6** Oxford, Elsevier, 391–432.

HERBERT, T. D. & SCHUFFERT, J. D. 1998. Alkenone unsaturation estimates of Late Miocene through Late Pliocene sea-surface temperatures at Site 958. *In*: FIRTH, J. V. (ed.) *Proceedings of the Ocean Drilling Program, Scientific Results*, **159T**, 17–21.

HERBERT, T. D., SCHUFFERT, J. D., THOMAS, D. *ET AL.* 1998. Depth and seasonality of alkenone production along the California margin inferred from a core top transect. *Paleoceanography*, **13**, 263–271.

HERBERT, T. D., SCHUFFERT, J. D., ANDREASEN, D. H. *ET AL.* 2001. Collapse of the California Current During Glacial Maxima Linked to Climate Change on Land. *Science*, **293**, 71–76.

HINRICHS, K. U., SCHNEIDER, R. R., MÜLLER, P. J. & RULLKÖTTER, J. 1999. A biomarker perspective on paleoproductivity variations in two Late Quaternary sediment sections from the Southeast Atlantic Ocean. *Organic Geochemistry*, **30**, 341–366.

HOEFS, M. J. L., VERSTEEGH, G. J. M., RIJPSTRA, W. I. C. *ET AL.* 1998. Postdepositional oxic degradation of alkenones: Implications for the measurement of palaeo sea surface temperatures. *Paleoceanography*, **13**, 42–49.

JORDAN, R. W. & GREEN, J. C. 1994. A check-list of the extant Haptophyta of the world. *Journal of Marine Biological Association of the United Kingdom*, **74**, 149–174.

KIENAST, M., STEINKE, S., STATTEGGER, K. & CALVERT, S. E. 2001. Synchronous tropical South China Sea SST change and Greenland warming during deglaciation. *Science*, **291**, 2132–2134.

KIRST, G. J., SCHNEIDER, R. R., MÜLLER, P. J. *ET AL.* 1999. Late Quaternary temperature variability in the Benguela Current System derived from alkenones. *Quaternary Research*, **52**, 92–103.

KORNILOVA, O. & ROSELL-MELÉ, A. 2003. Application of microwave-assisted extraction to the analysis of biomarker climate proxies in marine sediments. *Organic Geochemistry*, **34**, 1517–1523.

LAWRENCE, K. T. & HERBERT, T. D. 2005. Late Quaternary sea-surface temperatures in the western Coral Sea: Implications for the growth of the Australian Great Barrier Reef. *Geology*, **33**, 677–680.

LAWRENCE, K. T., LIU, Z. & HERBERT, T. D. 2006. Evolution of the eastern tropical Pacific through Plio-Pleistocene glaciation. *Science*, **312**, 79–83.

LEA, D., MASHIOTTA, T. A. & SPERO, H. J. 1999. Controls on magnesium and strontium uptake in planktonic foraminifera determined by live culturing. *Geochimica et Cosmochimica Acta*, **63**, 2369–2379.

LICHTFOUSE, E., LITTKE, R., DISKO, U. *ET AL.* 1992. Geochemistry and petrology of organic matter in

Miocene to Quaternary deep sea sediments from the Japan Sea. *Ocean Drilling Program, Scientific Results*, **127/128**, Ocean Drilling Program, 667–675.

LIU, Z. & HERBERT, T. D. 2004. High-latitude influence on the eastern equatorial Pacific climate in the early Pleistocene epoch. *Nature*, **427**, 720–723.

LYLE, M. W., PRAHL, F. G. & SPARROW, M. A. 1992. Upwelling and productivity changes inferred from a temperature record in the Central Equatorial Pacific. *Nature*, **355**, 812–815.

MARLOW, J. R., LANGE, C. B., WEFER, G. & ROSELL-MELÉ, A. 2000. Upwelling intensification as part of Pliocene-Pleistocene climate transition. *Science*, **290**, 2288–2291.

MARLOWE, I. T., BRASSELL, S. C., EGLINTON, G. & GREEN, J. C. 1984a. Long chain unsaturated ketones and esters in living algae and marine sediments. *In*: SCHENK, P. A. (ed.) *Advances in Organic Geochemistry*, **6**, 135–141.

MARLOWE, I. T., GREEN, J. C., NEAL, A. C. *ET AL.* 1984b. Long chain alkenones in the Prymnesiophyceae. Distribution of alkenones and other lipids and their taxonomic significance. *British Phycology Journal*, **19**, 203–216.

MARLOWE, I. T., BRASSELL, S. C., EGLINTON, G. & GREEN, J. C. 1990. Long-chain alkenones and alkyl alkenoates and the fossil coccolith record of marine sediments. *Chemical Geology*, **88**, 349–375.

McCAFFREY, M. A., FARRINGTON, J. W. & REPETA, D. J. 1990. The organic geochemistry of Peru Margin surface sediments. 1. A comparison of the C-37 Alkenone and historical El Niño records. *Geochimica et Cosmochimica Acta*, **54**, 1671–1682.

McCLYMONT, E. L. & ROSELL-MELÉ, A. 2005. Links between the onset of modern Walker circulation and the mid-Pleistocene climate transition. *Geology*, **33**, 389–392.

McCLYMONT, E. L., ROSELL-MELÉ, A., GIRAUDEAU, J. *ET AL.* 2005. Alkenone and coccolith records of the mid-Pleistocene in the south-east Atlantic: Implications for the $U^{K'}_{37}$ index and South African climate. *Quaternary Science Reviews*, **24**, 1559–1572.

McINTYRE, A. 1970. *Gephyrocapsa protohuxleyi* sp. n., a possible phyletic link and index fossil for the Pleistocene. *Deep-Sea Research*, **17**, 187–190.

MEDINA-ELIZALDE, M. & LEA, D. W. 2005. The mid-Pleistocene transition in the tropical Pacific. *Science*, **310**, 1009–1012.

MIX, A. C., LE, J. & SHACKLETON, N. J. 1995. Benthic foraminiferal stable isotope stratigraphy of Site 846:0–1.8 Ma. *In*: PISIAS, N. J., JANECEK, T. R., PALMER-JULSON, A. *ET AL.* (eds) *Proceedings of the Ocean Drilling Program, Scientific Results*, **138**, 839–854.

MOLFINO, B., KIPP, N. G. & MORLEY, J. J. 1982. Comparison of foraminiferal, coccolithophorid, and radiolarian paleotemperature equations: assemblage coherency and estimated concordancy. *Quaternary Research*, **17**, 279–313.

MOLLENHAUER, G., EGLINTON, T. I., OHKOUCHI, N. *ET AL.* 2003. Asynchronous alkenone and foraminifera records from the Benguela Upwelling System. *Geochimica et Cosmochimica Acta*, **67**, 2157–2171.

MOLLENHAUER, G., KIENAST, M., LAMY, F. *ET AL.* 2005. An evaluation of C-14 age relationships between co-occurring foraminifera, alkenones, and total organic carbon in continental margin sediments. *Paleoceanography*, **20**, PA1016, doi:10/1029/2004PA001103.

MOLNAR, P. & CANE, M. A. 2002. El Niño's tropical climate and teleconnections as a blueprint for pre-Ice Age climates. *Paleoceanography*, **17**, doi:10.1029/2001PA000663.

MORENO, A., CACHO, I., CANALS, M. *ET AL.* 2004. Millennial-scale variability in the productivity signal from the Alboran Sea record, Western Mediterranean Sea. *Palaeogeography, Palaeoclimatology, Palaeoecology*, **211**, 205–219.

MÜLLER, P. J., CEPEK, M., RUHLAND, G. & SCHNEIDER, R. R. 1997. Alkenone and coccolithophorid species changes in late Quaternary sediments from the Walvis Ridge: Implications for the alkenone paleotemperature method. *Palaeogeography, Palaeoclimatology, Palaeoecology*, **135**, 71–96.

MÜLLER, P. J., KIRST, G., RUHLAND, G. *ET AL.* 1998. Calibration of the alkenone paleotemperature index $U_{37}^{K'}$ on core-tops from the eastern South Atlantic and the global ocean (60° N–60° S). *Geochimica et Cosmochimica Acta*, **62**, 1757–1772.

MÜLLER, P. J. & FISCHER, G. 2001. A 4-year sediment trap record of alkenones from the filamentous upwelling region off Cape Blanc, NW Africa and a comparison with distributions in underlying sediments. *Deep-Sea Research Part I*, **48**, 1877–1903.

NIKOLAEV, S. D., OSKINA, N. S., BLYUM, N. S. & BUBENSHCHIKOVA, N. V. 1998. Neogene-Quaternary variations of the 'Pole-Equator' temperature gradient of the surface oceanic waters in the North Atlantic and North Pacific. *Global and Planetary Change*, **18**, 85–111.

NÜRNBERG, D. 1995. Magnesium in tests of *Neogloboquadrina pachyderma* sinistral from high northern and southern latitudes. *Journal of Foraminiferal Research*, **25**, 350–368.

NÜRNBERG, D., BIJMA, J. & HEMLEBEN, C. 1996. Assessing the reliability of magnesium in foraminiferal calcite as a proxy for water mass temperature. *Geochimica et Cosmochimica Acta*, **60**, 803–814.

NÜRNBERG, D., MÜLLER, A. & SCHNEIDER, R. R. 2000. Paleo-sea surface temperature calculations in the equatorial east Atlantic from Mg/Ca ratios in planktic foraminifera: A comparison to sea surface temperature estimates from U-37(K'), oxygen isotopes, and foraminiferal transfer function. *Paleoceanography*, **15**, 124–134.

OHKOUCHI, N., KAWAMURA, K., KAWAHATA, H. & OKADA, H. 1999. Depth ranges of alkenone production in the central Pacific Ocean. *Global Biogeochemical Cycles*, **13**, 695–704.

OHKOUCHI, N., EGLINTON, T. I., KEIGWIN, L. D. & HAYES, J. M. 2002. Spatial and temporal offsets between proxy records in a sediment drift. *Science*, **298**, 1224–1227.

OKADA, H. & HONJO, S. 1973. The distribution of oceanic coccolithophorids in the Pacific. *Deep-Sea Research*, **26**, 355–374.

OKADA, H. & MCINTYRE, A. 1979. Seasonal distribution of modern coccolithophores in the western North Atlantic Ocean. *Marine Biology*, **54**, 319–328.

SHIPBOARD SCIENTIFIC PARTY 1992. Site 846. *In*: MAYER, L. A., PISIAS, N. J., JANECEK, T. R. *ET AL.* (eds) *Proceedings of the Ocean Drilling Program, Initial Reports*, **138**. College Station, TX, Ocean Drilling Program, 265–333.

PEARSON, P. N., DITCHFIELD, P. W., SINGANO, J. *ET AL.* 2001. Warm tropical sea surface temperatures in the Late Cretaceous and Eocene epochs. *Nature*, **413**, 481–487.

PELEJERO, C., GRIMALT, J. O., HEILIG, S. *ET AL.* 1999. High-resolution $U_{37}^{K'}$ temperature reconstructions in the South China Sea over the past 220 kyr. *Paleoceanography*, **14**, 224–231.

PELEJERO, C. & CALVO, E. 2003. The upper end of the $U_{37}^{K'}$ temperature calibration revisited. *Geochemistry, Geophysics, Geosystems*, **4**, doi:10.1029/2002GC000431.

PFLAUMANN, U., DUPRAT, J., PUJOL, C. & LABEYRIE, L. 1996. SIMMAX: A modern analog technique to deduce Atlantic sea surface temperatures from planktonic foraminifera in deep-sea sediments. *Paleoceanography*, **11**, 15–35.

PHILANDER, S. G. & FEDOROV, A. V. 2003. Role of tropics in changing the response to Milankovich forcing some three million years ago. *Paleoceanography*, **18**, doi:10.1029/2002PA000837.

PRAHL, F. G. & WAKEHAM, S. G. 1987. Calibration of unsaturation patterns in long-chain ketone compositions for palaeotemperature assessment. *Nature*, **330**, 367–369.

PRAHL, F. G., MUEHLHAUSEN, L. A. & ZAHNLE, D. L. 1988. Further evaluation of long-chain alkenones as indicators of paleoceanographic conditions. *Geochimica et Cosmochimica Acta*, **52**, 2303–2310.

PRAHL, F. G., DE LANGE, G. J., LYLE, M. & SPARROW, M. A. 1989. Post-depositional stability of long-chain alkenones under contrasting redox conditions. *Nature*, **341**, 434–437.

PRAHL, F. G., COLLIER, R. B., DYMOND, J. *ET AL.* 1993*a*. A biomarker perspective on prymnesiophyte productivity in the northeast Pacific Ocean. *Deep-Sea Research*, **40**, 2061–2076.

PRAHL, F. G., COLLIER, R. B., DYMOND, J. *ET AL.* 1993*b*. A Biomarker Perspective on Prymnesiophyte Productivity in the Northeast Pacific-Ocean. *Deep-Sea Research Part I-Oceanographic Research Papers*, **40**, 2061–2076.

PRAHL, F. G., PILSKALN, C. H. & SPARROW, M. A. 2001. Seasonal record for alkenones in sedimentary particles from the Gulf of Maine. *Deep-Sea Research Part I-Oceanographic Research Papers*, **48**, 515–528.

PRAHL, F. G., WOLFE, G. V. & SPARROW, M. A. 2003. Physiological impacts on alkenone paleothermometry. *Paleoceanography*, **18**, doi:10.1029/2002PA000803.

PRELL, W. L. 1985. *The stability of low latitude sea surface temperatures An evaluation of the CLIMAP reconstruction with emphasis on the positive SST anomalies*: Washington, DC, Department of Energy, 1–60.

RAVELO, A. C., ANDREASEN, D. H., LYLE, M. *ET AL.* 2004. Regional climate shifts caused by gradual global cooling in the Pliocene epoch. *Nature*, **429**, 263–267.

RAVELO, A. C., DEKENS, P. S. & MCCARTHY, M. 2006. Evidence for El Niño-like conditions during the Pliocene. *GSA Today*, **16**, 4–11.

RECHKA, J. A. & MAXWELL, J. R. 1988. Characterisation of alkenone temperature indicators in sediments and organisms. *Organic Geochemistry*, **13**, 727–734.

RINNA, J., WARNING, B., MEYERS, P. A. *ET AL.* 2002. Combined organic and inorganic geochemical reconstruction of paleodepositional conditions of a Pliocene sapropel from the eastern Mediterranean Sea. *Geochimica et Cosmochimica Acta*, **66**, 1969–1986.

RONTANI, J. F., CUNY, P., GROSSI, V. & BEKER, B. 1997. Stability of long-chain alkenones in senescing cells of *Emiliania huxleyi*: effect of photochemical and aerobic microbial degradation on the alkenone unsaturation ratio $U^{K'}_{37}$. *Organic Geochemistry*, **26**, 503–509.

RONTANI, J. F., BEKER, B. & VOLKMAN, J. K. 2004. Long-chain alkenones and related compounds in benthic haptophye *Chrysotila lamellosa* Anand HAP 17. *Phytochemistry*, **65**, 117–126.

ROSELL-MELÉ, A., EGLINTON, G., PFLAUMANN, U. & SARNTHEIN, M. 1995. Atlantic core-top calibration of the $U^{K'}_{37}$ index as a sea-surface paleotemperature indicator. *Geochimica et Cosmochimica Acta*, **59**, 3099–3107.

ROSELL-MELÉ, A. 1998. Interhemispheric appraisal of the value of alkenone indices as temperature and salinity proxies in high-latitude locations. *Paleoceanography*, **13**, 694–703.

ROSELL-MELÉ, A., BARD, E., EMEIS, K. C. *ET AL.* 2001. Precision of the current methods to measure the alkenone proxy $U^{K'}_{37}$ and absolute alkenone abundance in sediments: Results of an interlaboratory comparison study. *Geochemistry, Geophysics, Geosystems*, **2**, doi:10.1029/2000GC000141.

ROSELL-MELÉ, A., BARD, E., EMEIS, K. C. *ET AL.* 2004. Sea surface temperature anomalies in the oceans at the LGM estimated from the alkenone- $U^{K'}_{37}$ index: comparison with GCMs. *Geophysical Research Letters*, **31**, doi:10.1029/2003GL018151.

ROSENTHAL, Y., BOYLE, E. A. & SLOWEY, N. 1997. Temperature control on the incorporation of magnesium, strontium, fluorine, and cadmium into benthic foraminiferal shells from Little Bahama Bank: Prospects from thermocline paleoceanography. *Geochimica et Cosmochimica Acta*, **61**, 3633–3643.

ROSENTHAL, Y., LOHMANN, G. P., LOHMANN, K. C. & SHERRELL, R. M. 2000. Incorporation and preservation of Mg in *Globigerinoides sacculifer*: Implications for reconstructing the temperature and $^{18}O/^{16}O$ of seawater. *Paleoceanography*, **15**, 135–145.

ROSENTHAL, Y. & LOHMANN, G. P. 2002. Accurate estimation of sea surface temperatures using dissolution-corrected calibrations of Mg/Ca paleothermometry. *Paleoceanography*, **17**, 1044, doi:10.1029/2001PA000749.

ROSTEK, F., RUHLAND, G., BASSINOT, F. C. *ET AL.* 1993. Reconstructing sea-surface temperature and salinity using $\delta^{18}O$ and alkenone records. *Nature*, **364**, 319–321.

ROSTEK, F., BARD, E., BEAUFORT, L. *ET AL.* 1997. Sea surface temperature and productivity records for the past 240 kyr in the Arabian Sea. *Deep-Sea Research Part II-Topical Studies in Oceanography*, **44**, 1461–1480.

RUDDIMAN, W. F. & JANECEK, T. R. 1989. Pliocene-Pleistocene Biogenic and Terrigenous Fluxes at Equatorial Atlantic Sites 662, 663 and 664. *In*: RUDDIMAN, W. F., SARNTHEIN, M., BALDAUF, J. G. *ET AL.* (eds) *Proceedings of the Ocean Drilling Program, Scientific Results*, **108**. College Station, TX, Ocean Drilling Program, 211–240.

RÜHLEMANN, C., MULITZA, S., MÜLLER, P. J. *ET AL.* 1999. Warming of the tropical Atlantic Ocean and slowdown of thermohaline circulation during the last deglaciation. *Nature*, **402**, 511–514.

SACHS, J. P. & ANDERSON, R. F. 2005. Increased productivity in the subantarctic ocean during Heinrich events. *Nature*, **434**, 1118–1121.

SAWADA, K., HANDA, N., SHIRAIWA, Y. *ET AL.* 1996. Long-chain alkenones and alkyl alkenoates in the coastal and pelagic sediments of the northwest North Pacific, with special reference to the reconstruction of *Emiliania huxleyi* and *Gephyrocapsa oceanica* ratios. *Organic Geochemistry*, **24**, 751–764.

SCHOUTEN, S., HOPMANS, E. C., SCHEFUSS, E. & SINNINGHE DAMSTÉ, J. S. 2002. Distributional variations in marine crenarchaeotal membrane lipids: a new tool for reconstructing ancient sea water temperatures? *Earth and Planetary Science Letters*, **204**, 265–274.

SCHOUTEN, S., HOPMANS, E. C., FORSTER, A. *ET AL.* 2003. Extremely high sea-surface temperatures at low latitudes during the middle Cretaceous as revealed by archael membrane lipids. *Geology*, **31**, 1069–1072.

SCHRAG, D. P., HAMPT, G. & MURRAY, D. W. 1996. Pore fluid constraints on the temperature and oxygen isotope composition of the glacial ocean. *Science*, **272**, 1930–1932.

SCHRAG, D. P. 1999. Effects of diagenesis on the isotopic record of late Paleogene tropical sea surface temperatures. *Chemical Geology*, **161**, 215–224.

SCHUBERT, C. J., VILLANUEVA, J., CALVERT, S. E. *ET AL.* 1998. Stable phytoplankton community structure in the Arabian Sea over the past 200,000 years. *Nature*, **394**, 563–566.

SEKI, O., ISHIWATARI, R. & MATSUMOTO, K. 2002. Millennial climate oscillations in NE Pacific surface waters. *Geophysical Research Letters*, **29**, 2144, doi: 10.1029/2002GL015200.

SEKI, O., IKEHARA, M., KAWAMURA, K. *ET AL.* 2004. Reconstruction of paleoproductivity in the Sea of Okhotsk over the last 30 kyr. *Paleoceanography*, **19**, PA1016, doi:10.1029/2002PA000808.

SEKI, O., KAWAMURA, K., SAKAMOTO, T. *ET AL.* 2005. Decreased surface salinity in the Sea of Okhotsk during the last glacial period estimated from alkenones. *Geophysical Research Letters*, **32**, LO8710, doi:10.1029/2004GL022177.

SHACKLETON, N. J. & OPDYKE, N. D. 1973. Oxygen Isotope and Palaeomagnetic Stratigraphy of Equatorial

Pacific Core V28-238: Oxygen Isotope Temperatures and Ice Volumes on a 10^5 and 10^6 Year Scale. *Quaternary Research*, **3**, 39–55.

SHACKLETON, N. J., HALL, M. A. & PATE, D. 1995. Pliocene stable isotope stratigraphy of Site 846. *In*: PISIAS, N. J., JANECEK, T. R., PALMER-JULSON, A. *ET AL.* (eds) *Proceedings of the Ocean Drilling Program, Scientific Results*, **138**. College Station, TX, Ocean Drilling Program, 337–355.

SICRE, M. A., TERNOIS, Y., PATERNE, M. *ET AL.* 2001. Climatic changes in the upwelling region off Cap Blanc, NW Africa, over the last 70 ky: a multi-biomarker approach. *Organic Geochemistry*, **32**, 981–990.

SICRE, M. A., BARD, E., EZAT, U. & ROSTEK, F. 2002. Alkenone distributions in the North Atlantic and Nordic sea surface waters. *Geochemistry, Geophysics, Geosystems*, **3**, doi:10.1029/2001GC000159.

SIKES, E. L. & KEIGWIN, L. D. 1994. Equatorial Atlantic sea-surface temperature for the last 30 kyr – a comparison of $U_{37}^{K'}$, $\delta^{18}O$ and foraminiferal assemblage temperature estimates. *Paleoceanography*, **9**, 31–45.

SIKES, E. L. & VOLKMAN, J. K. 1993. Calibration of alkenone unsaturation ratios ($U_{37}^{K'}$) for paleotemperature estimation in cold polar waters. *Geochimica et Cosmochimica Acta*, **57**, 1883–1889.

SIKES, E. L., VOLKMAN, J. K., ROBERTSON, L. G. & PICHON, J.-J. 1997. Alkenones and alkenes in surface waters and sediments of the Southern Ocean: Implications for paleotemperature estimation in polar regions. *Geochimica et Cosmochimica Acta*, **61**, 1495–1505.

SONZOGNI, C., BARD, E., ROSTEK, F. *ET AL.* 1997. Temperature and salinity effects on alkenone ratios measured in surface sediments from the Indian Ocean. *Quaternary Research*, **47**, 344–355.

SPRENGEL, C., BAUMANN, K. H. & NEUER, S. 2000. Seasonal and interannual variation of coccolithophore fluxes and species composition in sediment traps north of Gran Canaria (29 degrees N 15 degrees W). *Marine Micropaleontology*, **39**, 157–178.

STEINKE, S., KIENAST, M., PFLAUMANN, U. *ET AL.* 2001. A high-resolution sea-surface temperature record from the tropical South China Sea (16,500–3000 yr BP). *Quaternary Research*, **55**, 352–362.

STEPH, S., TIEDEMANN, R., GROENEVELD, J. *ET AL.* 2006. Pliocene Changes in Tropical East Pacific Upper Ocean Stratification: Response to Tropical Gateways? *In*: TIEDEMANN, R. (ed.) *Proceedings of the Ocean Drilling Program, Scientific Results*, **202**. College Station, TX, Ocean Drilling Program, 1–51.

SUN, D., CHU, G. Q., LI, S. Q. *ET AL.* 2004. Long-chain alkenones in sulfate lakes and its paleoclimatic implications. *Chinese Science Bulletin*, **49**, 2082–2086.

TEECE, M. A., GETLIFF, J. M., LEFTLEY, J. W. *ET AL.* 1998. Microbial degradation of the marine prymnesiophyte *Emiliania huxleyi* under oxic and anoxic conditions as a model for early diagenesis: long chain alkadienes, alkenones, and alkyl aklenoates. *Organic Geochemistry*, **29**, 863–880.

TERNOIS, Y., SICRE, M. A., BOIREAU, A. *ET AL.* 1997. Evaluation of long-chain alkenones as paleo-temperature indicators in the Mediterranean

Sea. *Deep-Sea Research Part I-Oceanographic Research Papers*, **44**, 271–286.

TERNOIS, Y., SICRE, M. A., BOIREAU, A. *ET AL.* 1998. Hydrocarbons, sterols and alkenones in sinking particles in the Indian Ocean sector of the Southern Ocean. *Organic Geochemistry*, **28**, 489–501.

THIERSTEIN, H. R., GEITZENAUER, K. R., MOLFINO, B. & SHACKLETON, N. 1977. Global synchroneity of late Quaternary coccolith datum levels: validation by oxygen isotopes. *Geology*, **5**, 400–404.

THOMPSON, P. R. 1976. Planktonic foraminiferal dissolution and the progress towards a Pleistocene Equatorial Pacific transfer function. *Journal of Foraminiferal Research*, **6**, 208–227.

THOMPSON, R. S. & FLEMING, R. F. 1996. Middle Pliocene vegetation: reconstructions, paleoclimatic inferences, and boundary conditions for climate modeling. *Marine Micropaleontology*, **27**, 27–49.

TREND-STAID, M. & PRELL, W. L. 2002. Sea surface temperature at the Last Glacial Maximum: A reconstruction using the modern analog technique. *Paleoceanography*, **17**, 1065, doi: 10.1029/2000 PA000506.

VERSTEEGH, G. J. M., RIEGMAN, R., DE LEEUW, J. W. & JANSEN, J. H.F. 2001. U_{37}^{K} values for *Isochrysis galbana* as a function of culture temperature, light intensity and nutrient concentrations. *Organic Geochemistry*, **32**, 785–794.

VILLANUEVA, J. & GRIMALT, J. O. 1996. Pitfalls in the chromatographic determination of the alkenone $U_{37}^{K'}$ index for paleotemperature estimation. *Journal of Chromatography A*, **723**, 285–291.

VILLANUEVA, J., GRIMALT, J. O., CORTIJO, E. *ET AL.* 1997. A biomarker approach to the organic matter deposited in the North Atlantic during the last climatic cycle. *Geochimica et Cosmochimica Acta*, **61**, 4633–4646.

VILLANUEVA, J., GRIMALT, J. O., LABEYRIE, L. D. *ET AL.* 1998. Precessional forcing of productivity in the North Atlantic Ocean. *Paleoceanography*, **13**, 561–571.

VILLANUEVA, J., CALVO, E., PELEJERO, C. *ET AL.* 2001. A latitudinal productivity band in the central North Atlantic over the last 270 kyr: An alkenone perspective. *Paleoceanography*, **16**, 617–626.

VILLANUEVA, J., FLORES, J. A. & GRIMALT, J. O. 2002. A detailed comparison of the $U_{37}^{K'}$ and coccolith records over the past 290 years: implications to the alkenone paleotemperature method. *Organic Geochemistry*, **33**, 897–905.

VOLKMAN, J. K. 2000. Ecological and environmental factors affecting alkenone distributions in seawater and sediments. *Geochemistry, Geophysics, Geosystems*, **1**, 2000GC000061.

VOLKMAN, J. K., EGLINTON, G., CORNER, E. D. S. & SARGENT, J. R. 1980. Novel unsaturated straight chain C_{37}-C_{39} methyl and ethyl ketones in marine sediments and a coccolithophore *Emiliania huxleyi*. *In*: DOUGLAS, A. G. & MAXWELL, J. R. (eds) *Advances in Organic Geochemistry*, Pergamon Press, 219–227.

VOLKMAN, J. K., JEFFREY, S. W., NICHOLS, P. D. *ET AL.* 1989. Fatty-acid and lipid-composition of 10 species of microalgae used in mariculture. *Journal of Experimental Marine Biology and Ecology*, **128**, 219–240.

VOLKMAN, J. K., BARRETT, S. M., BLACKBURN, S. I. & SIKES, E. L. 1995. Alkenones in *Gephyrocapsa oceanica*: Implications for studies of paleoclimate. *Geochimica et Cosmochimica Acta*, **59**, 513–520.

WARA, M. W., RAVELO, A. C. & DELANEY, M. L. 2005. Permanent El Niño-Like Conditions During the Pliocene Warm Period. *Science*, **309**, 758–761.

WINTER, A., JORDAN, R. & ROTH, P. 1994. Biogeography of living coccolithophores in ocean waters. *In*: WINTER, A. & SIESSER, W. G. (eds) *Coccolithophores*: Cambridge, UK, Cambridge University Press, 161–177.

ZACHOS, J., PAGANI, M., SLOAN, L. C. *ET AL.* 2001. Trends, rhythms, and aberrations in global climate 65 Ma to Present. *Science*, **292**, 686–693.

ZHAO, M. X., ROSELL, A. & EGLINTON, G. 1993. Comparison of 2 $U^{K'}_{37}$-sea surface-temperature records for the last climatic cycle at ODP Site 658 from the subtropical Northeast Atlantic. *Palaeogeography, Palaeoclimatology, Palaeoecology*, **103**, 57–65.

ZHAO, M. X., EGLINTON, G., HASLETT, S. K. *ET AL.* 2000. Marine and terrestrial biomarker records for the last 35,000 years at ODP site 658C off NW Africa. *Organic Geochemistry*, **31**, 919–930.

Onto the ice ages: proxy evidence for the onset of Northern Hemisphere glaciation

A. C. RAVELO[1], K. BILLUPS[2], P. S. DEKENS[3], T. D. HERBERT[4] & K. T. LAWRENCE[5]

[1]*Ocean Sciences Department, 1156 High Street, University of California, Santa Cruz, CA 95064, USA (e-mail: acr@es.ucsc.edu)*

[2]*College of Marine Studies, University of Delaware, 700 Pilottown Road, Lewes, DE 19958, USA*

[3]*Ocean Sciences Department, 1156 High Street, University of California, Santa Cruz, CA 95064, USA*

[4]*Dept. of Geological Sciences, Box 1846, Brown University, Providence RI 02912, USA*

[5]*Lafayette College, Deparment of Geology and Environmental Geosciences, 102 Van Wickle Hall, Easton, PA 18042, USA*

Abstract: Global climate change over the last five million years includes the early Pliocene warm period (*c.* 5–3 Ma) and a transition to the cool ice age climate of the latest *c.* 2 Ma. The growth of large ice sheets was accompanied by changes in deep and intermediate water mass characteristics, the end of El Niño-like mean conditions, and the development of cool tropical and subtropical upwelling regions. By considering how and when regional changes occurred, several possible causes of the warm to cold Pliocene climate transition are explored. There is intriguing evidence that, in addition to ice-albedo feedbacks that amplified cooling in high-latitude regions, the shoaling thermocline was a critical factor in the evolution of the Northern Hemisphere ice ages.

Cooling of global climate over the late Cenozoic culminated in the development of the large Northern Hemisphere glaciations of the Pleistocene epoch. Although records of ice-rafted debris indicate the presence of relatively small nascent Northern Hemisphere glaciations during the early Pliocene (*c.* 5 to 3 Ma), episodes of extremely cold climate, with ice-sheet expansion greater than the present-day size of the Greenland ice sheet, did not occur until after *c.* 3 Ma (Jansen & Sjoholm 1991). The transition from the globally warm average climate of the early Pliocene (*c.* 5 to 3 Ma) to the globally cool average climate of the late Pliocene and Pleistocene with large glaciations, is commonly referred to as the onset of Northern Hemisphere glaciation.

The onset of Northern Hemisphere glaciation was accompanied by changes in the amplitude and frequency of climate ('Milankovitch') cycles, which are presumably forced by changes in the latitudinal and seasonal distribution of solar heating on the Earth's surface as the Earth–Sun orbital parameters changed with time. This implies that the mean state of the climate could potentially influence the sensitivity of climate change to radiative forcing; as such, the characteristics of the warm Pliocene and the nature of the transition from a warm to a cold mean state may provide valuable information that can lead to a mechanistic understanding of long-term climate change.

The purpose of this study is to briefly present some proxy evidence for the onset of large Northern Hemisphere glaciations in the context of investigating processes that determine long-term climate change. This study will focus on the nature of the transition from warm to cold mean states by focusing on the expression of global cooling in different regions and water depths (Fig. 1 and Table 1), teleconnections that may have caused climate change in one region to dictate climate change in another, feedbacks that play a role in sustaining warm climatic states, and speculations about how long-term changes over the Pliocene and Pleistocene may have occurred.

Ice volume record

At present, the best proxy-records of high-latitude climate change over the last 5 Myr are benthic foraminiferal oxygen isotope records which were recently synthesized in a global stack (Lisiecki & Raymo 2005) (Fig. 2a). Although the oxygen isotope record is influenced by bottom-water temperature, it is a good first-order representation of general high-latitude cooling through the Pliocene–Pleistocene. Careful

From: WILLIAMS, M., HAYWOOD, A. M., GREGORY, F. J. & SCHMIDT, D. N. (eds) *Deep-Time Perspectives on Climate Change: Marrying the Signal from Computer Models and Biological Proxies.* The Micropalaeontological Society, Special Publications. The Geological Society, London, 563–573.
1747-602X/07/$15.00 © The Micropalaeontological Society 2007.

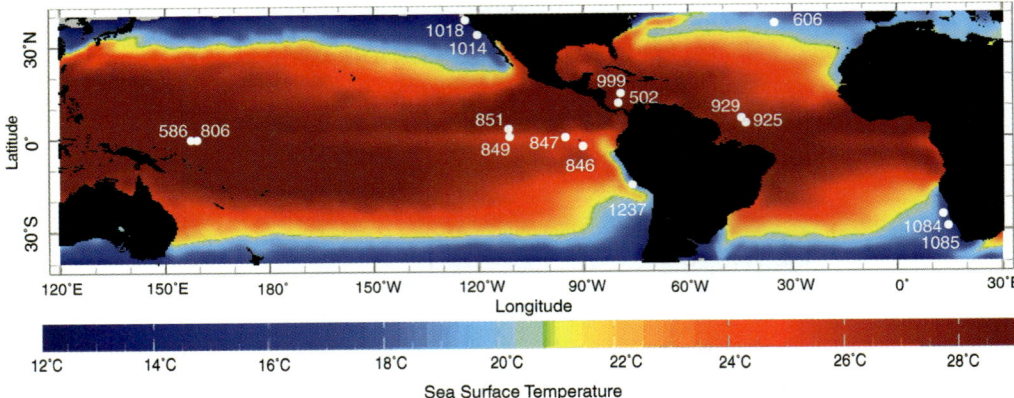

Fig. 1. March sea-surface temperature (Reynolds *et al.* 2002) with location of subtropical and tropical Ocean Drilling Program sites used in this study. High-latitude sites used in this study are not shown.

statistical analysis of the oxygen isotope record indicates that although small ice sheets were present in the Miocene and earliest Pliocene, the marked increase in global ice volume, or the onset of significant Northern Hemisphere glaciation, occurred gradually from 3.5–2.5 Ma (Mudelsee & Raymo 2005). Sea-surface temperature estimates support the idea that the North Atlantic and North Pacific were relatively warm at approximately 3 Ma (Dowsett *et al.* 1996) prior to the era of large glaciations.

Although the ice volume increase was relatively gradual, there were more abrupt climate changes in some regions, such as that inferred from ice-rafted debris records (Raymo *et al.* 1989; Haug *et al.* 1999). It could be that gradual cooling had a non-linear impact on regional climate; for example, a gradual shift in the subpolar front, once it passes over a site location, could result in an abrupt change in temperature or in ice-rafted debris deposition recorded in sediments from a core at that site (Fig. 2b). Alternatively, given the fact that abrupt changes in temperature, ice-rafted debris, and faunal assemblages are widespread and possibly synchronous (at *c.* 2.75 Ma), it could be that the

Table 1. *Site locations*

Site	Lat.	Long.	Water depth (m)
552 (North Atlantic)	56° N	23° W	2301
606 (North Atlantic)	37° N	35° W	3007
607 (North Atlantic)	41° N	33° W	3427
999 (Caribbean Sea)	13° N	79° W	2828[a]
502 (Caribbean Sea)	11° N	80° W	3051[b]
882 (North Pacific)	50° N	168° E	3244
925 (Tropical Atlantic)	4° N	43° W	3042
929 (Tropical Atlantic)	6° N	44° W	4361
1085 (Subtropical SE Atlantic)	29° S	14° E	1726
1084 (Subtropical SE Atlantic)	26° S	13° E	1192
704 (Subpolar SE Atlantic)	47° S	7° E	2532
586 (Tropical Pacific)	0°	158° E	2218
806 (Tropical Pacific)	0°	159° E	2520
1014 (NE Pacific)	33° N	120° W	1177[b]
1018 (NE Pacific)	37° N	123° W	2476
846 (Trop. Pacific)	3° S	91° W	3296
847 (Trop. Pacific)	0°	95° W	3373
849 (Trop. Pacific)	0°	111° W	3851
851 (Trop. Pacific)	3° N	111° W	3760
1237 (Subtropical SE Pacific)	16° S	76° W	3212

[a]Caribbean Sea sill depth is 1600–1900 m.
[b]Site 1014 is in Tanner Basin which has a sill depth of 1165 m.

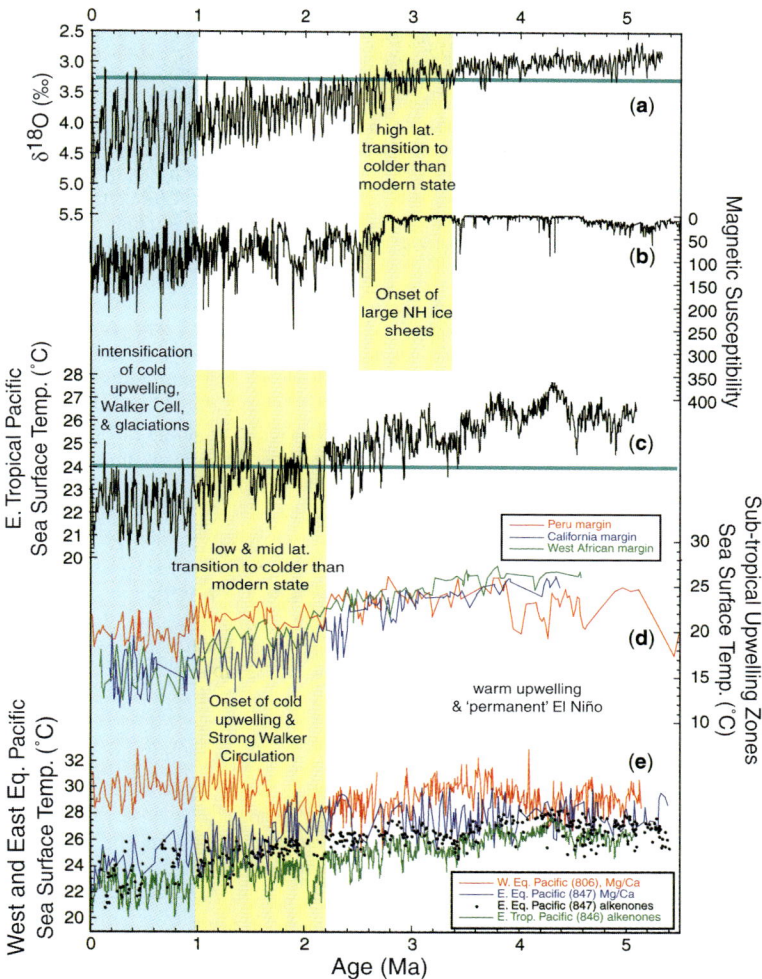

Fig. 2. Records of regional climate and ocean change over the last 5 Ma. (**a**) Stack of benthic oxygen isotope data (Lisiecki & Raymo 2005) represents the growth of high-latitude ice sheets. (**b**) The magnetic susceptibility record (Haug *et al.* 1999) indicates the concentration of ice-rafted debris in sediments in the North Pacific at ODP Site 882. (**c**) The Uk_{37}-derived sea-surface temperature record from eastern tropical pacific ODP Site 846 (Lawrence *et al.* 2006). (**d**) The Uk_{37}-derived sea-surface temperature records from three upwelling regions: the Peru margin (ODP Site 1237) and the California margin (ODP Site 1014) (Dekens *et al.* 2007), and the West African margin (ODP Site 1085) (Marlow *et al.* 2000). (**e**) The Mg/Ca and Uk_{37} sea-surface temperature records from the western tropical Pacific (ODP Site 806) compared to those from the eastern tropical Pacific (ODP Sites 847 and 846) showing the increase in west–east temperature difference across the equatorial Pacific as climate cooled.

benthic isotope record of a gradual increase in ice volume does not serve as an adequate indication of the average rate of high-latitude climate change through the Pliocene. Better constraints on the timing and rates of changes in different components of the subpolar and polar regions are needed to understand how well the ice volume (benthic oxygen isotope curve) represents high-latitude climate change.

Tropical and subtropical records

Records of cooling trends over the last 5 Myr, from middle- and low-latitude regions, are only available in a few locations. These records indicate that upwelling regions were relatively warm in the early Pliocene compared to today (Herbert & Schuffert 1998; Marlow *et al.* 2000; Lawrence *et al.* 2006); in some eastern boundary regions, sea-surface temperature was more than 8 °C warmer (Fig. 2c,d). Global reconstructions of sea-surface temperature during the early Pliocene indicate that regions to the west of upwelling zones were either the same or only moderately warmer than today (Dowsett *et al.* 2005), indicating that zonal sea-surface temperature gradients were reduced in middle- and low-latitude regions. In the tropical Pacific, the west-to-east sea-surface temperature difference, as recorded by planktonic foraminiferal magnesium to calcium ratios (Mg/Ca), was only about 1.5 °C in the early Pliocene and increased to the present-day difference of about 5 °C by the Pleistocene (Wara *et al.* 2005) (Fig. 2e).

In contrast, a recent study claims that sea-surface temperature in the east equatorial Pacific (at ODP Site 1241) was not warmer than today in the early Plio-cene, and concludes, therefore, that the east-to-west sea-surface temperature difference could not have been reduced (Groenveld *et al.* 2006). However, the Groenveld *et al.* (2006) study uses a different Mg/Ca-temperature calibration (Nurnberg *et al.* 1996) to estimate temperature at ODP Site 1241 than that used by Wara *et al.* (2005) to estimate temperatures at western Pacific ODP Site 806. In fact, the absolute Mg/Ca values at ODP Sites 1241 and 806 are nearly the same (average early Pliocene Mg/Ca values are 3.26 mmol/mol at Site 1241 and 3.54 mmol/mol at Site 806); if the same calibration is applied to both datasets, the resultant west-to-east temperature difference during the early Pliocene was <1 °C. Thus, the new Mg/Ca data from ODP Site 1241 (Groenveld *et al.* 2006) support previously published evidence for a reduced west-to-east temperature difference across the equatorial Pacific during the early Pliocene warm period (Wara *et al.* 2005; Ravelo *et al.* 2006).

In past studies, as in this one, the reduced west-east sea-surface temperature difference during the early Pliocene has been referred to as 'permanent El Niño-like conditions' (Wara *et al.* 2005; Fedorov *et al.* 2006; Ravelo *et al.* 2006); in fact, the sedimentary records that have been used to monitor changes in the west-east difference are of low resolution and cannot resolve individual El Niño events. The early Pliocene records indicate that the long-term *mean* pattern of sea-surface temp-erature resembled the pattern that occurs during short-lived El Niño events today. Hence, the term 'permanent El Niño-like conditions' refers to the mean state and does not preclude the possible exist-ence of interannual variability or El Niño events (although observational evidence for El Niño events in the Pliocene has not been found). The oceanic processes that cause changes in the long-term mean sea-surface temperature pattern through the Pliocene are thought to be different from the rapid air–sea processes that generate interannual variabil-ity and El Niño events in today's climate (Fedorov

et al. 2006). It is notable that the thermocline was on average deeper and/or warmer in the east equa-torial Pacific during the early Pliocene compared to today (Ravelo *et al.* 2006), as it is temporarily during El Niño events. It may be that the thermocline depth is an important factor in both short-lived El Niño events and long-term mean conditions in the tropical Pacific, but the adiabatic processes that dictate interannual fluctuations in the thermocline depth are different from the diabatic processes that influence long-term thermocline conditions (Boccaletti *et al.* 2004; Fedorov *et al.* 2006).

Overall, tropical and subtropical records indi-cate that sea-surface temperature cooled, especially in upwelling regions, suggesting that zonal sea-surface temperature gradients increased through the Pliocene. By about 3.0 Ma, the time-averaged 'permanent' El Niño-like conditions were ending. By just after 2.0 Ma, strong Walker circulation that characterizes the modern tropical Pacific climate system was established.

Deep-water records

Deep water is formed in high-latitude regions where surface water cools and sinks; thus, changes in the dis-tribution of deep-water masses reflects changes in high-latitude deep-water formation and deep-water mass mixing. Today, in the Atlantic Ocean, mixing between nutrient-depleted North Atlantic Deep Water flowing from north to south and denser nutrient-enriched Antarctic Bottom Water flowing from south to north results in a strong horizontal nutrient gradient in the deep Atlantic basin. In the Pacific Ocean, Antarctic Bottom Water enters the South Pacific and flows northward, becoming more nutrient-rich as it returns at mid-depth (*c.* 2 km). This mid-depth water, found throughout the North Pacific, is the most nutrient-enriched and oldest (radiocarbon-depleted) water in the global ocean. Deep-water nutrient distributions and gradients are reflected in the distribution of $\delta^{13}C$ composition of dissolved inorganic carbon ($\delta^{13}C_{DIC}$) (Fig. 3) because $\delta^{13}C_{DIC}$ is inversely related to nutrient and oxygen content of the deep ocean (Kroopnick 1985).

The $\delta^{13}C$ of benthic foraminifera has been used to reconstruct the $\delta^{13}C_{DIC}$ distribution in the deep ocean in the early Pliocene (Fig. 3) (Ravelo & Andreasen 2000). Deep-water mass distribution in the early Pliocene was qualitatively similar to today: the North Atlantic Deep Water was the most nutrient-depleted water mass and the mid-depth water of the North Pacific was the most nutrient-enriched water mass (Fig. 3). However, the $\delta^{13}C_{DIC}$ at sites within the mixing zone (*c.* 1.5 to 4.5 km) between the North Atlantic Deep Water and the Antarctic Bottom Water was higher than today, indicating that the deep Atlantic was

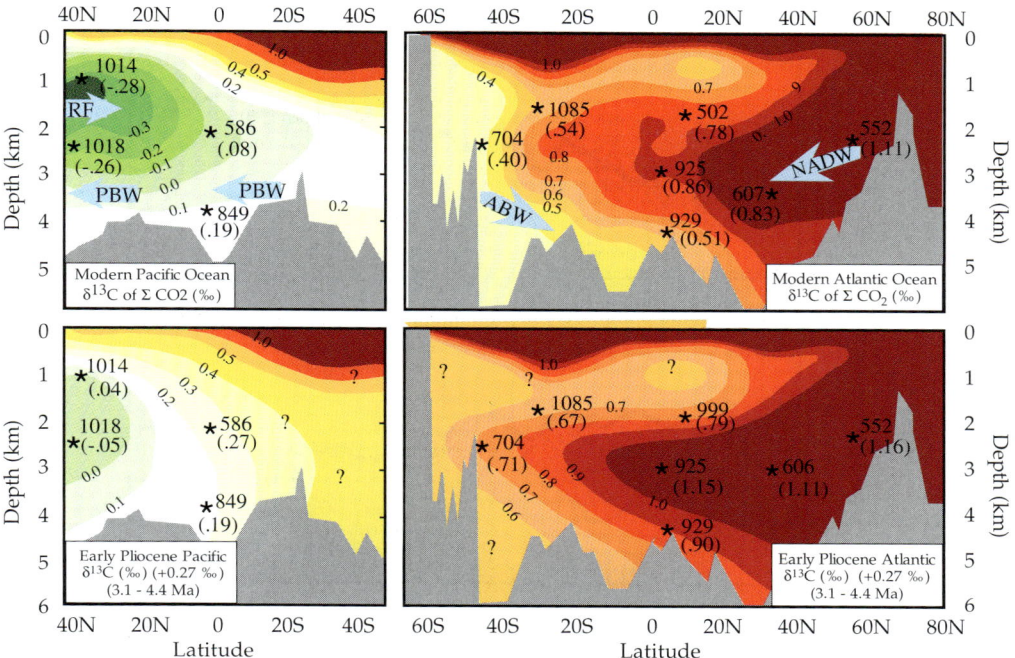

Fig. 3. Spatial distributions of the carbon isotopic composition of dissolved inorganic carbon ($\delta^{13}C_{DIC}$) and the carbon isotope composition of benthic foraminifera ($\delta^{13}C_{cal}$). In the upper panels, modern $\delta^{13}C_{DIC}$ is contoured (Kroopnick 1985), and discrete values are the average $\delta^{13}C_{cal}$ values for interglacial periods of the last 0.5 Ma plotted at the latitude and water depth of the site locations. In the lower panels, discrete average $\delta^{13}C_{cal}$ values for the Early Pliocene are contoured (after the global offset of $-0.27‰$ is subtracted). In the Early Pliocene, relatively high $\delta^{13}C_{cal}$ values in the deep Pacific and Atlantic Oceans indicates enhanced ventilation in both basins. This figure is from Ravelo & Andreasen (2000).

better ventilated (higher $\delta^{13}C_{DIC}$ is indicative of lower nutrient and higher oxygen concentrations). Greater ventilation could occur if the relative contribution of North Atlantic Deep Water to the South Atlantic was greater in the early Pliocene than today (Ravelo & Andreasen 2000) and/or if the composition of Antarctic Bottom Water was itself more ventilated and had a higher initial $\delta^{13}C_{DIC}$ signature that influenced the South Atlantic (Hodell & Venz-Curtis 2006). Similarly, the mid-depth water (c. 1 to 3 km) of the North Pacific had higher $\delta^{13}C_{DIC}$ values than today, indicating enhanced early Pliocene ventilation at those water depths in the North Pacific (Fig. 3). In all, the reduced $\delta^{13}C_{DIC}$ and inferred nutrient vertical gradients, in both the Atlantic and Pacific basins, indicate more rapid deep ocean ventilation and, possibly, thermohaline overturning during the early Pliocene warm period compared to today (Raymo *et al.* 1996; Ravelo & Andreasen 2000).

From c. 4.5 Ma to c. 3.5 Ma, North Atlantic Intermediate Water became relatively strong as indicated by increasing $\delta^{13}C_{DIC}$ in the Caribbean Sea (Haug & Tiedemann 1998) (Fig. 4b), which has a sill depth of 1600–1900 m and is filled with Atlantic Intermediate Water. This increase was approximately coincident with the effective closing of the Panama Seaway marked by the divergence in planktonic foraminiferal $\delta^{18}O$ records from the Caribbean and Eastern Tropical Pacific (Haug *et al.* 2001) (Fig. 4a). These observations are consistent with the idea that as flow through the seaway was restricted, northward heat flux increased (Berger & Wefer 1996) and was compensated for by increased southward transport of subsurface water (Haug & Tiedemann 1998). These changes have been used to argue that the closing of the Panama Seaway explains the intense warmth of the early Pliocene starting prior to 4.0 Ma, which may have delayed the onset of Northern Hemisphere glaciation (Berger & Wefer 1996). In contrast, it has also been suggested that these changes preconditioned the Earth system so that Northern Hemisphere glaciation could occur

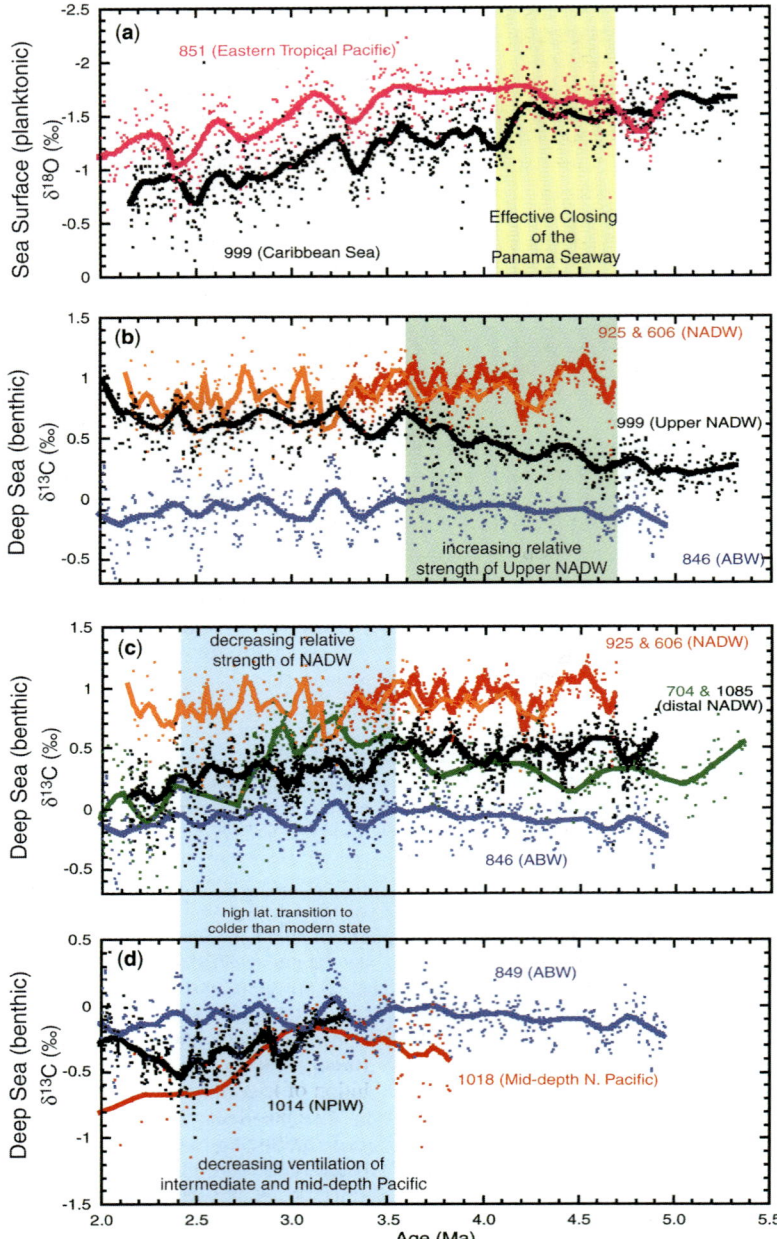

Fig. 4. The evolution of oceanic conditions from 5.5 to 2.0 Ma. (**a**) Records of $\delta^{18}O$ of *G. sacculifer* at ODP Sites 999 and 851 (Haug *et al.* 2001) diverge at about 4.2 Ma marking the effective closure of the Panama Seaway. (**b**) Records of $\delta^{13}C$ of benthic foraminifera at ODP Sites 925/606 and Site 846 reflect the nearly constant composition of the North Atlantic Deep Water and Antarctic Bottom Water respectively, while the record from ODP Site 999 reflects the increase in ventilation of North Atlantic Intermediate Water during and after the closure of the Panama Seaway. (**c**) The records from ODP Sites 704 and 1085 (located at the distal end of North Atlantic Deep Water) reflect the decrease in ventilation of the mid-depth South Atlantic during the onset of the ice ages. (**d**) The record at ODP Sites 1014 reflects nearly constant composition of North Pacific Intermediate Water while the record at Site 1018 reflects the increase in nutrient composition and decrease in ventilation of mid-depth water in the North Pacific during the onset of the ice ages. See Ravelo & Andreasen (2000) and references therein.

(Haug & Tiedemann 1998). These ideas will be discussed later after evidence for changes in thermocline conditions is presented.

While the relative strength of North Atlantic Intermediate Water increased from 4.5 to 3.5 Ma, possibly as a result of the closing of the Panama Seaway, as discussed above, North Atlantic Deep Water ventilation decreased much later. Specifically, $\delta^{13}C_{DIC}$ of the deep Central and South Atlantic decreased during the onset of Northern Hemisphere glaciation between 3.5 and 2.5 Ma (Fig. 4c) due to a reduction in the relative strength in North Atlantic Deep Water reaching the South Atlantic (Ravelo & Andreasen 2000) and/or to a reduction in Antarctic Bottom Water ventilation which impacted the $\delta^{13}C_{DIC}$ of the South Atlantic (Hodell & Venz-Curtis 2006).

In the Pacific, the mid-depth water became more nutrient-depleted during the onset of Northern Hemisphere glaciation (Fig. 4d); thus, like the deep Atlantic, the ventilation of the Pacific and the inferred turnover rate of the deep ocean were reduced as climate cooled through the Pliocene. Overall, the changes in the $\delta^{13}C_{DIC}$ of deep water indicate reduced deep-water ventilation in both the Pacific and Atlantic basins as climate cooled (Raymo 1994; Raymo *et al.* 1996; Ravelo & Andreasen 2000). The reduction of deep ocean ventilation during the onset of Northern Hemisphere glaciation may have been a response, not a cause, of global climate change. It is possible that high-latitude cooling affected conditions of deep-water formation in the Southern Ocean rather than in the North Atlantic (Kim & Crowley 2000), and/or that it led to increased stratification of polar regions which inhibited rapid thermohaline overturning and allowed greater carbon storage in the ocean interior, thereby acting as a positive feedback on the global cooling of the Pliocene (Sigman *et al.* 2004; Hodell & Venz-Curtis 2006).

The role of the thermocline

The onset of significant Northern Hemisphere glaciation and the establishment of the late Pliocene and Pleistocene ice-age climate included not just increasing ice volume, but also decreasing deep ocean ventilation, decreasing sea-surface temperature in upwelling regions, and an increasing west–east equatorial sea-surface temperature difference and Walker circulation, as described above. How did the climate in one region impact climate in another? In some cases, a close look at the regional timing of the transition to cold ice age climate can provide clues about the important processes that explain the warm Pliocene period and the mechanisms that cause long-term climate change.

Sea-surface temperature records from upwelling regions, the oxygen isotope record of ice volume changes, and deep ocean circulation records, indicate that the transition to cool ice age conditions and reduced deep ocean ventilation started at least 4 Myr ago, with marked cooling starting no later than 3.5 Ma (Figs 2 and 4). Some regions show a pronounced, relatively abrupt shift in conditions at *c.* 2.7 Ma, when critical high-latitude thresholds were crossed that resulted in, for example, increased iceberg discharge south of the present-day polar front (Fig. 2b) and reduced deep ocean ventilation (Fig. 4c, d), but these regional records do not accurately represent global cooling which appears to have been more gradual.

Although Pliocene cooling began in earnest by approximately 3.5 Ma in most regions, the timing of the establishment of 'ice age' conditions, or conditions that were as cold or colder than today, was different in high versus low latitudes. Specifically, mean ice volume has been greater than today for the last 2.5 Ma (Fig. 2a), while mean sea-surface temperature in upwelling regions has been cooler and the zonal SSTs difference across the Pacific (indicative of Walker circulation) has been greater than today for only the last *c.* 1.6 Ma (Fig. 2c–e). This implies that the cause of low-latitude cooling must be somewhat independent from changes in ice-sheet size since upwelling regions remained warmer than today even during the intense glaciations of the late Pliocene. For example, if the low-to-high-latitude temperature gradient and the resulting influence of that gradient on upwelling-favourable winds were the only factor controlling sea-surface temperature in upwelling regions, it would be expected that colder-than-modern high-latitude conditions and colder-than-modern upwelling regions would evolve at the same time. As this is not the case, other factors besides ice volume and high-latitude climate, and its possible influence on meridional temperature gradients and atmospheric forcing of upwelling, must have influenced the evolution of low-latitude climate through the Pliocene.

One possibility is that the sea-surface temperature of upwelling regions may have been influenced by changes in the conditions of subsurface water. For example, in the modern ocean, upwelling regions are cool only where the subsurface water is cool due to a relatively shallow thermocline; where the thermocline is deep, such as in the equatorial eastern Indian Ocean and the equatorial western and central Pacific Ocean, upwelled water is warm. Sea-surface temperature also varies temporally due to seasonal and interannual changes in the depth of the thermocline such as during El Niños when a deepening of the thermocline in the east equatorial Pacific prohibits cool water upwelling. Both warm upwelling regions and permanent El Niño-like conditions in the early

Pliocene could be explained by warmer or deeper mean state of the thermocline relative to today. Cooling of upwelling regions through the Pliocene (Marlow *et al.* 2000; Lawrence *et al.* 2006; Dekens *et al.* 2007) could therefore be explained by the shoaling of the thermocline (Fedorov *et al.* 2006) rather than solely by an increase in upwelling-favourable winds. There is limited evidence from the east equatorial Pacific that the permanent thermocline shoaled through the Pliocene (Chaisson & Ravelo 2000; Wara *et al.* 2005; Ravelo *et al.* 2006; Steph *et al.* 2006) although this should be verified at other locations.

That the conditions in the thermocline changed through the Pliocene is indirectly supported by records of palaeoproductivity and sea-surface temperature which indicate that the evolution of biological productivity and the cooling of upwelling zones throughout the world's oceans were decoupled (Lawrence *et al.* 2006; Dekens *et al.* 2007). As such, cooling of upwelling regions through the Pliocene are unlikely to be related simply to increasing wind strength, but rather there must have been changes in the subsurface source of upwelling water. Upwelling regions would have cooled once the thermocline became sufficiently shallow for winds to effectively mix cool subsurface water up to the surface. Consequently, stronger zonal sea-surface temperature and pressure gradients would have been established, which would have augmented upwelling-favourable winds. Such feedbacks are necessary to maintain strong Walker circulation in the tropical Pacific, and to enhance summertime subtropical high-pressure centres; through the Pliocene, these air–sea feedbacks must have come into play in eastern boundary upwelling regions around the globe.

In sum, the relative timing of climate change at high versus low latitude, the direct evidence for a shoaling thermocline in the east equatorial Pacific, and the decoupling between palaeoproductivity and sea-surface temperature changes in upwelling regions indicate that the far-field effects of ice-sheet growth itself cannot explain the cooling of upwelling regions and therefore the evolution of low-latitude climate. Rather, the cooling of low-latitude upwelling regions and the establishment of associated air–sea feedbacks must have been caused at least in part by changes in oceanic conditions such as the depth and/or temperature of the thermocline. It is possible that low-to-high teleconnections could therefore explain some of the characteristics of the warm Pliocene and the transition to the ice ages.

Causes of the onset of Northern Hemisphere glaciation

The general trends in climate over the entire Cenozoic are probably ultimately explained by decreasing atmospheric carbon dioxide concentration combined with changes in tectonic gateways and continental topography (Zachos *et al.* 2001), which influence large-scale ocean and atmospheric circulation. However, by the early Pliocene, atmospheric carbon dioxide concentrations were similar or only slightly (*c.* 30%) higher than pre-anthropogenic values (Kurschner *et al.* 1996; Raymo *et al.* 1996; Van der Burgh *et al.* 1993) while global temperature was about 2–3 °C warmer than today (Dowsett *et al.* 1996; Haywood & Valdes 2004; Sloan *et al.* 1996). In the absence of large changes in carbon dioxide concentrations, global cooling through the Pliocene and Pleistocene can only be explained by increasing albedo and/or decreasing greenhouse gases such as water vapour. Most climate models of future global warming predict that the sensitivity of global climate to a 30% increase in carbon dioxide would be significantly less than 2–3 °C (Intergovernmental Panel on Climate Change 2001); thus, the Pliocene provides an opportunity to constrain important factors, perhaps not captured by models, that would change the Earth's albedo and atmospheric water vapour content and amplify warming.

El Niño events in the modern climate system are temporary perturbations of mean tropical conditions, which have temporary global climate impacts. Could permanent El Niño-like conditions in the early Pliocene have had an impact on the mean global climate? A recent modelling study indicates that approximately 1 °C of global warming can be explained specifically by warmer sea-surface temperature in the east equatorial Pacific during the Pliocene, due to the accompanying reduction in highly reflective stratus clouds and increase in water vapour content of the atmosphere (Barreiro *et al.* 2005). A small amount of warming (0.6 °C) was also found to be attributable to a permanent reduction in the west–east temperature difference using a different model (Haywood *et al.* 2007). The end of permanent El Niño-like conditions (e.g. the development of a strong west-to-east SST difference and strong Walker Circulation in the tropical Pacific) may explain a portion of the cooling that accompanied the onset of the ice ages; unidentified processes or factors, either triggered by the end of El Niño-like conditions or not, must account for an additional *c.* 2 °C of global cooling through the late Pliocene.

Regional climate during the Pliocene could be related to the El Niño-like tropical conditions; for example, through teleconnections, a model predicts that a reduced tropical west-to-east SST gradient results in high-latitude North American warmth (Barreiro *et al.* 2005). However, a different climate model indicates that a reduced tropical west-to-east SST gradient results in cooler climate in North America

(Haywood *et al.* 2007), and that high-latitude North American warmth can result from Pliocene boundary conditions without permanent El Niño-like conditions (Haywood & Valdes 2004). Not surprisingly, different experimental designs and different climate models show that multiple factors can impact regional climate change. However, data-based reconstructions of Pliocene continental climate anomalies have the same pattern as those observed during El Niño events today (Molnar & Cane 2002), suggesting that low-to-high latitude teleconnections may impact North American climate. Assuming that the observations are solid evidence that the El Niño-like conditions of the early Pliocene had far-field effects, it follows that climate change through the Pliocene may be, at least in part, related mechanistically to the end of permanent El Niño-like conditions and associated feedbacks on regional climate temperature. But what ultimately forced the transition to occur?

One possible cause of global cooling through the Pliocene is the closing of the Panama Seaway. It has been suggested that the enhanced flow of warm water to the North Atlantic provided the water vapour and precipitation needed for the growth of large Northern Hemisphere ice sheets (Haug & Tiedemann 1998). As discussed above, the ventilation of North Atlantic Intermediate Water was enhanced starting around 4.5 Ma when the surface conditions in the Caribbean Sea and the eastern tropical Pacific diverged, signalling the effective closing of the Panama Seaway (Haug *et al.* 2001), providing evidence that North Atlantic conditions changed in the early Pliocene. However, the onset of Northern Hemisphere glaciation occurred *c.* 1 Myr after the effective closing of the Panama Seaway. To explain the delay in glaciation, it has been proposed that low-amplitude obliquity solar cycles from *c.* 4.0–3.0 Ma were not conducive to ice-sheet growth and that, once the amplitude of the obliquity cycle increased, around 3 Ma, large ice sheets developed (Haug & Tiedemann 1998). However, it is clear that global cooling began around 3.5 Ma or earlier when the amplitude of obliquity cycles was lowest: marked ice-sheet growth occurred starting around 3.5 Ma (Mudelsee & Raymo 2005) and upwelling regions began to cool prior to 3.5 Ma (Lawrence *et al.* 2006). Furthermore, a recent modelling study shows that the closing of the Panama Seaway does not intensify ice accumulation regardless of solar forcing (Klocker *et al.* 2004); large ice sheets were unlikely to develop while high-latitude temperatures were warm. Overall, it seems likely that increased warmth of high latitudes due to the closing of the Panama Seaway, if anything, inhibited or delayed the onset of the ice ages (Berger & Wefer 1996).

If the closing of the Panama Seaway contributed to Northern Hemisphere glaciation, then it may have done so through its impact on the thermocline conditions rather than through its impact on thermohaline circulation and North Atlantic conditions. There is evidence that the thermocline either shoaled or cooled starting between 4.5 and 4.0 Ma (Chaisson & Ravelo 2000; Wara *et al.* 2006; Ravelo *et al.* 2006; Steph *et al.* 2006), around the same time that the Caribbean and eastern tropical Pacific surface conditions became distinctly different (Fig. 4a). As the thermocline shoaled, cold upwelling conditions were established, highly reflective stratus clouds developed over upwelling regions, water vapour content decreased as the eastern equatorial Pacific cooled, and global climate cooled. Atmospheric teleconnections related to the demise of permanent El Niño conditions would have forced North America to cool and ice sheets to grow. Once ice sheets began to grow, then a series of feedbacks may have amplified global cooling: increased high-latitude albedo due to the ice sheets themselves, increased wind strength and upwelling (once the thermocline was shallow then increased winds due to ice age boundary conditions may have amplified global cooling), and increased ocean stratification and circulation which sequestered carbon dioxide from the atmosphere and stored it in the ocean interior. These ideas could begin to be tested with studies focused on how the closure of the Panamanian Seaway may have gradually impacted thermocline depth.

Alternatively, the restriction of flow through the Indonesian Seaways may have influenced thermocline conditions and tropical sea-surface temperature patterns; it is possible that the early Pliocene permanent El Niño conditions were due to the relatively high transport of warm thermocline water from the southern tropical Pacific to the Indian Ocean through the Indonesian Seaway (Cane & Molnar 2001). According to Cane and Molnar (2001), the tectonic movement of New Guinea caused cool thermocline water from the northern tropical Pacific to flow into the Indian Ocean, resulting in the cooling of upwelling regions in the Indian Ocean. These cool upwelling regions are linked to drier conditions in Africa and via atmospheric teleconnections to cooler conditions in North America, and, once established, allowed for the development of large Northern Hemisphere ice sheets. Although there is excellent supporting evidence for the drying and cooling of African climate (deMenocal 2004), more data are needed to confirm the development of cool upwelling regions in the Indian Ocean.

Future investigations of how the restriction of flow through the Panama or the Indonesian Seaways may have been responsible for the cooling of climate in the Pliocene and the onset of Northern Hemisphere glaciation should aim to explain how these tectonic changes could have influenced the depth of the thermocline, the cooling of upwelling regions, and the end of permanent El Niño-like conditions. Currently, it is difficult to envisage how regional tectonic events may have influenced the thermocline globally. If changes in flow through tropical gateways were not the sole cause of Northern Hemisphere glaciation, it may be that glaciation could have been the culmination of decreasing atmospheric carbon dioxide concentration through the Cenozoic, which was amplified dramatically in the Pliocene when the thermocline became shallow enough to cause upwelling regions to cool.

The thermocline depth is influenced by high-latitude conditions through adjustments in the oceanic heat budget. Today, the net positive flux of heat from the surface ocean to the atmosphere in the warm western boundaries of the ocean basins is balanced by a net flux of heat from the atmosphere to the surface ocean in low-latitude cold upwelling regions, and changes in the flux in one region cause changes in the other through adjustments in the thermocline (Boccaletti *et al.* 2004). For example, in a warm climate state when high-latitude air temperatures were warmer, the flux of heat to the atmosphere in western boundary regions would be relatively low and thus the flux of heat from the atmosphere to the surface ocean must be equally low in order to maintain a globally balanced heat budget. This would be achieved by decreasing the area of relatively cool upwelling water, thereby decreasing the intensity of heat flux into the ocean at low latitudes; such conditions require a deep thermocline (Philander & Fedorov 2003; Boccaletti *et al.* 2004).

Although atmospheric carbon dioxide concentrations decreased only slightly during the Pliocene, it is possible that the response of the climate system was amplified by adjustments of the thermocline (Fedorov *et al.* 2006). If the thermocline were at a critical threshold as the Miocene ended and the Pliocene began, then any number of perturbations (related to carbon cycle–climate feedbacks, gateway closures, volcanic eruptions, etc.) could have altered the ocean heat budget enough for the thermocline to shoal. This would have resulted in cooler upwelling regions, the end of permanent El Niño conditions and associated climate feedbacks that lead to the cooling of North America and the onset of Northern Hemisphere glaciation.

Much work needs to be done to verify many of the ideas presented in this study. Much of the insight into the evolution of climate in different regions is based on a small number of records that give us an incomplete picture of Pliocene climate change. Furthermore, the physical mechanisms that explain climate change through the Pliocene need to be explored through collaborative work between palaeoceanographers and theoreticians who are motivated by the same thing: to understand the causes of long-term climate change. If the onset of Northern Hemisphere glaciation is indeed related to crossing a critical threshold in which the thermocline shoaled enough for cold conditions to become 'normal', then future studies should be aimed at exploring whether future global warming may push the climate back across that threshold and into a warm, Pliocene-like, climatic state.

Many thanks to the National Science Foundation for supporting much of this research through grants (OCE-0081697, OCE-0623419 & ATM-0222383) to A.C.R. at the University of California, Santa Cruz. Data used in this study were generated by a large number of investigators over the years using samples collected by the Ocean Drilling Program.

References

BARREIRO, M., PHILANDER, G., PACANOWSKI, R. & FEDOROV, A. 2005. Simulations of warm tropical conditions with application to middle Pliocene atmospheres. *Climate Dynamics*, doi:10.1007/s00382-005-0086-4.

BERGER, W. H. & WEFER, G. 1996. Expeditions into the past: Paleoceanographic studies in the South Atlantic. *In*: WEFER, G. (ed.) *The South Atlantic: Present and Past Circulation*. New York, Springer, 363–410.

BOCCALETTI, G., PACANOWSKI, R. C., PHILANDER, S. G. & FEDOROV, A. V. 2004. The thermal structure of the upper ocean. *J. of Phys. Oceanogr*, **34**, 888–902.

CANE, M. A. & MOLNAR, P. 2001. Closing of the Indonesian seaway as a precursor to east African aridification around 3–4 million years ago. *Nature*, **411**, 157–162.

CHAISSON, W. P. & RAVELO, A. C. 2000. Pliocene development of the east–west hydrographic gradient in the equatorial Pacific. *Paleoceanography*, **15**, 497–505.

DEKENS, P. S., RAVELO, A. C. & MCCARTHY, M. D. 2007. Warm upwelling regions in the Pliocene warm period. *Paleoceanography*, doi: 10.1029/2006PA001394, in press.

DEMENOCAL, P. B. 2004. African climate change and faunal evolution during the Pliocene-Pleistocene. *Earth and Planetary Science Letters*, **220**, 3.

DOWSETT, H. J., BARRON, J. & POORE, R. 1996. Middle Pliocene sea surface temperatures: a global reconstruction. *Marine Micropaleontology*, **27**, 13–25.

DOWSETT, H. J., CHANDLER, M. A., CRONIN, T. M. & DWYER, G. S. 2005. Middle Pliocene sea surface temperature variability. *Paleoceanography*, **20**, doi:10.1029/2005P 2005PA001 A001133.

FEDOROV, A. V., DEKENS, P. S., MCCARTHY, M. *ET AL.* 2006. The Pliocene Paradox (Mechanisms for a Permanent El Niño). *Science*, **312**, 1485–1489.

GROENVELD, J., STEPH, S., TIEDEMANN, R., GARBE-SCHONBERG, D., NÜRNBERG, D. & STURM, A. 2006. Pliocene mixed-layer oceanography for Site 1241, using combined Mg/Ca and $\delta^{18}O$

analyses of *Globigerinoides sacculifer. In*: TIEDE-MANN, R., MIX, A. C., RICHTER, C. & RUDDIMAN, W. F., (eds) *Proceedings of the Ocean Drilling Program Scientific Results*, **202**, College Station. TX (Ocean Drilling Program), 1–27.

HAUG, G. H., SIGMAN, D. M., TIEDEMANN, R., PEDER-SEN, T. F. & SARNTHEIN, M. 1999. Onset of perma-nent stratification in the subarctic Pacific Ocean. *Nature*, **401**, 779–782.

HAUG, G. H. & TIEDEMANN, R. 1998. Effect off the for-mation of the Isthmus of Panama on Atlantic Ocean thermohaline circulation. *Nature*, **393**, 673–676.

HAUG, G. H., TIEDEMANN, R., ZAHN, R. & RAVELO, A. C. 2001. Role of Panama uplift on oceanic fresh-water balance. *Geology*, **29**, 207–210.

HAYWOOD, A. M. & VALDES, P. J. 2004. Modelling Plio-cene warmth: contribution of atmosphere, oceans and cryosphere. *Earth and Planetary Science Letters*, **218**, 363–377.

HAYWOOD, A. M., VALDES, P. J. & PECK, V. L. 2007. A Permanent El Niño-like state during the Pliocene? *Paleoceanography* **22**, PA1213, doi: 10.1029/2006 PA001323.

HERBERT, T. D. & SCHUFFERT, J. D. 1998. Alkenone unsaturation estimates of Late Miocene through Late Pliocene sea-surface temperatures at Site 958. *In*: FIRTH, J. V. (ed.) *Proceedings of the Ocean Drilling Program*, **159T**, College Station, TX, Ocean Drilling Program, 17–21.

HODELL, D. A. & VENZ-CURTIS, K. A. 2006. Late Neogene history of deepwater ventilation in the Southern Ocean. *Geochemistry, Geophysics, Geosys-tems*, **7**, doi:10.1029/2005GC001211.

INTERGOVERNMENTAL PANEL ON CLIMATE CHANGE 2001. *Climate Change: the scientific basis*, www.ipcc.ch.

JANSEN, E. & SJOHOLM, J. 1991. Reconstruction of gla-ciation over the past 6 Myr from ice-borne deposits in the Norwegian Sea. *Nature*, **349**, 600–603.

KIM, S.-J. & CROWLEY, T. J. 2000. Increased Pliocene North Atlantic Deep Water: cause or consequence of Pliocene warming? *Paleoceanography*, **15**, 451–455.

KLOCKER, A., PRANGE, M. & SCHULZ, M. 2004. Testing the influence of the Central American Seaway on orbitally forced Northern Hemisphere glaciation. *Geophysical Research Letters*, **32**, doi:10.1029/2004GL021564.

KROOPNICK, P. M. 1985. The distribution of ^{13}C of pCO_2 in the world oceans. *Deep-Sea Res.*, **32**, 57–84.

KURSCHNER, W. M., VAN DER BURGH, J., VISSCHER, H. & DILCHER, D. L. 1996. Oak leaves as biosensors of late Neogene and early Pleistocene paleoatmospheric CO_2 concentrations. *Marine Micropaleontology*, **27**, 299–312.

LAWRENCE, K. T., LIU, Z. & HERBERT, T. D. 2006. Evolution of the Eastern Tropical Pacific Through Plio-Pleistocene Glaciation. *Science*, **312**, 79–83.

LISIECKI, L. E. & RAYMO, M. E. 2005. A Pliocene-Pleistocene stack of 57 globally distributed benthic ^{18}O records. *Paleoceanography*, **20**, doi:10.1029/2004PA001071.

MARLOW, J. R., LANGE, C. B., WEFER, G. & ROSELL-MELE, A. 2000. Upwelling intensification as part of the Pliocene-Pleistocene climate transition. *Science*, **290**, 2288–2291.

MOLNAR, P. & CANE, M. A. 2002. El Niño's tropical climate and teleconnections as a blueprint for pre-Ice Age climates. *Paleoceanography*, **17**, doi:10.1029/2001PA000663.

MUDELSEE, M. & RAYMO, M. E. 2005. Slow dynamics of the Northern Hemisphere glaciation. *Paleoceano-graphy*, **20**, doi:10.1029/2005PA001153.

NURNBERG, D., BIJMA, J. & HEMLEBEN, C. 1996. Assessing the reliability of magnesium in forami-niferal calcite as a proxy for water mass tempera-tures. *Geochimica et Cosmochimica Acta*, **60**, 803–814.

PHILANDER, S. G. & FEDOROV, A. V. 2003. Role of tropics in changing the response to Milankovich forcing some three million years ago. *Paleoceanogra-phy*, **18**, doi:10.1029/2002PA000837.

RAVELO, A. C. & ANDREASEN, D. H. 2000. Enhanced circulation during a warm period. *Geophysical Research Letters*, **27**, 1001–1004.

RAVELO, A. C., DEKENS, P. S. & MCCARTHY, M. 2006. Evidence for El Niño-like conditions during the Pliocene. *GSA Today*, **16**, 4–11.

RAYMO, M. E. 1994. The Initiation of Northern Hemi-sphere Glaciation. *Annual Reviews of Earth and Planetary Sciences*, **22**, 353–383.

RAYMO, M. E., RUDDIMAN, W. F., BACKMAN, J., CLEMENT, B. M. & MARTINSON, D. G. 1989. Late Pliocene variation in northern hemisphere ice sheets and North Atlantic deepwater circulation. *Paleoceano-graphy*, **4**, 413–446.

RAYMO, M. E., GRANT, B., HOROWITZ, M. & RAU, G. H. 1996. Mid-Pliocene warmth: stronger green-house and stronger conveyor. *Marine Micropaleontol-ogy*, **27**, 313–326.

REYNOLDS, R. W., RAYNER, N. A., SMITH, T. M., STOKES, D. C. & WANG, W. 2002. An Improved in situ and satellite SST analysis for climate. *Journal of Climate*, **15**, 1609–1625.

SIGMAN, D. M., JACCARD, S. L. & HAUG, G. H. 2004. Polar ocean stratification in a cold climate. *Nature*, **428**, 59–63.

SLOAN, L. C., CROWLEY, T. J. & POLLARD, D. 1996. Modeling of middle Pliocene climate with the NCAR GENESIS general circulation model. *Marine Micropaleontology*, **27**, 51–61.

STEPH, S., TIEDEMANN, R., GROENEVELD, J., STURM, A. & NÜRNBERG, D. 2006. Pliocene changes in tropical East Pacific upper ocean stratification: Response to tropical gateways? *In*: TIEDEMANN, R., MIX, A. C., RICHTER, C. & RUDDIMAN, W. F. (eds) *Proceedings of the Ocean Drilling Program, Scientific Results*, **202**: College Station, TX (Ocean Drilling Program), 1–27.

VAN DER BURGH, J., VISSCHER, H., DILCHER, D. L. & KURSCHNER, W. M. 1993. Paleoatmospheric signa-tures in Neogene fossil leaves. *Science*, **260**, 1788–1790.

WARA, M. W., RAVELO, A. C. & DELANEY, M. L. 2005. Permanent El Niño-like conditions during the Pliocene warm period. *Science*, **309**, 758–761.

ZACHOS, J., PAGANI, M., SLOAN, L., THOMAS, E. & BILLUPS, K. 2001. Trends, rhythms, and aberrations in global climate 65 Ma to present. *Science*, **292**, 686–693.

Index

Page numbers in *italic* denote figures. Page numbers in **bold** denote tables.